PRINCIPLES OF NEURAL SCIENCE

Sixth Edition

神经科学原理

（英文版·原书第6版）

上册

埃里克·R. 坎德尔　　约翰·D. 凯斯特
(Eric R. Kandel)　　(John D. Koester)

[美]　　　　　　　　　　　　　　　　编著

萨拉·H. 麦克　　史蒂文·A. 西格尔鲍姆
(Sarah H. Mack)　　(Steven A. Siegelbaum)

机械工业出版社
CHINA MACHINE PRESS

图书在版编目（CIP）数据

神经科学原理：原书第 6 版 . 上册：英文 /（美）埃里克・R. 坎德尔（Eric R. Kandel）等编著 .—北京：机械工业出版社，2024.3

书名原文：Principles of Neural Science, Sixth Edition

ISBN 978-7-111-75299-8

I.①神… II.①埃… III.①神经科学－英文 IV.① Q189

中国国家版本馆 CIP 数据核字（2024）第 051540 号

机械工业出版社（北京市百万庄大街22 号　邮政编码100037）

策划编辑：向睿洋　　责任编辑：向睿洋　　欧阳智

责任校对：彭　萧　　责任印制：单爱军

保定市中画美凯印刷有限公司印刷

2025年1月第1版第1次印刷

214mm×275mm・46.25印张・1插页・1945千字

标准书号：ISBN 978-7-111-75299-8

定价：299.00 元

电话服务　　　　　　　　　网络服务

客服电话：010-88361066　　机　工　官　网：www.cmpbook.com

　　　　　010-88379833　　机　工　官　博：weibo.com/cmp1952

　　　　　010-68326294　　金　书　网：www.golden-book.com

封底无防伪标均为盗版　　机工教育服务网：www.cmpedu.com

PRINCIPLES OF NEURAL SCIENCE

Sixth Edition

Columns II (left) and IV (right) of the Edwin Smith Surgical Papyrus

This papyrus, transcribed in the Seventeenth Century B.C., is a medical treatise that contains the earliest reference to the brain anywhere in human records. According to James Breasted, who translated and published the document in 1930, the word brain 𓄹𓏛𓂧 occurs only 8 times in ancient Egyptian, 6 of them on these pages. The papyrus describes here the symptoms, diagnosis, and prognosis of two patients with compound fractures of the skull, and compares the surface of the brain to "those ripples that happen in copper through smelting, with a thing in it that throbs and flutters under your fingers like the weak spot of the crown of a boy before it becomes whole for him." The red ink highlights the patients' ailments and their prognoses.[⊖] (Reproduced, with permission, from the New York Academy of Medicine Library.)

⊖ 本书附有书配 U 盘，U 盘里有书中全部彩图，欢迎对照审阅。——编者注

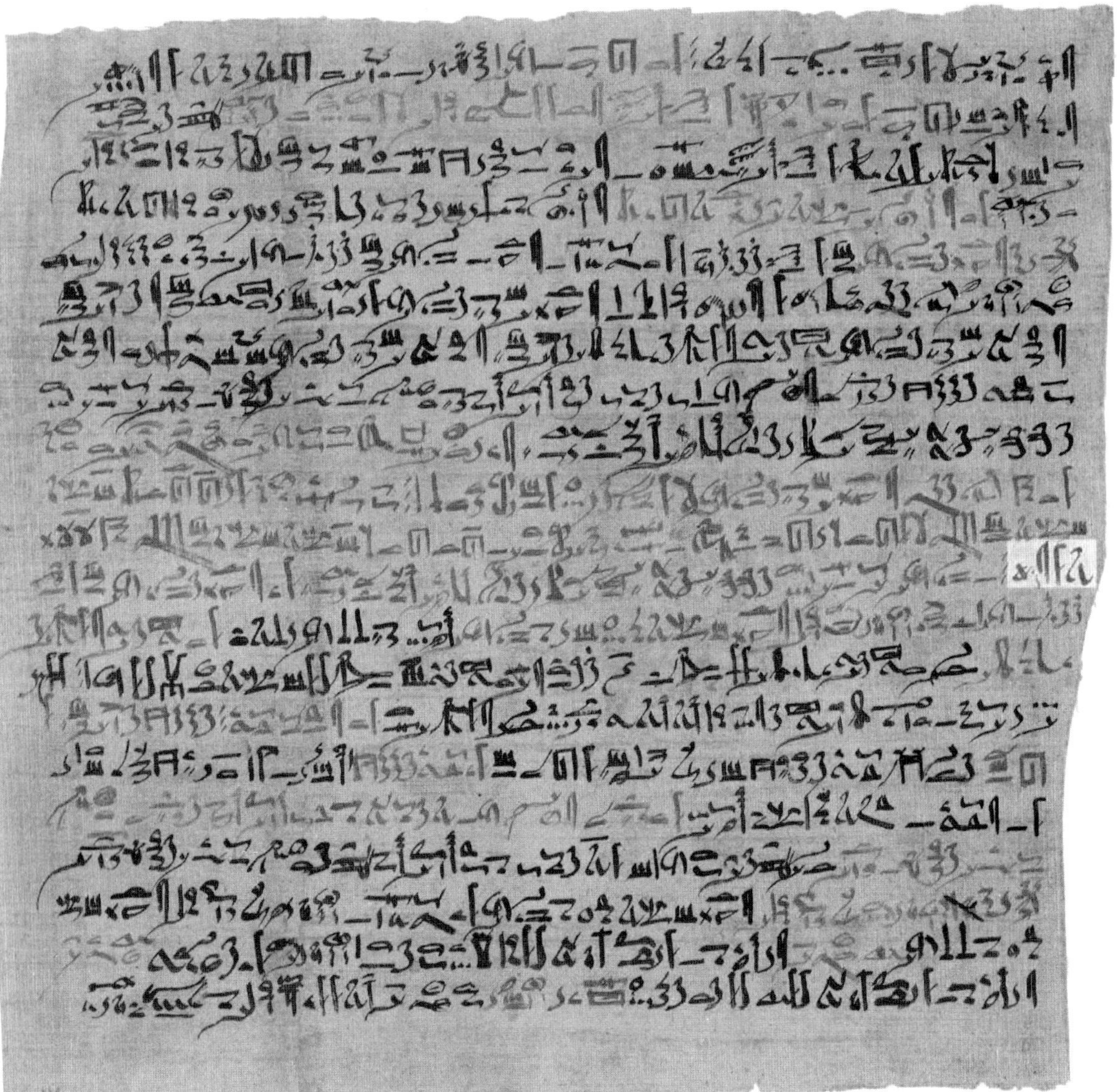

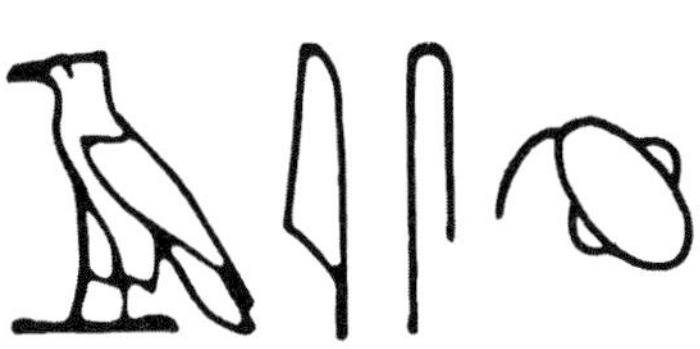

Men ought to know that from the brain, and
from the brain only, arise our pleasures, joys,
laughter and jests, as well as our sorrows,
pains, griefs and tears. Through it, in
particular, we think, see, hear, and distinguish
the ugly from the beautiful, the bad from the
good, the pleasant from the unpleasant. . . . It
is the same thing which makes us mad or
delirious, inspires us with dread and fear,
whether by night or by day, brings
sleeplessness, inopportune mistakes, aimless
anxieties, absent-mindedness, and acts that are
contrary to habit. These things that we suffer
all come from the brain, when it is not healthy,
but becomes abnormally hot, cold, moist, or
dry, or suffers any other unnatural affection to
which it was not accustomed. Madness comes
from its moistness. When the brain is
abnormally moist, of necessity it moves, and
when it moves neither sight nor hearing are
still, but we see or hear now one thing and
now another, and the tongue speaks in
accordance with the things seen and heard on
any occasion. But when the brain is still, a man
can think properly.

attributed to Hippocrates

Fifth Century, B.C.

Reproduced, with permission, from The Sacred Disease,
in *Hippocrates*, Vol. 2, page 175, translated by W.H.S.
Jones, London and New York: William Heinemann and
Harvard University Press. 1923.

Sarah H. Mack
1962–2020

WE DEDICATE THIS SIXTH EDITION OF *Principles of Neural Science* to our dear friends and colleagues, Thomas M. Jessell and Sarah H. Mack.

Sarah Mack, who contributed to and directed the art program of *Principles of Neural Science* during her more than 30-year tenure, passed away on October 2, 2020. She worked courageously and tirelessly to ensure that all the artwork for this edition met her high standards and could be completed while she still had the strength to continue.

After graduating from Williams College with honors in English literature in 1984, Sarah worked for five years in the field of social work, while taking courses at Columbia in studio art and computer graphics. She first contributed to the art program for the third edition of the book when she joined the Kandel lab as a graphic artist in 1989. Five years later, as the fourth edition went into the planning stage, Sarah, together with Jane Dodd as art editor, completely redesigned the art program, developing and converting hundreds of figures and introducing color. This monumental task required countless aesthetic decisions to develop a stylistic consistency for the various figure elements throughout the book. The result was a set of remarkably clear, didactic, and artistically pleasing diagrams and images. Sarah maintained and extended this high level of excellence as art editor of the fifth and sixth editions of the book. She has thus left an enduring mark on the thousands of students who over the years, as well as in years to come, have been introduced to neuroscience through her work.

Sarah was a most remarkable and gifted artist, who developed a deep understanding and appreciation of neuroscience during the many years she contributed to the book. In addition to her artistic contributions to the figures, she also edited the associated text and legends for maximum clarity. Because her contributions extended far beyond the preparation of the figures, Sarah was made co-editor of the current edition of the book. Sarah also had an amazing ability to juggle huge numbers of negotiations with dozens of authors simultaneously, all the while gently, but firmly, steering them to a final set of elegantly instructive images. She did this with such a spirit of generosity that her interactions with the authors, even those she never met in person, developed into warm friendships.

Over the past three editions, Sarah was the driving force that formed the basis for the aesthetic unifying vision running throughout the chapters of *Principles*. She will be greatly missed by us all.

Thomas M. Jessell
1951–2019

Tom Jessell was an extraordinary neuroscientist who made a series of pioneering contributions to our understanding of spinal cord development, the sensory-motor circuit, and the control of movement. Tom had a deep encyclopedic knowledge and understanding of all that came within his sphere of interest. Equally at home discussing a long-forgotten scientific discovery, quoting Shakespeare by heart, or enthusing about 20th-century British or Italian Renaissance art, Tom was a brilliant polymath.

Tom's interest in neuroscience began with his undergraduate studies of synaptic pharmacology at the University of London, from which he graduated in 1973. He then joined Leslie Iversen's laboratory at the Medical Research Council in Cambridge to pursue his PhD, where he investigated the mechanism by which the newly discovered neuropeptide substance P controls pain sensation. Tom made the pivotal observation that opioids inhibit transmission of pain sensation in the spinal cord by reducing substance P release. After receiving his doctoral degree in 1977, he continued to explore the role of substance P in pain processing as a postdoctoral fellow with Masanori Otsuka in Tokyo, solidifying his lifelong interest in spinal sensory mechanisms while managing to learn rudimentary Japanese. Tom then realized that deeper insights into spinal cord function might best be obtained through an understanding of neural development, prompting him to pursue research on the formation of a classic synapse, the neuromuscular junction, in Gerry Fischbach's laboratory at Harvard.

Tom then joined the faculty of Harvard's Department of Neurobiology as an Assistant Professor in 1981, where he explored the mechanisms of sensory synaptic transmission and the development of the somatosensory input to the spinal cord. In 1985 Tom was recruited to the position of Associate Professor and investigator of the Howard Hughes Medical Institute in the Center for Neurobiology and Behavior (now the Department of Neuroscience) and Department of Biochemistry and Molecular Biophysics at Columbia University's College of Physicians and Surgeons. Over the next 33 years, Tom, together with a remarkable group of students and collaborators, applied a multidisciplinary cellular, biochemical, genetic, and electrophysiological approach to identify and define spinal cord microcircuits that control sensory and motor behavior. His studies revealed the molecular and cellular mechanisms by which spinal neurons acquire their identity and by which spinal circuits are assembled and operate. He defined key concepts and principles of neural development and motor control, and his discoveries generated unprecedented insight into the neural principles that coordinate movement, paving the way for therapies for motor neuron disease.

Eric Kandel and Jimmy Schwartz, the initial editors of *Principles of Neural Science*, recruited Tom as co-editor as they began to plan the third edition of the book. Tom's role was to expand the treatment of developmental and molecular neural science. This proved to be a prescient choice as Tom's breadth of knowledge, clarity of thought, and precise, elegant style of writing helped shape and define the text for the next three editions. As co-authors of chapters in *Principles* during Tom's tenure, we can attest to the rigor of language and prose that he encouraged his authors to adopt.

In the last years of his life, Tom bravely faced a devasting neuro-degenerative disease that prevented him from actively participating in the editing of the current edition. Nonetheless Tom's vision remains in the overall design of *Principles* and its philosophical approach to providing a molecular understanding of the neural bases of behavior and neurological disease. Tom's towering influence on this and future editions of *Principles*, and on the field of neuroscience in general, will no doubt endure for decades to come.

Preface

As in previous editions, the goal of this sixth edition of *Principles of Neural Science* is to provide readers with insight into how genes, molecules, neurons, and the circuits they form give rise to behavior. With the exponential growth in neuroscience research over the 40 years since the first edition of this book, an increasing challenge is to provide a comprehensive overview of the field while remaining true to the original goal of the first edition, which is to elevate imparting basic principles over detailed encyclopedic knowledge.

Some of the greatest successes in brain science over the past 75 years have been the elucidation of the cell biological and electrophysiological functions of nerve cells, from the initial studies of Hodgkin, Huxley, and Katz on the action potential and synaptic transmission to our modern understanding of the genetic and molecular biophysical bases of these fundamental processes. The first three parts of this book delineate these remarkable achievements.

The first six chapters in Part I provide an overview of the broad themes of neural science, including the basic anatomical organization of the nervous system and the genetic bases of nervous system function and behavior. We have added a new chapter (Chapter 5) to introduce the principles by which neurons participate in neural circuits that perform specific computations of behavioral relevance. We conclude by considering how application of modern imaging techniques to the human brain provides a bridge between neuroscience and psychology. The next two parts of the book focus on the basic properties of nerve cells, including the generation and conduction of the action potential (Part II) and the electrophysiological and molecular mechanisms of synaptic transmission (Part III).

We then consider how the activity of neurons in the peripheral and central nervous systems gives rise to sensation and movement. In Part IV, we discuss the various aspects of sensory perception, including how information from the primary organs of sensation is transmitted to the central nervous system and how it is processed there by successive brain regions to generate a sensory percept. In Part V, we consider the neural mechanisms underlying movement, beginning with an overview of the field that is followed by a treatment ranging from the properties of skeletal muscle fibers to an analysis of how motor commands issued by the spinal cord are derived from activity in motor cortex and cerebellum. We include a new treatment that addresses how the basal ganglia regulate the selection of motor actions and instantiate reinforcement learning (Chapter 38).

In the latter parts of the book, we turn to higher-level cognitive processes, beginning in Part VI with a discussion of the neural mechanisms by which subcortical areas mediate homeostatic control mechanisms, emotions, and motivation, and the influence of these processes on cortical cognitive operations, such as feelings, decision-making, and attention. We then consider the development of the nervous system in Part VII, from early embryonic differentiation and the initial establishment of synaptic connections, to their experience-dependent refinement, to the replacement of neurons lost to injury or disease. Because learning and memory can be seen as a continuation of synaptic development, we next consider memory, together with language, and include a new chapter on decision-making and consciousness (Chapter 56) in Part VIII. Finally, in Part IX, we consider the neural mechanisms underlying diseases of the nervous system.

Since the last edition of this book, the field of neuroscience has continued to rapidly evolve, which is reflected in changes in this edition. The continued development of new electrophysiological and light microscopic–based imaging technologies has enabled the simultaneous recording of the activity of large populations of neurons in awake behaving animals. These large data sets have given rise to new computational and theoretical approaches to gain insight into how the activity of populations of neurons produce specific behaviors. Light microscopic imaging techniques using genetically encoded calcium sensors allow us to record the activity of hundreds or thousands of defined classes of neurons with subcellular resolution as an animal engages in defined behaviors. At the same time, the development of genetically encoded light-activated ion channels and ion pumps (termed optogenetics) or genetically engineered receptors activated by synthetic ligands (termed chemogenetics or pharmacogenetics) can be used to selectively activate or silence genetically defined populations of neurons to examine their causal role in such behaviors. In addition to including such material in chapters throughout the book, we introduce some of these developments in the new Chapter 5, which considers both the new experimental technologies as well as computational principles by which neural circuits give rise to behavior.

Over the past 20 years, there has also been an expansion of new technologies that enable noninvasive and invasive recordings from the human brain. These studies have narrowed the gap between neuroscience and psychology, as exemplified in the expanded discussion of different forms of human memory in Chapter 52. Noninvasive brain imaging methods have allowed scientists to identify brain areas in humans that are activated during cognitive acts. As discussed in a new chapter on the brain–machine interface (Chapter 39), the implantation of electrodes in the brains of patients permits both electrophysiological recordings and local neural stimulation, offering the promise of restoring some function to individuals with damage to the central or peripheral nervous system.

An understanding of basic and higher-order neural mechanisms is critical not only for our understanding of the normal function of the brain, but also for the insights they afford into a range of inherited and acquired neurological and psychiatric disorders.

With modern genetic sequencing, it is now clear that inherited or spontaneous mutations in neuronally expressed genes contribute to brain disease. At the same time, it is also clear that environmental factors interact with basic genetic mechanisms to influence disease progression. We now end the book with a new section, Part IX, which presents the neuroscientific principles underlying disorders of the nervous system. In previous editions, many of these chapters were dispersed throughout the book. However, we now group these chapters in their own part based on the increasing appreciation that the underlying causes of what appear to be separate diseases, including neurodegenerative diseases, such as Parkinson and Alzheimer disease, and neurodevelopmental disorders, such as schizophrenia and autism, share certain common principles. Finally, these chapters emphasize the historical tradition of how studies of brain disease provide deep insights into normal brain function, including memory and consciousness.

In writing this latest edition, it is our hope and goal that readers will emerge with an appreciation of the achievements of modern neuroscience and the challenges facing future generations of neuroscientists. By emphasizing how neuroscientists in the past have devised experimental approaches to resolve fundamental questions and controversies in the field, we hope that this textbook will also encourage readers to think critically and not shy away from questioning received wisdom, for every hard-won truth likely will lead to new and perhaps more profound questions in brain science. Thus, it is our hope that this sixth edition of *Principles of Neural Science* will provide the foundation and motivation for the next generation of neuroscientists to formulate and investigate these questions.

Acknowledgments

We were most fortunate to have had the creative editorial assistance of Howard P. Beckman, who passed away earlier this year after having finished his work on this edition. Following graduation from San Francisco State University with a BA in 1968, Howard began his distinguished career as a scientific editor. In 1997, he received a law degree from John F. Kennedy University and began a parallel career in environmental law. Howard has been an integral part of *Principles of Neural Science* since the third edition. Although he was not trained as a scientist, his logical thinking and rigorous intellect helped ensure that the book had a unified style of exposition. Howard's demand for clarity of writing has had an immeasurable impact on each edition of this book, and he will be greatly missed by all who worked with him over the years.

We owe an enormous debt of gratitude to Pauline Henick, who skillfully managed the editorial project with great care, keen intelligence, and the utmost diligence. Pauline somehow managed with good humor and understanding to keep all of the editors and authors of the book on track with their chapters through some very difficult circumstances. The timely publishing of the book would not have been possible without her stellar contributions.

We also wish to thank Jan Troutt of Troutt Visual Services for her superb technical and artistic contributions to the illustrations. We appreciate the artistic expertise and keen eye of Mariah Widman, who helped with the preparation of the figures.

We are indebted to our colleagues at McGraw Hill—Michael Weitz, Kim Davis, Jeffrey Herzich, and Becky Hainz-Baxter—for their invaluable help in the production of this edition. Anupriya Tyagi, Cenveo Publisher Services, did an outstanding job of overseeing the composition of the book, for which we are most grateful.

Many other colleagues have helped the editors by critically reading selected chapters of the book and have helped the authors with assistance in the research and writing of the chapters. We wish to acknowledge the contributions of Katherine W. Eyring to Chapter 15; Jeffrey L. Noebels, MD, PhD, to Chapter 58; and Gabriel Vazquez Velez, PhD, Maxime William C. Rousseaux, PhD, and Vicky Brandt to Chapter 63.

We also wish to acknowledge the important role of authors of chapters in previous editions of *Principles of Neural Science,* whose past contributions continue to be reflected in a number of chapters in the present edition. These legacy authors include Cori Bargmann, Uta Frith, James Gordon, A.J. Hudspeth, Conrad Gilliam, James E. Goldman, Thomas M. Jessell (deceased), Jane M. Macpherson, James H. Schwartz (deceased), Thomas Thach (deceased), and Stephen Warren.

We are especially indebted to the editors of the different sections (parts) of the book—Thomas D. Albright, Randy M. Bruno, Thomas M. Jessell (deceased), C. Daniel Salzman, Joshua R. Sanes, Michael N. Shadlen, Daniel M. Wolpert, and Huda Y. Zoghbi—who played a critical role in planning the overall organization of their sections and working with the authors to shape their chapters. Most importantly, we owe the greatest debt to the contributing authors of this edition.

We finally thank our spouses and families for their support and forbearance during the editorial process.

Contributors

Laurence F. Abbott, PhD
William Bloor Professor of Theoretical Neuroscience
Co-Director, Center for Theoretical Neuroscience
Zuckerman Mind Brain Behavior Institute
Department of Neuroscience, and Department of
Physiology and Cellular Biophysics
Columbia University College of Physicians and Surgeons

Ralph Adolphs, PhD
Bren Professor of Psychology, Neuroscience, and
Biology
Division of Humanities and Social Sciences
California Institute of Technology

Thomas D. Albright, PhD
Professor and Conrad T. Prebys Chair
The Salk Institute for Biological Studies

David G. Amaral, PhD
Distinguished Professor
Department of Psychiatry and Behavioral Sciences
The MIND Institute
University of California, Davis

Dora Angelaki, PhD
Center for Neural Science
New York University

Cornelia I. Bargmann, PhD
The Rockefeller University

Ben A. Barres, MD, PhD*
Professor and Chair, Department of Neurobiology
Stanford University School of Medicine

Allan I. Basbaum, PhD, FRS
Professor and Chair
Department of Anatomy
University California San Francisco

Amy J. Bastian, PhD
Professor of Neuroscience, Neurology, and Physical
Medicine and Rehabilitation
Department of Neuroscience
Johns Hopkins University
Director of the Motion Analysis Laboratory
Kennedy Krieger Institute

Bruce P. Bean, PhD
Department of Neurobiology
Harvard Medical School

Robert H. Brown, Jr, DPhil, MD
Professor of Neurology
Director, Program in Neurotherapeutics
University of Massachusetts Medical School

Randy M. Bruno, PhD
Associate Professor
Kavli Institute for Brain Science
Mortimer B. Zuckerman Mind Brain Behavior
Institute
Department of Neuroscience
Columbia University

Linda B. Buck, PhD
Professor of Basic Sciences
Fred Hutchinson Cancer Research Center
Affiliate Professor of Physiology and Biophysics
University of Washington

*Deceased

Stephen C. Cannon, MD, PhD
Professor and Chair of Physiology
Interim Chair of Molecular and Medical
Pharmacology
David Geffen School of Medicine
University of California, Los Angeles

David E. Clapham, MD, PhD
Aldo R. Castañeda Professor of Cardiovascular
Research, Emeritus
Professor of Neurobiology, Emeritus
Harvard Medical School
Vice President and Chief Scientific Officer
Howard Hughes Medical Institute

Rui M. Costa, DVM, PhD
Professor of Neuroscience and Neurology
Director/CEO Zuckerman Mind Brain Behavior
Institute
Columbia University

Aniruddha Das, PhD
Associate Professor
Department of Neuroscience
Mortimer B. Zuckerman Mind Brain Behavior
Institute
Columbia University

J. David Dickman, PhD
Vivian L. Smith Endowed Chair of Neuroscience
Department of Neuroscience
Baylor College of Medicine

Trevor Drew, PhD
Professor
Groupe de Recherche sur le Système Nerveux Central
(GRSNC)
Department of Neurosciences
Université de Montréal

Gammon M. Earhart, PT, PhD, FAPTA
Professor of Physical Therapy, Neuroscience, and
Neurology
Washington University in St. Louis

Joel K. Elmquist, DVM, PhD
Professor, Departments of Internal Medicine and
Pharmacology
Director, Center for Hypothalamic Research
Carl H. Westcott Distinguished Chair in Medical
Research
Maclin Family Professor in Medical Science
The University of Texas Southwestern Medical Center

Roger M. Enoka, PhD
Professor
Department of Integrative Physiology
University of Colorado

Gerald D. Fischbach, MD
Distinguished Scientist and Fellow, Simons
Foundation
Executive Vice President for Health and Biomedical
Sciences, Emeritus
Columbia University

Winrich Freiwald, PhD
Laboratory of Neural Systems
The Rockefeller University

Christopher D. Frith, PhD, FMedSci, FRS, FBA
Emeritus Professor of Neuropsychology, Wellcome
Centre for Human Neuroimaging
University College London
Honorary Research Fellow
Institute of Philosophy
School of Advanced Study
University of London

Daniel Gardner, PhD
Professor of Physiology and Biophysics
Departments of Physiology and Biophysics,
Neurology, and Neuroscience
Weill Cornell Medical College

Esther P. Gardner, PhD
Professor of Neuroscience and Physiology
Department of Neuroscience and Physiology
Member, NYU Neuroscience Institute
New York University Grossman School of Medicine

Charles D. Gilbert, MD, PhD
Arthur and Janet Ross Professor
Head, Laboratory of Neurobiology
The Rockefeller University

T. Conrad Gilliam, PhD
Marjorie I. and Bernard A. Mitchel Distinguished
Service Professor of Human Genetics
Dean for Basic Science
Biological Sciences Division and Pritzker School of
Medicine
The University of Chicago

Michael E. Goldberg, MD
David Mahoney Professor of Brain and Behavior
Departments of Neuroscience, Neurology, Psychiatry,
and Ophthalmology
Columbia University Vagelos College of Physicians
and Surgeons
Zuckerman Mind Brain Behavior Institute

Joshua A. Gordon, MD, PhD
Director, National Institute of Mental Health

David M. Holtzman, MD
Department of Neurology, Hope Center for
Neurological Disorders
Knight Alzheimer's Disease Research Center
Washington University School of Medicine

Fay B. Horak, PhD, PT
Professor of Neurology
Oregon Health and Science University

John P. Horn, PhD
Professor of Neurobiology
Associate Dean for Graduate Studies
Department of Neurobiology
University of Pittsburgh School of Medicine

Steven E. Hyman, MD
Stanley Center for Psychiatric Research
Broad Institute of MIT and Harvard University
Department of Stem Cell and Regenerative Biology
Harvard University

Jonathan A. Javitch, MD, PhD
Lieber Professor of Experimental Therapeutics in
Psychiatry
Professor of Pharmacology
Columbia University Vagelos College of Physicians
and Surgeons
Chief, Division of Molecular Therapeutics
New York State Psychiatric Institute

Thomas M. Jessell, PhD*
Professor (Retired)
Department of Neuroscience and Biochemistry and
Biophysics
Columbia University

John Kalaska, PhD
Professeur Titulaire
Département de Neurosciences
Faculté de Médecine
l'Université de Montréal

Eric R. Kandel, MD
University Professor
Kavli Professor and Director, Kavli Institute for
Brain Science
Co-Director, Mortimer B. Zuckerman Mind Brain
Behavior Institute
Senior Investigator, Howard Hughes Medical Institute
Department of Neuroscience
Columbia University

Ole Kiehn, MD, PhD
Professor, Department of Neuroscience
University of Copenhagen
Professor, Department of Neuroscience
Karolinska Institutet
Stockholm, Sweden

John D. Koester, PhD
Professor Emeritus of Clinical Neuroscience
Vagelos College of Physicians and Surgeons
Columbia University

Patricia K. Kuhl, PhD
The Bezos Family Foundation Endowed Chair in
Early Childhood Learning
Co-Director, Institute for Learning and Brain Sciences
Professor, Speech and Hearing Sciences
University of Washington

Joseph E. LeDoux, PhD
Henry And Lucy Moses Professor of Science
Professor of Neural Science and Psychology
Professor of Psychiatry and Child and Adolescent
Psychiatry
NYU Langone Medical School
Director of the Emotional Brain Institute
New York University and Nathan Kline Institute

---

*Deceased

Stephen G. Lisberger, PhD
Department of Neurobiology
Duke University School of Medicine

Attila Losonczy, MD, PhD
Professor, Department of Neuroscience
Mortimer B. Zuckerman Mind Brain Behavior
Institute
Kavli-Simons Fellow
Kavli Institute for Brain Science
Columbia University

Bradford B. Lowell, MD, PhD
Professor, Division of Endocrinology, Diabetes, and
Metabolism
Department of Medicine
Beth Israel Deaconess Medical Center
Program in Neuroscience
Harvard Medical School

Geoffrey A. Manley, PhD
Professor (Retired), Cochlear and Auditory Brainstem
Physiology
Department of Neuroscience
School of Medicine and Health Sciences
Cluster of Excellence "Hearing4all"
Research Centre Neurosensory Science
Carl von Ossietzky University
Oldenburg, Germany

Eve Marder, PhD
Victor and Gwendolyn Beinfield University Professor
Volen Center and Biology Department
Brandeis University

Pascal Martin, PhD
CNRS Research Director
Laboratoire Physico-Chimie Curie
Institut Curie, PSL Research University
Sorbonne Université
Paris, France

Markus Meister, PhD
Professor of Biology
Division of Biology and Biological Engineering
California Institute of Technology

Edvard I. Moser, PhD
Kavli Institute for Systems Neuroscience
Norwegian University of Science and Technology
Trondheim, Norway

May-Britt Moser, PhD
Kavli Institute for Systems Neuroscience
Norwegian University of Science and Technology
Trondheim, Norway

Eric J. Nestler, MD, PhD
Nash Family Professor of Neuroscience
Director, Friedman Brain Institute
Dean for Academic and Scientific Affairs
Icahn School of Medicine at Mount Sinai

Jens Bo Nielsen, MD, PhD
Professor, Department of Neuroscience
University of Copenhagen
The Elsass Foundation
Denmark

Donata Oertel, PhD*
Professor of Neurophysiology
Department of Neuroscience
University of Wisconsin

Franck Polleux, PhD
Professor, Columbia University
Department of Neuroscience
Mortimer B. Zuckerman Mind Brain Behavior
Institute
Kavli Institute for Brain Science

Peter Redgrave, PhD
University Professor, Emeritus
Department of Psychology
University of Sheffield
United Kingdom

Lewis P. Rowland, MD*
Professor of Neurology and Chair Emeritus
Department of Neurology
Columbia University

---

*Deceased

C. Daniel Salzman, MD, PhD
Professor, Departments of Neuroscience and
Psychiatry
Investigator, Mortimer B. Zuckerman Mind Brain
Behavior Institute
Investigator, Kavli Institute for Brain Science
Columbia University

Joshua R. Sanes, PhD
Jeff C. Tarr Professor of Molecular and Cellular
Biology
Paul J. Finnegan Family Director, Center for Brain
Science
Harvard University

Clifford B. Saper, MD, PhD
James Jackson Putnam Professor of Neurology and
Neuroscience
Harvard Medical School
Chairman, Department of Neurology
Beth Israel Deaconess Medical Center
Editor-in-Chief, *Annals of Neurology*

Nathaniel B. Sawtell, PhD
Associate Professor
Zuckerman Mind Brain Behavior Institute
Department of Neuroscience
Columbia University

Thomas E. Scammell, MD
Professor of Neurology
Beth Israel Deaconess Medical Center
Harvard Medical School

Daniel L. Schacter, PhD
William R. Kenan, Jr. Professor
Department of Psychology, Harvard University

Kristin Scott, PhD
Professor
University of California, Berkeley
Department of Molecular and Cell Biology

Stephen H. Scott, PhD
Professor and GSK Chair in Neuroscience
Centre for Neuroscience Studies
Department of Biomedical and Molecular Sciences
Department of Medicine
Queen's University
Kingston, Canada

Michael N. Shadlen, MD, PhD
Howard Hughes Medical Institute
Kavli Institute of Brain Science
Department of Neuroscience
Zuckerman Mind Brain Behavior Institute
Columbia University Irving Medical Center
Columbia University

Nirao M. Shah, MBBS, PhD
Department of Psychiatry and Behavioral Sciences
Department of Neurobiology
Stanford University

Krishna V. Shenoy, PhD
Investigator, Howard Hughes Medical Institute
Hong Seh and Vivian W. M. Lim Professor
Departments of Electrical Engineering, Bioengineer-
ing, and Neurobiology
Wu Tsai Neurosciences Institute and Bio-X Institute
Stanford University

Daphna Shohamy, PhD
Professor, Department of Psychology
Zuckerman Mind Brain Behavior Institute
Kavli Institute for Brain Science
Columbia University

Steven A. Siegelbaum, PhD
Chair, Department of Neuroscience
Gerald D. Fischbach, MD, Professor of Neuroscience
Professor of Pharmacology
Columbia University

Matthew W. State, MD, PhD
Oberndorf Family Distinguished Professor and Chair
Department of Psychiatry and Behavioral Sciences
Weill Institute for Neurosciences
University of California, San Francisco

Beth Stevens, PhD
Boston Children's Hospital
Broad Institute of Harvard and MIT
Howard Hughes Medical Institute

Thomas C. Südhof, MD
Avram Goldstein Professor in the School of Medicine
Departments of Molecular and Cellular Physiology
and of Neurosurgery
Howard Hughes Medical Institute
Stanford University

David Sulzer, PhD
Professor, Departments of Psychiatry, Neurology, and
Pharmacology
School of the Arts
Columbia University
Division of Molecular Therapeutics
New York State Psychiatric Institute

Larry W. Swanson, PhD
Department of Biological Sciences
University of Southern California

Carol A. Tamminga, MD
Professor and Chairman
Department of Psychiatry
UT Southwestern Medical School

Marc Tessier-Lavigne, PhD
President, Stanford University
Bing Presidential Professor
Department of Biology
Stanford University

Richard W. Tsien, DPhil
Druckenmiller Professor of Neuroscience
Chair, Department of Physiology and Neuroscience
Director, NYU Neuroscience Institute
New York University Medical Center
George D. Smith Professor Emeritus
Stanford University School of Medicine

Nicholas B. Turk-Browne, PhD
Professor, Department of Psychology
Yale University

Anthony D. Wagner, PhD
Professor, Department of Psychology
Wu Tsai Neurosciences Institute
Stanford University

Mark F. Walker, MD
Associate Professor of Neurology
Case Western Reserve University
Staff Neurologist, VA Northeast Ohio Healthcare
System

Xiaoqin Wang, PhD
Professor
Laboratory of Auditory Neurophysiology
Department of Biomedical Engineering
Johns Hopkins University

Gary L. Westbrook, MD
Senior Scientist, Vollum Institute
Dixon Professor of Neurology
Oregon Health and Science University

Daniel M. Wolpert, PhD, FMedSci, FRS
Department of Neuroscience
Mortimer B. Zuckerman Mind Brain Behavior
Institute
Columbia University

Robert H. Wurtz, PhD
Distinguished Investigator Emeritus
Laboratory of Sensorimotor Research
National Eye Institute
National Institutes of Health

Byron M. Yu, PhD
Department of Electrical and Computer Engineering
Department of Biomedical Engineering Neuroscience
Institute
Carnegie Mellon University

Rafael Yuste, MD, PhD
Columbia University
Professor of Biological Sciences
Director, Neurotechnology Center
Co-Director, Kavli Institute of Brain Sciences
Ikerbasque Research Professor
Donostia International Physics Center (DIPC)

Huda Y. Zoghbi, MD
Investigator, Howard Hughes Medical Institute
Professor, Baylor College of Medicine
Director, Jan and Dan Duncan Neurological Research
Institute
Texas Children's Hospital

Charles Zuker, PhD
Departments of Neuroscience, and Biochemistry and
Molecular Biophysics
Columbia University
Howard Hughes Medical Institute

Contents in Brief

Contents

Part II

Cell and Molecular Biology of Cells of the Nervous System

Part IV
Perception

28   Auditory Processing by the Central Nervous System 651

Donata Oertel, Xiaoqin Wang

29   Smell and Taste: The Chemical Senses 682

Linda Buck, Kristin Scott, Charles Zuker

Part I

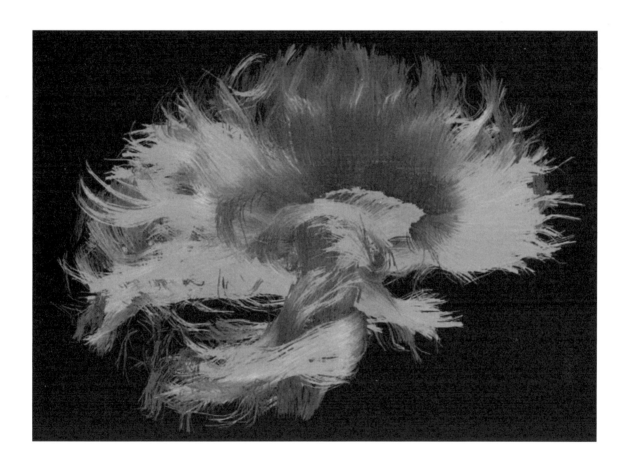

I   Overall Perspective

URING THE SECOND HALF OF THE 20TH CENTURY, the central focus of biology was on the gene. Now in the first half of the 21st century, the focus has shifted to neural science, and specifically to the biology of the mind. We wish to understand the processes by which we perceive, act, learn, and remember. How does the brain—an organ weighing only 1.5 kg—conceive of the infinite, discover new knowledge, and produce the remarkable individuality of human thoughts, feelings, and actions? How are these extraordinary mental capabilities distributed within the organ? What rules relate the anatomical organization and the cellular physiology of a region to its specific role in mentation? What do genes contribute to behavior, and how is gene expression in nerve cells regulated by developmental and learning processes? How does experience alter the way the brain processes subsequent events, and to what degree is that processing unconscious? Finally, what are the neural bases of neurological and psychiatric disorders? In this introductory section of *Principles of Neural Science*, we begin to address these questions. In so doing, we describe how neural science attempts to link the computational logic of neural circuitry to the mind—how the activities of nerve cells within defined neural circuits mediate complex mental processes.

Over the past several decades, technological advances have opened new horizons for the scientific study of the brain. Today, it is possible to link the cellular dynamics of interconnected circuits of neurons to the internal representations of perceptual and motor acts in the brain and to relate these internal mechanisms to observable behavior. New imaging techniques permit us to visualize the human brain in action—to identify specific regions of the brain associated with particular modes of thinking and feeling and their patterns of interconnections.

In the first part of this book, we consider the degree to which mental functions can be localized to specific regions of the brain. We also examine the extent to which such functions can be understood in terms of the properties of individual nerve cells, their molecular constituents, and their synaptic connections. In the later parts of the book, we examine in detail the mechanisms underlying cognitive and affective functions of the brain: perception, action, motivation, emotion, learning, and memory.

The human brain is a network of more than 80 billion individual nerve cells interconnected in systems—neural circuits—that construct our perceptions of the external world, fix our attention, guide our decisions, and implement our actions. A first step toward understanding the mind, therefore, is to learn how neurons are organized into signaling pathways and how they communicate by means of synaptic transmission. One of the chief ideas we shall develop in this book is that the specificity of the synaptic connections established during development and refined during experience underlie behavior. We must also understand both the innate and environmental determinants of behavior in which genes encode proteins that initially govern the development of the neural circuits that can then be modified by experience-dependent changes in gene expression.

A new science of mind is emerging through the application of modern cell and molecular biological techniques, brain imaging, theory, and clinical observation to the study of cognition, emotion, and behavior. Neural science has reinforced the idea first proposed by Hippocrates more than two millennia ago that the proper study of mind begins with study of the brain. Cognitive psychology and psychoanalytic theory have emphasized the diversity and complexity of human mental experience. These disciplines can now be enriched by insights into brain function from neural science. The task ahead is to produce a study of mental processes, grounded firmly in empirical neural science, concerned with questions of how internal representations and states of mind are generated.

Our goal is to provide not simply the facts but the principles of brain organization, function, and computation. The principles of neural science do not reduce the complexity of human thought to a set of molecules or mathematical axioms. Rather, they allow us to appreciate a certain beauty—a Darwinian elegance—in the complexity of the brain that accounts for mind and behavior. One might ask whether an idea gleaned from the detailed dissection of a more basic neural mechanism contains insight about higher brain function. Does the organization of a simple reflex bear on a volitional movement of the hand? Do the mechanisms that establish circuitry in the developing spinal cord bear on the mechanisms at play in storing a memory? Are the neural processes that awaken us from sleep similar to those that allow an unconscious process to pierce our conscious awareness? We hope readers will delight in the principles as they delve into their factual basis. No doubt, it is a work in progress.

Part Editors: Eric R. Kandel and Michael N. Shadlen

Part I

1

The Brain and Behavior

T HE LAST FRONTIER OF THE BIOLOGICAL SCIENCES— the ultimate challenge—is to understand the biological basis of consciousness and the brain processes by which we feel, act, learn, and remember. During the past few decades, a remarkable unification within the biological sciences has set the stage for addressing this great challenge. The ability to sequence genes and infer the amino acid sequences of the proteins they encode has revealed unanticipated similarities between proteins in the nervous system and those encountered elsewhere in the body. As a result, it has become possible to establish a general plan for the function of cells, a plan that provides a common conceptual framework for all of cell biology, including cellular neural science.

The current challenge in the unification within biology is the unification of psychology—the science of the mind—and neural science—the science of the brain. Such a unified approach, in which mind and body are not seen as separate entities, rests on the view that all behavior is the result of brain function. What we commonly call the mind is a set of operations carried out by the brain. Brain processes underlie not only simple motor behaviors such as walking and eating but also all the complex cognitive acts and behavior that we regard as quintessentially human—thinking, speaking, and creating works of art. As a corollary, all the behavioral disorders that characterize psychiatric illness—disorders of affect (feeling) and cognition (thought)—result from disturbances of brain function.

How do the billions of individual nerve cells in the brain produce behavior and cognitive states, and how are those cells influenced by the environment, which includes social experience? Explaining behavior in terms of brain activity is the important task of neural science, and the progress of neural science in this respect is a major theme of this book.

Neural science must continually confront certain fundamental questions. What is the appropriate level of biological description to understand a thought process, the movement of a limb, or the desire to make the movement? Why is a movement smooth or jerky or made unintentionally in certain neurological disease states? Answers to these questions might emerge from looking at the pattern of DNA expression in nerve cells and how this pattern regulates the electrical properties of neurons. However, we will also require knowledge of neural circuits comprising many neurons in specific brain areas and how the activity of specific circuits in many brain areas is coordinated.

Is there a level of biological description that is most apt? The short answer is, it depends. If one's goal is to understand and treat certain genetic epilepsy disorders, then DNA sequencing and measurements of electrical properties of individual neurons might be sufficient to produce an effective therapy.

If one is interested in learning, perception, and exploration, then an analysis of systems of circuits and brain regions is likely to be required.

The goal of modern neural science is to integrate all of these specialized levels into a coherent science. The effort forces us to confront new questions. If mental processes can be localized to discrete brain regions, what is the relationship between the functions of those regions and the anatomy and physiology of those regions? Is one kind of neural circuit required to process visual information, another type to parse speech, and yet another to sequence movements? Or do circuits with different functions share common organizational principles? Are the requisite neural computations best understood as operations on information represented by single neurons or populations of neurons? Is information represented in the electrical activity of individual nerve cells, or is it distributed over ensembles such that any one cell is no more informative than a random bit of computer memory? As we shall see, questions about levels of organization, specialization of cells, and localization of function recur throughout neural science.

To illustrate these points we shall examine how modern neural science describes language, a distinctive cognitive behavior in humans. In so doing, we shall focus broadly on operations in the cerebral cortex, the part of the brain that is most highly developed in humans. We shall see how the cortex is organized into functionally distinct regions, each made up of large groups of neurons, and how the neural apparatus of a highly complex behavior can be analyzed in terms of the activity of specific sets of interconnected neurons within specific regions. In Chapter 3, we describe how the neural circuit for a simple reflex behavior operates at the cellular level, illustrating how the interplay of sensory signals and motor signals leads to a motor act.

Two Opposing Views Have Been Advanced on the Relationship Between Brain and Behavior

Our views about nerve cells, the brain, and behavior emerged during the 20th century from a synthesis of five experimental traditions: anatomy, embryology, physiology, pharmacology, and psychology.

The 2nd century Greek physician Galen proposed that nerves convey fluid secreted by the brain and spinal cord to the body's periphery. His views dominated Western medicine until the microscope revealed the true structure of the cells in nervous tissue. Even so, nervous tissue did not become the subject of a special science until the late 1800s, when the Italian Camillo Golgi and the Spaniard Santiago Ramón y Cajal produced detailed, accurate descriptions of nerve cells but reached two quite different conclusions of how the brain functions.

Golgi developed a method of staining neurons with silver salts that revealed their entire cell structure under the microscope. Based on such studies, Golgi concluded that nerve cells are not independent cells isolated from one another but instead act together in one continuous web of tissue or syncytium. Using Golgi's technique, Ramón y Cajal observed that each neuron typically has a cell body and two types of processes: branching dendrites at one end and a long, cable-like axon at the other. Cajal concluded that nervous tissue is not a syncytium but a network of discrete cells. In the course of this work, Ramón y Cajal developed some of the key concepts and much of the early evidence for the *neuron doctrine*—the principle that individual neurons are the elementary building blocks and signaling elements of the nervous system.

In the 1920s the American embryologist Ross Harrison showed that the dendrites and axons grow from the cell body and do so even when each neuron is isolated from others in tissue culture. Harrison also confirmed Ramón y Cajal's suggestion that the tip of the axon gives rise to an expansion, the *growth cone*, which leads the developing axon to its target, either to other nerve cells or muscles. Both of these discoveries lent strong support to the neuron doctrine. The final definite evidence for the neuron doctrine came in the mid-1950s with the introduction of electron microscopy. A landmark study by Sanford Palay unambiguously demonstrated the existence of synapses, specialized regions of nerve cells that permit chemical or electrical signaling between them.

Physiological investigation of the nervous system began in the late 1700s when the Italian physician and physicist Luigi Galvani discovered that muscle and nerve cells produce electricity. Modern electrophysiology grew out of work in the 19th century by three German physiologists—Johannes Müller, Emil du Bois-Reymond, and Hermann von Helmholtz—who succeeded in measuring the speed of conduction of electrical activity along the axon of the nerve cell and further showed that the electrical activity of one nerve cell affects the activity of an adjacent cell in predictable ways.

Pharmacology made its first impact on our understanding of the nervous system and behavior at the end of the 19th century when Claude Bernard in France, Paul Ehrlich in Germany, and John Langley in England demonstrated that drugs do not act randomly on a cell, but rather bind to discrete receptors typically located in the cell membrane. This insight led to the discovery that nerve cells can communicate with each other by chemical means.

Psychological thinking about behavior dates back to the beginnings of Western science when the ancient Greek philosophers speculated about the causes of behavior and the relation of the mind to the brain. In subsequent centuries, two major views emerged. In the 17th century, René Descartes distinguished body and mind. In this *dualistic view*, the brain mediates perception, motor acts, memory, appetites, and passions—everything that can be found in the lower animals. But the mind—the higher mental functions, the conscious experience characteristic of human behavior—is not represented in the brain or any other part of the body but in the soul, a spiritual entity. Descartes believed that the soul communicated with the machinery of the brain by means of the pineal gland, a tiny structure in the midline of the brain. Descartes's position has had little sway in modern philosophy or neural science. Indeed, the underlying premise of neural science is that mind is a product of the brain and its neural activity. By this we do not mean that the aim of neural science is to explain *away* the mind by reduction to biological components, but rather to elucidate the biology of mind.

Attempts to join biological and psychological concepts in the study of behavior began as early as 1800, when Franz Joseph Gall, a Viennese physician and neuroanatomist, proposed a radically new idea of body and mind. He advocated that the brain is the organ of the mind and that all mental functions are embodied in the brain. He thus rejected the Cartesian idea that mind and body are separate entities. In addition, he argued that the cerebral cortex was not a unitary organ but contained within it many specialized organs, and that particular regions of the cerebral cortex control specific functions. Gall enumerated at least 27 distinct regions or organs of the cerebral cortex; later many more were added, each corresponding to a specific mental faculty (Figure 1–1). Gall assigned intellectual processes, such as the ability to evaluate causality, to calculate, and to sense order, to the front of the brain. Instinctive characteristics such as romantic love (*amativeness*) and combativeness were assigned to the back of the brain. Even the most abstract of human behaviors—generosity, secretiveness, and religiosity—were assigned a spot in the brain.

Although Gall's theory of the unity of body and mind and his idea that certain functions were localized to specific brain regions proved to be correct, the dominant view today is that many higher functions of mind are most likely highly distributed. Moreover, Gall's experimental approach to localization was extremely

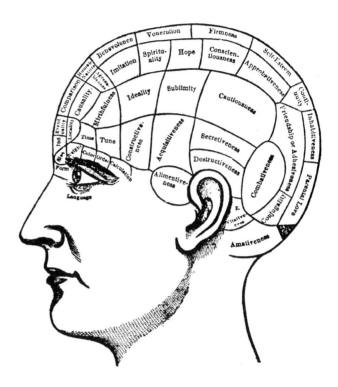

Figure 1–1 An early map of functional localization in the brain. According to the 19th century doctrine of phrenology, complex traits such as combativeness, spirituality, hope, and conscientiousness are controlled by specialized "organs," distinct areas of the cerebral cortex that expand as the traits develop. These enlargements of local areas of the brain were thought to produce characteristic bumps and ridges on the overlying skull, from which an individual's character could be determined. This map, taken from a drawing of the early 1800s, shows 42 intellectual and emotional "organs."

naive. Rather than locate functions empirically, by looking into the brain and correlating defects in mental attributes with lesions in specific regions following tumor or stroke, Gall spurned all evidence derived from studies of brain lesions, whether discovered through clinical examination or produced surgically in experimental animals. Influenced by physiognomy, the popular science based on the idea that facial features reveal character, Gall believed that the bumps and ridges on the skulls of people well endowed with specific cognitive faculties identified the centers for those faculties in the brain. He assumed that the size of an area of brain was related to the relative importance of the mental faculty represented in that area. Accordingly, exercise of a given mental faculty would cause the corresponding brain region to grow, and this growth in turn would cause the overlying skull to protrude.

Gall first had this idea as a young boy when he noticed that those of his classmates who excelled at memorizing school assignments had prominent eyes. He concluded that this was the result of an overdevelopment of regions in the front of the brain involved in verbal memory. He developed this idea further when, as a young physician, he was placed in charge of an asylum for the insane in Vienna. There he began to study patients suffering from monomania, a disorder characterized by an exaggerated interest in some key idea or a deep urge to engage in some specific behavior—theft, murder, eroticism, extreme religiosity. He reasoned that, because the patient functioned well in all other behaviors, the brain defect must be discrete and in principle could be localized by examining the skulls of these patients. Gall's studies of localized brain functions led to *phrenology*, a discipline concerned with determining personality and character based on the detailed shape of the skull.

In the late 1820s, Gall's ideas were subjected to experimental analysis by the French physiologist Pierre Flourens. Using experimental animals, Flourens destroyed some of Gall's functional centers in the brain, and in turn attempted to isolate the contribution of these "cerebral organs" to behavior. From these experiments, Flourens concluded that specific brain regions are not responsible for specific behaviors, but that all brain regions, especially the cerebral hemispheres of the forebrain, participate in every mental operation. Any part of a cerebral hemisphere, Flourens proposed, contributes to all the hemisphere's functions. Injury to any one area of the cerebral hemisphere should therefore affect all higher functions equally. Thus in 1823 Flourens wrote: "All perceptions, all volitions occupy the same seat in these (cerebral) organs; the faculty of perceiving, of conceiving, of willing merely constitutes therefore a faculty which is essentially one."

The rapid acceptance of this belief, later called the *holistic* view of the brain, was based only partly on Flourens's experimental work. It also represented a cultural reaction against the materialistic view that the human mind is a biological organ. It represented a rejection of the notion that there is no soul, that all mental processes can be reduced to activity within the brain, and that the mind can be improved by exercising it—ideas that were unacceptable to the religious establishment and landed aristocracy of Europe.

The holistic view was seriously challenged, however, in the mid-19th century by the French neurologist Paul Pierre Broca, the German neurologist Carl Wernicke, and the British neurologist Hughlings Jackson. For example, in his studies of focal epilepsy, a disease characterized by convulsions that begin in a particular part of the body, Jackson showed that different motor and sensory functions could be traced to specific parts

of the cerebral cortex. The regional studies by Broca, Wernicke, and Jackson were extended to the cellular level by Charles Sherrington and by Ramón y Cajal, who championed the view of brain function called *cellular connectionism*. According to this view, individual neurons are the signaling units of the brain; they are arranged in functional groups and connect to one another in a precise fashion. Wernicke's work and that of the French neurologist Jules Dejerine revealed that different behaviors are produced by different interconnected brain regions.

The first important evidence for localization emerged from studies of how the brain produces language. Before we consider the relevant clinical and anatomical studies, we shall first review the overall structure of the brain, including its major anatomical regions. This requires that we define some essential navigational terms used by neuroanatomists to describe the three-dimensional spatial relationships between parts of the brain and spinal cord. These terms are introduced in Box 1–1 and Figure 1–2.

The Brain Has Distinct Functional Regions

The central nervous system is a bilateral and largely symmetrical structure with two main parts, the spinal cord and the brain. The brain comprises six major structures: the medulla oblongata, pons, cerebellum, midbrain, diencephalon, and cerebrum (Box 1–2 and Figure 1–3). Each of these in turn comprise distinct groups of neurons with distinctive connectivity and developmental origin. In the medulla, pons, midbrain, and diencephalon, neurons are often grouped in distinct clusters termed nuclei. The surface of the cerebrum and cerebellum consists of a large folded sheet of neurons called the cerebral cortex and the cerebellar cortex, respectively, where neurons are organized in layers with stereotyped patterns of connectivity. The cerebrum also contains a number of structures located below the cortex (subcortical), including the basal ganglia and amygdala (Figure 1–4).

Modern brain imaging techniques make it possible to see activity in these structures in living people (see Chapter 6). Brain imaging is commonly used to evaluate the metabolic activity of discrete regions of the brain while people are engaged in specific tasks under controlled conditions. Such studies provide evidence that specific types of behavior recruit the activity of particular regions of the brain more than others. Brain imaging vividly demonstrates that cognitive operations rely primarily on the cerebral cortex, the furrowed gray matter covering the two cerebral hemispheres (Figure 1–5).

Box 1–1 Neuroanatomical Terms of Navigation

The location and orientation of components of the central nervous system within the body are described with reference to three axes: the rostral-caudal, dorsal-ventral, and medial-lateral axes (Figure 1–2). These terms allow the neuroanatomist to describe spatial relations between parts of the brain and spinal cord. They facilitate the comparison of brains of individuals of the same species as they develop or in the case of a disease. They also facilitate the comparison of brains from different species of animals, for example, to understand the brain's evolution.

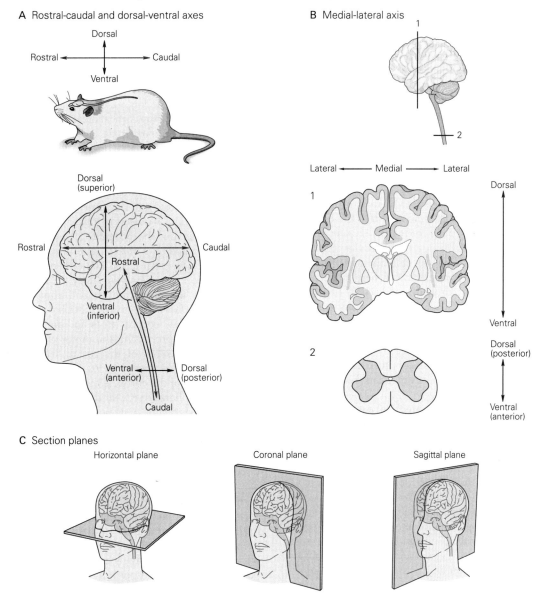

Figure 1–2 The central nervous system is described along three major axes. (Adapted, with permission, from Martin 2003.)

A. *Rostral* means toward the nose and *caudal* toward the tail. *Dorsal* means toward the back of the animal and *ventral* toward the belly. In lower mammals the orientations of these two axes are maintained through development into adult life. In humans and other higher primates, the longitudinal axis is flexed in the brain stem by approximately 110 degrees. Because of this flexure, the same positional terms have different meanings when referring to structures below and above the flexure. Below the flexure, in the spinal cord, rostral means toward the head, caudal means toward the coccyx (the lower end of the spinal column), ventral (anterior) means toward the belly, and dorsal (posterior) means toward the back. Above the flexure, rostral means toward the nose, caudal means toward the back of the head, ventral means toward the jaw, and dorsal means toward the top of the head. The term *superior* is often used synonymously with dorsal, and *inferior* means the same as ventral.

B. *Medial* means toward the middle of the brain and *lateral* toward the side.

C. When brains are sectioned for analysis, slices are typically made in one of three cardinal planes: horizontal, coronal, or sagittal.

Box 1–2 Anatomical Organization of the Central Nervous System

The Central Nervous System Has Seven Main Parts

The **spinal cord,** the most caudal part of the central nervous system, receives and processes sensory information from the skin, joints, and muscles of the limbs and trunk and controls movement of the limbs and the trunk. It is subdivided into cervical, thoracic, lumbar, and sacral regions (Figure 1–3A).

The spinal cord continues rostrally as the **brain stem**, which consists of the medulla oblongata, pons, and midbrain. The brain stem receives sensory information from the skin and muscles of the head and provides the motor control for the head's musculature. It also conveys information from the spinal cord to the brain and from the brain to the spinal cord, and regulates levels of arousal and awareness through the reticular formation.

The brain stem contains several collections of cell bodies, the cranial nerve nuclei. Some of these nuclei receive information from the skin and muscles of the head; others control motor output to muscles of the face, neck, and eyes. Still others are specialized to process information from three of the special senses: hearing, balance, and taste.

The **medulla oblongata,** directly rostral to the spinal cord, includes several centers responsible for vital autonomic functions, such as digestion, breathing, and the control of heart rate.

The **pons,** rostral to the medulla, conveys information about movement from the cerebral hemispheres to the cerebellum.

The **cerebellum,** behind the pons, modulates the force and range of movement and is involved in the learning of motor skills. It is functionally connected to the three main organs of the brain stem: the medulla oblongata, the pons, and the midbrain.

The **midbrain,** rostral to the pons, controls many sensory and motor functions, including eye movement and the coordination of visual and auditory reflexes.

The **diencephalon** lies rostral to the midbrain and contains two structures. The *thalamus* processes most of the information reaching the cerebral cortex from the rest of the central nervous system. The *hypothalamus* regulates autonomic, endocrine, and visceral functions.

The **cerebrum** comprises two cerebral hemispheres, each consisting of a heavily wrinkled outer layer (the *cerebral cortex*) and three deep-lying structures (components of the *basal ganglia*, the *hippocampus*, and *amygdaloid nuclei*). The basal ganglia, which include the caudate, putamen, and globus pallidus, regulate movement execution and motor- and habit-learning, two forms of memory that are referred to as implicit memory; the hippocampus is critical for storage of memory of people, places, things, and events, a form of memory that is referred to as explicit; and the amygdaloid nuclei coordinate the autonomic and endocrine responses of emotional states, including memory of threats, another form of implicit memory.

Each cerebral hemisphere is divided into four distinct lobes: frontal, parietal, occipital, and temporal (Figure 1–3B). These lobes are associated with distinct functions, although the cortical areas are all highly interconnected and can participate in a wide range of brain functions. The occipital lobe receives visual information and is critical for all aspects of vision. Information from the occipital lobe is then processed through two main pathways. The dorsal stream, connecting the occipital lobe to the parietal lobe, is concerned with the location and manipulation of objects in visual space. The ventral stream, connecting the occipital lobe to the temporal lobe, is concerned with object identity, including the recognition of individual faces. The temporal lobe is also important for processing auditory information (and also contains the hippocampus and amygdala buried beneath its surface). The frontal lobes are strongly interconnected with all cortical areas and are important for higher cognitive processing and motor planning.

About two-thirds of the cortex lies in the sulci, and many gyri are buried by overlying cortical lobes. The full extent of the cortex is made visible by separating the hemispheres to reveal the medial surface of the brain and by slicing the brain post mortem, for example in an autopsy (Figure 1–4). Much of this information can be visualized in the living brain through modern brain imaging (Figure 1–5; Chapter 6). These views also afford views of the white matter and subcortical gray matter.

Two important regions of cerebral cortex not visible on the surface include the cingulate cortex and insular cortex. The cingulate cortex lies dorsal to the corpus callosum and is important for regulation of emotion, pain perception, and cognition. The insular cortex, which lies buried within the overlying frontal, parietal, and temporal lobes, plays an important role in emotion, homeostasis, and taste perception. These internal views also afford examination of the *corpus callosum*, the prominent axon *fiber tract* that connects the two hemispheres.

The various brain regions described above are often divided into three broader regions: the *hindbrain* (comprising the medulla oblongata, pons, and cerebellum); *midbrain* (comprising the tectum, substantia nigra, reticular formation, and periaqueductal gray matter); and *forebrain* (comprising the diencephalon and cerebrum). Together the midbrain and hindbrain (minus the cerebellum) include the same structures as the brain stem. The anatomical organization of the nervous system is described in more detail in Chapter 4.

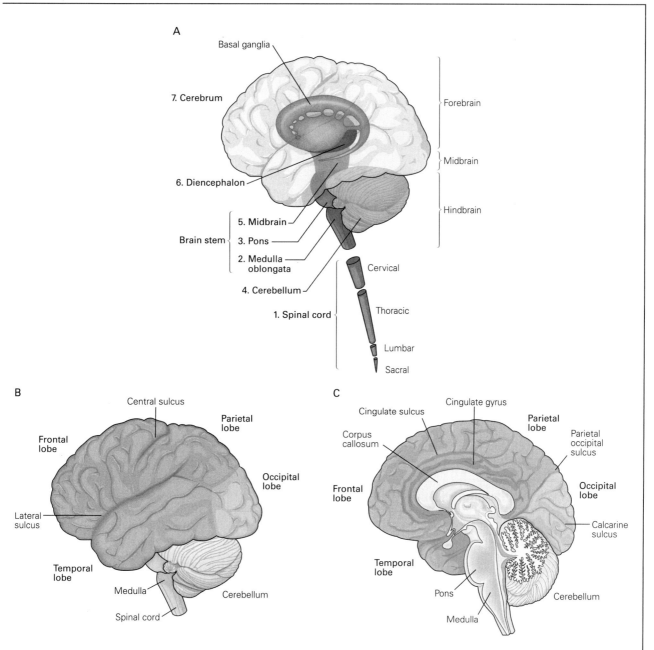

Figure 1–3 The divisions of the central nervous system.

A. The central nervous system can be divided into seven main regions, proceeding from the most caudal region, the spinal cord, to the brain stem (medulla, pons, and midbrain), to the diencephalon (containing the thalamus and hypothalamus), to the telencephalon or cerebrum (cerebral cortex, underlying white matter, subcortical nuclei, and the basal ganglia).

B. The four major lobes of the cerebrum are named for the parts of the cranium that cover them. This lateral view of the brain shows only the left cerebral hemisphere. The central

sulcus separates the frontal and parietal lobes. The lateral sulcus separates the frontal from the temporal lobes. The primary motor cortex occupies the gyrus immediately rostral to the central sulcus. The primary somatosensory cortex occupies the gyrus immediately caudal to the central sulcus.

C. Further divisions of the brain are visible when the hemispheres are separated in this medial view of the right hemisphere. The corpus collosum contains a large bundle of axons connecting the two hemispheres. The cingulate cortex is part of the cerebral cortex that surrounds the corpus collosum. The primary visual cortex occupies the calcarine sulcus.

(continued)

Box 1–2 Anatomical Organization of the Central Nervous System (continued)

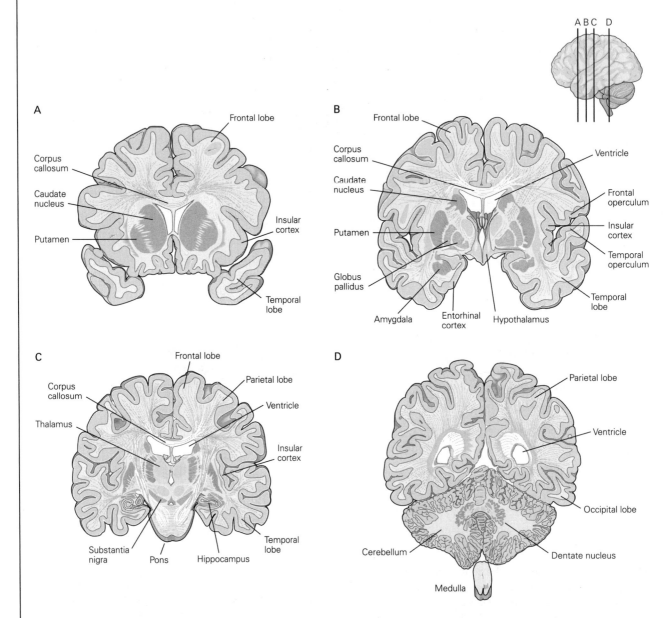

Figure 1–4 Major subcortical and deep cortical regions of the cerebral hemispheres are visible in drawings of brain slices from postmortem tissue.

Four sequential coronal sections (**A–D**) were made along the rostral-caudal axis indicated on the lateral view of the brain (top right, inset). The basal ganglia comprise the caudate nucleus, putamen, globus pallidus, substantia nigra, and subthalamic nucleus (not shown). The thalamus relays sensory information from the periphery to the cerebral cortex. The amygdala and hippocampus are regions of the cerebral cortex buried within the temporal lobe that are important for emotional responses and memory. The ventricles contain and produce the cerebrospinal fluid, which bathes the sulci, cisterns, and the spinal cord. (Adapted from Nieuwenhuys, Voogd, and van Huijzen 1988.)

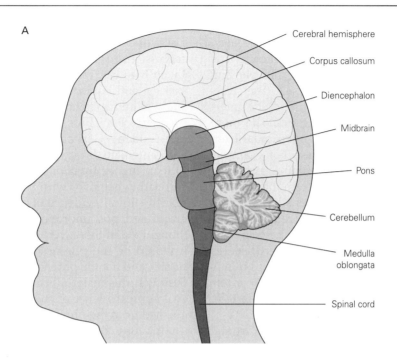

A

Cerebral hemisphere

Corpus callosum

Diencephalon

Midbrain

Pons

Cerebellum

Medulla oblongata

Spinal cord

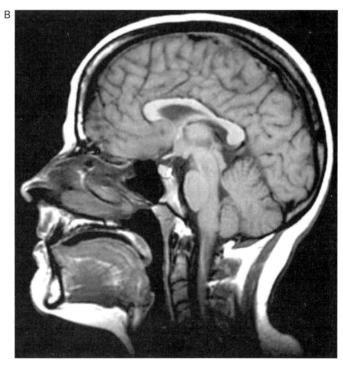

B

Figure 1–5 The main cortical and subcortical regions can be imaged in the brain of a living individual.

A. This schematic drawing shows, for reference, the major surface and deep regions of the brain, including the rostral end of the spinal cord.

B. The major brain divisions drawn in part **A** are evident in a magnetic resonance image of a living human brain.

In each of the hemispheres, the overlying cortex is divided into four lobes named for the skull bones that overlie them: *frontal, parietal, occipital*, and *temporal* (Figure 1–3B). Each lobe has several characteristic deep infoldings, an evolutionary strategy for packing a large sheet of cortex into a limited space. The crests of these convolutions are called *gyri*, and the intervening grooves are called *sulci* or *fissures*. The more prominent gyri and sulci, which are quite similar from person to person, bear specific names. For example, the central sulcus separates the precentral gyrus, an area concerned with motor function, from the postcentral gyrus, an area that deals with sensory function (Figure 1–3B). Several prominent gyri are only visible on the medial surface between the two hemispheres (Figure 1–3C), and others are deep within fissures and sulci and therefore only visible when the brain is sliced, either in postmortem tissue (Figure 1–4) or virtually, using magnetic resonance imaging (Figure 1–5), as explained in Chapter 6.

Each lobe has specialized functions. The frontal lobe is largely concerned with short-term memory, planning future actions, and control of movement; the parietal lobe mediates somatic sensation, forming a body image and relating it to extrapersonal space; the occipital lobe is concerned with vision; and the temporal lobe processes hearing, the recognition of objects and faces, and—through its deep structures, the hippocampus and amygdaloid nuclei—learning, memory, and emotion.

Two important features characterize the organization of the cerebral cortex. First, each hemisphere is concerned primarily with sensory and motor processes on the contralateral (opposite) side of the body. Thus sensory information that reaches the spinal cord from the left side of the body crosses to the right side of the nervous system on its way to the cerebral cortex. Similarly, the motor areas in the right hemisphere exert control over the movements of the left half of the body. The second feature is that the hemispheres, although similar in appearance, are not completely symmetrical in structure or function.

The First Strong Evidence for Localization of Cognitive Abilities Came From Studies of Language Disorders

The first areas of the cerebral cortex identified as important for cognition were areas concerned with language. These discoveries came from studies of *aphasia*, a language disorder that most often occurs when certain areas of brain tissue are destroyed by a stroke, the

occlusion or rupture of a blood vessel supplying a portion of a cerebral hemisphere. Many of the important discoveries in the study of aphasia occurred in rapid succession during the latter half of the 19th century. Taken together, these advances form one of the most exciting and important chapters in the neuroscientific study of human behavior.

Pierre Paul Broca, a French neurologist, was the first to identify specific areas of the brain concerned with language. Broca was influenced by Gall's efforts to map higher functions in the brain, but instead of correlating behavior with bumps on the skull, he correlated clinical evidence of aphasia with brain lesions discovered post mortem. In 1861 he wrote, "I had thought that if there were ever a phrenological science, it would be the phrenology of convolutions (*in the cortex*), and not the phrenology of bumps (*on the head*)." Based on this insight, Broca founded *neuropsychology*, an empirical science of mental processes that he distinguished from the phrenology of Gall.

In 1861 Broca described a patient, Leborgne, who as a result of a stroke could not speak, although he could understand language perfectly well. This patient had no motor deficits of the tongue, mouth, or vocal cords that would affect his ability to speak. In fact, he could utter isolated words, whistle, and sing a melody without difficulty. But he could not speak grammatically or create complete sentences, nor could he express ideas in writing. Postmortem examination of this patient's brain showed a lesion in a posterior inferior region of the left frontal lobe, now called *Broca's area* (Figure 1–6). Broca studied eight similar patients, all with lesions in this region, and in each case the lesion was located in the left cerebral hemisphere. This discovery led Broca to announce in 1864: "*Nous parlons avec l'hémisphère gauche!*" (We speak with the left hemisphere!).

Broca's work stimulated a search for cortical sites associated with other specific behaviors—a search soon rewarded. In 1870 Gustav Fritsch and Eduard Hitzig galvanized the scientific community when they showed that characteristic limb movements of dogs, such as extending a paw, could be produced by electrically stimulating discrete regions of the precentral gyrus. These regions were invariably located in the contralateral motor cortex. Thus the right hand, the one most used for writing and skilled movements, is controlled by the left hemisphere, the same hemisphere that controls speech. In most people, therefore, the left hemisphere is regarded as *dominant*.

The next step was taken in 1876 by Karl Wernicke, who at age 26 published a now-classic paper,

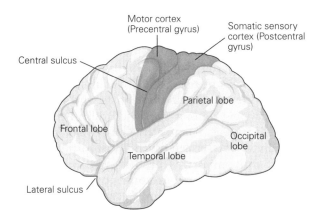

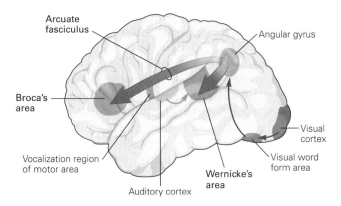

Figure 1–6 Language processing engages several regions of the left cerebral hemisphere.

Broca's area controls the production of speech. It lies near the region of the motor area that controls the mouth and tongue movements that form words. Wernicke's area processes auditory input for language and is important for understanding speech. It lies near the primary auditory cortex and the angular gyrus. The French neurologist Jules Dejerine proposed in the 1890s that a polymodal sensory area in the angular gyrus integrates information from vision and audition to represent words, but more recent studies implicate a more ventral occipitotemporal cortical area for processing of visual words. Wernicke's area communicates with Broca's area by a bidirectional pathway, part of which is made up of the arcuate fasciculus. (Adapted, with permission, from Geschwind 1979.)

"The Symptom-Complex of Aphasia: A Psychological Study on an Anatomical Basis." In it he described another type of aphasia, a failure of comprehension rather than speech: a *receptive* as opposed to an *expressive* malfunction. Whereas Broca's patients could understand language but not speak, Wernicke's patient could form words but could not understand language and produced senseless, yet grammatical, sentences. Moreover, the locus of this new type of aphasia was different from that described by Broca. The lesion occurred in the posterior part of the cerebral cortex where the temporal lobe meets the parietal lobe (Figure 1–6).

On the basis of this discovery, and the work of Broca, Fritsch, and Hitzig, Wernicke formulated a neural model of language that attempted to reconcile and extend the predominant theories of brain function at that time. Phrenologists and cellular connectionists argued that the cerebral cortex was a mosaic of functionally specific areas, whereas the holistic *aggregate-field* school claimed that every mental function involved the entire cerebral cortex. Wernicke proposed that only the most basic mental functions, those concerned with simple perceptual and motor activities, are mediated entirely by neurons in discrete local areas of the cortex. More complex cognitive functions, he argued, result from interconnections between several functional sites. By integrating the principle of localized function within a connectionist framework, Wernicke emphasized the idea that different components of a single behavior are likely to be processed in several regions of the brain. He was thus the first to advance the idea of *distributed processing*, now a central tenet of neural science.

Wernicke postulated that language involves separate motor and sensory programs, each governed by distinct regions of cortex. He proposed that the motor program that governs the mouth movements for speech is located in Broca's area, suitably situated in front of the region of the motor area that controls the mouth, tongue, palate, and vocal cords (Figure 1–6). He next assigned the sensory program that governs word perception to the temporal lobe area that he had discovered, now called *Wernicke's area*. This region is surrounded by the auditory cortex and by areas now known collectively as *association cortex*, integrating auditory, visual, and somatic sensations. According to Wernicke's model, the communication between these two language centers was mediated via a large bundle of axons known as the arcuate fasciculus.

Thus Wernicke formulated the first coherent neural model for language that is still useful today, with important modifications and elaborations described in Chapter 55. According to this model, the neural processing of spoken or written words begins in separate sensory areas of the cortex specialized for auditory or visual information. This information is then conveyed, via intermediate association areas that extract features suitable for recognition of spoken or written words, to Wernicke's area, where it is recognized as language and associated with meaning.

The power of Wernicke's model was not only its completeness but also its predictive utility. This model correctly predicted a third type of aphasia, one that results from disconnection. In this type, the receptive and expressive zones for speech are intact but the neuronal fibers that connect them (arcuate fasciculus) are destroyed. This *conduction aphasia*, as it is now called, is characterized by frequent, sound-based speech errors (*phonemic paraphasias*), repetition difficulties, and severe limitation in verbal working memory. Patients with conduction aphasia understand words that they hear and read, and have no motor difficulties when they speak. Yet they cannot speak coherently; they omit parts of words or substitute incorrect sounds and experience great difficulties in verbatim repetition of a multisyllabic word, phrase, or sentence that they hear or read or recall from memory. Although painfully aware of their own errors, their successive attempts at self-correction are often unsuccessful.

Inspired in part by Wernicke's advances and led by the anatomist Korbinian Brodmann, a new school of cortical localization arose in Germany at the beginning of the 20th century, one that distinguished functional areas of the cortex based on the shapes of cells and variations in their layered arrangement. Using this *cytoarchitectonic* method, Brodmann distinguished 52 anatomically and functionally distinct areas in the human cerebral cortex (Figure 1–7).

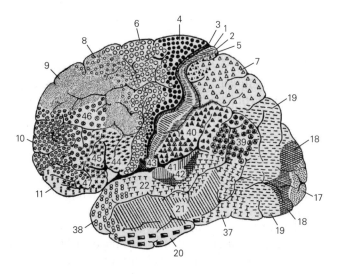

Figure 1–7 Early in the 20th century, the human cerebral cortex was classified into 52 discrete functional areas. The areas shown were identified by the anatomist Korbinian Brodmann on the basis of distinctive nerve cell structures and characteristic arrangements of cell layers. This scheme is still widely used today and is continually updated. Several areas defined by Brodmann have been found to control specific brain functions. For instance, area 4 is the motor cortex, responsible for voluntary movement. Areas 1, 2, and 3 constitute the primary somatosensory cortex, which receives sensory information primarily from the skin and joints. Area 17 is the primary visual cortex, which receives sensory signals from the eyes and relays them to other areas for further processing. Areas 41 and 42 constitute the primary auditory cortex. The drawing shows only areas visible on the outer surface of the cortex.

Even though the biological evidence for functionally discrete areas in the cortex was compelling, by the early 20th century, holistic views of the brain continued to dominate experimental thinking and clinical practice until 1950. This surprising state of affairs owed much to several prominent neural scientists who advocated the holistic view, among them the British neurologist Henry Head, the Russian behavioral physiologist Ivan Pavlov, and the American psychologist Karl Lashley.

Most influential was Lashley, who was deeply skeptical of the cytoarchitectonic approach to functional mapping of the cortex. "The 'ideal' architectonic map is nearly worthless," Lashley wrote. "The area subdivisions are in large part anatomically meaningless, and misleading as to the presumptive functional divisions of the cortex." His skepticism was reinforced by his studies of the effects of various brain lesions on the ability of rats to learn to run a maze. From these studies Lashley concluded that the severity of a learning defect depended on the size of the lesion, not on its precise location. Disillusioned, Lashley—and after him many other psychologists—concluded that learning and other higher mental functions have no special locus in the brain and consequently cannot be attributed to specific collections of neurons.

Based on his observations, Lashley reformulated the aggregate-field view by further minimizing the role of individual neurons, specific neuronal connections, and even specific brain regions in the production of specific behavior. According to Lashley's *theory of mass action*, it is the full mass of the brain, not its regional components, that is crucial for a function such as memory.

Lashley's experiments with rats have now been reinterpreted. A variety of studies have shown that the maze-learning used by Lashley is unsuited to the search for local cortical functions because it involves so many motor and sensory capabilities. Deprived of one sensory capability, say vision, a rat can still learn to run a maze using touch or smell. Besides, as we shall see later in the book, many mental functions are mediated by more than one region or neuronal pathway. Thus a given function may not be eliminated by a single lesion. This is especially germane when considering cognitive functions of the brain. For example, knowledge of space is supported by numerous parietal association areas that link vision to a potential shift of the gaze, turn of the head, reach of the hand, and so on. In principle, any one of these association areas can compensate for damage of another. It takes a large insult to the parietal lobe to produce obvious deficits of spatial knowledge (*spatial agnosia*) (Chapter 59). Such an observation would have seemed to support theories of mass action, but we now recognize that it is compatible with localization of function that incorporates the idea of redundancy of function.

Soon the evidence for localization of function became overwhelming. Beginning in the late 1930s, Edgar Adrian in England and Wade Marshall and Philip Bard in the United States discovered that touching different parts of a cat's body elicits electrical activity in distinct regions of the cerebral cortex. By systematically probing the body surface, they established a precise map of the body surface in specific areas of the cerebral cortex described by Brodmann. This result showed that functionally distinct areas of cortex could be defined unambiguously according to anatomical criteria such as cell type and cell layering, connections of cells, and—most importantly—behavioral function. As we shall see in later chapters, functional specialization is a key organizing principle in the cerebral cortex, extending even to individual columns of cells within an area of cortex. Indeed, the brain is divided into many more functional regions than Brodmann envisaged.

More refined methods made it possible to learn even more about the function of different brain regions involved in language. In the late 1950s Wilder Penfield, and later George Ojemann, reinvestigated the cortical areas that are essential for producing language. While locally anesthetized during brain surgery for epilepsy, awake patients were asked to name objects (or use language in other ways) while different areas of the exposed cortex were stimulated with small electrodes. If an area of the cortex was critical for language, application of the electrical stimulus blocked the patient's ability to name objects. In this way Penfield and Ojemann were able to confirm the language areas of the cortex described by Broca and Wernicke. As we shall learn in Chapter 55, the neural networks for language are far more extensive and complex than those described by Broca and Wernicke.

Initially almost everything known about the anatomical organization of language came from studies of patients with brain lesions. Today functional magnetic resonance imaging (fMRI) and other noninvasive methods allow analyses to be conducted on healthy people engaged in reading, speaking, and thinking (Chapter 6). fMRI not only has confirmed that reading and speaking activate different brain areas but also has revealed that just *thinking* about a word's meaning in the absence of sensory inputs activates a still different area in the left frontal cortex. Indeed, even within the traditional language areas, individual subregions are recruited to different degrees, depending on the way we think about words, express them, and resolve their meaning from the arrangement of other words (ie, syntax). The new imaging tools promise not only to teach us which areas are involved but also to expose the functional logic of their interconnection.

One of the great surprises emerging from modern methodologies is that so many areas of cortex are activated in language comprehension and production. These include the traditional language areas, identified by Broca, Wernicke, and Dejerine, in the left hemisphere; their homologs in the right hemisphere; and newly identified regions. Functional imaging tends to elucidate areas that are recruited differentially, whereas lesions from stroke, tumor, or injury distinguish brain areas that are essential for one or more functions. Thus it appears that Broca's area, once thought to be dedicated to language production, is in fact also involved in a variety of linguistic tasks including comprehension (Figure 1–6). In some cases, functional imaging invites refinement or revision of the critical areas identified by lesion studies. For example, reading is now thought to recruit specialized regions in the ventral occipitotemporal cortex in addition to the angular gyrus in the parietal cortex (shown in Figure 1–6).

Thus the processing of language in the brain exemplifies not only the principle of localized function but also the more sophisticated elaboration of this principle, that numerous distinct neural structures with specialized functions belong to systems. Perhaps this is the natural reconciliation of the controversy concerning localized and distributed processes—that is, a small number of distinct areas, each identified with a small set of functions and contributing through their interactions to the phenomenology of perception, action, and ideation. The brain may carve up a task differently than our intuition tells us. Who would have guessed that the neural analysis of the movement and color of an object would occur in different pathways rather than a single pathway mediating a unified percept of the object? Similarly, we might expect that the neural organization of language may not conform neatly to the axioms of a theory of universal grammar, yet support the very seamless functionality described by linguistic theory.

Studies of patients with brain damage continue to afford important insight into how the brain is organized for language. One of the most impressive results comes from the study of deaf people who have lost their ability to communicate through the use of a signed language (eg, British Sign Language [BSL] or American Sign Language [ASL]) after suffering cerebral damage. Signed languages use hand movements rather than vocalizations and are perceived by sight rather than sound, but have the same structural complexity as spoken languages. Sign language processing, as with spoken language processing, localizes to the left hemisphere. Damage to the left hemisphere can have quite specific consequences for signing just as for spoken language, affecting sign comprehension (following damage in Wernicke's area), grammar, or fluency (following damage in Broca's area). These clinical observations are supported by functional neuroimaging. Not surprisingly, production and comprehension of signed and spoken languages do not involve identical brain areas, but the overlap is truly remarkable (Figure 1–8). There is even evidence that processing the constituent parts of signs (eg, handshape used) involves some of the same brain regions involved when making rhyme judgements about speech.

These observations illustrate three points. First, language processing occurs primarily in the left hemisphere, independently of pathways that process the sensory and motor modalities used in language. Second, auditory input is not necessary for the emergence and operation of language capabilities in the left hemisphere. Third, spoken language is only one of a family of language skills mediated by the left hemisphere.

Investigations of other behaviors have provided additional support for the idea that the brain has distinct cognitive systems. These studies demonstrate that complex information processing requires many interconnected cortical and subcortical areas, each concerned with processing particular aspects of sensory stimuli or motor movement and not others. For example, perceptual awareness of an object's location, size, and shape relies on activity in numerous parietal association areas that link vision to potential actions, such as moving the eyes, orienting the head, reaching, and shaping the hand to grasp. The parietal areas do not initiate these actions but evaluate sensory information as evidence bearing on these potentialities. They receive information from the dorsal visual stream—sometimes referred to as the *where pathway*, but more aptly termed a *how pathway*—to construct a state of knowing (gnosia) about the location and other spatial properties of objects. The ventral visual stream, or *what pathway*, is also concerned with possible actions, but these are associated with socializing and foraging. These associations establish *gnosia* about the desirability of objects, faces, foods, and potential mates. In this sense, the *what pathway* might be a *how pathway* too.

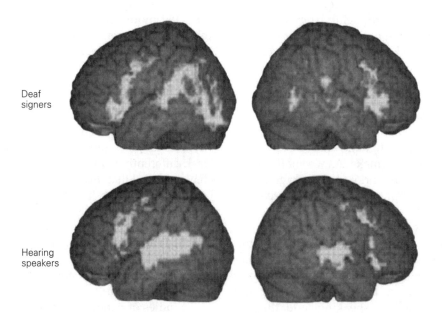

Figure 1–8 Deaf signing and hearing individuals share common language processing areas. Regions of the cortex involved in the recognition of a spoken or signed language, identified by functional magnetic resonance imaging (fMRI). **Yellow** highlight shows the areas of the left and right cerebral hemispheres (*left* and *right columns*, respectively) that were activated more when comprehending language than when performing a perceptual task. For the deaf signers (**top row**), the highlighted regions were more active during comprehension of British Sign Language than during the detection of a visual stimulus superimposed on the same motionless signer. For the hearing speakers (**bottom row**), highlighted regions were more active during comprehension of audio-visual speech than during the detection of a tone while viewing a motionless (silent) speaker. (Adapted, with permission, from MacSweeney et al., 2002. Copyright © 2002 Oxford University Press.)

Mental Processes Are the Product of Interactions Between Elementary Processing Units in the Brain

There are several reasons why the evidence for the localization of brain functions, which seems so obvious and compelling in retrospect, had been rejected so often in the past. Phrenologists introduced the idea of localization in an exaggerated form and without adequate evidence. They imagined each region of the cerebral cortex as an independent mental organ dedicated to a complete and distinct aspect of personality, much as the pancreas and the liver are independent digestive organs. Flourens's rejection of phrenology and the ensuing debate between proponents of the aggregate-field view (against localization) and the cellular connectionists (for localization) were responses to a theory that was simplistic and without adequate experimental evidence.

In the aftermath of Wernicke's discovery of the modular organization of language in the brain—interconnected nodes with distinctive functions—we now think that all cognitive abilities result from the interaction of many processing mechanisms distributed in several regions of the brain. That is, particular brain regions are not fully responsible for specific mental faculties but instead are *elementary processing units* that together have a role. Perception, movement, language, thought, and memory are all made possible by the interlinkage of serial and parallel processing in discrete brain regions—computational modules—within these regions. As a result, damage to a single area need not result in the complete loss of a cognitive function (or faculty) as many earlier neurologists believed. Even if a behavior initially disappears, it may partially return as undamaged parts of the brain reorganize their linkages. Further, when focal damage adversely affects a mental function it may do so indirectly by disrupting the function of other principal loci (*diaschisis*). Indeed, observations of this nature led Wernicke's student Kurt Goldstein to embrace the more holistic view.

Thus it is not accurate to think of a mental function as being mediated strictly by a chain of nerve cells and brain areas—each connected directly to the next—for in such an arrangement the entire process is disrupted when a single connection is damaged. A more realistic metaphor is that of a process consisting of several parallel pathways in a network of modules that interact and ultimately converge upon a common set of targets. The malfunction of a single pathway within a network may affect the information carried by that pathway without disrupting the entire system. The remaining parts of the network may be able to modify their performance to accommodate the breakdown of one pathway.

Modular processing in the brain was slow to be accepted because, until recently, it was difficult to demonstrate which components of a mental operation were mediated by a particular pathway or brain region. Nor is it easy to define mental operations in a manner that leads to testable hypotheses. Nevertheless, with the evolving convergence of modern cognitive psychology and brain science in recent decades, we have begun to appreciate that mental functions can successfully be broken down into subfunctions.

To illustrate this point, consider how we learn, store, and recall information about objects, people, and events. Simple introspection suggests that we store each piece of our knowledge as a single representation that can be recalled by memory-jogging stimuli or even by the imagination alone. Everything you know about an apple, for example, seems to be stored in one complete representation that is equally accessible whether you see a particular apple, a part of an apple, a red or green apple, the written word apple, or an apocryphal story about the discovery of gravity. Our experience, however, is not a faithful guide to how knowledge is stored in memory.

Knowledge about apples is not stored as a single coherent representation but rather is subdivided into distinct categories and stored separately. One region of the brain stores information about the way you would hold the apple, the way you would feel for softness (bearing on freshness), the color (bearing on preference or freshness), the way you might communicate the presence or taste of the apple to another person, as well as its semantic association with computers, physicists, worms, serpents, and biblical gardens. The concept "apple" entails each of these considerations and many more. A natural assumption is that a coherent concept comprising many details must exist in a single place in the brain; however, an equally valid assumption is that a unified concept like "apple" exists in the mind in the form of multiple links between a variety of neural structures, each with a particular kind of information, coordinated through the action of memory retrieval.

The most astonishing example of the modular organization of mental processes is the finding that our very sense of self—a self-aware being, the sum of what we mean when we say "I"—is achieved through the connection of independent circuits in our two cerebral hemispheres, each mediating its own sense of awareness. The remarkable discovery that even consciousness is not a unitary process was made by Roger Sperry, Michael Gazzaniga, and Joseph Bogen in the course of studying patients in whom the corpus callosum—the major tract connecting the two cerebral hemispheres—was severed as a treatment for epilepsy. They found that each hemisphere had a consciousness that functioned independently of the other.

Thus while one patient was reading a favorite book held in his left hand, the right hemisphere, which controls the left hand but plays only a minor role in language comprehension, found that the raw visual information it received from simply looking at the book was boring. The right hemisphere commanded the left hand to put the book down. Another patient would put on his clothes with the left hand while at the same time taking them off with the other. Each hemisphere has a mind of its own! In addition, the dominant hemisphere sometimes commented on the performance of the nondominant hemisphere, frequently manifesting a false sense of confidence regarding problems to which it could not know the solution, which was provided exclusively to the nondominant hemisphere.

Such findings have brought the study of consciousness, once the domain of philosophy and psychoanalysis, into the fold of neural science. As we shall see in later chapters, many of the issues described in this chapter reemerge in neural theories of consciousness. No one questions the idea that much information processing—perhaps the lion's share—does not reach conscious awareness. When sensory information, a plan of action, or an idea does become conscious, neural science seeks to explain the mechanisms that mediate this transition. While there is as yet no satisfactory explanation, some brain scientists would liken the process to a shift in the focus of attention, mediated by distinct groups of neurons, whereas others believe that awareness requires a qualitative change in the functional interaction between widely separated areas of the brain.

The main reason it has taken so long to understand which mental activities are mediated by which regions of the brain is that we are dealing with biology's deepest riddle: the neural mechanisms that account for consciousness and self-awareness. There is at present no satisfactory theory that explains why only some information that reaches our eyes leads to a state of subjective awareness of an item, person, or scene. We know that we are consciously aware of only a small fraction of our mental deliberations, and those thoughts that do pierce conscious awareness must arise from steps carried out by the brain unconsciously. As we propose in Chapter 56, some answers to the riddles of consciousness may be closer than imagined.

Meanwhile, the current gap in our understanding also poses practical, epistemological challenges for neural science. We cannot help but rely on our conscious experiences of the world, body, and ideation in our characterization of perception, behavior, and cognition. In doing so, however, we risk mischaracterizing many mental processes that do not pierce conscious awareness. For example, we tend to characterize the problem of perception in terms consistent with the subjective experience of sensory information, whereas even sophisticated but nonconscious knowledge of the content of perception may have greater resemblance to a behavioral utility (affordance), in effect an answer to whether this is something I might choose to eat, sit upon, or engage further. Similarly, cognitive processes, such as reasoning, strategizing, and decision making, are likely to be carried out by the brain in ways that only loosely resemble the steps we infer from conscious deliberation.

These cautionary notes have a bright corollary. The insight that many cognitive functions transpire without conscious awareness raises the possibility that principles of neural science revealed in the study of more rudimentary behaviors can furnish insight into more complex cognitive processes. Neural recordings from the brains of animals trained to perform complex tasks have led to an understanding of cognitive processes such as decision making, reasoning, planning, and allocating attention. These experimental models often extrapolate to human functions, and where they fall short, they inspire new hypotheses. For more often than not, there is inspiration if not insight to be gleaned from the gaps in our understanding.

To analyze how the brain gives rise to a specific mental process, we must determine not only which aspects of the process depend on which regions of the brain but also how the relevant information is represented, routed, and transformed. Modern neural science seeks to integrate such understanding across many scales. For example, studies at the level of both the single nerve cell and its molecular constituents elucidate the mechanisms underlying electrical excitability and synaptic connections. Studies of cells and simple circuits lend insights into neural computations, ranging from basic operations, like controlling net excitation, to more masterful feats of computation, such as the derivation of meaningful information from raw sensory data. Studies of the interactions between circuits and brain areas can explain how we coordinate widely separated muscle groups or express a belief in a proposition. Knowledge at all these levels is knit together by mathematical formalizations, computer simulation, and psychological theory. These conceptual tools can now be combined with modern physiological techniques and brain imaging methods, making it possible to track mental processes as they evolve in real time in living animals and humans. Indeed, the excitement evident in neural science today stems from the conviction that the biological principles that underlie human thought and behavior are within our grasp and may soon be harnessed to elucidate and improve the human condition.

Highlights

1. The neural sciences seek to understand the brain at multiple levels of organization, ranging from the cell and its constituents to the operations of the mind.
2. The fundamental principles of neural science bridge levels of time, complexity, and state—from cell to action and ideation, from development through learning to expertise and forgetting, from normal function to neurological deficits and recovery. As a first step, one must understand the building blocks—the electrical properties of the nerve cell and its connections to other nerve cells—and the organization of the nervous system from supporting cells to pathways.
3. The neuron doctrine states that individual nerve cells (neurons) are the elementary building blocks and signaling elements of the nervous system.
4. Neurons are organized into circuits with specialized functions. The simplest circuits mediate reflexes; more complex cognitive functions require more sophisticated circuits. This organizational principle extends the neuron doctrine to cellular connectionism.
5. Even within complex circuits, critical nodes can be identified as areas associated with a specific function. The first clear evidence for localization of brain function came from the study of a specific impairment of language production.
6. The two cerebral hemispheres receive information from the opposite side of the body and control the actions of the opposite side.
7. While the principle of localization of function in the brain is superior to its main historical alternatives—aggregate-field and the theory of mass action—it is constantly being refined. No area of the cerebral cortex functions independently of other cortical and subcortical structures.
8. A major refinement of localization is the principle of modular functional organization. The brain contains many representations of information organized by both the relevance of certain features for particular computations and by the variety of uses

to which such information may be put. This is a form of redundancy with respect to purpose or potential action.
9. The future of brain science will require integration of ideas that cross the boundaries of traditional disciplines. We must open our minds to a wide variety of sources to guide our intuitions and strategies for research, from the sublime—the nature of consciousness—to the seemingly mundane—what general anesthesia does to a calcium sensor in the ring of cells around the thalamus.

Eric R. Kandel
Michael N. Shadlen

Selected Reading

Churchland PS. 1986. *Neurophilosophy: Toward a Unified Science of the Mind-Brain.* Cambridge, MA: MIT Press.

Cooter R. 1984. *The Cultural Meaning of Popular Science: Phrenology and the Organization of Consent in Nineteenth-Century Britain.* Cambridge: Cambridge Univ. Press.

Cowan WM. 1981. Keynote. In: FO Schmitt, FG Worden, G Adelman, SG Dennis (eds). *The Organization of the Cerebral Cortex: Proceedings of a Neurosciences Research Program Colloquium*, pp. xi–xxi. Cambridge, MA: MIT Press.

Crick F, Koch C. 2003. A framework for consciousness. Nat Neurosci 6:119–126.

Dehaene S. 2009. *Reading in the Brain: The Science and Evolution of a Human Invention.* New York: Viking.

Ferrier D. 1890. *The Croonian Lectures on Cerebral Localisation.* London: Smith, Elder.

Geschwind N. 1974. *Selected Papers on Language and the Brain.* Dordrecht, Holland: Reidel.

Glickstein M. 2014. *Neuroscience. A Historical Introduction.* Cambridge, MA: MIT Press.

Gregory RL (ed). 1987. *The Oxford Companion to the Mind.* Oxford: Oxford Univ. Press.

Harrington A. 1987. *Medicine, Mind, and the Double Brain: A Study in Nineteenth-Century Thought.* Princeton, NJ: Princeton Univ. Press.

Harrison RG. 1935. On the origin and development of the nervous system studied by the methods of experimental embryology. Proc R Soc Lond B Biol Sci 118: 155–196.

Hickok G, Small S. 2015. *Neurobiology of Language.* Boston: Elsevier.

Jackson JH. 1884. The Croonian lectures on evolution and dissolution of the nervous system. Br Med J 1:591–593; 660–663; 703–707.

Kandel ER. 1976. The study of behavior: the interface between psychology and biology. In: *Cellular Basis of Behavior: An Introduction to Behavioral Neurobiology,* pp. 3–27. San Francisco: Freeman.

Ojemann GA. 1995. Investigating language during awake neurosurgery. In: RD Broadwell (ed). *Neuroscience, Memory, and Language.* Vol. 1, *Decade of the Brain,* pp. 117–131. Washington, DC: Library of Congress.

Petersen SE. 1995. Functional neuroimaging in brain areas involved in language. In: RD Broadwell (ed). *Neuroscience, Memory, and Language.* Vol. 1, *Decade of the Brain,* pp. 109–116. Washington DC: Library of Congress.

Shepherd GM. 1991. *Foundations of the Neuron Doctrine.* New York: Oxford Univ. Press.

Sperry RW. 1968. Mental unity following surgical disconnection of the cerebral hemispheres. Harvey Lect 62:293–323.

Young RM. 1990. *Mind, Brain and Adaptation in the Nineteenth Century.* New York: Oxford Univ. Press.

References

Adrian ED. 1941. Afferent discharges to the cerebral cortex from peripheral sense organs. J Physiol (Lond) 100:159–191.

Bernard C. 1878–1879. *Leçons sur les Phénomènes de la vie Communs aux Animaux et aux Végétaux.* Vols. 1, 2. Paris: Baillière.

Boakes R. 1984. *From Darwin to Behaviourism: Psychology and the Minds of Animals.* Cambridge: Cambridge Univ. Press.

Broca P. 1865. Sur le siége de la faculté du langage articulé. Bull Soc Anthropol 6:377–393.

Brodmann K. 1909. *Vergleichende Lokalisationslehre der Grosshirnrinde in ihren Prinzipien dargestellt auf Grund des Zeelenbaues.* Leipzig: Barth.

Darwin C. 1872. *The Expression of the Emotions in Man and Animals.* London: Murray.

Dejerine J. 1891. Sur un cas de cécité verbale avec agraphie suivi d'autopsie. Mémoires de la Société de Biologie 3:197–201.

Descartes R. [1649] 1984. *The Philosophical Writings of Descartes.* Cambridge: Cambridge Univ. Press.

Finger S., Koehler PJ, Jagella C. 2004. The Monakow concept of diaschisis: origins and perspectives. Arch Neurol 61:283–288.

Flourens P. [1824] 1953. Experimental research. P Flourens and JMD Olmsted. In: EA Underwood (ed). *Science, Medicine and History,* 2:290–302. London: Oxford Univ. Press.

Flourens P. 1824. *Recherches Expérimentales sur les Propriétés et les Fonctions du Système Nerveux, dans les Animaux Vertébrés.* Paris: Chez Crevot.

Fritsch G, Hitzig E. [1870] 1960. Electric excitability of the cerebrum. In: G. von Bonin (transl). *Some Papers on the Cerebral Cortex,* pp. 73–96. Springfield, IL: Thomas.

Gall FJ, Spurzheim G. 1810. *Anatomie et Physiologie du Système Nerveux en Général, et du Cerveau en Particulier, avec des Observations sur la Possibilité de Reconnoître Plusieurs Dispositions Intellectuelles et Morales de l'Homme et des Animaux, par la Configuration de leurs Têtes.* Paris: Schoell.

Galvani L. [1791] 1953. *Commentary on the Effect of Electricity on Muscular Motion.* RM Green (transl). Cambridge, MA: Licht.

Gazzaniga MS, LeDoux JE. 1978. *The Integrated Mind.* New York: Plenum.

Geschwind N. 1979. Specializations of the human brain. Sci Am 241:180–199.

Goldstein K. 1948. *Language and Language Disturbances: Aphasic Symptom Complexes and Their Significance for Medicine and Theory of Language.* New York: Grune & Stratton.

Golgi C. [1906] 1967. The neuron doctrine: theory and facts. In: *Nobel Lectures: Physiology or Medicine, 1901–1921,* pp. 189–217. Amsterdam: Elsevier.

Langley JN. 1906. On nerve endings and on special excitable substances in cells. Proc R Soc Lond B Biol Sci 78:170–194.

Lashley KS. 1929. *Brain Mechanisms and Intelligence: A Quantitative Study of Injuries to the Brain.* Chicago: Univ. Chicago Press.

Lashley KS, Clark G. 1946. The cytoarchitecture of the cerebral cortex of *Ateles:* a critical examination of architectonic studies. J Comp Neurol 85:223–305.

Loeb J. 1918. *Forced Movements, Tropisms and Animal Conduct.* Philadelphia: Lippincott.

MacSweeney M, Capek CM, Campbell R, Woll B. 2008. The signing brain: the neurobiology of sign language. Trends Cogn Sci 12:432–440.

MacSweeney M, Woll B, Campbell R, et al. 2002. Neural systems underlying British Sign Language and audio-visual English processing in native users. Brain 125:1583–1593.

Marshall WH, Woolsey CN, Bard P. 1941. Observations on cortical somatic sensory mechanisms of cat and monkey. J Neurophysiol 4:1–24.

Martin JH. 2003. *Neuroanatomy: Text and Atlas,* 3rd ed. New York: McGraw Hill.

McCarthy RA, Warrington EK. 1988. Evidence for modality-specific meaning systems in the brain. Nature 334:428–430.

Müller J. 1834–1840. *Handbuch der Physiologie des Menschen für Vorlesungen.* Vols. 1, 2. Coblenz: Hölscher.

Nieuwenhuys R, Voogd J, van Huijzen Chr. 1988. *The Human Central Nervous System: A Synopsis and Atlas,* 3rd rev. ed. Berlin: Springer.

Pavlov IP. 1927. *Conditioned Reflexes: An Investigation of the Physiological Activity of the Cerebral Cortex.* GV Anrep (transl). London: Oxford Univ. Press.

Penfield W. 1954. Mechanisms of voluntary movement. Brain 77:1–17.

Penfield W, Rasmussen T. 1950. *The Cerebral Cortex of Man: A Clinical Study of Localization of Function.* New York: Macmillan.

Penfield W, Roberts L. 1959. *Speech and Brain-Mechanisms.* Princeton, NJ: Princeton Univ. Press.

Ramón y Cajal S. [1892] 1977. A new concept of the histology of the central nervous system. DA Rottenberg (transl). In: DA Rottenberg, FH Hochberg (eds). *Neurological Classics in Modern Translation,* pp. 7–29. New York: Hafner. (See also historical essay by SL Palay, preceding Ramón y Cajal's paper.)

Ramón y Cajal S. [1906] 1967. The structure and connexions of neurons. In: *Nobel Lectures: Physiology or Medicine, 1901–1921*, pp. 220–253. Amsterdam: Elsevier.

Ramón y Cajal S. [1908] 1954. *Neuron Theory or Reticular Theory? Objective Evidence of the Anatomical Unity of Nerve Cells.* MU Purkiss, CA Fox (transl). Madrid: Consejo Superior de Investigaciones Científicas Instituto Ramón y Cajal.

Ramón y Cajal S. 1937. *1852–1934. Recollections of My Life.* EH Craigie (transl). Philadelphia: American Philosophical Society; reprinted 1989. Cambridge, MA: MIT Press.

Rose JE, Woolsey CN. 1948. Structure and relations of limbic cortex and anterior thalamic nuclei in rabbit and cat. J Comp Neurol 89:279–347.

Shadlen MN, Kiani R, Hanks TD, Churchland AK. 2008. Neurobiology of decision making: an intentional framework. In: C Engel, W Singer (eds.). *Better Than Conscious? Decision Making, the Human Mind, and Implications for Institutions*, pp. 71–102. Cambridge, MA: MIT Press.

Sherrington C. 1947. *The Integrative Action of the Nervous System*, 2nd ed. Cambridge: Cambridge Univ. Press.

Spurzheim JG. 1825. *Phrenology, or the Doctrine of the Mind*, 3rd ed. London: Knight.

von Helmholtz H. [1850] 1948. On the rate of transmission of the nerve impulse. In: W Dennis (ed). *Readings in the History of Psychology*, pp. 197–198. New York: Appleton-Century-Crofts.

Wandell BA, Rauschecker AM, Yeatman JD. 2012. Learning to see words. Annu Rev Psychol 63:31–53.

Wernicke C. 1908. The symptom-complex of aphasia. In: A Church (ed). *Diseases of the Nervous System*, pp. 265–324. New York: Appleton.

Yeatman JD, Rauschecker AM, Wandell BA. 2013. Anatomy of the visual word form area: adjacent cortical circuits and long-range white matter connections. Brain Lang 125:146–155.

2

Genes and Behavior

ALL BEHAVIORS ARE SHAPED BY THE interplay of genes and the environment. The most stereotypic behaviors of simple animals are influenced by the environment, while the highly evolved behaviors of humans are constrained by innate properties specified by genes. Genes do not control behavior directly, but the RNAs and proteins encoded by genes act at different times and at many levels to affect the brain. Genes specify the developmental programs that assemble the brain and are essential to the properties of neurons, glia, and synapses that allow neuronal circuits to function. Genes that are stably inherited over generations create the machinery by which new experiences can change the brain during learning.

In this chapter, we ask how genes contribute to behavior. We begin with an overview of the evidence that genes do influence behavior, and then review basic principles of molecular biology and genetic transmission. We then provide examples of the way that genetic influences on behavior have been documented. A deep understanding of the ways that genes regulate behavior has emerged from studies of worms, flies, and mice, animals whose genomes are accessible to experimental manipulation. Many persuasive links between genes and human behavior have emerged from the analysis of human brain development and function. Despite the formidable challenges inherent in studying complex traits in humans, recent progress has begun to reveal the genetic risk factors in neurodevelopmental and psychiatric syndromes such as autism, schizophrenia, and bipolar disorder, offering another important avenue to clarify the relationship between genes, brain, and behavior.

An Understanding of Molecular Genetics and Heritability Is Essential to the Study of Human Behavior

Many human psychiatric disorders and neurological diseases have a genetic component. The relatives of a patient are more likely than the general population to have the disease. The extent to which genetic factors account for traits in a population is called *heritability*. The strongest case for heritability is based on twin studies, first used by Francis Galton in 1883. Identical twins develop from a single fertilized egg that splits into two soon after fertilization; such monozygotic twins share all genes. In contrast, fraternal twins develop from two different fertilized eggs; these dizygotic twins, like normal siblings, share on average half their genetic information. Systematic comparisons over many years have shown that identical twins tend to be more similar (concordant) for neurological and psychiatric traits than fraternal twins, providing evidence of a heritable component of these traits (Figure 2–1A).

In a variation of the twin study model, the Minnesota Twin Study examined identical twins that were separated early in life and raised in different households. Despite sometimes great differences in their environment, twins shared predispositions for the same psychiatric disorders and even tended to share personality traits, like extraversion. This study provides considerable evidence that genetic variation contributes to normal human differences, not just to disease states.

Heritability for human diseases and behavioral traits is usually substantially less than 100%, demonstrating that the environment is an important factor in acquiring diseases or traits. Estimates of heritability for many neurological, psychiatric, and behavioral traits from twin studies are around 50%, but heritability can be higher or lower for particular traits (Figure 2–1B). Although studies of identical twins and other kinships provide support for the idea that human behavior has a hereditary component, they do not tell us which genes are important, let alone how specific genes influence behavior. These questions are addressed by studies in experimental animals in which genetic and environmental factors are strictly controlled and by modern methods of gene discovery that are now leading to the systematic, reliable identification of specific variations in DNA sequence and structure that contribute to human psychiatric and neurological phenotypes.

The Understanding of the Structure and Function of the Genome Is Evolving

The related fields of molecular biology and transmission genetics are central to our modern understanding of genes. Here we summarize some key ideas in these fields; a glossary at the end of the chapter defines commonly used terms.

Genes are made of DNA, and it is DNA that is passed on from one generation to the next. In most circumstances, exact copies of each gene are provided to all cells in an organism as well as to succeeding generations through DNA replication. The rare exceptions to this general rule—new (de novo) mutations that are introduced into the DNA of either germline or somatic cells and that play an important role in disease risk—are discussed later. DNA is made of two strands, each of which has a deoxyribose-phosphate backbone attached to a series of four subunits: the nucleotides adenine (A), guanine (G), thymine (T), and cytosine (C). The two strands are paired so that an A on one strand is always paired with a T on the complementary strand, and a G with a C (Figure 2–2). This complementarity ensures accurate copying of DNA during DNA replication and drives *transcription* of DNA into lengths of RNA called transcripts. Given that nearly all of the genome is double-stranded, bases or base pairs are used interchangeably as a unit of measurement. A segment of the genome encompassing a thousand base pairs is referred to as 1 kilobase (1 kb) or 1 kilobase pair (1 kbp), whereas a million base pairs are referred to as 1 megabase (1 Mb) or 1 megabase pairs (1 Mbp). RNA differs from DNA in that it is single-stranded, has a ribose rather than a deoxyribose backbone, and uses the nucleoside base uridine (U) in the place of thymine.

In the human genome, approximately 20,000 genes encode protein products, which are generated by *translation* of the linear messenger RNA (mRNA) sequence into a linear polypeptide (protein) sequence composed of amino acids. A typical protein-coding gene consists of a coding region, which is translated into the protein, and noncoding regions (Figure 2–3). The coding region is usually arranged in small coding segments called *exons*, which are separated by noncoding stretches called *introns*. The introns are deleted from the mRNA before its translation into protein.

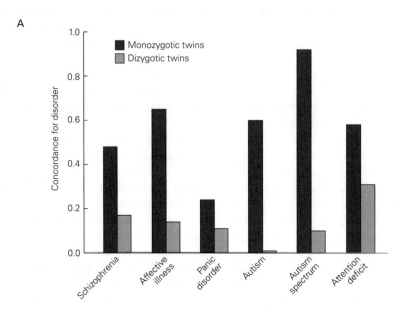

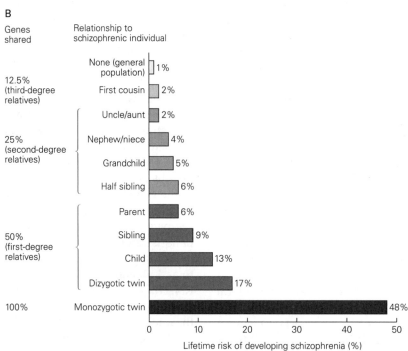

Figure 2–1 Familial risk of psychiatric disorders provides evidence of heritability.

A. Correlations between monozygotic twins for psychiatric disorders are considerably greater than those between dizygotic twins. Monozygotic twins share nearly all genes and have a high (but not 100%) risk of sharing the disease state. Dizygotic twins share 50% of their genetic material. A score of zero represents no correlation (the average result for two random people), whereas a score of 1.0 represents a perfect correlation. (Adapted from McGue and Bouchard 1998.)

B. The risk of developing schizophrenia is greater in close relatives of a schizophrenic patient. Like dizygotic twins, parents and children, as well as brothers and sisters, share 50% of their genetic material. If only a single gene accounted for schizophrenia, the risk should be the same for parents, siblings, children, and dizygotic twins of patients. The variation between family members shows that more complex genetic and environmental factors are in play. (Adapted, with permission, from Gottesman II. 1991.)

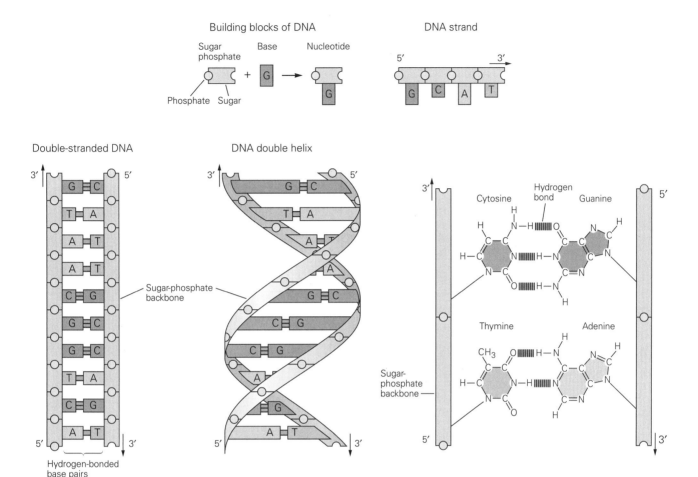

Figure 2–2 Structure of DNA. Four different nucleotide bases, adenine (**A**), thymine (**T**), cytosine (**C**), and guanine (**G**), are assembled on a sugar phosphate backbone in the double-stranded DNA helix. (Adapted from Alberts et al. 2002.)

Many functional RNA transcripts do not encode proteins. In fact, in the human genome, over 40,000 noncoding transcripts have been characterized as compared with approximately 20,000 protein-coding genes. Such genes include ribosomal RNAs (rRNAs) and transfer RNAs (tRNAs), essential components of the machinery for mRNA translation. Additional noncoding RNAs (ncRNA) include *long noncoding RNAs (lncRNAs)*, arbitrarily defined as longer than 200 bp in length, that do not encode proteins but can have roles in gene regulation; *small noncoding RNAs* of several types, including small nuclear RNAs (snRNAs), that guide mRNA splicing; and microRNAs (miRNAs) that pair with complementary sequences in specific mRNAs to inhibit their translation.

Each cell in the body contains the DNA for every gene but only expresses a specific subset of the genes as RNAs. The part of the gene that is transcribed into RNA is flanked by noncoding DNA regions that may be bound by other proteins, including *transcription factors*, to regulate gene expression. These sequence motifs include *promoters*, *enhancers*, *silencers*, and *insulators*, which together allow accurate expression of the RNA in the right cells at the right time. Promoters are typically found close to the beginning of the region to be transcribed; enhancers, silencers, and insulators may reside at a distance from the gene being regulated. Each type of cell has a unique complement of DNA-binding proteins that interact with promoters and other regulatory sequences to regulate gene expression and the resulting cellular properties.

The brain expresses a greater number of genes than any other organ in the body, and within the brain, diverse populations of neurons express different groups of genes. The selective gene expression controlled by promoters, other regulatory sequences, and the DNA-binding proteins that interact with them

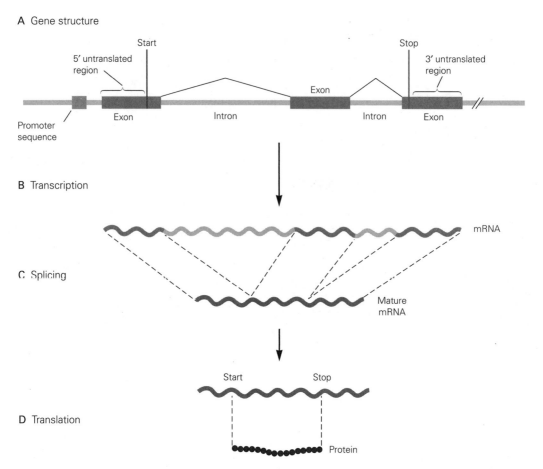

A Gene structure

5' untranslated region

Start

Stop

3' untranslated region

Promoter sequence

Exon

Intron

Exon

Intron

Exon

B Transcription

mRNA

C Splicing

Mature mRNA

D Translation

Start

Stop

Protein

Figure 2–3 Gene structure and expression.

A. A gene consists of coding regions (exons) separated by noncoding regions (introns). Its transcription is regulated by noncoding regions such as promoters and enhancers that are frequently found near the beginning of the gene.

B. Transcription leads to production of a primary single-stranded RNA transcript that includes both exons and introns.

C. Splicing removes introns from the immature transcript and ligates the exons into a mature messenger RNA (**mRNA**), which is exported from the nucleus of the cell.

D. Translation of the mature mRNA produces a protein product.

permits a fixed number of genes to generate a vastly larger number of neuronal cell types and connections in the brain.

Although genes specify the initial development and properties of the nervous system, the experience of an individual and the resulting activity in specific neural circuits can itself alter the expression of genes. In this way, environmental influences are incorporated into the structure and function of neural circuits. Some of the principal goals of genetic studies are to unravel the ways that individual genes affect biological processes, the ways that networks of genes influence each other's activity, and the ways that genes interact with the environment.

Genes Are Arranged on Chromosomes

The genes in a cell are arranged in an orderly fashion on long, linear stretches of DNA called *chromosomes*. Each gene in the human genome is reproducibly located at a characteristic position (locus) on a specific chromosome, and this genetic "address" can be used to associate biological traits with a gene's effects. Most multicellular animals (including worms, fruit flies, and mice, as well as humans) are *diploid*; every somatic cell carries two complete sets of chromosomes, one from the mother and the other from the father.

Humans have about 20,000 genes but only 46 chromosomes: 22 pairs of autosomes (chromosomes

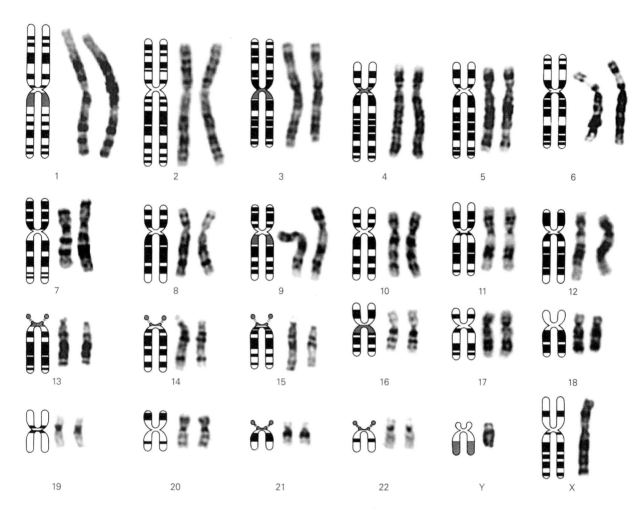

Figure 2–4 A map of normal human chromosomes at metaphase illustrates the distinctive morphology of each chromosome. Characteristic sizes and characteristic light and dark regions allow chromosomes to be distinguished from one another. (Adapted, with permission, from Watson, Tooze, and Kurtz 1983.)

that are present in both males and females) and two sex chromosomes (two X chromosomes in females, one X and one Y chromosome in males) (Figure 2–4). Each parent supplies one copy of each autosome to the diploid offspring. Each parent also supplies one X chromosome to female (XX) offspring, but XY males inherit their single X chromosome from their mothers and their single Y chromosome from their fathers. Sex-linked inheritance was discovered in fruit flies by Thomas Hunt Morgan in 1910. This pattern of sex-linked inheritance associated with the single X chromosome has been highly significant in human genetic studies, where certain X-linked genetic diseases are commonly observed only in males but are genetically transmitted from mothers to their sons.

In addition to the genes on chromosomes, a very small number of an organism's genes are transmitted through *mitochondria*, cytoplasmic organelles that carry out metabolic processes. Mitochondria in all children come from the ovum and therefore are transmitted from the mother to the child. Certain human disorders, including some neuromuscular degenerative diseases, some forms of intellectual disability, and some forms of deafness, are caused by mutations in the mitochondrial DNA.

The Relationship Between Genotype and Phenotype Is Often Complex

The two copies of a particular autosomal gene in an individual are called *alleles*. If the two alleles are identical, the individual is said to be *homozygous* at that locus. If the alleles vary because of mutations, the individual is *heterozygous* at that locus. Males are *hemizygous* for genes on the X chromosome. A population can have a large number of alleles of a gene; for example, a single gene that affects human eye color, called *OCA2*, can have alleles that encode shades of blue, green, hazel, or brown. Because of this variation, it is important to distinguish the *genotype* of an organism (its genetic makeup) and the *phenotype* (its appearance). In the broad sense, a genotype is the entire set of alleles forming the genome of an individual; in the narrow sense, it is the specific alleles of one gene. By contrast, a phenotype is a description of a whole organism, and is a result of the expression of the organism's genotype in a particular environment.

If a mutant phenotype is expressed only when both alleles of a gene are mutated, the resulting phenotype is called *recessive*. This can occur if individuals are homozygous for the mutant allele or if they are carrying a different damaging allele in a given gene on each of their chromosomes (so-called *compound heterozygote*). Recessive mutations usually result from loss or reduction of a functional protein. Recessive inheritance of mutant traits is commonly observed in humans and experimental animals.

If a mutant phenotype results from a combination of one mutant and one wild-type allele, the phenotypic trait and mutant allele are said to be *dominant.* Some mutations are dominant because 50% of the gene product is not enough for a normal phenotype (*haploinsufficiency*). Other dominant mutations lead to the production of an abnormal protein or to the expression of the wild-type gene product at an inappropriate time or place; if this acts antagonistically to the normal protein product, it is called a *dominant negative* mutation.

The difference between genotype and phenotype is evident when considering the consequences of having one normal (wild-type) allele and one mutant allele of the same gene. Recent progress in gene discovery in a range of neurodevelopmental disorders, including autism and epilepsy, has demonstrated that the human genome is more sensitive to haploinsufficiency than previously appreciated. However, while the complete inactivation of both copies of a gene typically has a reliable effect, the severity and manifestation of haploinsufficiency vary to a greater degree from individual to individual, a phenomenon known as *variable, partial,* or *incomplete penetrance.*

Genetic variations that disturb development, cell function, or behavior in humans fall on a continuum between common alleles (also referred to as

polymorphisms), which generally have small individual effects on biology and behavior, and rare variants, which may have larger biological effects (Box 2–1). While these categorizations are useful generalizations, there are nonetheless important cases in which common polymorphisms carry large disease risks; a common variation in the *APOE* gene, present in 16% of the population, results in a four-fold increase in the risk of late-onset Alzheimer disease.

Genes Are Conserved Through Evolution

The nearly complete nucleotide sequence of the human genome was reported in 2001, and the complete nucleotide sequences of many animal genomes have also been decoded. Comparisons between these genomes lead to a surprising conclusion: the unique human species did not result from the invention of unique new human genes.

Humans and chimpanzees are profoundly different in their biology and behavior, yet they share 99% of their protein-coding genes. Moreover, most of the approximately 20,000 genes in humans are also present in other mammals, such as mice, and over half of all human genes are very similar to genes in invertebrates such as worms and flies (Figure 2–5). The conclusion from this surprising discovery is that ancient genes that humans share with other animals are regulated in new ways to produce novel human properties, like the capacity to generate complex thoughts and language.

Because of this conservation of genes throughout evolution, insights from studies of one animal can often be applied to other animals with related genes, an important fact as animal experiments are often possible when experiments on humans are not. For example, a gene from a mouse that encodes an amino acid sequence similar to a human gene usually has a similar function to the *orthologous* human gene.

Approximately one-half of the human genes have functions that have been demonstrated or inferred from orthologous genes in other organisms (Figure 2–6). A set of genes shared by humans, flies, and even unicellular yeasts encodes the proteins for intermediary metabolism; synthesis of DNA, RNA, and protein; cell division; and cytoskeletal structures, protein transport, and secretion.

The evolution from single-cell organisms to multicellular animals was accompanied by an expansion of genes concerned with intercellular signaling and gene regulation. The genomes of multicellular

Box 2–1 Mutation: The Origin of Genetic Diversity

Although DNA replication generally is carried out with high fidelity, spontaneous errors called *mutations* do occur. Mutations can result from damage to the purine and pyrimidine bases in DNA, mistakes during the DNA replication process, and recombinations that occur during meiosis.

Changes in a single DNA base (also called a point mutation) within a coding region fall into five general categories:

1. A *silent mutation* changes a base but does not result in an obvious change in the encoded protein.
2. A *missense mutation* is a point mutation that results in one amino acid in a protein being substituted for another; increasingly these are being categorized using both informatics and empirical evidence into at least two subclasses: mutations that are damaging to protein function and those that may be functionally neutral.
3. A *nonsense mutation*, where a *codon* (a triplet of nucleotides) within the coding region specifying a specific amino acid is replaced by a stop codon, resulting in a shortened (truncated) protein product.
4. A *canonical splice site mutation* changes a nucleotide that specifies the exon/intron boundary.
5. A *frameshift mutation*, in which small insertions or deletions of nucleotides change the reading frame, leading to the production of a truncated or abnormal protein.

In the current literature, mutations falling into the latter four categories (including damaging missense mutations) are often referred to as *likely gene disrupting (LGD)* mutations.

The frequency of mutations greatly increases when the organism is exposed to chemical mutagens or ionizing radiation during experimental genetic studies. Chemical mutagens tend to induce *point mutations* involving changes in a single DNA base pair or the deletion of a few base pairs. Ionizing radiation can induce large insertions, deletions, or translocations.

In humans, point mutations occur at a low spontaneous rate in oocytes and sperm, leading to mutations present in the child but not in either parent, called de novo mutations. Each generation, between 70 and 90 single base changes are introduced into the entire genome (approximately 3 billion base pairs), of which one, on average, will cause a missense or nonsense mutation in a protein-coding gene. The number of de novo point mutations is increased in the children of older fathers, whereas the frequency of larger chromosome abnormalities is increased in the children of older mothers.

With the sequencing of the human genome in 2001 and increasingly high-resolution methods to detect genetic variation, it is also now clear that point mutations are not the only differences in sequence between humans. Certain sequences may be missing or repeated several times on a chromosome, and thus can have different numbers of copies in different individuals. When such variations encompass more than 1000 base pairs, they are called copy number variations (CNVs). Changes in more than a single base and less than 1000 base pairs are referred to as insertion/deletions (indels).

The contribution of any genetic variation to a disease or syndrome may be referred to as *simple* (or *Mendelian*) or *complex*. In general, a simple or Mendelian mutation is one that is sufficient to confer a phenotype without additional genetic risks. This does not imply that everyone with the mutation will show exactly the same phenotype. However, there is typically a highly reliable relationship between a specific disease allele and a phenotype, one that approaches a one-to-one relationship (as seen in sickle cell anemia or Huntington disease).

In contrast, a complex genetic disorder is one in which genetic risk factors change the probability of a disease but are not fully causal. This genetic contribution may involve rare mutations, common polymorphisms, or both, and is typically quite heterogeneous, with multiple different genes and alleles having the capacity to increase risk or play a protective role. Most complex disorders also involve a contribution from the environment.

---

animals, such as worms, flies, mice, and humans, typically encode thousands of transmembrane receptors, many more than are present in unicellular organisms. These transmembrane receptors are used in cell-to-cell communication in development, in signaling between neurons, and as sensors of environmental stimuli. The genome of a multicellular animal also encodes 1,000 or more different DNA-binding proteins that regulate the expression of other genes.

Many of the transmembrane receptors and DNA-binding proteins in humans are related to specific orthologous genes in other vertebrates and invertebrates. By enumerating the shared genetic heritage of the animals, we can infer that the basic molecular

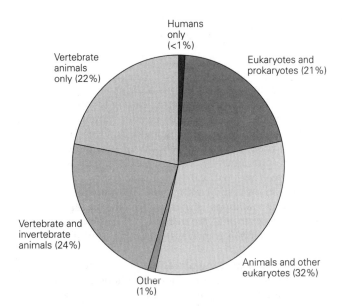

Figure 2–5 Most human genes are related to genes in other species. Less than 1% of human genes are specific to humans; other genes may be shared by all living things, by all eukaryotes, by animals only, or by vertebrate animals only. (Adapted, with permission, from Lander et al. 2001. Copyright © 2001 Springer Nature.)

pathways for neuronal development, neurotransmission, electrical excitability, and gene expression were present in the common ancestor of worms, flies, mice, and humans. Moreover, studies of animal and human genes have demonstrated that the most important genes in the human brain are those that are most conserved throughout animal phylogeny. Differences

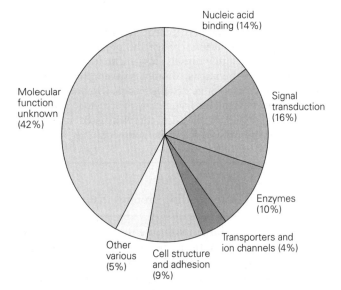

Figure 2–6 The predicted molecular functions of 26,383 human genes. (Adapted, with permission, from Venter et al. 2001.)

between mammalian genes and their invertebrate counterparts most often result from gene duplication in mammals or subtle changes in gene expression and function, rather than the creation of entirely new genes.

Genetic Regulation of Behavior Can Be Studied in Animal Models

Because of the evolutionary conservation between human and animal genes, studies in animal models of the relationships between the genes, proteins, and neural circuits that underlie behavior are likely to yield insight into these relationships in humans. Two important strategies have been applied with great success in the study of gene function.

In *classical genetic analysis*, organisms are first subjected to mutagenesis with a chemical or irradiation that induces random mutations and then screened for heritable changes that affect the behavior of interest, say, sleep. This approach does not impose a bias as to the kind of gene involved; it is a random search of all possible mutations that cause detectable changes. Genetic tracing of heritable changes allows the identification of the individual genes that are altered in the mutant organism. Thus the pathway of discovery in classical genetics moves from phenotype to genotype, from organism to gene. In *reverse genetics*, a specific gene of interest is targeted for alteration, a genetically modified animal is produced, and the animals with these altered genes are studied. This strategy is both focused and biased—one begins with a specific gene—and the pathway of discovery moves from genotype to phenotype, from gene to organism.

These two experimental strategies and their more subtle variations form the basis of experimental genetics. Gene manipulation by classical and reverse genetics is conducted in experimental animals, not in humans.

A Transcriptional Oscillator Regulates Circadian Rhythm in Flies, Mice, and Humans

The first large-scale studies of the influence of genes on behavior were initiated by Seymour Benzer and his colleagues around 1970. They used random mutagenesis and classical genetic analysis to identify mutations that affected learned and innate behaviors in the fruit fly *Drosophila melanogaster*: circadian (daily) rhythms, courtship behavior, movement, visual perception, and memory (Box 2–2 and Box 2–3). These induced mutations have had an immense influence on our understanding of the role of genes in behavior.

Box 2–2 Generating Mutations in Experimental Animals

Random Mutagenesis in Flies

Genetic analysis of behavior in the fruit fly (*Drosophila*) is carried out on flies in which individual genes have been mutated. Mutations can be made by chemical mutagenesis or by insertional mutagenesis, strategies that can affect any gene in the genome. Similar random mutagenesis strategies are used to create mutations in the nematode worm *Caenorhabditis elegans*, zebra fish, and mice.

Chemical mutagenesis, for example with ethyl methanesulfonate (EMS), typically creates random point mutations in genes. Insertional mutagenesis occurs when mobile DNA sequences called *transposable elements* randomly insert themselves into other genes.

The most widely used transposable elements in *Drosophila* are the P elements. P elements may be modified to carry genetic markers for eye color, which makes them easy to track in genetic crosses, and they may also be modified to alter expression of the gene into which they are inserted.

To cause P element transposition, *Drosophila* strains that carry P elements are crossed to those that do not. This genetic cross leads to destabilization and transposition of the P elements in the resulting offspring. The mobilization of the P element causes its transposition into a new location in a random gene.

Targeted Mutagenesis in Mice

Advances in molecular manipulation of mammalian genes have permitted precise replacement of a known normal gene with a mutant version. The process of generating a strain of mutant mice involves two separate manipulations. A gene on a chromosome is replaced by homologous recombination in a special cell line known as embryonic stem cells, and the modified cell line is incorporated into the germ cell population of the embryo (Figure 2–7).

The gene of interest must first be cloned. The gene is mutated, and a selectable marker, usually a drug-resistance gene, is then introduced into the mutated fragment. The altered gene is then introduced into embryonic stem cells, and clones of cells that incorporate the altered gene are isolated. DNA samples of each clone are tested to identify a clone in which the mutated gene has been integrated into the homologous (normal) site, rather than some other random site.

When a suitable clone has been identified, the cells are injected into a mouse embryo at the blastocyst stage (3 to 4 days after fertilization), when the embryo consists of approximately 100 cells. These embryos are then reintroduced into a female that has been hormonally prepared for implantation and allowed to come to term. The resulting embryos are chimeric mixtures between the stem cell line and the host embryo.

Embryonic stem cells in the mouse have the capability of participating in all aspects of development, including the germline. The injected cells can become germ cells and pass on the altered gene to future generations of mice. This technique has been used to generate mutations in various genes crucial to development or function in the nervous system.

Restricting Gene Knockout and Regulating Transgenic Expression

To improve the utility of gene knockout technology, methods have been developed that restrict deletions to cells in a specific tissue or at specific points in an animal's development. One method of regional restriction exploits the Cre/loxP system. The Cre/loxP system is a site-specific recombination system, derived from the P1 phage, in which the phage enzyme Cre recombinase catalyzes recombination between 34 bp loxP recognition sequences, which are normally not present in animal genomes.

The loxP sequences can be inserted into the genome of embryonic stem cells by homologous recombination such that they flank one or more exons of a gene of interest (called a floxed gene). When the stem cells are injected into an embryo, one can eventually breed a mouse in which the gene of interest is floxed and still functional in all cells of the animal.

A second line of transgenic mice can then be generated that expresses Cre recombinase under the control of a neural promoter sequence that is normally expressed in a restricted brain region. By crossing the Cre transgenic line of mice with the line of mice with the floxed gene of interest, the gene will only be deleted in those cells that express the Cre transgene.

In the example shown in Figure 2–8A, the gene encoding the NR1 (or GluN1) subunit of the *N*-methyl-D-aspartate (NMDA) glutamate receptor has been flanked with loxP elements and then crossed with a mouse line expressing Cre recombinase under control of the CaMKII promoter, which normally is expressed in forebrain neurons. In this particular line, expression was fortuitously limited to the CA1 region of the hippocampus, resulting in selective deletion of the NR1 subunit in this brain region (Figure 2–8B). Because the CaMKII promoter only activates gene transcription postnatally, early developmental changes are minimized by this strategy.

In addition to regional restriction of gene expression, control over the timing of gene expression gives the investigator an additional degree of flexibility and can exclude the possibility that any abnormality observed in the phenotype of the mature animal is the result of a developmental defect produced by the transgene. This can be done in mice by constructing a gene that can be turned on or off with a drug.

One starts by creating two lines of mice. Line 1 carries a particular transgene that is under control of the promoter tetO, which is ordinarily found only in bacteria. This promoter cannot by itself turn on the gene; it needs to be activated by a specific transcriptional regulator. Thus the second line of mice expresses a second transgene that encodes a hybrid transcription factor,

(continued)

Box 2–2 **Generating Mutations in Experimental Animals (continued)**

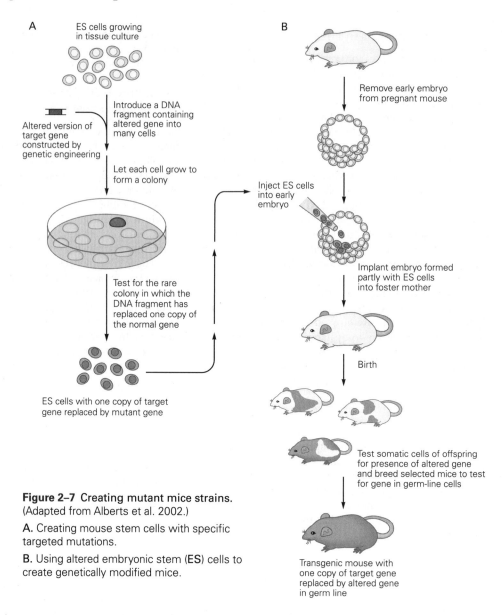

Figure 2–7 Creating mutant mice strains.
(Adapted from Alberts et al. 2002.)

A. Creating mouse stem cells with specific targeted mutations.

B. Using altered embryonic stem (**ES**) cells to create genetically modified mice.

the tetracycline transactivator (tTA), which recognizes and binds to the tetO promoter. Expression of tTA can be placed under the control of a promoter in the mouse genome that normally drives gene transcription in only specific classes of neurons or specific brain regions.

When the two lines of mice are mated, some of the offspring will carry both transgenes. In these mice, the tTA binds to the tetO promoter and activates the downstream transgene. What makes the tTA transcription factor particularly useful is that it becomes inactivated when it binds certain antibiotics, such as tetracycline, allowing transgene expression to be regulated by administering antibiotics to mice. One can also generate mice that express a mutant form of tTA called reverse tTA (rtTA). This transactivator will not bind to tetO unless the animal is fed doxycycline. In this case, the transgene is always turned off unless the drug is given (Figure 2–9).

Altering Gene Function by RNA Interference and CRISPR

Finally, genes can be inactivated by targeting them with modern molecular tools. One such method is RNA interference, which takes advantage of the fact that most double-stranded RNAs in eukaryotic cells are routinely destroyed; the whole RNA is destroyed even if only part of it is double-stranded. By introducing a short RNA sequence that artificially causes a selected mRNA to become double-stranded, researchers can activate this process to reduce the mRNA levels for specific genes.

Another experimental tool is CRISPR, a method in which components of a bacterial immune system are deployed in nonbacterial cells to directly attack a selected DNA sequence. To target a gene with CRISPR,

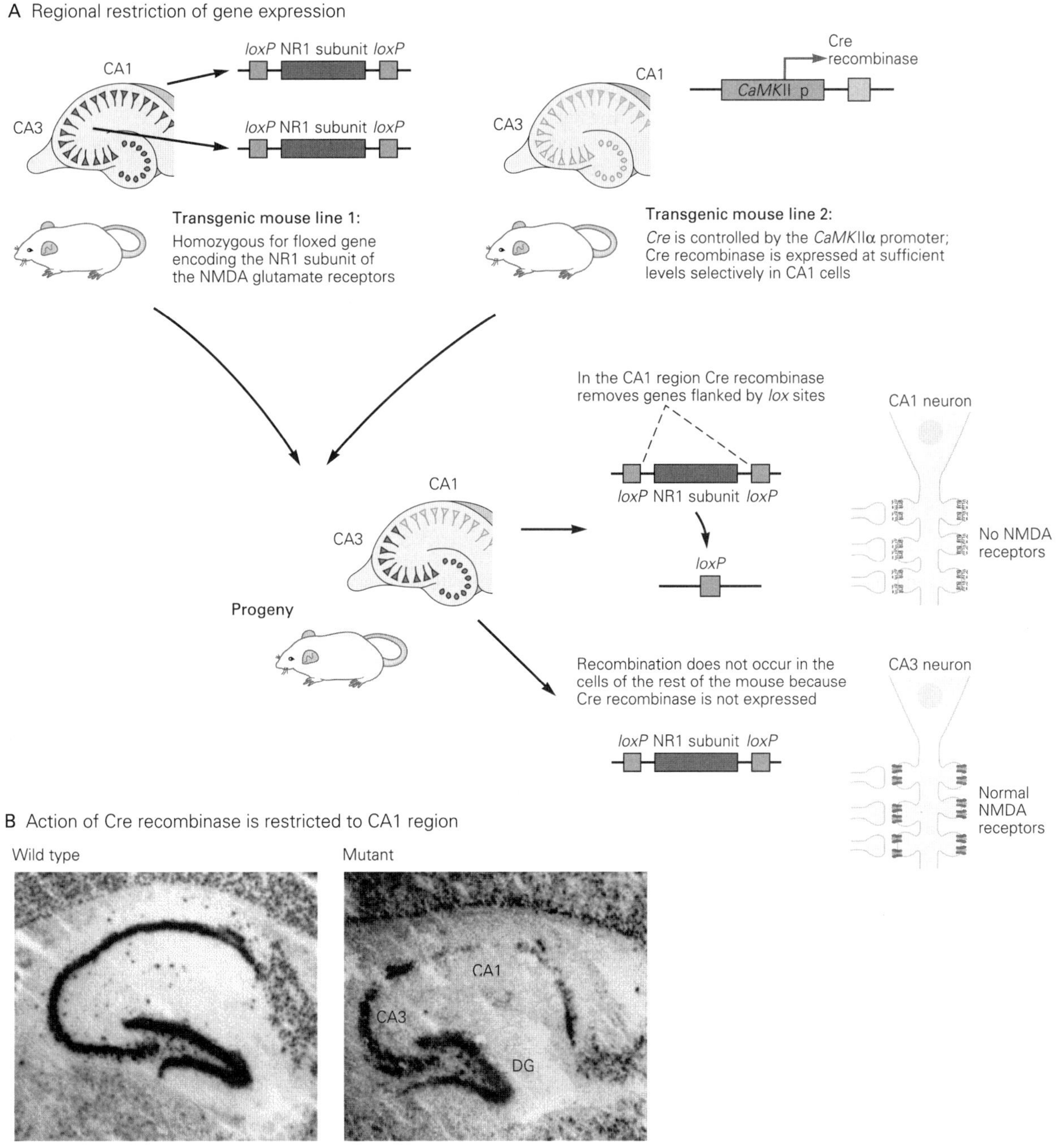

A Regional restriction of gene expression

Transgenic mouse line 1:
Homozygous for floxed gene encoding the NR1 subunit of the NMDA glutamate receptors

Transgenic mouse line 2:
Cre is controlled by the *CaMKIIα* promoter; Cre recombinase is expressed at sufficient levels selectively in CA1 cells

Progeny

In the CA1 region Cre recombinase removes genes flanked by *lox* sites

CA1 neuron

No NMDA receptors

Recombination does not occur in the cells of the rest of the mouse because Cre recombinase is not expressed

CA3 neuron

Normal NMDA receptors

B Action of Cre recombinase is restricted to CA1 region

Wild type

Mutant

Figure 2–8 The Cre/loxP system for gene knockout in selective regions.

A. A line of mice is bred in which the gene encoding the NR1 subunit of the NMDA receptor has been flanked by loxP genetic elements (transgenic mouse line 1). These so-called "floxed NR1" mice are then crossed with a second line of mice in which a transgene coding for Cre recombinase is placed under the control of a transcriptional promoter specific to a cell type or a tissue type (transgenic mouse line 2). In this example, the promoter from the *CaMKIIα* gene is used to drive expression of the *Cre* gene. In progeny that are homozygous for the floxed gene and that carry the *Cre*

recombinase transgene, the floxed gene will be deleted by Cre mediated loxP recombination only in cell type(s) in which the promoter driving Cre expression is active.

B. In situ hybridization is used to detect mRNA for the NR1 subunit in hippocampal slices from wild-type and mutant mice that contain two floxed NR1 alleles and express Cre recombinase under the control of the CaMKIIα promoter. In the mutant mice, expression of the mRNA for NR1 (dark staining) is greatly reduced in the CA1 region of the hippocampus but remains normal in CA3 and the dentate gyrus (**DG**). (Reproduced, with permission, from Tsien, Huerta, and Tonegawa 1996.)

Box 2–2 Generating Mutations in Experimental Animals (continued)

a bacterial protein (typically but not always a protein called CAS9) is produced together with an engineered guide RNA that has sequence similarity with the target gene. The CAS9-guide RNA complex seeks out and cleaves the target sequence in the genome of the cell of interest. That initial cleavage can induce point mutations, insertions, and deletions at that site, and can also facilitate desired recombination or genetic replacement events. CRISPR tools are increasing in their sophistication and

precision to the extent that they are now being considered for repair of hereditary mutations in people with severe inherited genetic diseases.

RNA interference and CRISPR have great potential to increase the power of genetic analysis because they can be used on any species in which DNA or RNA can be delivered to cells, including animals that are not now used in classical genetic analysis, such as long-lived birds, fish, and even primates.

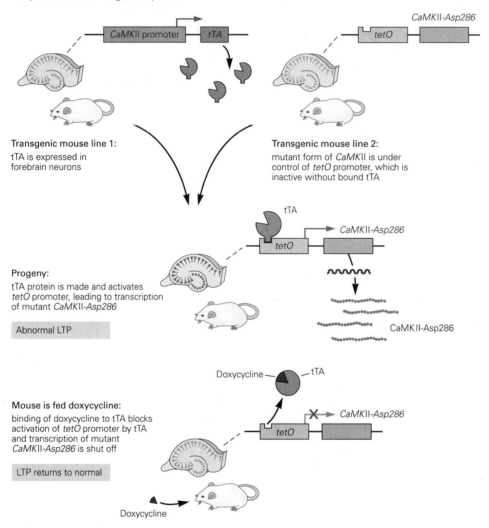

Figure 2–9 The tetracycline system for temporal and spatial regulation of transgene expression. Two independent lines of transgenic mice are bred. One line expresses, under the control of the CaMKIIα promoter, the tetracycline transactivator (**tTA**), an engineered protein incorporating a bacterial transcription factor that recognizes the bacterial tetO operon. The second line contains a transgene of interest—here encoding a constitutively active form of CaMKII (CaMKII–Asp286) that makes the kinase persistently active in the absence of Ca^{2+}—whose expression is under control of tetO. When these two lines are mated,

the offspring express the tTA protein in a pattern restricted to the forebrain. When the tTA protein binds to tetO, it will activate transcription of the downstream gene of interest. Tetracycline (or doxycycline) given to the offspring binds to the tTA protein and causes a conformational change that leads to the unbinding of the protein from tetO, blocking transgene expression. In this manner, mice will express CaMKII–Asp286 in the forebrain, and this expression can be turned off by administering doxycycline to the mice. (Reproduced, with permission, from Mayford et al. 1996.)

Box 2–3 Introducing Transgenes in Flies and Mice

Genes can be experimentally introduced in mice by injecting DNA into the nucleus of newly fertilized eggs (Figure 2–10). In some of the injected eggs, the new gene, or transgene, is incorporated into a random site on one of the chromosomes. Because the embryo is at the one-cell stage, the incorporated gene is replicated and ends up in all (or nearly all) of the animal's cells, including the germline.

Gene incorporation is illustrated with a coat color marker gene rescued by injecting a gene for pigment production into an egg obtained from an albino strain. Mice with patches of pigmented fur indicate successful expression of DNA. The transgene's presence is confirmed by testing a sample of DNA from the injected animals.

A similar approach is used in flies. The DNA to be injected is cloned into a transposable element (P element). When injected into the embryo, the DNA becomes inserted into the DNA of germ cell nuclei. P elements can be engineered to express genes at specific times and in specific cells. Transgenes may be wild-type genes that restore function to a mutant or *designer genes* that alter the expression of other genes or code for a specifically altered protein.

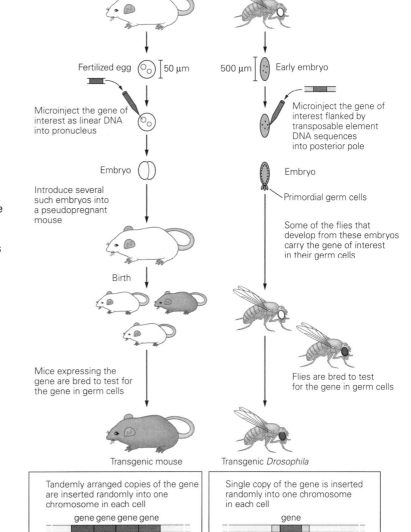

Figure 2–10 Generating transgenic mice and flies. Here the gene injected into the mouse causes a change in coat color, while the gene injected into the fly causes a change in eye color. In some transgenic animals of both species, the DNA is inserted at different chromosomal sites in different cells (see illustration at bottom). (Adapted from Alberts et al. 2002.)

We have a particularly complete picture of the genetic basis of the circadian control of behavior. An animal's circadian rhythm couples certain behaviors to a 24-hour cycle linked to the rising and setting of the sun. The core of circadian regulation is an intrinsic biological clock that oscillates over a 24-hour cycle. Because of the intrinsic periodicity of the clock, circadian behavior persists even in the absence of light or other environmental influences.

The circadian clock can be reset, such that changes in the day-night cycle eventually result in a matching shift in the intrinsic oscillator, a phenomenon familiar to any traveler recovering from jet lag. The clock is reset by light-driven signals transmitted by the eye to the brain. Finally, the clock drives effector pathways for specific behaviors, such as sleep and locomotion.

Benzer's group searched through thousands of mutant flies to look for rare flies that failed to follow circadian rhythms because of mutations in the genes that direct circadian oscillation. From this work emerged the first insight into the molecular machinery of the circadian clock. Mutations in the *period*, or *per*, gene affected all circadian behaviors generated by the fly's internal clock.

Interestingly, *per* mutations could change the circadian clock in several ways (Figure 2–11). Arrhythmic *per* mutant flies, which exhibited no discernible intrinsic rhythms in any behavior, lacked all function of the *per* gene, so *per* is essential for rhythmic behavior. *Per* mutations that maintained some function of the gene resulted in abnormal rhythms. Long-day alleles produced 28-hour behavioral cycles, whereas short-day alleles produced a 19-hour cycle. Thus *per* is not just an essential piece of the clock, it is actually a timekeeper whose activity can change the rate at which the clock runs.

The *per* mutant has no major adverse effects other than the change in circadian behavior. This observation is important because prior to the discovery of *per* many had questioned whether there could be true "behavior genes" that were not required for the physiological needs of an animal. *Per* does seem to be such a "behavior gene."

How does *per* keep time? The protein product PER is a transcriptional regulator that affects the expression of other genes. Levels of PER are regulated throughout the day. Early in the morning, PER and its mRNA are low. Over the course of the day, the PER mRNA and protein accumulate, reaching peak levels after dusk and during the night. The levels then decrease, falling before the next dawn. These observations provide an answer to the circadian rhythm puzzle—a central regulator appears and disappears throughout the day. But they are also unsatisfying because they only push the question back one step—what makes PER cycle? The answer to this question required the discovery of additional clock genes, which were discovered in flies and also in mice.

Emboldened by the success of the fly circadian rhythm screens, Joseph Takahashi began similar but far more labor-intensive genetic screens in mice in the 1990s. He screened hundreds of mutant mice for the rare individuals with alterations in their circadian locomotion period and found a single gene mutation that he called *clock*. When mice homozygous for the *clock* mutation are transferred to darkness, they initially experience extremely long circadian periods and later a complete loss of circadian rhythmicity (Figure 2–12). The *clock* gene therefore appears to regulate two fundamental properties of the circadian rhythm: the length of the circadian period and the persistence of rhythmicity in the absence of sensory input. These properties are conceptually identical to the properties of the *per* gene in flies.

The mouse *clock* gene, like the *per* gene in flies, encodes a transcriptional regulator whose activity oscillates through the day. The mouse CLOCK and fly PER proteins also shared a domain called a *PAS domain*, characteristic of a subset of transcriptional regulators. This observation suggests that the same molecular mechanism—oscillation of PAS-domain transcriptional regulation—controls circadian rhythm in flies and mice.

More significantly, parallel studies of flies and mice showed that similar groups of transcriptional regulators affect the circadian clock in both animals. After the mouse *clock* gene was cloned, a fly circadian rhythm gene was cloned and found to be even more closely related to mouse *clock*, than was *per*. In a different study, a mouse gene similar to fly *per* was identified and inactivated by reverse genetics. The mutant mouse had a circadian rhythm defect, like fly *per* mutants. In other words, both flies and mice use both *clock* and *per* genes to control their circadian rhythms. A group of genes, not one gene, are conserved regulators of the circadian clock.

Characterization of these genes has led to an understanding of the molecular mechanisms of circadian rhythm and a dramatic demonstration of the similarity of these mechanisms in flies and mice. In both flies and mice, the CLOCK protein is a transcriptional activator. Together with a partner protein, it controls the transcription of genes that determine behaviors such as locomotor activity levels. CLOCK and its partner also stimulate the transcription of the *per* gene. However, PER protein represses CLOCK's ability to stimulate *per* gene expression, so as PER protein

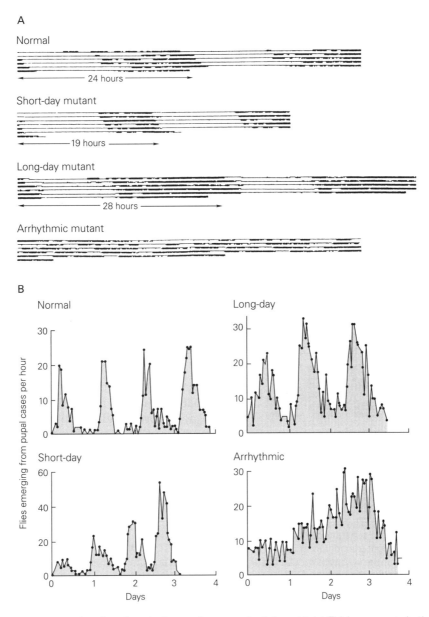

Figure 2–11 A single gene governs the circadian rhythms of behaviors in *Drosophila*. Mutations in the *period*, or *per*, gene affect all circadian behaviors regulated by the fly's internal clock. (Reproduced, with permission, from Konopka and Benzer 1971.)

A. Locomotor rhythms in normal *Drosophila* and three strains of *per* mutants: short-day, long-day, and arrhythmic. Flies were shifted from a cycle of 12 hours of light and 12 hours of dark into continuous darkness, and activity was then monitored

under infrared light. Thick segments in the record indicate activity.

B. Normal adult fly populations emerge from their pupal cases in cyclic fashion, even in constant darkness. The plots show the number of flies (in each of four populations) emerging per hour over a 4-day period of constant darkness. The arrhythmic mutant population emerges without any discernible rhythm.

accumulates, *per* transcription falls (Figure 2–13). The 24-hour cycle comes about because the accumulation and activation of PER protein is delayed by many hours after the transcription of *per*, a result of PER phosphorylation, PER instability, and interactions with other cycling proteins.

The molecular properties of *per*, *clock*, and related genes generate all properties essential for circadian rhythm.

1. The transcription of circadian rhythm genes varies with the 24-hour cycle: PER activity is high at night; CLOCK activity is high during the day.

The detailed elucidation of this molecular clock mechanism was recognized by the 2017 Nobel Prize in Physiology or Medicine, awarded to Jeffrey Hall, Michael Rosbash, and Michael Young.

The same genetic network controls circadian rhythm in humans. People with advanced sleep-phase syndrome have short 20-day cycles and an extreme early-to-bed, early-to-rise "morning lark" phenotype. Louis Ptáček and Ying-hui Fu found that these individuals have mutations in a human *per* gene. These results demonstrate that genes for behavior are conserved from insects to humans. Advanced sleep-phase syndrome is discussed in the chapter on sleep (Chapter 44).

Natural Variation in a Protein Kinase Regulates Activity in Flies and Honeybees

In the genetic studies of circadian rhythm described earlier, random mutagenesis was used to identify genes of interest in a biological process. All normal individuals have functional copies of *per*, *clock*, and the related genes; only after mutagenesis were different alleles generated. Another, more subtle question about the role of genes in behavior is to ask which genetic changes may be responsible for behavioral variation among normal individuals. Work by Marla Sokolowski and her colleagues led to the identification of the first gene associated with variation in behavior among normal individuals in a species.

Larvae of *Drosophila* vary in activity level and locomotion. Some larvae, called rovers, move over long distances (Figure 2–14). Others, called sitters, are relatively stationary. *Drosophila* larvae isolated from the wild can be either rovers or sitters, indicating that these are natural variations and not laboratory-induced mutations. These traits are heritable; rover parents have rover offspring and sitter parents have sitter offspring.

Sokolowski used crosses between different wild flies to investigate the genetic differences between rover and sitter larvae. These crosses showed that the difference between rover and sitter larvae lies in a single major gene, the *for* (forager) locus. The *for* gene encodes a signal transduction enzyme, a protein kinase activated by the cellular metabolite cGMP (cyclic guanosine 3′,5′-monophosphate). Thus this natural variation in behavior arises from altered regulation of signal transduction pathways. Many neuronal functions are regulated by protein kinases such as the cGMP-dependent kinase encoded by the *for* gene. Molecules such as protein kinases are particularly significant at transforming short-term neural signals into long-term changes in the property of a neuron or circuit.

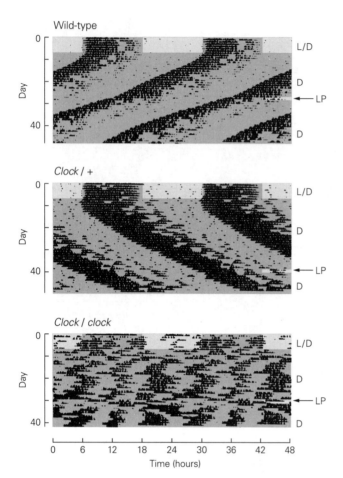

Figure 2–12 Circadian rhythm regulation by the *clock* gene in mice. The records show periods of locomotor activity by three animals: wild-type, heterozygous, and homozygous. All animals were kept on a light-dark (**L/D**) cycle of 12 hours for the first 7 days, then transferred to constant darkness (**D**). They later were exposed to a 6-hour light period (**LP**) to reset the rhythm. The circadian rhythm for the wild-type mouse has a period of 23.1 hours. The period for the heterozygous *clock*/+ mouse is 24.9 hours. The homozygous *clock*/*clock* mice experience a complete loss of circadian rhythmicity on transfer to constant darkness and transiently express a rhythm of 28.4 hours after the light period. (Reproduced, with permission, from Takahashi, Pinto, and Vitaterna. 1994. Copyright © 1994 AAAS.)

2. The circadian rhythm genes are transcription factors that affect each other's mRNA level, generating the oscillations. CLOCK activates *per* transcription and PER represses CLOCK function.

3. The circadian rhythm genes also control the transcription of other genes that in turn affect many downstream responses. For example, in flies, the neuropeptide gene *pdf* controls locomotor activity levels.

4. The oscillation of these genes can be reset by light.

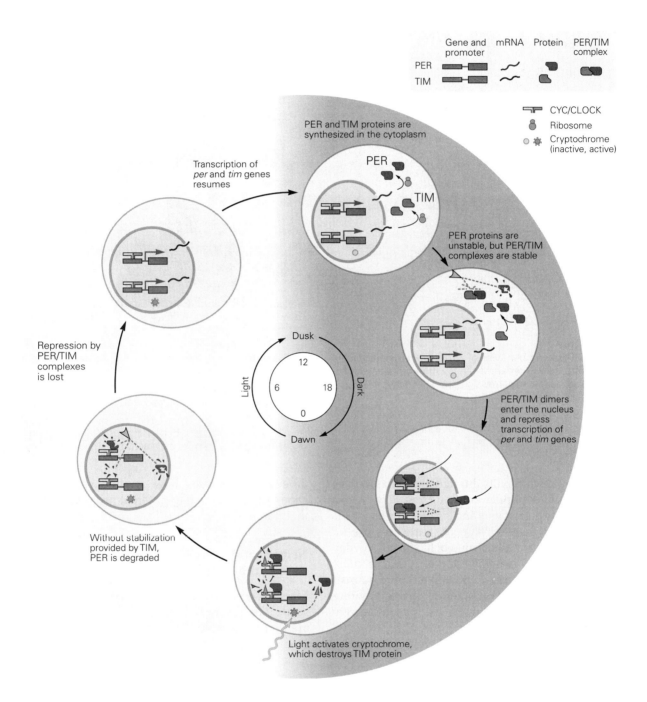

Figure 2–13 Molecular events that drive circadian rhythm.
The genes that control the circadian clock are regulated by two
nuclear proteins, PER and TIM. These proteins slowly accumulate and then bind to one another to form dimers. Once they
form dimers, they enter the nucleus and shut off the expression of circadian genes including their own. They do so by inhibiting CLOCK and CYCLE, which stimulate the transcription of
per and *tim* genes.

PER protein is highly unstable; most of it is degraded so quickly
that it never has a chance to repress CLOCK-dependent *per*
transcription. The degradation of PER is regulated by at least
two different phosphorylation events by different protein

kinases. When PER binds to TIM, PER is protected from degradation. As CLOCK drives more and more *per* and *tim* expression, enough PER and TIM eventually accumulate that the two
can bind and stabilize each other, at which point they enter
the nucleus where their own transcription is repressed. As a
result, *per* and *tim* mRNA levels fall; thereafter, PER and TIM
protein levels fall and CLOCK can (once again) drive expression
of *per* and *tim* mRNA. During daylight, TIM protein is degraded
by signaling pathways that are regulated by light (including
cryptochrome), so PER/TIM complexes form only at night. The
CLOCK protein induces PER and TIM expression but is inhibited by PER and TIM proteins.

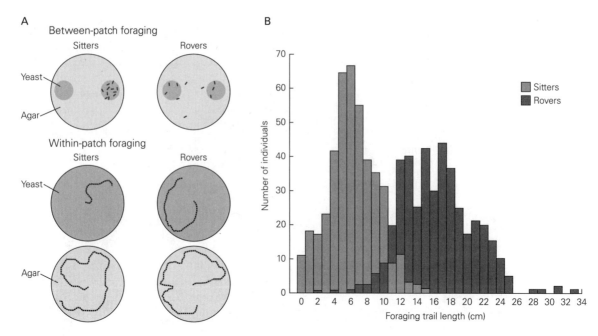

Figure 2–14 Foraging behavior of *Drosophila melanogaster* rover and sitter larvae differs while feasting on patches of yeast. (Reproduced, with permission, from Sokolowski. 2001. Copyright © 2001 Springer Nature.)

A. Rover-type larvae move from patch to patch, whereas sitter-type larvae stay put on a single patch for a long time. When foraging within a single patch, rover larvae move about more

than sitter larvae. On agar alone, rover and sitter larvae move about equally.

B. While foraging within a patch of food, rovers have longer trail lengths than sitters (trail lengths were measured over a period of 5 minutes).

This difference in foraging behavior maps to a single protein kinase gene, *for*, which varies in activity in different fly larvae.

Why would variability in signaling enzymes be preserved in wild populations of *Drosophila*, which typically include both rovers and sitters? The answer is that variations in the environment create pressure for balanced selection for alternative behaviors. Crowded environments favor the rover larva, which is more effective at moving to new, unexploited food sources in advance of competitors, whereas sparse environments favor the sitter larva, which exploits the current source more thoroughly.

The *for* gene is also found in honeybees. Honeybees exhibit different behaviors at different stages of their life; in general, young bees are nurses, while older bees become foragers that leave the hive. The *for* gene is expressed at high levels in the brains of active foraging honeybees and at low levels in the younger and more stationary nurse bees. Activation of cGMP signaling in young bees can cause them to enter the forager stage prematurely; this change could be induced by an environmental stimulus or the bee's advancing age.

Thus the same gene controls variation in a behavior in two different insects, but in different ways. In the fruit fly, variations in the behavior are expressed in different individuals, whereas in the honeybee, they are expressed in one individual at different ages. The

difference illustrates how an important regulatory gene can be recruited to different behavioral strategies in different species.

Neuropeptide Receptors Regulate the Social Behaviors of Several Species

Many aspects of behavior are associated with an animal's social interactions with other animals. Social behaviors are highly variable between species, yet have a large innate component within a species that is controlled genetically. A simple form of social behavior has been analyzed in the roundworm *Caenorhabditis elegans*. These animals live in soil and eat bacteria.

Different wild-type strains exhibit profound differences in feeding behavior. Animals from the standard laboratory strain are solitary, dispersing across a lawn of bacterial food and failing to interact with each other. Other strains have a social feeding pattern, joining large feeding groups of dozens or hundreds of animals (Figure 2–15). The difference between these strains is genetic, as both feeding patterns are stably inherited.

The difference between social and solitary worms is caused by a single amino acid substitution in a single gene, a member of a large family of genes involved in

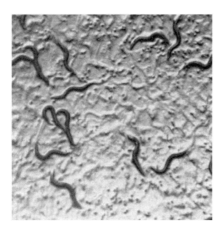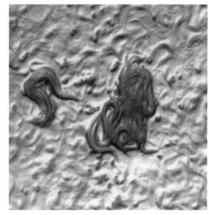

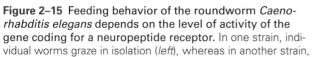

Figure 2–15 Feeding behavior of the roundworm *Caeno-rhabditis elegans* depends on the level of activity of the gene coding for a neuropeptide receptor. In one strain, individual worms graze in isolation (*left*), whereas in another strain, individuals mass together to feed. The difference is explained by a single amino acid substitution in the neuropeptide receptor gene. (Reproduced, with permission, from De Bono and Bargmann 1998.)

signaling between neurons. This gene, *npr-1*, encodes a neuropeptide receptor. Neuropeptides have long been appreciated for their roles in coordinating behaviors across networks of neurons. For example, a neuropeptide hormone of the marine snail *Aplysia* stimulates a complex set of movements and behavior patterns associated with a single behavior, egg laying. Mammalian neuropeptides have been implicated in feeding behavior, sleep, pain, and many other behaviors and physiological processes. The existence of a mutation in the neuropeptide receptor that alters social behavior suggests that this kind of signaling molecule is important both for generating the behavior and for generating the variation between individuals.

Neuropeptide receptors have also been implicated in the regulation of mammalian social behavior. The neuropeptides oxytocin and vasopressin stimulate mammalian affiliative behaviors such as pair bonding and parental bonding with offspring. In mice, oxytocin is required for social recognition, the ability to identify a familiar individual. Oxytocin and vasopressin have been studied in depth in prairie voles, rodents that form lasting pairs to raise their young. Oxytocin released in the brain of female prairie voles during mating stimulates pair-bond formation. Likewise, vasopressin released in the brain of male prairie voles during mating stimulates pair-bond formation and paternal behavior.

The extent of pair-bonding varies substantially between mammalian species. Male prairie voles form long-lasting pair-bonds with females and help them raise their offspring and are described as monogamous, but the closely related male montane voles breed widely

and do not engage in paternal behavior. The difference between the behaviors of males in these species correlates with differences in the expression of the V1a class of vasopressin receptors in the brain. In prairie voles, V1a vasopressin receptors are expressed at high levels in a specific brain region, the ventral pallidum (Figure 2–16). In montane voles, the levels are much lower in this region, although high in other brain regions.

The importance of oxytocin and vasopressin and their receptors has been confirmed and extended by reverse genetic studies in mice, which are easier than voles to manipulate genetically. Introducing the V1a vasopressin receptor gene from prairie voles into male mice, which behave more like montane voles, increases the expression of the V1a vasopressin receptor in the ventral pallidum and increases the affiliative behavior of the male mice toward females. Thus differences between species in the pattern of expression of the vasopressin receptor can contribute to differences in social behaviors.

The analysis of vasopressin receptors in different rodents provides insight into the mechanisms by which genes and behaviors can change during evolution. Thus evolutionary changes in the pattern of expression of the V1a vasopressin receptor in the ventral forebrain have altered the activity of a neural circuit, linking the function of the ventral forebrain to the function of the vasopressin-secreting neurons that are activated by mating. As a result, social behaviors are altered.

The importance of oxytocin and vasopressin in human social behavior is not known, but the central role of pair-bonding and pup rearing in mammalian species suggests that these molecules might play a role in our species as well.

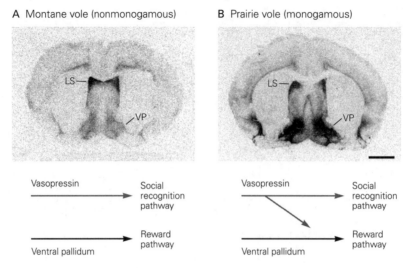

Figure 2–16 Distribution of vasopressin receptors (V1a) in two closely related rodent species. (Adapted, with permission, from Young et al. 2001. Copyright © 2001 Academic Press.)

A. Receptor expression is high in the lateral septum (**LS**) but low in the ventral pallidum (**VP**) in the montane vole, which does not form pair bonds.

B. Expression is high in the ventral pallidum of the monogamous prairie vole. Expression of the receptor in the ventral pallidum allows vasopressin to link the social recognition pathway to the reward pathway.

Studies of Human Genetic Syndromes Have Provided Initial Insights Into the Underpinnings of Social Behavior

Brain Disorders in Humans Result From Interactions Between Genes and the Environment

The first gene discovered for a neurological disease in humans clearly illustrates the interaction of genes and environment in determining cognitive and behavioral phenotypes. Phenylketonuria (PKU), described by Asbørn Følling in Norway in 1934, affects one in 15,000 children and results in severe impairment of cognitive function.

Children with this disease have two abnormal copies of the *PKU* gene that codes for phenylalanine hydroxylase, the enzyme that converts the amino acid phenylalanine to tyrosine. The mutation is recessive and heterozygous carrier individuals have no symptoms. Children who lack normal function in both copies of the gene accumulate high blood concentrations of phenylalanine from dietary proteins, which in turn leads to the production of toxic metabolites that interfere with neuronal function. The specific biochemical processes by which phenylalanine adversely affects the brain are still not understood.

The PKU phenotype (intellectual disability) results from the interaction of the genotype (the homozygous *pku* mutation) and the environment (the diet). The treatment for PKU is thus simple and effective: developmental delay can be prevented by a low-protein diet. The molecular and genetic analysis of gene function in PKU has led to dramatic improvements in the life of affected individuals. Since the early 1960s, the United States has instituted mandatory testing for PKU in newborns. Identifying children with the genetic disorder and modifying their diet before the disease appears prevents many aspects of the disorder.

Later chapters of this book describe many examples of single-gene traits that, like PKU, have led to insights into brain function and dysfunction. Certain themes have emerged from these studies. For example, a number of rare neurodegenerative disorders such as Huntington disease and spinocerebellar ataxia result from the pathological, dominant expansion of glutamate residues within proteins. The discovery of these polyglutamine repeat disorders highlighted the danger to the brain of unfolded and aggregated proteins. The discovery that epileptic seizures can be caused by a variety of mutations in ion channels led to the realization that these disorders are primarily disorders of neuronal excitability.

Rare Neurodevelopmental Syndromes Provide Insights Into the Biology of Social Behavior, Perception, and Cognition

Neurological and developmental disorders that manifest themselves in childhood have illuminated the importance and complexity of genetics in human brain function. Early evidence that genes affect specific cognitive and behavioral circuitry emerged from studies of a rare genetic condition known as Williams syndrome. Individuals with this disorder typically exhibit normal language as well as extreme sociability; early in development, they lack the reticence children typically have in the presence of strangers. At the same time, they are profoundly impaired in spatial processing, show overall intellectual disability, and have very high rates of anxiety (but rarely social anxiety disorder).

The patterns of impairments in Williams syndrome, as compared with for example autism spectrum disorders, suggest that language and social skills can be separated from some other brain functions. Brain areas concerned with language are impaired in children with autism but are active or accentuated in Williams syndrome. By contrast, general and spatial intelligence is more impaired in Williams syndrome than in about half of all children with autism spectrum disorder.

Williams syndrome is caused by a heterozygous deletion of the chromosome region 7q11.23, most often encompassing about 1.5 Mb and 27 genes. The simplest interpretation of this defect is that the level of expression of the genes within the interval is reduced because there is only one copy instead of two of each gene in the region. The precise genes in the interval that influence social communication and spatial processing are not yet known, but they are of great interest because of their potential to provide insight into the genetic regulation of human behavior.

A more recent discovery in studies of autism spectrum disorders has further highlighted the complex relationship between genetic variation and social and intellectual functioning first illuminated by Williams syndrome. Within about the past decade, advances in genomic technology have allowed for high-throughput methods to screen the genome for variations in chromosomal structure, and at much higher resolution than was allowed by the light microscope (see Box 2–1). Seminal studies in 2007 and 2008 demonstrated that individuals with autism spectrum disorder carry new (de novo) copy number variations much more often than unaffected individuals. These findings led to some of the first discoveries of specific genomic intervals contributing to common forms of the syndrome (ie, autism spectrum disorder without evidence of syndromal features, also known as *idiopathic* or *nonsyndromic* autism spectrum disorder).

In 2011, two simultaneous large-scale studies of de novo copy number variations in a very well-characterized cohort found that precisely the same region deleted

in Williams syndrome conferred substantial risk for autism spectrum disorder in an individual. However, in these cases, it was rare duplications (one excess copy of the region), and not deletions, that dramatically increased the risk for social disability. These findings, that losses and gains in the identical set of genes may lead to contrasting social phenotypes (while both typically lead to intellectual disability), further support the notion that domains of cognitive and behavioral functioning are separable but may share important molecular mechanisms.

Fragile X syndrome is another neurodevelopmental disorder of childhood that provides insight into the genetics of cognitive function; unlike Williams syndrome, it has been mapped to a single gene on the X chromosome. Fragile X syndrome varies in its presentation. Affected children may have intellectual disability, poor social cognition, high social anxiety, and repetitive behavior; about 30% of boys with fragile X syndrome meet diagnostic criteria for autism spectrum disorder. Fragile X syndrome is also associated with a broader range of traits, including physical characteristics such as an elongated face and protruding ears.

Fragile X syndrome has been shown to result from mutations that reduce expression of a gene called *fragile X mental retardation protein (FMRP)*. Because the gene falls on the X chromosome, males lose all expression of the gene when their one copy is mutated. FMRP protein regulates the translation of mRNAs into proteins in neurons, in a process that is itself regulated by neuronal activity. Regulated translation in neurons is an important component of the synaptic plasticity required for learning. The fragile X defect at the level of translation thus cascades up to affect neuronal function, learning, and higher-order cognitive processes. Interestingly, a large proportion of the other genes associated with increased risk for autism spectrum disorder as well as schizophrenia are regulated by the FMRP protein.

Another Mendelian disorder whose genetic basis is well understood is Rett syndrome (discussed in detail in Chapter 62). Rett syndrome is an X-linked, progressive neurodevelopmental disorder and one of the most common causes of intellectual disability in females. The disorder is almost always confined to females because canonical Rett mutations are very often lethal in the developing male embryo, which has a single X chromosome. Affected girls develop typically until they are 6 to 18 months of age, when they fail to acquire speech, show regression in intellectual functioning, and display compulsive, uncontrolled hand wringing instead of purposeful hand movement. In addition, girls with Rett syndrome often show a period of markedly impaired social interaction that may be indistinguishable from autism spectrum disorder, although it is thought that social functioning is largely preserved in later life. Huda Zoghbi and her colleagues found that the major cause of this syndrome results from mutations in the *methyl CpG binding protein 2 (MeCP2)* gene. Methylation of specific CpG sequences in DNA alters expression of nearby genes, and one of the established functions of *MeCP2* is that it binds methylated DNA as part of a process that regulates mRNA transcription.

Rare syndromes have also offered some of the first insights into the genetic substrates of schizophrenia (Chapter 60). For example, as first described by Robert Shprintzen and colleagues in 1978, deletions of chromosome 22q11 lead to a wide range of physical and behavioral symptoms, including psychosis, now often referred to as velocardiofacial syndrome (VCFS), DiGeorge syndrome, or 22q11 deletion syndrome. The initial descriptions by Shprintzen were met with some skepticism as a result of the extremely broad range of phenotypes associated with the identical deletion. It is now widely appreciated that the 22q11 deletion is the most common chromosomal abnormality associated with schizophrenia and childhood-onset schizophrenia. Moreover, chromosomal losses of the identical region have been found to be associated with large individual risks for autism. To date, the specific genes within the region responsible for the psychiatric phenotype(s) have not been definitively established. Moreover, recent evidence from the autism literature suggests that it is likely that a combination of multiple genes within this interval, each conferring relatively small individual effects, is responsible for the social disability phenotype.

Psychiatric Disorders Involve Multigenic Traits

As mentioned earlier, single-gene syndromes are rare compared to the total burden of neurodegenerative and psychiatric disease. Consequently, one might question the rationale for studying rare disorders if they represent just a fraction of the total disease burden. The reason is that rare conditions can provide insight into the biological processes involved in more common, complex forms of a disease. For example, among the prominent successes of human genetics has been the discovery of rare genetic variants that lead to early-onset Alzheimer disease or Parkinson disease. Individuals with these severe rare variants represent a tiny subset of all individuals with these conditions, but the identification of rare disease variants uncovered cellular processes that are also disrupted in the

larger patient pool, pointing to general therapeutic avenues. Similarly, pursuit of the pathophysiological mechanisms underlying Rett, fragile X, and other neurodevelopmental disorders has already led to some of the first attempts at rational drug development in psychiatric syndromes.

In the remainder of this chapter, we expand the discussion of the genetics of two complex neurodevelopmental and psychiatric phenotypes: autism spectrum disorders and schizophrenia. Compared to the rare Mendelian examples discussed earlier, the genetics of common forms of these conditions are indeed more diverse, varied, and heterogeneous, involving many different genes in different individuals as well as multiple risk genes conferring liability in combinations. Moreover, for both diagnoses, while the support for a genetic contribution is substantial, there is also compelling evidence for a contribution from environmental factors.

Progress in understanding these disorders came from the combination of rapidly advancing genomic technologies and statistical methods, a culture of open data sharing, and the consolidation of very large patient cohorts providing adequate power to detect very rare highly penetrant alleles as well as common genetic variants carrying small increments of risk. Importantly, recent successes in understanding both syndromes have provided a solid foundation for the pursuit of their biological consequences and the molecular, cellular, and circuit-level pathophysiology conveyed by these genetic risk factors.

Advances in Autism Spectrum Disorder Genetics Highlight the Role of Rare and De Novo Mutations in Neurodevelopmental Disorders

Autism spectrum disorders are a collection of developmental syndromes of varying severity affecting approximately 2% to 3% of the population and characterized by impairment in reciprocal social communication as well as stereotyped interests and repetitive behaviors. There is a significant male predominance; on average, three times as many boys as girls are affected. The clinical symptoms of autism spectrum disorders, by definition, emerge in the first 3 years of life, although highly reliable differences between affected and unaffected children are very often identifiable within the first months of life.

There is considerable phenotypic variability between those affected, leading to the development of the quite broad diagnostic classification of autism spectrum disorders. In addition, affected individuals have a higher frequency of seizures and cognitive problems than the general population and often have serious impairments in adaptive functioning. However, many autistic individuals are not as profoundly affected and lead highly successful lives.

Autism has a very strong heritable component (see Figure 2–1A), which is likely to account for its being among the first genetically complex neuropsychiatric syndromes to yield to modern gene discovery tools and methods. Autism spectrum disorder has broader significance because it provides insight into behaviors that are quintessentially human: language, complex intelligence, and interpersonal interactions. Importantly, the fact that the defects in social communication seen in autism spectrum disorders can coexist with normal intelligence and typical functioning in other domains suggests that the brain is to some degree modular with distinct cognitive functions that can vary independently.

While syndromic forms of autism spectrum disorder account for a small fraction of all cases, the first findings in more common so-called "idiopathic" or "nonsyndromic" forms of the disorder also demonstrated a role for rare mutations carrying large biological effects. For example, in 2003, the sequencing of genes within a region on the X chromosome deleted in a very small number of females with autistic features led to the discovery of rare, loss-of-function mutations in the gene *neuroligin 4X (NLGN4X)*, a gene encoding a synaptic adhesion molecule in excitatory neurons and found in several affected male family members. Soon thereafter, a linkage analysis of a large pedigree with intellectual disability and autism spectrum disorder showed that affected family members all carried a loss-of-function *NLGN4X* mutation.

De novo submicroscopic deletions and duplications in chromosomal structure may dramatically increase an individual's risk for autism spectrum disorder. These copy number variations (CNVs) cluster in specific regions of the genome, identifying specific risk intervals. The earliest reports using this approach showed that the de novo CNVs at chromosome 16p11.2, although present in only about 0.5% to 1% of affected individuals, carried substantial (greater than 10-fold) risk of autism spectrum disorder. Subsequent studies have now identified a dozen or more de novo CNVs that carry risk, including at chromosomes 16p11.2, 1q21, 15q11-13, and 3q29; deletions at 22q11, 22q13 (deleting the gene *SHANK3*), and 2p16 (deleting the gene *NXRN1*); and de novo duplications of 7q11.23 (the Williams syndrome region).

Interestingly, although these CNVs carry large risks for autism spectrum disorder, studies of other psychiatric disorders, including schizophrenia and bipolar disorder, have found that many of the same regions increase the risk for these conditions as well. Moreover, studies of individuals ascertained by genotype (eg, 16p11.2 deletions and duplications) have found a wide variety of associated behavioral phenotypes, ranging from specific language impairment to intellectual disability to schizophrenia. This "one-to-many" phenomenon presents important challenges to illuminating specific pathophysiological mechanisms in psychiatric illness and to conceptualizing the steps from gene discovery to therapies.

The widespread and replicable findings that de novo rare CNVs increase the risk for autism spectrum disorder and other developmental disorders immediately raised the question of whether rare de novo mutations in single genes might carry similar risks. Indeed, the development of technology for low-cost, high-throughput DNA sequencing, initially focused on the coding portion of the genome, led to the identification of an excess of de novo mutations deemed likely to disrupt gene function (LGD mutations) in affected individuals. The repeated occurrence of these mutations in close proximity among unrelated individuals has now been exploited as a means to identify specific risk genes for autism spectrum disorders.

Large-scale studies of de novo mutations in autism spectrum disorders have now identified more than 100 associated genes, with about 45 of these reaching the highest confidence level of statistical significance. These genes have a wide range of known functions, but analyses reveal a statistically significant overrepresentation of genes involved in synaptic formation and function, and in regulation of transcription. Moreover, there are a greater than expected number of risk genes that encode RNAs that are targets of fragile X mental retardation protein and/or proteins that are active in early brain development.

Identification of Genes for Schizophrenia Highlights the Interplay of Rare and Common Risk Variants

Schizophrenia affects about 1% of all young adults, causing a pattern of thought disorders and emotional withdrawal that profoundly impairs life. It is strongly heritable (see Figure 2–1B) and also has a strong environmental component that is associated with stress on a developing fetus. Children born just after the Dutch Hunger Winter famine of World War II had a greatly increased risk of schizophrenia many years later, and children whose mothers were infected with the rubella virus during pregnancy in the 1960s pandemic were also at considerably increased risk.

Genes, as well as the environment, contribute to schizophrenia. As with autism, the sequencing of the human genome, the development of inexpensive methods for genome-wide genotyping of common variants and detection of CNVs, and the consolidation of very large patient cohorts have all resulted in a transformation in the genetics of schizophrenia. First, essentially in parallel with the findings in autism spectrum disorders noted earlier, rare and de novo CNVs began to be implicated in the risk for schizophrenia by the early 2000s. A small percentage of cases are associated with chromosomal abnormalities that carry large risks, including, for example, deletions at chromosome 22q11. These chromosomal abnormalities overlap entirely, or nearly so, with those loci implicated in autism spectrum disorders, but the distribution of risk among deletions and duplications at these loci does not appear to be identical. For instance, although duplications and deletions of the 16p11.2 region are both associated with autism spectrum disorders and schizophrenia, duplications of the region are more likely to lead to schizophrenia, whereas deletions are more likely to be seen with autism spectrum disorders and intellectual disability.

With regard to schizophrenia, the most important development of the last decade and a half has been the emergence of common variant genome-wide association studies (GWASs). In contrast to studies of hypothesis-driven candidate genes described earlier, genome-wide association relies on assaying polymorphisms at every gene in the genome simultaneously. This hypothesis-free approach, when used with well-powered cohorts and appropriate correction for multiple comparisons, has proven to be a highly reliable and reproducible strategy for identifying common risk alleles in common disorders across all of medicine.

GWASs involving nearly 40,000 cases and 113,000 controls have resulted in the identification of 108 risk loci for schizophrenia. The effects attributable to any individual genetic variant in this set have been quite modest, typically accounting for a less than 25% increase in risk. Moreover, many of the genetic polymorphisms assayed in GWASs map to regions outside of the coding segment of the genome. Consequently, although 108 risk loci have been identified, it is not yet entirely clear which genes correspond to all of these risk variants. In some cases, the variations mapped sufficiently close to a single gene that such a relationship could be reasonably inferred; in other cases, this remains to be determined.

The genes implicated in schizophrenia risk provide a starting point for determining the biology underlying the disorder. For example, since the late 1990s,

evidence has pointed to the involvement of a region called the major histocompatibility complex (MHC) in schizophrenia risk. Accordingly, the MHC region has the strongest GWAS signal of any part of the human genome in the schizophrenia cohort. Detailed studies made possible by the very large number of patients in the cohort resolved this robust risk association signal in the MHC region into three different loci (and likely three different genes). Among these three loci, one gene, encoding the complement C4 factor, has a strong and definable effect on disease risk. Steven McCarroll and his colleagues showed that the complement C4 locus represents a natural case of CNV, that healthy individuals vary substantially in the number of copies of the gene they have, and that the level of expression of the C4A allele correlates with increasing schizophrenia risk. Subsequent follow-up studies showed that mice with knockouts of the C4 gene had a deficit in synaptic pruning during development, suggesting the hypothesis that excess C4A in humans might cause excessive synaptic pruning, a process that has long been of interest in the schizophrenia literature.

This finding represents an important demonstration of the ability to link genomics to a possible biological mechanism for increased disease risk. Even so, an individual with the highest-risk C4 haplotype who did not have a family history of schizophrenia would on average increase from having a 1% chance of being affected to approximately a 1.3% chance of being affected as a result of this allele. To get a sense of scale, having a first-degree relative with schizophrenia results in an approximately 10-fold increase in risk. This promising start and its limits reflect the challenges that geneticists and neurobiologists now face in moving from successful common variant gene discovery to the elaboration of the specific mechanisms leading to human pathology.

In addition to identifying numerous specific risk loci, GWASs in schizophrenia have repeatedly found that the small individual effects of many common alleles add up to increase risk. These results provide an additional, powerful avenue to study genotype-phenotype relationships in aggregate. Indeed, it is already clear that the number of risk alleles that an individual carries can have a significant (and additive) impact on the risk of developing the disorder. For instance, those in the top decile for a so-called polygenic risk score—a summary statistic relating to the overall amount of additive genetic risk an individual carries—are at 8- to 20-fold increased risk compared to the general population. Although the biology of the cumulative effect is not yet known, the observation nonetheless sets the stage for studying a series of interesting questions related to disease trajectory and treatment response and will almost certainly reinvigorate studies combining neuroimaing and genomics. These latter types of investigations, similar to early efforts at common variant discovery, have suffered from poor reliability due to the inherent limitations of studying selected, biologically plausible candidate genes.

Finally, high-throughput sequencing methods, similar to those employed in autism spectrum disorders, have begun to yield results in schizophrenia as well. Specifically, exome sequencing in search of rare and de novo risk alleles has been pursued with some success. However, such studies require much larger cohorts to identify statistically significant risks for LGD mutations compared to autism spectrum disorders, suggesting that the overall effect size of these types of variations is likely to be substantially less in schizophrenia. To date, these investigations have identified a handful of risk genes and implicated key neurobiological pathways. In particular, recent exome studies have pointed to the importance of the molecules within the activity-regulated cytoskeleton (ARC) complex, as well as the gene *set domain containing 1A (SETDIA)*, as relevant for schizophrenia pathogenesis.

Perspectives on the Genetic Bases of Neuropsychiatric Disorders

Genes affect many aspects of behavior. There are remarkable similarities in personality traits and psychiatric illnesses in human twins, even those raised separately. Domestic and laboratory animals can be bred for particular, stable behavioral traits; and increasingly, the contributions of a wide range of genetic variations for neurodevelopmental and psychiatric disorders are being discovered.

A series of parallel advances have ushered in an era of remarkable opportunity to understand the relationship between genes, brain, and behavior. The armamentarium available to manipulate and study model systems has been revolutionized. At the same time, progress in defining the genetic risk factors for human neuropsychiatric disorders has advanced considerably. Although the field remains in an early stage in this process, multiple examples of the value of successful gene discovery, and its application to deep biological understanding, have emerged.

Among the many striking findings from recent genetic studies of neurodevelopmental and psychiatric conditions is the overlap in genetic risks across a wide range of diagnostic boundaries. While it may not come as much of a surprise that biology does not

hew to categorical diagnostic criteria, it is nonetheless a formidable conceptual challenge to consider how the field will trace these effects and arrive at new therapeutic strategies.

In addition, it is worthwhile noting that for many other psychiatric conditions that have not yet seen the type of progress noted earlier, the calculus is straightforward: greater investment and larger sample sizes will lead to greater insight. For example, recent studies of de novo mutations in Tourette syndrome and obsessive-compulsive disorder clearly demonstrate that the rate-limiting factor for the identification of high confidence risk genes is the availability for sequencing of parent-child trios. In a similar vein, GWAS studies of major depression have only very recently reached sample sizes adequate to confirm statistically significant associated common variants. These studies have included hundreds of thousands of individuals and, not surprisingly, have identified risk alleles carrying very small individual effects.

This last point highlights the idea that one size does not fit all for the genomics of behavioral, developmental, and psychiatric disorders. From the investigations of model systems, to the illumination of rare Mendelian disorders, to the disentangling of both common and rare variants contributing to common disorders, the tools and opportunities available today are unprecedented. The coming years should see deep insights into the biology of psychiatric and neurodevelopmental disorders, and perhaps therapies with the potential to help patients and their families.

Highlights

1. Rare genetic syndromes such as fragile X syndrome, Rett syndrome, and Williams syndrome have provided important insights into the molecular mechanisms of complex human behaviors. Moreover, while considerable work remains to be done, the study of these syndromes has already challenged the notion that associated cognitive and behavioral deficits are immutable and has demonstrated the utility of a wide range of model systems in illuminating conserved biological mechanisms.

2. The sequencing of the human genome, the development of high-throughput genomic assays, and simultaneous computing and methodological advances have led to a profound change in the understanding of the genetics of human behavior and psychiatric illness. Several paradigmatic disorders, including schizophrenia and autism,

have seen dramatic progress, leading to the identification of dozens of definitive risk genes and chromosomal regions.

3. The maturation of the field of psychiatric genetics and genomics over the past decade has revealed the frailty of testing pre-specified candidate genes. These types of studies have now been supplanted by genome-wide scans of both common and rare alleles. Coupled with rigorous statistical frameworks and consensus statistical thresholds, these are yielding highly reliable and reproducible results.

4. At present, the cumulative evidence suggests that the full range of genetic variations underlies complex behavioral syndromes, including common and rare, transmitted and de novo, germline and somatic, and sequence and chromosomal structural variation. However, the relative contributions of these various types of genetic changes vary for a given disorder.

5. A striking finding from recent advances in the genetics of human behavior has been the overlap of genetic risks for syndromes with distinct symptoms and natural histories. Understanding how and why an identical mutation may lead to highly diverse phenotypic outcomes in different individuals will be a major challenge for the future.

6. Findings across common psychiatric disorders point to extremely high rates of genetic heterogeneity. This, coupled with the biological pleiotropy of the risk genes that have been identified to date, as well as the dynamism and complexity of human brain development, all point to important challenges ahead in moving from an understanding of risk genes to an understanding of behavior. Similarly, at present, an important distinction can be made between illuminating the biology of risk genes and unraveling the pathophysiology of behavioral syndromes.

Glossary[1]

Allele. Humans carry two sets of chromosomes, one from each parent. Equivalent genes in the two sets might be different, for example, because of single nucleotide polymorphisms. An allele is one of the two (or more) forms of a particular gene.

Centromere. Chromosomes contain a compact region known as a centromere, where sister chromatids (the two exact copies of each chromosome that are formed after replication) are joined.

Cloning. The process of generating sufficient copies of a particular piece of DNA to allow it to be sequenced or studied in some other way.

Complementary DNA (cDNA). A DNA sequence made from a messenger RNA molecule, using an enzyme called *reverse transcriptase*. cDNAs can be used experimentally to determine the sequence of messenger RNAs after their introns (non–protein-coding sections) have been spliced out.

Conservation of genes. Genes that are present in two distinct species are said to be conserved, and the two genes from the different species are called *orthologous genes*. Conservation can be detected by measuring the similarity of the two sequences at the base (RNA or DNA) or amino acid (protein) level. The more similarities there are, the more highly conserved the two sequences.

Copy number variation (CNV). A deletion or duplication of a limited genetic region that results in an individual having more or fewer than the usual two copies of some genes. Copy number variations are observed in some neurological and psychiatric disorders.

CRISPR (Clustered Regularly Interspaced Short Palindromic Repeats). An enzyme-RNA system in which the enzyme cleaves target sequences that match an RNA guide; the RNA guide can be engineered to recognize a desired gene or sequences within a cell for mutation.

Euchromatin. The gene-rich regions of a genome (see also heterochromatin).

Eukaryote. An organism with cells that have a complex internal structure, including a nucleus. Animals, plants, and fungi are all eukaryotes.

Genome. The complete DNA sequence of an organism.

Genotype. The set of genes that an individual carries; usually refers to the particular pair of alleles (alternative forms of a gene) that a person has at a given region of the genome.

Haplotype. A particular combination of alleles (alternative forms of genes) or sequence variations that are closely linked—that is, are likely to be inherited together—on the same chromosome.

Heterochromatin. Compact, gene-poor regions of a genome, which are enriched in simple sequence repeats.

Introns and exons. Genes are transcribed as continuous sequences, but only some segments of the resulting messenger RNA molecules contain information that encodes a protein product. These segments are called *exons*. The regions between exons are known as *introns* and are spliced from the RNA before the product is made.

---

[1]Based on Bork P, Copley R. 2001. Genome speak. Nature 409:815.

Long and short arms. The regions on either side of the centromere are known as arms. As the centromere is not in the center of the chromosome, one arm is longer than the other.

Messenger RNA (mRNA). Proteins are not synthesized directly from genomic DNA. Instead, an RNA template (a precursor mRNA) is constructed from the sequence of the gene. This RNA is then processed in various ways, including splicing. Spliced RNAs destined to become templates for protein synthesis are known as mRNAs.

Mutation. An alteration in a genome compared to some reference state. Mutations do not always have harmful effects.

Phenotype. The observable properties and physical characteristics of an organism.

Polymorphism. A region of the genome that varies between individual members of a population. To be called a polymorphism, a variant should be present in a significant number of people in the population.

Prokaryote. A single-celled organism with a simple internal structure and no nucleus. Bacteria and archaebacteria are prokaryotes.

Proteome. The complete set of proteins encoded by the genome.

Recombination. The process by which DNA is exchanged between pairs of equivalent chromosomes during egg and sperm formation. Recombination has the effect of making the chromosomes of the offspring distinct from those of the parents.

Restriction endonuclease. An enzyme that cleaves DNA at a particular short sequence. Different types of restriction endonuclease cleave at different sequences.

RNA interference (RNAi). A method for reducing the function of a specific gene by introducing into a cell small RNAs with complementarity to the targeted mRNA. Pairing of the mRNA with the small RNA leads to destruction of the endogenous mRNA.

Single nucleotide polymorphism (SNP). A polymorphism caused by the change of a single nucleotide. SNPs are often used in genetic mapping studies.

Splicing. The process that removes introns (noncoding portions) from transcribed RNAs. Exons (protein-coding portions) can also be removed. Depending on which exons are removed, different proteins can be made from the same initial RNA or gene. Different proteins created in this way are *splice variants* or *alternatively spliced.*

Transcription. The process of copying a gene into RNA. This is the first step in turning a gene into a protein, although not all transcripts lead to proteins.

Transcriptome. The complete set of RNAs transcribed from a genome.

Translation. The process of using a messenger RNA sequence to synthesize a protein. The messenger RNA serves as a template on which transfer RNA molecules, carrying amino acids, are lined up. The amino acids are then linked together to form a protein chain.

Matthew W. State
Cornelia I. Bargmann
T. Conrad Gilliam

Selected Reading

Alberts B, Johnson A, Lewis J, Raff M, Roberts K, Walter P. 2002. *Molecular Biology of the Cell,* 4th ed. New York: Garland Publishing. Also searchable at http://www.ncbi.nlm.nih.gov/entrez/query.fcgi?db=Books.

Allada R, Emery P, Takahashi JS, Rosbash M. 2001. Stopping time: the genetics of fly and mouse circadian clocks. Annu Rev Neurosci 24:1091–1119.

Bouchard TJ Jr, Lykken DT, McGue M, Segal NL, Tellegen A. 1990. Sources of human psychological differences: the Minnesota Study of Twins Reared Apart. Science 250:222–228.

Cong L, Ran FA, Cox D, et al. 2013. Multiplex genome engineering using CRISPR/Cas systems. Science 339:819–823.

Griffiths AJF, Gelbart WM, Miller JH, Lewontin RC. 1999. *Modern Genetic Analysis.* New York: Freeman. Also searchable at http://www.ncbi.nlm.nih.gov/entrez/query.fcgi?db=Books.

International Human Genome Sequencing Consortium. 2001. Initial sequencing and analysis of the human genome. Nature 409:860–921.

Jinek M, Chylinski K, Fonfara I, Hauer M, Doudna JA, Charpentier E. 2012. A programmable dual-RNA-guided DNA endonuclease in adaptive bacterial immunity. Science 337:816–821.

Online Mendelian Inheritance in Man, OMIM. McKusick-Nathans Institute of Genetic Medicine, Johns Hopkins University (Baltimore, MD) and National Center for Biotechnology Information, National Library of Medicine (Bethesda, MD). http://www.ncbi.nlm.nih.gov/omim/.

Venter JG, Adams MD, Myers EW, et al. 2001. The sequence of the human genome. Science 291:1304–1351.

References

Alberts B, Johnson A, Lewis J, Raff M, Roberts K, Walter P. 1998. *Molecular Biology of the Cell,* 3rd ed. New York: Garland Publishing.

Amir RE, Van den Veyver IB, Wan M, Tran CQ, Francke U, Zoghbi HY. 1999. Rett syndrome is caused by mutations in X-linked MECP2, encoding methyl-CpG-binding protein 2. Nat Genet 23:185–188.

Antoch MP, Song EJ, Chang AM, et al. 1997. Functional identification of the mouse circadian Clock gene by transgenic BAC rescue. Cell 89:655–667.

Arnold SE, Talbot K, Hahn CG. 2004. Neurodevelopment, neuroplasticity, and new genes for schizophrenia. Prog Brain Res 147:319–345.

Bear MF, Huber KM, Warren ST. 2004. The mGluR theory of fragile X syndrome. Trends Neurosci 27:370–377.

Bellugi U, Lichtenberger L, Jones W, Lai Z, St George M. 2000. I. The neurocognitive profile of Williams Syndrome: a complex pattern of strengths and weaknesses. J Cogn Neurosci 12:7–29. Suppl.

Ben-Shahar Y, Robichon A, Sokolowski MB, Robinson GE. 2002. Influence of gene action across different time scales on behavior. Science 296:741–744.

Caron H, van Schaik B, van der Mee M, et al. 2001. The human transcriptome map: clustering of highly expressed genes in chromosomal domains. Science 291:1289–1292.

De Bono M, Bargmann CI. 1998. Natural variation in a neuropeptide Y receptor homolog modifies social behavior and food responses in C. elegans. Cell 94:679–689.

De Rubeis S, He X, Goldberg AP, et al. 2014. Synaptic, transcriptional and chromatin genes disrupted in autism. Nature 515:209–215.

Fromer M, Pocklington AJ, Kavanagh DH, et al. 2014. De novo mutations in schizophrenia implicate synaptic networks. Nature 506:179–184.

Genovese G, Fromer M, Stahl EA, et al. 2016. Increased burden of ultra-rare protein-altering variants among 4,877 individuals with schizophrenia. Nat Neurosci 19:1433–1441.

Gottesman II. 1991. *Schizophrenia Genesis. The Origins of Madness*. New York: Freeman.

Iossifov I, O'Roak BJ, Sanders SJ, et al. 2014. The contribution of de novo coding mutations to autism spectrum disorder. Nature 515:216–221.

Jamain S, Quach H, Betancur C, et al. 2003. Mutations of the X-linked genes encoding neuroligins NLGN3 and NLGN4 are associated with autism. Nat Genet 34:27–29.

Kahler SG, Fahey MC. 2003. Metabolic disorders and mental retardation. Am J Med Genet C Semin Med Genet 117:31–41.

Khaitovich P, Muetzel B, She X, et al. 2004. Regional patterns of gene expression in human and chimpanzee brains. Genome Res 14:1462–1473.

Konopka RJ, Benzer S. 1971. Clock mutations of *Drosophila melanogaster*. Proc Natl Acad Sci U S A 68:2112–2116.

Lai CS, Fisher SE, Hurst JA, Vargha-Khadem F, Monaco AP. 2001. A forkhead-domain gene is mutated in a severe speech and language disorder. Nature 413:519–523.

Lander ES, Linton LM, Birren B, et al. 2001. Initial sequencing and analysis of the human genome. Nature 409:860–921.

Laumonnier F, Bonnet-Brilhault F, Gomot M, et al. 2004. X-linked mental retardation and autism are associated with a mutation in the NLGN4 gene, a member of the neuroligin family. Am J Hum Genet 74:552–557.

Lim MM, Wang Z, Olazabal DE, Ren X, Terwilliger EF, Young LJ. 2004. Enhanced partner preference in a promiscuous species by manipulating the expression of a single gene. Nature 429:754–757.

Mayford M, Bach ME, Huang Y-Y, Wang L, Hawkins RD, Kandel ER. 1996. Control of memory formation through regulated expression of a CaMKII transgene. Science 274:1678-1683.

McGue M, Bouchard TH Jr. 1998. Genetic and environmental influences on human behavioral differences. Ann Rev Neurosci 21:1–24.

Mendel G. 1866. Versuche über Pflanzen-hybriden. Verh Naturforsch 4:2–47. Translated in: C Stern, ER Sherwood (eds). *The Origin of Genetics: A Mendel Source Book*, 1966. San Francisco: Freeman.

Neale BM, Kou Y, Liu L, et al. 2012. Patterns and rates of exonic de novo mutations in autism spectrum disorders. Nature 485:242–245.

O'Roak BJ, Vives L, Girirajan S, et al. 2012. Sporadic autism exomes reveal a highly interconnected protein network of de novo mutations. Nature 485:246–250.

Sanders SJ, Ercan-Sencicek AG, Hus V, et al. 2011. Multiple recurrent de novo CNVs, including duplications of the 7q11.23 Williams syndrome region, are strongly associated with autism. Neuron 70:863–885.

Sanders SJ, He X, Willsey AJ, et al. 2015. Insights into autism spectrum disorder genomic architecture and biology from 71 risk loci. Neuron 87:1215–1233.

Sanders SJ, Murtha MT, Gupta AR, et al. 2012. De novo mutations revealed by whole exome sequencing are strongly associated with autism. Nature 485:237–241.

Satterstrom FK, Kosmicki JA, Wang J, et al. 2020. Large-scale exome sequencing study implicated both developmental and functional changes in the neurobiology of autism. Cell 180:568–594.

Schizophrenia Working Group of the Psychiatric Genomics Consortium. 2014. Biological insights from 108 schizophrenia-associated genetic loci. Nature 511:421–427.

Sebat J, Lakshmi B, Malhotra D, et al. 2007. Strong association of de novo copy number variation with autism. Science 316:445–449.

Sekar A, Bialas AR, de Rivera H, et al. 2016. Schizophrenia risk from complex variation of complement component 4. Nature 530:177–183.

Singh T, Kurki MI, Curtis D, et al. 2016. Rare loss-of-function variants in SETD1A are associated with schizophrenia and developmental disorders. Nat Neurosci 19:571–577.

Sokolowski MB. 1980. Foraging strategies of *Drosophila melanogaster*: a chromosomal analysis. Behav Genet 10: 291–302.

Sokolowski MB. 2001. *Drosophila*: genetics meets behavior. Nat Rev Genet 2:879–890.

Sztainberg Y, Zoghbi HY. 2016. Lessons learned from studying syndromic autism spectrum disorders. Nat Neurosci 19:1408–1417.

Takahashi JS, Pinto LH, Vitaterna MH. 1994. Forward and reverse genetic approaches to behavior in the mouse. Science 264:1724–1733.

Toh KL, Jones CR, He Y, et al. 2001. An hPer2 phosphorylation site mutation in familial advanced sleep phase syndrome. Science 291:1040–1043.

Tsien JZ, Huerta PT, Tonegawa S. 1996. The essential role of hippocampal CA1 NMDA receptor-dependent synaptic plasticity in spatial memory. Cell 87:1327–1338.

Walter J, Paulsen M. 2003. Imprinting and disease. Semin Cell Dev Biol 14:101–110.

Watson JD, Tooze J, Kurtz DT (eds). 1983. *Recombinant DNA: A Short Course*. New York: Scientific American.

Whitfield CW, Cziko AM, Robinson GE. 2003. Gene expression profiles in the brain predict behavior in individual honey bees. Science 302:296–299.

Young LJ, Lim MM, Gingrich B, Insel TR. 2001. Cellular mechanisms of social attachment. Horm Behav 40:132–138.

Zondervan KT, Cardon LR. 2004. The complex interplay among factors that influence allelic association. Nat Rev Genet 5:89–100.

3

Nerve Cells, Neural Circuitry, and Behavior

THE REMARKABLE RANGE OF HUMAN behavior depends on a sophisticated array of sensory receptors connected to the brain, a highly flexible neural organ that selects from among the stream of sensory signals those events in the environment and in the internal milieu of the body that are important for the individual. The brain actively organizes sensory information for perception, action, decision-making, aesthetic appreciation, and future reference—that is

to say, memory. It also ignores and discards information judiciously, one hopes, and reports to other brains about some of these operations and their psychological manifestations. All this is accomplished by interconnected nerve cells.

Individual nerve cells, or neurons, are the basic signaling units of the brain. The human brain contains a huge number of these cells, on the order of 86 billion neurons, that can be classified into at least a thousand different types. Yet this great variety of neurons is less of a factor in the complexity of human behavior than is their organization into anatomical circuits with precise functions. Indeed, one key organizational principle of the brain is that nerve cells with *similar* properties can produce different actions because of the way they are interconnected.

Because relatively few principles of organization of the nervous system give rise to considerable functional complexity, it is possible to learn a great deal about how the nervous system produces behavior by focusing on five basic features of the nervous system:

1. The structural components of individual nerve cells;
2. The mechanisms by which neurons produce signals within themselves and between each other;
3. The patterns of connection between nerve cells and between nerve cells and their targets (muscle and gland effectors);
4. The relationship of different patterns of interconnection to different types of behavior; and
5. How neurons and their connections are modified by experience.

The parts of this book are organized around these five major topics. In this chapter, we introduce these topics in turn in an overview of the neural control of behavior. We first consider the structure and function of neurons and the glial cells that surround and support them. We then examine how individual cells organize and transmit signals and how signaling between a few interconnected nerve cells produces a simple behavior, the knee-jerk reflex. We then extend these ideas to more complex behaviors, mediated by more complex and malleable circuits.

The Nervous System Has Two Classes of Cells

There are two main classes of cells in the nervous system: nerve cells, or neurons, and glial cells, or glia.

Nerve Cells Are the Signaling Units of the Nervous System

A typical neuron has four morphologically defined regions: (1) the cell body, (2) dendrites, (3) an axon, and (4) presynaptic terminals (Figure 3–1). As we shall see, each region has a distinct role in generating signals and communicating with other nerve cells.

The cell body or *soma* is the metabolic center of the cell. It includes the nucleus, which contains the genes of the cell, and the endoplasmic reticulum, an extension of the nucleus where the cell's proteins are synthesized. The cell body usually gives rise to two kinds of processes: several short *dendrites* and one long, tubular *axon*. Dendrites branch out in tree-like fashion and are the main apparatus for receiving incoming signals

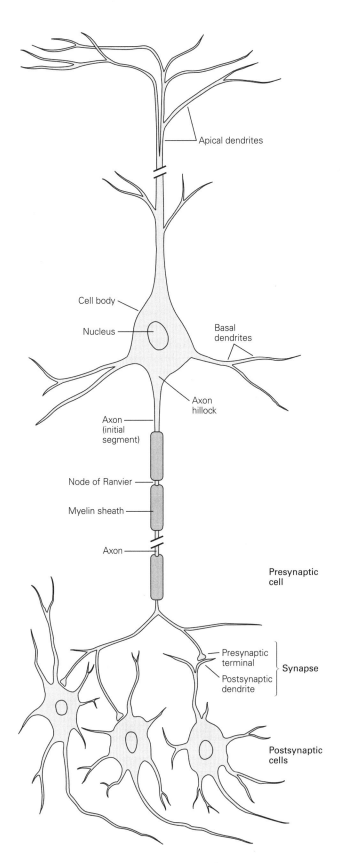

Figure 3–1 (Right) The structure of a neuron. Most neurons in the vertebrate nervous system have several main features in common. The cell body contains the nucleus, the storehouse of genetic information, and gives rise to two types of cell processes: axons and dendrites. Axons are the transmitting element of neurons; they vary greatly in length, some extending more than 1 m within the body. Most axons in the central nervous system are very thin (between 0.2 μm and 20 μm in diameter) compared with the diameter of the cell body (50 μm or more). Many axons are insulated by a sheath of fatty myelin that is regularly interrupted at gaps called the nodes of Ranvier. The action potential, the cell's conducting signal, is initiated at the initial segment of the axon and propagates to the synapse, the site at which signals flow from one neuron to another. Branches of the axon of the presynaptic neuron transmit signals to the postsynaptic cell. The branches of a single axon may form synapses with as many as 1,000 postsynaptic neurons. The apical and basal dendrites together with the cell body are the input elements of the neuron, receiving signals from other neurons.

from other nerve cells. The axon typically extends some distance from the cell body before it branches, allowing it to carry signals to many target neurons. An axon can convey electrical signals over distances ranging from 0.1 mm to 1 m. These electrical signals, or *action potentials,* are initiated at a specialized trigger region near the origin of the axon called the *initial segment* from which the action potentials propagate down the axon without failure or distortion at speeds of 1 to 100 m/s. The amplitude of an action potential traveling down the axon remains constant at 100 mV because the action potential is an all-or-none impulse that is regenerated at regular intervals along the axon (Figure 3–2).

Action potentials are the signals by which the brain receives, analyzes, and conveys information. These signals are highly stereotyped throughout the nervous system, even though they are initiated by a great variety of events in the environment that impinge on our bodies—from light to mechanical contact, from odorants to pressure waves. The physiological signals that convey information about vision are identical to those that carry information about odors. Here we see a key principle of brain function: the type of information conveyed by an action potential is determined not by the form of the signal but by the pathway the signal travels in the brain. The brain thus analyzes and interprets patterns of incoming electrical signals carried

over specific pathways, and in turn creates our sensations of sight, touch, taste, smell, and sound.

To increase the speed by which action potentials are conducted, large axons are wrapped in an insulating sheath of a lipid substance, myelin. The sheath is interrupted at regular intervals by the nodes of Ranvier, uninsulated spots on the axon where the action potential is regenerated. (Myelination is discussed in detail in Chapters 7 and 8 and action potentials in Chapter 10.)

Near its end, the axon divides into fine branches that contact other neurons at specialized zones of communication known as *synapses.* The nerve cell transmitting a signal is called the *presynaptic cell;* the cell receiving the signal is the *postsynaptic cell.* The presynaptic cell transmits signals from specialized enlarged regions of its axon's branches, called *presynaptic terminals* or *nerve terminals.* The presynaptic and postsynaptic cells are separated by a very narrow space, the *synaptic cleft.* Most presynaptic terminals end on the postsynaptic neuron's dendrites, but some also terminate on the cell body or, less often, at the beginning or end of the axon of the postsynaptic cell (see Figure 3–1). Some presynaptic neurons excite their postsynaptic target cells; other presynaptic neurons inhibit their target cells.

The neuron doctrine (Chapter 1) holds that each neuron is a discrete cell with distinctive processes arising from its cell body and that neurons are the signaling units of the nervous system. In retrospect, it is hard to appreciate how difficult it was for scientists to accept this elementary idea when first proposed. Unlike other tissues, whose cells have simple shapes and fit into a single field of the light microscope, nerve cells have complex shapes. The elaborate patterns of dendrites and the seemingly endless course of some axons made it extremely difficult to establish a relationship between these elements. Even after the anatomists Jacob Schleiden and Theodor Schwann put forward the cell theory in the early 1830s—and the idea that cells are the structural units of all living matter became a central dogma of biology—most anatomists did not accept that the cell theory applied to the brain, which they thought of as a continuous, web-like reticulum of very thin processes.

The coherent structure of the neuron did not become clear until late in the 19th century, when Ramón y Cajal began to use the silver-staining method introduced by Golgi. Still used today, this method has two advantages. First, in a random manner that is not understood, the silver solution stains only about 1% of the cells in any particular brain region, making it possible to examine a single neuron in isolation from

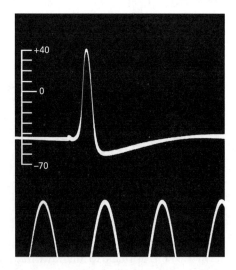

Figure 3–2 This historic tracing is the first published intracellular recording of an action potential. It was recorded in 1939 by Alan Hodgkin and Andrew Huxley from a squid giant axon, using glass capillary electrodes filled with sea water. The timing pulses (bottom) are separated by 2 ms. The vertical scale indicates the potential of the internal electrode in millivolts, the sea water outside being taken as zero potential. (Reproduced, with permission, from Hodgkin and Huxley 1939.)

its neighbors. Second, the neurons that do take up the stain are delineated in their entirety, including the cell body, axon, and full dendritic tree. The stain reveals that there is no cytoplasmic continuity between neurons, and Cajal concluded, prophetically and correctly, that there is no continuity even at synapses between two cells.

Ramón y Cajal applied Golgi's method to the embryonic nervous systems of many animals as well as humans. By examining the structure of neurons in almost every region of the nervous system, he could describe classes of nerve cells and map the precise connections between many of them. In this way, Ramón y Cajal deduced, in addition to the neuron doctrine, two other principles of neural organization that would prove particularly valuable in studying communication in the nervous system.

The first of these, the *principle of dynamic polarization*, states that electrical signals within a nerve cell flow in only one direction: from the postsynaptic sites of the neuron, usually the dendrites and cell body, to the trigger region at the axon. From there, the action potential is propagated along the entire length of the axon to its terminals. In most neurons studied to date, electrical signals in fact travel along the axon in one direction.

The second principle advanced by Ramón y Cajal, *connectional specificity*, states that nerve cells do not connect randomly with one another in the formation of networks but make specific connections—at particular contact points—with certain postsynaptic target cells and not with others. The principles of dynamic polarization and connectional specificity are the basis of the modern cellular-connectionist approach to studying the brain.

Ramón y Cajal was also among the first to realize that the feature that most distinguishes one type of neuron from another is form, specifically the number of the processes arising from the cell body. Neurons are thus classified into three large groups: unipolar, bipolar, and multipolar.

Unipolar neurons are the simplest because they have a single primary process, which usually gives rise to many branches. One branch serves as the axon; other branches function as receiving structures (Figure 3–3A). These cells predominate in the nervous systems of invertebrates; in vertebrates, they occur in the autonomic nervous system.

Bipolar neurons have an oval soma that gives rise to two distinct processes: a dendritic structure that receives signals from other neurons and an axon that carries information toward the central nervous system (Figure 3–3B). Many sensory cells are bipolar, including

those in the retina and olfactory epithelium of the nose. The receptor neurons that convey touch, pressure, and pain signals to the spinal cord develop initially as bipolar cells, but the two cell processes fuse into a single continuous structure that emerges from a single point in the cell body, and the dendrite is endowed with the specializations that render it an axon. In these so-called pseudo-unipolar cells, one axon transmits information from the sensory receptors in the skin, joints, and muscle toward the cell body, while the other carries this sensory information to the spinal cord (Figure 3–3C).

Multipolar neurons predominate in the nervous system of vertebrates. They typically have a single axon and many dendritic structures emerging from various points around the cell body (Figure 3–3D). Multipolar cells vary greatly in shape, especially in the length of their axons and in the extent, dimensions, and intricacy of their dendritic branching. Usually the extent of branching correlates with the number of synaptic contacts that other neurons make onto them. A spinal motor neuron with a relatively modest number of dendrites receives about 10,000 contacts—1,000 on the cell body and 9,000 on dendrites. In Purkinje cells in the cerebellum, the dendritic tree is much larger and bushier, receiving as many as a million contacts!

Nerve cells are also classified into three major functional categories: sensory neurons, motor neurons, and interneurons. *Sensory neurons* carry information from the body's peripheral sensors into the nervous system for the purpose of both perception and motor coordination. Some primary sensory neurons are called *afferent neurons*, and the two terms are used interchangeably. The term *afferent* (carried toward the central nervous system) applies to all information reaching the central nervous system from the periphery, whether or not this information leads to sensation. The term *sensory* designates those afferent neurons that convey information to the central nervous system from the sensory epithelia, from joint sensory receptors, or from muscle, but the concept has been expanded to include neurons in primary and secondary cortical areas that respond to changes in a sensory feature, such as displacement of an object in space, a shift in sound frequency, or the angular rotation of the head (via vestibular organs in the ear) or even something as complex as a face.

The term *efferent* applies to all information carried from the central nervous system toward the motor organs, whether or not this information leads to action. *Motor neurons* carry commands from the brain or spinal cord to muscles and glands (efferent information). The traditional definition of a *motor neuron* (or motoneuron) is a neuron that excites a muscle, but the designation of motor neuron now includes other

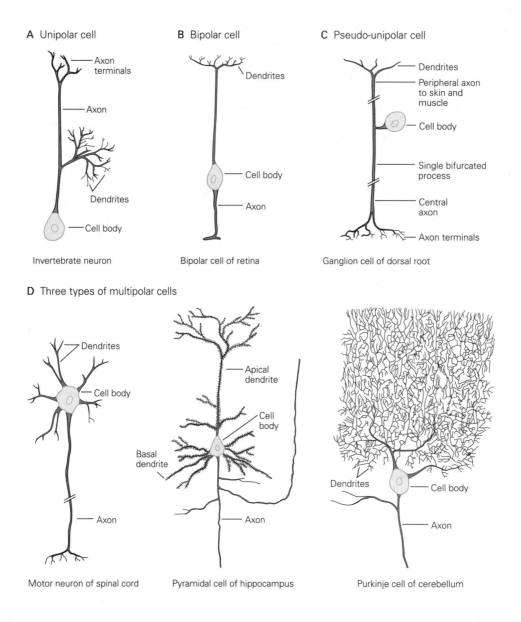

A Unipolar cell

Axon terminals

Axon

Dendrites

Cell body

Invertebrate neuron

B Bipolar cell

Dendrites

Cell body

Axon

Bipolar cell of retina

C Pseudo-unipolar cell

Dendrites

Peripheral axon to skin and muscle

Cell body

Single bifurcated process

Central axon

Axon terminals

Ganglion cell of dorsal root

D Three types of multipolar cells

Dendrites

Cell body

Axon

Motor neuron of spinal cord

Apical dendrite

Cell body

Basal dendrite

Axon

Pyramidal cell of hippocampus

Dendrites

Cell body

Axon

Purkinje cell of cerebellum

Figure 3–3 Neurons are classified as unipolar, bipolar, or multipolar according to the number of processes that originate from the cell body.

A. Unipolar cells have a single process emanating from the cell. Different segments serve as receptive surfaces or releasing terminals. Unipolar cells are characteristic of the invertebrate nervous system.

B. Bipolar cells have two types of processes that are functionally specialized. The dendrite receives electrical signals and the axon transmits signals to other cells.

C. Pseudo-unipolar cells, which are variants of bipolar cells, carry somatosensory information to the spinal cord. During development, the two processes of the embryonic bipolar cell fuse and emerge from the cell body as a single process that

has two functionally distinct segments. Both segments function as axons; one extends to peripheral skin or muscle, the other to the central spinal cord. (Adapted, with permission, from Ramón y Cajal 1933.)

D. Multipolar cells have a single axon and many dendrites. They are the most common type of neuron in the mammalian nervous system. Three examples illustrate the large diversity of these cells. Spinal motor neurons innervate skeletal muscle fibers. Pyramidal cells have a roughly triangular cell body; dendrites emerge from both the apex (the apical dendrite) and the base (the basal dendrites). Pyramidal cells are found in the hippocampus and throughout the cerebral cortex. Purkinje cells of the cerebellum are characterized by a rich and extensive dendritic tree that accommodates an enormous number of synaptic inputs. (Adapted, with permission, from Ramón y Cajal 1933.)

neurons that do not innervate muscle directly but that command action indirectly. A useful characterization of motor and sensory neurons alike is their temporal fidelity to matters outside the nervous system. Their activity keeps up with changes in external stimuli and dynamical forces exerted by the body musculature. Sensory neurons supply the brain with data, whereas motor neurons convert ideation into praxis. Together they compose our interface with the world.

Interneurons comprise the most numerous functional category and are subdivided into two classes: relay and local. Relay or projection interneurons have long axons and convey signals over considerable distances, from one brain region to another. Local interneurons have short axons because they form connections with nearby neurons in local circuits. Since almost every neuron can be regarded as an interneuron, the term is often used to distinguish between neurons that project to another neuron within a local circuit as opposed to neurons that project to a separate neural structure. The term is also sometimes used as shorthand for an inhibitory neuron, especially in studies of cortical circuits, but for clarity, the term *inhibitory interneuron* should be used when appropriate.

Each functional classification can be subdivided further. Sensory system interneurons can be classified according to the type of sensory stimuli to which they respond; these initial classifications can be broken down still further, according to location, density, and size as well as patterns of gene expression. Indeed, our view of neuronal complexity is rapidly evolving due to advances in mRNA sequence analysis that have enabled the molecular profiling of individual neurons. Such analyses have recently revealed a much greater heterogeneity of neuronal types than previously thought (Figure 3–4).

Glial Cells Support Nerve Cells

Glial cells greatly outnumber neurons—there are 2 to 10 times more glia than neurons in the vertebrate central nervous system. Although the name for these cells derives from the Greek for glue, glia do not commonly hold nerve cells together. Rather they surround the cell bodies, axons, and dendrites of neurons. Glia differ from neurons morphologically; they do not form dendrites and axons.

Glia also differ functionally. Although they arise from the same embryonic precursor cells, they do not have the same membrane properties as neurons and thus are not electrically excitable. Hence, they are not directly involved in electrical signaling, which is the function of nerve cells. Yet they play a role in allowing electrical signals to move quickly along the axons

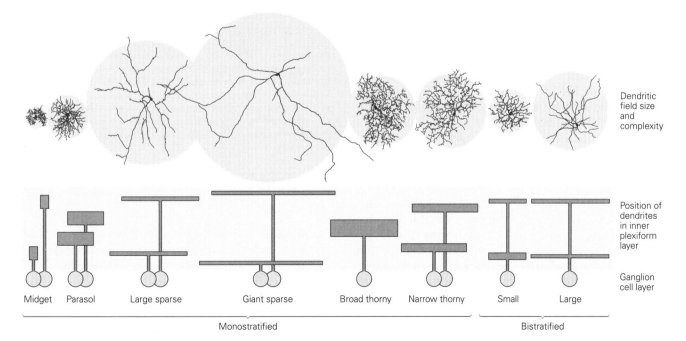

Figure 3–4 Sensory neurons can be subdivided into functionally distinct groups. For example, at least 13 types of retinal ganglion cells are distinguished based on the size and shape of their dendrites combined with the depth within the retina at which they receive their inputs. The inner plexiform layer contains the connections between interneurons of the retina (bipolar and amacrine cells) and the ganglion cells. (Reproduced, with permission, from Dacey et al. 2003. Copyright © 2003 Elsevier.)

of neurons, and they appear to play an important role in guiding connectivity during early development and stabilizing new or altered connections between neurons that occur through learning. Over the past decade, interest in the diverse functions of glia has accelerated, and their characterization has changed from support cells to functional partners of neurons (Chapter 7).

Each Nerve Cell Is Part of a Circuit That Mediates Specific Behaviors

Every behavior is mediated by specific sets of interconnected neurons, and every neuron's behavioral function is determined by its connections with other neurons. A simple behavior, the knee-jerk reflex, will illustrate this. The reflex is initiated when a transient imbalance of the body stretches the quadriceps extensor muscles of the leg. This stretching elicits sensory information that is conveyed to motor neurons, which in turn send commands to the extensor muscles to contract so that balance is restored.

This reflex is used clinically to test the integrity of the nerves as well as the cerebrospinal control of the reflex amplitude (or gain). The underlying mechanism is important because it maintains normal tone in the quadriceps and prevents our knees from buckling when we stand or walk. The tendon of the quadriceps femoris, an extensor muscle that moves the lower leg, is attached to the tibia through the tendon of the patella (kneecap). Tapping this tendon just below the patella stretches the quadriceps femoris. This stretch initiates reflex contraction of the quadriceps muscle to produce the familiar knee jerk. By increasing the tension of a select group of muscles, the stretch reflex changes the position of the leg, suddenly extending it outward (Figure 3–5).

The cell bodies of the sensory neurons involved in the knee-jerk reflex are clustered near the spinal cord in the dorsal root ganglia. They are pseudo-unipolar cells; one branch of each cell's axon runs to the quadriceps muscle at the periphery, while the other runs centrally into the spinal cord. The branch that innervates the quadriceps makes contact with stretch-sensitive receptors (muscle spindles) and is excited when the muscle is stretched. The branch reaching the spinal cord forms excitatory connections with the motor neurons that innervate the quadriceps and control its contraction. This branch also contacts local interneurons that *inhibit* the motor neurons controlling the opposing flexor muscles (Figure 3–5). Although these local interneurons are not involved in producing the stretch reflex itself, they increase the stability of the reflex by coordinating the actions of opposing muscle groups.

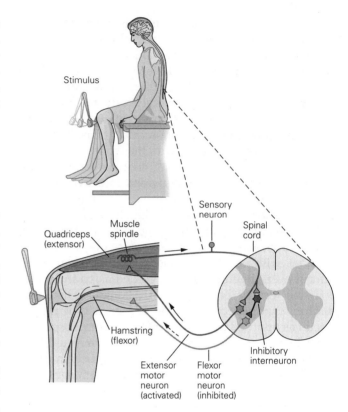

Figure 3–5 The knee-jerk reflex is controlled by a simple circuit of sensory and motor neurons. Tapping the kneecap with a reflex hammer pulls on the tendon of the quadriceps femoris, a muscle that extends the lower leg. When the muscle stretches in response to the pull of the tendon, information regarding this change in the muscle is conveyed to the central nervous system by sensory neurons. In the spinal cord, the sensory neurons form excitatory synapses with extensor motor neurons that contract the quadriceps, the muscle that was stretched. The sensory neurons act indirectly, through interneurons, to inhibit flexor motor neurons that would otherwise contract the opposing hamstring muscles. These actions combine to produce the reflex behavior. In the drawing, each extensor and flexor motor neuron represents a population of many cells.

Thus, the electrical signals that produce the stretch reflex carry four kinds of information:

1. Sensory information is conveyed to the central nervous system (the spinal cord) from muscle.
2. Motor commands from the central nervous system are issued to the muscles that carry out the knee jerk.
3. Inhibitory commands are issued to motor neurons that innervate opposing muscles.
4. Information about local neuronal activity related to the knee jerk is sent to higher centers of the central nervous system, permitting the brain to coordinate different behaviors simultaneously or in series.

In addition, the brain asserts context-dependent control of the reflex to adjust its gain. For example, when we run, the hamstring muscles flex the knee, thereby stretching the quadriceps. The brain and spinal cord suppress the stretch reflex to allow the quadriceps to relax. When these descending pathways are disrupted, as in some strokes, the reflex is exaggerated and the joint has stiffness.

The stretching of just one muscle, the quadriceps, activates several hundred sensory neurons, each of which makes direct contact with 45 to 50 motor neurons. This pattern of connection, in which one neuron activates many target cells, is called *divergence* (Figure 3–6A). It is especially common in the input stages of the nervous system; by distributing its signals to many target cells, a single neuron can exert wide and diverse influence. Conversely, a single motor cell in the knee-jerk circuit receives 200 to 450 input contacts from approximately 130 sensory cells. This pattern of connection is called *convergence* (Figure 3–6B). It is common at the output stages of the nervous system; a target motor cell that receives information from many sensory neurons is able to integrate information from many sources. Each sensory neuron input produces relatively weak excitation, so convergence also ensures that a motor neuron is activated only when a sufficient number of sensory neurons are activated together.

A stretch reflex such as the knee-jerk reflex is a simple behavior produced by two classes of neurons connecting at excitatory synapses. But not all important signals in the brain are excitatory. Many neurons produce inhibitory signals that reduce the likelihood of firing. Even in the simple knee-jerk reflex, the sensory neurons make both excitatory and inhibitory connections. Excitatory connections in the leg's extensor muscles cause these muscles to contract, whereas connections with inhibitory interneurons prevent the antagonist flexor muscles from contracting. This feature of the circuit is an example of *feedforward inhibition* (Figure 3–7A). In the knee-jerk reflex, feedforward

A Feedforward inhibition

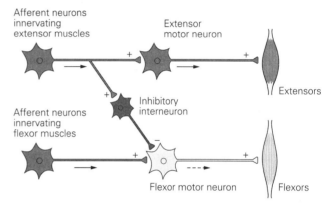

B Feedback inhibition

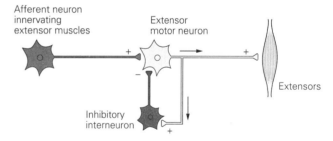

Figure 3–7 Inhibitory interneurons can produce either feedforward or feedback inhibition.

A. Feedforward inhibition enhances the effect of the active pathway by suppressing the activity of pathways mediating opposing actions. Feedforward inhibition is common in monosynaptic reflex systems. For example, in the knee-jerk reflex circuit (**Figure 3–5**) afferent neurons from extensor muscles excite not only the extensor motor neurons but also inhibitory interneurons that prevent the firing of the motor cells innervating the opposing flexor muscles.

B. Feedback inhibition is a self-regulating mechanism. Here extensor motor neurons act on inhibitory interneurons that in turn act on the extensor motor neurons themselves and thus reduce their probability of firing. The effect is to dampen activity within the stimulated pathway and prevent it from exceeding a certain critical level.

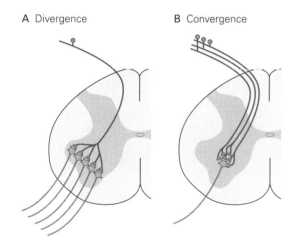

Figure 3–6 Diverging and converging neuronal connections are a key organizational feature of the brain.

A. In the sensory systems, each receptor neuron usually contacts several neurons that represent the second stage of processing. At subsequent processing stages, the incoming connections diverge even more. This allows sensory information from a single site to be distributed more widely in the spinal cord and brain.

B. By contrast, motor neurons are the targets of progressively converging connections. With this arrangement, input from many presynaptic cells is required to activate the motor neuron.

inhibition is *reciprocal*, ensuring that the flexor and extensor pathways always inhibit each other so that only muscles appropriate for the movement and not those opposed to it are recruited.

Some circuits provide *feedback inhibition*. For example, a motor neuron may have excitatory connections with both a muscle and an inhibitory interneuron that itself forms a connection with the motor neuron. When the inhibitory interneuron is excited by the motor neuron, the interneuron is able to limit the ability of the motor neuron to excite the muscle (Figure 3–7B). We will encounter many examples of feedforward and feedback inhibition when we examine more complex behaviors in later chapters.

Signaling Is Organized in the Same Way in All Nerve Cells

To produce a behavior, a stretch reflex for example, each participating sensory and motor nerve cell must generate four different signals in sequence, each at a different site within the cell. Despite variations in cell size and shape, transmitter biochemistry, or behavioral function, almost all neurons can be described by a model neuron that has four functional components

that generate the four types of signals: a receptive component for producing graded input signals, a summing or integrative component that produces a trigger signal, a conducting long-range signaling component that produces all-or-none conducting signals, and a synaptic component that produces output signals to the next neuron in line or to muscle or gland cells (Figure 3–8).

The different types of signals generated in a neuron are determined in part by the electrical properties of the cell membrane. Every cell, including a neuron, maintains a certain difference in the electrical potential on either side of the plasma membrane when the cell is at rest. This is called the *resting membrane potential*. In a typical resting neuron, the voltage of the inside of the cell is about 65 mV more negative than the voltage outside the cell. Because the voltage outside the membrane is defined as zero, we say the resting membrane potential is –65 mV. The resting potential in different nerve cells ranges from –40 to –80 mV; in muscle cells, it is greater still, about –90 mV. As described in detail in Chapter 9, the resting membrane potential results from two factors: the unequal distribution of electrically charged ions, in particular the positively charged Na^+ and K^+ ions, and the selective permeability of the membrane.

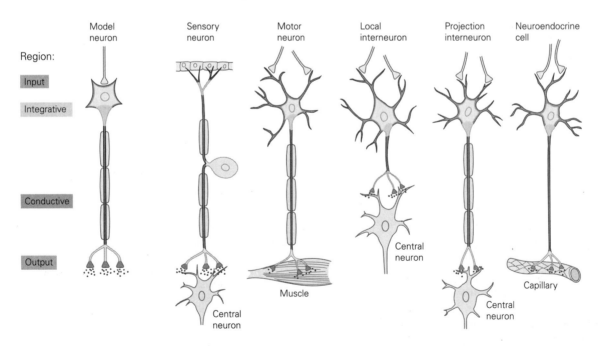

Figure 3–8 Most neurons have four functional regions in which different types of signals are generated. Thus, the functional organization of most neurons, regardless of type, can be represented schematically by a model neuron. This model neuron is the physiological expression of Ramón y Cajal's principle of dynamic polarization. The input, integrative, and conductive signals are all electrical and integral to the cell, whereas the output signal is a chemical substance ejected by the cell into the synaptic cleft. Not all neurons share all of these features; for example, some local interneurons lack a conductive component.

The unequal distribution of positively charged ions on either side of the cell membrane is maintained by two main mechanisms. Intracellular Na^+ and K^+ concentrations are largely controlled by a membrane protein that actively pumps Na^+ out of the cell and K^+ back into it. This *Na⁺-K⁺ pump*, about which we shall learn more in Chapter 9, keeps the Na^+ concentration in the cell low (about one-tenth the concentration outside the cell) and the K^+ concentration high (about 20 times the concentration outside). The extracellular concentrations of Na^+ and K^+ are maintained by the kidneys and the astroglial cells, also known as astrocytes.

The otherwise impermeable cell membrane contains proteins that form pores called *ion channels*. The channels that are active when the cell is at rest are highly permeable to K^+ but considerably less permeable to Na^+. The K^+ ions tend to leak out of these open channels, down the ion's concentration gradient. As K^+ ions exit the cell, they leave behind a cloud of unneutralized negative charge on the inner surface of the membrane, so that the net charge inside the membrane is more negative than that outside. With this state of affairs, the membrane potential is typically maintained at around –65 mV relative to outside of the neuron, and the neuron is said to be at rest.

The resting state is perturbed when the cell begins to take up Na^+ (or Ca^{2+}), which are at a higher concentration outside the cell. The inward movement of these positively charged ions (*inward current*) partially neutralizes the negative voltage inside the cell. We will say more about these events below. What happens next, however, holds the key to understanding what it is about neurons that makes signaling suitable for conveying information.

A cell, such as nerve and muscle, is said to be excitable when its membrane potential can be quickly and significantly altered. In many neurons, a 10-mV change in membrane potential (from –65 to –55 mV) makes the membrane much more permeable to Na^+ than to K^+. The resultant influx of Na^+ further neutralizes the negative charge inside the cell, leading to even more permeability to Na^+. The result is a brief and explosive change in membrane potential to +40 mV, the *action potential*. This potential is actively conducted down the cell's axon to the axon's terminal, where it initiates an elaborate chemical interaction with postsynaptic neurons or muscle cells. Since the action potential is actively propagated, its amplitude does not diminish by the time it reaches the axon terminal. An action potential typically lasts approximately 1 ms, after which the membrane returns to its resting state, with its normal separation of charges and higher permeability to K^+ than to Na^+.

The mechanisms underlying the resting potential and action potential are discussed in detail in Chapters 9 and 10. In addition to the long-distance signals represented by the action potential, nerve cells also produce local signals—receptor potentials and synaptic potentials—that are not actively propagated and that typically decay within just a few millimeters (see next section).

Changes in membrane potential that generate long-range and local signals can be either a decrease or an increase from the resting potential. That is, the resting membrane potential is the baseline from which all signaling occurs. A reduction in membrane potential, called *depolarization*, enhances a cell's ability to generate an action potential and is thus excitatory. In contrast, an increase in membrane potential, called *hyperpolarization*, makes a cell less likely to generate an action potential and is therefore inhibitory.

The Input Component Produces Graded Local Signals

In most neurons at rest, no current flows from one part of the cell to another, so the resting potential is the same throughout. In sensory neurons, current flow is typically initiated by a physical stimulus, which activates specialized receptor proteins at the neuron's receptive surface. In our example of the knee-jerk reflex, stretching of the muscle activates specific ion channels that open in response to stretch of the sensory neuron membrane, as we shall learn in Chapter 18. The opening of these channels when the cell is stretched permits the rapid influx of Na^+ ions into the sensory cell. This ionic current changes the membrane potential, producing a local signal called the *receptor potential*.

The amplitude and duration of a receptor potential depend on the intensity of the muscle stretch: The larger or longer-lasting the stretch, the larger or longer-lasting is the resulting receptor potential (Figure 3–9A). That is, receptor potentials are graded, unlike the all-or-none action potential. Most receptor potentials are depolarizing (excitatory); hyperpolarizing (inhibitory) receptor potentials are found in the retina.

The receptor potential is the first representation of stretch to be coded in the nervous system. However, because this depolarization spreads passively from the stretch receptor, it does not travel far. The distance is longer if the diameter of the axon is bigger, shorter if the diameter is smaller. Also, the distance is shorter if current can pass easily through the membrane, and longer if the membrane is insulated by myelin. The receptor potential from the stretch receptor therefore travels only 1 to 2 mm. In fact, just 1 mm away, the

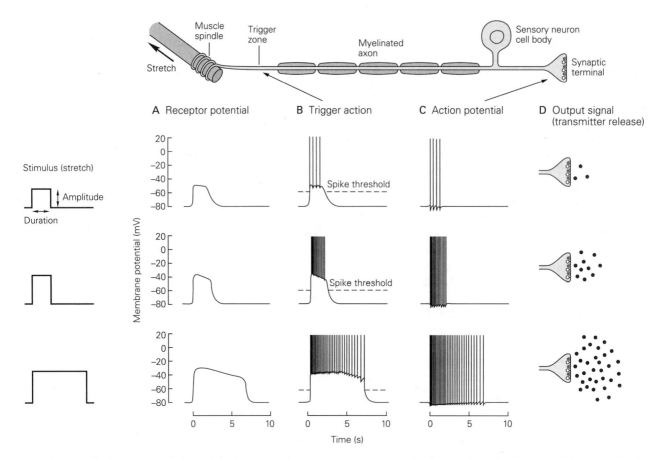

Figure 3–9 Each of the neuron's four signaling components produces a characteristic signal. The figure shows a sensory neuron activated by stretching of a muscle, which the neuron senses through a specialized receptor, the muscle spindle.

A. The input signal, called a receptor potential, is graded in amplitude and duration, proportional to the amplitude and duration of the stimulus.

B. The trigger zone sums the depolarization generated by the receptor potential. An action potential is generated only if the receptor potential exceeds a certain voltage threshold. Once this threshold is surpassed, any further increase in amplitude of the receptor potential can only increase the frequency with which the action potentials are generated, because action potentials have a constant amplitude. The duration of the

receptor potential determines the duration of the train of action potentials. Thus, the graded amplitude and duration of the receptor potential are translated into a frequency code in the action potentials generated at the trigger zone. All action potentials produced are propagated faithfully along the axon.

C. Action potentials are all-or-none. Because all action potentials have a similar amplitude and duration, the frequency and duration of firing encodes the information carried by the signal.

D. When the action potential reaches the synaptic terminal, it initiates the release of a neurotransmitter, the chemical substance that serves as the output signal. The frequency of action potentials in the presynaptic cell determines how much neurotransmitter is released by the cell.

amplitude of the signal is only about one-third what it was at the site of generation. To be carried successfully to the spinal cord, the local signal must be amplified—it must generate an action potential. In the knee-jerk reflex, if the receptor potential in the sensory neuron reaches the first node of Ranvier in the axon and is large enough, it will trigger an action potential (Figure 3–9B), which then propagates without failure to the axon terminals in the spinal cord (Figure 3–9C). At the synapse between the sensory neuron and a motor neuron, the action potential produces a chain of events that results in an input signal to the motor neuron.

In the knee-jerk reflex, the action potential in the presynaptic terminal of the sensory neuron initiates the release of a chemical substance, or neurotransmitter, into the synaptic cleft (Figure 3–9D). After diffusing across the cleft, the transmitter binds to receptor proteins in the postsynaptic membrane of the motor neuron, thereby directly or indirectly opening ion channels. The ensuing flow of current briefly alters the membrane potential of the motor cell, a change called the *synaptic potential*.

Like the receptor potential, the synaptic potential is graded; its amplitude depends on how much transmitter is released. In the same cell, the synaptic

Table 3–1 Comparison of Local (Passive) and Propagated Signals

Signal type	Amplitude (mV)	Duration	Summation	Effect of signal	Type of propagation
Local (passive) signals					
Receptor potentials	Small (0.1–10)	Brief (5–100 ms)	Graded	Hyperpolarizing or depolarizing	Passive
Synaptic potentials	Small (0.1–10)	Brief to long (5 ms–20 min)	Graded	Hyperpolarizing or depolarizing	Passive
Propagated (active) signals					
Action potentials	Large (70–110)	Brief (1–10 ms)	All-or-none	Depolarizing	Active

potential can be either depolarizing or hyperpolarizing depending on the type of receptor molecule that is activated. Synaptic potentials, like receptor potentials, spread passively. Thus, the change in potential will remain local unless the signal reaches beyond the axon's initial segment where it can give rise to an action potential. Some dendrites are not entirely passive but contain specializations that boost the synaptic potential, thereby increasing its efficacy to produce an action potential (Chapter 13). The features of receptor and synaptic potentials are summarized in Table 3–1.

The Trigger Zone Makes the Decision to Generate an Action Potential

Sherrington first pointed out that the function of the nervous system is to weigh the consequences of different types of information and then decide on appropriate responses. This *integrative* function of the nervous system is clearly seen in events at the trigger zone of the neuron, the initial segment of the axon.

Action potentials are generated by a sudden influx of Na^+ through channels in the cell membrane that open and close in response to changes in membrane potential. When an input signal (a receptor potential or synaptic potential) depolarizes an area of membrane, the local change in membrane potential opens local Na^+ channels that allow Na^+ to flow down its concentration gradient, from outside the cell where the Na^+ concentration is high to inside where it is low.

Because the initial segment of the axon has the highest density of voltage-sensitive Na^+ channels and therefore the lowest threshold for generating an action potential, an input signal spreading passively along the cell membrane is more likely to give rise to an action potential at the initial segment of the axon than at other sites in the cell. This part of the axon is therefore known as the *trigger zone*. It is here that the activity of all receptor (or synaptic) potentials is summed and

where, if the sum of the input signals reaches threshold, the neuron generates an action potential.

The Conductive Component Propagates an All-or-None Action Potential

The action potential is all-or-none: Stimuli below the threshold do not produce a signal, but stimuli above the threshold all produce signals of the same amplitude. Regardless of variation in intensity or duration of stimuli, the amplitude and duration of each action potential are pretty much the same, and this holds for each regenerated action potential at a node of Ranvier along a myelinated axon. In addition, unlike receptor and synaptic potentials, which spread passively and decrease in amplitude, the action potential, as we have seen, does not decay as it travels along the axon to its target—a distance that can be as great as 1 m—because it is periodically regenerated. This conducted signal can travel at rates as fast as 100 m/s. Indeed, the remarkable feature of action potentials is that they are highly stereotyped, varying only subtly (but in some cases importantly) from one nerve cell to another. This feature was demonstrated in the 1920s by Edgar Adrian, one of the first to study the nervous system at the cellular level. Adrian found that all action potentials have a similar shape or waveform (see Figure 3–2). The action potentials carried into the nervous system by a sensory axon often are indistinguishable from those carried out of the nervous system to the muscles by a motor axon.

Only two features of the conducting signal convey information: the number of action potentials and the time intervals between them (Figure 3–9C). As Adrian put it in 1928, summarizing his work on sensory fibers: "all impulses are very much alike, whether the message is destined to arouse the sensation of light, of touch, or of pain; if they are crowded together the sensation is intense, if they are separated by long intervals the sensation is correspondingly feeble." Thus, what determines the intensity of sensation or speed of movement is the frequency of the action potentials. Likewise, the duration of a sensation or movement is determined by the period over which action potentials are generated.

In addition to the frequency of the action potentials, the pattern of action potentials also conveys important information. For example, some neurons are spontaneously active in the absence of stimulation. Some spontaneously active nerve cells (beating neurons) fire action potentials regularly; others (bursting neurons) fire in brief bursts of action potentials. These diverse cells respond differently to the same excitatory synaptic input. An excitatory synaptic potential may initiate one or more action potentials in a cell that is not spontaneously active, whereas that same input to spontaneously active cells will simply increase the existing rate of firing.

An even more dramatic difference is seen when the input signal is inhibitory. Inhibitory inputs have little information value in a silent cell. By contrast, in spontaneously active cells, inhibition can have a powerful *sculpting* role. By establishing periods of silence in otherwise ongoing activity, inhibition can produce a complex pattern of alternating firing and silence where none existed. Such subtle differences in firing patterns may have important functional consequences for the information transfer between neurons. Mathematical modelers of neuronal networks have attempted to delineate neural codes in which information is also carried by the fine-grained pattern of firing—the exact timing of each action potential.

If signals are stereotyped and reflect only the most elementary properties of the stimulus, how can they carry the rich variety of information needed for complex behavior? How is a message that carries visual information about a bee distinguished from one that carries pain information about the bee's sting, and how are these sensory signals distinguished from motor signals for voluntary movement? The answer is simple and yet is one of the most important organizational principles of the nervous system: Interconnected neurons form anatomically and functionally distinct pathways—labeled lines—and it is these pathways of connected neurons, these labeled lines, not individual neurons, that convey information. The neural pathways activated by receptor cells in the retina that respond to light are completely distinct from the pathways activated by sensory cells in the skin that respond to touch.

The Output Component Releases Neurotransmitter

When an action potential reaches a neuron's terminal, it stimulates the release of chemical substances from the cell. These substances, called *neurotransmitters*, can be small organic molecules, such as L-glutamate and acetylcholine, or peptides like substance P or LHRH (luteinizing hormone–releasing hormone).

Neurotransmitter molecules are held in subcellular organelles called *synaptic vesicles*, which accumulate in the terminals of the axon at specialized release sites called *active zones*. To eject their transmitter substance into the synaptic cleft, the vesicles move up to and fuse with the neuron's plasma membrane, then burst open to release the transmitter into the synaptic cleft (the extracellular space between the pre- and postsynaptic cell) by a process known as *exocytosis*. The molecular machinery of neurotransmitter release is described in Chapters 14 and 15.

The released neurotransmitter molecules are the neuron's output signal. The output signal is thus graded according to the amount of transmitter released, which is determined by the number and frequency of the action potentials that reach the presynaptic terminals (Figure 3–9C,D). After release, the transmitter molecules diffuse across the synaptic cleft and bind to receptors on the postsynaptic neuron. This binding causes the postsynaptic cell to generate a synaptic potential. Whether the synaptic potential has an excitatory or inhibitory effect depends on the type of receptor in the postsynaptic cell, not on the particular chemical neurotransmitter. The same transmitter substance can have different effects at different receptors.

The Transformation of the Neural Signal From Sensory to Motor Is Illustrated by the Stretch-Reflex Pathway

As we have seen, the properties of a signal are transformed as the signal moves from one component of a neuron to another or between neurons. In the stretch reflex, when a muscle is stretched, the amplitude and duration of the stimulus are reflected in the amplitude and duration of the receptor potential generated in the sensory neuron (Figure 3–10A). If the receptor potential exceeds the threshold for an action potential in that cell, the graded signal is transformed at the trigger zone into an action potential. Although individual action potentials are all-or-none signals, the more the receptor potential exceeds threshold, the greater the depolarization and consequently the greater the frequency of action potentials in the axon. The duration of the input signal also determines the duration of the train of action potentials.

The information encoded by the frequency and duration of firing is faithfully conveyed along the axon to its terminals, where the firing of action potentials determines the amount of transmitter released.

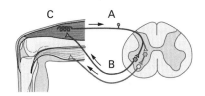

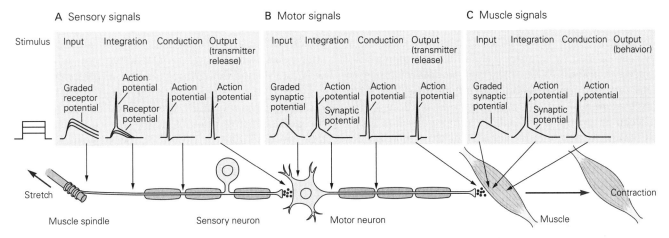

Figure 3–10 The sequence of signals that produces a reflex action.

A. The stretching of a muscle produces a receptor potential in the specialized receptor (the muscle spindle). The amplitude of the receptor potential is proportional to the intensity of the stretch. This potential spreads passively to the integrative or trigger zone at the first node of Ranvier. If the receptor potential is sufficiently large, it triggers an action potential that then propagates actively and without change along the axon to the axon terminal. At specialized sites in the terminal, the action potential leads to the release of a chemical neurotransmitter, the output signal. The transmitter diffuses across the synaptic cleft between the axon terminal and a target motor neuron that innervates the stretched muscle; it then binds to receptor molecules on the external membrane of the motor neuron.

B. This interaction initiates a synaptic potential that spreads passively to the trigger zone of the motor neuron's axon, where it initiates an action potential that propagates actively to the terminal of the motor neuron's axon. At the axon terminal, the action potential leads to release of a neurotransmitter near the muscle fiber.

C. The neurotransmitter binds receptors on the muscle fiber, generating a synaptic potential. The synaptic potential triggers an action potential in the muscle, which causes a contraction.

These stages of signaling have their counterparts in the motor neuron (Figure 3–10B) and in the muscle (Figure 3–10C).

Nerve Cells Differ Most at the Molecular Level

The model of neuronal signaling we have outlined is a simplification that applies to most neurons, but there are some important variations. For example, some neurons do not generate action potentials. These are typically local interneurons without a conductive component; they have no axon or such a short one that regeneration of the signal is not required. In these neurons, the input signals are summed and spread passively to the presynaptic terminal region near where transmitter is released. Neurons that are spontaneously active do not require sensory or synaptic inputs to fire action potentials because they have a special class of ion channels that permit Na$^+$ current flow even in the absence of excitatory synaptic input.

Even cells that are similar morphologically can differ importantly in molecular details. For example, they can have different combinations of ion channels. As we shall learn in Chapter 10, different ion channels provide neurons with various thresholds, excitability properties, and firing patterns. Such neurons can encode synaptic potentials into different firing patterns and thereby convey different information.

Neurons also differ in the chemical substances they use as transmitters and in the receptors that receive transmitter substances from other neurons. Indeed, many drugs that act on the brain do so by modifying the actions of specific chemical transmitters or receptors. Because of physiological differences among neurons, a disease may affect one class of neurons but not others. Certain diseases strike only motor neurons (amyotrophic lateral sclerosis and poliomyelitis), whereas others affect primarily sensory neurons (leprosy and tabes dorsalis, a late stage of syphilis). Parkinson disease, a disorder of voluntary movement, damages a small population of neurons that use dopamine as a neurotransmitter. Some diseases are selective even within the neuron, affecting only the receptive elements, the cell body, or the axon. In Chapter 57, we describe how research into myasthenia gravis, a disease caused by a faulty transmitter receptor in the muscle membrane, has provided important insights into synaptic transmission. Indeed, because the nervous system has so many cell types and variations at the molecular level, it is susceptible to more diseases (psychiatric as well as neurological) than any other organ system of the body.

Despite the morphological differences among nerve cells, the molecular mechanisms of electrical signaling are surprisingly similar. This simplicity is fortunate, for understanding the molecular mechanisms of signaling in one kind of nerve cell aids the understanding of these mechanisms in many other nerve cells.

The Reflex Circuit Is a Starting Point for Understanding the Neural Architecture of Behavior

The stretch reflex illustrates how interactions between just a few types of nerve cells can constitute a functional circuit that produces a simple behavior, even though the number of neurons involved is large (the stretch reflex circuit has perhaps a few hundred sensory neurons and a hundred motor neurons). Some invertebrate animals are capable of behavior as sophisticated as reflexes using far fewer neurons. Moreover, in some instances, just one critical command neuron can trigger a complex behavior such as the withdrawal of a body part from a noxious stimulus.

For more complex behaviors, especially in higher vertebrates, many neurons are required, but the basic neural structure of the simple reflex is often preserved. First, there is often an identifiable group of neurons whose firing rate changes in response to a particular type of environmental stimulus, such as a tone of a certain frequency, or the juxtaposition of light and dark at a particular angle. Just as the firing rate of the stretch receptor neurons encodes the degree of muscle tension, the firing rates of cortical neurons in sensory areas of the cortex encode the intensity of a sensory feature (eg, the degree of contrast of the contour). As we shall see in later chapters, it is possible to change

the features of a percept just by changing the firing rate of small groups of neurons.

Second, there is often an identifiable group of neurons whose firing rate changes before an animal performs a motor act. Just as the spike rate of motor neurons controls the magnitude of the contraction of the quadriceps muscle—hence the knee jerk—so does the firing rate of neurons in the motor cortex affect the latency and type of movement that will be performed. Exactly what aspect of the movement is encoded by such neurons remains an area of active inquiry, but it is well established that groups of neurons affect the ensuing action in a graded fashion by adjusting their firing rate. In other association areas of the cerebral cortex, the graded firing rates of neurons encode quantities that are essential for thought processes, such as the amount of evidence bearing on a choice (Chapter 56).

Although sophisticated mental operations are far more complicated than a simple stretch reflex, it may nevertheless prove useful to consider the extent to which cognitive functions are supported by neural mechanisms that are organized in any way that resembles a simple reflex. What types of elaborations might be required to mediate a sophisticated behavior and thought? Unlike a simple reflex, with a sophisticated behavior, activation of sensory neurons would not give rise to an immediate reflexive action. There is more contingency to the process. Although simple reflexes are modulated by context, mental functions are more deeply shaped by a complex repertoire of contingencies, allowing for many possible effects of any one stimulus and many possible precipitants of any one action. In light of these contingencies, we are forced to conceive of a flexible routing between the brain's data acquisition systems—not just the sensory systems but also the memory systems—and effector systems. As we shall see in later chapters, this is the role of the higher association areas of the cerebral cortex, acting in concert with several subcortical brain structures.

Perhaps a more salient difference between a complex mental function and a reflex is the timing of action. Once activated, a reflex circuit leads to action almost immediately after the sensory stimulus. Any delay depends mainly on the conduction velocity of the action potentials in the afferent and efferent limbs of the reflex (eg, the ankle jerk is slower than the knee jerk because the spinal cord is further from the stretch receptors of the calf muscles than it is from the thigh extensors). For more complex behaviors, action need not occur more or less instantaneously with the arrival of sensory information. It might be delayed to await additional information or be expressed only when specific circumstances occur.

Interestingly, the neurons in the association areas of the cortex of primates have the capacity to sustain graded firing rates for durations of many seconds. These neurons are abundant in the parts of the brain that mediate the flexible linkage between sensory and motor areas. They afford a freedom from the instantaneous nature of reflexive behavior and therefore may furnish the essential circuit properties that distinguish cognitive functions from more straightforward sensorimotor transformations like a reflex.

Neural Circuits Can Be Modified by Experience

Learning can result in behavioral changes that endure for years, even a lifetime. But even simple reflexes can be modified, albeit for a much briefer period of time. The fact that much behavior can be modified by learning raises an interesting question: How is it that behavior can be modified if the nervous system is wired so precisely? How can changes in the neural control of behavior occur when connections between the signaling units, the neurons, are set during early development?

Several solutions for this dilemma have been proposed. The proposal that has proven most farsighted is the *plasticity hypothesis*, first put forward at the turn of the 20th century by Ramón y Cajal. A modern form of this hypothesis was advanced by the Polish psychologist Jerzy Konorski in 1948.

The application of a stimulus leads to changes of a twofold kind in the nervous system … [T]he first property, by virtue of which the nerve cells *react* to the incoming impulse … we call *excitability*, and … changes arising … because of this property we shall call *changes due to excitability*. The second property, by virtue of which certain permanent functional transformations arise in particular systems of neurons as a result of appropriate stimuli or their combination, we shall call *plasticity* and the corresponding changes *plastic changes*.

There is now considerable evidence for functional plasticity at chemical synapses. These synapses often have a remarkable capacity for short-term physiological changes (lasting seconds to hours) that increase or decrease synaptic effectiveness. Long-term physiological changes (lasting days or longer) can give rise to anatomical alterations, including pruning of synapses and even growth of new ones. As we shall see in later chapters, chemical synapses are functionally and anatomically modified during critical periods of early development but also throughout life. This functional plasticity of neurons endows each of us with a characteristic manner of interacting with the surrounding world, both natural and social.

Highlights

1. Nerve cells are the signaling units of the nervous system. The signals are mainly electrical within the cell and chemical between cells. Despite variations in size and shape, nerve cells share certain common features. Each has specialized receptors or transducers that receive input from other nerve cells or from the senses respectively; a mechanism to convert input to electrical signals; a threshold mechanism to generate an all-or-none electrical impulse, the action potential, which can be regenerated along the axon that connects the nerve cell to its synaptic target (another nerve cell, a muscle, or gland); and the ability to produce the release of a chemical (neurotransmitter) that affects the target.

2. Glial cells support nerve cells. One type provides the insulation that speeds propagation of the action potential along the axon. Others help establish the chemical milieu for the nerve cells to operate, and still others couple nerve activity to the vascular supply of the nervous system.

3. Nerve cells differ in their morphology, the connections they make, and where they make them. This is clearest in specialized structures like the retina. Perhaps the largest difference between neurons is at the molecular level. Examples of molecular diversity include expression of different receptors, enzymes for synthesis of different neurotransmitters, and different expressions of ion channels. Differences in gene expression furnish the starting point for understanding why certain diseases affect some neurons and not others.

4. Each nerve cell is part of a circuit that has one or more behavioral functions. The stretch reflex circuit is an example of a simple circuit that produces a behavior in response to a stimulus. Its simplicity belies integrative functions, such as relaxation of muscles that oppose the stretched muscle.

5. Modern neural science aspires to explain mental processes far more complex than reflexes. A natural starting point is to understand the ways that circuits must be elaborated to support sensorymotor transformations, which unlike a reflex, are contingent, flexible and not beholden to the immediacy of sensory processing and movement control.

6. Neural connections can be modified by experience. In simple circuits, this process is a simple change in the strength of connections between neurons. A working hypothesis in modern neuroscience is that the "plastic" mechanisms at play

in simple circuits also play a critical role in the learning of more complex behavior and cognitive function.

Michael N. Shadlen
Eric R. Kandel

Selected Reading

Adrian ED. 1928. *The Basis of Sensation: The Action of the Sense Organs.* London: Christophers.

Jack JJB, Noble D, Tsien RW. 1983. *Electric Current Flow in Excitable Cells.* Oxford: Clarendon Press.

Jones EG. 1988. The nervous tissue. In: L Weiss (ed). *Cell and Tissue Biology: A Textbook of Histology,* 6th ed., pp. 277–351. Baltimore: Urban and Schwarzenberg.

Poeppel D, Mangun GR, Gazzaniga MS. 2020. *The Cognitive Neurosciences.* Cambridge, MA: The MIT Press.

Ramón y Cajal S. [1937] 1989. *Recollections of My Life.* EH Craigie (transl). Philadelphia: American Philosophical Society; 1989. Reprint. Cambridge, MA: MIT Press.

References

Adrian ED. 1932. *The Mechanism of Nervous Action: Electrical Studies of the Neurone.* Philadelphia: Univ. Pennsylvania Press.

Alberts B, Johnson A, Lewis J, Raff M, Roberts K, Walter JD. 2002. *Molecular Biology of the Cell,* 4th ed. New York: Garland.

Dacey DM, Peterson BB, Robinson FR, Gamlin PD. 2003. Fireworks in the primate retina: in vitro photodynamics reveals diverse LGN-projecting ganglion cell types. Neuron 37:15–27.

Erlanger J, Gasser HS. 1937. *Electrical Signs of Nervous Activity.* Philadelphia: Univ. Pennsylvania Press.

Hodgkin AL, Huxley AF. 1939. Action potentials recorded from inside a nerve fiber. Nature 144:710–711.

Kandel ER. 1976. The study of behavior: the interface between psychology and biology. In: *Cellular Basis of Behavior: An Introduction to Behavioral Neurobiology,* pp. 3–27. San Francisco: WH Freeman.

Konorski J. 1948. *Conditioned Reflexes and Neuron Organization.* Cambridge: Cambridge Univ. Press.

Martinez PFA. 1982. *Neuroanatomy: Development and Structure of the Central Nervous System.* Philadelphia: Saunders.

McCormick DA. 2004. Membrane potential and action potential. In: JH Byrne, JL Roberts (eds). *From Molecules to Networks: An Introduction to Cellular Neuroscience,* 2nd ed., p. 130. San Diego: Elsevier.

Newman EA. 1986. High potassium conductance in astrocyte endfeet. Science 233:453–454.

Nicholls JG, Wallace BG, Fuchs PA, Martin AR. 2001. *From Neuron to Brain,* 4th ed. Sunderland, MA: Sinauer.

Penfield W (ed). 1932. *Cytology & Cellular Pathology of the Nervous System,* Vol. 2. New York: Hoeber.

Ramón y Cajal S. 1933. *Histology,* 10th ed. Baltimore: Wood.

Sears ES, Franklin GM. 1980. Diseases of the cranial nerves. In: RN Rosenberg (ed). *The Science and Practice of Clinical Medicine.* Vol. 5, *Neurology,* pp. 471–494. New York: Grune & Stratton.

Sherrington C. 1947. *The Integrative Action of the Nervous System,* 2nd ed. Cambridge: Cambridge Univ. Press.

4

The Neuroanatomical Bases by Which Neural Circuits Mediate Behavior

T HE HUMAN BRAIN carries out actions in ways no current computer can begin to approach. Merely to see—to look onto the world and recognize a face or facial expression—entails amazing computational achievements. Indeed, all our perceptual abilities—seeing, hearing, smelling, tasting, and touching—are analytical triumphs. Similarly, all of our voluntary actions are triumphs of engineering. Sensation and movement, while wondrous in their own right, pale in comparison to complex cognitive behaviors such as forming memories or understanding social conventions.

The brain accomplishes these computational feats because its nerve cells are wired together in very precise functional circuits. The brain is hierarchically organized such that information processed at one level is passed to higher-level circuits for more complex and refined processing. In essence, the brain is a network of networks. Different brain areas work in an integrated fashion to accomplish purposeful behavior.

In this chapter, we outline the neuroanatomical organization of some of the circuits that enable the brain to process sensory input and produce motor output. We focus on touch as a sensory modality because the somatosensory system is particularly well understood and because touch clearly illustrates the interaction of sensory processing circuits at several levels, from the spinal cord to the cerebral cortex. Our purpose here is to illustrate the basic principles of how circuits control behavior. In the next chapter, we consider the functional properties of these circuits, including the computations by which they process information. In subsequent chapters, we consider in more detail the anatomy and function of the various sensory modalities and how sensory input regulates movement.

Finally, we provide a preview of the brain circuits that are instrumental in producing the memories of our daily lives, called explicit memory (see Chapters 52 and 54). We do this to make the point that while many of the neurons in the memory circuits are similar to those in the sensory and motor circuits, not all are. Moreover, the organization of the pathways between circuits is different in the memory system than it is in the motor and sensory systems. This highlights a basic neurobiological tenet that different circuits of the brain have evolved an organization to most efficiently carry out specific functions.

Comprehending the functional organization of the brain might at first seem daunting. But as we saw in the previous chapter, the organization of the brain is simplified by three anatomical considerations. First, there are relatively few types of neurons. Each of the many thousands of spinal motor neurons or millions of neocortical pyramidal cells has a similar anatomical structure and serves a similar function. Second, neurons in the brain and spinal cord are clustered in functional groups called nuclei or discrete areas of the cerebral cortex, which form networks or functional systems. Third, the discrete areas of the cerebral cortex are specialized for sensory, motor, or associative functions such as memory.

Local Circuits Carry Out Specific Neural Computations That Are Coordinated to Mediate Complex Behaviors

Neurons are interconnected to form functional circuits. Within the spinal cord, for example, simple reflex circuits receive sensory information from stretch receptors and send output to various muscle groups. For more complex behavioral functions, different stages of information processing are carried out in networks in different regions of the nervous system. Connections between neurons within the nervous system can be of different lengths.

Within a brain region, local connections, which may be excitatory or inhibitory, integrate many of the neurons into functional networks. Such local networks may then provide outputs to one or more other brain regions through longer projections. Many of these longer pathways have names. For example, projections from the lateral geniculate nucleus of the thalamus to the visual cortex are called the optic radiations. Connections from the neocortex—the region of the cerebral cortex nearest the surface of the brain—of one side of the brain to the other side of the brain form the corpus

callosum. Information carried by these long pathways integrates the output of many local circuits (Figure 4–1).

Consider the simple act of hitting a tennis ball (Figure 4–2). Visual information about the motion of the approaching ball is analyzed in the visual system, which is itself a hierarchically organized system extending from the retina to the lateral geniculate nucleus of the thalamus to dozens of cortical areas in the occipital and temporal lobes (Chapter 21). This information is combined in the motor cortex with proprioceptive information about the position of the arms, legs, and trunk to calculate the movement necessary to intercept the ball. Once the swing is initiated, many minor adjustments of the motor program are made by other brain regions dedicated to movement, such as the cerebellum, based on a steady stream of sensory information about the trajectory of the approaching ball and the position of the arm.

Like most motor behaviors, hitting a tennis ball is not hardwired into brain circuits but requires learning and memory. The memory for motor tasks, termed procedural or implicit memory, requires modifications to circuits in motor cortex, the basal ganglia, and the cerebellum. Finally, this entire act is accessible to consciousness and may elicit conscious recall of past similar experiences, termed explicit memory, and emotions. Explicit memory depends on circuits in the hippocampus (Chapters 52 and 54), whereas emotions are regulated by the amygdala (Chapters 42 and 53) and portions of the orbitofrontal, cingulate, and insular cortices. Of course, as the swing is being executed, the brain is also engaged in coordinating the player's heart rate, respiration, and other homeostatic functions through equally complex networks.

Sensory Information Circuits Are Illustrated in the Somatosensory System

Complex behaviors, such as differentiating the motor acts required to grasp a ball versus a book, require the integrated action of several nuclei and cortical regions. Information is processed in the brain in a hierarchical fashion. Thus, information about a stimulus is conveyed through a succession of subcortical and then cortical regions; at each level of processing, the information becomes increasingly more complex.

In addition, different types of information, even within a single sensory modality, are processed in several anatomically discrete pathways. In the somatosensory system, a light touch and a painful pin prick to the same area of skin are mediated by different sensory receptors in the skin that connect to distinct pathways

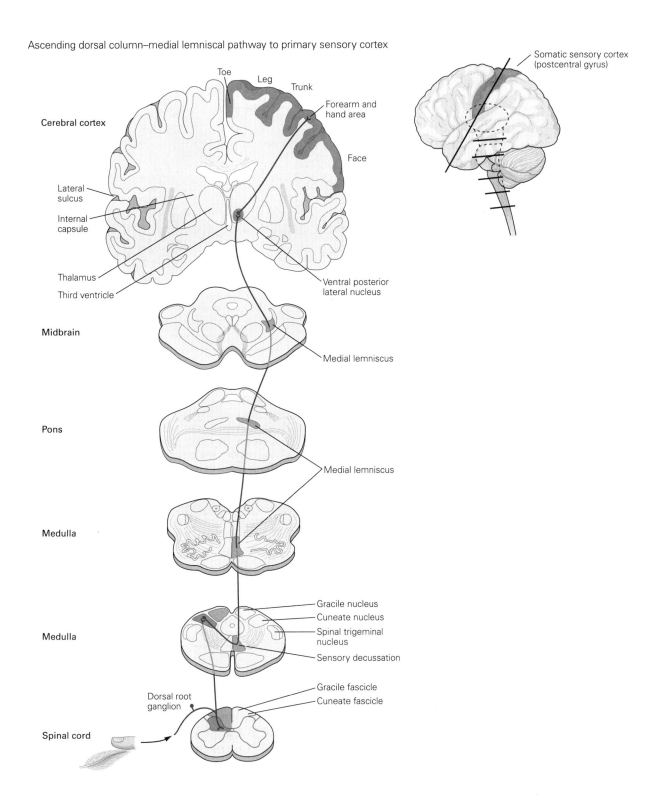

Figure 4–1 The dorsal column–medial lemniscal pathway is the major afferent pathway for somatosensory information. Somatosensory information enters the central nervous system through the dorsal root ganglion cells. The flow of information ultimately leads to the somatosensory cortex. Fibers that relay information from different parts of the body maintain an orderly relationship to each other and form a neural map of the body surface in their pattern of termination at each synaptic relay.

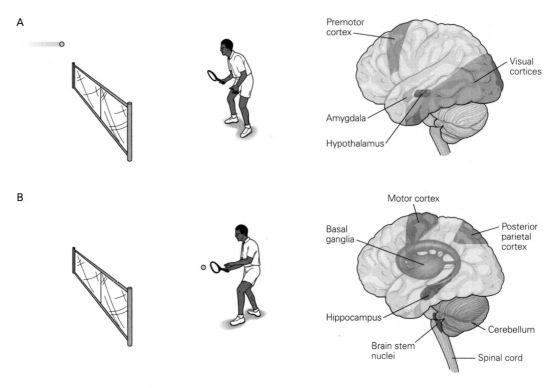

Figure 4–2 A simple behavior is mediated by many parts of the brain.

A. A tennis player watching an approaching ball uses the visual cortex to judge the size, direction, and velocity of the ball. The premotor cortex develops a motor program to return the ball. The amygdala acts in conjunction with other brain regions to adjust the heart rate, respiration, and other homeostatic mechanisms and also activates the hypothalamus to motivate the player to hit well.

B. To execute the shot, the player must use all of the structures illustrated in part **A** as well as others. The motor cortex sends signals to the spinal cord that activate and inhibit many muscles in the arms and legs. The basal ganglia become involved in initiating motor patterns and perhaps recalling learned movements to hit the ball properly. The cerebellum adjusts movements based on proprioceptive information from peripheral sensory receptors. The posterior parietal cortex provides the player with a sense of where his body is located in space and where his racket arm is located with respect to his body. Brain stem neurons regulate heart rate, respiration, and arousal throughout the movement. The hippocampus is not involved in hitting the ball but is involved in storing the memory of the return so that the player can brag about it later.

in the brain. The system for fine touch, pressure, and proprioception is called the epicritic system, whereas the system for pain and temperature is called the protopathic system.

Somatosensory Information From the Trunk and Limbs Is Conveyed to the Spinal Cord

All forms of somatosensory information from the trunk and limbs enter the spinal cord, which has a core H-shaped region of gray matter where neuronal cell bodies are located. The gray matter is surrounded by white matter formed by myelinated axons that make up both short and long connections. The gray matter on each side of the cord is divided into dorsal (or posterior) and ventral (or anterior) horns (Figure 4–3).

The dorsal horn contains groups of secondary sensory neurons (sensory nuclei) whose dendrites receive stimulus information from primary sensory neurons that innervate the body's skin, muscles, and joints. The ventral horn contains groups of motor neurons (motor nuclei) whose axons exit the spinal cord and innervate skeletal muscles. The spinal cord has circuits that mediate behaviors ranging from the stretch reflex to coordination of limb movements.

As we discussed in Chapter 3, when considering the knee-jerk reflex, interneurons of various types in the gray matter regulate the output of the spinal cord motor neurons (see Figure 3–5). Some of these interneurons are excitatory, whereas others are inhibitory. These interneurons modulate both sensory information flowing toward the brain and motor commands descending from the brain to the spinal motor neurons. Motor neurons can also adjust the output of other motor neurons via the interneurons. These

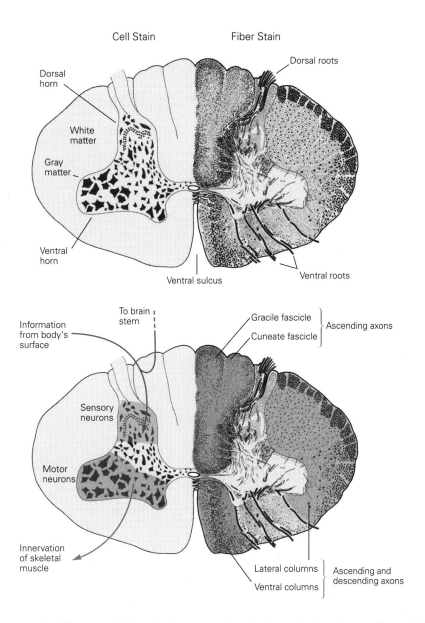

Figure 4–3 The major anatomical features of the spinal cord. The ventral horn (**green**) contains large motor neurons, whereas the dorsal horn (**orange**) contains smaller neurons. Fibers of the gracile fascicle carry somatosensory information from the lower limbs, whereas fibers of the cuneate fascicle carry somatosensory information from the upper body. Fiber bundles of the lateral and ventral columns include both ascending and descending fiber bundles.

circuits will be considered in more detail when we discuss the spinal cord in Chapter 32.

The white matter surrounding the gray matter contains bundles of ascending and descending axons that are divided into dorsal, lateral, and ventral columns. The dorsal columns, which lie between the two dorsal horns of the gray matter, contain only ascending axons that carry somatosensory information to the brain stem (Figure 4–1). The lateral columns include both ascending and descending axons from the brain

stem and neocortex that innervate spinal interneurons and motor neurons (Figure 4–3). This demonstrates a general principle about central nervous system connections. Processing tends to be hierarchical: Projections from a lower to a higher processing region are said to be feedforward, while descending projections can modulate spinal reflexes and are considered to be feedback. The motif in which region A projects to region B and, in turn, also receives return projections from B, is recapitulated throughout the nervous system. The ventral columns also include ascending and descending axons. The ascending somatosensory axons in the lateral and ventral columns constitute parallel pathways that convey information about pain and thermal sensation to higher levels of the central nervous system. The descending axons control axial muscles and posture.

The spinal cord is divided along its length into four major regions: cervical, thoracic, lumbar, and sacral (Figure 4–4). Connections arising from these regions

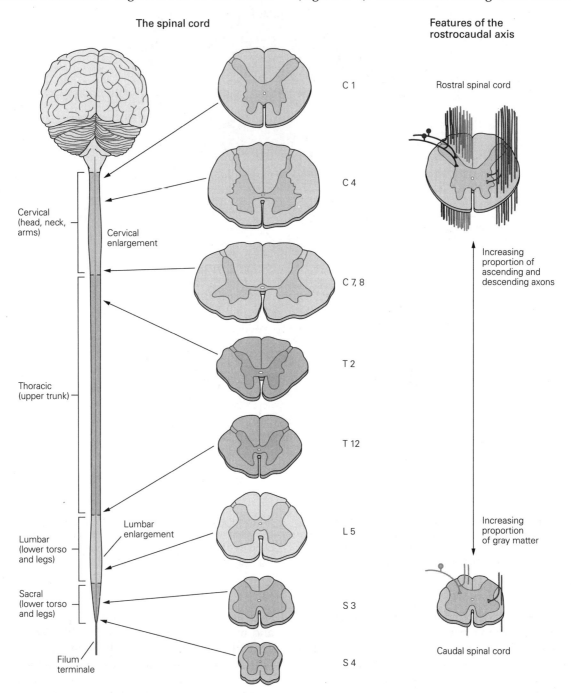

The spinal cord

Features of the rostrocaudal axis

Rostral spinal cord

C 1

C 4

Cervical (head, neck, arms)

Cervical enlargement

C 7, 8

Increasing proportion of ascending and descending axons

T 2

Thoracic (upper trunk)

T 12

Lumbar enlargement

L 5

Increasing proportion of gray matter

Lumbar (lower torso and legs)

Sacral (lower torso and legs)

S 3

S 4

Filum terminale

Caudal spinal cord

Figure 4–4 The internal and external appearances of the spinal cord vary at different levels. The proportion of gray matter (the H-shaped area within the spinal cord) to white matter is greater at sacral levels than at cervical levels. At sacral levels, very few incoming sensory axons have joined the spinal cord, whereas most of the motor axons have already terminated at higher levels of the spinal cord. The cross-sectional enlargements at the lumbar and cervical levels are regions where the large number of fibers innervating the limbs enter or leave the spinal cord.

are segregated according to the embryological somites from which muscles, bones, and other components of the body develop (Chapter 45). Axons projecting from the spinal cord to body structures that develop at the same segmental level join with axons entering the spinal cord in the intervertebral foramen to form spinal nerves. Spinal nerves at the cervical level are involved with sensory perception and motor function of the back of the head, neck, and arms; nerves at the thoracic level innervate the upper trunk; lumbar and sacral spinal nerves innervate the lower trunk, back, and legs.

Each of the four regions of the spinal cord contains multiple segments corresponding approximately to the different vertebrae in each region; there are 8 cervical segments, 12 thoracic segments, 5 lumbar segments, and 5 sacral segments. The actual substance of the mature spinal cord does not look segmented, but the segments of the four spinal regions are nonetheless defined by the number and location of the dorsal and ventral roots that enter or exit the spinal cord. The spinal cord varies in size and shape along its rostrocaudal axis because of two organizational features.

First, relatively few sensory axons enter the cord at the sacral level. The number of sensory axons entering the cord increases at progressively higher levels (lumbar, thoracic, and cervical). Conversely, most descending axons from the brain terminate at cervical levels, with progressively fewer descending to lower levels of the spinal cord. Thus, the number of fibers in the white matter is highest at cervical levels (where there are the highest numbers of both ascending and descending fibers) and lowest at sacral levels. As a result, sacral levels of the spinal cord have much less white matter than gray matter, whereas the cervical cord has more white matter than gray matter (Figure 4–4).

The second organizational feature is variation in the size of the ventral and dorsal horns. The ventral horn is larger at the levels where the motor nerves innervate the arms and legs. The number of ventral motor neurons dedicated to a body region roughly parallels the dexterity of movements of that region. Thus, a larger number of motor neurons is needed to innervate the greater number of muscles and to regulate the greater complexity of movement in the limbs as compared with the trunk. Likewise, the dorsal horn is larger where sensory nerves from the limbs enter the cord. Limbs have a greater density of sensory receptors to mediate finer tactile discrimination and thus send more sensory fibers to the cord. These regions of the cord are known as the lumbosacral and cervical enlargements (Figure 4–4).

The Primary Sensory Neurons of the Trunk and Limbs Are Clustered in the Dorsal Root Ganglia

The sensory neurons that convey information from the skin, muscles, and joints of the limbs and trunk to the spinal cord are clustered together in dorsal root ganglia within the vertebral column, immediately adjacent to the spinal cord (Figure 4–5). These neurons are pseudo-unipolar in shape; they have a bifurcated axon with central and peripheral branches. The peripheral branch innervates the skin, muscle, or other tissue as a free nerve ending or in association with specialized receptors for sensing touch, proprioception (stretch receptors), pain, and temperature.

The somatosensory system and its pathways from receptors to perception are more fully described in Chapters 17 to 20. Suffice it to say at this point that there are essentially two somatosensory pathways from the periphery that carry either touch and stretch (epicritic system) or pain and temperature (protopathic system). Epicritic fibers travel in the posterior column–medial lemniscal system (Figure 4–6). The centrally directed axons from neurons in the dorsal root ganglion ascend in the dorsal (or posterior) column white matter and terminate in the gracile nucleus or cuneate nucleus of the medulla. The centrally directed axons of the pain and temperature pathway form the spinothalamic

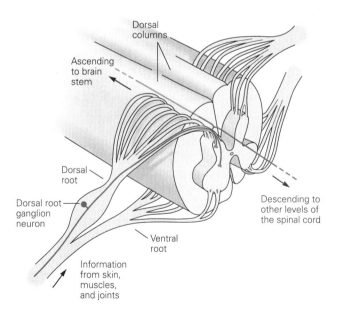

Figure 4–5 Dorsal root ganglia and spinal nerve roots. The cell bodies of neurons that bring sensory information from the skin, muscles, and joints lie in the dorsal root ganglia, clusters of cells that lie adjacent to the spinal cord. The axons of these neurons are bifurcated into peripheral and central branches. The central branch enters the dorsal portion of the spinal cord.

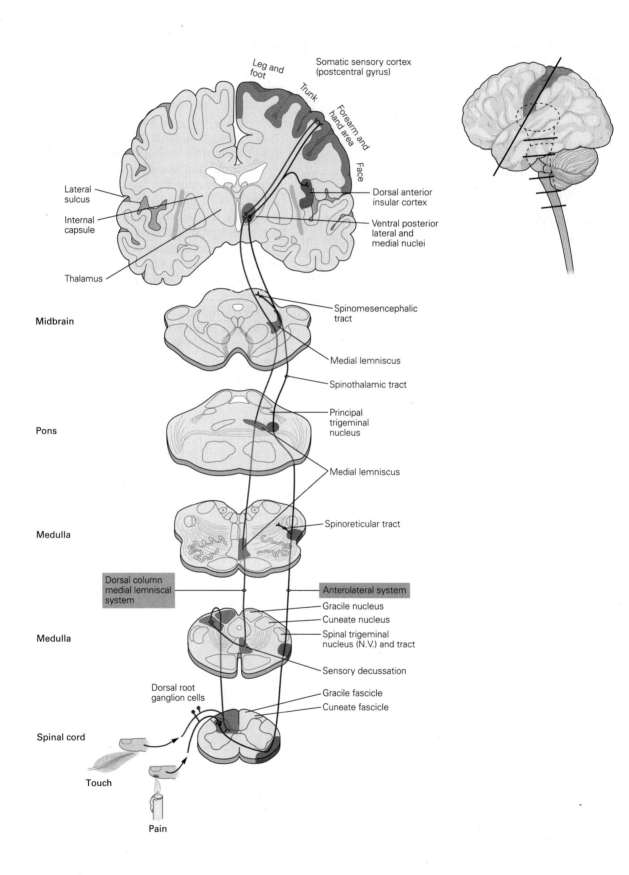

Somatic sensory cortex
(postcentral gyrus)

Leg and foot

Trunk

Forearm and hand area

Face

Lateral sulcus

Internal capsule

Dorsal anterior insular cortex

Ventral posterior lateral and medial nuclei

Thalamus

Midbrain

Spinomesencephalic tract

Medial lemniscus

Spinothalamic tract

Pons

Principal trigeminal nucleus

Medial lemniscus

Medulla

Spinoreticular tract

Dorsal column medial lemniscal system

Anterolateral system

Gracile nucleus

Cuneate nucleus

Medulla

Spinal trigeminal nucleus (N.V.) and tract

Sensory decussation

Dorsal root ganglion cells

Gracile fascicle

Cuneate fascicle

Spinal cord

Touch

Pain

pathway. They terminate within the gray matter of the dorsal horn of the spinal cord. Second-order neurons cross to the other side of the spinal cord and ascend in the anterior and lateral spinothalamic tracts (Figure 4–6). Both pathways ultimately terminate in the thalamus, which sends projections to the primary somatosensory area of the cerebral cortex. In the next section, we focus on the epicritic system.

The local and ascending branches from touch and proprioceptive sensory neurons provide two functional pathways for somatosensory information entering the spinal cord from dorsal root ganglion cells. The local branches can activate local reflex circuits that modulate motor output, while the ascending branches carry information into the brain, where this information is further processed in the thalamus and cerebral cortex.

The Terminals of Central Axons of Dorsal Root Ganglion Neurons in the Spinal Cord Produce a Map of the Body Surface

The manner in which the central axons of the dorsal root ganglion cells terminate in the spinal cord forms a neural map of the body surface. This orderly somatotopic distribution of inputs from different portions of the body surface is maintained throughout the entire ascending somatosensory pathway. This arrangement illustrates another important principle of neural organization. Neurons that make up neural circuits at any particular level are often connected in a systematic fashion and appear similar from individual to individual. Similarly, fiber bundles that connect different processing regions at different levels of the nervous system are also arranged in a highly organized and stereotypical fashion.

Axons that enter the cord in the sacral region ascend in the dorsal column near the midline, whereas those that enter at successively higher levels ascend at progressively more lateral positions within the dorsal columns. Thus, in the cervical cord, where axons from all portions of the body have already entered, sensory fibers from the lower body are located medially in the dorsal column, while fibers from the trunk, arm and shoulder, and finally the neck occupy progressively more lateral areas. In the cervical spinal cord, the axons forming the dorsal columns are divided into two bundles: a medially situated gracile fascicle and a more laterally situated cuneate fascicle (Figure 4–1).

Each Somatic Submodality Is Processed in a Distinct Subsystem From the Periphery to the Brain

The submodalities of somatic sensation—touch, pain, temperature, and position sense—are processed in the brain through different pathways that end in different brain regions. We illustrate the specificity of these parallel pathways by the path of information for the submodality of touch.

The primary afferent fibers that carry information about touch enter the ipsilateral dorsal column and ascend to the medulla. Fibers from the lower body run in the gracile fascicle and terminate in the gracile nucleus, whereas fibers from the upper body run in the cuneate fascicle and terminate in the cuneate nucleus. Neurons in the gracile and cuneate nuclei give rise to axons that cross to the other side of the brain and ascend to the thalamus in a long fiber bundle called the medial lemniscus (Figure 4–1).

As in the dorsal columns of the spinal cord, the fibers of the medial lemniscus are arranged somatotopically. Because the fibers carrying sensory information

Figure 4–6 (Opposite) Somatosensory information from the limbs and trunk is conveyed to the thalamus and cerebral cortex by two ascending pathways. Brain slices along the neuraxis from the spinal cord to the cerebrum illustrate the anatomy of the two principal pathways conveying somatosensory information to the cerebral cortex. The two pathways are separated until they reach the pons, where they are juxtaposed.

Dorsal column–medial lemniscal system (**orange**). Touch and limb proprioception signals are conveyed to the spinal cord and brain stem by large-diameter myelinated nerve fibers and transmitted to the thalamus in this system. In the spinal cord, the fibers for touch and proprioception divide, one branch going to the ipsilateral spinal gray matter and the other ascending in the ipsilateral dorsal column to the medulla. The second-order fibers from neurons in the dorsal column nuclei cross the

midline in the medulla and ascend in the contralateral medial lemniscus toward the thalamus, where they terminate in the lateral and medial ventral posterior nuclei. Thalamic neurons in these nuclei convey tactile and proprioceptive information to the primary somatosensory cortex.

Anterolateral system (**brown**). Pain, itch, temperature, and visceral information is conveyed to the spinal cord by small-diameter myelinated and unmyelinated fibers that terminate in the ipsilateral dorsal horn. This information is conveyed across the midline by neurons within the spinal cord and transmitted to the brain stem and the thalamus in the contralateral anterolateral system. Anterolateral fibers terminating in the brain stem compose the spinoreticular and spinomesencephalic tracts; the remaining anterolateral fibers form the spinothalamic tract.

cross the midline to the other side of the brain, the right side of the brain receives sensory information from the left side of the body, and vice versa. The fibers of the medial lemniscus end in a specific subdivision of the thalamus called the ventral posterior lateral nucleus (Figure 4–1). There the fibers maintain their somatotopic organization such that those carrying information from the lower body end laterally and those carrying information from the upper body end medially.

The Thalamus Is an Essential Link Between Sensory Receptors and the Cerebral Cortex

The thalamus is an egg-shaped structure that constitutes the dorsal portion of the diencephalon. It contains a class of excitatory neurons called thalamic relay cells that convey sensory input to the primary sensory areas of the cerebral cortex. However, the thalamus is not merely a relay. It acts as a gatekeeper for information to the cerebral cortex, preventing or enhancing the passage of specific information depending on the behavioral state of the organism.

The cerebral cortex has feedback projections that terminate, in part, in a special portion of the thalamus called the thalamic reticular nucleus. This nucleus

forms a thin sheet around the thalamus and is made up almost totally of inhibitory neurons that synapse onto the relay cells. It does not project to the neocortex at all. In addition to receiving feedback projections from the neocortex, the reticular nucleus receives input from axons leaving the thalamus en route to the neocortex, enabling the thalamus to modulate the response of its relay cells to incoming sensory information.

The thalamus is a good example of a brain region made up of several well-defined nuclei. As many as 50 thalamic nuclei have been identified (Figure 4–7). Some nuclei receive information specific to a sensory modality and project to a specific area of the neocortex. For example, cells in the ventral posterior lateral nucleus (where the medial lemniscus terminates) process somatosensory information, and their axons project to the primary somatosensory cortex (Figures 4–1 and 4–7). Projections from the retinal ganglion cells terminate in another portion of the thalamus called the lateral geniculate nucleus (Figure 4–7). Neurons in this nucleus project in turn to the visual cortex. Other portions of the thalamus participate in motor functions, transmitting information from the cerebellum and basal ganglia to the motor regions of the frontal lobe. Axons from cells of the thalamus that project to the neocortex travel in the corona radiata, a large

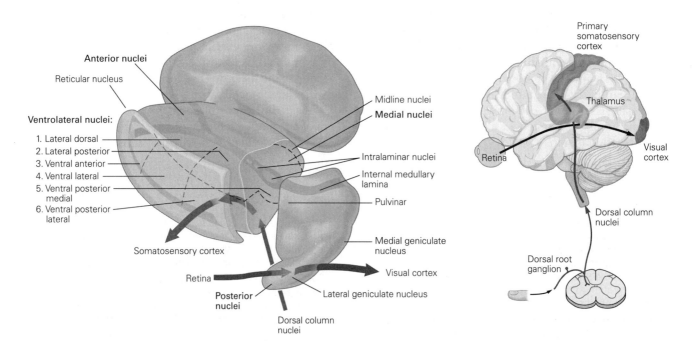

Figure 4–7 The major subdivisions of the thalamus. The thalamus is the critical relay for the flow of sensory information from peripheral receptors to the neocortex. Somatosensory information is conveyed from dorsal root ganglia to the ventral posterior lateral nucleus and from there to the primary

somatosensory cortex. Likewise, visual information from the retina reaches the lateral geniculate nucleus, from which it is conveyed to the primary visual cortex in the occipital lobe. Each of the sensory systems, except olfaction, has a similar processing step within a distinct region of the thalamus.

fiber bundle that carries most of the axons running to and from the cerebral hemispheres. Through its connections with the frontal lobe and hippocampus, the thalamus may play a role in cognitive functions, such as memory. Some nuclei that may play a role in attention project diffusely to large but distinctly different regions of cortex.

The nuclei of the thalamus are most commonly classified into four groups—anterior, medial, ventrolateral, and posterior—with respect to the internal medullary lamina, a sheet-like bundle of fibers that runs the rostrocaudal length of the thalamus (Figure 4–7). Thus, the medial group of nuclei is located medial to the internal medullary lamina, whereas the ventrolateral and posterior groups are located lateral to it. At the rostral pole of the thalamus, the internal medullary lamina splits and surrounds the anterior group. The caudal pole of the thalamus is occupied by the posterior group, dominated by the pulvinar nucleus. Groups of neurons are also located within the fibers of the internal medullary lamina and are collectively referred to as the intralaminar nuclei.

The *anterior group* receives its major input from the mammillary nuclei of the hypothalamus and from the presubiculum of the hippocampal formation. The role of the anterior group is uncertain, but because of its connections, it is thought to be related to memory and emotion. The anterior group is mainly interconnected with regions of the cingulate and frontal cortices.

The *medial group* consists mainly of the mediodorsal nucleus. This large thalamic nucleus has three subdivisions, each of which is connected to a particular portion of the frontal cortex. The nucleus receives inputs from portions of the basal ganglia, the amygdala, and midbrain and has been implicated in memory and emotional processing.

The nuclei of the *ventrolateral group* are named according to their positions within the thalamus. The ventral anterior and ventral lateral nuclei are important for motor control and carry information from the basal ganglia and cerebellum to the motor cortex. The ventral posterior nuclei convey somatosensory information to the neocortex. The ventroposterior lateral nucleus conveys information from the spinal cord tracts, as described earlier. The ventroposterior medial nucleus conveys information from the face, which enters the brain stem mainly through the trigeminal nerve (cranial nerve V).

The *posterior group* includes the medial and lateral geniculate nuclei, the lateral posterior nucleus, and the pulvinar. The medial geniculate nucleus is a component of the auditory system and is organized tonotopically based on the sound frequency information

carried by its inputs; it conveys auditory information to the primary auditory cortex in the superior temporal gyrus of the temporal lobe. The lateral geniculate nucleus receives information from the retina and conveys it to the primary visual cortex in the occipital lobe. Compared to rodents, the pulvinar is enlarged disproportionately in the primate brain, especially in the human brain, and its development seems to parallel the enlargement of the association regions of the parietal, occipital, and temporal cortices. It has been divided into at least three subdivisions and is extensively interconnected with widespread regions of the parietal, temporal, and occipital lobes, as well as with the superior colliculus and other nuclei of the brain stem related to vision.

As noted previously, the thalamus not only projects to the neocortex (feedforward connections) but also receives extensive return inputs back from the neocortex (feedback connections). For example, in the lateral geniculate nucleus, the number of synapses formed by axons from the feedback projection from the visual cortex is actually greater than the number of synapses that the lateral geniculate nucleus receives from the retina! This feedback is thought to play an important modulatory role in the processing of sensory information, although the exact function is not yet understood. Although this feedback is mainly from cortical neurons that are activated by both eyes, the neurons in the lateral geniculate nucleus are responsive to only one or the other eye. The implication is that they are primarily driven by input from the retina (which is from different eyes in different layers), not the feedback from the cortex, despite its numerical advantage. Most nuclei of the thalamus receive a similarly prominent return projection from the cerebral cortex, and the significance of these projections is one of the unsolved mysteries of neuroscience.

The thalamic nuclei described thus far are called the *relay* (or *specific*) *nuclei* because they have a specific and selective relationship with a particular portion of the neocortex. Other thalamic nuclei, called *nonspecific nuclei*, project to several cortical and subcortical regions. These nuclei are located either on the midline of the thalamus (the midline nuclei) or within the internal medullary lamina (the intralaminar nuclei). The largest of the midline nuclei are the paraventricular, paratenial, and reuniens nuclei; the largest of the intralaminar cell groups is the centromedian nucleus. The intralaminar nuclei project to medial temporal lobe structures, such as the amygdala and hippocampus, but also send projections to portions of the basal ganglia. These nuclei receive inputs from a variety of sources in the spinal cord, brain stem, and cerebellum and are thought to mediate cortical arousal.

The thalamus is an important step in the hierarchy of sensory processing, not a passive relay station where information is simply passed on to the neocortex. It is a complex brain region where substantial information processing takes place (Figure 4–1). To give but one example, the output of somatosensory information from the ventral posterior lateral nucleus is subject to four types of processing: (1) local processing within the nucleus; (2) modulation by brain stem inputs, such as from the noradrenergic and serotonergic systems; (3) inhibitory input from the reticular nucleus; and (4) modulatory feedback from the neocortex.

Sensory Information Processing Culminates in the Cerebral Cortex

Somatosensory information from the ventral posterior lateral nucleus is conveyed mainly to the primary somatosensory cortex (Figure 4–1). The neurons here are exquisitely sensitive to tactile stimulation of the skin surface. The somatosensory cortex, like earlier stages in tactile sensory processing, is somatotopically organized (Figure 4–8).

When the neurosurgeon Wilder Penfield stimulated the surface of the somatosensory cortex in patients undergoing brain surgery in the late 1940s and early 1950s, he found that sensation from the lower limbs is mediated by neurons located near the midline of the brain, whereas sensations from the upper body, hands and fingers, face, lips, and tongue are mediated by neurons located laterally. Penfield found that, although all parts of the body are represented in the cortex somatotopically, the amount of surface area of cortex devoted to each body part is not proportional to its mass. Instead, it is proportional to the fineness of discrimination in the body part, which in turn is related to the density of innervation of sensory fibers (Chapter 19). Thus, the area of cortex devoted to the fingers is larger than that for the arms. Likewise, the representation of the lips and tongue occupies more cortical surface than that of the remainder of the face (Figure 4–8). As we shall see in Chapter 53, the amount of cortex devoted to a particular body part is not fixed but can be modified by experience, as seen in concert violinists, where there is an expansion of the region of somatosensory cortex devoted to the fingers of the hand used to finger the strings. This illustrates

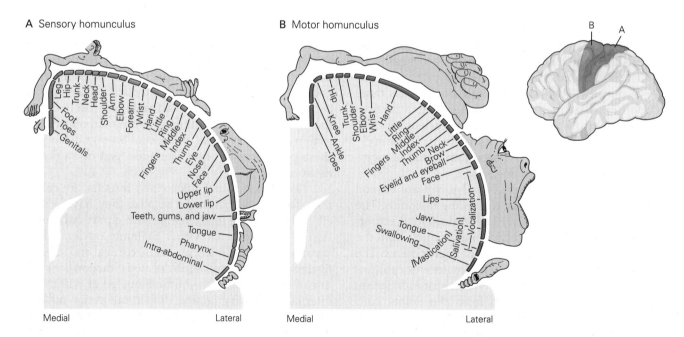

Figure 4–8 Homunculi illustrate the relative amounts of cortical area dedicated to sensory and motor innervation of individual parts of the body. The entire body surface is represented in an orderly array of somatosensory inputs in the cortex. (From Penfield and Rasmussen 1950. Reproduced by permission of the Osler Library of the History of Medicine, McGill University.)

A. The area of cortex dedicated to processing sensory information from a particular part of the body is not proportional to the mass of

the body part but instead reflects the density of sensory receptors in that part. Thus, sensory input from the lips and hands occupies more area of cortex than, say, that from the elbow.

B. Output from the motor cortex is organized in similar fashion. The amount of cortical surface dedicated to a part of the body is related to the degree of motor control of that part. Thus, in humans, much of the motor cortex is dedicated to controlling the muscles of the fingers and the muscles related to speech.

an important aspect of brain circuitry: It is capable of plastic changes in response to use or disuse. Such changes are important for various forms of learning, including the ability to recover function after a stroke.

The region of cerebral cortex nearest the surface of the brain is organized in layers and columns, an arrangement that increases its computational efficiency. The cortex has undergone dramatic expansion in evolution. The more recent neocortex comprises most of the cortex of mammals. In larger brains of primates and cetaceans, the neocortical surface is a sheet that is folded with deep wrinkles, thus allowing for three times more cortical surface to be packed into an only modestly enlarged head. Indeed, approximately two-thirds of the neocortex is along the deep wrinkles of the cortex, termed sulci. The remainder of neocortex

is at the external folds of the sheet, termed gyri. The neocortex receives inputs from the thalamus, other cortical regions on both sides of the brain, and other subcortical structures. Its output is directed to other regions of the cortex, basal ganglia, thalamus, pontine nuclei, and spinal cord.

These complex input–output relationships are efficiently organized in the orderly layering of cortical neurons; each layer contains different inputs and outputs. Many regions of the neocortex, in particular the primary sensory areas, contain six layers, numbered from the outer surface of the brain to the white matter (Figure 4–9).

Layer I, the molecular layer, is occupied by the dendrites of cells located in deeper layers and axons that travel through this layer to make connections in other areas of the cortex.

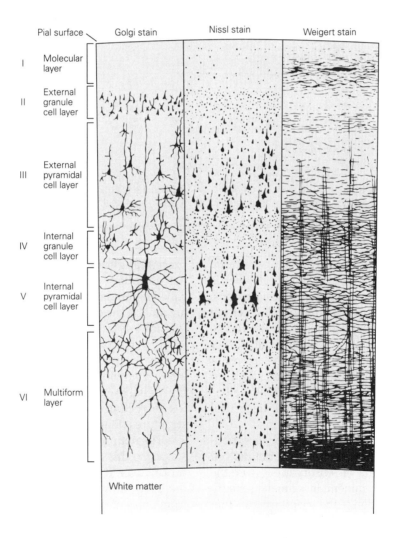

Figure 4–9 The neurons of the neocortex are arranged in distinctive layers. The appearance of the neocortex depends on what is used to stain it. The Golgi stain (*left*) reveals a subset of neuronal cell bodies, axons, and dendritic trees.

The Nissl method (*middle*) shows cell bodies and proximal dendrites. The Weigert stain (*right*) reveals the pattern of myelinated fibers. (Reproduced, with permission, from Heimer 1994.)

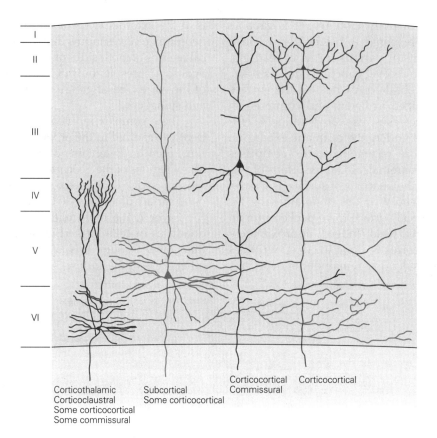

Corticothalamic
Corticoclaustral
Some corticocortical
Some commissural

Subcortical
Some corticocortical

Corticocortical
Commissural

Corticocortical

Figure 4–10 Neurons in different layers of neocortex project to different parts of the brain. Projections to all other parts of the neocortex, the so-called corticocortical or associational connections, arise primarily from neurons in layers II and III. Projections to subcortical regions arise mainly from layers V and VI. (Reproduced, with permission, from Jones 1986.)

Layers II and III contain mainly small pyramidal shaped cells. Layer II, the external granular cell layer, is one of two layers that contain small spherical neurons. Layer III is called the external pyramidal cell layer (an internal pyramidal cell layer lies at a deeper level). The neurons located deeper in layer III are typically larger than those located more superficially. The axons of pyramidal neurons in layers II and III project locally to other neurons within the same cortical area as well as to other cortical areas, thereby mediating intracortical communication (Figure 4–10).

Layer IV contains a large number of small spherical neurons and thus is called the internal granular cell layer. It is the main recipient of sensory input from the thalamus and is most prominent in primary sensory areas. For example, the region of the occipital cortex that functions as the primary visual cortex has an extremely prominent layer IV. Layer IV in this region is so heavily populated by neurons and so complex that it is typically divided into three sublayers. Areas with a prominent layer IV are called granular cortex. In contrast, the

precentral gyrus, the site of the primary motor cortex, has essentially no layer IV and is thus part of the so-called agranular frontal cortex. These two cortical areas are among the easiest to identify in histological sections (Figure 4–11).

Layer V, the internal pyramidal cell layer, contains mainly pyramidally shaped cells that are typically larger than those in layer III. Pyramidal neurons in this layer give rise to the major output pathways of the cortex, projecting to other cortical areas and to subcortical structures (Figure 4–9).

The neurons in layer VI are fairly heterogeneous in shape, so this layer is called the polymorphic or multiform layer. It blends into the white matter that forms the deep limit of the cortex and carries axons to and from areas of cortex.

The thickness of individual layers and the details of their functional organization vary throughout the cortex. An early student of the cerebral cortex, Korbinian Brodmann, used the relative prominence of the layers above and below layer IV, cell size, and packing

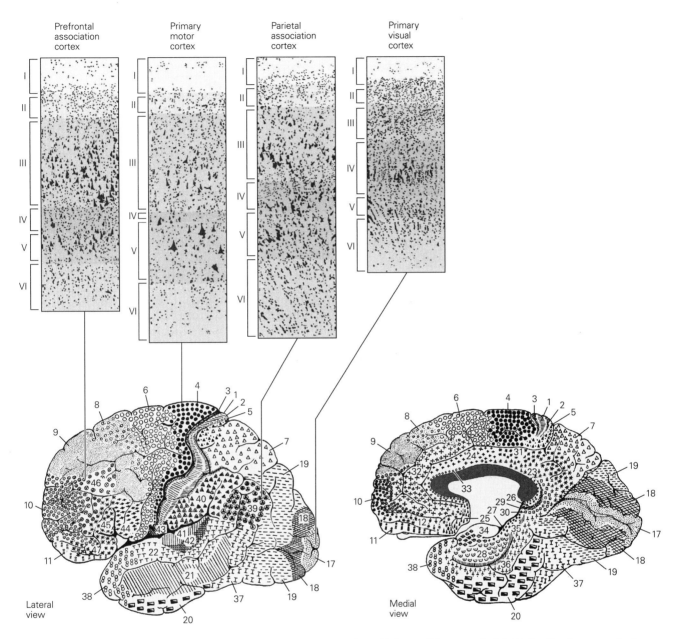

Figure 4–11 The extent of each cell layer of the neocortex varies throughout the cortex. Sensory areas of cortex, such as the primary visual cortex, tend to have a very prominent internal granular cell layer (layer IV), the site of sensory input. Motor areas of cortex, such as the primary motor cortex, have a very meager layer IV but prominent output layers, such as layer V. These differences led Korbinian Brodmann and others working at the turn of the 20th century to divide the cortex into various cytoarchitectonic regions. Brodmann's 1909 subdivision shown here is a classic analysis but was based on a single human brain. (Reproduced, with permission, from Martin 2012.)

characteristics to distinguish different areas of the neocortex. Based on such cytoarchitectonic differences, in 1909, Brodmann divided the cerebral cortex into 47 regions (Figure 4–11).

Although Brodmann's demarcation coincides in part with information on localized functions in the neocortex, the cytoarchitectonic method alone does not capture the subtlety or variety of function of all the distinct regions of the cortex. For example, Brodmann identified five regions (areas 17–21) as being concerned with visual function in the monkey. In contrast, modern connectional neuroanatomy and electrophysiology have identified more than 35 functionally distinct cortical regions within the five regions recognized by Brodmann.

Ascending

Descending

Superficial
layers (I–III)   Layer IV

Deep (Infragranular)
layers (V, VI)   Multilaminar

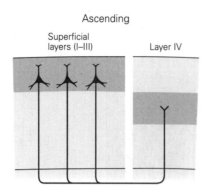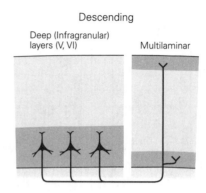

Figure 4–12 Ascending and descending cortical pathways are distinguished by the organization of their origins and terminations within the cortical layers. Ascending or feedforward pathways generally originate in superficial layers of the cortex and invariably terminate in layer IV. Descending or feedback pathways generally originate from deep layers and terminate in layers I and VI. (Adapted, with permission, from Felleman and Van Essen 1991.)

Within the neocortex, information passes from one synaptic relay to another using feedforward and feedback connections. In the visual system, for example, feedforward projections from the primary visual cortex to secondary and tertiary visual areas originate mainly in layer III and terminate mainly in layer IV of the target cortical area. In contrast, feedback projections to earlier stages of processing originate from cells in layers V and VI and terminate in layers I, II, and VI (Figure 4–12).

The cerebral cortex is organized functionally into columns of cells that extend from the white matter to the surface of the cortex. (This columnar organization is not particularly evident in standard histological preparations and was first discovered in electrophysiological studies.) Each column is about one-third of a millimeter in diameter. The cells in each column form a computational module with a highly specialized function. Neurons within a column tend to have very similar response properties, presumably because they form a local processing network. The larger the area of cortex dedicated to a function, the greater the number of computational columns that are dedicated to that function (Chapter 23). The highly discriminative sense of touch in the fingers is a result of many cortical columns in the large area of cortex dedicated to processing somatosensory information from the hand.

Beyond the identification of the cortical column, a second major insight from the early electrophysiological studies was that the somatosensory cortex contains not one but several somatotopic maps of the body surface. The primary somatosensory cortex (anterior parietal cortex) has four complete maps of the skin, one each in Brodmann areas 3a, 3b, 1, and 2. The thalamus sends, in parallel, a lot of deep receptor information

(eg, from muscles) to area 3a and most of its cutaneous information to areas 3b and 1. Area 2 receives input from these thalamorecipient cortical areas and may be responsible for our integrated perception of three-dimensional solid objects, termed stereognosis. Neurons in the primary somatosensory cortex project to neurons in adjacent areas, and these neurons in turn project to other adjacent cortical regions (Figure 4–13). At higher levels in the hierarchy of cortical connections, somatosensory information is used in motor control, eye–hand coordination, and memory related to touch.

The cortical areas involved in the early stages of sensory processing are concerned primarily with a single sensory modality. Such regions are called primary sensory or unimodal (sensory) association areas. Information from the unimodal association areas converges on multimodal association areas of the cortex concerned with combining sensory modalities (Figure 4–13). These multimodal association areas, which are heavily interconnected with the hippocampus, appear to be particularly important for two functions: (1) the production of a unified percept and (2) the representation of the percept in memory (we will return to this at the end of this chapter).

Thus, from the mechanical pressure on a receptor in the skin to the perception that a finger has been touched by a friend shaking your hand, somatosensory information is processed in a series of increasingly more complex circuits (networks) from the dorsal root ganglia to the somatosensory cortex, to unimodal association areas, and finally to multimodal association areas. One of the primary purposes of somatosensory information is to guide directed movement. As one might imagine, there is a close linkage between the somatosensory and motor functions of the cortex.

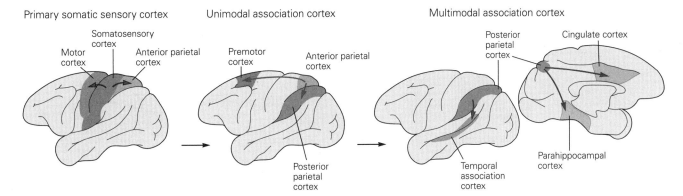

Figure 4–13 The processing of sensory information in the cerebral cortex begins with primary sensory areas, continues in unimodal association areas, and is further elaborated in multimodal association areas. Sensory systems also communicate with portions of the motor cortex. For example, the primary somatosensory cortex projects to the motor area in the frontal lobe and to the somatosensory association area in the parietal cortex. The somatosensory association area, in turn, projects to higher-order somatosensory association areas and to the premotor cortex. Information from different sensory systems converges in the multimodal association areas, which include the parahippocampal, temporal association, and cingulate cortices.

Voluntary Movement Is Mediated by Direct Connections Between the Cortex and Spinal Cord

As we shall see in Chapters 25 and 30, a major function of the perceptual systems is to provide the sensory information necessary for the actions mediated by the motor systems. The primary motor cortex is organized somatotopically like the somatosensory cortex (Figure 4–8B). Specific regions of the motor cortex influence the activity of specific muscle groups (Chapter 34).

The axons of neurons in layer V of the primary motor cortex provide the major output of the neocortex to control movement. Some layer V neurons influence movement directly through projections in the corticospinal tract to motor neurons in the ventral horn of the spinal cord. Others influence motor control by synapsing onto motor output nuclei in the medulla or onto striatal neurons in the basal ganglia. The human corticospinal tract consists of approximately one million axons, of which approximately 40% originate in the motor cortex. These axons descend through the subcortical white matter, the internal capsule, and cerebral peduncle in the midbrain (Figure 4–14). In the medulla, the fibers form prominent protuberances on the ventral surface called the medullary pyramids, and thus the entire projection is sometimes called the pyramidal tract.

Like the ascending somatosensory system, the descending corticospinal tract crosses to the opposite side of the spinal cord. Most of the corticospinal fibers cross the midline in the medulla at a location known as the pyramidal decussation. However, approximately 10% of the fibers do not cross until they reach the level of the spinal cord at which they will terminate. The corticospinal fibers make monosynaptic connections with motor neurons, connections that are particularly important for individuated finger movements. They also form synapses with both excitatory and inhibitory interneurons in the spinal cord, connections that are important for coordinating larger groups of muscles in behaviors such as reaching and walking.

The motor information carried in the corticospinal tract is significantly modulated by both sensory information and information from other motor regions. A continuous stream of tactile, visual, and proprioceptive information is needed to make voluntary movement both accurate and properly sequenced. In addition, the output of the motor cortex is under the substantial influence of other motor regions of the brain, including the cerebellum and basal ganglia, structures that are essential for smoothly executed movements. These two subcortical regions, which are described in detail in Chapters 37 and 38, provide feedback essential for the smooth execution of skilled movements and thus are also important for the improvement in motor skills through practice (Figure 4–15).

Modulatory Systems in the Brain Influence Motivation, Emotion, and Memory

Some areas of the brain are neither purely sensory nor purely motor, but instead modulate specific sensory or motor functions. Modulatory systems are often involved in behaviors that respond to a primary need such as hunger, thirst, or sleep. For example, sensory

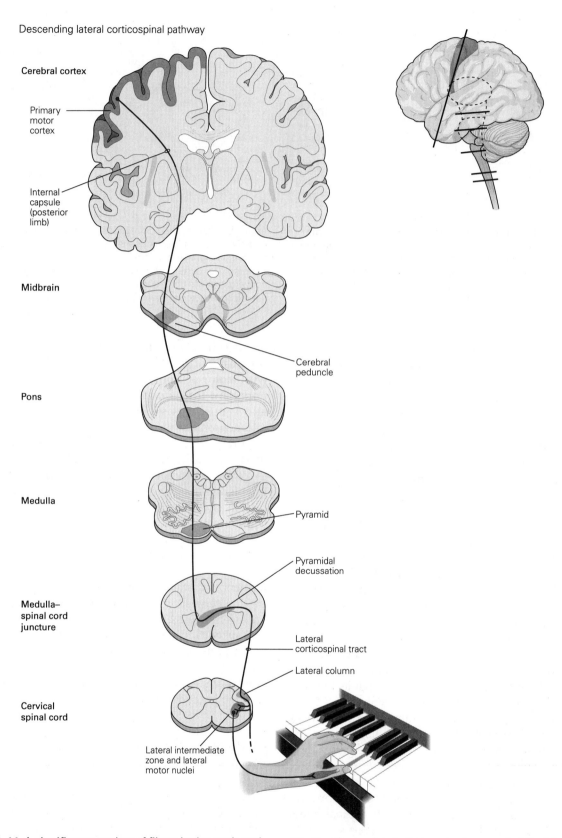

Figure 4–14 A significant number of fibers in the corticospinal tract originate in the primary motor cortex and terminate in the ventral horn of the spinal cord. The same axons are, at various points in their projections, part of the internal capsule, the cerebral peduncle, the medullary pyramid, and the lateral corticospinal tract.

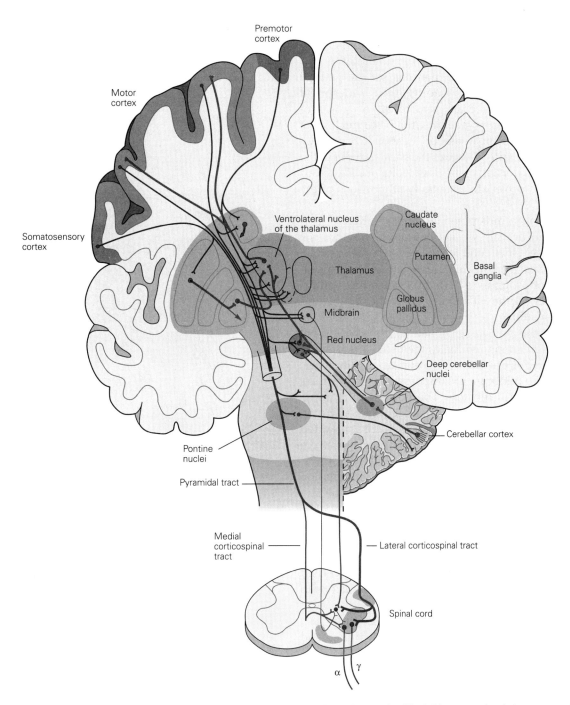

Figure 4–15 Voluntary movement requires coordination of all components of the motor system. The principal components are the motor cortex, basal ganglia, thalamus, midbrain, cerebellum, and spinal cord. The principal descending projections are shown in **green**; feedback projections and local connections are shown in **purple**. All of this processing is incorporated in the inputs to the motor neurons of the ventral horn of the spinal cord, the so-called "final common pathway" that innervates muscle and elicits movements. (This figure is a composite view made from sections of the brain taken at different angles.)

and modulatory systems in the hypothalamus determine blood glucose levels (Chapter 41). When blood sugar drops below a certain critical level, we feel hunger. To satisfy hunger, modulatory systems in the brain

focus vision, hearing, and smell on stimuli that are relevant to feeding.

Distinct modulatory systems within the brain stem modulate attention and arousal (Chapter 40). Small nuclei in the brain stem contain neurons that synthesize and release the modulatory neurotransmitter norepinephrine (the locus coeruleus) and serotonin (the dorsal raphe nucleus). Such neurons set the general arousal level of an animal through their widespread connections with forebrain structures. A group of cholinergic modulatory neurons, the basal nucleus of Meynert, is involved in arousal or attention (Chapter 40). This nucleus is located beneath the basal ganglia in the basal forebrain portion of the telencephalon. The axons of its neurons project to essentially all portions of the neocortex.

If a predator finds potential prey, a variety of cortical and subcortical systems determine whether the prey is edible. Once food is recognized, other cortical and subcortical systems initiate a comprehensive voluntary motor program to bring the animal into contact with the prey, capture it and place it in the mouth, and chew and swallow.

Finally, the physiological satisfaction the animal experiences in eating reinforces the behaviors that led to the successful predation. A group of dopaminergic neurons in the midbrain are important for monitoring reinforcements and rewards. The power of the dopaminergic modulatory systems has been demonstrated by experiments in which electrodes were implanted into the reward regions of rats and the animals were freely allowed to press a lever to electrically stimulate their brains. They preferred this self-stimulation to obtaining food or water, engaging in sexual behavior, or any other naturally rewarding activity. The role of the dopaminergic modulatory system in learning through reinforcement of exploratory behavior is described in Chapter 38.

How the brain's modulatory systems, concerned with reward, attention and motivation, interact with the sensory and motor systems is one of the most interesting questions in neuroscience, one that is also fundamental to our understanding of learning and memory storage (Chapter 40).

The Peripheral Nervous System Is Anatomically Distinct From the Central Nervous System

The peripheral nervous system supplies the central nervous system with a continuous stream of information about both the external environment and the internal environment of the body. It has somatic and autonomic divisions (Figure 4–16).

The *somatic division* includes the sensory neurons that receive information from the skin, muscles, and

Peripheral Nervous System

Somatic   Autonomic

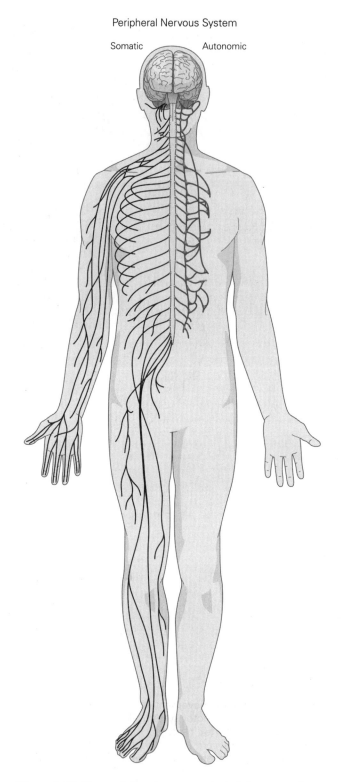

Figure 4–16 The peripheral nervous system has somatic and autonomic divisions. The somatic division carries information from the skin to the brain and from the brain to muscles. The autonomic division regulates involuntary functions, including activity of the heart and smooth muscles in the gut and glands.

joints. The cell bodies of these sensory neurons lie in the dorsal root ganglia and cranial ganglia. Receptors associated with these cells provide information about muscle and limb position and about touch and pressure at the body surface. In Part IV (Perception), we shall see how remarkably specialized these receptors are in transducing one or another type of physical energy (eg, deep pressure or heat) into the electrical signals used by the nervous system. In Part V (Movement), we shall see that sensory receptors in the muscles and joints are crucial to shaping coherent movement of the body.

The *autonomic division* of the peripheral nervous system mediates visceral sensation as well as motor control of the viscera, vascular system, and exocrine glands. It consists of the sympathetic, parasympathetic, and enteric systems. The sympathetic system participates in the body's response to stress, whereas the parasympathetic system acts to conserve body resources and restore homeostasis. The enteric nervous system, with neuronal cell bodies located in or adjacent to the viscera, controls the function of smooth muscle and secretions of the gut. The functional organization of the autonomic nervous system is described in Chapter 41 and its role in emotion and motivation in Chapter 42.

Memory Is a Complex Behavior Mediated by Structures Distinct From Those That Carry Out Sensation or Movement

Research over the past 50 years has provided a sophisticated view of memory systems in the brain. We now know that different forms of memory (eg, fear memory versus skill memory) are mediated by different brain regions. Here we contrast the organization of the system responsible for coding and storing our experiences of other individuals, places, facts, and episodes, a process called explicit memory.

We know that a structure called the hippocampus (or more properly the hippocampal formation, since it is several cortical regions) is a key component of a medial temporal lobe memory system that encodes and stores memories of our lives (Figure 4–17). This understanding is based largely on the analysis of the

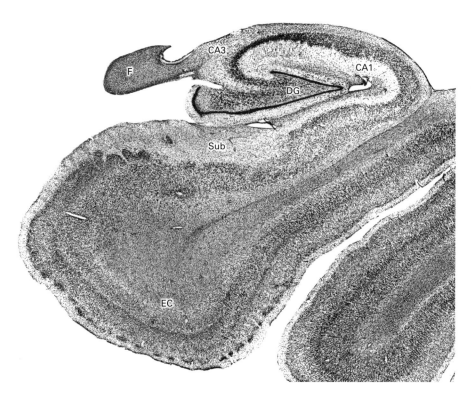

Figure 4–17 Coronal section of the human hippocampal formation stained by the Nissl methods to demonstrate cell bodies. The main cytoarchitectonic fields are shown in this section of the human hippocampal formation.

(Abbreviations: **CA3** and **CA1**, subdivisions of the hippocampus; **DG**, dentate gyrus; **EC**, entorhinal cortex; **F**, fimbria; **Sub**, subiculum.)

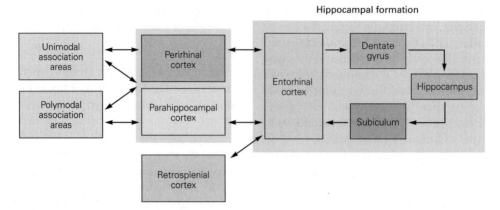

Figure 4–18 Hierarchical organization of connections to the hippocampal formation. The hippocampal formation receives highly processed sensory information, primarily through the entorhinal cortex, from multimodal association regions such as the perirhinal, parahippocampal, and retrosplenial cortices.

famous patient Henry Molaison (referred to as HM by the scientists who studied him during his life), who in the early 1950s had bilateral temporal lobe surgery to reduce his life-threatening epilepsy. In contrast to the six-layered neocortex, the hippocampus, along with olfactory cortex (piriform cortex), is a three-layered cortical structure referred to as archicortex, one of the phylogenetically older areas of cortex.

The reason we briefly describe the hippocampal formation in this chapter is to emphasize that not all brain circuits are alike. In fact, whether one talks about the olfactory bulb, where the sense of smell begins to be processed, or the cerebellum, where fine motor movements are refined, the general principle is that the structure of a circuit is specific to the function that it mediates. And the hippocampal circuit is as different from the circuits that mediate sensory perception or motor movement as one could imagine. Hippocampal circuitry of the brain will be dealt with in much more detail in later chapters. Chapter 5 introduces the idea that the hippocampus encodes information about an animal's spatial location in its environment and that the encoding of explicit memory (including spatial memory) requires plastic changes in synaptic function. Chapters 52 and 54 explore human memory function and the cellular and molecular bases of explicit memory and spatial representation, respectively.

The Hippocampal System Is Interconnected With the Highest-Level Polysensory Cortical Regions

Sensory systems are hierarchical and process progressively more complex stimuli at higher levels, particularly of the neocortex. Moreover, from the highest levels of each modality, circuits connect with polysensory cortical regions located at various places around the cortex, where information from many sensory modalities converges onto single neurons. The hippocampal system receives most of its input, the raw material with which it makes memories, from a few specific polysensory regions. These include the perirhinal and parahippocampal cortices, located in the medial temporal lobe, as well as the retrosplenial cortex, located in the caudal portion of the cingulate gyrus. These polysensory regions converge on the entry structure to the hippocampal system, the entorhinal cortex (Figure 4–18). The polysensory information that enters the entorhinal cortex can be thought of as providing summaries of immediate experience.

The Hippocampal Formation Comprises Several Different but Highly Integrated Circuits

The hippocampal formation is made up of a number of distinct cortical regions that are simpler in organization than the neocortex—at least they have fewer layers. The regions include the dentate gyrus, hippocampus, subiculum, and entorhinal cortex. Each of these regions is made up of subregions containing many neuronal cell types. The simplest subregion of the hippocampal formation is the dentate gyrus, which has a single principal neuron called the granule cell. The subregions of the hippocampus termed CA1, CA2, and CA3, consist of a single layer of pyramidal cells whose dendrites extend above and below the cell body layer and receive inputs from several regions. The subiculum (divided into subiculum, presubiculum and parasubiculum) is another region made up largely of pyramidal cells. Finally, the most complex part of the hippocampal formation is the entorhinal cortex, which has multiple layers but still has an organization distinctly different from the neocortex. For example, it lacks a layer IV and has a much more prominent layer II.

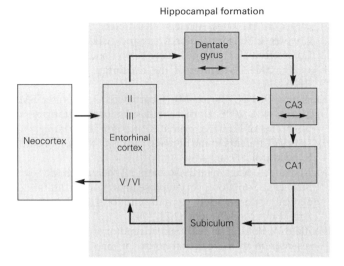

Hippocampal formation

Figure 4–19 Simplified diagram on internal connections within the hippocampal formation. The circuit begins from cells in layer II of the entorhinal cortex to the dentate gyrus, which then projects to the CA3 region of the hippocampus. The CA3 portion of the hippocampus projects to CA1, and CA1 then projects to the subiculum. The hippocampal circuit is closed when the subiculum projects to the deep layers of the entorhinal cortex. Not shown are the feedback pathways from entorhinal cortex to the same multimodal areas from which it receives sensory information.

The Hippocampal Formation Is Made Up Mainly of Unidirectional Connections

Here we describe the fundamental circuitry of the hippocampal formation. The circuitry is described in more detail in Chapter 54. The simplified version of the hippocampal circuit shown in Figure 4–19 emphasizes its stepwise serial processing of multimodal sensory information, with each hippocampal region contributing to the formation of explicit memories. This serial processing implies that damage to any one of the components of this system would lead to memory impairment. And, in fact, another famous patient, known by the initials R.B., did suffer profound memory impairment due to loss of cells in the CA1 region after an ischemic episode.

As it turns out, while the hippocampal formation is essential for the initial formation of memories of our lives, these memories are ultimately stored elsewhere in the brain. In patients such as HM, in whom the entorhinal cortex and much of the rest of the hippocampal system was removed, memories prior to the surgery were largely intact. Thus, to achieve creation and long-term storage of the memories of our lives, the hippocampus and entorhinal cortex must communicate with circuits

in the cerebral cortex. Where and precisely how that happens remain a mystery.

Highlights

1. Individual neurons are not able to carry out behavior. They must be incorporated into circuits that comprise different types of neurons that are interconnected by excitatory, inhibitory, and modulatory connections.

2. Sensory and motor information is processed in the brain in a variety of discrete brain regions that are active simultaneously.

3. A functional pathway is formed by the serial connection of identifiable brain regions, and each brain region's circuits process more complex or specific information than the preceding brain region.

4. The sensations of touch and pain are mediated by pathways that run between different circuits in the spinal cord, brain stem, thalamus, and neocortex.

5. All sensory and motor systems follow the pattern of hierarchical and reciprocal processing of information, whereas the hippocampal memory system is organized largely for serial processing of very complex, polysensory information. A general principle is that circuits in the brain have an organizational structure that is suited for the functions that they are carrying out.

6. Contrary to an intuitive analysis of our personal experience, perceptions are not precise copies of the world around us. Sensation is an abstraction, not a replication, of reality. The brain's circuits construct an internal representation of external physical events after first analyzing various features of those events. When we hold an object in the hand, the shape, movement, and texture of the object are simultaneously analyzed in different brain regions according to the brain's own rules, and the results are integrated in a conscious experience.

7. How sensation is integrated in a conscious experience—the *binding problem*—and how conscious experience emerges from the brain's analysis of incoming sensory information are two of the most intriguing questions in cognitive neuroscience (Chapter 56). An even more complex issue is how these conscious impressions are encoded into memories that are stored for decades.

David G. Amaral

Selected Reading

Brodal A. 1981. *Neurological Anatomy in Relation to Clinical Medicine,* 3rd ed. New York: Oxford Univ. Press.

Carpenter MB. 1991. *Core Text of Neuroanatomy,* 4th ed. Baltimore: Williams and Wilkins.

England MA, Wakely J. 1991. *Color Atlas of the Brain and Spinal Cord: An Introduction to Normal Neuroanatomy.* St. Louis: Mosby Year Book.

Martin JH. 2012. *Neuroanatomy: Text and Atlas,* 4th ed. New York: McGraw Hill.

Nieuwenhuys R, Voogd J, van Huijzen Chr. 1988. *The Human Central Nervous System: A Synopsis and Atlas,* 3rd rev. ed. Berlin: Springer-Verlag.

Peters A, Jones EG (eds). 1984. *Cerebral Cortex.* Vol. 1, *Cellular Components of the Cerebral Cortex.* New York: Plenum.

Peters A, Palay S, Webster H de F. 1991. *The Fine Structure of the Nervous System,* 3rd ed. New York: Oxford Univ. Press.

References

Brodmann K. 1909. *Vergleichende Lokalisationslehre der Grosshirnrinde in ihren Prinzipien dargestellt auf Grund des Zellenbaues.* Leipzig: Barth.

Felleman DJ, Van Essen DC. 1991. Distributed hierarchical processing in the primate cerebral cortex. Cereb Cortex 1: 1–47.

Heimer L. 1994. *The Human Brain and Spinal Cord: Functional Neuroanatomy and Dissection Guide,* 2nd ed. New York: Springer.

Jones EG. 1988. The nervous tissue. In: *Cell and Tissue Biology: A Textbook of Histology,* 6th ed., pp. 305–341. Baltimore: Urban & Schwarzenberg.

Jones EG. 1986. Connectivity of the primate sensory-motor cortex. In: EG Jones, A Peters (eds), Cerebral Cortex, Vol. 5, Chapter 4: Sensory-Motor Areas and Aspects of Cortical Connectivity, pp. 113–183. New York/London: Plenum.

Kaas JH. 2006. Evolution of the neocortex. Curr Biol 16: R910–914.

Kaas JH, Qi HX, Burish MJ, Gharbawie OA, Onifer SM, Massey JM. 2008. Cortical and subcortical plasticity in the brains of humans, primates, and rats after damage to sensory afferents in the dorsal columns of the spinal cord. Exp Neurol 209:407–416.

McKenzie AL, Nagarajan SS, Roberts TP, Merzenich MM, Byl NN. 2003. Somatosensory representation of the digits and clinical performance in patients with focal hand dystonia. Am J Phys Med Rehabil 82:737–749.

Penfield W, Boldrey E. 1937. Somatic motor and sensory representation in the cerebral cortex of man as studied by electrical stimulation. Brain 60:389–443.

Penfield W, Rasmussen T. 1950. *The Cerebral Cortex of Man: A Clinical Study of Localization of Function.* New York: Macmillan.

Ramón y Cajal S. 1995. *Histology of the Nervous System of Man and Vertebrates.* 2 vols. N Swanson, LW Swanson (transl). New York: Oxford Univ. Press.

Rockland KS, Ichinohe N. 2004. Some thoughts on cortical minicolumns. Exp Brain Res 158:265–277.

Zola-Morgan S, Squire LR, Amaral DG. 1986. Human amnesia and the medial temporal region: enduring memory impairment following a bilateral lesion limited to field CA1 of the hippocampus. J Neurosci 6:2950–2967.

5

The Computational Bases of Neural Circuits That Mediate Behavior

THE PREVIOUS CHAPTER focused on the neuroanatomy of the brain and the connections between different brain regions. An understanding of how these connections mediate behavior requires insight into how the information represented by the activity of different populations of neurons is communicated and processed. Much of this understanding has come from recordings of the minute electrical signals generated by individual neurons.

Although much has been learned by recording from just one or a few neurons at a time, advances in miniaturization and electronics technology now make it possible to record action potentials simultaneously from many hundreds of individual neurons across multiple brain areas, often in the context of a sensory, motor, or cognitive task (Box 5–1). Such advances, together with computational approaches for managing and making sense of large data sets, promise to revolutionize our understanding of neural function.

At the same time, modern genetic approaches based on mRNA sequencing from individual neurons are revealing the numerous types of cells that contribute to population activity. Genetic-based approaches also allow defined types of neurons to be activated or silenced during an experiment, supporting tests of causality (Box 5–2).

High-throughput anatomical methods, at the scales of both light and electron microscopy, are providing information about circuit wiring at an unprecedented level of detail and completeness. The complexity of neural circuits and the large data sets collected from them has motivated the development and application of statistical, computational, and theoretical methods for extracting, analyzing, modeling, and interpreting results. These methods are used to study a broad range of issues: experimental design, the extraction of signals from raw data, the analysis of large complex data sets, the construction and analysis of models simulating the data, and, finally and most importantly, building some form of understanding from the results.

Signal extraction is often done on the basis of a Bayesian approach, inferring the most likely signal

Box 5–1 Optical Neuroimaging

Optical imaging methods are a rapidly advancing field of technology for large-scale monitoring of neural circuit dynamics. Most of these approaches use fluorescent sensors—synthetic dyes or genetically engineered and encoded proteins—that signal changes in neural activity via changes in the magnitude or the wavelength of their emitted light following excitation. Various florescence imaging approaches have been developed, depending on the source of fluorescence excitation, including single-photon, multiphoton, and super-resolution fluorescent microscopic imaging.

The most commonly used fluorescence indicators signal changes in intracellular calcium levels as a proxy for the electrical activity of neurons. While the temporal resolution of fluorescence calcium imaging is generally lower than that of electrophysiology, fluorescent imaging with genetically encoded calcium indicators enables simultaneous monitoring of many thousands of individually identified neurons in the behaving animal over several days to weeks and months.

In addition to calcium imaging, synthetic and genetically encoded fluorescent indicators of electrical activity (eg, genetically encoded voltage indicators [GEVIs]), neurotransmitter concentration reporters (eg, glutamate-sensing fluorescent reporter [GluSnFR]), activity states of intracellular signaling molecules, and gene expression provide rapidly expanding and versatile techniques for monitoring neural activity on multiple spatial and temporal scales.

present in a noisy recording. Data analysis often consists of reducing the dimensionality of a large data set, not simply to make it more compact but to identify the essential components from which it is built.

Models of neural systems range from detailed simulations of the morphology and electrophysiology of individual neurons to more abstract models of large populations of neurons. Whatever the level of detail, the aim of models is to reveal how the measured features of a neuron or network of neurons contribute to the function of the neuron or neural circuit.

In addition, at the highest levels of functionality, such as identifying images, playing games, or performing tasks at human levels, ideas from machine learning are increasingly impacting neuroscience research.

In this chapter, we introduce ideas, techniques, and approaches that are used to characterize and interpret the activity of neural populations and circuits, with examples drawn from a number of areas of brain research. Many of these topics are covered in greater detail later in the book.

Neural Firing Patterns Provide a Code for Information

Sensory Information Is Encoded by Neural Activity

Animals and humans continually accumulate information about the world through their senses, make decisions on the basis of that information, and, when necessary, take action. In order for sensory information to be processed for decision making and action, it must be transformed into electrical signals that produce patterns of neural activity in the brain. Studying such neural representations and their relationship to external sensory cues, known collectively as neural coding, is a major area of neuroscience research. The process through which features of a stimulus are represented by neural activity is called encoding.

The structure of a neural representation plays an important role in how information is further processed by the nervous system. For example, visual information is initially encoded in the retina by photoreceptor responses to the color and light intensity over a small region of the visual field. This information is then transformed in the brain within the primary visual cortex to encode a visual scene on the basis of the edges and shapes that define the scene as well as on where these features are located. Further transformations occur in higher-order visual areas that extract complex shapes and further structure from the scene, including the identification of objects and even individual faces. In other brain areas, auditory encoding reflects the frequency spectrum of sounds, and touch is encoded in maps that represent the surface of the body. The sequence of action potentials fired by a neuron in response to a sensory stimulus represents how that stimulus changes over time. Research on neural coding aims to understand both the stimulus features that drive a neuron to respond and the temporal structure of the response and its relationship to changes in the external world.

Box 5–2 Optogenetic and Chemogenetic Manipulation of Neuronal Activity

Functional analysis of neural circuits relies on the ability to accurately manipulate identified circuit elements to elucidate their roles in physiology and behavior. Genetically encoded neural perturbation tools have been developed for remotely controlling neuron function using light (optogenetics) or small molecules (chemogenetics) that activate engineered receptors.

Genetically encoded foreign proteins can be expressed in molecularly, genetically, or spatially specified subsets of neurons using viruses or transgenic animals for subsequent selective perturbations of these cell populations. Optogenetic approaches involve the expression of light-sensitive proteins and subsequent light delivery to the resulting photosensitized neurons. Depending on the type of optogenetic actuator, light activation will then enhance neural activity (eg, light-gated

ion channels like channelrhodopsin) or suppress it (eg, light-gated ion pumps like halorhodopsin and archaerhodopsin) by depolarizing or hyperpolarizing the cell's membrane, respectively.

Alternatively, selected neuronal populations can be remotely activated or silenced using chemogenetic actuators, which are genetically engineered receptors that are targeted to defined neuronal populations using genetic methods; they can be activated via small-molecule synthetic ligands that selectively interact with these receptors upon delivery (eg, DREADDs [designer receptors exclusively activated by designer drugs]).

These optogenetic and chemogenetic tools offer precise spatiotemporal control over neuronal activity to probe the causal relationship between neuronal cell types, circuit physiology, and behavior.

Information Can Be Decoded From Neural Activity

Sensory neurons encode information by firing action potentials in response to sensory features. Other brain areas, such as those that lead to decisions or generate motor actions, must correctly interpret the meaning of action potential sequences that they receive from sensory areas in order to respond appropriately. The process by which information is extracted from neural activity is called decoding.

The decoding of neural signals can be done experimentally and in clinical contexts by neuroscientists. Such decoding can infer what an animal or a human is seeing or hearing from recordings of visual or auditory neurons, for example. In practice, only certain features of the stimulus are likely to be inferred, but the results can nevertheless be impressive. A large number of decoding procedures have been developed, ranging from simple weighted sums of neuronal firing rates to sophisticated statistical methods.

Decoding methods are central to the development of neuroprosthetics for people with various nervous system impairments that result in extensive paralysis (Chapter 39). To accomplish this, neurons are recorded in the parietal or motor cortices through implanted electrodes, and online decoding procedures are used to interpret the movement intentions that are represented by the recorded neural activity. The inferred intentions are then used to control a computer cursor or drive a robotic limb.

Decoding recorded neural activity also gives us a remarkable view of what is going on in a neural circuit, which in turn provides insight into memory

storage and retrieval, planning and decision making, and other cognitive functions. The following section illustrates these insights using a particularly interesting neural representation, the encoding of spatial location in the rodent hippocampus.

Hippocampal Spatial Cognitive Maps Can Be Decoded to Infer Location

One of the most complex cognitive challenges an animal faces is identifying and remembering its location in an environment relative to the location of other salient objects. For example, seed-caching birds can remember the location of hundreds of different places where they have stored food over a period of several months. The neural circuitry involved in formation of explicit memory—the memory of people, places, things, and events—was briefly introduced in the previous chapter. This form of memory requires the hippocampus, entorhinal cortex, and related structures in the temporal lobe. In 1971, John O'Keefe discovered physiological evidence of a neural representation of the spatial environment in the hippocampus. In 2014, he was awarded the Nobel Prize in Physiology or Medicine, together with May-Britt Moser and Edvard Moser, for their discoveries concerning the neuronal representation of space.

O'Keefe discovered that individual cells in the rat hippocampus, termed place cells, fire only when the animal traverses a particular area of the environment, termed the cell's place field (Figure 5–1). Subsequent research uncovered place cell–like activity in

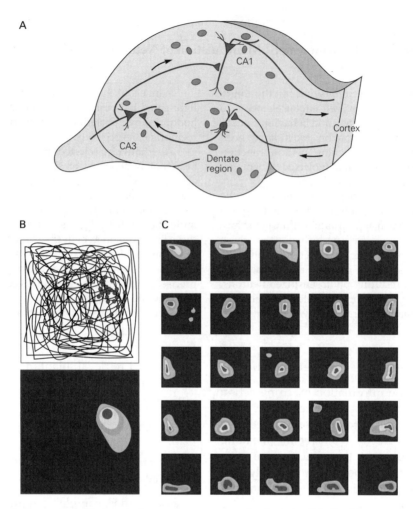

Figure 5–1 Hippocampal place cells and place cell maps.

A. Input–output transformations occur in the trisynaptic circuitry of the mammalian hippocampus, proceeding from the dentate gyrus input region, to the CA3 area, and to the CA1 output region, with principal excitatory neurons (**red**) in each region as primary processing units. Activity of principal cells is modulated by local circuit GABAergic interneurons (**gray**).

B. Place cell firing in the hippocampus. The path taken by a rat is shown in **black** as it traverses a square arena. Electrodes

were implanted within the hippocampus to record from individual cells. **Above:** A single place cell increases firing (each action potential represented by a red dot) at discrete locations in the environment. **Below:** A color-coded heat map of firing frequency of the schematic place cell. Lower wavelength colors (**yellow** and **red**) represent higher firing rates on a background of no activity (**dark blue**).

C. Color-coded heat maps showing the firing of 25 different place cells recorded simultaneously in the hippocampal CA1 region as the rat explores a square box.

the hippocampus of several other mammalian species, including bats, monkeys, and humans. Distinct sets of place cells are activated by distinct locations in a given environment. Consequently, although individual place cells represent relatively small spatial areas, the full diverse population of place cells in the hippocampus tiles the entire environment, and any given location is encoded by a unique ensemble of cells. The hippocampal place coding network provides an example of a cognitive map, initially postulated by the psychologist Edward Tolman, that enables an animal to successfully remember and then navigate its environment. The role

of the hippocampus in memory formation and the mechanisms by which the hippocampal spatial map is encoded are explored in detail in Chapters 52 and 54.

The electrophysiological methods available to O'Keefe in 1971 were limited to recording one place cell at a time, but subsequent advances allowed investigators to record dozens, and more recently hundreds, of place cells simultaneously. Critically, while single place cells encode only specific parts of the environment and are prone to occasional noisy firing outside of their place fields, entire populations of place cells provide more complete spatial coverage and the reliability of redundant place coding. These features of population coding have paved the way for new and powerful computational analyses. In particular, it is possible to decode the activity of populations of place cells and estimate an animal's location within an environment. This is accomplished by determining each cell's spatial selectivity and using this selectivity as a template to decode ongoing activity. In practice, this decoding is often performed by weighting each cell's contribution to the final estimate of the animal's position by a factor proportional to that cell's spatial coding reliability. Using this and similar techniques, one can reconstruct an animal's location from second to

second within room-sized environments with a precision of a few centimeters (Figure 5–1C).

Hippocampal function has been strongly implicated in spatial and declarative memory based on studies using spatial decoding techniques. During active exploration of an environment, hippocampal activity reflects place coding, but during immobile or resting behavior, the hippocampus enters a different regime in which neural activity is instead dominated by discrete semi-synchronous population bursts termed sharp-wave ripples (Figure 5–2A). These events are thought to be internally generated by circuitry within the hippocampus.

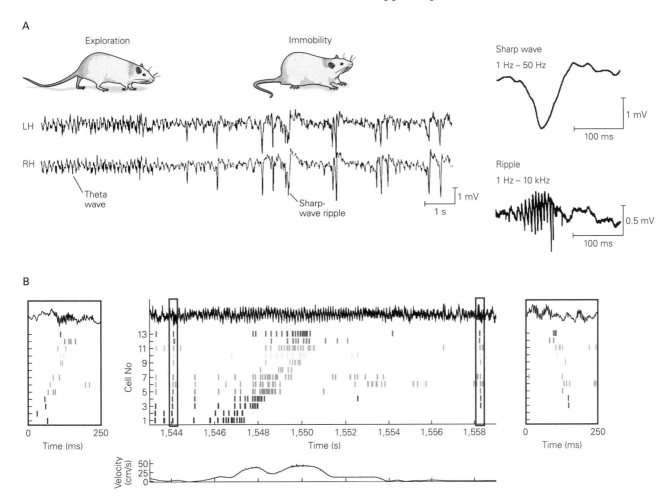

Figure 5–2 Hippocampal sharp-wave ripples and sequence replay.

A. *Left*: Behavior dependence of hippocampal local field potential activity (**LH** and **RH**, left and right hippocampus). Theta waves are present during exploration, and large negative sharp waves during immobility. *Right*: Sharp waves and ripples recorded from the hippocampal CA1 region. (Adapted, with permission, from Buzsaki 2015; and reproduced, with permission, from Buzsaki et al. 1992. Copyright © 1992 AAAS.)

B. Place cell sequences experienced during behavior (*middle*) are replayed in both forward (*left*) and reverse (*right*) direction during sharp-wave ripples. The rat moved from left to right on a familiar track. Spike trains for place fields of 13 CA3 pyramidal cells while the rat is on the track are shown before (forward replay; **red box**), during (*middle*), and after (reverse replay; **blue box**) a single traversal. The CA1 local field potential is shown on top (**black traces**), and the animal's velocity is shown below. (Adapted, with permission, from Diba and Buzsaki 2007. Copyright © 2007 Springer Nature.)

Notably, sharp-wave ripples are prominent during resting periods after recent learning, for example after exploration of an environment. Spatial decoding of the activity of place cells active within these short (50 to 500 ms) sharp-wave ripples reveals that hippocampal neurons recapitulate or replay discrete trajectories through the recently explored environment. Although these trajectories replicate paths taken through space, the replayed activity sequences differ from those observed during active exploration in several ways.

First, replayed sequences within sharp-wave ripples are time compressed, occurring about 10 to 20 times faster than during exploration (Figure 5–2B). Second, they can occur either in the same direction as behavioral spatial trajectories (forward replay) or in the opposite direction (reverse replay). Thus, decoding a single postexploration, 200-ms, sharp-wave ripple-replay event may reveal a virtual mental trajectory spanning 2 to 4 seconds of behavioral time replayed *backward* from how it was experienced. Replay is thought to represent a form of mental rehearsal by which certain memories are gradually consolidated and thus may be a crucial aspect of the role of the hippocampus in memory.

Neural Circuit Motifs Provide a Basic Logic for Information Processing

Neurons tend to be highly interconnected, both with nearby neurons and with neurons in distal brain areas. Knowledge of neuronal connections, called connectomics, is expanding rapidly due to a number of new methods for uncovering fine-scale anatomical structure. Patterns of neuronal interconnection come in several varieties.

Connections from one area to another, for example from the thalamus to primary visual cortex, are termed feedforward (Figure 5–3A). The forward direction is defined as extending from a more peripheral or primary area, such as the retina, thalamus, or primary visual cortex, to a higher area with more complex response properties, such as the visual areas that respond selectively to particular objects. In most cases, two areas that have feedforward connections also have feedback connections; for example, there are numerous connections from primary visual cortex back to the thalamus. Local connections often extend from one neuron to another, ultimately looping back onto the original neuron. Such looping connectivity is called recurrent. Many neurons are involved in all of these types of connectivity—feedforward, feedback, and recurrent—but it is useful to consider the functional implications of these different connectivity motifs separately.

Connections between neurons can be either excitatory or inhibitory. Normally, excitatory connections lead to increased neural firing and inhibitory connections lead to decreased neural firing. Many neural circuits receive strong excitatory drive from hundreds or thousands of synapses. If not checked by inhibition, this synaptic excitation would lead to unstable neural activity. A near balance of excitation and inhibition is a common feature of neural circuits that may enhance their computational capacity. However, this fine tuning may make the circuits prone to generating seizure activity if the balance between excitation and inhibition is not properly maintained, as occurs during epilepsy.

In mammals, visual information is processed in a series of brain areas that are often approximated as having feedforward circuitry. Feedforward circuits can process information in sophisticated ways, for example

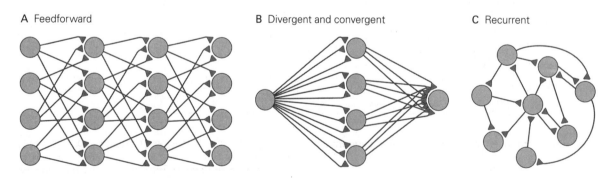

Figure 5–3 Four basic neural circuit motifs.

A. A feedforward circuit in which synaptic connections extend in a single direction from one processing level of neurons to another.

B. Divergent feedforward connections describe a small number of presynaptic neurons connecting to a larger number.

Convergent connections describe a large number of presynaptic neurons connecting to a smaller number.

C. In a recurrent network, synaptic connections occur in multiple directions between neurons, forming looping pathways through the circuit.

extracting and identifying objects from a complex visual scene, but they cannot produce ongoing, dynamic patterns of activity. For this purpose, recurrent circuitry is needed (Figure 5–3C).

Within feedforward circuitry, two submotifs can be identified: divergent and convergent connections (Figure 5–3B). In divergent connections, the number of neurons that receive a given type of input exceeds the number of neurons providing that input, so the information encoded in the presynaptic input neurons is expanded in the postsynaptic output neurons. In convergent connections, many presynaptic neurons send input to a smaller number of postsynaptic neurons.

The most prominent example of both divergent and convergent connectivity is provided by the cerebellum, as discussed later.

Visual Processing and Object Recognition Depend on a Hierarchy of Feed-Forward Representations

Visual information is processed within a large number of brain regions arranged hierarchically (Figure 5–4). Moving up the hierarchy from the primary sensory input generated by the retina, neurons respond to increasingly complex combinations of visual features, culminating in selectivity for complex objects, such as

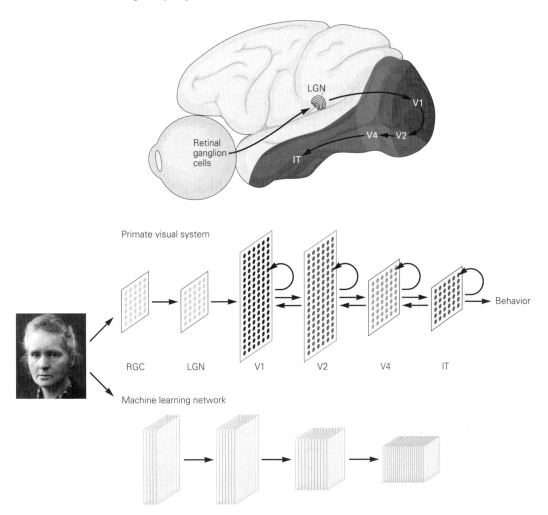

Figure 5–4 Comparison of biological and machine learning networks. In the visual system, multiple brain regions form a hierarchy in which neurons in series become progressively selective to more complex objects. The regions in the primate visual system pathway represent retinal ganglion cells (**RGC**), the lateral geniculate nucleus (**LGN**) of the thalamus, ventral stream visual areas (**V1**, **V2**, and **V4**), and the inferotemporal cortex (**IT**). The number of neurons per region varies (represented by the **colored dots**), but their selectivity steadily increases. The machine learning network pathway represents layers of a feedforward network trained to identify objects in images. Increased selectivity in the different regions of the machine learning network is indicated by the growing numbers of stacked sublayers, reflecting selectivity to a richer array of visual features. The hierarchy of response selectivities recorded in different visual areas resembles the activities seen in corresponding layers of the machine learning network. (Adapted, with permission, from Schrimpf et al. 2018.)

faces. Considerable research is devoted to identifying principles upon which the structure of the visual hierarchy is based. The development of artificial neural network models in machine vision has proven to be an instructive analogy for addressing this issue.

From the retina, to the thalamus, to the primary visual cortex, onto the highest visual areas associated with cognition in inferotemporal cortex, visual neurons respond selectively to particular patterns of light, dark, and color in regions of the visual field called their receptive fields. From the lowest to highest stages of visual processing, neurons have increasingly larger receptive fields and higher degrees of selectivity. At each stage, neurons with a particular type of selectivity tend to have receptive fields that tile the visual scene, providing full coverage for the selected feature. Moreover, the arrangement of the receptive fields in each visual brain area is topographically matched to the layout of the image of the external world on the retina, that is, the cortex forms a map of the visual field.

As receptive fields enlarge and selectivity increases, neural responses depend less on the precise location of the selected object or pattern and more on its overall features. In general, neurons in higher stages of visual processing respond more selectively to a larger portion of the visual field and depend less on features such as location, size, and orientation. This correlates with our ability to recognize objects independent of their location, size, and orientation in a scene. At the highest stages of the hierarchy, neurons can, for example, respond selectively to particular faces located across

the visual field, independent of the size of the face or its angular pose (ie, head direction).

The ideas of tiling, increased receptive field size, increased selectivity, and decreased dependence on view-dependent factors are central to the construction of artificial networks for machine vision. Such networks can reach human-level performance on some object recognition tasks. Furthermore, the pattern of errors that the machines make on difficult images matches, to some degree, the errors made by human subjects. Nonhuman primates can also perform these tasks at levels comparable to humans, and interestingly, recordings from different visual areas along the object recognition pathway correspond to activity seen in the artificial networks at similar stages in visual processing (Figure 5–4).

Diverse Neuronal Representations in the Cerebellum Provide a Basis for Learning

The most abundant class of neurons in our brains are the roughly 50 billion granule cells at the input stage of the cerebellum, composing more than half of all the neurons in the brain. The cerebellum is a hindbrain structure vital for motor coordination but also implicated in the adaptive regulation of autonomic, sensory, and cognitive functions (Figure 5–5). Malfunction of cerebellar circuits may contribute to various neurological disorders, including autism. In contrast to the thousands of inputs that most brain neurons receive, each granule cell receives just a handful of inputs (four on average).

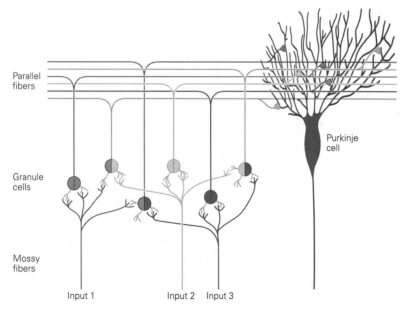

Figure 5–5 The cerebellum receives input from many regions of the brain and spinal cord. These inputs, known collectively as mossy fibers, are recoded in a vast number of granule cells, an example of divergent connectivity, allowing for many possible mixtures of the input signals. Dendrites of Purkinje cells receive convergent input from hundreds of thousands of granule cells relayed by their axons, known as parallel fibers. Parallel fiber to Purkinje cell synapses are modifiable, which is believed to be an important mechanism underlying motor and possibly other forms of learning.

Parallel fibers

Purkinje cell

Granule cells

Mossy fibers

Input 1   Input 2   Input 3

Recent experimental findings using neuroanatomical tracing and electrophysiological recording indicate that inputs converging onto a single granule cell often originate from distinct brain regions. As a result, the firing of individual granule cells may represent any one of an enormous range of combinations of stimuli or events. For example, a cell may fire only during the conjunction of a specific visual stimulus (such as a moving tennis ball) with the movement of a particular body part (such as the flexing of the wrist). Representations that combine different types of information in this way are called mixed.

Cerebellar granule cells provide an extreme example of divergent feedforward connectivity, with the information carried by approximately 200 million input fibers (called mossy fibers) mixed and expanded onto the 50 billion granule cells. Such a large representation is needed to handle the many different ways that multiple channels of information can be combined. For example, representing all possible combinations of 2 out of just 100 different input channels requires $100 \times 99/2$, or 4,950, different response types. Requiring a representation of all triplets pushes this number up over 150,000, and the number increases rapidly for four and more combinations. Because the large number of possible combinations would be difficult to specify genetically, it is generally thought that the assignment of mossy fibers to their granule cell targets is largely random.

This analysis suggests that the role of the cerebellar granule cells is to combine a large number of input channels in many possible ways. Such a representation clearly would be useful for making inferences and generating actions that depend on the co-occurrence of combinations of stimuli and actions. However, to be useful, this information must somehow be read out from the huge number of granule cells.

Read-out from the cerebellar cells is accomplished by Purkinje cells, the output neurons of the cerebellar cortex. In contrast to the highly divergent connectivity at the inputs to granule cells, connections between granule cells and Purkinje cells provide an extreme example of convergence. A single Purkinje cell receives input from over a hundred thousand granule cells. Theories of cerebellar function developed in the 1970s by David Marr and James Albus posited that this convergence allows Purkinje cells to extract useful information from the extremely rich representation provided by granule cells. By doing this, Purkinje cells may, for example, underlie the amazing human capacity to form the many complex associations required for motor skills, such as riding a bicycle or playing a musical instrument. However, to extract information that is useful for a number of purposes under a variety of conditions, the read-out provided by Purkinje cells must be adaptable. This adaptability is provided by the plasticity of the synapse between a granule cell and Purkinje cell synapse, as discussed in a later section.

Recurrent Circuitry Underlies Sustained Activity and Integration

Neurons are inherently forgetful. Transient synaptic input typically evokes a brief response that decays within a few tens of milliseconds. The time course of this decay is determined by an intrinsic property of neurons known as the membrane time constant (Chapter 9). How then do patterns of neural activity persist long enough to support cognitive operations such as memory or decision making that play out over seconds, minutes, or even longer periods of time?

Consider, for example, trying to detect whether you hear a familiar voice in a crowded room full of people talking loudly. As you listen, you may occasionally detect a bit of sound that resembles the voice you are searching for but that by itself is inconclusive. Nevertheless, over time, you may be able to accumulate enough evidence to arrive at a conclusion. This process of evidence accumulation requires integration, meaning that a running sum must be maintained and augmented as additional evidence is detected. Integration requires both a computation (addition) and memory to compute and maintain a running total (Chapter 56).

For a neural circuit to perform integration, a transient input must produce activity that is sustained at a constant level even after the input is gone. This sustained activity provides a memory of the transient input. As outlined in the previous paragraph, circuits that integrate can be useful for accumulating information, but they are also needed for noncognitive tasks such as maintaining the constant muscle tension required to hold a fixed body posture. One of the best studied neural integrators is the circuitry that allows humans and animals to maintain a constant gaze direction with their eyes, even in the dark. The fact that eye movements can be studied across a wide range of species, from fish to primates, has greatly facilitated progress. Moreover, the relative simplicity of the oculomotor system has fostered fruitful dialog between experimental and theoretical studies. (The oculomotor system is described in more detail in Chapter 35.)

The existence of integrator circuits in the oculomotor system was first suggested by a puzzling observation from neuronal recordings (Figure 5–6A). Oculomotor neurons that control the eye muscles increase action potential firing transiently to evoke

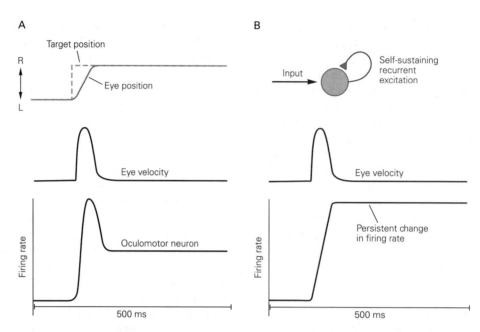

Figure 5–6 Recurrent circuitry and sustained neural activity are required for maintaining eye position.

A. *Above:* A saccadic eye movement consists of a rapid movement change in eye velocity to bring a target back to the center of gaze. This is followed by a sustained change in eye position to maintain the fovea on the target. The **dashed blue line** shows the location of the target, and the **gray line** shows the

eye movement and subsequent fixation on the target at its new position. *Below:* An oculomotor neuron exhibits a brief burst of activity related to eye velocity along with sustained activity related to eye position.

B. Recurrent excitation can explain how a brief pulse of input, such as an eye velocity signal, can lead to a persistent change in firing rate through a process akin to mathematical integration.

movements of the eye but also exhibit sustained action potential firing needed to hold the eye in fixed position. For example, a motor neuron that projects to an eye muscle that moves that eye to the left will fire at a high rate when gaze is maintained left of center and at a low rate when gaze is maintained right of center. The puzzle is that the premotor neurons in the superior colliculus and brain stem that project to the oculomotor neurons only fire transiently before eye movements. They do not show any sustained activity related to eye position. How then is this sustained activity generated?

An early conjecture, now strongly supported, is that steady eye position signals are computed by brain stem neurons that integrate the transient eye velocity signals. Such neurons receive velocity information and provide the steady output to the oculomotor neurons that maintain eye position. Lesions or inactivation of certain brain stem nuclei in monkeys, including the medial vestibular nucleus and the nucleus prepositus hypoglossi, result in a failure to maintain steady horizontal eye position following eye movements, suggesting that the neural integrator circuit lies within these structures. Damage to these brain stem structures in humans leads to the same problem, known clinically as gaze-evoked nystagmus (Chapter 35).

How do neural circuits perform integration? One possibility is that integration is supported by specialized intrinsic neuronal properties that effectively lengthen neuronal membrane time constants, allowing brief inputs to generate sustained output. A variety of candidate mechanisms have been described involving different voltage-activated ion channels. However, studies using intracellular recordings, which allow for direct control over the membrane voltage of the recorded neuron, have shown that sustained position-related signals persist even when the neuron's voltage-activated channels are blocked. A second possibility is that integration arises from interactions among a network of synaptically coupled neurons. Intracellular recordings in goldfish support this idea by showing that levels of synaptic input vary with eye position.

The question of what types of neural networks are capable of performing integration has been explored extensively in theoretical studies. One class of models that has been considered relies on recurrent connectivity, specifically a population of neurons that excite each other. A weakly coupled network of this type responds to an input pulse with activity that rapidly decays away. Increasing the strength of the recurrent excitation adds back some of the activity that would otherwise decay, lengthening the duration of the population response. If recurrent excitation is increased to the point where the recurrent excitation set up by a transient input precisely cancels the decay, the response can last indefinitely. This requires fine-tuning of network parameters.

In a perfectly tuned network, a transient pulse of input produces a change in firing rate that lasts forever in the absence of further input. Equivalently, such a population computes a running integral of the input it receives (Figure 5–6B). If the transient excitation in the network is not perfectly tuned but instead is slightly weaker, the input produces a change in firing rate that decays slowly. Eye position, in the dark, tends to drift back to the center in about 20 seconds, suggesting that the neural integrator is not tuned perfectly, but it is tuned well enough to extend the roughly 20-ms time constant of a typical neuron by a factor of about 1,000.

The fact that recurrent network models reproduce some of the core properties observed in biological integrator circuits has launched development of more detailed and realistic network models and testing the predictions of such models experimentally. These efforts also highlight the challenges involved in forging detailed links between the structure and function of neural circuits. Key questions remain even after decades of intensive study using a variety of systems and approaches.

For example, oculomotor integrator circuits typically contain two opposing classes of neurons, one increasing and the other decreasing their firing rates as eye position changes in a given direction. This arrangement is not restricted to oculomotor integrators but is also found in cortical regions implicated in decision making and working memory. Models have shown that mutual inhibition between these opposing populations can play a role in sustaining activity and integration. Although anatomical studies provide some support for this idea, studies in the goldfish showed that integration remains intact even when connections between the opposing populations are removed.

Another key question regards the mechanisms for tuning integrator networks. Experimental studies suggest that integrator networks are subject to modification via experience; in other words, they are tuneable. Although such tuning presumably occurs via changes in the strength of synaptic connections between neurons, direct evidence for this has yet to be obtained. In short, although much has been learned about how integration *could* be implemented, the details of the network architecture that actually support integration in any particular instance remain to be definitively established.

A detailed understanding of how we maintain the position of our eyes is an important end unto itself, with clinical relevance. However, as pointed out earlier, the solutions found here may apply equally to cognitive functions including short-term memory and decision making. Optical imaging of large populations of neurons along with temporally precise manipulations of their activity and detailed anatomical reconstructions, combined with theoretical models of network function, may soon provide the answers.

Learning and Memory Depend on Synaptic Plasticity

Experience can modify neural circuits to support memory and learning (Chapter 3). It is generally believed that experience-dependent changes responsible for learning and memory occur primarily at synapses. Multiple forms of synaptic plasticity have been identified, and each of these presumably supports a different set of functions.

Just as there are multiple forms of plasticity, there are multiple forms of learning. Different forms of learning can be defined based on the amount and type of information provided. In supervised learning, explicit instruction is given about the behavior needed to perform a task. In reinforcement learning, on the other hand, only a positive reward or a negative punishment is provided to indicate whether that task is being performed properly. Finally, unsupervised learning involves no instructive information at all, but rather organizes input data on the basis of its intrinsic structure without supervision. In the following sections, we discuss an example of unsupervised learning involving Hebbian plasticity and an example of reinforcement learning in the cerebellum. (The various types of learning and memory and their cellular and circuit mechanisms are described in detail in Chapters 52–54.)

Dominant Patterns of Synaptic Input Can be Identified by Hebbian Plasticity

Cortical neurons receive synaptic input from thousands of other neurons and combine this information in patterns of action potentials. The strength of synaptic transmission at each of the synapses determines how the information arriving from many inputs is combined to affect the firing of the neuron. Setting the strength of all the synapses to zero would obviously make for a noninformative neuron of no functional use. Similarly, setting them to nonzero values that extract a signal dominated by random noise would also not produce a signal of value. Instead, neurons can best serve a useful function by extracting the most interesting aspects of the information carried by their inputs. Theoretical analysis of a form of plasticity known as Hebbian indicates one way that this could happen in an unsupervised manner.

In 1949, Donald Hebb proposed that synapses should strengthen when a given presynaptic input to a neuron cooperates with a sufficient number of coactive inputs to cause that neuron to fire an action potential. Evidence for Hebbian synaptic plasticity has been obtained from many studies (Chapter 54). By itself, Hebbian plasticity would keep making synapses stronger and stronger, so some other form of plasticity must exist to prevent this from happening. Such compensatory forms of plasticity are called homeostatic, and experiments have revealed these forms of plasticity as well. Theoretical analysis indicates that a combination of Hebbian and homeostatic plasticity can adjust synapses, without any additional supervisory signal, so that they extract the combination of a neuron's inputs that is most highly modulated relative to other combinations (Figure 5–7). This is a reasonable candidate for the most interesting signal carried by those inputs, and thus, Hebbian plasticity provides a way for neurons to determine and extract such signals.

Synaptic Plasticity in the Cerebellum Plays a Key Role in Motor Learning

Although a detailed understanding of how the cerebellum contributes to complex human motor skills is lacking, a great deal is known about its role in simple forms of motor learning. Among the most thoroughly studied is a paradigm known as *delay eyeblink conditioning*, in which a neutral sensory stimulus such as a light or a tone is repeatedly paired with an aversive unconditioned stimulus (US) such as an air puff to the eye. After several days of such training, animals learn to close their eye in response to the previously neutral stimulus (the light or tone), known as the conditioned stimulus (CS), in anticipation of the US (the air puff). The timing of the eyelid closure is highly specific to the delay between the onset of the CS and the US.

Eyelid conditioning has been an extremely useful paradigm for understanding cerebellar function because it maps onto the structure of cerebellar circuitry in a particularly clear way (Figure 5–8). Information about the CS is first encoded by cerebellar granule cells and then relayed to Purkinje cells. The US is encoded by a completely separate input pathway, known as the olivocerebellar or climbing fiber system. In contrast to the many thousands of inputs from granule cells, each Purkinje cell receives a single powerful climbing fiber input from a brain stem nucleus known as the inferior olive. Electrophysiological recordings revealed that climbing fiber inputs to one particular region of the cerebellum signal the occurrence of the US, that is, a stimulus that is irritating to the cornea. This discovery was made possible by the fact that the climbing fiber evokes a distinct suprathreshold response in the Purkinje cell known as a complex spike.

A key to understanding how the cerebellum mediates learning was the discovery that the complex spike triggers plasticity at synapses between granule cells and Purkinje cells. Specifically, the co-occurrence of input from a presynaptic granule cell and a complex spike in the postsynaptic Purkinje cell results in a persistent weakening of the granule cell input, a form of plasticity known as cerebellar

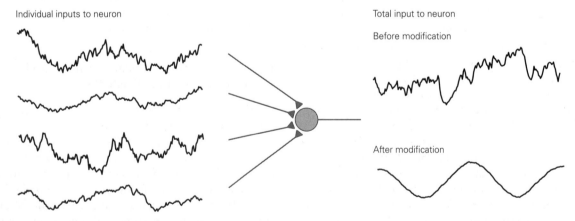

Figure 5–7 Hebbian plasticity can identify relevant input signals to a neuron. In this example, a neuron receives 100 inputs; firing rates for four of them are shown (*left*). Each of the input rates is noisy but contains, within the noise, a sinusoidal signal. The input rates are multiplied by synaptic strengths (**brown triangles**) and then summed to produce the total input to the neuron (*right*). Before Hebbian plasticity occurs, the synapses have random weights, resulting in the noisy trace; after modification, the total input reveals the underlying sinusoidal signal.

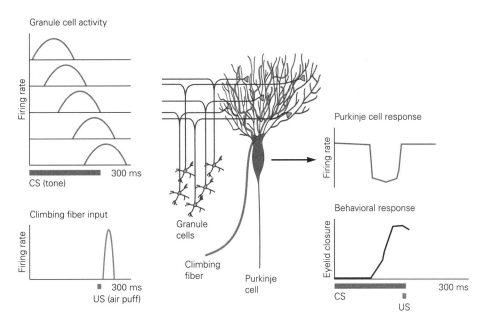

Figure 5–8 Hypothetical role of the cerebellum in eyeblink conditioning. Information about the conditioned stimulus (**CS**) and unconditioned stimulus (**US**) is relayed via mossy and climbing fiber pathways, respectively. Granule cell synapses active before presentation of the US are gradually weakened by long-term depression induced by climbing fiber input. This contributes to a pause in Purkinje cell firing that is precisely timed to occur just before the US. Since Purkinje cells are inhibitory, this pause excites downstream neurons in the cerebellar nucleus and red nucleus that drive eyelid closure.

long-term depression (Figure 5–8). Hence, for each occurrence of the US, the strength of granule–Purkinje cell synapses active immediately prior to the US is reduced. This plasticity leads to the gradual emergence of a learned pause in Purkinje cell firing due to the decrease in granule cell excitation just before the expected time of arrival of the US.

How does a decrease in Purkinje cell firing lead to a learned motor response? Purkinje cells are normally spontaneously active, and they inhibit their downstream targets. Purkinje cells in regions of the cerebellum receiving climbing fiber input related to noxious stimuli to the eye form synapses with neurons that indirectly activate the muscles that produce eyelid closure. Hence the learned pause in Purkinje cell firing causes the eyelid to close at just the right moment to protect the eye. Appropriate timing of the pause is thought to be mediated by a diversity of temporal response patterns in granule cells. Computer simulations have shown that learning of appropriately timed responses can be explained by plasticity in the granule–Purkinje cell synapse if individual granule cells are active at different delays after the CS or exhibit a variety of distinct, but repeatable, temporal patterns locked to the CS.

Due to technical challenges, direct evidence for such temporal representations has not yet been obtained for granule cells in the region of the mammalian cerebellum involved in eyeblink conditioning. However, a diversity of temporal patterns has been observed in granule cells in a structure analogous to the cerebellum in fish. More broadly, studies of the cerebellum, including those of eyeblink conditioning, provide a concrete illustration of how neural circuits can mediate learning though trial and error, even for learning more complex motor skills such as playing a musical instrument. Purkinje cells integrate a rich diversity of signals related both to the external world and internal state of the animal (conveyed by granule cells), with highly specific information about errors or unexpected events (conveyed by the climbing fibers). The climbing fiber acts as a teacher, weakening synapses that were active before, and hence could have contributed to errors. These changes in synaptic strength alter the firing patterns of Purkinje cells and, by virtue of specific wiring patterns, alter behavior such that errors are gradually reduced.

The cerebellum and cerebral cortex, including the hippocampal region, are foci of intense experimental and theoretical research on learning and memory. Technological advances are opening up new approaches for studying the contributions of synaptic actions, individual cells, and circuits to memory-related phenomena.

Highlights

1. Neural coding describes how stimulus features or intended actions are represented by neuronal activity. Decoding refers to the inverse process through which neural activity is interpreted to reveal the encoded signals. Mathematical decoding of neural responses can be used to interpret computations being performed by neural circuits and to drive prosthetic devices.

2. Neural circuits are highly interconnected, but a few basic motifs are used to characterize their functions and modes of operation. Feedforward circuits process information to extract structure and meaning from a sensory stream. Recurrent circuits can perform temporal processing and generate dynamic activity to drive motor responses.

3. Most neurons receive a finely tuned balance of excitatory and inhibitory inputs. Small changes in this balance in response to a sensory stimulus can evoke an action potential output.

4. Levels of neural activity must often be maintained for many seconds. Networks of recurrent excitation provide one mechanism to produce long-lasting changes in neural output.

5. Synaptic plasticity supports longer-lasting changes in neural circuits that underlie learning and memory. Hebbian plasticity can extract interesting signals from a complex set of inputs without the need for supervision (a "teacher"). Synaptic plasticity in the cerebellar cortex is driven by error signals (a form of supervision) and is used to tune motor responses and learn timing relationships.

Larry F. Abbott
Attila Losonczy
Nathaniel B. Sawtell

Selected Reading

Abbott LF. 2008. Theoretical neuroscience rising. Neuron 60:489–495.

Dayan P, Abbott LF. 2001. *Theoretical Neuroscience: Computational and Mathematical Modeling of Neural Systems.* Cambridge, MA: MIT Press.

Hebb DO. 1949. *The Organization of Behavior: A Neuropsychological Theory.* New York: Wiley.

LeCunn Y, Bengio Y, Hinton G. 2015. Deep learning. Nature 521:436–444.

Marr D. 1969. A theory of cerebellar cortex. J Physiol 202:437–470.

References

Buzsaki G. 2015. Hippocampal sharp wave-ripple: a cognitive biomarker for episodic memory and planning. Hippocampus 25:1073–1188.

Buzsaki G, Horváth Z, Urioste R, Hetke J, Wise K. 1992. High-frequency network oscillation in the hippocampus. Science 256:1025–1027.

Diba K, Buzsaki G. 2007. Forward and reverse hippocampal place-cell sequences during ripples. Nat Neurosci 10:1241–1242.

Fusi S, Miller EK, Rigotti M. 2016. Why neurons mix: high dimensionality for higher cognition. Curr Opin Neurobiol 37:66–74.

Litwin-Kumar A, Harris KD, Axel R, Sompolinsky H, Abbott LF. 2017. Optimal degrees of synaptic connectivity. Neuron 93:1153–1164.

Medina JF, Mauk MD. 2000. Computer simulation of cerebellar information processing. Nat Neurosci 3:1205–1211.

Miri A, Daie K, Arrenberg AB, Baier H, Aksay E, Tank DW. 2011. Spatial gradients and multidimensional dynamics in a neural integrator circuit. Nat Neurosci 14:1150–1159.

Oja E. 1982. A simplified neuron model as a principal component analyzer. J Math Biol 15:267–273.

O'Keefe J, Dostrovky J. 1971. The hippocampus as a spatial map. Preliminary evidence from unit activity in the freely-moving rat. Brain Res 34:171–175.

Schrimpf M, Kubilius J, Hong H, et al. 2018. Brain-Score: which artificial neural network for object recognition is most brain-like? bioRxiv doi:10.1101/407007.

Tolman EC. 1948. Cognitive maps in rats and men. Psychol Rev 55:189–208.

Wilson MA, McNaughton BL. 1994. Reactivation of hippocampal ensemble memories during sleep. Science 265:676–679.

Yamins DLK, DiCarlo JJ. 2016. Using goal-driven deep learning models to understand sensory cortex. Nat Neurosci 19:356–365.

6

Imaging and Behavior

T O EXPLAIN AN ORGANISM'S BEHAVIOR in biological terms, it is necessary to reconcile measures of biological processes (eg, action potentials, blood flow, release of neurotransmitters) with measures of cognitive and motor outputs. Relating biological and behavioral measures is challenging, however. Precise neural measurements and invasive techniques are possible in nonhuman animals, but many of these species have a relatively constrained behavioral repertoire. Moreover, it is far more difficult to directly measure or invasively manipulate neural activity in healthy humans, the species with the most advanced and varied behavior. Thus, a central effort of modern neuroscience has been to develop new methods for obtaining precise biological measures from the human brain and for modeling human behaviors in nonhuman animals.

The dominant approach in humans for measuring biological processes and linking them to behavior is functional magnetic resonance imaging (fMRI). Other imaging methods for measuring human brain function such as electroencephalography, positron emission tomography, and near-infrared spectroscopy have their own strengths. However, fMRI is particularly well suited for studying the neural underpinnings of human behavior for several reasons. First, it is non-invasive: It does not require surgery, ionizing radiation, or other disruptive intervention. Second, it can measure brain function over short periods of time (in seconds), which allows it to capture dynamic aspects of mental processes and behavior. Third, it measures activity across the whole brain simultaneously, providing the opportunity to examine how multiple brain regions interact to mediate complex behaviors. Thus, the focus of this chapter is fMRI.

We start by explaining the technicalities of how an fMRI experiment works and how the data are typically collected. We then explain how fMRI data are analyzed and how they provide insight into human behavior and thought. We then turn to a more conceptual overview of what has been learned from fMRI, using examples from the fields of perception, memory, and decision-making. Finally, we consider the strengths and limitations of fMRI and discuss what kinds of inferences about brain and behavior it can support.

Although the focus of this chapter is on imaging and behavior in the healthy brain, fMRI also has the potential to change the way we diagnose and treat psychiatric and neurological disorders. Virtually all such disorders (eg, autism, schizophrenia, depression, eating disorders) involve changes in large-scale circuit dynamics, in addition to the disruption of particular brain regions and cell types. Basic research into how healthy brain circuits mediate mental processes and behavior, combined with the ability to measure activity in these same circuits in clinical populations, holds tremendous promise for understanding disease and dysfunctional behavior.

Functional MRI Experiments Measure Neurovascular Activity

fMRI experiments enable investigators to track brain function based on changes in local blood oxygen levels that occur in response to neural activity. Like all forms of magnetic resonance imaging (MRI), fMRI requires both highly specialized equipment and sophisticated computer programs. In this section, we first consider the basic principles of how MRI can be used to image brain structure and then explain how fMRI extends this capability to image brain activity.

At the core of every MRI machine is a powerful magnet. The strength of the magnetic field is quantified in Tesla (T) units, and most modern MRI machines are 3T. The use of higher field strengths, such as 7T, offers some advantages, including the possibility of higher-resolution imaging of cortical layers. Such machines are not yet widespread, and layer-specific imaging is in its infancy, so we focus on the capabilities and configuration of 3T machines.

The outside of an MRI machine looks like a tunnel, known as the "bore" of the magnet. Subjects lie on a bed with their head in a helmet-like head coil, which receives signals from the brain. Visual stimuli are typically viewed through a mirror on the head coil angled toward a screen at the back of the bore. Auditory stimuli are presented through headphones. Behavior is typically measured in terms of manual responses with a button box and/or eye movements with an eye tracker. This apparatus constrains which experimental

tasks are possible. However, fMRI is flexible in other ways, including that it can be performed and repeated without harm in many different types of subjects, from children to the elderly, whether healthy or suffering from a disorder.

What does fMRI measure? There are two fundamental concepts that we will discuss in turn, first magnetic resonance and then neurovascular coupling (Figure 6–1).

fMRI Depends on the Physics of Magnetic Resonance

In general, MRI exploits the magnetic properties of hydrogen atoms, the dominant source of protons in the body, specifically the way each atom's proton interacts with a strong magnetic field. A key property of protons is that they intrinsically rotate around an axis. This *spin* gives protons angular momentum and a magnetic dipole along the axis, their own north and south poles. Under normal circumstances, the directions of these dipoles are random for different protons. When placed in a strong external magnetic field, however, a subset of the protons (how many is proportional to the field strength) align with the direction of this field, which extends from foot to head when lying in an MRI bore.

An important step toward measuring a signal from protons is to push them out of alignment with this main field. To understand why, it is helpful to think about a familiar object, the gyroscope. If a still gyroscope is tipped out of vertical balance, it will just fall over. However, if you spin the gyroscope before tipping it, inertial forces will prevent it from falling over. The axis around which the gyroscope is spinning will itself begin to rotate around the vertical axis. This *precession* occurs because gravity exerts a vertical torque on the tilted gyroscope, pulling its center of mass down so that it pivots around its bottom point and traces out a circle in the transverse plane (looking from above). Something similar happens to a proton that is tilted with respect to the strong magnetic field: The field applies a torque and the orientation of the rotational axis precesses around the field direction. The speed of precession, or the *resonant frequency*, is determined by the Larmor equation, according to the field strength and a *gyromagnetic ratio* specific to each type of atom. In the case of a 3T magnet and hydrogen atoms, this speed is in the radiofrequency (RF) range.

But how do protons get tipped out of alignment in the first place to enable precession? The answer depends upon the same principle of torque. A second, weaker magnetic field is applied in a perpendicular direction (eg, front to back of the head), introducing

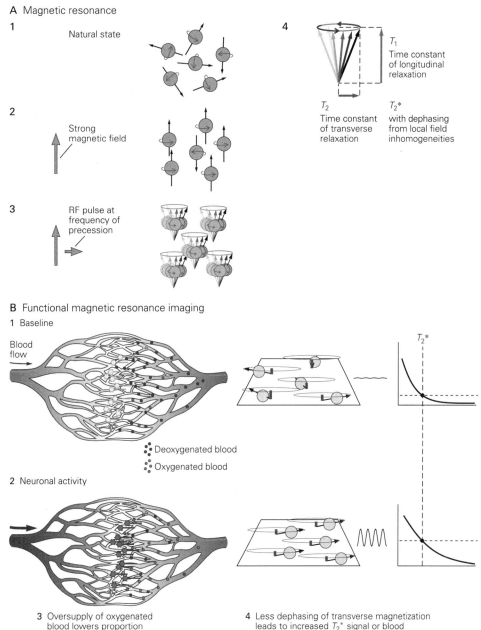

A Magnetic resonance

1 Natural state

2 Strong magnetic field

3 RF pulse at frequency of precession

4 T_1 Time constant of longitudinal relaxation

T_2 Time constant of transverse relaxation

T_2^* with dephasing from local field inhomogeneities

B Functional magnetic resonance imaging

1 Baseline

Blood flow

Deoxygenated blood

Oxygenated blood

T_2^*

2 Neuronal activity

3 Oversupply of oxygenated blood lowers proportion of deoxygenated blood

4 Less dephasing of transverse magnetization leads to increased T_2^* signal or blood oxygenation level–dependent contrast (BOLD)

Figure 6–1 How fMRI measures neural activity.

A. Outside of the MRI environment, protons in hydrogen atoms in the brain spin around axes that point in random directions (**1**). When a brain enters the strong magnetic field of the MRI bore, a subset of these axes aligns with this field, which is known as longitudinal magnetization (**2**). These protons can be measured by transmitting a radiofrequency (**RF**) pulse that induces a weaker magnetic field perpendicular to the strong field. This misaligns the protons with the strong field, which now acts as a torque, causing the proton spin axes to precess in an arc on the transverse plane. The frequency of the RF pulse is chosen to resonate with the precession rate of the protons, which in turn depends on the strength of the magnetic field (**3**). When the RF pulse is stopped, the protons initially continue to precess synchronously, inducing alternating current at the same frequency in receiver coils surrounding the head. These signals can be used to generate an image by applying magnetic gradients that adjust the field strength in orthogonal directions across the brain. This results in different resonant frequencies at different points in the brain, allowing the source of the received signals to be identified. The transverse magnetization

dissipates over time, and signal is lost. This relaxation occurs as the protons give off thermodynamic energy and their axes return to the longitudinal direction (**T$_1$**), and as the protons become desynchronized in the transverse plane from local interactions with other atoms and molecules (**T$_2$**), and because of inhomogeneities in the magnetic field (**T$_2$***) (**4**).

B. Magnetic resonance can be used to estimate neuronal activity in *functional* MRI because of the magnetic properties of blood. When a brain region is in a baseline state, there is a higher proportion of deoxygenated to oxygenated blood than when the region is active. Deoxygenated blood interacts with the magnetic field, causing local inhomogeneities that distort the rate of precession and disrupt the synchrony of protons in the transverse plane, leading to more rapid T_2^* decay and lower BOLD signal (**1**). Neuronal activity leads to metabolic demand (**2**), which in turn results in the delivery of excess oxygenated blood (**3**). Oxygenated blood does not interact with the magnetic field, and so the increased amount in active brain regions reduces field inhomogeneities. In turn, this reduces the dephasing of protons precessing in the transverse plane, leading to slower T_2^* decay and higher BOLD signal (**4**).

another torque that pulls protons away from alignment with the strong field. This misalignment causes precession about the direction of the strong field by allowing the strong field to act as a torque. Complicating matters, this precession makes protons a moving target for the weaker magnetic field that is needed to cause misalignment in the first place. This is solved by generating the second field using a transmit coil in the MRI machine, through which alternating current is passed to deliver an RF pulse at the resonant frequency of the protons. This induces a perpendicular magnetic field that rotates in lockstep with the precession. This RF pulse is sustained as long as needed to generate a specified change in the spin orientation of protons away from the strong field direction (eg, 90°). This change is known as the *flip angle* and is often chosen to maximize signal according to the Ernst equation.

Once the desired flip angle has been achieved, the RF pulse is stopped in order to measure the composition of tissue. At this point, protons are precessing around the strong magnetic field and tilted heavily into the transverse plane. This is akin to a bar magnet spinning on a table, where the north and south poles take turns passing any given location. If a coil is placed nearby, the spinning magnet induces a current in the wire that reverses as the poles alternate. This is what the receiver head coil in an MRI machine measures: alternating current induced by protons precessing synchronously (note: this is the same principle as described earlier for how the transmit coil works, just reversed). The amount of current indicates the concentration of precessing protons.

Critically, the frequency of these measured signals reflects the speed of precession, which in turn depends on the strength of the magnetic field experienced by the tissue. This can be used to generate three-dimensional images by imposing different gradients on the magnetic field (think of a staircase from higher to lower strength) that cause the Larmor frequency to vary systematically over space in the brain. During fMRI, one gradient is applied in a specific direction to select a slice of brain tissue. The RF pulse can be tailored to the resonant frequency for the exact field strength at this gradient step, such that only protons in this slice are excited. The same logic is used with additional gradients in orthogonal directions to impose a two-dimensional matrix on the selected slice, with each unit volume in the matrix or *voxel* having a unique frequency and phase. The head coil receives a composite signal with a mixture of these frequencies, but the signal can be decomposed to identify protons at every voxel in the slice.

There is another important property of precessing protons that contributes to MRI: The alternating current induced in the head coil begins to decay right after the RF pulse. There are different sources of decay. One source is that precessing protons give off thermodynamic energy (heat) to the surrounding tissue, just like a gyroscope will eventually lose energy to friction and topple over. As this occurs, the spin orientation of protons gradually relaxes back to the direction of the strong magnetic field, causing them to precess less in the transverse plane and thus generate less signal. This is called longitudinal relaxation and occurs with time constant T_1. A second type of decay occurs while protons are still precessing in the transverse plane. Individual protons are surrounded by a variable neighborhood of other atoms, which carry their own weak magnetic fields. This subtly changes the field strength the proton experiences, causing its Larmor frequency to vary unpredictably. Whereas right after the RF pulse protons precess in synchrony, these local interactions cause some protons to precess faster or slower. Because they get increasingly out of sync, the induced current alternates less reliably and signal is lost. This is called transverse relaxation and occurs with time constant T_2. This dephasing of protons can also result from inhomogeneities in the strong magnetic field itself, including how it is distorted by tissue placed into the field. The signal decay from both local interactions and field distortions has time constant T_2^* (pronounced "T2-star").

These different sources of decay are important because T_1 and T_2 time constants vary depending on tissue type. MRI can thus exploit signal decay to identify gray matter, white matter, fat, and/or cerebrospinal fluid. Depending on the configuration and timing of RF pulses, gradients, and other parameters set on the MRI machine (collectively known as a *pulse sequence*), the signals received from different voxels can highlight the contrast between tissues with different T_1 values (T_1-weighted image) and/or different T_2 values (T_2-weighted image). For example, white matter is brighter than gray matter in T_1-weighted images and vice versa for T_2-weighted images.

The standard pulse sequence for measuring brain function is the echo planar imaging (EPI) sequence. EPI has two desirable properties for fMRI: It is extremely fast, allowing an entire slice to be acquired from one RF pulse in less than 100 ms, and it is sensitive to T_2^*, which, as we will see later, is how MRI measures neural activity. When designing an fMRI study, several parameters of the EPI sequence need to be chosen, including how many slices to acquire in the brain volume (typically 30–90); how much time per volume (repetition time, typically 1–2 s); what voxel resolution to use (typically 2–3 mm in each dimension); and whether to use parallel acquisition (eg, acquire multiple parts of a slice and/or multiple slices at once). These choices are interdependent, imposing trade-offs between speed, precision, and signal-to-noise.

fMRI Depends on the Biology of Neurovascular Coupling

We have described general principles of magnetic resonance, but what about the second part of the story, neurovascular coupling? Active neurons consume energy obtained from oxygen in blood. Thus, when a brain area is active, blood oxygenation drops in that moment. To replenish these metabolic resources, the flow of blood to the local area increases over the next few seconds. Supply exceeds demand, and so, counterintuitively, there is a higher proportion of oxygenated (versus deoxygenated) blood in active brain areas.

To link this to magnetic resonance, remember that T_2^* decay reflects dephasing of protons caused by field inhomogeneities. Blood has different magnetic properties depending on oxygenation: Deoxygenated blood interacts with the magnetic field because the iron in hemoglobin is unbound, whereas oxygenated blood in which the iron is bound to oxygen does not. Deoxygenated blood thus causes faster T_2^* decay and reduces signal relative to oxygenated blood. This difference in signal is referred to as the *blood oxygenation level–dependent* (BOLD) contrast. Putting everything together, increased signal in a voxel measured with an EPI sequence indicates recent neuronal activity because of the relative increase in local blood oxygenation that accompanies such activity. The temporal profile of this BOLD response, known as the *hemodynamic response function*, looks like a bell curve with a long tail, peaking around 4 to 5 seconds after local neural activity and returning to baseline after 12 to 15 seconds.

There are many more details about the physics and biology of fMRI. In addition, our understanding of how it all works is still evolving. For example, it is unclear whether BOLD is more closely tied to the firing of individual neurons or to the activity of neural populations. Likewise, it may be difficult to distinguish whether increased blood oxygenation is caused by increases in local excitation or inhibition. More generally, the mechanisms of neurovascular coupling—how the brain knows when and where to deliver oxygenated blood—remain mysterious, with a growing focus on the functional role of astrocytes. There is also the possibility of obtaining better temporal and spatial resolution by measuring the initial consumption of oxygen at the precise site of neuronal activity (the "initial dip"), reflected in an immediate and focal rise in deoxygenated blood rather than the delayed and more diffuse oversupply of oxygenated blood. Nevertheless, even with an incomplete understanding, fMRI has utility as a tool to localize changes in neural activity in the human brain induced by mental operations.

Functional MRI Data Can Be Analyzed in Several Ways

When performing an fMRI experiment, researchers link the neurovascular measurements described earlier to cognitive tasks programmed into a computer script that a human subject performs. The script generally produces a series of *runs* that correspond to a continuous period of data collection (ie, several fMRI volumes in a row), typically lasting 5 to 10 minutes. Within each run, several *trials* are presented to the subject, often by showing a visual stimulus or playing an auditory stimulus. Depending on the task, the subject may, for example, passively view or listen to the stimulus, make a decision about it, or store it in memory. A button press or eye movement response is often collected as a behavioral index of cognitive processing on that trial. These trials are typically drawn from two or more task *conditions*, which determine the stimulus type, task difficulty, or other experimental parameters. In a basic subtraction design, trials are divided between an experimental condition and a control condition, which are identical but for one critical difference whose neural basis is being investigated. Trials usually last 2 to 10 seconds, often separated by a variable or "jittered" interval of several seconds. In all, such sessions typically last up to 2 hours.

Each fMRI session produces a large amount of raw data, with BOLD responses sampled thousands of times at hundreds of thousands of locations in the brain. How are these data translated into insights about cognition and behavior? Numerous approaches to fMRI analysis are possible (Box 6–1), but for the most part, they break down into three categories (Figure 6–2). We first describe preprocessing steps common to all three types and then explain how each is conducted and what it can tell us.

fMRI Data First Need to Be Prepared for Analysis by Following Preprocessing Steps

Before the data can be analyzed, they must be prepared for processing. This is accomplished with a series of

Box 6–1 Brain Imaging as Data Science

Compared to many areas of science, the basic methods of brain imaging have enjoyed remarkable standardization. A major reason for this has been the availability of widely adopted software packages since the earliest days of fMRI in the mid-1990s. These packages were created and released by research groups, and—before it was fashionable—most were open-source.

At first, they included tools for preprocessing, alignment, analysis models, and statistical corrections. They have since incorporated new tools developed by researchers, including nonlinear alignment, field map correction, nonparametric statistics, and parallelization.

As a result, virtually all fMRI researchers use one or more of these packages, at least for part of their analysis pipeline. The following are popular free software packages for fMRI analysis:

AFNI: https://afni.nimh.nih.gov

FSL: https://fsl.fmrib.ox.ac.uk

SPM: https://www.fil.ion.ucl.ac.uk/spm

Beyond these specialized packages, fMRI is increasingly being viewed through the more general lens of data science. There are two reasons for this. First, fMRI produces a huge amount of data, both within each session but also aggregated across the thousands of studies that have been conducted. Making sense of fMRI data can thus be considered a big-data problem. Second, the data are incredibly complex and noisy, and the cognitive signals of interest are weak and hard to find. This creates a data mining challenge that has inspired many computer scientists.

The most concrete manifestation of this trend is the rise of machine learning in fMRI analysis. Other points of contact with data science include the challenges associated with the real-time analysis of streaming data, the application of network analysis and graph theoretic approaches, the use of high-performance computing clusters and cloud systems, and the growing practice of researchers publicly sharing data (eg, https://openneuro.org), code (on services such as GitHub), and educational materials (eg, https://brainiak.org/tutorials). Thus, the field of brain imaging will continue to benefit from advances in computer science, engineering, applied math, and statistics.

---

steps referred to as *preprocessing*. Preprocessing seeks to remove known sources of noise in the data, caused by either the subject or the MRI machine. Standard practice includes five basic steps known as motion correction, slice-time correction, temporal filtering, spatial smoothing, and anatomical alignment.

Motion correction seeks to address inevitable noise in the data due to a subject's head movement. Even the best subjects move their heads a few millimeters over the course of a scan, such that the voxels across three-dimensional brain volumes become somewhat misaligned. This movement can be corrected for using a spatial interpolation algorithm that lines up all of the volumes within each run. This algorithm quantifies the amount of movement at each point during the scan, including the translation in the x, y, and z dimensions, and the amount of rotation about these axes (*pitch, roll,* and *yaw*, respectively). These six time courses can later be included in the data analysis as *regressors*, to further remove motion artifacts.

Slice-time correction is applied to deal with differences in the timing of the acquisition of samples across different slices. EPI sequences collect the slices that make up each brain volume sequentially, often in an interleaved order to avoid contamination of adjacent slices. Thus, there is a large difference in the timing of the first- and last-acquired slices of the same volume, which are closer in time to the preceding and subsequent volumes, respectively, than to each other. Correcting for this difference in the timing of the slices can be accomplished with temporal interpolation to estimate what the signal would have been if all slices were acquired simultaneously.

Temporal filtering and *spatial smoothing* aim to increase the signal-to-noise ratio. Temporal filtering removes components of the time course in each voxel that are highly likely to be noise rather than meaningful variance, such as very low frequencies (>100-second period) that typically result from scanner drift. Spatial smoothing applies a kernel (typically 4–8 mm wide) to blur individual volumes, averaging out noise across adjacent voxels and improving the odds that functions will overlap across subjects after anatomical alignment.

This *anatomical alignment* is accomplished by registering data across runs and subjects, usually with simple transformations (eg, shift, rotate, scale), to a standard template such as Montreal Neurological Institute or Talairach space. Typically, fMRI data are

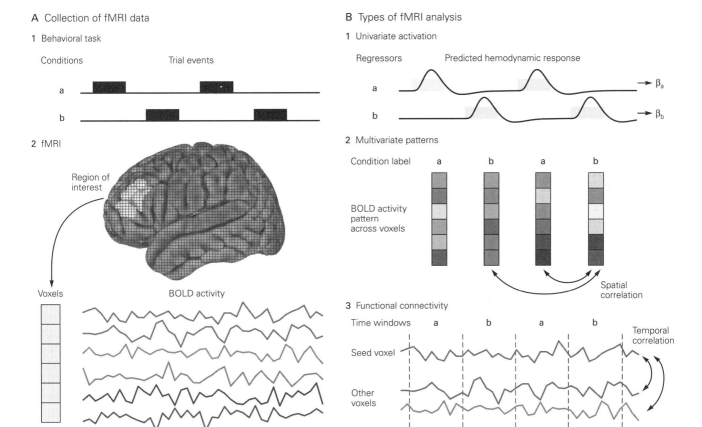

A Collection of fMRI data

1 Behavioral task

2 fMRI

Region of interest

Voxels   BOLD activity

Time →

B Types of fMRI analysis

1 Univariate activation

2 Multivariate patterns

3 Functional connectivity

Figure 6–2 Collecting and analyzing fMRI data.

A. An fMRI experiment typically involves subjects performing a behavioral task while BOLD activity is measured from the brain.

1. The example task consists of two conditions (**a**, **b**) that alternate in time, each with two events depicted (**black rectangles**).

2. The time course of BOLD activity in six example voxels (different colors) from a region of interest (**ROI**) during the task. Analysis often focuses on an ROI or other subset of voxels in the brain to reduce the number of statistical tests performed. When all voxels in the brain are analyzed, statistical corrections are applied to reduce the number of false positives. The results of such analyses are often overlaid on a structural MRI as a color-coded heat map. The map is the result of extensive preprocessing and analysis and does not directly reflect neuronal activity or even blood oxygenation. Rather, voxels are colored to indicate that they have passed the threshold of being considered significant in a statistical test.

B. Three analysis approaches are often used in fMRI experiments such as the one depicted in **A.**

1. Univariate activation analysis attempts to explain the BOLD activity of each individual voxel in terms of what happened in the task. This is accomplished using a statistical model that contains a regressor for each task condition specifying the predicted hemodynamic response (**bell curves**) for trial events from that condition (**gray rectangles**). The result of fitting the model to BOLD activity is a beta value for each regressor in every voxel, quantifying the average response of the voxel to trials of that condition. The beta values for a voxel can be subtracted to measure whether there is a greater response in one condition than another. To determine statistical significance, this difference in activation between conditions in each voxel is compared across subjects.

2. Multivariate pattern analysis considers the pattern of BOLD activity across voxels. These spatial patterns are extracted for

each trial from a subset of voxels (six depicted) and at a particular moment in time, often the peak of the predicted hemodynamic response (color saturation indicates amplitude of BOLD activity in each voxel on that trial). There are two common ways of analyzing these patterns. The first (shown) involves calculating the spatial correlation of patterns from a pair of trials to explore how similarly voxels responded to the trials. If a brain region represents different information across conditions, this pattern similarity should be higher for pairs of trials from the same versus different conditions. The second type of multivariate pattern analysis (not shown) uses a type of machine learning known as pattern classification. Some of the patterns and their corresponding condition labels are used to train a classifier model, assigning weights to voxels based on how useful they are at distinguishing between conditions. The model is then tested with other patterns on which it was not trained. If a brain region represents different information across conditions, the model should be able to correctly guess from which condition the patterns were extracted. To determine statistical significance, spatial correlations or classification accuracies in a region are compared across subjects.

3. Functional connectivity analysis examines how BOLD activity is correlated between voxels over time. Typically, a seed voxel or ROI is chosen and its time course (**pink curve**) is correlated with the time courses of other voxels (two shown here). This can be performed while the subject is resting, resulting in a correlation value for every voxel that can be used to identify brain networks in a baseline state. Functional connectivity can also be calculated in different time windows of a task (**dashed lines**), resulting in a correlation value for each trial that can be used to understand the dynamics of these networks. To determine statistical significance, temporal correlations for each voxel are compared across subjects between conditions or against zero.

first aligned to a structural scan from the same subject, and then this structural scan is aligned to the standard template.

Once these five steps are completed, the data are ready for analysis.

fMRI Can Be Used to Localize Cognitive Functions to Specific Brain Regions

The first kind of fMRI analysis seeks to localize functions in the brain and to determine what brain regions are associated with a behavior. This is based on having subjects complete a task during fMRI and then examining the relationship between different phases of the experiment and changes in BOLD activity in different parts of the brain. Based on researchers' knowledge of what happened at different times in the experiment, the function of the regions can be inferred.

A series of statistical analyses are performed to quantify this relationship and to determine its significance. Typically, this is accomplished using a statistical regression method known as a *general linear model* (GLM). The GLM attempts to explain observed data (here, the time course of the BOLD activity in each voxel) as a linear combination of regressors that reflect independent variables (eg, task conditions) and covariates (eg, movement parameters).

The regressors that model task conditions serve as a hypothesis about how a voxel should respond if involved in the cognitive function manipulated by that task. The regressor for each condition is generated by marking the onset and duration of each trial of that condition in the experimental time line, corresponding to the expected neuronal activity, and then accounting for the delayed hemodynamic response. All regressors are fit simultaneously to the fMRI activity in each voxel, and the result is a parameter estimate (or "beta") for each condition and voxel, reflecting how much of the temporal variance of the voxel is uniquely explained by that condition's trials on average.

To localize a function, betas from two or more conditions are compared in a contrast. The most basic form of contrast is to subtract one beta (eg, control condition) from another (eg, experimental condition). Contrasts are typically averaged over runs within each subject and then entered into a t-test to assess reliability across subjects. Because statistics are calculated for every voxel, there is a high risk of false positives, and a correction for multiple comparisons is required (eg, by giving voxels more credence if they cluster together with other significant voxels). Alternatively, a more constrained analysis can be performed, focusing on a limited number of regions of interest (ROIs) that are

defined a priori. Contrast values can then be averaged over the voxels in an ROI to produce regional estimates, rather than examining all voxels in the brain, thereby reducing the number of comparisons.

This general family of approaches is often described as measuring *univariate activation*— "univariate" because each voxel or region is treated independently and "activation" because the result is a measure of the relative activity evoked by one condition versus another. This kind of analysis is typically used to localize a cognitive function to a set of voxels or regions in the brain.

However, univariate activation can be used for more than localization. For example, a GLM can make quantitative predictions about BOLD activity by assigning a continuous weight, rather than a categorical one, to each trial in a regressor based on an experimental parameter (eg, working memory load), behavioral measurement (eg, response time), or computational model (eg, prediction error in reinforcement learning). The resulting beta reflects how much a voxel correlates with the variable of interest.

Another use of univariate activation is for measuring changes in BOLD activity as a function of repeating a stimulus. Such studies take advantage of *adaptation* (or *repetition suppression*)—the tendency of stimulus-selective neurons to respond less to repeated versus new stimuli. This fact allows the tuning of a brain region to be inferred by conducting an experiment in which related and unrelated stimuli are presented sequentially. In some trials, one stimulus is followed by a near-repetition of the same stimulus, but with a feature changed (eg, its location or size). A univariate analysis tests whether BOLD activity in voxels from the region is lower on these trials compared to other trials in which either (1) the first stimulus is followed by an unrelated second stimulus or (2) the changed stimulus is preceded by an unrelated stimulus. If such a BOLD reduction is observed, the region can be interpreted as not tuned for the changed feature (eg, the region could be considered location or size invariant).

fMRI Can Be Used to Decode What Information Is Represented in the Brain

The second category of fMRI analysis seeks to characterize what kinds of information are represented in different regions of the brain to guide behavior. Rather than analyze voxels independently or average over voxels within an ROI, these analyses examine the information carried by spatial *patterns* of BOLD activity over multiple voxels. This is typically referred to as *multivariate pattern analysis* (MVPA). There are two types of MVPA, based on the similarity or classification of activity patterns.

Similarity-based MVPA tries to understand what information is contained or "represented" in a brain region. This is accomplished by examining how similarly the region processes different conditions or stimuli in an experiment. This similarity is calculated from the pattern of activation across voxels in an ROI, defined as either the pattern of beta values from a GLM or the pattern of raw BOLD activity from preprocessed data. Once these patterns have been defined for multiple conditions or stimuli, the correlation or distance of each pair of patterns is calculated. This produces a matrix of the pairwise similarities between conditions or stimuli within the ROI. With this matrix, it is possible to infer to what information the ROI is most sensitive. For example, if subjects are shown photos of different objects (eg, a banana, canoe, taxi), a matrix of distances between the activity patterns evoked by these objects can be computed for different brain regions. An ROI in which there is less distance between banana and canoe than between either of them and taxi could be interpreted to mean that the region represents shape (ie, concavity); another region in which the lowest distance is between banana and taxi might represent color (ie, yellow); or one with the lowest distance between canoe and taxi might be interpreted as representing function (ie, transportation).

Neural similarity from fMRI can also be compared with similarity calculated in other ways for the same conditions or stimuli, including from human judgments, computational models, or neural measures in other species. For example, if human subjects rate a large set of stimuli in terms of how similar they look to each other, a brain region with a matching similarity structure could be considered a candidate source of this behavior. This approach of calculating *second-order* correlations between neural and behavioral similarity matrices, or between neural similarity matrices from two sources, is called *representational similarity analysis* (RSA).

Classifier-based MVPA uses techniques from machine learning (discussed in Chapter 5) to decode what information is present in a brain region. The first step is to train a classifier model on a subset of the fMRI data to discriminate between conditions or stimulus classes from patterns of BOLD activity across voxels in an ROI. These patterns are usually obtained from individual trials, and each is labeled according to the condition or stimulus on the corresponding trial. This training set thus contains several brain pattern examples of each class. Classifier training can use many different algorithms, the two

most common being support vector machine and regularized logistic regression. The result is typically a weight for each voxel reflecting how activity in that voxel contributes to classification collectively with the other voxels. The second step after training is to test the classifier by examining how well it can decode patterns from a held-out and independent subset of fMRI data (eg, from a different run or subject). The pattern of BOLD activity on each test trial is multiplied by the learned classifier weights and summed to produce a guess about how the pattern should be labeled. Classification accuracy is quantified as the proportion of these guesses that match the correct labels. Importantly, this approach can be used to understand how different brain regions give rise to behavior, such as by attempting to classify which action was performed, which decision was made, or which memory was retrieved.

fMRI Can Be Used to Measure Correlated Activity Across Brain Networks

The third category of fMRI analysis seeks to understand the organization of the brain as a network. Knowing what brain regions do individually does not fully explain how the brain as a whole generates behavior. It is additionally critical to know how brain regions relate to each other—that is, where do the inputs to a region come from and where do the outputs go? This requires an understanding of which regions communicate with each other and when and how they transmit information. This is difficult to determine definitively with fMRI but can be estimated by measuring the correlation of BOLD activity between voxels or regions over time. If two parts of the brain have correlated activity, they may be sharing the same information or participating in the same process. Such correlations are interpreted as measures of *functional connectivity*.

One way to study functional connectivity with fMRI is to measure BOLD correlations in a resting state. Subjects are scanned while they lie still without performing a task, and then the time course of BOLD activity from one "seed" ROI is extracted and correlated with the time courses from other ROIs or from all voxels in the brain. Alternatively, clustering or component analyses can be used without a seed to identify collections of voxels with similar temporal profiles. Resting functional connectivity defined in these ways has helped reveal that the brain contains several large-scale networks of regions. The most widely studied of these networks is referred to as the default mode network, which includes the posterior medial cortex, lateral parietal cortex, and medial prefrontal cortex.

By definition, resting connectivity cannot be linked to concurrent behavior. Nor is it static, as telling subjects not to do anything does not restrict what they think about. Nevertheless, resting connectivity can be linked to behavior indirectly by examining how it goes awry in disease or disorders and how it relates to cognitive differences between people.

Functional connectivity can be linked more directly to behavior if it is measured during tasks rather than at rest. One difficulty in interpreting such correlations between regions is that two regions might be correlated during a task not because they are communicating with each other, but because of a third variable. For example, the regions might be responding independently but coincidentally to the same stimulus. Thus, task-based functional connectivity is typically calculated after removing, or otherwise accounting for, BOLD responses evoked by stimuli. This approach allows functional connectivity to be manipulated experimentally and compared across task conditions. These comparisons provide insight into how the involvement and interaction of brain regions in a network change dynamically to support different behaviors. This has proven useful for understanding cognitive functions such as attention, motivation, and memory, which depend on some brain regions modulating others.

Functional connectivity can also be viewed as a pattern (of correlations rather than activity) and submitted to MVPA. Correlation patterns are larger in scale than activity patterns: If there are n voxels in an activity pattern, there are on the order of n^2 voxel pairs in a correlation pattern. Thus, it can be helpful to summarize the properties of correlation patterns using graph theory, where individual voxels or regions are treated as the nodes in the graph and the functional connectivity between these nodes determines the edge strengths.

Functional MRI Studies Have Led to Fundamental Insights

Functional MRI has changed our understanding of the basic neurobiological building blocks of human behavior. Combining experimental manipulations and computational models from cognitive psychology with precise neurobiological measurements has expanded existing theories of the mind and brain and has stimulated new ideas. Discoveries from fMRI have impacted not just our understanding of behaviors presumed to be uniquely human, but also behaviors that have long been investigated in animals.

In this section, we review three examples of this progress. The study of face perception reveals how human fMRI studies have inspired research in animals. The study of memory illustrates how fMRI has challenged theories from cognitive psychology and systems neuroscience. The study of decision-making shows how animal studies and computational models have advanced fMRI research.

fMRI Studies in Humans Have Inspired Neurophysiological Studies in Animals

Our understanding of how the brain perceives faces has grown tremendously over the past two decades (Chapter 24). The advances described below provide an example of how findings from fMRI in humans inspired follow-up studies with neuronal recordings and causal interventions in nonhuman primates. This synergy across species and techniques led to a more complete understanding of the fundamental process by which faces are recognized.

Some classes of stimuli are more important for survival than others. Does the brain have dedicated machinery for the processing of such stimuli? Faces are an obvious case in humans. The development of fMRI combined with careful and systematic experimental designs led to important insights into how and where faces are processed in the human brain. One region in the fusiform gyrus, often referred to as the fusiform face area (FFA), was found to show robust and selective BOLD activity when humans view faces.

Early fMRI studies that led to this discovery relied on simple designs in which subjects were presented with a series of different types of visual stimuli. To measure the face selectivity of brain areas, the BOLD response to faces was compared with the BOLD responses for the other categories (eg, places, objects). An area of the lateral fusiform gyrus, most reliably in the right hemisphere, was strongly activated by faces. These findings fit with earlier findings of individual neurons in nonhuman primates that respond to faces, but inspired a new wave of animal studies to examine a larger-scale network of brain regions. These newer animal studies, borrowing experimental designs from the human studies, first used fMRI to find orthologs of the FFA. The resulting face patches were then probed invasively with neuronal recording and stimulation. This revealed insights into the distributed neural circuitry for face processing in primates.

In addition to responding selectively to face stimuli, does the FFA contribute to the behavior of face recognition? This question has been addressed using stimulus variations that are known to affect face recognition (eg, presenting faces that are inverted or presenting parts of faces). Initial fMRI studies using simple comparisons of stimulus categories (inverted versus upright faces) produced weak and mixed results. Follow-up studies used an adaptation design to determine how BOLD activity changes when a face is repeated intact or altered. The findings suggested that the FFA represents intact faces differently than when the same visual features are reconfigured in a way that disrupts behavioral recognition.

Another way to examine the behavioral significance of a region is to study patients who have behavioral deficits—in this case, an impairment of face recognition known as prosopagnosia. Surprisingly, some fMRI studies found an intact FFA in these patients, casting doubt on its necessity for face perception. However, here too follow-up studies using an adaptation design proved informative: The otherwise intact FFA of prosopagnosics did not adapt when the same face was repeated. This suggests that the FFA responds differently in people with prosopagnosia, consistent with its importance to face recognition.

The finding that visual categories, or mental processes more generally, can be mapped to one or a small number of regions like the FFA was important for thinking about the relationship between mind and brain. Whether specific functions are localized or broadly distributed has been a central question regarding brain organization throughout the history of neuroscience (Chapter 1). The discovery of the FFA and the face patch system provided new evidence of localization, and encouraged researchers to pursue the hypothesis that other complex cognitive functions might be localized in specific brain areas or small sets of nodes, but also to question whether localization is the right way to think about brain organization. For example, further studies showed that faces produce widely distributed responses over visual cortex and that the FFA can be co-opted for recognition of other kinds of objects with which we have expertise. These debates reflect the transformative nature of this original work, both for studies of the human brain and for related questions in animal models.

fMRI Studies Have Challenged Theories From Cognitive Psychology and Systems Neuroscience

Many theoretical models from cognitive psychology were originally agnostic about the brain. However, there are now several examples of fMRI findings that changed our understanding of the organization and mechanisms of cognition.

One prominent example is the study of memory. The overall goal of memory research, beginning in the 19th century, has been to understand how a memory is created, retrieved, and used, and whether these processes differ across types of memory. A key discovery came from research on patient H.M. and the realization that damage to the hippocampus causes a loss of the ability to form new autobiographical memories but does not impact the ability to learn certain skills (Chapter 52). These findings led to the idea that memory can be divided into two broad classes, conscious versus unconscious (also known as declarative versus procedural or explicit versus implicit). In the tradition of localization, these and other types of memory were mapped onto distinct brain regions, based on where in the brain a patient had damage and which behavioral symptoms they exhibited.

Later fMRI studies of the healthy human brain helped reveal that this dichotomy was oversimplified. First, several studies using what came to be known as the *subsequent memory task* showed that regions beyond the hippocampus are implicated in the successful formation of declarative memory. In such studies, subjects are presented with a series of stimuli (pictures or words) while being scanned. Later, usually outside of the MRI machine, their memory for these stimuli is tested. The BOLD responses from when a stimulus was initially encoded are then sorted based on whether it was subsequently remembered or forgotten. These conditions are contrasted to reveal which brain regions show more (or less) activity during successful memory formation. In addition to finding such differences in the hippocampus and surrounding medial temporal lobe, BOLD activity in prefrontal and parietal cortices is also predictive of later memory. By measuring the whole brain of healthy individuals, fMRI revealed that declarative memory is served by more than one brain system—processes linked to prefrontal cortex (eg, semantic elaboration) and parietal cortex (eg, selective attention) are also involved in encoding.

The traditional taxonomy of memory organization was challenged in another way by fMRI studies. fMRI revealed that a wide range of tasks that were previously assumed to not involve the hippocampus (or declarative memory) in fact do consistently engage this region. These studies often use learning tasks that would classically be considered unconscious, in which subjects have the opportunity to learn but are never asked to report their memories and, in some instances, are unable to do so if prompted. For example, in the *probabilistic classification task*, subjects learn by trial and error to sort visual cues into categories, even when the relationship between cues and categories is sometimes unreliable. BOLD activity during such learning trials is estimated and compared to a baseline task that does not involve trial-and-error learning (eg, studying cues with their categories provided). Such comparisons generally reveal activation in the striatum, but also reliably in the hippocampus (see Chapter 52).

In summary, fMRI studies of tasks thought to rely on declarative memory often recruit regions outside of the hippocampus, and tasks thought to rely on procedural memory can recruit the hippocampus. In both cases, these discoveries were serendipitous and made possible only because data were obtained from the whole brain with fMRI. Although these began as unexpected results, they led to systematic follow-up studies that have updated our understanding of the organization of memory. Chiefly, they challenged the original emphasis on conscious awareness as the defining characteristic of hippocampal processing. This in turn helped relate the findings from human studies to those from animal studies, where the notion of conscious memory is less central and where tasks that engage the hippocampus often involve spatial navigation. Thus, fMRI findings in humans have been transformative for our understanding of theoretical models of memory, in terms of both neural structures and cognitive behaviors.

fMRI Studies Have Tested Predictions From Animal Studies and Computational Models

The integration of computational models with fMRI has been an important development in cognitive neuroscience. One example of this comes from studies of how the brain learns to predict and obtain rewards, combined with models of reinforcement learning that formalize this process. These models co-evolved with studies of reward-based decision-making in animals, which also inspired later human studies.

Central to these studies and theories, midbrain dopaminergic neurons increase their firing in response to unexpected rewards, such as juice (Chapter 43). Once a predictive cue has been reliably paired with a reward, the neurons shift their response in time to this predictive cue. If a predicted reward fails to occur, firing decreases. This pattern of responses suggests that midbrain dopaminergic neurons signal the difference between expected and actual rewards. This difference is commonly known as *reward prediction error* and has been modeled using equations based on reinforcement learning theory. When this model is applied to human tasks involving rewards, hypothesized reward prediction errors can be estimated on a trial-by-trial basis. These estimates can then be used to predict BOLD activity

and identify voxels and regions that may be involved in reinforcement learning in the human brain.

In a typical study of this type, subjects perform a learning task during fMRI, making a series of choices about visual cues to predict possible rewards. They learn the outcome immediately after each choice. For example, a subject might view two shapes (eg, circle, triangle), choose one by pressing a button, and then learn whether the choice led to a monetary reward. The key feature of such tasks is that the association between shapes and rewards is probabilistic and changes over the course of the experiment. Because of this noisy relationship, subjects must learn to track the likelihood of reward for each shape. Reward prediction error can be calculated on each trial based on the history of the subject's choices and rewards and then included in the analysis of their fMRI data. Many studies using this approach have found that trial-by-trial reward prediction error correlates with BOLD activity in the ventral striatum, an area that receives input from midbrain dopaminergic neurons.

Other computational models, such as *deep neural networks*, which integrate cognitive psychology, computer science, and neuroscience, have also served an important theoretical purpose by generating novel hypotheses about brain activity. Because these models are often inspired by the architecture and functions of the brain, they help bridge levels of analysis, from physiological recordings in animals to fMRI in humans. They also serve a useful purpose in data analysis by simulating variables of psychological and neurobiological interest that can be sought in the brain, an approach often referred to as model-based analysis.

Functional MRI Studies Require Careful Interpretation

The examples provided earlier illustrate how fMRI can improve our understanding of the links between brain and behavior. At the interface with psychology, fMRI can complement purely behavioral measurements. Many complex human behaviors (eg, memory recall, decision-making) depend on multiple processing stages and components. Measuring these processes with fMRI can provide richer and more mechanistic explanations of behavior than those based on simple behavioral measurements such as accuracy or response time alone. At the interface with systems neuroscience, fMRI complements direct neuronal recordings. Most brain areas (eg, hippocampus) support multiple behaviors and do so in concert with other regions. The ability to image the whole brain with fMRI makes it possible to arrive at a more complete understanding of neural mechanisms at the network level.

What does it mean then to find BOLD activity in a region during a task? The multiplicity of mappings between brain and behavior poses serious challenges to interpretation of fMRI results (Figure 6–3). One fundamental consideration is the type of inference. Most fMRI studies use *forward inference*, in which an experiment compares BOLD activity between task conditions that manipulate the engagement of a particular mental process (eg, comparing the effects of face versus nonface stimuli to study face recognition). Brain regions that differ between these conditions can be inferred to take part in the manipulated process. Forward inference relies on a task manipulation and therefore allows a researcher to infer that differences in brain activity are related to the mental process of interest.

With *reverse inference*, differences in neural activity are the basis for inferring which specific mental process is active, even when the conditions that gave rise to the differences were not designed to manipulate that process. For example, in the previous face versus nonface contrast, a researcher might interpret differential activity in the striatum as evidence that faces are rewarding. This kind of reverse inference is often unjustified, as reward was neither measured nor manipulated—the interpretation is based on other studies that manipulated reward and found striatal activity. The problem arises because each brain region generally supports more than one function, meaning that it is unclear from the observation of activity alone which function(s) were engaged. Indeed, the striatum is also strongly implicated in movement, so perhaps faces are engaging motor rather than reward processes? The logically sound conclusion in this example, reflecting forward inference, is that the striatum is involved in some (as yet unresolved) aspect of face recognition.

One solution therefore is not to use reverse inference in fMRI studies. However, there are some situations in which reverse inference can be desirable or even necessary. For example, reverse inference can allow researchers to perform exploratory analyses and generate new hypotheses, even from data that were collected for other purposes. This may be especially important for getting the most out of fMRI data that are hard to collect, such as from children, the elderly, and patients (Box 6–2). Motivated by this need, statistical tools have been developed to support reverse inference. For example, the web-based tool Neurosynth uses a large database of published studies to assign a probability that a specific mental process (eg, reward) is involved given that BOLD activity has been observed in a particular region (eg, striatum).

It is also important to make a distinction between a correlation of brain activity with behavior versus a cause-and-effect relationship between the brain activity and the behavior. If a brain region is selectively and consistently involved in a specific mental process, this correlation does not license the conclusion that it plays a necessary or sufficient role in the process. With respect to sufficiency, the brain region might (and likely does) work with one or more other brain regions to accomplish the process. With respect to necessity, activity in the region might be a secondary by-product of processing elsewhere.

One approach to bolstering the interpretation of an fMRI study is to evaluate how the findings converge with those from more invasive methods, such as electrical stimulation in epilepsy patients. Because every tool has limitations, including other correlational measures such as neuronal recordings, this principle of converging evidence is central to advancing understanding of how the brain supports behavior. In addition to converging evidence across studies and tools, there are also efforts to manipulate brain function simultaneously with fMRI, using transcranial magnetic stimulation or real-time neurofeedback.

Future Progress Depends on Technological and Conceptual Advances

Functional MRI is the best technology we have so far for probing the healthy human brain. It allows measurement of the whole brain at reasonably high resolution as well as many aspects of the mind in large subject samples without harm. However, in other ways, it is far from what we ultimately need if we are to obtain a deeper and more precise understanding of how the brain works. When compared to tools available in animals, fMRI provides relatively noisy, slow, and indirect measurements of neuronal activity and circuit dynamics.

Efforts are underway to address these limitations, both technically and biologically. On the technical front, multiband imaging sequences can enhance the temporal and spatial resolution of fMRI data by enabling the acquisition of multiple slices through the brain in parallel. However, faster measurements are inherently limited by the slow speed of the hemodynamic response and smaller voxels still average across hundreds of thousands of neurons.

On the biological front, we have a rudimentary understanding of how BOLD activity emerges from physiological mechanisms in the brain, such as single neuron activity, population activity, the function of

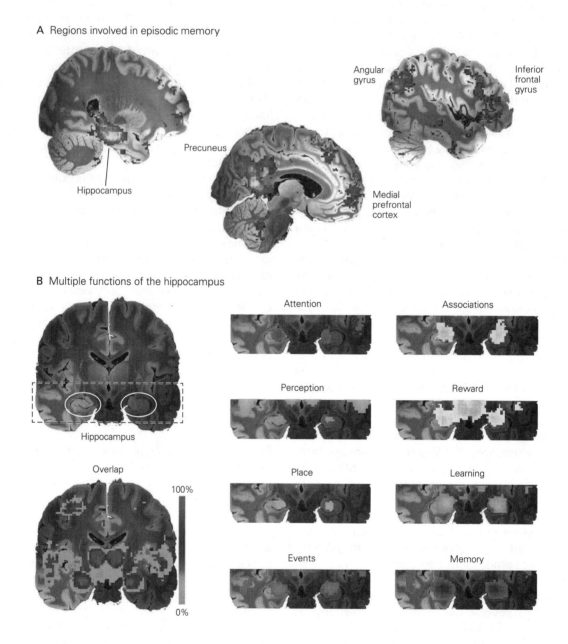

A Regions involved in episodic memory

Hippocampus

Precuneus

Angular gyrus

Inferior frontal gyrus

Medial prefrontal cortex

B Multiple functions of the hippocampus

Hippocampus

Overlap

100%

0%

Attention

Associations

Perception

Reward

Place

Learning

Events

Memory

Figure 6–3 Challenges of mapping mind and brain. Any interpretation of data from fMRI must consider the complexity of the relationship between cognitive functions and brain regions. This complexity is illustrated here with a meta-analysis from a database containing more than 14,000 published fMRI studies. (Data retrieved in 2019 from http://neurosynth.org, displayed on brain from Edlow et al. 2019; figure updated and adapted from Shohamy and Turk-Browne 2013 by Tristan Yates.)

A. This map shows that multiple brain regions are engaged by episodic memory—that is, encoding and retrieval of specific events from one's past. Colored voxels indicate a high probability of the term "episodic" in studies that reported activation in

these voxels (reverse inference). This example illustrates how a single cognitive function can be associated with multiple brain regions (one-to-many mapping).

B. These maps show that multiple cognitive functions engage the hippocampus (circled in white in each hemisphere). Colored voxels in each inset brain indicate a high probability that these voxels were activated in studies that examined the corresponding term (forward inference). The overlap map shows the percentage of these terms that activated each voxel. This example illustrates how a single brain region can be associated with multiple cognitive functions and behaviors (many-to-one mapping).

Box 6–2 Brain Imaging in the Real World

The ability to image the human brain with noninvasive tools and to measure internal mental processes has led to interest in applying fMRI to a variety of real-world problems, such as clinical diagnosis and treatment, law and justice, artificial intelligence, marketing and economics, and politics.

In the clinical realm, one interesting direction is the use of fMRI to examine patients in a vegetative state. Studies suggest that some such patients exhibit brain activity that reflects mental processing. For example, a patient might appear comatose—unconscious, noncommunicative, and unreactive to external stimuli—yet exhibit neural activity in the motor cortex when asked to think of an action or in category-specific visual regions when asked to imagine specific visual cues. Such findings could influence the prognosis and treatment of patients by clinicians.

Another potential real-world application of fMRI is lie detection. The ability to accurately distinguish truth from lies based on brain activity could have significant value in the courtroom. Some laboratory studies have reported differential brain activity when groups of subjects are instructed to lie repeatedly. To be useful, however, fMRI would need to provide highly reliable evidence about whether an *individual* person is lying about a *specific* event, in a way that is immune from strategies or countermeasures. This is not possible at present, and indeed, fMRI evidence is generally inadmissible in court.

These and other applications of fMRI raise ethical and privacy concerns. For example, authorities could use fMRI data to justify consequential decisions (eg, guilt or innocence), exploiting the public's bias to believe biological explanations, even when the underlying science is not settled. More troubling, humans currently have autonomy over whether we share our internal thoughts and feelings, but devices that sense this information could change that. As a result, an important challenge for neuroscientists when considering practical applications is to accurately convey that fMRI is powerful but has limitations and that our understanding of the human brain is a work in progress.

---

astrocytes and other glia, neuromodulatory systems, and the vascular system. A better understanding of the relationship between BOLD activity and these processes is essential for knowing when and why measurements of different types align and diverge. Although some experimental conditions lead to an increase in both neuronal activity and BOLD activity, others do not. For example, although the presentation of a visual cue increases both blood flow in the visual cortex and neuronal firing, if this visual cue is highly expected but not presented, blood flow can still increase but without an increase in neuronal activity. This suggests that there are important nuances to the coupling of neural and vascular activity that may have functional significance and that the vascular signals themselves may be more complex than previously appreciated.

As the history of fMRI shows, scientific discoveries in one field can lead to unexpected breakthroughs in other fields. The discovery of MRI in the 1970s (which 20 years later led to fMRI) came from physics and chemistry, and was so profound and far-reaching that it was recognized by the Nobel Prize in Physiology or Medicine in 2003 to Paul Lauterbur and Peter Mansfield. This in turn was made possible by the discovery of nuclear magnetic resonance decades earlier, which resulted in the Nobel Prize in Physics in 1944

to Isidor Rabi and in 1952 to Felix Bloch and Edward Purcell. These discoveries had no initial connection to neuroscience but came to spark a revolution in the study of mind, brain, and behavior.

Highlights

1. Functional brain imaging methods in cognitive neuroscience seek to record activity in the human brain associated with mental processes as they unfold in the human mind, linking biological and behavioral measures. Currently, the dominant technique is fMRI.

2. fMRI is based on two main concepts: the physics of magnetic resonance and the biology of neurovascular coupling. Combined, they allow fMRI to measure the BOLD response to neuronal activity. When human subjects perform cognitive tasks during fMRI, measurements of BOLD activity can be linked to particular mental processes and behaviors over time.

3. The link between BOLD activity and behavior is inferred through a series of preprocessing steps and statistical analyses. These analyses can answer

a range of questions, such as which brain regions are active during specific tasks, what information is coded in the spatial pattern of activity within a region, and how regions interact with each other over time as part of a network.

4. Human brain imaging has led to fundamental insights about the neural mechanisms of behavior across many domains. Some prominent examples are understanding how the human brain processes faces, how memories are stored and retrieved, and how we learn from trial and error. Across these domains, data from fMRI have converged with findings from neuronal recordings in animals and with theoretical predictions from computational models, providing a more complete picture of the relationship between brain and mind.

5. fMRI records brain activity but does not directly modify activity. Therefore, it does not support inferences about whether a brain region is necessary for a behavior, but rather whether the region is involved in that behavior. Most studies support forward inferences about this involvement, whereby activity in the brain can be linked to a mental process because the experiment manipulates that process.

6. fMRI provides an opportunity to study the function of the human brain as it engages in a variety of mental processes, in both health and disease. This technology and analyses of the data it generates are undergoing continual development to improve the temporal and spatial resolution of biological measurements and to clarify the links between these measurements, mental processes, and behavior.

Daphna Shohamy
Nick Turk-Browne

Suggested Reading

Bandettini PA. 2012. Twenty years of functional MRI: the science and the stories. NeuroImage 62:575–588.

Bullmore E, Sporns O. 2009. Complex brain networks: graph theoretical analysis of structural and functional systems. Nat Rev Neurosci 10:186–198.

Cohen JD, Daw N, Engelhardt B, et al. 2017. Computational approaches to fMRI analysis. Nat Neurosci 20:304–313.

Daw ND, Doya K. 2006. The computational neurobiology of learning and reward. Curr Opin Neurobiol 16:199–204.

Kanwisher N. 2010. Functional specificity in the human brain: a window into the functional architecture of the mind. Proc Natl Acad Sci U S A 107:11163–11170.

Poldrack RA, Mumford JA, Nichols TE. 2011. *Handbook of Functional MRI Data Analysis.* Cambridge: Cambridge Univ. Press.

Shohamy D, Turk-Browne NB. 2013. Mechanisms for widespread hippocampal involvement in cognition. J Exp Psychol Gen 142:1159–1170.

References

Aly M, Ranganath C, Yonelinas AP. 2013. Detecting changes in scenes: the hippocampus is critical for strength-based perception. Neuron 78:1127–1137.

Aly M, Turk-Browne NB. 2016. Attention stabilizes representations in the human hippocampus. Cereb Cortex 26:783–796.

Brewer JB, Zhao Z, Desmond JE, Glover GH, Gabrieli JD. 1998. Making memories: brain activity that predicts how well visual experience will be remembered. Science 281:1185–1187.

Dricot L, Sorger B, Schiltz C, Goebel R, Rossion B. 2008. The roles of "face" and "non-face" areas during individual face perception: evidence by fMRI adaptation in a brain-damaged prosopagnosic patient. NeuroImage 40:318–332.

Edlow BL, Mareyam A, Horn A, et al. 2019. 7 Tesla MRI of the ex vivo human brain at 100 micron resolution. Sci Data 30:244.

Farah MJ, Hutchinson JB, Phelps EA, Wagner AD. 2014. Functional MRI-based lie detection: scientific and societal challenges. Nat Rev Neurosci 15:123–131.

Foerde K, Shohamy D. 2011. Feedback timing modulates brain systems for learning in humans. J Neurosci 31:13157–13167.

Fox MD, Raichle ME. 2007. Spontaneous fluctuations in brain activity observed with functional magnetic resonance imaging. Nat Rev Neurosci 8:700–711.

Friston KJ, Holmes AP, Worsley KJ, Poline JP, Frith CD, Frackowiak, RS. 1994. Statistical parametric maps in functional imaging: a general linear approach. Hum Brain Mapp 2:189–210.

Grill-Spector K, Henson R, Martin A. 2006. Repetition and the brain: neural models of stimulus-specific effects. Trends Cogn Sci 10:14–23.

Hannula DE, Ranganath C. 2009. The eyes have it: hippocampal activity predicts expression of memory in eye movements. Neuron 63:592–599.

Huettel SA, Song AW, McCarthy G. 2014. *Functional Magnetic Resonance Imaging.* Sunderland, Massachusetts: Sinauer Associates, Inc.

Kanwisher N, McDermott J, Chun MM. 1997. The fusiform face area: a module in human extrastriate cortex specialized for face perception. J Neurosci 17:4302–4311.

Kim H. 2011. Neural activity that predicts subsequent memory and forgetting: a meta-analysis of 74 fMRI studies. NeuroImage 54:2446–2461.

Kriegeskorte N, Mur M, Bandettini P. 2008. Representational similarity analysis—connecting the branches of systems neuroscience. Front Syst Neurosci 2:4.

McCarthy G, Puce A, Gore JC, Allison T. 1997. Face-specific processing in the human fusiform gyrus. J Cog Neurosci 9:605–610.

Moeller S, Freiwald WA, Tsao DY. 2008. Patches with links: a unified system for processing faces in the macaque temporal lobe. Science 320:1355–1359.

Norman KA, Polyn SM, Detre GJ, Haxby JV. 2006. Beyond mind-reading: multi-voxel pattern analysis of fMRI data. Trends Cogn Sci 10:424–430.

O'Doherty JP, Hampton A, Kim H. 2007. Model-based fMRI and its application to reward learning and decision making. Ann NY Acad Sci 1104:35–53.

Owen AM, Coleman MR, Boly M, Davis MH, Laureys S, Pickard JD. 2006. Detecting awareness in the vegetative state. Science 313:1402.

Schapiro AC, Kustner LV, Turk-Browne NB. 2012. Shaping of object representations in the human medial temporal lobe based on temporal regularities. Curr Biol 22:1622–1627.

Sirotin Y, Das A. 2009. Anticipatory haemodynamic signals in sensory cortex not predicted by local neuronal activity. Nature 457:475–479.

Squire LR. 1992. Memory and the hippocampus: a synthesis from findings with rats, monkeys, and humans. Psychol Rev 99:195–231.

Turk-Browne NB. 2013. Functional interactions as big data in the human brain. Science 342:580–584.

Wagner AD, Schacter DL, Rotte M, et al. 1998. Building memories: remembering and forgetting of verbal experiences as predicted by brain activity. Science 281:1188–1191.

Wimmer GE, Shohamy D. 2012. Preference by association: how memory mechanisms in the hippocampus bias decisions. Science 338:270–273.

Yarkoni T, Poldrack RA, Nichols TE, Van Essen DC, Wager TD. 2011. Large-scale automated synthesis of human functional neuroimaging data. Nat Methods 8:665–670.

Part II

II Cell and Molecular Biology of Cells of the Nervous System

IN ALL BIOLOGICAL SYSTEMS, FROM THE MOST primitive to the most advanced, the basic building block is the cell. Cells are often organized into functional modules that are repeated in complex biological systems. The vertebrate brain is the most complex example of a modular system. Complex biological systems have another basic feature: They are architectonic—that is, their anatomy, fine structure, and dynamic properties all reflect a specific physiological function. Thus, the construction of the brain and the cell biology, biophysics, and biochemistry of its component cells reflect its fundamental function, which is to mediate behavior.

The nervous system is made up of glial cells and nerve cells. Earlier views of glia as purely structural elements have been supplanted by our current understanding that there are several types of glial cells, each of which is specialized to regulate one or more particular aspects of neuronal function. Different varieties of glial cells play essential roles in enabling and guiding neural development, insulating axonal processes, controlling the extracellular milieu, supporting synaptic transmission, facilitating learning and memory, and modulating pathological processes within the nervous system. Some glial cells have receptors for neurotransmitters and voltage-gated ion channels that enable them to communicate with one another and with neurons to support neuronal signaling.

In contrast to glial cells, the great diversity of nerve cells—the fundamental units from which the modules of the nervous systems are assembled—are variations on one basic cell plan. Four features of this plan give nerve cells the unique ability to communicate with one another precisely and rapidly over long distances. First, the neuron is polarized, possessing receptive dendrites on one end and communicating axons with presynaptic terminals at the other. This polarization of functional properties restricts the predominant flow of voltage impulses to one direction. Second, the neuron is electrically excitable. Its cell membrane contains specialized proteins—ion channels and receptors—that permit the influx and efflux of specific inorganic ions, thus creating electrical currents that generate the voltage signals across the membrane. Third, the neuron contains proteins and organelles that endow it with specialized secretory properties that allow it to release neurotransmitters at synapses. Fourth, this system for rapid signaling over the long distances between the cell body and its terminals is enabled by a cytoskeletal structure that mediates, on a slower time scale, efficient transport of various proteins, mRNAs, and organelles between the two compartments.

In this part of the book, we shall be concerned with the distinctive cell biological properties that allow neurons and glia to fulfill their various specialized functions. A major emphasis will be on properties of ion channels that endow neurons with the ability to generate and propagate electrical signals in the form of action potentials. We begin the discussion of neurons by considering general properties shared by ion channels—the ability to select and conduct ions, and to gate between open and closed conformations. Neurons use four major classes of channels for signaling: (1) resting channels generate the resting potential and underlie the passive electrical properties of neurons that determine the time course of synaptic potentials, their spread along dendrites, and the threshold for firing an action potential; (2) sensory receptor channels respond to certain sensory stimuli to generate local receptor potentials; (3) ligand-gated channels open in response to neurotransmitters, generating local synaptic potentials; and (4) voltage-gated channels produce the currents that generate self-propagating action potentials. In this part, we focus mainly on resting and voltage-gated channels. In Part III, we consider in more detail ligand-gated channels, and the neurotransmitters and second messengers that control their activity. The channels that are activated by sensory stimuli will be examined in Part IV.

Part Editors: John D. Koester and Steven A. Siegelbaum

Part II

7

The Cells of the Nervous System

THE CELLS OF THE NERVOUS SYSTEM— neurons and glia—share many characteristics with cells in general. However, neurons are specially endowed with the ability to communicate precisely and rapidly with other cells at distant sites in the body. Two features give neurons this ability.

First, they have a high degree of morphological and functional asymmetry: Neurons have receptive dendrites at one end and a transmitting axon at the other. This arrangement is the structural basis for unidirectional neuronal signaling.

Second, neurons are both electrically and chemically excitable. The cell membrane of neurons contains specialized proteins—ion channels and receptors—that facilitate the flow of specific inorganic ions, thereby redistributing charge and creating electrical currents that alter the voltage across the membrane. These changes in charge can produce a wave of depolarization in the form of action potentials along the axon, the usual way a signal travels within the neuron. Glia are less excitable, but their membranes contain transporter proteins that facilitate the uptake of ions as well as proteins that remove neurotransmitter molecules from the extracellular space, thus regulating neuronal function.

There are hundreds of distinct types of neurons depending on their dendritic morphology, pattern of axonal projections, and electrophysiological properties. This structural and functional diversity is largely specified by the genes expressed by each neuronal cell type. Although neurons all inherit the same complement of genes, each expresses a restricted set and thus produces only certain molecules—enzymes, structural proteins, membrane constituents, and secretory products—and not others. In large part, this expression depends on the cell's developmental history. In essence, each cell *is* the set of molecules it expresses.

There are also many kinds of glial cells that can be identified based on their unique morphological, physiological, and biochemical features. The diverse morphologies of glial cells suggest that glia are probably as heterogeneous as neurons. Nonetheless, glia in the vertebrate nervous system can be divided into two major classes: macroglia and microglia. There are three main types of macroglia: oligodendrocytes, Schwann cells, and astrocytes. In the human brain, about 90% of all glial cells are macroglia. Of these, approximately half are myelin-producing cells (oligodendrocytes and Schwann cells) and half are astrocytes. *Oligodendrocytes* provide the insulating myelin sheaths of the axons of some neurons in the central nervous system (CNS) (Figure 7–1A). *Schwann cells* myelinate the axon of neurons in the peripheral nervous system (Figure 7–1B); nonmyelinating Schwann cells have other functions, including promoting development, maintenance, and repair at the neuromuscular synapse. *Astrocytes* owe their name to their irregular, roughly star-shaped cell bodies and large numbers of processes; they support neurons and modulate neuronal signaling in several ways (Figure 7–1C). *Microglia* are the brain's resident immune cells and phagocytes, but also have homeostatic functions in the healthy brain.

Neurons and Glia Share Many Structural and Molecular Characteristics

Neurons and glia develop from common neuroepithelial progenitors of the embryonic nervous system and share many structural characteristics (Figure 7–2). The boundaries of these cells are defined by the cell membrane or *plasmalemma*, which has the asymmetric bilayer structure of all biological membranes and provides a hydrophobic barrier impermeable to most water-soluble substances. Cytoplasm has two main components: cytosol and membranous organelles.

Cytosol is the aqueous phase of cytoplasm. In this phase, only a few proteins are actually free in solution. With the exception of some enzymes that catalyze metabolic reactions, most proteins are organized into functional complexes. A recent subdiscipline called *proteomics* has determined that these complexes can consist of many distinct proteins, none of which are covalently linked to another. For example, the cytoplasmic tail of the N-methyl-D-aspartate (NMDA)-type glutamate receptor, a membrane-associated protein that mediates excitatory synaptic transmission in the

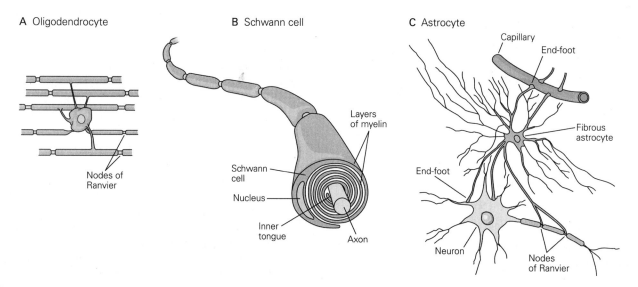

Figure 7–1 The principal types of glial cells are oligodendrocytes and astrocytes in the central nervous system and Schwann cells in the peripheral nervous system.

A. Oligodendrocytes are small cells with relatively few processes. In the white matter of the brain, as shown here, they provide the myelin sheaths that insulate axons. A single oligodendrocyte can wrap its membranous processes around many axons. In the gray matter, perineural oligodendrocytes surround and support the cell bodies of neurons.

B. Schwann cells furnish the myelin sheaths for axons in the peripheral nervous system. During development, several Schwann cells are positioned along the length of a single axon.

Each cell forms a myelin sheath approximately 1 mm long between two nodes of Ranvier. The sheath is formed as the inner tongue of the Schwann cell turns around the axon several times, wrapping the axon in layers of membrane. In actuality, the layers of myelin are more compact than what is shown here. (Adapted from Alberts et al. 2002.)

C. Astrocytes, a major class of glial cells in the central nervous system, are characterized by their star-like shape and the broad end-feet on their processes. Because these end-feet put the astrocyte into contact with both capillaries and neurons, astrocytes are thought to have a nutritive function. Astrocytes also play an important role in maintaining the blood–brain barrier (described later in the chapter).

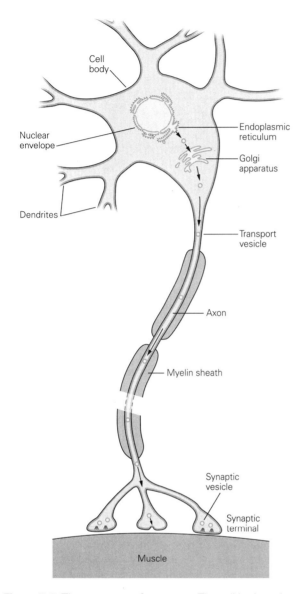

Figure 7–2 The structure of a neuron. The cell body and nucleus of a spinal motor neuron are surrounded by a double-layered membrane, the nuclear envelope, which is continuous with the endoplasmic reticulum. The space between the two membrane layers that constitutes the nuclear envelope is continuous with the lumen of the endoplasmic reticulum. Dendrites emerge from the basal aspect of the neuron, the axon from the apical aspect. (Adapted, with permission, from Williams et al. 1989.)

CNS, is anchored in a large complex of more than 100 scaffold proteins and protein-modifying enzymes. (Many cytosolic proteins involved in second-messenger signaling, discussed in later chapters, are embedded in the cytoskeletal matrix immediately below the plasmalemma.) *Ribosomes*, the organelle on which messenger RNA (mRNA) molecules are translated, are made up of several protein subunits. *Proteasomes*,

large multi-enzyme organelles that degrade ubiquitinated proteins (a process described later in this chapter), are also present throughout the cytosol of neurons and glia.

Membranous organelles, the second main component of cytoplasm, include mitochondria and peroxisomes as well as a complex system of tubules, vesicles, and cisternae called the *vacuolar apparatus*. Mitochondria and peroxisomes process molecular oxygen. Mitochondria generate adenosine triphosphate (ATP), the major molecule by which cellular energy is transferred or spent, whereas peroxisomes prevent accumulation of the strong oxidizing agent hydrogen peroxide. Mitochondria, which are derived from symbiotic archeobacteria that invaded eukaryotic cells early in evolution, are not functionally continuous with the vacuolar apparatus. Mitochondria also play other essential roles in Ca^{2+} homeostasis and lipid biogenesis.

The vacuolar apparatus includes the smooth endoplasmic reticulum, the rough endoplasmic reticulum, the Golgi complex, secretory vesicles, endosomes, lysosomes, and a multiplicity of transport vesicles that interconnect these various compartments (Figure 7–3). Their lumen corresponds topologically to the outside of the cell; consequently, the inner leaflet of their lipid bilayer corresponds to the outer leaflet of the plasmalemma.

The major subcompartments of this system are anatomically discontinuous but functionally connected because membranous and lumenal material is moved from one compartment to another by means of transport vesicles. For example, proteins and phospholipids synthesized in the rough endoplasmic reticulum (the portion of the reticulum studded with ribosomes) and the smooth endoplasmic reticulum are transported to the Golgi complex and then to secretory vesicles, which empty their contents when the vesicle membrane fuses with the plasmalemma (a process called *exocytosis*). This secretory pathway adds membranous components to the plasmalemma and also releases the contents of these secretory vesicles into the extracellular space.

Conversely, components of cell membrane are taken into the cell through endocytic vesicles (*endocytosis*). These are incorporated into early endosomes, sorting compartments that are concentrated at the cell's periphery. The endocytosed membrane, which typically contains specific proteins such as transmembrane receptors, can be either directed back to the plasma membrane by maturing into recycling endosomes or can mature into late endosomes which are targeted for degradation by fusion with lysosomes. (Exocytosis and endocytosis are discussed in detail later in this

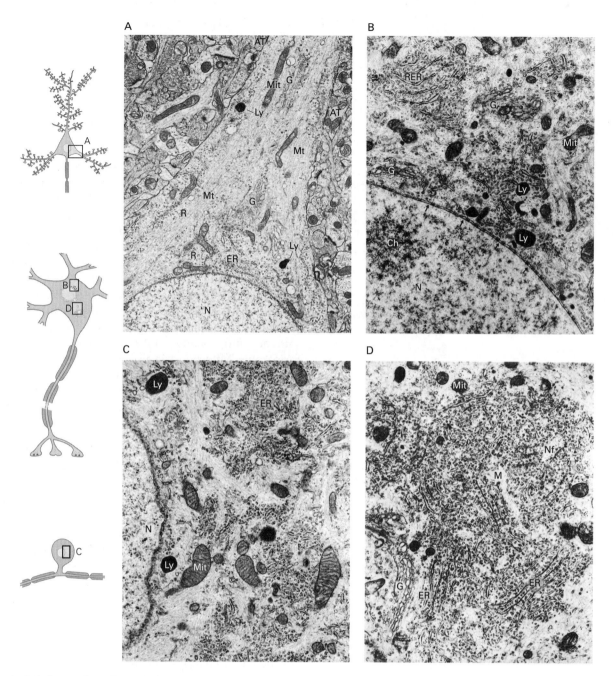

Figure 7–3 Organelles of the neuron. Electron micrographs show cytoplasm in four different regions of the neuron. (Adapted, with permission, from Peters et al. 1991.)

A. A dendrite emerges from a pyramidal neuron's cell body, which includes the endoplasmic reticulum (**ER**) above the nucleus (**N**) and a portion of the Golgi complex (**G**) nearby. Some Golgi cisternae have entered the dendrite, as have mitochondria (**Mit**), lysosomes (**Ly**), and ribosomes (**R**). Microtubules (**Mt**) are prominent cytoskeletal filaments in the cytosol. Axon terminals (**AT**) making contact with the dendrite are seen at the top and right.

B. Some components of a spinal motor neuron that participate in the synthesis of macromolecules. The nucleus (**N**) contains masses of chromatin (**Ch**) and is bounded by the nuclear envelope, which contains many nuclear pores (**arrows**). The mRNA leaves the nucleus through these pores and attaches to ribosomes that either remain free in the cytoplasm or attach to the membranes of the endoplasmic reticulum to form the rough endoplasmic reticulum (**RER**). Regulatory proteins synthesized in the cytoplasm are imported into the nucleus through the

pores. Several parts of the Golgi apparatus (**G**) are seen, as are lysosomes (**Ly**) and mitochondria (**Mit**).

C, D. Micrographs of a dorsal root ganglion cell (**C**) and a motor neuron (**D**) show the organelles in the cell body that are chiefly responsible for synthesis and processing of proteins. The mRNA enters the cytoplasm through the nuclear envelope and is translated into proteins. Free polysomes, strings of ribosomes attached to a single mRNA, generate cytosolic proteins and proteins to be imported into mitochondria (**Mit**) and peroxisomes. Proteins destined for the endoplasmic reticulum are formed after the polysomes attach to the membrane of the endoplasmic reticulum (**ER**). The particular region of the motor neuron shown here also includes membranes of the Golgi apparatus (**G**), in which membrane and secretory proteins are further processed. Some of the newly synthesized proteins leave the Golgi apparatus in vesicles that move down the axon to synapses; other membrane proteins are incorporated into lysosomes (**Ly**) and other membranous organelles. The microtubules (**M**) and neurofilaments (**Nf**) are components of the cytoskeleton.

chapter.) The smooth endoplasmic reticulum also acts as a regulated internal Ca^{2+} store throughout the neuronal cytoplasm (see the discussion of Ca^{2+} release in Chapter 14).

A specialized portion of the rough endoplasmic reticulum forms the *nuclear envelope*, a spherical flattened cisterna that surrounds chromosomal DNA and its associated proteins (histones, transcription factors, polymerases, and isomerases) and defines the nucleus (Figure 7–3). Because the nuclear envelope is continuous with other portions of the endoplasmic reticulum and other membranes of the vacuolar apparatus, it is presumed to have evolved as an invagination of the plasmalemma to ensheathe eukaryotic chromosomes. The nuclear envelope is interrupted by nuclear pores, where fusion of the inner and outer membranes of the envelope results in the formation of hydrophilic channels through which proteins and RNA are exchanged between the cytoplasm proper and the nuclear cytoplasm.

Even though nucleoplasm and cytoplasm are continuous domains of cytosol, only molecules with molecular weights less than 5,000 can pass through the nuclear pores freely by diffusion. Larger molecules need help. Some proteins have special nuclear localization signals, domains that are composed of a sequence of basic amino acids (arginine and lysine) that are recognized by soluble proteins called *nuclear import receptors* (importins). At a nuclear pore, this complex is guided into the nucleus by another group of proteins called *nucleoporins*.

The cytoplasm of the nerve cell body extends into the dendritic tree without functional differentiation. Generally, all organelles in the cytoplasm of the cell body are also present in dendrites, although the densities of the rough endoplasmic reticulum, Golgi complex, and lysosomes rapidly diminish with distance from the cell body. In dendrites, the smooth endoplasmic reticulum is prominent at the base of thin processes called *spines* (Figures 7–4 and 7–5), the receptive portion of excitatory synapses. Concentrations of polyribosomes in dendritic spines mediate local protein synthesis (see below).

In contrast to the continuity of the cell body and dendrites, a sharp functional boundary exists between the cell body at the axon hillock, where the axon emerges. The organelles that compose the main biosynthetic machinery for proteins in the

Figure 7–4 Golgi and endoplasmic reticulum membranes extend from the cell body into dendrites.

A. The Golgi complex (**solid arrow**) appears under the light microscope as several filaments that extend into the dendrites (**open arrow**) but not into the axon. The **arrowheads** at the bottom indicate the axon hillock. For this micrograph, a large neuron of the brain stem was immunostained with antibodies specifically directed against the Golgi complex. (Reproduced, with permission, from De Camilli et al. 1986. Copyright © 1986 Rockefeller University Press.)

B. Smooth endoplasmic reticulum (**arrowhead**) extends into the neck of a dendritic spine, while another membrane compartment sits at the origin of the spine (**arrow**). (Reproduced, with permission, from Cooney et al. 2002. Copyright © 2002 Society for Neuroscience.)

A
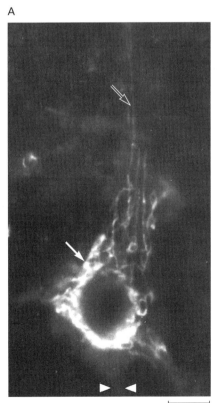
10 μm

B
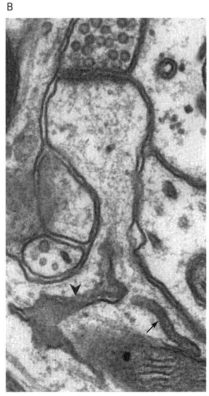
0.5 μm

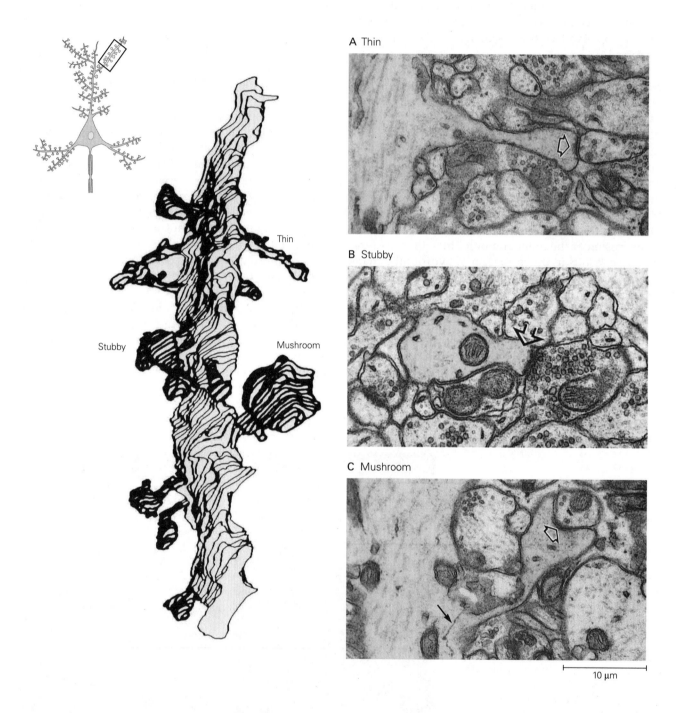

A Thin

B Stubby

C Mushroom

10 μm

Figure 7–5 Types of dendritic spines. Three types of dendritic spine shapes are shown in a mature dendrite in a pyramidal cell in the CA1 region of the hippocampus. The drawing at left is based on a series of electron micrographs. (Drawing reproduced, with permission, from Harris and Stevens 1989; A, B, and C are reproduced, with permission, from Sorra and Harris 1993. Copyright © 1993 Society for Neuroscience.)

A. In this thin dendritic spine, the thickened receptive surface (**arrow**), located across from the presynaptic axon, contains synaptic receptors. The tissue shown here and in **B** and **C** is from the hippocampus of a postnatal day-15 rat brain.

B. Stubby spines containing postsynaptic densities (**arrow**) are both small and rare in the mature hippocampus. Their larger counterparts (not shown) predominate in the immature brain.

C. Mushroom-shaped spines have a larger head. The immature spine shown here contains flat cisternae of smooth endoplasmic reticulum, some with a beaded appearance (**solid arrow**). The postsynaptic density is indicated by the **open arrow**.

neuron—ribosomes, rough endoplasmic reticulum, and the Golgi complex—are generally excluded from axons (Figure 7–4), as are lysosomes and certain proteins. However, axons are rich in smooth endoplasmic reticulum, individual synaptic vesicles, and their precursor membranes.

The Cytoskeleton Determines Cell Shape

The cytoskeleton determines the shape of a cell and is responsible for the asymmetric distribution of organelles within the cytoplasm. It includes three filamentous structures: microtubules, neurofilaments, and microfilaments. These filaments and associated proteins account for approximately a quarter of the total protein in the cell.

Microtubules form long scaffolds that extend from one end of a neuron to the other and play a key role in developing and maintaining cell shape. A single microtubule can be as long as 0.1 mm. Microtubules consist of protofilaments, each of which consists of multiple pairs of *α*- and *β-tubulin* subunits arranged longitudinally along the microtubule (Figure 7–6A). Tubulin subunits bind to neighboring subunits along the protofilament and also laterally between adjacent protofilaments. Microtubules are polarized with a plus end (or growing end) and a minus end (where microtubules can be depolymerized). Interestingly, microtubule orientations differ between axons and dendrites. In the axon, microtubules display a single orientation with the plus end directed away from the cell body. In proximal dendrites, microtubules can be oriented both ways, with a plus end oriented toward or away from the cell body.

Microtubules grow by addition of guanosine triphosphate (GTP)-bound tubulin dimers at their plus end. Shortly after polymerization, the GTP is hydrolyzed to guanosine diphosphate (GDP). When a microtubule stops growing, its positive end is capped by a GDP-bound tubulin monomer. The low affinity of the GDP-bound tubulin for the polymer would lead to catastrophic depolymerization, were not for the fact that the microtubules are stabilized by interaction with other proteins.

In fact, while microtubules undergo rapid cycles of polymerization and depolymerization in dividing cells, a phenomenon referred to as *dynamic instability*, in mature dendrites and axons, they are more stable. This stability is thought to be caused by *microtubule-associated proteins* (MAPs) that promote the oriented polymerization and assembly of the tubulin polymers. MAPs in axons differ from those in dendrites.

For example, MAP2 is present in dendrites but not in axons, where tau proteins (see Box 7–1) and MAP1b are present. Furthermore, microtubule stability is also tightly regulated by many different types of reversible tubulin posttranslational modifications such as acetylation, detyrosination, and polyglutamylation. In Alzheimer disease and some other degenerative disorders, tau proteins are modified and abnormally polymerized, forming a characteristic lesion called the *neurofibrillary tangle* (Box 7–1).

Tubulins are encoded by a multigene family. At least six genes code the α- and β-subunits. Because of the expression of the different genes or posttranscriptional modifications, more than 20 isoforms of tubulin are present in the brain.

Neurofilaments, 10 nm in diameter, are the bones of the cytoskeleton (Figure 7–6B). Neurofilaments are related to intermediate filaments of other cell types, including the cytokeratins of epithelial cells (hair and nails), glial fibrillary acidic protein in astrocytes, and desmin in muscle. Unlike microtubules, neurofilaments are stable and almost totally polymerized in the cell.

At 3 to 7 nm in diameter, *microfilaments* are the thinnest of the three main types of fibers that make up the cytoskeleton (Figure 7–6C). Like thin filaments of muscle, microfilaments are made up of two strands of polymerized globular actin monomers, each bearing an ATP or adenosine diphosphate (ADP), wound into a double-stranded helix. Actin is a major constituent of all cells, perhaps the most abundant animal protein in nature. There are several closely related molecular forms: the α actin of skeletal muscle and at least two other molecular forms, β and γ. Each is encoded by a different gene. Neural actin in higher vertebrates is a mixture of the β and γ species, which differ from muscle actin by a few amino acid residues. Most actin molecules are highly conserved, not only in different cell types of a species but also in organisms as distantly related as humans and protozoa.

Unlike microtubules and neurofilaments, actin filaments are short. They are concentrated at the cell's periphery in the cortical cytoplasm just beneath the plasmalemma, where they form a dense network with many actin-binding proteins (eg, spectrin-fodrin, ankyrin, talin, and actinin). This matrix plays a key role in the dynamic function of the cell's periphery, such as the motility of growth cones (the growing tips of axons) during development, generation of specialized microdomains at the cell surface, and the formation of pre- and postsynaptic morphological specializations.

Like microtubules, microfilaments undergo cycles of polymerization and depolymerization. At any one

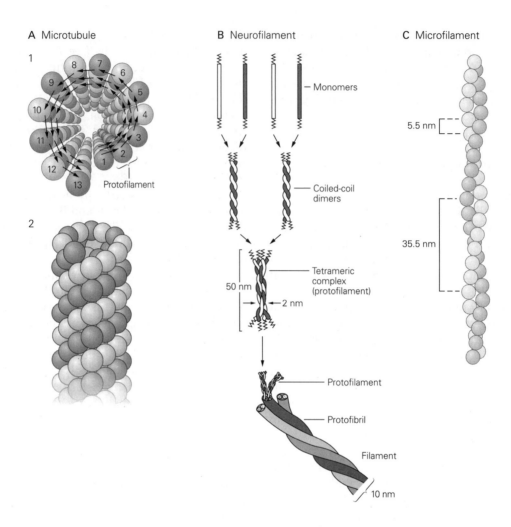

Figure 7–6 Atlas of fibrillary structures.

A. Microtubules, the largest-diameter fibers (25 nm), are helical cylinders composed of 13 protofilaments, each 5 nm in width. Each protofilament is made up of a column of alternating α- and β-tubulin subunits; each subunit has a molecular weight of approximately 50,000 Da. Adjacent subunits bind to each other along the longitudinal protofilaments and laterally between subunits of adjacent protofilaments.

A tubulin molecule is a heterodimer consisting of one α- and one β-tubulin subunit. **1.** View of a microtubule. The **arrows** indicate the direction of the right-handed helix. **2.** A side view of a microtubule shows the alternating α- and β-subunits.

B. Neurofilaments are built with fibers that twist around each other to produce coils of increasing thickness. The thinnest units are monomers that form coiled-coil heterodimers. These dimers form a tetrameric complex that becomes the protofilament. Two protofilaments become a protofibril, and three protofibrils are helically twisted to form the 10-nm diameter neurofilament. (Adapted, with permission, from Bershadsky and Vasiliev 1988.)

C. Microfilaments, the smallest-diameter fibers (approximately 7 nm), are composed of two strands of polymerized globular actin (G-actin) monomers arranged in a helix. At least six different (but closely related) actins are found in mammals; each variant is encoded by a separate gene. Microfilaments are polar structures because the globular monomers are asymmetric.

time, approximately half of the total actin in a cell can exist as unpolymerized monomers. The state of actin is controlled by binding proteins, which facilitate assembly and limit polymer length by capping the rapidly growing end of the filament or by severing it. Other binding proteins crosslink or bundle actin filaments.

The dynamic state of microtubules and microfilaments permits a mature neuron to retract old axons and dendrites and extend new ones. This structural plasticity is thought to be a major factor in changes of synaptic connections and efficacy and, therefore, cellular mechanisms of long-term memory and learning.

Box 7–1 Abnormal Accumulations of Proteins Are Hallmarks of Many Neurological Disorders

Tau is a microtubule-binding protein normally present in nerve cells. In Alzheimer disease, abnormal aggregates of tau are visible in the light microscope in neurons and glia as well as in the extracellular space. Highly phosphorylated tau molecules arranged in long, thin polymers wind around one another to form paired helical filaments (Figure 7–7A and Chapter 64). Bundles of the polymers, known as *neurofibrillary tangles*, accumulate in neuronal cell bodies, dendrites, and axons (Figure 7–7A).

In normal neurons, tau is either bound to microtubules or free in the cytosol. In the tangles, it is not bound to microtubules but is highly insoluble. The tangles form at least in part because tau is not proteolytically degraded. The accumulations disturb the polymerization of tubulin and therefore interfere with axonal transport. Consequently, the shape of the neuron is not maintained.

Tau accumulations are also found in neurons of patients with progressive supranuclear palsy, a movement disorder, and in patients with frontotemporal dementias, a group of neurodegenerative disorders that affect the frontal and temporal lobes (Chapter 63). The familial forms of frontotemporal dementias are caused by mutations in the *tau* gene. Abnormal aggregates are also found in glial cells, both astrocytes and oligodendrocytes, in progressive supranuclear palsy, cortico-basoganglionic degeneration, and frontotemporal dementias.

The peptide *β-amyloid* also accumulates in the extracellular space in Alzheimer disease (Figure 7–7B and Chapter 64). It is a small proteolytic product of a much larger integral membrane protein, amyloid precursor protein, which is normally processed by several proteolytic enzymes associated with intracellular membranes.

The proteolytic pathway that generates β-amyloid requires the enzyme β-secretase.

For unknown reasons, in Alzheimer disease, abnormal amounts of the amyloid precursor are processed by β-secretase. Some patients with early-onset familial Alzheimer disease either have mutations in the amyloid precursor gene or in the genes coding for the membrane proteins presenilin 1 and 2, which are closely associated with secretase activity.

In Parkinson disease, abnormal aggregates of α-synuclein accumulate in cell bodies of neurons. Like tau, *α-synuclein* is a normal soluble constituent of the cell. But in Parkinson disease, it becomes insoluble, forming spherical inclusions called *Lewy bodies* (Figure 7–7C and Chapter 63).

These abnormal inclusions also contain ubiquitin. Because ubiquitin is required for proteasomal degradation of proteins, its presence suggests that affected neurons have attempted to target α-synuclein or other molecular constituents for proteolysis. Apparently, degradation does not occur, possibly because of misfolding or the abnormal aggregation of the proteins or because of faulty proteolytic processing in the cell.

Do these abnormal protein accumulations affect the physiology of the neurons and glia? On the one hand, the accumulations may form in response to altered posttranslational processing of the proteins and serve to isolate the abnormal proteins, permitting normal cell activities. On the other hand, the accumulations may disrupt cellular activities such as membrane trafficking, axonal and dendritic transport, and the maintenance of synaptic connections between specific classes of neurons. In addition, the altered proteins themselves, aside from the aggregations, may have deleterious effects. With β-amyloid, there is evidence that the peptide itself is toxic.

(continued)

In addition to serving as the cytoskeleton, microtubules and actin filaments act as tracks along which organelles and proteins are rapidly driven by molecular motors. The motors used by the actin filaments, the *myosins*, also mediate other types of cell motility, including extension of the cell's processes and translocation of membranous organelles from the bulk cytoplasm to the region adjacent to the plasma membrane. (Actomyosin is responsible for muscle contraction.) Because the microtubules and actin filaments are polar, each motor drives its organelle cargo in only one direction.

As already mentioned, microtubules are arranged in parallel in the axon with plus ends pointing away from the cell body and minus ends facing the cell body. This regular orientation allows some organelles to move toward and others to move away from nerve endings, the direction being determined by the specific type of molecular motor, thus maintaining the distinctive distribution of axonal organelles (Figure 7–8). In dendrites, however, microtubules with opposite polarities are mixed together, explaining why the organelles of the cell body and dendrites are similar.

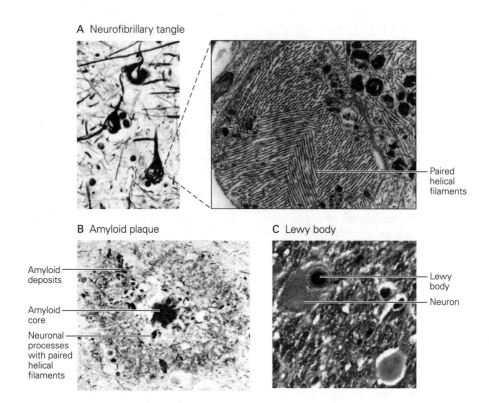
Protein Particles and Organelles Are Actively Transported Along the Axon and Dendrites

In neurons, most proteins are made in the cell body from mRNAs. Important examples are neurotransmitter biosynthetic enzymes, synaptic vesicle membrane components, and neurosecretory peptides. Because axons and terminals often lie at great distances from the cell body, sustaining the function of these remote regions presents a challenge. Passive diffusion would be far too slow to deliver vesicles, particles, or even single macromolecules over this great distance.

The axon terminal, the site of secretion of neurotransmitters, is particularly distant from the cell body. In a motor neuron that innervates a muscle of the leg in humans, the distance of the nerve terminal from the cell body can exceed 10,000 times the cell body diameter. Thus, membrane and secretory products formed in the cell body must be actively transported to the end of the axon (Figure 7–9).

In 1948, Paul Weiss first demonstrated axonal transport when he tied off a sciatic nerve and observed that axoplasm in the nerve accumulated with time on the proximal side of the ligature. He concluded that axoplasm

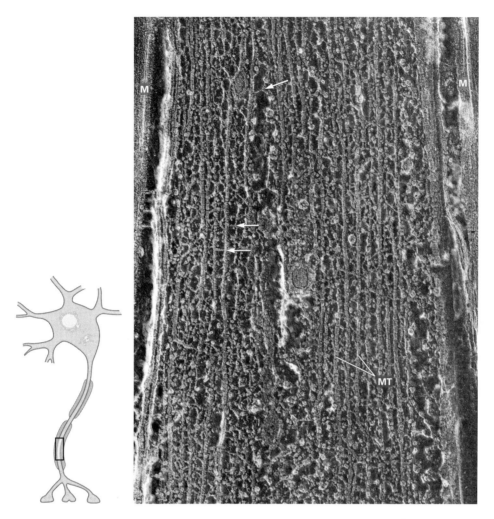

Figure 7–8 The cytoskeletal structure of an axon. The micrograph shows the dense packing of microtubules and neurofilaments linked by cross bridges (**arrows**). Organelles are transported in both the anterograde and retrograde directions in the microtubule-rich domains. Visualization in the micrograph was achieved by quick freezing and deep etching. **M**, myelin sheath; **MT**, microtubules. ×105,000. (Adapted, with permission, from Schnapp and Reese 1982. Copyright © 1982 Rockefeller University Press.)

moves at a slow, constant rate from the cell body toward terminals in a process he called *axoplasmic flow*. Today we know that the flow Weiss observed consists of two discrete mechanisms, one fast and the other slow.

Membranous organelles move toward axon terminals (anterograde direction) and back toward the cell body (retrograde direction) by *fast axonal transport*, a form of transport that is up to 400 mm per day in warm-blooded animals. In contrast, cytosolic and cytoskeletal proteins move only in the anterograde direction by a much slower form of transport, *slow axonal transport*. These transport mechanisms in neurons are adaptations of processes that facilitate intracellular movement of organelles in all secretory cells. Because all these mechanisms operate along axons, they have been used by neuroanatomists to trace the course of individual axons as well as interconnections between neurons (Box 7–2).

Fast Axonal Transport Carries Membranous Organelles

Large membranous organelles are carried to and from the axon terminals by fast transport (Figure 7–11). These organelles include synaptic vesicle precursors, large dense-core vesicles, mitochondria, elements of the smooth endoplasmic reticulum, and protein particles carrying RNAs. Direct microscopic analysis reveals that fast transport occurs in a stop-and-start (saltatory) fashion along linear tracks of microtubules aligned with the main axis of the axon. The saltatory nature of the movement results from the periodic dissociation

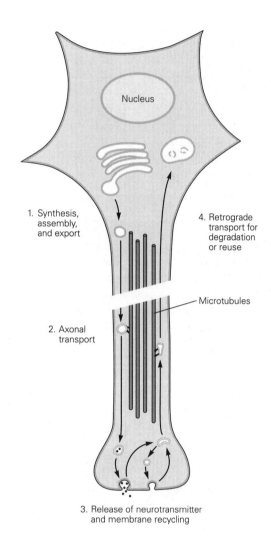

Figure 7–9 Membrane trafficking in the neuron. 1. Proteins and lipids of secretory organelles are synthesized in the endoplasmic reticulum and transported to the Golgi complex, where large dense-core vesicles (peptide-containing secretory granules) and synaptic vesicle precursors are assembled. 2. Large dense-core vesicles and transport vesicles carry synaptic vesicle proteins down the axon via axonal transport. 3. At the nerve terminals, the synaptic vesicles are assembled and loaded with nonpeptide neurotransmitters. Synaptic vesicles and large dense-core vesicles release their contents by exocytosis. 4. Following exocytosis, large dense-core vesicle membranes are returned to the cell body for reuse or degradation. Synaptic vesicle membranes undergo many cycles of local exocytosis and endocytosis in the presynaptic terminal.

of an organelle from the track or from collisions with other particles.

Early experiments on dorsal root ganglion cells showed that anterograde fast transport depends critically on ATP, is not affected by inhibitors of protein synthesis (once the injected labeled amino acid is incorporated), and does not depend on the cell body,

because it occurs in axons severed from their cell bodies. In fact, active transport can occur in reconstituted cell-free axoplasm.

Microtubules provide an essentially stationary track on which specific organelles can be moved by molecular motors. The idea that microtubules are involved in fast transport emerged from the finding that certain alkaloids that disrupt microtubules and block mitosis, which depends on microtubules, also interfere with fast transport.

Molecular motors were first visualized in electron micrographs as cross bridges between microtubules and moving particles (Figure 7–8). More advanced fluorescence time-lapse microscopy techniques are able to visualize the dynamics of axon transport for specific cargos such as mitochondria and synaptic vesicles. The motor molecules for anterograde transport are plus-end-directed motors called *kinesin* and a variety of kinesin-related proteins. The kinesins represent a large family of adenosine triphosphatases (ATPase), each of which transports different cargos. Kinesin is a heterotetramer composed of two heavy chains and two light chains. Each heavy chain has three domains: (1) a globular head (the ATPase domain) that acts as the motor when attached to microtubules, (2) a coiled-coil helical stalk responsible for dimerization with the other heavy chain, and (3) a fan-like carboxyl-terminus that interacts with the light chains. This end of the complex attaches indirectly to the organelle that is moved through specific families of proteins referred to as cargo adapters.

Fast retrograde transport primarily moves endosomes generated by endocytic activity at nerve endings, mitochondria, and elements of the endoplasmic reticulum. Many of these components are degraded through fusion with lysosomes. Fast retrograde transport also delivers signals that regulate gene expression in the neuron's nucleus. For example, activated growth factor receptors at nerve endings are taken up into vesicles and carried back along the axon to the nucleus. Transport of transcription factors informs the gene transcription apparatus in the nucleus of conditions in the periphery. Retrograde transport of these molecules is especially important during nerve regeneration and axon regrowth (Chapter 47). Certain toxins (tetanus toxin) as well as pathogens (herpes simplex, rabies, and polio viruses) are also transported toward the cell body along the axon.

The rate of retrograde fast transport is approximately one-half to two-thirds that of anterograde fast transport. As in anterograde transport, particles move along microtubules during retrograde flow. The motor molecules for retrograde axonal transport are minus-end-directed motors called *dyneins*, similar to those found in cilia and

Box 7–2 Neuroanatomical Tracing Makes Use of Axonal Transport

Neuroanatomists typically locate axons and terminals of specific nerve cell bodies by microinjection of dyes; expression of fluorescent proteins; or autoradiographically tracing specific proteins soon after administering radioactively labeled amino acids, certain labeled sugars (fucose or amino sugars, precursors of glycoprotein), or specific transmitter substances.

Similarly, particles, proteins, or dyes that are readily taken up at nerve terminals by endocytosis and transported back to cell bodies are used to identify the cell bodies. The enzyme horseradish peroxidase has been most widely used for this type of study because it readily undergoes retrograde transport and its reaction product is conveniently visualized histochemically.

Axonal transport is also used by neuroanatomists to label material exchanged between neurons, making it possible to identify neuronal networks (Figure 7–10).

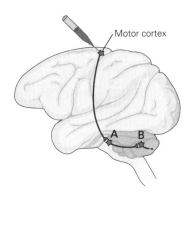

Motor cortex

A Pontine nuclei

B Cerebellar cortex

1 mm

30 µm

Figure 7–10 Axonal transport of the herpes simplex virus (HSV) is used to trace cortical pathways in monkeys. Depending on the strain, the virus moves in the anterograde or retrograde direction by axonal transport. In either direction, it enters a neuron with which the infected cell makes synaptic contact. Here the projections of cells in the primary motor cortex to the cerebellum in monkeys were traced using an anterograde-moving strain (HSV-1 [H129]). Monkeys were injected in the region of the primary motor cortex that controls the arm. After 4 days, the brain was sectioned and immunostained for viral antigen. Micrographs show the virus was transported from the primary motor cortex to second-order neurons in pontine nuclei (**A**) and then to third-order neurons in the cerebellar cortex (**B**). (Reproduced, with permission, from P.L. Strick.)

flagella of nonneuronal cells. They consist of a multimeric ATPase protein complex with two globular heads on two stalks connected to a basal structure. The globular heads attach to microtubules and act as motors, moving toward the minus end of the polymer. As with kinesin, the other end of the complex attaches to the transported organelle through specialized cargo adapters.

Microtubules also mediate anterograde and retrograde transport of mRNAs and ribosomal RNA carried in particles formed with RNA-binding proteins. These proteins have been characterized in both vertebrate and invertebrate nervous systems and include the cytoplasmic polyadenylation element binding protein (CPEB), the fragile X protein, Hu proteins, NOVA, and Staufen. The activities of these proteins are critical. For example, CPEB keeps select mRNAs dormant during transport from the cell body to nerve endings; once there (upon stimulation), the binding protein can facilitate the local translation of the RNA by mediating polyadenylation and activation of the messenger. Both CPEB and Staufen were discovered in *Drosophila*, where they maintain maternal mRNAs dormant in

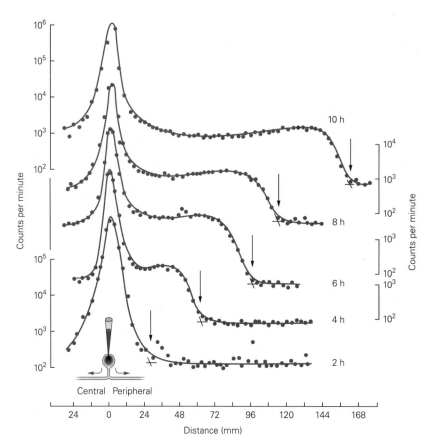

Figure 7–11 Early experiments on anterograde axonal transport used radioactive labeling of proteins. In the experiment illustrated here, the distribution of radioactive proteins along the sciatic nerve of the cat was measured at various times after injection of [³H]-leucine into dorsal root ganglia in the lumbar region of the spinal cord. To show transport curves for various times (2, 4, 6, 8, and 10 hours after the injection) in one figure, several ordinate scales (in logarithmic units) are used. Large amounts of labeled protein stay in the ganglion cell bodies but, with time, move out along axons in the sciatic nerve, so the advancing front of the labeled protein is progressively farther from the cell body (**arrows**). The velocity of transport can be calculated from the distance of the front at the various times. From experiments of this kind, Sidney Ochs found that the rate of fast axonal transport is constant at 410 mm per day at body temperature. (Adapted, with permission, from Ochs 1972. Copyright © 1972 AAAS.)

unfertilized eggs and, upon fertilization, distribute and localize mRNA to various regions of the dividing embryo. Loss-of-function mutations in the fragile X (*FMR1*) gene lead to a severe form of mental retardation.

Proteins, ribosomes, and mRNA are concentrated at the base of some dendritic spines (Figure 7–12). Only a select group of mRNAs are transported into the dendrites from the soma. These include mRNAs that encode actin- and cytoskeletal-associated proteins, MAP2, and the α-subunit of the Ca²⁺/calmodulin-dependent protein kinase. They are translated in the dendrites in response to activity in a presynaptic neuron. This local protein synthesis is thought to be important in sustaining the molecular changes at the synapse that underlie long-term memory and learning.

Likewise, the mRNA for myelin basic protein is transported to the distant ends of the oligodendrocytes, where it is translated as the myelin sheath grows, as discussed later in this chapter.

Slow Axonal Transport Carries Cytosolic Proteins and Elements of the Cytoskeleton

Cytosolic proteins and cytoskeletal proteins are moved from the cell body by slow axonal transport. Slow transport occurs only in the anterograde direction and consists of at least two kinetic components that carry different proteins at different rates.

The slower component travels at 0.2 to 2.5 mm per day and carries the proteins that make up the fibrillar elements of the cytoskeleton: the subunits of

A

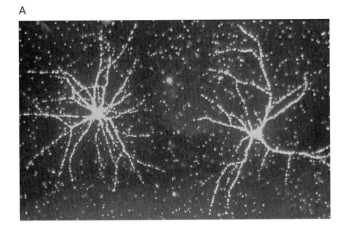

B

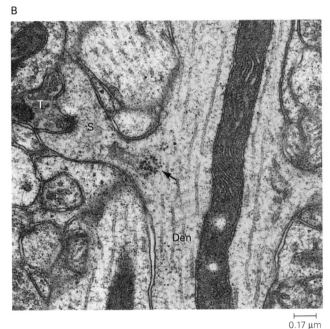

0.17 μm

Figure 7–12 Ribosomes in the dendritic arbor. (Images reproduced, with permission, from Oswald Steward.)

A. Some ribosomes are dispatched from the cell body to dendrites where they are used in local protein synthesis. This autoradiograph shows the distribution of ribosomal RNA (rRNA) in hippocampal neurons in low-density cultures using in situ hybridization. The image was made with dark field illumination, in which silver grains reflect light and thus appear as bright spots. Silver grains, denoting the rRNA, are heavily concentrated over cell bodies and dendrites but are not detectable over the axons that crisscross among the dendrites.

B. Ribosomes in dendrites are selectively concentrated at the junction of the spine and main dendritic shaft (**arrow**), where the spine contacts the axon terminal of a presynaptic neuron. This electron micrograph shows a mushroom-shaped spine of a neuron in the hippocampal dentate gyrus. Note the absence of ribosomes in the dendritic shaft. **S**, spine head; **T**, presynaptic terminal; **Den**, main shaft of the dendrite containing a long mitochondrion. ×60,000.

neurofilaments and α- and β-tubulin subunits of microtubules. These fibrous proteins constitute approximately 75% of the total protein moved in the slower component. Microtubules are transported in polymerized form by a mechanism involving microtubule sliding in which relatively short preassembled microtubules move along existing microtubules. Neurofilament monomers or short polymers move passively together with the microtubules because they are crosslinked by protein bridges.

The other component of slow axonal transport is approximately twice as fast as the slower component. It carries clathrin, actin, and actin-binding proteins as well as a variety of enzymes and other proteins.

Proteins Are Made in Neurons as in Other Secretory Cells

Secretory and Membrane Proteins Are Synthesized and Modified in the Endoplasmic Reticulum

The mRNAs for secretory and membrane proteins are translated through the membrane of the rough endoplasmic reticulum, and their polypeptide products are processed extensively within the lumen of the endoplasmic reticulum. Most polypeptides destined to become proteins are translocated across the membrane of the rough endoplasmic reticulum during synthesis, a process called *cotranslational transfer.*

Transfer is possible because ribosomes, the site where proteins are synthesized, attach to the cytosolic surface of the reticulum (Figure 7–13). Complete transfer of the polypeptide chain into the lumen of the reticulum produces a secretory protein (recall that the inside of the reticulum is related to the outside of the cell). Important examples are the neuroactive peptides. If the transfer is incomplete, an integral membrane protein results. Because a polypeptide chain can thread its way through the membrane many times during synthesis, several membrane-spanning configurations are possible depending on the primary amino acid sequence of the protein. Important examples are neurotransmitter receptors and ion channels (Chapter 8).

Some proteins transported into the endoplasmic reticulum remain there. Others are moved to other compartments of the vacuolar apparatus or to the plasmalemma, or are secreted into the extracellular space. Proteins that are processed in the endoplasmic reticulum are extensively modified. One important modification is the formation of intramolecular disulfide linkages (Cys-S-S-Cys) caused by oxidation of pairs of free sulfhydryl side chains, a process that cannot occur

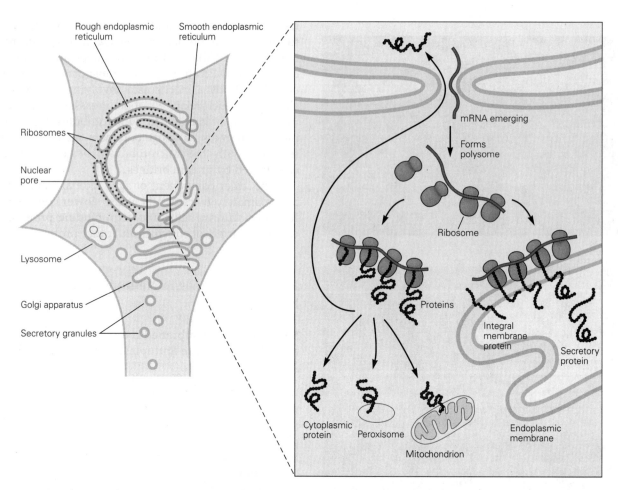

Figure 7–13 Protein synthesis in the endoplasmic reticulum. Free and membrane-bound polysomes translate mRNA that encodes proteins with a variety of destinations. Messenger RNA, transcribed from genomic DNA in the neuron's nucleus, emerges into the cytoplasm through nuclear pores to form polyribosomes (see enlargement). The polypeptides that become secretory and membrane proteins are translocated across the membrane of the rough endoplasmic reticulum.

in the reducing environment of the cytosol. Disulfide linkages are crucial to the tertiary structure of these proteins.

Proteins may be modified by cytosolic enzymes either during synthesis (cotranslational modification) or afterward (posttranslational modification). One example is *N*-acylation, the transfer of an acyl group to the N-terminus of the growing polypeptide chain. Acylation by a 14-carbon fatty acid myristoyl group permits the protein to anchor in membranes through the lipid chain.

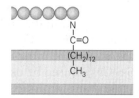

Other fatty acids can be conjugated to the sulfhydryl group of cysteine, producing a thioacylation:

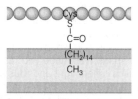

Isoprenylation is another posttranslational modification important for anchoring proteins to the cytosolic side of membranes. It occurs shortly after synthesis of the protein is completed and involves a series of enzymatic steps that result in thioacylation by one of two long-chain hydrophobic polyisoprenyl moieties (farnesyl, with 15 carbons, or geranyl-geranyl, with 20) of the sulfhydryl group of a cysteine at the C-terminus of proteins.

Some posttranslational modifications are readily reversible and thus used to regulate the function of a protein transiently. The most common of these modifications is the phosphorylation at the hydroxyl group in serine, threonine, or tyrosine residues by protein kinases. Dephosphorylation is catalyzed by protein phosphatases. (These reactions are discussed in Chapter 14.) As with all posttranslational modifications, the sites to be phosphorylated are determined by particular sequences of amino acids around the residue to be modified. Phosphorylation can alter physiological processes in a reversible fashion. For example, protein phosphorylation–dephosphorylation reactions regulate the kinetics of ion channels, the activity of transcription factors, and the assembly of the cytoskeleton.

Still another important posttranslational modification is the addition of *ubiquitin*, a highly conserved protein with 76 amino acids, to the ϵ-amino group of specific lysine residues in the protein molecule. Ubiquitination, which regulates protein degradation, is mediated by three enzymes. E1 is an activating enzyme that uses the energy of ATP. The activated ubiquitin is next transferred to a conjugase, E2, which then transfers the activated moiety to a ligase, E3. The E3, alone or together with E2, transfers the ubiquitinyl group to the lysine residue in a protein. Specificity arises because a given protein molecule can only be ubiquinated by a specific E3 or combination of E3 and E2. Some E3s also require special cofactors—ubiquitination occurs only in the presence of E3 and a cofactor protein.

Monoubiquitination tags a protein for degradation in the endosomal–lysosomal system. This is especially important in endocytosis and recycling of surface receptors. Ubiquitinyl monomers are successively linked to the ϵ-amino group of a lysine residue in the previously added ubiquitin moiety. Addition of more than five ubiquitins to the multiubiquitin chain tags the protein for degradation by the proteasome, a large complex containing multifunctional protease subunits that cleave proteins into short peptides.

The ATP–ubiquitin–proteasome pathway is a mechanism for the selective and regulated proteolysis of proteins that operates in the cytosol of all regions of the neuron—dendrites, cell body, axon, and terminals. Until recently, this process was thought to be primarily directed to poorly folded, denatured, or aged and damaged proteins. We now know that ubiquitin-mediated proteolysis can be regulated by neuronal activity and plays specific roles in many neuronal processes, including synaptogenesis and long-term memory storage.

Another important protein modification is glycosylation, which occurs on amino groups of asparagine residues (N-linked glycosylation) and results in the addition en bloc of complex polysaccharide chains. These chains are then trimmed within the endoplasmic reticulum by a series of reactions controlled by chaperones, including heat shock proteins, calnexin, and calreticulin. Because of the great chemical specificities of oligosaccharide moieties, these modifications can have important implications for cell function. For example, cell-to-cell interactions that occur during development rely on molecular recognition between glycoproteins on the surfaces of the two interacting cells. Also, because a given protein can have somewhat different oligosaccharide chains, glycosylation can diversify the function of a protein. It can increase the hydrophilicity of the protein (useful for secretory proteins), fine-tune its ability to bind macromolecular partners, and delay its degradation.

An interesting posttranslational modification of mRNA is RNA interference (RNAi), the targeted destruction of double-stranded RNAs. This mechanism, which is believed to have arisen to protect cells against viruses and other rogue fragments of nucleic acids, shuts down the synthesis of any targeted protein. Double-stranded RNAs are taken up by an enzyme complex that cleaves the molecule into oligomers. The RNA sequences are retained by the complex. As a result, any homologous hybridizing RNA strands, either double- or single-stranded, will be destroyed. The process is regenerative: The complex retains a hybridizing fragment and goes on to destroy another RNA molecule until none remain in the cell. Although the physiological role of RNAi in normal cells is unclear, transfection or injection of RNAi into cells is of great research and clinical importance (Chapter 2).

Secretory Proteins Are Modified in the Golgi Complex

Proteins from the endoplasmic reticulum are carried in transport vesicles to the Golgi complex where they are modified and then moved to synaptic terminals and other parts of the plasma membrane. The Golgi complex appears as a grouping of membranous sacks aligned with one another in long ribbons.

The mechanism by which vesicles are transported between stations of the secretory and endocytic pathways has been remarkably conserved from simple unicellular prokaryotes (yeast) to neurons and glia of multicellular organisms. Transport vesicles develop from membrane, beginning with the assembly of proteins that form *coats*, or coat proteins, at selected patches of the cytosolic surface of the membrane. A coat has two functions. It forms rigid cage-like structures that produce evagination of the membrane into a bud shape, and it selects the protein cargo to be incorporated into the vesicles.

There are several types of coats. *Clathrin coats* assist in evaginating Golgi complex membrane and plasmalemma during endocytosis. Two other coats, COPI and COPII, cover transport vesicles that shuttle between the endoplasmic reticulum and the Golgi complex. Coats usually are rapidly dissolved once free vesicles have formed. The fusion of vesicles with the target membrane is mediated by a cascade of molecular interactions, the most important of which is the reciprocal recognition of small proteins on the cytosolic surfaces of the two interacting membranes: vesicular soluble *N*-ethylmaleimide–sensitive factor attachment protein receptors (v-SNAREs) and t-SNAREs (target-membrane SNAREs). The role of SNARE proteins in neurotransmitter release through synaptic vesicle fusion with the plasma membrane is discussed in Chapter 15.

Vesicles from the endoplasmic reticulum arrive at the *cis* side of the Golgi complex (the aspect facing the nucleus) and fuse with its membranes to deliver their contents into the Golgi complex. These proteins travel from one Golgi compartment (cisterna) to the next, from the *cis* to the *trans* side, undergoing a series of enzymatic reactions. Each Golgi cisterna or set of cisternae is specialized for a particular type of reaction. Several types of protein modifications, some of which begin in the endoplasmic reticulum, occur within the Golgi complex proper or within the transport station adjacent to its *trans* side, the *trans-Golgi network* (the aspect of the complex typically facing away from the nucleus toward the axon hillock). These modifications include the addition of N-linked oligosaccharides, O-linked (on the hydroxyl groups of serine and threonine) glycosylation, phosphorylation, and sulfation.

Both soluble and membrane-bound proteins that travel through the Golgi complex emerge from the trans-Golgi network in a variety of vesicles that have different molecular compositions and destinations. Proteins transported from the trans-Golgi network include secretory products as well as newly synthesized components for the plasmalemma, endosomes, and other membranous organelles (see Figure 7–2). One class of vesicles carries newly synthesized plasmalemmal proteins and proteins that are continuously secreted (*constitutive secretion*). These vesicles fuse with the plasma membrane in an unregulated fashion. An important type of these vesicles delivers lysosomal enzymes to late endosomes.

Still other classes of vesicles carry secretory proteins that are released by an extracellular stimulus (*regulated secretion*). One type stores secretory products, primarily neuroactive peptides, in high concentrations. Called *large dense-core vesicles* because of their electron-dense (osmophilic) appearance in the electron microscope, these vesicles are similar in function and biogenesis to peptide-containing granules of endocrine cells. Large dense-core vesicles are targeted primarily to axons but can be seen in all regions of the neuron. They accumulate in the cytoplasm just beneath the plasma membrane and are highly concentrated at axon terminals, where their contents are released through Ca^{2+}-regulated exocytosis.

Recent work has demonstrated that *small synaptic vesicles*—the electron-lucent vesicles responsible for the rapid release of neurotransmitter at axon terminals—are actively transported toward the synaptic terminals as individual cargoes. It is thought that protein components of small synaptic vesicles originate from large precursor vesicles from the trans-Golgi network. These synaptic vesicles already incorporate most of the proteins that enable their fusion at the presynaptic active zone. The neurotransmitter molecules stored in these synaptic vesicles are released by exocytosis regulated by Ca^{2+} influx through channels close to the release site. The vesicles can then undergo cycles of recycling/exocytosis as described in Chapter 15. Importantly, these vesicles are refilled through specialized transporters called vesicular transporters that are specific for each neurotransmitter (eg, glutamate, γ-aminobutyric acid [GABA], acetylcholine).

Surface Membrane and Extracellular Substances Are Recycled in the Cell

Vesicular traffic toward the cell surface is continuously balanced by *endocytic traffic* from the plasma membrane to internal organelles. This traffic is essential for maintaining the area of the membrane in a steady state. It can alter the activity of many important regulatory molecules on the cell surface (eg, by removing receptors and adhesion molecules). It also removes nutrients and molecules, such as expendable receptor ligands and damaged membrane proteins, to the degradative compartments of the cells. Finally, it serves to recycle synaptic vesicles at nerve terminals (Chapter 15).

A significant fraction of endocytic traffic is carried in clathrin-coated vesicles. The clathrin coat interacts selectively through transmembrane receptors with extracellular molecules that are to be taken up into the cell. For this reason, clathrin-mediated uptake is often referred to as *receptor-mediated endocytosis*. The vesicles eventually shed their clathrin coats and fuse with the early endosomes, in which proteins to be recycled to the cell surface are separated from those destined for other intracellular organelles. Patches of the plasmalemma can also be recycled through larger, uncoated vacuoles that also fuse with early endosomes (*bulk endocytosis*).

Glial Cells Play Diverse Roles in Neural Function

Ramón y Cajal recognized the close association of glia with neurons and synapses in the brain (Figure 7–14). Although their function was at that time a mystery, he predicted that glia must do more than hold neurons together. Indeed, it is now clear that glial cells are critical players in brain development, function, and disease.

Glia Form the Insulating Sheaths for Axons

A major function of oligodendrocytes and Schwann cells is to provide the insulating material that allows rapid conduction of electrical signals along the axon. These cells produce thin sheets of myelin that wrap concentrically, many times, around the axon. CNS myelin, produced by oligodendrocytes, is similar, but not identical, to peripheral nervous system myelin, produced by Schwann cells.

Both types of glia produce myelin only for segments of axons. This is because the axon is not continuously wrapped in myelin, a feature that facilitates propagation of action potentials (Chapter 9). One Schwann cell produces a single myelin sheath for one segment of one axon, whereas one oligodendrocyte produces myelin sheaths for segments of as many as 30 axons (Figures 7–1 and 7–15).

The number of layers of myelin on an axon is proportional to the diameter of the axon—larger axons have thicker sheaths. Axons with very small diameters are not myelinated; nonmyelinated axons conduct action potentials much more slowly than do myelinated axons because of their smaller diameter and lack of myelin insulation (Chapter 9).

The regular lamellar structure and biochemical composition of the sheath are consequences of how

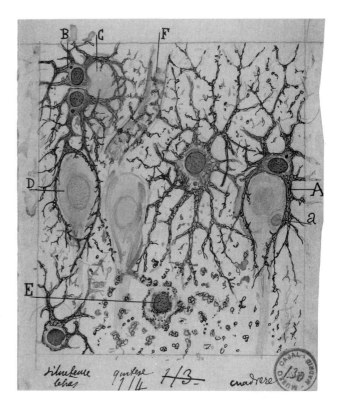

Figure 7–14 Astrocytes interact with neurons and synapses in the brain. This drawing by Ramón y Cajal (based on tissue stained with the sublimated gold chloride method) shows astrocytes of the pyramidal layer and stratum radiatum of Ammon's horn in the human brain. **(A)** A large astrocyte ensheathes a pyramidal neuron. **(B)** Twin astrocytes form a nest around a nerve cell body **(C)**. One of the astrocytes sends two branches to form another nest **(D)**. **(E)** A cell shows signs of autolysis. **(F)** Capillary vessel. (Reproduced, with permission, from the Instituto Cajal, Madrid, Spain.)

myelin is formed from the glial plasma membrane. In the development of the peripheral nervous system, before myelination takes place, the axon lies within a trough formed by Schwann cells. Schwann cells line up along the axon at regular intervals that become the myelinated segments of axon. The external membrane of each Schwann cell surrounds the axon to form a double membrane structure called the *mesaxon*, which elongates and spirals around the axon in concentric layers (Figure 7–15C). As the axon is ensheathed, the cytoplasm of the Schwann cell is squeezed out to form a compact lamellar structure.

The regularly spaced segments of myelin sheath are separated by unmyelinated gaps, called *nodes of Ranvier*, where the plasma membrane of the axon is exposed to the extracellular space for approximately 1 µm (Figure 7–16). This arrangement greatly increases the speed at which nerve impulses are conducted

A Myelination in the central nervous system

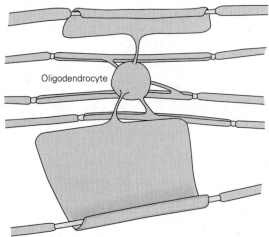

Oligodendrocyte

B Myelination in the peripheral nervous system

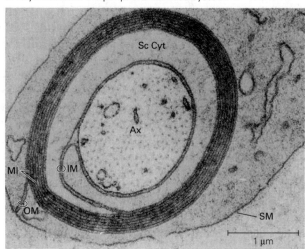

Sc Cyt

Ax

MI

IM

OM

SM

1 µm

C Development of myelin sheath in the peripheral nervous system

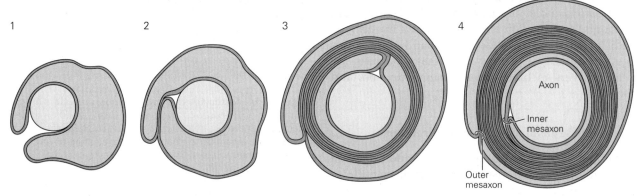

1

2

3

4

Axon

Inner
mesaxon

Outer
mesaxon

Figure 7–15 Glial cells produce the myelin that insulates the axons of central and peripheral neurons.

A. Axons in the central nervous system are wrapped in several layers of myelin produced by oligodendrocytes. Each oligodendrocyte can myelinate many axons. (Adapted from Raine 1984.)

B. This electron micrograph of a transverse section through an axon (**Ax**) in the sciatic nerve of a mouse shows the origin of a sheet of myelin (**MI**) at a structure called the inner mesaxon (**IM**). The myelin arises from the surface membrane (**SM**) of a Schwann cell, which is continuous with the outer mesaxon (**OM**). In this image, the Schwann cell cytoplasm (**Sc Cyt**) still surrounds the axon; eventually it is squeezed out and the

myelin layers become compact, as shown in part **C**. (Reproduced, with permission, from Dyck et al. 1984.)

C. A peripheral nerve fiber is myelinated by a Schwann cell in several stages. In stage 1, the Schwann cell surrounds the axon. In stage 2, the outer aspects of the plasma membrane have become tightly apposed in one area. This membrane fusion reflects early myelin membrane formation. In stage 3, several layers of myelin have formed because of continued rotation of the Schwann cell cytoplasm around the axon. In stage 4, a mature myelin sheath has formed; much of the Schwann cell cytoplasm has been squeezed out of the inner-most loop. (Adapted, with permission, from Williams et al. 1989.)

(up to 100 m/s in humans) because the signal jumps from one node to the next, a mechanism called *saltatory conduction* (Chapter 9). Nodes are easily excited because the density of Na⁺ channels, which generate the action potential, is approximately 50 times greater in the axon membrane at the nodes than in myelin-sheathed regions of membrane. Cell adhesion

molecules around nodes keep the myelin boundaries stable.

In the human femoral nerve, the primary sensory axon is approximately 0.5 m long and the internodal distance is 1 to 1.5 mm; thus, approximately 300 to 500 nodes of Ranvier occur along a primary afferent fiber between the thigh muscle and the cell body in the

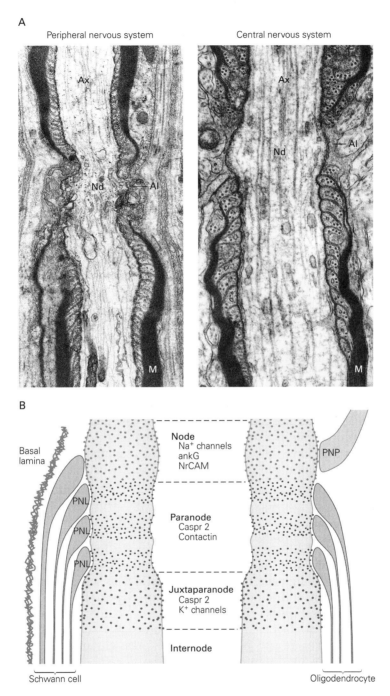

Figure 7–16 The myelin sheath of axons has regular gaps called the *nodes of Ranvier.*

A. Electron micrographs show the region of nodes in axons from the peripheral nervous system and spinal cord. The axon (**Ax**) runs vertically in both micrographs. The layers of myelin (**M**) are absent at the nodes (**Nd**), where the axon's membrane (axolemma, **Al**) is exposed. (Reproduced, with permission, from Peters et al. 1991.)

B. Regions on both sides of a node of Ranvier are rich in stable contacts between myelinating cells and the axon, to ensure that the nodes do not move or change in size and to restrict the localization of K^+ and Na^+ channels in the axon.

Potassium-permeable channels and the adhesion protein Caspr2 are concentrated in the juxtaparanode. Paranodal loops (**PNL**) of Schwann cell or oligodendrocyte cytoplasm form a series of stable junctions with the axon. The paranode region is rich with adhesion proteins such as Caspr2, contactin, and neu-rofascin (NF155). At the nodes in central axons, perinodal astro-glial processes (**PNP**) contact the axonal membrane, which is enormously enriched with Na^+ channels. This localization of Na^+ permeability is a major basis for the saltatory conduction in myelinated axons. The membrane-cytoskeletal linker ankyrin G (**ankG**) and the cell adhesion molecules NrCAM and NF186 are also concentrated at the nodes. (Reproduced, with permission, from Peles and Salzer 2000. Copyright © 2000 Elsevier.)

dorsal root ganglion. Because each internodal segment is formed by a single Schwann cell, as many as 500 Schwann cells participate in the myelination of each peripheral sensory axon.

Myelin has bimolecular layers of lipid interspersed between protein layers. Its composition is similar to that of the plasmalemma, consisting of 70% lipid and 30% protein with high concentrations of cholesterol and phospholipid. In the CNS, myelin has two major proteins: myelin basic protein, a small, positively charged protein that is situated on the cytoplasmic surface of compact myelin, and proteolipid protein, a hydrophobic integral membrane protein. Presumably, both provide structural stability for the sheath.

Both have also been implicated as important autoantigens against which the immune system can react to produce the demyelinating disease multiple sclerosis. In the peripheral nervous system, myelin contains a major protein, P_0, as well as the hydrophobic protein PMP22. Autoimmune reactions to these proteins produce a demyelinating peripheral neuropathy, the *Guillain-Barré syndrome*. Mutations in myelin protein genes also cause a variety of demyelinating diseases in both peripheral and central axons (Box 7–3). Demyelination slows down, or even stops, conduction of the action potential in an affected axon, because it allows electrical current to leak out of the axonal membrane. Demyelinating diseases thus have devastating effects on neuronal circuits in the central and peripheral nervous systems (Chapter 57).

Astrocytes Support Synaptic Signaling

Astrocytes are found in all areas of the brain; indeed, they constitute nearly half the number of brain cells. They play important roles in nourishing neurons and in regulating the concentrations of ions and neurotransmitters in the extracellular space. But astrocytes and neurons also communicate with each other to modulate synaptic signaling in ways that are still poorly understood. Astrocytes are generally divided into two main classes, which are distinguished by morphology, location, and function. *Protoplasmic astrocytes* are found in gray matter, and their processes are closely associated with synapses as well as blood vessels. *Fibrillary (or fibrous) astrocytes* in white matter contact axons and nodes of Ranvier. In addition, specialized astrocytes include Bergmann glia in the cerebellum and Müller glia in the retina.

Astrocytes have large numbers of thin processes that enfold all the blood vessels of the brain and ensheathe synapses or groups of synapses. By their intimate physical association with synapses, often closer than 1 µm, astrocytes are positioned to regulate extracellular concentrations of ions, neurotransmitters, and other molecules (Figure 7–19). In fact, astrocytes express many of the same voltage-gated ion channels and neurotransmitter receptors that neurons do and are thus well equipped to receive and transmit signals that could affect neuronal excitability and synaptic function.

How do astrocytes regulate axonal conduction and synaptic activity? The first recognized physiological role was that of K^+ buffering. When neurons fire action potentials, they release K^+ ions into the extracellular space. Because astrocytes have high concentrations of K^+ channels in their membranes, they can act as *spatial buffers*: They take up K^+ at sites of neuronal activity, mainly synapses, and release it at distant contacts with blood vessels. Astrocytes can also accumulate K^+ locally within their cytoplasmic processes along with Cl^- ions and water. Unfortunately, accumulation of ions and water in astrocytes can contribute to severe brain swelling after head injury.

Astrocytes also regulate neurotransmitter concentrations in the brain. For example, high-affinity transporters located in the astrocyte's plasma membrane rapidly clear the neurotransmitter glutamate from the synaptic cleft (Figure 7–19C). Once within the glial cell, glutamate is converted to glutamine by the enzyme glutamine synthetase. Glutamine is then transferred to neurons, where it serves as an immediate precursor of glutamate (Chapter 16). Interference with these uptake mechanisms results in high concentrations of extracellular glutamate that can lead to the death of neurons, a process termed excitotoxicity. Astrocytes also degrade dopamine, norepinephrine, epinephrine, and serotonin.

Astrocytes sense when neurons are active because they are depolarized by the K^+ released by neurons and have neurotransmitter receptors similar to those of neurons. For example, Bergmann glia in the cerebellum express glutamate receptors. Thus, the glutamate released at cerebellar synapses affects not only postsynaptic neurons but also astrocytes near the synapse. The binding of these ligands to glial receptors increases the intracellular free Ca^{2+} concentration, which has several important consequences. The processes of one astrocyte connect to those of neighboring astrocytes through intercellular aqueous channels called gap junctions (Chapter 11), allowing transfer of ions and small molecules between many cells. An increase in free Ca^{2+} within one astrocyte increases Ca^{2+} concentrations in adjacent astrocytes. This spread of Ca^{2+} through the astrocyte network occurs over hundreds of micrometers. It is likely that this Ca^{2+} wave modulates

Box 7–3 Defects in Myelin Proteins Disrupt Conduction of Nerve Signals

Because in myelinated axons normal conduction of the nerve impulse depends on the insulating properties of the myelin sheath, defective myelin can result in severe disturbances of motor and sensory function.

Many diseases that affect myelin, including some animal models of demyelinating disease, have a genetic basis. The *shiverer* (or *shi*) mutant mice have tremors and frequent convulsions and tend to die young. In these mice, the myelination of axons in the central nervous system is greatly deficient and the myelination that does occur is abnormal.

The mutation that causes this disease is a deletion of five of the six exons of the gene for myelin basic protein, which in the mouse is located on chromosome 18. The mutation is recessive; a mouse develops the disease only if it has inherited the defective gene from both parents. *Shiverer* mice that inherit both defective genes have only approximately 10% of the myelin basic protein (MBP) found in normal mice (Figure 7–17A).

When the wild-type gene is injected into fertilized eggs of the *shiverer* mutant with the aim of rescuing the mutant, the resulting transgenic mice express the wild-type gene but produce only 20% of the normal amounts of MBP. Nevertheless, myelination of central neurons in the transgenic mice is much improved. Although they still have occasional tremors, the transgenic mice do not have convulsions and have a normal life span (Figure 7–17B).

In both the central and peripheral nervous systems, myelin contains a protein termed *myelin-associated glycoprotein* (MAG). MAG belongs to the immunoglobulin superfamily that includes several important cell surface proteins thought to be involved in cell-to-cell recognition, eg, the major histocompatibility complex of antigens, T-cell surface antigens, and the neural cell adhesion molecule (NCAM).

A

Normal mouse has abundant myelination

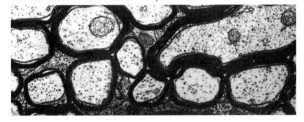

Shiverer mutant has scant myelination

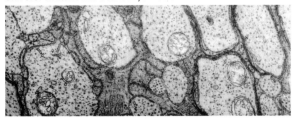

Transfected normal gene improves myelination

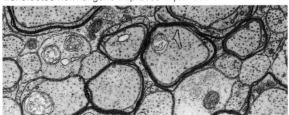

B

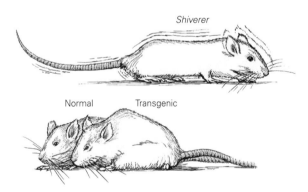

Figure 7–17 A genetic disorder of myelination in mice can be partially cured by transfection of the normal gene that encodes myelin basic protein.

A. Electron micrographs show the state of myelination in the optic nerve of a normal mouse, a *shiverer* mutant, and a *shiverer* mutant with the transfected gene for myelin basic protein.

B. The *shiverer* mutant exhibits poor posture and weakness. Injection of the wild-type gene into the fertilized egg of the mutant improves myelination; the treated mutant looks as perky as a normal mouse. (Reproduced, with permission, from Readhead et al. 1987.)

(continued)

Box 7–3 Defects in Myelin Proteins Disrupt Conduction of Nerve Signals (continued)

In the peripheral nervous system, MAG is expressed by Schwann cells early during production of myelin and eventually becomes a component of mature (compact) myelin. Its early expression, subcellular location, and structural similarity to other surface recognition proteins suggest that it is an adhesion molecule important for the initiation of the myelination process. Two isoforms of MAG are produced from a single gene through alternative RNA splicing.

The major protein in mature peripheral myelin, *myelin protein zero* (MPZ or P$_0$), spans the plasmalemma of the Schwann cell. It has a basic intracellular domain and, like MAG, is a member of the immunoglobulin superfamily. The glycosylated extracellular part of the protein, which contains the immunoglobulin domain, functions as a homophilic adhesion protein during myelin-ensheathing by interacting with identical domains on the surface of the apposed membrane. Genetically engineered mice in which the function of P$_0$ has been eliminated have poor motor coordination, tremors, and occasional convulsions.

Observation of *trembler* mouse mutants led to the identification of *peripheral myelin protein 22* (PMP22). This Schwann cell protein spans the membrane four times and is normally present in compact myelin. PMP22 is altered by a single amino acid in the mutants. A similar protein is found in humans, encoded by a gene on chromosome 17.

Mutations of the *PMP22* gene on chromosome 17 produce several hereditary peripheral neuropathies,

while a duplication of this gene causes one form of *Charcot-Marie-Tooth disease* (Figure 7–18). This disease is the most common inherited peripheral neuropathy and is characterized by progressive muscle weakness, greatly decreased conduction in peripheral nerves, and cycles of demyelination and remyelination. Because both duplicated genes are active, the disease results from *increased* production of PMP22 (a two- to three-fold increase in gene dosage). Mutations in a number of genes expressed by Schwann cells can produce inherited peripheral neuropathies.

In the central nervous system, more than half of the protein in myelin is the proteolipid protein (PLP), which has five membrane-spanning domains. Proteolipids differ from lipoproteins in that they are insoluble in water. Proteolipids are soluble only in organic solvents because they contain long chains of fatty acids that are covalently linked to amino acid residues throughout the proteolipid molecule. In contrast, lipoproteins are noncovalent complexes of proteins with lipids and often serve as soluble carriers of the lipid moiety in the blood.

Many mutations of PLP are known in humans as well as in other mammals, eg, the *jimpy* mouse. One example is Pelizaeus-Merzbacher disease, a heterogeneous X-linked disease in humans. Almost all PLP mutations occur in a membrane-spanning domain of the molecule. Mutant animals have reduced amounts of (mutated) PLP, hypomyelination, and degeneration and death of oligodendrocytes. These observations suggest that PLP is involved in the compaction of myelin.

A

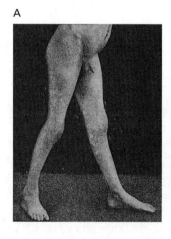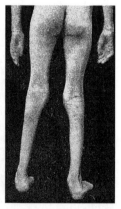

Figure 7–18 Charcot-Marie-Tooth disease (type 1A) results from increased production of peripheral myelin protein 22.

A. A patient with Charcot-Marie-Tooth disease shows impaired gait and deformities. (Reproduced, with permission, from Charcot's original description of the disease, Charcot and Marie 1886.)

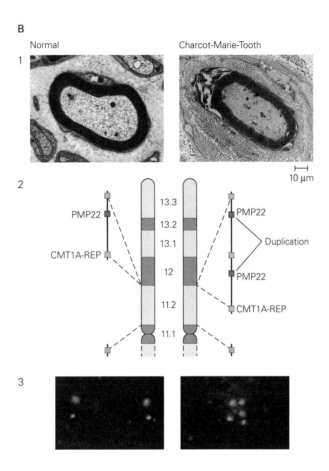

B

Normal

Charcot-Marie-Tooth

1

10 µm

2

PMP22

CMT1A-REP

13.3

13.2

13.1

12

11.2

11.1

PMP22

Duplication

PMP22

CMT1A-REP

3

B. The disordered myelination in Charcot-Marie-Tooth disease (type 1A) results from increased production of peripheral myelin protein 22 (PMP22).

1. Sural nerve biopsies from a normal individual (reproduced, with permission, from A.P. Hays) and from a patient with Charcot-Marie-Tooth disease (reproduced, with permission, from Lupski and Garcia 1992). In the patient's biopsy, the myelin sheath is slightly thinner than normal and is surrounded by concentric rings of Schwann cell processes. These changes are typical of the recurrent demyelination and remyelination seen in this disorder.

2. The increase in PMP22 is caused by a duplication of a normal 1.5-Mb region of the DNA on the short arm of chromosome 17 at 17p11.2-p12. The *PMP22* gene is flanked by two similar repeat sequences (CMT1A-REP), as shown in the normal chromosome 17 on the *left*. Normal individuals have two normal chromosomes. In patients with the disease (*right*), the duplication results in two functioning *PMP22* genes, each flanked by a repeat sequence. The normal and duplicated regions are shown in the expanded diagrams indicated by the **dashed lines**. (The repeats are

thought to have given rise to the original duplication, which was then inherited. The presence of two similar flanking sequences with homology to a transposable element is believed to increase the frequency of unequal crossing over in this region of chromosome 17 because the repeats enhance the probability of mispairing of the two parental chromosomes in a fertilized egg.)

3. Although a large duplication (3 Mb) cannot be detected in routine examination of chromosomes in the light microscope, evidence for the duplication can be obtained using fluorescence in situ hybridization. The *PMP22* gene is detected with an oligonucleotide probe tagged with the dye Texas Red. An oligonucleotide probe that hybridizes with DNA from region 11.2 (indicated by the green segment close to the centromere) is used for in situ hybridization on the same sample. A nucleus from a normal individual (*left*) shows a pair of chromosomes, each with one red site (*PMP22* gene) for each green site. A nucleus from a patient with the disease (*right*) has one extra red site, indicating that one chromosome has one *PMP22* gene and the other has two *PMP22* genes. (Adapted, with permission, from Lupski et al. 1991.)

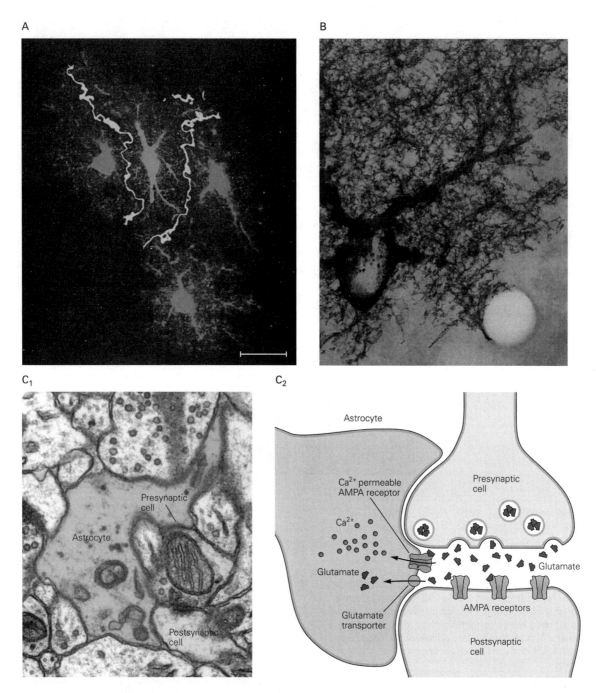

Figure 7–19 Astrocyte processes are intimately associated with synapses.

A. Astrocytes occupy discrete volumes. The central astrocyte (green) is shown to occupy a volume distinct from its three neighbors (red), with only a small overlap (yellow) at the ends of their processes, which are interconnected by gap junctions Bar = 20 μm. (Reproduced, with permission, from Bushong et al. 2002. Copyright © 2002 Society for Neuroscience.)

B. This high-voltage electron micrograph shows several thick processes emanating from the cell body of an astrocyte and branching into extraordinarily fine processes. The typical envelopment of a blood vessel is shown at lower right. (Reproduced, with permission, from Hama et al. 1994. Copyright © 1994 Wiley.)

C. The processes of astrocytes are intimately associated with both presynaptic and postsynaptic elements. **1.** The close association between astrocyte processes and synapses is seen in this electron micrograph of hippocampal cells. (Reproduced, with permission, from Ventura and Harris 1999. Copyright © 1999 Society for Neuroscience.) **2.** Glutamate released from the presynaptic neuron activates not only receptors on the postsynaptic neuron but also AMPA-type (α-amino-3-hydroxy-5-methylisoxazole-4-propionate) receptors on astrocytes. Astrocytes remove glutamate from the synaptic cleft by uptake through high-affinity transporters. (Adapted from Gallo and Chittajallu 2001.)

nearby neuronal activity by triggering the release of nutrients and regulating blood flow. An increase in Ca^{2+} in astrocytes leads to the secretion of signals that enhance synaptic function and even behavior. Thus astrocyte–neuron signaling contributes to normal neural circuit functioning.

Astrocytes also are important for the development of synapses. Their appearance at synapses in the postnatal brain coincides with periods of synaptogenesis and synapse maturation. Astrocytes prepare the surface of the neuron for synapse formation and stabilize newly formed synapses. For example, astrocytes secrete several synaptogenic factors, including thrombospondins, hevin, and glycipans, that promote the formation of new synapses. Astrocytes can also help remodel and eliminate excess synapses during development by phagocytosis (Chapter 48). In the adult CNS, astrocytes continue to phagocytose synapses, and as this phagocytosis is dependent on neuronal activity, it is possible that this remodeling of synapses contributes to learning and memory. In pathological states, such as chromatolysis produced by axonal damage, astrocytes and presynaptic terminals temporarily retract from the damaged postsynaptic cell bodies. Astrocytes release neurotrophic and gliotrophic factors that promote the development and survival of neurons and oligodendrocytes. They also protect other cells from the effects of oxidative stress. For example, the glutathione peroxidase in astrocytes detoxifies toxic oxygen free radicals released during hypoxia, inflammation, and neuronal degeneration.

Finally, astrocytes ensheathe small arterioles and capillaries throughout the brain, forming contacts between the ends of astrocyte processes and the basal lamina around endothelial cells. The CNS is sequestered from the general circulation so that macromolecules in the blood do not passively enter the brain and spinal cord (the *blood–brain barrier*). The barrier is largely the result of tight junctions between endothelial cells and cerebral capillaries, a feature not shared by capillaries in other parts of the body. Nevertheless, endothelial cells have a number of transport properties that allow some molecules to pass through them into the nervous system. Because of the intimate contacts of astrocytes and blood vessels, the transported molecules, such as glucose, can be taken up by astrocyte end-feet.

Following brain injury and disease, astrocytes undergo a dramatic transformation called *reactive astrocytosis*, which involves changes in gene expression, morphology, and signaling. The functions of reactive astrocytes are complex and poorly understood, as they both hinder and support CNS recovery. Recent studies have found evidence for at least two kinds of reactive astrocytes; one type helps to promote repair and recovery, whereas another is harmful, actively contributing to the death of neurons after acute CNS injury; however there are likely other subtypes. These neurotoxic reactive astrocytes are prominent in patients with Alzheimer disease and other neurodegenerative diseases and thus are an attractive target for new therapies. An interesting question is why the brain ever generates a neurotoxic reactive astrocyte. Quite possibly, removal of injured or sick neurons allows synapses to reorganize to help preserve neural circuit function. In addition, removal of virally infected neurons could help limit the spread of viral infections.

Microglia Have Diverse Functions in Health and Disease

Microglia compose about 10% of glia in the CNS and exist in multiple morphological states in the healthy and damaged brain. Despite being described by Rio Hortega over 100 years ago, the functions of microglia are poorly understood compared to other cell types. Unlike neurons, astrocytes, and oligodendrocytes, microglia do not belong to the neuroectodermal lineage. Long thought to derive from the bone marrow, recent fate mapping studies reveal that microglia are in fact derived from myeloid progenitors in the yolk sac.

Microglia colonize brain very early in embryonic development and reside in all regions of the brain throughout life (Figure 7–20). During development, microglia help sculpt developing neural circuits by engulfing pre- and postsynaptic structures (Figure 7–21), and emerging evidence suggests microglia may modulate other aspects of brain development and brain homeostasis. Recent in vivo imaging studies have revealed dynamic interactions between microglia and neurons. In the healthy adult cerebral cortex, microglia processes continuously survey their surrounding extracellular environment and contact neurons and synapses, but the functional significance of this activity remains unknown.

Following injury and disease, microglia undergo a dramatic increase in the motility of their processes and changes in morphology and gene expression and can be rapidly recruited to sites of damage where they can have beneficial roles. For example, they serve to bring lymphocytes, neutrophils, and monocytes into the CNS and expand the lymphocyte population, important immunological activities in infection, stroke, and immunologic demyelinating disease. They also protect the brain by phagocytosing debris as well as unwanted

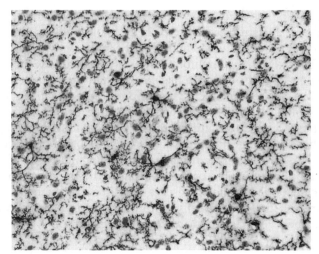

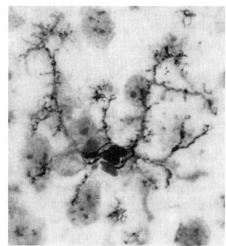

Figure 7–20 Large numbers of microglia reside in the mammalian central nervous system. The micrograph on the *left* shows microglia (in **brown**, immunocytochemistry) in the cerebral cortex of an adult mouse. The **blue** blobs are the nuclei of nonmicroglial cells. The microglial cells have fine, lacy processes, as shown in the higher magnification micrograph on the *right*. (Reproduced, with permission, from Berry et al. 2002.)

and dying cells and toxic proteins, actions that are critical for preventing further damage and maintaining brain homeostasis. Although critical for the immune response to infection or trauma, microglia also contribute to pathological neuroinflammation by releasing cytokines and neurotoxic proteins and by inducing neurotoxic reactive astrocytes. They also contribute to synapse loss and dysfunction in models of Alzheimer disease and neurodegenerative disease.

Choroid Plexus and Ependymal Cells Produce Cerebrospinal Fluid

The function of neurons and glia is tightly regulated by the extracellular environment of the CNS. *Interstitial fluid* (ISF) fills spaces between neurons and glia in the parenchyma. *Cerebrospinal fluid* (CSF) bathes the brain's ventricles, the subarachnoid space of the brain and spinal cord, and the major cisterns of the CNS. The ISF and CSF deliver nutrients to cells in the CNS, maintain ion homeostasis, and serve as a removal system for metabolic waste products. In conjunction with the meningeal layers that surround the brain and spinal cord, the CSF provides a cushion that protects CNS tissues from mechanical damage. The fluid environment of the CNS is maintained by endothelial cells of the blood–brain barrier and choroid plexus epithelial cells of the blood–CSF barrier. These barriers not only serve to regulate the extracellular environment of the brain and spinal cord but also relay critical information between the CNS and the periphery.

The cells of the *choroid plexus* and the *ependymal layer* contribute to CSF production, composition, and dynamics. The choroid plexuses appear as epithelial invaginations soon after neural tube closure where the lateral, third, and fourth ventricles will eventually

Figure 7–21 Microglia interact with and sculpt synaptic elements in the healthy brain. Two-photon imaging in the olfactory bulb of adult mice shows microglial processes expressing a fractalkine receptor–GFP fusion (CX3CR1-GFP) (**green**) connecting to tdTomato-labeled neurons (**red**). (Reproduced, with permission, from Hong and Stevens 2016.)

form. Through embryonic development, the choroid plexuses mature, each forming a ciliated cuboidal epithelial layer that encapsulates a stromal and immune cell network and an extensive capillary bed. The ependyma is a single layer of ciliated cuboidal cells, a type of glia cell that lines the ventricles of the brain. At several places in the lateral and fourth ventricles, specialized ependymal cells form the epithelial layer that surrounds the choroid plexus (Figure 7–22B).

The choroid plexus produces most of the CSF that bathes the brain. Loose junctions between ependymal cells provide access for CSF to the brain's interstitial space. Ciliary motion in the ependymal cells helps to move CSF through the ventricular system (Figure 7–22A), facilitating long-range delivery of molecules to other cells in the CNS and transport of waste from the CNS to the periphery.

The choroid plexus transports fluid and solutes from the serum into the CNS to generate CSF. The fenestrated capillaries that traverse the choroid plexus allow free passage of water and small molecules from the blood into the stromal space of the choroid plexus. The choroid plexus epithelial cells, however, form tight junctions, preventing further unregulated movement of these molecules into the brain. Instead, import of water, ions, metabolites, and protein mediators that compose the CSF is tightly regulated by transporters and channels in the choroid plexus epithelium. Active transport mechanisms in the epithelium are bidirectional, additionally mediating the flux of molecules from the CSF back into the peripheral circulation.

The choroid plexus epithelial cells also synthesize and secrete many proteins into the CSF. In the healthy

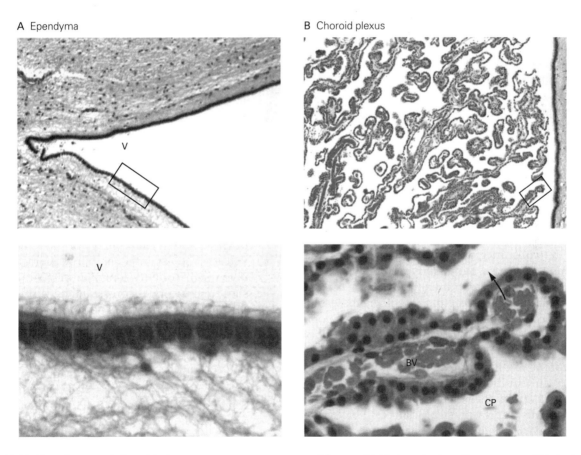

A Ependyma

B Choroid plexus

Figure 7–22 Ependyma and choroid plexus.

A. The ependyma is a single layer of ciliated, cuboidal cells lining the cerebral ventricles (V). The lower image, a high magnification of the ependymal lining (rectangle in upper image), shows the cilia on the ventricular side of the ependymal cells.

B. The choroid plexus is continuous with the ependyma but projects into the ventricles, where it covers thin blood vessels

and forms a highly branched papillary structure. This is the site of cerebrospinal fluid formation. High magnification (lower image) shows the blood vessel core (BV) and overlying choroid plexus (CP). The arrow denotes the direction of fluid flow from capillary into ventricle during the formation of cerebrospinal fluid.

embryonic and postnatal brain, these proteins modulate development of neural stem cells and may regulate processes such as cortical plasticity. The choroid plexus epithelial cell secretome can also be altered by inflammatory signals from the periphery or from within the brain, with consequences for neuronal function during infection and in aging. Functional roles for other choroid plexus–derived factors in the healthy and diseased brain—including microRNAs, long noncoding RNAs, and extracellular vesicles—are beginning to emerge, further underlining the important contribution of this structure to brain development and homeostasis.

Highlights

1. The morphology of neurons is elegantly suited to receive, conduct, and transmit information in the brain. Dendrites provide a highly branched, elongated surface for receiving signals. Axons conduct electrical impulses rapidly over long distances to their synaptic terminals, which release neurotransmitters onto target cells.

2. Although all neurons conform to the same basic cellular architecture, different subtypes of neurons vary widely in their specific morphological features, functional properties, and molecular identities.

3. Neurons in different locations differ in the complexity of their dendritic trees, extent of axon branching, and the number of synaptic terminals that they form and receive. The functional significance of these morphological differences is plainly evident. For example, motor neurons must have a more complex dendritic tree than sensory neurons, as even simple reflex activity requires integration of many excitatory and inhibitory inputs. Different types of neurons use different neurotransmitters, ion channels, and neurotransmitter receptors. Together, these biochemical, morphological, and electrophysiological differences contribute to the great complexity of information processing in the brain.

4. Neurons are among the most highly polarized cells in our body. The considerable size and complexity of their dendritic and axonal compartments represent significant cell biological challenges for these cells, including transport of various organelles, proteins, and mRNA over long distances (up to a meter for some axons). Most neuronal proteins are synthesized in the cell body, but some synthesis occurs in dendrites and axons. The newly synthesized proteins are folded with the assistance of chaperones, and their final structure is often modified by permanent or reversible posttranslational modifications.

The final destination of a protein in the neuron depends on signals encoded in its amino acid sequence.

5. Transport of proteins and mRNA occurs with great specificity and results in the vectorial transport of selected membrane components. The cytoskeleton provides an important framework for the transport of organelles to different intracellular locations in addition to controlling axonal and dendritic morphology.

6. All these fundamental cell biological processes are profoundly modifiable by neuronal activity, which produces the dramatic changes in cell structure and function by which neural circuits adapt to experience (learning).

7. The nervous system also contains several types of glial cells. Oligodendrocytes and Schwann cells produce the myelin insulation that enables axons to conduct electrical signals rapidly. Astrocytes and nonmyelinating Schwann cells ensheathe other parts of the neuron, particularly synapses. Astrocytes control extracellular ion and neurotransmitter concentrations and actively participate in the formation and function of synapses. Microglia resident immune cells and phagocytes dynamically interact with neurons and glial cells and have diverse roles in health and disease.

8. The cells of the choroid plexus and the ependymal layer contribute to CSF production, composition, and dynamics.

9. New advances in genomics and single-cell RNA sequencing are beginning to define the immense diversity of cell types, not only among neurons but also among glial cells.

10. Recent progress in genetics, cell biology, and in vivo microscopy (two-photon microscopy, light-sheet microscopy) is providing new insights into the unique mechanisms by which neurons establish and maintain their polarity throughout an individual's life span.

11. These new insights provide important clues into the cell biological steps, including for example defects in axon transport, that trigger neurodegenerative diseases such as Huntington, Parkinson, and Alzheimer disease.

Beth Stevens
Franck Polleux
Ben A. Barres

Selected Reading

Alberts B, Johnson A, Lewis J, Raff M, Roberts K, Walter P (eds). 2002. *Molecular Biology of the Cell,* 4th ed. New York: Garland.

Chung WS, Allen NJ, Eroglu C. 2015. Astrocytes control synapse formation, function, and elimination. Cold Spring Harb Perspect Biol 7:a020370.

Damkier HH, Brown P, Praetorius J. 2013. Cerebrospinal fluid secretion by the choroid plexus. Physiol Rev 93:1847–1892.

Dyck PJ, Thomas PK, Griffin JW, Low PA, Poduslo JF (eds). 1993. *Peripheral Neuropathy,* 3rd ed. Philadelphia: Saunders.

Dyck PJ, Thomas PK, Lambert EH, Bunge R (eds). 1984. *Peripheral Neuropathy,* 2nd ed., Vols. 1, 2. Philadelphia: Saunders.

Glickman MH, Ciechanover A. 2002. The ubiquitin-proteasome proteolytic pathway: destruction for the sake of construction. Physiol Rev 82:373–428.

Hartl FU. 1996. Molecular chaperones in cellular protein folding. Nature 381:571–579.

Kapitein LC, Hoogenraad CC. 2015. Building the neuronal microtubule cytoskeleton. Neuron 87:492–506.

Kelly RB. 1993. Storage and release of neurotransmitters. Cell 72:43–53.

Kreis T, Vale R (eds). 1999. *Guidebook to the Cytoskeletal and Motor Proteins,* 2nd ed. Oxford: Oxford Univ. Press.

Lun MP, Monuki ES, Lehtinen MK. 2015. Development and functions of the choroid plexus-cerebral fluid system. Nature Rev Neurosci 16:445–457.

Nigg EA. 1997. Nucleocytoplasmic transport: signals, mechanisms and regulation. Nature 386:779–787.

Pemberton LF, Paschal BM. 2005. Mechanisms of receptor-mediated nuclear import and nuclear export. Traffic 6:187–198.

Rothman JE. 2002. Lasker Basic Medical Research Award: the machinery and principles of vesicle transport in the cell. Nat Med 8:1059–1062.

Schafer DP, Stevens B. 2015. Microglia function in central nervous system development and plasticity. Cold Spring Harb Perspect Biol 7:a020545.

Schatz G, Dobberstein B. 1996. Common principles of protein translocation across membranes. Science 271:1519–1526.

Schwartz JH. 2003. Ubiquitination, protein turnover, and long-term synaptic plasticity. Sci STKE 190:26.

Siegel GJ, Albers RW, Brady S, Price DL (eds). 2005. *Basic Neurochemistry: Molecular, Cellular, and Medical Aspects,* 7th ed. Amsterdam: Elsevier.

Signor D, Scholey JM. 2000. Microtubule-based transport along axons, dendrites and axonemes. Essays Biochem 35:89–102.

St Johnston D. 2005. Moving messages: the intracellular localization of mRNAs. Nat Rev Mol Cell Biol 6:363–375.

Stryer L. 1995. *Biochemistry,* 4th ed. New York: Freeman.

Tahirovic S, Bradke F. 2009. Neuronal polarity. Cold Spring Harb Perspect Biol 1:a001644.

Zhou L, Griffin JW. 2003. Demyelinating neuropathies. Curr Opin Neurol 16:307–313.

References

Barnes AP, Polleux F. 2009. Establishment of axon-dendrite polarity in developing neurons. Ann Rev Neurosci 32:347–381.

Berry M, Butt AM, Wilkin G, Perry VH. 2002. Structure and function of glia in the central nervous system. In: Graham DI and Lantos PL (eds). *Greenfield's Neuropathology.* 7th ed., pp. 104–105. London: Arnold.

Bershadsky AD, Vasiliev JM. 1988. *Cytoskeleton.* New York: Plenum.

Brendecke SM, Prinz M. 2015. Do not judge a cell by its cover—diversity of CNS resident, adjoining and infiltrating myeloid cells in inflammation. Semin Immunopathol 37:591–605.

Bushong EA, Martone ME, Jones YZ, Ellisman MH. 2002. Protoplasmic astrocytes in CA1 stratum radiatum occupy separate anatomical domains. J Neurosci 22:183–192.

Charcot J-M, Marie P. 1886. Sur une forme particulière d'atrophie musculaire progressive, souvent familiale, débutant par les pieds et les jambes et atteignant plus tard les mains. Rev Med 6:97–138.

Christopherson KS, Ullian EM, Stokes CC, et al. 2005. Thrombospondins are astrocyte-secreted proteins that promote CNS synaptogenesis. Cell 120:421–433.

Chung WS, Clarke LE, Wang GX, et al. 2013. Astrocytes mediate synapse elimination through MEGF10 and MERTK pathways. Nature 504:394–400.

Ciechanover A, Brundin P. 2003. The ubiquitin proteasome system in neurodegenerative diseases: sometimes the chicken, sometimes the egg. Neuron 40:427–446.

Cooney JR, Hurlburt JL, Selig DK, Harris KM, Fiala JC. 2002. Endosomal compartments serve multiple hippocampal dendritic spines from a widespread rather than a local store of recycling membrane. J Neurosci 22:2215–2224.

De Camilli P, Moretti M, Donini SD, Walter U, Lohmann SM. 1986. Heterogeneous distribution of the cAMP receptor protein RII in the nervous system: evidence for its intracellular accumulation on microtubules, microtubule-organizing centers, and in the area of the Golgi complex. J Cell Biol 103:189–203.

Divac I, LaVail JH, Rakic P, Winston KR. 1977. Heterogeneous afferents to the inferior parietal lobule of the rhesus monkey revealed by the retrograde transport method. Brain Res 123:197–207.

Duxbury MS, Whang EE. 2004. RNA interference: a practical approach. J Surg Res 117:339–344.

Esiri MM, Hyman BT, Beyreuther K, Masters C. 1997. Ageing and dementia. In: DI Graham, PL Lantos (eds). *Greenfield's Neuropathology,* 6th ed. Vol II. London: Arnold.

Gallo V, Chittajallu R. 2001. Neuroscience. Unwrapping glial cells from the synapse: what lies inside? Science 292:872–873.

Giraudo CG, Hu C, You D, et al. 2005. SNAREs can promote complete fusion and hemifusion as alternative outcomes. J Cell Biol 170:249–260.

Goldberg AL. 2003. Protein degradation and protection against misfolded or damaged proteins. Nature 426:895–899.

Görlich D, Mattaj IW. 1996. Nucleocytoplasmic transport. Science 271:1513–1518.

Hama K, Arii T, Kosaka T. 1994. Three-dimensional organization of neuronal and glial processes: high voltage electron microscopy. Microsc Res Tech 29:357–367.

Harris KM, Jensen FE, Tsao B. 1992. Three-dimensional structure of dendritic spines and synapses in rat hippocampus (CA1) at postnatal day 15 and adult ages: implications for the maturation of synaptic physiology and long-term potentiation. J Neurosci 12:2685–2705.

Harris KM, Stevens JK. 1989. Dendritic spines of CA1 pyramidal cells in the rat hippocampus: serial electron microscopy with reference to their biophysical characteristics. J Neurosci 9:2982–2997.

Hirokawa N. 1997. The mechanisms of fast and slow transport in neurons: identification and characterization of the new Kinesin superfamily motors. Curr Opin Neurobiol 7:605–614.

Hirokawa N, Pfister KK, Yorifuji H, Wagner MC, Brady ST, Bloom GS. 1989. Submolecular domains of bovine brain kinesin identified by electron microscopy and monoclonal antibody decoration. Cell 56:867–878.

Hoffman PN, Lasek RJ. 1975. The slow component of axonal transport: identification of major structural polypeptides of the axon and their generality among mammalian neurons. J Cell Biol 66:351–366.

Hong S, Stevens B. 2016. Microglia: phagocytosing to clear, sculpt and eliminate. Dev Cell 38:126–128.

Ko CO, Robitaille R. 2015. Perisynaptic Schwann cells at the neuromuscular synapse: adaptable, multitasking glial cells. Cold Spring Harb Perspect Biol 7:a020503.

Lemke G. 2001. Glial control of neuronal development. Annu Rev Neurosci 24:87–105.

Liddelow SA, Guttenplan KA, Clarke LE, et al. 2016 Neurotoxic reactive astrocytes are induced by activated microglia. Nature 541:481–487.

Lupski JR, de Oca-Luna RM, Slaugenhaupt S, et al. 1991. DNA duplication associated with Charcot-Marie-Tooth disease type 1A. Cell 66:219–232.

Lupski JR, Garcia CA. 1992. Molecular genetics and neuropathology of Charcot-Marie-Tooth disease type 1A. Brain Pathol 2:337–349.

Ma Z, Stork T, Bergles DE, Freeman MR. 2016. Neuromodulators signal through astrocytes to alter neural circuit activity and behaviour. Nature 539:428–432.

Maday S, Twelvetrees AE, Moughamian AJ, Holzbaur EL. 2014. Axonal transport: cargo-specific mechanisms of motility and regulation. Neuron 84:292–309.

McNew JA, Goodman JM. 1996. The targeting and assembly of peroxisomal proteins: some old rules do not apply. Trends Biochem Sci 21:54–58.

Mirra SS, Hyman BT. 2002. Aging and dementia. In: DI Graham, PL Lantos (eds). Greenfield's Neuropathology, 7th ed., Vol. 2, p. 212. London: Arnold.

Ochs S. 1972. Fast transport of materials in mammalian nerve fibers. Science 176:252–260.

Peles E, Salzer JL. 2000. Molecular domains of myelinated axons. Curr Opin Neurobiol 10:558–565.

Peters A, Palay SL, Webster H de F. 1991. The Fine Structure of the Nervous System, 3rd ed. New York: Oxford University Press.

Raine CS. 1984. Morphology of myelin and myelination. In: P Morell (ed). Myelin. New York: Plenum Press.

Ransohoff RM, Cardona AE. 2010. The myeloid cells of the central nervous system parenchyma. Nature 468:253–262.

Ramón y Cajal S. [1901] 1988. Studies on the human cerebral cortex. IV. Structure of the olfactory cerebral cortex of man and mammals. In: J DeFelipe, EG Jones (eds, transl). Cajál on the Cerebral Cortex, pp. 289–362. New York: Oxford Univ. Press.

Ramón y Cajal S. [1909] 1995. Histology of the Nervous System of Man and Vertebrates. N Swanson, LW Swanson (transl). Vols. 1, 2. New York: Oxford Univ. Press.

Readhead C, Popko B, Takahashi N, et al. 1987. Expression of a myelin basic protein gene in transgenic Shiverer mice: correction of the dysmyelinating phenotype. Cell 48:703–712.

Roa BB, Lupski JR. 1994. Molecular genetics of Charcot-Marie-Tooth neuropathy. Adv Human Genet 22:117–152.

Schafer DP, Lehrman EK, Kautzman AG, et al. 2012. Microglia sculpt postnatal neural circuits in an activity and complement-dependent manner. Neuron 74:691–705.

Schnapp BJ, Reese TS. 1982. Cytoplasmic structure in rapid-frozen axons. J Cell Biol 94:667–679.

Silva-Vargas V, Maldonado-Soto AR, Mizrak D, Codega P, Doetsch F. 2016. Age-dependent niche signals from the choroid plexus regulate adult neural stem cells. Cell Stem Cell 19:643–652.

Sorra KE, Harris KM. 1993. Occurrence and three-dimensional structure of multiple synapses between individual radiatum axons and their target pyramidal cells in hippocampal area CA1. J Neurosci 13:3736–3748.

Sossin W. 1996. Mechanisms for the generation of synapse specificity in long-term memory: the implications of a requirement for transcription. Trends Neurosci 19:215–218.

Takei K, Mundigl O, Daniell L, De Camilli P. 1996. The synaptic vesicle cycle: a single vesicle budding step involving clathrin and dynamin. J Cell Biol 1335:1237–1250.

Ventura R, Harris KM. 1999. Three-dimensional relationships between hippocampal synapses and astrocytes. J Neurosci 19:6897–6906.

Weiss P, Hiscoe HB. 1948. Experiments on the mechanism of nerve growth. J Exp Zool 107:315–395.

Wells DG, Richter JD, Fallon JR. 2000. Molecular mechanisms for activity-regulated protein synthesis in the synapto-dendritic compartment. Curr Opin Neurobiol 10:132–137.

Williams PL, Warwick R, Dyson M, Bannister LH (eds). 1989. Gray's Anatomy, 37th ed., pp 859–919. Edinburgh: Churchill Livingstone.

Zemanick MC, Strick PL, Dix RD. 1991. Direction of transneuronal transport of herpes simplex virus 1 in the primate motor system is strain-dependent. Proc Natl Acad Sci U S A 88:8048–8051.

8

Ion Channels

S IGNALING IN THE BRAIN DEPENDS on the ability of nerve cells to respond to very small stimuli with rapid and large changes in the electrical potential difference across the cell membrane. In sensory cells, the membrane potential changes in response to minute physical stimuli: Receptors in the eye respond to a single photon of light; olfactory neurons detect a single molecule of odorant; and hair cells in the inner ear respond to tiny movements of atomic dimensions. These sensory responses ultimately lead to the firing of an action potential during which the membrane potential changes at up to 500 volts per second.

The rapid changes in membrane potential that underlie signaling throughout the nervous system are mediated by specialized pores or openings in the membrane called ion channels, a class of integral membrane proteins found in all cells of the body. The ion channels of nerve cells are optimally tuned to respond to specific physical and chemical signals. They are also heterogeneous—in different parts of the nervous system different types of channels carry out specific signaling tasks.

Because of their key roles in electrical signaling, malfunctioning of ion channels can cause a wide variety of neurological diseases (Chapters 57 and 58). Diseases caused by ion channel malfunction are not limited to the brain; for example, cystic fibrosis, skeletal muscle disease, and certain types of cardiac arrhythmia are also caused by ion channel malfunction. Moreover, ion channels are often the site of action of drugs, poisons, or toxins. Thus, ion channels have crucial roles in both the normal physiology and pathophysiology of the nervous system.

In addition to ion channels, nerve cells contain a second important class of proteins specialized for moving ions across cell membranes, the ion transporters and pumps. These proteins do not participate in rapid neuronal signaling but rather are important for establishing and maintaining the concentration gradients of physiologically important ions between the inside and outside of the cell. As we will see in this and the next chapters, ion transporters and pumps differ in important aspects from ion channels, but also share certain common features.

Ion channels have three important properties: (1) They recognize and select specific ions; (2) they open and close in response to specific electrical, chemical, or mechanical, signals; and (3) they conduct ions across the membrane. The channels in nerve and muscle conduct ions across the cell membrane at extremely rapid rates, thereby providing a large flow of electric charge. Up to 100 million ions can pass through a single channel each second. This current causes the rapid changes in membrane potential required for signaling (Chapter 10). The fast flow of ions through channels is comparable to the turnover rate of the fastest enzymes, catalase and carbonic anhydrase, which are limited by diffusion of substrate. (The turnover rates of most other enzymes are considerably slower, ranging from 10 to 1,000 per second.)

Despite such an extraordinary rate of ion flow, channels are surprisingly selective for the ions they allow to permeate. Each type of channel allows only one or a few types of ions to pass. For example, the negative resting potential of nerve cells is largely determined by a class of K^+ channels that are 100-fold more permeable to K^+ than to Na^+. In contrast, generation of the action potential involves a class of Na^+ channels that are 10- to 20-fold more permeable to Na^+ than to K^+. Thus, a key to the great versatility of neuronal signaling is the regulated activation of different classes of ion channels, each of which is selective for specific ions.

Many channels open and close in response to a specific event: Voltage-gated channels are regulated by changes in membrane potential, ligand-gated channels by binding of chemical transmitters, and mechanically gated channels by membrane stretch. Other channels are normally open when the cell is at rest. The ion flux through these "resting" channels largely determines the resting potential.

The flux of ions through ion channels is passive, requiring no expenditure of metabolic energy by the channels. Ion channels are limited to catalyzing the passive movement of ions down their thermodynamic concentration and electrical gradients. The direction of this flux is determined not by the channel itself, but rather by the electrostatic and diffusional driving forces across the membrane. For example, Na^+ ions flow into a cell through voltage-gated Na^+ channels during an action potential because the external Na^+ concentration is much greater than the internal concentration; the open channels allow Na^+ to diffuse into the cell down its concentration gradient.

With such passive ion movement, the Na^+ concentration gradient would eventually dissipate were it not for ion pumps. Different types of ion pumps maintain the concentration gradients for Na^+, K^+, Ca^{2+} and other ions.

These pumps differ from ion channels in two important details. First, whereas open ion channels have a continuous water-filled pathway through which ions flow unimpeded from one side of the membrane to the other, each time a pump moves an ion or group of ions across the membrane, it must undergo a series of conformational changes. As a result, the rate of ion flow through pumps is 100 to 100,000 times slower than through channels. Second, pumps that maintain ion gradients use chemical energy, often in the form of adenosine triphosphate (ATP), to transport ions against their electrical and chemical gradients. Such ion movements are termed *active transport*. The function and structure of ion pumps and transporters are considered in detail at the end of this chapter and in Chapter 9.

In this chapter, we examine six questions: Why do nerve cells have channels? How can channels conduct ions at such high rates and still be selective? How are channels gated? How are the properties of these channels modified by various intrinsic and extrinsic conditions? How does channel structure explain function? Finally, how do ion movements through channels differ from ion movements through transporters? In succeeding chapters, we consider how resting channels and pumps generate the resting potential (Chapter 9), how voltage-gated channels generate the action potential (Chapter 10), and how ligand-gated channels produce synaptic potentials (Chapters 11, 12, and 13).

Ion Channels Are Proteins That Span the Cell Membrane

To appreciate why nerve cells use channels, we need to understand the nature of the plasma membrane and the physical chemistry of ions in solution. The plasma membrane of all cells, including nerve cells, is approximately 6 to 8 nm thick and consists of a mosaic of lipids and proteins. The core of the membrane is formed by a double layer of phospholipids approximately 3 to 4 nm thick. Embedded within this continuous lipid sheet are integral membrane proteins, including ion channels.

The lipids of the membrane do not mix with water—they are hydrophobic. In contrast, the ions within the cell and those outside strongly attract water molecules—they are hydrophilic (Figure 8–1). Ions attract water because water molecules are dipolar: Although the net charge on a water molecule is zero, charge is separated within the molecule. The oxygen atom in a water molecule tends to attract electrons and so bears a small net negative charge, whereas the hydrogen atoms tend to lose electrons and therefore carry a small net positive charge. As a result of this unequal distribution of charge, positively charged ions (cations) are strongly attracted electrostatically to the oxygen atoms of water, and negatively charged ions (anions) are attracted to the hydrogen atoms. Similarly, ions attract water; they become surrounded by electrostatically bound *waters of hydration* (Figure 8–1).

An ion cannot move from water into the uncharged hydrocarbon tails of the lipid bilayer in the membrane unless a large amount of energy is expended to overcome the attraction between the ion and the surrounding water molecules. For this reason, it is extremely unlikely that an ion will move from solution into the lipid bilayer, and therefore, the bilayer itself is almost completely impermeable to ions. Rather, ions cross the membrane through ion channels, where the energetics favor ion movement.

Although their molecular nature has been known with certainty for only approximately 35 years, the idea of ion channels dates to the work of Ernst Brücke at the end of the 19th century. Physiologists had long known that, despite the fact that the cell membrane acts as a barrier, cell membranes are nevertheless permeable to water and many small solutes, including some ions. To explain osmosis, the flow of water across biological membranes, Brücke proposed that membranes contain channels or pores that allow water but not larger solutes to flow. Over 100 years later, Peter Agre found that a family of proteins termed *aquaporins* form channels with a highly selective permeability to water. At the beginning of the 20th century, William Bayliss suggested that water-filled channels would permit ions to cross the cell membrane easily, as the ions would not need to be stripped of their waters of hydration.

The idea that ions move through channels leads to a question: How can a water-filled channel conduct ions at high rates and yet be selective? How, for instance, does a channel allow K^+ ions to pass while excluding Na^+ ions? Selectivity cannot be based solely on the diameter of the ion because K^+, with a crystal radius of 0.133 nm, is larger than Na^+ (crystal radius of 0.095 nm). One important factor that determines ion selectivity is the size of an ion's shell of waters of hydration, because the ease with which an ion moves in solution (its mobility) depends on the size of the ion together with the shell of water surrounding it. The

smaller an ion, the more highly localized is its charge and the stronger its electric field. As a result, smaller ions attract water more strongly. Thus, as Na^+ moves through solution, its stronger electrostatic attraction for water causes it to have a larger water shell, which tends to slow it down relative to K^+. Because of its larger water shell, Na^+ behaves as if it is larger than K^+. The smaller an ion, the lower its mobility in solution. Therefore, we can construct a model of a channel that selects K^+ rather than Na^+ simply on the basis of the interaction of the two ions with water in a water-filled channel (Figure 8–1).

Although this model explains how a channel can select K^+ and exclude Na^+, it does not explain how a channel could select Na^+ and exclude K^+. This problem led many physiologists in the 1930s and 1940s to abandon the channel theory in favor of the idea that ions cross cell membranes by first binding to a specific carrier protein, which then shuttles the ion through the membrane. In this carrier model, selectivity is based on the chemical binding between the ion and the carrier protein, not on the mobility of the ion in solution.

Even though we now know that ions can cross membranes by means of a variety of transport macromolecules, the Na^+-K^+ pump being a well-characterized example (Chapter 9), many properties of membrane ion permeability do not fit the carrier model. Most important is the rapid rate of ion transfer across membranes. An example is provided by the transmembrane current that is initiated when the neurotransmitter acetylcholine (ACh) binds its receptor in the postsynaptic membrane of the nerve–muscle synapse. As described later, the current conducted by a single ACh receptor is 12.5 million ions per second. In contrast, the Na^+-K^+ pump transports at most 100 ions per second.

If the ACh receptor acted as a carrier, it would have to shuttle an ion across the membrane in 0.1 μs (one ten-millionth of a second), an implausibly fast rate. The 100,000-fold difference in rates between the Na^+-K^+ pump and ACh receptor strongly suggests that the ACh receptor (and other ligand-gated receptors) must conduct ions through a channel. Later measurements in many voltage-gated pathways selective for K^+, Na^+, and Ca^{2+} also demonstrated large currents carried by single macromolecules, indicating that they too are channels.

But we are still left with the problem of what makes a channel selective. To explain selectivity, Bertil Hille extended the pore theory by proposing that channels have narrow regions that act as molecular sieves. At this *selectivity filter*, an ion must shed most of its waters of hydration to traverse the channel; in their place, weak chemical bonds (electrostatic interactions) form

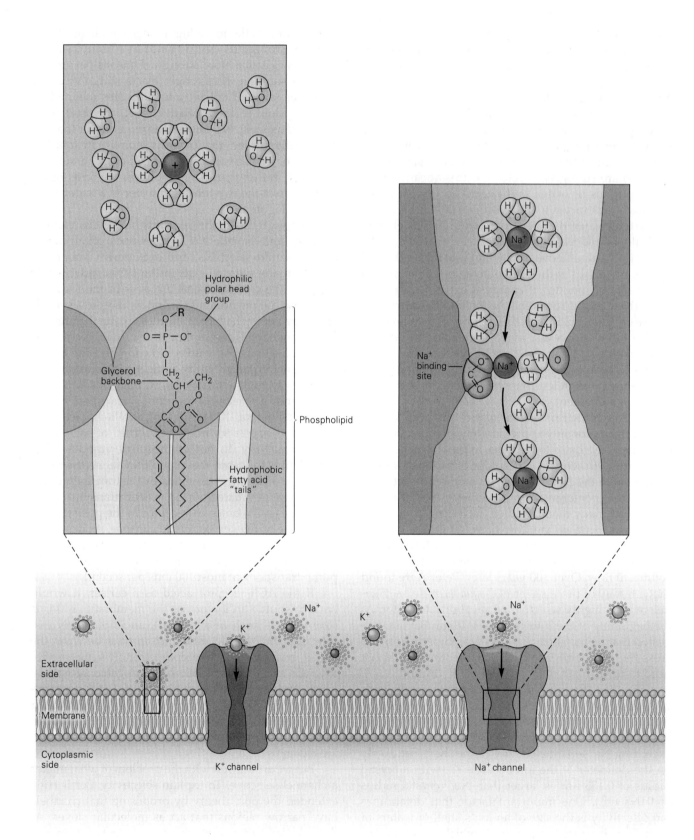

with polar (charged) amino acid residues that line the walls of the channel (Figure 8–1). Because shedding its waters of hydration is energetically unfavorable, the ion will traverse a channel only if its energy of interaction with the selectivity filter compensates for the loss of the energy of interaction with its waters of hydration. Ions traversing the channel are normally bound to the selectivity filter for only a short time (less than 1 μs), after which electrostatic and diffusional forces propel the ion through the channel. In channels where the pore diameter is large enough to accommodate several water molecules, an ion need not be stripped completely of its water shell.

How is this chemical recognition and specificity established? One theory was developed in the early 1960s by George Eisenman to explain the properties of ion-selective glass electrodes. According to this theory, a binding site with high negative field strength—for example, one formed by negatively charged carboxylic acid groups of glutamate or aspartate—will bind Na^+ more tightly than K^+. This selectivity results because the electrostatic interaction between two charged groups, as governed by Coulomb's law, depends inversely on the distance between the two groups.

Because it has a smaller crystal radius than K^+, Na^+ can approach a binding site with a high negative field strength more closely than K^+ can and thus will derive a more favorable free-energy change on binding. This compensates for the requirement that Na^+ lose some of its waters of hydration in order to traverse the narrow selectivity filter. In contrast, a binding site with a low negative field strength—one that is composed, for example, of polar carbonyl or hydroxyl oxygen

atoms—would select K^+ over Na^+. At such a site, the binding of Na^+ would not provide a sufficient free-energy change to compensate for the loss of the ion's waters of hydration, which Na^+ holds strongly. However, such a site would be able to compensate for the loss of a K^+ ion's waters of hydration since the larger K^+ ions interact more weakly with water. It is currently thought that ion channels are selective both because of such specific chemical interactions and because of molecular sieving based on pore diameter.

Ion Channels in All Cells Share Several Functional Characteristics

Most cells are capable of local signaling, but only nerve and muscle cells are specialized for rapid signaling over long distances. Although nerve and muscle cells have a particularly rich variety and high density of membrane ion channels, their channels do not differ fundamentally from those of other cells in the body. Here we describe the general properties of ion channels in a wide variety of cells determined by recording current flow through channels under various experimental conditions.

Currents Through Single Ion Channels Can Be Recorded

Studies of ion channels were originally limited to recording the total current through the entire population of a class of ion channels, an approach that obscures some details of channel function. Later developments

Figure 8–1 (Opposite) The permeability of the cell membrane to ions is determined by the interaction of ions with water, the membrane lipid bilayer, and ion channels. Ions in solution are surrounded by a cloud of water molecules (waters of hydration) that are attracted by the net charge of the ion. This cloud is carried along by the ion as it diffuses through solution, increasing the effective size of the ion. It is energetically unfavorable, and therefore improbable, for the ion to leave this polar environment to enter the nonpolar environment of the lipid bilayer formed from phospholipids.

Phospholipids have a hydrophilic head and a hydrophobic tail. The hydrophobic tails join to exclude water and ions, whereas the polar hydrophilic heads face the aqueous environments of the extracellular fluid and cytoplasm. The phospholipid is composed of a backbone of glycerol in which two –OH groups are linked by ester bonds to fatty acid molecules. The third –OH group of glycerol is linked to phosphoric acid. The phosphate group is further linked to one of a variety of small, polar alcohol head groups (**R**).

Ion channels are integral membrane proteins that span the lipid bilayer, providing a pathway for ions to cross the membrane. The channels are selective for specific ions.

Potassium channels have a narrow pore that excludes Na^+. Although a Na^+ ion is smaller than a K^+ ion, in solution, the effective diameter of Na^+ is larger because its local field strength is more intense, causing it to attract a larger cloud of water molecules. The K^+ channel pore is too narrow for the hydrated Na^+ ion to permeate.

Sodium channels have a selectivity filter that weakly binds Na^+ ions. According to the hypothesis developed by Bertil Hille and colleagues, a Na^+ ion binds transiently at an active site as it moves through the filter. At the binding site, the positive charge of the ion is stabilized by a negatively charged amino acid residue on the channel wall and also by a water molecule that is attracted to a second polar amino acid residue on the other side of the channel wall. It is thought that a K^+ ion, because of its larger diameter, cannot be stabilized as effectively by the negative charge and therefore will be excluded from the filter. (Adapted from Hille 1984.)

Box 8–1 Recording Current in Single Ion Channels: The Patch Clamp

The *patch-clamp* technique was developed in 1976 by Erwin Neher and Bert Sakmann to record current from single ion channels. It is a refinement of the original voltage clamp technique (see Box 10–1).

A small fire-polished glass micropipette with a tip diameter of approximately 1 μm is pressed against the membrane of a skeletal muscle fiber. A metal electrode in contact with the electrolyte in the micropipette connects the pipette to a special electrical circuit that measures the current through channels in the membrane under the pipette tip (Figure 8–2A).

In 1980, Neher discovered that applying a small amount of suction to the patch pipette greatly increased the tightness of the seal between the pipette and the membrane. The result was a seal with extremely high resistance between the inside and the outside of the pipette. The seal lowered the electronic noise and extended the usefulness of the patch-clamp technique to the whole range of ion channels. Since this discovery, the patch-clamp technique has been used to study all the major classes of ion channels in a variety of neurons and other cells (Figure 8–2B).

Christopher Miller independently developed a method for incorporating ion channels from biological membranes into artificial lipid bilayers. He first homogenized the membranes in a blender; using centrifugation of the homogenate, he then separated out a fraction composed only of membrane vesicles. He studied the functional components of these vesicles using a technique developed by Paul Mueller and Donald Rudin in the 1960s. They discovered how to create an artificial lipid bilayer by painting a thin drop of phospholipid over a hole in a nonconducting barrier separating two salt solutions. Miller found that under appropriate ionic conditions his membrane vesicles fused with the planar phospholipid membrane, incorporating any ion channel in the vesicle into the planar membrane.

This technique has two experimental advantages. First, it allows recording from ion channels in regions of cells that are inaccessible to patch clamp; for example, Miller has successfully studied a K⁺ channel isolated from the internal membrane of skeletal muscle (the sarcoplasmic reticulum). Second, it allows researchers to study how the composition of the membrane lipids influences channel function.

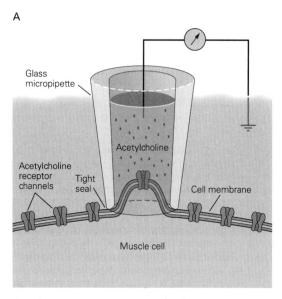

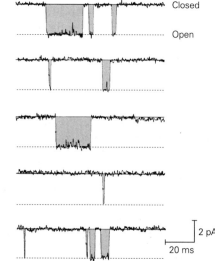

Figure 8–2 Patch-clamp setup and recording.

A. A pipette containing a low concentration of acetylcholine (**ACh**) in saline solution is used to record current through ACh receptor channels in skeletal muscle. (Adapted from Alberts et al. 1994.)

B. Patch-clamp recording of the current through a single ACh receptor channel as the channel switches between closed and open states. (Reproduced, with permission, from B. Sakmann.)

have made it possible to obtain much higher resolution by recording the current through single ion channels. The first direct recordings of individual ion channels in biological membranes were obtained by Erwin Neher and Bert Sakmann in 1976. A glass micropipette containing ACh—the neurotransmitter that activates the ACh receptors in the membrane of skeletal muscle—was pressed tightly against a muscle membrane. Small unitary current pulses representing the opening and closing of a single ACh receptor channel were recorded from the membrane under the pipette tip. The current pulses all had the same amplitude, indicating that the channels open in an all-or-none fashion (Box 8–1).

The pulses measured 2 pA (2×10^{-12} A) at a membrane potential of –80 mV, which according to Ohm's law ($I = V/R$) indicates that the channels had a resistance of 5×10^{11} ohms. In dealing with ion channels, it is more useful to speak of conductance, the reciprocal of resistance ($\gamma = 1/R$), as this provides an electrical measure related to ion permeability. Thus, Ohm's law for a single ion channel can be expressed as $i = \gamma \times V$. The conductance of the ACh receptor channel is approximately 25×10^{-12} siemens (S), or 25 picosiemens (pS), where 1 S equals 1/ohm.

The Flux of Ions Through a Channel Differs From Diffusion in Free Solution

The kinetic properties of ion permeation are best described by the channel's conductance, which is determined by measuring the current (ion flux) through the open channel in response to an electrochemical driving force. The net electrochemical driving force is determined by two factors: the electrical potential difference across the membrane and the concentration gradients of the permeant ions across the membrane. Changing either one can change the net driving force (Chapter 9).

In some open channels, the current varies linearly with driving force—that is, the channels behave as simple resistors. In others, the current is a nonlinear function of driving force. This type of channel behaves as a rectifier—it conducts ions more readily in one direction than in the other because of asymmetry in the channel's structure or ionic environment (Figure 8–3).

The rate of net ion flux (current) through a channel depends on the concentration of the permeant ions in the surrounding solution. At low concentrations, the current increases almost linearly with concentration. At higher concentrations, the current tends to reach a point at which it no longer increases. At this point, the current is said to *saturate*. This saturation effect indicates that ion flux across the cell membrane is not like electrochemical diffusion in free solution but involves the binding of ions to specific polar sites within the pore of the channel. A simple electrodiffusion model would predict that the ionic current should increase in proportion to increases in concentration.

The relation between current and ionic concentration for a wide range of ion channels is well described by a simple chemical binding equation, identical to the Michaelis-Menten equation for enzymes, suggesting that a single ion binds within the channel during permeation. The ionic concentration at which current reaches half its maximum defines the *dissociation constant*, the concentration at which half of the channels will be occupied by a bound ion. The dissociation constant in plots of current versus concentration is typically quite high, approximately 100 mM, indicating weak binding. (In typical interactions between enzymes and substrates, the dissociation constant is below 1 μM.) The rapid rate at which an ion unbinds is necessary for the channel to achieve the very high conduction rates responsible for the rapid changes in membrane potential during signaling.

Figure 8–3 Current–voltage relations. In many ion channels, the relation between current (i) through the open channel and membrane voltage (V_m) is linear (left plot). Such channels are said to be "ohmic" because they follow Ohm's law, $i = V_m/R$ or $V_m \times \gamma$, where γ is conductance. In other channels, the relation between current and membrane potential is nonlinear. This kind of channel is said to "rectify," in the sense that it conducts current more readily in one direction than the other. The right plot shows an outwardly rectifying channel for which positive current (right side) is larger than the negative current (left side) for a given absolute value of voltage.

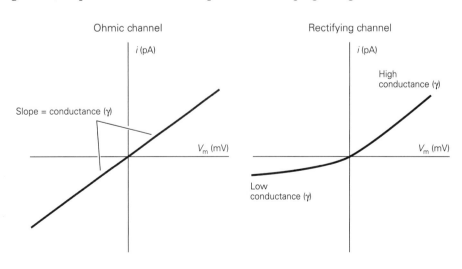

Some ion channels can be blocked by certain free ions or molecules in the cytoplasm or extracellular fluid that bind either to the mouth of the aqueous pore or somewhere within the pore. If the blocker is an ion that binds to a site within the pore, it will be influenced by the membrane electric field as it enters the channel. For example, if a positively charged blocker enters the channel from outside the membrane, then making the cytoplasmic side of the membrane more negative will drive the blocker into the channel by electrostatic attraction, increasing the block. Although some blockers are toxins or drugs that originate outside the body, others are common ions normally present in the cell or its environment. Physiological blockers of certain classes of channels include Mg^{2+}, Ca^{2+}, Na^+, and polyamines such as spermine.

The Opening and Closing of a Channel Involve Conformational Changes

In ion channels that mediate electrical signaling, the channel protein has two or more conformational states that are relatively stable. Each conformation represents a different functional state. For example, each ion channel has at least one open state and one or two closed states. The transition of a channel between these different states is called *gating*.

The molecular mechanisms of gating are only partially understood. In some cases, such as the voltage-gated Cl^- channel described later in the chapter, a local conformational change along the channel lumen gates the channel (Figure 8–4A). In most cases, channel gating involves widespread changes in the channel's conformation (Figure 8–4B). For example, concerted movements, such as twisting, bending, or tilting, of the subunits that line the channel pore mediate the opening and closing of some ion channels (see Figure 8–14 and Chapters 11 and 12). The molecular rearrangements that occur during the transition from closed to open states appear to enhance ion conduction through the channel not only by creating a wider lumen, but also by positioning relatively more polar amino acid constituents at the surface that lines the aqueous pore. In other cases (eg, inactivation of K^+ channels described in Chapter 10), part of the channel protein acts as a particle that can close the channel by blocking the pore (Figure 8–4C).

Three major transduction mechanisms have evolved to control channel opening in neurons. Certain channels are opened by the binding of chemical ligands, known as agonists (Figure 8–5A). Some ligands bind directly to the channel either at an extracellular or intracellular site; transmitters bind

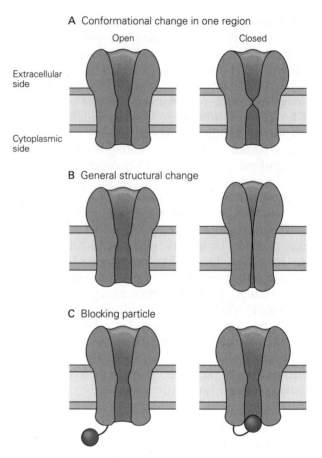

Figure 8–4 Three physical models for the opening and closing of ion channels.

A. A localized conformational change occurs in one region of the channel.

B. A generalized structural change occurs along the length of the channel.

C. A blocking particle swings into and out of the channel mouth.

at extracellular sites, whereas certain cytoplasmic constituents, such as Ca^{2+}, cyclic nucleotides, and GTP-binding proteins, bind at intracellular sites, as do certain dynamically regulated mobile lipid components of the membrane (Chapter 14). Other ligands activate intracellular second messenger signaling cascades that can covalently modify channel gating through protein phosphorylation (Figure 8–5B). Many ion channels are regulated by changes in membrane potential (Figure 8–5C). Some voltage-gated channels act as temperature sensors; changes in temperature shift their voltage gating to higher or lower membrane potentials, giving rise to heat- or cold-sensitive channels. Finally, some channels are regulated by mechanical force (Figure 8–5D).

A Ligand gating

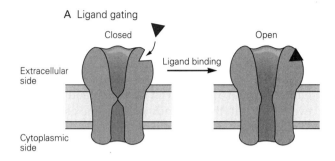

B Phosphorylation gating

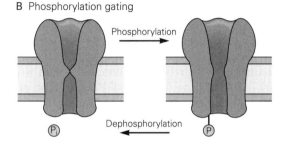

C Voltage gating

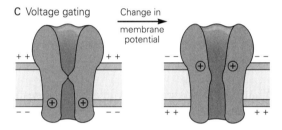

D Stretch or pressure gating

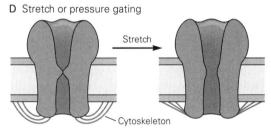

Figure 8–5 Several types of stimuli control the opening and closing of ion channels.

A. A ligand-gated channel opens when a ligand binds a receptor site on the external surface of the channel protein. The energy from ligand binding drives the channel toward an open state.

B. Some channels are regulated by protein phosphorylation and dephosphorylation. The energy for channel opening comes from the transfer of the high-energy phosphate, P_i.

C. Voltage-gated channels open and close with changes in the electrical potential difference across the membrane. The change in membrane potential causes a local conformational change by acting on a region of the channel that has a net charge.

D. Some channels open and close in response to membrane stretch or pressure. The energy for gating may come from mechanical forces that are passed to the channel either directly by distortion of the membrane lipid bilayer or by protein filaments attached to the cytoskeleton or surrounding tissues.

The rapid gating actions necessary for moment-to-moment signaling may also be influenced by certain long-term changes in the metabolic state of the cell. For example, the gating of some K^+ channels is sensitive to intracellular levels of ATP. Some channel proteins contain a subunit with an integral oxidoreductase catalytic domain that is thought to alter channel gating in response to the redox state of the cell.

These regulators control the entry of a channel into one of three functional states: closed and activatable (resting), open (active), or closed and nonactivatable (*inactivated* or *refractory*). A change in the functional state of a channel requires energy. In voltage-gated channels, the energy is provided by the movement of a charged region of the channel protein through the membrane's electric field. This region, the *voltage sensor*, contains a net electric charge, called a *gating charge*, resulting from the presence of basic (positively charged) or acidic (negatively charged) amino acids. The movement of the charged voltage sensor through the electric field in response to a change in membrane potential imparts a change in free energy to the channel that alters the equilibrium between the closed and open states of the channel. For most voltage-gated channels, channel opening is favored by making the inside of the membrane more positive (depolarization).

In transmitter-gated channels, gating is driven by the change in chemical free energy that results when the transmitter binds to a receptor site on the channel protein. Finally, mechanosensitive channels are gated by force transmitted by the distortion of the surrounding lipid bilayer or by protein tethers.

The stimuli that gate the channel also control the rates of transition between the open and closed states of a channel. For voltage-gated channels, the rates are steeply dependent on membrane potential. Although the time scale can vary from several microseconds to a minute, the transition tends to require a few milliseconds on average. Thus, once a channel opens, it stays open for a few milliseconds, and after closing, it stays closed for a few milliseconds before reopening. Once the transition between open and closed states begins, it proceeds virtually instantaneously (in less than 10 µs, the present limit of experimental measurements), thus giving rise to abrupt, all-or-none, step-like changes in current through the channel (Figure 8–2 in Box 8–1).

Ligand-gated and voltage-gated channels enter refractory states through different processes. Ligand-gated channels can enter the refractory state when their exposure to the agonist is prolonged, a process called *desensitization*—an intrinsic property of the interaction between ligand and channel.

Many, but not all, voltage-gated channels enter a refractory state after opening, a process termed *inactivation*. In the inactivated state, the channel is closed and can no longer be opened by positive voltages. Rather, the membrane potential must be returned to its initial negative resting level before the channel can recover from inactivation so that it can again open in response to depolarization. Inactivation of voltage-gated Na^+ and K^+ channels is thought to result from a conformational change, controlled by a subunit or region of the channel separate from that which controls activation. In contrast, the inactivation of certain voltage-gated Ca^{2+} channels is thought to require Ca^{2+} influx. An increase in cytoplasmic Ca^{2+} concentration inactivates the Ca^{2+} channel by binding to the regulatory molecule calmodulin, which is permanently associated with the Ca^{2+} channel protein (Figure 8–6).

Some mechanically gated ion channels that mediate touch sensation inactivate in response to a prolonged stimulus or to a train of brief stimuli. Although the molecular mechanism of this inactivation is not known, it is thought to be an intrinsic property of the channel.

Exogenous factors, such as drugs and toxins, can also affect the gating control sites of an ion channel. Most of these agents tend to inhibit channel opening, but a few facilitate opening. *Competitive antagonists* interfere with normal gating by binding to the same site at which the endogenous agonist normally binds. Antagonist binding, which does not open the channel, blocks access of agonist to the binding site, thereby preventing channel opening. The antagonist binding can be weak and reversible, as in the blockade of the nicotinic ACh-gated channel in skeletal muscle by the plant alkaloid curare, a South American arrow poison (Chapters 11 and 12), or it can be strong and virtually irreversible, as in the blockade of the same channel by the snake venom α-bungarotoxin.

Some exogenous substances modulate gating in a noncompetitive manner, without directly interacting with the transmitter-binding site. For example, binding of the drug diazepam (Valium) to a regulatory site on Cl^- channels that are gated by γ-aminobutyric acid (GABA), an inhibitory neurotransmitter, enhances the frequency with which the channels open in response to GABA binding (Figure 8–7B). This type of indirect, allosteric modulatory effect is found in some voltage- and mechanically gated channels as well.

The Structure of Ion Channels Is Inferred From Biophysical, Biochemical, and Molecular Biological Studies

What do ion channels look like? How does the channel protein span the membrane? What happens to the structure of the channel when it opens and closes? Where along the length of the channel protein do drugs and transmitters bind?

Figure 8–6 Voltage-gated channels are inactivated by two mechanisms.

A. Many voltage-gated channels enter a refractory (inactivated) state after briefly opening in response to depolarization of the membrane. They recover from the refractory state and return to the resting state only after the membrane potential is restored to its resting value.

B. Some voltage-dependent Ca^{2+} channels become inactivated when the internal Ca^{2+} level increases following channel opening. The internal Ca^{2+} binds to calmodulin (**CaM**), a specific regulatory protein associated with the channel.

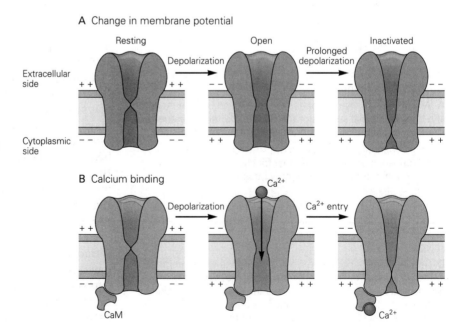

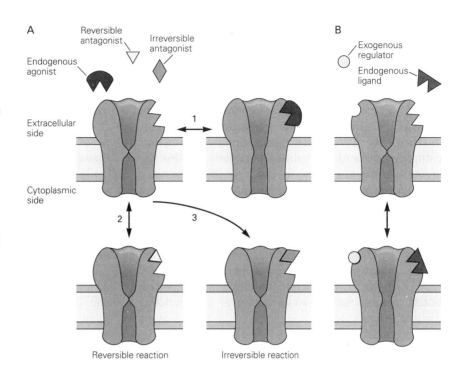

Figure 8–7 Exogenous ligands, such as drugs, can bias an ion channel toward an open or closed state.

A. In channels that are normally opened by the binding of an endogenous ligand (**1**), a drug or toxin may block the binding of the agonist by means of a reversible (**2**) or irreversible (**3**) reaction.

B. Some exogenous agents can bias a channel toward the open state by binding to a regulatory site, distinct from the ligand-binding site that normally opens the channel.

Biophysical, biochemical, and molecular biological studies have provided a basic understanding of channel structure and function. More recent studies using X-ray crystallography and cryo-electron microscopy have provided information about the structure of an increasing number of channels at the atomic level. All ion channels are large integral-membrane proteins with a core transmembrane domain that contains a central aqueous pore spanning the entire width of the membrane. The channel protein often contains carbohydrate groups attached to its external surface. The pore-forming region of many channels is made up of two or more subunits, which may be identical or different. In addition, some channels have auxiliary subunits that may have a variety of effects, including facilitating cell surface expression of the channel, targeting the channel to its appropriate location on the cell surface, and modifying gating properties of the channel. These subunits may be attached to the cytoplasmic end or embedded in the membrane (Figure 8–8).

The genes for all the major classes of ion channels have been cloned and sequenced. The amino acid

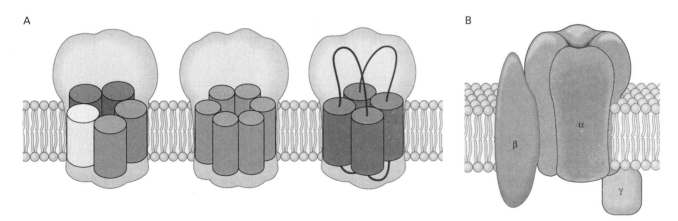

Figure 8–8 Ion channels are integral membrane proteins composed of several subunits.

A. Ion channels can be constructed as hetero-oligomers from distinct subunits (*left*), as homo-oligomers from a single type of subunit (*middle*), or from a single polypeptide chain organized into repeating motifs, where each motif functions as the equivalent of one subunit (*right*).

B. In addition to one or more α-subunits that form a central pore, some channels contain auxiliary subunits (β or γ) that modulate pore gating, channel expression, and membrane localization.

sequence of a channel, inferred from its DNA sequence, can be used to create a structural model of the channel protein. Regions of secondary structure—the arrangement of the amino acid residues into α-helixes and β-sheets—as well as regions that are likely to correspond to membrane-spanning domains of the channel are then predicted based on the structures of related proteins that have been experimentally determined using electron and X-ray diffraction analysis. This type of analysis has identified, for example, the presence of four hydrophobic regions, each around 20 amino acids in length, in the amino acid sequence of a subunit of the ACh receptor channel. Each of these regions is thought to form an α-helix that spans the membrane (Figure 8–9).

Comparing the amino acid sequences of the same type of channel from different species provides additional insights into channel structure and function.

Regions that show a high degree of sequence similarity (that is, have been highly conserved throughout evolution) are likely to be important in determining the structure and function of the channel. Likewise, conserved regions in different but related channels are likely to serve a common biophysical function.

The functional consequences of changes in a channel's primary amino acid sequence can be explored through a variety of techniques. One particularly versatile approach is to use genetic engineering to construct channels with parts that are derived from the genes of different species—so-called *chimeric channels*. This technique takes advantage of the fact that the same type of channel can have somewhat different properties in different species. For example, the bovine ACh receptor channel has a slightly greater single-channel conductance than the ACh receptor channel in electric fish. By comparing the properties of a chimeric channel

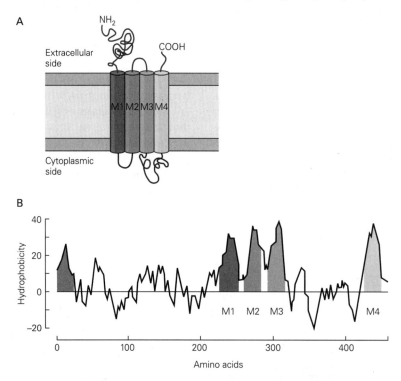

Figure 8–9 The secondary structure of membrane-spanning proteins.

A. A proposed secondary structure for a subunit of the nicotinic acetylcholine (ACh) receptor channel present in skeletal muscle. Each cylinder (**M1–M4**) represents a putative membrane-spanning α-helix comprised of approximately 20 hydrophobic amino acid residues. The membrane segments are connected by cytoplasmic or extracellular segments (loops) of hydrophilic residues. The amino terminus (**NH₂**) and carboxyl terminus (**COOH**) of the protein lie on the extracellular side of the membrane.

B. The membrane-spanning regions of an ion channel protein can be identified using a hydrophobicity plot. The amino acid

sequence of the α-subunit of the nicotinic ACh receptor was inferred from the nucleotide sequence of the cloned receptor subunit gene. Then a running average of hydrophobicity was plotted for the entire amino acid sequence of the subunit. Each point in the plot represents an average hydrophobic index of a sequence of 19 amino acids and corresponds to the midpoint of the sequence. Four of the hydrophobic regions (M1–M4) correspond to the membrane-spanning segments. The hydrophobic region at the far left in the plot is the signal sequence, which positions the hydrophilic amino terminus of the protein on the extracellular surface of the cell during protein synthesis. The signal sequence is cleaved from the mature protein. (Reproduced, with permission, from Schofield et al. 1987.)

to those of the two original channels, one can assess the functions of specific regions of the channel. This technique has been used to identify the membrane-spanning segment that forms the lining of the pore of the ACh receptor channel (Chapter 12).

The roles of different amino acid residues or stretches of residues can be tested using *site-directed mutagenesis*, a type of genetic engineering in which specific amino acid residues are substituted or deleted. Finally, one can exploit the naturally occurring mutations in channel genes. A number of inherited and spontaneous mutations in the genes that encode ion channels in nerve or muscle produce changes in channel function that can underlie certain neurological diseases. Many of these mutations are caused by localized changes in single amino acids within channel proteins, demonstrating the importance of that region for channel function. The detailed functional changes in such channels can then be examined in an artificial expression system.

Ion Channels Can Be Grouped Into Gene Families

The great diversity of ion channels in a multicellular organism is illustrated by the human genome. Our genome contains nine genes encoding variants of voltage-gated Na^+ channels, 10 genes for different Ca^{2+} channels, 80 genes for K^+ channels, 70 genes for ligand-gated channels, and more than a dozen genes for Cl^- channels. Fortunately, the evolutionary relationships between the genes that encode ion channels provide a relatively simple framework with which to categorize them.

Most of the ion channels that have been described in nerve and muscle cells fall into a few gene superfamilies. Members of each gene superfamily have similar amino acid sequences and transmembrane topology and, importantly, related functions. Each superfamily is thought to have evolved from a common ancestral gene by gene duplication and divergence. Several superfamilies can be further subdivided into families of genes encoding channels with more closely related structure and function.

One superfamily encodes ligand-gated ion channels that are receptors for the neurotransmitters ACh, GABA, glycine, or serotonin (Chapter 12). All of these receptors are composed of five subunits, each of which has four transmembrane α-helixes (Figure 8–10A). In addition, the N-terminal extracellular domain that forms the receptor for the ligand contains a conserved loop of 13 amino acids flanked by a pair of cysteine residues that form a disulfide bond. As a result, this receptor superfamily is referred to as the *cys-loop receptors*. Ligand-gated channels can be classified by

their ion selectivity in addition to their agonist. The genes that encode glutamate receptor channels belong to a separate gene family.

Gap-junction channels, which bridge the cytoplasm of two cells at electrical synapses (Chapter 11), are encoded by a separate gene superfamily. A gap-junction channel is composed of two hemi-channels, one from each connected cell. A hemi-channel has six identical subunits, each of which has four membrane-spanning segments (Figure 8–10B).

The genes that encode the voltage-gated ion channels responsible for generating the action potential belong to another superfamily (Chapter 10). These channels are selective for Ca^{2+}, Na^+, or K^+. Comparative DNA sequence data suggest that most voltage-sensitive cation channels stem from a common ancestral channel—perhaps a K^+ channel—that can be traced to a single-cell organism living more than 1.4 billion years ago, before the evolution of separate plant and animal kingdoms.

All voltage-gated cation channels have a similar four-fold symmetric architecture, with a core motif composed of six transmembrane α-helical segments termed S1–S6. A seventh hydrophobic region, the *P-region*, connects the S5 and S6 segments by dipping into and out of the extracellular side of the membrane (Figures 8–10C and 8–11A); it forms the channel's selectivity filter. Voltage-gated Na^+ and Ca^{2+} channels are composed of a large subunit that contains four repeats of this basic motif (Figure 8–10C). Voltage-gated K^+ channels are tetramers, with each separate subunit containing one copy of the basic motif (Figure 8–11A). Each subunit contributes one P-region to the pore of the fully assembled channel. This structural configuration is also shared by other, more distantly related channel families described later and in Chapter 10.

The S4 segment is thought to play a particularly important role in voltage gating. It contains an unusual pattern of amino acids in which every third position contains a positively charged arginine or lysine residue. This region was originally proposed to be the voltage sensor because, according to fundamental biophysical principles, voltage-gating must involve the movement of intramembrane gating charges within the membrane electric field. Additional evidence implicating S4 as the voltage sensor comes from the finding that this pattern of positive charges is highly conserved in all voltage-gated cation-selective channels but is absent in channels that are not voltage-gated. Further support comes from site-directed mutagenesis experiments showing that neutralization of these positive charges in S4 decreases the voltage sensitivity of channel activation.

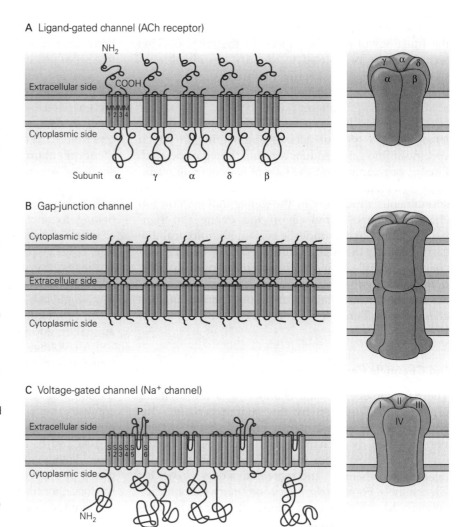

Figure 8–10 Three superfamilies of ion channels.

A. Members of a large family of ligand-gated channels, such as the acetylcholine receptor channel, are composed of five identical or closely related subunits, each of which contains four transmembrane α-helixes (M1–M4). Each cylinder in the figure represents a single transmembrane α-helix.

B. The gap-junction channel is formed from a pair of hemichannels, one each in the pre- and postsynaptic cell membranes, that join in the space between two cells. Each hemichannel is made of six identical subunits, each containing four transmembrane α-helixes. Gap-junction channels serve as conduits between the cytoplasm of the pre- and postsynaptic cells at electrical synapses (Chapter 11).

C. The voltage-gated Na^+ channel is formed from a single polypeptide chain that contains four homologous domains or repeats (motifs I–IV), each with six membrane-spanning α-helixes (**S1–S6**). The S5 and S6 segments are connected by an extended strand of amino acids, the **P-region**, which dips into and out of the external surface of the membrane to form the selectivity filter of the pore. Voltage-gated Ca^{2+} channels share the same general structural pattern, although the amino sequences are different.

A Ligand-gated channel (ACh receptor)

B Gap-junction channel

C Voltage-gated channel (Na^+ channel)

The major gene family encoding the voltage-gated K^+ channels is related to three additional families of K^+ channels, each with distinctive properties and structure. One family includes genes encoding three types of channels activated by either intracellular Na^+ or Ca^{2+} or by intracellular Ca^{2+} plus depolarization. A second family consists of the genes encoding inward-rectifying K^+ channels. Because they are open at the resting potential and rapidly occluded by cytosolic cations during depolarization, they conduct ions more readily in the inward than in the outward direction. Each channel subunit has only two transmembrane segments connected by a pore-forming P-region. A third family of genes encodes K2P channels composed of subunits with two repeated pore-forming segments (Figure 8–11). Various members are regulated by temperature, mechanical force, and intracellular ligands.

These channels may also contribute to the K^+ permeability of the resting membrane.

The sequencing of the genomes of a variety of species, from bacteria to humans, has led to the identification of additional ion channel gene families. These include channels with related P-regions but that are only very distantly related to the family of voltage-gated channels. An example is the excitatory postsynaptic glutamate-gated channel, in which the P-region is inverted, entering and leaving the internal surface of the membrane (Figure 8–11D).

Finally, the transient receptor potential (TRP) family of nonselective cation channels (named after a mutant *Drosophila* strain in which light evokes a brief receptor potential in photoreceptors) comprises a very large and diverse group of tetrameric channels that contain P-regions. Like the voltage-gated K^+ channels,

Figure 8–11 Four related families of ion channels with P-regions.

A. Voltage-gated K⁺ channels are composed of four subunits, each of which corresponds to one repeated domain of voltage-gated Na⁺ or Ca²⁺ channels, with six transmembrane segments and a pore-forming P-region (see Figure 8–10C).

B. Inward-rectifying K⁺ channels are composed of four subunits, each of which has only two transmembrane segments connected by a P-region.

C. The K2P K⁺ channels are composed of subunits that contain two repeats similar to the inward-rectifying K⁺ channel subunit, with each repeat containing a P-region. Two of these subunits combine to form a channel with four P-regions.

D. Glutamate receptors constitute a distinct family of tetrameric channels with P-regions. Their pore regions are nonselectively permeable to cations. In these receptors, the amino terminus is extracellular and the P-region enters and exits the cytoplasmic side of the membrane. The distantly related bacterial GluR0 K⁺-permeable glutamate receptor has four subunits, which contain two transmembrane segments (*left*); in higher organisms, the subunits contain three (*right*).

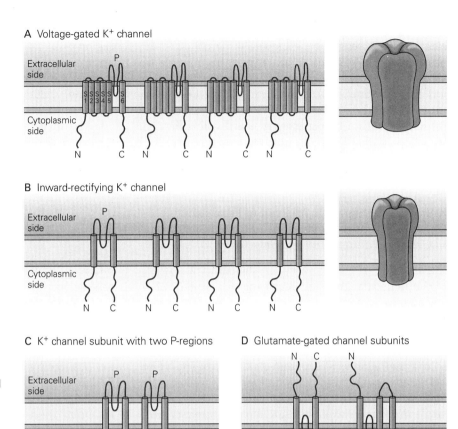

A Voltage-gated K⁺ channel

B Inward-rectifying K⁺ channel

C K⁺ channel subunit with two P-regions

D Glutamate-gated channel subunits

TRP channels also contain six transmembrane segments, but are in most cases gated by intracellular or intramembrane ligands. TRP channels are important for Ca²⁺ metabolism in all cells, visual signaling in insects, and pain, heat, and cold sensation in the nervous system of higher animals (Chapter 18). TRP channels have been implicated in osmoreception and certain taste sensations in mammals.

A number of other families of channels have been identified, structurally unrelated to those considered earlier. These include CLC Cl⁻ channels that help set the resting potential of skeletal muscle cells and certain neurons, nonspecific cation-permeable Piezo channels that are activated by mechanical stimuli (Chapter 18), Na⁺ channels that are activated by H⁺ ions released during inflammation, and ligand-gated cation channels that are activated by ATP, which functions as a neurotransmitter at certain excitatory synapses. With the completion of the human genome project, it is likely that nearly all of the major classes of ion channel genes have now been identified.

The diversity of ion channels is even greater than the large number of ion channel genes. Because most channels in a subfamily are composed of multiple subunits, each type of which may be encoded by a family of closely related genes, combinatorial permutations of these subunits can generate a diverse array of heteromultimeric channels with different functional properties. Additional diversity can be produced by posttranscriptional and posttranslational modifications. These subtle variations in structure and function presumably allow channels to perform highly specific functions. As with enzyme isoforms, variants of a channel with slightly different properties may be expressed at distinct stages of development, in different cell types throughout the brain, and even in different regions within a cell. Changes in neuronal activity can also lead to changes in ion channel expression patterns (Chapter 10).

Biochemical, biophysical, and molecular biological approaches have been important in defining structure–function relationships among the large variety of ion channels. The use of X-ray crystallography and cryo-electron microscopy to define the structure of channels at atomic resolution provides a framework for achieving greater understanding of the mechanisms of ion channel function and malfunction due to disease-causing mutations. Combining a wide array of data from these various approaches makes possible the construction of detailed molecular models, which can be tested by further experiments, as well as by theoretical approaches such as molecular dynamics simulation.

X-Ray Crystallographic Analysis of Potassium Channel Structure Provides Insight Into Mechanisms of Channel Permeability and Selectivity

The first high-resolution X-ray crystallographic analysis of the molecular architecture of the pore region of an ion-selective channel was provided by Rod MacKinnon and his colleagues. To overcome the difficulties inherent in obtaining crystals of large integral membrane proteins, they initially focused on a non–voltage-gated K$^+$ channel, termed KcsA, from a bacterium. This channel is advantageous for crystallography as it can be expressed at high levels for purification, is relatively small, and has a simple transmembrane topology similar to that of the inward-rectifying K$^+$ channel in higher organisms, including mammals (Figure 8–11B).

The crystal structure of the KcsA protein provides several important insights into the mechanisms by which the channel facilitates the movement of K$^+$ ions across the hydrophobic lipid bilayer. The channel is made up of four identical subunits arranged symmetrically around a central pore (Figure 8–12A). Each subunit has two membrane-spanning α-helixes, an inner and outer helix. They are connected by the P-loop, which forms the selectivity filter of the channel. The amino acid sequence of these subunits is homologous to that of the S5-P-S6 region of vertebrate voltage-gated K$^+$ channels. The two α-helixes of each subunit tilt away from the central axis of the pore such that the structure resembles an inverted tepee (Figure 8–12B,C).

The four inner α-helixes from each of the subunits line the cytoplasmic end of the pore. At the intracellular mouth of the channel, these four helixes cross, forming a very narrow opening—the "smoke hole" of the tepee. Because this hole is too small to allow passage of K$^+$ ions, the crystal structure is presumed to represent the channel in the closed state. The inner helixes are homologous to the S6 membrane-spanning segment of voltage-gated K$^+$ channels (Figure 8–11A). At the extracellular end of the channel, the transmembrane helixes in each subunit are connected by a region consisting of three elements: (1) a chain of amino acids that surrounds the mouth of the channel (the turret region), (2) an abbreviated α-helix (the pore helix) approximately 10 amino acids in length that projects toward the central axis of the pore, and (3) a stretch of 5 amino acids near the C-terminal end of the P-region that forms the selectivity filter.

The shape and structure of the pore determine its ion-conducting properties. Both the inner and outer mouths of the pore are lined with acidic amino acids whose negative charges help attract cations from the bulk solution. Going from inside to outside, the pore consists of a medium-wide tunnel, 18 Å in length, which leads into a wider (10 Å diameter) spherical inner chamber. This chamber is lined predominantly by the side chains of hydrophobic amino acids. These relatively wide regions are followed by the very narrow selectivity filter, only 12 Å in length, which is rate-limiting for the passage of K$^+$ ions. A high K$^+$ ion throughput rate is ensured by the fact that the inner 28 Å of the pore, from the cytoplasmic entrance to the selectivity filter, lacks polar groups that could delay ion passage by binding and unbinding the ion (Figure 8–12C,D).

An ion passing from the polar solution through the nonpolar lipid bilayer encounters the least energetically favorable region in the middle of the bilayer. The large energy difference between these two regions for a K$^+$ ion is minimized by two details of channel structure. The inner chamber is filled with water, which provides a highly polar environment, and the pore helixes provide dipoles whose electronegative carboxyl ends point toward this inner chamber (Figure 8–12C).

The high energetic cost incurred as a K$^+$ ion sheds its waters of hydration is partially compensated by the presence of 20 electronegative oxygen atoms that line the walls of the selectivity filter and form favorable electrostatic interactions with the permeant ions. Each of the four subunits contributes four main-chain carbonyl oxygen atoms from the protein backbone and one side-chain hydroxyl oxygen atom to form a total of four binding sites for K$^+$ ions. Each bound K$^+$ ion is thus stabilized by interactions with a total of eight oxygen atoms, which lie in two planes above and below the bound cation. In this way, the channel is able to compensate for the loss of the K$^+$ ion's waters of hydration. The selectivity filter is stabilized at a critical width, such that it provides optimal electrostatic interactions with K$^+$ ions as they pass but is too wide for the smaller Na$^+$ ions to interact effectively with all eight oxygen atoms at any point along the length of the filter (Figure 8–12C).

In light of the extensive interactions between a K$^+$ ion and the channel, how does the KcsA channel

A Looking down the channel

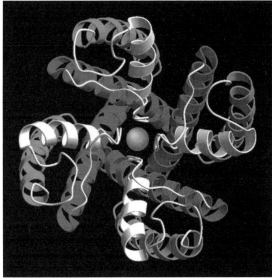

B Cross-section

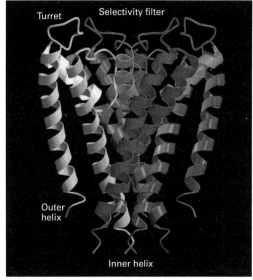

C K⁺ ion binding sites

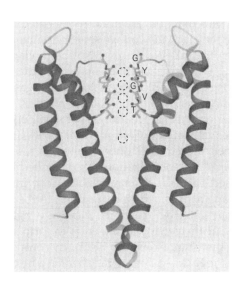

D K⁺ ion movements

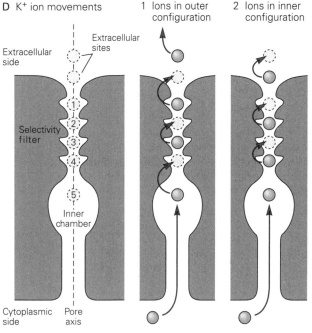

Figure 8–12 The X-ray crystal structure of a bacterial potassium channel. (Reproduced, with permission, from Doyle et al. 1998. Copyright © 1998 AAAS.)

A. The view is looking down the channel pore from outside the cell. Each of the four subunits of the KcsA K⁺ channel contributes two membrane-spanning helixes, an outer helix (**blue**) and an inner helix (**red**). The P-region (**white**) lies near the extracellular surface of the channel pore and consists of a short α-helix (pore helix) and a loop that forms the selectivity filter of the channel. In the center of the pore is a K⁺ ion (**pink**).

B. The channel is seen in a side view within a cross section of the membrane. The four subunits are shown in different colors.

C. Another view in the same orientation as in **part B** shows only two of the four subunits. The channel contains five K⁺ binding sites (**dashed**). Four of the sites lie in the selectivity filter (**yellow**), while the fifth site lies in an inner chamber near the center of the channel. The four K⁺ binding sites of the selectivity filter are formed by successive rings of oxygen atoms (**red**) from five amino acid residues per subunit. Four of the rings are formed by carbonyl oxygen atoms from the main chain backbone of four consecutive amino acid residues—glycine (**G**), tyrosine

(**Y**), glycine (**G**), and valine (**V**). A fifth ring of oxygen near the internal end of the selectivity filter is formed by the side-chain hydroxyl oxygen of threonine (**T**). Each ring contains four oxygen atoms, one from each subunit. Only the oxygen atoms from two of the four subunits are shown in this view. (Reproduced, with permission, from Morais-Cabral et al. 2001. Copyright © 2001 Springer Nature.)

D. A view of K⁺ ion permeation through the channel illustrates the sequence of changes in occupancy of the various K⁺ binding sites. A pair of ions hops in concert between a pair of binding sites in the selectivity filter. In the initial state, the "outer configuration," a pair of ions is bound to sites 1 and 3. As an ion enters the inner mouth of the channel, the ion in the inner chamber jumps to occupy the innermost binding site of the selectivity filter (site 4). This causes the pair of ions in the outer configuration to hop outward, expelling an ion from the channel. The two ions now in the inner configuration (sites 2 and 4) can hop to binding sites 1 and 3, returning the channel to its initial state (the outer configuration), from which it can conduct a second K⁺ ion. (Adapted, with permission, from Miller 2001. Copyright © 2001 Springer Nature.)

manage its high rate of conduction? Although the channel contains a total of five potential binding sites for K^+ ions, X-ray analysis shows that the channel can be occupied by at most three K^+ ions at any instant. One ion is normally present in the wide inner chamber, and two ions occupy two of the four binding sites within the selectivity filter (Figure 8–12D).

These structural data led to the following hypothesis. Because of electrostatic repulsion, two K^+ ions never simultaneously occupy adjacent binding sites within the selectivity filter; rather, a water molecule is always interposed between K^+ ions. During conduction, a pair of K^+ ions within the selectivity filter hop in tandem between pairs of binding sites. If only one ion were in the selectivity filter, it would be bound rather tightly, and the throughput rate for ion permeation would be compromised. But the mutual electrostatic repulsion between two K^+ ions occupying nearby sites ensures that the ions will linger only briefly, thus resulting in a high overall rate of K^+ conduction.

The form of the KcsA selectivity filter appears to be highly conserved among various types of mammalian voltage-gated K^+ channels. However, more recent studies by MacKinnon and colleagues have revealed how variations in geometric and surface charge features below the selectivity filter of this canonical pore can cause some voltage-gated K^+ channels to differ markedly in single-channel conductance and in affinity for various open channel blockers.

The snug fit between the K^+ channel selectivity filter and K^+ ions that helps explain the unusually high selectivity of these channels is not representative of all channel types. As we shall see in later chapters, in many channels pore diameters are significantly wider than the principal permeating ion, contributing to a lower degree of selectivity.

X-Ray Crystallographic Analysis of Voltage-Gated Potassium Channel Structures Provides Insight into Mechanisms of Channel Gating

As described earlier, the S4 segment of voltage-gated ion channels is thought to be the voltage sensor that detects changes in membrane potential. How do the positive charges in S4 move through the membrane electric field in response to a change in membrane potential? How is S4 movement coupled to gating? What is the relationship of the voltage-sensing region to the pore-forming region of the channel? What is the configuration of the open channel? Some answers to these questions have come from X-ray crystallographic analyses of mammalian voltage-gated K^+ channels, as well as from a number of studies using mutagenesis

and other biophysical approaches. MacKinnon and colleagues studied the mammalian voltage-gated Kv1.2 K^+ channel, as well as a closely related chimera Kv1.2-Kv2.1, which yielded higher-resolution images.

Their analysis of X-ray crystal structures of the Kv1.2 channel and the Kv1.2-2.1 chimera showed that a K^+ channel subunit has two domains. The S1–S4 segments form a voltage-sensing domain at the periphery of the channel, whereas the S5-P–S6 segments form the pore domain at the central axis of the channel. The two domains are linked at their intracellular ends by the short S4–S5 coupling helix (Figure 8–13). The idea that the S1–S4 voltage sensor is a separate domain is supported by the fact that certain bacterial proteins contain S1–S4 domains but lack a pore domain. One such protein is a voltage-sensitive phosphatase, while another forms a voltage-gated proton channel. Conversely, the inward-rectifying K^+ channels (Figure 8–11B) have a high K^+ selectivity but are not directly gated by voltage because they lack the voltage sensor domain.

The crystal structures also help clarify what happens when the channel opens. Studies by Clay Armstrong in the 1960s suggested that a gate exists at the intracellular mouth of voltage-gated K^+ channels of higher organisms. He found that small organic cations such as tetraethylammonium (TEA) can enter and block the channel only when this internal gate is opened by depolarization. As described earlier, in the closed bacterial K^+ channels the four inner transmembrane helixes, which correspond to the S6 helixes in voltage-gated K^+ channels, meet at a tight bundle crossing at their cytoplasmic ends to form the closed gate of the channel (Figure 8–12). In contrast, the S6 helix of the Kv1.2–2.1 chimera is bent at a flexible three-amino-acid hinge (proline-valine-proline), causing the inner end of the helix to bend outward. This configuration results in an open channel conformation with an internal orifice that is dilated to 12 Å in diameter, wide enough to pass fully hydrated K^+ ions as well as larger cations such as TEA (Figures 8–13 and 8–14C). Once inside the channel lumen, TEA blocks K^+ permeation because it is too large to pass through the selectivity filter. It is not surprising that the Kv channel is in the open state, as there is no voltage gradient across the channel in the crystals. This is similar to the situation in a membrane that has been depolarized to 0 mV, a voltage at which the channels are normally open. This opening mechanism is likely to be a general one because the inner helixes of many K^+ channels in bacteria and higher organisms also have a flexible hinge at this position.

One long-standing question concerns the placement and movement of the gating charges on the S4

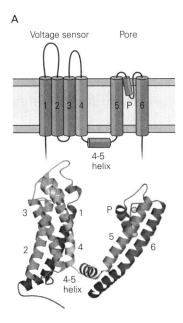

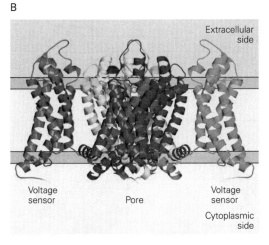

Figure 8–13 X-ray crystal structure of a voltage-gated potassium channel. (Adapted, with permission, from Long et al. 2007. Copyright © 2007 Springer Nature.)

A. *Top*: In addition to its six transmembrane α-helixes (**S1–S6**), a voltage-gated K⁺ channel subunit contains a short α-helix (the **P** helix) that is part of the P-region selectivity filter, as well as an α-helix on the cytoplasmic side of the membrane that connects transmembrane helixes S4 and S5 (4–5 coupling helix). *Bottom*: An X-ray structural model of a single subunit shows the positions of the six membrane-spanning helixes, the P helix, and the 4–5 coupling helix. The S1–S4 voltage-sensing region and S5-P–S6 pore-forming regions are localized in separate domains. Two K⁺ ions bound in the pore are shown in **pink**.

B. In this side view of the channel, each subunit is a different color. Subunit 6 (**red**) is in the same orientation as in **part A.**

segment voltage sensor. As mentioned earlier, movement of these charges within the plane of the membrane in response to changes in membrane potential is thought to couple membrane depolarization to channel gating. However, placement of charges within the hydrophobic membrane results in an unfavorable energy state, as discussed earlier for free ions in solutions. How do ion channels compensate for this unfavorable free energy?

The crystal structure provides some answers to this question. Mutagenesis studies indicate that four positively charged arginine residues in the external half of the S4 segment are likely to carry most of the gating charge. In the open state, the four positive charges face outward toward the extracellular side of the membrane, where they may undergo energetically favorable interactions with water or with the negatively charged head groups of the phospholipid bilayer. Positive charges on other S4 residues that lie more deeply within the lipid bilayer are stabilized by interactions with negatively charged acidic residues on the S1–S3 transmembrane helixes.

At present, a crystal structure for the closed state of the Kv1.2–Kv2.1 chimera is lacking. However, MacKinnon and colleagues have proposed a plausible model for voltage gating based on the structures of the open voltage-gated K⁺ channel and the closed bacterial K⁺ channel KcsA (Figure 8–14). According to this model, a negative voltage inside the cell exerts a force on the positively charged S4 helix that causes it to move inward by about 1.0 to 1.5 nm. As a result, the four positively charged S4 residues, which in the depolarized state face the external environment and sense the extracellular potential, now face the intracellular side of the membrane and sense the intracellular potential. In this manner, movement of each S4 segment will translocate 3 to 4 gating charges across the membrane electric field as the channel transitions between the closed and open states, for a total of 12 to 16 charges moved per tetrameric channel. This number is similar to the total gating charge movement determined from biophysical measurements (Chapter 10).

How are S4 movements coupled to the gate of the channel? According to the model, when the membrane voltage becomes negative, the resulting inward movement of the S4 segment exerts a downward force on the S4–S5 coupling helix. This helix, which lies roughly parallel to the membrane at its cytoplasmic surface, rests on the inner end of the S6 helix gate in the open state. As the S4–S5 helix moves downward, it acts as a lever, applying force to S6 and closing the gate. Thus, voltage-gating is thought to rely on the electromechanical coupling between the voltage-sensing domain and the

Figure 8–14 Model for voltage gating based on X-ray crystal structures of two potassium channels. In each part of the figure, the drawing on the *left* shows the actual structure of the open voltage-gated Kv1.2-2.1 chimera, while the drawing on the *right* shows the hypothetical structure of a closed voltage-gated K$^+$ channel, based in part on the structure of the pore region of the bacterial K$^+$ channel KcsA in the closed state. (Adapted, with permission, by Yu-hang Chen from Long et al. 2007. Copyright © 2007 Springer Nature.)

A. A view looking down on the open and closed channel from outside the cell. The central pore is constricted in the closed state, preventing K$^+$ flow through the channel.

B. A view of the S1–S4 voltage-sensing domain from the side, parallel to the plane of the membrane. Positively charged residues in S4 are shown as **blue sticks**. In the open state, when the membrane is depolarized, four positive charges on the S4 helix are located in the external half of the membrane, facing the external solution. The positive charges in the interior of the membrane are stabilized by interactions with negatively charged residues in S1 and S2 (**red sticks**). In the closed state, when the membrane potential is negative, the S4 region moves inward so that its positive charges now lie in the inner half of the membrane. The inward movement of S4 causes the cytoplasmic S4–S5 coupling helix (**orange**) to move downward.

C. The putative conformational change in the channel pore upon voltage gating. A side view of the tetrameric S5-P–S6 pore region of the channel shows the S4–S5 coupling helix. Membrane repolarization causes the downward movement of the S4–S5 helix, applying force to the S6 inner helix of the pore (**blue**). This causes the S6 helix to bend at its pro-val-pro hinge, thereby closing the gate of the channel.

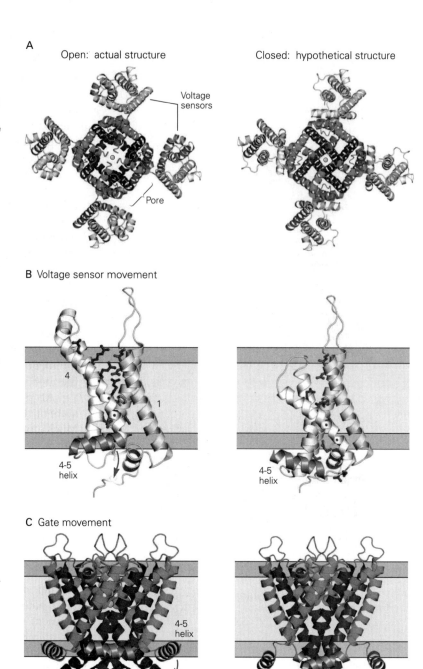

pore domain of the channel. Although this electromechanical coupling model provides a satisfying picture of how changes in membrane voltage lead to channel gating, a definitive answer to this key problem will require resolution of the structure of the closed state of a related voltage-gated mammalian K$^+$ channel.

Such a direct coupling between the sensing element of a channel (S4/S4–S5 linker) and the pore gate (S6) is found in most voltage-gated K$^+$ channels, as well

as in voltage-gated Na$^+$ and Ca^{2+} channels. However, in many cases, the element of a channel that responds directly to the gating signal is not in direct contact with the channel gate, and instead, an allosteric mechanism propagates the response indirectly by more remote conformational changes. For example, in the voltage-gated K$^+$ channel Kv10, the S4–S5 linker is not in a position to act as a lever on S6. Rather, the inward movement of S4 in response to a negative potential closes the S6 gate indirectly, by laterally compressing the S5 helix, which is packed against the S6 helix. Additional cases of allosteric gating mechanisms are discussed later in the context of inactivation of voltage-gated Na⁺ channels (Chapter 10) and activation of ligand-gated (Chapters 12 and 13) and mechanically gated channels (Chapter 18).

The Structural Basis of the Selective Permeability of Chloride Channels Reveals a Close Relation Between Channels and Transporters

Ions move across cell membranes by active transport by ion transporters or pumps, as well as by passive diffusion through ion channels. Ion transporters are distinguished from ion channels because (1) they use a source of energy to actively transport ions against their electrochemical gradients, and (2) they transport ions at rates much lower than ion channels, too low to support fast neuronal signaling. Nevertheless, some types of transporters and certain ion channels have similar structures, according to studies of the CLC family of proteins.

The CLC proteins expressed in vertebrates consist of a family of Cl⁻ channels and a closely related family of Cl⁻-H⁺ cotransporters. Cotransporters use the electrochemical gradient of one ion to move another ion against its electrochemical gradient. The CLC Cl⁻-H⁺ cotransporters, which are expressed in intracellular organelles, transfer two Cl⁻ ions across the membrane in exchange for one proton. This type of transporter is termed an ion exchanger.

The human voltage-gated ClC-1 channels are important for maintaining the resting potential in skeletal muscle (Chapter 57).

The crystal structures of the human ClC-1 channel and the homologous *E. coli* CLC exchangers have been determined by MacKinnon and colleagues. They found a close similarity in amino acid sequence to be reflected in a marked overall structural resemblance. Both types of CLC proteins consist of a homodimer composed of two identical subunits. Each subunit forms a separate ion pathway, and the two subunits function independently (Figure 8–15). The structures of the CLC proteins are quite different from those of K⁺ channels. Unlike the pore of a K⁺ channel, which is widest in the central region, each pore of the CLC protein has an hourglass profile. The neck of the hourglass, a tunnel 12 Å in length which forms the selectivity filter, is just wide enough to contain three fully dehydrated Cl⁻ ions.

Although the ion permeation pathways of the CLC proteins and the K⁺ channel differ in significant respects, they have evolved four similar features that are critical to their function. First, their selectivity filters contain multiple sequential binding sites for the permeating ion. Multi-ion occupancy creates a metastable state that facilitates rapid ion passage. Second, the ion binding sites are formed by polar, partially charged atoms, not by fully ionized atoms. The resultant relatively weak binding energy ensures that the permeating ions do not become too tightly bound. Third, permeant ions are stabilized in the center of the membrane by the positively polarized ends of two α-helixes. Fourth, wide, water-filled vestibules at either end of the selectivity filter allow ions to approach the filter in a partially hydrated state. Thus although the K⁺ channels and CLC proteins differ fundamentally in amino acid sequence and overall structure, strikingly

Figure 8–15 The vertebrate CLC family of chloride channels and transporters are double-barrel channels with two identical pores.

A. Recordings of current through a single vertebrate Cl⁻ channel show three levels of current: both pores closed (**0**), one pore open (**1**), and both pores open (**2**). (Adapted from Miller 1982.)

B. The channel is shown from the side (*left*) and looking down on the membrane from outside the cell (*right*). Each subunit contains its own ion transport pathway and gate. In addition, the dimer has a gate shared by both subunits (not shown).

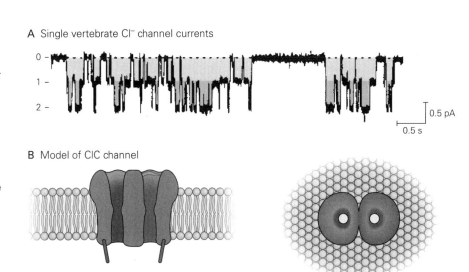

A Single vertebrate Cl⁻ channel currents

B Model of ClC channel

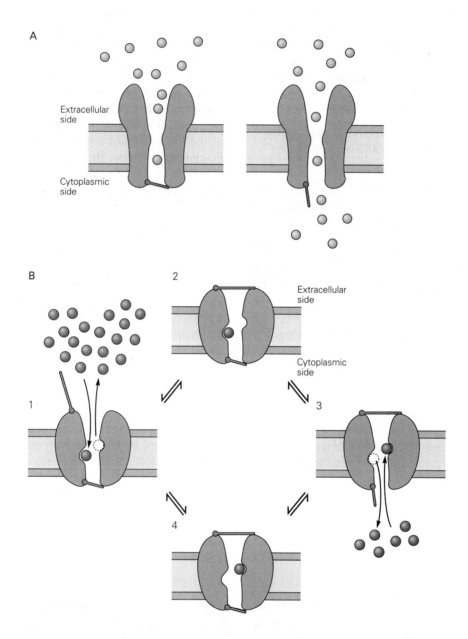

Figure 8–16 The functional difference between ion channels and transporters or pumps. (Adapted, with permission, from Gadsby 2004. Copyright © 2004 Springer Nature.)

A. Ion channels have a continuous aqueous pathway for ion conduction across the membrane. This pathway can be occluded by the closing of a gate.

B. Ion pumps and transporters have two gates in series that control ion flux. The gates are never open simultaneously, but both can close to trap one or more ions in the pore. The type of transporter illustrated here moves two different types of ions in opposite directions and is termed an *exchanger* or *antiporter*. Ion movement is tightly linked to a cycle of opening and closing of the two gates. When the external gate is open, one type of

ion leaves while the other type enters the pore (**1**). This triggers a conformational change, causing the external gate to shut, thereby trapping the incoming ion (**2**). A second conformational change then causes the internal gate to open, allowing the trapped ion to leave and a new ion to enter (**3**). A further conformational change closes the internal gate, allowing the cycle to continue (**4**). With each cycle, one type of ion is transported from the outside to the inside of the cell, whereas the other type is transported from the inside to the outside of the cell. By coupling the movements of two or more ions, an exchanger can use the energy stored in the electrochemical gradient of one ion to actively transport another ion against its electrochemical gradient.

similar functional features have evolved in these two classes of membrane proteins that promote both a high degree of ion selectivity and efficient throughput. These features have been conserved with surprising fidelity from prokaryotes through humans.

More detailed structural studies will be needed to understand how some CLC proteins function as Cl^--H^+ exchangers whereas others act as conventional channels. Most exchangers and pumps, such as the Na^+-K^+ pump (Chapter 9), are thought to have two gates, one external and one internal, which are never open simultaneously. Instead, ion movements and gate movements are presumed to be part of a tightly coupled reaction cycle (Figure 8–16). CLC exchangers apparently have two gates that control Cl^- flux, coupled to the protonation-deprotonation cycle of a flexible glutamate residue in the selectivity filter that shuttles protons across the membrane. The resultant conformational changes enable transport of Cl^- against its concentration gradient, driven by the flux of H^+ ions down their electrochemical gradient. CLC channels are apparently built on a very similar structure as the transporters, but with modified gates and small structural changes in the ion transport pathway that allow much more rapid movement of Cl^- down its electrochemical gradient.

Highlights

1. Ions cross cell membranes through two main classes of integral membrane proteins—ion channels and ion pumps or transporters.

2. Ion channels act as catalysts for the passive flux of ions across the membrane. Channels have a central water-filled pore that substitutes for the polar environment on either side of the membrane. It allows the electrically charged ions to rapidly cross the nonpolar environment of the cell membrane, driven by the ion's electrochemical gradient.

3. Most ion channels are selectively permeable to certain ions. The portion of the channel pore called the selectivity filter determines which ions can penetrate based on the ion's charge, size, and physicochemical interactions with the amino acids that line the wall of the pore.

4. Ion channels have gates that open and close in response to different signals. In the open state, channels generate ionic currents that produce rapid voltage signals that carry information in the nervous system and in other excitable cells.

5. Most ion channels have three states: open, closed, and inactivated or desensitized. Transitioning

between these states is termed gating. Depending on the type of channel, gating is controlled by various factors, including membrane voltage, ligand binding, mechanical force, phosphorylation state, and temperature.

6. The most common ion channels in nerve and muscle cells belong to three major gene superfamilies, the members of which are related by gene sequence homology and, in most cases, by functional properties.

7. Most ion channels are composed of multiple subunits. Combinatorial permutations of these subunits can generate a diverse array of channels with different functional properties. Posttranscriptional modifications generate additional diversity.

8. The various types of ion channels are differentially expressed in different types of neurons and different regions of neurons, contributing to the functional complexity and computational power of the nervous system. The expression patterns of some ion channels and transporters change during development and in response to changes in neuronal activity patterns.

9. The rich variety of ion channels in different types of neurons has stimulated intensive efforts to develop drugs that can activate or block specific channel types in nerve and muscle cells. Such drugs would, in principle, maximize therapeutic effectiveness with minimal side effects.

10. Structure-function and X-ray crystallographic studies of voltage-gated ion channels have provided key insights into the molecular and atomic-level details of K^+ channel conduction, selectivity, and gating. Recent technical advances in single-particle cryo-electron microscopy have led to rapid progress in studies of a wide range of ion channels.

11. Active transport, which is mediated by integral membrane proteins called transporters or pumps, enables ions to move across the membrane against their electrochemical gradient. The driving force that generates active ion fluxes comes either from chemical energy (the hydrolysis of ATP) or from the favorable electrochemical potential difference for a cotransported ion.

12. Most ion transporters and pumps do not provide a continuous pathway for ions. Rather, they undergo conformational changes for the different phases of the transport cycle, thereby providing alternating access of the molecule's central lumen to the two sides of the membrane. Because these conformational changes are relatively slow,

they are much less efficient than ion channels in mediating ion fluxes.

John D. Koester
Bruce P. Bean

Selected Reading

Hille B. 2001. *Ion Channels of Excitable Membranes,* 3rd ed. Sunderland, MA: Sinauer.

Isacoff EY, Jan LY, Minor DL. 2013. Conduits of life's spark: a perspective on ion channel research since the birth of *Neuron*. Neuron 80:658–674.

Jentsch TJ, Pusch M. 2018. CLC chloride channels and transporters: structure, function, physiology and disease. Physiol Rev 98:1493–1590.

Miller C. 1987. How ion channel proteins work. In: LK Kaczmarek, IB Levitan (eds). *Neuromodulation: The Biological Control of Neuronal Excitability,* pp. 39–63. New York: Oxford Univ. Press.

Yu FH, Yarov-Yarovoy V, Gutman GA, Catterall WA. 2005. Overview of molecular relationships in the voltage-gated ion channel superfamily. Pharmacol Rev 57:387–395.

References

Accardi A, Miller C. 2004. Secondary active transport mediated by a prokaryotic homologue of ClC Cl⁻ channels. Nature 427:803–807.

Alberts B, Bray D, Lewis J, Raff M, Roberts K, Watson JD. 1994. *Molecular Biology of the Cell,* 3rd ed. New York: Garland.

Armstrong CM. 1971. Interaction of tetraethylammonium ion derivatives with the potassium channels of giant axons. J Gen Physiol 58:413–437.

Basilio D, Noack K, Picollo A, Accardi A. 2014. Conformational changes required for H^+/Cl^- exchange mediated by a CLC transporter. Nat Struct Mol Biol 21:456–464.

Bayliss WM. 1918. *Principles of General Physiology,* 2nd ed., rev. New York: Longmans, Greene.

Boscardin E, Alijevic O, Hummler E, Frateschi S, Kellenberger S. 2016. The function and regulation of acid-sensing ion channels (ASICs) and the epithelial Na+ channel (ENaC): IUPHAR Review 19. Br J Pharmacol. 173:2671–2701.

Brücke E. 1843. Beiträge zur Lehre von der Diffusion tropfbarflüssiger Korper durch poröse Scheidenwände. Ann Phys Chem 58:77–94.

Coste B, Xiao B, Santos JS, et al. 2012. Piezo proteins are poreforming subunits of mechanically activated channels. Nature 483:176–181.

Doyle DA, Cabral JM, Pfuetzner RA, et al. 1998. The structure of the potassium channel: molecular basis of K^+ conduction and selectivity. Science 280:69–77.

Eisenman G. 1962. Cation selective glass electrodes and their mode of operation. Biophys J 2:259–323. Suppl 2.

Enyedi P, Gabor G. 2010. Molecular background of leak K^+ currents: two-pore domain potassium channels. Physiol Rev 90:550–605.

Feng L, Campbell EB, MacKinnon R. 2012. Molecular mechanism of proton transport in CLC Cl^-/H^+ exchange transporters. Proc Natl Acad Sci U S A 109:11699–11704.

Gadsby DC. 2004. Ion transport: spot the difference. Nature 427:795–797.

Hamill OP, Marty A, Neher E, Sakmann B, Sigworth FJ. 1981. Improved patch-clamp techniques for high-resolution current recording from cells and cell-free membrane patches. Pflugers Arch 391:85–100.

Hansen SB. 2015. Lipid antagonism: the PIP2 paradigm of ligand-gated ion channels. Biochim Biophys Acta 1851:620–628.

Henderson R, Unwin PNT. 1975. Three-dimensional model of purple membrane obtained by electron microscopy. Nature 257:28–32.

Hille B. 1973. Potassium channels selective permeability to small cations. J Gen Physiol 61:669–686.

Hille B. 1984. *Ion Channels of Excitable Membranes,* Sunderland, MA: Sinauer.

Isom LL, DeJongh KS, Catterall WA. 1994. Auxiliary subunits of voltage-gated ion channels. Neuron 12:1183–1194.

Kaczmarek LK. 2013. Slack, slick, and sodium-activated potassium channels. ISRN Neurosci 2013:354262.

Katz B, Thesleff S. 1957. A study of the "desensitization" produced by acetylcholine at the motor end-plate. J Physiol (Lond) 138:63–80.

Kyte J, Doolittle RF. 1982. A simple method for displaying the hydropathic character of a protein. J Mol Biol 157:105–132.

Lau C, Hunter MJ, Stewart A, Perozo E, and Vandenberg JI. 2019. Never at rest: insights into the conformational dynamics of ion channels from cryo-electron microscopy. J Physiol 596:1107–1119.

Lewis AH, Cui AF, McDonald MF, Grandl J. 2017. Transduction of repetitive mechanical stimuli by Piezo1 and Piezo2 ion channels. Cell Rep 19:2572–2585.

Long SB, Tao X, Campbell EB, MacKinnon R. 2007. Atomic structure of a voltage-dependent K^+ channel in a lipid membrane-like environment. Nature 450:376–382.

Miller C. 1982. Open-state substructure of single chloride channels from *Torpedo electroplax*. Philos Trans R Soc Lond B Biol Sci 299:401–411.

Miller C (ed). 1986. *Ion Channel Reconstitution.* New York: Plenum.

Miller C. 2001. See potassium run. Nature 414:23–24.

Morais-Cabral JH, Zhou Y, MacKinnon R. 2001. Energetic optimization of ion conduction rate by the K^+ selectivity filter. Nature 414:37–42.

Moran Y, Barzilai MG, Liebeskind BJ, Zakon HH. 2015. Evolution of voltage-gated ion channels at the emergence of Metazoa. J Exp Biol 218:515–525.

Murata Y, Iwasaki H, Sasaki M, Inaba K, Okamura Y. 2005. Phosphoinositide phosphatase activity coupled to an intrinsic voltage sensor. Nature 435:1239–1243.

Neher E, Sakmann B. 1976. Single-channel currents recorded from membrane of denervated frog muscle fibres. Nature 260:799–802.

Nishida M, MacKinnon R. 2002. Structural basis of inward rectification: cytoplasmic pore of the G protein-gated inward rectifier GIRK1 at 1.8 Å resolution. Cell 111:957–965.

Noda M, Takahashi H, Tanabe T, et al. 1983. Structural homology of *Torpedo californica* acetylcholine receptor subunits. Nature 302:528–532.

Noda M, Shimizu S, Tanabe T, et al. 1984. Primary structure of *Electrophorus electricus* sodium channel deduced from cDNA sequence. Nature 312:121–127.

Park E, MacKinnon R. 2018. Structure of the CLC chloride channel from *Homo sapiens*. eLife 7:36629.

Payandeh J, Scheuer T, Zheng N, Catterall WA. 2011. The crystal structure of a voltage-gated sodium channel. Nature 475:353–359.

Peterson BZ, DeMaria CD, Yue DT. 1999. Calmodulin is the Ca^{2+} sensor for Ca^{2+}-dependent inactivation of L-type calcium channels. Neuron 22:549–558.

Pongs O, Schwarz JR. 2010. Ancillary subunits associated with voltage-dependent K^+ channels. Physiol Rev 90:755–796.

Prager-Khoutorsky M, Arkady Khoutorsky A, Bourque CW. 2014. Unique interweaved microtubule scaffold mediates osmosensory transduction via physical interaction with TRPV1. Neuron 83:866–878.

Preston GM, Agre P. 1992. Appearance of water channels in *Xenopus* oocytes expressing red cell CHIP28 protein. Science 256:385–387.

Ramsey IS, Moran MM, Chong JA, Clapham DE. 2006. A voltage-gated proton-selective channel lacking the pore domain. Nature 440:1213–1216.

Rogers CJ, Twyman RE, MacDonald RL. 1994. Benzodiazepine and β-carboline regulation of single $GABA_A$ receptor channels of mouse spinal neurones in culture. J Physiol (Lond) 475:69–82.

Schofield PR, Darlison MG, Fujita N, et al. 1987. Sequence and functional expression of the $GABA_A$ receptor shows a ligand-gated receptor super-family. Nature 328:221–227.

Tao X, Hite RK, MacKinnon R. 2017. Cryo-EM structure of the open high-conductance Ca^{2+}-activated K^+ channel. Nature 541:46–51.

Voets T, Droogmans T, Wissenbach U, Jannssens A, Flockerzi V, Nillus B. 2004. The principle of temperature-dependent gating in cold- and heat-sensitive TRP channels. Nature 430:748–754.

Wang W, MacKinnon R. 2017. Cryo-EM structure of the open human ether-à-go-go-related K^+ channel hERG. Cell 169:422–430.

Wu LJ, Sweet TB, Clapham DE. 2010. Current progress in the mammalian TRP ion channel family. Pharmacol Rev 62:381–404.

Yang N, George AL Jr, Horn R. 1996. Molecular basis of charge movement in voltage-gated sodium channels. Neuron 16:113–122.

Zhou Y, Morais-Cabral JH, Kaufman A, MacKinnon R. 2001. Chemistry of ion coordination and hydration revealed by a K^+ channel-Fab complex at 2.0 Å resolution. Nature 414:43–48.

9

Membrane Potential and the Passive Electrical Properties of the Neuron

INFORMATION IS CARRIED WITHIN neurons and from neurons to their target cells by electrical and chemical signals. Transient electrical signals are particularly important for carrying time-sensitive information rapidly and over long distances. These transient electrical signals—receptor potentials, synaptic potentials, and action potentials—are all produced by temporary changes in the electric current into and out of the cell, changes that drive the electrical potential across the cell membrane away from its resting value. This current represents the flow of negative and positive ions through ion channels in the cell membrane.

Two types of ion channels—resting and gated—have distinctive roles in neuronal signaling. Resting channels are primarily important in maintaining the resting membrane potential, the electrical potential across the membrane in the absence of signaling. Some types of resting channels are constitutively open and are not gated by changes in membrane voltage; other types are gated by changes in voltage but are also open at the negative resting potential of neurons. Most voltage-gated channels, in contrast, are closed when the membrane is at rest and require membrane depolarization to open.

In this and the next several chapters, we consider how transient electrical signals are generated in the neuron. We begin by discussing how particular ion channels establish and maintain the membrane potential when the membrane is at rest and briefly describe the mechanism by which the resting potential can be perturbed, giving rise to transient electrical signals such as the action potential. We then consider how the passive electrical properties of neurons—their resistive and capacitive characteristics—contribute to the integration and local propagation of synaptic and receptor potentials within the neuron. In Chapter 10 we examine the detailed mechanisms by which voltage-gated Na^+, K^+, and Ca^{2+} channels generate the action potential, the electrical signal conveyed along the axon. Synaptic potentials are considered in Chapters 11 to 14, and receptor potentials are discussed in Part IV in connection with the actions of sensory receptors.

The Resting Membrane Potential Results From the Separation of Charge Across the Cell Membrane

The neuron's cell membrane has thin clouds of positive and negative ions spread over its inner and outer surfaces. At rest, the extracellular surface of the membrane has an excess of positive charge and the cytoplasmic surface an excess of negative charge (Figure 9–1). This

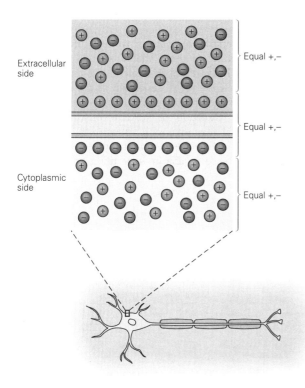

Figure 9–1 The cell membrane potential results from the separation of net positive and net negative charges on either side of the membrane. The excess of positive ions outside the membrane and negative ions inside the membrane represents a small fraction of the total number of ions inside and outside the cell at rest.

separation of charge is maintained because the lipid bilayer of the membrane is a barrier to the diffusion of ions (Chapter 8). The charge separation gives rise to the *membrane potential* (V_m), a difference of electrical potential, or voltage, across the membrane defined as

$$V_m = V_{in} - V_{out},$$

where V_{in} is the potential on the inside of the cell and V_{out} the potential on the outside.

The membrane potential of a cell at rest, the *resting membrane potential* (V_r), is equal to V_{in} since by convention the potential outside the cell is defined as zero. Its usual range is –60 mV to –70 mV. All electrical signaling involves brief changes away from the resting membrane potential caused by electric currents across the cell membrane.

The electric current is carried by ions, both positive (cations) and negative (anions). The direction of current is conventionally defined as the direction of *net movement of positive charge*. Thus, in an ionic solution, cations move in the direction of the electric current and anions move in the opposite direction. In the nerve cell at rest, there is no net charge movement across the membrane. When there is a net flow of cations or anions into or out of the cell, the charge separation across the resting membrane is disturbed, altering the electrical potential of the membrane. A reduction or reversal of charge separation, leading to a less negative membrane potential, is called *depolarization*. An increase in charge separation, leading to a more negative membrane potential, is called *hyperpolarization*.

Changes in membrane potential that do not lead to the opening of gated ion channels are passive responses of the membrane and are called *electrotonic potentials*. Hyperpolarizing responses are almost always passive, as are small depolarizations. However, when depolarization approaches a critical level, or threshold, the cell responds actively with the opening of voltage-gated ion channels, which produces an all-or-none *action potential* (Box 9–1).

The Resting Membrane Potential Is Determined by Nongated and Gated Ion Channels

The resting membrane potential is the result of the passive flux of individual ion species through several classes of resting channels. Understanding how this passive ionic flux gives rise to the resting potential enables us to understand how the gating of different

Box 9–1 Recording the Membrane Potential

Reliable techniques for recording the electrical potential across cell membranes were developed in the late 1940s. These techniques allow accurate recordings of both the resting membrane potential and action potentials (Figure 9–2).

Glass micropipettes filled with a concentrated salt solution serve as electrodes and are placed on either side of the cell membrane. Wires inserted into the back ends of the pipettes are connected via an amplifier to an oscilloscope, which displays the amplitude of the membrane potential in volts. Because the diameter of such a *microelectrode* tip is minute (<1 µm), it can be inserted into a cell with relatively little damage to the cell membrane (Figure 9–2A).

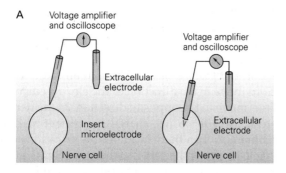

Figure 9–2A Recording setup.

When both electrodes are outside the cell, no electrical potential difference is recorded. But as soon as one microelectrode is inserted into the cell, the oscilloscope shows a steady voltage, the resting membrane potential. In most nerve cells at rest, the membrane potential is approximately –65 mV (Figure 9–2B).

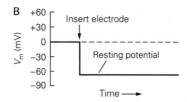

Figure 9–2B Oscilloscope display.

The membrane potential can be experimentally changed using a current generator connected to a second pair of electrodes—one intracellular and one extracellular. When the intracellular electrode is made positive with respect to the extracellular one, a pulse of positive current from the current generator causes positive charge to flow into the neuron from the intracellular electrode. This current returns to the extracellular electrode by flowing outward across the membrane.

As a result, the inside of the membrane becomes more positive while the outside of the membrane becomes more negative. This decrease in the separation of charge is called *depolarization*.

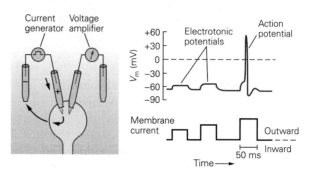

Figure 9–2C Depolarization.

Small depolarizing current pulses evoke purely electrotonic (passive) potentials in the cell—the size of the change in potential is proportional to the size of the current pulses. However, a sufficiently large depolarizing current triggers the opening of voltage-gated ion channels. The opening of these channels leads to the action potential, which differs from electrotonic potentials in the way in which it is generated as well as in magnitude and duration (Figure 9–2C).

Reversing the direction of current—making the intracellular electrode negative with respect to the extracellular electrode—makes the membrane potential more negative. This increase in charge separation is called *hyperpolarization*.

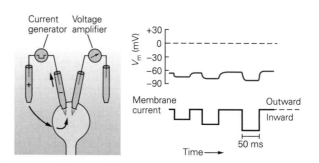

Figure 9–2D Hyperpolarization.

Hyperpolarization does not trigger an active response in the cell. The responses of the cell to hyperpolarization are usually purely electrotonic. As the size of the current pulse increases, the hyperpolarization increases proportionately (Figure 9–2D).

Table 9–1 Distribution of the Major Ions Across a Neuronal Membrane at Rest: The Giant Axon of the Squid

Species of ion	Concentration in cytoplasm (mM)	Concentration in extracellular fluid (mM)	Equilibrium potential[1] (mV)
K^+	400	20	–75
Na^+	50	440	+55
Cl^-	52	560	–60
A^- (organic anions)	385	None	None

[1]The membrane potential at which there is no net flux of the ion species across the cell membrane.

types of ion channels generates the action potential, as well as the receptor and synaptic potentials.

No single ion species is distributed equally on the two sides of a nerve cell membrane. Of the four most abundant ions found on either side of the cell membrane, Na^+ and Cl^- are concentrated outside the cell and K^+ and A^- (organic anions, primarily amino acids and proteins) inside. Table 9–1 shows the distribution of these ions inside and outside of one particularly well-studied nerve cell process, the giant axon of the squid, whose extracellular fluid has a salt concentration similar to that of seawater. Although the absolute values of the ionic concentrations for vertebrate nerve cells are two- to three-fold lower than those for the squid giant axon, the *concentration gradients* (the ratio of the external to internal ion concentration) are similar.

The unequal distribution of ions raises several important questions. How do ionic gradients contribute to the resting membrane potential? What prevents the ionic gradients from dissipating by diffusion of ions across the membrane through the resting channels? These questions are interrelated, and we shall answer them by considering two examples of membrane permeability: the resting membranes of glial cells, which are permeable to only one species of ion, and the resting membranes of nerve cells, which are permeable to three. For the purposes of this discussion, we shall only consider the resting channels that are not gated by voltage and thus are always open.

Open Channels in Glial Cells Are Permeable to Potassium Only

The permeability of a cell membrane to a particular ion species is determined by the relative proportions of the various types of ion channels that are open. The simplest case is that of the glial cell, which has a resting potential of approximately –75 mV. Like most cells, a glial cell has high concentrations of K^+ and A^- on the

inside and high concentrations of Na^+ and Cl^- on the outside. However, most resting channels in the membrane are permeable only to K^+.

Because K^+ ions are present at a high concentration inside the cell, they tend to diffuse across the membrane from the inside to the outside of the cell down their chemical concentration gradient. As a result, the outside of the membrane accumulates a net positive charge (caused by the slight excess of K^+) and the inside a net negative charge (because of the deficit of K^+ and the resulting slight excess of anions). Because opposite charges attract each other, the excess positive charges on the outside and the excess negative charges on the inside collect locally on either surface of the membrane (Figure 9–1).

The flux of K^+ out of the cell is self-limiting. The efflux of K^+ gives rise to an electrical potential difference—positive outside, negative inside. The greater the flow of K^+, the more charge is separated and the greater is the potential difference. Because K^+ is positive, the negative potential inside the cell tends to oppose the further efflux of K^+. Thus, K^+ ions are subject to two forces driving them across the membrane: (1) a *chemical driving force,* a function of the concentration gradient across the membrane, and (2) an *electrical driving force,* a function of the electrical potential difference across the membrane.

Once K^+ diffusion has proceeded to a certain point, the electrical driving force on K^+ exactly balances the chemical driving force. That is, the outward movement of K^+ (driven by its concentration gradient) is equal to the inward movement of K^+ (driven by the electrical potential difference across the membrane). This potential is called the K^+ *equilibrium potential*, E_K (Figure 9–3). In a cell permeable only to K^+ ions, E_K determines the resting membrane potential, which in most glial cells is approximately –75 mV.

The equilibrium potential for any ion X can be calculated from an equation derived in 1888 from basic

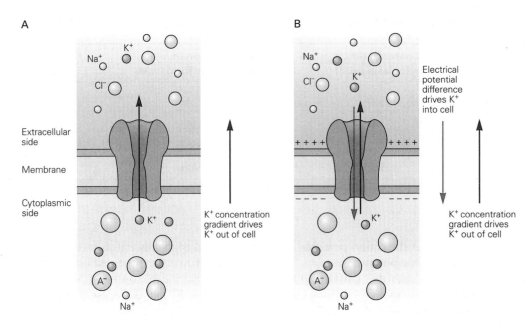

Figure 9–3 The flux of K⁺ across a cell membrane is determined by both the K⁺ concentration gradient and the membrane potential.

A. In a cell permeable only to K⁺, the resting potential is generated by the efflux of K⁺ down its concentration gradient.

B. The continued efflux of K⁺ builds up an excess of positive charge on the outside of the cell and leaves behind an excess

of negative charge inside the cell. This buildup of charge leads to a potential difference across the membrane that impedes the further efflux of K⁺, so eventually an equilibrium is reached: The electrical and chemical driving forces are equal and opposite, so as many K⁺ ions move in as move out.

thermodynamic principles by the German physical chemist Walter Nernst:

$$E_x = \frac{RT}{zF} \ln \frac{[X]_o}{[X]_i}, \qquad \textbf{Nernst Equation}$$

where R is the gas constant, T the temperature (in degrees Kelvin), z the valence of the ion, F the Faraday constant, and $[X]_o$ and $[X]_i$ the concentrations of the ion outside and inside the cell. (To be precise, chemical activities rather than concentrations should be used.)

Since RT/F is 25 mV at 25°C (77°F, room temperature), and the constant for converting from natural logarithms to base 10 logarithms is 2.3, the Nernst equation can also be written as follows:

$$E_x = \frac{58\,\text{mV}}{z} \log \frac{[X]_o}{[X]_i}.$$

Thus, for K⁺, since $z = +1$ and given the concentrations inside and outside the squid axon in Table 9–1:

$$E_K = \frac{58\,\text{mV}}{1} \log \frac{[20]}{[400]} = -75\,\text{mV}.$$

The Nernst equation can be used to find the equilibrium potential of any ion that is present on both sides of a membrane permeable to that ion (the potential is sometimes called the *Nernst potential*). The equilibrium potentials for the distributions of Na⁺, K⁺, and Cl⁻ ions across the squid giant axon are given in Table 9–1.

In our discussion so far, we have treated the generation of the resting potential as a passive mechanism—the diffusion of ions down their chemical gradients—one that does not require the expenditure of energy by the cell. However, energy from hydrolysis of adenosine triphosphate (ATP) is required to set up the initial concentration gradients and to maintain them in neurons, as we shall see below.

Open Channels in Resting Nerve Cells Are Permeable to Three Ion Species

Unlike glial cells, nerve cells at rest are permeable to Na⁺ and Cl⁻ ions in addition to K⁺ ions. Of the abundant ion species in nerve cells, only the large organic anions (A⁻) are unable to permeate the cell membrane. How are the concentration gradients for the three permeant ions (Na⁺, K⁺, and Cl⁻) maintained across the membrane of a single cell, and how do these three gradients interact to determine the cell's resting membrane potential?

To answer these questions, it is easiest to examine first only the diffusion of K^+ and Na^+. Let us return to the simple example of a cell having only K^+ channels, with concentration gradients for K^+, Na^+, Cl^-, and A^- as shown in Table 9–1. Under these conditions, the resting membrane potential V_r is determined solely by the K^+ concentration gradient and is equal to E_K (–75 mV) (Figure 9–4A).

Now consider what happens if a few resting Na^+ channels are added to the membrane, making it slightly permeable to Na^+. Two forces drive Na^+ into the cell: Na^+ tends to flow into the cell down its chemical concentration gradient, and it is driven into the cell by the negative electrical potential difference across the membrane (Figure 9–4B). The influx of Na^+ depolarizes the cell, but only slightly from the K^+ equilibrium potential (–75 mV). The new membrane potential does not come close to the Na^+ equilibrium potential of +55 mV because there are many more resting K^+ channels than Na^+ channels in the membrane.

As soon as the membrane potential begins to depolarize from the value of the K^+ equilibrium potential, K^+ flux is no longer in equilibrium across the membrane. The reduction in the electrical force driving K^+ into the cell means that there is now a net flow of K^+ out of the cell, tending to counteract the Na^+ influx. The more the membrane potential is depolarized and driven away from the K^+ equilibrium potential, the greater is the net electrochemical force driving K^+ out of the cell and consequently the greater the net K^+ efflux. Eventually the membrane potential reaches a new resting level at which the increased outward movement of K^+ just balances the inward movement of Na^+ (Figure 9–4C). This balance point (usually approximately –65 mV) is far from the Na^+ equilibrium potential (+55 mV) and is only slightly more positive than the K^+ equilibrium potential (–75 mV).

To understand how this balance point is determined, bear in mind that the magnitude of the flux of an ion across a cell membrane is the product of its *electrochemical driving force* (the sum of the electrical and chemical driving forces) and the conductance of the membrane to the ion:

$$\text{ion flux} = (\text{electrical driving force}$$
$$+ \text{ chemical driving force})$$
$$\times \text{ membrane conductance.}$$

In a resting nerve cell, relatively few Na^+ channels are open, so the membrane conductance of Na^+ is quite low. Thus, despite the large chemical and electrical forces driving Na^+ into the cell, the influx of Na^+ is small. In contrast, many K^+ channels are open in the membrane of a resting cell so that the membrane conductance of K^+ is relatively large. Because of the high conductance of K^+ relative to Na^+ in the cell at rest, the small net outward force acting on K^+ is enough to produce a K^+ efflux equal to the Na^+ influx.

The Electrochemical Gradients of Sodium, Potassium, and Calcium Are Established by Active Transport of the Ions

As we have seen, the passive movement of K^+ out of the resting cell through open channels balances the passive movement of Na^+ into the cell. However, this steady leakage of ions cannot be allowed to continue unopposed for any appreciable length of time because the Na^+ and K^+ gradients would eventually run down, reducing the resting membrane potential.

Dissipation of ionic gradients is prevented by the sodium-potassium pump (*Na^+-K^+ pump*), which moves Na^+ and K^+ *against* their electrochemical gradients: It extrudes Na^+ from the cell while taking in K^+. The pump therefore requires energy, and the energy comes from hydrolysis of ATP. Thus, at the resting membrane potential, the cell is not in equilibrium but rather in a *steady state*: There is a continuous passive influx of Na^+ and efflux of K^+ through resting channels that is exactly counterbalanced by the Na^+-K^+ pump.

As we saw in the previous chapter, pumps are similar to ion channels in that they catalyze the movement of ions across cell membranes. However, they differ in two important respects. First, whereas ion channels are passive conduits that allow ions to move down their electrochemical gradient, pumps require a source of chemical energy to transport ions against their electrochemical gradient. Second, ion transport is much faster in channels: Ions typically flow through channels at a rate of 10^7 to 10^8 per second, whereas pumps operate at speeds more than 10,000 times slower.

The Na^+-K^+ pump is a large membrane-spanning protein with catalytic binding sites for Na^+ and ATP on its intracellular surface and for K^+ on its extracellular surface. With each cycle, the pump hydrolyzes one molecule of ATP. (Because the Na^+-K^+ pump hydrolyzes ATP, it is also referred to as the Na^+-K^+ ATPase.) It uses this energy of hydrolysis to extrude three Na^+ ions from the cell and bring in two K^+ ions. The unequal flux of Na^+ and K^+ ions causes the pump to generate a net outward ionic current. Thus, the pump is said to be *electrogenic*. This pump-driven efflux of positive charge tends to set the resting potential a few millivolts more negative than would be achieved by the passive

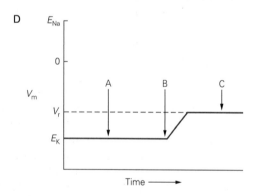

A K⁺ channels only

K⁺ driving forces
Chemical Electrical

B K⁺ and Na⁺ channels

Na⁺ driving forces
Chemical Electrical

Na⁺

K⁺

C Steady state

Na⁺

K⁺

Net driving forces		Net currents	
K⁺	Na⁺	K⁺	Na⁺

D

E_{Na}

0

V_m

V_r

E_K

A   B   C

Time ⟶

Figure 9–4 The resting potential of a cell is determined by the proportions of different types of ion channels that are open, together with the value of their equilibrium potentials. The channels in the figures represent the entire complement of K⁺ or Na⁺ channels in this hypothetical cell membrane. The lengths of the arrows within the channels represent the relative amplitudes of the electrical (**red**) and chemical (**blue**) driving forces acting on Na⁺ or K⁺. The lengths of the arrows in the diagram on the right denote the relative sizes of the net driving force (the sum of the electrical and chemical driving forces) for Na⁺ and K⁺ and the net ion currents. Three hypothetical situations are illustrated.

A. In a resting cell in which only K⁺ channels are present, K⁺ ions are in equilibrium and $V_m = E_K$.

B. Adding a few Na⁺ channels to the resting membrane allows Na⁺ ions to diffuse into the cell, and this influx begins to depolarize the membrane.

C. The resting potential settles at a new level (V_r), where the influx of Na⁺ is balanced by the efflux of K⁺. In this example, the aggregate conductance of the K⁺ channels is much greater than that of the Na⁺ channels because the K⁺ channels are more numerous. As a result, a relatively small net driving force for K⁺ drives a current equal and opposite to the Na⁺ current driven by the much larger net driving force for Na⁺. This is a steady-state condition, in which neither Na⁺ nor K⁺ is in equilibrium but the net flux of charge is null.

D. Membrane voltage changes during the hypothetical situations illustrated in parts **A**, **B**, and **C**.

diffusion mechanisms discussed earlier. During periods of intense neuronal activity, the increased influx of Na^+ leads to an increase in Na^+-K^+ pump activity that generates a prolonged outward current, leading to a hyperpolarizing after-potential that can last for several minutes, until the normal Na^+ concentration is restored. The Na^+-K^+ pump is inhibited by ouabain or digitalis plant alkaloids, an action that is important in the treatment of heart failure.

The Na^+-K^+ pump is a member of a large family of pumps known as *P-type ATPases* (because the phosphoryl group of ATP is temporarily transferred to the pump). P-type ATPases include a *Ca^{2+} pump* that transports Ca^{2+} across cell membranes (Figure 9–5A). All cells normally maintain a very low cytoplasmic Ca^{2+} concentration, between 50 and 100 nM. This concentration is more than four orders of magnitude lower than the external concentration, which is approximately 2 mM

in mammals. Calcium pumps in the plasma membrane transport Ca^{2+} out of the cell; other Ca^{2+} pumps located in internal membranes, such as the smooth endoplasmic reticulum, transport Ca^{2+} from the cytoplasm into these intracellular Ca^{2+} stores. Calcium pumps are thought to transport two Ca^{2+} ions for each ATP molecule that is hydrolyzed, with two protons transported in the opposite direction.

The Na^+-K^+ pump and Ca^{2+} pump have similar structures. They are formed from 110 kD α-subunits, whose large transmembrane domain contains 10 membrane-spanning α-helixes (Figure 9–5A). In the Na^+-K^+ pump, an α-subunit associates with an obligatory β-subunit that is required for proper assembly and membrane expression of the pump. In humans, four genes encode highly related Na^+-K^+ pump α-subunits (*ATP1A1, ATP1A2, ATP1A3, ATP1A4*). Mutations in *ATP1A2* result in familial hemiplegic migraine, a form

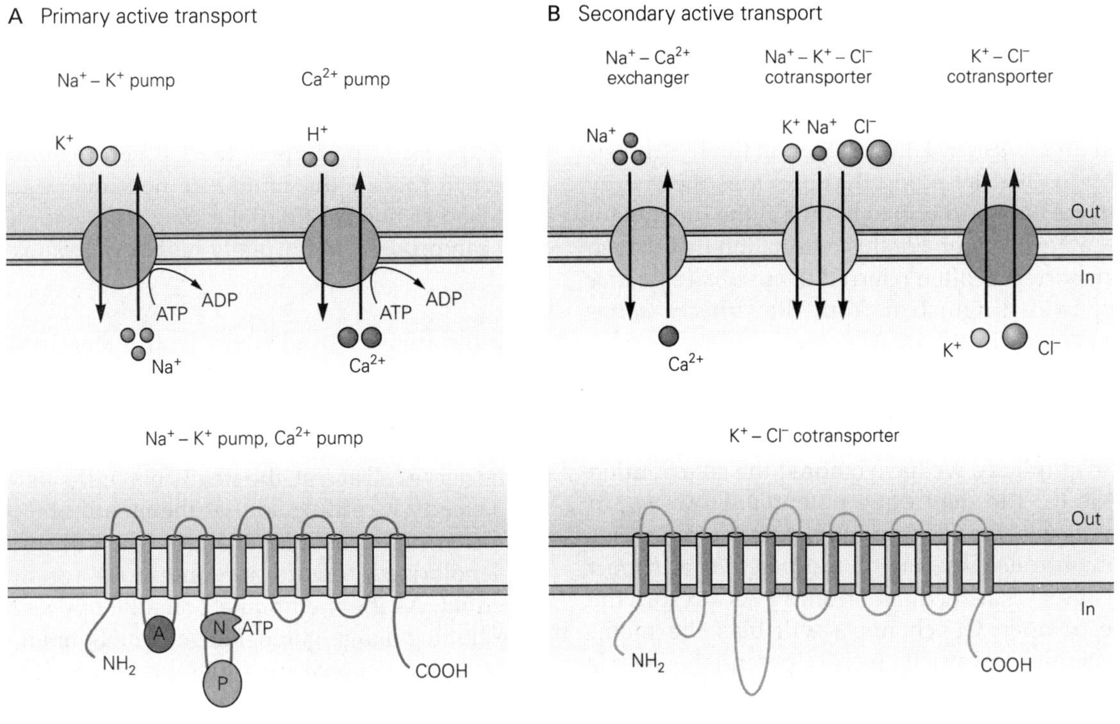

A Primary active transport

B Secondary active transport

Figure 9–5 Pumps and transporters regulate the chemical concentration gradients of Na^+, K^+, Ca^{2+}, and Cl^- ions.

A. The Na-K^+ pump and Ca^{2+} pump are two examples of active transporters that use the energy of adenosine triphosphate (ATP) hydrolysis to transport ions against their concentration gradient. The α-subunit of a Na-K^+ pump or homologous Ca^{2+} pump (**below**) has 10 transmembrane segments, a cytoplasmic amino terminus, and a cytoplasmic carboxyl terminus. There are also cytoplasmic loops important for binding ATP (**N**), ATP hydrolysis and phosphorylation of the pump (**P**), and transducing phosphorylation to transport (**A**). The Na^+-K^+ pump

also contains a smaller β-subunit with a single transmembrane domain plus a small accessory integral membrane protein **FXYD**, which modulates pump kinetics (not shown).

B. The Na^+-Ca^{2+} exchanger uses the potential energy of the electrochemical gradient of Na^+ to transport Ca^{2+} out of a cell. The Na^+-Ca^{2+} exchanger contains nine transmembrane segments, two reentrant membrane loops important for ion transport, and a large cytoplasmic regulatory loop. Chloride ions are transported into the cell by the Na^+-K^+-Cl^- cotransporter and out of the cell by the K^+-Cl^- cotransporter. These transporters are members of a family of Cl^- transport proteins with 12 transmembrane segments (**below**).

of migraine associated with an aura and muscle weakness. Certain mutations in the neuron-specific *ATP1A3* isoform lead to rapid-onset dystonia parkinsonism, a movement disorder that first occurs in late adolescence or early adulthood. A different set of mutations lead to a distinct neurological disorder, alternating hemiplegia of childhood, a paralysis that affects one side of the body and develops in children under the age of 2.

Most neurons have relatively few Ca^{2+} pumps in the plasma membrane. Instead, Ca^{2+} is transported out of the cell primarily by the *Na^+-Ca^{2+} exchanger* (Figure 9–5B). This membrane protein is not an ATPase but a different type of molecule called a *cotransporter*. Cotransporters move one type of ion against its electrochemical gradient by using the energy stored in the electrochemical gradient of a second ion. (The CLC Cl^--H^+ cotransporter discussed in Chapter 8 is a type of exchanger.) In the case of the Na^+-Ca^{2+} exchanger, the electrochemical gradient of Na^+ drives the efflux of Ca^{2+}. The exchanger transports three or four Na^+ ions into the cell (down the electrochemical gradient for Na^+) for each Ca^{2+} ion it removes (against the electrochemical gradient of Ca^{2+}). Because Na^+ and Ca^{2+} are transported in opposite directions, the exchanger is termed an *antiporter*. Ultimately, it is the hydrolysis of ATP by the Na^+-K^+ pump that provides the energy (stored in the Na^+ gradient) to maintain the function of the Na^+-Ca^{2+} exchanger. For this reason, ion flux driven by cotransporters is often referred to as *secondary active transport*, to distinguish it from the *primary active transport* driven directly by ATPases.

Chloride Ions Are Also Actively Transported

So far, for simplicity, we have ignored the contribution of chloride (Cl^-) to the resting potential. However, in most nerve cells, the Cl^- gradient across the cell membrane is controlled by one or more active transport mechanisms so that E_{Cl} differs from V_r. As a result, the presence of open Cl^- channels will bias the membrane potential toward its Nernst potential. *Chloride transporters* typically use the energy stored in the gradients of other ions—they are cotransporters.

Cell membranes contain a number of different types of Cl^- cotransporters (Figure 9–5B). Some transporters increase intracellular Cl^- to levels greater than those that would be passively reached if the Cl^- Nernst potential was equal to the resting potential. In such cells, E_{Cl} is positive to V_r so that the opening of Cl^- channels depolarizes the membrane. An example of this type of transporter is the Na^+-K^+-Cl^- cotransporter. This protein transports two Cl^- ions into the cell together with one Na^+ and one K^+ ion. As a result, the transporter is electroneutral. The Na^+-K^+-Cl^- cotransporter differs from the Na^+-Ca^{2+} exchanger in that the former transports all three ions in the same direction—it is a *symporter*.

In most neurons, the Cl^- gradient is determined by cotransporters that move Cl^- out of the cell. This action lowers the intracellular concentration of Cl^- so that E_{Cl} is typically more negative than the resting potential. As a result, the opening of Cl^- channels leads to an influx of Cl^- that hyperpolarizes the membrane. The K^+-Cl^- cotransporter is an example of such a transport mechanism; it moves one K^+ ion out of the cell for each Cl^- ion it exports.

Interestingly, in early neuronal development, cells tend to express primarily the Na^+-K^+-Cl^- cotransporter. As a result, at this stage the neurotransmitter γ-aminobutyric acid (GABA), which activates ligand-gated Cl^- channels, typically has an excitatory (depolarizing) effect. As neurons develop, they begin to express the K^+-Cl^- cotransporter, such that in most mature neurons GABA typically hyperpolarizes the membrane and thus acts as an inhibitory neurotransmitter. In some pathological conditions in adults, such as certain types of epilepsy or chronic pain syndromes, the expression pattern of the Cl^- cotransporters may revert to that of the immature nervous system. This will lead to aberrant depolarizing responses to GABA that can produce abnormally high levels of excitation.

The Balance of Ion Fluxes in the Resting Membrane Is Abolished During the Action Potential

In the nerve cell at rest, the steady Na^+ influx is balanced by a steady K^+ efflux, so that the membrane potential is constant. This balance changes when the membrane is depolarized toward the threshold for an action potential. As the membrane potential approaches this threshold, voltage-gated Na^+ channels open rapidly. The resultant increase in membrane conductance to Na^+ causes the Na^+ influx to exceed the K^+ efflux once threshold is exceeded, creating a net influx of positive charge that causes further depolarization. The increase in depolarization causes still more voltage-gated Na^+ channels to open, resulting in a greater influx of Na^+, which accelerates the depolarization even further.

This regenerative, positive feedback cycle develops explosively, driving the membrane potential rapidly toward the Na^+ equilibrium potential of +55 mV:

$$E_{Na} = \frac{RT}{F} \ln \frac{[Na]_o}{[Na]_i} = 58 \text{ mV} \times \log \frac{[440]}{[50]} = +55 \text{ mV}.$$

However, the membrane potential never quite reaches E_{Na} because K^+ efflux continues throughout the depolarization. A slight influx of Cl^- into the cell also counteracts the depolarizing effect of the Na^+ influx. Nevertheless, so many voltage-gated Na^+ channels open during the rising phase of the action potential that the cell membrane's Na^+ conductance is much greater than the conductance of either Cl^- or K^+. Thus, at the peak of the action potential, the membrane potential approaches the Na^+ equilibrium potential, just as at rest (when permeability to K^+ is predominant), the membrane potential tends to approach the K^+ equilibrium potential.

The Contributions of Different Ions to the Resting Membrane Potential Can Be Quantified by the Goldman Equation

Although K^+, Na^+, and Cl^- fluxes set the value of the resting potential, V_m is not equal to E_K, E_{Na}, or E_{Cl} but lies at some intermediate value. As a general rule, when V_m is determined by two or more species of ions, the contribution of one species is determined not only by the concentrations of the ion inside and outside the cell but also by the ease with which the ion crosses the membrane.

One convenient measure of how readily the ion crosses the membrane is the *permeability* (P) of the membrane to that ion, which has units of velocity (cm/s). This measure is similar to that of a diffusion constant, which determines the rate of solute movement in solution driven by a local concentration gradient. The dependence of membrane potential on ionic permeability and concentration is given by the Goldman equation:

$$V_m = \frac{RT}{F} \ln \frac{P_K[K^+]_o + P_{Na}[Na^+]_o + P_{Cl}[Cl^-]_i}{P_K[K^+]_i + P_{Na}[Na^+]_i + P_{Cl}[Cl^-]_o}.$$

Goldman Equation

This equation applies only when V_m is not changing. It states that the greater the concentration of an ion species and the greater its membrane permeability, the greater is its contribution to determining the membrane potential. In the limit, when permeability to one ion is exceptionally high, the Goldman equation reduces to the Nernst equation for that ion. For example, if $P_K \gg P_{Cl}$ or P_{Na}, as in glial cells, the equation becomes as follows:

$$V_m \cong \frac{RT}{F} \ln \frac{[K^+]_o}{[K^+]_i}.$$

Alan Hodgkin and Bernard Katz used the Goldman equation to analyze changes in membrane potential in the squid giant axon. They measured the variations in membrane potential in response to systematic changes in the extracellular concentrations of Na^+, Cl^-, and K^+. They found that if V_m is measured shortly after the extracellular concentration is changed (before the internal ionic concentrations are altered), $[K^+]_o$ has a strong effect on the resting potential, $[Cl^-]_o$ has a moderate effect, and $[Na^+]_o$ has little effect. The data for the membrane at rest could be fit accurately by the Goldman equation using the following permeability ratios:

$$P_K : P_{Na} : P_{Cl} = 1.0 : 0.04 : 0.45.$$

At the peak of the action potential, there is an instant in time when V_m is not changing and the Goldman equation is applicable. At that point, the variation of V_m with external ionic concentrations is fit best if a quite different set of permeability ratios is assumed:

$$P_K : P_{Na} : P_{Cl} = 1.0 : 20 : 0.45.$$

For these values of permeability, the Goldman equation approaches the Nernst equation for Na^+:

$$V_m \cong \frac{RT}{F} \ln \frac{[Na^+]_o}{[Na^+]_i} = +55 \, mV.$$

Thus, at the peak of the action potential, when the membrane is much more permeable to Na^+ than to any other ion, V_m approaches E_{Na}. However, the finite permeability of the membrane to K^+ and Cl^- results in K^+ efflux and Cl^- influx that partially counterbalance Na^+ influx, thereby preventing V_m from quite reaching E_{Na}.

The Functional Properties of the Neuron Can Be Represented as an Electrical Equivalent Circuit

The utility of the Goldman equation is limited because it cannot be used to determine how membrane potential changes with time or distance within a neuron in response to a local change in permeability. It is also inconvenient for determining the magnitude of the individual Na^+, K^+, and Cl^- currents. This information can be obtained using a simple mathematical model derived from electric circuit theory. The model, called an *equivalent circuit*, represents all of the important electrical properties of the neuron by a circuit consisting of conductors or resistors, batteries, and capacitors. Equivalent circuits provide us with an intuitive understanding as well as a quantitative description of how current caused by the movement of ions generates electrical signals in nerve cells.

The first step in developing an equivalent circuit is to relate the membrane's discrete physical properties to its electrical properties. The lipid bilayer endows the membrane with electrical *capacitance*, the ability of an electrical nonconductor (insulator) to separate electrical charges on either side of it. The nonconducting phospholipid bilayer of the membrane separates the cytoplasm and extracellular fluid, both of which are highly conductive environments. The separation of charges on the inside and outside surfaces of the cell membrane (the capacitor) gives rise to the electrical potential difference across the membrane. The electrical potential difference or voltage across a capacitor is

$$V = Q/C,$$

where Q is the net excess positive or negative charge on each side of the capacitor and C is the capacitance.

Capacitance is measured in units of farads (F), and charge is measured in coulombs (where 96,500 coulombs of a univalent ion is equivalent to 1 mole of that ion). A charge separation of 1 coulomb across a capacitor of 1 F produces a potential difference of 1 volt. A typical value of membrane capacitance for a nerve cell is approximately 1 µF per cm^2 of membrane area. Very few charges are required to produce a significant potential difference across such a capacitance. For example, the excess of positive and negative charges separated by the membrane of a spherical cell body with a diameter of 50 µm and a resting potential of −60 mV is 29×10^6 ions. Although this number may seem large, it represents only a tiny fraction (1/200,000) of the total number of positive or negative charges in solution within the cytoplasm. The bulk of the cytoplasm and the bulk of the extracellular fluid are electroneutral.

The membrane is a *leaky capacitor* because it is studded with ion channels that can conduct charge. Ion channels endow the membrane with conductance and with the ability to generate an electrical potential difference. The lipid bilayer itself has effectively zero conductance or infinite resistance. However, because ion channels are highly conductive, they provide pathways of finite electrical resistance for ions to cross the membrane. Because neurons contain many types of channels selective for different ions, we must consider each class of ion channel separately.

In an equivalent circuit we can represent each K^+ channel as a resistor or conductor of ionic current with a single-channel conductance γ_K (remember, conductance = 1/resistance) (Figure 9–6A). If there were no K^+ concentration gradient, the current through a single K^+

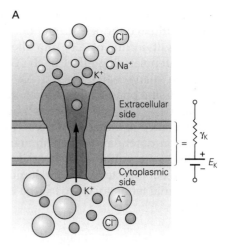

A

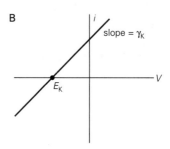

B

Figure 9–6 Chemical and electrical forces contribute to current through an ion channel.

A. A concentration gradient for K^+ gives rise to an electromotive force, which has a value equal to E_K, the Nernst potential for K^+. This can be represented by a battery. In this circuit, the battery E_K is in series with the conductor γ_K, representing the conductance of the K^+ channel.

B. The current-voltage relation for a K^+ channel in the presence of both electrical and chemical driving forces. The membrane potential at which the current is zero is equal to the K^+ Nernst potential.

channel would be given by Ohm's law: $i_K = \gamma_K \times V_m$. However, normally there is a K^+ concentration gradient, and thus also a chemical force driving K^+ across the membrane, represented in the equivalent circuit by a battery. (A source of electrical potential is called an *electromotive force* [EMF], and an electromotive force generated by a difference in chemical potentials is called a battery.) The electromotive force of this battery is given by E_K, the Nernst potential for K^+ (Figure 9–6).

In the absence of voltage across the membrane, the normal K^+ concentration gradient causes an outward K^+ current. According to our convention for current, an outward movement of positive charge across the membrane corresponds to a positive current. According to the Nernst equation, when the concentration gradient for a positively charged ion, such as K^+, is directed outward (ie, the K^+ concentration inside the cell is higher than outside), the equilibrium potential for that ion is negative. Thus, the K^+ current that flows solely because of its concentration gradient is given by $i_K = -\gamma_K \times E_K$ (the negative sign is required because a negative equilibrium potential produces a positive current at 0 mV).

Finally, for a real neuron that has both a membrane potential and a K^+ concentration gradient, the net K^+ current is given by the sum of the currents caused by the electrical and chemical driving forces:

$$i_K = (\gamma_K \times V_m) - (\gamma_K \times E_K) = \gamma_K \times (V_m - E_K). \quad \textbf{(9–1)}$$

The factor $(V_m - E_K)$ is called the *electrochemical driving force*. It determines the direction of ionic current and (along with the conductance) its magnitude. This equation is a modified form of Ohm's law that takes into account the fact that ionic current through a membrane is determined not only by the voltage across the membrane but also by the ionic concentration gradients.

A cell membrane has many resting K^+ channels, all of which can be combined into a single equivalent circuit element consisting of a conductor in series with a battery. In this equivalent circuit, the total conductance of all the K^+ channels (g_K), ie, the K^+ conductance of the cell membrane in its resting state, is equal to the number of resting K^+ channels (N_K) multiplied by the conductance of an individual K^+ channel (γ_K):

$$g_K = N_K \times \gamma_K.$$

Because the battery in this equivalent circuit depends solely on the concentration gradient for K^+ and is independent of the number of K^+ channels, its value is the equilibrium potential for K^+, E_K.

Like the population of resting K^+ channels, all the resting Na^+ channels can be represented by a single conductor in series with a single battery, as can the resting Cl^- channels. Because the K^+, Na^+, and Cl^- channels account for the bulk of the passive ionic current through the membrane in the cell at rest, we can calculate the resting potential by incorporating these three pathways into a simple equivalent circuit of a neuron (Figure 9–7).

To complete this circuit, we first connect the elements representing each type of channel at their two ends with elements representing the extracellular fluid and cytoplasm. The extracellular fluid and cytoplasm are both good conductors (compared with the membrane) because they have relatively large cross-sectional areas and many ions available to carry charge. In a small region of a neuron, the extracellular

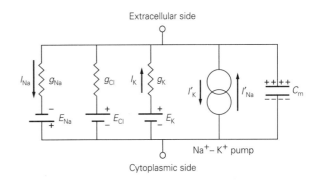

Figure 9–7 An equivalent circuit of passive and active current in a resting neuron. The total K^+ conductance represented by the symbol g_K is the product of $\gamma_K \times N$, the total number of open K^+ channels in the resting membrane. The total conductances for Na^+ and Cl^- channels are determined in a similar fashion. Under steady-state conditions, the passive Na^+ and K^+ currents are balanced by active Na^+ and K^+ fluxes (I'_{Na} and I'_K) driven by the Na-K$^+$ pump. The active Na^+ flux (I'_{Na}) is 50% greater than the active K^+ flux (I'_K) because the Na^+-K^+ pump transports three Na^+ ions out for every two K^+ ions it transports into the cell. As a result, for the cell to remain in a steady state, I_{Na} must be 50% greater than I_K (arrow size is proportional to current magnitude). There is no current through the Cl^- channels because in this example V_m is at E_{Cl}, the Cl^- equilibrium potential.

and cytoplasmic resistances can be approximated by a *short circuit*—a conductor with zero resistance. The membrane capacitance (C_m) is determined by the insulating properties of the lipid bilayer and its area.

Finally, the equivalent circuit can be made complete by incorporating the active ion fluxes driven by the Na^+-K^+ pump, which extrudes three Na^+ ions from the cell for every two K^+ ions it pumps in. This electrogenic ATP-dependent pump, which keeps the ionic batteries charged, is represented in the equivalent circuit by the symbol for a current generator (Figure 9–7). The use of the equivalent circuit to analyze neuronal properties quantitatively is illustrated in Box 9–2, where the equivalent circuit is used to calculate the resting potential.

The Passive Electrical Properties of the Neuron Affect Electrical Signaling

Once an electrical signal is generated in part of a neuron, for example in response to a synaptic input on a branch of a dendrite, it is integrated with the other inputs to the neuron and then propagated to the axon initial segment, the site of action potential generation. When synaptic potentials, receptor potentials, or action potentials are generated in a neuron, the membrane potential changes rapidly.

Box 9–2 Using the Equivalent Circuit Model to Calculate Resting Membrane Potential

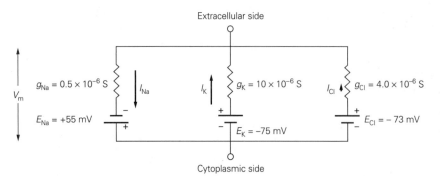

Figure 9–8 The electrical equivalent circuit used to calculate the resting membrane potential. In this example, it is assumed that the Cl⁻ cotransporter maintains intracellular Cl⁻ at a relatively low value. As a result, the Cl⁻ equilibrium potential is more negative than the resting potential.

An equivalent circuit model of the resting membrane can be used to calculate the resting potential (Figure 9–8). To simplify the calculation, we ignore the electrogenic influence of the Na⁺-K⁺ pump because it is small. We also ignore membrane capacitance because V_m is unchanging, so the charge on the capacitance is also not changing.

Because there are more resting channels for K⁺ than for Na⁺, the membrane conductance for K⁺ is much greater than that for Na⁺. In the equivalent circuit in Figure 9–8, g_K (10×10^{-6} S) is 20 times higher than g_{Na} (0.5×10^{-6} S). For most nerve cells, the value of g_{Cl} ranges from one-fourth to one-half of g_K. In this example, g_{Cl} equals 4.0×10^{-6} S. Given these values and the values of E_K, E_{Cl}, and E_{Na}, we can calculate V_m as follows.

Since the membrane potential is constant, there is no net current through the three sets of ion channels:

$$I_K + I_{Cl} + I_{Na} = 0. \qquad (9\text{–}2)$$

We can easily calculate each current in two steps. First, we add up the separate potential differences across each branch of the circuit. For example, in the K⁺ branch,

the total potential difference is the sum of the the battery E_K and the voltage drop across g_K given by Ohm's law ($V_m = I_K/g_K$):[*]

$$V_m = E_K + I_K/g_K$$

Similarly, for the Na⁺ and Cl⁻ conductance branches:

$$V_m = E_{Cl} + I_{Cl}/g_{Cl}$$

$$V_m = E_{Na} + I_{Na}/g_{Na}$$

Next, we rearrange and solve for the ionic current I in each branch:

$$I_{Na} = g_{Na} \times (V_m - E_{Na}) \qquad (9\text{–}3a)$$

$$I_K = g_K \times (V_m - E_K) \qquad (9\text{–}3b)$$

$$I_{Cl} = g_{Cl} \times (V_m - E_{Cl}). \qquad (9\text{–}3c)$$

These equations are similar to Equation 9–1, in which the net current through a single ion channel is derived from the currents caused by the individual

---

[*]Because we have defined V_m as $V_{in} - V_{out}$, the following convention must be used for these equations. Outward current (in this case I_K) is positive and inward current is negative. Batteries whose positive pole is directed toward the inside of the membrane (eg, E_{Na}) are given positive values in the equations. The reverse is true for batteries whose negative pole is directed toward the inside, such as the K⁺ battery.

What determines the rate of change in potential with time or distance? What determines whether a stimulus will or will not produce an action potential? Here we consider the neuron's passive electrical properties and geometry and how these relatively constant properties affect the cell's electrical signaling. The actions of the gated channels and the ionic currents that change the membrane potential are described in the next five chapters.

Neurons have three passive electrical properties that are important for electrical signaling. We have already described the resting membrane conductance or resistance ($g_r = 1/R_r$) and the membrane capacitance, C_m. A third important property that determines signal propagation along dendrites or axons is their intracellular axial resistance (r_a). Although the resistivity of cytoplasm is much lower than that of the membrane, the axial resistance along the entire length of an

driving forces. As these equations illustrate, the ionic current through each conductance branch is equal to the conductance of that branch multiplied by the net electrochemical driving force. Thus, for the K^+ current, the conductance is proportional to the number of open K^+ channels, and the driving force is equal to the difference between V_m and E_K. If V_m is more positive than E_K (−75 mV), the driving force is positive and the current is outward; if V_m is more negative than E_K, the driving force is negative and the current is inward.

Similar equations are used in a variety of contexts throughout this book to relate the magnitude of a particular ionic current to its membrane conductance and driving force.

As we saw in Equation 9–2, $I_{Na} + I_K + I_{Cl} = 0$. If we now substitute Equations 9–3a,b,c for I_{Na}, I_K, and I_{Cl} in Equation 9–2, multiply through, and rearrange, we obtain the following expression:

$$V_m \times (g_{Na} + g_K + g_{Cl}) = (E_{Na} \times g_{Na}) + (E_K \times g_K)$$
$$+ (E_{Cl} \times g_{Cl}).$$

Solving for V_m, we obtain an equation for the resting membrane potential that is expressed in terms of membrane conductances g and batteries E:

$$V_m = \frac{(E_{Na} \times g_{Na}) + (E_K \times g_K) + (E_{Cl} \times g_{Cl})}{g_{Na} + g_K + g_{Cl}}. \qquad (9\text{–}4)$$

From this equation, using the values in our equivalent circuit (Figure 9–8), we calculate that $V_m = -70$ mV.

Equation 9–4 states that V_m approaches the value of the ionic batteries that have the greater conductance. This principle can be illustrated by considering what happens during the action potential. At the peak of the action potential, g_K and g_{Cl} are essentially unchanged from their resting values, but g_{Na} increases as much as 500-fold. This increase in g_{Na} is caused by the opening of voltage-gated Na^+ channels. In the equivalent circuit in Figure 9–8, a 500-fold increase would change g_{Na} from 0.5×10^{-6} S to 250×10^{-6} S.

If we substitute this new value of g_{Na} into Equation 9–4 and solve for V_m, we obtain +48 mV.

V_m is closer to E_{Na} than to E_K at the peak of the action potential because g_{Na} is now 25-fold greater than g_K and 62.5-fold greater than g_{Cl}, so that the Na^+ battery becomes much more important than the K^+ and Cl^- batteries in determining V_m.

Equation 9–4 is similar to the Goldman equation in that the contribution to V_m of each ionic battery is weighted in proportion to the conductance of the membrane for that particular ion. In the limit, if the conductance for one ion is much greater than that for the other ions, V_m approaches the value of that ion's Nernst potential.

The equivalent circuit can be further simplified by lumping the conductance of all the resting channels that contribute to the resting potential into a single conductance, g_r, and replacing the battery for each conductance channel with a single battery E_r, whose value is given by Equation 9–4 (Figure 9–9). Here the subscript r stands for the resting channel pathway. Because the resting channels provide a pathway for the steady leakage of ions across the membrane, they are sometimes referred to as *leakage channels* (Chapter 10). This consolidation of resting pathways will prove useful when we consider the effects on membrane voltage of current through voltage-gated and ligand-gated channels in later chapters.

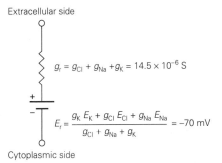

Extracellular side

$g_r = g_{Cl} + g_{Na} + g_K = 14.5 \times 10^{-6}$ S

$E_r = \dfrac{g_K E_K + g_{Cl} E_{Cl} + g_{Na} E_{Na}}{g_{Cl} + g_{Na} + g_K} = -70$ mV

Cytoplasmic side

Figure 9–9 The Na^+, K^+, and Cl^- resting channels can be simplified to a single conductance and battery. For an equivalent circuit model of the resting membrane (eg, Figure 9–8), the total membrane conductance (g_r) is calculated from the sum of the Na^+, K^+, and Cl^- conductances, and the value of the resting potential battery (E_r) is calculated from Equation 9–4.

extended thin neuronal process can be considerable. Because these three elements provide the return pathway to complete the electrical circuit when active ionic currents flow into or out of the cell, they determine the time course of the change in synaptic potential generated by the synaptic current. They also determine whether a synaptic potential generated in a dendrite will depolarize the trigger zone at the axon initial segment enough to fire an action potential. Finally, the

passive properties influence the speed at which an action potential is conducted.

Membrane Capacitance Slows the Time Course of Electrical Signals

The steady-state change in a neuron's voltage in response to subthreshold current resembles the behavior of a simple resistor, but the *time course* of the change

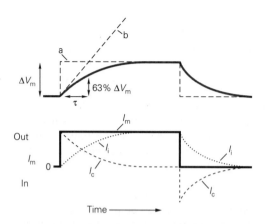

Figure 9–10 The rate of change in the membrane potential is slowed by the membrane capacitance. The **upper** plot shows the response of the membrane potential (ΔV_m) to a step current pulse (I_m). The shape of the actual voltage response (**red line**) combines the properties of a purely resistive element (**dashed line a**) and a purely capacitive element (**dashed line b**). The time taken to reach 63% of the final voltage defines the membrane time constant, τ. The **lower** plot shows the two elements of the total membrane current (I_m) during the current pulse: the ionic current (I_i) across the resistive elements of the membrane (ion channels) and the capacitive current (I_c).

does not. A true resistor responds to a step change in current with a similar step change in voltage, but the neuron's membrane potential rises and decays more slowly than the step change in current because of its *capacitance* (Figure 9–10).

To understand how the capacitance slows down the voltage response, recall that the voltage across a capacitor is proportional to the charge stored on the capacitor. To alter the voltage, charge Q must be added to or removed from the capacitor C:

$$\Delta V = \Delta Q/C.$$

To change the charge across the capacitor (the membrane lipid bilayer), there must be current across the capacitor (I_c). Since current is the flow of charge per unit time ($I_c = \Delta Q/\Delta t$), the change in voltage across a capacitor is a function of the magnitude and duration of the current:

$$\Delta V = I_c \cdot \Delta t/C.$$

Thus, the magnitude of the change in voltage across a capacitor in response to a current pulse depends on the duration of the current, because time is required to deposit and remove charge from the capacitor.

If the membrane had only resistive properties, a step pulse of outward current would change the membrane potential instantaneously. Conversely, if the membrane had only capacitive properties, the membrane potential would change linearly with time

in response to the same step of current. Because the membrane has both capacitive and resistive properties in parallel, the actual change in membrane potential combines features of the two pure responses. The initial slope of the change reflects a purely capacitive element, whereas the final slope and amplitude reflect a purely resistive element (Figure 9–10, upper plot).

In the simple case of the spherical cell body of a neuron, the time course of the potential change is described by the following equation:

$$\Delta V_m(t) = I_m R_m (1 - e^{-t/\tau}),$$

where e is the base of the system of natural logarithms with a value of approximately 2.72, and τ is the *membrane time constant*, given by the product of the membrane resistance and capacitance ($R_m C_m$). The time constant can be measured experimentally as the time it takes the membrane potential to rise to $1 - 1/e$, or approximately 63% of its steady-state value (Figure 9–10, upper plot). Typical values of τ for neurons range from 20 to 50 ms. We shall return to the time constant in Chapter 13 where we consider the temporal summation of synaptic inputs in a cell.

Membrane and Cytoplasmic Resistance Affect the Efficiency of Signal Conduction

So far, we have considered the effects of the passive properties of neurons on signaling only within the cell body. Distance is not a factor in the propagation of a signal in the neuron's soma because the cell body can be approximated as a sphere whose membrane voltage is uniform. However, a subthreshold voltage signal traveling along extended structures (dendrites, axons, and muscle fibers) decreases in amplitude with distance from the site of initiation because some charge leaks out of the resting membrane conductance as it flows along the dendrite or axon. To show how this attenuation occurs, we will consider how the geometry of a neuron influences the distribution of current.

If current is injected into a dendrite at one point, how will the membrane potential change along its length? For simplicity, consider how membrane potential varies with distance after a constant-amplitude current pulse has been on for some time ($t \gg \tau$). Under these conditions, the membrane capacitance is fully charged, so membrane potential reaches a steady value. The variation of the potential with distance depends on the fraction of charge that leaks out of the dendrite compared to the fraction that flows inside the dendrite towards the soma. Since charge flows along the path of least resistance, this depends on the relative values of the *membrane resistance* in a unit length of dendrite r_m

A

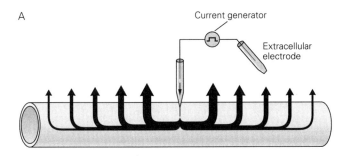

B

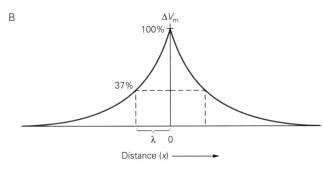

Figure 9–11 The change in membrane potential along a neuronal process during electrotonic conduction decreases with distance.

A. Current injected into a neuronal process by a microelectrode follows the path of least resistance to the return electrode in the extracellular fluid. (The thickness of the arrows represents the magnitude of membrane current.)

B. The change in V_m decays exponentially with distance from the site of current injection. The distance at which ΔV_m has decayed to 37% of its value at the point of current injection defines the length constant, λ.

(units of $\Omega \cdot cm$) and the *axial resistance* per unit length of dendrite r_a (units of Ω/cm). The change in membrane potential along the dendrite becomes smaller with distance from the current electrode (Figure 9–11A). This decay with distance is exponential and expressed by

$$\Delta V(x) = \Delta V_0\, e^{-x/\lambda},$$

where λ is the membrane *length constant*, x is the distance from the site of current injection, and ΔV_0 is the change in membrane potential produced by the current at the site of injection ($x = 0$). The length constant is the distance along the dendrite to the site where ΔV_m has decayed to $1/e$, or 37% of its initial value (Figure 9–11B). It is a measure of the efficiency of electrotonic conduction—the passive spread of voltage changes along the neuron—and is determined by the values of membrane and axial resistance as follows:

$$\lambda = \sqrt{(r_m/r_a)}.$$

The better the insulation of the membrane (that is, the greater r_m) and the better the conducting properties

of the inner core (the lower r_a), the greater the length constant of the dendrite. That is because current is able to spread farther along the inner conductive core of the dendrite before leaking across the membrane at some point x to alter the local membrane potential:

$$\Delta V(x) = i(x) \cdot r_m.$$

The length constant is also a function of the diameter of the neuronal process. Neuronal processes vary greatly in diameter, from as much as 1 mm for the squid giant axon to 1 μm for fine dendritic branches in the mammalian brain. For neuronal processes with similar ion channel surface densities (number of channels per unit membrane area) and cytoplasmic composition, thicker axons and dendrites have longer length constants than do narrower processes and hence can transmit passive electrical signals for greater distances. Typical values for neuronal length constants for unmyelinated axons range from about 0.5 to 1.0 mm. Myelinated axons have longer length constants—up to about 1.5 mm—because the insulating properties of myelin lead to an increase in the effective r_m of the axon.

To understand how the diameter of a process affects the length constant, we must consider how the diameter (or radius) affects r_m and r_a. Both r_m and r_a are measures of resistance for a unit length of a neuronal process of a given radius. The axial resistance r_a of the process depends inversely on the number of charge carriers (ions) in a cross section of the process. Therefore, given a fixed cytoplasmic ion concentration, r_a depends inversely on the cross-sectional area of the process $1/(\pi \cdot radius^2)$. The resistance of a unit length of membrane r_m depends inversely on the total number of channels in a unit length of the neuronal process.

Channel density, the number of channels per μm^2 of membrane, is often similar among different-sized processes. As a result, the number of channels per unit length of a neuronal process increases in direct proportion to increases in membrane area, which depends on the circumference of the process times its length; therefore, r_m varies as $1/(2 \cdot \pi \cdot radius)$. Because r_m/r_a varies in direct proportion to the radius of the process, the length constant is proportional to the square root of the radius. In this analysis, we have assumed that dendrites have only passive electrical properties. As discussed in Chapter 13, however, voltage-gated ion channels endow most dendrites with active properties that modify their purely passive length constants.

The efficiency of electrotonic conduction has two important effects on neuronal function. First, it influences spatial summation, the process by which synaptic potentials generated in different regions of the neuron are added together at the trigger zone of the axon (Chapter 13). Second, electrotonic conduction is a factor in the propagation of the action potential. Once the membrane at any point along an axon has been depolarized beyond threshold, an action potential is generated in that region. This local depolarization spreads passively down the axon, causing successive adjacent regions of the membrane to reach the threshold for generating an action potential (Figure 9–12). Thus, the depolarization spreads along the length of the axon by local current driven by the difference in potential between the active and resting regions of the axon membrane. In axons with longer length constants, local current spreads a greater distance down the axon, and therefore, the action potential propagates more rapidly.

Large Axons Are More Easily Excited Than Small Axons

The influence of axonal geometry on action potential conduction plays an important role in a common neurological exam. In the examination of a patient for

diseases of peripheral nerves, the nerve often is stimulated by passing current between a pair of external cutaneous electrodes placed over the nerve, and the population of resulting action potentials (the *compound action potential*) is recorded farther along the nerve by a second pair of cutaneous voltage-recording electrodes. In this situation, the total number of axons that generate action potentials varies with the amplitude of the current pulse (Chapter 57).

To drive a cell to threshold, a stimulating current from the positive electrode must pass through the cell membrane into the axon. There it travels along the axoplasmic core, eventually exiting the axon into the extracellular fluid through the membrane to reach the second (negative) electrode. However, most of the stimulating current does not even enter the axon, moving instead through neighboring axons or through the low-resistance pathway of the extracellular fluid. Thus, the axons into which current enters most easily are the ones most excitable.

In general, axons with the largest diameter have the lowest threshold for such excitation. The greater the

Figure 9–12 Electrotonic conduction contributes to propagation of the action potential.

A. An action potential propagating from right to left causes a difference in membrane potential between two adjacent regions of the axon. The difference creates a local-circuit current that causes the depolarization to spread passively. Current spreads from the more positive active region (**2**) to the less positive resting region *ahead* of the action potential (**1**), as well as to the less positive area *behind* the action potential (**3**). However, because there is also an increase in membrane K$^+$ conductance in the wake of the action potential (Chapter 10), the buildup of positive charge along the inner side of the membrane in area 3 is more than balanced by the local efflux of K$^+$, allowing this region of membrane to repolarize.

B. A short time later, the action potential has traveled down the axon and the process is repeated.

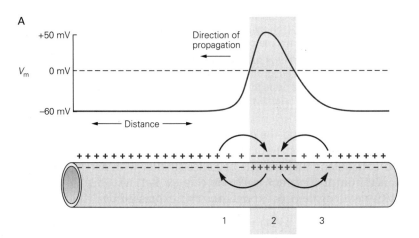

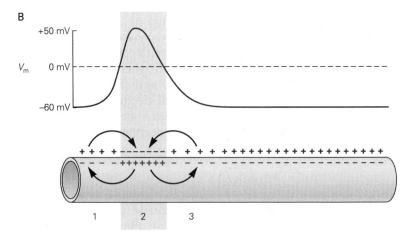

diameter of the axon, the lower is the axial resistance to the flow of current down the axon because the number of charge carriers (ions) per unit length of the axon is greater. Because more current enters the larger axon, the axon is depolarized more efficiently than a smaller axon. For these reasons, larger axons are recruited at low values of current; axons with smaller diameter are recruited only at relatively greater current strengths.

The fact that larger axons conduct more rapidly and have a lower current threshold for excitation aids in the interpretation of clinical nerve-stimulation tests. Neurons that convey different types of information (eg, motor versus sensory) often differ in axon diameter and thus conduction velocity (Chapter 18). In addition, a specific disease may preferentially affect certain functional classes of axons. Thus, using conduction velocity as a criterion to determine which classes of axons have defective conduction properties can help one infer the neuronal basis for the neurological deficit.

Passive Membrane Properties and Axon Diameter Affect the Velocity of Action Potential Propagation

The passive spread of depolarization during conduction of the action potential is not instantaneous. In fact, electrotonic conduction is a rate-limiting factor in the propagation of the action potential. We can understand this limitation by considering a simplified equivalent circuit of two adjacent segments of axon membrane connected by a segment of axoplasm.

An action potential generated in one segment of membrane supplies depolarizing current to the adjacent membrane, causing it to depolarize gradually toward threshold (Figure 9–12). According to Ohm's law, the larger the axoplasmic resistance, the smaller is the current between adjacent membrane segments ($I = V/R$) and thus the longer it takes to change the charge on the membrane capacitance of the adjacent segment.

Recall that, since $\Delta V = \Delta Q/C$, the membrane potential changes slowly if the current is small because ΔQ, equal to the magnitude of the current multiplied by time, changes slowly. Similarly, the larger the membrane capacitance, the more charge must be deposited on the membrane to change the potential across the membrane, so the current requires a longer time to produce a given depolarization. Therefore, the time it takes for depolarization to spread along the axon is determined by both the axial resistance r_a and the capacitance per unit length of the axon c_m (units F/cm). The rate of passive spread of charge varies inversely with the product $r_a c_m$. If this product is reduced, the rate of passive spread increases and the action potential propagates faster.

Rapid propagation of the action potential is functionally important, and two adaptations have evolved to increase it. One is an increase in the diameter of the axon core. Because r_a decreases in proportion to the square of axon diameter, whereas c_m increases in direct proportion to diameter, the net effect of an increase in diameter is a decrease in $r_a c_m$. This adaptation has been carried to an extreme in the giant axon of the squid, which can reach a diameter of 1 mm. No larger axons have evolved, presumably because of the competing need to keep neuronal size small so that many cells can be packed into a limited space.

The second adaptation that increases conduction velocity is the wrapping of a myelin sheath around an axon (Chapter 7). This process is functionally equivalent to increasing the thickness of the axonal membrane by as much as 100-fold. Because the capacitance of a parallel-plate capacitor such as the membrane is inversely proportional to the thickness of the insulation, myelination decreases c_m and thus $r_a c_m$. Each layer of myelin is extremely thin—only 80 Å. Therefore, myelination results in a proportionally much greater decrease in $r_a c_m$ than does the same increase in the diameter of a bare axon core, because the many layers of membrane in the myelin sheath produce a large decrease in c_m with a relatively small increase in overall axon diameter. For this reason, conduction in myelinated axons is faster than in nonmyelinated axons of the same diameter.

In a neuron with a myelinated axon, the action potential is triggered at the nonmyelinated initial segment of the axon. The inward current through this region of membrane is available to discharge the capacitance of the myelinated axon ahead. Even though the capacitance of the axon is quite small (because of the myelin insulation), the amount of current down the core of the axon from the trigger zone is not enough to discharge the capacitance along the entire length of the myelinated axon.

To prevent the action potential from dying out, the myelin sheath is interrupted every 1 to 2 mm by the nodes of Ranvier, bare patches of axon membrane approximately 1 μm in length (Chapter 7). Although the area of membrane at each node is quite small, the nodal membrane is rich in voltage-gated Na^+ channels and thus can generate an intense depolarizing inward Na^+ current in response to the passive spread of depolarization down the axon. These regularly distributed nodes thus periodically boost the amplitude of the action potential, preventing it from decaying with distance.

The action potential, which spreads quite rapidly along the internodal region because of the low capacitance of the myelin sheath, slows down as it crosses the high-capacitance region of each bare node. Consequently, as the action potential moves down the axon, it jumps quickly from node to node (Figure 9–13A). For this

A Normal axon

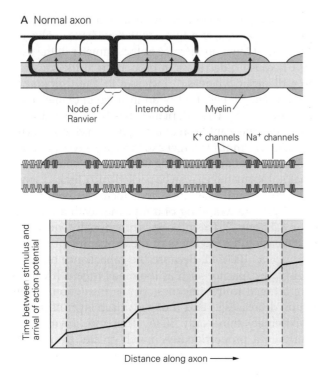

B Demyelinated axon

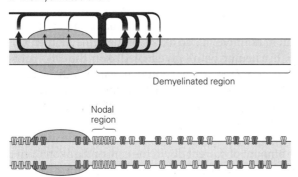

Figure 9–13 Action potentials in myelinated nerves are regenerated at the nodes of Ranvier.

A. The densities of capacitive and ionic membrane currents (membrane current per unit area of membrane) are much higher at the nodes of Ranvier than in the myelin-insulated internodal regions. (The density of membrane current at any point along the axon is represented by the thickness of the arrows.) Because of the higher capacitance of the axon membrane at the nodes, the action potential slows down as it approaches each node and thus appears to skip rapidly from node to node as it propagates from left to right.

B. In regions of the axon that have lost their myelin, the spread of the action potential is slowed down or blocked. The local-circuit currents must discharge a larger membrane capacitance, and because of the shorter length constant (caused by the low membrane resistance in demyelinated stretches of axon), they do not spread as well down the axon. In response to demyelination, additional voltage-gated Na^+ and K^+ ion channels are inserted into the membrane that is normally myelinated.

reason, the action potential in a myelinated axon is said to move by *saltatory conduction* (from the Latin *saltare*, to jump). Because ions flow across the membrane only at the nodes in myelinated fibers, saltatory conduction is also favorable from a metabolic standpoint. Less energy must be expended by the Na^+-K^+ pump to restore the Na^+ and K^+ concentration gradients, which tend to run down as the action potential is propagated.

The distribution of conduction velocities varies widely among neurons and even between different branches of an axon, depending on axon diameter and degree of myelination. Additional geometric features of myelinated axons, such as internodal length and nodal diameter, can also affect velocity. Evolution has adapted conduction velocities to optimize the behavioral functions of each neuron. In general, axons that are involved in rapid sensory and motor computations generally have high rates of conduction. More specifically, in certain neural circuits in the auditory system, an optimal behavioral response depends on the precise temporal relationship of presynaptic action potentials in two pathways that converge on the same postsynaptic neuron (Chapter 28). In such cases, values of the geometrical parameters of myelinated axons in the two input pathways can result in different conduction velocities that compensate for the differences in the input path lengths.

Various diseases of the nervous system are caused by demyelination, such as multiple sclerosis and Guillain-Barré syndrome. As an action potential goes from a myelinated region to a bare stretch of demyelinated axon, it encounters a region of relatively high c_m and low r_m. The inward current generated at the node just before the demyelinated segment may be too small to provide the capacitive current required to depolarize the segment of demyelinated membrane to threshold. In addition, this local-circuit current does not spread as far as it normally would because it encounters a segment of axon that has a relatively short length constant resulting from its low r_m (Figure 9–13B). These two factors can combine to slow, and in some cases actually block, the conduction of action potentials, causing devastating effects on behavior (Chapter 57).

Highlights

1. When the cell is at rest, passive fluxes of ions into and out of the cell through ion channels are balanced, such that the charge separation across the membrane remains constant and the membrane potential is maintained at its resting value.

2. The permeability of the cell membrane for an ion species is proportional to the number of open channels that allow passage of that ion. According to the Goldman equation, the value of the resting membrane potential in nerve cells is determined by resting channels that conduct K^+, Cl^-, and Na^+; the membrane potential is closest to the equilibrium (Nernst) potential of the ion or ions with the greatest membrane permeability.

3. Changes in membrane potential that generate neuronal electrical signals (action potentials, synaptic potentials, and receptor potentials) are caused by changes in the membrane's relative permeabilities to these three ions and to Ca^{2+} ions.

4. Although the changes in permeability caused by the opening of gated ion channels change the net charge separation across the membrane, they typically produce only negligible changes in the bulk concentrations of ions.

5. The functional properties of a neuron can be described by an electrical equivalent circuit, which includes the membrane capacitance, the ionic conductances, the EMF–generating properties of ion channels, and cytoplasmic resistance. In this model, membrane potential is determined by the ion or ions with the greatest membrane conductances.

6. Ion pumps prevent the ionic batteries from running down due to passive fluxes through the ion channels. The Na^+-K^+ pump uses the chemical energy of one molecule of ATP to exchange three intracellular Na^+ ions for two extracellular K^+ ions, an example of primary active transport. Secondary active transport by cotransporters is powered by coupling the downhill ionic gradients of one or two types of ions to drive the uphill transport of another ion. The coupling may take the form of symtransport (in the same direction) or antitransport (opposite directions).

7. The Na^+-Ca^{2+} antitransporter exchanges internal Ca^{2+} ion for external Na^+ ions. There are two types of Cl^- cotransporters in the cell membrane. The Cl^--K^+ symtransporter, which transports Cl^- and K^+ out of the cell, maintains E_{Cl} at a relatively negative potential, and is the most common variant of Cl^- transporter found in mature neurons. The Cl^--Na^+-K^+ symtransporter, which transports Cl^-, Na^+, and K^+ into the cell, generates an E_{Cl} that is relatively positive. It is expressed in immature neurons and in certain adult neurons.

8. The details of the molecular transitions during primary and secondary active transport are an area of active investigation.

9. The nerve cell membrane has a relatively high capacitance per unit of membrane area. As a result, when a channel opens and ions begin to flow, the membrane potential changes more slowly than the membrane current.

10. The currents that change the charge on the membrane capacitance along the length of an axon or dendrite pass through a relatively poor conductor—a thin column of cytoplasm. These two factors combine to slow down the conduction of voltage signals. Moreover, the various ion channels that are open at rest and that give rise to the resting potential also degrade the signaling function of the neuron, as they make the cell leaky and limit how far a signal can travel passively.

11. To overcome the physical constraints on long-distance signaling, neurons use sequential transient opening of voltage-gated Na^+ and K^+ channels to generate action potentials. The action potential is continually regenerated along the axon and thus conducted without attenuation.

12. For pathways in which rapid signaling is particularly important, conduction of the action potential is enhanced by myelination of the axon, an increase in axon diameter, or both. Conduction velocities can vary between or within axons in ways that optimize the timing of neuronal signals within a neuronal circuit.

John D. Koester
Steven A. Siegelbaum

Selected Reading

Clausen MV, Hilbers F, Poulsen H. 2017. The structure and function of the Na,K-ATPase isoforms in health and disease. Front Physiol 8:371.

Hille B. 2001. *Ionic Channels of Excitable Membranes*, 3rd ed. Sunderland, MA: Sinauer.

Hodgkin AL. 1964. Saltatory conduction in myelinated nerve. In: *The Conduction of the Nervous Impulse. The Sherrington Lecture, VII*, pp. 47–55. Liverpool: Liverpool University Press.

Jack JB, Noble D, Tsien RW. 1975. *Electric Current Flow in Excitable Cells*, pp. 1–4, 83–97, 131–224, 276–277. Oxford: Clarendon.

Johnston D, Wu M-S. 1995. Functional properties of dendrites. In: *Foundations of Cellular Neurophysiology*, pp. 55–120. Cambridge, MA: MIT Press.

Koch C. 1999. *Biophysics of Computation*, pp. 25–48. New York: Oxford Univ. Press.

References

Debanne D, Campanac E, Bialowas A, Carlier E, Alcaraz G. 2011. Axon physiology. Physiol Rev 91:555–602.

Ford MC, Alexandrova O, Cossell L, et al. 2015. Tuning of Ranvier node and internode properties in myelinated axons to adjust action potential timing. Nat Commun 6:8073.

Friedrich T, Tavraz NN, Junghans C. 2016. ATP1A2 mutations in migraine: seeing through the facets of an ion pump onto the neurobiology of disease. Front Physiol 7:239.

Gadsby DC. 2009. Ion channels versus ion pumps: the principal difference, in principle. Nat Rev Mol Cell Biol 10:344–352.

Goldman DE. 1943. Potential, impedance, and rectification in membranes. J Gen Physiol 27:37–60.

Hodgkin AL, Katz B. 1949. The effect of sodium ions on the electrical activity of the giant axon of the squid. J Physiol 108:37–77.

Hodgkin AL, Rushton WAH. 1946. The electrical constants of a crustacean nerve fibre. Proc R Soc Lond Ser B 133:444–479.

Huxley AF, Stämpfli R. 1949. Evidence for saltatory conduction in peripheral myelinated nerve fibres. J Physiol 108:315–339.

Jorgensen PL, Hakansson KO, Karlish SJ. 2003. Structure and mechanism of Na,K-ATPase: functional sites and their interactions. Annu Rev Physiol 65:817–849.

Kaila K, Price T, Payne J, Puskarjov M, Voipio J. 2014. Cation-chloride cotransporters in neuronal development, plasticity and disease. Nat Rev Neurosci 15:637–654.

Lytton J. 2007. Na^+/Ca^{2+} exchangers: three mammalian gene families control Ca^{2+} transport. Biochem J 406:365–382.

Moore JW, Joyner RW, Brill MH, Waxman SD, Najar-Joa M. 1978. Simulations of conduction in uniform myelinated fibers: relative sensitivity to changes in nodal and internodal parameters. Biophys J 21:147–160.

Nernst W. [1888] 1979. Zur Kinetik der in Lösung befindlichen Körper. [On the kinetics of substances in solution.] Z Physik Chem 2:613–622, 634–637. English translation in: GR Kepner (ed). 1979. *Cell Membrane Permeability and Transport*, pp. 174–183. Stroudsburg, PA: Dowden, Hutchinson & Ross.

Ren D. 2011. Sodium leak channels in neuronal excitability and rhythmic behaviors. Neuron 72:899–911.

Seidl AH, Rubel EW, Barría A. 2014. Differential conduction velocity regulation in ipsilateral and contralateral collaterals innervating brainstem coincidence detector neurons. J Neurosci 34:4914–4919.

Stokes DL, Green NM. 2003. Structure and function of the calcium pump. Annu Rev Biophys Biomol Struct 32:445–468.

Toyoshima C, Flemming F. 2013. New crystal structures of PII-type ATPases: excitement continues. Curr Op Struct Biol 23:507–514.

10

Propagated Signaling: The Action Potential

NERVE CELLS ARE ABLE TO CARRY electrical signals over long distances because the long-distance signal, the action potential, is continually regenerated and thus does not attenuate as it moves down the axon. In Chapter 9, we saw how an action potential arises from sequential changes in the membrane's permeability to Na^+ and K^+ ions and how the membrane's passive properties influence the speed at which the action potential is conducted. In this chapter, we describe in detail the voltage-gated ion channels that are critical for generating and propagating action potentials and consider how these channels are responsible for important features of a neuron's electrical excitability.

Action potentials have four properties important for neuronal signaling. First, they can be initiated only when the cell membrane voltage reaches a *threshold*. As we saw in Chapter 9, in many nerve cells, the membrane behaves as a simple resistor in response to small hyperpolarizing or depolarizing current steps. The membrane voltage changes in a graded manner as a function of the size of the current step according to Ohm's law, $\Delta V = \Delta I \cdot R$ (in terms of conductance, $\Delta V = \Delta I / G$). However, as the size of the depolarizing current increases, the membrane voltage will eventually reach a threshold, typically at around -50 mV, at which an action potential can be generated (see Figure 9–2C). Second, the action potential is an all-or-none event. The size and shape of an action potential initiated by a large depolarizing current is the same as that of an action potential evoked by a current that just surpasses the threshold.[1] Third, the action potential is conducted without decrement. It has a self-regenerative feature that keeps the amplitude constant, even when it is conducted over great distances. Fourth, the action potential is followed by a *refractory period*. For a brief time after an action potential is generated, the neuron's ability to fire a second action potential is suppressed. The refractory period limits the frequency at which a nerve can fire action potentials, and thus limits the information-carrying capacity of the axon.

These four properties of the action potential— initiation threshold, all-or-none amplitude, conduction without decrement, and refractory period—are unusual for biological processes, which typically respond in a graded fashion to changes in the environment. Biologists were puzzled by these properties for almost 100 years after the action potential was first recorded in the mid-1800s. Finally, in the late 1940s and early 1950s, studies of the membrane properties of the giant axon of the squid by Alan Hodgkin, Andrew Huxley, and Bernard Katz provided the first quantitative insight into the mechanisms underlying the action potential.

The Action Potential Is Generated by the Flow of Ions Through Voltage-Gated Channels

An important early insight into how action potentials are generated came from an experiment performed by Kenneth Cole and Howard Curtis that predated the studies by Hodgkin, Huxley, and Katz. While recording from the giant axon of the squid, they found that the conductance of the membrane increases dramatically during the action potential (Figure 10–1). This discovery provided evidence that the action potential results from a dramatic increase in the ion permeability of the cell membrane. It also raised two central questions: Which ions are responsible for the action potential, and how is the permeability of the membrane regulated?

Hodgkin and Katz provided a key insight into this problem by demonstrating that the amplitude of the action potential is reduced when the external Na^+

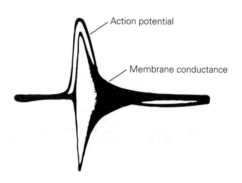

Figure 10–1 The action potential results from an increase in ion conductance of the axon membrane. This historic recording from an experiment conducted in 1939 by Kenneth Cole and Howard Curtis shows the oscilloscope record of an action potential superimposed on a simultaneous record of axonal membrane conductance.

concentration is lowered, indicating that Na^+ influx is responsible for the rising phase of the action potential. They proposed that depolarization of the cell above the threshold for an action potential causes a brief increase in the cell membrane's Na^+ conductance, during which the Na^+ conductance overwhelms the K^+ conductance that predominates in the cell at rest, thereby driving the membrane potential towards E_{Na}. Their data also suggested that the falling phase of the action potential was caused by a later increase in K^+ permeability.

Sodium and Potassium Currents Through Voltage-Gated Channels Are Recorded With the Voltage Clamp

This insight of Hodgkin and Katz raised a further question. What mechanism is responsible for regulating the changes in the Na^+ and K^+ permeabilities of the membrane? Hodgkin and Andrew Huxley reasoned that the Na^+ and K^+ permeabilities were regulated directly by the membrane voltage. To test this hypothesis, they systematically varied the membrane potential in the squid giant axon and measured the resulting changes in the conductance of voltage-gated Na^+ and K^+ channels. To do this, they made use of a new apparatus, the voltage clamp, developed by Kenneth Cole.

Prior to the availability of the voltage-clamp technique, attempts to measure Na^+ and K^+ conductance as a function of membrane potential had been limited by the strong interdependence of the membrane potential and the gating of Na^+ and K^+ channels. For example, if the membrane is depolarized sufficiently to open some voltage-gated Na^+ channels, the influx of Na^+ through these channels causes further depolarization. The additional depolarization causes still more

---

[1]The all-or-none property describes an action potential that is generated under a specific set of conditions. The size and shape of the action potential *can* be affected by changes in membrane properties, ion concentrations, temperature, and other variables, as discussed later in the chapter. The shape can also be affected slightly by the current that is used to evoke it, if measured near the point of stimulation.

Na$^+$ channels to open and consequently induces more inward Na$^+$ current:

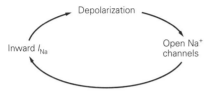

This positive feedback cycle drives the membrane potential to the peak of the action potential, making it impossible to achieve a stable membrane potential.

The voltage clamp interrupts the interaction between the membrane potential and the opening and closing of voltage-gated ion channels. It does so by adding or withdrawing a current from the axon that is equal to the current through the voltage-gated membrane channels. In this way, the voltage clamp prevents the membrane potential from changing. Thus, the amount of current that must be generated by the voltage clamp to keep the membrane potential constant provides a direct measure of the current through the voltage-gated channels (Box 10–1). Using the voltage-clamp technique, Hodgkin and Huxley were able to completely describe the ionic mechanisms underlying the action potential.

One advantage of the voltage clamp is that it readily allows the ionic and capacitive components of membrane current to be analyzed separately. As described in Chapter 9, the membrane potential V_m is proportional to the charge Q_m on the membrane capacitance C_m. When V_m is not changing, Q_m is constant, and no capacitive current ($\Delta Q_m / \Delta t$) flows. Capacitive current flows *only* when V_m is changing. Therefore, when the membrane potential changes in response to a commanded depolarizing step, capacitive current flows only at the beginning and end of the step. Because the capacitive current is essentially instantaneous, the ionic currents that subsequently flow through the voltage-gated channels can be analyzed separately.

Measurements of these ionic currents can be used to calculate the voltage and time dependence of changes in membrane conductance caused by the opening and closing of Na$^+$ and K$^+$ channels. This information provides insights into the properties of these two types of channels.

A typical voltage-clamp experiment starts with the membrane potential clamped at its resting value. When a small (10 mV) depolarizing step is applied, a very brief outward current instantaneously discharges the membrane capacitance by the amount required for a 10 mV depolarization. This capacitive current (I_c) is followed by a smaller outward current that persists for the duration of the voltage step. This steady ionic current flows through the nongated resting ion channels of the membrane, which we refer to here as *leakage channels* (see Box 9–2). The current through these channels is called the *leakage current*, I_l, and the total conductance of this population of channels is called the *leakage conductance* (g_l). At the end of the step, a brief inward capacitive current repolarizes the membrane to its initial voltage and the total membrane current returns to zero (Figure 10–3A).

If a large depolarizing step is commanded, the current record is more complicated. The capacitive and leakage currents both increase in amplitude. In addition, shortly after the end of the capacitive current and the start of the leakage current, an inward (negative) current develops; it reaches a peak within a few milliseconds, declines, and gives way to an outward current. This outward current reaches a plateau that is maintained for the duration of the voltage step (Figure 10–3B).

A simple interpretation of these results is that the depolarizing voltage step sequentially turns on two types of voltage-gated channels, each selective for a distinct ion species. One type of channel conducts ions that generate a rapidly rising inward current, while the other conducts ions that generate a more slowly rising outward current. Because these two oppositely directed currents partially overlap in time, the most difficult task in analyzing voltage-clamp experiments is to determine their separate time courses.

Hodgkin and Huxley achieved this separation by changing ions in the bathing solution. By replacing Na$^+$ with a larger, impermeant cation (choline · H$^+$), they eliminated the inward Na$^+$ current. Later the task of separating inward and outward currents was made easier by the discovery of drugs or toxins that selectively block the different classes of voltage-gated channels. Tetrodotoxin, a poison from a certain Pacific puffer fish, blocks the voltage-gated Na$^+$ channel with a very high potency in the nanomolar range of concentration. (Ingestion of only a few milligrams of tetrodotoxin from improperly prepared puffer fish, consumed as the Japanese sashimi delicacy *fugu*, can be fatal.) The cation tetraethylammonium (TEA) specifically blocks some voltage-gated K$^+$ channels.

When TEA is applied to the axon to block the K$^+$ channels, the total membrane current (I_m) consists of I_c, I_l, and I_{Na}. The leakage conductance g_l is constant; it does not vary with V_m or with time. Therefore, the leakage current I_l can be readily calculated and subtracted from I_m, leaving I_{Na} and I_c. Because I_c occurs only briefly at the beginning and end of the pulse, it is easily isolated by visual inspection, revealing the pure I_{Na}.

Box 10–1 Voltage-Clamp Technique

The voltage clamp permits the experimenter to "clamp" the membrane potential at predetermined levels, preventing changes in membrane current from influencing the membrane potential. By controlling the membrane potential, one can measure the effect of changes in membrane potential on the membrane conductance of individual ion species.

The voltage clamp is connected to a pair of electrodes (one intracellular and one extracellular) used to measure the membrane potential and another pair of electrodes used to pass current across the membrane (Figure 10–2A). Through the use of a negative feedback

amplifier, the voltage clamp is able to pass the correct amount of current across the cell membrane to rapidly step the membrane to a constant predetermined potential.

Depolarization opens voltage-gated Na$^+$ and K$^+$ channels, initiating movement of Na$^+$ and K$^+$ across the membrane. This change in membrane current ordinarily would change the membrane potential, but the voltage clamp maintains the membrane potential at the predetermined (commanded) level.

When Na$^+$ channels open in response to a moderate depolarizing voltage step, an inward ionic current

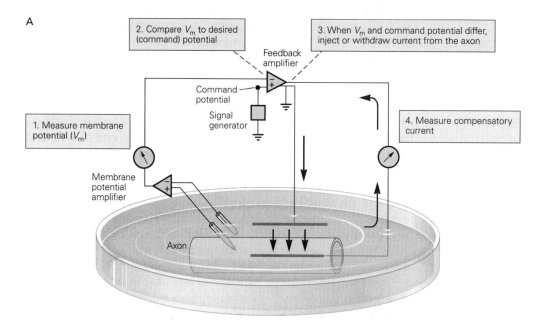

Figure 10–2 The negative feedback mechanism of the voltage clamp.

A. Membrane potential (V_m) is measured by two electrodes, one intracellular and one in the bath, connected to an amplifier. The membrane potential signal is displayed on an oscilloscope and also fed into the negative terminal of the feedback amplifier. The command potential, which is selected by the experimenter and can be of any desired amplitude and waveform, is fed into the positive terminal of the feedback amplifier. The feedback amplifier subtracts the membrane potential from the command potential and amplifies any difference between these two signals. The voltage output of the amplifier is connected to an internal current electrode, a thin wire that runs the length of the

axon core. The negative feedback ensures that the voltage output of the amplifier will drive a current across the resistance of the current electrode that eliminates any difference between V_m and the command potential. To accurately measure the current–voltage relationship of the cell membrane, the membrane potential must be uniform along the entire surface of the axon. This is made possible by the highly conductive internal current electrode, which short-circuits the axoplasmic resistance, reducing axial resistance to near zero (see Chapter 9). This low-resistance pathway eliminates all variations in electrical potential along the axon core.

develops because Na$^+$ ions are driven through these channels by their electrochemical driving force. This Na$^+$ influx would normally depolarize the membrane by increasing the positive charge on the inside of the membrane and reducing the positive charge on the outside.

The voltage clamp intervenes in this process by simultaneously withdrawing positive charges from the cell and depositing them in the external solution. By generating a current that is equal and opposite to the ionic current through the membrane, the voltage-clamp circuit automatically prevents the ionic current from changing the membrane potential from the commanded value. As a result, the *net* amount of charge separated by the membrane does not change, and therefore, no significant change in V_m occurs.

The voltage clamp is a negative feedback system, a type of system in which the value of the output of the system (V_m in this case) is fed back as the input to the system and compared to a reference value (the command signal). Any difference between the command signal and the output signal activates a "controller" (the feedback amplifier in this case) that automatically reduces the difference. Thus, the actual membrane potential automatically and precisely follows the command potential.

For example, assume that an inward Na$^+$ current through the voltage-gated Na$^+$ channels ordinarily causes the membrane potential to become more positive than the command potential. The input to the feedback amplifier is equal to ($V_{command} - V_m$). The amplifier generates an output voltage equal to this error signal multiplied by the gain of the amplifier. Thus, both the input and the resulting output voltage at the feedback amplifier will be negative.

This negative output voltage will make the internal current electrode negative, withdrawing net positive charge from the cell through the voltage-clamp circuit. As the current flows around the circuit, an equal amount of net positive charge will be deposited into the external solution through the other current electrode.

Today, most voltage-clamp experiments use a patch-clamp amplifier. The patch-clamp technique uses a feedback amplifier to control the voltage in a saline-filled micropipette and measures the current flowing through a patch of membrane to which the pipette is sealed. This allows the functional properties of single ion channels to be analyzed (see Box 8–1 and Figure 10–9).

If the pipette is sealed onto a cell and the patch under the membrane is ruptured by a pulse of suction, the result is a "whole-cell patch clamp" recording in which the intracellular voltage of the cell is controlled by the patch-clamp amplifier and the current flowing through the entire cell membrane is measured (Figure 10–2B). Whole-cell patch clamp recording allows voltage-clamp measurements in small cell bodies of neurons and is widely used to study the electrophysiological properties of neurons in cell culture, in brain slice preparations, and, recently, in vivo.

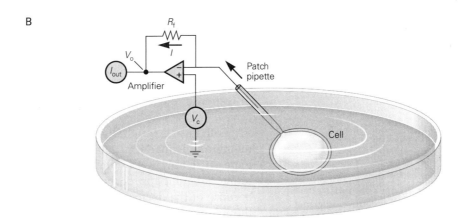

B. Voltage clamp of a neuronal cell body using the whole-cell mode of a patch-clamp amplifier. The patch pipette is sealed onto the cell and the membrane under the pipette is ruptured, providing electrical continuity between the inside of the cell and the pipette. An electrode in the pipette controls V_m, with an amplifier providing current (*I*) through a feedback resistor (*R$_f$*) to clamp the electrode (and therefore the pipette solution and the inside of the cell) to the command voltage (*V$_c$*), which is applied to the other amplifier input. The voltage on the output of the amplifier (*V$_o$*) is proportional to current flowing through the electrode and through the membrane.

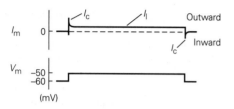

A Currents from small depolarization

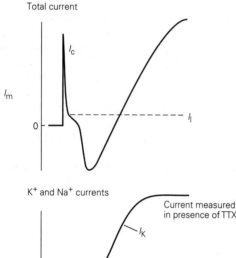

B Currents from large depolarization

Total current

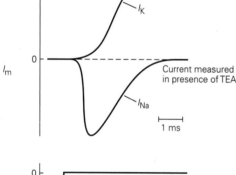

K⁺ and Na⁺ currents

Current measured
in presence of TTX

Current measured
in presence of TEA

1 ms

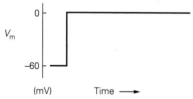

Figure 10–3 A voltage-clamp experiment demonstrates the sequential activation of voltage-gated sodium and potassium channels.

A. A small depolarization (10 mV) elicits capacitive and leakage currents (I_c and I_l, respectively), the components of the total membrane current (I_m).

B. A larger depolarization (60 mV) results in larger capacitive and leakage currents, plus a time-dependent inward ionic current followed by a time-dependent outward ionic current.

Top: Total (net) current in response to the depolarization. *Middle*: Individual Na⁺ and K⁺ currents. Depolarizing the cell in the presence of tetrodotoxin (**TTX**), which blocks the Na⁺ current, or in the presence of tetraethylammonium (**TEA**), which blocks the K⁺ current, reveals the pure K⁺ and Na⁺ currents (I_K and I_{Na}, respectively) after subtracting I_c and I_l. *Bottom*: Voltage step.

Similarly, I_K can be measured when the Na⁺ channels are blocked by tetrodotoxin (Figure 10–3B).

By stepping the membrane through a wide range of potentials, Hodgkin and Huxley were able to measure the Na⁺ and K⁺ currents over the entire voltage extent of the action potential (Figure 10–4). They found that the Na⁺ and K⁺ currents vary as a graded function of the membrane potential. As the membrane voltage is made more positive, the outward K⁺ current becomes larger. The inward Na⁺ current also becomes larger with increases in depolarization, up to a certain extent. However, as the voltage becomes more and more positive, the Na⁺ current eventually declines in amplitude. When the membrane potential is +55 mV, the Na⁺ current is zero. Positive to +55 mV, the Na⁺ current reverses direction and becomes outward.

Hodgkin and Huxley explained this behavior by a simple model in which the size of the Na⁺ and K⁺ currents is determined by two factors. The first is the magnitude of the Na⁺ or K⁺ conductance, g_{Na} or g_K, which reflects the number of Na⁺ or K⁺ channels open at any instant (Chapter 9). The second factor is the electrochemical driving force on Na⁺ ions ($V_m - E_{Na}$) or K⁺ ions ($V_m - E_K$). The model is thus expressed as:

$$I_{Na} = g_{Na} \times (V_m - E_{Na})$$

$$I_K = g_K \times (V_m - E_K).$$

According to this model, the amplitudes of I_{Na} and I_K change as the voltage is made more positive because there is an increase in g_{Na} and g_K. The conductances increase because the opening of the Na⁺ and K⁺ channels is voltage-dependent. The currents also change in response to changes in the electrochemical driving forces.

Both I_{Na} and I_K initially increase in amplitude as the membrane is made more positive because g_{Na} and g_K increase steeply with voltage. However, as the membrane potential approaches E_{Na} (+55 mV), I_{Na} declines because of the decrease in inward driving force, even though g_{Na} is large. That is, the positive membrane voltage now opposes the influx of Na⁺ down its chemical concentration gradient. At +55 mV the chemical and electrical driving forces are in balance so there is no net I_{Na}, even though g_{Na} is quite large. As the membrane is made positive to E_{Na}, the driving force on Na⁺ becomes positive. That is, the electrical driving force pushing Na⁺ out is now greater than the chemical driving force pulling Na⁺ in, so I_{Na} becomes outward. The behavior of I_K is simpler; E_K is quite negative (–75 mV), so in addition to an increase in g_K, the outward driving force on K⁺ also becomes larger as the membrane

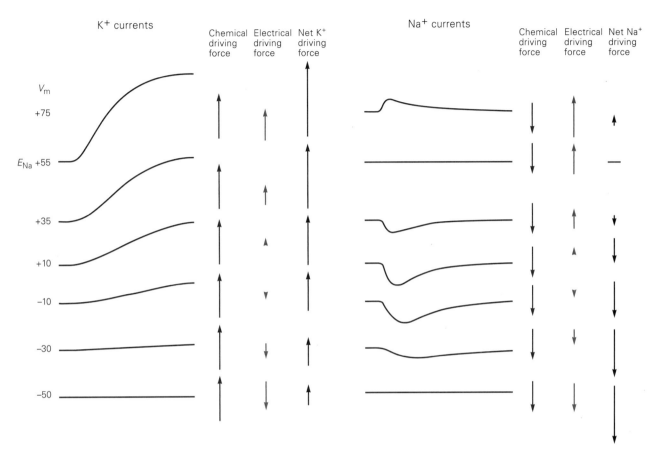

Figure 10–4 The magnitude and polarity of the sodium and potassium membrane currents vary with the amplitude of membrane depolarization. *Left*: With progressive depolarization, the voltage-clamped membrane K⁺ current increases monotonically, because both g_k and $(V_m - E_K)$, the driving force for K⁺, increase with increasing depolarization. The voltage during the depolarization is indicated at left. The direction and magnitude of the chemical (E_K) and electrical driving force on K⁺, as well as the net driving force, are given by arrows at the right of each trace. (**Up arrows** = outward force; **down arrows** = inward force.) *Right*: At first, the Na⁺ current becomes increasingly inward with greater depolarization due to the increase in g_{Na}. However, as the membrane potential approaches E_{Na} (+55 mV), the magnitude of the inward Na⁺ current begins to decrease due to the decrease in inward driving force $(V_m - E_{Na})$. Eventually, I_{Na} goes to zero when the membrane potential reaches E_{Na}. At depolarizations positive to E_{Na}, the sign of $(V_m - E_{Na})$ reverses and I_{Na} becomes outward.

is made more positive, thereby increasing the outward K⁺ current.

Voltage-Gated Sodium and Potassium Conductances Are Calculated From Their Currents

From the two preceding equations, Hodgkin and Huxley were able to calculate g_{Na} and g_K by dividing measured Na⁺ and K⁺ currents by the known Na⁺ and K⁺ electrochemical driving forces. Their results provided direct insight into how membrane voltage controls channel opening because the values of g_{Na} and g_K reflect the number of open Na⁺ and K⁺ channels (Box 10–2).

Measurements of g_{Na} and g_K at various levels of membrane potential reveal two functional similarities and two differences between the Na⁺ and K⁺ channels.

Both types of channels open in response to depolarization. Also, as the size of the depolarization increases, the extent and rate of opening increase for both types of channels. The Na⁺ and K⁺ channels differ, however, in the rate at which they open and in their responses to prolonged depolarization. At all levels of depolarization, the Na⁺ channels open more rapidly than K⁺ channels (Figure 10–6). When the depolarization is maintained for some time, the Na⁺ channels begin to close, leading to a decrease of inward current. The process by which Na⁺ channels close during a prolonged depolarization is termed *inactivation*.

Thus, depolarization causes Na⁺ channels to switch between three different states—resting, activated, or inactivated—which represent three different conformations of the Na⁺ channel protein (see Figure 8–6).

Box 10–2 Calculation of Membrane Conductances From Voltage-Clamp Data

Membrane conductances can be calculated from voltage-clamp currents using equations derived from an equivalent circuit (Figure 10–5) that includes the membrane capacitance (C_m); the leakage conductance (g_l), representing the conductance of all of the resting (nongated) K^+, Na^+, and Cl^- channels (Chapter 9); and g_{Na} and g_K, the conductances of the voltage-gated Na^+ and K^+ channels.

In the equivalent circuit, the ionic battery of the leakage channels, E_l, is equal to the resting membrane potential, and g_{Na} and g_K are in series with their appropriate ionic batteries.

The current through each class of voltage-gated channel can be calculated from a modified version of Ohm's law that takes into account the electrical driving force (V_m) and chemical driving forces (E_{Na} or E_K) on Na^+ and K^+ (Chapter 9):

$$I_K = g_K \times (V_m - E_K)$$
$$I_{Na} = g_{Na} \times (V_m - E_{Na}).$$

Rearranging and solving for g gives two equations that can be used to compute the conductances of the active Na^+ and K^+ channel populations:

$$g_K = \frac{I_K}{(V_m - E_K)}$$

$$g_{Na} = \frac{I_{Na}}{(V_m - E_{Na})}.$$

In these equations, the independent variable V_m is set by the experimenter. The dependent variables I_K and I_{Na} can be calculated from the records of voltage-clamp experiments (see Figure 10–4). The parameters E_K and E_{Na} can be determined empirically by finding the values of V_m at which I_K and I_{Na} reverse their polarities, that is, their *reversal potentials*.

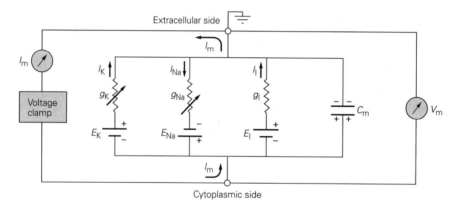

Figure 10–5 Equivalent circuit of a voltage-clamped neuron. The voltage-gated conductance pathways (g_K and g_{Na}) are represented by the symbol for a variable conductance—a conductor (resistor) with an arrow through it. The conductance is variable because of its dependence on time and voltage. These conductances are in series with batteries representing the chemical gradients for Na^+ and K^+. In addition, there are parallel pathways for leakage current (g_l and E_l) and capacitive current (C_m). Arrows indicate current flow during a depolarizing step that has activated g_K and g_{Na}.

In contrast, squid axon K^+ channels do not inactivate; they remain open as long as the membrane is depolarized, at least for voltage-clamp depolarizations lasting up to tens of milliseconds (Figure 10–6).

In the inactivated state, the Na^+ channel cannot be opened by further membrane depolarization. The inactivation can be reversed only by repolarizing the membrane to its negative resting potential, whereupon membrane to its negative resting potential, whereupon

the channel switches to the resting state. This switch takes some time.

These variable, time-dependent effects of depolarization on g_{Na} are determined by the kinetics of two gating mechanisms in Na^+ channels. Each Na^+ channel has an *activation gate* that is closed while the membrane is at the resting potential and opened by depolarization. An *inactivation gate* is open at the resting potential

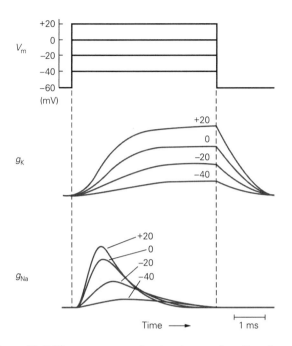

Figure 10–6 The responses of potassium and sodium ion channels to prolonged depolarization. Increasing depolarizations elicit graded increases in K⁺ and Na⁺ conductance (g_{Na} and g_K), which reflect the proportional opening of thousands of voltage-gated K⁺ and Na⁺ channels. The Na⁺ channels open more rapidly than the K⁺ channels. During a maintained depolarization, the Na⁺ channels close after opening because of the closure of an inactivation gate. The K⁺ channels remain open because they lack a fast inactivation process. At very positive V_m, the K⁺ and Na⁺ conductances approach a maximal value because the depolarization is sufficient to open nearly all available channels.

and closes after the channel opens in response to depolarization. The channel conducts Na⁺ only for the brief period during depolarization when *both* gates are open.

The Action Potential Can Be Reconstructed From the Properties of Sodium and Potassium Channels

Hodgkin and Huxley were able to fit their measurements of membrane conductance to a set of empirical equations that completely describe the Na⁺ and K⁺ conductances as a function of membrane potential and time. Using these equations and measured values for the passive properties of the axon, they computed the shape and conduction velocity of the action potential. Remarkably, these equations also provided insights into the biophysical bases for voltage-gating that were confirmed over 50 years later when the structure of certain voltage-gated channels was elucidated through X-ray crystallography.

The calculated waveform of the action potential matched the waveform recorded in the unclamped

axon almost perfectly. This close agreement indicates that the mathematical model developed by Hodgkin and Huxley accurately describes the properties of the channels that are responsible for generating and propagating the action potential. More than a half-century later, the Hodgkin-Huxley model stands as the most successful quantitative model in neural science if not in all of biology.

According to the model, an action potential involves the following sequence of events. Depolarization of the membrane causes Na⁺ channels to open rapidly (an increase in g_{Na}), resulting in an inward Na⁺ current. This current, by discharging the membrane capacitance, causes further depolarization, thereby opening more Na⁺ channels, resulting in a further increase in inward current. This regenerative process drives the membrane potential toward E_{Na}, causing the rising phase of the action potential. The depolarization limits the duration of the action potential in two ways: (1) It gradually inactivates the voltage-gated Na⁺ channels, thus reducing g_{Na}, and (2) it opens, with some delay, the voltage-gated K⁺ channels, thereby increasing g_K. Consequently, the inward Na⁺ current is followed by an outward K⁺ current that tends to repolarize the membrane (Figure 10–7).

Two features of the action potential predicted by the Hodgkin-Huxley model are its threshold and all-or-none behavior. A fraction of a millivolt can be the difference between a subthreshold stimulus and a stimulus that generates a full-sized action potential.

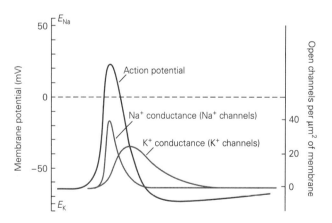

Figure 10–7 The sequential opening of voltage-gated Na⁺ and K⁺ channels generates the action potential. One of Hodgkin and Huxley's great achievements was to dissect the change in conductance during an action potential into separate components attributable to the opening of Na⁺ and K⁺ channels. The shape of the action potential and the underlying conductance changes can be calculated from the properties of the voltage-gated Na⁺ and K⁺ channels. (Adapted, with permission, from Hille 2001.)

This all-or-none phenomenon may seem surprising when one considers that Na^+ conductance increases in a strictly *graded* manner as depolarization increases (Figure 10–6). Each increment of depolarization increases the number of voltage-gated Na^+ channels that open, thereby gradually increasing Na^+ current. How then can there be a discrete threshold for generating an action potential?

Although a small subthreshold depolarization increases the inward I_{Na}, it also increases two *outward* currents, I_K and I_l, by increasing the electrochemical driving forces acting on K^+ and Cl^-. In addition, the depolarization augments K^+ conductance by gradually opening more voltage-gated K^+ channels (Figure 10–6). As the outward K^+ and leakage currents increase with depolarization, they tend to repolarize the membrane and thereby resist the depolarizing action of the Na^+ influx. However, because of the high voltage sensitivity and more rapid kinetics of activation of the Na^+ channels, the depolarization eventually reaches the point at which the increase in inward I_{Na} exceeds the increase in outward I_K and I_l. At this point, there is a net inward ionic current. This produces a further depolarization, opening even more Na^+ channels, so that the depolarization becomes regenerative, rapidly driving the membrane potential V_m all the way to the peak of the action potential. The specific value of V_m at which the net ionic current ($I_{Na} + I_K + I_l$) changes from outward to inward, depositing a positive charge on the inside of the membrane capacitance, is the threshold.

Early experiments with extracellular stimulation of nerve fibers showed that, for a short time after an action potential (typically a few milliseconds), it is impossible to generate another action potential. This *absolute refractory period* is followed by a period when it is possible to stimulate another action potential, but only with a stimulus larger than what was needed for the first. This *relative refractory period* typically lasts 5 to 10 ms.

The Hodgkin-Huxley analysis provided a mechanistic explanation of two factors underlying the refractory period. In the immediate aftermath of an action potential, it is impossible to evoke another one, even with a very strong stimulus, because the Na^+ channels remain inactivated. After repolarization, Na^+ channels recover from inactivation and reenter the resting state, a transition that takes several milliseconds (Figure 10–8). The relative refractory period corresponds to partial recovery from inactivation.

The relative refractory period is also influenced by a residual increase in K^+ conductance that follows the action potential. It takes several milliseconds for all of the K^+ channels that open during the action potential to return to their closed state. During this period, when the K^+ conductance remains somewhat elevated, V_m is slightly more negative than its normal resting value, as V_m approaches E_K (Figure 10–7, Equation 9–4). This *afterhyperpolarization* and residual increase in g_K contribute to the increase in depolarizing current required to drive V_m to threshold during the relative refractory period.

The Mechanisms of Voltage Gating Have Been Inferred From Electrophysiological Measurements

The empirical equations derived by Hodgkin and Huxley are quite successful in describing how the flow of ions through the Na^+ and K^+ channels generates the action potential. However, these equations describe the process of excitation in terms of changes in membrane conductance and current. They tell little about the mechanisms that activate or inactivate channels in response to changes in membrane potential or about channel selectivity for specific ions.

We now know that the voltage-dependent conductances described by Hodgkin and Huxley are generated by ion channels that open in a voltage- and time-dependent manner. Patch-clamp recordings from a variety of nerve and muscle cells have provided detailed information about the properties of the voltage-dependent Na^+ channels that generate the action potential. Recordings of single voltage-gated Na^+ channels show that, in response to a depolarizing step, each channel opens in an all-or-none fashion, conducting brief current pulses of constant amplitude but variable duration.

Each channel opening is associated with a current of about 1 pA (at voltages near −30 mV), and the open state is rapidly terminated by inactivation. Each channel behaves stochastically, opening after a variable time and staying open for a variable time before inactivating. If the openings of all the channels in a cell membrane in response to a step depolarization are summed or the openings of a single channel to multiple trials of the same depolarization are summed (Figure 10–9), the result is an averaged current with the same time course as the macroscopic Na^+ current recorded in voltage-clamp experiments (see Figure 10–4B).

To explain how changes in membrane potential lead to an increase in Na^+ conductance, Hodgkin and Huxley deduced from basic thermodynamic considerations that a conformational change in some membrane component that regulates the conductance must move charged particles through the membrane electric field. As a result, membrane depolarization would exert a force causing the charged particles to move, thereby opening the channel. For a channel with positively charged mobile particles, the depolarization

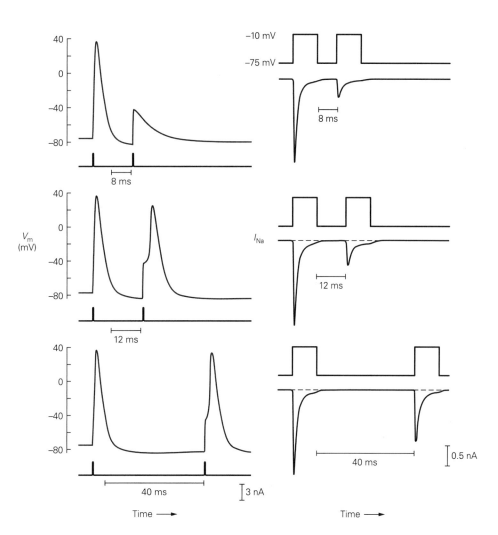

Figure 10–8 The refractory period is associated with recovery of the sodium channels from inactivation. *Left:* The voltage response of a mouse dorsal root ganglion neuron to two current pulses (**bottom traces**). The first triggers an action potential; the second triggers a variable voltage response depending on the delay between the current pulses. *Right:* Sodium currents recorded under voltage clamp in the same cell evoked by two depolarizing voltage pulses separated by the intervals indicated in the records on the left. In this neuron, the refractory period corresponds to the time needed for recovery of about 20% of the sodium channels. (Data from Pin Liu and Bruce Bean.)

would produce an outward movement of charge that should precede the opening of the channel. Upon membrane repolarization the charge would move in the opposite direction, closing the channel. Because the mobile charge movement is confined to the membrane, it is a type of capacitive current. This *gating charge* movement was predicted to generate a small outward current (or *gating current*), which was later confirmed when the membrane current was examined using very sensitive techniques. Blocking the inward ionic current with tetrodotoxin revealed a small outward capacitive current during the time the channels were activating (Figure 10–10A). In later experiments, this gating current (I_g) was progressively reduced by mutating the positively charged lysine and arginine residues in the four S4 transmembrane regions of the Na$^+$ channel to neutral residues. Thus, the gating current is produced by outward movement of the positively charged residues in the S4 regions through the membrane electric field (Chapter 8). Voltage-gated K$^+$ and Ca^{2+} channels also generate gating currents during channel opening.

Recent experiments have shown that the four S4 transmembrane regions of the Na$^+$ channel move with different time courses. Movement of the S4 regions of the first three domains (DI, DII, DIII) occurs first and is associated with channel activation. Movement of the S4 region of domain IV occurs more slowly and is associated with inactivation. Inactivation of Na$^+$ channels likely involves a series of conformational changes whereby outward movement of the S4 region of domain IV enables binding of the cytoplasmic linker connecting domains III and IV to a binding site near the intracellular ends of the pore-forming S6 helices, stabilizing a nonconducting inactivated state of the pore (Figure 10–10B,C).

Control of channel activation by gating charges results in a characteristic feature of voltage-dependent channels: The conductance change occurs over a relatively narrow range of V_m with a saturating value for larger depolarizations. When peak I_{Na} is measured over a wide range of V_m and then converted to a conductance, as illustrated in Figure 10–6, the voltage dependence of peak conductance has a sigmoid shape (Figure 10–11). Activation of voltage-dependent

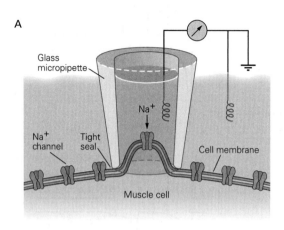

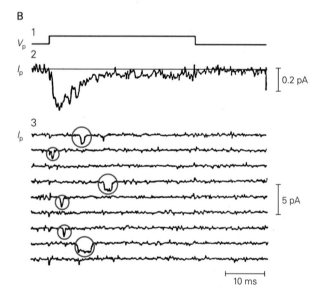

Figure 10–9 Individual voltage-gated ion channels open in an all-or-none fashion.

A. A small patch of membrane containing a single voltage-gated Na⁺ channel is electrically isolated from the rest of the cell by the patch electrode. The Na⁺ current that enters the cell through the channel is recorded by a monitor connected to the patch electrode (see Box 8–1).

B. Recordings of single Na⁺ channels in cultured muscle cells of rats. **(1)** Time course of a 10 mV depolarizing voltage step applied across the isolated patch of membrane (V_p = potential difference across the patch). **(2)** The sum of the inward current through the Na⁺ channel in the patch during 300 trials (I_p = current through the patch). The trace was obtained by blocking the K⁺ channels with tetraethylammonium and subtracting the leakage and capacitive currents electronically. **(3)** Nine individual trials from the set of 300, showing six openings of the channel (**circled**). These data demonstrate that the total Na⁺ current recorded in a conventional voltage-clamp record (see Figure 10–3B) can be accounted for by the statistical nature of the all-or-none opening and closing of a large number of Na⁺ channels. (Reproduced, with permission, from Sigworth and Neher 1980.)

sodium conductance begins at about –50 mV (near the threshold for action potential firing), reaches a midpoint near –25 mV, and saturates at about 0 mV. The saturation of conductance occurs when the S4 regions of the entire population of Na⁺ channels have moved to the activated conformation.

The relationship of conductance to voltage can be approximately fit by the Boltzmann function, an equation from statistical mechanics that describes the distribution of a population of molecules that can exist in distinct states with different potential energies. In the case of Na⁺ channels, the channels move between closed and open states that differ in potential energy because of the work done when the S4 region gating charges move through the electric field of the membrane (Figure 10–10C). The two parameters of the fitted Boltzmann curve, the midpoint and the slope factor, provide a convenient characterization of the voltage dependence with which channels open. The curve is steeper if more gating charge moves as channels convert between closed and open states. The voltage dependence of activation and inactivation of many other types of voltage-dependent channels can also be approximated by Boltzmann curves with characteristic midpoints and slopes.

Voltage-Gated Sodium Channels Select for Sodium on the Basis of Size, Charge, and Energy of Hydration of the Ion

In Chapter 8, we saw how the structure of the K⁺ channel pore could explain how such channels are able to select for K⁺ over Na⁺ ions. The narrow diameter of the K⁺ channel selectivity filter (around 0.3 nm) requires that a K⁺ or Na⁺ ion must shed nearly all of its waters of hydration to enter the channel, an energetically unfavorable event.

The energetic cost of dehydration of a K⁺ ion is well compensated by its close interaction with a cage of electronegative carbonyl oxygen atoms contributed by the peptide backbones of the four subunits of the K⁺ channel selectivity filter. Because of its smaller radius, a Na⁺ ion has a higher local electric field than does a K⁺ ion and therefore interacts more strongly with its waters of hydration than does K⁺. On the other hand, the small diameter of the Na⁺ ion precludes close interaction with the cage of carbonyl oxygen atoms in the selectivity filter; the resultant high energetic cost of dehydrating the Na⁺ ion excludes it from entering the channel.

How then does the selectivity filter of the Na⁺ channel select for Na⁺ over K⁺ ions? Bertil Hille deduced a model for the Na⁺ channel's selectivity mechanism from measurements of the channel's relative

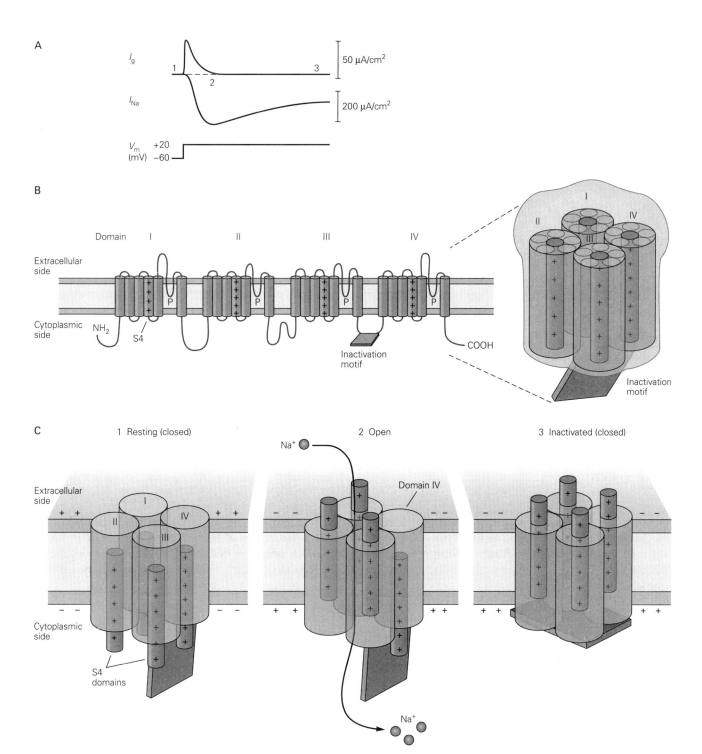

Figure 10–10 The opening and closing of the sodium channel are associated with a redistribution of charges.

A. When the membrane is depolarized, the Na$^+$ current (I_{Na}) is activated and then inactivated. The activation of the Na$^+$ current is preceded by a brief capacitive *gating current* (I_g), reflecting the outward movement of positive charges within the walls of the Na$^+$ channels. To detect this small gating current, it is necessary to block the flow of ionic current through the Na$^+$ and K$^+$ channels and subtract the capacitive current that depolarizes the lipid bilayer. (Adapted, with permission, from Armstrong and Gilly 1979.)

B. Secondary structure of the α-subunit of mammalian sodium channels showing location of gating charges. The sodium channel α-subunit is a single polypeptide consisting of four repeated domains, each containing six transmembrane regions. The fourth transmembrane region (S4 region) of each domain contains positively charged arginine and lysine residues that form

the gating charge of the channels. (Adapted, with permission, from Ahern et al. 2016. Permission conveyed through Copyright Clearance Center, Inc.)

C. Diagrams depict the redistribution of gating charge and positions of the activation and inactivation gates when the channel is at rest, open, and inactivated. The **red** cylinders represent the S4 regions containing the positive gating charges. Depolarization of the cell membrane from the resting state causes the gating charge to move outward. Outward movement of the S4 regions of domains I, II, and III is associated with activation (**1–2**), whereas slower movement of the S4 region of domain IV is associated with inactivation (**2–3**). The movement of the S4 region of domain IV allows an intracellular loop between domains III and IV (depicted as a **green** rectangle) to bind to a docking site near the S6 helices at the inside of the pore, allosterically stabilizing an inactivated closed state. (Adapted from Ahern et al. 2016.)

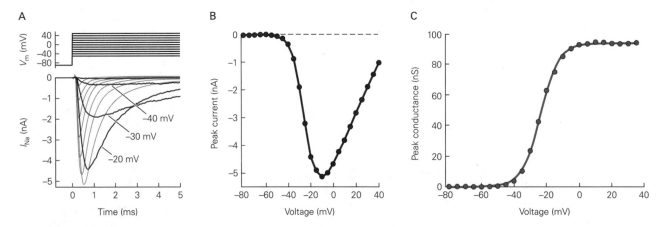

Figure 10–11 The voltage dependence of sodium channel activation is determined by the number of gating charges.

A. Whole-cell patch-clamp recordings of voltage-gated Na⁺ currents were made in a dissociated hippocampal pyramidal neuron. Sodium channel currents were isolated by blocking currents through K⁺ and Ca²⁺ channels and then subtracting the capacitive and leakage currents that remained after blocking Na⁺ currents.

B. Current–voltage curve for peak Na⁺ current.

C. Peak Na⁺ conductance versus membrane potential. Increases in peak g_{Na} in response to a series of depolarizing voltage steps were calculated from peak current, as in Box 10–2. The experimental data points are fit by the Boltzmann relation with the form $g_{Na}/g_{Na(max)} = 1/[1 + \exp - (V_m - V_h)/k]$, where $V_h = -24$ mV is the midpoint of the activation curve and $k = 5.5$ is the "slope factor," with units of mV, and $g_{Na(max)}$ is the maximal sodium conductance at positive voltages. The greater the number of gating charges that must move to open the channel, the smaller is the slope factor. The voltage dependence of most voltage-gated channels can be fit by similar Boltzmann curves. (Data from Indira M. Raman.)

permeability to a range of organic and inorganic cations. As we learned in Chapter 8, the channel behaves as if it contains a selectivity filter, or recognition site, which selects partly on the basis of size, thus acting as a molecular sieve (see Figure 8–1). Based on the size and hydrogen-bonding characteristics of the largest organic cation that could readily permeate the channel, Hille deduced that the selectivity filter has rectangular dimensions of 0.3 × 0.5 nm. This cross section is just large enough to accommodate one Na⁺ ion contacting one water molecule (see Figure 8–1). Because a K⁺ ion in contact with one water molecule is larger than the size of the selectivity filter, it cannot readily permeate.

According to Hille's model, negatively charged carboxylic acid groups of glutamate or aspartate residues at the outer mouth of the pore perform the first step in the selection process by attracting cations and repelling anions. The negative carboxylic acid groups, as well as other oxygen atoms that line the pore, can substitute for waters of hydration, but the degree of effectiveness of this substitution varies among ion species. The negative charge of a carboxylic acid is able to form a stronger coulombic interaction with the smaller Na⁺ ion than with the larger K⁺ ion. Because the Na⁺ channel is large enough to accommodate a cation in contact with several water molecules, the energetic cost of dehydration is not as great as it is in a K⁺ channel. As a result of these two features, the Na⁺ channel is able

to select for Na⁺ over K⁺, but not perfectly, with P_{Na}/P_K ~ 12/1. Structures of bacterial and vertebrate voltage-gated Na⁺ channels obtained by X-ray crystallography and cryo-electron microscopy have confirmed many of the key features of Hille's model.

Individual Neurons Have a Rich Variety of Voltage-Gated Channels That Expand Their Signaling Capabilities

The basic mechanism of electrical excitability identified by Hodgkin and Huxley in the squid giant axon is common to most excitable cells: Voltage-gated channels conduct an inward Na⁺ current followed by an outward K⁺ current. However, we now know that the squid axon is unusually simple in expressing only two types of voltage-gated ion channels. In contrast, the genomes of both vertebrates and invertebrates include large families of voltage-gated Na⁺, K⁺, and Ca²⁺ channels encoded by sub-families of related genes that are widely expressed in different kinds of nerve and muscle cells.

A neuron in the mammalian brain typically expresses a dozen or more different types of voltage-gated ion channels. The voltage dependence and kinetic properties of various Na⁺, Ca²⁺, and K⁺ channels can differ widely. Moreover, the distribution of these channels varies between different types of neurons and even between different regions of a single neuron. The great variety of voltage-gated channels in the membranes of most neurons enables a neuron to fire action potentials with a much greater range of frequencies and patterns than is possible in the squid axon, and thus allows much more complex information-processing abilities and modulatory control than is possible with just two types of channels.

The Diversity of Voltage-Gated Channel Types Is Generated by Several Genetic Mechanisms

The conservative mechanism by which evolution proceeds—creating new structural or functional entities by duplicating, modifying, shuffling, and recombining existing gene-coding sequences—is illustrated by the diversity and modular design of the members of the extended gene superfamily that encodes the voltage-gated Na^+, K^+, and Ca^{2+} channels. This family also includes genes that encode calcium-activated K^+ channels, the hyperpolarization-activated HCN nonselective cation channels, and a voltage-independent cyclic nucleotide-gated cation channel important for phototransduction and olfaction.

The functional differences between these channels are produced by differences in amino acid sequences in their core transmembrane domains as well as by the addition of regulatory elements in cytoplasmic domains. For example, some K^+ channels have a mechanism of inactivation mediated by a tethered plug formed by the cytoplasmic N-terminus of the channel protein, which binds to the inner mouth of the channel when the activation gate opens. The C-terminal cytoplasmic end of the channel proteins is a particularly rich locus for regulatory elements, including domains that bind either Ca^{2+} or cyclic nucleotides, enabling these agents to regulate channel gating. Inward-rectifying K^+ channels, which are tetramers of subunits with only a P-region and flanking transmembrane regions, have an internal cation-binding site that produces rectification. When the cell is depolarized, cytoplasmic Mg^{2+} or positively charged polyamines (small organic molecules that are normal constituents of the cytoplasm) are electrostatically driven to this binding site from the cytoplasm, plugging the channel (Figure 10–12).

Figure 10–12 represents large families of channels within which there is considerable structural and functional diversity. Five different mechanisms contribute to diversity in voltage-gated channels.

1. Multiple genes encode related principal subunits within each class of channel. For example,

in mammalian neurons and muscle, at nine different genes encode voltage-gated Na^+ channel α-subunits.
2. The four α-subunits that form a voltage-gated K^+ channel (Figure 8–11) can be encoded by different genes. After translation, the α-subunits are in some cases mixed and matched in various combinations, thus forming different subclasses of heteromeric channels.
3. A single gene product may be alternatively spliced, resulting in variations in the messenger RNA (mRNA) molecules that encode the α-subunit.
4. The mRNA that encodes an α-subunit may be edited by chemical modification of a single nucleotide, thereby changing the composition of a single amino acid in the channel subunit.
5. Principal pore-forming α-subunits of all channel types are commonly combined with different accessory subunits to form functionally different channel types.

These accessory subunits (often termed β-, γ-, or δ-subunits) may be either cytoplasmic or membrane-spanning and can produce a wide range of effects on channel function. For example, some β-subunits enhance the efficiency with which the channel protein is transported from the rough endoplasmic reticulum to the membrane, as well determining its final destination on the cell surface. Other subunits can regulate the voltage sensitivity or kinetics of channel gating. In contrast to the α-subunits, there is no known homology among the β-, γ-, and δ-subunits from the three major subfamilies of voltage-gated channels.

These various sources of channel diversity also vary widely between different areas of the nervous system, between different types of neurons, and within different subcellular compartments of a given neuron. A corollary of this regional differentiation is that mutations or epigenetic mechanisms that alter voltage-gated channel function can have very selective effects on neuronal or muscular function. The result is a large array of neurological diseases called channelopathies (Chapters 57 and 58).

Voltage-Gated Sodium Channels

The α-subunits of mammalian voltage-dependent Na^+ channels are encoded by nine genes. Three of the α-subunits encoded by these genes (Nav1.1, Nav1.2, and Nav1.6) are widely expressed in neurons in the mature mammalian brain, while four others have more restricted expression in neurons. Nav1.3 is strongly expressed early in development, with little expression in the mature brain, but can be reexpressed in injured

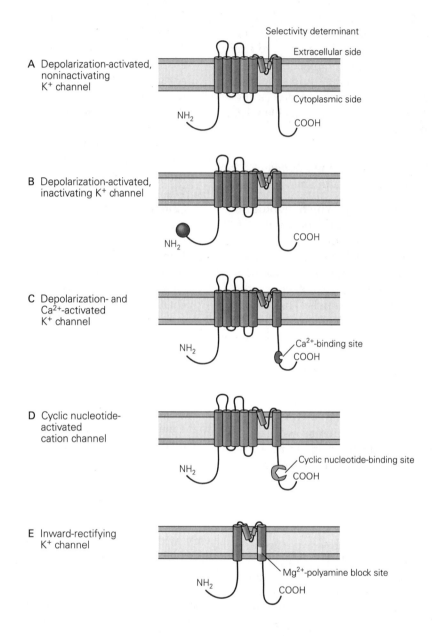

Figure 10–12 The extended gene family of voltage-gated channels produces variants of a common molecular design.

A. The basic transmembrane topology of an α-subunit of a voltage-gated K$^+$ channel. The S4 membrane-spanning α-helix is labeled in **red**.

B. Many K$^+$ channels that are first activated and then inactivated by prolonged depolarization have a ball-and-chain segment at their N-terminal end that inactivates the channel by plugging its inner mouth.

C. Some K$^+$ channels that require both depolarization and an increase in intracellular Ca^{2+} to activate have a Ca^{2+}-binding sequence attached to the C-terminal end of the channel.

D. Cation channels gated by cyclic nucleotides have a cyclic nucleotide-binding domain attached to the C-terminal end. One

subclass of such channels includes the voltage-independent, cyclic nucleotide-gated channels important in the transduction of olfactory and visual sensory signals. Another subclass consists of the hyperpolarization-activated cyclic nucleotide-gated (HCN) channels important for pacemaker activity (see Figure 10–15D). The P loops in these channels lack key amino acid residues required for K$^+$ selectivity. As a result, these channels do not show a high degree of discrimination between Na$^+$ and K$^+$.

E. Inward-rectifying K$^+$ channels, which are gated by blocking particles available in the cytoplasm, are formed from a truncated version of the basic building block, with only two membrane-spanning regions and a P-region.

tissue, for example following spinal cord injury. Nav1.7 channels are confined to autonomic and sensory neurons in the peripheral nervous system. Nav1.8 and Nav1.9 channels are largely restricted to a subset of peripheral sensory neurons, with particularly prominent expression in pain-sensing primary sensory neurons (nociceptors). Nav1.1, Nav1.2, Nav1.3, Nav1.6, and Nav1.7 channels have generally similar voltage dependence and relatively fast activation and inactivation kinetics compared to Nav1.8 and Nav1.9 channels. Nav1.4 channels in skeletal muscle fibers and Nav1.5 channels in cardiac muscle conduct the voltage-gated Na^+ current that generates action potentials in these tissues.

Although Nav1.1, Nav1.2, and Nav1.6 are all widely expressed in mammalian central neurons, they are expressed in different proportions in different types of neurons. Nav1.1 channels are particularly strongly expressed in some inhibitory GABAergic interneurons, and some loss-of-function mutations in Nav1.1 channels can lead to epilepsy, as in Dravet syndrome, perhaps reflecting greater loss of excitability of inhibitory neurons relative to excitatory neurons.

Voltage-Gated Calcium Channels

Virtually all neurons contain voltage-gated Ca^{2+} channels that open in response to membrane depolarization. A strong electrochemical gradient drives Ca^{2+} into the cell, so these channels give rise to an inward current that helps depolarize the cell.

A single neuron typically expresses at least four or five different types of voltage-gated Ca^{2+} channels with different voltage dependence, kinetic properties, and subcellular localization. Calcium channels that are widely expressed in central and peripheral neurons include Cav1.2 and Cav1.3 channels (collectively known as L-type channels), Cav2.1 (P/Q-type channels), Cav2.2 (N-type channels), and Cav2.3 (R-type channels). The various Cav1 and Cav2 family channels are collectively known as *high-threshold* or *high-voltage activated* (HVA) Ca^{2+} channels because activation generally requires relatively large depolarizations.

Members of the Cav3 family, collectively known as *T-type* or *low-voltage activated* (LVA) channels, are more selectively expressed in certain neurons. They are opened by small depolarizations (as negative as −65 mV) and undergo inactivation over tens of milliseconds. At normal resting potentials, Cav3 channels are typically inactivated. Hyperpolarization of the membrane voltage (as by inhibitory synaptic input) removes resting inactivation, which enables transient activation of the LVA channels following the hyperpolarization as the membrane voltage moves back toward its resting level. This activation can produce a regenerative depolarization that triggers a burst of Na^+ action potentials, which terminates when the Cav3 channels inactivate. Such postinhibitory burst firing is common in some regions of the thalamus and can help drive synchronized burst firing in neural circuits (see Figure 44–2). Similar rebound activation of Cav3 channels following the hyperpolarizing phase of a slow pacemaker potential contributes to spontaneous rhythmic bursting in some thalamocortical neurons (see Figure 10–15).

Voltage-Gated Potassium Channels

Voltage-dependent K^+ channels comprise an especially varied group of channels that differ in their kinetics of activation, voltage-activation range, and sensitivity to various ligands. Mammalian neurons typically express members of at least five families of voltage-dependent K^+ channels: Kv1, Kv2, Kv3, Kv4, and Kv7 (Figure 10–13). Each family consists of multiple gene products, with each channel composed of four α-subunits. For example, there are eight closely-related genes encoding the members of the Kv1 gene family (Kv1.1–Kv1.8).

The subunits can be of the same type (a homomeric channel) or of different gene products from within the same Kv family (a heteromeric channel). For example, Kv1 channels can be formed by heteromeric combinations of at least five different gene products that have wide expression in central neurons, with each combination possessing different properties of kinetics and voltage dependence. The possible functional variation provided by different combinations of α-subunits in heteromeric channels is immense, although not all possible combinations actually occur.

One way of distinguishing different components of voltage-dependent K^+ currents in a neuron is by the presence or absence of inactivation. Non-inactivating K^+ current, like that described in the squid axon by Hodgkin and Huxley, is called *delayed rectifier* K^+ current. In the squid axon, delayed rectifier current flows through a single Kv1 family channel type. In most mammalian neurons, delayed rectifier current includes multiple components from Kv1, Kv2, and Kv3 family channels, each with different kinetics and voltage dependence. Kv3 channels are unusual in requiring large depolarizations to become activated and also in having very rapid kinetics of activation. As a result, Kv3 channels are not activated until the action potential is near its peak, but they activate quickly enough to help terminate the action potential.

In addition to delayed rectifier current, many neurons also have a component of inactivating K^+ current known as A-type current. In cell bodies and dendrites,

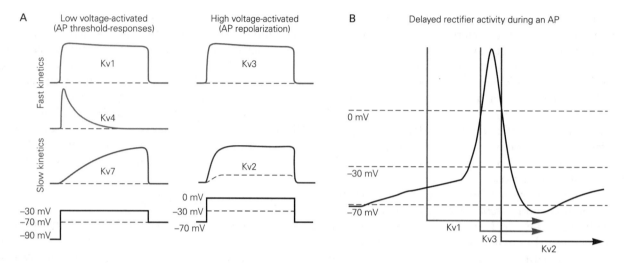

Figure 10–13 Different voltage dependence and kinetics of major classes of mammalian voltage-activated potassium channels.

A. Simplified generalization of the voltage dependence and kinetics of the major voltage-gated K⁺ families. Because Kv1, Kv4, and Kv7 channels can be activated by relatively small depolarizations, they often help control action potential (**AP**) threshold. Kv2 and Kv3 channels require larger depolarizations to be activated. Kv1, Kv3, and Kv4 channels are activated relatively rapidly, whereas Kv7 and Kv2 channels are activated more slowly.

B. Simplified generalization of the differing activation times of the major components of delayed rectifier K⁺ channels during an action potential. Kv1 channels require small depolarizations and are activated rapidly, sometimes significantly in advance of the action potential. Kv3 channels require large depolarizations and are activated late in the rising phase of the action potential and deactivated very rapidly thereafter. Kv2 channels are activated relatively slowly during the falling phase of the action potential and remain open during the afterhyperpolarization. (Adapted, with permission, from Johnston et al. 2010.)

A-type current is formed primarily by Kv4 family α-subunits, which form channels that inactivate over a range of time scales from a few milliseconds to tens of milliseconds. Kv1 channels that include Kv1.4 subunits or the auxiliary subunit Kvβ1 also mediate an inactivating component of current, which is highly expressed in some nerve terminals as well as some cell bodies.

As is the case for Na⁺ channels and Cav3 family channels, A-type K⁺ current not only inactivates during large depolarizations but is also subject to steady-state inactivation by small depolarizations from rest, providing a mechanism by which its amplitude can be modulated by small voltage changes around resting potential (see Figure 10–15B).

Kv7 subunits form non-inactivating channels that require only small depolarizations from rest to be activated and can even be activated significantly at the resting potential. In some neurons, Kv7 channels are downregulated by the transmitter acetylcholine acting through muscarinic G protein–coupled receptors (thus the origin of an alternative name of "M-current"). Kv7 channels typically activate relatively slowly, over tens of milliseconds, and provide little current during a single action potential but tend to suppress firing of subsequent action potentials (Chapter 14).

The KCNH gene family consists of three subfamilies of voltage-gated K⁺ channels (Kv10, Kv11, and Kv12), which are also expressed in the brain. They influence resting potential, action potential threshold, and frequency and pattern of firing.

Voltage-Gated Hyperpolarization-Activated Cyclic Nucleotide-Gated Channels

Many neurons have cation channels that are slowly activated by hyperpolarization. This sensitivity to hyperpolarization is enhanced when intracellular cyclic nucleotides bind to the channel. Because these hyperpolarization-activated cyclic nucleotide-gated (HCN) channels have only two of the four negative binding sites found in the selectivity filter of K⁺ channels, they are permeable to both K⁺ and Na⁺ and have a reversal potential around −40 to −30 mV. As a result, hyperpolarization from rest, as during strong synaptic inhibition or following an action potential, opens the channels to generate an inward depolarizing current referred to as I_h (see Figure 10–15D).

Gating of Ion Channels Can Be Controlled by Cytoplasmic Calcium

In a typical neuron, the opening and closing of certain ion channels can be modulated by various cytoplasmic factors, thus affording the neuron's excitability properties greater flexibility. Changes in the levels of such cytoplasmic factors can result from the activity of the neuron itself or from the influences of other neurons (Chapters 14 and 15).

Intracellular Ca^{2+} concentration is one important factor that modulates ion channel activity. Although ionic currents through membrane channels during an action potential generally do not result in appreciable changes in the intracellular concentrations of most ion species, calcium is a notable exception to this rule. The concentration of free Ca^{2+} in the cytoplasm of a resting cell is extremely low, about 10^{-7} M, several orders of magnitude below the external Ca^{2+} concentration, which is approximately 2 mM. Thus, intracellular Ca^{2+} concentration may increase many-fold above its resting value as a result of voltage-gated Ca^{2+} influx.

Intracellular Ca^{2+} concentration controls the gating of a number of channels. Several kinds of channels are activated by increases in intracellular Ca^{2+}. For example, the Ca^{2+}-activated *BK channels* (named for their big single-channel conductance), which are widely expressed in neurons, are voltage-dependent K^+ channels that require a very large, nonphysiological depolarization to open in the absence of Ca^{2+}. The binding of Ca^{2+} to a site on the cytoplasmic surface of the channel shifts its voltage gating to allow the channel to open at more negative potentials. With the influx of Ca^{2+} during an action potential, BK channels can open and help repolarize the action potential. Another family of calcium-activated K^+ channels, the *SK channels* (named for their small single-channel conductance), are not voltage dependent but open only in response to increases in intracellular Ca^{2+}. SK channels can open in response to relatively small changes in intracellular Ca^{2+} but gate slowly, so their activation gradually builds up as more Ca^{2+} enters the cell during repeated action potential firing. Some Ca^{2+} channels are themselves sensitive to levels of intracellular Ca^{2+}, becoming inactivated when intracellular Ca^{2+} increases as a result of entry through the channel itself.

As is described in later chapters, changes in the intracellular concentration of Ca^{2+} can also influence a variety of cellular metabolic processes, as well as neurotransmitter release and gene expression.

Excitability Properties Vary Between Types of Neurons

Through the expression of a distinct complement of ion channels, the electrical properties of different neuronal types have evolved to match the dynamic demands of information processing. Thus, the function of a neuron is defined not only by its synaptic inputs and outputs but also by its intrinsic excitability properties.

Different types of neurons in the mammalian nervous system generate action potentials that have different shapes and fire in different characteristic patterns, reflecting different expression of voltage-gated channels. For example, cerebellar Purkinje neurons and GABAergic cortical interneurons are associated with high levels of expression of Kv3 channels. The rapid activation of these channels produces narrow action potentials. In dopaminergic and other monoaminergic neurons, there is a high level of expression of voltage-activated Ca^{2+} channels that open during the falling phase of the action potential. The inward Ca^{2+} current from these channels slows repolarization, resulting in broader action potentials.

In the squid axon, the action potential is followed by an *afterhyperpolarization* (see Figure 10–7). In some mammalian neurons, the afterhyperpolarization has slow components lasting tens or even hundreds of milliseconds, generated by calcium-activated K^+ channels or voltage-activated K^+ channels with slow deactivation kinetics. Slow afterhyperpolarizations mediated by SK channels can be enhanced by repeated action potentials, reflecting buildup in intracellular Ca^{2+} concentrations.

In many pyramidal neurons in the cortex and hippocampus, the action potential is followed by an *afterdepolarization*, a transient depolarization that sometimes follows an earlier faster afterhyperpolarization. If the afterdepolarization is large enough, it can trigger a second action potential, resulting in all-or-none burst firing. The afterdepolarization can be caused by a variety of ionic currents, including Na^+ and Ca^{2+} currents from a number of voltage-dependent channels.

The shape of the action potential in a neuron is not always invariant. In some cases it can be dynamically regulated either intrinsically (eg, by repetitive firing) or extrinsically (eg, by synaptic modulation) (Chapter 15).

The pattern of action potential firing evoked by depolarization varies widely between neurons. The input-output function of a neuron can be characterized by the frequency and pattern of action potential firing in response to a series of current injections of different magnitudes. In the mammalian cerebral cortex, glutamatergic pyramidal neurons typically fire rapidly at the beginning of the current pulse followed by progressive slowing of firing, a pattern known as *adaptation* (Figure 10–14A). In contrast, many GABAergic interneurons fire with very little change in frequency (Figure 10–14B). Other neurons have more complex firing patterns. Some pyramidal neurons in the cerebral cortex tend to fire with an initial burst of action potentials (Figure 10–14C); "chattering" cells respond with repetitive brief bursts of high-frequency firing (Figure 10–14D).

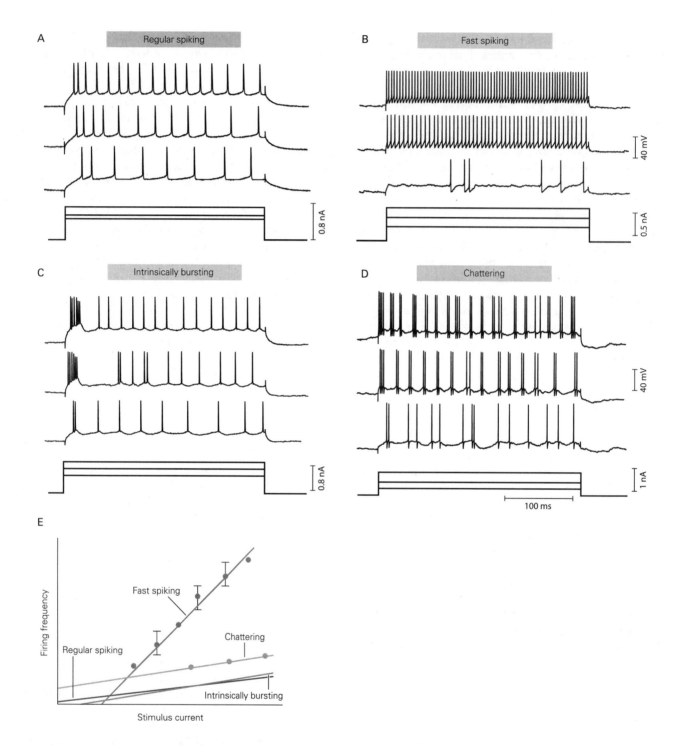

Figure 10–14 Different firing patterns in four types of corti-cal neurons. Three steps of depolarizing current, each of differ-ent amplitude, were injected into each cell to evoke firing. (Adapted, with permission, from Nowak et al. 2003.)

A. A cortical neuron with a firing pattern typical of many gluta-matergic cortical pyramidal neurons, illustrating characteristic adaptation.

B. A firing pattern typical of many GABAergic interneurons, illustrating maintained high-frequency repetitive firing.

C. Firing in an intrinsically bursting neuron, a subtype of pyrami-dal neuron in cortical layer II/III.

D. Firing in a chattering cell, a subtype of pyramidal neuron in cortical layer V.

E. Firing frequency versus stimulus current for these four cell types, showing their different sensitivities to increasing stimu-lus strength.

The sensitivity of these four classes of neurons to excitatory input can also be characterized by their frequency–current relationships. The fast spiking neurons are the most sensitive to increases in depolarizing excitatory current.

Some neurons can sustain repetitive firing at high frequencies up to 500 Hz. Such *fast-spiking* neurons occur throughout the mammalian central nervous system, including many principal neurons in the auditory system, where neurons must respond to sound waves of very high frequencies. The ability to fire repetitively at high frequencies is correlated with high expression levels of Kv3 family channels, which produce rapid repolarization and close extremely rapidly following repolarization, resulting in a minimal afterhyperpolarization and a brief refractory period.

The different firing patterns of neurons can be understood in terms of the expression and gating properties of particular channels. For example, adaptation of firing frequency during a maintained current pulse can be produced by activation of particular Kv1 family channels, which are strongly activated following an action potential and thus impede firing of a subsequent spike (Figure 10–15A). Because many channels are controlled by a process of inactivation that regulates their availability for activation, synaptic inputs that produce small voltage changes around the resting potential can greatly modify the cell's excitability. For example, in some neurons, a steady hyperpolarizing synaptic input makes the cell less excitable by reducing the extent of inactivation of the A-type K^+ channels at the normal resting potential of the cell (Figure 10–15B). In other neurons, such a steady hyperpolarization makes the cell *more* excitable because it reduces the inactivation of Cav3 voltage-gated Ca^{2+} channels (Figure 10–15C).

A surprisingly large number of neurons in the mammalian brain fire spontaneously in the absence of any synaptic input. When such activity is regular and rhythmic, it is often referred to as "pacemaking," by analogy to the rhythmic spontaneous firing of the cardiac pacemaker in the sinoatrial node of the heart. Many neurons that release modulatory neurotransmitters, such as dopamine, serotonin, norepinephrine, and acetylcholine, fire spontaneously, typically at frequencies of 0.5 to 5 Hz, resulting in constant tonic release of transmitter in the target areas of the neuron.

One mechanism causing spontaneous firing is exemplified by neurons in the suprachiasmatic nucleus of the hypothalamus, which helps control the circadian rhythm of overall metabolism and the sleep–wake cycle. These neurons fire spontaneously, with faster firing during the daytime than the nighttime (Chapter 44). Pacemaking in these cells is driven in part by subthreshold *persistent Na^+ current*, a small voltage-dependent current which flows through Na^+ channels at voltages as negative as –70 mV. This current can slowly depolarize the neuron to the point where a fast action potential fires (Figure 10–15E). In the same neurons, there are non–voltage-dependent channels that conduct Na^+ "leak" current, which depolarizes the cells into the voltage range where voltage-dependent persistent Na^+ current is activated. The higher expression level in the cell membrane of such Na^+ leakage channels during the daytime leads to the day–night difference in firing rate.

In dopaminergic neurons of the substantia nigra, pacemaking is unusual in being driven partly by voltage-dependent Ca^{2+} currents. The continual entry of Ca^{2+} during the lifetime of the neurons may contribute to metabolic stress associated with death of these neurons in Parkinson disease (Chapter 63).

Excitability Properties Vary Between Regions of the Neuron

Different regions of a neuron have different types of ion channels that support the specialized functions of each region. The axon, for example, functions as a relatively simple relay line. In contrast, the input, integrative, and output regions of a neuron typically perform more complex processing of the information they receive before passing it along (Chapter 3).

The trigger zone at the axon initial segment has the lowest threshold for action potential generation, in part because it has an exceptionally high density of voltage-gated Na^+ channels. In addition, it typically has voltage-gated ion channels that are sensitive to relatively small deviations from the resting potential. These channels thus play a critical role in transforming graded synaptic or receptor potentials into a train of all-or-none action potentials. Channels highly expressed at the axon initial segment of many neurons include Nav1.6, Kv1, Kv7, and low voltage-activated T-type Ca^{2+} channels.

Dendrites in many types of neurons have voltage-gated ion channels, including Ca^{2+}, K^+, HCN, and Na^+ channels (Chapter 13). When activated, these channels help shape the amplitude, time course, and propagation of the synaptic potentials to the cell body. In some neurons, the density of voltage-gated channels in the dendrites is sufficient to support local action potentials, typically with relatively high threshold voltages.

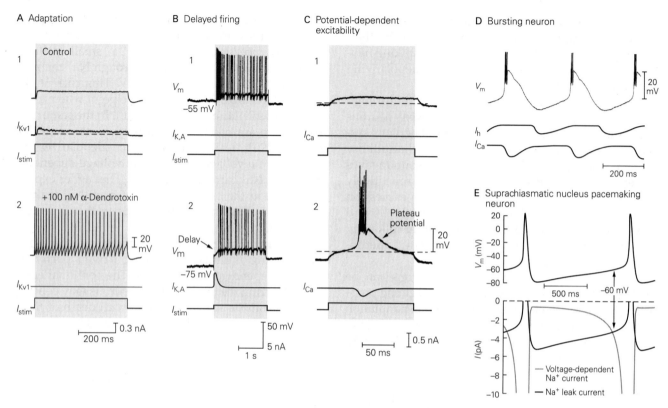

Figure 10–15 Regulation of firing pattern by a variety of voltage-gated channels.

A. Activation of Kv1 channels normally prevents firing of a second action potential by increasing spike threshold in a mouse dorsal root ganglion neuron (1). Blocking Kv1 channels with the snake toxin α-dendrotoxin changes the firing pattern from strong adaptation to maintained repetitive firing in response to a steady stimulating current (I_{stim}) (2). (Data from Pin Liu.)

B. Injection of a depolarizing current pulse (I_{stim}) into a neuron in the nucleus tractus solitarius normally triggers an immediate train of action potentials (1). If the cell is first held at a hyperpolarized membrane potential, the spike train is delayed (2). The delay is caused when A-type K⁺ channels are activated by the depolarizing current pulse. The channels generate a transient outward K⁺ current, $I_{K,A}$, that briefly slows the approach of V_m to threshold. These channels typically are inactivated at the resting potential (–55 mV), but steady hyperpolarization removes the inactivation. (Adapted with permission, from Dekin and Getting 1987.)

C. A small depolarizing current pulse injected into a thalamic neuron at rest generates a subthreshold depolarization (1). If the membrane potential is held at a hyperpolarized level, the same current pulse triggers a burst of action potentials (2). The effectiveness of the current pulse is enhanced because the hyperpolarization causes a type of voltage-gated Ca²⁺ channel to recover from inactivation. Depolarizing inward current through these Ca²⁺ channels (I_{Ca}) generates a plateau potential of about 20 mV that triggers a burst of action potentials. The **dashed line** indicates the level of the normal resting potential. (Adapted with permission, from Llinás and Jahnsen 1982.)

   The data in parts **B** and **C** demonstrate that steady hyperpolarization, such as might be produced by inhibitory synaptic input to a neuron, can profoundly affect the spike train pattern of a neuron. This effect varies greatly among cell types and depends on the presence or absence of particular types of voltage-gated Ca²⁺ and K⁺ channels.

D. In the absence of synaptic input, thalamocortical relay neurons can fire spontaneously in brief bursts of action potentials. These bursts are produced by current through two types of voltage-gated ion channels. The gradual depolarization that leads to a burst is driven by inward current (I_h) through HCN channels, which open in response to hyperpolarization. The burst is triggered by an inward Ca²⁺ current through voltage-gated Cav3 channels, which are activated at relatively low levels of depolarization. This Ca²⁺ influx generates sufficient depolarization to reach threshold and drive a brief burst of Na⁺-dependent action potentials. The strong depolarization during the burst causes the HCN channels to close and inactivates the Ca²⁺ channels, allowing hyperpolarization to develop between bursts of firing. This hyperpolarization then opens the HCN channels, initiating the next cycle in the rhythm. (Adapted, with permission, from McCormick and Huguenard 1992.)

E. Neurons from the suprachiasmatic nucleus of the hypothalamus generate spontaneous pacemaker potentials. Following an action potential, the neuron spontaneously depolarizes, first slowly and then more rapidly, resulting in another action potential. The depolarization is driven by two inward Na⁺ currents during the interspike interval. One is "persistent Na⁺ current," which flows through voltage-dependent sodium channels sensitive to block by tetrodotoxin, probably the same population of channels that underlie the much larger sodium current during the upstroke of the action potential. The second current flows through non–voltage-activated sodium leak nonselective (*NALCN*) channels, which provide a steady conductance pathway for Na⁺ and K⁺ ions. At negative voltages, Na⁺ driving force is large and K⁺ driving force is small, so the leak current is carried predominantly by Na⁺ ions. This inward current depolarizes the neuron to the point at which voltage-dependent persistent Na⁺ current becomes dominant (around –60 mV). (Adapted, with permission, from Jackson et al. 2004. Copyright © 2004 Society for Neuroscience.)

With moderate synaptic stimulation, full-blown action potentials are first generated at the trigger zone at the initial segment of the axon and then propagate back into the dendrites, serving as a signal to the synaptic regions that the cell has fired.

Conduction of the action potential down the axon is mediated primarily by voltage-gated Na^+ and K^+ channels that function much like those in the squid axon. In myelinated axons, the nodes of Ranvier have a high density of Na^+ channels but a low density of voltage-activated K^+ channels. There is a higher density of voltage-activated K^+ channels under the myelin sheath near the two ends of each internodal segment. The normal function of these K^+ channels is to suppress generation of action potentials in portions of the axon membrane under the myelin sheath. In demyelinating diseases, these channels become exposed and thus may inhibit the ability of the bare axon to conduct action potentials (Chapters 9 and 57).

Presynaptic nerve terminals at chemical synapses have a high density of voltage-gated Ca^{2+} channels, most commonly Cav2.1 (P/Q-type) channels, Cav2.2 (N-type) channels, or a mixture of the two. Arrival of an action potential in the terminal opens these channels, causing an influx of Ca^{2+} that triggers transmitter release (Chapter 15).

Neuronal Excitability Is Plastic

The expression, localization, and functional state of voltage-gated ion channels controlling the rate and pattern of action potential firing in a particular neuron are not always fixed, but can change in response to changes in the neuron's synaptic input, its activity, or its environment, as well as in response to injury or disease. For example, synaptic input that causes channel phosphorylation via second messenger pathways can lead to transient changes in a channel's functional properties, which in turn modulate cell excitability (Chapter 14). Plasticity can also occur on a longer time scale, such as when increased activity of a neuronal network as a whole leads to decreased excitability of individual neurons—a homeostatic feedback system. In some cases, activity-induced structural changes, such as change in length of the axon initial segment or its migration relative to the soma, can also affect excitability. The molecular mechanisms of homeostatic changes in neuronal excitability are not well understood but likely involve intracellular calcium signaling pathways that control transcription or cellular trafficking of specific ion channels. Dysfunction of such

regulatory pathways may underlie some types of epilepsy and hyperexcitability associated with conditions such as neuropathic pain.

Highlights

1. An action potential is a transient depolarization of membrane voltage lasting about 1 ms that is produced when ions move across the cell membrane through voltage-gated channels and thereby change the charge separation across the membrane.

2. The depolarizing phase of the action potential results from rapid, regenerative opening of voltage-activated Na^+ channels. Repolarization is due to inactivation of Na^+ channels and activation of K^+ channels.

3. The sharp threshold for action potential generation occurs at a voltage at which inward Na^+ channel current just exceeds outward currents through leak channels and voltage-gated K^+ channels.

4. The refractory period reflects Na^+ channel inactivation and K^+ channel activation continuing after the action potential. The refractory period limits the action potential firing rate.

5. The conformational changes of channel proteins underlying voltage-dependent activation and inactivation are not yet completely understood, but key regions involved in channel gating have been identified.

6. Voltage-gated sodium channels select for sodium on the basis of size, charge, and energy of hydration of the ion.

7. Most neurons express multiple kinds of voltage-gated Na^+, Ca^{2+}, K^+, HCN, and Cl^- channels, with especially large diversity in the properties of K^+ channels.

8. The diversity of voltage-dependent channels reflects the expression of multiple genes, formation of heteromeric channels from multiple gene products, alternative splicing of gene transcripts, mRNA editing, and combination of pore-forming subunits with a variety of accessory proteins.

9. Activity of some voltage-gated ion channels can be modulated by cytoplasmic Ca^{2+}.

10. The diversity in expression of voltage-gated ion channels results in differences in excitability properties of different types of neurons and in different regions of the same neuron.

11. The regional expression and functional state of ion channels can be regulated in response to cell

activity, changes in cell environment, or pathological processes, resulting in plasticity of the intrinsic excitability of neurons.

Bruce P. Bean
John D. Koester

Selected Reading

Ahern CA, Payandeh J, Bosmans F, Chanda B. 2016. The hitchhiker's guide to the voltage-gated sodium channel galaxy. J Gen Physiol 147:1–24.

Armstrong CM, Hille B. 1998. Voltage-gated ion channels and electrical excitability. Neuron 20:371–380.

Bezanilla F. 2008. How membrane proteins sense voltage. Nat Rev Mol Cell Biol 9:323–332.

Duménieu M, Oulé M, Kreutz MR, Lopez-Rojas J. 2017. The segregated expression of voltage-gated potassium and sodium channels in neuronal membranes: functional implications and regulatory mechanisms. Front Cell Neurosci 11:115.

Hille B. 2001. *Ion Channels of Excitable Membranes*, 3rd ed. Sunderland, MA: Sinauer.

Hodgkin AL. 1992. *Chance & Design: Reminiscences of Science in Peace and War*. Cambridge: Cambridge Univ. Press.

Johnston J, Forsythe ID, Kopp-Scheinpflug C. 2010. Going native: voltage-gated potassium channels controlling neuronal excitability. J Physiol 588:3187–3200.

Llinás RR. 1988. The intrinsic electrophysiological properties of mammalian neurons: insights into central nervous system function. Science 242:1654–1664.

Rudy B, McBain C. 2001. Kv3 channels: voltage-gated K+ channels designed for high-frequency repetitive firing. Trends Neurosci 24:517–526.

Turrigiano GG, Nelson SB. 2004. Homeostatic plasticity in the developing nervous system. Nat Rev Neurosci 5:97–107.

Vacher H, Mohapatra DP, Trimmer JS. 2008. Localization and targeting of voltage-dependent ion channels in mammalian central neurons. Physiol Rev 88:1407–1447.

References

Aplizar SA, Cho H, Hoppa M. 2019. Subcellular control of membrane excitability in the axon. Curr Opin Neurobiol 57:117–125.

Armstrong CM, Gilly WF. 1979. Fast and slow steps in the activation of sodium channels. J Gen Physiol 59:691–711.

Battefeld A, Tran BT, Gavrilis J, Cooper EC, Kole MH. 2014. Heteromeric Kv7.2/7.3 channels differentially regulate action potential initiation and conduction in neocortical myelinated axons. J Neurosci 34:3719–3732.

Bauer CK, Schwarz JR. 2018. Ether-à-go-go K+ channels: effective modulators of neuronal excitability. J Physiol (Lond) 596:769–783.

Bender KJ, Trussell LO. 2012. The physiology of the axon initial segment. Annu Rev Neurosci 35:249–265.

Capes DL, Goldschen-Ohm MP, Arcisio-Miranda M, Bezanilla F, Chanda B. 2013. Domain IV voltage-sensor movement is both sufficient and rate limiting for fast inactivation in sodium channels. J Gen Physiol 142:101–112.

Carrasquillo Y, Nerbonne JM. 2014. I$_A$ channels: diverse regulatory mechanisms. Neuroscientist 20:104–111.

Catterall WA. 1988. Structure and function of voltage-sensitive ion channels. Science 242:50–61.

Catterall WA. 2010. Ion channel voltage sensors: structure, function, and pathophysiology. Neuron 67:915–928.

Catterall WA. 2011. Voltage-gated calcium channels. Cold Spring Harb Perspect Biol 3:a003947.

Catterall WA, Few AP. 2008. Calcium channel regulation and presynaptic plasticity. Neuron 59:882–8901.

Cole KS, Curtis HJ. 1939. Electric impedance of the squid giant axon during activity. J Gen Physiol 22:649–670.

Dekin MS, Getting PA. 1987. In vitro characterization of neurons in the vertical part of the nucleus tractus solitarius. II. Ionic basis for repetitive firing patterns. J Neurophysiol 58:215–229.

Erisir A, Lau D, Rudy B, Leonard CS. 1999. Function of specific K+ channels in sustained high-frequency firing of fast-spiking neocortical interneurons. J Neurophysiol 82:2476–2489.

Flourakis M, Kula-Eversole E, Hutchison AL, et al. 2015. A conserved bicycle model for circadian clock control of membrane excitability. Cell 162:836–848.

Hodgkin AL, Huxley AF. 1952. A quantitative description of membrane current and its application to conduction and excitation in nerve. J Physiol 117:500–544.

Hodgkin AL, Katz B. 1949. The effect of sodium ions on the electrical activity of the giant axon of the squid. J Physiol 108:37–77.

Isom LL, DeJongh KS, Catterall WA. 1994. Auxiliary subunits of voltage-gated ion channels. Neuron 12:1183–1194.

Jackson AC, Yao GL, Bean BP. 2004. Mechanism of spontaneous firing in dorsomedial suprachiasmatic nucleus neurons. J Neurosci 24:7985–7998.

Joseph A, Turrigiano GG. 2017. All for one but not one for all: excitatory synaptic scaling and intrinsic excitability are coregulated by CaMKIV, whereas inhibitory synaptic scaling is under independent control. J Neurosci 37:6778–6785.

Kaczmarek LK. 2012. Gradients and modulation of K+ channels optimize temporal accuracy in networks of auditory neurons. PLoS Comput Biol 8:e1002424.

Kole MH, Ilschner SU, Kampa BM, Williams SR, Ruben PC, Stuart GJ. 2008. Action potential generation requires a high sodium channel density in the axon initial segment. Nat Neurosci 11:178–186.

Lee C-H, MacKinnon R. 2017. Structures of the human HCN1 hyperpolarization-activated channel. Cell 168:111–120.

Llinás R, Jahnsen H. 1982. Electrophysiology of mammalian thalamic neurones in vitro. Nature 297:406–408.

McCormick DA, Connors BW, Lighthall JW, Prince DA. 1985. Comparative electrophysiology of pyramidal and sparsely spiny stellate neurons of the neocortex. J Neurophysiol 54:782–806.

McCormick DA, Huguenard JR. 1992. A model of electrophysiological properties of thalamocortical relay neurons. J Neurophysiol 68:1384–1400.

Nowak LG, Azouz R, Sanchez-Vives MV, Gray CM, McCormick DA. 2003. Electrophysiological classes of cat primary visual cortical neurons in vivo as revealed by quantitative analyses. J Neurophysiol 89:1541–1566.

Pan X, Li Z, Zhou Q, et al. 2018. Structure of the human voltage-gated sodium channel Na(v)1.4 in complex with β1. Science 362:eaau2486.

Payandeh J, Scheuer T, Zheng N, Catterall WA. 2011. The crystal structure of a voltage-gated sodium channel. Nature 475:353–359.

Proft J, Weiss N. 2015. G protein regulation of neuronal calcium channels: back to the future. Mol Pharmacol 87:890–906.

Puopolo M, Raviola E, Bean BP. 2007. Roles of subthreshold calcium current and sodium current in spontaneous firing of mouse midbrain dopamine neurons. J Neurosci 27:645–656.

Sigworth FJ, Neher E. 1980. Single Na^+ channel currents observed in cultured rat muscle cells. Nature 287:447–449.

Tai C, Abe Y, Westenbroek RE, Scheuer T, Catterall WA. 2014. Impaired excitability of somatostatin- and parvalbumin-expressing cortical interneurons in a mouse model of Dravet syndrome. Proc Natl Acad Sci U S A 111:E3139–E3148.

Tateno T, Harsch A, Robinson HP. 2004. Threshold firing frequency-current relationships of neurons in rat somatosensory cortex: type 1 and type 2 dynamics. J Neurophysiol 92:2283–2294.

Vassilev PM, Scheuer T, Catterall WA. 1988. Identification of an intracellular peptide segment involved in sodium channel inactivation. Science 241:1658–1661.

Yamada R, Kuba H. 2016, Structural and functional plasticity at the axon initial segment. Front Cell Neurosci 10:250.

Yang N, George AL Jr, Horn R. 1996. Molecular basis of charge movement in voltage-gated sodium channels. Neuron 16:113–122.

Yu FH, Catterall WA. 2003. Overview of the voltage-gated sodium channel family. Genome Biol 4:207.

Part III

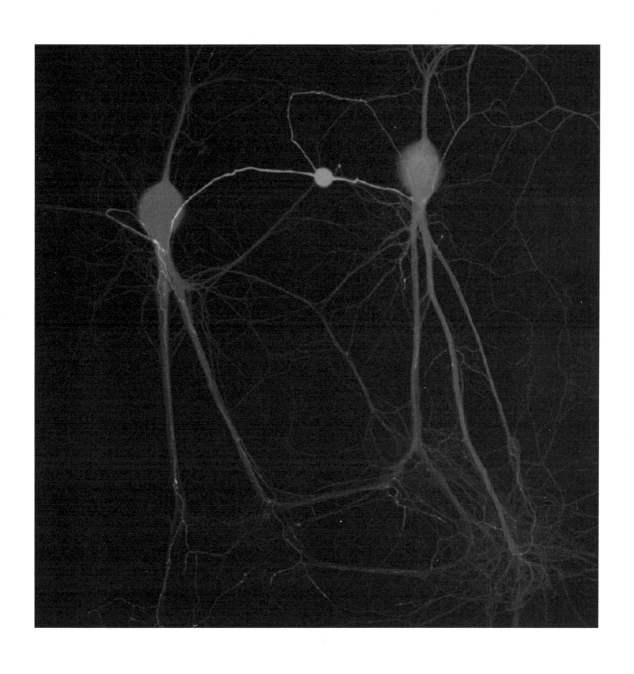

III Synaptic Transmission

I N PART II, WE EXAMINED HOW ELECTRICAL signals are initiated and propagated within an individual neuron. We now turn to synaptic transmission, the process by which one nerve cell communicates with another.

With some exceptions, the synapse consists of three components: (1) the terminals of the presynaptic axon, (2) a target on the postsynaptic cell, and (3) a zone of apposition. Based on the structure of the apposition, synapses are categorized into two major groups: electrical and chemical. At electrical synapses, the presynaptic terminal and the postsynaptic cell are in very close apposition at regions termed *gap junctions*. The current generated by an action potential in the presynaptic neuron directly enters the postsynaptic cell through specialized bridging channels called *gap junction channels*, which physically connect the cytoplasm of the presynaptic and postsynaptic cells. At chemical synapses, a cleft separates the two cells, and the cells do not communicate through bridging channels. Rather, an action potential in the presynaptic cell leads to the release of a chemical transmitter from the nerve terminal. The transmitter diffuses across the synaptic cleft and binds to receptor molecules on the postsynaptic membrane, which regulates the opening and closing of ion channels in the postsynaptic cell. This leads to changes in the membrane potential of the postsynaptic neuron that can either excite or inhibit the firing of an action potential.

Receptors for transmitters can be classified into two major groups depending on how they control ion channels in the postsynaptic cell. One type, the ionotropic receptor, is an ion channel that opens when the transmitter binds. The second type, the metabotropic receptor, acts indirectly on ion channels by activating a biochemical second-messenger cascade within the postsynaptic cell. Both types of receptors can result in excitation or inhibition. The sign of the signal depends not on the identity of the transmitter but on the properties of the receptor with which the transmitter interacts. Most transmitters are low-molecular-weight molecules, but certain peptides also can act as messengers at synapses. The methods of electrophysiology, biochemistry, and molecular biology have been used to characterize the receptors in postsynaptic cells that respond to these various chemical messengers. These methods have also clarified how second-messenger pathways transduce signals within cells.

In this part of the book, we consider synaptic transmission in its most elementary forms. We first compare and contrast the two major classes of synapses, chemical and electrical (see Chapter 11). We then focus on a model chemical synapse in the peripheral nervous system, the neuromuscular junction between a presynaptic motor neuron and a postsynaptic skeletal muscle fiber (see Chapter 12). Next we examine chemical synapses between neurons in the central nervous system, focusing on the postsynaptic cell and its integration of synaptic signals from multiple presynaptic inputs acting on both ionotropic receptors (see Chapter 13) and metabotropic receptors (see Chapter 14). We then turn to the presynaptic terminal and consider the mechanisms by which neurons release transmitter from their presynaptic terminals, how transmitter release can be regulated by neural activity (see Chapter 15), and the chemical nature of the neurotransmitters (see Chapter 16). Because the molecular architecture of chemical synapses is complex, many inherited and acquired diseases can affect chemical synaptic transmission, which we consider in detail later in Chapter 57.

One key theme running throughout the chapters of this section, and indeed throughout the book, is the concept of plasticity. At all synapses, the strength of a synaptic connection is not fixed but can be modified in various ways by behavioral context or experience, through a variety of mechanisms referred to as synaptic plasticity. Some modifications result from the activity of the synapse itself (homosynaptic plasticity). Other modifications depend on extrinsic factors, often due to the release of a modulatory transmitter (heterosynaptic plasticity). In Chapters 53 and 54, we will see how such modifications provide a cellular substrate for different forms of memory storage that range in duration from seconds to a lifetime. In the chapters of Part IX, we will see how dysfunction in synaptic plasticity can contribute to a variety of neurological and psychiatric disorders.

Part Editor: Steven A. Siegelbaum

Part III

11

Overview of Synaptic Transmission

W HAT GIVES NERVE CELLS THEIR SPECIAL ABILITY to communicate with one another rapidly and with such great precision? We have already seen how signals are propagated *within* a neuron, from its dendrites and cell body to its axonal terminals. With this chapter, we begin to consider the signaling *between* neurons through the process of synaptic transmission. Synaptic transmission is fundamental to the neural functions we consider in this book, such as perception, voluntary movement, and learning.

Neurons communicate with one another at a specialized site called a *synapse*. The average neuron forms several thousand synaptic connections and receives a similar number of inputs. However, this number can vary widely depending on the particular type of

neuron. Whereas the Purkinje cell of the cerebellum receives up to 100,000 synaptic inputs, the neighboring granule neurons, the most numerous class of neurons in the brain, receive only around four excitatory inputs. Although many of the synaptic connections in the central and peripheral nervous systems are highly specialized, all neurons make use of one of the two basic forms of synaptic transmission: electrical or chemical. Moreover, the strength of both forms of synaptic transmission is not fixed, but can be enhanced or diminished by neuronal activity. This synaptic *plasticity* is crucial for memory and for other higher brain functions.

Electrical synapses are employed primarily to send rapid and stereotyped depolarizing signals. In contrast, chemical synapses are capable of more variable signaling and thus can produce more complex interactions. They can produce either excitatory or inhibitory actions in postsynaptic cells and initiate changes in the postsynaptic cell that last from milliseconds to hours. Chemical synapses also serve to amplify neuronal signals, so even a small presynaptic nerve terminal can alter the response of large postsynaptic cells. Because chemical synaptic transmission is so central to understanding brain and behavior, it is examined in detail in the next four chapters.

Synapses Are Predominantly Electrical or Chemical

The term *synapse* was introduced at the beginning of the 20th century by Charles Sherrington to describe the specialized zone of contact at which one neuron communicates with another. This site had first been

Table 11–1 Distinguishing Properties of Electrical and Chemical Synapses

Type of synapse	Distance between pre- and postsynaptic cell membranes	Cytoplasmic continuity between pre- and postsynaptic cells	Ultrastructural components	Agent of transmission	Synaptic delay	Direction of transmission
Electrical	4 nm	Yes	Gap-junction channels	Ion current	Virtually absent	Usually bidirectional
Chemical	20–40 nm	No	Presynaptic vesicles and active zones; postsynaptic receptors	Chemical transmitter	Significant: at least 0.3 ms, usually 1–5 ms or longer	Unidirectional

described histologically at the level of light microscopy by Ramón y Cajal in the late 19th century.

All synapses were initially thought to operate by means of electrical transmission. In the 1920s, however, Otto Loewi discovered that a chemical compound, most likely acetylcholine (ACh), conveys signals from the vagus nerve to slow the beating heart. Loewi's discovery provoked considerable debate in the 1930s over whether chemical signaling existed at the fast synapses between motor nerve and skeletal muscle as well as synapses in the brain.

Two schools of thought emerged, one physiological and the other pharmacological. Each championed a single mechanism for all synaptic transmission. Led by John Eccles (Sherrington's student), the physiologists argued that synaptic transmission is electrical, that the action potential in the presynaptic neuron generates a current that flows passively into the postsynaptic cell. The pharmacologists, led by Henry Dale, argued that transmission is chemical, that the action potential in the presynaptic neuron leads to the release of a chemical substance that in turn initiates current in the postsynaptic cell. When physiological and ultrastructural techniques improved in the 1950s and 1960s, it became clear that both forms of transmission exist. While most neurons initiate electrical signaling with a chemical transmitter, many others produce an electrical signal directly in the postsynaptic cell.

Once the fine structure of synapses was made visible with the electron microscope, chemical and electrical synapses were found to have different structures. At chemical synapses, the presynaptic and postsynaptic neurons are completely separated by a small space, the synaptic cleft; there is no continuity between the cytoplasm of one cell and the next. In contrast, at electrical synapses, the pre- and postsynaptic cells communicate through special channels that directly connect the cytoplasm of the two cells.

The main functional properties of the two types of synapses are summarized in Table 11–1. The most important difference can be observed by injecting a positive current into the presynaptic cell to elicit a depolarization. At both types of synapses, outward current across the presynaptic cell membrane deposits positive charge on the inside of its membrane, thereby depolarizing the cell (Chapter 9). At electrical synapses, some of the current will enter the postsynaptic cell through the gap-junction channels, depositing a positive charge on the inside of the membrane and depolarizing it. The current leaves the postsynaptic cell across the membrane through resting channels (Figure 11–1A). If the depolarization exceeds threshold, voltage-gated ion channels in the postsynaptic cell open and generate an action potential. By contrast, at chemical synapses, there is no direct low-resistance pathway between the pre- and postsynaptic cells. Instead, the action potential in the presynaptic neuron initiates the release of a chemical transmitter, which diffuses across the synaptic cleft and binds with receptors on the membrane of the postsynaptic cell (Figure 11–1B).

Electrical Synapses Provide Rapid Signal Transmission

During excitatory synaptic transmission at an electrical synapse, voltage-gated ion channels in the presynaptic cell generate the current that depolarizes the postsynaptic cell. Thus, these channels not only depolarize the presynaptic cell above the threshold for an action potential but also generate sufficient ionic current to produce a change in potential in the postsynaptic cell. To generate such a large current, the presynaptic terminal must be big enough for its membrane to contain many ion channels. At the same time, the postsynaptic cell must be relatively small. This is because a small cell has a higher input resistance (R_{in}) than a large cell

A Current pathways at electrical synapses   B Current pathways at chemical synapses

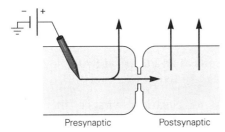 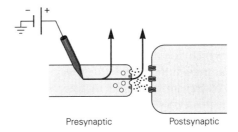

Presynaptic   Postsynaptic   Presynaptic   Postsynaptic

Figure 11–1 Functional properties of electrical and chemical synapses.

A. At an electrical synapse, some current injected into the presynaptic cell escapes through resting (nongated) ion channels in the cell membrane. However, some current also enters the postsynaptic cell through gap-junction channels that connect the cytoplasm of the pre- and postsynaptic cells and that provide a low-resistance (high-conductance) pathway for electrical current.

B. At chemical synapses, all current injected into the presynaptic cell escapes into the extracellular fluid. However, the resulting depolarization of the presynaptic cell membrane can produce an action potential that causes the release of neurotransmitter molecules that bind receptors on the postsynaptic cell. This binding opens ion channels that initiate a change in membrane potential in the postsynaptic cell.

and, according to Ohm's law ($\Delta V = I \times R_{in}$), undergoes a greater voltage change (ΔV) in response to a given presynaptic current (I).

Electrical synaptic transmission was first described by Edwin Furshpan and David Potter in the giant

motor synapse of the crayfish, where the presynaptic fiber is much larger than the postsynaptic fiber (Figure 11–2A). An action potential generated in the presynaptic fiber produces a depolarizing postsynaptic potential that often exceeds the threshold to fire an action

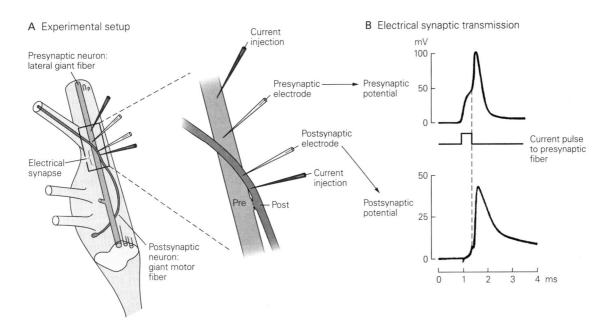

Figure 11–2 Electrical synaptic transmission was first demonstrated at the giant motor synapse in the crayfish. (Adapted, with permission, from Furshpan and Potter 1957 and 1959.)

A. The lateral giant fiber running down the nerve cord is the presynaptic neuron. The giant motor fiber, which projects from the cell body in the ganglion to the periphery, is the postsynaptic

neuron. Electrodes for passing current and for recording voltage are placed within the pre- and postsynaptic cells.

B. Transmission at an electrical synapse is virtually instantaneous—the postsynaptic response follows presynaptic stimulation in a fraction of a millisecond. The **dashed line** shows how the responses of the two cells correspond in time. At chemical synapses, there is a delay (the synaptic delay) between the pre- and postsynaptic potentials (see Figure 11–8).

potential. At electrical synapses, the synaptic delay—the time between the presynaptic spike and the postsynaptic potential—is remarkably short (Figure 11–2B).

Such a short latency is not possible with chemical transmission, which requires several biochemical steps: release of a transmitter from the presynaptic neuron, diffusion of transmitter molecules across the synaptic cleft to the postsynaptic cell, binding of transmitter to a specific receptor, and subsequent gating of ion channels (all described in this and the next chapter). Only current passing directly from one cell to another can produce the near-instantaneous transmission observed at the giant motor electrical synapse.

Another feature of electrical transmission is that the change in potential of the postsynaptic cell is directly related to the size and shape of the change in potential of the presynaptic cell. Even when a weak subthreshold depolarizing current is injected into the presynaptic neuron, some current enters the postsynaptic cell and depolarizes it (Figure 11–3). In contrast, at a chemical synapse, the current in the presynaptic cell must reach the threshold for an action potential before it can release transmitter and elicit a response in the postsynaptic cell.

Most electrical synapses can transmit both depolarizing and hyperpolarizing currents. A presynaptic action potential with a large hyperpolarizing afterpotential produces a biphasic (depolarizing-hyperpolarizing) change in potential in the postsynaptic cell. Signal transmission at electrical synapses is similar to the passive propagation of subthreshold electrical signals along axons (Chapter 9) and therefore is also referred to as *electrotonic transmission*. At some specialized gap junctions, the channels have voltage-dependent gates that permit them to conduct depolarizing current in only one direction, from the presynaptic cell to the postsynaptic cell. These junctions are called *rectifying synapses*. (The crayfish giant motor synapse is an example.)

Cells at an Electrical Synapse Are Connected by Gap-Junction Channels

At an electrical synapse, the pre- and postsynaptic components are apposed at the *gap junction*, where the separation between the two neurons (4 nm) is much less than the normal nonsynaptic space between neurons (20 nm). This narrow gap is bridged by *gap-junction channels*, specialized protein structures that conduct ionic current directly from the presynaptic to the postsynaptic cell.

A gap-junction channel consists of a pair of *hemichannels*, or *connexons*, one in the presynaptic and the other in the postsynaptic cell membrane. These hemichannels thus form a continuous bridge between the two cells (Figure 11–4). The pore of the channel has a large diameter of approximately 1.5 nm, much larger than the 0.3- to 0.5-nm diameter of ion-selective ligand-gated or voltage-gated channels. The large pore of gap-junction channels does not discriminate among inorganic ions and is even wide enough to permit small organic molecules and experimental markers such as fluorescent dyes to pass between the two cells.

Each connexon is composed of six identical subunits, called *connexins*. Connexins in different tissues are encoded by a large family of 21 separate but related genes. In mammals, the most common connexon in neurons is formed from the product of *connexin 36*. Connexin genes are named for their predicted molecular weight, in kilodaltons, based on their primary amino acid sequence. All connexin subunits have an intracellular N- and C-terminus with four interposed α-helices that span the cell membrane (Figure 11–4C).

Many gap-junction channels in different cell types are formed by the products of different connexin genes and thus respond differently to modulatory factors that control their opening and closing. For example, although most gap-junction channels close in response to lowered cytoplasmic pH or elevated cytoplasmic Ca^{2+}, the sensitivity of different channel isoforms to these factors varies widely. The closing of gap-junction channels in response to pH and Ca^{2+} plays an important role in the decoupling of damaged cells from healthy cells, as damaged cells contain elevated Ca^{2+} levels and a high concentration of protons. Finally, neurotransmitters released from nearby chemical synapses can modulate the opening of gap-junction channels through intracellular metabolic reactions (Chapter 14).

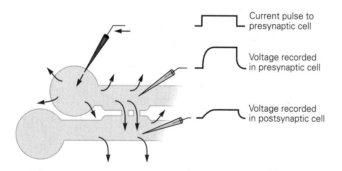

Figure 11–3 Electrical transmission is graded. It occurs even when the current in the presynaptic cell is below the threshold for an action potential. As demonstrated by single-cell recordings, a subthreshold depolarizing stimulus causes a passive depolarization in the presynaptic and postsynaptic cells. (Depolarizing or outward current is indicated by an upward deflection.)

Current pulse to presynaptic cell

Voltage recorded in presynaptic cell

Voltage recorded in postsynaptic cell

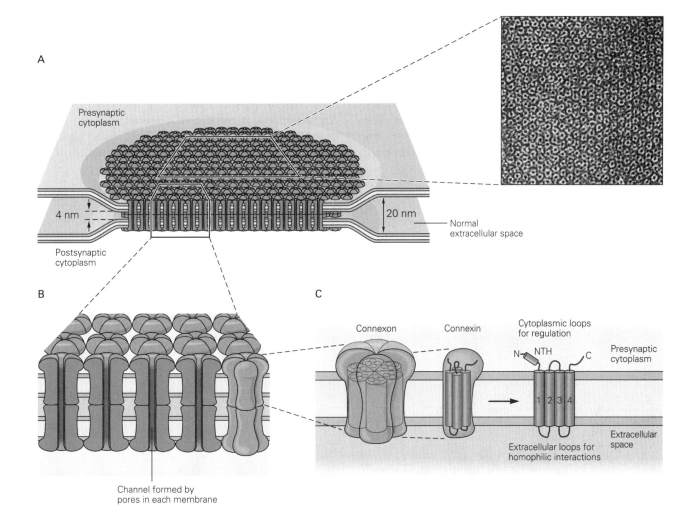

A

Presynaptic
cytoplasm

4 nm

Postsynaptic
cytoplasm

20 nm — Normal
extracellular space

B

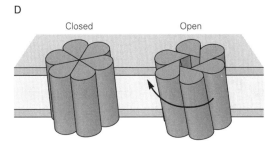

Channel formed by
pores in each membrane

C

Connexon   Connexin   Cytoplasmic loops
for regulation

N   NTH   C   Presynaptic
                                    cytoplasm

                    1 2 3 4

Extracellular loops for   Extracellular
homophilic interactions   space

D

Closed   Open

Figure 11–4 A three-dimensional model of the gap-junction channel, based on X-ray and electron diffraction studies.

A. The electrical synapse, or gap junction, is composed of numerous specialized channels that span the membranes of the pre- and postsynaptic neurons. These gap-junction channels allow current to pass directly from one cell to the other. The array of channels in the electron micrograph was isolated from the membrane of a rat liver cell that had been negatively stained, a technique that darkens the area around the channels and in the pores. Each channel appears hexagonal in outline. Magnification ×307,800. (Reproduced, with permission, from N. Gilula.)

B. A gap-junction channel is actually a pair of hemichannels, one in each apposite cell that connects the cytoplasm of the two cells. (Adapted from Makowski et al. 1977.)

C. Each hemichannel, or connexon, is made up of six identical subunits called connexins. Each connexin is approximately

7.5 nm long and spans the cell membrane. A single connexin has intracellular N- and C-terminals, including a short intracellular N-terminal α-helix (**NTH**), and four membrane-spanning α-helixes (**1–4**). The amino acid sequences of gap-junction proteins from many different kinds of tissue have regions of similarity that include the transmembrane helixes and the extracellular regions, which are involved in the homophilic matching of apposite hemichannels.

D. The connexins are arranged in such a way that a pore is formed in the center of the structure. The resulting connexon, with a pore diameter of approximately 1.5 to 2 nm, has a characteristic hexagonal outline, as shown in the photograph in part A. In some gap-junction channels, the pore is opened when the subunits rotate approximately 0.9 nm at the cytoplasmic base in a clockwise direction. (Reproduced, with permission, from Unwin and Zampighi 1980. Copyright © 1980 Springer Nature.)

The three-dimensional structure of a gap-junction channel formed by the human connexin 26 subunit has been determined by X-ray crystallography. This structure shows how the membrane-spanning α-helixes assemble to form the central pore of the channel and how the extracellular loops connecting the transmembrane helixes interdigitate to connect the two hemichannels (Figure 11–5). The pore is lined with polar residues that facilitate the movement of ions. An N-terminal α-helix may serve as the voltage gate of the connexin 26 channel, plugging the cytoplasmic mouth of the pore in the closed state. A separate gate

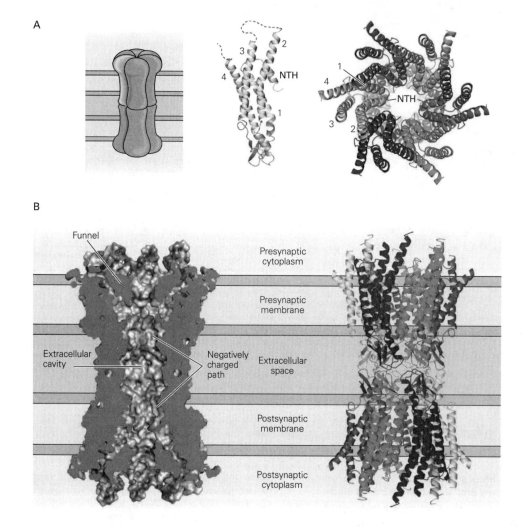

Figure 11–5 High-resolution three-dimensional structure of a gap-junction channel. All structures were determined by X-ray crystallography of gap-junction channels formed by the human connexin 26 subunit. (Reproduced, with permission, from Maeda et al. 2009. Copyright © 2009 Springer Nature.)

A. *Left:* Diagram of an intact gap-junction channel showing the pair of apposed hemichannels. *Middle:* This high-resolution structure of a single connexin subunit shows four transmembrane α-helixes (1–4) and a short N-terminal helix (**NTH**). The orientation of the subunit corresponds to that of the yellow subunit in the diagram to the right. *Right:* Bottom-up view looking into a hemichannel from the cytoplasm. Each of the six subunits has a different color. The helixes of the yellow subunit

are numbered. The orientation corresponds to that of the yellow hemichannel in the diagram at left, following a 90-degree rotation toward the viewer.

B. Two side views of the gap-junction channel in the plane of the membrane show the two apposed hemichannels. The orientation is the same as in part A. *Left:* Cross section through the channel shows the internal surface of the channel pore. **Blue** indicates positively charged surfaces; **red** indicates negatively charged surfaces. The **green mass** inside the pore at the cytoplasmic entrance (funnel) is thought to represent the channel gate formed by the N-terminal helix. *Right:* A side view of the channel shows each of the six connexin subunits in the same color scheme as in part A. The entire gap-junction channel is approximately 9 nm wide by 15 nm tall.

at the extracellular side of the channel, formed by the extracellular loop connecting the first two membrane helixes, has been inferred from functional studies. This loop gate is thought to close isolated hemichannels that are not docked to a hemichannel partner in the apposing cell.

Electrical Transmission Allows Rapid and Synchronous Firing of Interconnected Cells

How are electrical synapses useful? As we have seen, electrical synaptic transmission is rapid because it results from the direct passage of current between cells. Speed is important for escape responses. For example, the tail-flip response of goldfish is mediated by a giant neuron in the brain stem (known as the Mauthner cell), which receives sensory input at electrical synapses. These electrical synapses rapidly depolarize the Mauthner cell, which in turn activates the motor neurons of the tail, allowing rapid escape from danger.

Electrical transmission is also useful for orchestrating the actions of groups of neurons. Because current crosses the membranes of all electrically coupled cells

at the same time, several small cells can act together as one large cell. Moreover, because of the electrical coupling between the cells, the effective resistance of the network is smaller than the resistance of an individual cell. Thus, from Ohm's law, the synaptic current required to fire electrically coupled cells is larger than that necessary to fire an individual cell. That is, electrically coupled cells have a higher firing threshold. Once this high threshold is surpassed, however, electrically coupled cells fire synchronously because voltage-activated Na^+ currents generated in one cell are very rapidly conducted to other cells.

Thus, a behavior controlled by a group of electrically coupled cells has an important adaptive advantage: It is triggered explosively. For example, when seriously perturbed, the marine snail *Aplysia* releases massive clouds of purple ink that provide a protective screen. This stereotypic behavior is mediated by three electrically coupled motor cells that innervate the ink gland. Once the action potential threshold is exceeded in these cells, they fire synchronously (Figure 11–6). In certain fish, rapid eye movements (called saccades) are also mediated by electrically coupled motor neurons

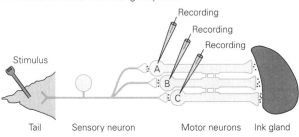

A Neural circuit of the inking response

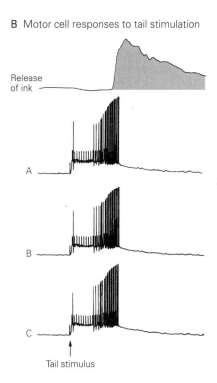

B Motor cell responses to tail stimulation

Figure 11–6 Electrically coupled motor neurons firing together can produce synchronous behaviors. (Adapted, with permission, from Carew and Kandel 1976.)

A. In the marine snail *Aplysia*, sensory neurons from the tail ganglion form synapses with three motor neurons that

innervate the ink gland. The motor neurons are interconnected by electrical synapses.

B. A train of stimuli applied to the tail produces a synchronized discharge in all three motor neurons that results in the release of ink.

firing together. Gap junctions are also important in the mammalian brain, where the synchronous firing of electrically coupled inhibitory interneurons generates synchronous 40- to 100-Hz (gamma) oscillations in large populations of cells.

In addition to providing speed or synchrony in neuronal signaling, electrical synapses also can transmit metabolic signals between cells. Because of their large-diameter pore, gap-junction channels conduct a variety of inorganic cations and anions, including the second messenger Ca^{2+}, and even conduct moderate-sized organic compounds (<1 kDa molecular weight) such as the second messengers inositol 1,4,5-trisphosphate (IP_3), cyclic adenosine monophosphate (cAMP), and even small peptides.

Gap Junctions Have a Role in Glial Function and Disease

Gap junctions are formed between glial cells as well as between neurons. In glia, the gap junctions mediate both intercellular and intracellular signaling. In the brain, individual astrocytes are connected to each other through gap junctions forming a glial cell network. Electrical stimulation of neuronal pathways in brain slices can release neurotransmitters that trigger a rise in intracellular Ca^{2+} in certain astrocytes. This produces a wave of Ca^{2+} that propagates from astrocyte to astrocyte at a rate of approximately 1–20 μm/s, about a million-fold slower than the propagation of an action potential (10–100 m/s). Although the precise function of the waves is unknown, their existence suggests that glia may play an active role in intercellular signaling in the brain.

Gap-junction channels also enhance communication *within* certain glial cells, such as the Schwann cells that produce the myelin sheath of axons in the peripheral nervous system. Successive layers of myelin formed by a single Schwann cell are connected by gap junctions. These gap junctions may help to hold the layers of myelin together and promote the passage of small metabolites and ions across the many layers of myelin. The importance of the Schwann cell gap-junction channels is underscored by certain genetic diseases. For example, the X chromosome–linked form of Charcot-Marie-Tooth disease, a demyelinating disorder, is caused by single mutations that disrupt the function of *connexin 32*, the gene expressed in Schwann cells. Inherited mutations that prevent the function of a connexin in the cochlea (connexin 26), which normally forms gap-junction channels that are important for fluid secretion in the inner ear, underlie up to half of all instances of congenital deafness.

Chemical Synapses Can Amplify Signals

In contrast to electrical synapses, at chemical synapses, there is no structural continuity between presynaptic and postsynaptic neurons. In fact, the separation between the two cells at a chemical synapse, the synaptic cleft, is usually slightly wider (20–40 nm) than the nonsynaptic intercellular space (20 nm). Chemical synaptic transmission depends on a neurotransmitter, a chemical substance that diffuses across the synaptic cleft and binds to and activates receptors in the membrane of the target cell. At most chemical synapses, transmitter is released from specialized swellings of the presynaptic axon—synaptic boutons—that typically contain 100 to 200 synaptic vesicles, each of which is filled with several thousand molecules of neurotransmitter (Figure 11–7).

The synaptic vesicles are clustered at specialized regions of the synaptic bouton called *active zones*. During a presynaptic action potential, voltage-gated Ca^{2+} channels at the active zone open, allowing Ca^{2+} to enter the presynaptic terminal. The rise in intracellular Ca^{2+} concentration triggers a biochemical reaction that causes the vesicles to fuse with the presynaptic membrane and release neurotransmitter into the synaptic cleft, a process termed *exocytosis*. The transmitter molecules then diffuse across the synaptic cleft and bind

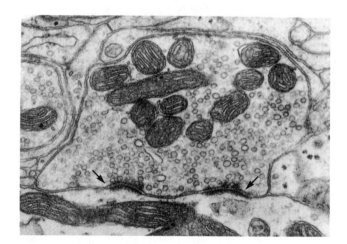

Figure 11–7 The fine structure of a presynaptic terminal. This electron micrograph shows an axon terminal in the cerebellum. The large dark structures are mitochondria. The many small round bodies are vesicles that contain neurotransmitter. The fuzzy dark thickenings along the presynaptic membrane (**arrows**) are the active zones, specialized areas that are thought to be docking and release sites for synaptic vesicles. The synaptic cleft is the space separating the pre- and postsynaptic cell membranes. (Reproduced, with permission, from Heuser and Reese 1977.)

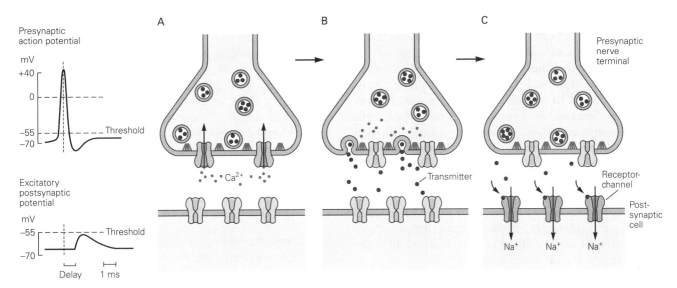

Figure 11–8 Synaptic transmission at chemical synapses involves several steps. The complex process of chemical synaptic transmission accounts for the delay between an action potential in the presynaptic cell and the synaptic potential in the postsynaptic cell, as compared with the virtually instantaneous transmission of signals at electrical synapses (see Figure 11–2B).

A. An action potential arriving at the terminal of a presynaptic axon causes voltage-gated Ca^{2+} channels at the active zone to open. The **gray filaments** represent the docking and release sites of the active zone.

B. The Ca^{2+} channel opening produces a high concentration of intracellular Ca^{2+} near the active zone, causing vesicles containing neurotransmitter to fuse with the presynaptic cell membrane and release their contents into the synaptic cleft (a process termed *exocytosis*).

C. The released neurotransmitter molecules then diffuse across the synaptic cleft and bind specific receptors on the postsynaptic membrane. These receptors cause ion channels to open (or close), thereby changing the membrane conductance and membrane potential of the postsynaptic cell.

to their receptors on the postsynaptic cell membrane. This in turn activates the receptors, leading to the opening or closing of ion channels. The resulting flux of ions alters the membrane conductance and potential of the postsynaptic cell (Figure 11–8).

These several steps account for the synaptic delay at chemical synapses. Despite its biochemical complexity, the release process is remarkably efficient—the synaptic delay is usually only 1 ms or less. Although chemical transmission lacks the immediacy of electrical synapses, it has the important property of *amplification.* Just one synaptic vesicle releases several thousand molecules of transmitter that together can open thousands of ion channels in the target cell. In this way, a small presynaptic nerve terminal, which generates only a weak electrical current, can depolarize a large postsynaptic cell.

The Action of a Neurotransmitter Depends on the Properties of the Postsynaptic Receptor

Chemical synaptic transmission can be divided into two steps: a transmitting step, in which the presynaptic cell releases a chemical messenger, and a receptive step, in which the transmitter binds to and activates the receptor molecules in the postsynaptic cell. The transmitting process in neurons resembles endocrine hormone release. Indeed, chemical synaptic transmission can be seen as a modified form of hormone secretion. Both endocrine glands and presynaptic terminals release a chemical agent with a signaling function, and both are examples of regulated secretion (Chapter 7). Similarly, both endocrine glands and neurons are usually some distance from their target cells.

There is one important difference, however, between endocrine and synaptic signaling. Whereas the hormone released by a gland travels through the blood stream until it interacts with all cells that contain an appropriate receptor, a neuron usually communicates only with the cells with which it forms synapses. Because the presynaptic action potential triggers the release of a chemical transmitter onto a target cell across a distance of only 20 nm, the chemical signal travels only a small distance to its target. Therefore, neuronal signaling has two special features: It is fast and it is precisely directed.

In most neurons, this directed or focused release is accomplished at the active zones of synaptic boutons.

In presynaptic neurons neurons without active zones, the distinction between neuronal and hormonal transmission becomes blurred. For example, the neurons in the autonomic nervous system that innervate smooth muscle reside at some distance from their postsynaptic cells and do not have specialized release sites in their terminals. Synaptic transmission between these cells is slower and relies on a more widespread diffusion of transmitter. Furthermore, the same transmitter substance can be released differently from different cells. A substance can be released from one cell as a conventional transmitter acting directly on neighboring cells. From other cells, it can be released in a less focused way as a modulator, producing a more diffuse action; and from still other cells, it can be released into the blood stream as a neurohormone.

Although a variety of chemicals serve as neurotransmitters, including both small molecules and peptides (Chapter 16), the action of a transmitter depends on the properties of the postsynaptic receptors that recognize and bind the transmitter, not the chemical properties of the transmitter. For example, ACh can excite some postsynaptic cells and inhibit others, and at still other cells, it can produce both excitation and inhibition. It is the receptor that determines the action of ACh, including whether a cholinergic synapse is excitatory or inhibitory.

Within a group of closely related animals, a transmitter substance binds conserved families of receptors and is often associated with specific physiological functions. In vertebrates, ACh acts on excitatory ACh receptors at all neuromuscular junctions to trigger contraction while also acting on inhibitory ACh receptors to slow the heart.

The distinction between the transmitting and receptive processes is not absolute; many presynaptic terminals contain transmitter receptors that can modify the release process. In some instances, these presynaptic receptors are activated by the transmitter released from the same presynaptic terminal. In other instances, the presynaptic terminal can be contacted by presynaptic terminals from other classes of neurons that release distinct neurotransmitters.

The notion of a receptor was introduced in the late 19th century by the German bacteriologist Paul Ehrlich to explain the selective action of toxins and other pharmacological agents and the great specificity of immunological reactions. In 1900, Ehrlich wrote: "Chemical substances are only able to exercise an action on the tissue elements with which they are able to establish an intimate chemical relationship ... [This relationship] must be specific. The [chemical] groups must be adapted to one another ... as lock and key."

In 1906, the English pharmacologist John Langley postulated that the sensitivity of skeletal muscle to curare and nicotine was caused by a "receptive molecule." A theory of receptor function was later developed by Langley's students (in particular, A.V. Hill and Henry Dale), a development that was based on concurrent studies of enzyme kinetics and cooperative interactions between small molecules and proteins. As we shall see in the next chapter, Langley's "receptive molecule" has been isolated and characterized as the ACh receptor of the neuromuscular junction.

All receptors for chemical transmitters have two biochemical features in common:

1. They are membrane-spanning proteins. The region exposed to the external environment of the cell recognizes and binds the transmitter from the presynaptic cell.
2. They carry out an effector function within the target cell. The receptors typically influence the opening or closing of ion channels.

Activation of Postsynaptic Receptors Gates Ion Channels Either Directly or Indirectly

Neurotransmitters control the opening of ion channels in the postsynaptic cell either directly or indirectly. These two classes of transmitter actions are mediated by receptor proteins derived from different gene families.

Receptors that gate ion channels directly, such as the ACh receptor at the neuromuscular junction, are composed of four or five subunits that form a single macromolecule. Such receptors contain both an extracellular domain that forms the binding site for the transmitter and a membrane-spanning domain that forms an ion-conducting pore (Figure 11–9A). This kind of receptor is often referred to as *ionotropic* because the receptor directly controls ion flux. Upon binding neurotransmitter, the receptor undergoes a conformational change that opens the ion channel. The actions of ionotropic receptors, also called *receptor-channels* or *ligand-gated channels*, are discussed in detail in Chapters 12 and 13.

Receptors that gate ion channels indirectly, like the several types of receptors for norepinephrine or dopamine in neurons of the cerebral cortex, are normally composed of one or at most two subunits that are distinct from the ion channels they regulate. These receptors, which commonly have seven membrane-spanning α-helices, act by altering intracellular metabolic reactions and are often referred to as *metabotropic receptors*. Activation of these receptors often stimulates

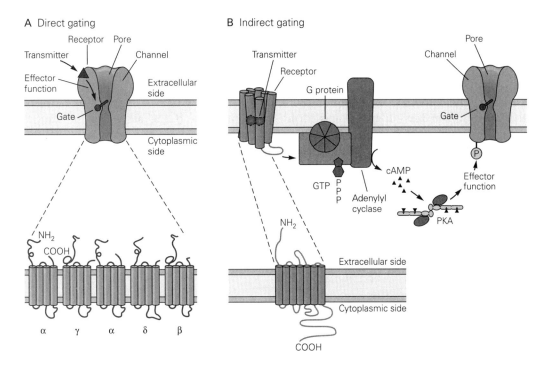

Figure 11–9 Neurotransmitters open postsynaptic ion channels either directly or indirectly.

A. A receptor that directly opens ion channels is an integral part of the macromolecule that also forms the channel. Many such ligand-gated channels are composed of five subunits, each of which is thought to contain four membrane-spanning α-helical regions.

B. A receptor that indirectly opens an ion channel is a distinct macromolecule separate from the channel it regulates. In one large family of such receptors, the receptors are composed of a single subunit with seven membrane-spanning α-helical regions that bind the ligand within the plane of the membrane. These receptors activate a guanosine triphosphate (**GTP**)–binding protein (**G protein**), which in turn activates a second-messenger cascade that modulates channel activity. In the cascade illustrated here, the G protein stimulates adenylyl cyclase, which converts adenosine triphosphate (**ATP**) to cyclic adenosine monophosphate (**cAMP**). The cAMP activates the cAMP-dependent protein kinase (**PKA**), which phosphorylates the channel (**P**), leading to a change in opening.

the production of second messengers, small freely diffusible intracellular metabolites such as cAMP or diacylglycerol. Many of these second messengers activate protein kinases, enzymes that phosphorylate different substrate proteins. In many instances, the protein kinases directly phosphorylate ion channels, including gap-junction channels and ionotropic receptors, modulating their opening or closing (Figure 11–9B). The actions of metabotropic receptors are examined in detail in Chapter 14.

Ionotropic and metabotropic receptors have different functions. The ionotropic receptors produce relatively fast synaptic actions lasting only milliseconds. These are commonly found at synapses in neural circuits that mediate rapid behaviors, such as the stretch receptor reflex. The metabotropic receptors produce slower synaptic actions, lasting hundreds of milliseconds to minutes. These slower actions can modulate a behavior by altering the excitability of neurons and the strength of the synaptic connections in the neural circuit that mediates the behavior. Such modulatory synaptic actions often act as crucial reinforcing pathways in the process of learning.

Electrical and Chemical Synapses Can Coexist and Interact

As we now realize, both Henry Dale and John Eccles were correct about the existence of chemical and electrical synapses, respectively. Furthermore, we now know that both forms of synaptic transmission can coexist in the same neuron and that electrical and chemical synapses can modify each other's efficacy. For example, during development, many neurons are initially connected by electrical synapses, whose presence helps in the formation of chemical synapses. As chemical synapses begin to form, they often initiate the down-regulation of electrical transmission.

Both types of synapses also can coexist in neurons in the mature nervous system. The role of these two types of synapses is perhaps best understood in the circuitry of the retina. There, rod and cone photoreceptors release the neurotransmitter glutamate and form chemical synapses on a class of interneurons called bipolar cells. Each bipolar cell extends its dendrites horizontally, receiving chemical synaptic input from a number of overlying rods and cones that respond to light from a very small region of the visual field. The receptive field of a bipolar neuron, however, extends about twice as far as the receptive field of the photoreceptors from which it receives chemical synaptic input. This is a result of electrical synapses formed between neighboring bipolar cells and between bipolar cells and a second type of interneuron, the amacrine cell (Chapter 22).

Finally, the efficacy of gap junctions can be regulated by phosphorylation through different protein kinases, which generally enhances gap-junction coupling. For example, dopamine and other transmitters can increase or decrease gap-junction coupling by acting on metabotropic G protein–coupled receptors to regulate levels of cAMP and thereby enhance or decrease channel phosphorylation. Such complex signaling loops are a hallmark of many neural circuits and greatly expand their computational powers.

Highlights

1. Neurons communicate by two major mechanisms: electrical and chemical synaptic transmission.

2. Electrical synapses are formed at regions of tight apposition called gap junctions, which provide a direct pathway for charge to flow between the cytoplasm of communicating neurons. This results in very rapid synaptic transmission that is suited for synchronizing the activity of populations of neurons.

3. Neurons at electrical synapses are connected through gap-junction channels, which are formed from a pair of hemichannels, called connexons, one each contributed by the presynaptic and postsynaptic cells. Each connexon is a hexamer, composed of six subunits termed connexins.

4. At chemical synapses, a presynaptic action potential triggers the release of a chemical transmitter from the presynaptic cell through the process of exocytosis. Transmitter molecules then rapidly diffuse across the synaptic cleft to bind to and activate transmitter receptors in the postsynaptic cell.

5. Although slower than electrical synaptic transmission, chemical transmission allows for amplification of the presynaptic action potential through the release of tens of thousands of molecules of transmitter and the activation of hundreds to thousands of receptors in the postsynaptic cell.

6. There are two major classes of transmitter receptors. Ionotropic receptors are ligand-gated ion channels. Binding of transmitter to an extracellular binding site triggers a conformational change that opens the channel pore, generating an ionic current that excites (depolarizes) or inhibits (hyperpolarizes) the postsynaptic cell, depending on the receptor. Ionotropic receptors underlie fast chemical synaptic transmission that mediates rapid signaling in the nervous system.

7. Metabotropic receptors are responsible for the second major class of chemical synaptic actions. These receptors activate intracellular metabolic signaling pathways, often leading to the synthesis of second messengers, such as cAMP, that regulate levels of protein phosphorylation. Metabotropic receptors underlie slow, modulatory synaptic actions that contribute to changes in behavioral state and arousal.

Steven A. Siegelbaum
Gerald D. Fischbach

Selected Reading

Bennett MV, Zukin RS. 2004. Electrical coupling and neuronal synchronization in the mammalian brain. Neuron 19:495–511.

Colquhoun D, Sakmann B. 1998. From muscle endplate to brain synapses: a short history of synapses and agonist-activated ion channels. Neuron 20:381–387.

Cowan WM, Kandel ER. 2000. A brief history of synapses and synaptic transmission. In: MW Cowan, TC Südhof, CF Stevens (eds). *Synapses*, pp. 1–87. Baltimore and London: The Johns Hopkins Univ. Press.

Curti S, O'Brien J. 2016. Characteristics and plasticity of electrical synaptic transmission. BMC Cell Biol 17:13. Suppl 1.

Eccles JC. 1976. From electrical to chemical transmission in the central nervous system. The closing address of the

Sir Henry Dale Centennial Symposium. Notes Rec R Soc Lond 30:219–230.

Furshpan EJ, Potter DD. 1959. Transmission at the giant motor synapses of the crayfish. J Physiol 145:289–325.

Goodenough DA, Paul DL. 2009. Gap junctions. Cold Spring Harb Perspect Biol 1:a002576.

Jessell TM, Kandel ER. 1993. Synaptic transmission: a bidirectional and a self-modifiable form of cell-cell communication. Cell 72:1–30.

Nakagawa S, Maeda S, Tsukihara T. 2010. Structural and functional studies of gap junction channels. Curr Opin Struct Biol 20:423–430.

Pereda AE. 2014. Electrical synapses and their functional interactions with chemical synapses. Nat Rev Neurosci 15:250–263.

References

Beyer EC, Paul DL, Goodenough DA. 1987. Connexin 43: a protein from rat heart homologous to a gap junction protein from liver. J Cell Biol 105:2621–2629.

Bruzzone R, White TW, Scherer SS, Fischbeck KH, Paul DL. 1994. Null mutations of connexin 32 in patients with X-linked Charcot-Marie-Tooth disease. Neuron 13:1253–1260.

Carew TJ, Kandel ER. 1976. Two functional effects of decreased conductance EPSP's: synaptic augmentation and increased electrotonic coupling. Science 192:150–153.

Cornell-Bell AH, Finkbeiner SM, Cooper MS, Smith SJ. 1990. Glutamate induces calcium waves in cultured astrocytes: long-range glial signaling. Science 247:470–473.

Dale H. 1935. Pharmacology and nerve-endings. Proc R Soc Lond 28:319–332.

Eckert R. 1988. Propagation and transmission of signals. In: Animal Physiology: Mechanisms and Adaptations, 3rd ed., pp. 134–176. New York: Freeman.

Ehrlich P. 1900. On immunity with special reference to cell life. Croonian Lect Proc R Soc Lond 66:424–448.

Furshpan EJ, Potter DD. 1957. Mechanism of nerve-impulse transmission at a crayfish synapse. Nature 180:342–343.

Harris AL. 2009. Gating on the outside. J Gen Physiol 133:549–553.

Heuser JE, Reese TS. 1977. Structure of the synapse. In: ER Kandel (ed). Handbook of Physiology: A Critical, Comprehensive Presentation of Physiological Knowledge and Concepts, Sect. 1. The Nervous System, Vol. 1 Cellular Biology of Neurons, Part 1, pp. 261–294. Bethesda, MD: American Physiological Society.

Jaslove SW, Brink PR. 1986. The mechanism of rectification at the electrotonic motor giant synapse of the crayfish. Nature 323:63–65.

Langley JN. 1906. On nerve endings and on special excitable substances in cells. Proc R Soc Lond B Biol Sci 78:170–194.

Loewi O, Navratil E. [1926] 1972. On the humoral propagation of cardiac nerve action. Communication X. The fate of the vagus substance. English translation in: I Cooke, M Lipkin Jr (eds). Cellular Neurophysiology: A Source Book, pp. 4711–485. New York: Holt, Rinehart and Winston.

Maeda S, Nakagawa S, Suga M, et al. 2009. Structure of the connexin 26 gap junction channel at 3.5 Å resolution. Nature 458:597–602.

Makowski L, Caspar DL, Phillips WC, Goodenough DA. 1977. Gap junction structures. II. Analysis of the X-ray diffraction data. J Cell Biol 74:629–645.

Pappas GD, Waxman SG. 1972. Synaptic fine structure: morphological correlates of chemical and electronic transmission. In: GD Pappas, DP Purpura (eds). Structure and Function of Synapses, pp. 1–43. New York: Raven.

Ramón y Cajal S. 1894. La fine structure des centres nerveux. Proc R Soc Lond 55:444–468.

Ramón y Cajal S. 1911. Histologie du Système Nerveux de l'Homme & des Vertébrés, Vol. 2. L Azoulay (transl). Paris: Maloine, 1955. Reprint. Madrid: Instituto Ramón y Cajal.

Sherrington C. 1947. The Integrative Action of the Nervous System, 2nd ed. New Haven: Yale Univ. Press.

Unwin PNT, Zampighi G. 1980. Structure of the junction between communicating cells. Nature 283:545–549.

Whittington MA, Traub RD. 2003. Interneuron diversity series: inhibitory interneurons and network oscillations in vitro. Trends Neurosci 26:676–682.

12

Directly Gated Transmission:
The Nerve-Muscle Synapse

MUCH OF OUR UNDERSTANDING of the principles that govern chemical synapses in the brain is based on studies of synapses formed by motor neurons on skeletal muscle cells. The landmark work of Bernard Katz and his colleagues over three decades beginning in 1950 defined the basic parameters of synaptic transmission and opened the door to modern molecular analyses of synaptic function. Therefore, before we examine the complexities of synapses in the central nervous system, we will examine the basic features of chemical synaptic transmission at the simpler nerve-muscle synapse.

The early studies capitalized on several experimental advantages offered by nerve-muscle preparations of various species. Muscles and attached motor axons are easy to dissect and maintain for several hours in vitro. Muscle cells are large enough to be penetrated with two or more fine-tipped microelectrodes, enabling precise analyses of synaptic potentials and underlying ionic currents. In most species, innervation is restricted to one site, the motor endplate, and in adult animals that site is innervated by only one motor axon. In contrast, central neurons receive many convergent inputs that are distributed throughout the dendritic arbor and the soma, and thus the impact of single inputs is more difficult to discern.

Most important, the chemical transmitter that mediates synaptic transmission between nerve and muscle, acetylcholine (ACh), was identified early in the 20th century. We now know that signaling at the nerve-muscle synapse involves a relatively simple mechanism: Neurotransmitter released from the presynaptic nerve binds to a single type of receptor in the postsynaptic membrane, the nicotinic ACh receptor.[1] Binding of transmitter to the receptor directly opens an ion channel; both the receptor and channel are components of the same macromolecule. Synthetic and natural agents that activate or inhibit nicotinic ACh receptors have proven useful in analyzing not only the ACh receptors in muscle, but also cholinergic synapses in peripheral ganglia and in the brain. Moreover, such ligands can be useful therapeutic agents, including the treatment of inherited and acquired neurological diseases resulting from alterations in ACh receptor function or genetic mutations.

The Neuromuscular Junction Has Specialized Presynaptic and Postsynaptic Structures

As the motor axon approaches the *end-plate*, the site of contact between nerve and muscle (also known as the *neuromuscular junction*), it loses its myelin sheath and divides into several fine branches. At their ends, these fine branches form multiple expansions or varicosities called *synaptic boutons* (Figure 12–1) from which the motor axon releases its transmitter. Although myelin ends some distance from the sites of transmitter release, Schwann cells cover and partially enwrap the nerve terminal. A terminal "arbor" defines the area of the motor end-plate. In different species, end-plates range from compact elliptical structures about 20 µm across to linear arrays more than 100 µm in length.

The nerve terminals lie in grooves, the primary folds, along the muscle surface. The membrane under each synaptic bouton is further invaginated to form a series of secondary or junctional folds (Figure 12–1). The muscle cytoplasm beneath the nerve terminals contains many round muscle nuclei that likely are involved in synthesis of synapse-specific molecules. They are different from the flat nuclei located away from the synapse along the length of the muscle fiber.

Action potentials in the axon are conducted to the tips of the fine branches where they trigger the release of ACh. The synaptic boutons contain all the machinery required to synthesize and release ACh. This includes the synaptic vesicles containing the transmitter ACh and the active zones where the synaptic vesicles are clustered. In addition, each active zone

contains voltage-gated Ca^{2+} channels that conduct Ca^{2+} into the terminal with each action potential. This influx of Ca^{2+} triggers the fusion of the synaptic vesicles with the plasma membrane at the active zones, releasing the contents of the synaptic vesicles into the synaptic cleft by the process of exocytosis (Chapter 15).

The distribution of ACh receptors can be studied using α-bungarotoxin (αBTX), a peptide isolated from the venom of the snake *Bungarus multicinctus* that binds tightly and specifically to the ACh receptors at the neuromuscular junction (Figure 12–2B). Quantitative autoradiography of iodinated BTX (^{125}I-αBTX) showed that the ACh receptors are packed at the crests of the secondary folds at a surface density in excess of $10,000/\mu m^2$ (Figure 12–3). The factors responsible for localizing the receptor are discussed in Chapter 48, where we consider the development of synaptic connections.

Presynaptic and postsynaptic membranes at the neuromuscular junction are separated by a cleft approximately 100 nm wide. Although such a gap was postulated by Ramón y Cajal in the last years of the 19th century, it was not visualized until synapses were examined by electron microscopy more than 50 years later! A basement membrane (basal lamina) composed of collagen and other extracellular matrix proteins is present throughout the synaptic cleft. Acetylcholinesterase (AChE), an enzyme that rapidly hydrolyzes ACh, is anchored to collagen fibrils within the basement membrane. The ACh released into the synaptic cleft must run a "gauntlet" of AChE before it reaches the ACh receptors in the muscle membrane. As AChE is inhibited by high concentrations of ACh, most molecules get through. Nevertheless, the enzyme limits the action of ACh to "one hit" because AChE hydrolyzes transmitter as soon as it dissociates from its receptor in the postsynaptic membrane.

The Postsynaptic Potential Results From a Local Change in Membrane Permeability

Once ACh is released from a synaptic terminal, it rapidly binds to and opens the ACh receptor-channels in the end-plate membrane. This produces a dramatic increase in the permeability of the muscle membrane to cations, which leads to the entry of positive charge into the muscle fiber and a rapid depolarization of the end-plate membrane. The resulting excitatory postsynaptic potential (EPSP) is very large; stimulation of a single motor axon produces an EPSP of approximately 75 mV. At the nerve-muscle synapse, the EPSP is also referred to as the *end-plate potential*.

This change in membrane potential usually is large enough to rapidly activate the voltage-gated Na^+

---

[1]There are two basic types of receptors for ACh: nicotinic and muscarinic, so called because the alkaloids nicotine and muscarine bind exclusively to and activate one or the other type of ACh receptor. The nicotinic ACh receptor is ionotropic, whereas the muscarinic receptor is metabotropic. We shall learn more about muscarinic ACh receptors in Chapter 14.

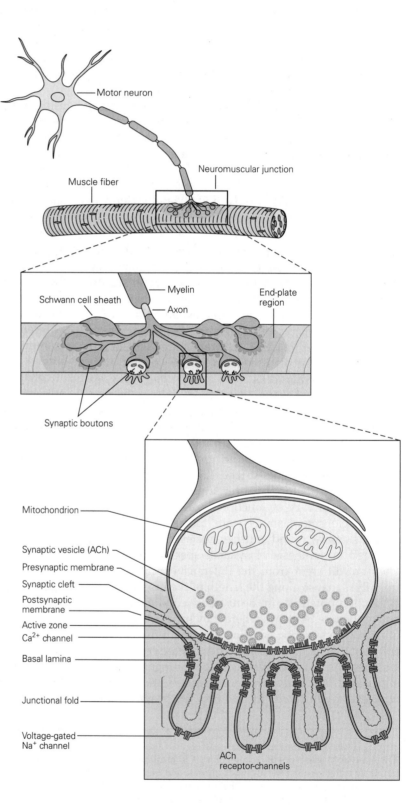

Figure 12–1 The neuromuscular junction is an ideal site for studying chemical synaptic signaling. At the muscle, the motor axon ramifies into several fine branches approximately 2 μm thick. Each branch forms multiple swellings called *synaptic boutons*, which are covered by a thin layer of Schwann cells. The boutons contact a specialized region of the muscle fiber membrane, the *end-plate*, and are separated from the muscle membrane by a 100-nm synaptic cleft. Each bouton contains mitochondria and synaptic vesicles clustered around *active zones*, where the neurotransmitter acetylcholine (**ACh**) is released. Immediately under each bouton in the end-plate are several junctional folds, the crests of which contain a high density of ACh receptors.

The muscle fiber and nerve terminal are covered by a layer of connective tissue, the basal lamina, consisting of collagen and glycoproteins. Unlike the cell membrane, the basal lamina is freely permeable to ions and small organic compounds, including the ACh transmitter. Both the presynaptic terminal and the muscle fiber secrete proteins into the basal lamina, including the enzyme acetylcholinesterase, which inactivates the ACh released from the presynaptic terminal by breaking it down into acetate and choline. The basal lamina also organizes the synapse by aligning the presynaptic boutons with the postsynaptic junctional folds. (Adapted from McMahan and Kuffler 1971.)

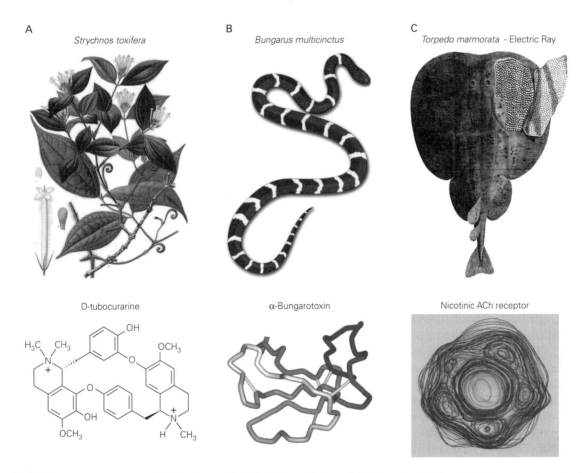

A   *Strychnos toxifera*

B   *Bungarus multicinctus*

C   *Torpedo marmorata* - Electric Ray

D-tubocurarine

α-Bungarotoxin

Nicotinic ACh receptor

Figure 12–2 Poisons, venoms, and high-voltage electric fish help elucidate the structure and function of the nicotinic ACh receptor.

A. Curare is a mixture of toxins extracted from the leaves of *Strychnos toxifera* and is used by South American indigenous people on arrowheads to paralyze their quarry. The active compound, D-tubocurarine, is a complex multiring structure with positively charged amine groups that bear some similarity to ACh. It binds tightly to the ACh binding site on the nicotinic receptor, where it acts as a competitive antagonist for ACh. (Reproduced from Pabst, G (ed). 1898. *Köhler's Medizinal-Pflanzen*, Vol. 3, Plate 45. Gera-Untermhaus, Germany: Franz Eugen Köhler.)

B. The toxin α-bungarotoxin is obtained from the venom of the banded krait, *Bungarus*. It is a 74-amino acid polypeptide

that contains five disulfide bonds (**yellow lines**), producing a rigid structure (From https://en.wikipedia.org/wiki/Alpha-bungarotoxin. Adapted from Zeng et al. 2001.). The toxin binds extremely tightly to the ACh binding site and acts as an irreversible, noncompetitive antagonist of ACh.

C. The electric ray *Torpedo marmorata* has a specialized structure, the electric organ, which consists of a large number of small, flat, muscle-like cells, or electroplaques, arranged in series like a stack of batteries. When a motor nerve releases ACh, a large current is generated by the opening of a very large number of nicotinic ACh receptor-channels, which produces a very large voltage drop of up to 200 V outside the fish, thereby stunning nearby prey. The electroplaques provide a rich source of ACh receptors for biochemical purification and characterization. (From Walsh 1773.)

channels in the muscle membrane, converting the end-plate potential into an action potential, which then propagates along the muscle fiber. The threshold for generating an action potential in the muscle is particularly low at the end-plate, owing to a high density of voltage-gated Na$^+$ channels in the bottom of the junctional folds. The combination of a very large EPSP and low threshold results in a high safety factor for triggering an action potential in the muscle fiber. In contrast, in the central nervous system, most presynaptic neurons produce postsynaptic potentials less than 1 mV

in amplitude, such that inputs from many presynaptic neurons are needed to generate an action potential in most central neurons.

The end-plate potential was first studied in detail in the 1950s by Paul Fatt and Bernard Katz using intracellular voltage recordings. Fatt and Katz were able to isolate the end-plate potential by applying the drug curare (Figure 12–2A) to reduce the amplitude of the postsynaptic potential below the threshold for the action potential (Figure 12–4). At the end-plate, the synaptic potential rises within 1 to 2 ms but decays more slowly.

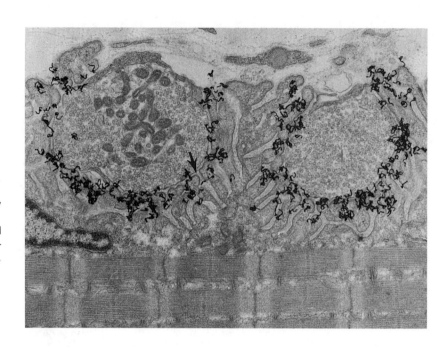

Figure 12–3 Acetylcholine receptors in the vertebrate neuromuscular junction are concentrated at the top one-third of the junctional folds. This receptor-rich region is characterized by an increased density of the postjunctional membrane (**arrow**). The autoradiograph shown here was made by first incubating the membrane with radiolabeled α-bungarotoxin, which binds to the ACh receptor. Radioactive decay results in the emittance of a particle that causes overlaid silver grains to become fixed along its trajectory (**black grains**). Magnification ×18,000. (Reproduced, with permission, from Salpeter 1987.)

By recording at different points along the muscle fiber, Fatt and Katz found that the EPSP is maximal at the end-plate and decreases progressively with distance (Figure 12–5). In addition, the time course of the EPSP slows progressively with distance.

From this, Fatt and Katz concluded that the end-plate potential is generated by an inward ionic current that is confined to the end-plate and then spreads passively away. (An inward current corresponds to an influx of positive charge, which depolarizes the inside

of the membrane.) Inward current is confined to the end-plate because the ACh receptors are concentrated there, opposite the presynaptic terminal from which transmitter is released. The decrease in amplitude and slowing of the EPSP as a function of distance is a result of the passive cable properties of the muscle fiber.

Electrophysiological evidence that the ACh receptors are localized to the end-plate was provided by Stephen Kuffler and his colleagues, who applied ACh to precise points on the muscle membrane

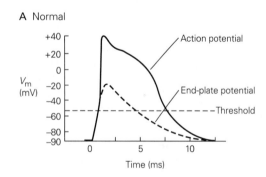

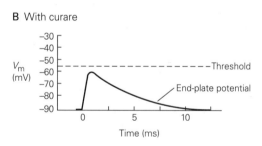

Figure 12–4 The end-plate potential can be isolated pharmacologically for study.

A. Under normal circumstances, stimulation of the motor axon produces an action potential in a skeletal muscle cell. The **dashed curve** in the plot shows the inferred time course of the end-plate potential that triggers the action potential. The lighter **dashed line** shows the action potential threshold.

B. Curare blocks the binding of ACh to its receptor and so prevents the end-plate potential from reaching the threshold

for an action potential. In this way, the currents and channels that contribute to the end-plate potential, which are different from those producing an action potential, can be studied. The end-plate potential shown here was recorded in the presence of a low concentration of curare, which blocks only a fraction of the ACh receptors. The values for the resting potential (–90 mV), end-plate potential, and action potential in these intracellular recordings are typical of a vertebrate skeletal muscle.

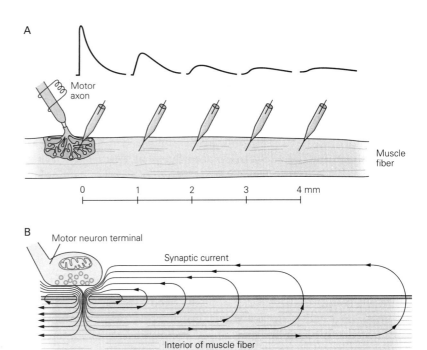

Figure 12–5 The end-plate potential decreases with distance as it passively propagates away from the end-plate. (Adapted, with permission, from Miles 1969.)

A. The amplitude of the postsynaptic potential decreases and the time course of the potential slows with distance from the site of initiation in the end-plate.

B. The decay results from leakiness of the muscle fiber membrane. Because charge must flow in a complete circuit, the inward synaptic current at the end-plate gives rise to a return outward current through resting channels and across the lipid bilayer (the capacitor). This return outward flow of positive charge depolarizes the membrane. Because current leaks out all along the membrane, the outward current and resulting depolarization decreases with distance from the end-plate.

using a technique called micro-iontophoresis. In this approach, the positively charged ACh is ejected from an ACh-filled extracellular microelectrode by applying a positive voltage to the inside of the electrode. Exposing the end-plate region to proteolytic enzymes allows the nerve terminal to be pulled away from the muscle surface and the ACh to be applied directly to the postsynaptic membrane directly under the tip of the small microelectrode. Using this technique, Kuffler

found that the postsynaptic depolarizing response to ACh declined steeply within a few micrometers of the synaptic terminal.

Voltage-clamp experiments have revealed that the end-plate current rises and decays more rapidly than the resultant end-plate potential (Figure 12–6). The time course of the end-plate current is directly determined by the rapid opening and closing of the ACh receptor-channels. Because it takes time for an ionic

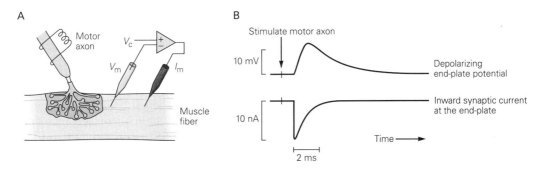

Figure 12–6 The end-plate current increases and decays more rapidly than the end-plate potential.

A. The membrane at the end-plate is voltage-clamped by inserting two microelectrodes into the muscle near the end-plate. One electrode measures membrane potential (V_m), and the second passes current (I_m). Both electrodes are connected to a negative feedback amplifier, which ensures that sufficient current (I_m) is delivered so that V_m will remain clamped at the command potential V_c. The synaptic current evoked by stimulating

the motor nerve can then be measured at constant V_m, for example, –90 mV (see Box 10–1).

B. The end-plate potential (measured when V_m is not clamped) changes relatively slowly and lags behind the more rapid inward synaptic current (measured under voltage-clamp conditions). This is because synaptic current must first alter the charge on the muscle membrane capacitance before the muscle membrane can be depolarized.

current to charge or discharge the muscle membrane capacitance, and thus alter the membrane voltage, the EPSP lags behind the synaptic current (see Figure 9–10 and the Postscript at the end of this chapter).

The Neurotransmitter Acetylcholine Is Released in Discrete Packets

During their first microelectrode recordings at frog motor end-plates in the 1950s, Fatt and Katz observed small spontaneous depolarizing potentials (0.5–1.0 mV) that occurred at an average rate of about 1/s. Such spontaneous potentials were restricted to the end-plate, exhibited the same time course as stimulus-evoked EPSPs, and were blocked by curare. Hence, they were named "miniature" end-plate potentials (mEPPs, or mEPSPs in our current terminology).

What could account for the small, fixed size of the miniature end-plate potential? Del Castillo and Katz tested the possibility that an mEPSP represents the action of a *single* ACh molecule. This hypothesis was quickly dismissed, because applying very small amounts of ACh to the end-plate could elicit depolarizing responses that were much smaller than the 1.0-mV mEPSP. The low doses of ACh did produce an increase in baseline fluctuations or "noise." Later analysis of the statistical components of this noise led to estimates that the underlying unitary postsynaptic response was a depolarization of 0.3 μV in amplitude and 1.0 ms in duration. This was the first hint of the electrical signaling properties of a single ACh receptor-channel (described later).

Del Castillo and Katz concluded that each mEPSP must represent the action of a multimolecular packet or "quantum" of transmitter. Further, they suggested that the large, stimulus-evoked EPSP was made up of an integral number of quanta. Evidence for this quantal hypothesis is presented in Chapter 15.

Individual Acetylcholine Receptor-Channels Conduct All-or-None Currents

What are the properties of the ACh receptor-channels that produce the inward current that generates the depolarizing end-plate potential? Which ions move through the channels to produce this inward current? And what does the current carried by a single ACh receptor-channel look like?

In 1976, Erwin Neher and Bert Sakmann obtained key insights into the biophysical nature of ACh receptor-channel function from recordings of the current conducted by single ACh receptor-channels in skeletal

muscle cells, the unitary or elementary current. They found that the opening of an individual channel generates a very small rectangular step of ionic current (Figure 12–7A). At a given resting potential, each channel opening generates the same-size current pulse. At –90 mV, the current steps are approximately –2.7 pA in amplitude. Although this is a very small current, it corresponds to a flow of approximately 17 million ions per second!

Whereas the amplitude of the current through a single ACh receptor-channel is constant for every opening, the duration of each opening and the time between openings vary considerably. These variations occur because channel openings and closings are stochastic; they obey the same statistical law that describes the exponential time course of radioactive decay. Because channels and ACh undergo random thermal motions and fluctuations, it is impossible to predict exactly how long it will take any one channel to bind ACh or how long that channel will stay open before the ACh dissociates and the channel closes. However, the average length of time a particular type of channel stays open is a well-defined property of that channel, just as the half-life of radioactive decay is an invariant property of a particular isotope. The mean open time for ACh receptor-channels is approximately 1 ms. Thus, each channel opening permits the movement of approximately 17,000 ions. Once a channel closes, the ACh molecules dissociate and the channel remains closed until it binds ACh again.

The Ion Channel at the End-Plate Is Permeable to Both Sodium and Potassium Ions

Once a receptor-channel opens, which ions flow through the channel, and how does this lead to depolarization of the muscle membrane? One important means of identifying the ion (or ions) responsible for the synaptic current is to measure the value of the chemical driving force (the chemical battery) propelling ions through the channel. Remember, the current through a single open channel is given by the product of the single-channel conductance and the electrochemical driving force on the ions conducted through the channel (Chapter 9). Thus, the current generated by a single ACh receptor-channel is given by:

$$I_{EPSP} = \gamma_{EPSP} \times (V_m - E_{EPSP}), \qquad (12\text{–}1)$$

where I_{EPSP} is the amplitude of current through one channel, γ_{EPSP} is the conductance of a single open channel, E_{EPSP} is membrane potential at which the net flux of ions through the channel is zero, and $V_m - E_{EPSP}$ is the electrochemical driving force for ion flux. The current steps change in

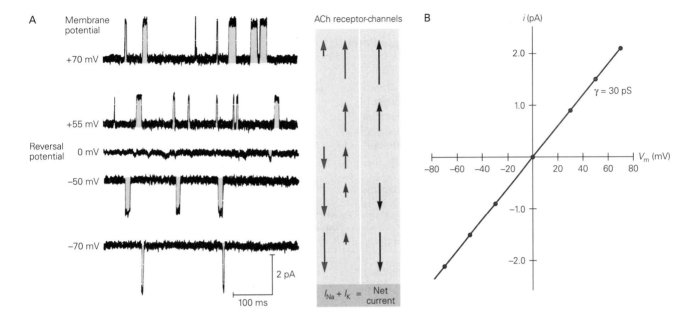

Figure 12–7 Individual acetylcholine (ACh) receptor-channels conduct an all-or-none elementary current.

A. The patch-clamp technique is used to record currents from single ACh receptor-channels. The patch electrode is filled with salt solution that contains a low concentration of ACh and is then brought into close contact with the surface of the muscle membrane (see Box 8–1). At a fixed membrane potential, each time a channel opens, it generates a relatively constant elementary current. At the resting potential of –90 mV, the current is approximately –2.7 pA (1 pA = 10^{-12} A). As the voltage across a patch of membrane is systematically varied, the resultant current varies in amplitude as a result of changes in driving force. The current is inward at voltages negative to 0 mV and outward at voltages positive to 0 mV, thus defining 0 mV as the reversal potential. The arrows on the right side of the traces illustrate the individual sodium and potassium fluxes and resultant net current as a function of voltage.

B. The linear relation between current through a single ACh receptor-channel and membrane voltage shows that the channel behaves as a simple resistor having a single-channel conductance (γ) of about 30 pS.

size as the membrane potential changes because of the change in driving force. For the ACh receptor-channels, the relationship between I_{EPSP} and membrane voltage is linear, indicating that the single-channel conductance is constant and does not depend on membrane voltage; that is, the channel behaves as a simple ohmic resistor. From the slope of this relation, the channel is found to have a conductance of 30 pS (Figure 12–7B). As we saw in Chapter 9, the total conductance, g, due to the opening of a number of receptor-channels (n) is given by:

$$g = n \times \gamma.$$

The current–voltage relation for a single channel shows that the reversal potential for ionic current through ACh receptor-channels, obtained from the intercept of the membrane voltage axis, is 0 mV, which is not equal to the equilibrium potential for Na^+ or any of the other major cations or anions. This is due to the fact that this chemical potential is produced not by a single ion species but by a combination of two species: The ligand-gated channels at the end-plate are almost equally permeable to both major cations, Na^+ and K^+.

Thus, during the end-plate potential, Na^+ flows into the cell and K^+ flows out. The reversal potential is at 0 mV because this is a weighted average of the equilibrium potentials for Na^+ and K^+ (Box 12–1). At the reversal potential, the influx of Na^+ is balanced by an equal efflux of K^+ (Figure 12–7A).

The ACh receptor-channels at the end-plate are not selective for a single ion species, as are the voltage-gated Na^+ or K^+ channels, because the diameter of the pore of the ACh receptor-channel is substantially larger than that of the voltage-gated channels. Electrophysiological measurements suggest that it may be 0.6 nm in diameter, an estimate based on the size of the largest organic cation that can permeate the channel. For example, tetramethylammonium (TMA) is approximately 0.6 nm in diameter and yet still permeates the channel. In contrast, the voltage-gated Na^+ channel is only permeant to organic cations that are smaller than 0.5 × 0.3 nm in cross section, and voltage-gated K^+ channels will only conduct ions less than 0.3 nm in diameter.

The relatively large diameter of the ACh receptor-channel pore is thought to provide a water-filled

Box 12–1 Reversal Potential of the End-Plate Potential

The reversal potential of a membrane current carried by more than one ion species, such as the end-plate current through the ACh receptor-channels, is determined by two factors: (1) the relative conductance for the permeant ions (g_{Na} and g_K in the case of the end-plate current) and (2) the equilibrium potentials of the ions (E_{Na} and E_K).

At the reversal potential for the ACh receptor-channel current, inward current carried by Na^+ is balanced by outward current carried by K^+:

$$I_{Na} + I_K = 0. \qquad (12\text{--}2)$$

The individual Na^+ and K^+ currents can be obtained from

$$I_{Na} = g_{Na} \times (V_m - E_{Na}) \qquad (12\text{--}3a)$$

and

$$I_K = g_K \times (V_m - E_K). \qquad (12\text{--}3b)$$

We can substitute Equations 12–3a and 12–3b for I_{Na} and I_K in Equation 12–2, replacing V_m with E_{EPSP} (because at the reversal potential $V_m = E_{EPSP}$):

$$g_{Na} \times (E_{EPSP} - E_{Na}) + g_K \times (E_{EPSP} - E_K) = 0. \qquad (12\text{--}4)$$

Solving this equation for E_{EPSP} yields

$$E_{EPSP} = \frac{(g_{Na} \times E_{Na}) + (g_K \times E_K)}{g_{Na} + g_K}. \qquad (12\text{--}5)$$

This equation can also be used to solve for the ratio g_{Na}/g_K if one knows E_{EPSP}, E_K, and E_{Na}. Thus, rearranging Equation 12–5 yields

$$\frac{g_{Na}}{g_K} = \frac{E_{EPSP} - E_K}{E_{Na} - E_{EPSP}}. \qquad (12\text{--}6)$$

At the neuromuscular junction, $E_{EPSP} = 0$ mV, $E_K = -100$ mV, and $E_{Na} = +55$ mV. Thus, from Equation 12–6, g_{Na}/g_K has a value of approximately 1.8, indicating that the conductance of the ACh receptor-channel for Na^+ is slightly higher than for K^+. A comparable approach can be used to analyze the reversal potential and the movement of ions during excitatory and inhibitory synaptic potentials in central neurons (Chapter 13).

---

environment that allows cations to diffuse through the channel relatively unimpeded, much as they would in free solution. This explains why the pore does not discriminate between Na^+ and K^+ and why even divalent cations, such as Ca^{2+}, are able to pass through. Anions are excluded by the presence of fixed negative charges in the channel, as described later in this chapter. Recent X-ray crystallographic data have provided a direct view of the large pore of the ACh receptor-channel (see Figure 12–12).

Four Factors Determine the End-Plate Current

How do the rectangular current steps carried by single ACh receptor-channels produce the large synaptic current at the end-plate in response to motor nerve stimulation? Stimulation of a motor nerve releases a large quantity of ACh into the synaptic cleft. The ACh rapidly diffuses across the cleft and binds to the ACh receptors, causing more than 200,000 receptor-channels to open almost simultaneously. (This number is obtained by comparing the total end-plate current, approximately −500 nA,

with the current through a single channel, approximately −2.7 pA).

The rapid opening of so many channels causes a large increase in the total conductance of the end-plate membrane, g_{EPSP}, and produces the fast rising phase of the end-plate current. As the ACh in the cleft decreases rapidly to zero (in <1 ms), because of enzymatic hydrolysis and diffusion, the channels begin to close randomly. Although each closure produces only a small step-like decrease in end-plate current, the random closing of large numbers of small unitary currents causes the total end-plate current to appear to decay smoothly (Figure 12–8).

The Acetylcholine Receptor-Channels Have Distinct Properties That Distinguish Them From the Voltage-Gated Channels That Generate the Muscle Action Potential

The ACh receptors that produce the end-plate potential differ in two important ways from the voltage-gated channels that generate the action potential in

A Idealized time course of opening of six ion channels

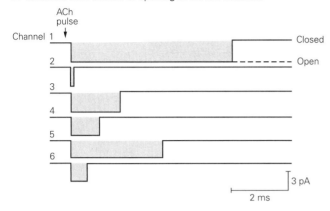

B Total current of the six channels

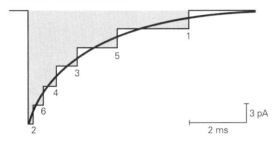

Figure 12–8 The time course of the total current at the end-plate reflects the summation of contributions of many individual acetylcholine receptor-channels. (Reproduced, with permission, from Colquhoun 1981. Copyright © 1981 Elsevier.)

A. Individual ACh receptor-channels open in response to a brief pulse of ACh. In this idealized example, the membrane contains six ACh receptor-channels, all of which open rapidly and nearly simultaneously. The channels remain open for varying times and close independently.

B. The stepped trace shows the sum of the six single-channel current records in part A. It represents the current during the sequential closing of each channel (the number indicates which channel has closed). In the final period of current, only channel one is open. In a current record from a whole muscle fiber, with thousands of channels, individual channel closings are not detectable because the scale needed to display the total end-plate current (hundreds of nanoamperes) is so large that the contributions of individual channels cannot be resolved. As a result, the total end-plate current appears to decay smoothly.

muscle. First, the action potential is generated by sequential activation of two distinct classes of voltage-gated channels, one selective for Na^+ and the other for K^+. In contrast, a single type of ion channel, the ACh receptor-channel, generates the end-plate potential by allowing both Na^+ and K^+ to pass with nearly equal permeability.

Second, the Na^+ flux through voltage-gated channels is regenerative: By increasing the depolarization of the cell, the Na^+ influx opens more voltage-gated

Na^+ channels. This regenerative feature is responsible for the all-or-none property of the action potential. In contrast, the number of ACh receptor-channels opened during the synaptic potential is fixed by the amount of ACh available. The depolarization produced by Na^+ influx through the ACh-gated channels does not lead to the opening of more ACh receptor-channels and cannot produce an action potential. To trigger an action potential, a synaptic potential must recruit neighboring voltage-gated Na^+ channels (Figure 12–9).

As might be expected from these two differences in physiological properties, the ACh receptor-channels and voltage-gated channels are formed by different macromolecules that exhibit different sensitivities to drugs and toxins. Tetrodotoxin, which blocks the voltage-gated Na^+ channel, does not block the influx of Na^+ through the nicotinic ACh receptor-channels. Similarly, α-bungarotoxin binds tightly to the nicotinic receptors and blocks the action of ACh but does not interfere with voltage-gated Na^+ or K^+ channels.

Transmitter Binding Produces a Series of State Changes in the Acetylcholine Receptor-Channel

Each ACh receptor has two binding sites for ACh; both must be occupied by transmitter for the channel to open efficiently. However, during prolonged applications of ACh, the channel enters a desensitized state where it no longer conducts. The time course of desensitization of the muscle nicotinic receptor is too slow to contribute to the time course of the EPSP under normal conditions, where ACh is present in the synaptic cleft for only a very brief period of time. However, desensitization can play a more important role in determining the time course of the postsynaptic response at certain neuronal synapses, where the transmitter may persist in the synaptic cleft for more prolonged times or where the postsynaptic receptors undergo more rapid desensitization.

For example, the persistence of ACh in the synaptic cleft at cholinergic synapses in the brain may lead to significant desensitization of certain subtypes of neuronal nicotinic receptors. Heavy smokers can build up sufficient levels of nicotine to desensitize receptors in the brain. Desensitization also plays a role in the action of the drug succinylcholine, a dimer of ACh that is resistant to acetylcholinesterase and is used during general anesthesia to produce muscle relaxation. Succinylcholine does so through its ability to produce both receptor desensitization and prolonged depolarization, which blocks muscle action potentials by inactivating voltage-gated Na^+ channels.

Figure 12–9 The end-plate potential resulting from the opening of acetylcholine receptor-channels opens voltage-gated sodium channels. The end-plate potential is normally large enough to open a sufficient number of voltage-gated Na⁺ channels to exceed the threshold for an action potential. (Adapted from Alberts et al. 1989.)

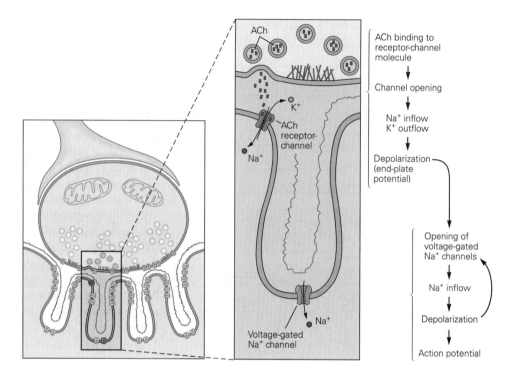

A minimal reaction model, first proposed by Katz and his colleagues, captures many (but not all) of the key steps of ACh receptor-channel function, in which a closed receptor-channel (R) successively binds two molecules of ACh (A) prior to undergoing a rapid conformational change to an open state (R*). This is followed by a slower conformational change to the nonconducting desensitized state (D). The model also incorporates the finding that there is a small probability that an individual receptor may enter the desensitized state even in the absence of ACh. These binding and gating reactions can be summarized by the following scheme:

$$A + R \longleftrightarrow AR + A \longleftrightarrow A_2R \longleftrightarrow A_2R^*$$
$$\updownarrow \qquad\qquad \updownarrow \qquad\qquad \updownarrow \qquad\qquad \updownarrow$$
$$A + D \longleftrightarrow AD + A \longleftrightarrow A_2D \longleftrightarrow A_2D^*$$

X-ray crystal structure models have now been obtained for all three states of the ACh receptor (described later).

The Low-Resolution Structure of the Acetylcholine Receptor Is Revealed by Molecular and Biophysical Studies

The nicotinic ACh receptor at the nerve-muscle synapse is part of a single macromolecule that includes the pore in the membrane through which ions flow. Where in the molecule is the binding site located? How is the

pore of the channel formed? How is ACh binding coupled to channel gating?

Insights into these questions have been obtained from molecular and biophysical studies of the ACh receptor proteins and their genes, beginning with the purification of the macromolecule from the electric ray *Torpedo marmorata* (Figure 12–2). Using different biochemical approaches, Arthur Karlin and Jean Pierre Changeux purified the receptor from electroplaques, specialized muscle-like cells whose stack-like packing enables their individual EPSPs to summate in series to generate the large voltages (>100 V) used by the electric ray to stun its prey. Their studies indicate that the mature nicotinic ACh receptor is a membrane glycoprotein formed from five subunits of similar molecular weight: two α-subunits and one β-, one γ-, and one δ-subunit (Figure 12–10).

Karlin and his colleagues identified two extracellular binding sites for ACh on each receptor protein in the clefts between each α-subunit and its neighboring γ- or δ-subunit. One molecule of ACh must bind at each of the two sites for the channel to open efficiently (Figure 12–10). Because α-bungarotoxin binds remarkably tightly to the same binding site on the α-subunit as does ACh, the toxin acts as an irreversible transmitter antagonist.

Further insights into the structure of the ACh receptor-channel come from the analysis of the primary amino acid sequence of the receptor's four different

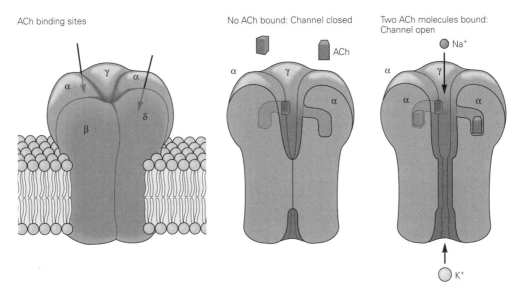

Figure 12–10 The nicotinic ACh receptor-channel is a pentameric macromolecule. The receptor and channel are components of a single macromolecule consisting of five subunits: two identical α-subunits and one each of β-, γ-, and δ-subunits. The subunits form a pore through the cell membrane. When two molecules of ACh bind to the extracellular binding sites—formed at the interfaces of the two α-subunits and their neighboring γ- and δ-subunits—the conformation of the receptor-channel molecule changes (see Figure 12–12). This change opens the pore through which K⁺ and Na⁺ flow down their electrochemical gradients.

subunits and from biophysical studies. Molecular cloning by Shosaku Numa and colleagues demonstrated that the four subunits are encoded by distinct but related genes. Sequence comparison of the subunits shows a high degree of similarity—one-half of the amino acid residues are identical or conservatively substituted—which suggests that all subunits have a similar structure. Furthermore, all four of the genes for the subunits are homologous; that is, they are derived from a common ancestral gene. Nicotinic ACh receptors in neurons are encoded by a set of distinct but related genes. All of these receptors are pentamers; however, their subunit composition and stoichiometry vary. Whereas most neuronal receptors are composed of two α-subunits and three β-subunits, some neuronal receptors are composed of five identical α-subunits (the α7 isoform) and so can bind five molecules of ACh.

All nicotinic ACh receptor subunits contain a highly conserved sequence near the extracellular binding site for ACh consisting of two disulfide-bonded cysteine (cys) residues with 13 intervening amino acids. The resultant 15-amino acid loop forms a signature sequence both for nicotinic ACh receptor subunits and for related receptors for other transmitters in neurons. The *cys-loop receptor family*, also known as *pentameric ligand-gated ion channels* (pLGIC), includes

receptors for the neurotransmitters γ-aminobutyric acid (GABA), glycine, and serotonin.

The distribution of the polar and nonpolar amino acids of the subunits provided the first clues as to how the subunits are threaded through the membrane bilayer. Each subunit contains four hydrophobic regions of approximately 20 amino acids called M1 to M4, each of which forms an α-helix that spans the membrane (Figure 12–11A). The amino acid sequences of the subunits suggest that the subunits are arranged such that they create a central pore through the membrane (Figure 12–11B).

The walls of the channel pore are formed by the M2 membrane-spanning segment and by the loop connecting M2 to M3. Three rings of negative charges that flank the external and internal boundaries of the M2 segment play an important role in the channel's selectivity for cations. Certain local anesthetic drugs block the channel by interacting with one ring of polar serine residues and two rings of hydrophobic residues in the central region of the M2 helix, midway through the membrane.

Three-dimensional models of the entire receptor-channel complex were initially proposed by Karlin based on low-resolution neutron scattering and by Nigel Unwin based on electron diffraction images. The complex is divided into three regions: a large

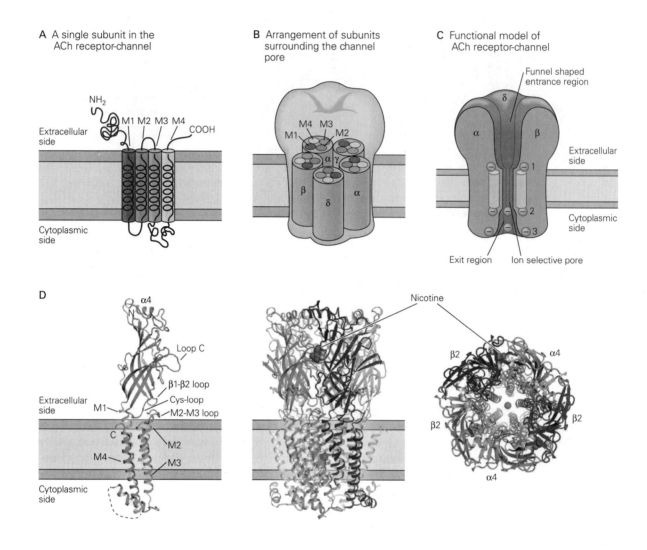

A A single subunit in the ACh receptor-channel

NH₂

Extracellular side

M1 M2 M3 M4

COOH

Cytoplasmic side

B Arrangement of subunits surrounding the channel pore

M4 M3
M1   M2
α γ
β
δ   α

C Functional model of ACh receptor-channel

Funnel shaped entrance region

δ

α   β

Extracellular side

1

2

Cytoplasmic side

3

Exit region   Ion selective pore

D

α4
N

Loop C

β1-β2 loop
Cys-loop
M2-M3 loop

Extracellular side
M1

C

M2
M4
M3

Cytoplasmic side

Nicotine

β2   α4

β2   β2

α4

Figure 12–11 The ACh receptor subunits are homologous membrane-spanning proteins.

A. Each subunit contains a large extracellular N-terminus, four membrane-spanning α-helixes (M1–M4), and a short extracellular C-terminus. The N-terminus contains the ACh-binding site, and the membrane helixes form the pore.

B. The five subunits are arranged such that they form a central aqueous channel, with the M2 segment of each subunit forming the lining of the pore. The γ-subunit lies between the two α-subunits. (Dimensions are not to scale.)

C. Negatively charged amino acids on each subunit form three rings of negative charge around the pore. As an ion traverses the channel, it encounters these rings of charge. The rings at the external and internal surfaces of the cell membrane (**1, 3**) may serve as prefilters that help repel anions and form divalent cation blocking sites. The central ring near the cytoplasmic side of the membrane bilayer (**2**) may contribute more importantly to establishing the specific cation selectivity of the selectivity filter, which is the narrowest region of the pore.

D. A high-resolution X-ray crystal structure model of a human neuronal nicotinic ACh receptor-channel. *Right:* A top-down view of the open channel, which is composed of two α₄-subunits and three β₂-subunits arranged around the central pore. These subunits are closely related variants of the α- and β-subunits of the muscle receptor. Two molecules of nicotine (atoms shown as **red** spheres) are bound to the receptor. A permeating cation is shown as a **pink** sphere. *Center:* A side view of the receptor showing the location of the phospholipid bilayer of the membrane and bound nicotine. *Left:* A side view of a single α₄-subunit in the plane of the membrane. The amino-terminus of the subunit consists of a large extracellular domain. Loop C helps form the ligand-binding site. The β₁-β₂ and cys-loops at the interface between the extracellular domain and the M1–M4 membrane-spanning α-helixes transmit a conformational change from the ligand-binding site to the pore to open the channel. (Reproduced, with permission, from Morales-Perez et al. 2016. Copyright © 2016 Springer Nature.)

extracellular portion that contains the ACh binding site, a narrow transmembrane pore selective for cations, and a large exit region at the internal membrane surface (Figure 12–11C). The extracellular region is surprisingly large, approximately 6 nm in length. In addition, the extracellular end of the pore has a wide mouth approximately 2.5 nm in diameter. Within the bilayer of the membrane, the pore gradually narrows.

The autoimmune disorder myasthenia gravis results from the production of antibodies that bind to the extracellular domain of the ACh receptor, leading to a decrease in the number or function of the nicotinic ACh receptors at the neuromuscular junction. If the change is severe enough, this can decrease the EPSP below the threshold for triggering an action potential, resulting in debilitating weakness. Several congenital forms of myasthenia result from mutations in nicotinic ACh receptor subunits that can also alter receptor number or channel function. For example, a mutation in an amino acid residue in the M2 segment leads to a prolonged channel open time, termed the slow channel syndrome, which results in excessive postsynaptic excitation that leads to degeneration of the end-plate (Chapter 57).

The High-Resolution Structure of the Acetylcholine Receptor-Channel Is Revealed by X-Ray Crystal Studies

A deeper understanding of the fine details of the ACh binding site initially came from high-resolution X-ray crystallographic studies of a molluscan ACh-binding protein, which is homologous to the extracellular amino terminus of nicotinic ACh receptor subunits. Remarkably, unlike typical ACh receptors, the molluscan ACh-binding protein is a soluble protein secreted by glial cells into the extracellular space. At cholinergic synapses in snails, it acts to reduce the size of the EPSP, perhaps by buffering the free concentration of ACh in the synaptic cleft.

Further insights into the structure of the complete receptor-channel have come from X-ray crystal structures of related pentameric ligand-gated channels from bacteria and multicellular animals, culminating with a recent X-ray crystal structure of a human neuronal nicotinic ACh receptor in complex with nicotine. Combined with knowledge of structures of related proteins, we now have a remarkably detailed knowledge of the structure and mechanisms underlying ligand binding, channel gating, and ion permeation of the ACh receptor-channel and related ligand-gated channels.

In the neuronal ACh receptor, two α-subunits combine with three β-subunits to form the pentamer (Figure 12–11D). The large extracellular domain of the receptor contains two ACh binding sites and forms a pentameric ring that surrounds a large central vestibule, which presumably funnels ions toward the narrow transmembrane domain of the receptor. Each α-subunit binds one molecule of nicotine at a site located at the interface with a neighboring β-subunit. Electron diffraction data from Nigel Unwin's higher-resolution structures of related cys-loop receptors and from the high-resolution structure of the desensitized state of the neuronal nicotinic receptor show that the four transmembrane segments of each subunit are indeed α-helices that traverse the 3-nm length of the lipid bilayer (Figure 12–12). In the desensitized state, the M2 segments from the five subunits form a narrow constriction near the intracellular side of the membrane, preventing ion permeation.

Our picture of the transmembrane region of the nicotinic ACh receptor-channel in the open and closed state is still incomplete. However, by comparison with structures of related pLGICs, a coherent picture of the receptor is beginning to emerge. In the closed state, the pore-lining M2 segments lie roughly parallel to each other, forming a narrow central pore. The pore is further constricted to a diameter of 0.3 to 0.4 nm by a ring of highly conserved hydrophobic leucine residues near the middle of the M2 segment (Figure 12–12). This hydrophobic constriction is thought to provide a high-energy barrier that restricts the passage of hydrated cations whose diameter is greater than the constriction in the pore. At present the discrepancy in pore diameter inferred from electrophysiological measuements (0.6 nm) and the narrower value from the crystal structure remains unresolved.

In the open state, the M2 segments are thought to tilt outward and rotate, widening the constriction of the leucine residues in the middle of M2, thus enabling ion permeation. The narrowest constriction in the open pore lies near the intracellular mouth of the channel, where the electronegative hydroxyl side chains from one ring of threonine residues (serine and threonine residues in the muscle ACh receptor) and a second ring of negatively charged glutamate residues are thought to form the selectivity filter. In the desensitized state the M2 segments tilt further, causing the selectivity filter to constrict even more, preventing ion permeation.

A detailed picture of how ligand binding leads to channel opening is now emerging based on various structural and functional studies. Binding of ligand is thought to promote the closure of the cleft between neighboring subunits, leading to the tightening of the extracellular domain of the pentamer, similar to the closing of the petals of a flower. This results in a twisting motion that causes the bottom of the extracellular domain of the receptor to push on the M1 segment and extracellular loop connecting the M2 and

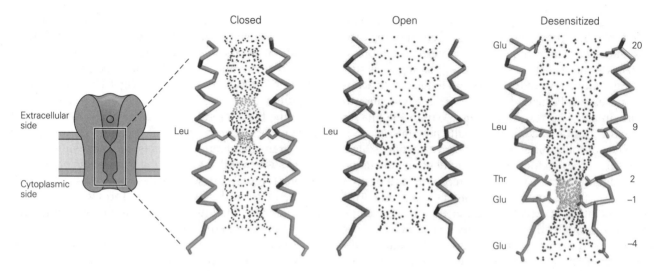

Figure 12–12 A high-resolution three-dimensional structural model of a neuronal nicotinic ACh receptor-channel. High-resolution models of the pentameric family of ligand-gated channels are shown for the closed, open, and desensitized states of the receptor-channel. Two out of five of the M2 α-helices are shown. The desensitized structure is from the human neuronal ACh receptor. The closed and open states are based on structures of neuronal glycine receptors, which are closely related in amino acid sequence to ACh receptor subunits. Key amino acid side chains are illustrated for the desensitized ACh receptor with position numbering on the right and amino acid abbreviations on the left. According to convention, position 0 is near the intracellular surface of the phospholipid bilayer; other positions are labeled according to relative position in the primary amino acid sequence. A conserved leucine in the middle of the M2 segment (position 9) forms a gate that constricts the pore in the closed state. Ligand binding causes the subunits to tilt outward and twist, opening up the leucine gate. A further conformational change during desensitization causes the subunits to tilt inward near the bottom, constricting the pore near the intracellular side of the channel and thereby producing a nonconducting state. The negatively charged glutamates at positions 20, –1, and –4 correspond to the external (**1**), middle (**2**), and internal (**3**) rings of charge in Figure 12–11C. The negatively charged glutamate at position –1 and the electronegative threonine at position 2 form the selectivity filter of the channel. (Reproduced, with permission, from Morales-Perez et al. 2016. Copyright © 2016 Springer Nature.)

M3 transmembrane segments. This motion exerts a force on the M2 segment that leads to its rotation and tilting, thereby opening up the hydrophobic leucine gate in the middle of the pore and allowing ion permeation. Although future studies will no doubt refine our understanding of the structural bases for nicotinic receptor-channel and function, these recent advances give us an unprecedented molecular understanding of one of the most fundamental processes in the nervous system: synaptic transmission and, specifically, the signaling of information from nerve to muscle.

Highlights

1. The terminals of motor neurons form synapses with muscle fibers at specialized regions in the muscle membrane called end-plates. When an action potential reaches the terminals of a presynaptic motor neuron, it causes the release of ACh.
2. ACh diffuses across the narrow (100-nm) synaptic cleft in a matter of microseconds and binds to nicotinic ACh receptors in the end-plate membrane.

The energy of binding is translated into a conformational change that opens a cation-selective channel in the protein, allowing Na⁺, K⁺, and Ca²⁺ to flow across the postsynaptic membrane. The net effect, due largely to the influx of Na⁺ ions, produces a depolarizing synaptic potential called the end-plate potential.

3. Because the ACh receptor-channels are concentrated at the end-plate, the opening of these channels produces a local depolarization. This local depolarization is large enough (75 mV) to exceed the threshold for action potential generation by a factor of three to four.
4. It is important that the safety factor of nerve-muscle transmission be at a high level, as it determines our ability to move, breath, and escape from danger. Decreases in ACh receptor number or function as a result of autoimmune disease or genetic mutations can contribute to neurological disorders.
5. Patch-clamp recordings have revealed the step-like increase and decrease in current in response to the opening and closing of single ACh receptor-channels. A typical excitatory postsynaptic current

at the neuromuscular junction is generated by the opening of approximately 200,000 individual channels.

6. The biochemical structure of the muscle nicotinic ACh receptor has been determined. The receptor is a pentamer composed of two α-subunits and one β-γ-, and δ-subunit. The four genes encoding the subunits are closely related, and more distantly related to the genes encoding other pentameric ligand-gated channels for other transmitters.

7. Higher-resolution structures have provided a detailed view of the ACh ligand-binding pocket and the pore of the channel and further insight into how ligand binding leads to conformational changes associated with receptor-channel opening and desensitization gating reactions.

Postscript: The End-Plate Current Can Be Calculated From an Equivalent Circuit

The current through a population of ACh receptor-channels can be described by Ohm's law. However, to describe how the current generates the end-plate potential, the conductance of the resting channels in the surrounding membrane must also be considered. We must also take into consideration the capacitive properties of the membrane and the ionic batteries determined by the distribution of Na$^+$ and K$^+$ inside and outside the cell.

The dynamic relationships between these various components can be explained using the same rules we used in Chapter 9 to analyze the current in passive electrical devices that consist only of resistors, capacitors, and batteries. We can represent the end-plate region with an equivalent circuit that has three parallel current paths: (1) one for the synaptic current through the transmitter-gated channels, (2) one for the return current through resting channels (the nonsynaptic membrane), and (3) one for the capacitive current across the lipid bilayer (Figure 12–13). For simplicity, we ignore the voltage-gated channels in the surrounding nonsynaptic membrane.

Because the end-plate current is carried by both Na$^+$ and K$^+$ flowing through the same ion channel, we combine the Na$^+$ and K$^+$ current pathways into a single conductance (g_{EPSP}) representing the ACh receptor-channels. The conductance of this pathway is proportional to the number of channels opened, which in turn depends on the concentration of transmitter in the synaptic cleft. In the absence of transmitter, no channels are open and the conductance is zero. When a presynaptic action potential causes the release

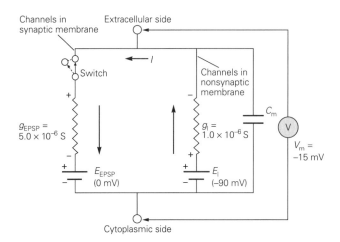

Figure 12–13 The equivalent circuit of the end-plate. The circuit has three parallel current pathways. One conductance pathway carries the end-plate current and consists of a battery (E_{EPSP}) in series with the conductance of the ACh receptor-channels (g_{EPSP}). Another conductance pathway carries current through the nonsynaptic membrane and consists of a battery representing the resting potential (E_l) in series with the conductance of the resting channels (g_l). In parallel with both of these conductance pathways is the membrane capacitance (C_m). The voltmeter (V) measures the potential difference between the inside and the outside of the cell.

When no ACh is present, the ACh receptor-channels are closed and carry no current. This state is depicted as an open electrical circuit in which the synaptic conductance is not connected to the rest of the circuit. The binding of ACh opens the synaptic channels. This event is electrically equivalent to throwing the switch that connects the gated conductance pathway (g_{EPSP}) with the resting pathway (g_l). In the steady state, an inward current through the ACh receptor-channels is balanced by an outward current through the resting channels. With the indicated values of conductances and batteries, the membrane will depolarize from –90 mV (its resting potential) to –15 mV (the peak of the end-plate potential).

of ACh, the conductance of this pathway increases to approximately 5×10^{-6} S, which is about five times the conductance of the parallel branch representing the resting (leakage) channels (g_l).

The end-plate conductance is in series with a battery (E_{EPSP}) with a value given by the reversal potential for synaptic current (0 mV) (Figure 12–13). This value is the weighted algebraic sum of the Na$^+$ and K$^+$ equilibrium potentials (see Box 12–1). The current during the excitatory postsynaptic potential (I_{EPSP}) is given by

$$I_{EPSP} = g_{EPSP} \times (V_m - E_{EPSP}).$$

Using this equation and the equivalent circuit of Figure 12–13, we can now analyze the EPSP in terms of its components (Figure 12–14).

At the onset of the EPSP (the dynamic phase), an inward current (I_{EPSP}) flows through the ACh

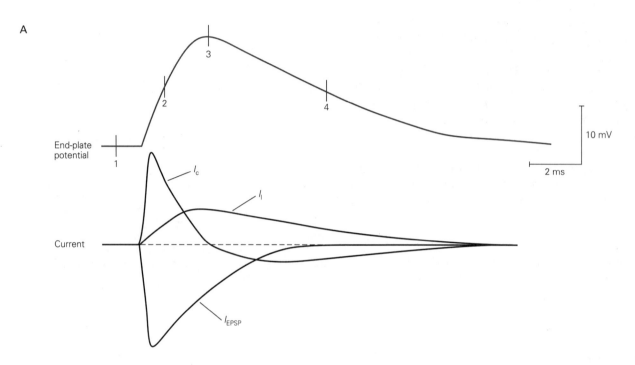

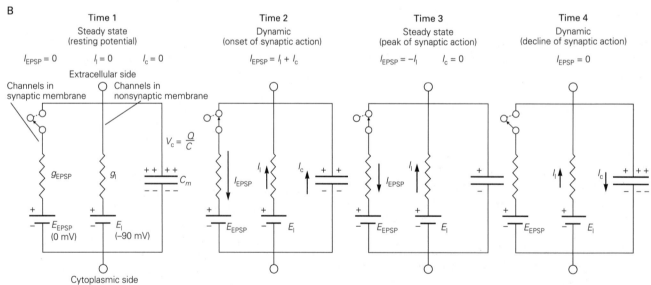

Figure 12–14 The time course of the end-plate potential is determined by both the ACh-gated synaptic conductance and the passive membrane properties of the muscle cell.

A. The time course of the end-plate potential and the component currents through the ACh receptor-channels (I_{EPSP}), the resting (or leakage) channels (I_l), and the capacitor (I_c). There is a capacitive current only when the membrane potential is

changing. In the steady state, such as at the peak of the end-plate potential, the inward flow of positive charge through the ACh receptor-channels is exactly balanced by the outward ionic current across the resting channels, and there is no capacitive current.

B. Equivalent circuits for the current at times 1, 2, 3, and 4 shown in part A. (The relative magnitude of a current is represented by the arrow length.)

receptor-channels because of the increased conductance to Na^+ and K^+ and the large inward driving force on Na^+ at the resting potential of -90 mV (Figure 12–14B, time 2). Because charge flows in a closed loop, the inward synaptic current leaves the cell as outward current through two parallel pathways: a pathway for ionic current (I_l) through the resting (or leakage) channels and a pathway for capacitive current (I_c) across the lipid bilayer. Thus,

$$I_{EPSP} = -(I_l + I_c).$$

During the earliest phase of the EPSP, the membrane potential, V_m, is still close to its resting value, E_l. As a result, the outward driving force on current through the resting channels ($V_m - E_l$) is small. Therefore, most of the outward current leaves the cell as capacitive current and the membrane depolarizes rapidly (Figure 12–14B, time 2). As the cell depolarizes, the outward driving force on current through the resting channels increases, while the inward driving force on synaptic current through the ACh receptor-channels decreases. Concomitantly, as the concentration of ACh in the synapse decreases, the ACh receptor-channels begin to close, and eventually the inward current through the gated channels is exactly balanced by outward current through the resting channels ($I_{EPSP} = -I_l$). At this point, no charge flows into or out of the capacitor ($I_c = 0$). Because the rate of change of membrane potential is directly proportional to I_c,

$$I_c / C_m = \Delta V_m / \Delta t,$$

the membrane potential will have reached a peak or new steady-state value, $\Delta V_m / \Delta t = 0$ (Figure 12–14B, time 3).

As the ACh receptor-channels close, I_{EPSP} decreases further. Now I_{EPSP} and I_l are no longer in balance and the membrane potential starts to repolarize, because the outward current through leak channels (I_l) becomes larger than the inward synaptic current. During most of the declining phase of the synaptic action, the ACh receptor-channels carry no current because they are all closed. Instead, current is conducted through the membrane only as outward current carried by resting channels, balanced by inward capacitive current (Figure 12–14B, time 4).

When the EPSP is at its peak or steady-state value, $I_c = 0$, and therefore the value of V_m can be easily calculated. The inward current through the ACh receptor-channels (I_{EPSP}) must be exactly balanced by outward current through the resting channels (I_l):

$$I_{EPSP} + I_l = 0. \tag{12–7}$$

The current through the ACh receptor-channels (I_{EPSP}) and resting channels (I_l) is given by Ohm's law:

$$I_{EPSP} = g_{EPSP} \times (V_m - E_{EPSP}),$$

and

$$I_l = g_l \times (V_m - E_l).$$

By substituting these two expressions into Equation 12–7, we obtain

$$g_{EPSP} \times (V_m - E_{EPSP}) + g_l \times (V_m - E_l) = 0.$$

Solving for V_m, we obtain

$$V_m = \frac{(g_{EPSP} \times E_{EPSP}) + (g_l \times E_l)}{g_{EPSP} + g_l}. \tag{12–8}$$

This equation is similar to that used to calculate the resting and action potentials (Chapter 9). According to Equation 12–8, the peak voltage of the EPSP is a weighted average of the electromotive forces of the two batteries for the ACh receptor-channels and the resting (leakage) channels. The weighting factors are given by the relative magnitude of the two conductances. Since g_l is a constant, the greater the value of g_{EPSP} (ie, the more ACh channels are open), the more closely V_m will approach the value of E_{EPSP}.

We can now calculate the peak EPSP for the specific case shown in Figure 12–13, where $g_{EPSP} = 5 \times 10^{-6}$ S, $g_l = 1 \times 10^{-6}$ S, $E_{EPSP} = 0$ mV, and $E_l = -90$ mV. Substituting these values into Equation 12–8 yields

$$V_m =$$

$$\frac{[(5 \times 10^{-6}\ S) \times (0\ mV)] + [(1 \times 10^{-6}\ S) \times (-90\ mV)]}{(5 \times 10^{-6}\ S) + (1 \times 10^{-6}\ S)}.$$

or

$$V_m = \frac{(1 \times 10^{-6}\ S) \times (-90\ mV)}{(6 \times 10^{-6}\ S)} = -15\ mV.$$

The peak amplitude of the EPSP is then

$$\Delta V_{EPSP} = V_m - E_l = -15\ mV - (-90\ mV) = 75\ mV.$$

Gerald D. Fischbach
Steven A. Siegelbaum

Selected Reading

Fatt P, Katz B. 1951. An analysis of the end-plate potential recorded with an intracellular electrode. J Physiol 115:320–370.

Heuser JE, Reese TS. 1977. Structure of the synapse. In: ER Kandel (ed). *Handbook of Physiology: A Critical, Comprehensive Presentation of Physiological Knowledge and Concepts*, Sect. 1 *The Nervous System*, Vol. 1 *Cellular Biology of Neurons*, Part 1, pp. 261–294. Bethesda, MD: American Physiological Society.

Hille B. 2001. *Ion Channels of Excitable Membranes*, 3rd ed., pp. 169–199. Sunderland, MA: Sinauer.

Imoto K, Busch C, Sakmann B, et al. 1988. Rings of negatively charged amino acids determine the acetylcholine receptor-channel conductance. Nature 335:645–648.

Karlin A. 2002. Emerging structure of the nicotinic acetylcholine receptors. Nat Rev Neurosci 3:102–114.

Neher E, Sakmann B. 1976. Single-channel currents recorded from membrane of denervated frog muscle fibres. Nature 260:799–802.

Nemecz Á, Prevost MS, Menny A, Corringer PJ. 2016. Emerging molecular mechanisms of signal transduction in pentameric ligand-gated ion channels. Neuron 90:452–470.

References

Akabas MH, Kaufmann C, Archdeacon P, Karlin A. 1994. Identification of acetylcholine receptor-channel lining residues in the entire M2 segment of the α-subunit. Neuron 13:919–927.

Alberts B, Bray D, Lewis J, Raff M, Roberts K, Watson JD. 1989. *Molecular Biology of the Cell*, 2nd ed. New York: Garland.

Brejc K, van Dijk WJ, Klaassen RV, et al. 2001. Crystal structure of an ACh-binding protein reveals the ligand-binding domain of nicotinic receptors. Nature 411:269–276.

Charnet P, Labarca C, Leonard RJ, et al. 1990. An open channel blocker interacts with adjacent turns of α-helices in the nicotinic acetylcholine receptor. Neuron 4:87–95.

Claudio T, Ballivet M, Patrick J, Heinemann S. 1983. Nucleotide and deduced amino acid sequences of *Torpedo californica* acetylcholine receptor γ-subunit. Proc Natl Acad Sci U S A 80:1111–1115.

Colquhoun D. 1981. How fast do drugs work? Trends Pharmacol Sci 2:212–217.

Dwyer TM, Adams DJ, Hille B. 1980. The permeability of the endplate channel to organic cations in frog muscle. J Gen Physiol 75:469–492.

Fertuck HC, Salpeter MM. 1974. Localization of acetylcholine receptor by ^{125}I-labeled α-bungarotoxin binding at mouse motor endplates. Proc Natl Acad Sci U S A 71:1376–1378.

Heuser JE, Salpeter SR. 1979. Organization of acetylcholine receptors in quick-frozen, deep-etched, and rotary-replicated Torpedo postsynaptic membrane. J Cell Biol 82:150–173.

Ko C-P. 1984. Regeneration of the active zone at the frog neuromuscular junction. J Cell Biol 98:1685–1695.

Kuffler SW, Nicholls JG, Martin AR. 1984. *From Neuron to Brain: A Cellular Approach to the Function of the Nervous System*, 2nd ed. Sunderland, MA: Sinauer.

McMahan UJ, Kuffler SW. 1971. Visual identification of synaptic boutons on living ganglion cells and of varicosities in postganglionic axons in the heart of the frog. Proc R Soc Lond B Biol Sci 177:485–508.

Miles FA. 1969. *Excitable Cells*. London: Heinemann.

Miyazawa A, Fujiyoshi Y, Unwin N. 2003. Structure and gating mechanism of the acetylcholine receptor pore. Nature 424:949–955.

Morales-Perez CL, Noviello CM, Hibbs RE. 2016. X-ray structure of the human α4β2 nicotinic receptor. Nature 538:411–415.

Noda M, Furutani Y, Takahashi H, et al. 1983. Cloning and sequence analysis of calf cDNA and human genomic DNA encoding α-subunit precursor of muscle acetylcholine receptor. Nature 305:818–823.

Noda M, Takahashi H, Tanabe T, et al. 1983. Structural homology of *Torpedo californica* acetylcholine receptor subunits. Nature 302:528–532.

Palay SL. 1958. The morphology of synapses in the central nervous system. Exp Cell Res 5:275–293. Suppl.

Revah F, Galzi J-L, Giraudat J, Haumont PY, Lederer F, Changeux J-P. 1990. The noncompetitive blocker [3H] chlorpromazine labels three amino acids of the acetylcholine receptor gamma subunit: implications for the alpha-helical organization of regions MII and for the structure of the ion channel. Proc Natl Acad Sci U S A 87:4675–4679.

Salpeter MM (ed). 1987. *The Vertebrate Neuromuscular Junction*, pp. 1–54. New York: Liss.

Takeuchi A. 1977. Junctional transmission. I. Postsynaptic mechanisms. In: ER Kandel (ed). *Handbook of Physiology: A Critical, Comprehensive Presentation of Physiological Knowledge and Concepts*, Sect. 1 *The Nervous System*, Vol. 1 *Cellular Biology of Neurons*, Part 1, pp. 295–327. Bethesda, MD: American Physiological Society.

Verrall S, Hall ZW. 1992. The N-terminal domains of acetylcholine receptor subunits contain recognition signals for the initial steps of receptor assembly. Cell 68:23–31.

Villarroel A, Herlitze S, Koenen M, Sakmann B. 1991. Location of a threonine residue in the alpha-subunit M2 transmembrane segment that determines the ion flow through the acetylcholine receptor-channel. Proc R Soc Lond B Biol Sci 243:69–74.

Walsh J. 1773. Of the electric property of the torpedo. Phil Trans 63(1773):480.

Zeng H, Moise L, Grant MA, Hawrot E. 2001. The solution structure of the complex formed between alpha-bungarotoxin and an 18-mer cognate peptide derived from the alpha 1 subunit of the nicotinic acetylcholine receptor from Torpedo californica. J Biol Chem 276: 22930–22940.

13

Synaptic Integration in the Central Nervous System

LIKE SYNAPTIC TRANSMISSION at the neuromuscular junction, most rapid signaling between neurons in the central nervous system involves ionotropic receptors in the postsynaptic membrane. Thus, many principles that apply to the synaptic connection between the motor neuron and skeletal muscle fiber at the neuromuscular junction also apply in the central nervous system. Nevertheless, synaptic transmission between central neurons is more complex for several reasons.

First, although most muscle fibers are typically innervated by only one motor neuron, a central nerve cell (such as pyramidal neurons in the neocortex) receives connections from thousands of neurons. Second, muscle fibers receive only excitatory inputs, whereas central neurons receive both excitatory and inhibitory inputs. Third, all synaptic actions on muscle fibers are mediated by one neurotransmitter, acetylcholine (ACh), which activates only one type of receptor (the ionotropic nicotinic ACh receptor). A single central neuron, however, can respond to many different types of inputs, each mediated by a distinct transmitter that activates a specific type of receptor.

These receptors include ionotropic receptors, where binding of transmitter directly opens an ion channel, and metabotropic receptors, where transmitter binding indirectly regulates a channel by activating second messengers. As a result, unlike muscle fibers, central neurons must integrate diverse inputs into a single coordinated action.

Finally, the nerve–muscle synapse is a model of efficiency—every action potential in the motor neuron produces an action potential in the muscle fiber. In comparison, connections made by a presynaptic neuron onto a central neuron are only modestly effective—in many cases at least 50 to 100 excitatory neurons must fire together to produce a synaptic potential large enough to trigger an action potential in postsynaptic neurons.

The first insights into synaptic transmission in the central nervous system came from experiments by John Eccles and his colleagues in the 1950s on the synaptic inputs onto spinal motor neurons that control the stretch reflex, the simple behavior we considered in Chapter 3. The spinal motor neurons have been particularly useful for examining central synaptic mechanisms because they have large, accessible cell bodies and, most important, they receive both excitatory and inhibitory connections and therefore allow us to study the integrative action of the nervous system at the cellular level.

Central Neurons Receive Excitatory and Inhibitory Inputs

To analyze the synapses that mediate the stretch reflex, Eccles activated a large population of axons of the sensory cells that innervate the stretch receptor organs in the quadriceps (extensor) muscle (Figure 13–1A,B). Nowadays the same experiments can be done by stimulating a single sensory neuron.

Passing sufficient current through a microelectrode into the cell body of a stretch-receptor sensory neuron that innervates the extensor muscle generates an action potential. This in turn produces a small excitatory postsynaptic potential (EPSP) in the motor neuron that innervates precisely the same muscle (in this case the quadriceps) monitored by the sensory neuron (Figure 13–1B, upper panel). The EPSP produced by one sensory cell, the unitary EPSP, depolarizes the extensor motor neuron by less than 1 mV, often only 0.2 to 0.4 mV, far below the threshold for generating an action potential. Typically, a depolarization of 10 mV or more is required to reach threshold.

The generation of an action potential in a motor neuron thus requires the near-synchronous firing of a number of sensory neurons. This can be observed in an experiment in which a population of sensory neurons is stimulated by passing current through an extracellular electrode. As the strength of the extracellular stimulus is increased, more sensory afferent fibers are excited, and the depolarization produced by the EPSP becomes larger. The depolarization eventually becomes large enough to bring the membrane potential of the motor neuron axon initial segment (the region with the lowest threshold) to the threshold for an action potential.

In addition to the EPSP produced in the extensor motor neuron, stimulation of extensor stretch-receptor neurons also produces a small inhibitory postsynaptic potential (IPSP) in the motor neuron that innervates the flexor muscle, which is antagonistic to the extensor muscle (Figure 13–1B, lower panel). This hyperpolarizing action is generated by an inhibitory interneuron, which receives excitatory input from the sensory neurons of the extensor muscle and in turn makes synapses with the motor neurons that innervate the flexor muscle. In the laboratory, a single interneuron can be stimulated intracellularly to directly elicit a small unitary IPSP in the motor neuron. Extracellular activation of an entire population of interneurons elicits a larger IPSP. If strong enough, IPSPs can counteract the EPSP and prevent the membrane potential from reaching threshold.

Excitatory and Inhibitory Synapses Have Distinctive Ultrastructures and Target Different Neuronal Regions

As we learned in Chapter 11, the effect of a synaptic potential—whether it is excitatory or inhibitory—is determined not by the type of transmitter released from the presynaptic neuron but by the type of ion channels in the postsynaptic cell activated by the transmitter. Although some transmitters can produce both EPSPs and IPSPs, by acting on distinct classes of ionotropic receptors at different synapses, most transmitters produce a single predominant type of synaptic response; that is, a transmitter is usually inhibitory or excitatory. For example, in the vertebrate central nervous system, neurons that release glutamate typically act on receptors that produce excitation; neurons that release γ-aminobutyric acid (GABA) or glycine act on receptors that produce inhibition.

The synaptic terminals of excitatory and inhibitory neurons can be distinguished by their ultrastructure.

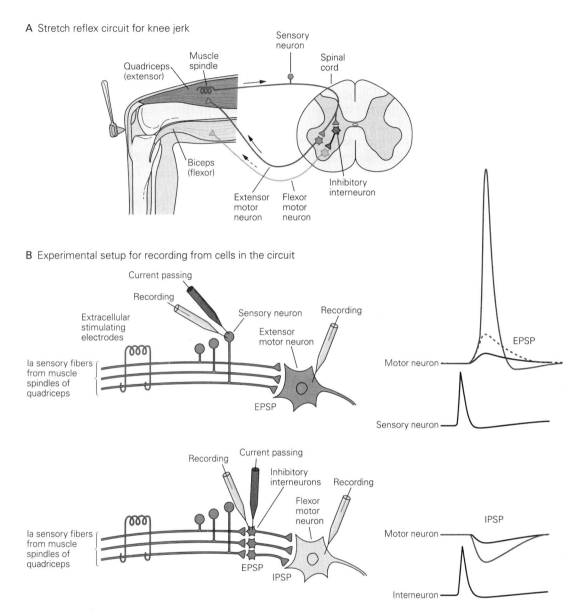

Figure 13–1 The combination of excitatory and inhibitory synaptic connections mediating the stretch reflex of the quadriceps muscle is typical of circuits in the central nervous system.

A. A sensory neuron activated by a stretch receptor (muscle spindle) at the extensor (quadriceps) muscle makes an excitatory connection with an extensor motor neuron in the spinal cord that innervates this same muscle group. It also makes an excitatory connection with an interneuron, which in turn makes an inhibitory connection with a flexor motor neuron that innervates the antagonist (biceps femoris) muscle group. Conversely, an afferent fiber from the biceps (not shown) excites an interneuron that makes an inhibitory synapse on the extensor motor neuron.

B. This idealized experimental setup shows the approaches to studying the inhibition and excitation of motor neurons in the pathway illustrated in part A. **Upper panel:** Two alternatives for eliciting excitatory postsynaptic potentials (**EPSPs**) in

the extensor motor neuron. A single presynaptic axon can be stimulated by inserting a current-passing electrode into the sensory neuron cell body. An action potential in the sensory neuron stimulated in this way triggers a small EPSP in the extensor motor neuron (**black trace**). Alternatively, the whole afferent nerve from the quadriceps can be stimulated electrically with an extracellular electrode. The excitation of many afferent neurons through the extracellular electrode generates a synaptic potential (**dashed trace**) large enough to initiate an action potential (**red trace**). **Lower panel:** The setup for eliciting and measuring inhibitory potentials in the flexor motor neuron. Intracellular stimulation of a single inhibitory interneuron receiving input from the quadriceps pathway produces a small inhibitory (hyperpolarizing) postsynaptic potential (**IPSP**) in the flexor motor neuron (**black trace**). Extracellular stimulation recruits numerous inhibitory neurons and generates a larger IPSP (**red trace**). (Action potentials in the sensory neuron and interneuron appear smaller because they were recorded at lower amplification than those in the motor neuron.)

Two morphological types of synapses are common in the brain: Gray types I and II (named after E. G. Gray, who described them using electron microscopy). Most type I synapses are glutamatergic and excitatory, whereas most type II synapses are GABAergic and inhibitory. Type I synapses have round synaptic vesicles, an electron-dense region (the *active zone*) on the presynaptic membrane, and an even larger electron-dense region in the postsynaptic membrane opposed to the active zone (known as the *postsynaptic density*), which gives type I synapses an asymmetric appearance. Type II synapses have oval or flattened synaptic vesicles and less obvious presynaptic membrane specializations and postsynaptic densities, resulting in a more symmetric appearance (Figure 13–2). (Although type I synapses are mostly excitatory and type II inhibitory, the two morphological types have proved to be only a first approximation to transmitter biochemistry. Immunocytochemistry affords much more reliable distinctions between transmitter types, as discussed in Chapter 16).

Although dendrites are normally postsynaptic and axon terminals presynaptic, all four regions of the nerve cell—axon, presynaptic terminals, cell body, and dendrites—can be presynaptic or postsynaptic sites of chemical synapses. The most common types of contact, illustrated in Figure 13–2, are axodendritic, axosomatic, and axo-axonic (by convention, the presynaptic element is identified first). Excitatory synapses are typically axodendritic and occur mostly on dendritic spines. Inhibitory synapses are normally formed on dendritic shafts, the cell body, and the axon initial segment. Dendrodendritic and somasomatic synapses are also found, but they are rare.

As a general rule, the proximity of a synapse to the axon initial segment is thought to determine its effectiveness. A given postsynaptic current generated at a site near the cell body will produce a greater change in membrane potential at the trigger zone of the axon initial segment, and therefore have a greater influence on action potential output than an equal current generated at more remote sites in the dendrites. This is because some of the charge entering the postsynaptic membrane at a remote site will leak out of the dendritic membrane as the synaptic potential propagates to the cell body (Chapter 9). Some neurons compensate for this effect by placing

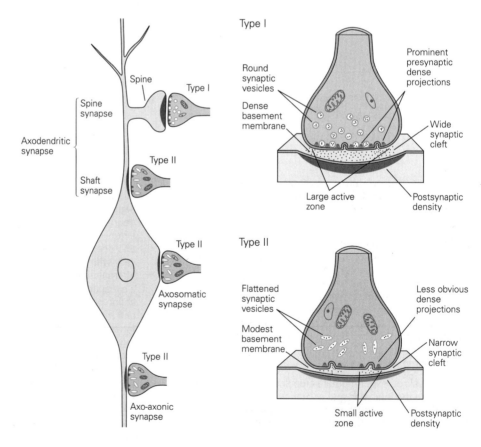

Figure 13–2 The two most common morphological types of synapses in the central nervous system are Gray type I and type II. Type I is usually excitatory, whereas type II is usually inhibitory. Differences include the shape of vesicles, the prominence of presynaptic densities, total area of the active zone, width of the synaptic cleft, and presence of a dense basement membrane. Type I synapses typically contact specialized dendritic projections, called spines, and less commonly contact the shafts of dendrites. Type II synapses contact the cell body (axosomatic), dendritic shaft (axodendritic), axon initial segment (axo-axonic), and presynaptic terminals of another neuron (not shown).

more glutamate receptors at distal synapses than at proximal synapses, ensuring that inputs at different locations along the dendritic tree will have a more equivalent influence at the initial segment. In contrast to axodendritic and axosomatic input, most axo-axonic synapses have no direct effect on the trigger zone of the postsynaptic cell. Instead, they affect neural activity by controlling the amount of transmitter released from the presynaptic terminals (Chapter 15).

Excitatory Synaptic Transmission Is Mediated by Ionotropic Glutamate Receptor-Channels Permeable to Cations

The excitatory transmitter released from the presynaptic terminals of the stretch-receptor sensory neurons is the amino acid L-glutamate, the major excitatory transmitter in the brain and spinal cord. Eccles and his colleagues discovered that the EPSP in spinal motor cells results from the opening of ionotropic glutamate receptor-channels, which are permeable to both Na^+ and K^+. This ionic mechanism is similar to that produced by ACh at the neuromuscular junction described in Chapter 12. Like the ACh receptor-channels, glutamate receptor-channels conduct both Na^+ and K^+ with nearly equal permeability. As a result, the reversal potential for current flow through these channels is 0 mV (see Figure 12–7).

Glutamate receptors can be divided into two broad categories: ionotropic receptors and metabotropic receptors (Figure 13–3). There are three major types of ionotropic glutamate receptors: *AMPA*, *kainate*, and *NMDA*, named according to the types of pharmacological agonists that activate them (α-amino-3-hydroxy-5-methylisoxazole-4-propionic acid, kainate, and N-methyl-D-aspartate, respectively). These receptors are also differentially sensitive to antagonists. The NMDA receptor is selectively blocked by the drug *APV* (2-amino-5-phosphono-valeric acid). The AMPA and kainate receptors are not affected by APV, but both are blocked by the drug *CNQX* (6-cyano-7-nitroquinoxaline-2,3-dione). Because of this shared pharmacological sensitivity, these two types are sometimes called the *non-NMDA receptors*. Another important distinction between NMDA and non-NMDA receptors is that the NMDA receptor channel is highly permeable to Ca^{2+}, whereas most non-NMDA receptors are not. There are several types of metabotropic glutamate receptors, most of which can be activated by *trans-(1S,3R)-1-amino-1,3-cyclopentanedicarboxylic acid (ACPD)*.

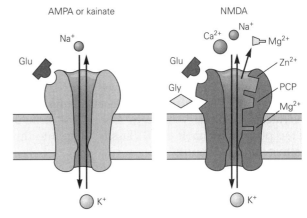

A Ionotropic glutamate receptor

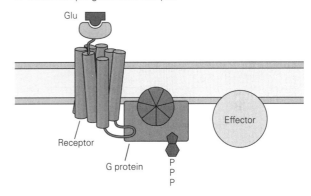

B Metabotropic glutamate receptor

Figure 13–3 Different classes of glutamate receptors regulate excitatory synaptic actions in neurons in the spinal cord and brain.

A. Ionotropic glutamate receptors directly gate ion channels permeable to cations. The AMPA and kainate types bind the glutamate agonists AMPA or kainate, respectively; these receptors contain a channel that is permeable to Na^+ and K^+. The NMDA receptor binds the glutamate agonist NMDA; it contains a channel permeable to Ca^{2+}, K^+, and Na^+. It has binding sites for glutamate, glycine, Zn^{2+}, phencyclidine (**PCP**, or angel dust), MK801 (an experimental drug), and Mg^{2+}, each of which regulates the functioning of the channel differently.

B. Binding of glutamate (**Glu**) to metabotropic glutamate receptors indirectly gates ion channels by activating a GTP-binding protein (**G protein**), which in turn interacts with effector molecules that alter metabolic and ion channel activity (Chapter 11).

The action of all ionotropic glutamate receptors is excitatory or depolarizing because the reversal potential of their ionic current is near zero, causing channel opening to produce a depolarizing inward current at negative membrane potentials. In contrast, metabotropic receptors can produce either excitation or inhibition, depending on the reversal potential of the ionic currents that they regulate and whether they promote channel opening or channel closing.

The Ionotropic Glutamate Receptors Are Encoded by a Large Gene Family

Over the past 30 years, a large variety of genes coding for the subunits of all the major neurotransmitter receptors have been identified. In addition, many of these subunit genes are alternatively spliced, generating further diversity. This molecular analysis demonstrates evolutionary linkages among the structure of receptors that enable us to classify them into three distinct families (Figure 13–4).

The ionotropic glutamate receptor family includes the AMPA, kainate, and NMDA receptors. The genes encoding the AMPA and kainate receptors are more closely related to one another than are the genes encoding the NMDA receptors. Surprisingly, the glutamate receptor family bears little resemblance to the two other gene families that encode ionotropic receptors (one of which encodes the nicotinic ACh, GABA, and glycine receptors, and the other the ATP receptors, as described later).

The AMPA, kainate, and NMDA receptors are tetramers composed of two or more types of related subunits, with all four subunits arranged around a central pore. The AMPA receptor subunits are encoded by four separate genes (*GluA1–GluA4*), whereas the kainate receptor subunits are encoded by five different genes (*GluK1–GluK5*). Autoantibodies to the GluA3 subunit of the AMPA receptor are thought to play an important role in some forms of epilepsy. These antibodies actually mimic glutamate by activating GluA3-containing receptors, resulting in excessive excitation and seizures. NMDA receptors, on the other hand, are encoded by a family consisting of five genes that fall

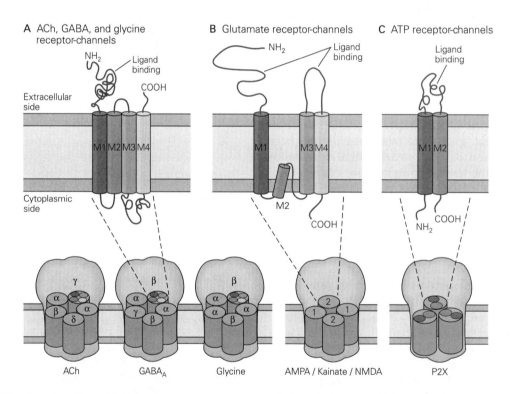

Figure 13–4 The three families of ionotropic receptors.

A. The nicotinic ACh, GABA_A, and glycine receptor-channels are all pentamers composed of several types of related subunits. As shown here, the ligand-binding domain is formed by the extracellular amino-terminal region of the protein. Each subunit has a membrane domain with four membrane-spanning α-helixes (M1–M4) and a short extracellular carboxyl terminus. The M2 helix lines the channel pore.

B. The glutamate receptor-channels are tetramers, often composed of two different types of closely related subunits (here denoted 1 and 2). The subunits have a large extracellular amino terminus, a membrane domain with three membrane-spanning

α-helixes (M1, M3, and M4), a large extracellular loop connecting the M3 and M4 helixes, and an intracellular carboxyl terminus. The M2 segment forms a loop that dips into and out of the cytoplasmic side of the membrane, contributing to the selectivity filter of the channel. The glutamate binding site is formed by residues in the extracellular amino terminus and in the M3–M4 extracellular loop.

C. The adenosine triphosphate (**ATP**) receptor-channels (or purinergic P2X receptors) are trimers. Each subunit possesses two membrane-spanning α-helixes (M1 and M2) and a large extracellular loop that binds ATP. The M2 helix lines the pore.

into two groups: The *GluN1* gene encodes one type of subunit, whereas four distinct *GluN2* genes (*A–D*) encode a second type. Each NMDA receptor contains two GluN1 subunits and two GluN2 subunits.

Glutamate Receptors Are Constructed From a Set of Structural Modules

All ionotropic glutamate receptor subunits share a common architecture with similar motifs. Eric Gouaux and colleagues have provided important insights into the structure of the ionotropic glutamate receptors, initially through an X-ray crystallographic model of an AMPA receptor composed of four GluA2 subunits. The subunits have a large extracellular amino-terminal domain, which is followed in the primary amino acid sequence by an extracellular ligand-binding domain and a transmembrane domain (Figures 13–4B and 13–5). The transmembrane domain contains three transmembrane α-helices (M1, M3, and M4) and a loop (M2) between the M1 and M3 helices that dips into and out of the cytoplasmic side of the membrane. This M2 loop resembles the pore-lining P loop of K^+ channels and helps form the selectivity filter of the channel (see Figure 8–12).

Both extracellular domains are homologous to bacterial amino acid binding protein domains. The ligand-binding domain is a bi-lobed clamshell-like structure (Figure 13–5A), whereas the amino-terminal domain is homologous to the glutamate-binding domain of metabotropic glutamate receptors but does not bind glutamate. Instead, in the ionotropic glutamate receptors, this domain is involved in subunit assembly, the modulation of receptor function by ligands other than glutamate, and/or the interaction with other synaptic proteins to regulate synapse development.

The ligand-binding domain is formed by two distinct regions in the linear sequence of the protein. One region comprises the end of the amino-terminal domain up to the M1 transmembrane helix; the second region is formed by the large extracellular loop connecting the M3 and M4 helices (Figure 13–5A). In the ionotropic receptors, the binding of a molecule of glutamate within the clamshell triggers the closure of the lobes of the clamshell; competitive antagonists also bind to the clamshell but fail to trigger clamshell closure. This suggests that the conformational change associated with clamshell closure is important for opening the ion channel.

In addition to the core subunits that form the receptor-channel, AMPA receptors contain additional (or auxiliary) subunits that regulate receptor trafficking to the membrane and function. One important class of auxiliary subunits comprises the *transmembrane AMPA receptor regulatory proteins* (TARPs). A TARP subunit

has four transmembrane domains, and its association with the pore-forming AMPA receptor subunits enhances the surface membrane trafficking, synaptic localization, and gating of the AMPA receptors. The first TARP family member to be identified was stargazin, which was isolated through a genetic screen in the *stargazer* mutant mouse, so named because these animals have a tendency to tip their heads backward and stare upward. Loss of stargazin leads to a complete loss of AMPA receptors from cerebellar granule cells, which results in cerebellar ataxia and frequent seizures. Other members of the TARP family are similarly required for AMPA receptor trafficking to the surface membrane in other types of neurons.

High-resolution cryo-electron microscopy has revealed the structure of TARP subunits in association with the AMPA receptor subunits (Figure 13–5D,E). These studies suggest that interactions between a TARP subunit and the ligand-binding domain clamshell of an AMPA receptor can stabilize the receptor in the glutamate-bound open state, thereby enhancing the channel open time, single-channel conductance, and affinity for glutamate.

Given the homology among the various subtypes of glutamate receptors, it is not surprising that the overall structure of the kainate and NMDA receptors is similar to that of the homomeric GluA2 receptor. However, there are some important differences that give rise to the distinct physiological functions of the different receptors. The high permeability of the NMDA receptor-channels to Ca^{2+} has been localized to a single amino acid residue in the pore-forming M2 loop. All NMDA receptor subunits contain the neutral residue asparagine at this position in the pore. In most types of AMPA receptor subunits, the residue at this position is the uncharged amino acid glutamine; in the GluA2 subunit, however, the corresponding M2 residue is arginine, a positively charged basic amino acid. Inclusion of even a single GluA2 subunit prevents the AMPA receptor-channel from conducting Ca^{2+} (Figure 13–6B), most likely as a result of strong electrostatic repulsion by the arginine. The opening of AMPA receptor-channels in cells that lack the GluA2 subunit can produce a significant Ca^{2+} influx because the pores of these receptors lack the positively charged arginine residue.

Interestingly, the DNA of the *GluA2* gene does not encode an arginine residue at this position in the M2 loop but rather codes for a glutamine residue. After transcription, the codon for glutamine in the *GluA2* mRNA is replaced with one for arginine through an enzymatic process termed RNA editing (Figure 13–6A). The importance of this RNA editing was investigated using a genetically modified mouse whose *GluA2* gene was engineered so that the relevant nucleotide in the

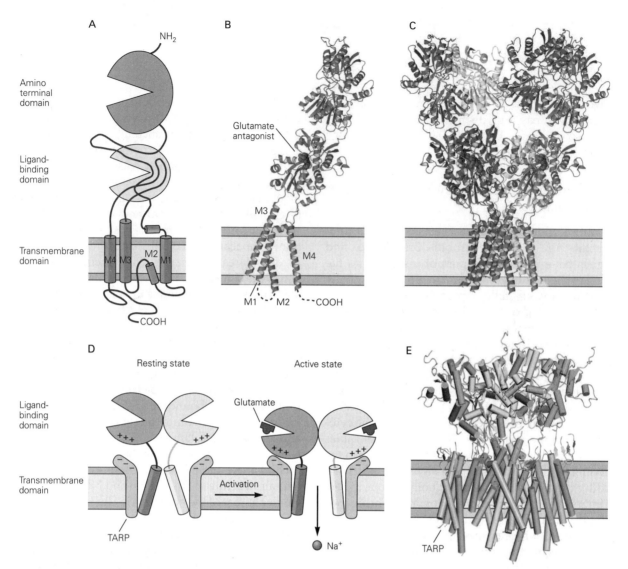

Figure 13–5 Atomic structure of an ionotropic glutamate receptor.

A. Schematic organization of the ionotropic glutamate receptors. The receptors contain a large extracellular amino terminus, a transmembrane domain containing three membrane-spanning α-helixes (M1, M3, and M4), and a loop that dips into the cytoplasmic side of the membrane (M2). The ligand-binding domain is formed by the extracellular region of the receptor on the amino-terminal side of the M1 segment and by the extracellular loop connecting M3 and M4. These two regions intertwine to form a clamshell-like structure that binds glutamate and various pharmacological agonists and competitive antagonists. A similar structure is formed at the extreme amino terminus of the receptor. In ionotropic glutamate receptors, this amino-terminal domain does not bind glutamate but is thought to modulate receptor function and synapse development. (Reproduced, with permission, from Armstrong et al. 1998.)

B. Three-dimensional X-ray crystal structure of a single AMPA receptor GluA2 subunit. This side view shows the amino-terminal, ligand-binding, and transmembrane domains (compare to panel A). The M1, M3, and M4 transmembrane α-helixes are indicated, as is a short α-helix in the M2 loop. A molecule of a competitive antagonist of glutamate bound to the ligand-binding domain is shown (**red space-filling representation**). The cytoplasmic loops connecting the membrane α-helixes were not resolved in the structure and have been drawn as **dashed lines**. (Reproduced, with permission, from Sobolevsky, Rosconi, and Gouaux 2009.)

C. This side view shows the structure of a receptor assembled from four identical GluA2 subunits (the subunits are colored differently for illustrative purposes). The subunits associate through their extracellular domains as a pair of dimers (two-fold symmetry). In the amino-terminal domain, one dimer is formed by the **blue** and **yellow** subunits, the other dimer by the **red** and **green** subunits. In the ligand-binding domain, the subunits change partners. In one dimer, the blue subunit associates with the red subunit, whereas in the other dimer, the yellow subunit associates with the green subunit. In the transmembrane region, the subunits associate as a four-fold symmetric tetramer. The significance of this highly unusual subunit arrangement is not fully understood. (Reproduced, with permission, from Sobolevsky, Rosconi, and Gouaux 2009.)

D. Cartoon side view of auxiliary TARP subunits (**blue**) associated with pore-forming GluA2 subunits. For simplicity, only the transmembrane and ligand-binding domain of two of the four GluA2 subunits is shown. Two of four TARP subunits are also shown. Binding of glutamate causes the clamshell-like ligand-binding domain to close, leading to a conformational change in the transmembrane domain that opens the pore. An electrostatic interaction between TARP and GluA2 stabilizes the receptor in the open state. (Adapted, with permission, from Mayer 2016. Copyright © 2016 Elsevier Ltd.)

E. Three-dimensional structure of the TARP-GluA2 complex. The α-helices are shown as cylinders. The four TARP subunits are shown in **blue**. Transmembrane and ligand-binding domains of GluA2 subunits are shown in **yellow** and **green**. (Adapted, with permission, from Mayer 2016. Copyright © 2016 Elsevier Ltd.)

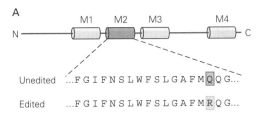

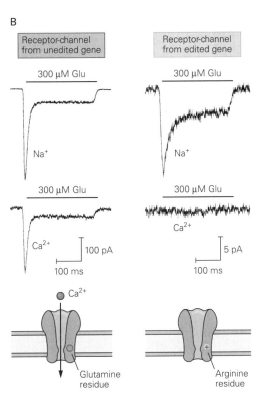

Figure 13–6 Determinants of calcium ion permeability of the AMPA receptor-channel.

A. Comparison of amino acid sequences in the M2 region of the AMPA receptor-channel coded by the *GluA2* gene before and after RNA editing. The unedited transcript codes for the polar residue glutamine (**Q**, the single-letter amino acid notation), whereas the edited transcript codes for the positively charged residue arginine (**R**). In adults, the GluA2 protein exists almost exclusively in the edited form.

B. AMPA receptor-channels expressed from unedited transcripts conduct Ca²⁺ (*left traces*), whereas those expressed from edited transcripts do not (*right traces*). The traces show currents elicited by glutamate with either extracellular Na⁺ (*top*) or Ca²⁺ (*bottom*) as the predominant permeant cation. (Reproduced, with permission, from Sakmann 1992. Copyright © 1992 Elsevier.)

glutamine codon could no longer be changed to arginine. Such mice develop seizures and die within a few weeks after birth, presumably because the high Ca²⁺ permeability of all the AMPA receptors results in an excess of intracellular Ca²⁺.

NMDA and AMPA Receptors Are Organized by a Network of Proteins at the Postsynaptic Density

How are the different glutamate receptors localized and arranged at excitatory synapses? Like most ionotropic receptors, glutamate receptors are normally clustered at postsynaptic sites in the membrane, precisely opposed to glutamatergic presynaptic terminals. The vast majority of excitatory synapses in the mature nervous system contain both NMDA and AMPA receptors, whereas in early development, synapses containing only NMDA receptors are common. The pattern of receptor localization and expression at individual synapses depends on a large number of regulatory proteins that constitute the postsynaptic density and help organize the three-dimensional structure of the postsynaptic cell membrane.

The postsynaptic density (PSD) is a remarkably stable structure, permitting its biochemical isolation, purification, and characterization. Electron microscopic studies of intact and isolated PSDs provide a strikingly detailed view of their structure (Figure 13–7A). By using gold-labeled antibodies, it is possible to identify specific protein components of the postsynaptic membrane, including the location and number of glutamate receptors. A typical PSD is around 350 nm in diameter and contains about 20 NMDA receptors, which tend to be localized near the center of the PSD, and 10 to 50 AMPA receptors, which are less centrally localized. The metabotropic glutamate receptors are located on the periphery, outside the main area of the PSD. All three receptor types interact with a wide array of cytoplasmic and membrane proteins to ensure their proper localization (Figure 13–7C).

One of the most prominent proteins in the PSD important for the clustering of glutamate receptors is PSD-95 (PSD protein of 95 kD molecular weight). PSD-95 is a membrane-associated protein that contains three repeated regions—the so-called PDZ domains—important for protein–protein interactions. (The PDZ domains are named after the three proteins in which they were first identified: PSD-95, the DLG tumor suppressor protein in *Drosophila*, and a protein termed ZO-1.) The PDZ domains bind to specific sequences at the carboxy terminus of a number of proteins. In PSD-95, the PDZ domains bind the NMDA receptor and Shaker-type voltage-gated K⁺ channels, thereby localizing and concentrating these channels at postsynaptic sites. PSD-95 also interacts with the postsynaptic membrane protein neuroligin, which contacts the presynaptic membrane protein neurexin in the synaptic cleft, an interaction important for synapse development. Mutations in neuroligin are thought to contribute to some cases of autism.

A Purified postsynaptic densities

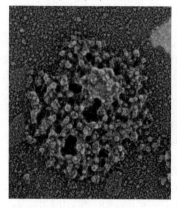
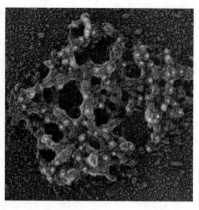

B Distribution of receptors

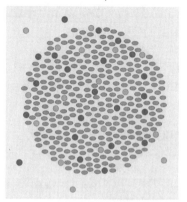

● AMPA receptors
● NMDA receptors
● PSD-95

C Molecular organization of synapse at dendritic spine

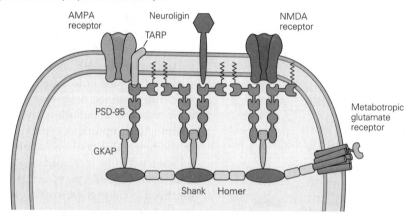

Figure 13–7 The postsynaptic cell membrane is organized into a macromolecular complex at excitatory synapses. Proteins containing PDZ domains help organize the distribution of AMPA and NMDA glutamate receptors at the postsynaptic density. (Reproduced, with permission, from Sheng and Hoogenrad 2007. Micrographs provided by Thomas S. Reese and Xiaobing Chen; National Institutes of Health, USA.)

A. Electron microscope images of biochemically purified post-synaptic densities, showing organization of the protein network. The membrane lipid bilayer is no longer present. *Left:* View of postsynaptic density from what would normally be the outside of the cell. This image consists of the extracellular domains of various receptors and membrane proteins. *Right:* View of a postsynaptic density from what would normally be the cytoplasmic side of the membrane. **White dots** show immunolabeled guanylate kinase anchoring protein, an important component of the postsynaptic density.

B. The distribution of NMDA receptors, AMPA receptors, and PSD-95, a prominent postsynaptic density protein, at a synapse.

C. The network of receptors and their interacting proteins in the postsynaptic density. PSD-95 contains three PDZ domains at its amino terminus and two other protein-interacting motifs at its carboxyl terminus, an SH3 domain and guanylate kinase (GK) domain. Certain PDZ domains of PSD-95 bind to the carboxyl terminus of the GluN2 subunit of the NMDA receptor. PSD-95 does not directly interact with AMPA receptors but binds to the carboxyl terminus of the TARP family of membrane proteins, which interact with the AMPA receptors as auxiliary subunits. PSD-95 also acts as a scaffold for various cytoplasmic proteins by binding to GK-associated protein (**GKAP**), which interacts with Shank, a large protein that associates into a meshwork linking the various components of the postsynaptic density. PSD-95 also interacts with the cytoplasmic region of neuroligin. The metabotropic glutamate receptor is localized on the periphery of the synapse where it interacts with the protein Homer, which in turn binds to Shank.

Although PSD-95 does not directly bind to AMPA receptors, it does interact with the TARP subunits. The proper localization of AMPA receptors in the postsynaptic membrane depends on the interaction between the carboxy terminus of the TARP subunit and PSD-95.

AMPA receptors also bind to a distinct PDZ domain protein called GRIP, and metabotropic glutamate receptors interact with yet another PDZ domain protein called Homer. In addition to interacting with receptors, proteins with PDZ domains interact with many other cellular proteins, including proteins that bind to the actin cytoskeleton, providing a scaffold around which a complex of postsynaptic proteins is constructed. Indeed, a biochemical analysis of the PSD has identified dozens of proteins that participate in NMDA or AMPA receptor complexes.

NMDA Receptors Have Unique Biophysical and Pharmacological Properties

The NMDA receptor has several interesting properties that distinguish it from AMPA receptors. As mentioned earlier, NMDA receptors have a distinctively high permeability to Ca^{2+}. In addition, the NMDA receptor is unique among ligand-gated channels thus far characterized because its opening depends on membrane voltage as well as transmitter binding.

The voltage dependence is caused by a mechanism that is quite different from that employed by the voltage-gated channels that generate the action

potential. In the latter, changes in membrane potential are translated into conformational changes in the channel by an intrinsic voltage sensor. In the NMDA receptors, however, depolarization removes an extrinsic plug from the channel. At the resting membrane potential (–65 mV), extracellular Mg^{2+} binds tightly to a site in the pore of the channel, blocking ionic current. But when the membrane is depolarized (for example, by the opening of AMPA receptor-channels), Mg^{2+} is expelled from the channel by electrostatic repulsion, allowing Na^+, K^+, and Ca^{2+} to flow (Figure 13–8). The NMDA receptor has the further interesting property of being inhibited by the hallucinogenic drug *phencyclidine* (PCP, also known as angel dust) and the experimental compound MK801. Both drugs bind to a site in the pore of the channel that is distinct from the Mg^{2+} binding site (Figure 13–3A).

At most glutamatergic central synapses, the postsynaptic membrane contains both NMDA and AMPA receptors. The relative contributions of current through

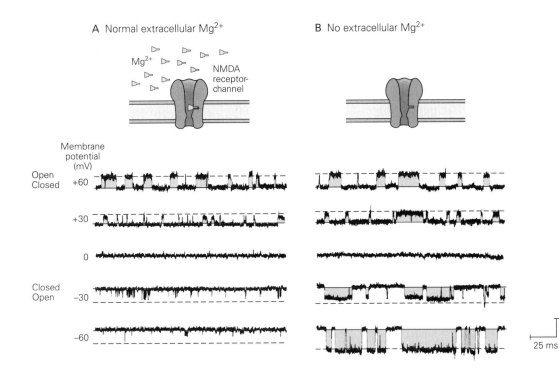

Figure 13–8 Opening of individual NMDA receptor-channels depends on membrane potential in addition to glutamate. These patch-clamp recordings are from individual NMDA receptor-channels (from rat hippocampal cells in culture). **Downward** deflections indicate pulses of inward (negative) current; **upward** deflections indicate outward (positive) current. (Reproduced, with permission, from J. Jen and C.F. Stevens.)

A. When Mg^{2+} is present in normal concentration in the extracellular solution (1.2 mM), the channel is largely blocked at the resting potential (–60 mV). At negative membrane potentials,

only brief, flickering, inward currents are seen upon channel opening because of the Mg^{2+} block. Substantial depolarization to voltages positive to the reversal potential of 0 mV (to +30 mV or +60 mV) relieves the Mg^{2+} block, permitting longer-lasting pulses of outward current through the channel.

B. When Mg^{2+} is removed from the extracellular solution, the opening and closing of the channel do not depend on voltage. The channel is open at the resting potential of –60 mV, and the synaptic current reverses near 0 mV, like the total synaptic current (see Figure 13–9B).

NMDA and AMPA receptors to the total excitatory postsynaptic current (EPSC) can be quantified using pharmacological antagonists in a voltage-clamp experiment (Figure 13–9). Since NMDA receptors are largely inhibited by Mg^{2+} at the normal resting potential of most neurons, the EPSC is predominantly determined by charge flow through the AMPA receptors. This current has very rapid rising and decay phases. However, as a neuron becomes depolarized and Mg^{2+} is driven out of the mouth of the NMDA receptors, more charge flows through them. Thus, the NMDA receptor-channel conducts current maximally when two conditions are met: Glutamate is present, and the cell is depolarized. That is, the NMDA receptor acts as a molecular "coincidence detector," opening during the concurrent activation of the presynaptic and postsynaptic cells. In addition, because of its intrinsic kinetics of ligand gating, the current through the NMDA receptor-channel rises and decays with a much slower time course than the current through AMPA receptor-channels. As a result, the NMDA receptors contribute to a late, slow phase of the EPSC and EPSP.

As most glutamatergic synapses contain AMPA receptors that are capable of triggering an action potential by themselves, what is the function of the NMDA receptor? At first glance, the function of these receptors is even more puzzling because their intrinsic channel is normally blocked by Mg^{2+} at the resting potential. However, the high permeability of the NMDA receptor-channels to Ca^{2+} endows them with the special ability to produce a marked rise in intracellular $[Ca^{2+}]$ that can activate various calcium-dependent signaling cascades, including several different protein kinases (Chapters 15 and 53). Thus, NMDA receptor activation can translate electrical signals into biochemical ones. Some of these biochemical reactions lead to long-lasting changes in synaptic strength through a set of processes called long-term synaptic plasticity, which are important for refining synaptic connections during early development and regulating neural circuits in the adult brain, including circuits critical for long-term memory.

The Properties of the NMDA Receptor Underlie Long-Term Synaptic Plasticity

In 1973, Tim Bliss and Terje Lomo found that a brief period of high-intensity and high-frequency synaptic stimulation (known as a tetanus) leads to *long-term potentiation* (LTP) of excitatory synaptic transmission in the hippocampus, a region of the mammalian brain required for many forms of long-term memory (Figure 13–10; see Chapters 53 and 54). Subsequent studies demonstrated that LTP requires Ca^{2+} influx through

the NMDA receptor-channels, which open in response to the combined effect of glutamate release and strong postsynaptic depolarization during the tetanic stimulation. LTP is blocked if the tetanus is delivered in the presence of APV, which blocks the NMDA receptors, or if the postsynaptic neuron is injected with a compound that chelates intracellular Ca^{2+}.

The rise of Ca^{2+} in the postsynaptic cell is thought to potentiate synaptic transmission by activating postsynaptic biochemical cascades that trigger the insertion of additional AMPA receptors into the postsynaptic membrane. Under some circumstances, postsynaptic Ca^{2+} can trigger production of a retrograde messenger, a chemical signal that enhances transmitter release from the presynaptic terminal (Chapter 14). As we will discuss later, the Ca^{2+} accumulation and biochemical activation are largely restricted to the individual spines that are activated by the tetanic stimulation. As a result, LTP is input-specific; only those synapses that are activated during the tetanic stimulation are potentiated.

The prolonged high-frequency presynaptic firing required to induce LTP is unlikely to be achieved under physiological conditions. However, a more physiologically relevant form of plasticity, termed spike-timing-dependent plasticity (STDP), can be induced if a single presynaptic stimulus is paired at low frequency with the triggered firing of one or more postsynaptic action potentials, providing sufficient depolarization to relieve Mg^{2+} block of the NMDA receptor pore. The presynaptic activity must precede postsynaptic firing, following a rule proposed in 1949 by the psychologist Donald Hebb for how individual neurons could become grouped together into functional assemblies during associative memory storage. A number of lines of evidence now suggest that LTP, STDP, or related processes provide an important cellular mechanism for memory storage (Chapters 53 and 54) and fine-tuning synaptic connections during development (Chapter 49).

NMDA Receptors Contribute to Neuropsychiatric Disease

Unfortunately, there is also a downside to recruiting Ca^{2+} through the NMDA receptors. Excessively high concentrations of glutamate are thought to result in an overload of Ca^{2+} in the postsynaptic neurons, a condition that can be toxic to neurons. In tissue culture, even a brief exposure to high concentrations of glutamate can kill many neurons, an action called *glutamate excitotoxicity*. High concentrations of intracellular Ca^{2+} are thought to activate calcium-dependent proteases and phospholipases and lead to the production of free radicals that are toxic to the cell.

A Early and late components of synaptic current

B Current-voltage relationship of the synaptic current

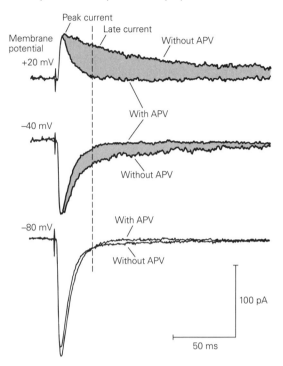

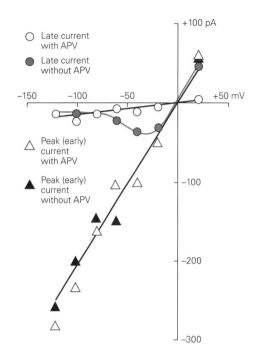

Figure 13–9 The contributions of the AMPA and NMDA receptor-channels to the excitatory postsynaptic current. These voltage-clamp current records are from a cell in the rat hippocampus. Similar receptor-channels are present in motor neurons and throughout the brain. (Adapted, with permission, from Hestrin et al. 1990.)

A. The drug APV selectively binds to and blocks the NMDA receptor. Shown here is the excitatory postsynaptic current (EPSC) before and during application of 50 µM APV at three different membrane potentials. The difference between the traces (**blue region**) represents the contribution of the NMDA receptor-channel to the EPSC. The current that remains in the presence of APV is the contribution of the AMPA receptor-channels. At −80 mV, there is no current through the NMDA receptor-channels because of pronounced Mg^{2+} block (see Figure 13–8). At −40 mV, a small late inward current through NMDA receptor-channels is evident. At +20 mV, the late component is more prominent and has reversed to become an outward current. The time 25 ms after the peak of the synaptic current (**dashed line**) is used for the calculations of late current in part B.

B. The postsynaptic currents through the NMDA and AMPA receptor-channels differ in their dependence on the membrane potential. The current through the AMPA receptor-channels contributes to the early phase of the synaptic current (**filled triangles**). The early phase is measured at the peak of the synaptic current and plotted here as a function of membrane potential. The current through the NMDA receptor-channels contributes to the late phase of the synaptic current (**filled circles**). The late phase is measured 25 ms after the peak of the synaptic current, a time at which the AMPA receptor component has decayed almost to zero (see part A). Note that the AMPA receptor-channels behave as simple resistors; current and voltage have a linear relationship. In contrast, current through the NMDA receptor-channels is nonlinear and increases as the membrane is depolarized from −80 to −40 mV, owing to progressive relief of the Mg^{2+} block. The reversal potential of both receptor-channel types is at 0 mV. The components of the synaptic current in the presence of 50 µm APV are indicated by the **unfilled circles** and **triangles**. Note how APV blocks the late (NMDA receptor) component of the EPSC but not the early (AMPA receptor) component.

Glutamate toxicity may contribute to cell damage after stroke, to the cell death that occurs with episodes of rapidly repeated seizures experienced by patients who have status epilepticus, and to degenerative diseases such as Huntington disease. Agents that selectively block the NMDA receptor may protect against the toxic effects of glutamate and have been tested clinically. The hallucinations that accompany NMDA receptor blockade have so far limited the usefulness of such compounds. A further complication of attempts

to control excitotoxicity by blocking NMDA receptor function is that physiological levels of NMDA receptor activation may actually protect neurons from damage and cell death.

Not all of the physiological and pathophysiological effects mediated by the NMDA receptor may result from Ca^{2+} influx. There is increasing evidence that binding of glutamate to the NMDA receptor may cause a conformational change in the receptor that activates downstream intracellular signaling pathways independently of ion flux. Such metabotropic functions of the NMDA receptor may contribute to long-term depression, a form of synaptic plasticity in which low-frequency synaptic activity produces a long-lasting decrease in glutamatergic synaptic transmission, the opposite of LTP. Metabotropic actions of the NMDA receptor may also contribute to the effect of β-amyloid, the peptide fragment implicated in Alzheimer disease, in depressing synaptic function.

A number of lines of evidence implicate NMDA receptor malfunction in schizophrenia. Pharmacological blockade of NMDA receptors with drugs such as phencyclidine or the general anesthetic ketamine, a derivative of PCP, produces symptoms that resemble the hallucinations associated with schizophrenia; in contrast, certain antipsychotic drugs enhance current through the NMDA receptor-channels. A particularly

A Schaffer collateral pathway LTP

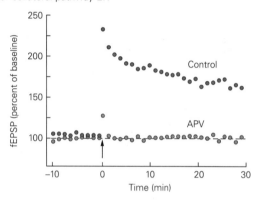

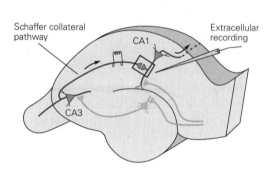

B Mechanism of LTP

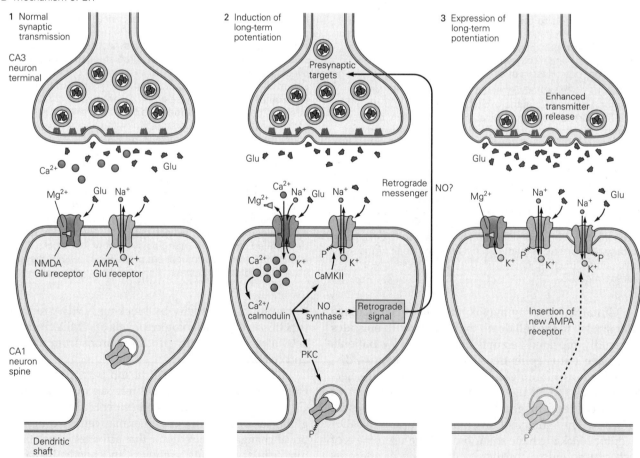

striking link with schizophrenia is seen in anti-NMDA receptor encephalitis, an autoimmune disorder in which the production of antibodies to the NMDA receptor reduces levels of the receptor in the membrane. Individuals with this disorder often experience severe seizures, most likely a result of the loss of inhibitory tone because of a reduction in NMDA receptor excitation in GABAergic interneurons, as well as psychoses, including hallucinations and other symptoms resembling schizophrenia. Treatments that reduce antibody levels often lead to complete remission of these symptoms. The idea that a decrease in NMDA receptor function may contribute to the symptoms of schizophrenia is further supported by recent genome-wide linkage analysis suggesting an association between the *NR2A* gene and schizophrenia. One additional link between the NMDA receptor and neuropsychiatric disorders is provided by the finding that low doses of ketamine exert a rapid and powerful antidepressant action.

Fast Inhibitory Synaptic Actions Are Mediated by Ionotropic GABA and Glycine Receptor-Channels Permeable to Chloride

Although glutamatergic excitatory synapses account for the vast majority of synapses in the brain, inhibitory synapses play an essential role in the nervous system both by preventing too much excitation and by regulating the firing patterns of networks of neurons. IPSPs in spinal motor neurons and most central neurons are generated by the amino acid neurotransmitters GABA and glycine.

GABA acts on both ionotropic and metabotropic receptors. The $GABA_A$ receptor is an ionotropic receptor that directly opens a Cl^- channel. The $GABA_B$ receptor is a metabotropic receptor that activates a second-messenger cascade, which often indirectly activates a K^+ channel (Chapter 15). Glycine, a less common inhibitory transmitter in the brain, also activates ionotropic receptors that directly open Cl^- channels. Glycine is the major transmitter released in the spinal cord by the interneurons that inhibit antagonist motorneurons.

Ionotropic Glutamate, GABA, and Glycine Receptors Are Transmembrane Proteins Encoded by Two Distinct Gene Families

The individual subunits that form the $GABA_A$ and glycine receptors are encoded by two distinct but closely related sets of genes. More surprisingly, these receptor subunits are structurally related to the nicotinic ACh receptor subunits, even though the latter select for cations and are therefore excitatory. Thus, as we saw above (Figure 13–4), the three types of receptor subunits are members of one large gene family.

Figure 13–10 (Opposite) NMDA receptor-dependent long-term potentiation of synaptic transmission at Schaffer collateral synapses.

A. Tetanic stimulation of the Schaffer collateral pathway for 1 second (**arrow**) induces LTP at the synapses between the presynaptic terminals of CA3 pyramidal neurons and the postsynaptic dendritic spines of CA1 pyramidal neurons. The graph shows the size of the synaptic response (extracellular field EPSP or **fEPSP**) as a percentage of the initial response prior to induction of LTP. At these synapses, LTP requires activation of the NMDA receptor-channels in the CA1 neurons; LTP is completely blocked when the tetanus is delivered in the presence of the NMDA receptor antagonist APV. (Adapted from Morgan and Teyler 2001.)

B. A model for the mechanism of long-term potentiation at Schaffer collateral synapses.

1. During normal, low-frequency synaptic transmission, glutamate (**Glu**) released from the terminals of CA3 Schaffer collateral axons binds to both NMDA and AMPA receptors in the postsynaptic CA1 neurons (specifically at the postsynaptic membrane of dendritic spines, the site of excitatory input). Sodium and potassium ions flow through the AMPA receptors but not through the NMDA receptor-channels, because their pores are blocked by Mg^{2+} at negative membrane potentials.

2. During a high-frequency tetanus, the large depolarization of the postsynaptic membrane (caused by the large amount of glutamate release resulting in strong activation of the AMPA receptors) relieves the Mg^{2+} blockade of the NMDA receptor-channels, allowing Ca^{2+}, Na^+, and K^+ to flow through these channels. The resulting increase of Ca^{2+} in the dendritic spine activates calcium-dependent protein kinases—calcium/calmodulin–dependent kinase (**CaMKII**) and protein kinase C (**PKC**)—leading to induction of LTP.

3. Second-messenger cascades activated during induction of LTP have two main effects on synaptic transmission. Phosphorylation through activation of protein kinases, including PKC, enhances current through the AMPA receptor-channels, in part by causing insertion of new receptors into the postsynaptic CA1 neuron. In addition, the postsynaptic cell releases (in ways that are still not understood) retrograde messengers that diffuse to the presynaptic terminal to enhance subsequent transmitter release. One such retrograde messenger may be nitric oxide (**NO**), produced by the enzyme NO synthase (shown in panel B-2).

Like the nicotinic ACh receptor-channels, the GABA_A and glycine receptor-channels are pentamers. The GABA_A receptors are usually composed of two α-, two β-, and one γ- or δ-subunit and are activated by the binding of two molecules of GABA in clefts formed between the two α- and β-subunits. The glycine receptors are composed of three α- and two β-subunits and require the binding of up to three molecules of ligand to open. The transmembrane topology of each GABA_A and glycine receptor subunit is similar to that of a nicotinic ACh receptor subunit, consisting of a large extracellular ligand-binding domain followed by four hydrophobic transmembrane α-helices (labeled M1, M2, M3, and M4), with the M2 helix forming the lining of the channel pore (Figure 13–4A). However, the amino acids flanking the M2 domain are strikingly different from those of the nicotinic ACh receptor. As discussed in Chapter 12, the pore of the ACh receptor contains rings of negatively charged acidic residues that help the channel select for cations over anions. In contrast, the GABA and glycine receptor-channels contain either neutral or positively charged basic residues at the homologous positions, which contribute to the selectivity of these channels for anions.

Most of the major classes of receptor subunits are encoded by multiple related genes. Thus, there are six types of GABA_A α-subunits (α1–α6), three β-subunits (β1–β3), three γ-subunits (γ1–γ3), and one δ-subunit. The genes for these different subtypes are often differentially expressed in different types of neurons, endowing their inhibitory synapses with distinct properties. The possible combinatorial arrangements of these subunits in a fully assembled pentameric receptor provides an enormous potential diversity of receptors.

The GABA_A and glycine receptors play important roles in disease and in the actions of drugs. GABA_A receptors are the target for several types of drugs that are clinically important and socially abused, including general anesthetics, benzodiazepines, barbiturates, and alcohol. General anesthetics, either gases or injectable compounds, induce loss of consciousness and are therefore widely used during surgery. Benzodiazepines are antianxiety agents and muscle relaxants that include diazepam (Valium), lorazepam (Ativan), and clonazepam (Klonopin). Zolpidem (Ambien) is a benzodiazepine compound that promotes sleep. The barbiturates comprise a distinct group of hypnotics that includes phenobarbital and secobarbital.

The different classes of compounds—GABA, general anesthetics, benzodiazepines, barbiturates, and alcohol—bind to different sites on the receptor but act similarly to increase the opening of the GABA receptor-channel. For example, whereas GABA binds to a cleft between the α- and β-subunits, benzodiazepines bind to a cleft between the α- and γ-subunits. In addition, the binding of any one of these classes of drug influences the binding of the others. For example, a benzodiazepine (or a barbiturate) binds more strongly to the receptor-channel when GABA also is bound, and this tight binding helps stabilize the channel in the open state. In this manner, the various compounds all enhance inhibitory synaptic transmission.

How do these different compounds, all acting on GABA_A receptors to promote channel opening, produce such diverse behavioral and psychological effects, for example, reducing anxiety versus promoting sleep? It turns out that many of these compounds bind selectively to specific subunit types, which can be expressed in different types of neurons in different regions of the brain. For example, zolpidem binds selectively to GABA_A receptors containing the α_1- subunit. In contrast, the anxiolytic effect of benzodiazepines requires binding to the α_2- and γ-subunits.

In addition to being important pharmacological targets, the GABA_A and glycine receptors are targets of disease and poisons. Missense mutations in the α-subunit of the glycine receptor underlie an inherited neurological disorder called *familial startle disease* (or *hyperekplexia*), characterized by abnormally high muscle tone and exaggerated responses to noise. These mutations decrease the opening of the glycine receptor and so reduce the normal levels of inhibitory transmission in the spinal cord. The poison strychnine, a plant alkaloid compound, causes convulsions by blocking the glycine receptor and decreasing inhibition. Nonsense mutations that result in truncations of GABA_A receptor α- and γ-subunits have been implicated in congenital forms of epilepsy.

Chloride Currents Through GABA_A and Glycine Receptor-Channels Normally Inhibit the Postsynaptic Cell

The function of GABA receptors is intimately linked to their biophysical properties. Eccles and his colleagues determined the ionic mechanism of the IPSP in spinal motor neurons by systematically changing the level of the resting membrane potential in a motor neuron while stimulating a presynaptic inhibitory interneuron (Figure 13–11).

When the motor neuron membrane is held at the normal resting potential (–65 mV), a small hyperpolarizing potential is generated when the presynaptic interneuron is stimulated. When the motor neuron membrane is held at –70 mV, no change in potential

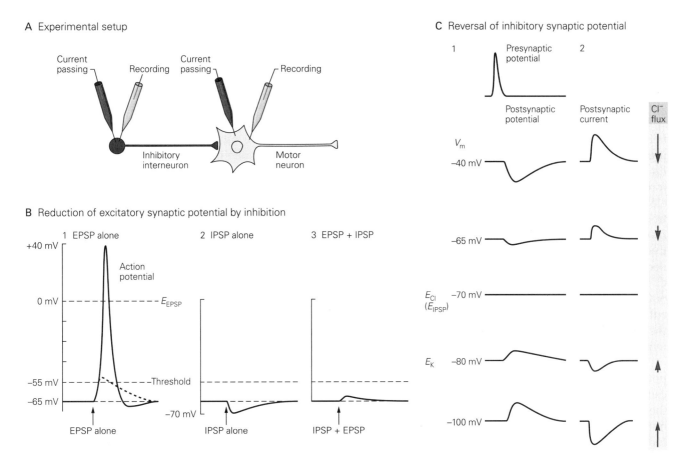

A Experimental setup

B Reduction of excitatory synaptic potential by inhibition

C Reversal of inhibitory synaptic potential

Figure 13–11 Inhibitory actions at chemical synapses result from the opening of ion channels selective for chloride.

A. In this hypothetical experiment, two electrodes are placed in the presynaptic interneuron and two in the postsynaptic motor neuron. The current-passing electrode in the presynaptic cell is used to produce an action potential; in the postsynaptic cell, it is used to alter the membrane potential systematically prior to the presynaptic input.

B. Inhibitory actions counteract excitatory actions. **1.** A large EPSP occurring alone depolarizes the membrane toward E_{EPSP} and exceeds the threshold for generating an action potential. **2.** An IPSP alone moves the membrane potential away from the threshold toward E_{Cl}, the equilibrium potential for Cl^- (−70 mV). **3.** When inhibitory and excitatory synaptic potentials occur together, the effectiveness of the EPSP is reduced and prevented from reaching the threshold for an action potential.

C. The IPSP and inhibitory synaptic current reverse at E_{Cl}. **1.** A presynaptic spike produces a hyperpolarizing IPSP at the resting membrane potential (−65 mV). The IPSP is larger when the membrane potential is set at −40 mV due to the increased inward driving force on Cl^-. When the membrane potential is set at −70 mV the IPSP is nullified. This reversal potential for the IPSP occurs at E_{Cl}. With further hyperpolarization of the membrane, the IPSP is inverted to a depolarizing postsynaptic potential (at −80 and −100 mV) because the membrane potential is negative to E_{Cl}. **2.** The reversal potential of the inhibitory postsynaptic current measured under voltage clamp. An inward (negative) current flows at membrane potentials negative to the reversal potential (corresponding to an efflux of Cl^-), and an outward (positive) current flows at membrane potentials positive to the reversal potential (corresponding to an influx of Cl^-). (**Up arrows** = efflux; **down arrows** = influx.)

is recorded when the interneuron is stimulated. But at potentials more negative than −70 mV, the motor neuron generates a *depolarizing* response following stimulation of the inhibitory interneuron. This reversal potential of −70 mV corresponds to the Cl^- equilibrium potential in spinal motor neurons (the extracellular concentration of Cl^- is much greater than the intracellular concentration). Thus, at −70 mV, the tendency of

Cl^- to diffuse into the cell down its chemical concentration gradient is balanced by the electrical force (the negative membrane potential) that opposes Cl^- influx. Replacement of extracellular Cl^- with an impermeant anion reduces the size of the IPSP and shifts the reversal potential to more positive values in accord with the predictions of the Nernst equation. Thus, the IPSP results from an increase in Cl^- conductance.

The currents through single GABA and glycine receptor-channels, the unitary currents, have been measured using the patch-clamp technique. Both transmitters activate Cl⁻ channels that open in an all-or-none manner, similar to the opening of ACh and glutamate-gated channels. The inhibitory effect of GABA and glycine on neuronal firing depends on two related mechanisms. First, in a typical neuron, the resting potential of −65 mV is slightly more positive than E_{Cl} (−70 mV). At this resting potential, the chemical force driving Cl⁻ into the cell is slightly greater than the electrical force opposing Cl⁻ influx—that is, the electrochemical driving force on Cl⁻ ($V_m − E_{Cl}$) is positive. As a result, the opening of Cl⁻ channels leads to a positive current, based on the relation $I_{Cl} = g_{Cl} (V_m − E_{Cl})$. Because the charge carrier is the negatively charged Cl⁻ ion, the positive current corresponds to an influx of Cl⁻ into the neuron, down its electrochemical gradient. This causes a net increase in the negative charge on the inside of the membrane—the membrane becomes hyperpolarized.

However, some central neurons have a resting potential that is approximately equal to E_{Cl}. In such cells, an increase in Cl⁻ conductance does not change the membrane potential—the cell does not become hyperpolarized—because the electrochemical driving force on Cl⁻ is nearly zero. However, the opening of Cl⁻ channels in such a cell still inhibits the cell from firing an action potential in response to a near-simultaneous EPSP. This is because the depolarization produced by an excitatory input depends on a weighted average of the batteries for all types of open channels—that is, the batteries for the excitatory and inhibitory synaptic conductances and the resting conductances—with the weighting factor equal to the total conductance for a particular type of channel (see Chapter 12, Postscript). Because the battery for Cl⁻ channels lies near the resting potential, opening these channels helps hold the membrane near its resting potential during the EPSP by increasing the weighting factor for the Cl⁻ battery.

The effect that the opening of Cl⁻ channels has on the magnitude of an EPSP can also be described in terms of Ohm's law. Accordingly, the amplitude of the depolarization during an EPSP, ΔV_{EPSP} is given by:

$$\Delta V_{EPSP} = I_{EPSP}/g_l$$

where I_{EPSP} is the excitatory synaptic current and g_l is the conductance from all other channels open in the membrane, including resting channels and transmitter-gated Cl⁻ channels. Because the opening of the Cl⁻ channels increases the resting conductance, ie, makes the neuron more leaky, the depolarization during the EPSP decreases. This consequence of synaptic inhibition is called the *short-circuiting* or *shunting* effect.

By counteracting synaptic excitation, synaptic inhibition can exert powerful control over action potential firing in neurons that are spontaneously active because of the presence of intrinsic pacemaker channels. This function, called the *sculpting* role of inhibition, shapes the pattern of firing in such cells (Figure 13–12). In fact, this sculpting role of inhibition likely happens in all neurons, leading to the temporal patterning of neuronal spiking and to the control of the synchronization of neural circuits.

The different biophysical properties of synaptic conductances can be understood as distinct mathematical operations carried out by the postsynaptic neuron. Thus, inhibitory inputs that hyperpolarize the cell perform a *subtraction* on the excitatory inputs, whereas the shunting effect of the conductance increase performs a *division*. Adding excitatory inputs (or removing nonshunting inhibitory inputs) results in *summation*. Finally, the combination of an excitatory input with the removal of an inhibitory shunt produces a *multiplication*. These arithmetic effects, however, are often mixed and can vary with time as the membrane potential of neurons constantly varies, leading to changes in the driving force on Cl⁻ through GABA$_A$ receptor-channels.

In some cells, such as those with metabotropic GABA$_B$ receptors, inhibition is caused by the opening of K⁺ channels. Because the K⁺ equilibrium potential of neurons ($E_K = −80$ mV) is always negative to the resting potential, opening K⁺ channels inhibits the cell even

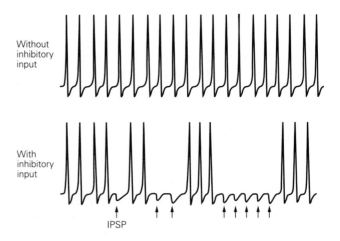

Without inhibitory input

With inhibitory input

IPSP

Figure 13–12 Inhibition can shape the firing pattern of a spontaneously active neuron. Without inhibitory input, the neuron fires continuously at a fixed interval. With inhibitory input (**arrows**), some action potentials are inhibited, resulting in a distinctive pattern of impulses.

more profoundly than opening Cl^- channels (assuming a similar-size synaptic conductance), generating a more "subtractive" inhibition. $GABA_B$ responses turn on more slowly and persist for a longer time compared with $GABA_A$ responses.

Paradoxically, under some conditions, the activation of $GABA_A$ receptors in brain cells can cause excitation. This is because the influx of Cl^- after intense periods of stimulation can be so great that the intracellular Cl^- concentration increases substantially. It may even double. As a result, the Cl^- equilibrium potential may become more positive than the resting potential. Under these conditions, the opening of Cl^- channels leads to Cl^- efflux and depolarization of the neuron. Such depolarizing Cl^- responses occur normally in many neurons in newborn animals, where the intracellular Cl^- concentration tends to be high even at rest. This is because the K^+-Cl^- cotransporter responsible for maintaining low intracellular Cl^- is expressed at low levels during early development (Chapter 9). Depolarizing Cl^- responses may also occur in the distal dendrites of more mature neurons and perhaps also at their axon initial segment. Such excitatory $GABA_A$ receptor actions in adults may contribute to epileptic discharges in which large, synchronized, and depolarizing GABA responses are observed.

Some Synaptic Actions in the Central Nervous System Depend on Other Types of Ionotropic Receptors

A minority of fast excitatory synaptic actions in the brain are mediated by the neurotransmitter serotonin (5-HT) acting at the $5-HT_3$ class of ionotropic receptor-channels. These pentameric receptors, which are made up of subunits with four transmembrane segments, are structurally similar to nicotinic ACh receptors. Like the ACh receptor-channels, $5-HT_3$ receptor-channels are permeable to monovalent cations and have a reversal potential near 0 mV.

Ionotropic receptors for adenosine triphosphate (ATP) serve an excitatory function at other selected synapses and constitute a third family of transmitter-gated ion channels. These so-called purinergic receptors (named for the purine ring in adenosine) occur on smooth muscle cells innervated by sympathetic neurons of the autonomic ganglia as well as on certain central and peripheral neurons. At these synapses, ATP activates an ion channel that is permeable to both monovalent cations and Ca^{2+}, with a reversal potential near 0 mV. Several genes coding for this family of ionotropic ATP receptors (termed the *P2X receptors*) have

been identified. The amino acid sequence and subunit structure of these ATP receptors are different from the other two ligand-gated channel families. An X-ray crystal structure of the P2X receptor reveals that it has an exceedingly simple organization in which three subunits, each containing only two transmembrane segments, surround a central pore (Figure 13–4C).

Excitatory and Inhibitory Synaptic Actions Are Integrated by Neurons Into a Single Output

Each neuron in the central nervous system is constantly bombarded by an array of synaptic inputs from many other neurons. A single motor neuron, for example, may be the target of as many as 10,000 different presynaptic terminals. Some are excitatory, others inhibitory; some are strong, others weak. Some inputs contact the motor cell on the tips of its apical dendrites, others on proximal dendrites, some on the dendritic shaft, others on the soma. The different inputs can reinforce or cancel one another. How does a given neuron integrate these signals into a coherent output?

As we saw earlier, the synaptic potentials produced by a single presynaptic neuron typically are not large enough to depolarize a postsynaptic cell to the threshold for an action potential. The EPSPs produced in a motor neuron by most stretch-sensitive afferent neurons are only 0.2 to 0.4 mV in amplitude. If the EPSPs generated in a single motor neuron were to sum linearly, at least 25 afferent neurons would have to fire together and release transmitter to depolarize the trigger zone by the 10 mV required to reach threshold. But at the same time the postsynaptic cell is receiving excitatory inputs, it may also be receiving inhibitory inputs that prevent the firing of action potentials by either a subtractive or shunting effect.

The net effect of the inputs at any individual excitatory or inhibitory synapse will therefore depend on several factors: the location, size, and shape of the synapse; the proximity and relative strength of other synergistic or antagonistic synapses; and the resting potential of the cell. And, in addition, all of this is exquisitely dependent on the timing of the excitatory and inhibitory input. Inputs are coordinated in the postsynaptic neuron by a process called *neuronal integration*. This cellular process reflects the task that confronts the nervous system as a whole. A cell at any given moment has two options: to fire or not to fire an action potential. Charles Sherrington described the brain's ability to choose between competing alternatives as the *integrative action of the nervous system*.

He regarded this decision making as the brain's most fundamental operation (see Chapter 56).

Synaptic Inputs Are Integrated at the Axon Initial Segment

In most neurons, the decision to initiate an action potential output is made at one site: the axon initial segment. Here, the cell membrane has a lower threshold for action potential generation than at the cell body or dendrites because it has a higher density of voltage-dependent Na^+ channels (Figure 13–13). With each increment of membrane depolarization, more Na^+ channels open, providing a higher density of inward current (per unit area of membrane) at the axon initial segment than elsewhere in the cell.

At the initial segment, the depolarization increment required to reach the threshold for an action potential (–55 mV) is only 10 mV from the resting level of –65 mV. In contrast, the membrane of the cell body must be depolarized by 30 mV before reaching its threshold (–35 mV). Therefore, synaptic excitation first discharges the region of membrane at the initial segment, also called the *trigger zone*. The action potential generated at this site then depolarizes the membrane of the cell body to threshold and at the same time is propagated along the axon .

Because neuronal integration involves the summation of synaptic potentials that spread to the trigger zone, it is critically affected by two passive membrane properties of the neuron (Chapter 9). First, the membrane time constant helps determine the time course of the synaptic potential in response to the EPSC, thereby controlling *temporal summation*, the process by which consecutive synaptic potentials are added together in the postsynaptic cell. Neurons with a large membrane time constant have a greater capacity for temporal summation than do neurons with a shorter time constant (Figure 13–14A). As a result, the longer the time constant, the greater is the likelihood that two consecutive inputs will summate to bring the cell membrane to its threshold for an action potential.

Second, the *length* constant of the cell determines the degree to which the EPSP decreases as it spreads

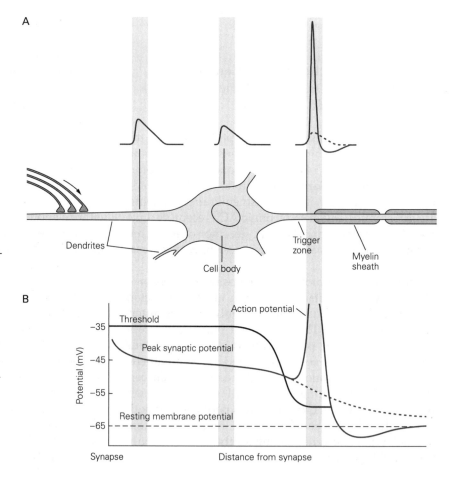

Figure 13–13 A synaptic potential arising in a dendrite can generate an action potential at the axon initial segment. (Adapted, with permission, from Eckert et al. 1988.)

A. An excitatory synaptic potential originating in the dendrites decreases with distance as it propagates passively to the soma. Nevertheless, an action potential can be initiated at the trigger zone (the axon initial segment) because the density of the Na^+ channels in this region is high and thus the threshold for an action potential is low.

B. Comparison of the threshold for initiation of the action potential at different sites in the neuron (corresponding to drawing A). An action potential is generated when the amplitude of the synaptic potential exceeds the threshold. The **dashed line** shows the decay of the synaptic potential if no action potential is generated at the axon initial segment.

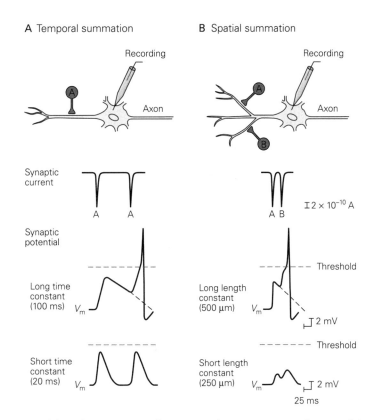

Figure 13–14 Central neurons are able to integrate a variety of synaptic inputs through temporal and spatial summation of synaptic potentials.

A. Temporal summation. The time constant of a postsynaptic cell (see Figure 9–10) affects the amplitude of the depolarization caused by consecutive EPSPs produced by a single presynaptic neuron (cell **A**). Here the synaptic current generated by the presynaptic neuron is nearly the same for both EPSPs. In a cell with a *long* time constant, the first EPSP does not fully decay by the time the second EPSP is triggered. In that instance, the depolarizing effects of both potentials are additive, bringing the membrane potential above the threshold and triggering an action potential. In a cell with a *short* time constant, the first EPSP decays to the resting potential before the second EPSP is triggered, and in that instance, the second EPSP alone does not cause enough depolarization to trigger an action potential.

B. Spatial summation. The length constant of a postsynaptic cell (see Figure 9–11B) affects the amplitudes of two

excitatory postsynaptic potentials produced by two presynaptic neurons (cells **A** and **B**). For illustrative purposes, both synapses are the same distance (500 μm) from the postsynaptic cell's trigger zone, and the current produced by each synaptic contact is the same. If the distance between the site of synaptic input and the trigger zone in the postsynaptic cell is only one length constant (that is, the postsynaptic cell has a long length constant of 500 μm), the synaptic potentials produced by each of the two presynaptic neurons will decrease to 37% of their original amplitude by the time they reach the trigger zone. Summation of the two potentials results in enough depolarization to exceed threshold, triggering an action potential. If the distance between the synapse and the trigger zone is equal to two length constants (ie, the postsynaptic cell has a short length constant of 250 μm), each synaptic potential will be less than 15% of its initial amplitude, and summation will not be sufficient to trigger an action potential.

passively from a synapse along the length of the dendrite to the cell body and axon initial segment (the trigger zone). In cells with a longer length constant, signals spread to the trigger zone with minimal decrement; in cells with a short length constant, the signals decay rapidly with distance. Because the depolarization produced by one synapse is almost never sufficient to trigger an action potential at the trigger zone, the inputs from many presynaptic neurons acting at different sites on the postsynaptic neuron must be added together. This process is called *spatial summation.* Neurons with

a large length constant are more likely to be brought to threshold by inputs arising from different sites than are neurons with a short length constant (Figure 13–14B).

Subclasses of GABAergic Neurons Target Distinct Regions of Their Postsynaptic Target Neurons to Produce Inhibitory Actions With Different Functions

In contrast to the relatively few types of glutamatergic pyramidal neurons, the mammalian central nervous system has a large variety of GABAergic inhibitory interneurons that differ in developmental origin, molecular composition, morphology, and connectivity (Figure 13–15). Up to 20 different subtypes of GABAergic neurons have been identified in one subregion of the hippocampus alone. The different types of GABAergic interneurons form extensive synaptic connections

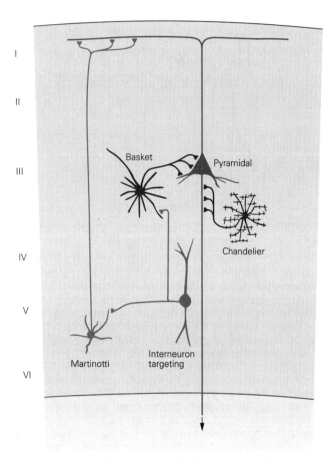

Figure 13–15 Different GABAergic inhibitory neurons target different regions of a postsynaptic cell. A diverse array of interneurons can be distinguished by their morphology, expression of different molecular markers, and their preferred site of targeting of postsynaptic neurons. *Basket cells* send their axons to form synapses on the cell body and proximal dendrites of postsynaptic neurons. The dendrites of the basket cells are shown as short lines radiating from the soma. *Axo-axonic cells*, also called *chandelier cells*, send their axons to form clusters of synapses along the axon initial segment of their targets. Both basket cells and chandelier cells express the calcium-binding protein parvalbumin. *Dendrite-targeting cells*, also called *Martinotti cells*, send their axons to form synapses on the distal dendrites of pyramidal cells. These cells also release the neuropeptide somatostatin. Other classes of GABAergic neurons selectively form synapses onto other inhibitory interneurons. These interneuron-targeting inhibitory neurons often release neuropeptide Y in addition to GABA.

with their neighboring excitatory and inhibitory neurons. Thus, even though only 20% of all neurons are inhibitory, the overall levels of inhibition and excitation tend to be nearly balanced in most brain regions. This results in the tuning of neural circuits to respond to only the most salient excitatory information. While the diversity of interneurons is challenging to understand, it is clear that different types of interneurons selectively target different regions of their postsynaptic neurons.

This selective targeting is important because the location of inhibitory inputs in relation to excitatory synapses is critical in determining the effectiveness of inhibition (Figure 13–16). Inhibition of action potential output in response to excitatory input is more effective when inhibition is initiated at the cell body or near the axon trigger zone. The depolarization produced by an excitatory current from a dendrite must pass along the cell body membrane as it moves toward the axon. Inhibitory actions at the cell body or axon initial segment open Cl⁻ channels, thus increasing Cl⁻ conductance and reducing (by shunting) much of the depolarization produced by the spreading excitatory current. In addition, the size of the hyperpolarization at the cell body in response to an IPSP is largest when the inhibitory input targets the cell body, not a dendrite, owing to the attenuation of the dendritic IPSP by the cable properties of the dendrite.

Two classes of inhibitory neurons, basket cells and chandelier cells, exert strong control over neuronal output by specifically targeting the soma and axon initial segment, respectively (Figure 13–15). Basket cells often express the calcium-binding protein parvalbumin and are the most common type of inhibitory neuron in the brain. Chandelier cells, which also express parvalbumin, have axonal arbors with a branching pattern and clustering of synaptic terminals that resemble the numerous candles of a chandelier. Under some circumstances, the chandelier cells may paradoxically enhance neuronal firing because the Cl⁻ reversal potential in some axons can be positive to the threshold for action potential firing.

A third class of interneurons, the Martinotti cells, specifically targets distal dendrites and spines. These common interneurons release the neuropeptide somatostatin in addition to GABA. Inhibitory actions at a remote part of a dendrite act to decrease the local depolarization produced by a nearby excitatory input, with less of an effect on EPSPs generated on other dendritic branches. Somatostatin-positive interneurons activate slowly in response to an excitatory input and generate IPSPs that increase in size with repetitive activation (synaptic facilitation). In contrast, parvalbumin-expressing interneurons fire rapidly and generate

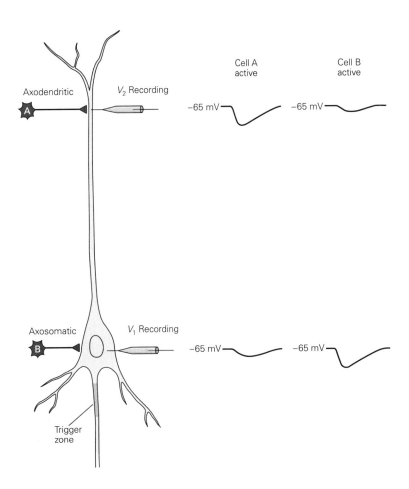

Figure 13–16 The effect of an inhibitory current in the postsynaptic neuron depends on the distance the current travels from the synapse to the cell's trigger zone. In this hypothetical experiment, the inputs from inhibitory axosomatic and axodendritic synapses are compared by recording from both the cell body (V_1) and a dendrite (V_2) of the postsynaptic cell. Stimulating cell B (the axosomatic synapse) produces a large IPSP in the cell body. The IPSP decays as it propagates up the dendrite, producing only a small hyperpolarization at the site of dendritic recording. Stimulating cell A activates an axodendritic synapse, producing a large local IPSP in the dendrite but only a small IPSP in the cell body, because the synaptic potential decays as it propagates down the dendrite. Thus, the axosomatic IPSP is more effective than the axodendritic IPSP in inhibiting action potential firing in the postsynaptic cell, whereas the axodendritic IPSP is more effective in preventing local dendritic depolarization.

IPSPs that decrease in size with repetitive activation (synaptic depression). These properties allow the somatostatin and parvalbumin interneurons to control the spread through neural circuits of later and earlier phases of neural signals, respectively.

A fourth major type of inhibitory interneuron expresses the neuropeptide vasoactive intestinal peptide (VIP). These interneurons selectively target other interneurons and thus serve to decrease the level of inhibition in a neural circuit, thereby enhancing overall excitation, a process termed disinhibition.

Dendrites Are Electrically Excitable Structures That Can Amplify Synaptic Input

Propagation of signals in dendrites was originally thought to be purely passive. However, intracellular recordings from the cell body of neurons in the 1950s and from dendrites beginning in the 1970s demonstrated that dendrites could produce action potentials. Indeed, we now know that the dendrites of most neurons contain voltage-gated Na+, K+, and Ca2+ channels in addition to ligand-gated channels and resting

channels. In fact, the rich diversity of dendritic conductances suggests that central neurons rely on a sophisticated repertory of electrophysiological properties to integrate synaptic inputs.

One function of the voltage-gated Na+ and Ca2+ channels in dendrites is to amplify the EPSP. In some neurons, there is a sufficient density of voltage-gated channels in the dendritic membrane to serve as a local trigger zone. This can produce nonlinear electrical responses that enhance the depolarization generated by excitatory inputs that arrive at remote parts of the dendrite. When a cell has several dendritic trigger zones, each one sums the local excitation and inhibition produced by nearby synaptic inputs; if the net input is above threshold, a dendritic action potential may be generated, usually by voltage-gated Na+ or Ca2+ channels (Figure 13–17A). Nevertheless, the number of voltage-gated Na+ or Ca2+ channels in the dendrites is usually not sufficient to support all-or-none regenerative propagation of an action potential to the cell body. Rather, action potentials generated in the dendrites are usually local events that spread electrotonically to the cell body and axon initial segment, producing a

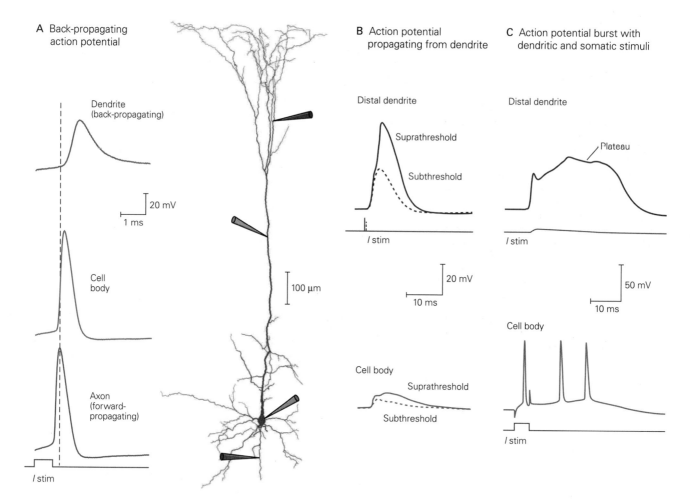

A Back-propagating action potential

Dendrite (back-propagating)

20 mV

1 ms

Cell body

Axon (forward-propagating)

I stim

100 µm

B Action potential propagating from dendrite

Distal dendrite

Suprathreshold

Subthreshold

I stim

20 mV

10 ms

Cell body

Suprathreshold

Subthreshold

C Action potential burst with dendritic and somatic stimuli

Distal dendrite

Plateau

I stim

50 mV

10 ms

Cell body

I stim

Figure 13–17 Active properties of dendrites can amplify synaptic inputs and support propagation of electrical signals to and from the axon initial segment. The figure illustrates an experiment in which several electrodes are used to record membrane voltage and pass stimulating current in the axon, cell body, and at several locations along the dendritic tree. The recording electrodes and corresponding voltage traces are matched by color. Stimulating current pulses are also indicated (*I* stim). (Panels A and B adapted from Stuart et al. 2016.)

A. An action potential initiated in the axon initial segment can propagate to the dendrites. Such backpropagation depends on activation of voltage-gated Na⁺ channels in the dendrites. Unlike the nondecrementing action potential that is continually regenerated along an axon, the amplitude of a back-propagating action potential decreases as it travels along a dendrite due to its relatively low density of voltage-gated Na⁺ channels.

B. A strong depolarizing EPSP at a dendrite can generate a dendritic action potential that travels to the cell body. Such action potentials are often generated by dendritic voltage-gated Ca^{2+} channels and have a high threshold. They propagate relatively slowly and attenuate with distance, often failing to reach the cell body. The **solid blue line** shows a suprathreshold response generated in the dendrite in response to a large depolarizing current pulse, and the **dotted blue line** shows a subthreshold response to a weaker current stimulus. The **solid and dotted orange lines** show the corresponding voltage responses recorded in the cell body.

C. Near simultaneous injection of a subthreshold stimulating current resembling a weak EPSC into the dendrite and a strong brief suprathreshold stimulating current into the cell body (which by itself evokes a single somatic action potential) triggers a long-lasting plateau potential in the dendrite and the firing of a burst of action potentials in the cell body. (Adapted, with permission from Larkum et al. 1999. Copyright © 1999 Springer Nature.)

subthreshold somatic depolarization that is integrated with other input signals in the cell.

Dendritic voltage-gated channels also permit action potentials generated at the axon initial segment to propagate backward into the dendritic tree (Figure 13–17B). These *backpropagating* action potentials are largely generated by dendritic voltage-gated Na⁺ channels. Although the precise role of these action potentials is unclear, they may provide a temporally precise mechanism for enhancing current through NMDA receptor-channels by providing the depolarization necessary to remove the Mg^{2+} block, thereby contributing to the induction of synaptic plasticity (Figure 13–10).

NMDA receptors are able to mediate another type of nonlinear integration in dendrites as a result of their voltage dependence. Moderate synaptic stimuli are able to activate a sufficient number of AMPA receptors to produce an intermediate level of depolarization that is able to lead to expulsion of Mg^{2+} from a fraction of NMDA receptors. As these receptors begin to conduct cations into the postsynaptic dendrite, they produce a further depolarization that leads to even greater unblocking of Mg^{2+}, increasing further the size of the NMDA receptor EPSC, resulting in even greater depolarization. In some instances, this leads to a local regenerative depolarization, referred to as an NMDA spike. Such NMDA spikes are purely local events—they cannot propagate actively in the absence of synaptic stimulation because they require glutamate release. NMDA spikes have been implicated in different forms of synaptic plasticity and in the enhancement of dendritic integration of synaptic inputs.

Under what conditions do active conductances influence dendritic integration? There is now evidence that dendrites may switch between passive and active integration depending on the precise timing and strength of synaptic inputs. One interesting example of such a switch is the way some cortical neurons respond to inputs arriving at their distal and proximal dendrites. In many neurons, inputs from relatively nearby neurons arrive at more proximal regions of the dendrites, closer to the cell body. Inputs from more distant brain areas arrive at the distal tips of dendrites. Although excitatory synaptic inputs to the distal dendrites usually produce only a very small depolarizing response at the soma, due to electronic decay along the dendritic cable, these inputs can significantly enhance spike firing when paired with excitatory inputs to more proximal regions of the dendrites. Thus, a single strong EPSP at a proximal site (or a single brief somatic current pulse) normally produces a single action potential at the axon initial segment, which can then backpropagate into the dendrites. However, when a distal stimulus is paired with a proximal stimulus, the backpropagating spike summates with the distal EPSP to trigger a long-lasting type of dendritic spike called a plateau potential, which depends on activation of voltage-gated Ca^{2+} channels and NMDA receptors. When the plateau potential arrives at the cell body, it can trigger a brief burst of three or more spikes at rates as high as 100 Hz (Figure 13–17C). These spike bursts are thought to provide a very potent means of inducing long-term synaptic plasticity and releasing transmitter as the burst propagates to the presynaptic terminal.

A more localized form of synaptic integration occurs in dendritic spines. Even though some excitatory inputs occur on dendritic shafts, close to 95% of all excitatory inputs in the brain terminate on spines, surprisingly avoiding dendritic shafts (see Figure 13–2). Although the function of spines is not completely understood, their thin necks provide a barrier to diffusion of various signaling molecules from the spine head to the dendritic shaft. As a result, a relatively small Ca^{2+} current through the NMDA receptors can lead to a relatively large increase in $[Ca^{2+}]$ that is localized to the head of the individual spine that is synaptically activated (Figure 13–18A). Moreover, because action potentials can backpropagate from the cell body to the dendrites, spines also serve as sites at which information about presynaptic and postsynaptic activity is integrated.

Indeed, when a backpropagating action potential is paired with presynaptic stimulation, the spine Ca^{2+} signal is greater than the linear sum of the individual Ca^{2+} signals from synaptic stimulation alone or action potential stimulation alone. This "supralinearity" is specific to the activated spine and occurs because depolarization during the action potential causes Mg^{2+} to be expelled from the NMDA receptor-channel, allowing it to conduct Ca^{2+} into the spine. The resultant Ca^{2+} accumulation thus provides, at an individual synapse, a biochemical detector of the near simultaneity of the input (EPSP) and output (backpropagating action potential), which is thought to be a key requirement of memory storage (Chapter 54).

Because the thin spine neck restricts, at least partly, the rise in Ca^{2+} and, thus, long-term plasticity to the spine that receives the synaptic input, spines also ensure that activity-dependent changes in synaptic function, and thus memory storage, are restricted to the synapses that are activated. The ability of spines to implement such synapse-specific local learning rules may be of fundamental importance for the ability of neural networks to store meaningful information (Chapter 54). Finally, in some spines, local synaptic potentials are filtered as they propagate through the spine neck and enter the dendrite, such that the size of the EPSP at the cell body is reduced. The regulation of this electrical filtering could provide another means of controlling the efficacy with which a given synaptic conductance is able to excite the cell body.

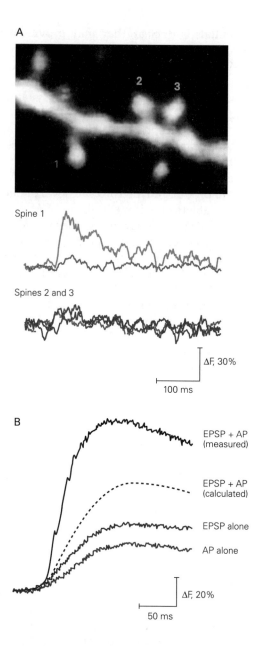

Figure 13–18 Dendritic spines compartmentalize calcium influx through NMDA receptors.

A. This fluorescence image of a hippocampal CA1 pyramidal neuron filled with a calcium-sensitive dye shows the outline of a dendritic shaft with several spines. When the dye binds Ca^{2+}, its fluorescence intensity increases. The traces plot fluorescence intensity over the time following the extracellular stimulation of the presynaptic axon. Spine 1 shows a large, rapid increase in fluorescence (ΔF) in response to synaptic stimulation (**red trace**), reflecting Ca^{2+} influx through the NMDA receptors. In contrast, there is little change in the fluorescence intensity in the neighboring dendrite shaft (**gray trace**), showing that Ca^{2+} accumulation is restricted to the head of the spine. Spines 2 and 3 show little increase in fluorescence in response to synaptic stimulation because their presynaptic axons were not activated. (Reproduced, with permission, from Lang et al. 2004. Copyright © 2004 National Academy of Sciences.)

B. Calcium accumulation is greatest in spines when synaptic stimulation is paired with postsynaptic action potentials. The Ca^{2+} signal generated when an EPSP and a backpropagating action potential are evoked at the same time is greater than the expected sum of the individual Ca^{2+} signals when either an EPSP or a backpropagating action potential (**AP**) alone is evoked. (Adapted, with permission, from Yuste and Denk 1995.)

Highlights

1. A typical central neuron integrates a large number of excitatory and inhibitory synaptic inputs. The amino acid transmitter glutamate is responsible for most excitatory synaptic actions in the central nervous system, with the inhibitory amino acids GABA and glycine mediating inhibitory synaptic actions.

2. Glutamate activates families of ionotropic and metabotropic receptors. The three major classes of ionotropic glutamate receptors—AMPA, NMDA, and kainate—are named for the chemical agonists that activate them.

3. The ionotropic glutamate receptors are tetramers composed of subunits encoded by homologous genes. Each subunit has a large extracellular amino terminus, with three membrane-spanning segments and a large cytoplasmic tail. A pore-forming loop dips into and out of the membrane between the first and second transmembrane segments.

4. Binding of glutamate to all three ionotropic receptors opens a nonselective cation channel equally permeable to Na^+ and K^+. The NMDA receptor-channel also has a high permeability to Ca^{2+}.

5. The NMDA receptor acts as a coincidence detector. It is normally blocked by extracellular Mg^{2+}

lodged in its pore; it only conducts when gluta-mate is released *and* the postsynaptic membrane is sufficiently depolarized to expel the Mg^{2+} ion by electrostatic repulsion.

6. Calcium influx through the NMDA receptor dur-ing strong synaptic activation can trigger intra-cellular signaling cascades, leading to long-term synaptic plasticity, which can potentiate synap-tic transmission for a period of hours to days, providing a potential mechanism for memory storage.

7. Inhibitory synaptic actions in the brain are medi-ated by the binding of GABA to both ionotropic ($GABA_A$) and metabotropic ($GABA_B$) receptors. The $GABA_A$ receptors are pentamers, whose subunits are homologous to those of the nicotinic ACh receptors. Glycine ionotropic receptors are structurally similar to $GABA_A$ receptors and are largely confined to inhibitory synapses in the spi-nal cord.

8. Binding of GABA or glycine to its receptor acti-vates a Cl^- selective channel. In most cells, the Cl^- equilibrium potential is slightly negative to the resting potential. As a result, inhibitory synaptic actions hyperpolarize the cell membrane away from threshold for firing an action potential.

9. The decision as to whether a neuron fires an action potential depends on spatial and temporal summation of the various excitatory and inhibi-tory inputs and is determined by the size of the resulting depolarization at the axon initial seg-ment, the region of the neuron with the lowest threshold.

10. Dendrites also have voltage-gated channels, ena-bling them to fire local action potentials in some circumstances. This can amplify the size of the local EPSP to produce a larger depolarization at the cell body.

Rafael Yuste
Steven A. Siegelbaum

Selected Reading

Arundine M, Tymianski M. 2004. Molecular mechanisms of glutamate-dependent neurodegeneration in ischemia and traumatic brain injury. Cell Mol Life Sci 61:657–668.

Basu J, Siegelbaum SA. 2015. The corticohippocampal circuit, synaptic plasticity, and memory. Cold Spring Harb Per-spect Biol 7:11.

Colquhoun D, Sakmann B. 1998. From muscle endplate to brain synapses: a short history of synapses and agonist-activated ion channels. Neuron 20:381–387.

Granger AJ, Gray JA, Lu W, Nicoll RA. 2011. Genetic analy-sis of neuronal ionotropic glutamate receptor subunits. J Physiol 589:4095–4101.

Herring BE, Nicoll RA. 2016. Long-term potentiation: from CaMKII to AMPA receptor trafficking. Annu Rev Physiol 78:351–365.

Karnani M, Agetsuma M, Yuste R. 2014. A blanket of inhibi-tion: functional inferences from dense inhibitory connec-tivity. Curr Opin Neurobiol 26:96–102.

Martenson JS, Tomita S. 2015. Synaptic localization of neuro-transmitter receptors: comparing mechanisms for AMPA and $GABA_A$ receptors. Curr Opin Pharmacol 20:102–108.

Mayer ML. 2016. Structural biology of glutamate receptor ion channel complexes. Curr Opin Struct Biol 41:119–127.

Olsen RW, Sieghart W. 2009. $GABA_A$ receptors: subtypes pro-vide diversity of function and pharmacology. Neurophar-macology 56:141–148.

Peters A, Palay SL, Webster HD. 1991. *The Fine Structure of the Nervous System.* New York: Oxford Univ. Press.

Sheng M, Hoogenraad CC. 2007. The postsynaptic architec-ture of excitatory synapses: a more quantitative view. Ann Rev Biochem 76:823–847.

Stuart GJ, Spruston N. 2015. Dendritic integration: 60 years of progress. Nat Neurosci 18:1713–1721.

Valbuena S, Lerma J. 2016. Non-canonical signaling, the hid-den life of ligand-gated ion channels. Neuron 92:316–329.

References

Araya R, Vogels T, Yuste R. 2014. Activity-dependent den-dritic spine neck changes are correlated with synaptic strength. Proc Natl Acad Sci U S A 111:E2895–E2904.

Armstrong N, Sun Y, Chen GQ, Gouaux E. 1998. Structure of a glutamate-receptor ligand-binding core in complex with kainate. Nature 395:913–917.

Bormann J, Hamill O, Sakmann B. 1987. Mechanism of anion permeation through channels gated by glycine and γ-aminobutyric acid in mouse cultured spinal neurones. J Physiol 385:243–286.

Cash S, Yuste R. 1999. Linear summation of excitatory inputs by CA1 pyramidal neurons. Neuron 22:383–394.

Coombs JS, Eccles JC, Fatt P. 1955. The specific ionic conduct-ances and the ionic movements across the motoneuronal membrane that produce the inhibitory post-synaptic potential. J Physiol 130:326–373.

Eccles JC. 1964. *The Physiology of Synapses.* New York: Academic.

Eckert R, Randall D, Augustine G. 1988. Propagation and transmission of signals. In: *Animal Physiology: Mechanisms and Adaptations,* 3rd ed., pp. 134–176. New York: Freeman.

Finkel AS, Redman SJ. 1983. The synaptic current evoked in cat spinal motoneurones by impulses in single group Ia axons. J Physiol 342:615–632.

Gray EG. 1963. Electron microscopy of presynaptic organelles of the spinal cord. J Anat 97:101–106.

Grenningloh G, Rienitz A, Schmitt B, et al. 1987. The strychnine-binding subunit of the glycine receptor shows homology with nicotinic acetylcholine receptors. Nature 328:215–220.

Hamill OP, Bormann J, Sakmann B. 1983. Activation of multiple-conductance state chloride channels in spinal neurones by glycine and GABA. Nature 305:805–808.

Hestrin S, Nicoll RA, Perkel DJ, Sah P. 1990. Analysis of excitatory synaptic action in pyramidal cells using whole-cell recording from rat hippocampal slices. J Physiol 422:203–225.

Heuser JE, Reese TS. 1977. Structure of the synapse. In: ER Kandel (ed), *Handbook of Physiology: A Critical, Comprehensive Presentation of Physiological Knowledge and Concepts*, Sect. 1 *The Nervous System.* Vol. 1, *Cellular Biology of Neurons*, Part 1, pp. 261–294. Bethesda, MD: American Physiological Society.

Hollmann M, O'Shea-Greenfield A, Rogers SW, Heinemann S. 1989. Cloning by functional expression of a member of the glutamate receptor family. Nature 342:643–648.

Jia H, Rochefort NL, Chen X, Konnerth A. 2010. Dendritic organization of sensory input to cortical neurons in vivo. Nature 464:1307–1312.

Kayser MS, Dalmau J. 2016. Anti-NMDA receptor encephalitis, autoimmunity, and psychosis. Schizophr Res 176:36–40.

Lang C, Barco A, Zablow L, Kandel ER, Siegelbaum SA, Zakharenko SS. 2004. Transient expansion of synaptically connected dendritic spines upon induction of hippocampal long-term potentiation. Proc Natl Acad Sci U S A 101:16665–16670.

Larkum ME, Zhu JJ, Sakmann B. 1999. A new cellular mechanism for coupling inputs arriving at different cortical layers. Nature 398:338–341.

Llinas R. 1988. The intrinsic electrophysiological properties of mammalian neurons: insights into central nervous system function. Science 242:1654–1664.

Llinas R, Sugimori M. 1980. Electrophysiological properties of in vitro Purkinje cell dendrites in mammalian cerebellar slices. J Physiol 305:197–213.

Markram H, Lubke J, Frotscher M, Sakmann B. 1997. Regulation of synaptic efficacy by coincidence of postsynaptic APs and EPSPs. Science 275:213–215.

Morgan SL, Teyler TJ. 2001. Electrical stimuli patterned after the theta-rhythm induce multiple forms of LTP. J Neurophysiol 86:1289–1296.

Palay SL. 1958. The morphology of synapses in the central nervous system. Exp Cell Res Suppl 5:275–293.

Pfeffer CK, Xue M, He M, Huang ZJ, Scanziani M. 2013. Inhibition of inhibition in visual cortex: the logic of connections between molecularly distinct interneurons. Nat Neurosci 16:1068–1076.

Pritchett DB, Sontheimer H, Shivers BD, et al. 1989. Importance of a novel GABA$_A$ receptor subunit for benzodiazepine pharmacology. Nature 338:582–585.

Redman S. 1979. Junctional mechanisms at group Ia synapses. Prog Neurobiol 12:33–83.

Sakmann B. 1992. Elementary steps in synaptic transmission revealed by currents through single ion channels. Neuron 8:613–629.

Schiller J, Schiller Y. 2001. NMDA receptor-mediated dendritic spikes and coincident signal amplification. Curr Opin Neurobiol 11:343–348.

Sheng M, Hoogenraad C. 2007. The postsynaptic architecture of excitatory synapses: a more quantitative view. Ann Rev Biochem 76:823–847.

Sherrington CS. 1897. The central nervous system. In: M Foster (ed). *A Text Book of Physiology*, 7th ed. London: Macmillan.

Sobolevsky AI, Rosconi MP, Gouaux E. 2009. X-ray structure, symmetry and mechanism of an AMPA-subtype glutamate receptor. Nature 462:745–756.

Sommer B, Köhler M, Sprengel R, Seeburg PH. 1991. RNA editing in brain controls a determinant of ion flow in glutamate-gated channels. Cell 67:11–19.

Stuart G, Spruston N, Häuser M (eds). 2016. *Dendrites*, 3rd ed. Oxford, England, and New York: Oxford Univ. Press.

Yuste R. 2010. *Dendritic Spines*. Cambridge, MA and London, England: MIT Press.

Yuste R, Denk W. 1995. Dendritic spines as basic functional units of neuronal integration. Nature 375:682–684.

14

Modulation of Synaptic Transmission and Neuronal Excitability: Second Messengers

T HE BINDING OF NEUROTRANSMITTER to postsynaptic receptors produces a postsynaptic potential either directly, by opening ion channels, or indirectly, by altering ion channel activity through changes in the postsynaptic cell's biochemical state. As we saw in Chapters 11 to 13, the type of postsynaptic action depends on the type of receptor. Activation of an *ionotropic receptor* directly opens an ion channel that is part of the receptor macromolecule itself. In contrast, activation of *metabotropic receptors* regulates the opening of ion channels indirectly through biochemical signaling pathways; the metabotropic receptor and the ion channels regulated by the receptor are distinct macromolecules (Figure 14–1).

Whereas the action of ionotropic receptors is fast and brief, metabotropic receptors produce effects that begin slowly and persist for long periods, ranging from hundreds of milliseconds to many minutes. The two types of receptors also differ in their functions. Ionotropic receptors underlie fast synaptic signaling that is the basis of all behaviors, from simple reflexes to complex cognitive processes. Metabotropic receptors

A Direct gating

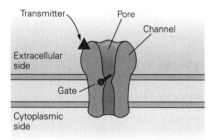

B Indirect gating

1 G protein–coupled receptor

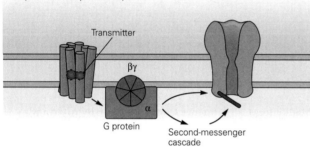

2 Receptor tyrosine kinase

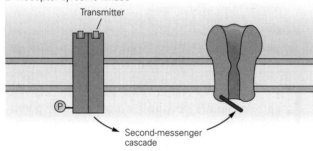

Figure 14–1 Neurotransmitter actions can be divided into two groups according to the way receptor and effector functions are coupled.

A. Direct transmitter actions are produced by the binding of transmitter to *ionotropic receptors*, ligand-gated channels in which the receptor and ion channel are domains within a single macromolecule. The binding of transmitter to the receptor on the extracellular aspect of the receptor-channel protein directly opens the ion channel embedded in the cell membrane.

B. Indirect transmitter actions are caused by binding of transmitter to *metabotropic receptors*, macromolecules that are separate from the ion channels they regulate. There are two families of these receptors. **1.** G protein–coupled receptors activate guanosine triphosphate (GTP)-binding proteins that engage a second-messenger cascade or act directly on ion channels. **2.** Receptor tyrosine kinases initiate a cascade of protein phosphorylation reactions, beginning with autophosphorylation (**P**) of the kinase itself on tyrosine residues.

modulate behaviors; they modify reflex strength, activate motor patterns, focus attention, set emotional states, and contribute to long-lasting changes in neural circuits that underlie learning and memory. Metabotropic receptors are responsible for many of the actions of transmitters, hormones, and growth factors. The actions of these neuromodulators can produce remarkable and dramatic changes in neuronal excitability and synaptic strength and, in so doing, can profoundly alter the state of activity in an entire circuit important for behavior.

Ionotropic receptors change the membrane potential quickly. As we have seen, this change is local at first but is propagated as an action potential along the axon if the change in membrane potential is suprathreshold. Activation of metabotropic receptors also begins as a local action that can spread to a wider region of the cell. The binding of a neurotransmitter with a metabotropic receptor activates proteins that in turn activate effector enzymes. The effector enzymes then often produce second-messenger molecules that can diffuse within a cell to activate still other enzymes that catalyze modifications of a variety of target proteins, greatly changing their activities.

There are two major families of metabotropic receptors: G protein–coupled receptors and receptor tyrosine kinases. We first describe the G protein–coupled receptor family and later discuss the receptor tyrosine kinase family.

The G protein–coupled receptors are coupled to an effector by a trimeric guanine nucleotide-binding protein, or G protein (Figure 14–1B). This receptor family comprises α- and β-adrenergic receptors for norepinephrine, muscarinic acetylcholine (ACh) receptors, γ-aminobutyric acid B (GABA$_B$) receptors, certain glutamate and serotonin receptors, all receptors for dopamine, receptors for neuropeptides, odorant receptors, rhodopsin (the protein that reacts to light, initiating visual signals; see Chapter 22), and many others. Many of these receptors are thought to be involved in neurological and psychiatric diseases and are key targets for the actions of important classes of therapeutic drugs.

G protein–coupled receptors activate a variety of effectors. The typical effector is an enzyme that produces a diffusible second messenger. These second messengers in turn trigger a biochemical cascade, either by activating specific protein kinases that phosphorylate the hydroxyl group of specific serine or threonine residues in various proteins or by mobilizing Ca^{2+} ions from intracellular stores, thereby initiating reactions that change the cell's biochemical state. In some instances, the G protein or the second messenger acts directly on an ion channel.

The Cyclic AMP Pathway Is the Best Understood Second-Messenger Signaling Cascade Initiated by G Protein–Coupled Receptors

The adenosine 3′,5′-cyclic monophosphate (cyclic AMP or cAMP) pathway is a prototypic example of a G protein–coupled second-messenger cascade. It was the first second-messenger pathway to be discovered, and our conception of other second-messenger pathways is based on it.

The binding of transmitter to receptors linked to the cAMP cascade first activates a specific G protein, G_s (named for its action to *stimulate* cAMP synthesis). In its resting state, G_s, like all G proteins, is a trimeric protein consisting of an α-, β-, and γ-subunit. The α-subunit is only loosely associated with the membrane and is usually the agent that couples the receptor to its primary effector enzyme. The β- and γ-subunits form a strongly bound complex that is more tightly associated with the membrane. As described later in this chapter, the βγ complex of G proteins can regulate the activity of certain ion channels directly.

In the resting state, the α-subunit binds a molecule of guanosine diphosphate (GDP). Upon the binding of ligand, a G protein–coupled receptor undergoes a conformational change that enables it to bind to the α-subunit, thereby promoting the exchange of GDP with a molecule of guanosine triphosphate (GTP). This leads to a conformational change that causes the α-subunit to dissociate from the βγ complex, thereby activating the α-subunit.

The particular class of α-subunit that is coupled to the cAMP cascade is termed $α_s$, which stimulates the integral membrane protein adenylyl cyclase to catalyze the conversion of adenosine triphosphate (ATP) to cAMP. When associated with the cyclase, $α_s$ also acts as a GTPase, hydrolyzing its bound GTP to GDP. When GTP is hydrolyzed, $α_s$ becomes inactive. It dissociates from adenylyl cyclase and reassociates with the βγ complex, thereby stopping the synthesis of cAMP (Figure 14–2A). A G_s protein typically remains active for a few seconds before its bound GTP is hydrolyzed.

Once a G protein–coupled receptor binds a ligand, it can interact sequentially with more than one G protein macromolecule. As a result, the binding of relatively few molecules of transmitter to a small number of receptors can activate a large number of cyclase complexes. The signal is further amplified in the next step in the cAMP cascade, the activation of the protein kinase.

The major target of cAMP in most cells is the cAMP-dependent protein kinase (also called protein kinase A or PKA). This kinase, identified and characterized by Edward Krebs and colleagues, is a heterotetrameric enzyme consisting of a dimer of two regulatory (R) subunits and two catalytic (C) subunits. In the absence of cAMP, the R subunits bind to and inhibit the C subunits. In the presence of cAMP, each R subunit binds two molecules of cAMP, leading to a conformational change that causes the R and C subunits to dissociate (Figure 14–2B). Dissociation frees the C subunits to transfer the γ-phosphoryl group of ATP to the hydroxyl groups of specific serine and threonine residues in substrate proteins. The action of PKA is terminated by phosphoprotein phosphatases, enzymes that cleave the phosphoryl group from proteins, producing inorganic phosphate.

Protein kinase A is distantly related through evolution to other serine and threonine protein kinases that we shall consider: the calcium/calmodulin-dependent protein kinases and protein kinase C. These kinases also have regulatory and catalytic domains, but both domains are within the same polypeptide molecule (see Figure 14–4).

In addition to blocking enzymatic activity, the regulatory subunits of PKA also target the catalytic subunits to distinct sites within cells. Human PKA has two types of R subunits, R_I and R_{II}, each with two subtypes: $R_{Iα}$, $R_{Iβ}$, $R_{IIα}$, and $R_{IIβ}$. The genes for each derive from a common ancestor but have different properties. For example, type II PKA (containing R_{II}-type subunits) is targeted to the membrane by *A kinase attachment proteins* (AKAPs). One type of AKAP targets PKA to the *N*-methyl-D-aspartate (NMDA)-type glutamate receptor by binding both PKA and the postsynaptic density protein PSD-95, which binds to the cytoplasmic tail of the NMDA receptor (Chapter 13). In addition, this AKAP also binds a protein phosphatase, which removes the phosphate group from substrate proteins. By localizing PKA and other signaling components near their substrate, AKAPs form local signaling complexes that increase the specificity, speed, and efficiency of second-messenger cascades. Because AKAPs have only a weak affinity for R_I subunits, most type I PKA is free in the cytoplasm.

Kinases can only phosphorylate proteins on serine and threonine residues that are embedded within a context of specific *phosphorylation consensus sequences* of amino acids. For example, phosphorylation by PKA usually requires a sequence of two contiguous basic amino acids—either lysine or arginine—followed by any amino acid, and then by the serine or threonine residue that is phosphorylated (for example, Arg-Arg-Ala-Thr).

Several important protein substrates for PKA have been identified in neurons. These include voltage-gated

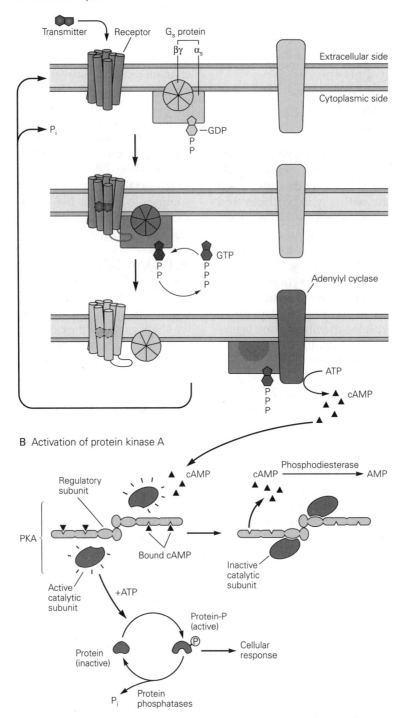

A The cAMP cycle

B Activation of protein kinase A

Figure 14–2 Activation of G protein–coupled receptors stimulates cyclic adenosine monophosphate (cAMP) production and protein kinase A. (Adapted from Alberts et al. 1994.)

A. The binding of a transmitter to certain receptors activates the stimulatory G protein (G_s), consisting of α_s-, β-, and γ-subunits. When activated, the α_s-subunit exchanges its bound guanosine diphosphate (**GDP**) for guanosine triphosphate (**GTP**), causing α_s to dissociate from the $\beta\gamma$ complex. Next, α_s associates with an intracellular domain of adenylyl cyclase, thereby stimulating the enzyme to produce cAMP from adenosine triphosphate (**ATP**). The hydrolysis of GTP to GDP and inorganic phosphate (P_i) leads to dissociation of α_s from the cyclase and its reassociation with the $\beta\gamma$ complex. The cyclase then stops producing the second messenger. As transmitter

dissociates from the receptor, the three subunits of the G protein reassociate, and the guanine nucleotide-binding site on the α-subunit is occupied by GDP.

B. Four cAMP molecules bind to the two regulatory subunits of protein kinase A (**PKA**), liberating the two catalytic subunits, which are then free to phosphorylate specific substrate proteins on certain serine or threonine residues, thereby regulating protein function to produce a given cellular response. Two kinds of enzymes regulate this pathway. Phosphodiesterases convert cAMP to adenosine monophosphate (which is inactive), and protein phosphatases remove phosphate groups (**P**) from the substrate proteins, releasing inorganic phosphate, P_i. Phosphatase activity is, in turn, decreased by the protein inhibitor-1 (not shown), when it is phosphorylated by PKA.

and ligand-gated ion channels, synaptic vesicle proteins, enzymes involved in transmitter biosynthesis, and proteins that regulate gene transcription. As a result, the cAMP pathway has widespread effects on the electrophysiological and biochemical properties of neurons. We shall consider some of these actions later in this chapter.

The Second-Messenger Pathways Initiated by G Protein–Coupled Receptors Share a Common Molecular Logic

Approximately 3.5% of genes in the human genome code for G protein–coupled receptors. Although many of these are odorant receptors in olfactory neurons (Chapter 29), many others are receptors for well-characterized neurotransmitters used throughout the nervous system. Despite their enormous diversity, all G protein–coupled receptors consist of a single polypeptide with seven characteristic membrane-spanning regions (serpentine receptors) (Figure 14–3A). Recent results from X-ray crystallography have provided detailed insights into the three-dimensional structure of these receptors in contact with their respective G proteins (Figure 14–3B).

The number of substances that act as second messengers in synaptic transmission is much fewer than the number of transmitters. More than 100 substances serve as transmitters; each can activate several types of receptors present in different cells. The few second messengers that have been well characterized fall into two categories, intracellular and transcellular. Intracellular messengers are molecules whose actions are confined to the cell in which they are produced. Transcellular messengers are molecules that can readily cross the cell membrane and thus can leave the cell in which they are produced to act as intercellular signals, or first messengers, on neighboring cells.

A Family of G Proteins Activates Distinct Second-Messenger Pathways

Approximately 20 types of α-subunits have been identified, 5 types of β-subunits, and 12 types of γ-subunits. G proteins with different α-subunits couple different classes of receptors and effectors and therefore have different physiological actions. For example, the inhibitory G_i *proteins*, which contain the α_i-subunit, inhibit adenylyl cyclase and decrease cAMP levels. Other G proteins ($G_{q/11}$ proteins, which contain α_q- or α_{11}-subunits) activate phospholipase C and probably other signal transduction mechanisms not yet identified. The

G_o protein, which contains the α_o-subunit, is expressed at particularly high levels in the brain, but its exact targets are not known. Compared with other organs of the body, the brain contains an exceptionally large variety of G proteins. Even so, because of the limited number of classes of G proteins compared to the much larger number of receptors, one type of G protein can often be activated by different classes of receptors.

The number of known effector targets for G proteins is even more limited than the types of G proteins. Important effectors include certain ion channels that are activated by the $\beta\gamma$ complex, adenylyl cyclase in the cAMP pathway, phospholipase C in the diacylglycerol-inositol polyphosphate pathway, and phospholipase A_2 in the arachidonic acid pathway. Each of these effectors (except for the ion channels) initiates changes in specific target proteins within the cell, either by generating second messengers that bind to the target protein or by activating a protein kinase that phosphorylates it.

Hydrolysis of Phospholipids by Phospholipase C Produces Two Important Second Messengers, IP₃ and Diacylglycerol

Many important second messengers are generated through the hydrolysis of phospholipids in the inner leaflet of the plasma membrane. This hydrolysis is catalyzed by three enzymes—phospholipase C, D, and A_2—named for the ester bonds they hydrolyze in the phospholipid. The phospholipases each can be activated by different G proteins coupled to different receptors.

The most commonly hydrolyzed phospholipid is *phosphatidylinositol 4,5-bisphosphate* (PIP₂), which typically contains the fatty acid stearate esterified to the glycerol backbone in the first position and the unsaturated fatty acid arachidonate in the second. Activation of receptors coupled to G_q or G_{11} stimulates *phospholipase C*, which leads to the hydrolysis of PIP₂ (specifically the phosphodiester bond that links the glycerol backbone to the polar head group) and production of two second messengers, *diacylglycerol* (DAG) and *inositol 1,4,5-trisphosphate* (IP₃).

Diacylglycerol, which is hydrophobic, remains in the membrane when formed, where it recruits the cytoplasmic protein kinase C (PKC). Together with DAG and certain membrane phospholipids, PKC forms an active complex that can phosphorylate many protein substrates in the cell, both membrane-associated and cytoplasmic (Figure 14–4A). Activation of some isoforms of PKC requires elevated levels of cytoplasmic Ca^{2+} in addition to DAG.

A Typical G protein–coupled receptor

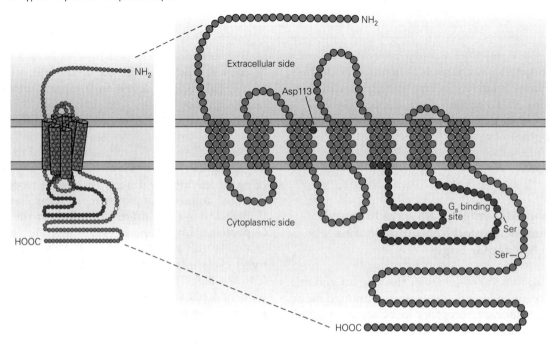

B Interaction of receptor and G protein

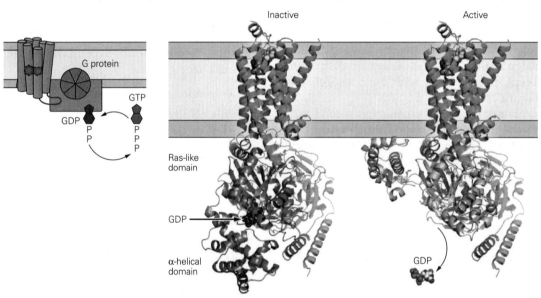

Figure 14–3 G protein–coupled receptors contain seven membrane-spanning domains.

A. The β_2-adrenergic receptor shown here is representative of G protein–coupled receptors, including the β_1-adrenergic and muscarinic acetylcholine (ACh) receptors and rhodopsin. It consists of a single subunit with an extracellular amino terminus, intracellular carboxy terminus, and seven membrane-spanning α-helixes. The binding site for the neurotransmitter lies in a cleft in the receptor formed by the transmembrane helixes. The amino acid residue aspartic acid (**Asp**)-113 participates in binding. The part of the receptor indicated in **brown** associates with G_s protein α-subunits. Two serine (**Ser**) residues in the intracellular carboxy-terminal tail are sites for phosphorylation by specific receptor kinases, which helps

inactivate the receptor. (Adapted, with permission, from Frielle et al. 1989.)

B. Models based on X-ray crystal structures of the β_2-adrenergic receptor (**blue**) interacting with the G_s protein in the inactive guanosine diphosphate (**GDP**)-bound state and the active guanosine triphosphate (**GTP**)-bound state. A high-affinity synthetic agonist is bound in the transmembrane region near the extracellular surface of the membrane (space-filling model). The α_s-, β-, and γ-subunits of the inactive G_s protein are shown in **brown**, **cyan**, and **purple**, respectively. In the active state, α_s (**gold**) undergoes a conformational change that enables it to interact with adenylyl cyclase. (Adapted, with permission, from Kobilka 2013. Copyright © 2013 Wiley-VCH Verlag GmbH & Co. KGaA, Weinheim.)

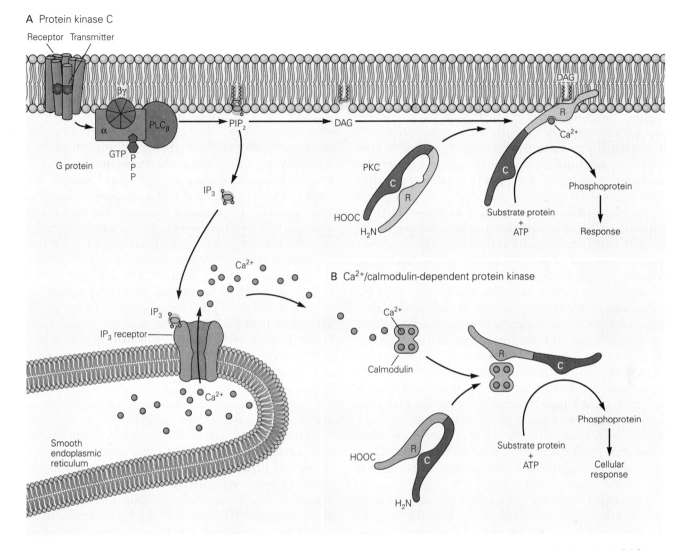

Figure 14–4 Hydrolysis of phospholipids in the cell membrane activates three major second-messenger cascades.

A. The binding of transmitter to a receptor activates a G protein that activates phospholipase C_β (**PLC$_\beta$**). This enzyme cleaves phosphatidylinositol 4,5-bisphosphate (**PIP$_2$**) into the second messengers inositol 1,4,5-trisphosphate (**IP$_3$**) and diacylglycerol (**DAG**). IP$_3$ is water soluble and diffuses into the cytoplasm, where it binds to the IP$_3$ receptor-channel on the smooth endoplasmic reticulum, thereby releasing Ca^{2+} from internal stores. DAG remains in the membrane, where it recruits and activates protein kinase C (**PKC**). Membrane phospholipid is also a necessary cofactor for PKC activation. Some isoforms of PKC also require Ca^{2+} for activation. PKC is composed of a single protein

molecule that has both a regulatory domain that binds DAG and a catalytic domain that phosphorylates proteins on serine or threonine residues. In the absence of DAG the regulatory domain inhibits the catalytic domain.

B. The calcium/calmodulin-dependent protein kinase is activated when Ca^{2+} binds to calmodulin and the calcium/calmodulin complex then binds to a regulatory domain of the kinase. The kinase is composed of many similar subunits (only one of which is shown here), each having both regulatory and catalytic functions. The catalytic domain phosphorylates proteins on serine or threonine residues. (**ATP**, adenosine triphosphate; **C**, catalytic subunit; **COOH**, carboxy terminus; **H$_2$N**, amino terminus; **R**, regulatory subunit.)

The second product of the phospholipase C pathway, IP$_3$, stimulates the release of Ca^{2+} from intracellular membrane stores in the lumen of the smooth endoplasmic reticulum. The membrane of the reticulum contains a large integral membrane macromolecule, the IP$_3$ receptor, which forms both a receptor for

IP$_3$ on its cytoplasmic surface and a Ca^{2+} channel that spans the membrane of the reticulum. When this macromolecule binds IP$_3$, the channel opens, releasing Ca^{2+} into the cytoplasm (Figure 14–4A).

The increase in intracellular Ca^{2+} triggers many biochemical reactions and opens calcium-gated channels in the plasma membrane. Calcium can also act as a second messenger to trigger the release of additional Ca^{2+} from internal stores by binding to another integral protein in the membrane of the smooth endoplasmic reticulum, the *ryanodine receptor* (so called because it binds the plant alkaloid ryanodine, which inhibits the receptor; in contrast, caffeine opens the ryanodine receptor). Like the IP_3 receptor to which it is distantly related, the ryanodine receptor forms a Ca^{2+} channel that spans the reticulum membrane; however, cytoplasmic Ca^{2+}, not IP_3, opens the ryanodine receptor-channel.

Calcium often acts by binding to the small cytoplasmic protein calmodulin. An important function of the calcium/calmodulin complex is to activate *calcium/calmodulin-dependent protein kinase* (CaM kinase). This enzyme is a complex of many similar subunits, each containing both regulatory and catalytic domains within the same polypeptide chain. When the calcium/calmodulin complex is absent, the C-terminal regulatory domain of the kinase binds and inactivates the catalytic portion. Binding to the calcium/calmodulin complex causes conformational changes of the kinase molecule that unfetter the catalytic domain for action (Figure 14–4B). Once activated, CaM kinase can phosphorylate itself through intramolecular reactions at many sites in the molecule. Autophosphorylation has an important functional effect: It converts the enzyme into a form that is independent of calcium/calmodulin and therefore persistently active, even in the absence of Ca^{2+}.

Persistent activation of protein kinases is a general and important mechanism for maintaining biochemical processes that underlie long-term changes in synaptic function associated with certain forms of memory. In addition to the persistent activation of calcium/calmodulin-dependent protein kinase, PKA can also become persistently active following a prolonged increase in cAMP because of a slow enzymatic degradation of free regulatory subunits through the ubiquitin pathway. The decline in regulatory subunit concentration results in the long-lasting presence of free catalytic subunits, even after cAMP levels have declined, leading to the continued phosphorylation of substrate proteins. PKC can also become persistently active through proteolytic cleavage of its regulatory and catalytic domains or through the expression of a PKC isoform that lacks a regulatory domain. Finally, the duration of phosphorylation can be enhanced by certain proteins that act to inhibit the activity of phosphoprotein phosphatases. One such protein, inhibitor-1, inhibits phosphatase activity only when the inhibitor is itself phosphorylated by PKA.

Receptor Tyrosine Kinases Compose the Second Major Family of Metabotropic Receptors

The *receptor tyrosine kinases* represent a distinct family of receptors from the G protein–coupled receptors. The receptor tyrosine kinases are integral membrane proteins composed of a single subunit with an extracellular ligand-binding domain connected to a cytoplasmic region by a single transmembrane segment. The cytoplasmic region contains a protein kinase domain that phosphorylates both itself (autophosphorylation) and other proteins on tyrosine residues (Figure 14–5A). This phosphorylation results in the activation of a large number of proteins, including other kinases that are capable of acting on ion channels.

Receptor tyrosine kinases are activated when bound by peptide hormones, including epidermal growth factor (EGF), fibroblast growth factor (FGF), nerve growth factor (NGF), brain-derived neurotrophic factor (BDNF), and insulin. Cells also contain important nonreceptor cytoplasmic tyrosine kinases, such as the protooncogene *src*. These nonreceptor tyrosine kinases are often activated by interactions with receptor tyrosine kinases and are important in regulating growth and development.

Many (but not all) of the receptor tyrosine kinases exist as monomers in the plasma membrane in the absence of ligand. Ligand binding causes two monomeric receptor subunits to form a dimer, thereby activating the intracellular kinase. Each monomer phosphorylates its counterpart at a tyrosine residue, an action that enables the kinase to phosphorylate other proteins. Like the serine and threonine protein kinases, tyrosine kinases regulate the activity of neuronal proteins they phosphorylate, including the activity of certain ion channels. Tyrosine kinases also activate an isoform of phospholipase C, phospholipase Cγ, which like PLCβ cleaves PIP_2 into IP_3 and DAG.

Receptor tyrosine kinases initiate cascades of reactions involving several adaptor proteins and other protein kinases that often lead to changes in gene transcription. The mitogen-activated protein kinases (MAP kinases) are an important group of serine-threonine kinases that can be activated by a signaling cascade initiated by receptor tyrosine kinase. MAP kinases are activated by cascades of protein-kinase reactions (kinase kinases), each cascade specific to one of three types of MAP kinase: extracellular signal-regulated kinase (ERK), p38 MAP kinase, and *c-Jun* N-terminal kinase (JNK). Activated MAP kinases have several important actions. They translocate to the nucleus where they

Figure 14–5 Receptor tyrosine kinases.

A. Receptor tyrosine kinases are monomers in the absence of a ligand. The receptor contains a large extracellular binding domain that is connected by a single transmembrane segment to a large intracellular region that contains a catalytic tyrosine kinase domain. Ligand binding to the receptor often causes two receptor subunits to form dimers, enabling the enzyme to phosphorylate itself on various tyrosine residues on the cytoplasmic side of the membrane.

B. After the receptor is autophosphorylated, several downstream signaling cascades become activated through the binding of specific adaptor proteins to the receptor phosphotyrosine residues (**P**). *Left:* Activation of mitogen-activated protein kinase (**MAPK**). A series of adaptor proteins recruits the small guanosine triphosphate (GTP)-binding protein Ras, which activates a protein kinase cascade, leading to the dual phosphorylation of MAP kinase on nearby threonine and tyrosine residues. The activated MAP kinase then phosphorylates substrate proteins on serine and threonine residues, including ion channels and transcription factors. *Center:* Phospholipase C$_\gamma$ (**PLC$_\gamma$**) becomes activated on binding to a different phosphotyrosine residue, providing a mechanism for producing inositol 1,4,5-trisphosphate (**IP$_3$**) and diacylglycerol (**DAG**) that does not rely on G proteins. *Right:* Activation of the **Akt** protein kinase (also called PKB). Adaptor proteins first activate phosphoinositide 3-kinase (**PI3K**), which adds a phosphate group to PIP$_2$, yielding PIP$_3$, which then enables Akt activation.

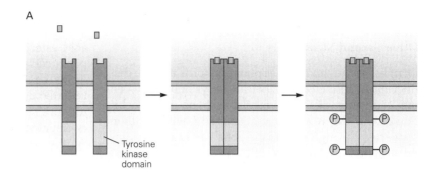

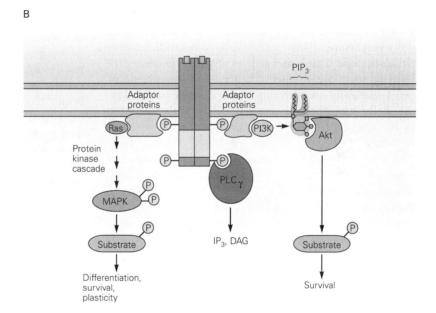

turn on gene transcription by phosphorylating certain transcription factors. This action is thought to be important in stabilizing long-term memory formation (Chapters 53 and 54). MAP kinases also phosphorylate cytoplasmic and membrane proteins to produce short-term modulatory actions (Figure 14–5B).

Several Classes of Metabolites Can Serve as Transcellular Messengers

The metabolic products we have considered so far in response to metabotropic receptor actions do not readily cross the cell membrane. As a result, they act as true intracellular second messengers: They only affect the cell that produces them. However, cells can also synthesize metabolites that are lipid soluble and so can both act on the cell that produces them and diffuse across the plasma membrane to affect neighboring cells. We refer to such molecules as transcellular messengers.

Although these molecules have some functional resemblance to neurotransmitters, they differ in a number of important ways. They are not contained within vesicles and are not released at specialized synaptic contacts. They often do not act on membrane receptors but cross the plasma membrane of neighboring cells to reach intracellular targets. And their release and actions are much slower than those at fast synapses. We will consider three broad classes of transcellular messengers: the cyclooxygenase and lipoxygenase metabolites of the lipid molecule arachidonic acid, the endocannabinoids, and the gas nitric oxide.

Hydrolysis of Phospholipids by Phospholipase A₂ Liberates Arachidonic Acid to Produce Other Second Messengers

Phospholipase A₂ hydrolyzes phospholipids that are distinct from PIP_2, cleaving the fatty acyl bond between the 2′ position of the glycerol backbone and arachidonic acid. This releases *arachidonic acid*, which is then converted through enzymatic action to one of a family of active metabolites called *eicosanoids*, so called because of their 20 (Greek *eicosa*) carbon atoms.

Three types of enzymes metabolize arachidonic acid: (1) cyclooxygenases, which produce prostaglandins and thromboxanes; (2) several lipoxygenases, which produce a variety of other metabolites; and (3) the cytochrome P450 complex, which oxidizes arachidonic acid itself as well as cyclooxygenase and lipoxygenase metabolites (Figure 14–6). Synthesis of prostaglandins and thromboxanes in the brain is dramatically increased by nonspecific stimulation such as electroconvulsive shock, trauma, or acute cerebral ischemia (localized absence of blood flow). These metabolites can all be released by the cell that synthesizes them and thus act as transcellular signals. Many of the actions of prostaglandins are mediated by acting in the plasma membrane on a family of G protein–coupled receptors. The members of this receptor family can, in turn, activate or inhibit adenylyl cyclase or activate phospholipase C.

Endocannabinoids Are Transcellular Messengers That Inhibit Presynaptic Transmitter Release

In the early 1990s, researchers identified two types of G protein–coupled receptors, CB1 and CB2, which bind with high affinity the active compound in marijuana, Δ^9-tetrahydrocannabinol (THC). Both classes of receptors are coupled to G_i and G_o types of G proteins. The CB1 receptors are the most abundant type of G protein–coupled receptor in the brain and are found predominantly on axons and presynaptic terminals in both the central and peripheral nervous systems. Activation of these receptors inhibits release of several types of neurotransmitters, including both GABA and glutamate. The CB2 receptors are found mainly on lymphocytes, where they modulate the immune response.

The identification of the cannabinoid receptors led to the purification of their endogenous ligands, the *endocannabinoids*. Two major endocannabinoids have been identified; both contain an arachidonic acid moiety and bind to both CB1 and CB2 receptors. *Anandamide* (Sanskrit *ananda*, bliss) consists of arachidonic acid

coupled to ethanolamine (arachidonyl-ethanolamide); *2-arachidonylglycerol* (2-AG) consists of arachidonic acid esterified at the 2 position of glycerol. Both are produced by the enzymatic hydrolysis of phospholipids containing arachidonic acid, a process that is initiated either when certain G protein–coupled receptors are stimulated or the internal Ca^{2+} concentration is elevated (Figure 14–6). However, whereas 2-AG is synthesized in nearly all neurons, the sources of anandamide are less well characterized.

Because the endocannabinoids are lipid metabolites that can diffuse through the membrane, they function as transcellular signals that act on neighboring cells, including presynaptic terminals. Production of these metabolites is often stimulated in postsynaptic neurons by the increase in intracellular Ca^{2+} that results from postsynaptic excitation. Once produced, the endocannabinoids diffuse through the cell membrane to nearby presynaptic terminals, where they bind to CB1 receptors and inhibit transmitter release. In this manner, the postsynaptic cell can control activity of the presynaptic neuron. There is now intense interest in understanding how the activation of these receptors in the brain leads to the various behavioral effects of marijuana.

The Gaseous Second Messenger Nitric Oxide Is a Transcellular Signal That Stimulates Cyclic GMP Synthesis

Nitric oxide (NO) acts as a transcellular messenger in neurons as well as in other cells of the body. The modulatory function of NO was discovered through its action as a local hormone released from the endothelial cells of blood vessels, causing relaxation of the smooth muscle of vessel walls. Like the metabolites of arachidonic acid, NO readily passes through cell membranes and can affect nearby cells without acting on a surface receptor. Nitric oxide is a free radical and so is highly reactive and short-lived.

Nitric oxide produces many of its actions by stimulating the synthesis of guanosine 3′,5′-cyclic monophosphate (cyclic GMP or cGMP), which like cAMP is a cytoplasmic second messenger that activates a protein kinase. Specifically, NO activates guanylyl cyclase, the enzyme that converts GTP to cGMP. There are two types of guanylyl cyclase. One is an integral membrane protein with an extracellular receptor domain and an intracellular catalytic domain that synthesizes cGMP. The other is cytoplasmic (soluble guanylyl cyclase) and is the isoform activated by NO. In some instances, NO is thought

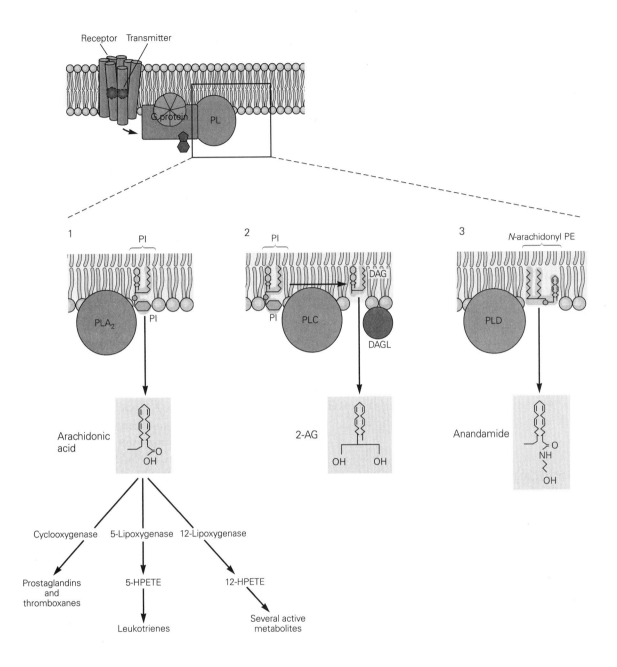

Figure 14–6 Three phospholipases generate distinct second messengers by hydrolysis of phospholipids containing arachidonic acid.

Pathway 1. Stimulation of G protein–coupled receptors leads to activation of phospholipase A_2 (**PLA$_2$**) by the free βγ-subunit complex. Phospholipase A_2 hydrolyzes phosphatidylinositol (**PI**) in the plasma membrane, leading to the release of arachidonic acid, a 20-carbon fatty acid with four double bonds that is a component of many phospholipids. Once released, arachidonic acid is metabolized through several pathways, three of which are shown. The 12- and 5-lipoxygenase pathways both produce several active metabolites; the cyclooxygenase pathway produces prostaglandins and thromboxanes. Cyclooxygenase is inhibited by indomethacin, aspirin, and other nonsteroidal anti-inflammatory drugs. Arachidonic acid and many of its metabolites modulate the activity of certain ion channels. (**HPETE,** hydroperoxyeicosatetraenoic acid.)

Pathway 2. Other G proteins activate phospholipase C (**PLC**), which hydrolyzes PI in the membrane to generate DAG (see Figure 14–4). Hydrolysis of DAG by a second enzyme, diacylglycerol lipase (**DAGL**), leads to production of 2-arachidonyl-glycerol (**2-AG**), an endocannabinoid that is released from neuronal membranes and then activates G protein–coupled endocannabinoid receptors in the plasma membrane of other neighboring neurons.

Pathway 3. Elevation of intracellular Ca^{2+} activates phospholipase D (**PLD**), which hydrolyzes phospholipids that have an unusual polar head group containing arachidonic acid (N-arachidonylphosphatidylethanolamine [**N-arachidonyl PE**]). This action generates a second endocannabinoid termed anandamide (arachidonylethanolamide).

to act directly by modifying sulfhydryl groups on cysteine residues of various proteins, a process termed nitrosylation.

Cyclic GMP has two major actions. It acts directly to open cyclic nucleotide-gated channels (important for phototransduction and olfactory signaling, as described in Chapters 22 and 29, respectively), and it activates the *cGMP-dependent protein kinase* (PKG), which like PKA phosphorylates substrate proteins on certain serine or threonine residues. PKG differs from the PKA in that it is a single polypeptide with both regulatory (cGMP-binding) and catalytic domains, which are homologous to regulatory and catalytic domains in other protein kinases. It also phosphorylates a distinct set of substrates from PKA.

Cyclic GMP–dependent phosphorylation of proteins is prominent in Purkinje cells of the cerebellum, large neurons with copiously branching dendrites. There, the cGMP cascade is activated by NO produced and released from the presynaptic terminals of granule cell axons (the parallel fibers) that make excitatory synapses onto the Purkinje cells. This increase in cGMP in the Purkinje neuron reduces the response of the AMPA receptors to glutamate, thereby depressing fast excitatory transmission at the parallel fiber synapse.

The Physiological Actions of Metabotropic Receptors Differ From Those of Ionotropic Receptors

Second-Messenger Cascades Can Increase or Decrease the Opening of Many Types of Ion Channels

The functional differences between metabotropic and ionotropic receptors reflect the differences in their properties. For example, metabotropic receptor actions are much slower than ionotropic ones (Table 14–1). The

physiological actions of the two classes of receptors also differ.

Ionotropic receptors are channels that function as simple on-off switches; their main job is either to excite a neuron to bring it closer to the threshold for firing or inhibit the neuron to decrease its likelihood to fire. Because these channels are normally confined to the postsynaptic region of the membrane, the action of ionotropic receptors is local. Metabotropic receptors, on the other hand, because they activate diffusible second messengers, can act on channels some distance from the receptor. Moreover, metabotropic receptors regulate a variety of channel types, including resting channels, ligand-gated channels, and voltage-gated channels that generate action potentials, underlie pacemaker potentials, and provide Ca^{2+} influx for neurotransmitter release.

Finally, whereas transmitter binding leads to an increase in the opening of ionotropic receptor-channels, the activation of metabotropic receptors can lead to an increase or decrease in channel opening. For example, MAP kinase phosphorylation of an inactivating (A-type) K^+ channel in the dendrites of hippocampal pyramidal neurons decreases channel opening and, thus, K^+ current magnitude, thereby enhancing dendritic action potential firing.

The binding of transmitter to metabotropic receptors can greatly influence the electrophysiological properties of a neuron (Figure 14–7). Metabotropic receptors in a presynaptic terminal can alter transmitter release by regulating either Ca^{2+} influx or the efficacy of the synaptic release process itself (Figure 14–7A). Metabotropic receptors in the postsynaptic cell can influence the strength of a synapse by modulating the ionotropic receptors that mediate the postsynaptic potential (Figure 14–7B). By acting on resting and voltage-gated channels in the postsynaptic neuron's cell body, dendrites, and axon, metabotropic receptor actions can also alter the resting potential, membrane resistance, length and time constants, threshold potential, action potential

Table 14–1 Comparison of Synaptic Excitation Produced by the Opening and Closing of Ion Channels

	Ion channels involved	Effect on total membrane conductance	Contribution to action potential	Time course	Second messenger	Nature of synaptic action
EPSP caused by opening of channels	Nonselective cation channel	Increase	Triggers action potential	Usually fast (milliseconds)	None	Mediating
EPSP caused by closing of channels	K^+ channel	Decrease	Modulates action potential	Slow (seconds or minutes)	Cyclic AMP (or other second messengers)	Modulating

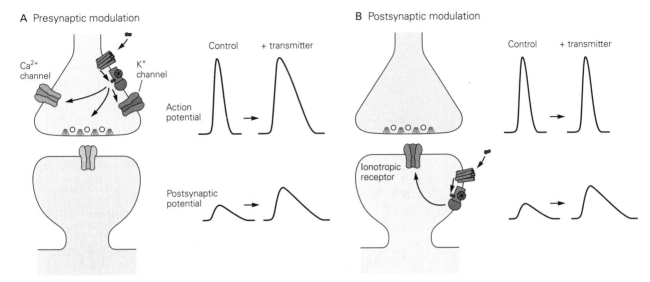

Figure 14–7 The modulatory actions of second messengers can regulate fast synaptic transmission by acting at two synaptic sites.

A. In the presynaptic terminal, second messengers can regulate the efficacy of transmitter release and thus the size of the fast postsynaptic potential mediated by ionotropic receptors. This can occur by altering presynaptic Ca²⁺ influx, either directly by modulating presynaptic voltage-gated Ca²⁺ channels or indirectly by modulating presynaptic K⁺ channels, which alters Ca²⁺ influx by controlling action potential duration as illustrated (and thereby the length of time Ca²⁺ channels remain open). Some modulatory transmitters act to directly modulate the efficacy of the release machinery.

B. In the postsynaptic terminal, second messengers can alter directly the amplitude of postsynaptic potentials by modulating ionotropic receptors.

duration, and repetitive firing characteristics. Such modulation of the intrinsic excitability of neurons can play an important role in regulating information flow through neuronal circuits to alter behavior.

The distinction between direct and indirect regulation of ion channels is nicely illustrated by cholinergic synaptic transmission in autonomic ganglia of the peripheral nervous system. Stimulation of the presynaptic nerve releases ACh from the nerve terminals, directly opening nicotinic ACh receptor-channels in the postsynaptic neuron, thereby producing a fast excitatory postsynaptic potential (EPSP). The fast EPSP is followed by a slow EPSP that takes approximately 100 ms to develop but then lasts for several seconds. The slow EPSP is produced by an action of ACh on metabotropic muscarinic receptors that leads to the closing of a delayed-rectifier K⁺ channel called the muscarine-sensitive (or M-type) K⁺ channel (Figure 14–8A). These voltage-gated channels, which are formed by members of the KCNQ gene family, are partially activated when the cell is at rest; as a result, the current they carry helps determine the cell resting potential and membrane resistance.

The M-type K⁺ channel differs from other delayed-rectifier K⁺ channels by its much slower activation. It requires several hundred milliseconds to fully activate on depolarization. Because M-type channels are partially open at the resting potential, their closure in response to muscarinic stimulation causes a decrease in resting K⁺ conductance, thus depolarizing the cell (Figure 14–8B). How far will the membrane depolarize? This can be calculated using the equivalent circuit form of the Goldman equation (Chapter 9) by decreasing the g_K term from its initial value. As the change in g_K due to closure of M-type K⁺ channels is relatively modest, the depolarization at the peak of the slow EPSP is small, only a few millivolts. Nonetheless, M-type K⁺ channel closure by ACh can lead to a striking increase in action potential firing in response to a depolarizing input.

What are the special properties of M-type K⁺ channel closure that dramatically enhance excitability? First, the depolarization resulting from the reduction in resting g_K drives the membrane closer to threshold. Second, the increase in membrane resistance decreases the amount of excitatory current necessary to depolarize the cell to a given voltage. Third, the reduction in the delayed K⁺ current enables the cell to produce a more sustained firing of action potentials in response to a prolonged depolarizing stimulus.

In the absence of ACh, a ganglionic neuron normally fires only one or two action potentials and then stops firing in response to prolonged excitatory stimulation that is just above threshold. This process, termed *spike-frequency adaptation*, results in part from

A Fast and slow synaptic transmission

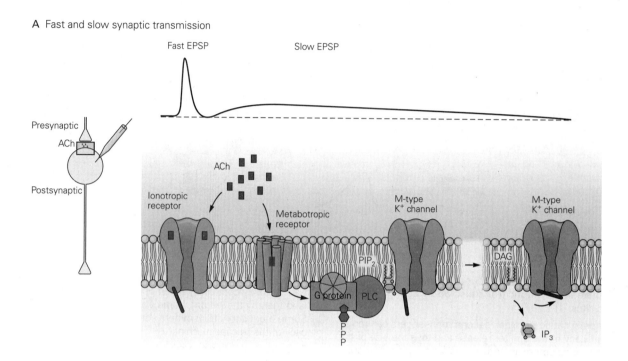

B The effect of muscarine on the M-type K⁺ current

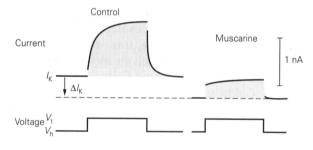

C M-type K⁺ current inhibition reduces spike adaptation

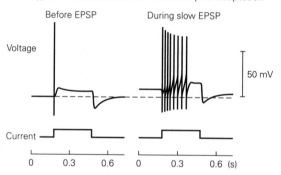

Figure 14–8 Fast ionotropic and slow metabotropic synaptic actions at autonomic ganglia.

A. The release of ACh onto a postsynaptic neuron in autonomic ganglia produces a fast EPSP followed by a slow EPSP. The fast EPSP is produced by activation of ionotropic nicotinic ACh receptors, the slow EPSP by activation of metabotropic muscarinic ACh receptors. The metabotropic receptor stimulates PLC to hydrolyze PIP_2, yielding IP_3 and DAG. The decrease in PIP_2 causes the closure of M-type delayed-rectifier K⁺ channels.

B. Voltage-clamp recordings from an autonomic ganglion neuron indicate that ACh decreases the magnitude of the current carried by the voltage-gated M-type K⁺ channels. In this experiment, the cell is initially clamped at a holding potential (V_h) near the resting potential in the absence of ACh (typically –60 mV). At this potential, the M-type K⁺ channels are partially open, leading to a steady outward K⁺ current. The voltage is then stepped for 1 second to a more positive test potential (V_t, typically –40 mV), which normally causes a slow increase in outward K⁺ current (I_K) as the M-type K⁺ channels respond

to the more positive voltage by increasing their opening (control). Application of muscarine, a plant alkaloid that selectively stimulates the muscarinic ACh receptor, causes a fraction of the M-type K⁺ channels to close. This decreases the outward K⁺ current at the holding potential (note the shift in baseline current, ΔI_K), by closing the M-type K⁺ channels that are open at rest, and decreases the magnitude of the slowly activating K⁺ current in response to the step depolarization. (Adapted from Adams et al. 1986.)

C. In the absence of muscarinic ACh receptor stimulation, the neuron fires only a single action potential in response to a prolonged depolarizing current stimulus, a process termed spike-frequency adaptation (*left*). This is because the slow activation of the M-type K⁺ channel during the depolarization generates an outward current that repolarizes the membrane below threshold. When the same current stimulus is applied during a slow EPSP, when a large fraction of M-type channels are now unable to open, the neuron fires a more sustained train of action potentials (*right*). (Adapted from Adams et al. 1986.)

the increase in M-type K^+ current in response to the prolonged depolarization, which helps repolarize the membrane below threshold. As a result, if the same prolonged stimulus is applied during a slow EPSP (when the M-type K^+ channels are closed), the neuron remains depolarized above threshold during the entire stimulus and thus fires a prolonged burst of impulses (Figure 14–8C). As this modulation by ACh illustrates, the M-type K^+ channels do more than help set the resting potential—they also control excitability.

Although it has been known for some time that muscarinic receptor actions in autonomic ganglia result in the activation of PLC and the production of DAG and IP_3, the precise mechanism by which this signaling cascade produces M-type channel closure remained mysterious. However, it is now clear that M-channel closure upon muscarinic receptor activation is not due to the production of a second messenger. Rather, the M-channels, as well as a number of other types of channels (eg, see Figure 14–10), bind membranous PIP_2 as a cofactor for their proper functioning. Thus, muscarinic receptor activation closes M-type channels by activating PLC, and thereby decreasing the levels of PIP_2 in the membrane due to hydrolysis by PLC. We shall next discuss the mechanisms by which other signaling cascades are capable of modulating other types of ion channels. We start by describing the simplest mechanism, the direct gating of ion channels by G proteins, and then consider a more complex mechanism dependent on protein phosphorylation by PKA.

G Proteins Can Modulate Ion Channels Directly

The simplest mechanism for the indirect gating of a channel occurs when transmitter binding to a metabotropic receptor releases a G protein subunit that directly interacts with the channel to modify its opening. This mechanism is used to gate two kinds of ion channels: the *G protein–gated inward-rectifier K^+* channels (GIRK1–4; encoded by the *KCNJ1–4* genes) and a voltage-dependent Ca^{2+} channel. With both kinds of channels, it is the G protein's βγ complex that binds to and regulates channel opening (Figure 14–9A).

The GIRK channel, like other inward-rectifier channels, passes current more readily in the inward than the outward direction, although in physiological situations, K^+ current is always outward. Inward-rectifier channels resemble a truncated voltage-gated K^+ channel in having two transmembrane regions connected by a P-region loop that forms the selectivity filter in the channel (see Figure 8–11).

In the 1920s, Otto Loewi described how the release of ACh in response to stimulation of the vagus nerve slows the heart rate (Figure 14–9B). We now know that ACh activates muscarinic receptors to stimulate G protein activity, which directly opens the GIRK channel. For many years, this transmitter action was puzzling because it has properties of both ionotropic and metabotropic receptor actions. The time course of activation of the K^+ current following release of ACh is slower (50- to 100-ms rise time) than that of ionotropic receptors (rise time <1 ms). However, the rate of GIRK channel activation is much faster than that of second-messenger-mediated actions that depend on protein phosphorylation (which can take many seconds to turn on). Although biochemical and electrophysiological studies clearly demonstrated that a G protein was required for this action, patch-clamp experiments showed that the G protein did not trigger production of a diffusible second messenger (Figure 14–9C). These findings were reconciled when it was found that the GIRK channel was activated directly by the G protein's βγ-subunit complex, which becomes available to interact with the GIRK channel when it dissociates from the G protein α-subunit upon activation of the muscarinic receptors.

The mechanism by which the βγ-subunits activate the GIRK channel was recently elucidated at the atomic resolution through the solving of the X-ray crystal structure of the GIRK channel in a complex with the βγ-subunits. Each of the four GIRK channel subunits binds a single βγ-subunit complex, which interacts with the cytoplasmic surface of the channel, leading to a conformational change that promotes channel opening (Figure 14–10).

Activation of GIRK channels hyperpolarizes the membrane in the direction of E_K (–80 mV). In certain classes of spontaneously active neurons, the outward K^+ current through these channels acts predominantly to decrease the neuron's intrinsic firing rate, opposing the slow depolarization caused by excitatory pacemaker currents carried by the hyperpolarization-activated, cyclic nucleotide-regulated channels, which are encoded by the *HCN* gene family (Chapter 10). Because GIRK channels are activated by neurotransmitters, they provide a means for synaptic modulation of the firing rate of excitable cells. These channels are regulated in a wide variety of neurons by a large number of transmitters and neuropeptides that act on different G protein–coupled receptors to activate either G_i or G_o, thereby releasing the βγ-subunits.

Several G protein–coupled receptors also act to inhibit the opening of certain voltage-gated Ca^{2+} channels, again as a result of the direct binding of the βγ complex of G_i or G_o to the channel. Because Ca^{2+} influx through voltage-gated Ca^{2+} channels normally has a depolarizing effect, the dual action of G protein βγ-subunits—Ca^{2+} channel inhibition and K^+ channel activation—strongly inhibits neuronal firing. As we will see in Chapter 15, inhibition

A Direct opening of the GIRK channel by a G protein

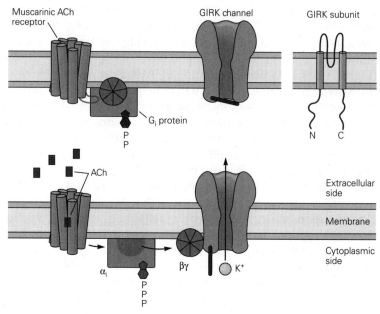

Figure 14–9 Some G proteins can open ion channels directly without employing second messengers.

A. An inward-rectifying K$^+$ channel (**GIRK**) is opened directly by a G protein. Binding of ACh to a muscarinic receptor causes the G$_i$ protein and α_i-$\beta\gamma$ complex to dissociate; the free $\beta\gamma$-subunits bind to a cytoplasmic domain of the channel, causing the channel to open.

B. Stimulation of the parasympathetic vagus nerve releases ACh, which acts at muscarinic receptors to open GIRK channels in cardiac muscle cell membranes. The current through the GIRK channel hyperpolarizes the cells, thus slowing the heart rate. (Adapted from Toda and West 1967.)

C. Three single-channel records show that opening of GIRK channels does not involve a freely diffusible second messenger. In this experiment, the pipette contained a high concentration of K$^+$, which makes E_K less negative. As a result, when GIRK channels open, they generate brief pulses of inward (downward) current. In the absence of ACh, channels open briefly and infrequently (**top record**). Application of ACh in the bath (outside the pipette) does not increase channel opening in the patch of membrane under the pipette (**middle record**). This is because the free $\beta\gamma$-subunits, released by the binding of ACh to its receptor, remain tethered to the membrane near the receptor and can only activate nearby channels. The subunits are not free to diffuse to the channels under the patch pipette. The ACh must be in the pipette to activate the channel (**bottom record**). (Reproduced, with permission, from Soejima and Noma 1984. Copyright © 1984 Springer Nature.)

B Opening of GIRK channels by ACh hyperpolarizes cardiac muscle cells

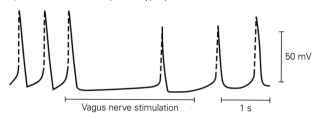

C Opening of GIRK channels by ACh does not require second messengers

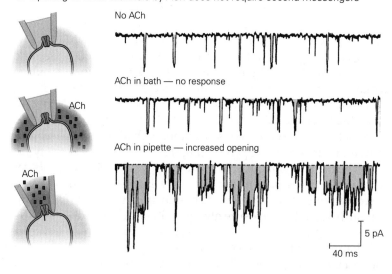

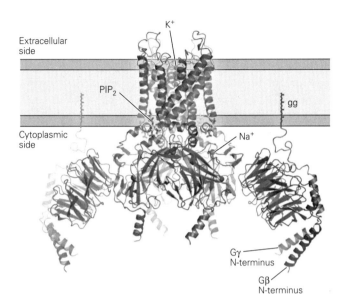

Figure 14–10 G protein βγ-subunits can directly bind and activate GIRK channels. A high-resolution structure of a GIRK channel (**green**) interacting with the G protein β-subunit (**Gβ, cyan**) and γ-subunit (**Gγ, purple**). A geranylgeranyl lipid molecule (**gg**) is attached to the C-terminus of Gγ. The structure illustrates that Na^+ ions and the phospholipid PIP_2 also bind to the channel, thereby enhancing channel opening. The **pink spheres** inside the channel represent K^+ ions. (Adapted with permission from Whorton and MacKinnon 2013. Copyright © 2013 Springer Nature.)

of voltage-gated Ca^{2+} channels in presynaptic terminals can suppress the release of neurotransmitter.

Cyclic AMP–Dependent Protein Phosphorylation Can Close Potassium Channels

In the marine mollusk *Aplysia*, a group of mechanoreceptor sensory neurons initiates defensive withdrawal reflexes in response to tactile stimuli through fast excitatory synapses with motor neurons. Certain interneurons form serotonergic synapses with these sensory neurons, and the serotonin released by the interneurons sensitizes the withdrawal reflex, enhancing the animal's response to a stimulus and thus producing a simple form of learning (Chapter 53).

The modulatory action of serotonin depends on its binding to a G protein–coupled receptor that activates a G_s protein, which elevates cAMP and thus activates PKA. This leads to the direct phosphorylation and subsequent closure of the serotonin-sensitive (or S-type) K^+ channel that acts as a resting channel (Figure 14–11). Like the closing of the M-type K^+ channel by ACh, closure of the S-type K^+ channel decreases K^+ efflux from the cell, thereby depolarizing the cell and decreasing its resting membrane conductance. Conversely, the opening of the same S-type K^+

channels can be enhanced by the neuropeptide FMR-Famide, acting through 12-lipoxygenase metabolites of arachidonic acid. This enhanced channel opening leads to a slow hyperpolarizing inhibitory postsynaptic potential (IPSP) associated with an increase in resting membrane conductance.

Thus, a single channel can be regulated by distinct second-messenger pathways that produce opposite effects on neuronal excitability. Likewise, a resting K^+ channel with two pore-forming domains in each subunit (the TREK-1 channel) in mammalian neurons is dually regulated by PKA and arachidonic acid in a manner very similar to the dual regulation of the S-type channel in *Aplysia*.

Second Messengers Can Endow Synaptic Transmission with Long-Lasting Consequences

So far, we have described how synaptic second messengers alter the biochemistry of neurons for periods lasting seconds to minutes. Second messengers can also produce long-term changes lasting days to weeks as a result of alterations in a cell's expression of specific genes (Figure 14–12). Such changes in gene expression result from the ability of second-messenger cascades to control the activity of transcription factors, regulatory proteins that control mRNA synthesis.

Some transcription factors can be directly regulated by phosphorylation. For example, the cAMP response element-binding protein (CREB) is activated when phosphorylated by PKA, calcium/calmodulin-dependent protein kinases, PKC, or MAP kinases. Once activated, CREB enhances transcription by binding to specific DNA sequences, the cAMP response elements or CRE, and recruiting a component of the transcription machinery, the CREB-binding protein (CBP). CBP activates transcription by recruiting RNA polymerase II and by functioning as a histone acetylase, adding acetyl groups to certain histone lysine residues. The acetylation weakens the binding between histones and DNA, thus opening up the chromatin structure and enabling specific genes to be transcribed. The changes in transcription and chromatin structure are important for regulating neuronal development, as well as for long-term learning and memory (Chapters 53 and 54).

Modulators Can Influence Circuit Function by Altering Intrinsic Excitability or Synaptic Strength

Most of this chapter has been devoted to understanding the cellular mechanisms and signal transduction pathways that allow neuromodulator-activated

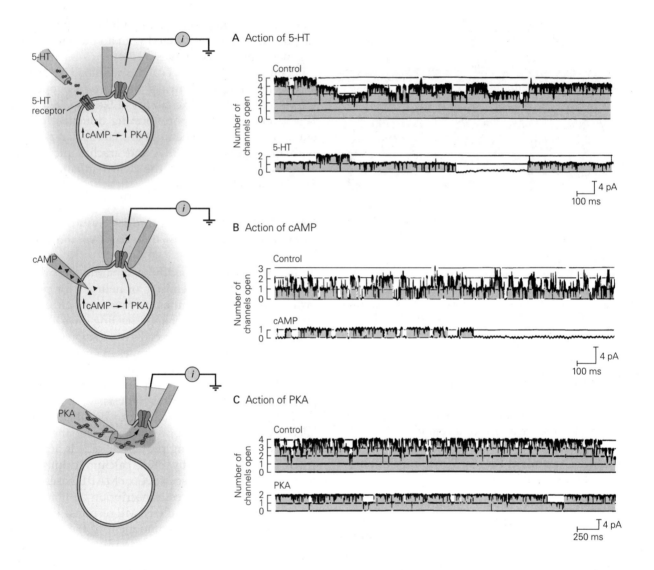

Figure 14–11 Serotonergic interneurons close a K⁺ channel through the diffusible second-messenger cAMP. Serotonin (5-HT) produces a slow EPSP in *Aplysia* sensory neurons by closing the serotonin-sensitive or S-type K⁺ channels. The 5-HT receptor is coupled to G_s, which stimulates adenylyl cyclase. The increase in cAMP activates cAMP-dependent protein kinase A (**PKA**), which phosphorylates the S-type channel, leading to its closure. Single-channel recordings illustrate the actions of 5-HT, cAMP, and PKA on the S-type channels.

A. Addition of 5-HT to the bath closes three of five S-type K⁺ channels active in this cell-attached patch of membrane. The experiment implicates a diffusible messenger, as the 5-HT applied in the bath has no direct access to the S-type channels in the membrane under the pipette. Each

channel opening contributes an outward (positive) current pulse. (Adapted, with permission, from Siegelbaum, Camardo, and Kandel 1982.)

B. Injection of cAMP into a sensory neuron through a microelectrode closes all three active S-type channels in this patch. The bottom trace shows the closure of the final active channel in the presence of cAMP. (Adapted, with permission, from Siegelbaum, Camardo, and Kandel 1982.)

C. Application of the purified catalytic subunit of PKA to the cytoplasmic surface of the membrane closes two out of four active S-type K⁺ channels in this cell-free patch. ATP was added to the solution bathing the inside surface of the membrane to provide the source of phosphate for protein phosphorylation. (Adapted, with permission, from Shuster et al. 1985.)

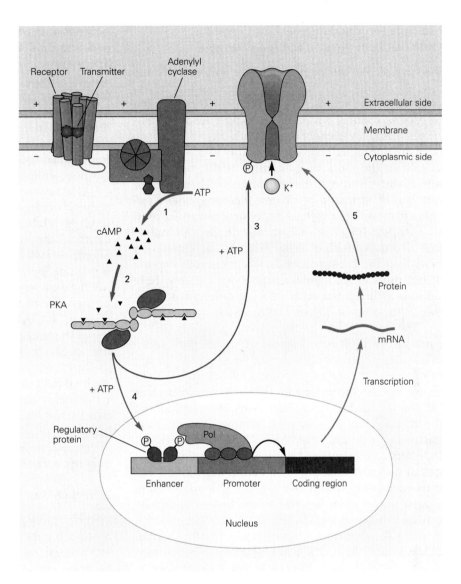

Figure 14–12 A single neurotransmitter can have either short-term or long-term effects on an ion channel. In this example, a short exposure to transmitter activates the cAMP second-messenger system (**1**), which in turn activates PKA (**2**). The kinase phosphorylates a K⁺ channel; this leads to a synaptic potential that lasts for several minutes and modifies the excitability of the neuron (**3**). With sustained activation of the receptor, the kinase translocates to the nucleus, where it phosphorylates one or more transcription factors that turn on gene expression (**4**). As a result of the new protein synthesis, the synaptic actions are prolonged—closure of the channel and changes in neuronal excitability last days or longer (**5**). (**Pol,** polymerase.)

pathways to alter the activity of ion channels, receptors, and synapses in individual neurons. However, in the intact brain, modulatory transmitters released either from diffuse projections over large areas of the brain (Chapter 16) or from more locally targeted connections can alter the dynamics of brain circuits in a number of important ways. In this section, we examine one well-studied example of modulatory control of circuit function—the control of crustacean feeding behavior by the neurons of the stomatogastric ganglion to illustrate the following general properties.

1. Modulatory projection neurons or neurohormones can coordinately influence the properties of large numbers of neurons to change the state of a neural circuit or of the entire animal. For example,

modulators released from a relatively small number of neurons are important in the control of the transitions between sleep and wakefulness (Chapter 44).

2. Neuromodulators act over intermediate time scales, ranging from many milliseconds to hours. Fast synaptic transmission and rapid action potential propagation are well suited for rapid computation of all kinds of processes important for behavior. Nevertheless, modulators that act over longer time scales can bias a circuit's dynamics to expand its dynamic range or to adapt it to the behavioral needs of the animal. For example, many sensory processes will evoke very different responses depending on the behavioral state of the animal, and modulators that alter synaptic strength and intrinsic excitability are often involved in such actions.

Multiple Neuromodulators Can Converge Onto the Same Neuron and Ion Channels

We have seen in our discussion of the *Aplysia* S-channel how the same ion channel can be regulated by different modulatory agents. This is a common theme, as the M-type K$^+$ channel is modulated by acetylcholine, substance P, and a variety of other peptides.

One particularly striking example of convergence is seen in the modulatory control of the neurons of the crustacean stomatogastric ganglion. There, a large number of structurally diverse neuropeptides converge to modulate a voltage-dependent inward current (I_{MI}). Although I_{MI} is a small current, it plays an important role in regulating excitability and the generation of plateau and burst potentials. Many neurons express a large number of different receptor types, giving these cells the ability to respond flexibly to different modulatory inputs during different brain states.

The crustacean stomatogastric ganglion (STG) contains 26 to 30 neurons and generates two rhythmic motor patterns important for feeding—the gastric rhythm and the pyloric rhythm. One set of STG neurons generates the pyloric rhythm, which is important for filtering food and is continuously active throughout the animal's life. Another set of neurons generates the gastric mill rhythm, which moves three teeth inside of the stomach that are used to chew and grind food. The gastric mill rhythm is activated in response to food and is therefore only intermittently active in vivo. Whether a particular rhythm is active at any time is under the control of a variety of neuromodulators, some of which activate the pyloric and gastric mill rhythms, while others inhibit them. These modulators can be released at specific synaptic contacts or can act diffusely as neurohormones. Interestingly, modulators can also cause individual neurons to switch between these two circuits, thereby increasing the computational power that this small number of neurons can achieve.

The fundamental circuit (the kernel) that serves as the pacemaker of the STG pyloric rhythm consists of a single anterior burster (AB) neuron and two pyloric dilator (PD) neurons. Both types of neurons make inhibitory synaptic connections with a third type of neuron, the pyloric (PY) neuron. During bursting, a slowly depolarizing pacemaker potential (slow wave) triggers a burst of action potentials in both AB and PD neurons. As these neurons are strongly coupled by electrical (gap-junction) synapses, they depolarize and synchronously fire bursts of action potentials, resulting in transient inhibition of the downstream PY neuron (Figure 14–13A).

Dopamine, which functions both as a fast neurotransmitter and as a neurohormone in crustaceans, influences feeding behavior by acting on many neurons and synapses to influence synaptic strength and neuronal and muscle excitability. For example, application of dopamine decreases the slow-wave amplitude in the PD neurons but increases the amplitude of the slow wave in the AB neurons. Ron Harris-Warrick found that dopamine modulates different sets of membrane currents in the two neurons, providing a clear example of how a single modulatory transmitter can exert distinct actions in different postsynaptic cells (Figure 14–13B).

Dopamine also alters the relative timing of the activity of these neurons. Although the PY neuron receives inhibitory input from both the AB and PD neurons, the inhibitory synaptic action from the AB neuron is faster than that from the PD neuron. Thus, dopamine, by inhibiting the PD neuron and suppressing the slow component of the IPSP, acts to speed the time course of the combined IPSP in the PY neurons (Figure 14–13A), contributing to a change in the timing of the activity of the PY neurons relative to that of the pacemaker group. Dopamine also enhances firing in the PY neuron by modulating its intrinsic *excitability*, by decreasing the transient A-type K$^+$ current ($I_{K,A}$) while increasing the *excitatory* slow inward current carried by the HCN channels (I_h) (Figure 14–13B). Thus, the effects of a modulator on the circuit result from its selective actions on a number of voltage-gated channels and synapses in distributed circuit elements.

Why So Many Modulators?

We now know that the STG is the direct target of 50 or more different neuromodulatory substances, including biogenic amines, amino acids, NO, and a host of neuropeptides that are released from descending modulatory projection neurons and sensory neurons and that circulate as hormones in the hemolymph. Many of these modulators are released as cotransmitters from the terminals of certain descending fibers that are activated by sensory neurons. Many neuromodulators are both released synaptically in the STG neuropil and also function as neurohormones.

Why should a small ganglion composed of only 26 to 30 neurons be modulated by so many substances? At first, it was thought that the richness of the modulatory innervation was important for producing different behaviorally relevant motor outputs. This remains true, but it is now also evident that some modulators may be used exclusively in special circumstances, such as molting, and that different modulators with similar effects ensure that important functions are preserved even if one modulatory system is lost. Thus, diverse modulators may be used in the service of both plasticity and stability.

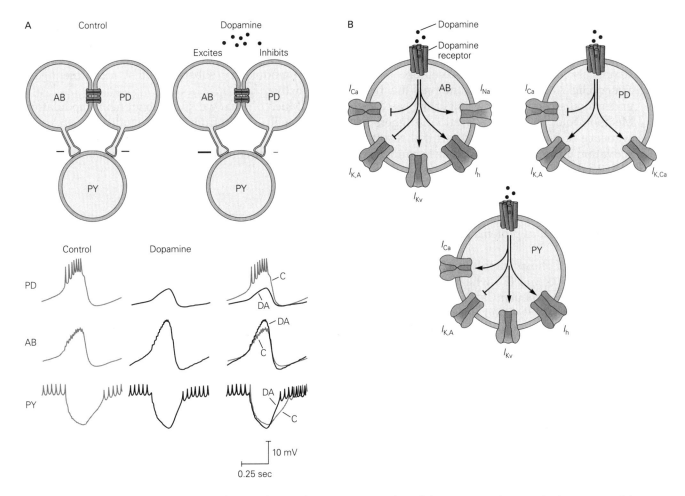

Figure 14–13 The modulatory action of dopamine on the pyloric rhythm of the lobster stomatogastric ganglion results from numerous actions.

A. A circuit diagram shows the interactions between three of the pyloric circuit neurons. The anterior burster (**AB**) and pyloric dilator (**PD**) neurons are strongly electrically coupled by gap-junction channels. Both the AB and PD neurons form inhibitory synapses with the pyloric (**PY**) neuron that generate inhibitory postsynaptic potentials (IPSPs) in this cell. Intracellular voltage recordings illustrate phases of pyloric rhythm from PD, AB, and PY neurons without dopaminergic input (control) and with dopamine. On the *right*, the voltage traces from control cells (**C**) and cells with dopaminergic input (**DA**) are overlaid. Dopamine enhances the amplitude of the slow-wave burst in the AB neuron (in this neuron, axonal action potentials are highly attenuated by the cable

properties of the neuron and appear in the soma as faint ripples) but hyperpolarizes and decreases the amplitude of the slow wave in the PD neurons. These combined actions result in a shorter IPSP in the PY neuron, enabling it to fire earlier relative to the PD neurons. (Adapted, with permission, from Eisen and Marder 1984.)

B. Dopamine modulates a number of different voltage-dependent channels in the AB, PD, and PY neurons. These include Ca^{2+} currents (I_{Ca}), a calcium-activated K^+ current ($I_{K,Ca}$), an inactivating K^+ current ($I_{K,A}$), a delayed rectifier K^+ current (I_{Kv}), the hyperpolarization-activated cation current (I_h), and a persistent Na^+ current (I_{Na}). **Lines with arrowheads** indicate current increase, **lines ending in short line segment** indicate current decrease. (Adapted, with permission, from Marder and Bucher 2007. For effects of dopamine on the complete pyloric circuit, see Harris-Warrick, 2011.)

Highlights

1. Neuromodulators are substances that bind to receptors, most of which are metabotropic, to alter the excitability of neurons, the likelihood of transmitter release, or the functional state of receptors on postsynaptic neurons.

2. When neuromodulators activate second-messenger pathways, the modulator can influence

the properties of ion channels and other targets at some distance from the site of release.

3. Some neuromodulatory systems have widespread and pronounced actions over many neurons and many brain areas.

4. There are two major families of metabotropic receptors: G protein–coupled receptors and receptor tyrosine kinases. Many important brain signaling molecules, such as norepinephrine,

ACh, GABA, glutamate, serotonin, dopamine, and many diverse neuropeptides, activate metabotropic receptors; many of these same substances also activate ionotropic receptors.

5. The cyclic AMP pathway is among the best-understood second-messenger signaling cascades. Metabotropic receptor activation triggers a sequence of biochemical reactions that result in activation of adenylyl cyclase, which synthesizes cAMP, which in turn activates protein kinase A. The kinase then phosphorylates target proteins, altering their functional state. Important targets for PKA include voltage- and ligand-gated ion channels as well as proteins important in vesicle release.

6. Hydrolysis of phospholipids by phospholipase C produces DAG and IP_3, which plays an important role in intracellular Ca^{2+} handling. Endocannabinoids are synthesized from lipid precursors and can act across synapses as retrograde messengers. Another generalized signaling molecule is the gas nitric oxide, which diffuses across membranes and stimulates cyclic GMP synthesis.

7. The receptor tyrosine kinases also gate ion channels indirectly in response to binding a variety of peptide hormones.

8. Neuromodulators can close ion channels, thus producing decreases in membrane conductance. The M-type current is a slowly activating voltage-gated K^+ current that underlies action potential adaptation. ACh and several neuropeptides decrease M-type current amplitude, thereby producing a slow depolarization and decreasing adaptation. The S-type K^+ channel contributes to the resting K^+ conductance of certain neurons, including a class of sensory neurons mediating the *Aplysia* gill withdrawal reflex. Closure of the channel by serotonin, acting through a cAMP signaling cascade, depolarizes the resting membrane, increases excitability, and enhances transmitter release from sensory neuron terminals. Prolonged exposure to serotonin can alter gene transcription to produce long-term changes in synaptic strength.

9. Modulators can alter the output of neuronal circuits by acting on numerous circuit targets.

10. Given that all brain neurons and synapses are likely to be modulated by one or more substances, it is remarkable that brain circuits are only rarely "overmodulated" so that they lose their function. Much additional research is needed to understand the rules that allow robust and stable network performance in the face of the modulators that allow network plasticity.

11. Except in a few notable cases such as small ganglia or the retina, it is likely that we still have only a partial catalog of the total number of neuromodulatory substances that are present and active.

12. Much of what we know about neuromodulatory actions comes from in vitro studies. Much less is known about how neuromodulatory concentrations are controlled in behaving animals.

Steven A. Siegelbaum
David E. Clapham
Eve Marder

Selected Reading

Berridge MJ. 2016. The inositol trisphosphate/calcium signaling pathway in health and disease. Physiol Rev 96:1261–1296.

Greengard P. 2001. The neurobiology of slow synaptic transmission. Science 294:1024–1030.

Hille B, Dickson EJ, Kruse M, Vivas O, Suh BC. 2015. Phosphoinositides regulate ion channels. Biochim Biophys Acta 1851:844–856.

Kobilka B. 2013. The structural basis of G-protein-coupled receptor signaling (Nobel Lecture). Angew Chem Int Ed Engl 52:6380–6388.

Levitan IB. 1999. Modulation of ion channels by protein phosphorylation. How the brain works. Adv Second Messenger Phosphoprotein Res 33:3–22.

Lu HC, Mackie K. 2016. An introduction to the endogenous cannabinoid system. Biol Psychiatry 79:516–525.

Marder E. 2012. Neuromodulation of neuronal circuits: back to the future. Neuron 76:1–11.

Schwartz JH. 2001. The many dimensions of cAMP signaling. Proc Natl Acad Sci U S A 98:13482–13484.

Syrovatkina V, Alegre KO, Dey R, Huang XY. 2016. Regulation, signaling, and physiological functions of G-proteins. J Mol Biol 428:3850–3868.

Takemoto-Kimura S, Suzuki K, Horigane SI, et al. 2017. Calmodulin kinases: essential regulators in health and disease. J Neurochem 141:808–818.

References

Adams PR, Jones SW, Pennefather P, Brown DA, Koch C, Lancaster B. 1986. Slow synaptic transmission in frog sympathetic ganglia. J Exp Biol 124:259–285.

Alberts B, Bray D, Lewis J, Raff M, Roberts K, Watson JD. 1994. *Molecular Biology of the Cell*, 3rd ed. New York: Garland.

Eisen JS, Marder E. 1984. A mechanism for the production of phase shifts in a pattern generator. J Neurophysiol 51:1375–1393.

Fantl WJ, Johnson DE, Williams LT. 1993. Signalling by receptor tyrosine kinases. Annu Rev Biochem 62:453–481.

Frielle T, Kobilka B, Dohlman H, Caron MG, Lefkowitz RJ. 1989. The β-adrenergic receptor and other receptors coupled to guanine nucleotide regulatory proteins. In: S Chien (ed). *Molecular Biology in Physiology,* pp. 79–91. New York: Raven.

Halpain S, Girault JA, Greengard P. 1990. Activation of NMDA receptors induces dephosphorylation of DARPP-32 in rat striatal slices. Nature 343:369–372.

Harris-Warrick, RM. 2011. Neuromodulation and flexibility in central pattern generating networks. Curr Opin Neurobiol 21:685-692.

Logothetis DE, Kurachi Y, Galper J, Neer EJ, Clapham DE. 1987. The βγ subunits of GTP-binding proteins activate the muscarinic K$^+$ channel in heart. Nature 325:321–326.

Marder E, Bucher D. 2007. Understanding circuit dynamics using the stomatogastric nervous system of lobsters and crabs. Annu Rev Physiol 69:291–316.

Nusbaum MP, Blitz DM, Marder E. 2017. Functional consequences of neuropeptide/small molecule cotransmission. Nature Rev Neurosci 18:389–403.

Osten P, Valsamis L, Harris A, Sacktor TC. 1996. Protein synthesis-dependent formation of protein kinase Mzeta in long-term potentiation. J Neurosci 16:2444–2451.

Pfaffinger PJ, Martin JM, Hunter DD, Nathanson NM, Hille B. 1985. GTP-binding proteins couple cardiac muscarinic receptors to a K channel. Nature 317:536–538.

Phillis JW, Horrocks LA, Farooqui AA. 2006. Cyclooxygenases, lipoxygenases, and epoxygenases in CNS: their role and involvement in neurological disorders. Brain Res Rev 52:201–243.

Shuster MJ, Camardo JS, Siegelbaum SA, Kandel ER. 1985. Cyclic AMP-dependent protein kinase closes the serotonin-sensitive K$^+$ channels of *Aplysia* sensory neurones in cell-free membrane patches. Nature 313:392–395.

Siegelbaum SA, Camardo JS, Kandel ER. 1982. Serotonin and cyclic AMP close single K$^+$ channels in *Aplysia* sensory neurones. Nature 299:413–417.

Soejima M, Noma A. 1984. Mode of regulation of the ACh-sensitive K-channel by the muscarinic receptor in rabbit atrial cells. Pflugers Arch 400:424–431.

Tedford HW, Zamponi GW. 2006. Direct G protein modulation of Cav2 calcium channels. Pharmacol Rev 58: 837–862.

Toda N, West TC. 1967. Interactions of K, Na, and vagal stimulation in the S-A node of the rabbit. Am J Physiol 212:416–423.

Whorton MR, MacKinnon R. 2013. X-ray structure of the mammalian GIRK2-betagamma G-protein complex. Nature 498:190–197.

Zeng L, Webster SV, Newton PM. 2012. The biology of protein kinase C. Adv Exp Med Biol 740:639–661.

15

Transmitter Release

SOME OF THE BRAIN'S MOST remarkable abilities, such as learning and memory, are thought to emerge from the elementary properties of chemical synapses, where the presynaptic cell releases chemical transmitters that activate receptors in the membrane of the postsynaptic cell. At most central synapses, transmitter is released from the presynaptic cell at presynaptic boutons, varicosities along the axon (like beads on a string) filled with synaptic vesicles and other organelles that contact postsynaptic targets. At other synapses, including the neuromuscular junction, transmitter is released from presynaptic terminals at the end of the axon. For convenience, we will refer to both types of release sites as presynaptic terminals. In the last three chapters, we saw how postsynaptic receptors control ion channels that generate the postsynaptic potential. Here we consider how electrical and biochemical events in the presynaptic terminal lead to the rapid release of small-molecule neurotransmitters, such as acetylcholine (ACh), glutamate, and γ-aminobutyric acid (GABA), that underlie fast synaptic transmission. In the next chapter, we examine the chemistry of the neurotransmitters themselves as well as the biogenic amines (serotonin, norepinephrine, and dopamine) and neuropeptides, which underlie slower forms of intercellular signaling.

Transmitter Release Is Regulated by Depolarization of the Presynaptic Terminal

What event at the presynaptic terminal leads to the release of transmitter? Bernard Katz and Ricardo Miledi first demonstrated the importance of depolarization of the presynaptic membrane. For this purpose, they used the giant synapse of the squid, a synapse large enough to permit insertion of electrodes into both pre- and postsynaptic structures. Two electrodes are inserted into the presynaptic terminal—one for stimulating and one for recording—and one electrode is inserted into the postsynaptic cell for recording the excitatory postsynaptic potential (EPSP), which provides an index of transmitter release (Figure 15–1A).

After the presynaptic neuron is stimulated and fires an action potential, an EPSP large enough to trigger an action potential is recorded in the postsynaptic cell. Katz and Miledi then asked how the presynaptic action potential triggers transmitter release. They found that as voltage-gated Na⁺ channels are blocked by application of tetrodotoxin, successive action potentials become progressively smaller. As the action potential is reduced in size, the EPSP decreases accordingly (Figure 15–1B). When the Na⁺ channel blockade becomes so profound as to reduce the amplitude of the presynaptic spike below 40 mV (positive to the resting potential), the EPSP disappears altogether. Thus, the amount of transmitter release (as measured by the size of the postsynaptic depolarization) is a steep function of the amount of presynaptic depolarization (Figure 15–1C).

Katz and Miledi next investigated how presynaptic depolarization triggers transmitter release. The action potential is produced by an influx of Na⁺ and an efflux of K⁺ through voltage-gated channels. To determine

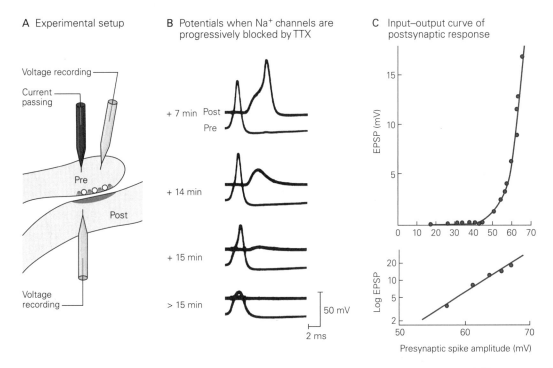

Figure 15–1 Transmitter release is triggered by changes in presynaptic membrane potential. (Adapted, with permission, from Katz and Miledi 1967a.)

A. Voltage recording electrodes are inserted in both the pre- and postsynaptic fibers of the giant synapse in the stellate ganglion of a squid. A current-passing electrode is also inserted presynaptically to elicit a presynaptic action potential.

B. Tetrodotoxin (**TTX**) is added to the solution bathing the cell to block the voltage-gated Na⁺ channels that underlie the action potential. The amplitudes of both the presynaptic action potential and the excitatory postsynaptic potential (**EPSP**) gradually decrease as more and more Na⁺ channels are blocked. After 7 minutes, the presynaptic action potential can still produce a suprathreshold EPSP that triggers an action potential in the postsynaptic cell. After about 14 to 15 minutes, the presynaptic spike gradually becomes smaller and produces smaller postsynaptic depolarizations. When the presynaptic spike is reduced to

40 mV or less, it fails to produce an EPSP. Thus, the size of the presynaptic depolarization (here provided by the action potential) controls the magnitude of transmitter release.

C. The dependence of the amplitude of the EPSP on the amplitude of the presynaptic action potential is the basis for the input–output curve for transmitter release. This relation is obtained by stimulating the presynaptic nerve during the onset of the blockade by TTX of the presynaptic Na⁺ channels, when there is a progressive reduction in the amplitude of the presynaptic action potential and postsynaptic depolarization. The upper plot demonstrates that a 40-mV presynaptic action potential is required to produce a postsynaptic potential. Beyond this threshold, there is a steep increase in amplitude of the EPSP in response to small increases in the amplitude of the presynaptic potential. The relationship between the presynaptic spike and the EPSP is logarithmic, as shown in the lower plot. A 13.5-mV increase in the presynaptic spike produces a 10-fold increase in the EPSP.

whether Na$^+$ influx or K$^+$ efflux is required to trigger transmitter release, Katz and Miledi first blocked the Na$^+$ channels with tetrodotoxin. They then asked whether direct depolarization of the presynaptic membrane, by current injection, would still trigger transmitter release. Indeed, depolarization of the presynaptic membrane beyond a threshold of about 40 mV positive to the resting potential elicits an EPSP in the postsynaptic cell even with the Na$^+$ channels blocked. Beyond that threshold, progressively greater depolarization leads to progressively greater amounts of transmitter release. This result shows that presynaptic Na$^+$ influx is not necessary for release; it is important only insofar as it depolarizes the membrane enough for transmitter release to occur (Figure 15–2B).

To examine the contribution of K$^+$ efflux to transmitter release, Katz and Miledi blocked the voltage-gated K$^+$ channels with tetraethylammonium at the same time that they blocked the voltage-sensitive Na$^+$ channels with

tetrodotoxin. They then injected a depolarizing current into the presynaptic terminals and found that the EPSPs were of normal size, indicating that normal transmitter release occurred (Figure 15–2C). Thus, neither Na$^+$ nor K$^+$ flux is required for transmitter release.

In the presence of tetraethylammonium, the current pulse elicits presynaptic depolarization throughout the duration of the pulse because the K$^+$ current that normally repolarizes the presynaptic membrane is blocked. As a result, transmitter release is sustained throughout the current pulse as reflected in the prolonged depolarization of the postsynaptic cell (Figure 15–2C). Quantification of the sustained depolarization was used by Katz and Miledi to determine a complete input–output curve relating presynaptic depolarization to transmitter release (Figure 15–2D). They confirmed the steep dependence of transmitter release on presynaptic depolarization. In the range of depolarization over which transmitter release increases (40–70 mV

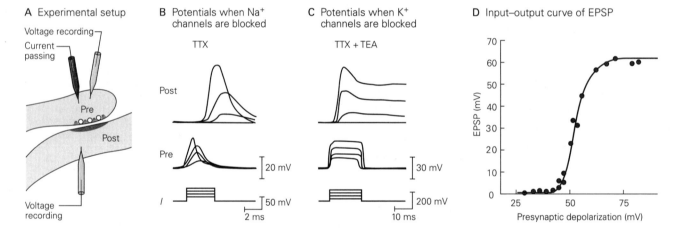

Figure 15–2 Transmitter release is not directly triggered by the opening of presynaptic voltage-gated Na$^+$ or K$^+$ channels. (Adapted, with permission, from Katz and Miledi 1967a.)

A. Voltage recording electrodes are inserted in both the pre- and postsynaptic fibers of the giant synapse in the stellate ganglion of a squid. A current-passing electrode has also been inserted into the presynaptic cell.

B. Depolarizing the presynaptic terminal with direct current injection through a microelectrode can trigger transmitter release even after the voltage-gated Na$^+$ channels are completely blocked by adding tetrodotoxin (**TTX**) to the cell-bathing solution. Three sets of traces represent (from bottom to top) the depolarizing current pulse (*I*) injected into the presynaptic terminal, the resulting potential in the presynaptic terminal (**Pre**), and the EPSP generated by the release of transmitter onto the postsynaptic cell (**Post**). Progressively stronger current pulses in the presynaptic cell produce correspondingly greater depolarizations of the presynaptic terminal. The greater the presynaptic depolarization, the larger is the EPSP. The presynaptic depolarizations are not maintained throughout the duration of

the depolarizing current pulse because delayed activation of the voltage-gated K$^+$ channels causes repolarization.

C. Transmitter release occurs even after the voltage-gated Na$^+$ channels have been blocked with TTX *and* the voltage-gated K$^+$ channels have been blocked with tetraethylammonium (**TEA**). In this experiment, TEA was injected into the presynaptic terminal. The three sets of traces represent the same measurements as in part B. Because the presynaptic K$^+$ channels are blocked, the presynaptic depolarization is maintained throughout the current pulse. The large sustained presynaptic depolarization produces large sustained EPSPs.

D. Blocking both the Na$^+$ and K$^+$ channels permits accurate control of presynaptic voltage and the determination of a complete input–output curve. Beyond a certain threshold (40 mV positive to the resting potential), there is a steep relationship between presynaptic depolarization and transmitter release, as measured from the size of the EPSP. Depolarizations greater than a certain level do not cause any additional release of transmitter. The initial presynaptic resting membrane potential was approximately –70 mV.

positive to the resting level), a 10-mV increase in presynaptic depolarization produces as much as a 10-fold increase in transmitter release. Depolarization of the presynaptic membrane above an upper limit produces no further increase in the postsynaptic potential.

Release Is Triggered by Calcium Influx

Katz and Miledi next turned their attention to Ca^{2+} ions. Earlier, Katz and José del Castillo had found that increasing the extracellular Ca^{2+} concentration enhanced transmitter release, whereas lowering the concentration reduced and ultimately blocked synaptic transmission. Because transmitter release is an intracellular process, these findings implied that Ca^{2+} must enter the cell to influence transmitter release.

Previous work on the squid giant axon membrane had identified a class of voltage-gated Ca^{2+} channels, the opening of which results in a large Ca^{2+} influx because of the large inward electrochemical driving force on Ca^{2+}. The extracellular Ca^{2+} concentration, approximately 2 mM in vertebrates, is normally four orders of magnitude greater than the intracellular concentration, approximately 10^{-7} M at rest. However, because these Ca^{2+} channels are sparsely distributed along the axon, they cannot, by themselves, provide enough current to produce a regenerative action potential.

Katz and Miledi found that the Ca^{2+} channels were much more abundant at the presynaptic terminal. There, in the presence of tetraethylammonium and tetrodotoxin, a depolarizing current pulse was sometimes able to trigger a regenerative depolarization that required extracellular Ca^{2+}, a *calcium spike*. Katz and Miledi therefore proposed that Ca^{2+} serves dual functions. It is a carrier of depolarizing charge during the action potential (like Na^+), and it is a special chemical signal—a second messenger—conveying information about changes in membrane potential to the intracellular machinery responsible for transmitter release. Calcium ions are able to serve as an efficient chemical signal because of their low intracellular resting concentration, approximately 10^5-fold lower than the resting concentration of Na^+. As a result, the small amount of Ca^{2+} ions that enter or leave a cell during an action potential can lead to large percentage changes in intracellular Ca^{2+} that can trigger various biochemical reactions. Proof of the importance of Ca^{2+} channels in release has come from more recent experiments showing that specific toxins that block Ca^{2+} channels also block release.

The properties of the voltage-gated Ca^{2+} channels at the squid presynaptic terminal were measured by Rodolfo Llinás and his colleagues. Using a voltage

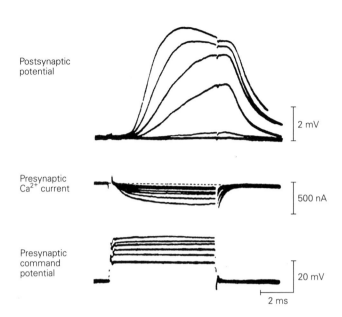

Figure 15–3 Transmitter release is regulated by Ca^{2+} influx into the presynaptic terminals through voltage-gated Ca^{2+} channels. The voltage-sensitive Na^+ and K^+ channels in a squid giant synapse were blocked by tetrodotoxin and tetraethylammonium. The membrane of the presynaptic terminal was voltage-clamped and membrane potential stepped to six different command levels of depolarization (**bottom**). The amplitude of the postsynaptic depolarization (**top**) varies with the size of the presynaptic inward Ca^{2+} current (**middle**) because the amount of transmitter release is a function of the concentration of Ca^{2+} in the presynaptic terminal. The notch in the postsynaptic potential trace is an artifact that results from turning off the presynaptic command potential. (Adapted, with permission, from Llinás and Heuser 1977.)

clamp, Llinás depolarized the terminal while blocking the voltage-gated Na^+ channels with tetrodotoxin and the K^+ channels with tetraethylammonium. He found that graded depolarizations activated a graded inward Ca^{2+} current, which in turn resulted in graded release of transmitter (Figure 15–3). The Ca^{2+} current is graded because the Ca^{2+} channels are voltage-dependent like the voltage-gated Na^+ and K^+ channels. Calcium ion channels in squid terminals differ from Na^+ channels, however, in that they do not inactivate quickly but stay open as long as the presynaptic depolarization lasts.

Calcium channels are largely localized in presynaptic terminals at *active zones*, the sites where neurotransmitter is released, exactly opposite the postsynaptic receptors (Figure 15–4). This localization is important as Ca^{2+} ions do not diffuse long distances from their site of entry because free Ca^{2+} ions are rapidly buffered by Ca^{2+}-binding proteins. As a result, Ca^{2+} influx creates a sharp local rise in Ca^{2+} concentration at the active zones. This rise in Ca^{2+} in the presynaptic terminals can be visualized using Ca^{2+}-sensitive

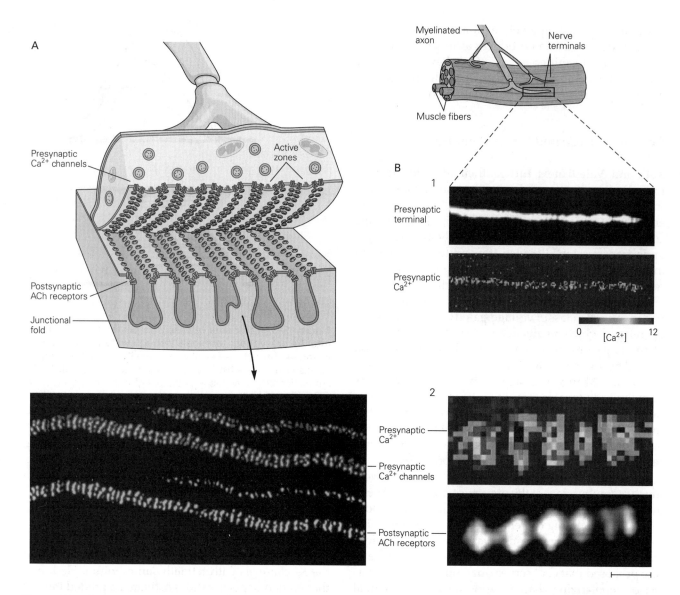

Figure 15–4 Calcium flowing into the presynaptic nerve terminal during synaptic transmission at the neuromuscular junction is concentrated at the active zone. Calcium channels in presynaptic terminals at the end-plate are concentrated opposite clusters of nicotinic acetylcholine (**ACh**) receptors on the postsynaptic muscle membrane. Two drawings show the frog neuromuscular junction.

A. The enlarged view shows the microanatomy of the neuromuscular junction with the presynaptic terminal peeled back. A fluorescent image shows the presynaptic Ca^{2+} channels (labeled with a Texas red-coupled marine snail toxin that binds to Ca^{2+} channels) and postsynaptic ACh receptors (labeled with fluorescently tagged α-bungarotoxin, which binds selectively to ACh receptors). The two images are normally superimposed but have been separated for clarity. The patterns of labeling with both probes are in almost precise register, indicating that the active zone of the presynaptic neuron is in almost perfect alignment with the postsynaptic membrane containing the high concentration of ACh receptors. (Reproduced, with permission, from Robitaille, Adler, and Charlton 1990.)

B. Calcium influx in presynaptic terminals is localized at active zones. Calcium can be visualized using Ca^{2+}-sensitive fluorescent dyes. **1.** A presynaptic terminal at a neuromuscular junction filled with the dye fura-2 under resting conditions is shown in the black and white image. The fluorescence intensity of the dye changes as it binds Ca^{2+}. In the color image, color-coded fluorescence intensity changes show local hotspots of intracellular Ca^{2+} in response to a single presynaptic action potential. **Red** indicates regions with a large increase in Ca^{2+}; **blue** indicates regions with little increase in Ca^{2+}. Regular peaks in Ca^{2+} concentration are seen along the terminal, corresponding to the localization of Ca^{2+} channels at the active zones.
2. The color image shows a high-magnification view of the peak increase in terminal Ca^{2+} levels. The corresponding black and white image shows fluorescence labeling of nicotinic ACh receptors in the postsynaptic membrane, illustrating the close spatial correspondence between areas of presynaptic Ca^{2+} influx and areas of postsynaptic receptors. Scale bar = 2 μm. (Reproduced, with permission, from Wachman et al. 2004. Copyright © 2004 Society for Neuroscience.)

fluorescent dyes (Figure 15–4B). One striking feature of transmitter release at all synapses is its steep and nonlinear dependence on Ca^{2+} influx; a 2-fold increase in Ca^{2+} influx can increase the amount of transmitter released by more than 16-fold. This relationship indicates that at some regulatory site, the *calcium sensor*, the cooperative binding of several Ca^{2+} ions is required to trigger release.

The Relation Between Presynaptic Calcium Concentration and Release

How much Ca^{2+} is necessary to induce release of neurotransmitters? To address this question, Bert Sakmann and Erwin Neher and their colleagues measured synaptic transmission in the calyx of Held, a large synapse in the mammalian auditory brain stem, composed of axons from the cochlear nucleus to the medial nucleus of the trapezoid body. This synapse is specialized for very rapid and reliable transmission to allow for precise localization of sound in the environment.

The calyx forms a cup-like presynaptic terminal that engulfs a postsynaptic cell body (Figure 15–5A). The calyx synapse includes almost a thousand active zones that function as independent release sites. This enables a presynaptic action potential to release a large amount of transmitter that results in a reliably large postsynaptic depolarization. In contrast, individual synaptic boutons of a typical neuron in the brain contain only a single active zone. Because the calyx terminal is large, it is possible to insert electrodes into both the pre- and postsynaptic structures, much as with the squid giant synapse, and directly measure the synaptic coupling between the two compartments. This paired recording allows a precise determination of the time course of activity in the presynaptic and postsynaptic cells (Figure 15–5B).

These recordings revealed a brief lag of 1 to 2 ms between the onset of the presynaptic action potential and the EPSP, which accounts for what Sherrington termed the *synaptic delay*. Because Ca^{2+} channels open more slowly than Na^+ channels, and the inward Ca^{2+} driving force increases as the neuron repolarizes, Ca^{2+} does not begin to enter the presynaptic terminal in full force until the membrane has begun to repolarize. Surprisingly, once Ca^{2+} enters the terminal, transmitter is rapidly released with a delay of only a few hundred microseconds. Thus, the synaptic delay is largely attributable to the time required to open Ca^{2+} channels. The astonishing speed of Ca^{2+} action indicates that, prior to Ca^{2+} influx, the biochemical machinery underlying the release process must already exist in a primed and ready state. Such rapid kinetics are vital for neuronal

information processing and require elegant molecular mechanisms that we shall consider later.

A presynaptic action potential normally produces only a brief rise in presynaptic Ca^{2+} concentration because the Ca^{2+} channels open only for a short time. In addition, Ca^{2+} influx is localized at the active zone. These two properties contribute to a concentrated local pulse of Ca^{2+} that induces a burst of transmitter release (Figure 15–5B). As we shall see later in this chapter, the duration of the action potential regulates the amount of Ca^{2+} that flows into the terminal and thus the amount of transmitter release.

To determine how much Ca^{2+} is needed to trigger release, the Neher and Sakmann groups introduced into the presynaptic terminal an inactive form of Ca^{2+} complexed within a light-sensitive *chemical cage*. They also loaded the terminals with a Ca^{2+}-sensitive fluorescent dye to assay the intracellular free Ca^{2+} concentration. By uncaging the Ca^{2+} ions with a flash of light, they could trigger transmitter release by a uniform and quantifiable increase in Ca^{2+} concentration. These experiments revealed that a rise in Ca^{2+} concentration of less than 1 μM is sufficient to induce release of some transmitter, but approximately 10 to 30 μM Ca^{2+} is required to release the amount normally observed during an action potential. Here again, the relationship between Ca^{2+} concentration and transmitter release is highly nonlinear, consistent with a model in which at least four or five Ca^{2+} ions must bind to the Ca^{2+} sensor to trigger release (Figure 15–5C,D).

Several Classes of Calcium Channels Mediate Transmitter Release

Calcium channels are found in all nerve cells and in many nonneuronal cells. In skeletal and cardiac muscle cells, they are important for excitation-contraction coupling; in endocrine cells, they mediate release of hormones. Neurons contain five broad classes of voltage-gated Ca^{2+} channels: the L-type, P/Q-type, N-type, R-type, and T-type, which are encoded by distinct but closely related genes that can be divided into three gene families based on amino acid sequence similarity. L-type channels are encoded by the Ca_V1 family. Members of the Ca_V2 family comprise P/Q- ($Ca_V2.1$), N- ($Ca_V2.2$), and R-type ($Ca_V2.3$) channels. Finally, T-type channels are encoded by the Ca_V3 gene family. Each channel type has specific biophysical and pharmacological properties and physiological functions (Table 15–1).

Calcium channels are multimeric proteins whose distinct properties are determined by their poreforming subunit, the α_1-subunit. The α_1-subunit is homologous to the α-subunit of the voltage-gated Na^+

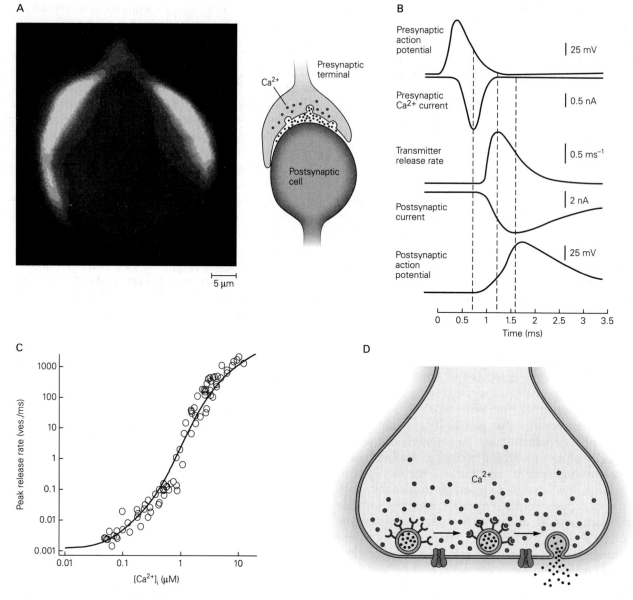

Figure 15–5 The precise relation between presynaptic Ca²⁺ and transmitter release at a central synapse has been measured. (Reproduced, with permission, from Meinrenken, Borst, and Sakmann 2003, and Sun et al. 2007. Parts A and B: Copyright © 2003 John Wiley and Sons.)

A. The large presynaptic terminal of the calyx of Held in the mammalian brain stem engulfs a postsynaptic cell body. The fluorescence image at left shows a calyx filled with a Ca²⁺-sensitive dye.

B. Time courses for several synaptic events. The **dashed lines** indicate the timing of the peak responses for the Ca²⁺ current, transmitter release, and postsynaptic current.

C. Transmitter release is steeply dependent on the Ca²⁺ concentration in the presynaptic terminal. The calyx was loaded with a caged Ca²⁺ compound that releases its bound Ca²⁺ in

response to a flash of ultraviolet light and with a Ca²⁺-sensitive dye that allows the intracellular Ca²⁺ concentration to be measured. By controlling the intensity of light, one can regulate the increase in Ca²⁺ in the presynaptic terminal. The plot, on a logarithmic scale, shows the relation between the rate of vesicle release and intracellular Ca²⁺ concentration. The **blue line** depicts a fit of the data by a model that assumes that release is triggered by a major Ca²⁺ sensor that binds five Ca²⁺ ions, resulting in a Ca²⁺ cooperativity of five. Due to the nonlinear relationship between Ca²⁺ and release, small increments in Ca²⁺ at concentrations of more than 1 μm cause massive increases in release.

D. The release of transmitter from a vesicle requires the binding of five Ca²⁺ ions to a Ca²⁺-sensing synaptic vesicle protein. In the figure, Ca²⁺ ions bind to five sensors present on a single vesicle; in reality, each sensor molecule binds multiple Ca²⁺ ions.

Table 15–1 Voltage-Gated Ca^{2+} Channels of Neurons

Channel	Former name	Ca^{2+} channel type	Tissue	Blocker	Voltage dependence[1]	Function
Ca$_V$1.1–1.4	$\alpha_{1C,D,F,S}$	L	Muscle, neurons	Dihydropyridines	HVA	Contraction, slow and some limited fast release
Ca$_V$2.1	α_{1A}	P/Q	Neurons	ω-Agatoxin (spider venom)	HVA	Fast release +++
Ca$_V$2.2	α_{1B}	N	Neurons	ω-Conotoxin (cone snail venom)	HVA	Fast release ++
Ca$_V$2.3	α_{1E}	R	Neurons	SNX-482 (tarantula venom)	HVA	Fast release +
Ca$_V$3.1–3.3	$\alpha_{1G,H,I}$	T	Muscle, neurons	Mibefradil (limited selectivity)	LVA	Pacemaker firing

[1]HVA, high voltage activated; LVA, low voltage activated.

channel, comprised of four repeats of a domain with six membrane-spanning segments that includes the S4 voltage-sensor and pore-lining P-region (see Figure 8–10). Calcium channels also have auxiliary subunits (termed α_2, β, γ, and δ) that modify the properties of the channel formed by the α_1-subunit. The subcellular localization in neurons of different types of calcium channels also varies. The N- and P/Q-type Ca^{2+} channels are found predominantly in the presynaptic terminal, whereas L-, R-, and T-type channels are found largely in the soma and dendrites.

Four of the types of voltage-gated Ca^{2+} channels—the L-type, P/Q-type, N-type, and R-type—generally require fairly strong depolarization to be activated (voltages positive to –40 to –20 mV are required) and thus are sometimes loosely referred to as *high-voltage-activated* Ca^{2+} channels (Table 15–1). In contrast, T-type channels open in response to small depolarizations around the threshold for generating an action potential (–60 to –40 mV) and are therefore called *low-voltage-activated* Ca^{2+} channels. Because they are activated by small changes in membrane potential, the T-type channels help control excitability at the resting potential and are an important source of the excitatory current that drives the rhythmic pacemaker activity of certain cells in both the brain and heart.

In neurons, the rapid release of conventional transmitters during fast synaptic transmission is mediated mainly by P/Q-type and N-type Ca^{2+} channels, the channel types most concentrated at the active zone. The localization of N-type Ca^{2+} channels at the frog neuromuscular junction has been visualized using a fluorescence-labeled snail toxin that binds selectively to these channels (see Figure 15–4A). The L-type channels are not found in the active zone and thus do not normally contribute to the fast release of conventional transmitters such as ACh and glutamate. However, Ca^{2+} influx through L-type channels is important for slower forms of release that do not occur at specialized active zones, such as the release of neuropeptides from neurons and of hormones from endocrine cells. As we shall see later, regulation of Ca^{2+} influx into presynaptic terminals controls the amount of transmitter release and hence the strength of synaptic transmission.

Mutations in voltage-gated Ca^{2+} channels are responsible for certain acquired and genetic diseases. Timothy syndrome, a developmental disorder characterized by a severe form of autism with impaired cognitive function and a range of other pathophysiological changes, results from a mutation in the α_1-subunit of L-type channels that alters their voltage-dependent gating, thereby affecting dendritic integration. Different point mutations in the P/Q-type channel α_1-subunit

give rise to hemiplegic migraine or epilepsy. Patients with Lambert-Eaton syndrome, an autoimmune disease associated with muscle weakness, make antibodies to the P/Q-type channel α_1-subunit that decrease total Ca^{2+} current (Chapter 57).

Transmitter Is Released in Quantal Units

How does the influx of Ca^{2+} trigger transmitter release? Katz and his colleagues provided a key insight into this question by showing that transmitter is released in discrete amounts they called *quanta*. Each quantum of transmitter produces a postsynaptic potential of fixed size, called the *quantal synaptic potential*. The total postsynaptic potential is made up of a large number of quantal potentials. EPSPs seem smoothly graded in amplitude only because each quantal (or unit) potential is small relative to the total potential.

Katz and Fatt obtained the first clue as to the quantal nature of synaptic transmission in 1951 when they observed spontaneous postsynaptic potentials of approximately 0.5 mV at the nerve-muscle synapse of the frog. Like end-plate potentials evoked by nerve stimulation, these small depolarizing responses are largest at the site of nerve-muscle contact and decay electrotonically with distance (see Figure 12–5). Small spontaneous potentials have since been observed in mammalian muscle and in central neurons. Because postsynaptic potentials at vertebrate nerve-muscle synapses are called end-plate potentials, Fatt and Katz called these spontaneous potentials *miniature end-plate potentials*.

Several results convinced Fatt and Katz that the miniature end-plate potentials represented responses to the release of small amounts of ACh, the neurotransmitter used at the nerve-muscle synapse. The time course of the miniature end-plate potentials and the effects of various drugs on them are indistinguishable from the properties of the end-plate potential. Like the end-plate potentials, the miniature end-plate potentials are enhanced and prolonged by prostigmine, a drug that blocks hydrolysis of ACh by acetylcholinesterase. Conversely, they are abolished by agents that block the ACh receptor, such as curare. The miniature end-plate potentials represent responses to small packets of transmitter that are spontaneously released from the presynaptic nerve terminal in the absence of an action potential. Their frequency can be increased by a small depolarization of the presynaptic terminal. They disappear if the presynaptic motor nerve degenerates and reappear when a new motor synapse is formed.

What could account for the small, fixed size of the miniature end-plate potential? Del Castillo and Katz first tested the possibility that each event represents a response to the opening of a *single* ACh receptor-channel. However, application of very small amounts of ACh to the frog muscle end-plate elicited depolarizing postsynaptic responses that were much smaller than the 0.5 mV response of a miniature end-plate potential. This finding made it clear that the miniature end-plate potential represents the opening of more than one ACh receptor-channel. In fact, Katz and Miledi were later able to estimate the voltage response to the elementary current through a single ACh receptor-channel as being only approximately 0.3 μV (Chapter 12). Based on this estimate, a miniature end-plate potential of 0.5 mV would represent the summation of the elementary currents of approximately 2,000 channels. Later work showed that a miniature end-plate potential is the response to the synchronous release of approximately 5,000 molecules of ACh.

What is the relationship of the large end-plate potential evoked by nerve stimulation and the small, spontaneous miniature end-plate responses? This question was first addressed by del Castillo and Katz in a study of synaptic signaling at the nerve-muscle synapse bathed in a solution low in Ca^{2+}. Under this condition, the end-plate potential is reduced markedly, from the normal 70 mV to about 0.5 to 2.5 mV. Moreover, the amplitude of each successive end-plate potential now varies randomly from one stimulus to the next; often, no response can be detected at all (termed *failures*). However, the minimum response above zero—the unit end-plate potential in response to a presynaptic action potential—is identical in amplitude (approximately 0.5 mV) and shape to the spontaneous miniature end-plate potentials. Importantly, the amplitude of each end-plate potential is an integral multiple of the unit potential (Figure 15–6).

Now del Castillo and Katz could ask: How does the rise of intracellular Ca^{2+} that accompanies each action potential affect the release of transmitter? They found that increasing the external Ca^{2+} concentration does not change the amplitude of the unit synaptic potential. However, the proportion of failures decreases and the incidence of higher-amplitude responses (composed of multiple quantal units) increases. These observations show that an increase in external Ca^{2+} concentration does not enhance the *size* of a quantum of transmitter (that is, the number of ACh molecules in each quantum) but rather acts to increase the average number of quanta that are released in response to a presynaptic action potential. The greater the Ca^{2+} influx into the terminal, the larger the number of transmitter quanta released.

Thus, three findings led del Castillo and Katz to conclude that transmitter is released in packets with a fixed amount of transmitter, a quantum: The amplitude of the end-plate potential varies in a stepwise manner at low levels of ACh release, the amplitude of each step increase is an integral multiple of the unit potential, and the unit potential has the same mean amplitude and shape as that of the spontaneous miniature end-plate potentials. Moreover, by analyzing the statistical distribution of end-plate potential amplitudes, del Castillo and Katz and other subsequent researchers were able to show that a single action potential produced a transient increase in the probability that a given quantum of transmitter is released according to a random process, similar to that governing the outcome of a coin toss (Box 15–1).

In the absence of an action potential, the rate of quantal release is low—only one quantum per second is released spontaneously at the end-plate. In contrast, the firing of an action potential releases approximately 150 quanta, each approximately 0.5 mV in amplitude, resulting in a large end-plate potential. Thus, the influx of Ca^{2+} into the presynaptic terminal during an action potential dramatically increases the rate of quantal release by a factor of 150,000, triggering the synchronous release of about 150 quanta in about 1 ms.

Transmitter Is Stored and Released by Synaptic Vesicles

What morphological features of the cell might account for the quantal release of transmitter? The physiological observations indicating that transmitter is released in fixed quanta coincided with the discovery, through electron microscopy, of accumulations of small clear vesicles in the presynaptic terminal. Del Castillo and Katz speculated that the vesicles were organelles for the storage of transmitter, each vesicle stored one quantum of transmitter (amounting to several thousand molecules), and each vesicle released its entire contents into the synaptic cleft in an all-or-none manner at sites specialized for release.

The sites of release, the active zones, contain a cloud of synaptic vesicles that cluster above a fuzzy electron-dense material attached to the internal face of the presynaptic membrane (see Figure 15–4A). At all synapses, the vesicles are typically clear, small, and ovoid, with a diameter of approximately 40 nm (in distinction with the large dense-core vesicles described in Chapter 16). Although most synaptic vesicles do not contact the active zone, some are physically bound. These are called the *docked* vesicles and are thought to

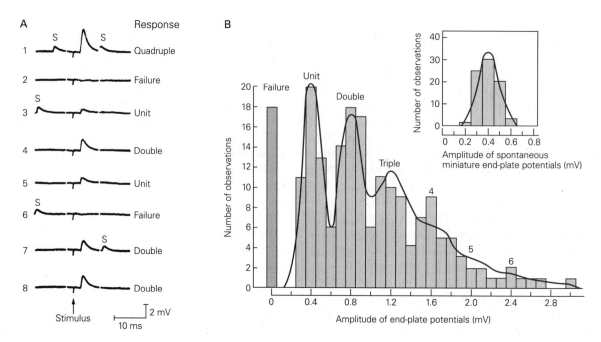

Figure 15–6 Neurotransmitter is released in fixed increments. Each increment or quantum of transmitter produces a unit end-plate potential of fixed amplitude. The amplitude of the response evoked by nerve stimulation is thus equal to the amplitude of the unit end-plate potential multiplied by the number of quanta of transmitter released.

A. Intracellular recordings from a muscle fiber at the end-plate show the change in postsynaptic potential when eight consecutive stimuli of the same size are applied to the motor nerve. To reduce transmitter release and to keep the end-plate potentials small, the tissue is bathed in a Ca^{2+}-deficient (and magnesium-rich) solution. The postsynaptic responses to the nerve stimulus vary. Two of the eight presynaptic stimuli elicit no EPSP (failures), two produce unit potentials, and the others produce EPSPs that are approximately two to four times the amplitude of the unit potential. Note that the spontaneous miniature end-plate potentials (**S**), which occur at random intervals in the traces, are the same size as the unit potential. (Adapted, with permission, from Liley 1956.)

B. After many end-plate potentials are recorded, the number of responses with a given amplitude is plotted as a function of this amplitude in the histogram shown here. The distribution of responses falls into a number of peaks. The first peak, at 0 mV, represents failures. The first peak of responses, at 0.4 mV, represents the unit potential, the smallest elicited response. The unit response has the same amplitude as the spontaneous

miniature end-plate potentials (inset), indicating that the unit response is caused by the release of a single quantum of transmitter. The other peaks in the histogram are integral multiples of the amplitude of the unit potential; that is, responses are composed of two, three, four, or more quantal events.

The number of responses under each peak divided by the total number of events in the entire histogram is the probability that a single presynaptic action potential triggers the release of the number of quanta that comprise the peak. For example, if there are 30 events in the peak corresponding to the release of two quanta out of a total of 100 events recorded, the probability that a presynaptic action potential releases exactly two quanta is 30/100 or 0.3. This probability follows a Poisson distribution (**red curve**). This theoretical distribution is composed of the sum of several Gaussian functions. The spread of the unit peak (standard deviation of the Gaussian function) reflects the fact that the amount of transmitter in a quantum, and hence the amplitude of the quantal postsynaptic response, varies randomly about a mean value. The successive Gaussian peaks widen progressively because the variability (or variance) associated with each quantal event increases linearly with the number of quanta per event. The distribution of amplitudes of the spontaneous miniature potentials (inset) is fit by a Gaussian curve whose width is identical to that of the Gaussian curve for the unit synaptic responses. (Adapted, with permission, from Boyd and Martin 1956.)

be the ones immediately available for release (sometimes referred to as the *readily releasable pool*). At the neuromuscular junction, the active zones are linear structures (see Figure 15–4), whereas in central synapses, they are disc-shaped structures approximately 0.1 μm² in area with dense projections pointing into the cytoplasm. Active zones are generally found in precise apposition to the postsynaptic membrane

patches that contain the neurotransmitter receptors (see Figure 13–2). Thus, presynaptic and postsynaptic specializations are functionally and morphologically attuned to each other, sometimes precisely aligned in structural "nanocolumns." As we shall learn later, several key active zone proteins involved in transmitter release have now been identified and characterized.

Box 15–1 Synaptic Strength Depends on the Probability of Transmitter Release and Other Quantal Parameters

The mean size of a synaptic response E evoked by an action potential has often been described as the product of the total number of releasable quanta (n), the probability that an individual quantum of transmitter is released (p), and the size of the response to a quantum (a):

$$E = n \cdot p \cdot a.$$

These parameters are statistical terms, useful for describing the size and variability of the postsynaptic response. At some but not all central synapses, they can also be assigned to biological processes. We begin by focusing on synapses of the kind envisioned by Katz and colleagues, where the interpretation of the parameters is most straightforward. At these synapses, the presynaptic terminal typically contains multiple active zones, and each active zone releases at most a single vesicle in response to an action potential (*univesicular release*).

We then consider another kind of synapse that requires a different interpretation. At these synapses, each active zone can release multiple vesicles in response to a single action potential (*multivesicular release*), leading to very high concentrations of transmitter in the synaptic cleft that can cause the postsynaptic receptors to become saturated with transmitter.

Univesicular Release at Multiple Active Zones

In the simplest case, the parameter a is the response of the postsynaptic membrane to the release of a single vesicle's contents of transmitter. It is assumed that transmitter is packaged in synaptic vesicles, that release of the contents of a vesicle is a stereotyped, all-or-none event, and that single release events occur in physical isolation from each other. Quantal size depends on the amount of transmitter in a vesicle and on the properties of the postsynaptic cell, such as the membrane resistance and capacitance (which can be independently estimated) and the responsiveness of the postsynaptic membrane to the transmitter substance. This can also be measured experimentally by the

postsynaptic membrane's response to the application of a known amount of transmitter.

The parameter n describes the maximum number of quantal units that can be released in response to a single action potential if the probability p reaches 1.0. At some central synapses, this maximum may be imposed by the number of release sites (active zones) in the terminals of a presynaptic neuron that contact a given postsynaptic neuron. Multiple studies have found that for this kind of connection n corresponds with the number of release sites determined by electron microscopy, as if those sites obeyed a rough rule wherein a presynaptic action potential triggers the exocytosis of at most one vesicle per active zone.

The parameter p represents the likelihood of vesicle release. This likelihood encompasses a series of events necessary for a particular release site to contribute a quantal event: (1) The active zone must be loaded with at least one releasable vesicle (a process referred to as vesicle mobilization); (2) the presynaptic action potential must evoke Ca^{2+} influx in sufficient quantity and proximity to the vesicle; and (3) the Ca^{2+}-sensitive synaptotagmin and SNARE machinery must cause the vesicle to fuse and discharge its contents.

Here, we focus mainly on the determinants of p. We can treat quantal release at a single active zone as a random event with only two possible outcomes in response to an action potential—the quantum of transmitter is or is not released. Because the quantal responses from different active zones are thought to occur independently of each other in some situations, this is similar to tossing a set of n coins in the air and counting the number of heads or tails. The equivalent of individual coin flips (Bernoulli trials) are then totaled up in a binomial distribution, where p stands for the average probability of success (that is, the probability that any given quantum will be released) and q (equal to $1 - p$) stands for the mean probability of failure.

Both the average probability (p) that an individual quantum will be released and the maximal number (n)

(continued)

Quantal transmission has been demonstrated at all chemical synapses so far examined. Nevertheless, the efficacy of transmitter release from a single presynaptic cell onto a single postsynaptic cell varies widely in the nervous system and depends on several factors: (1) the number of individual synapses

between a pair of presynaptic and postsynaptic cells (that is, the number of presynaptic boutons that contact the postsynaptic cell); (2) the number of active zones in an individual synaptic terminal; and (3) the probability that a presynaptic action potential will trigger release of one or more quanta of transmitter

Box 15–1 Synaptic Strength Depends on the Probability of Transmitter Release and Other Quantal Parameters (continued)

of releasable quanta are assumed to be constant. (Any reduction in the store of vesicles is assumed to be quickly replenished after each stimulus.) The product of n and p yields an estimate m of the mean number of quanta that will be released. This mean is called the *quantal content* or *quantal output*.

Calculation of the probability of transmitter release can be illustrated with the following example. Consider a terminal that has a releasable store of five quanta ($n = 5$). Assuming $p = 0.1$, then the probability that an individual quantum will not be released from the terminals (q) is $1 - p$, or 0.9. We can now determine the probability that a stimulus will release no quanta (failure), a single quantum, or any other number of quanta (up to n).

The probability that none of the five available quanta will be released by a given stimulus is the product of the individual probabilities that each quantum will not be released: $q^5 = (0.9)^5$, or 0.59. We would thus expect to see 59 failures in a hundred stimuli. The probabilities of observing zero, one, two, three, four, or five quanta are represented by the successive terms of the binomial expansion:

$$(q + p)^5 = q^5 \text{ (failures)} + 5\,q^4p \text{ (1 quantum)}$$
$$+ 10\,q^3p^2 \text{ (2 quanta)} + 10\,q^2p^3 \text{ (3 quanta)}$$
$$+ 5\,qp^4 \text{ (4 quanta)} + p^5 \text{ (5 quanta)}.$$

Thus, in 100 stimuli, the binomial expansion would predict 33 single unit responses, 7 double responses, 1 triple response, and 0 quadruple or quintuple responses.

Values for the quantal output m vary from approximately 100 to 300 at the vertebrate nerve-muscle synapse, the squid giant synapse, and *Aplysia* central synapses, to as few as 1 to 4 in the synapses of the sympathetic ganglion and spinal cord of vertebrates. The probability of release p also varies, ranging from as high as 0.7 at the neuromuscular junction in the frog and 0.9 in the crab down to around 0.1 at some mammalian central synapses. Estimates for n range from as much as 1,000

(at the vertebrate nerve-muscle synapse) to 1 (at single terminals of mammalian central neurons).

This numerical example illustrates a characteristic feature of synapses with simple binomial features— their substantial variability. This holds just as strongly whether p is high or low. For example, for $p = 0.9$ and 100 stimuli, the binomial expansion predicts 0 failures, 0 single unit responses, 1 double response, 7 triple responses, 33 quadruple responses, and 59 quintuple responses, the mirror-image of the distribution for $p = 0.1$. Even if each sequential event that supports vesicle release is highly likely, the aggregate strength of the synapse will vary widely.

Multivesicular Release with Receptor Saturation

One well-studied mechanism for achieving high synaptic reliability is through the release of multiple vesicles onto a single postsynaptic site. In the extreme, this can release sufficient amounts of transmitter in the synaptic cleft to cause the postsynaptic receptor binding sites to become fully occupied by transmitter (receptor saturation).

Under these conditions, the postsynaptic response will reach a maximal amplitude. Further release of transmitter, for example in response to a modulatory neurotransmitter, would fail to increase the postsynaptic response. Variability in response size would shrink greatly if, say, three to five vesicles worth of transmitter activated the same number of receptors as a single vesicle. The postsynaptic response would be highly stereotyped (it would appear to result from release of a single quantum of transmitter) even though the presynaptic terminal was releasing multiple vesicles. However, the binomial treatment could still retain some usefulness as a way of adding up the contributions of multiple synapses of this kind, so long as each synapse released transmitter simultaneously and independently. But in such a case, n, p, and a would take on biological meanings different from those in which only a single vesicle could be released per synapse.

at an active zone. As we will see later, release probability can be powerfully regulated as a function of neuronal activity.

In the central nervous system, most presynaptic boutons have only a single active zone where an action potential usually releases at most a single quantum

of transmitter in an all-or-none manner. However, at some central synapses, such as the calyx of Held, transmitter is released from a large presynaptic terminal that may contain many active zones and thus can release a large number of quanta in response to a single presynaptic action potential. Central neurons also vary in the number of synapses that a typical presynaptic cell forms with a typical postsynaptic cell. Whereas most central neurons form only a few synapses with any one postsynaptic cell, a single climbing fiber from neurons in the inferior olive forms up to 10,000 terminals on a single Purkinje neuron in the cerebellum! Finally, the mean probability of transmitter release from a single active zone also varies widely among presynaptic terminals, from less than 0.1 (that is, a 10% chance that a presynaptic action potential will trigger release of a vesicle) to greater than 0.9. This wide range of probabilities can even be seen among the boutons at individual synapses between a specific type of presynaptic cell and a specific type of postsynaptic cell.

Thus, central neurons vary widely in the efficacy and reliability of synaptic transmission. Synaptic *reliability* is defined as the probability that an action potential in a presynaptic cell leads to some measurable response in the postsynaptic cell—that is, the probability that a presynaptic action potential will release one or more quanta of transmitter. *Efficacy* refers to the mean amplitude of the synaptic response, which depends on both the reliability of synaptic transmission and on the mean size of the response when synaptic transmission does occur.

Most central neurons communicate at synapses that have a low probability of transmitter release. The high failure rate of release at most central synapses (that is, their low release probability) is not a *design defect* but serves a purpose. As we discuss later, this feature allows transmitter release to be regulated over a wide dynamic range, which is important for adapting neural signaling to different behavioral demands. In synaptic connections where a low probability of release is deleterious for function, this limitation can be overcome by simply having many active zones in one synapse, as is the case at the calyx of Held and the nerve-muscle synapse. Both contain hundreds of independent active zones, so an action potential reliably releases 150 to 250 quanta, ensuring that a presynaptic signal is always followed by a postsynaptic action potential. Reliable transmission at the neuromuscular junction is essential for survival. An animal would not survive if its ability to move away from a predator was hampered by a low-probability response. Another strategy for increasing reliability is to use multivesicular release, the simultaneous fusion of multiple vesicles at a single active zone, to ensure that postsynaptic receptors are consistently exposed to a saturating concentration of neurotransmitter (see Box 15–1).

Not all chemical signaling between neurons depends on the synaptic machinery described earlier. Some substances, such as certain lipid metabolites and the gas nitric oxide (Chapter 14), can diffuse across the lipid bilayer of the membrane. Others can be moved out of nerve endings by carrier proteins if their intracellular concentration is sufficiently high. Plasma membrane transporters for glutamate or GABA normally take up transmitter into a cell from the synaptic cleft following a presynaptic action potential (Chapter 13). However, in some glial cells of the retina, the direction of glutamate transport can be reversed under certain conditions, causing glutamate to leave the cell through the transporter into the synaptic cleft. Still other substances simply leak out of nerve terminals at a low rate. Surprisingly, approximately 90% of the ACh that leaves the presynaptic terminals at the neuromuscular junction does so through continuous leakage. This leakage is ineffective, however, because it is diffuse and not targeted to receptors at the end-plate region and because it is continuous and low level rather than synchronous and concentrated.

Synaptic Vesicles Discharge Transmitter by Exocytosis and Are Recycled by Endocytosis

The quantal hypothesis of del Castillo and Katz has been amply confirmed by direct experimental evidence that synaptic vesicles do indeed package neurotransmitter and that they release their contents by directly fusing with the presynaptic membrane, a process termed *exocytosis*.

Forty years ago, Victor Whittaker discovered that the synaptic vesicles in the motor nerve terminals of the electric organ of the electric fish *Torpedo* contain a high concentration of ACh. Later, Thomas Reese and John Heuser and their colleagues obtained electron micrographs that caught vesicles in the act of exocytosis. To observe the brief exocytotic event, they rapidly froze the nerve-muscle synapse by immersing it in liquid helium at precisely defined intervals after the presynaptic nerve was stimulated. In addition, they increased the number of quanta of transmitter discharged with each nerve impulse by applying the drug 4-aminopyridine, a compound that blocks certain voltage-gated K^+ channels, thus increasing the duration of the action potential and enhancing Ca^{2+} influx. (The spike broadening produced by this pharmacological intervention resembles spike broadening resulting from cumulative inactivation of K^+ channels during repetitive firing; see Figure 15–15C.) In both cases, prolonged action potentials evoke greater opening of presynaptic Ca^{2+} channels.

These techniques provided clear images of synaptic vesicles at the active zone during exocytosis. Using a technique called *freeze-fracture electron microscopy*,

Reese and Heuser noted deformations of the presynaptic membrane along the active zone immediately after synaptic activity, which they interpreted as invaginations of the cell membrane caused by fusion of synaptic vesicles. These deformations lay along one or two rows of unusually large intramembranous particles, visible along both margins of the presynaptic density. Many of these particles are now thought to be voltage-gated Ca^{2+} channels (Figure 15–7). The particle density (approximately 1,500 per μm^2) is similar to the Ca^{2+} channel density that is thought to be present in the presynaptic plasma membrane at the active zone. Moreover, the proximity of the particles to the release site is consistent with the short time interval between the onset of the Ca^{2+} current and the release of transmitter.

Finally, Heuser and Reese found that these deformations are transient; they occur only when vesicles are discharged and do not persist after transmitter has been released. Thin-section electron micrographs revealed a number of omega-shaped (Ω) structures with the appearance of synaptic vesicles that have just fused with the membrane, prior to the complete collapse of the vesicle membrane into the plasma membrane (Figure 15–7B). Heuser and Reese confirmed this idea by showing that the number of Ω-shaped structures is directly correlated with the size of the EPSP when they varied the concentration of 4-aminopyridine to alter the amount of transmitter release. These morphological studies provide striking evidence that transmitter is released from synaptic vesicles by means of exocytosis.

Following exocytosis, the excess membrane added to the presynaptic terminal is retrieved. In images of presynaptic terminals made 10 to 20 seconds after stimulation, Heuser and Reese observed new structures at the plasma membrane, the coated pits, which are formed by the protein *clathrin* that helps mediate membrane retrieval through the process of endocytosis (Figure 15–7C). Several seconds later, the coated pits are seen to pinch off from the membrane and appear as coated vesicles in the cytoplasm. As we will see later, endocytosis through coated pit formation represents one of several means of vesicle membrane retrieval.

Capacitance Measurements Provide Insight Into the Kinetics of Exocytosis and Endocytosis

In certain neurons with large presynaptic terminals, the increase in surface area of the plasma membrane during exocytosis can be detected in electrical measurements as increases in membrane capacitance. As we saw in Chapter 9, the capacitance of the membrane is proportional to its surface area. Erwin Neher discovered that one could use measurements of capacitance to monitor exocytosis in secretory cells.

In adrenal chromaffin cells (which release epinephrine and norepinephrine) and in mast cells of the rat peritoneum (which release histamine and serotonin), individual dense-core vesicles are large enough to permit measurement of the increase in capacitance associated with fusion of a single vesicle. Release of transmitter in these cells is accompanied by stepwise increases in capacitance, followed somewhat later by stepwise decreases, which reflect the retrieval and recycling of the excess membrane (Figure 15–8).

In neurons, the changes in capacitance caused by fusion of single, small synaptic vesicles are usually too small to resolve. In certain favorable synaptic preparations that release large numbers of vesicles (such as the giant presynaptic terminals of bipolar neurons in the retina), membrane depolarization triggers a transient smooth rise and fall in the total capacitance of the terminal as a result of the exocytosis and retrieval of the membrane from hundreds of individual synaptic vesicles (Figure 15–8C). These results provide direct measurements of the rates of membrane fusion and retrieval.

Exocytosis Involves the Formation of a Temporary Fusion Pore

Morphological studies of mast cells using rapid freezing suggest that exocytosis depends on the formation of a temporary fusion pore that spans the membranes of the vesicle and plasma membranes. In electrophysiological studies of capacitance increases in mast cells, a channel-like fusion pore was detected in the electrophysiological recordings prior to complete fusion of vesicles and cell membranes. This fusion pore starts out with a single-channel conductance of approximately 200 pS, similar to that of gap-junction channels, which also bridge two membranes. During exocytosis, the pore rapidly dilates, probably from around 5 to 50 nm in diameter, and the conductance increases dramatically (Figure 15–9A).

The fusion pore is not just an intermediate structure leading to exocytosis of transmitter, as transmitter can be released through the pore prior to pore expansion and vesicle collapse. This was first shown by amperometry, a method that uses an extracellular carbon-fiber electrode to detect certain amine neurotransmitters, such as serotonin, based on an electrochemical reaction between the transmitter and the electrode that generates an electrical current proportional to the local transmitter concentration. Firing of an action potential in serotonergic cells leads to a large transient increase in electrode

Cytoplasmic half of presynaptic membrane (freeze fracture)

Presynaptic membrane (thin section)

A Cell membrane at synapse

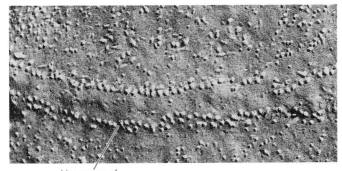

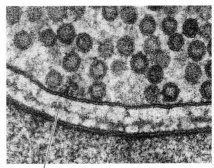

Linear array of
intramembranous particles

Synaptic cleft

B Exocytosis

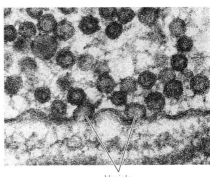

Vesicle
fusions

Vesicle
fusions

C Endocytosis

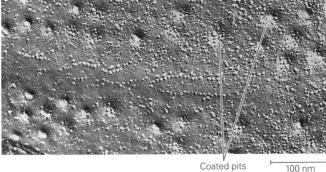

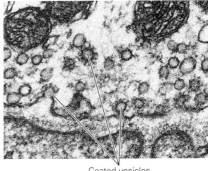

Coated pits

100 nm

Coated vesicles
and pits

Figure 15–7 Synaptic vesicles release transmitter by exocytosis and are retrieved by endocytosis. The images on the left are freeze-fracture electron micrographs at a neuromuscular junction. The freeze-fracture technique exposes the intramembranous area to view by splitting the membrane along the hydrophobic interior of the lipid bilayer. The views shown are of the cytoplasmic leaflet of the bilayer presynaptic membrane looking up from the synaptic cleft (see Figure 15–4A). Conventional thin-section electron micrographs on the right show cross-section views of the presynaptic terminal, synaptic cleft, and postsynaptic muscle membrane. (Reproduced, with permission, from Heuser and Reese 1981. Permission conveyed through Copyright Clearance Center, Inc.)

A. Parallel rows of intramembranous particles arrayed on either side of an active zone are thought to be the voltage-gated Ca^{2+} channels essential for transmitter release (see Figure 15–4A). The thin-section image at right shows the synaptic vesicles adjacent to the active zone.

B. Synaptic vesicles release transmitter by fusing with the plasma membrane (exocytosis). Here, synaptic vesicles are caught in the act of fusing with the plasma membrane by rapid freezing of the tissue within 5 ms after a depolarizing stimulus. Each depression in the plasma membrane represents the fusion of one synaptic vesicle. In the micrograph at right, fused vesicles are seen as Ω-shaped structures.

C. After exocytosis, synaptic vesicle membrane is retrieved by endocytosis. Within approximately 10 seconds after fusion of the vesicles with the presynaptic membrane, coated pits form. After another 10 seconds, the coated pits begin to pinch off by endocytosis to form coated vesicles. These vesicles store the membrane proteins of the original synaptic vesicle and also molecules captured from the extracellular medium. The vesicles are recycled at the terminals or are transported to the cell body, where the membrane constituents are degraded or recycled (see Chapter 7).

A Mast cell before and after exocytosis of secretory vesicles

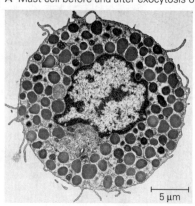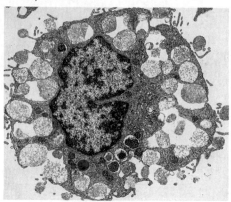

5 µm

B Membrane capacitance during and after exocytosis of mast cell vesicles

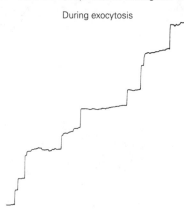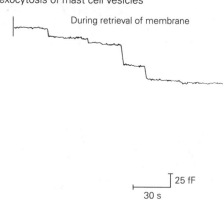

During exocytosis

During retrieval of membrane

25 fF

30 s

C Retinal bipolar neuron terminal

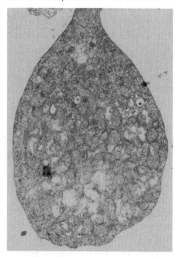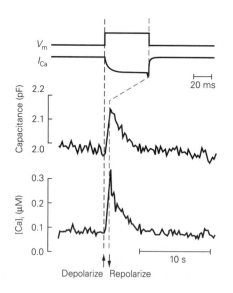

V_m

I_{Ca}

20 ms

Capacitance (pF)

2.2

2.1

2.0

$[Ca]_i$ (µM)

0.3

0.2

0.1

0.0

Depolarize   Repolarize

10 s

current, corresponding to the exocytosis of the contents of a single dense-core vesicle. In some instances, these large transient increases are preceded by smaller, longer-lasting current signals that reflect leakage of transmitter through a fusion pore that flickers open and closed several times prior to complete fusion (Figure 15–9B).

It is possible that transmitter can also be released solely through transient fusion pores that fleetingly connect vesicle lumen and extracellular space without full collapse of the vesicle membrane into the plasma membrane. Capacitance measurements for exocytosis of large dense-core vesicles in neuroendocrine cells show that the fusion pore can open and close rapidly and reversibly. The reversible opening and closing of a fusion pore represents a very rapid method of membrane retrieval. The circumstances under which the small clear vesicles at fast synapses discharge transmitter through a fusion pore, as opposed to full membrane collapse, are uncertain.

The Synaptic Vesicle Cycle Involves Several Steps

When firing at high frequency, a typical presynaptic neuron is able to maintain a high rate of transmitter release. This can result in the exocytosis of a large number of vesicles over time, more than the number morphologically evident within the presynaptic terminal. To prevent the supply of vesicles from being rapidly depleted during fast synaptic transmission, used vesicles are rapidly retrieved and recycled. Because nerve terminals are usually some distance from the cell body, replenishing vesicles by synthesis in the cell body and transport to the terminals would be too slow to be practical at fast synapses.

Synaptic vesicles are released and reused in a simple cycle. Vesicles fill with neurotransmitter and cluster in the nerve terminal. They then dock at the active zone where they undergo a complex *priming* process that makes vesicles competent to respond to the Ca^{2+} signal that triggers the fusion process (Figure 15–10A). Numerous mechanisms exist for retrieving the synaptic vesicle membrane following exocytosis, each with a distinct time course (Figure 15–10B).

The first, most rapid mechanism involves the reversible opening and closing of the fusion pore, without the full fusion of the vesicle membrane with the plasma membrane. In the *kiss-and-stay* pathway, the vesicle remains at the active zone after the fusion pore closes, ready for a second release event. In the *kiss-and-run* pathway, the vesicle leaves the active zone after the fusion pore closes, but is competent for rapid rerelease. These pathways are thought to be used preferentially during stimulation at low frequencies.

Jorgensen and colleagues have described a second pathway of *ultrafast* clathrin-independent endocytosis that is 200 times faster than the classical clathrin-mediated pathway. Beginning just 50 ms after exocytosis, ultrafast endocytosis occurs just outside of the active zone.

Stimulation at higher frequencies recruits a third, slower recycling pathway that uses clathrin to retrieve the vesicle membrane after fusion with the plasma membrane. Clathrin forms a lattice-like structure that surrounds the membrane during endocytosis, giving rise to the appearance of a coat around the coated pits observed by Heuser and Reese. In this pathway, the retrieved vesicular membrane must be recycled through an endosomal compartment before the vesicles can be reused. Clathrin-mediated recycling requires up

Figure 15–8 (Opposite) Changes in capacitance reveal the time course of exocytosis and endocytosis.

A. Electron micrographs show a mast cell before (*left*) and after (*right*) exocytosis. Mast cells are secretory cells of the immune system that contain large dense-core vesicles filled with the transmitters histamine and serotonin. Exocytosis of the secretory vesicles is normally triggered by the binding of antigen complexed to an immunoglobulin (IgE). Under experimental conditions, massive exocytosis can be triggered by the inclusion of a nonhydrolyzable analog of guanosine triphosphate (GTP) in an intracellular recording electrode. (Reproduced, with permission, from Lawson et al. 1977. Permission conveyed through Copyright Clearance Center, Inc.)

B. Stepwise increases in capacitance reflect the successive fusion of individual secretory vesicles with the mast cell membrane. The step increases are unequal because of variability in the membrane area of the vesicles. After exocytosis, the membrane added through fusion is retrieved through endocytosis. Endocytosis of individual vesicles gives rise to the stepwise decreases in membrane capacitance. In this way, the

cell maintains a constant size. (Units are femtofarads, where 1 fF = 0.1 μm^2 of membrane area.) (Adapted, with permission, from Fernandez, Neher, and Gomperts 1984.)

C. The giant presynaptic terminals of bipolar neurons in the retina are more than 5 μm in diameter, permitting direct patch-clamp recordings of membrane capacitance and Ca^{2+} current. A brief depolarizing voltage-clamp step in membrane potential (V_m) elicits a large sustained Ca^{2+} current (I_{Ca}) and a rise in the cytoplasmic Ca^{2+} concentration, $[Ca]_i$. This results in the fusion of several thousand small synaptic vesicles with the cell membrane, leading to an increase in total membrane capacitance. The increments in capacitance caused by fusion of individual vesicles are too small to resolve. As the internal Ca^{2+} concentration falls back to its resting level upon repolarization, the extra membrane area is retrieved and capacitance returns to its baseline value. The increases in capacitance and Ca^{2+} concentration outlast the brief depolarization and Ca^{2+} current (note different time scales) because of the relative slowness of endocytosis and Ca^{2+} metabolism. (Micrograph reproduced, with permission, from Zenisek et al. 2004. Copyright © 2004 Society for Neuroscience.)

A Electrical events associated with opening of fusion pore

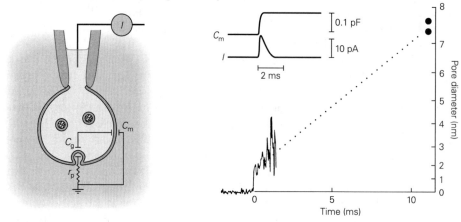

B Transmitter release through fusion pore

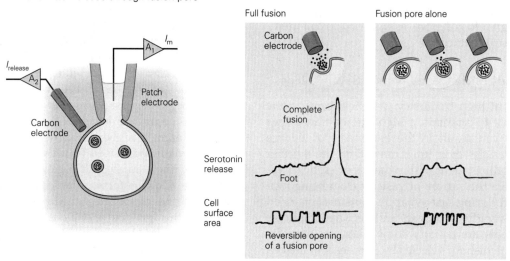

Figure 15–9 Reversible opening and closing of fusion pores.

A. A whole cell patch clamp is used to record membrane current associated with the opening of a fusion pore. As a vesicle fuses with the plasma membrane, the capacitance of the vesicle (C_g) is initially connected to the capacitance of the rest of the cell membrane (C_m) through the high resistance of the fusion pore (r_p). Because the membrane potential of the vesicle (lumenal side negative) is normally much more negative than the membrane potential of the cell, charge flows from the vesicle to the cell membrane during fusion. This transient current (I) is associated with the increase in membrane capacitance (C_m).

The magnitude of the conductance of the fusion pore (g_p) can be calculated from the time constant of the transient current according to $\tau = C_g r_p = C_g / g_p$. The pore diameter can be calculated from the pore conductance, assuming that the pore spans two lipid bilayers and is filled with a solution whose resistivity is equal to that of the cytoplasm. The plot on the right shows the pore has an initial conductance of approximately 200 pS, similar to the conductance of a gap-junction channel, corresponding to a pore diameter of approximately 2 nm. The pore diameter and conductance rapidly increase as the pore dilates to approximately 7 to 8 nm in 10 ms (**filled circles**). (Reproduced, with permission, from Monck and Fernandez 1992. Permission conveyed through Copyright Clearance Center, Inc; and adapted, with permission, from Spruce et al. 1990.)

B. Transmitter release is measured by amperometry. A cell is voltage-clamped with a whole cell patch electrode while an extracellular carbon fiber is pressed against the cell surface. A large voltage applied to the tip of the carbon electrode oxidizes certain amine transmitters (such as serotonin or norepinephrine). This oxidation of one molecule generates one or more free electrons, which results in an electrical current that is proportional to the amount of transmitter release. The current can be recorded through an amplifier (A_2) connected to the carbon electrode. Membrane current and capacitance are recorded through the patch electrode amplifier (A_1). Recordings of serotonin release (**top traces**) and capacitance measurements (**bottom traces**) from mast cell secretory vesicles are shown at the *right*. The records indicate that serotonin may be released through the reversible opening and closing of the fusion pore prior to full fusion (*traces on left*). During these brief openings, small amounts of transmitter escape through the pore, resulting in a low-level signal (a *foot*) that precedes a large spike of transmitter release upon full fusion. During the foot, the cell surface area (proportional to membrane capacitance) undergoes reversible step-like changes as the fusion pore opens and closes. Sometimes the reversible opening and closing of the fusion pore are not followed by full fusion (*traces on right*). (Adapted, with permission, from Neher 1993.)

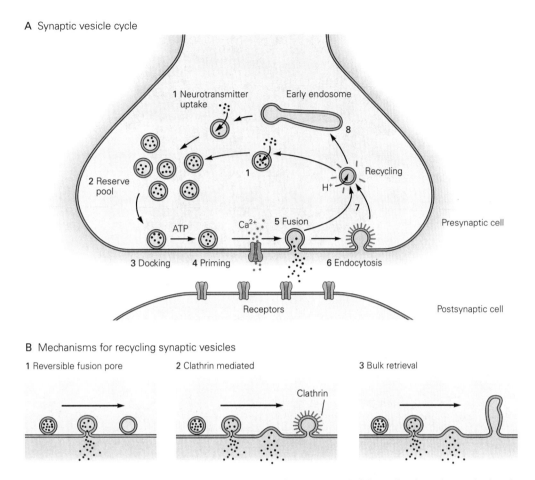

A Synaptic vesicle cycle

B Mechanisms for recycling synaptic vesicles

Figure 15–10 The synaptic vesicle cycle.

A. Synaptic vesicles are filled with neurotransmitters by active transport (**step 1**) and join the vesicle cluster that may represent a reserve pool (**step 2**). Filled vesicles dock at the active zone (**step 3**) where they undergo an ATP-dependent priming reaction (**step 4**) that makes them competent for Ca²⁺-triggered fusion (**step 5**). After discharging their contents, synaptic vesicles are recycled through one of several routes (see part B). In one common route, vesicle membrane is retrieved via clathrin-mediated endocytosis (**step 6**) and recycled directly (**step 7**) or via endosomes (**step 8**).

B. Retrieval of vesicles after transmitter discharge is thought to occur via three mechanisms, each with distinct kinetics. **1.** A reversible fusion pore is the most rapid mechanism for reusing vesicles. The vesicle membrane does not completely fuse with the plasma membrane, and transmitter is released through the

fusion pore. Vesicle retrieval requires only the closure of the fusion pore and thus can occur rapidly, in tens to hundreds of milliseconds. This pathway may predominate at lower to normal release rates. The spent vesicle may either remain at the membrane (kiss-and-stay) or relocate from the membrane to the reserve pool of vesicles (kiss-and-run). **2.** In the classical pathway, excess membrane is retrieved through endocytosis by means of clathrin-coated pits. These pits are found throughout the axon terminal except at the active zones. This pathway may be important at normal to high rates of release. **3.** In the bulk retrieval pathway, excess membrane reenters the terminal by budding from uncoated pits. These uncoated cisternae are formed primarily at the active zones. This pathway may be used only after high rates of release and not during the usual functioning of the synapse. (Adapted, with permission, from Schweizer, Betz, and Augustine 1995; Südhof 2004.)

to a minute for completion and also appears to shift from the active zone to the membrane surrounding the active zone (see Figure 15–7). A fourth mechanism operates after prolonged high-frequency stimulation. Under these conditions, large membranous invaginations into the presynaptic terminal are visible, which are thought to reflect membrane recycling through a process called *bulk retrieval*.

Exocytosis of Synaptic Vesicles Relies on a Highly Conserved Protein Machinery

Many key proteins of synaptic vesicles as well as their interacting partners in the plasma membrane have been isolated and purified. Proteomic analysis of isolated synaptic vesicles has provided a census of the many types of proteins they contain (Figure 15–11).

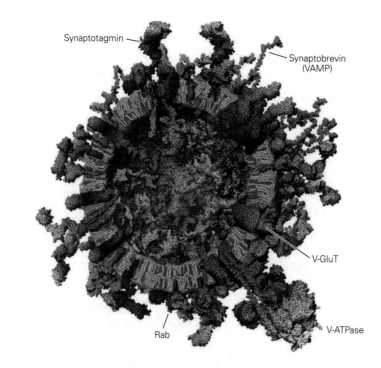

A
Synaptotagmin

Synaptobrevin
(VAMP)

V-GluT

V-ATPase

Rab

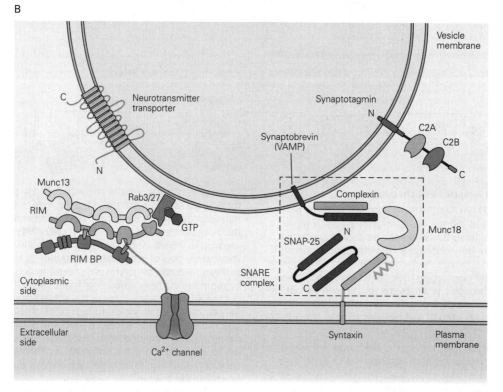

B

Vesicle
membrane

C

Neurotransmitter
transporter

N

Synaptotagmin

N

C2A

C2B

C

Synaptobrevin
(VAMP)

Munc13

RIM

Rab3/27

GTP

Complexin

Munc18

RIM BP

SNAP-25

N

Cytoplasmic
side

C

SNARE
complex

Extracellular
side

Ca²⁺ channel

Syntaxin

Plasma
membrane

Figure 15–11 Molecular components of exocytosis.

A. Depiction of protein constituents of a glutamatergic synaptic vesicle (and their approximate copy numbers). Proteins are shown embedded in a synaptic vesicle, drawn to scale. Components include the vesicular ATPase (**V-ATPase**; 1–2 per vesicle), vesicular glutamate transporter (**V-GluT**; ~10 per vesicle), synaptobrevin/VAMP (~70 per vesicle), synaptotagmin (~15 per vesicle) and the small GTPases Rab3 and/or Rab27. Estimates are obtained as an average over many vesicles. (Reproduced from Takamori et al. 2006. Copyright © 2006 Elsevier.)

B. The molecular machinery mediating Ca²⁺-triggered vesicle fusion with the presynaptic cell membrane. This depiction of a portion of a docked synaptic vesicle and the presynaptic active zone illustrates the interactions of several key functional proteins of the neurotransmitter release machinery. *Right:* The dotted

box shows the core fusion machine, which is comprised of the SNARE proteins synaptobrevin/VAMP, syntaxin-1, and SNAP-25, along with Munc18-1. The Ca²⁺ sensor synaptotagmin-1 functions in coordination with complexin (shown bound to the SNARE complex). *Left:* The active zone protein complex also contains RIM, Munc13, and RIM-BP and a Ca²⁺ channel in the presynaptic plasma membrane. RIM plays a central role in this complex, coordinating multiple functions of the active zone by binding to specific target proteins: (1) vesicular Rab proteins (Rab3 and Rab27) to mediate vesicle docking; (2) Munc13 to activate vesicle priming; and (3) the Ca²⁺ channel, both directly and indirectly via RIM-BP, to tether Ca²⁺ channels within 100 nm of docked vesicles. The active zone protein complex puts into close proximity key elements that enable vesicles to dock, prime, and fuse rapidly in response to action potential–triggered Ca²⁺ entry near the docked vesicle. (Reproduced from Südhof 2013.)

Two of the most abundant proteins, *synaptobrevin* and *synaptotagmin-1*, are involved in vesicle fusion and are discussed later. Another key class of vesicle proteins are the neurotransmitter transporters (Chapter 16). These transmembrane proteins (exemplified by the *glutamate transporter* v-GluT) harness energy stored in the electrochemical gradient for protons to pump transmitter molecules against their concentration gradient from the cytoplasm into the vesicle. The proton-motive force is generated by a vesicular H^+ pump, the V-ATPase, that pumps protons into the lumen of the vesicle from the cytoplasm, leading to an acidic vesicular pH of around 5.0.

Other synaptic vesicle proteins direct vesicles to their release sites, participate in the discharge of transmitter by exocytosis, and mediate recycling of the vesicle membrane. The protein machinery involved in these three steps has been conserved throughout evolution, in species ranging from worms to humans, and forms the basis for the regulated release of neurotransmitter. We consider each of these steps in turn.

The Synapsins Are Important for Vesicle Restraint and Mobilization

The vesicles outside the active zone represent a reserve pool of transmitter. Paul Greengard discovered a family of proteins, *synapsins*, that are thought to be important regulators of the reserve pool of vesicles. Synapsins are peripheral membrane proteins that are bound to the cytoplasmic surface of synaptic vesicles. Synapsins contain a conserved central ATPase domain that accounts for most of their structure, but whose function remains unknown. In addition, synapsin-1 binds actin.

The synapsins are substrates for both protein kinase A and Ca^{2+}/calmodulin-dependent protein kinase II. When the nerve terminal is depolarized and Ca^{2+} enters, the synapsins become phosphorylated by the kinase and are thus released from the vesicles. Strikingly, stimulation of synapsin phosphorylation, genetic deletion of synapsins or intracellular injection of a synapsin antibody leads to a decrease in the number of synaptic vesicles in the nerve terminal and a resulting decrease in the ability of a terminal to maintain a high rate of transmitter release during repetitive stimulation.

SNARE Proteins Catalyze Fusion of Vesicles With the Plasma Membrane

Because a membrane bilayer is a stable structure, fusion of the synaptic vesicle and plasma membrane must overcome a large unfavorable activation energy. This is accomplished by a family of fusion proteins now referred to as *SNAREs* (soluble *N*-ethylmaleimide–sensitive factor attachment receptors) (Figure 15–12).

SNAREs are universally involved in membrane fusion, from yeast to humans. They mediate both constitutive membrane trafficking during the movement of proteins from the endoplasmic reticulum to the Golgi apparatus to the plasma membrane, as well as synaptic vesicle trafficking important for regulated exocytosis. SNAREs have a conserved protein sequence, the SNARE motif, that is 60 residues long. They come in two forms. Vesicle SNAREs, or v-SNAREs (also referred to as R-SNAREs because they contain an important central arginine residue), reside in the vesicle membranes. Target-membrane SNAREs, or t-SNAREs (also referred to as Q-SNAREs because they contain an important glutamine residue), are present in target membranes, such as the plasma membrane.

Each synaptic vesicle contains a v-SNARE called *synaptobrevin* (also called vesicle-associated membrane protein or VAMP). By contrast, the presynaptic active zone contains two types of t-SNARE proteins, *syntaxin* and *SNAP-25*. (Synaptobrevin and syntaxin have one SNARE motif; SNAP-25 has two.) The first clue that synaptobrevin, syntaxin, and SNAP-25 are all involved in fusion of the synaptic vesicle with the plasma membrane came from the finding that all three proteins are substrates for botulinum and tetanus toxins, bacterial proteases that are potent inhibitors of transmitter release. James Rothman then provided the crucial insight that these three proteins interact in a tight biochemical complex. In experiments using purified v-SNAREs and t-SNAREs in solution, four SNARE motifs bind tightly to each other to form an α-helical coiled-coil complex (Figure 15–12B).

How does formation of the SNARE complex drive synaptic vesicle fusion? During exocytosis, the SNARE motif of synaptobrevin on the synaptic vesicle forms a tight complex with the SNARE motifs of SNAP-25 and syntaxin on the plasma membrane (Figure 15–12B). The crystal structure of the SNARE complex suggests that this complex draws the membranes together. The ternary complex of synaptobrevin, syntaxin, and SNAP-25 is extraordinarily stable. The energy released in its assembly is thought to draw the negatively charged phospholipids of the vesicle and plasma membranes in close apposition, forcing them into a prefusion intermediate state (Figure 15–12). Such an unstable state may start the formation of the fusion pore and generate the rapid opening and closing (flickering) of the fusion pore observed in electrophysiological measurements.

However, the SNAREs do not fully account for fusion of the synaptic vesicle and plasma membranes. Reconstitution experiments with purified proteins in

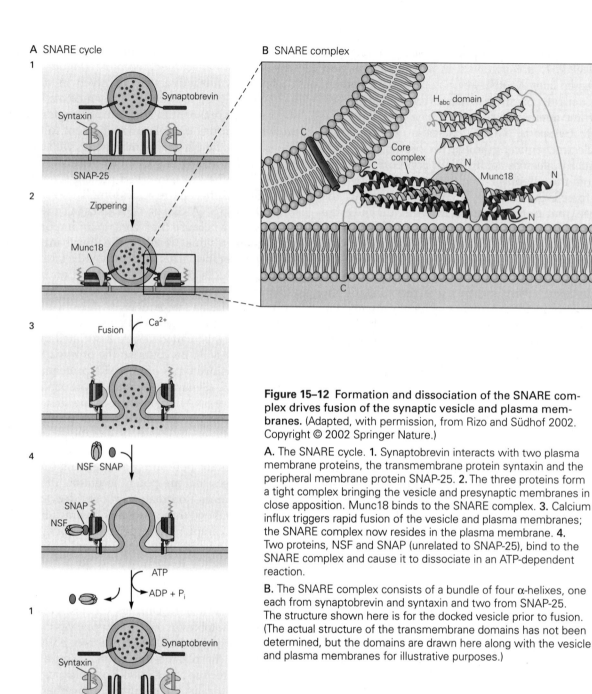

A SNARE cycle

1

Synaptobrevin

Syntaxin

SNAP-25

2

Zippering

Munc18

3

Fusion   Ca²⁺

4

NSF SNAP

SNAP

NSF

ATP

ADP + Pᵢ

1

Synaptobrevin

Syntaxin

SNAP-25

B SNARE complex

H_abc domain

C

Core
complex

C

N

Munc18

N

C

N

Figure 15–12 Formation and dissociation of the SNARE complex drives fusion of the synaptic vesicle and plasma membranes. (Adapted, with permission, from Rizo and Südhof 2002. Copyright © 2002 Springer Nature.)

A. The SNARE cycle. **1.** Synaptobrevin interacts with two plasma membrane proteins, the transmembrane protein syntaxin and the peripheral membrane protein SNAP-25. **2.** The three proteins form a tight complex bringing the vesicle and presynaptic membranes in close apposition. Munc18 binds to the SNARE complex. **3.** Calcium influx triggers rapid fusion of the vesicle and plasma membranes; the SNARE complex now resides in the plasma membrane. **4.** Two proteins, NSF and SNAP (unrelated to SNAP-25), bind to the SNARE complex and cause it to dissociate in an ATP-dependent reaction.

B. The SNARE complex consists of a bundle of four α-helixes, one each from synaptobrevin and syntaxin and two from SNAP-25. The structure shown here is for the docked vesicle prior to fusion. (The actual structure of the transmembrane domains has not been determined, but the domains are drawn here along with the vesicle and plasma membranes for illustrative purposes.)

lipid vesicles indicate that synaptobrevin, syntaxin, and SNAP-25 can catalyze fusion, but the in vitro reaction shows little regulation by Ca²⁺, and the reaction is much slower and less efficient than vesicle fusion in a real synapse. One important additional protein required for exocytosis of synaptic vesicles is Munc18 (mammalian unc18 homolog). Homologs of Munc18, referred to as SM proteins (sec1/Munc18-like proteins), are essential for all SNARE-mediated intracellular fusion reactions. Munc18

binds to syntaxin before the SNARE complex assembles. Deletion of Munc18 prevents all synaptic fusion in neurons. The core fusion machinery is thus composed of SNARE and SM proteins that are modulated by various accessory factors specific for particular fusion reactions. Finally, the synaptic SNARE complex also interacts with a small soluble protein called *complexin*, which suppresses the spontaneous release of transmitter but enhances Ca²⁺-dependent evoked release.

After fusion, the SNARE complex must be disassembled for efficient vesicle recycling to occur. Rothman discovered that a cytoplasmic ATPase called *NSF* (*N*-ethylmaleimide-sensitive fusion protein) binds to SNARE complexes via an adaptor protein called *SNAP* (soluble NSF-attachment protein, not related to the SNARE protein SNAP-25). NSF and SNAP use the energy of ATP hydrolysis to dissociate SNARE complexes, thereby regenerating free SNARE (Figure 15–12A). SNAREs and NSF also participate in the cycling of postsynaptic AMPA-type glutamate receptors in dendritic spines.

Calcium Binding to Synaptotagmin Triggers Transmitter Release

Because fusion of synaptic vesicles with the plasma membrane must occur within a fraction of a millisecond, it is thought that most proteins responsible for fusion are assembled prior to Ca^{2+} influx. According to this view, once Ca^{2+} enters the presynaptic terminal, it binds a Ca^{2+} sensor on the vesicle, triggering immediate fusion of the membranes.

Members of a family of closely related proteins, the synaptotagmins, have been identified as the major Ca^{2+} sensors that trigger fusion of synaptic vesicles. Synaptotagmins are membrane proteins with a single N-terminal transmembrane region that anchors them to the synaptic vesicle (Figure 15–13A,B). The cytoplasmic region of each synaptotagmin protein is largely composed of two domains, the C2 domains, which are a common protein motif homologous to the Ca^{2+} and phospholipid-binding C2 domain of protein kinase C. The finding that the C2 domains bind not only Ca^{2+} but also phospholipids is consistent with their importance in Ca^{2+}-dependent exocytosis. Synaptotagmin-1, -2, and -9 have been identified as Ca^{2+} sensors for fast and synchronous vesicle fusion. Each exhibits distinct Ca^{2+} binding affinities and kinetics, endowing different synapses with distinct release properties on the basis of the particular synaptotagmin isoform that is expressed. In contrast, synaptotagmin-7 mediates a slower form of Ca^{2+}-triggered exocytosis that is important for synaptic transmission during prolonged periods of activity periods of repeated firing of action potential. All of these synaptotagmins also function as Ca^{2+} sensors in other forms of exocytosis, such as exocytosis in endocrine cells and the insertion of AMPA-type glutamate receptors into the postsynaptic cell membrane from a pool of intracellular vesicles during NMDA-receptor-dependent long-term potentiation.

Studies with mutant mice in which synaptotagmin-1 is deleted or in which its Ca^{2+} affinity is altered through genetic engineering provide important evidence that synaptotagmin is the physiological Ca^{2+} sensor. When the affinity of synaptotagmin for Ca^{2+} is decreased two-fold, the Ca^{2+} required for transmitter release is changed by the same amount. When synaptotagmin-1 is deleted in mice, flies, or worms, an action potential is no longer able to trigger fast synchronous release. However, Ca^{2+} is still capable of stimulating a slower form of transmitter release referred to as asynchronous release (Figure 15–13A), mediated by synaptotagmin-7. Thus, nearly all Ca^{2+}-triggered neurotransmitter release depends on the synaptotagmins.

How does Ca^{2+} binding to synaptotagmin trigger vesicle fusion? The two C2 domains bind a total of five Ca^{2+} ions, the same minimal number of Ca^{2+} ions required to trigger release of a quantum of transmitter (Figure 15–13B). However, as multiple synaptotagmins may be engaged to trigger release, more than five bound Ca^{2+} ions may be distributed among the multiple synaptotagmin molecules on a single vesicle.

The binding of the Ca^{2+} ions to synaptotagmin is thought to act as a switch, promoting the interaction of the C2 domains with phospholipids. The C2 domains of synaptotagmin also interact with SNARE proteins and complexin. Crystal structures of synaptotagmin reveal a conserved primary interface with the associated SNARE complex. In addition, a second molecule of synaptotagmin forms a tripartite interaction with the same SNARE complex and complexin. Brunger and colleagues found that both the primary SNARE complex/synaptotagmin interface and the tripartite SNARE complex/synaptotagmin/complexin interface are essential for fast Ca^{2+}-triggered fusion. These findings have led to the hypothesis that: (1) at rest, synaptotagmin of primed vesicles exists in a complex with partially pre-zippered SNARE proteins and complexin; (2) upon action potential–triggered Ca^{2+} influx, Ca^{2+} binds to synaptotagmin. This triggers an interaction between synaptotagmin and the plasma membrane that causes the complex to rotate en bloc, which induces complexin to partly dissociate from the SNARE complex; (3) this rotation causes a dimpling of the plasma membrane, rearrangement of its cytoplasm-facing lipids, and ultimately the fusion of plasma and vesicle membranes (Figure 15–13C). In this way, the energy of the favorable interaction of synaptotagmin, Ca^{2+}, and the membrane can be harnessed to both relieve the complexin-mediated lock on fusion and promote the energetically unfavorable merging of a vesicle membrane with the plasma membrane.

The Fusion Machinery Is Embedded in a Conserved Protein Scaffold at the Active Zone

As we have seen, a defining feature of fast synaptic transmission is that neurotransmitters are released by

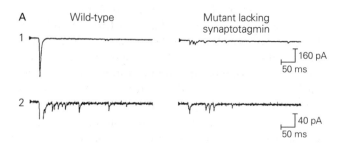

A   Wild-type   Mutant lacking synaptotagmin

1

160 pA
50 ms

2

40 pA
50 ms

B1 Calcium-bound synaptotagmin

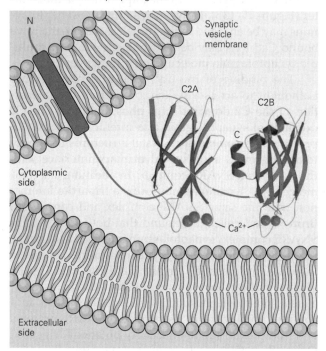

N

Synaptic
vesicle
membrane

C2A   C2B

C

Cytoplasmic
side

Ca²⁺

Extracellular
side

B2 Synaptotagmin/SNARE complex

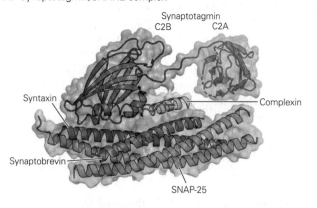

Synaptotagmin
C2B   C2A

Syntaxin

Complexin

Synaptobrevin

SNAP-25

C

Locked

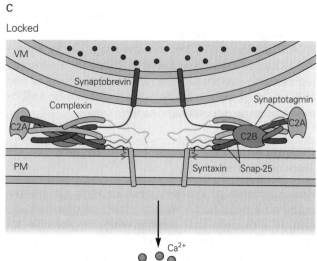

VM

Synaptobrevin

Complexin

Synaptotagmin

C2A   C2A

C2B

PM

Syntaxin   Snap-25

Ca²⁺

Unlocking and triggering

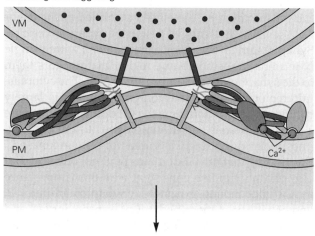

VM

PM

Ca²⁺

Fusion pore formation

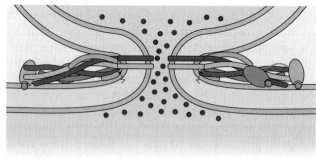

exocytosis at the active zone. Other types of exocytosis, such as that which occurs in the adrenal medulla, do not require a specialized domain of the plasma membrane. The active zone is thought to coordinate and regulate the docking and priming of synaptic vesicles to enable the speed and tight regulation of release. This is accomplished through a conserved set of proteins that form one large macromolecular structure at active zones.

An exquisitely detailed view of the active zone at the frog neuromuscular junction was obtained by Jack MacMahan using a powerful ultrastructural technique called electron microscopic tomography. This technique has shown how synaptic vesicles are tethered to the membrane by a series of distinctive structural entities, termed *ribs* and *beams*, that attach to defined sites on the vesicles and to particles (*pegs*) in the presynaptic membrane that may correspond to voltage-gated Ca^{2+} channels (Figure 15–14).

A key goal in understanding how the various synaptic vesicle and active zone proteins are coordinated during exocytosis is to match up the various proteins that have been identified with elements of this electron microscopic structure. Several cytoplasmic proteins have been identified that are thought to be components of a structural matrix at the active zone. These include three large cytoplasmic multidomain proteins, *Munc13* (not related to the Munc18 protein discussed earlier), *RIM*, and *RIM-binding proteins* (RIM-BPs), which form a tight complex with each other and may comprise part of the ribs and beams. The binding of synaptic vesicles to RIM and Munc13 is essential for priming the vesicles for exocytosis. Phosphorylation of RIM by cAMP-dependent protein kinase is implicated in the enhancement of transmitter release associated with certain forms of long-term synaptic plasticity that may contribute to learning and memory. As we will see later, regulation of Munc13 by second messengers is involved in shorter-term forms of synaptic plasticity.

RIM binds the synaptic vesicle proteins *Rab3* and *Rab27*, members of the family of low-molecular-weight guanosine triphosphatases (GTPases). Rab3 and Rab27 proteins transiently associate with synaptic vesicles as a GTP-bound Rab3 complex (Figure 15–11B). The binding of RIM to Rab3 or Rab27 is thought to tether synaptic vesicles to the active zone during the vesicle cycle prior to SNARE-complex assembly. Moreover, RIM and RIM-BP together mediate the recruitment of Ca^{2+} channels to the active zone, allowing tight coupling of Ca^{2+} influx to vesicle release. This general machinery is conserved through evolution and is present in invertebrates, although with modifications.

At the *Drosophila* neuromuscular junction, Sigrist and colleagues identified another protein, Bruchpilot, as a major component of the electron-dense active zone "T-bar" structure; Bruchpilot is associated with the fly homolog of RIM-binding protein, which also serves to recruit Ca^{2+} channels to active zones in *Drosophila*. As coordinators of both presynaptic Ca^{2+} channels and synaptic vesicles, these proteins act as essential regulators of presynaptic release in the fly. In *Caenorhabditis elegans*, RIM plays a central role for the same processes.

Figure 15–13 (Opposite) Synaptotagmin mediates Ca^{2+}-dependent transmitter release by forming a protein complex that favors vesicle fusion.

A. Fast Ca^{2+}-triggered transmitter release is absent in mutant mice lacking synaptotagmin-1. Recordings show excitatory postsynaptic currents evoked in vitro by stimulation of cultured hippocampal neurons from wild-type mice and from mutant mice in which synaptotagmin-1 has been deleted by homologous recombination (1). Neurons from wild-type mice show large, fast excitatory postsynaptic currents evoked by presynaptic action potentials, reflecting the fact that synaptic transmission is dominated by the rapid synchronous release of transmitter from a large number of synaptic vesicles. In the **bottom trace** (2), where the synaptic current is shown at a highly expanded scale, one can see that a small, prolonged phase of asynchronous release of transmitter follows the fast phase of synchronous release. During this slow phase, there is a prolonged increase in frequency of individual quantal responses. In neurons from a mutant mouse, a presynaptic action potential triggers only the slow asynchronous phase of release; the rapid synchronous phase has been abolished. (Reproduced, with permission, from Geppert et al., 1994.)

B. The X-ray crystal structure synaptotagmin. **B1.** A ribbon diagram shows that the C2A domain binds three Ca^{2+} ions and the C2B domain two Ca^{2+} ions. The **blue arrows** show β-strands. There are two short α-helixes (**orange**) at the C-terminus of the C2B domain. The structures of the other regions of synaptotagmin have not yet been determined and are drawn here for illustrative purposes. The membrane and structures are drawn to scale. (Adapted, with permission, from Fernandez et al., 2001.) **B2.** The X-ray crystal structure of synaptotagmin (**light blue**) bound to the SNARE complex (synaptobrevin, syntaxin and SNAP-25) and complexin. The transmembrane domain of synaptotagmin is not shown. (Adapted, with permission, from Zhou et al, 2017.)

C. Zippering of the synaptotagmin-complexin-SNARE complex mediates vesicle fusion. **Top**, in the absence of Ca^{2+}, the α-helixes of the SNARE complex and complexin, with the bound synaptotagmin, are only partially zippered. **Middle**, binding of Ca^{2+} to the C2A and C2B domains of synaptotagmin allows them to interact with the plasma membrane, applying force to bring the vesicle and plasma membranes closer together. **Bottom**, synaptotagmin-mediated proximity and the final zippering of the complexin-SNARE-synaptotagmin complex triggers membrane fusion. (Adapted, with permission, from Zhou et al., 2017.)

A

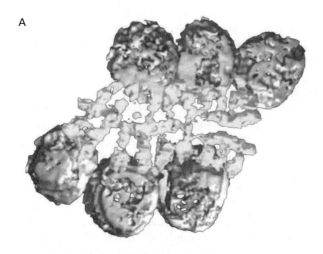

B

C

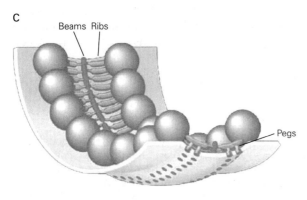

Figure 15–14 Synaptic vesicles at the active zone. The images are obtained from electron microscopic tomography. (Reproduced, with permission, from Harlow et al. 2001. Copyright © 2001 Springer Nature.)

A. Vesicles are tethered to filamentous proteins of the active zone. Three distinct filamentous structures are resolved: pegs, ribs, and beams. Ribs protruding from the vesicles are attached to long horizontal beams, which are anchored to the membrane by vertical pegs.

B. Ribs and beams superimposed on a freeze fracture view of intramembranous particles at the active zone show how the ribs are aligned with the particles, some of which are presumed to be voltage-gated Ca^{2+} channels. Scale bar = 100 nm.

C. A model for the structure of the active zone shows the relation between synaptic vesicles, pegs, ribs, and beams.

Modulation of Transmitter Release Underlies Synaptic Plasticity

The effectiveness of chemical synapses can be modulated dramatically and rapidly—by several-fold in a matter of seconds—and this change can be maintained for seconds, to hours, or even days or longer, a property called *synaptic plasticity.*

Synaptic strength can be modified presynaptically, by altering the release of neurotransmitter, postsynaptically, by modulating the response to transmitter (as discussed in Chapter 13), or both. Long-term changes in presynaptic and postsynaptic mechanisms are crucial for the refinement of synaptic connections during development (Chapter 49) and for storing information during learning and memory (Chapters 53 and 54). Here, we focus on how synaptic strength can be changed through modulation of the amount of transmitter released. In principle, changes in transmitter release can be mediated by two different mechanisms: changes in Ca^{2+} influx or changes in the amount of transmitter released in response to a given Ca^{2+} concentration. As we will see later, both types of mechanisms contribute to different forms of plasticity.

Synaptic strength is often altered by the pattern of activity of the presynaptic neuron. Trains of action potentials produce successively larger postsynaptic currents at some synapses and successively smaller currents in others (Figure 15–15A). A decrease in the size of the postsynaptic response to repeated stimulation is referred to as *synaptic depression* (Figure 15–15A, upper); the opposite, enhancement of transmission with repeated stimulation, is called *synaptic facilitation* or *potentiation* (Figure 15–15A, lower, 15–15E). Various synapses exhibit these disparate forms of *short-term synaptic plasticity*—sometimes overlapping and sometimes with one predominating—resulting in characteristic patterns of short-term dynamics in individual synapse types (Figure 15–15A).

Whether a synapse facilitates or depresses often is determined by the probability of release in response to the first action potential of a train. Synapses with an initial high probability of release normally undergo depression because the high rate of release transiently depletes docked vesicles at the active zone. Synapses with an initial low probability of release undergo synaptic facilitation, in part because the buildup in intracellular Ca^{2+} during the train increases the probability of release (see later). The importance of release probability in controlling the sign of plasticity can be seen by the effect of genetic mutations. Synapses formed by hippocampal neurons in cell culture have an initially high release probability and so normally depress in response to 20-Hz stimulation. However, a mutation that reduces by approximately two-fold the Ca^{2+}-binding affinity of synaptotagmin-1, thus reducing the initial probability of release, converts the depressing synapse into a facilitating one (Figure 15–15B).

Mechanisms that affect the concentration of free Ca^{2+} in the presynaptic terminal also affect the amount of transmitter released. For example, the buildup of inactivation of certain voltage-gated K^+ channels during high-frequency firing leads to a gradual increase in the duration of the action potential. Prolongation of the action potential increases the time that voltage-gated Ca^{2+} channels stay open, which leads to enhanced entry of Ca^{2+} and a subsequent increase in transmitter release, resulting in a larger postsynaptic potential (Figure 15–15C).

Most studies of the functional implications of short-term synaptic dynamics have been performed in vitro or are based on computational results. However, recent in vivo experiments are beginning to shed light on the behavioral importance of short-term plasticity. For example, in vivo recordings in rodents from thalamocortical synapses have suggested that synaptic depression may contribute to sensory adaptation during repeated whisker stimulation. The time course of this sensory adaptation parallels the attenuation of cortical spiking to whisker stimulation and the synaptic depression of EPSPs at thalamocortical synapses (Figure 15–15D).

High-frequency stimulation of the presynaptic neuron, which in some cells can generate up to 500 to 1,000 action potentials per second, is called *tetanic stimulation*. Such intense stimulation can cause dramatic changes in synaptic strength. The increase in size of the EPSP during tetanic stimulation is called *potentiation*; the increase that persists after tetanic stimulation is called *posttetanic potentiation* (Figure 15–15E). In contrast to synaptic facilitation, which lasts milliseconds to seconds, posttetanic potentiation usually lasts several minutes, but it can persist for an hour or more at some synapses.

Synapses utilize a complex containing Munc13 and RIM, two of the active zone proteins discussed earlier, to counteract vesicle depletion during high-frequency stimulation. The rise in presynaptic Ca^{2+} during tetanic stimulation activates phospholipase C, which produces inositol 1,4,5-trisphosphate (IP_3) and diacylglycerol. Diacylglycerol directly interacts with a protein domain on Munc13 called the C1 domain (homologous to the diacylglycerol-binding domain in protein kinase C but distinct from the C2 domain of synaptotagmin), thereby accelerating the rate of synaptic vesicle recycling. At the same time, IP_3 causes additional release of Ca^{2+} from intracellular stores, and the increase in Ca^{2+} further activates Munc13 by binding to its C2 domain,

which resembles the C2 domain of synaptotagmin but acts as an agent of short-term synaptic plasticity.

Activity-Dependent Changes in Intracellular Free Calcium Can Produce Long-Lasting Changes in Release

Several Ca^{2+}-dependent mechanisms contribute to longer-lasting changes in transmitter release that persist after a high-frequency tetanus is terminated. Normally the rise in Ca^{2+} in the presynaptic terminal in response to an action potential is rapidly buffered by cytoplasmic Ca^{2+}-binding proteins and mitochondria. Calcium ions are also actively transported out of the neuron by pumps and transporters. However, during tetanic stimulation, so much Ca^{2+} flows into the axon terminals that the Ca^{2+} buffering and clearance systems can become saturated.

This leads to a temporary excess of Ca^{2+} called *residual* Ca^{2+}. The residual free Ca^{2+} enhances synaptic transmission for many minutes or longer by activating certain enzymes that are sensitive to enhanced levels of resting Ca^{2+}, including the Ca^{2+}/calmodulin-dependent protein kinases. Activation of such Ca^{2+}-dependent enzymatic pathways is thought to increase the priming of synaptic vesicles in the terminals. Here, then, is the simplest kind of cellular memory! The presynaptic cell can store information about the history of its activity in the form of residual free Ca^{2+} in its terminals (or residual Ca^{2+} bound to sensor proteins).

This Ca^{2+} acts by multiple pathways that have different half-times of decay. In Chapter 13, we saw how posttetanic potentiation at certain synapses is followed by an even longer-lasting process (also initiated by Ca^{2+} influx), called *long-term potentiation*, which can last for many hours or even days. The importance of long-term potentiation for learning and memory will be considered in Chapters 53 and 54.

Axo-axonic Synapses on Presynaptic Terminals Regulate Transmitter Release

Synapses are formed on axon terminals as well as the cell body and dendrites of neurons (see Chapter 13). Although axosomatic synaptic actions affect all branches of the postsynaptic neuron's axon (because they affect the probability that the neuron will fire an action potential), axo-axonic actions selectively control individual terminals of the axon. One important action of axo-axonic synapses is to increase or decrease Ca^{2+} influx into the presynaptic terminals of the postsynaptic cell, thereby enhancing or depressing transmitter release, respectively.

As we saw in Chapter 13, when one neuron releases transmitter that hyperpolarizes the cell body (or

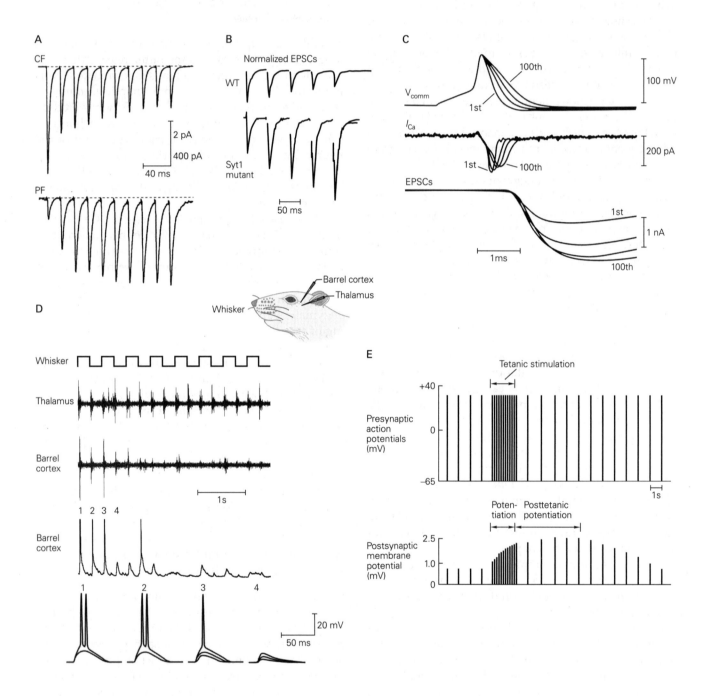

dendrites) of another, it decreases the likelihood that the postsynaptic cell will fire; this action is called *postsynaptic inhibition*. In contrast, when a neuron forms synapses on the axon terminal of another cell, it can reduce the amount of transmitter that will be released by the postsynaptic cell onto a third cell; this action is called *presynaptic inhibition* (Figure 15–16A). Other axo-axonic synaptic actions can increase the amount of transmitter released by the postsynaptic cell; this action is called *presynaptic facilitation* (Figure 15–16B). Both presynaptic inhibition and facilitation can occur in response to activation of ionotropic or metabotropic receptors in the membrane of the presynaptic terminals.

The best-analyzed mechanisms of presynaptic inhibition and facilitation are found in invertebrate neurons and vertebrate mechanoreceptor neurons (whose axons project to neurons in the spinal cord). Three mechanisms for presynaptic inhibition have been identified in these cells. One depends on the activation of inhibitory interneurons that form axo-axonic synapses on the sensory neuron presynaptic terminals, where they activate ionotropic GABA$_A$ receptor-channels. Because the Cl$^-$ reversal potential in the presynaptic terminals is relatively positive, the increased Cl$^-$ conductance resulting from activation of GABA$_A$ channels depolarizes the presynaptic terminal. This voltage change, termed the primary afferent depolarization, is thought to inactivate voltage-gated Na$^+$ channels, reducing the amplitude of the presynaptic action potential, which decreases the activation of voltage-gated Ca^{2+} channels and thereby decreases the amount of transmitter release.

The other two mechanisms for presynaptic inhibition both result from the activation of presynaptic G protein–coupled metabotropic receptors. One type of action results from the modulation of ion channels. As we saw in Chapter 14, the βγ-subunit complex of G proteins can simultaneously close voltage-gated Ca^{2+} channels and open K$^+$ channels. This decreases the influx of Ca^{2+} and enhances repolarization of the presynaptic terminal following an action potential, thus diminishing transmitter release. A second type of G protein–dependent action depends on a direct action by the βγ-subunit complex on the release machinery itself, independent of any changes in ion channel activity or Ca^{2+} influx. This second action is thought to involve a decrease in the Ca^{2+} sensitivity of the release machinery.

In contrast, presynaptic facilitation can be caused by enhanced influx of Ca^{2+}. In certain molluscan neurons, serotonin acts through cAMP-dependent protein phosphorylation to close K$^+$ channels in the presynaptic

Figure 15–15 (Opposite) Diversity of short-term plasticity in the central nervous system.

A. Excitatory postsynaptic currents (**EPSCs**) were recorded from a cerebellar Purkinje neuron under voltage clamp in response to repetitive stimulation of either the climbing fiber (**CF**) or parallel fiber (**PF**) inputs to the Purkinje cells. In both cases EPSCs were recorded while afferents were stimulated 10 times at 50 Hz. Note that the CF EPSC depresses whereas the PF EPSC facilitates during repetitive stimulation. (Reproduced, with permission, from Dittman et al. 2000. Copyright © 2000 Society for Neuroscience.)

B. EPSCs were recorded from hippocampal neurons in culture during stimulation at 20 Hz. EPSC size was normalized by dividing each response by peak amplitude of first EPSC in each individual train. The EPSC depresses in neurons cultured from wild-type mice (**WT**) whereas the EPSC facilitates in neurons from mice harboring a mutated form of synaptotagmin-1 that reduces its Ca^{2+} binding affinity (**Syt1 mutant**, R233Q). (Reproduced, with permission, from Fernandez-Chacon et al. 2001. Copyright © 2001 Springer Nature.)

C. The action potential recorded at the presynaptic terminals of dentate gyrus granule neurons broadens progressively during a 2-s long train of 50 Hz stimulation. This results in enhanced synaptic transmission from the granule neurons onto their CA3 postsynaptic target. The 1st, 25th, 50th, and 100th action potentials are shown. These action potential waveforms were then used as the command waveforms (**Vcomm, top**) to voltage clamp the presynaptic nerve terminal ("action potential clamp"), eliciting the voltage-gated Ca^{2+} current (I_{Ca}) recorded in the terminal (**middle**) and the EPSCs in a postsynaptic CA3 neuron (**bottom**). As the action potential, waveform increases in duration, the duration of I_{Ca} increases, increasing the amplitude of the EPSCs. (Adapted,

with permission, from Geiger and Jonas, 2000. Copyright © 2000 Cell Press.)

D. Simultaneous extracellular multiunit recordings of action potentials from thalamus and barrel cortex (*second* and *third* traces from top) during a train of 4-Hz mechanical stimulation of the primary whisker (**top trace**). Cortical and thalamic responses both depress during stimulation, although cortical responses depress faster. Intracellular voltage responses of a cortical neuron in a whisker barrel to 4-Hz stimulation of the primary whisker (**bottom two traces**). The first of the two traces shows responses to successive whisker stimuli in one train. Time scale same as top traces. The bottom trace shows an expanded view of the first four responses to whisker stimulation in three separate trains. Note there is variability from trial to trial in the spiking responses to the second and third stimuli in the train, likely due to the probabilistic nature of transmitter release. (Reproduced from Chung et al. 2002. Copyright © 2002 Cell Press.)

E. A brief burst of high-frequency stimulation leads to sustained enhancement in transmitter release. The time scale of the experimental records here has been compressed (each presynaptic and postsynaptic potential appears as a single line indicating its amplitude). A stable excitatory postsynaptic potential (EPSP) of around 1 mV is produced when the presynaptic neuron is stimulated at a relatively low rate of one action potential per second. The presynaptic neuron is then stimulated for a few seconds at a higher rate of 50 action potentials per second. During this *tetanic stimulation*, the EPSP increases in size because of enhanced transmitter release, a phenomenon known as *potentiation*. After several seconds of stimulation, the presynaptic neuron is returned to the initial rate of stimulation (one per second). However, the EPSPs remain enhanced for minutes and, in some cells, for several hours. This persistent increase is called *posttetanic potentiation*.

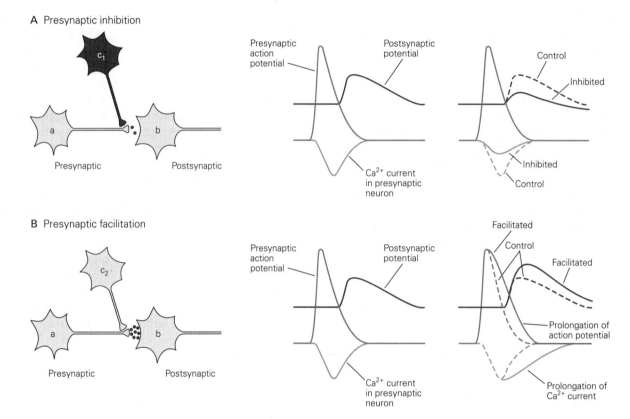

Figure 15–16 Axo-axonic synapses can inhibit or facilitate transmitter release by the presynaptic cell.

A. An inhibitory neuron (c_1) forms a synapse on an axon terminal of neuron **a**. Release of transmitter by cell c_1 activates metabotropic receptors on the terminal, thus inhibiting the Ca^{2+} current in the terminal and reducing the amount of transmitter released by cell a onto cell **b**. The reduction of transmitter release from cell a in turn reduces the amplitude of the excitatory postsynaptic potential in cell b, a process termed presynaptic inhibition.

B. A facilitating neuron (c_2) forms a synapse on an axon terminal of neuron **a**. Release of transmitter by cell c_2 activates metabotropic receptors on the terminal, thus decreasing a K^+ current in the terminals and thereby prolonging the action potential and increasing Ca^{2+} influx into cell a. This increases transmitter release from cell a onto cell **b**, thereby increasing the size of the EPSP in cell b, a process termed presynaptic facilitation.

terminal (including the *Aplysia* S-type K^+ channel discussed in Chapter 14). This action increases the duration of the presynaptic action potential, thereby increasing Ca^{2+} influx by enabling the voltage-dependent Ca^{2+} channels to remain open for a longer period. In other cells, activation of presynaptic ionotropic receptors increases transmitter release. Activation of presynaptic Ca^{2+}-permeable ionotropic receptor-channels, including presynaptic NMDA-type glutamate receptors, can increase release by directly enhancing Ca^{2+} influx. Activation of presynaptic ionotropic receptor-channels that are not permeable to Ca^{2+} can indirectly increase presynaptic Ca^{2+} levels by depolarizing the terminal and activating voltage-gated Ca^{2+} channels.

Thus, presynaptic terminals are endowed with a variety of mechanisms that allow for the fine-tuning of the strength of synaptic transmission. Although we know a fair amount about the mechanisms of short-term changes in synaptic strength—changes that last

seconds, minutes, and hours—we are only beginning to learn about their functional roles. The mechanisms that support changes that persist for days, weeks, and longer also remain mysterious. These long-term changes often require alterations in gene expression and growth of presynaptic and postsynaptic structures in addition to alterations in Ca^{2+} influx and enhancement of transmitter release from existing terminals. We will discuss how such changes may contribute to different forms of long-term learning and memory in Chapters 53 and 54.

Highlights

1. Chemical neurotransmission is the primary mechanism by which neurons communicate, and process information; it occurs throughout the nervous system. Release of neurotransmitter is

stimulated by a series of electrical and biochemical processes in the presynaptic nerve terminal.

2. Neurotransmitter release is steeply dependent on depolarization of the presynaptic terminal. While the action potential is controlled by sodium and potassium conductances, it is the depolarization itself, rather than opening of either voltage-gated sodium or potassium channels, that triggers release.

3. Depolarization of the presynaptic terminal opens voltage-gated Ca^{2+} channels (VGCCs), resulting in Ca^{2+} influx. These channels are concentrated at presynaptic "active zones," very close to the sites at which release occurs. The relationship between Ca^{2+} influx and neurotransmitter release is tightly coupled and steeply nonlinear. The peak Ca^{2+} entry lags slightly behind the peak of the action potential and quickly produces a marked rise in the rate of transmitter release.

4. VGCCs are heterogeneous—five classes have been described with distinct biophysical, biochemical, and pharmacological properties. Multiple classes of VGCCs can contribute to neurotransmitter release at individual nerve terminals and are targets of disease-causing mutations. P/Q- and N-type Ca^{2+} channels are particularly prominent at active zones in the central nervous system.

5. Chemical transmission generally involves the release of quantal packets of neurotransmitter, with a quantum corresponding to the contents of a single synaptic vesicle. Under conditions that decrease transmitter release, such as lowered extracellular Ca^{2+}, a presynaptic action potential triggers the probabilistic release of a few quanta, which produce postsynaptic responses of variable amplitude that are integral multiples of the unitary response to a single quantum, interspersed by complete failures of transmission.

6. The unitary events are driven by the fusion of individual synaptic vesicles of relatively homogeneous size and transmitter content. Visualized as small, clear, spherical membrane organelles, single vesicles contain thousands of small-molecule neurotransmitters. Other neurotransmitters, including the biogenic amines and neuropeptides, are packaged into a distinct class of larger, dense-core vesicles that mediate slower forms of synaptic transmission. A typical presynaptic terminal in the mammalian central nervous system involved in fast synaptic transmission contains 100 to 200 vesicles. A small number of vesicles dock along the presynaptic membrane of the active zone and are the most ready to fuse.

7. At many synaptic connections, the amplitude of a postsynaptic potential can be described as a product of multiple factors: (1) the number of presynaptic sites occupied by a readily releasable vesicle (n), (2) the release probability of individual sites (p), and (3) the size of the postsynaptic response to the release of a single vesicle (a). On individual trials, the number of vesicles released can be described by a binomial distribution reflecting the likelihood of release of zero, one, two, or more vesicles, as if we were to count the number of heads when n coins were being flipped.

8. Exocytosis, the process by which vesicles fuse with the presynaptic membrane, and endocytosis, the process that retrieves vesicles, occur in rapid succession in nerve terminals and other secretory structures. These events are evident in morphological studies and are studied in real time by electrical measurements of membrane surface area.

9. Exocytosis is mediated by evolutionarily conserved SNARE proteins. Together, the presynaptic plasma membrane proteins syntaxin and SNAP-25 and the synaptic vesicle membrane protein synaptobrevin contribute to the SNARE complex, a set of four helical domains. Formation of this complex is critical for vesicle fusion as shown by the ability of various neurotoxins to block transmitter release through the cleavage of SNARE proteins. SNARE complex assembly is modulated by a family of SM proteins, exemplified by Munc18.

10. Synaptotagmins, such as synaptotagmin-1 (syt1), are abundant vesicular proteins that act as Ca^{2+} sensors for regulation of vesicle release. Syt1 binds multiple Ca^{2+} ions and thus forms a close association with the plasma membrane following Ca^{2+} influx. By binding to the SNARE complex even before the rise in presynaptic Ca^{2+}, it may enable that complex to cause fusion quickly.

11. Synaptic vesicle exocytosis is exquisitely precise and rapid because its molecular machinery is embedded in an active zone protein scaffold consisting of RIM, RIM-BP, and Munc-13. The complex: 1. Tethers vesicles to the plasma membrane through binding of RIM to Rab3 and Rab27 vesicle proteins; 2. Recruits calcium channels to the vicinity of tethered vesicles via binding to RIM and RIM-BP; and 3. Facilitates SNARE complex assembly via interaction with Munc13. The active zone complex also mediates many forms of short- and long-term synaptic plasticity.

12. Rapid endocytosis of vesicle membranes after release enables fast recycling of vesicles for a continuous supply during prolonged stimulation.

13. Synaptic terminals are diverse and vary in their release properties. Active zone scaffolding proteins differ across synapses and species, as does

expression of presynaptic Ca^{2+} channels and synaptotagmins. At some synapses, vesicles and Ca^{2+} channels appear aligned by an intricate structural network.

14. Transmitter release can be modulated intrinsically or extrinsically as an aspect of synaptic plasticity. Synaptic strength can be strongly influenced intrinsically by the pattern of firing in phenomena known as "depression" and "facilitation." In addition, extrinsic neuromodulators can alter the dynamics of release by regulation of Ca^{2+} channels or events downstream of Ca^{2+} entry.

Steven A. Siegelbaum
Thomas C. Südhof
Richard W. Tsien

Selected Reading

Katz B. 1969. *The Release of Neural Transmitter Substances.* Springfield, IL: Thomas.

Lonart G. 2002. RIM1: an edge for presynaptic plasticity. Trends Neurosci 25:329–332.

Meinrenken CJ, Borst JG, Sakmann B. 2003. Local routes revisited: the space and time dependence of the Ca^{2+} signal for phasic transmitter release at the rat calyx of Held. J Physiol 547:665–689.

Reid CA, Bekkers JM, Clements JD. 2003. Presynaptic Ca^{2+} channels: a functional patchwork. Trends Neurosci 26:683–687.

Stevens CF. 2003. Neurotransmitter release at central synapses. Neuron 40:381–388.

Südhof TC. 2014. The molecular machinery of neurotransmitter release (Nobel lecture). Angew Chem Int Ed Engl 53:12696–12717.

References

Acuna C, Liu X, Südhof TC. 2016. How to make an active zone: unexpected universal functional redundancy between RIMs and RIM-BPs. Neuron 91:792–807.

Akert K, Moor H, Pfenninger K. 1971. Synaptic fine structure. Adv Cytopharmacol 1:273–290.

Bacaj T, Wu D, Yang X, et al. 2013. Synaptotagmin-1 and -7 trigger synchronous and asynchronous phases of neurotransmitter release. Neuron 80:947–959.

Baker PF, Hodgkin AL, Ridgway EB. 1971. Depolarization and calcium entry in squid giant axons. J Physiol 218:709–755.

Bollmann JH, Sakmann B, Gerard J, Borst G. 2000. Calcium sensitivity of glutamate release in a calyx-type terminal. Science 289:953–957.

Borst JG, Sakmann B. 1996. Calcium influx and transmitter release in a fast CNS synapse. Nature 383:431–434.

Boyd IA, Martin AR. 1956. The end-plate potential in mammalian muscle. J Physiol 132:74–91.

Chung S, Li X, Nelson SB. 2002. Short-term depression at thalamocortical synapses contributes to rapid adaptation of cortical sensory responses in vivo. Neuron 34:437–446.

Couteaux R, Pecot-Dechavassine M. 1970. Vésicules synaptiques et poches au niveau des "zones actives" de la jonction neuromusculaire. C R Acad Sci Hebd Seances Acad Sci D 271:2346–2349.

Del Castillo J, Katz B. 1954. The effect of magnesium on the activity of motor nerve endings. J Physiol 124:553–559.

Dittman JS, Kreitzer AC, Regehr WG. 2000. Interplay between facilitation, depression and residual calcium at three presynaptic terminals. J Neurosci 20:1374–1385.

Enoki R, Hu YL, Hamilton D, Fine A. 2009. Expression of long-term plasticity at individual synapses in hippocampus is graded, bidirectional, and mainly presynaptic: optical quantal analysis. Neuron 62:242–253.

Fatt P, Katz B. 1952. Spontaneous subthreshold activity at motor nerve endings. J Physiol 117:109–128.

Fawcett DW. 1981. *The Cell,* 2nd ed. Philadelphia: Saunders.

Fernandez I, Araç D, Ubach J, et al. 2001. Three-dimensional structure of the synaptotagmin 1 C2B-domain: synaptotagmin 1 as a phospholipid binding machine. Neuron 32:1057–1069.

Fernandez JM, Neher E, Gomperts BD. 1984. Capacitance measurements reveal stepwise fusion events in degranulating mast cells. Nature 312:453–455.

Fernandez-Chacon R, Konigstorfer A, Gerber SH, et al. 2001. Synaptotagmin I functions as a calcium regulator of release probability. Nature 410:41–49.

Geiger JR, Jonas P. 2000. Dynamic control of presynaptic Ca(2+) inflow by fast-inactivating K(+) channels in hippocampal mossy fiber boutons. Neuron 28:927–939.

Geppert M, Goda Y, Hammer RE, et al. 1994. Synaptotagmin I: a major Ca^{2+} sensor for transmitter release at a central synapse. Cell 79:717–727.

Harlow LH, Ress D, Stoschek A, Marshall RM, McMahan UJ. 2001. The architecture of active zone material at the frog's neuromuscular junction. Nature 409:479–484.

Heuser JE, Reese TS. 1981. Structural changes in transmitter release at the frog neuromuscular junction. J Cell Biol 88:564–580.

Hille B. 2001. *Ionic Channels of Excitable Membranes,* 3rd ed. Sunderland, MA: Sinauer.

Kaeser PS, Deng L, Wang Y, et al. 2011. RIM proteins tether Ca^{2+} channels to presynaptic active zones via a direct PDZ-domain interaction. Cell 144:282–295.

Kandel ER. 1981. Calcium and the control of synaptic strength by learning. Nature 293:697–700.

Katz B, Miledi R. 1967a. The study of synaptic transmission in the absence of nerve impulses. J Physiol 192:407–436.

Katz B, Miledi R. 1967b. The timing of calcium action during neuromuscular transmission. J Physiol 189:535–544.

Klein M, Shapiro E, Kandel ER. 1980. Synaptic plasticity and the modulation of the Ca^{2+} current. J Exp Biol 89:117–157.

Kretz R, Shapiro E, Connor J, Kandel ER. 1984. Post-tetanic potentiation, presynaptic inhibition, and the modulation of the free Ca^{2+} level in the presynaptic terminals. Exp Brain Res Suppl 9:240–283.

Kuffler SW, Nicholls JG, Martin AR. 1984. *From Neuron to Brain: A Cellular Approach to the Function of the Nervous System,* 2nd ed. Sunderland, MA: Sinauer.

Lawson D, Raff MC, Gomperts B, Fewtrell C, Gilula NB. 1977. Molecular events during membrane fusion. A study of exocytosis in rat peritoneal mast cells. J Cell Biol 72:242–259.

Liley AW. 1956. The quantal components of the mammalian end-plate potential. J Physiol 133:571–587.

Llinás RR. 1982. Calcium in synaptic transmission. Sci Am 247:56–65.

Llinás RR, Heuser JE. 1977. Depolarization-release coupling systems in neurons. Neurosci Res Program Bull 15:555–687.

Llinás R, Steinberg IZ, Walton K. 1981. Relationship between presynaptic calcium current and postsynaptic potential in squid giant synapse. Biophys J 33:323–351.

Lynch G, Halpain S, Baudry M. 1982. Effects of high-frequency synaptic stimulation on glumate receptor binding studied with a modified in vitro hippocampal slice preparation. Brain Res 244:101–111.

Magee JC, Johnston D. 1997. A synaptically controlled, associative signal for Hebbian plasticity in hippocampal neurons. Science 275:209–213.

Magnus CJ, Lee PH, Atasoy D, Su HH, Looger LL, Sternson SM. 2011. Chemical and genetic engineering of selective ion channel-ligand interactions. Science 333:1292–1296.

Martin AR. 1977. Junctional transmission. II. Presynaptic mechanisms. In: ER Kandel (ed). *Handbook of Physiology: A Critical, Comprehensive Presentation of Physiological Knowledge and Concepts,* Sect. 1 *The Nervous System,* Vol. 1 *Cellular Biology of Neurons,* Part 1, pp. 329–355. Bethesda, MD: American Physiological Society.

Monck JR, Fernandez JM. 1992. The exocytotic fusion pore. J Cell Biol 119:1395–1404.

Neher E. 1993. Cell physiology. Secretion without full fusion. Nature 363:497–498.

Nicoll RA. 1982. Neurotransmitters can say more than just "yes" or "no." Trends Neurosci 5:369–374.

Park M, Penick EC, Edwards JG, Kauer JA, Ehlers MD. 2004. Recycling endosomes supply AMPA receptors for LTP. Science 305:1972–1975.

Peters A, Palay SL, Webster H deF. 1991. *The Fine Structure of the Nervous System: Neurons and Supporting Cells,* 3rd ed. Philadelphia: Saunders.

Redman S. 1990. Quantal analysis of synaptic potentials in neurons of the central nervous system. Physiol Rev 70:165–198.

Rhee JS, Betz A, Pyott S, et al. 2002. Beta phorbol ester- and diacylglycerol-induced augmentation of transmitter release is mediated by Munc13s and not by PKCs. Cell 108:121–133.

Rizo J, Südhof TC. 2002. Snares and Munc18 in synaptic vesicle fusion. Nat Rev Neurosci 3:641–653.

Robitaille R, Adler EM, Charlton MP. 1990. Strategic location of calcium channels at transmitter release sites of frog neuromuscular synapses. Neuron 5:773–779.

Schneggenburger R, Neher E. 2000. Intracellular calcium dependence of transmitter release rates at a fast central synapse. Nature 406:889–893.

Schoch S, Castillo PE, Jo T, et al. 2002. RIM1alpha forms a protein scaffold for regulating neurotransmitter release at the active zone. Nature 415:321–326.

Schoch S, Deak F, Konigstorfer A, et al. 2001. SNARE function analyzed in synaptobrevin/VAMP knockout mice. Science 294:1117–1122.

Schweizer FE, Betz H, Augustine GJ. 1995. From vesicle docking to endocytosis: intermediate reactions of exocytosis. Neuron 14:689–696.

Smith SJ, Augustine GJ, Charlton MP. 1985. Transmission at voltage-clamped giant synapse of the squid: evidence for cooperativity of presynaptic calcium action. Proc Natl Acad Sci U S A 82:622–625.

Söllner T, Whiteheart SW, Brunner M, et al. 1993. SNAP receptors implicated in vesicle targeting and fusion. Nature 362:318–324.

Spruce AE, Breckenridge LJ, Lee AK, Almers W. 1990. Properties of the fusion pore that forms during exocytosis of a mast cell secretory vesicle. Neuron 4:643–654.

Sudhof TC. 2004. The synaptic vesicle cycle. Annu Rev Neurosci. 27:509–547.

Südhof TC. 2013. Neurotransmitter release: the last millisecond in the life of a synaptic vesicle. Neuron 80:675–690.

Sun J, Pang ZP, Qin D, Fahim AT, Adachi R, Südhof TC. 2007. A dual-Ca2+-sensor model for neurotransmitter release in a central synapse. Nature 450:676–682.

Sun JY, Wu LG. 2001. Fast kinetics of exocytosis revealed by simultaneous measurements of presynaptic capacitance and postsynaptic currents at a central synapse. Neuron 30:171–182.

Sutton RB, Fasshauer D, Jahn R, Brunger AT. 1998. Crystal structure of a SNARE complex involved in synaptic exocytosis at 2.4 A resolution. Nature 395:347–353.

Takamori S, Holt M, Stenius K, et al. 2006. Molecular anatomy of a trafficking organelle. Cell 127:831–846.

von Gersdorff H, Matthews G. 1994. Dynamics of synaptic vesicle fusion and membrane retrieval in synaptic terminals. Nature 367:735–739.

Wachman ES, Poage RE, Stiles JR, Farkas DL, Meriney SD. 2004. Spatial distribution of calcium entry evoked by single action potentials within the presynaptic active zone. J Neurosci 24:2877–2885.

Wernig A. 1972. Changes in statistical parameters during facilitation at the crayfish neuromuscular junction. J Physiol 226:751–759.

Whittaker VP. 1993. Thirty years of synaptosome research. J Neurocytol. 22:735–742.

Zenisek D, Horst NK, Merrifield C, Sterling P, Matthews G. 2004. Visualizing synaptic ribbons in the living cell. J Neurosci 24:9752–9759.

Zhou Q, Zhou P, Wang AL, et al. 2017. The primed SNARE-complexin-synaptotagmin complex for neuronal exocytosis. Nature. 548:420–425.

Zucker RS. 1973. Changes in the statistics of transmitter release during facilitation. J Physiol 229:787–810.

16

Neurotransmitters

CHEMICAL SYNAPTIC TRANSMISSION can be divided into four steps: (1) synthesis and storage of a transmitter substance, (2) release of the transmitter, (3) interaction of the transmitter with receptors at the postsynaptic membrane, and (4) removal of the transmitter from the synapse. In the previous chapters, we considered steps 2 and 3. We now turn to the initial and final steps of chemical synaptic transmission: the synthesis and storage of transmitter molecules and their removal from the synaptic cleft after synaptic action.

A Chemical Messenger Must Meet Four Criteria to Be Considered a Neurotransmitter

Before considering the biochemical processes involved in synaptic transmission, it is important to make clear what is meant by a chemical transmitter. The concept is empirical and has changed over the years with increased understanding of synaptic transmission and a corresponding expansion of signaling agents. The concept that a released chemical could act as a transmitter was introduced by the British physician George Oliver and his colleague Edward Albert Schaefer, who in 1894 reported that injection of an adrenal gland extract increases blood pressure (Sir Henry Dale claimed that Oliver discovered this by injecting the extract into his own son). The constituent responsible was independently identified by three laboratories in 1897, and competing claims for priority provide one reason that this transmitter has 38 different names in the Merck Index, including adrenaline (as it was obtained from the adrenal gland) and epinephrine.

Experiments reported in 1904 by Thomas Elliott, a student in the lab of the physiologist John Langley, are generally credited as the first report of chemical neurotransmission. Elliott concluded that "adrenaline might then be the chemical stimulant liberated on each occasion when the impulse arrives at the periphery." Not incidentally, Elliott also proposed as early as 1914 that nerves could accumulate transmitter by an uptake system, suggesting that adrenal gland signaling might "depend on what could be picked up from the circulating blood and stored in its nerve endings," although uptake mechanisms were not demonstrated until more than 40 years later.

In 1913, Arthur Ewins, working with Henry Dale, discovered acetylcholine (ACh) as a component of the ergot fungus. In 1921, Otto Loewi demonstrated that stimulation of the vagus nerve terminals in frog hearts released "vagustoff," which was later shown to be ACh. Dale and Loewi later shared the Nobel Prize in 1946. The terms *cholinergic* and *adrenergic* were introduced to indicate that a neuron makes and releases ACh or norepinephrine (or epinephrine), respectively, the two substances first recognized as neurotransmitters. The term *catecholaminergic*, encompassing dopamine and the adrenergic transmitters, was derived from one of many natural sources, the catechu tree of India. Since that time, many other substances have been identified as transmitters.

The first secretory vesicles shown to accumulate and release neurotransmitters were the chromaffin vesicles of the adrenal gland, named in 1902 by Alfred Kohn due to their colorimetric reaction with chromate. William Cramer later showed that these organelles accumulate epinephrine. More recently, Mark Wightman provided direct evidence that these vesicles released epinephrine using carbon fiber electrodes as an electrochemical detector to measure catecholamine molecules released following fusion of chromaffin vesicles with the plasma membrane.

As a first approximation, a neurotransmitter can be defined as a substance that is released by a neuron that affects a specific target in a specific manner. A target can be either another neuron or an effector organ, such as muscle or gland. As with many other operational concepts in biology, the concept of a transmitter is not precise. Although the actions of hormones and neurotransmitters are quite similar, neurotransmitters usually act on targets that are close to the site of transmitter release, whereas hormones are released into the bloodstream to act on distant targets.

Neurotransmitters typically act on a target other than the releasing neuron itself, whereas substances termed autacoids act on the cell from which they are released. Nevertheless, at many synapses, transmitters activate not only postsynaptic receptors but also autoreceptors at the presynaptic release site. Autoreceptors usually modulate synaptic transmission that is in progress, for example, by limiting further release of transmitter or inhibiting subsequent transmitter synthesis. Receptors can also exist on presynaptic release sites that receive synaptic input from another neuron. These receptors function as heteroreceptors that regulate presynaptic excitability and transmitter release (Chapters 13 and 15).

Following release, the interaction of neurotransmitters with receptors is typically transient, lasting for periods ranging from less than a millisecond to several seconds. Nevertheless, neurotransmitter actions can result in long-term changes within target cells lasting hours or days, often by activating gene transcription. Moreover, nonneural cells, including astrocytes and microglia, can also synthesize, store, and release neurotransmitters, as well as express receptors that modulate their own function.

A limited number of substances of low molecular weight are generally accepted as classical neurotransmitters, and these exclude many neuropeptides, as well as other substances that are not released by exocytosis. Even so, it is often difficult to demonstrate that a specific neurotransmitter operates at a particular synapse, particularly given the diffusion and rapid reuptake or degradation of transmitters at the synaptic cleft.

A classical neurotransmitter is considered to meet four criteria:

1. It is synthesized in the presynaptic neuron.
2. It is accumulated within vesicles present in presynaptic release sites and is released via exocytosis in amounts sufficient to exert a defined action on the postsynaptic neuron or effector organ.
3. When administered exogenously in reasonable concentrations, it mimics the action of the endogenous transmitter (for example, it activates the same ion channels or second-messenger pathway in the postsynaptic cell).
4. A specific mechanism usually exists for removing the substance from the extracellular environment. This may be the synaptic cleft in the case of "wired" or "private" neurotransmission (in which the action of the substance is limited to a single synapse) or the extrasynaptic space in the case of "volume" or "social" neurotransmission (in which the substance diffuses to multiple synapses).

The nervous system makes use of two main classes of chemical substances that fit these criteria for signaling: small-molecule transmitters and neuropeptides. Neuropeptides are short polymers of amino acids processed in the Golgi apparatus, where they are packaged in large dense-core vesicles (approximately 70–250 nm in diameter). Small-molecule transmitters are packaged in small vesicles (~40 nm in diameter) that are usually electron-lucent. Vesicles are closely associated with specific Ca^{2+} channels at active zones and release their contents through exocytosis in response to a rise in intracellular Ca^{2+} evoked by an action potential (Chapter 15). Vesicle membrane is retrieved through endocytosis and recycled locally in the axon to produce new synaptic vesicles. Large dense-core vesicles can contain both small-molecule transmitters and neuropeptides and do not undergo local recycling following full fusion with the plasma membrane.

Both types of vesicles are found in most neurons but in different proportions. Small synaptic vesicles are characteristic of neurons that use ACh, glutamate, γ-aminobutyric acid (GABA), and glycine as transmitters, whereas neurons that use catecholamines and serotonin as transmitters often have both small and large dense-core vesicles. The adrenal medulla—the tissue in which most discoveries on secretion were made and still widely used as a model for studying exocytosis—contains only large dense-core vesicles that contain both catecholamines and neuroactive peptides.

Only a Few Small-Molecule Substances Act as Transmitters

A relatively small number of low-molecular-weight substances are generally accepted as neurotransmitters. These include ACh, the excitatory amino acid glutamate, the inhibitory amino acids GABA and glycine, amine containing amino acid derivatives, and adenosine triphosphate (ATP) and its metabolites (Table 16–1). A

Table 16–1 Small-Molecule Neurotransmitter Substances and Their Precursors

Transmitter	Precursor
Acetylcholine	Choline
Biogenic amines	
Dopamine	Tyrosine
Norepinephrine	Tyrosine via dopamine
Epinephrine	Tyrosine via norepinephrine
Octopamine	Tyrosine via tyramine
Serotonin	Tryptophan
Histamine	Histidine
Melatonin	Tryptophan via serotonin
Amino acids	
Aspartate	Oxaloacetate
γ-Aminobutyric acid	Glutamine
Glutamate	Glutamine
Glycine	Serine
Adenosine triphosphate (ATP)	Adenosine diphosphate (ADP)
Adenosine	ATP
Endocannabinoids	Phospholipids
Nitric oxide	Arginine

small set of small molecules, such as the gas nitric oxide lipid metabolites are not released from vesicles and tend to break all of the classical rules (Chapter 14).

The amine messengers share many biochemical similarities. All are charged small molecules that are formed in relatively short biosynthetic pathways and synthesized either from essential amino acids or from precursors derived from the major carbohydrate substrates of intermediary metabolism. Like other pathways of intermediary metabolism, synthesis of these neurotransmitters is catalyzed by enzymes that, with the notable exception of dopamine β-hydroxylase, are cytosolic. ATP, which originates in mitochondria, is abundantly present throughout the cell.

As in any biosynthetic pathway, the overall synthesis of amine transmitters typically is regulated at one rate-limiting enzymatic reaction. The rate-limiting step often is characteristic of one type of neuron and usually is absent in other types of mature neurons. The classical small-molecule neurotransmitters released from a particular neuron are thus determined by their presence in the cytosol due to synthesis and reuptake and to the selectivity of the vesicular transporter.

Acetylcholine

Acetylcholine is the only low-molecular-weight aminergic transmitter substance that is not an amino acid or derived directly from one. The biosynthetic pathway for ACh has only one enzymatic reaction, catalyzed by choline acetyltransferase (step 1 below):

$$\text{Acetyl CoA + choline}$$
$$(1) \Big\updownarrow$$
$$CH_3 - \overset{\overset{\displaystyle O}{\|}}{C} - O - CH_2 - CH_2 - \overset{+}{N} - (CH_3)_3 + CoA$$
$$\text{Acetylcholine}$$

This transferase is the characteristic and limiting enzyme in ACh biosynthesis. Nervous tissue cannot synthesize choline, which is derived from the diet and delivered to neurons through the bloodstream. The co-substrate, acetyl coenzyme A (acetyl CoA), participates in many general metabolic pathways and is not restricted to cholinergic neurons.

Acetylcholine is released at all vertebrate neuromuscular junctions by spinal motor neurons (Chapter 12). In the autonomic nervous system, it is the transmitter released by all preganglionic neurons and by parasympathetic postganglionic neurons (Chapter 41). Cholinergic neurons form synapses throughout the brain; those in the nucleus basalis have particularly widespread projections to the cerebral cortex. Acetylcholine (together with a noradrenergic component) is a principal neurotransmitter of the reticular activating system, which modulates arousal, sleep, wakefulness, and other critical aspects of human consciousness.

Biogenic Amine Transmitters

The terms *biogenic amine* or *monoamine*, although chemically imprecise, have been used for decades to designate certain neurotransmitters. This group includes the catecholamines and serotonin. Histamine, an imidazole, is also often included with biogenic amine transmitters, although its biochemistry is remote from the catecholamines and the indolamines.

Catecholamine Transmitters

The catecholamine transmitters—dopamine, norepinephrine, and epinephrine—are all synthesized from the essential amino acid tyrosine in a biosynthetic pathway containing five enzymes: tyrosine hydroxylase, pteridine reductase, aromatic amino acid decarboxylase, dopamine β-hydroxylase, and phenylethanolamine-N-methyl transferase. Catecholamines contain a catechol nucleus, a 3,4-dihydroxylated benzene ring.

The first enzyme, tyrosine hydroxylase (step 1 below), is an oxidase that converts tyrosine to L-dihydroxyphenylalanine (L-DOPA):

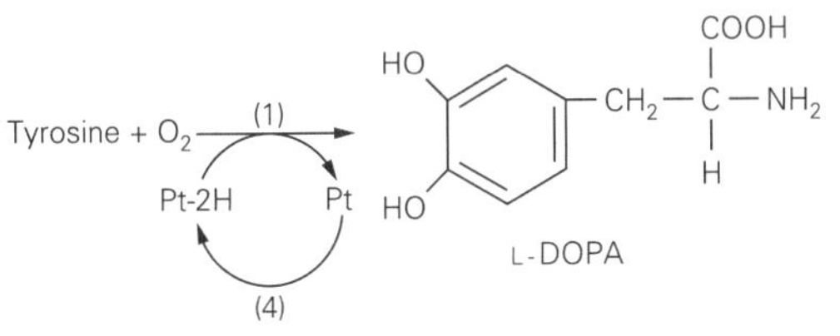

This enzyme is rate-limiting for the synthesis of both dopamine and norepinephrine. A distinct pathway is used to synthesize L-DOPA for production of the melanin pigments found throughout the plant and animal kingdoms, while the neuromelanin pigment found in some dopamine and norepinephrine neurons are metabolites of the oxidized neurotransmitters.

L-DOPA is present in all cells producing catecholamines, and its synthesis requires a reduced pteridine cofactor, Pt-2H, which is regenerated from pteridine (Pt) by another enzyme, pteridine reductase, which uses nicotinamide adenine dinucleotide (NADH) (step 4 above). This reductase is not specific to neurons.

Based on the finding that individuals with Parkinson disease have lost dopamine neurons of the substantia nigra, L-DOPA has been used to restore dopamine and motor function in patients. L-DOPA, whether exogenous or produced by tyrosine hydroxylase, is decarboxylated by a widespread enzyme known as aromatic amino acid decarboxylase, also called L-DOPA decarboxylase (step 2 below), to yield dopamine and carbon dioxide:

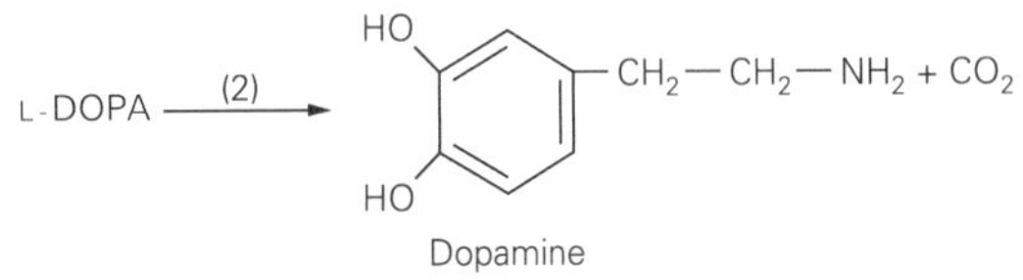

Interestingly, dopamine was initially thought to be present in neurons only as a precursor to norepinephrine. That dopamine also functions as a neurotransmitter itself was demonstrated in 1957 by Aarvid Carlsson, who found that rabbits treated with the synaptic vesicle dopamine uptake blocker, reserpine, exhibited floppy ears, but that L-DOPA, under conditions that produced dopamine but not norepinephrine, restored the normal erect ear posture.

In adrenergic neurons, the third enzyme in the sequence, dopamine β-hydroxylase (step 3 below), further converts dopamine to norepinephrine:

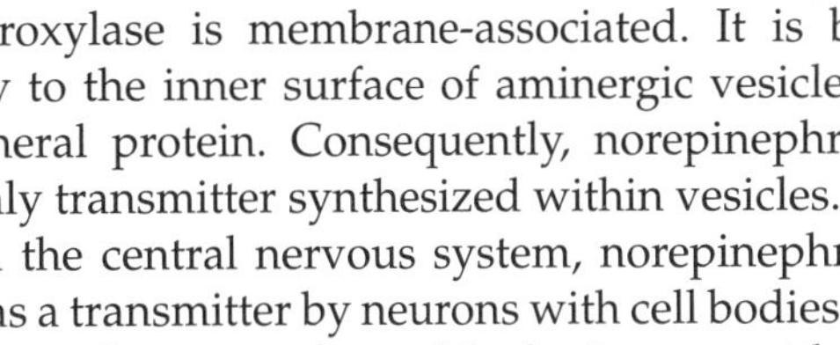

Unlike all other enzymes in the biosynthetic pathways of small-molecule neurotransmitters, dopamine β-hydroxylase is membrane-associated. It is bound tightly to the inner surface of aminergic vesicles as a peripheral protein. Consequently, norepinephrine is the only transmitter synthesized within vesicles.

In the central nervous system, norepinephrine is used as a transmitter by neurons with cell bodies in the locus coeruleus, a nucleus of the brain stem with many complex modulatory functions (Chapter 40). Although these adrenergic neurons are relatively few in number, they project widely throughout the cortex, cerebellum, hippocampus, and spinal cord. In many cases, neurons that release norepinephrine can also release the precursor dopamine, and thus can act at neurons expressing receptors for dopamine or norepinephrine. In the peripheral nervous system, norepinephrine is the transmitter of the postganglionic neurons in the sympathetic nervous system (Chapter 41).

In addition to these four catecholaminergic biosynthetic enzymes, a fifth enzyme, phenylethanolamine-N-methyltransferase (step 5 below), methylates norepinephrine to form epinephrine (adrenaline) in the adrenal medulla:

Epinephrine

This reaction requires *S*-adenosyl-methionine as a methyl donor. The transferase is a cytoplasmic enzyme. Thus, for epinephrine to be formed, its immediate precursor norepinephrine must exit from vesicles into the cytoplasm. For epinephrine to be released, it must then be taken back up into vesicles. Only a small number of neurons in the brain use epinephrine as a transmitter.

The production of these catecholamine neurotransmitters is controlled by feedback regulation of the first enzyme in the pathway, tyrosine hydroxylase (Box 16–1). Not all cells that release catecholamines express all five biosynthetic enzymes, although cells that release epinephrine do. During development, the expression of the genes encoding these synthetic enzymes is independently regulated and the particular catecholamine that is produced by a cell is determined by which enzyme(s) in the step-wise pathway is not

---

Box 16–1 Catecholamine Production Varies With Neuronal Activity

Norepinephrine neurotransmission is far more active during awake states than sleep or anesthesia, with locus coeruleus noradrenergic neurons nearly silenced during rapid eye movement (REM) sleep. The production of catecholamine is able to keep up with wide variations in neuronal activity because catecholamine synthesis is highly regulated. Circadian changes in extracellular dopamine in the striatum have been suggested to result from altered activity of the dopamine uptake transporter.

In autonomic ganglia, the amount of norepinephrine in postganglionic neurons is regulated transsynaptically. Activity in the presynaptic neurons, which are both cholinergic and peptidergic, first induces short-term changes in second messengers in the postsynaptic adrenergic cells. These changes increase the supply of norepinephrine through the cAMP-dependent phosphorylation of tyrosine hydroxylase, the first enzyme in the catecholamine biosynthetic pathway.

Phosphorylation enhances the affinity of the hydroxylase for the pteridine cofactor and diminishes feedback inhibition by end products such as norepinephrine. Phosphorylation of tyrosine hydroxylase lasts only as long as cAMP remains elevated, as the phosphorylated hydroxylase is quickly dephosphorylated by protein phosphatases.

If presynaptic activity is sufficiently prolonged, however, other changes in the production of norepinephrine will occur. Severe stress to an animal results in intense presynaptic activity and persistent firing of the postsynaptic adrenergic neuron, placing a greater demand on transmitter synthesis. To meet this challenge, the tyrosine hydroxylase gene is induced to increase transcription and thus production of the protein. Elevated amounts of tyrosine hydroxylase are observed in the cell body within hours after stimulation and at nerve endings days later.

This induction of increased levels of tyrosine hydroxylase begins with the persistent release of chemical transmitters from the presynaptic neurons and prolonged activation of the cAMP pathway in postsynaptic adrenergic cells, which activates the cAMP-dependent protein kinase (PKA). This kinase phosphorylates not only existing tyrosine hydroxylase molecules but also the transcription factor, cAMP response element binding protein (CREB).

Once phosphorylated, CREB binds a specific DNA enhancer sequence called the cAMP-recognition element (CRE), which lies upstream (5′) of the gene for the hydroxylase. Binding of CREB to CRE facilitates the binding of RNA polymerase to the gene's promoter, increasing tyrosine hydroxylase transcription. Induction of tyrosine hydroxylase was the first known example of a neurotransmitter altering gene expression.

Based on similarity in portions of the amino acid and nucleic acid sequences encoding three of the biosynthetic enzymes—tyrosine hydroxylase, dopamine β-hydroxylase, and phenylethanolamine-*N*-methyltransferase—it has been suggested that the three enzymes may have arisen from a common ancestral protein. Moreover, long-term changes in the synthesis of these enzymes are coordinately regulated in adrenergic neurons.

At first, this discovery suggested that the genes encoding these enzymes might be located sequentially along the same chromosome and be controlled by the same promoter, like genes in a bacterial operon. But in humans, the genes for the biosynthetic enzymes for norepinephrine are not located on the same chromosome. Therefore, coordinate regulation is likely achieved by parallel activation through similar but independent transcription activator systems.

expressed. Thus, neurons that release norepinephrine do not express the methyltransferase, and neurons that release dopamine do not express the transferase or dopamine β-hydroxylase. Some neurons that express tyrosine hydroxylase, and thus produce dopamine, do not express the vesicular monoamine transporter (VMAT), the transporter that accumulates dopamine in synaptic vesicles, and thus do not appear to release dopamine as a transmitter.

Of the four major dopaminergic nerve tracts, three arise in the midbrain (Chapters 40 and 43). Dopaminergic neurons in the substantia nigra that project to the striatum are important for the control of movement and are affected in Parkinson disease and other disorders of movement, but projections to the associative striatum have also been implicated more recently in dopamine dysfunction in schizophrenia. The mesolimbic and mesocortical tracts are critical for affect, emotion, attention, and motivation and are implicated in drug addiction and schizophrenia. A fourth dopaminergic tract, the tuberoinfundibular pathway, originates in the arcuate nucleus of the hypothalamus and projects to the pituitary gland, where it regulates secretion of hormones (Chapter 41).

The synthesis of biogenic amines is highly regulated and can be rapidly increased. As a result, the amounts of transmitter available for release can keep up with wide variations in neuronal activity. Mechanisms for regulating both the synthesis of catecholamine transmitters and the production of enzymes in the step-wise catecholamine pathway are discussed in Box 16–1.

Trace amines, naturally occurring catecholamine derivatives, may also serve as transmitters. In invertebrates, the tyrosine derivatives tyramine and octopamine (so called because it was originally identified in the octopus salivary gland) play key roles in numerous physiological processes including behavioral regulation. Trace amine receptors also have been identified in mammals, where their functional role is still being characterized. In particular, trace amine-associated receptor 1 (TAAR1) has been shown to modulate aspects of biogenic amine neurotransmission as well as to play a role in the immune system.

Serotonin

Serotonin (5-hydroxytryptamine or 5-HT) and the essential amino acid tryptophan from which it is derived belong to a group of aromatic compounds called indoles, with a five-member ring containing nitrogen joined to a benzene ring. Two enzymes are needed to synthesize serotonin: tryptophan (Trp) hydroxylase (step 1 below), an oxidase similar to

tyrosine hydroxylase, and aromatic amino acid decarboxylase, also called 5-hydroxytryptophan (5-HTP) decarboxylase (step 2 below):

As with the catecholamines, the limiting reaction in serotonin synthesis is catalyzed by the first enzyme in the pathway, tryptophan hydroxylase. Tryptophan hydroxylase is similar to tyrosine hydroxylase not only in catalytic mechanism but also in amino acid sequence. The two enzymes are thought to stem from a common ancestral protein by gene duplication because the two hydroxylases are encoded by genes close together on the same chromosome (tryptophan hydroxylase, 11p15.3-p14; tyrosine hydroxylase, 11p15.5). The second enzyme in the pathway, 5-hydroxytryptophan decarboxylase, is identical to L-DOPA decarboxylase. Enzymes with similar activity, L-aromatic amino acid decarboxylases, are present in nonnervous tissues as well.

The cell bodies of serotonergic neurons are found in and around the midline raphe nuclei of the brain stem and are involved in regulating affect, attention, and other cognitive functions (Chapter 40). These cells, like the noradrenergic cells in the locus coeruleus, project widely throughout the brain and spinal cord. Serotonin and the catecholamines norepinephrine and dopamine are implicated in depression, a major mood disorder. Antidepressant medications inhibit the uptake of serotonin, norepinephrine, and dopamine, thereby increasing the magnitude and duration of the action of these transmitters, which in turn leads to altered cell signaling and adaptations (Chapter 61).

Histamine

Histamine, derived from the essential amino acid histidine by decarboxylation, contains a characteristic five-member ring with two nitrogen atoms. It has long been recognized as an autacoid, active when released from mast cells in the inflammatory reaction and in the control of vasculature, smooth muscle, and exocrine glands (eg, secretion of highly acidic gastric juice). Histamine is a transmitter in both invertebrates and vertebrates. It is concentrated in the hypothalamus, one of the brain centers for regulating the secretion of hormones (Chapter 41). The decarboxylase catalyzing its synthesis (step 1 below), although not extensively analyzed, appears to be characteristic of histaminergic neurons.

Histidine —(1)→ [imidazole ring] CH₂—CH₂—NH₂ + CO₂

$$\text{Histidine} \xrightarrow{(1)} \text{Histamine} + CO_2$$

Histamine

As described in the next section, the biogenic amines are loaded into synaptic and secretory vesicles by two transporters, VMAT1, mostly in peripheral cells, and VMAT2, mostly in the central nervous system. As the transporters are not selective for a given biogenic amine, a mixture of transmitters can be present. Some neurons co-release dopamine with norepinephrine, whereas secretory vesicles from the adrenal medulla can co-release epinephrine and norepinephrine.

Amino Acid Transmitters

In contrast to acetylcholine and the biogenic amines, which are not intermediates in general metabolic pathways and are produced only in certain neurons, the amino acids glutamate and glycine are not only neurotransmitters but also universal cellular constituents. Because they can be synthesized in neurons and other cells, neither is an essential amino acid.

Glutamate, the neurotransmitter most frequently used at excitatory synapses throughout the central nervous system, is produced from α-ketoglutarate, an intermediate in the tricarboxylic acid cycle of intermediary metabolism. After it is released, glutamate is taken up from the synaptic cleft by specific transporters in the membrane of both neurons and glia (see later). The glutamate taken up by astrocytes is converted to glutamine by the enzyme glutamine synthase. This glutamine is transported back into neurons that use glutamate as a transmitter, where it is hydrolyzed to glutamate by the enzyme glutaminase. Cytoplasmic glutamate is then loaded into synaptic vesicles by the vesicular glutamate transporter, VGLUT.

Glycine is the major transmitter used by inhibitory interneurons of the spinal cord. It is also a necessary cofactor for activation of the N-methyl-D-aspartate (NMDA) glutamate receptors (Chapter 13). Glycine is synthesized from serine by the mitochondrial form of the serine hydroxymethyltransferase. The amino acid GABA is synthesized from glutamate in a reaction catalyzed by glutamic acid decarboxylase (step 1 below):

COOH   COOH
| |
CH₂   CH₂
| |
CH₂   (1)   CH₂   + CO₂
| → |
H₂N—CH   H₂N—CH₂
|
COOH

Glutamate   GABA

GABA is present at high concentrations throughout the central nervous system and is detectable in other tissues. It is used as a transmitter by an important class of inhibitory interneurons in the spinal cord. In the brain, GABA is the major transmitter of a wide array of inhibitory neurons and interneurons. Both GABA and glycine are loaded into synaptic vesicles by the same transporter, VGAT, and thus can be co-released from the same vesicles.

ATP and Adenosine

ATP and its degradation products (eg, adenosine) act as transmitters at some synapses by binding to several classes of G protein–coupled receptors (the P1 and P2Y receptors). ATP can also produce excitatory actions by binding to ionotropic P2X receptors. Caffeine's stimulatory effects depend on its inhibition of adenosine binding to the P1 receptors. Adenine and guanine and their sugar-containing derivatives are called purines; the evidence for transmission at purinergic receptors is especially strong for autonomic neurons that innervate the vas deferens, bladder, and muscle fibers of the heart; for nerve plexuses on smooth muscle in the gut; and for some neurons in the brain. Purinergic transmission is particularly important for nerves mediating pain (Chapter 20).

ATP released by tissue damage acts to transmit pain sensation through one type of ionotropic purine receptor present on the terminals of peripheral axons of dorsal root ganglion cells that act as nociceptors. ATP released from terminals of the central axons of these dorsal root ganglion cells excites another type of ionotropic purine receptor on neurons in the dorsal horn of the spinal cord. ATP and other nucleotides also act at the family of P2Y G protein–coupled receptors to modulate various downstream signaling pathways.

Small-Molecule Transmitters Are Actively Taken Up Into Vesicles

Common amino acids act as transmitters in some neurons but not in others, indicating that the presence of a substance in a neuron, even in substantial amounts, is not in itself sufficient evidence that the substance is used as a transmitter. For example, at the neuromuscular junction of the lobster (and other arthropods), GABA is inhibitory and glutamate is excitatory. The concentration of GABA is approximately 20 times greater in inhibitory cells than in excitatory cells, supporting the idea that GABA is the inhibitory transmitter at the lobster neuromuscular junction. In contrast, the concentration of the excitatory transmitter glutamate is similar in both excitatory and inhibitory cells. Glutamate therefore must be compartmentalized within these neurons; that is, *transmitter* glutamate must be kept separate from *metabolic* glutamate. In fact, transmitter glutamate is compartmentalized in synaptic vesicles.

Although the presence of a specific set of biosynthetic enzymes can determine whether a small molecule can be used as a transmitter, the presence of the enzymes does not mean that the molecule will be used. Before a substance can be released as a transmitter, it usually must first be concentrated in synaptic vesicles. Transmitter concentrations within vesicles are high, on the order of several hundred millimolar. Neurotransmitter substances are concentrated in vesicles by transporters that are specific to each type of neuron and energized by a vacuolar-type H^+-ATPase (V-ATPase) that is found not only in synaptic and secretory vesicles but also in all organelles in the secretory pathway, including endosomes and lysosomes.

Using the energy generated by the hydrolysis of cytoplasmic ATP, the V-ATPase creates an H^+ electrochemical gradient by promoting the influx of protons into the vesicle. Transporters use this proton gradient to drive transmitter molecules into the vesicles against their concentration gradient through a proton-antiport mechanism. A number of different vesicular transporters in mammals are responsible for concentrating different transmitter molecules in vesicles (Figure 16–1). These proteins span the vesicle membrane 12 times and are distantly related to a class of bacterial transporters that mediate drug resistance. (Vesicular transporters differ structurally and mechanistically from the transporters in the plasma membrane, as discussed later.)

Transmitter molecules are classically modeled to be taken up into a vesicle by vesicular transporters in exchange for the transport of two protons out of the vesicle. Because the maintenance of the pH gradient requires the hydrolysis of ATP, the uptake of transmitter into vesicles is energy-dependent. Vesicular transporters can concentrate some neurotransmitters such as dopamine up to 100,000-fold relative to their concentration in the cytoplasm. Uptake of transmitters by the transporters is rapid, enabling vesicles to be quickly refilled after they release their transmitter and are retrieved by endocytosis; this is important for maintaining the supply of releasable vesicles during periods of rapid nerve firing (Chapter 15).

The specificity of transporters for substrate is quite variable. The vesicular ACh transporter (VAChT) does not transport choline or other transmitters. Likewise, the vesicular glutamate transporters, for which there are three types (VGLUT1, 2, and 3) that are differentially expressed in the CNS, carry negligible amounts of the other acidic amino acid, aspartate. However, VMAT2 can transport all of the biogenic amines as well as drugs including amphetamine and even some neurotoxic compounds such as N-methyl-4-phenylpyridinium (MPP^+). 1-Methyl-4-phenyl-1,2,3,6-tetrahydropyridine (MPTP), a contaminant of a synthetic opiate drug of abuse, is metabolized to MPP^+ by the enzyme monoamine oxidase (MAO) type B. In fact, VMAT1 was cloned by Robert Edwards and colleagues based on the ability of the transporter to protect cells from the neurotoxic effects of MPP^+; cells expressing VMAT were able to sequester the toxin in vesicle-like compartments, thereby lowering its cytoplasmic concentration and promoting cell survival. By expressing genes obtained from a cDNA library from adrenal pheochromocytoma cells in a mammalian cell line sensitive to MPP^+, Edwards was able to identify cells that expressed VMAT1 based on their selective survival. VMAT2 was subsequently identified by homology cloning, as well as directly by a number of other groups.

Transporters and V-ATPases are present in the membranes of both small synaptic vesicles and large dense-core vesicles. Vesicular transporters are the targets of several important pharmacological agents. Reserpine and tetrabenazine inhibit uptake of amine transmitters by binding to the vesicular monoamine transporter. The psychostimulants amphetamine, methamphetamine, and 3,4-methylenedioxy-N-methylamphetamine (MDMA or ecstasy) act to deplete vesicles of amine transmitter molecules, but also cause their efflux from the cytoplasm into the extracellular space via the plasma membrane biogenic amine transporters (see below). These compounds accumulate inside vesicles through proton-antiport–driven transport mediated by VMAT, which diminishes the proton gradient necessary for loading amine transmitters into vesicles.

Drugs that are sufficiently similar to the normal transmitter substance can act as *false transmitters*. These are packaged in vesicles and released by exocytosis as if they were true transmitters, but they often bind only weakly or not at all to the postsynaptic receptor for the natural transmitter, and thus their release decreases the efficacy of transmission. Several drugs historically used to treat hypertension, such as α-methyldopa and guanethidine, are taken up into adrenergic synapses

A Monoamines

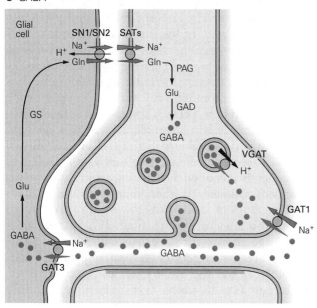

B Acetylcholine

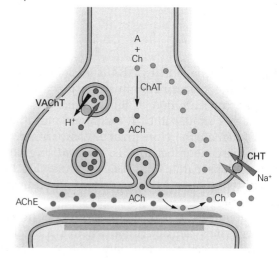

C GABA

D Glutamate

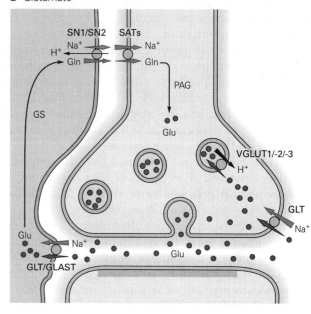

Figure 16–1 Small-molecule transmitters are transported from the cytosol into vesicles or from the synaptic cleft to the cytosol by transporters. Most small-molecule neurotransmitters are released by exocytosis from the nerve terminal and act on specific postsynaptic receptors. The signal is terminated and transmitter recycled by specific transporter proteins located at the nerve terminal or in surrounding glial cells. Transport by these proteins (**orange circles**) is driven by the electrochemical gradients of H$^+$ (**black arrows**) or Na$^+$ (**red arrows**). (Adapted, with permission, from Chaudhry et al. 2008. Copyright © 2008 Springer-Verlag.)

A. Three distinct transporters mediate reuptake of monoamines across the plasma membrane. The dopamine transporter (**DAT**), norepinephrine transporter (**NET**), and serotonin transporter (**SERT**) are responsible for the reuptake (**dark blue arrows**) of their cognate transmitters. The vesicular monoamine transporter (**VMAT2**) transports all three monoamines into synaptic vesicles for subsequent exocytotic release.

B. Cholinergic signaling is terminated by metabolism of acetylcholine (**ACh**) to the inactive choline and acetate by acetylcholinesterase (**AChE**), which is located in the synaptic cleft (**green bar**). Choline (**Ch**) is transported by the choline transporter (**CHT**) back into the nerve terminal (**light blue arrow**) where choline acetyltransferase (**ChAT**) subsequently catalyzes acetylation of

choline to reconstitute ACh. The ACh is transported into the vesicle by the vesicular ACh transporter (**VAChT**).

C. At GABAergic and glycinergic nerve terminals, the GABA transporter (**GAT1**) and glycine transporter (**GLYT2**, not shown) mediate reuptake of GABA and glycine (**gray arrow**), respectively. GABA may also be taken up by surrounding glial cells (eg, by GAT3). In the glial cells, GABA is first converted to glutamate (**Glu**) by GAD. Glu is then is converted by glial glutamine synthetase (**GS**) to glutamine (**Gln**). Glutamine is transported back to the nerve terminal by the concerted action of the system N transporter (**SN1/SN2**) and system A transporter (**SAT**) (**brown arrows**). In the nerve terminal, phosphate-activated glutaminase (**PAG**) converts glutamine to glutamate, which is converted to GABA by glutamate decarboxylase (**GAD**). VGAT then transports GABA into vesicles. The glial transporter GLYT1 (not shown) also contributes to the clearance of glycine.

D. After release from excitatory neuronal terminals, the majority of glutamate is taken up by surrounding glial cells (eg, by **GLT** and **GLAST**) for conversion to glutamine, which is subsequently transported back to the nerve terminals by SN1/SN2 and a type of SAT (**SATx**) (**brown arrows**). Reuptake of glutamate at glutamatergic terminals also has been demonstrated for a GLT isoform (**purple arrows**). Glutamate is transported into vesicles by **VGLUT**.

(and converted into α-methyldopamine in the case of α-methyldopa) and replace norepinephrine in synaptic vesicles. When released, these drugs fail to stimulate postsynaptic adrenergic receptors, thereby relaxing vascular smooth muscle by inhibiting adrenergic tone. Tyramine, which is found in high quantities in dietary red wine and cheese, also acts as a false transmitter; however, it also can act as a stimulant by releasing biogenic amines through a mechanism akin to amphetamine. Another false transmitter, 5-hydroxydopamine, can produce an electron-dense reaction product and has been used to identify synaptic vesicles that acquire biogenic amines.

More recently, several fluorescent false neurotransmitters have been designed, enabling researchers to use imaging methods to monitor the uptake and release of neurotransmitter derivatives during synaptic activity in rodent and fly brain (see Figure 16–5 in Box 16–2).

An unexpected finding is that dopamine can be released from dendrites as well as from axons, despite the lack of synaptic vesicles in dendrites. Organelles that express VMAT2 seem likely to be the source of the release, albeit with different requirements for intracellular Ca^{2+} than classical neurotransmission at presynaptic terminals. For technical reasons, this phenomenon has been studied mostly in dendrites of dopaminergic neurons of the substantia nigra: dopamine can be measured directly by electrochemical techniques, and the dendrites are well separated from the cell bodies. However, it is possible that dendritic neurotransmitter release occurs more widely throughout the nervous system.

Many Neuroactive Peptides Serve as Transmitters

The enzymes that catalyze the synthesis of the low-molecular-weight neurotransmitters, with the exception of dopamine β-hydroxylase, are found in the cytoplasm. These enzymes are synthesized on free polysomes in the cell body and probably in dendrites and are distributed throughout the neuron by axoplasmic flow. Thus, small-molecule transmitter substances can be formed in all parts of the neuron; most importantly, they can be synthesized at axonal presynaptic sites from which they are released.

In contrast, neuroactive peptides are derived from secretory proteins that are formed in the cell body. More than 50 short peptides are produced by neurons or neuroendocrine cells and exert physiological actions (Table 16–2). Some act as hormones

Table 16–2 Neuroactive Mammalian Peptides

Category	Peptide
Hypothalamic neuropeptides	Thyrotropin-releasing hormone Gonadotropin-releasing hormone Corticotropin-releasing factor (CRF) Growth hormone–releasing hormone Melanocyte-stimulating hormone Melanocyte-inhibiting factor Somatostatin β-Endorphin Dynorphin Galanin Neuropeptide Y Orexin Oxytocin Vasopressin
Neurohypophyseal neuropeptides	Oxytocin Vasopressin
Pituitary peptides	Adrenocorticotropic hormone β-Endorphin α-MSH Prolactin Luteinizing hormone Growth hormone Thyrotropin
Pineal hormones	Melatonin
Basal ganglia	Substance P Enkephalin Dynorphin Neuropeptide Y Neurotensin Cholecystokinin Glucagon-like peptide-1 Cocaine- and amphetamine-regulated transcript (CART)
Gastrointestinal peptides	Vasoactive intestinal polypeptide Cholecystokinin Gastrin Substance P Neurotensin Methionine-enkephalin Leucine-enkephalin Insulin Glucagon Bombesin Secretin Somatostatin Thyrotropin-releasing hormone Motilin
Heart	Atrial natriuretic peptide
Other	Angiotensin II Bradykinin Calcitonin Calcitonin gene-related peptide (CGRP) Galanin Leptin Sleep peptide(s) Substance K (neurokinin A)

on targets outside the brain (eg, angiotensin and gastrin) or are products of neuroendocrine secretion (eg, oxytocin, vasopressin, somatostatin, luteinizing hormone, and thyrotropin-releasing hormone). In addition, many neuropeptides act as neurotransmitters when released close to a target neuron, where they can cause inhibition, excitation, or both.

Neuroactive peptides have been implicated in modulating sensory perception and affect. Some peptides, including substance P and the enkephalins, are preferentially located in regions of the central nervous system involved in the perception of pain. Other neuropeptides regulate complex responses to stress; these peptides include γ-melanocyte-stimulating hormone, corticotropin-releasing hormone (CRH), adrenocorticotropin (ACTH), dynorphin, and β-endorphin.

Although the diversity of neuroactive peptides is enormous, as a class these chemical messengers share a common cell biology. A striking generality is that neuroactive peptides are grouped in families with members that have similar sequences of amino acid residues. At least 10 have been identified; the seven main families are listed in Table 16–3.

Several different neuroactive peptides can be encoded by a single continuous messenger RNA (mRNA), which is translated into one large polyprotein precursor (Figure 16–2). Polyproteins can serve as a mechanism for amplification by providing more than one copy of the same peptide from the one precursor. For example, the precursor of glucagon contains two copies of the hormone. Polyproteins also generate diversity by producing several distinct peptides cleaved from one precursor, as in the case of the opioid peptides. The opioid peptides are derived from polyproteins encoded by three distinct genes. These peptides are endogenous ligands for a family of G protein–coupled receptors. In addition to endogenous agonists, the mu opioid receptor also binds drugs with analgesic and addictive properties, such as morphine and synthetic derivatives, including heroin and oxycodone.

The processing of more than one functional peptide from a single polyprotein is not unique to neuroactive peptides. The mechanism was first described for proteins encoded by small RNA viruses. Several viral polypeptides are produced from the same viral polyprotein, and all contribute to the generation of new virus particles. As with the virus, where the different proteins obviously serve a common biological purpose (formation of new viruses), a neuronal polypeptide will in many instances yield peptides that work together to serve a common physiological goal. Sometimes the biological functions appear to be more complex, as peptides with related or antagonistic activities can be generated from the same precursor.

A particularly striking example of this form of synergy is the group of peptides formed from the precursor of egg-laying hormone (ELH), a set of neuropeptides that govern diverse reproductive behaviors in the marine mollusk *Aplysia*. Egg-laying hormone can act as a hormone causing the contraction of oviduct muscles; it can also act as a neurotransmitter to alter the firing of several neurons involved in producing behaviors, as do the other peptides cut from the polyprotein.

The processing of neuroactive peptides takes place within the neuron's intracellular membrane system and in vesicles. Several peptides are produced from a single polyprotein by limited and specific proteolytic cleavage, catalyzed by proteases within these internal membrane systems. Some of these enzymes are serine proteases, a class that also includes the pancreatic enzymes trypsin and chymotrypsin. As with trypsin, the cleavage site of the peptide bond is determined by basic amino acid residues (lysine and arginine) in the substrate protein. Although cleavage is most common at dibasic residues, it can also occur at single basic residues, and polyproteins sometimes are cleaved at other peptide bonds.

Table 16–3 The Main Families of Neuroactive Peptides

Family	Peptide members
Opioids	Opiocortin, enkephalins, dynorphin, FMRFamide (Phe-Met-Arg-Phe-amide)
Neurohypophyseal neuropeptides	Vasopressin, oxytocin, neurophysins
Tachykinins	Substance P, physalaemin, kassinin, uperolein, eledoisin, bombesin, substance K
Secretins	Secretin, glucagon, vasoactive intestinal peptide, gastric inhibitory peptide, growth hormone–releasing factor, peptide histidine isoleucine amide
Insulins	Insulin, insulin-like growth factors I and II
Somatostatins	Somatostatins, pancreatic polypeptide
Gastrins	Gastrin, cholecystokinin

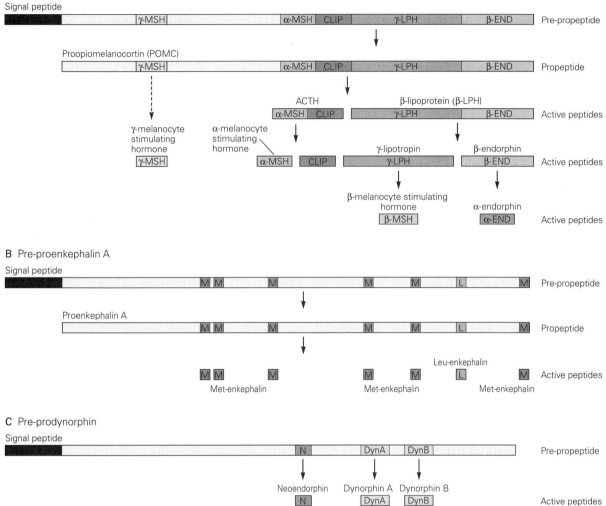

Figure 16–2 Hormone and neuropeptide precursors are processed differentially: The opioid family of neuropeptides. The opioid neuropeptides are derived from larger precursor molecules that require multiple rounds of protease-mediated cleavage. These precursors are processed differentially to yield their specific peptide products. Transport of these precursors through the membrane of the endoplasmic reticulum is initiated by a hydrophobic signal sequence. Internal cleavages often occur at basic residues within the polypeptide. Moreover, these precursors have key cysteine residues and sugar moieties that play roles in their processing and function. Generally, the first iteration of processing begins with the newly synthesized polyprotein precursor (known as the pre-propeptide form). Cleavage of an amino-terminal signal sequence generates a smaller molecule, the propeptide. Three major opioid peptide precursor proteins are encoded by three genes: *proopiomelanocortin* (*POMC*), *proenkephalin* (*PENK*), and *prodynorphin* (*PDYN*) (not shown). Differential processing of the three resultant pre-propeptides gives rise

to the major opioid peptides—endorphins, enkephalins, and dynorphins.

A. The POMC precursor is processed differently in different lobes of the pituitary gland, resulting in α-melanocyte-stimulating hormone (α-MSH) and γ-MSH, corticotropin-like intermediate lobe peptide (CLIP), and β-lipotropin (β-LPH). β-LPH is cleaved to yield γ-LPH and β-endorphin (β-END), which themselves yield β-melanocyte-stimulating hormone (β-MSH) and α-endorphin (α-END), respectively. The endoproteolytic cleavages within adrenocorticotropic hormone (ACTH) and β-LPH take place in the intermediate lobe but not the anterior lobe.

B. Similar principles are evident in the processing of the enkephalin precursor, which gives rise to six Met-enkephalin peptides and one Leu-enkephalin peptide.

C. The dynorphin precursor is cleaved into at least three peptides that are related to Leu-enkephalin since the amino-terminal sequences of all three peptides contain the sequence of Leu-enkephalin.

Other types of peptidases also catalyze the limited proteolysis required for processing neuroactive peptides. Among these are thiol endopeptidases (with catalytic mechanisms like that of pepsin), amino peptidases (which remove the N-terminal amino acid of the peptide), and carboxy-peptidase B (an enzyme that removes an amino acid from the C-terminal end of the peptide if it is basic).

Different neurons that produce the same polyprotein may release different neuropeptides because of differences in the way the polyprotein is processed. An example is proopiomelanocortin (POMC), one of the three branches of the opioid family. POMC is found in neurons of the anterior and intermediate lobes of the pituitary, in the hypothalamus, and in several other regions of the brain, as well as in the placenta and gut. The same mRNA for POMC is found in all of these tissues, but different peptides are produced from POMC in different tissues in a regulated manner. One possibility is that two neurons that process the same polyprotein might differently express proteases with different specificities within the lumina of the endoplasmic reticulum, Golgi apparatus, or vesicles. Alternatively, the two neurons might contain the same processing proteases, but each cell might glycosylate the common polyprotein at different sites, thereby protecting different regions of the polypeptide from cleavage.

Peptides and Small-Molecule Transmitters Differ in Several Ways

Large dense-core vesicles are homologous to the secretory granules of nonneuronal cells. These vesicles are formed in the trans-Golgi network, where they are loaded with neuropeptides and other proteins that enable formation of the dense core. The dense-core vesicles are then transported from the soma to presynaptic sites in axons. In addition to containing neuropeptides, these vesicles often contain small molecule transmitters due to their expression of vesicular transporters. After large dense-core vesicles release their contents through exocytosis, the membrane is not recycled to form new large dense core vesicles. Rather the vesicles must be replaced by transport from the soma. In contrast, mature small synaptic vesicles are not synthesized in the soma. Rather, their protein components are delivered to release sites by transport of large dense-core precursor vesicles. To form a mature small synaptic vesicle, the precursor vesicles must first fuse with the

plasma membrane. Following endocytosis, mature synaptic vesicles are then produced by local processing. Once their contents are released by exocytosis, synaptic vesicles can be rapidly recycled to maintain their local concentration during periods of sustained neural firing.

Although both types of vesicles contain many similar proteins, dense-core vesicles lack several proteins needed for release at the active zones. The membranes from dense-core vesicles are used only once; new dense-core vesicles must be synthesized in the cell body and transported to the axonal terminals by anterograde transport. Moreover, no uptake mechanisms exist for neuropeptides. Thus, once a peptide is released, a new supply must arrive from the cell body. Although there is evidence for local protein synthesis in some axons, it has not been shown that this provides new peptides for release.

The large dense-core vesicles release their contents by an exocytotic mechanism that is not specialized to nerve cells and does not require active zones; release can thus take place anywhere along the membrane of the axon that has the appropriate fusion machinery. As in other examples of regulated secretion, exocytosis of the dense-core vesicles depends on a general elevation of intracellular Ca^{2+} through voltage-gated Ca^{2+} channels that are not localized to the site of release. As a result, this form of exocytosis is slow and requires high stimulation frequencies to raise Ca^{2+} to levels sufficient to trigger release. This is in contrast to the rapid exocytosis of synaptic vesicles following a single action potential, which initiates the large, rapid increase in Ca^{2+} through voltage-gated Ca^{2+} channels tightly clustered at the active zone (Chapter 15).

Peptides and Small-Molecule Transmitters Can Be Co-released

Neuroactive peptides, small-molecule transmitters, and other neuroactive molecules coexist in the same dense-core vesicles of some neurons (Chapters 7 and 15). In mature neurons, the combination usually consists of one of the small-molecule transmitters and one or more peptides derived from a polyprotein. For example, ACh and vasoactive intestinal peptide (VIP) can be released together and work synergistically on the same target cells.

Another example is calcitonin gene–related peptide (CGRP), which in most spinal motor neurons is packaged together with ACh, the transmitter used at the neuromuscular junction. CGRP activates adenylyl cyclase, raising cyclic adenosine monophosphate (cAMP) levels and cAMP-dependent protein phosphorylation in the target muscles (Chapter 14). Increased protein phosphorylation results in an increase in the force of contraction. Another example is the co-release of glutamate and dynorphin in neurons of the hippocampus, where glutamate is excitatory and dynorphin inhibitory. Because postsynaptic target cells have receptors for both chemical messengers, all of these examples of co-release are also examples of cotransmission.

As already described, the dense-core vesicles that release peptides differ from the small clear vesicles that release only small-molecule transmitters. The peptide-containing vesicles may or may not contain small-molecule transmitter, but both types of vesicles contain ATP. As a result, ATP is released by exocytosis of both large dense-core vesicles and synaptic vesicles. Moreover, it appears that ATP may be stored and released in a number of distinct ways: (1) ATP is co-stored and co-released with transmitters, (2) ATP release is simultaneous but independent of transmitter release, and (3) ATP is released alone. Co-release of ATP (which after release can be degraded to adenosine) may be an important illustration that coexistence and co-release do not necessarily signify cotransmission. ATP, like many other substances, can be released from neurons but still not be involved in signaling if there are no receptors nearby.

As mentioned earlier, one criterion for judging whether a particular substance is used as a transmitter is that the substance is present in high concentrations in a neuron. Identification of transmitters in specific neurons has been important in understanding synaptic transmission, and a variety of histochemical methods are used to detect chemical messengers in neurons (Box 16–2).

The glutamate synaptic vesicle transporters VGLUT2 and VGLUT3 are expressed in neurons that release other classes of neurotransmitter, particularly cholinergic, serotonergic, and catecholaminergic neurons. An interesting example of co-release of two small-molecule transmitters is that of glutamate and dopamine by neurons projecting to the ventral striatum, cortex, and elsewhere. This co-release may have important implications for modulation of motivated behaviors and for establishing the patterns of axonal projections. In some cases, glutamate is released together with dopamine in response to different patterns of dopaminergic neuron firing. While there is a controversy regarding whether the same synaptic vesicles can accumulate both neurotransmitters, in isolated synaptic vesicles, glutamate uptake enhances vesicular monoamine storage by increasing the pH gradient that drives vesicular monoamine transport, providing a presynaptic mechanism to regulate quantal size.

Removal of Transmitter From the Synaptic Cleft Terminates Synaptic Transmission

Timely removal of transmitters from the synaptic cleft is critical to synaptic transmission. If transmitter molecules released in one synaptic action were allowed to remain in the cleft after release, this would impede the normal spatial and temporal dynamics of signaling, initially boosting the signal but preventing new signals from getting through. The synapse would ultimately become refractory, mainly because of receptor desensitization resulting from continued exposure to transmitter.

Transmitter substances are removed from the cleft by three mechanisms: diffusion, enzymatic degradation, and reuptake. Diffusion removes some fraction of all chemical messengers, but in brain regions with very high innervation and thus a high requirement for neurotransmitter release, diffusion can play a relatively small role in tapering signaling. In contrast, in regions of low innervation, diffusion is a major means by which signaling is decreased.

At cholinergic synapses, the dominant means of clearing ACh is enzymatic degradation of the transmitter by acetylcholinesterase. At the neuromuscular junction, the active zone of the presynaptic nerve terminal is situated just above the junctional folds of the muscle membrane. The ACh receptors are located at the surface of the muscle facing the release sites and do not extend deep into the folds (see Figure 12–1), whereas acetylcholinesterase is anchored to the basement membrane within the folds. This anatomical arrangement of transmitter, receptor, and degradative enzyme serves two functions.

First, on release, ACh reacts with its receptor; after dissociation from the receptor, the ACh diffuses into the cleft and is hydrolyzed to choline and acetate by acetylcholinesterase. As a result, the transmitter molecules are used only once. Thus, one function of the esterase is to punctuate the synaptic message. Second, the choline that otherwise might be lost by diffusion away from the synaptic cleft is recaptured. Once hydrolyzed by the esterase, the choline lingers in the reservoir provided by the junctional folds and is taken back up into cholinergic nerve endings by a high-affinity choline transporter. (Unlike the biogenic amines, there is no uptake mechanism for ACh itself at the plasma membrane.) In addition to acetylcholinesterase, ACh is also degraded by another esterase,

Box 16–2 Detection of Chemical Messengers and Their Processing Enzymes Within Neurons

Powerful histochemical techniques are available for detecting both small-molecule transmitter substances and neuroactive peptides in histological sections of nervous tissue.

Catecholamines and serotonin, when reacted with formaldehyde vapor, form fluorescent derivatives. In an early example of transmitter histochemistry, the Swedish neuroanatomists Bengt Falck and Nils Hillarp found that the reaction can be used to locate transmitters with fluorescence microscopy under properly controlled conditions.

Because individual vesicles are too small to be resolved by the light microscope, the exact position of the vesicles containing the transmitter was inferred by

comparing the fluorescence under the light microscope with the position of vesicles under the electron microscope. A number of fluorescent false transmitters, particularly those that mimic catecholamines, are substrates for plasma membrane and/or vesicular transporters, enabling their use to label vesicles and assess their turnover in living tissue. In addition, a variety of genetically expressed neurotransmitter reporters based on green fluorescent protein can be used to detect extracellular levels of neurotransmitters.

Histochemical analysis can be extended to the ultrastructure of neurons under special conditions. Fixation of nervous tissue in the presence of potassium permanganate, chromate, or silver salts, or the dopamine analog

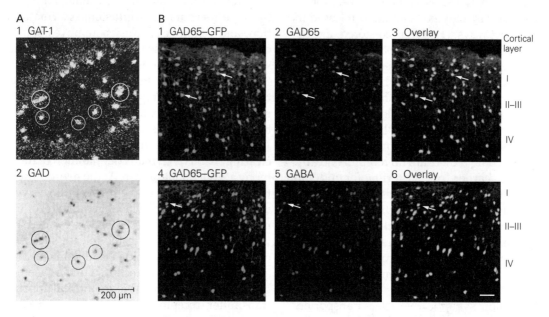

Figure 16–3 Techniques for visualizing chemical messengers.

A. A light-microscope section of the hippocampus of a rat. **1.** In situ hybridization using a probe for the mRNA encoding GAT-1, a GABA transporter. The probe was end-labeled with α-^{35}S-dATP and visualized by clusters of silver grains in the overlying autoradiographic photographic emulsion. **2.** In situ hybridization of the mRNA for glutamic acid decarboxylase (**GAD**), the specific biosynthetic enzyme for GABA, was carried out with an oligonucleotide probe linked to the enzyme alkaline phosphatase. The GAD probe was visualized by accumulation of colored alkaline phosphatase reaction product in the cytoplasm. Neurons expressing both GAT-1 and GAD transcripts were labeled

by silver grains and the phosphatase reaction, respectively, and are indicated by circles enclosing cells bodies that contain both labels. (Used with permission of Sarah Augood.)

B. Images of neocortex from a GAD65-GFP transgenic mouse in which green fluorescent protein (**GFP**) is expressed under the control of the GAD65 promotor. GFP is co-localized with GAD65 (1–3) and GABA (4–6) (both detected by indirect immunofluorescence) in neurons in the different layers. Most of the GFP-positive neurons are immunopositive for GAD65 and GABA (**arrows** show selected examples). Scale bar = 100 μm. (Adapted, with permission, from López-Bendito et al. 2004. Copyright © 2004 Oxford University Press.)

5-hydroxdopamine, which forms an electron-dense product, intensifies the electron density of vesicles containing biogenic amines and thus reveals the large number of dense-core vesicles that are characteristic of aminergic neurons.

It is also possible to identify neurons that express the gene for a particular transmitter enzyme or peptide precursor. Many methods for detecting specific mRNAs depend on nucleic acid hybridization. One such method is in situ hybridization.

Two single strands of a nucleic acid polymer will pair if their sequence of bases is complementary. With in situ hybridization, the strand of noncoding DNA (the negative or antisense strand or its corresponding RNA) is applied to tissue sections under conditions suitable for hybridizing with endogenous (sense) mRNA. If the probes are radiolabeled, autoradiography reveals the locations of neurons that contain the complex formed by the labeled complementary nucleic acid strand and the mRNA.

Hybrid oligonucleotides synthesized with nucleotides containing base analogs tagged chemically, fluorescently, or with antibodies can be detected histochemically. Multiple labels can be used at the same time (Figure 16–3A). RNAscope, a more recent mRNA hybridization method, allows for simultaneous detection of different mRNAs with lower background and

single-molecule sensitivity. Another approach to detecting the synthetic proteins involves viral or transgenic expression of proteins fused to variants of green fluorescent protein (Figure 16–3B).

Transmitter substances can also be detected using immunohistochemical techniques. Amino acid transmitters, biogenic amines, and neuropeptides have a primary amino group that becomes covalently fixed within the neurons; this group becomes cross-linked to proteins by aldehydes, the usual fixatives used in microscopy for immunohistochemical techniques.

Specific antibodies against the transmitter substances are necessary. Antibodies specific to serotonin, histamine, and many neuroactive peptides can be detected by a second antibody (in a technique called *indirect immunofluorescence*). As an example, if the first antibody is rabbit-derived, the second antibody can be goat antibody raised against rabbit immunoglobulin.

These commercially available secondary antibodies are tagged with fluorescent dyes and used under the fluorescence microscope to locate antigens in regions of individual neurons—cell bodies, axons, and presynaptic release sites (Figure 16–3).

Immunohistochemical techniques are also used with electron microscopy to locate chemical transmitters in the ultrastructure of neurons. Such techniques

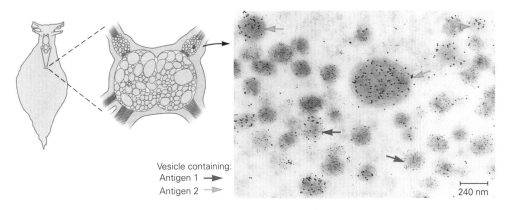

Vesicle containing:
Antigen 1 ➡
Antigen 2 ➡

240 nm

Figure 16–4 Electron-opaque gold particles linked to antibody are used to locate antigens in tissue at the ultrastructural level. The electron micrograph shows a section through the cell body of an *Aplysia* bag cell. Bag cells control reproductive behavior by releasing a group of neuropeptides cleaved from the egg-laying hormone (ELH) precursor. The cells contain several kinds of dense-core vesicles. The cell shown here was treated with two antibodies against different amino acid sequences contained in different regions of the ELH precursor. One antibody

was raised in rabbits and the other in rats. These antibodies were detected with anti-rabbit or anti-rat immunoglobulins (secondary antibodies) raised in goats. Each secondary antibody was coupled to colloidal gold particles of a distinct size. Vesicles identified by antigen 1 (labeled with the smaller gold particles) are smaller than vesicles identified by antigen 2 (labeled with the larger gold particles), indicating that the specific fragments cleaved from the precursor are localized in different vesicles. (Reproduced, with permission, from Fisher et al. 1988.)

(*continued*)

Box 16–2 Detection of Chemical Messengers and Their Processing Enzymes Within Neurons (continued)

usually involve a peroxidase-antiperoxidase system that produces an electron-dense reaction product. Another method is to use antibodies linked to electron-dense gold particles. Spheres of colloidal gold can be generated with precise diameters in the nanometer range. When coated with an appropriate antibody, these gold particles can be used to detect proteins and peptides with high resolution. This technique has the additional useful feature that more than one specific antibody can be examined in the same tissue section if each antibody is linked to gold particles of different sizes (Figure 16–4).

A number of fluorescent vesicular transporter substrates have been used as fluorescent false neurotransmitters (FFNs) to monitor transmitter release in mouse brain slice or whole fly brain (Figure 16–5). This approach allows visualization of nerve terminals in which synaptic vesicles have been loaded with FFNs; release can then be monitored optically in real time in response to either depolarization, which leads to exocytosis and synaptic vesicle emptying, or amphetamine, which leads to nonexocytic release of vesicular contents into the cytoplasm in response to vesicle deacidification.

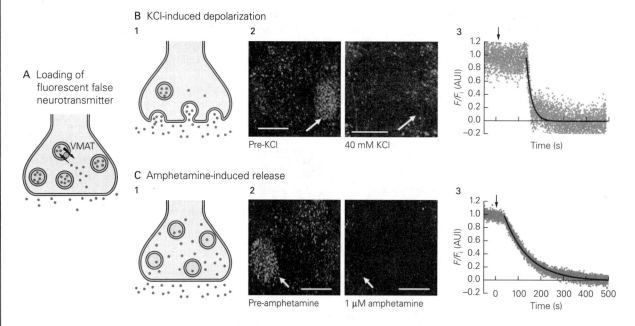

Figure 16–5 Fluorescent false neurotransmitter (FFN) labeling permits optical monitoring of neurotransmitter release.

A. FFN (**blue dots**) is transported by VMAT into synaptic vesicles in dopamine nerve terminals. Vesicles at steady state are acidic, as indicated by **yellow shading**.

B. 1. Raising the extracellular KCl concentration leads to depolarization and release of vesicles through exocytosis, resulting in the loss of fluorescent label (destaining). **2.** KCl (40 mM) depolarization caused rapid FFN206 destaining in presynaptic dopamine nerve terminals. Whole fly brains were loaded to steady state with FFN206 (300 nM) and treated with KCl. Projected image stacks of the neuropil before (*left*) and after (*right*) KCl-induced depolarization.

3. Kinetics of fluorescence decay from representative experiments. **Black arrow** indicates initiation of KCl addition. Scale bar = 25 µm.

C. 1. Amphetamine leads to deacidification of vesicles (**loss of yellow shading**) and their destaining through a nonexocytic mechanism discussed in the text. **2.** Amphetamine (1 µM) caused FFN206 destaining. Whole fly brains were loaded to steady state with FFN206 (300 nM) and treated with amphetamine. Projected image stacks of neuropil before (*left*) and after (*right*) treatment. **3.** Kinetics of fluorescence decay from representative experiments. **Black arrow** indicates initiation of drug addition. Scale bar = 25 µm. (Reproduced, with permission, from Freyberg et al. 2016.)

butyrylcholinesterase, which can also degrade other molecules including cocaine and the paralytic drug succinylcholine. However, the precise functions of butyrylcholinesterase are not fully understood.

Many other enzymatic pathways that degrade released transmitters are not directly involved in terminating synaptic transmission but are important for controlling the concentration of the transmitter within the neuron or for inactivating transmitter molecules that have diffused away from the synaptic cleft. Many of these degradative enzymes are important clinically— they provide sites for drug action and serve as diagnostic indicators. For example, inhibitors of MAO, an intracellular enzyme that degrades amine transmitters, are used to treat depression and Parkinson disease. Catechol-O-methyltransferase (COMT) is another cytoplasmic enzyme that is important for degrading biogenic amines. Measurement of its metabolites provides a useful clinical index of the efficacy of drugs that affect the synthesis or degradation of the biogenic amines in nervous tissue. COMT is thought to play a particularly critical role in regulating cortical dopamine levels because of the low levels of the dopamine uptake transporter. The relevance of this enzyme is underscored by the finding that a functional polymorphism in the COMT gene has been related to cognitive performance.

Neuropeptides are removed relatively slowly from the synaptic cleft by slow diffusion and proteolysis by extracellular peptidases. In contrast, small-molecule transmitters are removed more quickly from the synaptic cleft and extrasynaptic space. The critical mechanism for inactivation of most small molecule neurotransmitters is reuptake at the plasma membrane. This mechanism serves the dual purposes of terminating the synaptic action of the transmitter as well as recapturing transmitter molecules for subsequent reuse. Although Elliott had hypothesized in 1914 that uptake transporters might exist, their discovery waited until 1958 when F. Barbara Hughes and Benjamin Brodie found that blood platelets accumulated norepinephrine and serotonin, which could compete with each other for uptake. Julius Axelrod,[*] also a member of Brodie's group, soon afterward characterized norepinephrine uptake into neurons using a radiolabeled substrate.

High-affinity uptake is mediated by transporter molecules in the membranes of nerve terminals and glial cells. Unlike vesicular transporters, which are powered by the H^+ electrochemical gradient in an antiport mechanism, plasma membrane transporters are driven by the Na^+ electrochemical gradient through a symport mechanism in which Na^+ ions and transmitter move in the same direction.

Each type of neuron has its own characteristic uptake mechanism. For example, noncholinergic neurons do not take up choline with high affinity. Certain powerful psychotropic drugs can block uptake processes. For example, cocaine blocks the uptake of dopamine, norepinephrine, and serotonin; the tricyclic antidepressants block uptake of serotonin and norepinephrine. The selective serotonin reuptake inhibitors, such as fluoxetine (Prozac), were an important therapeutic innovation and are generally better tolerated than tricyclic antidepressants, although treatment-resistant depression remains a critical problem. The application of appropriate drugs that block transporters can prolong and enhance synaptic signaling by the biogenic amines and GABA. In some instances, drugs act both on transporters on the neuron's surface and on vesicular transporters within the cell. For example, amphetamines are actively taken up by the dopamine or other biogenic amine transporters on the external membrane of the neuron as well as by VMAT2.

Transporter molecules for neurotransmitters belong to two distinct groups that are different in both structure and mechanism. High-resolution structures of bacterial homologs from each of these families have been solved, which has greatly advanced our understanding of transporter mechanisms.

One group of transporters is the neurotransmitter sodium symporters (NSS), a superfamily of transmembrane proteins that thread through the plasma membrane 12 times (11 times for many prokaryotic homologs). These proteins are comprised of a pseudo-symmetric inverted repeat in which membrane-spanning segments 1 to 5 are homologous to membrane-spanning segments 6 to 10. The NSS family includes the transporters of GABA, glycine, norepinephrine, dopamine, serotonin, osmolytes, and amino acids. Crystal structures for the human serotonin transporter and the fly dopamine transporter, which share the same structure and general mechanism as the bacterial homologs crystallized previously, have recently been solved.

The second family consists of transporters of glutamate. These proteins traverse the plasma membrane eight times and contain two helical hairpins that are

---

[*]Axelrod received a bachelor's degree in chemistry and wrote many of his celebrated papers as a technician in Brodie's lab before entering graduate school and receiving a PhD 21 years later. He was awarded a share of the Nobel Prize in 1970 for his co-discovery of neuronal norepinephrine uptake and his discovery of COMT. His co-recipients of the prize that year were Bernard Katz, who described quantal neurotransmission, and Ulf von Euler, who also studied vesicular uptake and epinephrine release.

thought to serve a role in gating access of substrate from each side of the membrane (see Figure 8–16). Each group includes several transporters for each transmitter substance; for example, there are multiple GABA, glycine, and glutamate transporters, each with somewhat different localization, function, and pharmacology.

The two groups can be distinguished functionally. Although both are driven by the electrochemical potential provided by the Na^+ gradient, transport of glutamate requires the countertransport of K^+, whereas transport by NSS proteins often requires the cotransport of Cl^- (or H^+ antiport in the case of prokaryotic homologs). During transport of glutamate, one negatively charged molecule of the transmitter is imported with three Na^+ ions and one proton (symport) in exchange for the export of one K^+. This leads to a net influx of two positive charges for each transport cycle, generating an inward current. As a result of this charge transfer, the negative resting potential of the cell generates a large inward driving force that results in an enormous gradient of glutamate across the cell membrane. In contrast, the NSS proteins transport one to three Na^+ ions and one Cl^- ion together with their substrates. While under most conditions the electrochemical driving force is sufficient for NSS transporters to carry transmitter into the cell, thereby increasing the cytoplasmic transmitter concentration, the concentration of transmitter in the cytoplasm is quite low and ultimately determined by the action of vesicular transporters to load transmitter into synaptic vesicles.

A fascinating aspect of the function of the NSS proteins is the ability of these transporters to run backward, allowing them to generate transmitter efflux. This is best characterized for the neurotransmitter dopamine, as amphetamine and related analogs lead to massive release of dopamine through a nonexocytotic mechanism. As discussed earlier, at pharmacological doses, amphetamine is actively transported by both the plasma membrane dopamine transporter (DAT) and the vesicular VMAT2; the latter effect dissipates the vesicular H^+ gradient, leading to the escape of dopamine to the cytoplasm. This dopamine then moves "backward", out of the cell, through DAT, a process that requires phosphorylation of its N-terminus. While essential for amphetamine function, the normal physiological role of this phosphorylation remains a mystery, as it does not seem essential for dopamine uptake. Computational studies suggest that phosphorylation-regulated interactions of the N-terminus with acidic lipids in the inner leaflet play a role in modulating transporter

function. Nevertheless, the ultimate answer may require atomic resolution structures that include the N-terminal domain coupled with biophysical data on N-terminal dynamics.

Highlights

1. Information carried by a neuron is encoded in electrical signals that travel along its axon to a synapse, where these signals are transformed and carried across the synaptic cleft by one or more chemical messengers.

2. Two major classes of chemical messengers, small-molecule transmitters and neuroactive peptides, are packaged in vesicles within the presynaptic neuron. After their synthesis in the cytoplasm, small-molecule transmitters are taken up and highly concentrated in vesicles, where they are protected from degradative enzymes in the cytoplasm.

3. Synaptic vesicles in the periphery are highly concentrated in nerve endings and, in the brain, tend to be at varicosities along the axon at presynaptic sites. Classical excitatory synapses with ionotropic glutamate receptors are examples of "private" synapses that communicate with a closely apposed postsynaptic structure such as a dendritic spine. In contrast, the dopamine system exemplifies "social" synapses that can interact with extrasynaptic receptors on many neurons.

4. To prevent depletion of small molecule transmitters during rapid synaptic transmission, most are synthesized locally at terminals.

5. The protein precursors of neuroactive peptides are synthesized only in the cell body, the site of transcription and translation. The neuropeptides are packaged in secretory granules and vesicles that are carried from the cell body to the terminals by axoplasmic transport. Unlike the vesicles that contain small-molecule transmitters, these vesicles are not refilled at the terminal.

6. The enzymes that regulate transmitter biosynthesis are under tight regulatory control, and changes in neuronal activity can produce homeostatic changes in the levels and activity of these enzymes. This regulation can occur both posttranslationally in the cytoplasm, as a result of phosphorylation and dephosphorylation reactions, as well as by transcriptional control in the nucleus.

7. Precise mechanisms for terminating transmitter actions represent a key step in synaptic

transmission that is nearly as important as transmitter synthesis and release. Some released transmitter is lost as a result of simple diffusion out of the synaptic cleft. However, for the most part, transmitter actions are terminated by specific molecular reactions.

8. Acetylcholine is rapidly hydrolyzed by acetylcholinesterase to choline and acetate. Glutamate, GABA, glycine, and the biogenic amines are taken up into presynaptic terminals and/or glia by specific transporters at the plasma membrane that are driven by the Na^+ gradient.

9. Some of the most potent psychoactive compounds act at neurotransmitter transporters. The psychostimulatory effects of cocaine result from its action to prevent reuptake of dopamine, thereby increasing its extracellular levels. In contrast, amphetamine and its derivatives promote nonexocytotic release of dopamine through a mechanism involving both the plasma membrane DAT and the vesicular transporter VMAT2.

10. The first step in understanding the molecular strategy of chemical transmission usually involves identifying the contents of synaptic vesicles. Except for those instances in which transmitter is released by transporter molecules or by diffusion through the membrane (in the case of gases and lipid metabolites, see Chapter 14), only molecules suitably packaged in vesicles can be released from a neuron's terminals. However, not all molecules released by a neuron are chemical messengers—only those that bind to appropriate receptors and initiate functional changes in that target neuron can usefully be considered neurotransmitters.

11. Information is transmitted when transmitter molecules bind to receptor proteins in the membrane of another cell, causing them to change conformation, leading either to increased ion conductance in the case of ligand-gated ion channels or to alterations in downstream signaling pathways in the case of G protein–coupled receptors.

12. The co-release of several neuroactive substances onto appropriate postsynaptic receptors permits great diversity of information to be transferred in a single synaptic action.

Jonathan A. Javitch
David Sulzer

Selected Reading

Alberts B, Johnson A, Lewis J, Raff M, Roberts K, Walter P. 2002. Membrane transport of small molecules and the electrical properties of membranes. In: *Molecular Biology of the Cell*, 4th ed. New York and Oxford: Garland Science.

Axelrod J. 2003. Journey of a late blooming biochemical neuroscientist. J Biol Chem 278:1–13.

Burnstock G. 1986. Purines and cotransmitters in adrenergic and cholinergic neurones. Prog Brain Res 68:193–203.

Chaudhry FA, Boulland JL, Jenstad M, Bredahl MK, Edwards RH. 2008. Pharmacology of neurotransmitter transport into secretory vesicles. Handb Exp Pharmacol 184:77–106.

Cooper JR, Bloom FE, Roth RH. 2003. *The Biochemical Basis of Neuropharmacology*, 8th ed. New York: Oxford Univ. Press.

Dale H. 1935. Pharmacology and nerve endings. Proc R Soc Med (Lond) 28:319–332.

Edwards RH. 2007. The neurotransmitter cycle and quantal size. Neuron 55:835–858.

Falck B, Hillarp NÅ, Thieme G, Torp A. 1982. Fluorescence of catecholamines and related compounds condensed with formaldehyde. Brain Res Bull 9:11–15.

Fatt P, Katz B. 1950. Some observations on biological noise. Nature 166:597–598.

Jiang J, Amara SG. 2011. New views of glutamate transporter structure and function: advances and challenges. Neuropharmacology 60:172–181.

Johnson RG. 1988 Accumulation of biological amines into chromaffin granules: a model for hormone and neurotransmitter transport. Physiol Rev 68:232–307.

Katz B. 1966. *Nerve, Muscle, and Synapse*. New York: McGraw-Hill.

Koob GF, Sandman CA, Strand FL (eds). 1990. A decade of neuropeptides: past, present and future. Ann N Y Acad Sci 579:1–281.

Loewi O. 1960. An autobiographic sketch. Perspect Biol Med 4:3–25.

Lohr KM, Masoud ST, Salahpour A, Miller GW. 2017. Membrane transporters as mediators of synaptic dopamine dynamics: implications for disease. Eur J Neurosci 45:20–33.

Pereira D, Sulzer D. 2012. Mechanisms of dopamine quantal size regulation. Front Biosci 17:2740–2767.

Sames D, Dunn M, Karpowicz RJ Jr, Sulzer D. 2013. Visualizing neurotransmitter secretion at individual synapses. ACS Chem Neurosci 4:648–651.

Siegel GJ, Agranoff BW, Albers RW, Molinoff PB (eds). 1998. *Basic Neurochemistry: Molecular, Cellular, and Medical Aspects*, 6th ed. Philadelphia: Lippincott.

Snyder SH, Ferris CD. 2000. Novel neurotransmitters and their neuropsychiatric relevance. Am J Psychiatry 157:1738–1751.

Sossin WS, Fisher JM, Scheller RH. 1989. Cellular and molecular biology of neuropeptide processing and packaging. Neuron 2:1407–1417.

Sulzer D, Cragg SJ, Rice ME. 2016. Striatal dopamine neurotransmission: regulation of release and uptake. Basal Ganglia 6:123–148.

Sulzer D, Pothos EN. 2000. Regulation of quantal size by pre-synaptic mechanisms. Rev Neurosci 11:159–212.

Toei M, Saum R, Forgac M. 2010. Regulation and isoform function of the V-ATPases. Biochemistry 49:4715–4723.

Torres GE, Amara SG. 2007. Glutamate and monoamine transporters: new visions of form and function. Curr Opin Neurobiol 17:304–312.

Van der Kloot W. 1991 The regulation of quantal size. Prog Neurobiol 36:93–130.

Weihe E, Eiden LE. 2000. Chemical neuroanatomy of the vesicular amine transporters. FASEB J 15:2435–2449.

References

Augood SJ, Herbison AE, Emson PC. 1995. Localization of GAT-1 GABA transporter mRNA in rat striatum: cellular coexpression with GAD_{67} mRNA, GAD_{67} immunoreactivity, and parvalbumin mRNA. J Neurosci 15:865–874.

Coleman JA, Green EM, Gouaux E. 2016. X-ray structures and mechanism of the human serotonin transporter. Nature 532:334–339.

Danbolt NC, Chaudhry FA, Dehnes Y, et al. 1998. Properties and localization of glutamate transporters. Prog Brain Res 116:23–43.

Fisher JM, Sossin W, Newcomb R, Scheller RH. 1988. Multiple neuropeptides derived from a common precursor are differentially packaged and transported. Cell 54:813–822.

Freyberg Z, Sonders MS, Aguilar JI, et al. 2016. Mechanisms of amphetamine action illuminated through in vivo optical monitoring of dopamine synaptic vesicles. Nat Commun 7:10652.

Hnasko TS, Chuhma N, Zhang H, et al. 2010. Vesicular glutamate transport promotes dopamine storage and glutamate corelease in vivo. Neuron 65:643–656.

Khoshbouei, H, Sen N, Guptaroy B, et al. 2004. N-terminal phosphorylation of the dopamine transporter is required for amphetamine-induced efflux. PLoS Biol 2(3): E78.

Krieger DT. 1983. Brain peptides: what, where, and why? Science 222:975–985.

Liu Y, Kranz DE, Waites C, Edwards RH. 1999. Membrane trafficking of neurotransmitter transporters in the regulation of synaptic transmission. Trends Cell Biol 9:356–363.

Lloyd PE, Frankfurt M, Stevens P, Kupfermann I, Weiss KR. 1987. Biochemical and immunocytological localization of the neuropeptides FMRFamide SCPA, SCPB, to neurons involved in the regulation of feeding in *Aplysia*. J Neurosci 7:1123–1132.

López-Bendito G, Sturgess K, Erdélyi F, Szabó G, Molnár Z, Paulsen O. 2004. Preferential origin and layer destination of GAD65-GFP cortical interneurons. Cereb Cortex 14:1122–1133.

Okuda T, Haga T. 2003. High-affinity choline transporter. Neurochem Res 28:483–488.

Otsuka M, Kravitz EA, Potter DD. 1967. Physiological and chemical architecture of a lobster ganglion with particular reference to γ-aminobutyrate and glutamate. J Neurophysiol 30:725–752.

Palay SL. 1956. Synapses in the central nervous system. J Biophys Biochem Cytol 2:193–202.

Pereira DB, Schmitz Y, Mészáros J, et al. 2016. Fluorescent false neurotransmitter reveals functionally silent dopamine vesicle clusters in the striatum. Nat Neurosci 19:578–586.

Rubin RP. 2007. A brief history of great discoveries in pharmacology: in celebration of the centennial anniversary of the founding of the American Society of Pharmacology and Experimental Therapeutics. Pharmacol Rev 59:289–359.

Yamashita A, Singh SK, Kawate T, Jin Y, Gouaux E. 2005. Crystal structure of a bacterial homologue of Na^+/Cl^--dependent neurotransmitter transporters. Nature 437:205–223.

Yernool D, Boudker O, Jin Y, Gouaux E. 2004. Structure of a glutamate transporter homologue from *Pyrococcus horikoshii*. Nature 431:811–818.

Part IV

Preceding Page

In Plato's "Allegory of the Cave," which addresses the origin of knowledge, his early insight into the constructive nature of perception offers illuminating metaphors for the process. The parable begins with the premise that a group of prisoners has never seen the outside world. Their experience is limited to shadows cast upon the wall of the cave by objects passing before a fire. The causes of those shadows—even the fact that they are shadows—is unknown to the prisoners. Nonetheless, over time, the shadows become imbued with meaning in the prisoners' minds. Metaphorically, the shadows represent sensations, which are fleeting and incoherent. The assignment of meaning represents the construction of intelligible percepts. The prisoner turning the corner of the wall has been freed to witness the larger world of causes, which he reports back to those still imprisoned. In a novel metaphorical take on this ancient story, this returning prisoner represents the field of modern neuroscience, which sheds light on the relationship between our shadowy sensations and our rich perceptual experience of the world. (Plato's Cave, 1604. Jan Pietersz Saenredam, after Cornelis Cornelisz van Haarlem. National Gallery, Washington D.C.)

IV   Perception

> I understood that the world was nothing: a mechanical chaos of
> casual, brute enmity on which we stupidly impose our hopes and
> fears. I understand that, finally and absolutely, I alone exist. All
> the rest, I saw, is merely what pushes me, or what I push against,
> blindly—as blindly as all that is not myself pushes back. I create
> the whole universe, blink by blink.... Nevertheless, something will
> come of all this.[1]

JOHN GARDNER'S HEARTRENDING TALE OF THE TORMENTED
MONSTER Grendel's perspective on life captures the fundamental nature of perceptual experience: It is a construct that we alone
impose. Or, as Grendel keenly observes, "The mountains are what I
define them as." Isolated and tortured by loneliness, Grendel sees the
world as do the shackled prisoners in Plato's Cave, where mere shadows are what is sensed, but those shadows are imbued with meaning,
utility, agency, beauty, joy, and sadness, all through the constructive
process of perception: "What I see, I inspire with usefulness . . . and all
that I do not see is useless, void."

Like the prisoner who escapes from Plato's Cave to view a larger
world of causes, or the all-knowing dragon who fills Grendel with
ideas from another dimension—"But dragons, my boy, have a whole
different kind of mind. . . . We see from the mountaintop: all time, all
space"—modern neuroscience promises a mountaintop understanding of perceptual experience, an understanding not simply of the
things we construct from our shadowy sensations, but how we do
so, and for what purpose.

This section on perception offers that expansive mountaintop
view. For each of the sensory modalities in turn, these chapters
begin by examining environmental stimuli—light, sound, gravity,
touch, and chemicals—that are the origins of human experience and
knowledge of the world. In hierarchical fashion, the chapters survey
the mechanisms that enable stimulus detection and discrimination,
the perceptual processes that fill evanescent sensations with meaning, and the operations that support attention, decision, and action,
based on what is perceived.

Vision—a sense both particularly well understood and heavily
utilized by humans—acquires information through properties of
light. Light reflected from objects in the environment varies in wavelength and intensity and fluctuates over space and time, and through

---

[1]Gardner J. 1971. *Grendel*. Alfred A. Knopf, New York.

those physical properties conveys evidence of the world around us. Cast as patterned images upon the retina, luminous energy is transduced into neuronal signals by dedicated receptor cells. The evidential properties of these images are detected by a collection of specialized neuronal systems that sense forms of contrast and convey this information to the rest of the brain.

Similarly, the auditory system acquires information about the world through the simple compression and rarefaction of air, as caused by spoken language, music, or environmental sounds. This sensory evidence is detected—even in minute quantities and with incredibly precise timing—by an extraordinarily intricate amplification system consisting of small drums, levers, tubes, and hair cells, whose bendable stereocilia transduce mechanical energy into neuronal signals. Similar motion-detecting hair cells serve the vestibular senses of balance, acceleration, and rotation of the head.

The somatosensory system acquires information about physical stimuli impinging on the body in the form of pressure, vibration, and temperature—and, in the extreme, pain—as would be caused by touch, movement of the skin across a textured surface, or contact with a source of heat. The peripheral nerve endings of a variety of specialized detector neurons embedded in skin, viscera, and muscle transduce this mechanical energy into neuronal signals, which are carried via the spinal cord and cranial nerves to the brain.

Finally, the senses of taste and smell acquire information about the chemical composition of the world, in the form of form of food, drink, and airborne molecules. From one of the most exciting and rapidly developing areas of sensory biology today, we now know that there are hundreds of olfactory receptors that have unique patterns of affinity for airborne molecules, which accounts for the human ability to detect and discriminate a staggering number and diversity of odors.

All of these receptive systems serve as filters, characterized by neural "receptive fields" that highlight certain forms of information and restrict others. These selective filters are tunable over different timescales, enhancing attention to salient stimuli and adapting to the statistics of the sensory world. This flexibility accommodates variations in both behavioral goals and environmental conditions.

Like the shackled prisoners in Plato's cave, our sensory systems initially convey simple filtered representations of sensory input, which are fundamentally ambiguous, noisy, and incomplete. Alone they have no meaning. Quite remarkably, our brain enables us to ultimately experience this sensory information as the environmental objects and events that *cause* those patterns. The constructive transition from a world of sensory evidence to one of meaning lies at the heart of perception and has long been one of the most engaging mysteries of human cognition. The 19th-century English philosopher John Stuart Mill wrote "perception reflects the permanent possibilities of sensation," and in doing so reclaims from transient sensory events the enduring structural and relational properties of the world.

This section reveals how perception overcomes the vagaries of sensory evidence to develop hypotheses or inferences about the causes of sensation, by reference to past knowledge. Much of this happens through the machinery of the cerebral cortex, where sensory signals are linked both within and between modalities and to feedback from the memory store. Like a detective viewing a crime scene, informed by memory and context, the activity of cortical neurons begins to yield what William James aptly called the "perception of probable things."

With this perceptual transformation also comes the ability to recognize objects familiar to us. We readily generalize across different sensory manifestations of the same or similar objects, in the form of perceptual constancies and categorical percepts, and we link these with other meaningful events. The sound of the coffee grinder in the morning, the smell of a lover's perfume or the sight of her face expands our experience beyond the immediate to a realm of recall and imagination. The chapters in this collection review the brain structures and computations that underlie these associative functions, which include highly specialized neuronal systems for recognizing and interpreting complex and behaviorally significant objects, such as faces.

Perceptual experience of the world around us is a prerequisite for meaningful interaction with that world. Decisions are made based on the accumulation of sensory evidence in support of one percept versus another. Is that my suitcase on the carousel? Is this where we turn? Was that aria from Wagner or Strauss? Is that the fragrance of jasmine or gardenia? Cortical neurons form salience maps, which represent the outcome of these perceptual decisions with respect to behavioral goals and rewards and prioritize actions accordingly.

Perception is generally treated—as it is herein—as a distinct sub-discipline of neuroscience. Increasingly we see this compartmentalization breaking down. The relationship of perception to other brain functions—learning, memory, emotion, motor control, language, development—is ever clearer with the explosive growth of new concepts and experimental methods for monitoring and manipulating brain structure and function, and for revealing the extensive anatomical and functional neural connections between seemingly distinct brain regions. Thus have we begun to fully appreciate how the brain's system for acquiring and interpreting information, for becoming aware of and understanding the world—for *perceiving*—is the functional centerpiece of human cognition and behavior.

Part Editors: Thomas D. Albright and Randy M. Bruno

Part IV

17

Sensory Coding

O UR SENSES ENLIGHTEN AND EMPOWER US. Through sensation, we form an immediate and relevant picture of the world and our place in

it, informed by our past experience and preparing us for probable futures. Sensation provides immediate answers to three ongoing and essential questions: *Is something there? What is it?* and *What's changed?* To answer these questions, all sensory systems perform two fundamental functions: *detection* and *discrimination*. Because our world and our needed responses to it change with time, sensory systems can both *preferentially respond* and *adapt* to changing stimuli in the short term, and also *learn* to modify our responses to stimuli as our needs and circumstances change.

Since ancient times, humans have been fascinated by the nature of sensory experience. Aristotle defined five senses—vision, hearing, touch, taste, and smell— each linked to specific sense organs in the body: eyes, ears, skin, tongue, and nose. Pain was not considered to be a specific sensory modality but rather an affliction of the soul. Intuition, often referred to colloquially as a "sixth sense," was not yet understood to depend upon the experience of the classic sensory systems. Today, neurobiologists recognize intuition as inferences derived from previous experience and thus the result of cognitive as well as sensory processes.

In this chapter, we consider the organizational principles and coding mechanisms that are universal to all sensory systems. *Sensory information* is defined as neural activity originating from stimulation of receptor cells in specific parts of the body. Our senses include the classic five senses plus a variety of modalities not recognized by the ancients but essential to bodily function: the *somatic* sensations of pain, itch, temperature, and proprioception (posture and movement of our own body); *visceral* sensations (both conscious and unconscious) necessary for homeostasis; and the *vestibular* senses of balance (the position of the body in the gravitational field) and head movement.

Sensation informs and enriches all life, and the fundamentals of sensory processing have been conserved throughout vertebrate evolution. Specialized receptors in each of the sensory systems provide the first neural representation of the external and internal world, transforming a specific type of stimulus energy into electrical signals (Figure 17–1). All sensory information is then transmitted to the central nervous system by trains of action potentials that represent particular aspects of the stimulus. This information flows centrally to regions of the brain involved in the processing of individual senses, multisensory integration, and cognition.

The sensory pathways have both serial and parallel components, consisting of fiber tracts with

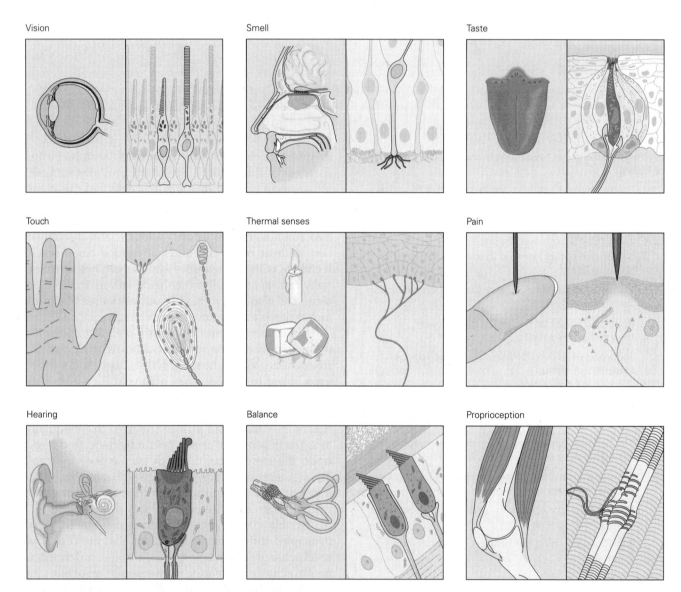

Figure 17–1 The major sensory modalities in humans are mediated by distinct classes of receptor neurons located in specific sense organs. Each class of receptor cell transforms one type of stimulus energy into electrical signals that are encoded as trains of action potentials (see Figure 17–4). The principal receptor cells include photoreceptors (vision), chemoreceptors (smell, taste, and pain), thermal receptors, and mechanoreceptors (touch, hearing, balance, and proprioception). The classic five senses—vision, smell, taste, touch, and hearing—and the sense of balance are mediated by receptors in the eye, nose, mouth, skin, and inner ear, respectively. The other somatosensory modalities—thermal senses, pain, visceral sensations, and proprioception—are mediated by receptors distributed throughout the body.

thousands or millions of axons linked by synapses that both transmit and transform information. Relatively simple forms of neural coding of stimuli by receptors are modulated by complex mechanisms in the brain to form the basis of cognition. Sensory pathways are also controlled by higher centers in the brain that modify and regulate incoming sensory signals by feeding information back to earlier stages of processing. Thus, perception is the product not simply of "raw" physical sensory information but also cognition and experience.

Both scientists and philosophers have examined the extent to which the sensations we experience accurately reflect the stimuli that produce them, and how they are altered by our inherently subjective and imprecise knowledge of the world. In prior centuries, the interest of European philosophers in sensation and perception was related to the question of human nature itself. Two schools of thought eventually dominated: empiricism, represented by John Locke, George Berkeley, and David Hume, and idealism, represented by René Descartes, Immanuel Kant, and Georg Wilhelm Friedrich Hegel.

Locke, the preeminent empiricist, advanced the idea that the mind at birth is a blank slate, or *tabula rasa*, void of any ideas. Knowledge, he asserted, is obtained only through sensory experience—what we see, hear, feel, taste, and smell. Berkeley extended this topic by questioning whether there was any sensory reality beyond the experiences and knowledge acquired through the senses. He famously asked: Does a falling tree make a sound if no one is near enough to hear it?

Idealists argued that the human mind possesses certain innate abilities, including logical reasoning itself. Kant classified the five senses as categories of human understanding. He argued that perceptions were not direct records of the world around us but rather were products of the brain and thus depended on the architecture of the nervous system. Kant referred to these brain properties as *a priori knowledge*.

Thus, in Kant's view, the mind was not the passive receiver of sense impressions envisaged by the empiricists. Rather, it had evolved to conform to certain universal conditions such as space, time, and causality. These conditions were independent of any physical stimuli detected by the body. For Kant and other idealists, this meant that knowledge is based not only on sensory stimulation alone but also on our ability to organize and interpret sensory experience. If sensory experience is inherently subjective and personal, they said, it may not be subject to empirical analysis. As the empirical investigation of perception matured, both schools proved partially correct.

Psychophysics Relates Sensations to the Physical Properties of Stimuli

The modern study of sensation and perception began in the 19th century with the emergence of experimental psychology as a scientific discipline. The first scientific psychologists—Ernst Weber, Gustav Fechner, Hermann Helmholtz, and Wilhelm Wundt—focused their experimental study of mental processes on sensation, which they believed was the key to understanding the mind. Their findings gave rise to the fields of psychophysics and sensory physiology.

Psychophysics describes the relationship between the physical characteristics of a stimulus and attributes of the sensory experience. *Sensory physiology* examines the neural consequences of a stimulus—how the stimulus is transduced by sensory receptors and processed in the brain. Some of the most exciting advances in our understanding of perception have come from merging these two approaches in both human and animal studies. For example, functional magnetic resonance imaging (fMRI) and positron emission tomography (PET) have been used in controlled experiments to identify regions of the human brain involved in the perception of pain or the identification of specific types of objects or particular persons and places.

Psychophysics Quantifies the Perception of Stimulus Properties

Early scientific studies of the mind focused not on the perception of complex qualities such as color or taste but on phenomena that could be isolated and measured precisely: the size, shape, amplitude, velocity, and timing of stimuli. Weber and Fechner developed simple experimental paradigms to study how and under what conditions humans are able to distinguish between two stimuli of different amplitudes. They quantified the intensity of sensations in the form of mathematical laws that allowed them to predict the relationship between the magnitude of a stimulus and its detectability, including the ability to discriminate between different stimuli.

In 1953, Stanley S. Stevens demonstrated that the subjective experience of the intensity (I) of a stimulus (S) is best described by a power function. Stevens's law states that,

$$I = K(S - S_0)^n,$$

where the *sensory threshold* (S_0) is the lowest stimulus strength a subject can detect, and K is a constant. For some sensations, such as the sense of pressure on the hand, the relationship between the stimulus magnitude and its perceived intensity is linear, that is, a power function with a unity exponent ($n = 1$).

All sensory systems have a threshold, and thresholds have two essential functions. First, by asking if a sensation is large enough to have a high enough probability of being of interest or relevance, they reduce unwanted responses to noise. Second, the specific nonlinearity introduced by thresholds aids encoding and processing, even if the rest of the primary sensory response scales linearly with the stimulus. Sensory thresholds are a feature, not a bug. Thresholds are normally determined statistically by presenting a subject with a series of stimuli of random amplitude. The percentage of times the subject reports detecting the stimulus is plotted as a function of stimulus amplitude, forming a relation called the *psychometric function* (Figure 17–2). By convention, threshold is defined as the stimulus amplitude detected in half of the trials.

The measurement of sensory thresholds is a useful technique for diagnosing sensory function in individual modalities. An elevated threshold may signal an abnormality in sensory receptors (such as loss of hair cells in the inner ear caused by aging or exposure to very loud noise), deficits in nerve conduction properties (as in multiple sclerosis), or a lesion in sensory-processing areas of the brain. Sensory thresholds may also be altered by emotional or psychological factors related to the conditions in which stimulus detection

is measured. Thresholds can also be determined by the method of limits, in which the subject reports the intensity at which a progressively decreasing stimulus is no longer detectable or an increasing stimulus becomes detectable. This technique is widely used in audiology to measure hearing thresholds.

Subjects can also provide nonverbal responses in sensory detection or discrimination tasks using levers, buttons, or other devices that allow accurate measurement of decision times. Experimental animals can be trained to respond to controlled sensory stimuli using such devices, allowing neuroscientists to investigate the underlying neural mechanisms by combining electrophysiological and behavioral studies in the same experiment. Methods for quantifying responses to stimuli are summarized in Box 17–1.

Stimuli Are Represented in the Nervous System by the Firing Patterns of Neurons

Psychophysical methods provide objective techniques for analyzing sensations evoked by stimuli. These quantitative measures have been combined with neurophysiological techniques to study the neural mechanisms that transform sensory neural signals into percepts. The goal of sensory neuroscience is to follow the flow of sensory information from receptors toward the cognitive centers of the brain, to understand the processing mechanisms that occur at successive synapses, and to decipher how this shapes our internal representation of the external world. The neural coding of sensory information is better understood at the early stages of processing than at later stages in the brain.

This approach to the *neural coding problem* was pioneered in the 1960s by Vernon Mountcastle, who showed that single-cell recordings of spike trains from peripheral and central sensory neurons provide a statistical description of the neural activity evoked by a physical stimulus. He then investigated which quantitative aspects of neural responses might correspond to the psychophysical measurements of sensory tasks and, just as important, which do not.

The study of neural coding of information is fundamental to understanding how the brain works. A neural code describes the relationship between the activity in a specified neural population and its functional consequences for perception or action. Sensory systems are ideal for the study of neural coding because both the physical properties of the stimulus input and the neural or behavioral output of these systems can be precisely defined and quantified in a controlled setting.

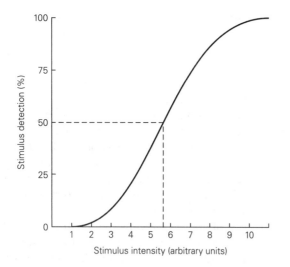

Figure 17–2 The psychometric function. The psychometric function plots the percentage of stimuli detected by a human observer as a function of the stimulus magnitude. Threshold is defined as the stimulus intensity detected on 50% of the trials, which in this example would be about 5.5 (arbitrary units). Psychometric functions are also used to measure the *just noticeable difference* (JND) between stimuli that differ in intensity, frequency, or other parametric properties.

Box 17–1 Signal Detection Theory: Quantifying Detection and Discrimination

Two major functions of our sensory systems are to tell us if something is there and what it is. To test our ability and the ability of our sensory systems to answer these questions, experimental protocols, tools, and methods have been developed to quantify the response of sensory systems to stimuli. These include *decision theory* and *signal detection theory*. Each uses statistical methods to quantify the variability of subjects' responses.

In an "Is something there?" task, for example, subjects or experimental animals can correctly detect a specific stimulus (a "hit" or "true positive"), respond incorrectly in the absence of that stimulus ("false positive" or "false alarm"), fail to respond to a true stimulus ("miss"), or correctly decline to respond in the absence of the stimulus ("true negative" or "correct rejection"). With repeated presentations, these choices can be tabulated in a four-cell stimulus–response matrix (Figure 17–3A).

This quantifies *sensitivity*, defined as the number of true positives divided by the number of stimuli presented, and *specificity*, defined as the number of true negatives divided by the number of presentations without a stimulus.

In 1927, L. L. Thurstone proposed that the variability of sensations evoked by stimuli could be represented as normal or Gaussian probability functions, equating the physical distance between the amplitudes of two stimuli to a psychological scale value of inferred intensity called the *discrimination index* or *d'*.

Decision theory methods were first applied to psychophysical studies in 1954 by the psychologists Wilson Tanner and John Swets. They developed a series of experimental protocols for stimulus detection that allowed accurate calculation of *d'* as well as techniques for quantitative analyses of sensations in both human and animal subjects. Such studies can be designed to measure not just "Is something there?" as in the earlier

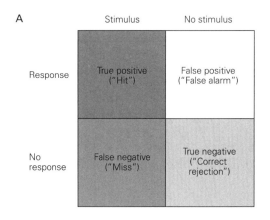

Figure 17–3A The stimulus–response matrix for data collected during a yes–no stimulus detection task ("Is a particular stimulus there?"). Each trial updates one of the four totals. For example, correct detection of the stimulus would update the count of true positives (hits), but an incorrect positive response in the absence of the stimulus would count as a false positive. From such a table, important measures such as the sensitivity and false-positive rate can be calculated.

example, but also comparative judgments of a physical property of a stimulus such as its intensity, size, or temporal frequency, thereby measuring a *two-alternative forced-choice* analog of "What is it?"

When subjects are asked to report whether the second stimulus is stronger or weaker, higher or lower, larger or smaller, or same or different than the first stimulus, responses in each trial can again be tabulated in a four-cell stimulus–response matrix similar to the one in Figure 17–3A, but with the terms "stimulus" or "no stimulus" replaced by the two distinct stimuli.

(continued)

By recording neuronal activity at various stages of sensory processing, neuroscientists attempt to decipher the mechanisms used by various sensory modalities to represent information and the transformations needed to convey these signals to the brain encoded by sequences of action potentials. Additional analyses are performed of the transformation of signals by neural networks along pathways to and within the cerebral cortex. Neuroscientists can also modify activity within sensory circuits by direct stimulation with electrical pulses, chemical neurotransmitters, and modulators,

or can use genetically encoded light-activated ion channels (optogenetics) to depolarize or hyperpolarize sensory neurons. How sensory stimuli are encoded by neurons may lead to insight into the coding principles that underlie cognition.

It is often said that the power of the brain lies in the millions of neurons processing information in parallel. That formulation, however, does not capture the essential difference between the brain and all the other organs of the body. In the kidney or a muscle, most cells do similar things; if we understand typical

Box 17–1 Signal Detection Theory: Quantifying Detection and Discrimination (continued)

B

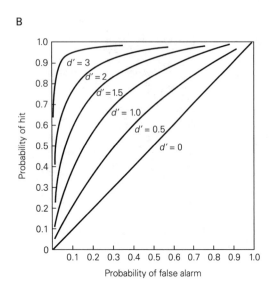

Figure 17–3B A receiver operating characteristic (ROC) plot displays the results of sets of trials, each collected in matrices such as those in Figure 17–3A. The vertical axis plots the fraction or probability of hits as a function of fraction or probability of false alarms on the horizontal axis. It is also common to label the vertical axis TPR (true-positive rate), or sensitivity, and the horizontal axis FPR (false-positive rate), or (1 − specificity). A set of trials in which yes or no responses are delivered randomly (discriminability [d'] = 0) plots as a straight line from the origin to the upper right corner. The area under such an ROC curve (AUC) would be 0.5. A perfect set of trials, in which observers accurately detect the presence of every stimulus and fail to be fooled by any trials without stimuli ($d' > 3$), would rise sharply along the left axis, and the AUC would be 1.0. AUC values are increasingly quoted as single-number measures of confidence. The (theoretical) curves shown demonstrate how higher values of d' result in larger AUC. (Adapted, with permission, from Swets 1973. Copyright © 1973 AAAS.)

Discriminability (d') in these studies is measured with *receiver operating characteristic* (ROC) analyses that compare the neural firing rates or choice probability evoked by pairs of stimuli that differ in some property. The assumption is that one of the two stimuli evokes higher responses than the other. ROC graphs of neural or psychophysical data plot the proportion of trials judged correctly (hits) and incorrectly (false positives) when the decision criteria are set at various firing levels or choice rates (Figure 17–3B). The area under the ROC curve provides an accurate estimate of d' for each stimulus pair.

Signal detection methods have been applied by William Newsome, Michael Shadlen, and J. Anthony Movshon in studies of neural responses to visual stimuli that differ in orientation, spatial frequency, or coherence of motion in order to correlate changes in neural firing rates with sensory processing. The neurometric function, plotting neural discriminability as a function of stimulus differences, corresponds closely to the psychometric function obtained in forced-choice paradigms testing the same stimuli, thereby providing a physiological basis for the observed behavioral responses.

Many of these tools, developed in part to study sensory systems, have been generalized to apply broadly beyond neuroscience. ROC curves, sensitivity, and specificity are essential in quantification of diagnosis and treatment of disease. The area under an ROC curve, or AUC, is today used much more than d'. Values of AUC close to 1 characterize high sensitivity and high specificity. The *false positive rate* (1 − specificity, or the number of false positives divided by the number of presentations without a stimulus) is, for many experiments or clinical investigations in which true positive findings are rare, a more meaningful measure than the classical p value.

muscle cells, we essentially understand how whole muscles work. In the brain, millions of cells each do something *different*. To understand the brain, we need to understand how its tasks are organized in networks of neurons.

Sensory Receptors Respond to Specific Classes of Stimulus Energy

Functional differences between sensory systems arise from two features: the different stimulus energies that

drive them and the discrete pathways that compose each system. Each neuron performs a specific task, and the train of action potentials it produces has a specific functional significance for all postsynaptic neurons in that circuit. This basic idea was expressed in the theory of specificity set forward by Charles Bell and Johannes Müller in the 19th century, and remains one of the cornerstones of sensory neuroscience.

When analyzing sensory experience, it is important to realize that our conscious sensations differ qualitatively from the physical properties of stimuli because, as Kant and the idealists predicted, the nervous system extracts only certain features of each stimulus while ignoring others. It then interprets this information within the constraints of the brain's intrinsic structure and previous experience. Thus, we *receive* electromagnetic waves of different frequencies, but we *see* them as colors. We receive pressure waves from objects vibrating at different frequencies but we hear sounds, words, and music. We encounter chemical compounds floating in the air or water but we experience them as odors and tastes. Colors, tones, odors, and tastes are mental creations constructed by the brain out of sensory experience. They do not exist as such outside the brain but are linked to specific physical properties of stimuli.

The richness of sensory experience begins with millions of highly specific sensory receptors. Sensory receptors are found in specialized epithelial structures called sense organs, principally the eye, ear, nose, tongue, and skin. Each receptor responds to a specific kind of energy at specific locations in the sense organ and sometimes only to energy with a particular temporal or spatial pattern. The receptor transforms the stimulus energy into electrical energy; thus, all sensory systems use a common signaling mechanism. The amplitude and duration of the electrical signal produced by the receptor, termed the *receptor potential*, are related to the intensity and time course of stimulation of the receptor. The process by which a specific stimulus energy is converted into an electrical signal is called *stimulus transduction*.

Sensory receptors are morphologically specialized to transduce specific forms of energy, and each receptor has a specialized anatomical region within the sense organ where stimulus transduction occurs (Figure 17–4). Most receptors are optimally selective for a single type of stimulus energy, a property termed

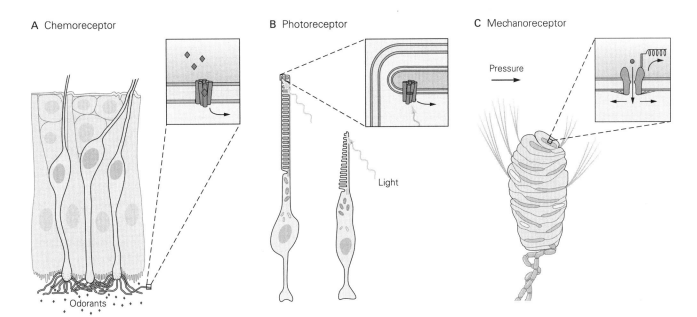

A Chemoreceptor   B Photoreceptor   C Mechanoreceptor

Odorants

Light

Pressure

Figure 17–4 Sensory receptors are specialized to transduce a particular type of stimulus energy into electrical signals. Sensory receptors are classified as chemoreceptors, photoreceptors, or mechanoreceptors depending on the class of stimulus energy that excites them. They transform that energy into an electrical signal that is transmitted along pathways that serve one sensory modality. The inset in each panel illustrates the location of the ion channels that are activated by stimuli.

A. The olfactory hair cell responds to chemical molecules in the air. The olfactory cilia on the mucosal surface bind specific odorant molecules and depolarize the sensory nerve through a second-messenger system. The firing rate signals the concentration of odorant in the inspired air.

B. Rod and cone cells in the retina respond to light. The outer segment of both receptors contains the photopigment rhodopsin, which changes configuration when it absorbs light of particular wavelengths. Stimulation of the chromophore by light reduces the concentration of cyclic guanosine 3′,5′-monophosphate (cGMP) in the cytoplasm, closing cation channels and thereby hyperpolarizing the photoreceptor. (Adapted from Shepherd 1994.)

C. Meissner's corpuscles respond to mechanical pressure. The fluid-filled capsule (**pale blue**) surrounding the sensory nerve endings (**pink**) is linked by collagen fibers to the fingerprint ridges. Pressure or motion on the skin opens stretch-sensitive ion channels in the nerve fiber endings, thus depolarizing them. (Adapted, with permission, from Andres and von Düring 1973.)

receptor specificity. We see particular colors, for example, because we have receptors that are selectively sensitive to photons with specific ranges of wavelengths, and we smell particular odors because we have receptors that bind specific odorant molecules.

In all sensory systems, each receptor encodes the type of energy applied to its receptive field, the local stimulus magnitude, and how it changes with time. For example, photoreceptors in the retina encode the hue, brightness, and duration of light striking the retina from a specific location in the visual field. Hair cell receptors in the cochlea encode the tonal frequency, loudness, and duration of sound-pressure waves hitting the ear. The neural representation of an object, sound, or scene is therefore composed of a mosaic of individual receptors that collectively signal its size, contours, texture, temporal frequency, color, and temperature.

The arrangement of receptors in the sense organ allows further specialization of function within each sensory system. Mammalian sensory receptors are classified as mechanoreceptors, chemoreceptors, photoreceptors, or thermoreceptors (Table 17–1). Mechanoreceptors and chemoreceptors are the most widespread and the most varied in form and function.

Four different kinds of *mechanoreceptors* that sense skin deformation, motion, stretch, and vibration are responsible for the sense of touch in the human hand and elsewhere (Chapters 18 and 19). Muscles contain three kinds of mechanoreceptors that signal muscle length, velocity, and force, whereas other mechanoreceptors in the joint capsule signal joint angle (Chapter 31). Hearing is based on two kinds of mechanoreceptors, inner and outer hair cells, that transduce motion of the basilar membrane in the inner ear (Chapter 26). Other hair cells in the vestibular labyrinth sense motion and acceleration of the fluids of the inner ear to signal head motion and orientation (Chapter 27). Visceral mechanoreceptors detect the distension of internal organs such as the bowel and bladder. Osmoreceptors in the brain, which sense the state of hydration, are activated when a cell swells. Certain mechanoreceptors report extreme distortion that threatens to damage tissue; their signals reach pain centers in the brain (Chapter 20).

Table 17–1 Classification of Sensory Receptors

Sensory system	Modality	Stimulus	Receptor class	Receptor cells
Visual	Vision	Light (photons)	Photoreceptor	Rods and cones
Auditory	Hearing	Sound (pressure waves)	Mechanoreceptor	Hair cells in cochlea
Vestibular	Head motion	Gravity, acceleration, and head motion	Mechanoreceptor	Hair cells in vestibular labyrinths
Somatosensory				Cranial and dorsal root ganglion cells with receptors in:
	Touch	Skin deformation and motion	Mechanoreceptor	Skin
	Proprioception	Muscle length, muscle force, and joint angle	Mechanoreceptor	Muscle spindles, Golgi tendon organs, and joint capsules
	Pain	Noxious stimuli (thermal, mechanical, and chemical stimuli)	Thermoreceptor, mechanoreceptor, and chemoreceptor	All tissues except central nervous system
	Itch	Histamine, pruritogens	Chemoreceptor	Skin
	Visceral (not pain)	Wide range (thermal, mechanical, and chemical stimuli)	Thermoreceptor, mechanoreceptor, and chemoreceptor	Cardiovascular, gastrointestinal tract, urinary bladder, and lungs
Gustatory	Taste	Chemicals	Chemoreceptor	Taste buds, intraoral thermal, and chemoreceptors
Olfactory	Smell	Odorants	Chemoreceptor	Olfactory sensory neurons

Chemoreceptors are responsible for olfaction, gustation, itch, pain, and many visceral sensations. A significant part of pain is due to chemoreceptors that detect molecules spilled into the extracellular fluid by tissue injury and molecules that are part of the inflammatory response. Several kinds of *thermoreceptors* in the skin sense skin warming and cooling. Another thermoreceptor, which monitors blood temperature in the hypothalamus, is mainly responsible for whether we feel warm or cold.

Vision is mediated by five kinds of *photoreceptors* in the retina. The light sensitivities of these receptors define the visible spectrum. The photopigments in rods and cones detect electromagnetic energy of wavelengths that span the range of 390 to 670 nm (Figure 17–5A), the principal wavelengths of sunlight and moonlight reaching the earth and informing our visual world. Unlike some other species, such as birds or reptiles, humans do not detect ultraviolet light or infrared radiation because we lack receptors that detect the appropriate short or long wavelengths. Likewise, we do not perceive radio waves and microwave energy bands because we have not evolved receptors for these wavelengths.

Multiple Subclasses of Sensory Receptors Are Found in Each Sense Organ

Each major sensory system has several *submodalities*. For example, taste can be sweet, sour, salty, savory, or bitter; visual objects have qualities of color, shape, and pattern; and touch includes qualities of temperature, texture, and rigidity. Some submodalities are mediated by discrete subclasses of receptors that respond to limited ranges of stimulus energies of that modality; others are derived by combining information from different receptor types.

The receptor behaves as a filter for a narrow range or bandwidth of energy. For example, an individual photoreceptor is not sensitive to all wavelengths of light but only to a small part of the spectrum. We say that a receptor is *tuned* to an optimal or best stimulus, the *preferred* stimulus that activates the receptor at low energy and evokes the strongest neural response. As a result, we can plot a *tuning curve* for each receptor based on physiological experiments (see the light absorbance curves for photoreceptors in Figure 17–5A). The tuning curve shows the range of sensitivity of the receptor, including its preferred stimulus. For example, blue cone cells in the retina are most sensitive to light of 430 to 440 nm, green cone cells respond best to 530 to 540 nm, and red cone cells respond most vigorously to light of 560 to 570 nm. Responses of the three cone cells to other wavelengths of light are weaker as the incident wavelengths differ from these optimal ranges (Chapter 22).

Each rod and cone cell thus responds to a wide spectrum of colors. The graded sensitivity of photoreceptors encodes specific wavelengths by the amplitude of the evoked receptor potential. However, this amplitude also depends upon the intensity or brightness of the light, so a green cone responds similarly to bright orange or dimmer green light. How are these distinguished? Stronger stimuli activate more photoreceptors than do weaker ones, and the resulting population code of multiple receptors, combined with receptors of different wavelength preferences, distinguishes intensity from hue. Such neural ensembles enable individual visual neurons to multiplex signals of color and brightness in the same pathway.

Additionally, because the tuning curve of a photoreceptor is roughly symmetric around the best frequency, wavelengths of greater or lesser values may evoke similar responses. For example, red cones respond similarly to light of 520 and 600 nm. How does the brain interpret these signals? The answer again lies with multiple receptors, in this case the green and blue cones. Green cones respond very strongly to light of 520 nm, as it is close to their preferred wavelength, but respond weakly to 600 nm light. Blue cones do not respond to 600 nm light and are barely activated at 520 nm. As a result, 520 nm light is perceived as green, whereas 600 nm is seen as orange. Thus, through varying combinations of photoreceptors, we are able to perceive a spectrum of colors.

Similarly, the complex flavors we perceive when eating are a result of combinations of chemoreceptors with different affinities for natural ligands. The broad tuning curves of a large number of distinct olfactory and gustatory receptors afford many combinatorial possibilities.

The existence of submodalities points to an important principle of sensory coding, namely that the range of stimulus energies—such as the wavelength of light—is deconstructed into smaller, simpler components whose intensity is monitored over time by specialized receptors that transmit information in parallel to the brain. The brain eventually integrates these diverse components of the stimulus to convey an ensemble representation of the sensory event. The ensemble hypothesis is even more important when we examine the representation of sensory events in the central nervous system. Although most studies of sensory processing have examined how individual neurons respond to temporally varying stimuli, the current challenge is to decipher how sensory information is

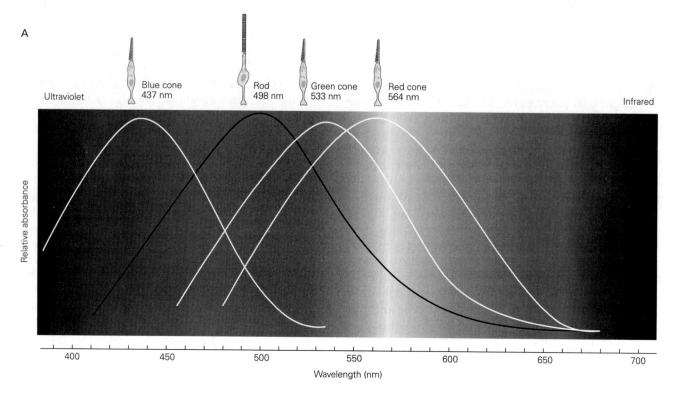

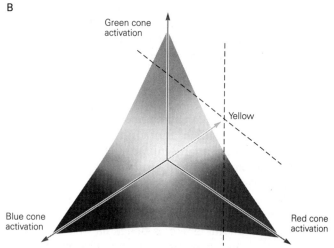

Figure 17–5 Human perception of colors results from the simultaneous activation of three different classes of photoreceptors in the retina.

A. The visible spectrum of light spans wavelengths of 390 to 670 nm. Individual photoreceptors are sensitive to a broad range of wavelengths, but each is most responsive to light in a particular spectral band. Thus, cone cells are classified as red, green, or blue type photoreceptors. Changes in the relative activation of each of the three cone types account for the perception of specific colors. (Adapted from Dowling 1987.)

B. The neural coding of color and brightness in the retina can be portrayed as a three-dimensional vector in which the strength of activation of each cone type is plotted along one of the three axes. Each point in the vector space represents a unique pattern of activation of the three cone types. Direction in the vector indicates the relative activity of each cone type and the color seen. In the example shown here, strong activation of **red cones** along with moderate stimulation of **green cones** and weak activation of **blue cones** produces the perception of **yellow**. The length of the vector from the origin to the point represents the intensity or brightness of light in that region of the retina.

distributed across populations of neurons responding to the same event at the same time.

Receptor Population Codes Transmit Sensory Information to the Brain

The receptor potential generated by an adequate stimulus produces a local depolarization or hyperpolarization of the sensory receptor neuron whose amplitude is proportional to the stimulus intensity. However, the sense organs are located at distances far enough from the central nervous system that passive propagation of receptor potentials is insufficient to transmit signals there. To communicate sensory information to the brain, a second step in neural coding must occur. The receptor potential produced by the stimulus must be transformed into sequences of action potentials that can be propagated along axons. The analog signal of stimulus magnitude in the receptor potential is transformed into a digital pulse code in which the frequency of action potentials is proportional to the intensity of the stimulus (Figure 17–6A). This is spike train *encoding*.

The recognition of an analog-to-digital transformation dates back to 1925 when Edgar Adrian and Yngve Zotterman discovered the all-or-none properties of the action potential in sensory neurons. Despite the simple recording instruments available at that time, Adrian and Zotterman discovered that the frequency of firing—the number of action potentials per second—varies with the strength of the stimulus and its duration; stronger stimuli evoke larger receptor potentials that generate a greater number and a higher frequency of action potentials. This signaling mechanism is termed *rate coding*.

In later years, as recording technology improved and digital computers allowed precise quantification of the timing of action potentials, Vernon Mountcastle and his colleagues demonstrated a precise correlation between sensory thresholds and neural responses, as well as the parametric relationship between neural firing rates and the perceived intensity of sensations (Figure 17–6B). They also found that the intensity of a stimulus is represented in the brain by all active neurons in the receptor population. This type of *population code* depends on the fact that individual receptors in a

A Neural code of stimulus magnitude

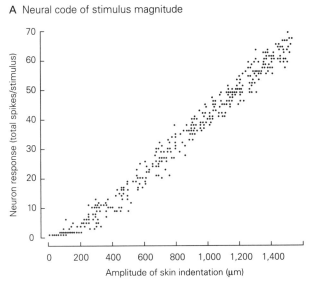

B Perceived sensation intensity

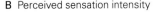
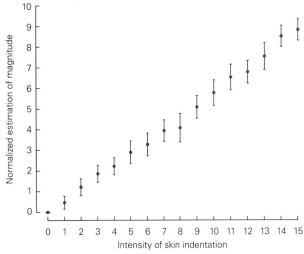

Figure 17–6 The firing rates of sensory neurons encode the stimulus magnitude. The two plots indicate that the neural coding of stimulus intensity is faithfully transmitted from peripheral receptors to cortical centers that mediate conscious sensation. (Adapted, with permission, from Mountcastle, Talbot, and Kornhuber 1966.)

A. The number of action potentials per second recorded from a touch receptor in the hand is proportional to the amplitude of skin indentation. Each dot represents the response of the receptor to pressure applied by a small probe. The relationship

between the neural firing rate and the pressure stimulus is linear. This receptor does not respond to stimuli weaker than 200 µm, its touch threshold.

B. Estimates made by human subjects of the magnitude of sensation produced by pressure on the hand increase linearly as a function of skin indentation. The relation between a subject's estimate of the intensity of the stimulus and its physical strength resembles the relation between the discharge frequency of the sensory neuron and the stimulus amplitude.

sensory system differ in their sensory thresholds or in their affinity for particular molecules.

Most sensory systems have low- and high-threshold receptors. When stimulus intensity changes from weak to strong, low-threshold receptors are first recruited, followed by high-threshold receptors. For example, rod cells in the retina are activated by very low light levels and reach their maximal receptor potentials and firing rates in dim daylight. Cone cells do not respond in very dim light but do report differences in daylight brightness. The combination of the two types of photoreceptors allows us to perceive light intensity over several orders of magnitude. Parallel processing by low- and high-threshold receptors thus extends the dynamic range of a sensory system.

Distributed patterning of firing in neural ensembles allows the use of vector algebra to quantify how stimulus properties are distributed across populations of active neurons. For example, although humans possess only three types of cone cells in the retina, we can clearly identify colors across the entire spectrum of visible light. In Figure 17–5B, we see that the color yellow can be synthesized in the mind by specific combinations of activity in red, green, and blue cone cells (Figure 17–5B). Likewise, the color magenta results from other combinations of the same photoreceptor classes. Mathematically, the perceived hue can be represented in a three-dimensional vector space in which the strengths of activation of each receptor class are combined to yield a unique sensation.

High-dimensional multineuronal representation of stimuli across large populations of neurons is beginning to be analyzed as new techniques are developed for simultaneous recording and imaging of activity in neural ensembles. Ideally, the firing rates of each neuron in a population can be plotted in a coordinate system with multiple axes such as modality, location, intensity, and time. The neural components along these axes combine to form a vector that represents the population's activity. The vector interpretation is useful because it makes available powerful analytic techniques.

The possibilities for information coding through *temporal patterning* within and between neurons in a population are enormous. For example, the timing of action potentials in a presynaptic neuron can determine whether the postsynaptic cell fires. Two action potentials that arrive near synchronously will alter the postsynaptic neuron's probability of firing more than would action potentials arriving at different times. The relative timing of action potentials between neurons also has a profound effect on mechanisms of learning and synaptic plasticity, including long-term potentiation and depression at synapses (Chapter 54).

Sequences of Action Potentials Signal the Temporal Dynamics of Stimuli

The instantaneous firing patterns of sensory neurons are as important to sensory perception as the total number of spikes fired over long periods. Steady rhythmic firing in nerves innervating the hand is perceived as steady pressure or vibration depending upon which touch receptors are activated (Chapter 19). Bursting patterns may be perceived as motion. The patterning of spike trains plays an important role in encoding temporal fluctuations of the stimulus, such as the frequency of vibration or auditory tones, or changes in rate of movement. Humans can report changes in sensory experience that correspond to alterations within a few milliseconds in the firing patterns of sensory neurons.

Sensory systems detect *contrasts*, changes in the temporal and spatial patterns of stimulation. If a stimulus persists unchanged for several minutes without a change in position or amplitude, the neural response and corresponding sensation diminishes, a condition called *receptor adaptation*. Receptor adaptation is thought to be an important neural basis of perceptual adaptation, whereby a constant stimulus fades from consciousness. Receptors that respond to prolonged and constant stimulation—known as *slowly adapting* receptors—encode stimulus duration by generating action potentials throughout the period of stimulation (Figure 17–7A). In contrast, *rapidly adapting* receptors respond only at the beginning and end of a stimulus; they *cease* firing in response to constant amplitude stimulation and are active only when the stimulus intensity increases or decreases (Figure 17–7B). Rapidly and slowly adapting sensors illustrate another important principle of sensory coding: Neurons signal important properties of stimuli not only when they fire but also when they slow or stop firing.

The temporal properties of a changing stimulus are encoded as changes in the firing pattern, including the *interspike intervals*, of sensory neurons. For example, the touch receptors illustrated in Figure 17–7 fire at higher rates when a probe initially contacts the skin than when the pressure is maintained. The time interval between spikes is shorter when the skin is indented rapidly than when pressure is applied gradually. The firing rate of these neurons is proportional to both the speed at which the skin is indented and the total amount of pressure applied. During steady pressure, the firing rate slows to a level proportional to skin indentation (Figure 17–7A) or ceases entirely (Figure 17–7B). Firing of both neurons stops after the probe is retracted.

A Slowly adapting receptor

B Rapidly adapting receptor

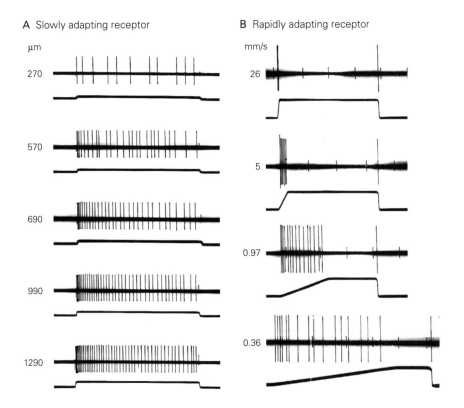

Figure 17–7 Firing patterns of sensory neurons convey information about the stimulus intensity and time course. These records illustrate responses of two different classes of touch receptors to a probe pressed into the skin. The stimulus amplitude and time course are shown in the lower trace of each pair; the upper trace shows the action potentials recorded from the sensory nerve fiber in response to the stimulus.

A. A slowly adapting mechanoreceptor responds as long as pressure is applied to the skin. The total number of action potentials discharged during the stimulus is proportional to the amount of pressure applied to the skin. The firing rate is higher

at the beginning of skin contact than during steady pressure, as this receptor also detects how rapidly pressure is applied to the skin. When the probe is removed from the skin, the spike activity ceases. (Adapted, with permission, from Mountcastle, Talbot, and Kornhuber 1966.)

B. A rapidly adapting mechanoreceptor responds at the beginning and end of the stimulus, signaling the rate at which the probe is applied and removed; it is silent when pressure is maintained at a fixed amplitude. Rapid motion evokes a brief burst of high-frequency spikes, whereas slow motion evokes a longer-lasting, low-frequency spike train. (Adapted, with permission, from Talbot et al. 1968.)

The Receptive Fields of Sensory Neurons Provide Spatial Information About Stimulus Location

The position of a sensory neuron's input terminals in the sense organ is a major component of the specific information conveyed by that neuron. The skin area, location in the body, retinal area, or tonal domain in which stimuli can activate a sensory neuron is called its *receptive field* (Figure 17–8). The region from which a sensation is perceived to arise is called the neuron's *perceptive field*. The two usually coincide.

The dimensions of receptive fields play an important role in the ability of a sensory system to encode detailed spatial information. The objects that we see with our eyes or hold in our hands are much larger than the receptive field of an individual sensory neuron,

and therefore stimulate groups of adjacent receptors. The size of the stimulus therefore determines the total number of receptors that are activated. In this manner, the spatial distribution of active and silent receptors provides a neural image of the size and contours of the stimulus.

The spatial resolution of a sensory system depends on the total number of receptor neurons and the distribution of receptive fields across the area covered. The projection neurons for regions of the body with a high density of receptors, such as the retinal ganglion cells representing the central retina (the fovea), have small receptive fields because they receive inputs from a small number of bipolar cells, each of which receives input from a few closely packed photoreceptors. Because of the high density of receptors in the

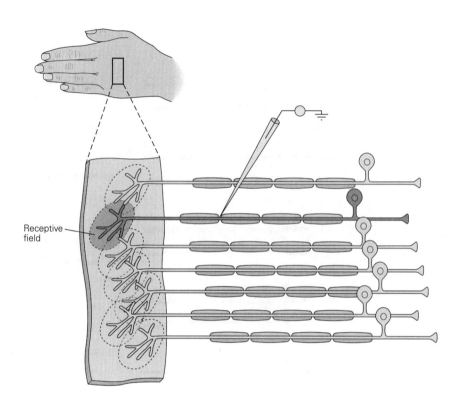

Figure 17–8 The receptive field of a sensory neuron. The receptive field of a touch-sensitive neuron denotes the region of skin where gentle tactile stimuli evoke action potentials in that neuron. It encompasses all of the receptive endings and terminal branches of the sensory nerve fiber. If the fiber is stimulated electrically with a microelectrode, the subject experiences touch localized on the skin. The area from which the sensation is perceived to arise is called the *perceptive field*. A patch of skin contains many overlapping receptive fields, allowing sensations to shift smoothly from one sensory neuron to the next in a continuous sweep. The axon terminals of sensory neurons in the central nervous system are arranged somatotopically, providing an orderly map of the innervated region of the body.

Receptive field

fovea, the population of neurons transmits a very detailed representation of the visual scene. Ganglion cells in the periphery of the retina have larger receptive fields because the receptor density is much lower. The dendrites of these ganglion cells receive information from a wider area of the retina, and thus integrate light intensity over a greater portion of the visual field. This arrangement yields a less detailed image of the visual scene (Figure 17–9). Similarly, the region of the body most often used to touch objects is the hand. Not surprisingly, mechanoreceptors for touch are concentrated in the fingertips, and the receptive fields on the hand are smaller than those on the arm or trunk.

Central Nervous System Circuits Refine Sensory Information

The central connections of a sensory neuron determine how that neuron's signals influence our sensory experience. Action potentials in nerve fibers of the cochlea, for example, evoke the sensation of a tone whether they are initiated by sound waves acting on hair cells or by electrical stimulation with a neural prosthesis.

The parcellation of a stimulus into its components, each encoded by an individual type of sensory receptor or projection neuron, is an initial step in sensory processing. These components are integrated into a representation of an object or scene by neural networks in the brain. This process allows the brain to select certain abstract features of an object, person, scene, or external event from the detailed input of many receptors. As a result, the representation formed in the brain may enhance the saliency of features that are important at the moment while ignoring others. In this sense, our percepts are not merely reflections of environmental events, but also constructs of the mind.

How we experience the sensations reported by primary receptors is also subject to modification or learning. Initially aversive odors and tastes, for example, can become attractive over time because of familiarity or changes in context or association. The pleasure elicited by photos of a respected baseball player can be converted to disdain should he subsequently appear in the uniform of a rival team.

In the early stages of sensory information processing in the central nervous system, each class of peripheral receptors provides input to clusters of neurons in relay nuclei that are dedicated to one sensory modality. That is, each sensory modality is represented by an ensemble of central neurons connected to a specific class of receptors. Such ensembles are referred to as *sensory systems*, and include the somatosensory, visual, auditory, vestibular, olfactory, and gustatory systems (see Table 17–1).

A 20 × 20 pixels   B 60 × 60 pixels   C 400 × 400 pixels

Figure 17–9 The visual resolution of scenes and objects depends on the density of photoreceptors that mediate the image. The resolution of detail is inversely correlated with the area of the receptive field of individual neurons. Each square or pixel in these images represents a receptive field. The gray scale in each pixel is proportional to the average light intensity in the corresponding receptive field. If there are a small number of neurons, and each spans a large area of the image, the result is a very schematic representation of the scene (**A**). As the density of neurons increases, and the size of each receptive field decreases, the spatial detail becomes clearer (**B, C**). The increased spatial resolution comes at the cost of the larger number of neurons required to transmit the information. (Photographs reproduced, with permission, from Daniel Gardner.)

The brain has evolved to process and respond to this rich ensemble of sensory information. The activation of sensory, cognitive, and motor systems in the human brain can be visualized in real time with fMRI techniques. Maurizio Corbetta, Marcus Raichle, and colleagues discovered coherent fluctuations in low-frequency (0.01–0.1 Hz) components of the blood oxygen level–dependent (BOLD) signal during the "resting" state in brain areas that are anatomically connected and activated together during specific behaviors. Figure 17–10 highlights three functionally specialized networks of brain areas that respond to auditory (in red), somatomotor (in green), and visual (in blue) inputs. Other areas are multisensory, integrating information from several different modalities. Spontaneous correlation of firing of these networks in the absence of direct sensory stimuli or performance of motor tasks suggests that excitability within resting state sensory or motor networks may signal readiness to process information for future sensation or action. Deficits in sensory, cognitive, or motor function following local brain injury may result not just from impairment of one specific area, or node, but rather disruption of the circuit or circuits that include that node.

Synapses in sensory pathways provide an opportunity to modify the signals from receptors. Most neurons in relay nuclei receive convergent excitatory inputs from many presynaptic neurons (Figure 17–11A), integrate those inputs, combine them with inhibitory and top-down signals, and transmit the processed information to higher brain areas. Horace Barlow proposed that sensory systems demonstrate *efficient coding*, which includes sensory relays recoding sensory messages so that their redundancy is reduced, but comparatively little information is lost. Likewise, each receptor neuron excites multiple postsynaptic relay neurons.

Convergent excitatory networks provide a mechanism for spatial summation of inputs, strengthening signals of functional importance. One example of how such circuits are used is detection of synchronous inputs from multiple nearby locations but not others, thereby providing the first step toward *orientation tuning* of central neurons. Relay neurons are also interconnected with their neighbors, forming recurrent excitatory connections that amplify sensory signals. Such *recurrent networks* are also a feature of some deep learning algorithms used by artificial neural networks to classify sensory patterns.

A relay neuron's receptive field is also shaped by inhibitory input. The inhibitory region of a receptive field provides an important mechanism for enhancing the contrast between stimuli, giving the sensory system additional power to resolve spatial detail. Inhibitory interneurons modulate the excitability of neurons in relay nuclei, thereby regulating the amount of sensory information transmitted to higher levels of

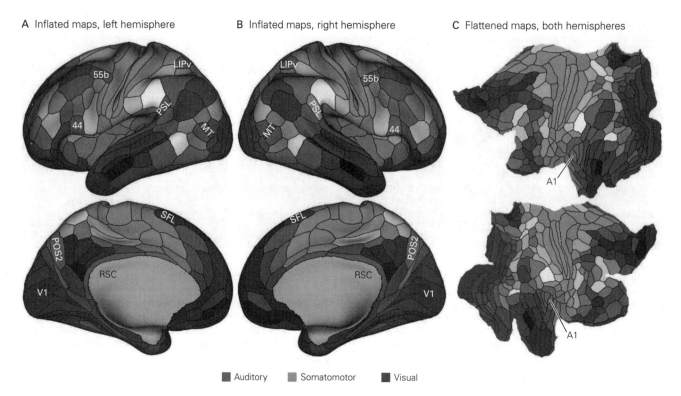

A Inflated maps, left hemisphere   B Inflated maps, right hemisphere   C Flattened maps, both hemispheres

■ Auditory   ■ Somatomotor   ■ Visual

Figure 17–10 Distinct regions of the human brain process information for individual sensory modalities, multisensory systems, motor activity, or cognitive function. The human cerebral cortex has been divided into 180 functional areas by the Human Connectome Project based largely on a variety of fMRI techniques and neuroanatomy. Early auditory areas (**red**), somatosensory and motor areas (**green**), and visual areas (**blue**) are shaded in primary colors. Mixed colors indicate multisensory areas: visual and somatosensory/motor (**blue-green, LIPv, MT**); or visual and auditory (**pink to purple, POS2, RSC**). Language networks include areas 55b, 44, SFL, and PSL in both hemispheres. Gray-scaled regions serve cognitive functions; they comprise the anticorrelated "task-positive" (**light shading**) and "default mode" (**dark shading**) networks. The maps show brain regions located on the surface gyri and within adjacent cortical sulci. Note the similarity of brain organization

between the two hemispheres. Data available at https://balsa.wustl.edu/study/RVVG. (Reproduced, with permission, from Glasser et al. 2016. Copyright © 2016 Springer Nature.)

A. Inflated maps of the left hemisphere. The top map is a lateral view and the bottom map is a medial view.

B. Similar maps of the right hemisphere.

C. Flattened maps show the functional organization of both hemispheres (left at top, right at bottom).

(Abbreviations: **A1**, primary auditory cortex; **LIPv**, lateral intraparietal area, ventral portion; **MT**, middle temporal area; **POS2**, parieto-occipital sulcus area 2; **PSL**, perisylvian language area; **RSC**, retrosplenial complex; **SFL**, superior frontal language area; **V1**, primary visual cortex; **Area 55b**, newly identified language area; **Area 44**, part of Broca's area.)

a network (Figure 17–11B). Inhibitory circuits are also useful for suppressing irrelevant information during goal-directed behaviors, thereby focusing attention on specific task-related inputs. Additionally, inhibitory networks allow the context of a stimulus to modify the strength of excitation evoked by that stimulus, an important process called *normalization*.

The responses of central neurons to sensory stimuli are more variable from trial to trial than those of peripheral receptors. Central sensory neurons also fire irregularly before and after stimulation and during periods when no stimuli are present. The variability of the evoked central responses is a result of several factors: the subject's state of alertness, whether attention

is engaged (Figure 17–12), previous experience of that stimulus, and recent activation of the pathway by similar stimuli. Similarly, the context of stimulus presentation, subjective intentions, motor plans that may require feedback, or intrinsic oscillations of the neuron's membrane potential can all modify incoming sensory information.

The Receptor Surface Is Represented Topographically in the Early Stages of Each Sensory System

The axons of sensory projection neurons terminate in the brain in an orderly manner that retains their spatial arrangement in the receptor sheet. Sensory neurons for

A Typical neural circuit for sensory processing

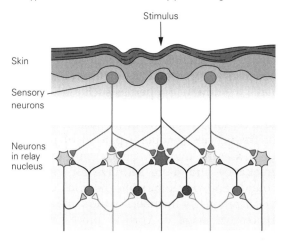

B Spatial distribution of excitation and inhibition among
   relay neurons

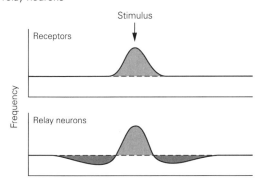

C Types of inhibition in relay nuclei

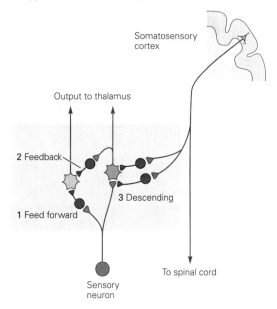

**Figure 17–11 Relay neurons in sensory systems integrate a
variety of inputs that shape stimulus information.**

A. Sensory information is transmitted in the central nervous
system through hierarchical processing networks. Neural signaling initiated by a stimulus to the skin reaches a large group of
postsynaptic neurons in relay nuclei in the brain stem and thalamus and is most strong in neurons in the center of the array
of postsynaptic cells (**red neuron**). (Adapted, with permission,
from Dudel 1983.)

B. Inhibition (**gray areas**) mediated by local interneurons (**gray**)
confines excitation (**orange area**) to the central zone in the
array of relay neurons where stimulation is strongest. This pattern of inhibition within the relay nucleus enhances the contrast
between strongly and weakly stimulated relay neurons.

C. Inhibitory interneurons in a relay nucleus are activated by
three distinct excitatory pathways. **1. Feed-forward** inhibition

is initiated by the afferent fibers of sensory neurons that
terminate on the inhibitory interneurons. **2. Feedback** inhibition is initiated by recurrent collateral axons of neurons in
the output pathway from the nucleus that project back to
interneurons in the source nucleus. The interneurons in turn
inhibit nearby output neurons, creating sharply defined zones
of excitatory and inhibitory activity in the relay nucleus. In this
way, the most active relay neurons reduce the output of adjacent, less active neurons, thus ensuring that only one of two
or more active neurons will send out signals. **3. Descending**
inhibition is initiated by neurons in other brain regions such as
the cerebral cortex. The descending commands allow cortical
neurons to control the afferent relay of sensory information,
providing a mechanism by which attention can select sensory
inputs.

touch in adjacent regions of the skin project to neighboring neurons in the central nervous system, and
this topographic arrangement of receptive fields is
preserved throughout the early somatosensory pathways. Each primary sensory area in the brain thus
contains a topographic, spatially organized map of the
sense organ. This topography extends to all levels of

a sensory system. Within these maps, specificity—the
qualities to which neurons are most narrowly tuned—
provides clues to the functional organization of that
region of the brain.

In the first and subsequent relay nuclei of the
somatosensory, visual, and auditory systems, adjacent neurons represent adjacent areas of the body,

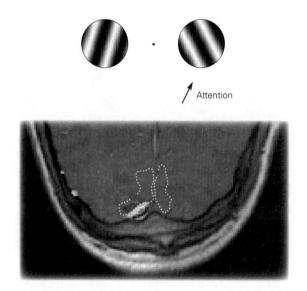

Attention

Figure 17–12 Attention to a visual stimulus alters responses of neurons in visual cortical areas. When we pay attention to a stimulus, we select certain sensory inputs for cognitive processing and ignore or suppress other information. Functional MRI is used in this study to measure the effects of attention to visual stimuli on neural responses in human primary visual (V1) cortex (**white dashed lines** on the brain anatomical section, **lower panel**). Moving grating stimuli (**upper panel**) were presented simultaneously to the right and left visual fields while subjects stared at a central fixation point (**black dot**). The subjects performed a motion discrimination task, attending (without moving their eyes) to one of the two oriented gratings. When stimuli were attended in the right visual field, neural activity (**red**) increased significantly in the left hemisphere, but not in the right hemisphere, even though the stimuli were presented to both eyes. When the subject attended to the grating in the left visual field, a similar focus of activity occurred in the right V1 cortex, and activity dropped in the left hemisphere (not shown). (Adapted from Gandhi, Heeger, and Boynton, 1999.)

retina, and cochlea, respectively. The organization of these nuclei is thus said to be somatotopic, retinotopic, or tonotopic. Nuclei in the auditory system are tonotopic because the cochlear hair cells of the inner ear are arranged to create an orderly shift in frequency sensitivity from cell to cell (Figure 26–2). Neurons in the primary sensory areas of the cerebral cortex maintain these location-specific features of a stimulus, and the functional maps of these early cortical areas are likewise somatotopic, retinotopic, or tonotopic.

Sensory information flows serially through hierarchical pathways, including multiple levels of the cerebral cortex, before ending in brain regions that are concerned with cognition and action. Forming the percepts that inform these regions requires integration of lower-level inputs that report only information from

small areas of the sense organ. Neurons in the cerebral cortex are specialized to integrate and so detect specific features of stimuli beyond merely their location in the sense organ. Such neurons are said to be *tuned* to combined stimulus features represented by ensembles of sensory receptors. These neurons respond preferentially to stimulus properties such as the orientation of edges (eg, simultaneous activation of specific groups of receptors), direction of motion, or tonal sequences of frequencies (temporal pattern of receptor activation). Central auditory neurons are less selective for frequency and more selective for certain kinds of sound. For example, some neurons are specific for vocalizations by members of the same species. In each successive stage of cortical processing, the spatial organization of stimuli is progressively lost as neurons become less concerned with the descriptive features of stimuli and more concerned with properties of behavioral importance. Details of these central sensory transformations are presented in succeeding chapters that describe specific sensory systems.

Sensory Information Is Processed in Parallel Pathways in the Cerebral Cortex

Distributed spatial coding is ubiquitous in sensory systems for two reasons. First, it takes advantage of the parallel architecture of the nervous system. There are approximately 100 million neurons in each primary sensory area of the cerebral cortex, and the possible number of combinatorial patterns of neural activity far exceeds the number of atoms in the universe. Second, each neuron codes the intensity and timing of a stimulus as well as its location in the receptor sheet. It fires only when many of its excitatory synapses receive action potentials and most of the inhibitory synapses do not, firing in response to specific patterns of stimulation but not to others. Since many cortical neurons receive input from 1,000 to 10,000 synapses, the information coding potential is enormous.

One of the most important insights into feature detection in the cortex arose from combined physiological and anatomical studies of the cortical visual pathways by Mortimer Mishkin and Leslie Ungerleider in the early 1980s. They discovered that sensory information arriving in the primary visual areas is divided in two parallel pathways.

One pathway carries information needed for classification of images, while the other conveys information needed for immediate action. Visual features that identify *what* an object is are transmitted in a *ventral pathway* to the temporal lobe and eventually to the hippocampus and entorhinal cortex. Visual information

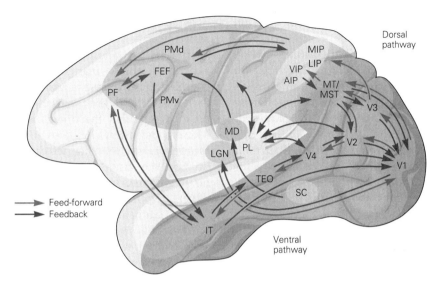

Figure 17–13 Visual stimuli are processed by serial and parallel networks in the cerebral cortex. When you read this text, the spatial pattern of the letters is sent to the cerebral cortex through successive synaptic links comprising photoreceptors, bipolar cells of the retina, retinal ganglion cells, cells in the lateral geniculate nucleus (**LGN**) of the thalamus, and neurons of the primary visual cortex (**V1**). Within the cortex, there is a gradual divergence to successive processing areas called ventral and dorsal streams that are neither wholly serial nor parallel. The ventral stream in the temporal lobe (**red shading**) analyzes and encodes information about the form and structure of the visual scene and objects within it, delivering this information to the parahippocampal cortex (not shown) and prefrontal cortex (**PF**). The dorsal stream in the parietal lobe (**blue shading**) analyzes and represents

information about stimulus location and motion and delivers this information to motor areas of the frontal cortex that control movements of the eyes, hand, and arm. The anatomical connections between these areas are reciprocal, involving both feedforward and feedback circuits. The zone of overlap (**purple**) shows that both pathways originate from the same source in **V1**. Connections to subcortical structures in the thalamus and midbrain are defined in Figure 21–7B. (Abbreviations: **V1, V2, V3,** and **V4,** occipital visual areas; **MT,** middle temporal; **MST,** medial superior temporal; **AIP, VIP, LIP,** and **MIP,** anterior, ventral, lateral, and medial intraparietal; **TEO,** temporal-occipital; **IT,** inferior temporal; **PMd** and **PMv,** dorsal and ventral premotor; **FEF,** frontal eye fields.) (Adapted from Albright and Stoner 2002.)

about *where* an object is located, its size and shape, and *how* it might be acquired and used is transmitted in a more *dorsal pathway* to the parietal lobe and eventually to the motor areas of frontal cortex (Figure 17–13).

Ventral and dorsal streams are evident in other sensory systems as well. In the auditory system, acoustic information from speech is transmitted to Wernicke's area in the temporal lobe, which has a strong role in language comprehension, and to Broca's area in the frontal cortex, which is involved in speech production. In the somatosensory system, information about an object's size and shape is transmitted to ventral areas of parietal cortex for object recognition. Tactile information about object size, weight, and texture is also communicated to posterior parietal and frontal motor areas, where it is needed to plan the handling of the object.

Ventral and dorsal streams of sensory information also contribute to two major forms of memory: semantic (also called explicit) memory, which we use to talk about objects or persons, and procedural (also

called implicit) memory, which we use to interact with objects, persons, or the immediate environment.

Ventral stream information generates *nouns* that we use to identify and classify persons, places, and objects, such as spheres, bricks, and cars. Dorsal stream information motivates *verbs* enabling the actions performed based on sensory inputs and subjective intentions, such as grasping, lifting, or driving.

Feedback Pathways From the Brain Regulate Sensory Coding Mechanisms

Sensory systems are not simply automated assembly lines that reassemble fragmented neural representations of environmental events (eg, light, sound, odor) into more coherent percepts. We have enormous control over our own experience of sensation and perception, and even our conscious attention.

We can to some extent control which sensations reach our consciousness. We may, for example, watch television to take our minds off the pain of a sprained ankle.

Direct, volitional control of the sensory information that reaches consciousness can be readily demonstrated by suddenly directing your attention to a body part, such as the fingers of your left hand, to which you were initially oblivious as you were attending to this text. Sensations from the fingers flood consciousness until attention is redirected to the text. Neural recordings in somatosensory and visual cortex confirm that neurons change their sensitivity, as reflected in their firing rates, much more so than their selectivity for particular stimuli. At a more abstract level, for example, we can switch our attention from the subject matter of a painting to the artist's technique.

Each primary sensory area of cortex has extensive projections back to its principal afferent relay nucleus in the thalamus. In fact, the number of feedback axons exceeds the number of afferent axons from the thalamus to the cortex. These projections have an important function that is not yet clear. One possibility is that they modulate the activity of certain neurons when attention and vigilance change or during motor tasks.

Centers in the brain are also able to modulate the responsiveness of sensory receptors. For example, neurons in the motor cortex can alter the sensitivity of sensory receptors in skeletal muscle that signal muscle length. Activation of gamma motor neurons by corticospinal pathways enhances the sensory responses of muscle spindle afferents to stretch. Neurons in the brain stem can directly modulate the frequency sensitivity of hair cells in the cochlea. Thus, information about a stimulus sent from peripheral sensory neurons to the brain is conditioned by the entire organism.

Top-Down Learning Mechanisms Influence Sensory Processing

What we perceive is always some combination of the sensory stimulus itself and the memories it both evokes and builds upon. The relationship between perception and memory was originally developed by empiricists, particularly the associationist philosophers James and John Stuart Mill. Their idea was that sensory and perceptual experiences that occur together or in close succession, particularly those that do so repeatedly, become associated so that thereafter the one triggers the other. Association is a powerful mechanism, and much of learning consists of forging associations through repetition.

Contemporary neuroscientists using multineuronal recordings discovered that sensory events evoke sequences of neuronal activation. These patterns of neural activity are believed to trigger memories of previous experiences of such stimulation patterns. For example, as we hear a work of music over and over again, the circuits of our auditory system are modified by the experience, and we learn to anticipate what comes next, completing the phrase before it occurs. Familiarity with the phrasing and harmonies used by a composer allows us to distinguish the operas of Verdi from those of Mozart, and the symphonies of Bruckner from those of Brahms. Likewise, when we drive to an unknown destination, our visual system is initially overwhelmed by new landmarks, as we assess which are important and which are not. With repeated trips, the journey becomes second nature and seems to take less time.

Percepts are uniquely subjective. When we look at a work of art, we superimpose our personal experience on the view; what we see is not just the image projected on the retina, but its contextual meaning to us as individuals. For example, when we view a historic photograph of important events in our lives, or persons we admired or detested, we recall not only the event in the image but also the words spoken and our emotional reactions in the past. The emotional response is muted or absent if we did not experience a direct connection to the event or person illustrated.

How can a network of neurons "recognize" a specific pattern of inputs from a population of presynaptic neurons? One potential mechanism is called *template matching*. Each neuron in the target population has a pattern of excitatory and inhibitory presynaptic connections. If the pattern of arriving action potentials fits the postsynaptic neuron's pattern of synaptic connections even approximately—activating many of its excitatory synapses but mostly avoiding activating its inhibitory synapses—the target neuron fires. The codes may also be combinatorial: the overall activity of a region remains the same with different stimuli, but the specific subset of neurons that are active when a particular input is presented constitute a "tag" specifying that input.

Charles F. Stevens has identified these in very different sensory systems and noted that such *maximum entropy* codes are highly efficient, able to represent many different stimuli for a set number of neurons. Refining our understanding of efficient coding, the Carandini and Harris labs have recently shown that the neural code in mouse visual cortex is indeed efficient and preserves fine detail, but in a manner that retains the ability to generalize by responding similarly to closely related visual stimuli. Such computational or algorithmic views have great promise for our understanding of sensory systems. *Artificial neural networks*, simulated using computers, can be trained on images and taught to "see." Daniel L. Yamins and James J.

DiCarlo have pointed out that as these artificial networks evolve the ability to recognize objects and faces, the properties of neuron-like "units" in particular layers begin to resemble the distribution of activity seen in corresponding cortical areas. Such artificial neural networks are trained by machine learning algorithms that modify the connection strength between units, similar to neuronal learning with repetition and synapse modification.

Precisely how the brain solves the recognition problem is uncertain. There is currently much evidence that the neural representation of a stimulus in the initial pathways of sensory systems is an isomorphic representation of the stimulus. Successive synaptic regions transform these initial representations into abstractions of our environment that we are beginning to decipher. In contrast, we barely understand the top-down mechanisms by which incoming sensory information invokes memories of past occurrences and activates our prejudices and opinions.

One view of these processes is Bayesian: Our experience and understanding of the world inform a top-down *sensory prior* that describes our likely environment. The primary insight of Bayes's rule is that decisions are made by the likelihood ratio of current evidence from a test stimulus and the subject's previous experience of similar stimuli (priors), all modified by the task contingencies (rewards and hazards). Ongoing sensory information contributes immediate data, and the two combine to form an up-to-the-moment *posterior* estimate of our surroundings and our place in them. When we do understand these neural codes and the algorithms and mechanisms that generate and interpret them, it is likely that we will be on the verge of understanding cognition, the way in which information is coded in our memory and our understanding. That is what makes the study of neural coding so challenging and exciting.

Highlights

1. Our sensory systems provide the means by which we perceive the external world, remain alert, form a body image, and regulate our movements. Sensations arise when external stimuli interact with some of the billion sensory receptors that innervate every organ of the body. The information detected by these receptors is conveyed to the brain as trains of action potentials traveling along individual sensory axons.

2. All sensory systems respond to four elementary features of stimuli—modality, location, intensity, and duration. The diverse sensations we experience—the sensory modalities—reflect different forms of energy that are transformed by receptors into depolarizing or hyperpolarizing electrical signals called receptor potentials. Receptors specialized for particular forms of energy, and sensitive to particular ranges of the energy bandwidth, allow humans to sense many kinds of mechanical, thermal, chemical, and electromagnetic events.

3. The intensity and duration of stimulation are represented by the amplitude and time course of the receptor potential and by the total number of receptors activated. In order to transmit sensory information over long distances, the receptor potential is transformed into a digital pulse code, sequences of action potentials whose frequency of firing is proportional to the strength of the stimulus. The pattern of action potentials in peripheral nerves and in the brain gives rise to sensations whose qualities can be measured directly using a variety of psychophysical paradigms such as magnitude estimation, signal detection methods, and discrimination tasks. The temporal features of a stimulus, such as its duration and changes in magnitude, are signaled by the dynamics of the spike train.

4. The location and spatial dimensions of a stimulus are conveyed through each receptor's receptive field, the precise area in the sensory domain in which stimulation activates the receptor. The identity of the active sensory neurons therefore signals not only the modality of a stimulus but also the place where it occurs.

5. These messages are analyzed centrally by several million sensory neurons performing different, specific functions in parallel. Each sensory neuron extracts highly specific and localized information about the external or internal environment, and in turn has a specific effect on sensation and cognition because it projects to specific places in the brain that have specific sensory, motor, or cognitive functions. To maintain the specificity of each modality within the nervous system, receptor axons are segregated into discrete anatomical pathways that terminate in unimodal nuclei.

6. Sensory information in the central nervous system is processed in stages, in the sequential relay nuclei of the spinal cord, brain stem, thalamus, and cerebral cortex. Each nucleus integrates sensory inputs from adjacent receptors and, using networks of inhibitory neurons, emphasizes the strongest signals. After about a dozen synaptic steps in each sensory system, neural activity

converges on neuronal groups whose function is multisensory and more directly cognitive.

7. Processing of sensory information in the cerebral cortex occurs in multiple cortical areas in parallel and is not strictly hierarchical. Feedback connections from areas of the brain involved in cognition, memory, and motor planning control the incoming stream of sensory information, allowing us to interpret sensory stimulation in the context of past experience and current goals.

8. The richness of sensory experience—the complexity of sounds in a Mahler symphony, the subtle layering of color and texture in views of the Grand Canyon, or the multiple flavors of a salsa—requires the activation of large ensembles of receptors acting in parallel, each one signaling a particular aspect of a stimulus. The neural activity in a set of thousands or millions of neurons should be thought of as coordinated activity that conveys a "neural image" of specific properties of the external world.

9. Our sensory systems are increasingly appreciated as computational and algorithmic encoders, processors, and decoders of information. Insights from machine learning, information theory, artificial neural networks, and Bayesian inference continue to inform our understanding of what we perceive in our bodies and from the world around us.

Esther P. Gardner
Daniel Gardner

Selected Reading

Basbaum AI, Kaneko JH, Shepherd GM, Westheimer G (eds). 2008. *The Senses: A Comprehensive Reference* (6 vols). Oxford: Elsevier.

Dowling JE. 1987. *The Retina: An Approachable Part of the Brain.* Cambridge, MA: Belknap.

Gerstein GL, Perkel DH, Dayhoff JE. 1985. Cooperative firing activity in simultaneously recorded populations of neurons: detection and measurement. J Neurosci 5:881–889.

Green DM, Swets JA. 1966. *Signal Detection Theory and Psychophysics.* New York: Wiley. (Reprinted 1974, Huntington, NY: Robert E. Krieger.)

Kandel ER. 2016. *Reductionism in Art and Brain Science: Bridging the Two Cultures.* New York: Columbia Univ. Press.

Moore GP, Perkel DH, Segundo JP. 1966. Statistical analysis and functional interpretation of neuronal spike data. Annu Rev Physiol 28:493–522.

Mountcastle VB. 1998. *Perceptual Neuroscience: The Cerebral Cortex.* Cambridge, MA: Harvard Univ. Press.

Singer W. 1999. Neuronal synchrony: a versatile code for the definition of relations? Neuron 24:49–65.

Stevens SS. 1961. The psychophysics of sensory function. In: WA Rosenblith (ed). *Sensory Communication*, pp. 1–33. Cambridge, MA: MIT Press.

Stevens SS. 1975. *Psychophysics: Introduction to Its Perceptual, Neural, and Social Prospects.* New York: Wiley.

References

Adrian ED, Zotterman Y. 1926. The impulses produced by sensory nerve-endings. Part 2. The response of a single end-organ. J Physiol (Lond) 61:151–171.

Albright TD, Stoner GR. 2002. Contextual influences on visual processing. Annu Rev Neurosci 25:339–379.

Andres KH, von Düring M. 1973. Morphology of cutaneous receptors. In: Iggo A (ed). *Handbook of Sensory Physiology,* Vol. 2, *Somatosensory System*, pp. 3–28. Berlin: Springer-Verlag.

Barch DM, Burgess GC, Harms MP, et al. 2013. Function in the human connectome: task-fMRI and individual differences in behavior. Neuroimage 80:169–189.

Berkeley G. [1710] 1957. *A Treatise Concerning the Principles of Human Knowledge.* K Winkler (ed). Indianapolis: Bobbs-Merrill.

Britten KH, Shadlen MN, Newsome WT, Movshon JA. 1992. The analysis of visual motion: a comparison of neuronal and psychophysical performance. J Neurosci 12:4745–4768.

Carandini M, Heeger DJ. 2011. Normalization as a canonical neural computation. Nat Rev Neurosci 13:51–62.

Chang L, Tsao DY. 2017. The code for facial identity in the primate brain. Cell 169:1013–1028.

Colquhoun D. 2014. An investigation of the false discovery rate and the misinterpretation of p-values. Royal Soc Open Sci 1:140216.

DiCarlo JJ, Zoccolan D, Rust NC. 2012. How does the brain solve visual object recognition? Neuron 73:415–434.

Dudel J. 1983. General sensory physiology. In: RF Schmitt, G Thews (eds). *Human Physiology*, pp. 177–192. Berlin: Springer-Verlag.

Gandhi SP, Heeger DJ, Boynton GM. 1999. Spatial attention affects brain activity in human primary visual cortex. Proc Natl Acad Sci U S A 96:3314–3319.

Gazzaniga MS (ed). 2009. *The Cognitive Neurosciences*, 4th ed. Cambridge, MA: MIT Press.

Glasser MF, Coalson TS, Robinson EC, et al. 2016. A multimodal parcellation of human cerebral cortex. Nature 536:171–178.

Hubel DH, Wiesel TN. 1968. Receptive fields and functional architecture of monkey striate cortex. J Physiol 195:215–243.

Hume D. [1739] 1984. *A Treatise of Human Nature.* EC Mossner (ed). New York: Penguin.

Johansson RS, Vallbo AB. 1979. Detection of tactile stimuli. thresholds of afferent units related to psychophysical thresholds in the human hand. J Physiol 297:405–422.

Johnson KO, Hsiao SS, Yoshioka T. 2002. Neural coding and the basic law of psychophysics. Neuroscientist 8:111–121.

Kant I. [1781/1787] 1961. *Critique of Pure Reason*. NK Smith (transl.). London: Macmillan.

Kirkland KL, Gerstein GL. 1999. A feedback model of attention and context dependence in visual cortical networks. J Comput Neurosci 7:255–267.

LaMotte RH, Mountcastle VB. 1975. Capacities of humans and monkeys to discriminate between vibratory stimuli of different frequency and amplitude: a correlation between neural events and psychophysical measurements. J Neurophysiol 38:539–559.

Li HH, Rankin J, Rinzel J, Carrasco M, Heeger DJ. 2017. Attention model of binocular rivalry. Proc Natl Acad Sci U S A 114:E6192-E6201.

Livingstone MS, Hubel DH. 1987. Psychophysical evidence for separate channels for the perception of form, color, movement, and depth. J Neurosci 7:3416–3468.

Locke J. 1690. *An Essay Concerning Human Understanding: In Four Books*, Book 2, Chapter 1. London.

Mountcastle VB, Talbot WH, Kornhuber HH. 1966. The neural transformation of mechanical stimuli delivered to the monkey's hand. In: AVS de Reuck, J Knight (eds). *Ciba Foundation Symposium: Touch, Heat and Pain*, pp. 325–351. London: Churchill.

Ochoa J, Torebjörk E. 1983. Sensations evoked by intraneural microstimulation of single mechanoreceptor units innervating the human hand. J Physiol 342:633–654.

Raichle ME. 2011. The restless brain. Brain Connect 1:3–12.

Roy A, Steinmetz PN, Hsiao SS, Johnson KO, Niebur E. 2007. Synchrony: a neural correlate of somatosensory attention. J Neurophysiol 98:1645–1661.

Shepherd GM. 1994. *Neurobiology*, 3rd ed. New York: Oxford Univ. Press.

Smith SM, Beckmann CF, Andersson J, et al. 2013. Resting-state fMRI in the Human Connectome Project. Neuroimage 80:144–168.

Stevens CF. 2015. What the fly's nose tells the fly's brain. Proc Natl Acad Sci U S A 112:9460–9465.

Stevens CF. 2018. Conserved features of the primate face code. Proc Natl Acad Sci U S A 115:584–588.

Stringer C, Pachitariu M, Steinmetz N, Carandini M, Harris KD. 2019. High-dimensional geometry of population responses in visual cortex. Nature 571:361–365.

Swets JA. 1973. The relative operating characteristic in psychology: a technique for isolating effects of response bias finds wide use in the study of perception and cognition. Science 182:990–1000.

Swets JA. 1986. Indices of discrimination or diagnostic accuracy: their ROCs and implied models. Psychol Bull 99:100–117.

Talbot WH, Darian-Smith I, Kornhuber HH, Mountcastle VB. 1968. The sense of flutter-vibration: comparison of the human capacity with response patterns of mechanoreceptive afferents from the monkey hand. J Neurophysiol 31:301–334.

Tanner WP, Swets JA. 1954. A decision-making theory of visual detection. Psychol Rev 61:401–409.

Thurstone LL. 1927. A law of comparative judgment. Psychol Rev 34:273–286.

Ungerleider LG, Mishkin M. 1982. Two cortical visual systems. In: DG Ingle, MA Goodale, RJW Mansfield (eds). *Analysis of Visual Behavior*, pp. 549–586. Cambridge, MA: MIT Press.

Yamins DLK, DiCarlo JJ. 2016. Using goal-driven deep learning models to understand sensory cortex. Nat Neurosci 19:356–365.

18

Receptors of the Somatosensory System

NEUROPHYSIOLOGICAL STUDIES OF THE INDIVID-
UAL sensory modalities were first conducted
in the somatosensory system (Greek *soma*, the
body), the system that transmits information coded
by receptors distributed throughout the body. Charles
Sherrington, one of the earliest investigators of these
bodily senses, noted that the somatosensory system

serves three major functions: proprioception, exteroception, and interoception.

Proprioception is the sense of oneself (Latin *proprius*,
one's own). Receptors in skeletal muscle, joint capsules, and the skin enable us to have conscious awareness of the posture and movements of our own body,
particularly the four limbs and the head. Although one
can move parts of the body without sensory feedback
from proprioceptors, the movements are often clumsy,
poorly coordinated, and inadequately adapted to complex tasks, particularly if visual guidance is absent.

Exteroception is the sense of direct interaction with
the external world as it impacts the body. The principal mode of exteroception is the sense of *touch*, which
includes sensations of contact, pressure, stroking,
motion, and vibration, and is used to identify objects.
Some touch involves an active motor component—
stroking, tapping, grasping, or pressing—whereby a
part of the body is moved against another surface or
organism. The sensory and motor components of touch
are intimately connected anatomically in the brain and
are important in guiding behavior.

Exteroception also includes the *thermal senses*
of heat and cold. Thermal sensations are important
controllers of behavior and homeostatic mechanisms
needed to maintain the body temperature near 37°C
(98.6°F). Finally, exteroception includes the sense
of *pain*, or nociception, a response to external events
that damage or harm the body. Nociception is a prime
motivator of actions necessary for survival, such as
fight or flight.

The third component of somatic sensation, *interoception*, is the sense of the function of the major
organ systems of the body and its internal state.

The information conveyed by receptors in the viscera is crucial for regulating autonomic functions, particularly in the cardiovascular, respiratory, digestive, and renal systems, although most of the stimuli registered by these receptors do not lead to conscious sensations. Interoceptors are primarily chemoreceptors that monitor organ function through such indicators as blood gases and pH, and mechanoreceptors that sense tissue distention, which may be perceived as painful.

This diverse group of sensory functions may seem an unlikely combination to form a sensory system. We treat all of the somatic senses in one introductory chapter because they are mediated by one class of sensory neurons, the dorsal root ganglion (DRG) neurons. Somatosensory information from the skin, muscles, joint capsules, and viscera is conveyed by DRG neurons innervating the limbs and trunk or by trigeminal sensory neurons that innervate cranial structures (the face, lips, oral cavity, conjunctiva, and dura mater). These sensory neurons perform two major functions: the transduction and encoding of stimuli into electrical signals and the transmission of those signals to the central nervous system.

The study of somatic sensation has been revolutionized in the past 10 years by three important advances. First, the development of transgenic mice with fluorescent reporters of gene expression in DRG neurons has allowed neuroscientists to assess the physiological responses of specific receptor classes and their anatomical projections to sensory receptors in the body and in the central nervous system. Functional imaging of individual DRG neurons expressing genetically encoded calcium sensors such as GCaMP6 enables simultaneous optical recordings of activity from populations of receptor neurons innervating a specific region of the body, thereby providing a useful tool for analyzing ensemble responses to somatosensory stimuli. Second, studies of isolated DRG neurons in vitro, or in reduced skin-nerve preparations, enable biophysical assessment of receptor responses and characterization of ion channels expressed in individual somatosensory neurons. Third, the identification of Piezo protein ion channels as the molecular transducers of touch and proprioception in mammalian mechanoreceptors has provided a novel system for assessing the role of these channels in the senses of touch, proprioception, and visceral function.

In this chapter, we consider the principles common to all DRG neurons and those that distinguish their individual sensory function. We begin with a description of the peripheral nerves and their organization, followed by a survey of the receptor classes responsible for each of the major bodily senses. We also examine the sensory transduction mechanisms that convert various stimulus energies into electrical signals. We then describe the integration of information by the parent axon from multiple receptors in its receptive field and conclude with a discussion of the central processing centers for each submodality in the spinal cord and brain stem. Higher-order processing of touch, pain, proprioception, and autonomic regulation of viscera is described in later chapters.

Dorsal Root Ganglion Neurons Are the Primary Sensory Receptor Cells of the Somatosensory System

The cell body of a DRG neuron lies in a ganglion on the dorsal root of a spinal or cranial nerve. Dorsal root ganglion neurons originate from the neural crest and are intimately associated with the nearby segment of the spinal cord. Individual neurons in a DRG respond selectively to specific types of stimuli because of morphological and molecular specializations of their peripheral terminals.

Dorsal root ganglion neurons are a type of bipolar cell, called pseudo-unipolar cells. The axon of a DRG neuron has two branches, one projecting to the periphery and one projecting to the central nervous system (Figure 18–1). The peripheral terminals of individual DRG neurons innervate the skin, muscle, joint capsules, or viscera and contain receptors specialized for particular kinds of stimuli. The region of the body innervated by these sensory endings is called a *dermatome* (see Figure 18-13). Sensory peripheral nerve endings differ in receptor morphology and stimulus selectivity, allowing them to detect mechanical, thermal, or chemical events. The central branches terminate in the spinal cord or brain stem, forming the first synapses in somatosensory pathways. Thus, the axon of each DRG cell serves as a single transmission line with one polarity between the receptor terminal and the central nervous system. This axon is called the *primary afferent fiber*.

Individual primary afferent fibers innervating a particular region of the body, such as the thumb or fingers, are grouped together into bundles or fascicles of axons forming the *peripheral nerves*. They are guided during development to a specific location in the body by various trophic factors such as brain-derived neurotrophic factor (BDNF), neurotrophin-3 (NT3), neurotrophin-4 (NT4), or nerve growth factor (NGF). The peripheral nerves also include motor axons innervating nearby muscles, blood vessels, glands, or viscera.

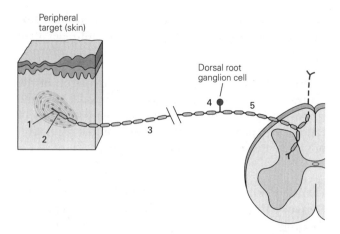

Figure 18–1 The dorsal root ganglion neuron is the primary sensory cell of the somatosensory system. The cell body is located in a dorsal root ganglion (**DRG**) adjacent to the spinal cord. The axon has two branches, one projecting to the body, where its specialized terminal contains receptors for a particular form of stimulus energy, and one projecting to the spinal cord or brain stem, where the afferent signals are processed. All DRG neurons contain five functional zones:
1. The *distal terminals* in skin, muscle, or viscera contain specialized receptor-channels that convert specific types of stimulus energy (mechanical, thermal, or chemical) into a depolarizing receptor potential. DRG neurons typically have multiple sensory endings. 2. The *spike generation site* contains voltage-gated Na^+ and K^+ channels (Na_V and K_V) that are located near the initial segment of the axon within the receptor capsule; they convert the receptor potential into a stream of action potentials. 3. The *peripheral nerve fiber* transmits action potentials from the spike initiation site to the DRG cell body. 4. The *cell body* of the DRG neuron is contained within a ganglion adjacent to the spinal cord or brain stem. 5. A *spinal or cranial nerve* connects the DRG or trigeminal neuron to the ipsilateral spinal cord or brain stem.

Damage to peripheral nerves or their targets in the brain may produce sensory deficits in more than one somatosensory submodality or motor deficits in specific muscle groups. Knowledge of where somatosensory modalities overlap morphologically, and where they diverge, facilitates diagnosis of neurological disorders and malfunction.

Each DRG neuron can be subdivided into five functional zones: the receptive zone, the spike generation site, the peripheral nerve fiber, the DRG cell body, and the spinal or cranial nerve (Figure 18–1). The receptive zone, at the distal end of the DRG axon, contains specialized receptor proteins that sense mechanical force, thermal events, or chemicals in the local environment and translate these signals into a local depolarization of the axonal terminals, called the *receptor potential* (see Figure 3–9A). This local depolarization spreads passively toward the central axon where action potentials

are generated, usually at the initial segment (distal to the first node of Ranvier in myelinated fibers) (see Figure 3–10A). Stimuli of sufficient strength produce action potentials that are transmitted along the peripheral nerve fiber, through the cell soma, and into the central branch that terminates in the spinal cord or brain stem.

The soma of a DRG neuron contains the cell nucleus. Sensory receptor proteins are expressed in the soma, providing a convenient expression system for characterizing their conductance properties in vitro. Isolated DRG neurons have been widely used in patch-clamp studies of sensory receptor currents and voltage-gated action potential channels.

DRG neurons differ in the size of their cell soma, gene expression profile, conduction velocity of their axons, sensory transduction molecule(s), innervation pattern in the body, and physiological function. For example, DRGs that innervate mechanoreceptors that sense touch and proprioception have the largest cell bodies and large myelinated axons; they express proteins such as Npy2r or parvalbumin (PV) (Figure 18–2). In contrast, DRG neurons that sense temperature or irritant chemicals have small cell bodies and unmyelinated axons; they express calcitonin gene-related peptide (CGRP) or the lectin IB4 (Figure 18–2C,D). As these fluorescent molecular labels extend through the axons to their peripheral endings in the body and in the central nervous system, David Ginty and colleagues were able to characterize the pattern of somatosensory nerve endings in the body (Figure 18–2H) and trace their central projections to the spinal cord (Figure 18–2G) and brain stem.

Peripheral Somatosensory Nerve Fibers Conduct Action Potentials at Different Rates

The peripheral nerves that transmit spike trains from the site of spike generation to the central nervous system have classically served as the primary recording sites for neurophysiological studies of somatosensory receptor mechanisms. Individual peripheral nerve fibers in animals are typically dissected from the main axon bundle and placed on fine wires that serve as recording electrodes. Microelectrodes—manufactured from sharpened tungsten or platinum wires—have also been inserted through the skin into the peripheral nerves of humans (a technique known as *microneurography*) to measure sensory responses to various somatic stimuli (Chapter 19).

Peripheral nerve fibers are classified into functional groups based on properties related to axon diameter

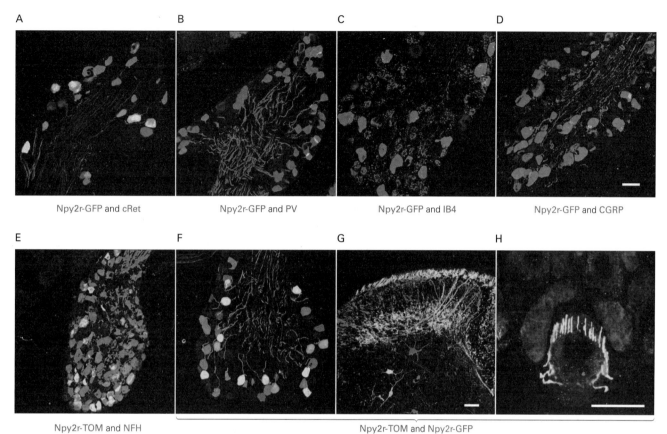

A   B   C   D

Npy2r-GFP and cRet   Npy2r-GFP and PV   Npy2r-GFP and IB4   Npy2r-GFP and CGRP

E   F   G   H

Npy2r-TOM and NFH   Npy2r-TOM and Npy2r-GFP

Figure 18–2 Dorsal root ganglion neurons differ in size, gene expression, and skin innervation patterns. (Reproduced, with permission, from Li et al. 2011. Copyright © 2011 Elsevier Inc.)

Panels **A–F** show double immunostaining of histological sections through a thoracic dorsal root ganglion. Individual dorsal root ganglion (DRG) neurons in these sections express genetic markers for specific classes of somatosensory nerve fibers. The G protein–coupled receptor Npy2r-GFP (**green**) or Npy2r-TOM (**red**) labels physiologically identified Aβ rapidly adapting low-threshold mechanoreceptors (Aβ RA-LTMRs). These fibers also express neurofilament heavy polypeptide (NFH), a marker of heavily myelinated axons (**E**), form longitudinal lanceolate (comb-like) endings surrounding individual guard hairs or awl/auchene hairs in hairy skin (**H**), and terminate in laminae III to V of the dorsal horn (**G**). Double-labeled neurons or fibers are stained **yellow**.

A. Aβ RA-LTMRs express the receptor tyrosine kinase *Ret* early in development (named early *Ret* and stained **red**). A majority of these neurons also express Npy2r-GFP (**green**); neurons that express both markers are stained **yellow**. Aβ RA-LTMRs have medium-sized cell bodies.

B. Aβ RA-LTMRs (**green**) have smaller cell bodies than proprioceptors such as muscle spindle afferents and Golgi tendon organs that express parvalbumin (PV, **red**).

C, D. Aβ RA-LTMRs (Npy2r-GFP, **green**) have larger cell bodies than unmyelinated purinergic C fibers that release ATP as cotransmitters (**IB4, red**) and peptidergic Aδ LTMRs that express calcitonin gene-related peptide (CGRP, **red**).

E. Heavily myelinated peripheral nerve fibers with large cell bodies express neurofilament heavy polypeptide (NFH, **green**). These include group Ia and Ib muscle afferents, Aβ SA-LTMRs, and Aβ RA-LTMRs (also labeled with Npy2r-tdTom [**red**]). Only Aβ RA-LTMRs express both markers and are stained **yellow**.

F–H. Double immunostaining with Npy2r-GFP (**green**) and Npy2r-tdTomato (**red**) of thoracic DRG neurons (**F**), their central processes in lamina III through V in the spinal cord dorsal horn (**G**), and their peripheral lanceolate endings at hair follicles in hairy skin sections (**H**) shows that the labeled peripheral and central Aβ RA-LTMR neurons largely overlap with each other (**yellow**) and that such genetic markers are useful for tracing sensory nerve endings.

and myelination, conduction velocity, and whether they are sensory or motor. The first nerve classification scheme was devised in 1894 by Charles Sherrington, who measured the diameter of myelin-stained axons in sensory nerves, and subsequently codified by David Lloyd (Table 18–1). They found two or three overlapping groups of axonal diameters (Figure 18–3). It was later discovered that these anatomical groupings are functionally important. Group I axons in *muscle* nerves innervate muscle spindle receptors and Golgi tendon

Table 18–1 Classification of Sensory Fibers in Peripheral Nerves[1]

	Muscle nerve	Cutaneous nerve[2]	Fiber diameter (μm)	Conduction velocity (m/s)
Myelinated				
Large diameter	I	Aα	12–20	72–120
Medium diameter	II	Aβ	6–12	36–72
Small diameter	III	Aδ	1–6	4–36
Unmyelinated	IV	C	0.2–1.5	0.4–2.0

[1]Sensory fibers from muscle are classified according to their diameter, whereas those from the skin are classified by conduction velocity.
[2]The types of receptors innervated by each type of fiber are listed in Table 18–2.

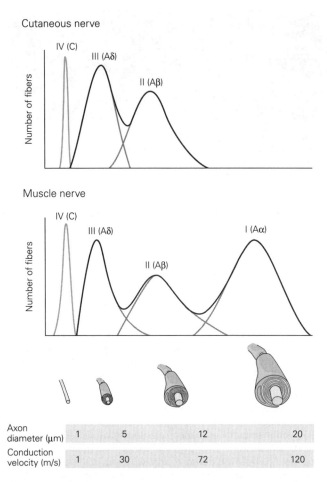

Axon diameter (μm): 1 5 12 20
Conduction velocity (m/s): 1 30 72 120

Figure 18–3 Classification of mammalian peripheral nerve fibers. The histograms illustrate the distribution of axon diameter for four groups of sensory nerve fibers innervating skeletal muscle and the skin. Each group has a characteristic axon diameter and conduction velocity (see Table 18–1). **Light blue lines** mark the boundaries of fiber profiles in each group in the zones of overlap. The conduction velocity (m/s) of myelinated peripheral nerve fibers is approximately six times the fiber diameter (μm). (Adapted, with permission, from Boyd and Davey 1968.)

organs, which signal muscle length and contractile force. Group II fibers innervate secondary spindle endings and receptors in joint capsules; these receptors also mediate proprioception. Group III fibers, the smallest myelinated muscle afferents, and the unmyelinated group IV afferents signal trauma or injuries in muscles and joints that are sensed as painful.

Nerves that innervate the skin contain two sets of myelinated fibers: Group II fibers innervate cutaneous mechanoreceptors that respond to touch, and group III fibers mediate thermal and noxious stimuli, as well as light touch in hairy skin. Unmyelinated group IV cutaneous afferents, like those in muscle, also mediate thermal and noxious stimuli.

Another method for classifying peripheral nerve fibers is based on electrical stimulation of whole nerves. In this widely used diagnostic technique, nerve conduction velocities are measured between pairs of stimulating and recording electrodes placed on the skin above a peripheral nerve. When studying conduction in the median or ulnar nerve, for example, the stimulation electrode might be placed at the wrist and the recording electrode on the upper arm. Brief electrical pulses applied through the stimulating electrode evoke action potentials in the nerve. The neural signal recorded a short time later in the arm represents the summed action potentials of all of the nerve fibers excited by the stimulus pulse and is called the *compound action potential* (Chapter 9). It increases in amplitude as more nerve fibers are stimulated; the summed activity is roughly proportional to the total number of active nerve fibers.

Electrical stimuli of increasing strength evoke action potentials first in the largest axons, because they have the lowest electrical resistance, and then progressively in smaller axons (Figure 18–4). Large-diameter fibers conduct action potentials more rapidly because

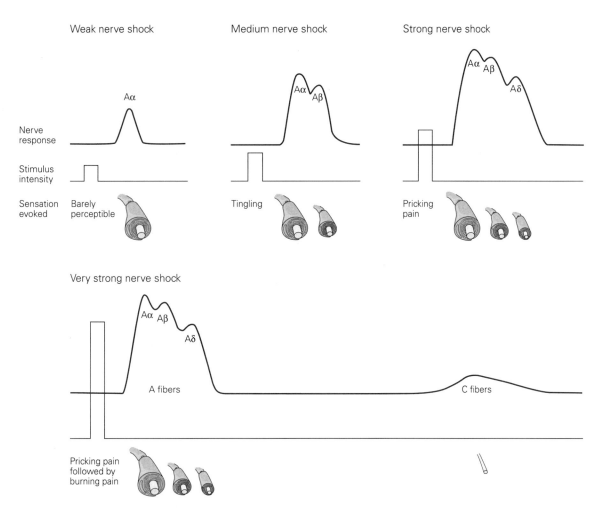

Figure 18–4 Conduction velocities of peripheral nerves are measured clinically from compound action potentials. Electrical stimulation of a peripheral nerve at varying intensities activates different types of nerve fibers. The action potentials of all the nerves stimulated by a particular amount of current are summed to create the compound action potential. The distinct conduction velocities of different classes of sensory and motor axons produce multiple peaks. (Adapted from Erlanger and Gasser 1938.)

the internal resistance to current flow along the axon is low, and the nodes of Ranvier are widely spaced along its length (Chapter 9). The conduction velocity of large myelinated fibers (in meters per second) is approximately six times the axon diameter (in micrometers), whereas thinly myelinated fibers conduct at five times the axon diameter. For unmyelinated fibers, the factor for converting axon diameter to conduction velocity is 1.5 to 2.5.

Following the stimulus artifact, the earliest neural signal recorded in the compound action potential occurs in fibers with conduction velocities greater than 90 m/s. Called the Aα wave (Figure 18–4), this signal reflects the action potentials generated in group I fibers and in motor neurons innervating skeletal muscle.

The sensation is barely perceived by the subject in the region innervated.

As more large fibers are recruited, a second signal, the Aβ wave, appears. This component corresponds to group II fibers in skin or muscle nerves that innervate mechanoreceptors mediating touch and proprioception and becomes larger as the shock intensity is increased. At higher voltages, when axons in the smaller Aδ range are recruited, the stimulus becomes painful, resembling an electric shock produced by static electricity. Voltages sufficient to activate unmyelinated C fibers evoke sensations of burning pain. As we shall learn later in this chapter, some Aδ and C fibers also respond to light touch on hairy skin, but such gentle tactile stimuli are masked by concurrent activation of pain fibers when whole nerves are stimulated electrically. Stimulation of motor neurons innervating the intrafusal fibers of muscle spindles (see Figure 18–9) evokes an intermediate wavelet called the Aγ wave, but this is usually difficult to discern because the conduction velocities of these motor neurons overlap those of Aβ and Aδ sensory axons. These differences in fiber diameter and conduction velocity of peripheral nerves allow signals of touch and proprioception to reach the spinal cord and higher brain centers earlier than noxious or thermal signals.

The clinician takes advantage of the known distribution of the conduction velocities of the various afferent fibers to diagnose diseases that result in sensory-fiber degeneration or motor neuron loss. In certain conditions, the loss of peripheral nerve axons is selective; in the neuropathy characteristic of diabetes, for example, the large-diameter sensory fibers degenerate. Such a selective loss is reflected in a reduction in the appropriate peak of the compound action potential, a slowing of nerve conduction, and a corresponding diminution of sensory capacity. Similarly, in multiple sclerosis, degeneration of the myelin sheath of large-diameter afferent fibers in the central nervous system results in slowing and, if severe enough, failure of nerve conduction.

A Variety of Specialized Receptors Are Employed by the Somatosensory System

The functional specialization of individual DRG neurons is determined by the molecular mechanisms of sensory transduction that occur at the distal nerve terminals in the body. When a somatic receptor is activated by an appropriate stimulus, its sensory terminal is typically depolarized. The amplitude and time course of the depolarization reflect the strength of the stimulus and its duration (see Figure 3–9A). Stimuli of sufficient strength produce action potentials that are transmitted along the peripheral branch of the DRG neuron's axon and into the central branch that terminates in the spinal cord or brain stem.

The sensory neurons that mediate touch and proprioception terminate in a nonneural capsule (Figure 18–1) or form morphologically distinctive endings surrounding hair follicles (Figure 18–2H) or intrafusal muscle fibers (see Figure 18–9A). They sense mechanical stimuli that indent or stretch their receptive surface. In contrast, the peripheral axons of neurons that detect noxious, thermal, or chemical events have unsheathed endings with multiple branches that terminate in the epidermis or in the viscera.

Several different morphologically specialized receptors underlie the various somatosensory submodalities. For example, the median nerve that innervates the skin of the hand and some of the muscles controlling the hand contains tens of thousands of nerve fibers that can be classified into 30 functional types. Of these, 22 types are afferent fibers (sensory axons conducting impulses toward the spinal cord), and eight types are efferent fibers (motor axons conducting impulses away from the spinal cord to skeletal muscle, blood vessels, and sweat glands). The afferent fibers convey signals from eight kinds of cutaneous mechanoreceptors that are sensitive to different kinds of skin deformation; five kinds of proprioceptors that signal information about muscle force, muscle length, and joint angle; four kinds of thermoreceptors that report the temperatures of objects touching the skin; and four kinds of nociceptors that signal potentially injurious stimuli. The major receptor groups within each submodality are listed in Table 18–2.

Mechanoreceptors Mediate Touch and Proprioception

A mechanoreceptor senses physical deformation of the tissue surrounding it. Mechanical distension—such as pressure on the skin, stretch of muscles, suction applied directly to cell membranes, or osmotic swelling of tissue—is transduced into electrical energy by the physical action of the stimulus on mechanoreceptor ion channels in the membrane. Mechanical stimulation deforms the receptor protein, thus opening stretch-sensitive ion channels and increasing nonspecific cation conductances that depolarize the receptor neuron (see Figure 3–9A). Removal of the stimulus relieves mechanical stress on the receptor and allows stretch-sensitive channels to close.

Various mechanisms for activation of mechanoreceptor ion channels have been proposed. Some mechanoreceptors appear to respond to forces conveyed through tension or deformation of the lipids of the plasma membrane, a mechanism called *force from lipids* (Figure 18–5A). Here, deformation of membrane lipids changes the cell surface curvature, exposing hydrophobic residues in the receptor protein to the membrane phospholipids, thereby opening the channel pore to cation flow. This may be the mechanism for detection of cellular swelling, which plays an important role in osmoregulation, or changes in shear stress on the walls of blood vessels due to altered fluid flow.

Another postulated mechanism for activation of mechanoreceptors involves linking the channel protein to the surrounding tissue through structural proteins,

Table 18–2 Receptor Types Active in Somatic Sensory Processing

Receptor type	Fiber group[1]	Fiber name	Receptor	Marker(s)	Modality
Cutaneous mechanoreceptors					Touch
Meissner corpuscle	Aα,β	RA1	Piezo2	*cRet/Npy2r/* NFH	Stroking, flutter
Merkel disk receptor	Aα,β	SA1	Piezo2	Troma1/ Keratin8/*Npy2r*	Pressure, texture
Pacinian corpuscle[2]	Aα,β	RA2	Piezo2	*cRet/Npy2r/* NFH	Vibration
Ruffini ending	Aα,β	SA2	Piezo2		Skin stretch
Hair (guard)	Aα,β	Aβ RA-LTMR	Piezo2	*cRet/Npy2r/* NFH	Stroking, hair movement
Hair (awl/auchene)	Aδ	Aδ-LTMR	Piezo2	TrkB	Light stroking, air puff
Field receptor (circumferential endings)	Aβ	Aβ Field-LTMR	Piezo2	NFH	Skin stretch
Hair (zigzag)	C	C-LTMR		TH	Slow stroking, gentle touch
Thermal receptors					Temperature
Cool receptors	Aδ	III	TRPM8		Skin cooling (<25°C)
Warm receptors	C	IV	TRPV3		Skin warming (>35°C)
Heat nociceptors	Aδ	III	TRPV1/ TRPV2		Hot temperature (>45°C)
Cold nociceptors	C	IV	TRPA1/ TRPM8		Cold temperature (<5°C)
Nociceptors					Pain
Mechanical	Aδ	III		CGRP	Sharp, pricking pain
Thermal-mechanical (heat)	Aδ	III	TRPV2		Burning pain
Thermal-mechanical (cold)	C	IV	TRPV1/ TRPA1	IB4	Freezing pain
Polymodal	C	IV	TRPV1/ TRPA1		Slow, burning pain
Muscle and skeletal mechanoreceptors					Limb proprioception
Muscle spindle primary	Aα	Ia	Piezo2	PV/NFH	Muscle length and speed
Muscle spindle secondary	Aβ	II	Piezo2	PV/NFH	Muscle stretch
Golgi tendon organ	Aα	Ib	Piezo2	PV/NFH	Muscle contraction
Joint capsule receptors	Aβ	II			Joint angle
Stretch-sensitive free endings	Aδ	III			Excess stretch or force

[1]See Table 18–1.
[2]Pacinian corpuscles are also located in the mesentery, between layers of muscle, and on interosseous membranes.

a mechanism termed *force from filaments* (Figure 18–5B). In this arrangement, mechanical force applied to the skin or muscle by direct pressure or lateral stretch of the tissue distorts the extracellular matrix or intracellular cytoskeletal proteins (actin, integrins, microtubules). These tethering molecules interact with the receptor-channel proteins, change their conformation, and open cation channels. The extracellular linkage to the channel proteins is elastic and often represented as a spring-loaded gate. Direct channel gating in this model may be produced by forces that stretch the extracellular linkage protein. The channel closes when the force is removed. This type of direct channel gating is used by hair cells of the inner ear and by some touch receptors in the skin.

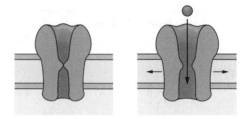

A Direct activation through lipid tension

Figure 18–5 Ion channels in mechanoreceptor nerve terminals are activated by mechanical stimuli that stretch or deform the cell membrane. Mechanical displacement leads to channel opening, permitting the influx of cations. (Adapted, with permission, from Lin and Corey 2005. Copyright © 2005 Elsevier Ltd.)

A. Force from lipids. Channels can be directly activated by forces conveyed through lipid tension in the cell membrane, such as changes in blood pressure.

B. Force from filaments. Forces conveyed through structural proteins linked to the ion channel can also directly activate mechanosensory channels. The linking structural proteins may be extracellular (attached to the surrounding tissue) or intracellular (bound to the cytoskeleton) or both.

B Direct activation through structural proteins

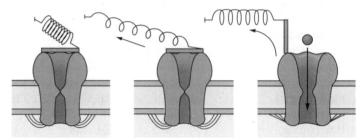

It is remarkable that although the receptor end organs for touch in the skin were first studied by Edgar Adrian and Yngve Zotterman in the 1920s and receptor potentials were recorded from isolated touch receptors from the mesentery (Pacinian corpuscles) in the 1960s, there was little consensus about the molecular biology of mechanosensation in mammalian touch. The leading candidates were derived from invertebrate model organisms such as the nematode worm *Caenorhabditis elegans* whose touch receptors were identified as members of the degenerin superfamily of ion channels and are similar to vertebrate epithelial Na^+ channels (DEG/ENac channels). Other candidate molecules included TRPV4 receptors (members of the transient receptor potential [TRP] receptors that are also involved in thermal senses), and NOMPC, a *Drosophila* member of the TRPN family. However, these molecules are not expressed in mammalian DRG neurons.

The Piezo protein family of transmembrane ion channels was recently identified by Ardem Patapoutian and colleagues as molecular mediators of mechanoreception in mammals. Piezo1 proteins are composed of approximately 2,500 amino acids, with at least 26 transmembrane α-helices (Figure 18–6A). The ion channel is a trimer formed from three identical Piezo protein subunits, with two pore-forming α-helices at the C-terminal end of each Piezo protein. The N-terminals of the subunits form a propeller-like structure (Figure 18–6B), which is thought to be involved in coupling mechanical stimuli to channel gating. Piezo proteins

form nonspecific cation-permeable channels that conduct excitatory depolarizing current.

Two different isoforms of the Piezo proteins serve as mechanosensors: Piezo1 is found primarily in nonneural tissue, such as epithelia in blood vessels, the kidney, and bladder, and in red blood cells. Piezo2 is expressed in mechanosensory DRG and trigeminal neurons that mediate the senses of touch and proprioception and in vagal afferents innervating smooth muscle of the lung, where they mediate the Hering-Breuer reflex by sensing lung stretch (Chapter 32).

Specialized End Organs Contribute to Mechanosensation

In addition to the molecular composition of the ion channels expressed in the distal nerve endings, components of surrounding tissue such as epithelial cells or muscle fibers play a significant role in mechanotransduction. The specialized nonneural end organs that surround the nerve terminals of a DRG neuron must be deformed in specific ways to excite the fiber. For example, individual mechanoreceptors respond selectively to pressure or motion, and thereby detect the direction of force applied to the skin, joints, or muscle fibers. The end organ can also amplify or modulate the sensitivity of the receptor axon to mechanical displacement.

Specialized epithelial cells in the skin—such as Merkel cells, the epithelium lining hair follicles, and the papillary ridges that form the fingerprints of glabrous

A Molecular organization of the Piezo1 protein

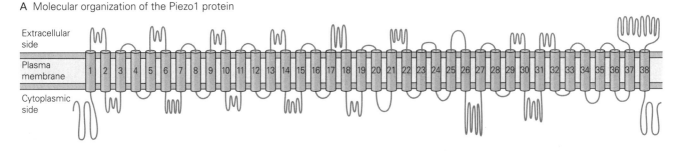

B Structure of the Piezo1 ion channel

1 Side view

2 Top-down view

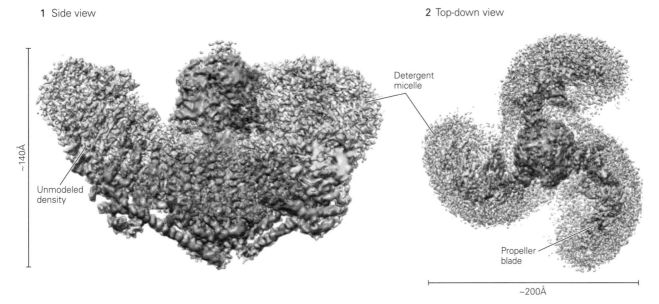

Figure 18–6 Structure and molecular organization of Piezo1 ion channels.

A. Piezo1 and Piezo2 have homologous protein structures, containing approximately 2,500 amino acids, with at least 26 putative transmembrane segments. Combined as trimers, they form the largest membrane ion channels in mammals. (Adapted, with permission, from Murthy, Dubin, and Patapoutian 2017. Copyright © 2017 Springer Nature.)

B. Putative structure of the Piezo1 ion channel deduced from cryo-electron microscopy. **1.** Side view, cytoplasmic surface down. **2.** Top-down view from extracellular side. The receptor is a triskelion made up of three identical Piezo1 subunits. The C-terminals of the three Piezo proteins form a central extracellular cap tethered to the extracellular surface of the transmembrane pore, which extends beyond the membrane into a cytoplasmic tail domain. The aqueous pore through the channel extends through the central axis of the cap, the transmembrane pore, and the cytoplasmic tail domain. The N-terminals of the three protein subunits are arrayed peripherally, forming a propeller-like helical structure. **Blue** indicates area of high-resolution modeling. (Adapted, with permission, from Saotome et al. 2018. Copyright © 2018 Springer Nature.)

skin—play important auxiliary roles in the sense of touch. The best studied of these end organs are Merkel cells—sensory epithelial cells that form close contacts with the terminals of large-diameter (Aβ) sensory nerve axons at the epidermal–dermal junction, forming Merkel cell–neurite complexes. Merkel cells cluster in swellings of the epidermis in hairy skin called touch domes (Figure 18–7A) and near the center of the fingerprint ridges in glabrous skin (see Figure 19–3). When a probe contacts the touch dome, the sensory

nerve responds with a train of action potentials whose frequency is proportional to the velocity and amplitude of pressure applied to the skin (Figure 18–7A2). These spike trains typically last throughout the period of stimulation and are termed *slowly adapting* because the firing persists for periods of up to 30 minutes. Likewise, the sensory nerve is called an *SA1 fiber* (slowly adapting type 1 fiber).

Merkel cells serve a similar receptive function in the sense of touch as auditory hair cells in the cochlea

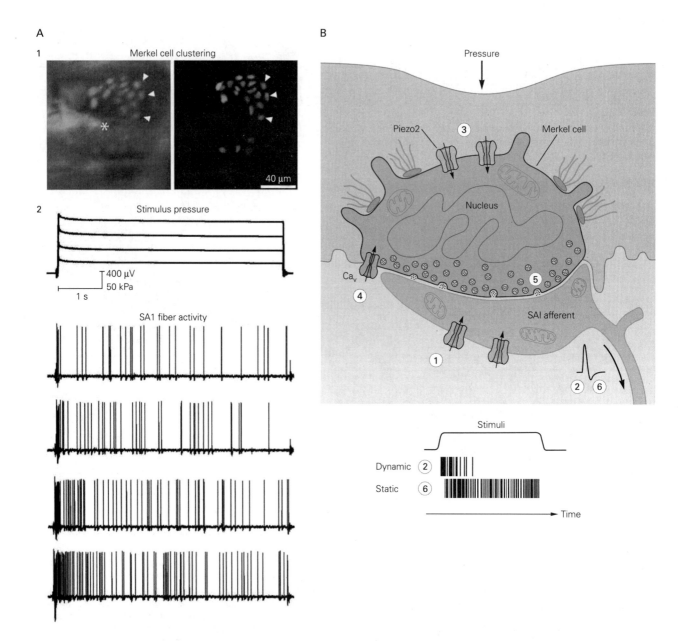

Figure 18–7 Afferent fibers innervating Merkel cells respond continuously to pressure on the skin.

A. 1. An individual slowly adapting type 1 (**SA1**) mechanoreceptive fiber innervates a cluster of 22 Merkel cells (each labeled with enhanced green fluorescent protein [eGFP]) in a touch dome of the hairy skin. *Left*: In vivo epifluorescent images of the isolated skin-nerve recording preparation. Asterisk (*) indicates the location of the associated guard hair within the touch dome. *Right*: Confocal z-series projections of the entire touch dome innervated by the SA1 fiber. **Arrowheads** are used to align Merkel cells in the two images. **2.** The SA1 fiber responds to 5-second duration steps of pressure (measured in kilopascals [kPa]) applied over the touch dome (upper records) with irregular, slowly adapting spike trains (recorded extracellularly), whose mean frequency of firing is proportional to the applied force (lower records). The neuron fires at its highest rate at the

start of stimulation and fires fewer spikes during maintained pressure. (Reproduced, with permission, from Wellnitz et al. 2010.)

B. A model of sensory transduction in SA1 mechanoreceptors. Pressure on the skin opens Piezo2 channels (**blue**) in the Merkel cell and in the peripheral neurite of the SA1 fiber that receives synaptic input from the Merkel cell. Piezo2 channels in the neurite open at the onset of stimulation (**1**) generating the initial dynamic response to touch (**2**). Skin deformation simultaneously activates Piezo2 channels in the Merkel cell (**3**), depolarizing it and allowing voltage-gated Ca_V channels in the Merkel cell (**4**) to open and release neurotransmitter continuously (**5**). Binding of the neurotransmitter further depolarizes the SA1 neurite, producing sustained firing in the principal axon (**6**). (Reproduced, with permission, from Maksimovic et al. 2014. Copyright © 2014 Springer Nature.)

(Chapter 26) and taste cells in the tongue (Chapter 32). Merkel cells studied in vitro respond to mechanical force such as pressure or suction with depolarizing currents that are similar in time course and conductance to those evoked in isolated DRG neurons. They express synaptic release proteins and contain vesicles that release excitatory neurotransmitters during sustained pressure. Merkel cells express Piezo2 proteins and show increased cytoplasmic Ca^{2+} levels when stimulated by pressure.

The importance of Merkel cells for physiological responses to touch is seen in mice that fail to develop Merkel cells in the epidermis (*Atoh1* conditional knockout mice). The firing rates of SA1 fibers in these animals are reduced in amplitude and duration compared to wild-type. These experiments indicate that Merkel cells are responsible for the sustained response to static touch. Recently, Ellen Lumpkin and colleagues used optogenetic stimulation of Merkel cells rather than direct pressure on the skin to demonstrate that SA1 fibers innervating touch domes use a dual-mechanism to sense pressure on the skin (Figure 18–7B). The initial dynamic response to touch is generated primarily by current flow through Piezo2 channels in the SA1 nerve terminal. The subsequent static response results from excitatory synaptic transmission from Merkel cells that express Piezo2 channels and continuously release neurotransmitter during sustained pressure on the skin.

Hairs that protrude from the surface of the skin provide another important set of touch end organs. Sensory hair fibers are extremely sensitive to motion. Deflection of hairs by light breezes or air puffs evokes one or more action potentials from hair follicle afferent fibers. Humans can perceive motion of individual hairs and localize the sensation to the base of the hair, where it emerges from the skin. Sensory hairs serve an important protective function as they detect objects, other organisms, or obstacles in the environment at a distance before they impact the body. Hairs or sensory antennae detect important object features such as texture, curvature, and rigidity that aid recognition as friend or foe. These neurons are named *rapidly adapting low-threshold mechanoreceptors* (RA-LTMRs) because they respond to gentle touch or hair movement with brief bursts of spikes when the hair is moved by external forces.

Hairs are embedded in skin invaginations called hair follicles. Three types of hairs are found in mammalian skin (Figure 18–8A). The largest, longest, and stiffest hairs (named guard hairs) are the first to emerge from the skin during development. *Guard hairs* are innervated by the largest-diameter and fastest-conducting

sensory nerve fibers (type Aβ); these fibers form lanceolate (comb-like) endings in the epidermis of the follicle surrounding the hair (Figure 18–2H). Aβ RA-LTMR nerve fibers also innervate intermediate-sized hairs (called *awl/auchene hairs*) with lanceolate endings. Awl/auchene hairs are triply innervated: They provide inputs to fast-conducting (Aβ) myelinated fibers; smaller-diameter, slower-conducting myelinated (Aδ) fibers; and unmyelinated C fibers. The smallest and most numerous hairs (called *zigzag* or *down hairs*) are also innervated by Aδ and C fibers.

Until recently, Aδ and C fibers were thought to mediate only thermal or painful sensations. However, microneurography studies in humans by Johan Wessberg, Håkan Olausson, and Åke Vallbo demonstrated that hairy skin is also innervated by unmyelinated C-LTMR fibers that respond to slowly moving tactile stimuli and are thought to mediate social or pleasurable touch. They may also play a role in pain inhibition in the spinal cord dorsal horn.

The innervation pattern of hair follicles in the skin illustrates two important principles of sensory innervation of the body: convergence and divergence. Each individual hair follicle in the skin provides input to multiple sensory afferent fibers. This pattern of overlap provides redundancy of sensory input from a small patch of skin. Shared lines of communication innervate each hair follicle, rather than a single labeled line. Tactile information from the skin is therefore transmitted in parallel by an ensemble of sensory neurons.

The skin area innervated by the sensory nerve terminals of a DRG neuron defines the cell's *receptive field*, the region of the body that can excite the cell. Each sensory nerve fiber collects information from a wide area of skin because its distal terminals have multiple branches that can be activated independently. This morphology enables each afferent fiber to provide unique patterns of sensory input to the brain.

The diversity of receptive field sizes and territories encompassed by individual classes of DRG neurons is illustrated in Figure 18–8B. Because of the large size of tactile receptive fields, gentle touch excites many different sensory fibers at the site of contact, each conveying a specific sensory message. The smallest tactile receptive fields are the touch domes innervated by SA1 fibers (see Figure 19–8B Aβ SA1). An individual SA1 fiber innervates all of the Merkel cells in a touch dome and typically collects information from one to three touch domes in adjacent skin regions. Hair follicles innervated by individual RA fibers are spread further apart and the sensory endings differ somewhat in size, with the largest-diameter Aβ fibers encompassing the

A Hairy skin innervation

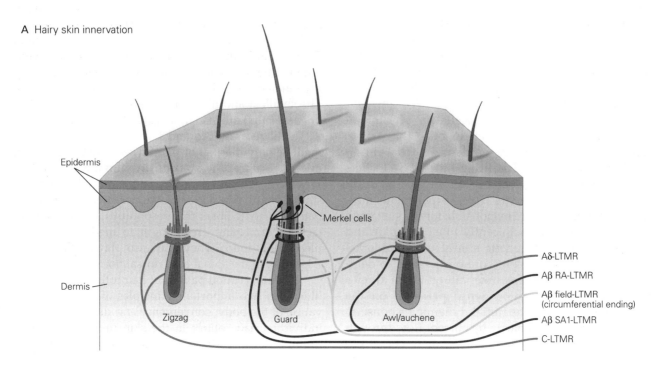

Epidermis

Merkel cells

Dermis

Zigzag   Guard   Awl/auchene

Aδ-LTMR

Aβ RA-LTMR

Aβ field-LTMR
(circumferential ending)

Aβ SA1-LTMR

C-LTMR

B Receptive fields of low-threshold mechanoreceptors

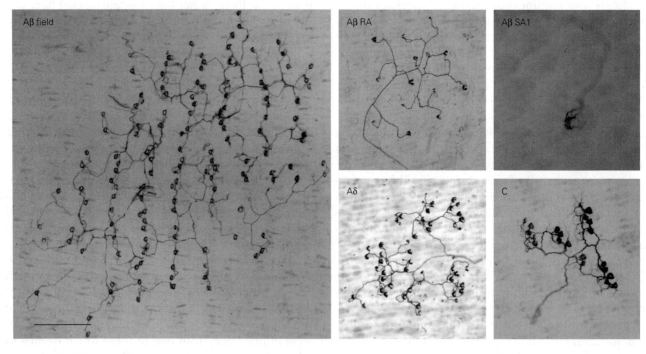

Aβ field

Aβ RA

Aβ SA1

Aδ

C

smallest hair follicle receptive fields (see Figure 19–8B Aβ RA). The largest receptive fields in the skin are those of Aβ field receptors (see Figure 19–8B Aβ Field). These fibers form circumferential endings around hair follicles but do not respond to hair movement or air puffs. Instead, field receptors respond to stroking or stretching of the skin in their receptive fields. Field receptors are also excited by painful stimuli such as pulling hairs or strong pressure, suggesting that they may also mediate sensations of mechanical pain.

Proprioceptors Measure Muscle Activity and Joint Positions

Mechanoreceptors in muscles and joints convey information about the posture and movements of the body and thereby play an important role in proprioception and motor control. Mechanical coupling of sensory nerve terminals to skeletal muscle, tendons, joint capsules, and the skin is thought to underlie proprioception. These receptors include two types of muscle-length sensors, the type Ia and II muscle spindle endings; one muscle force sensor, the Golgi tendon organ; joint-capsule receptors, which transduce tension in the joint capsule; and Ruffini endings that sense skin stretch over joints.

The muscle spindle consists of a bundle of thin muscle fibers, or intrafusal fibers, that are aligned parallel to the larger fibers of the muscle and enclosed within a capsule (Figure 18–9A). The intrafusal fibers are entwined by a pair of sensory axons that detect muscle stretch because of mechanoreceptive ion channels in the nerve terminals. Intrafusal muscles also receive inputs from motor axons that regulate contractile force and receptor sensitivity. (See Box 32–1 for details on muscle spindles.)

Although the receptor potential and firing rates of muscle spindle afferent fibers are proportional to muscle length (Figure 18–9B), these responses can be modulated by higher centers in the brain that regulate contraction of intrafusal muscles. Spindle afferent fibers are thus able to encode the amplitude and speed of internally generated voluntary movements as well as passive limb displacement by external forces (Chapter 32).

Golgi tendon organs, located at the junction between skeletal muscle and tendons, measure the forces generated by muscle contraction. (See Box 32–4 for details on Golgi tendon organs.) Although these receptors play an important role in reflex circuits modulating muscle force, they appear to contribute little to conscious sensations of muscle activity. Psychophysical experiments in which muscles are fatigued or partially paralyzed have shown that perceived muscle force is mainly related to centrally generated effort rather than to actual muscle force.

Recent studies by Ardem Patapoutian and colleagues suggest that Piezo2 mediates the signals transmitted by afferent fibers from muscle spindles and Golgi tendon organs, as these fibers express the Piezo2 protein in their distal terminals and cell body.

Joint receptors play little if any role in postural sensations of joint angle. Instead, perception of the angle of proximal joints such as the elbow or knee

---

Figure 18–8 (Opposite) Innervation of the hairy skin by low-threshold mechanoreceptors.

A. Hairy skin of mammals is innervated by specific combinations of low-threshold mechanoreceptors (**LTMRs**); these multiple classes of nerve fibers allow touch information to be transmitted along multiple parallel nerve fibers to the central nervous system. Touch domes of Merkel cells are located at the epidermal–dermal boundary surrounding large-diameter guard hairs. The axons of Merkel cells are classified as Aβ SA1-LTMRs, and they compose approximately 3% of sensory fibers innervating hairy skin. Guard hair follicles are innervated by rapidly adapting touch fibers, classified as Aβ RA-LTMRs, which form longitudinal lanceolate (comb-like) endings surrounding the hair follicle. They compose another 3% of sensory fibers innervating hairy skin. Aβ RA-LTMR fibers also form lanceolate endings on medium size awl/auchene hairs; each fiber innervates multiple hair follicles in neighboring regions of skin. Awl/auchene hairs are innervated by lanceolate endings from three different classes of sensory fibers; Aβ RA-LTMRs (**blue**), Aδ-LTMRs (**red**, 7% of fibers), and C-LTMRs (**green**, 15%–27% of fibers). Zigzag, or down hairs, are the most numerous type;

they are innervated by the smallest-diameter, slowest-conducting peripheral nerve fibers (Aδ- and C-LTMRs). All three types of hair follicles are also innervated by circumferential endings (**yellow**). (Reproduced with permission from Zimmerman, Bai, and Ginty 2014. Copyright © 2014 AAAS.)

B. Whole mount sections of skin illustrate the spread of LTMR sensory nerve terminals in hairy skin and the skin region that can activate an individual sensory fiber. All five classes innervate multiple hair follicles and have branched sensory nerve endings. Scale bar (which applies to all images) = 500 μm. Firing rates in each of these axons reflect inputs from multiple receptor end organs in the skin. Aβ field-LTMRs form circumferential endings around all classes of hair follicles; they have the largest receptive fields in hairy skin, innervating up to 180 hair follicles/fibers and spanning areas up to 6 mm². Aβ SA1-LTMRs have the smallest receptive fields but innervate all of the Merkel cells within a touch dome; each touch dome is innervated by only a single Aβ SA1-LTMR. Aβ RA- Aδ-, and C-LTMRs form lanceolate endings enclosing up to 40 individual hair follicles and span skin areas of 0.5 to 4 mm². (Reproduced, with permission, from Bai et al. 2015. Copyright © 2015 Elsevier Inc.)

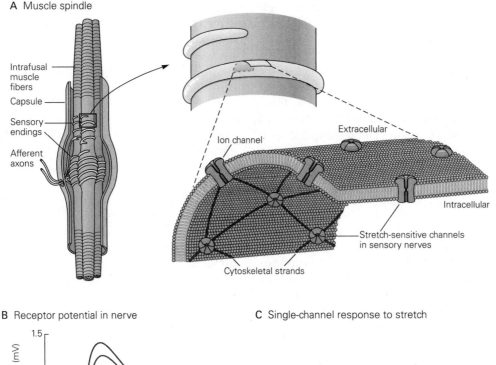

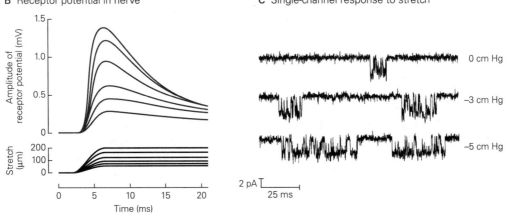

Figure 18–9 The muscle spindle is the principal receptor for proprioception.

A. The muscle spindle is located within skeletal muscle and is excited by stretch of the muscle. It consists of a bundle of thin (intrafusal) muscle fibers entwined by a pair of sensory axons. It is also innervated by several motor axons (not shown) that produce contraction of the intrafusal muscle fibers. Stretch-sensitive ion channels in the sensory nerve terminals are linked to the cytoskeleton by the protein spectrin. (Adapted, with permission, from Sachs 1990.)

B. The depolarizing receptor potential recorded in a group Ia fiber innervating the muscle spindle is proportional to both the velocity and amplitude of muscle stretch parallel to the myofilaments. When stretch is maintained at a fixed length, the receptor potential decays to a lower value. (Adapted, with permission, from Ottoson and Shepherd 1971.)

C. Patch-clamp recordings of a single stretch-sensitive channel in myocytes. Pressure is applied to the receptor cell membrane by suction. At rest (0 cm Hg) the channel opens sporadically for short time intervals. As the pressure applied to the membrane increases, the channel opens more often and remains in the open state longer. This allows more current to flow into the receptor cell, resulting in higher levels of depolarization. (Adapted, with permission, from Guharay and Sachs 1984. Copyright © 1984 The Physiological Society.)

depends on afferent signals from muscle spindle receptors and efferent motor commands. Additionally, conscious sensations of finger position and hand shape depend on cutaneous stretch receptors as well as muscle spindles.

Thermal Receptors Detect Changes in Skin Temperature

Although the size, shape, and texture of objects held in the hand can be apprehended visually as well as by touch, the thermal qualities of objects are uniquely somatosensory. Humans recognize four distinct types of thermal sensation: cold, cool, warm, and hot. These sensations result from differences between the normal skin temperature of approximately 32°C (90°F) and the external temperature of the air or of objects contacting the body. Temperature sense, like the other *protopathic* modalities of pain and itch, is mediated by a *combinatorial code* of multiple receptor types, transmitted by small-diameter afferent fibers.

Although humans are exquisitely sensitive to sudden changes in skin temperature, we are normally unaware of the wide swings in skin temperature that occur as our cutaneous blood vessels expand or contract to discharge or conserve body heat. If skin temperature changes slowly, we are unaware of changes in the range 31° to 36°C (88–97°F). Below 31°C (88°F), the sensation progresses from cool to cold and, finally,

beginning at 10° to 15°C (50–59°F), to pain. Above 36°C (97°F), the sensation progresses from warm to hot and then, beginning at 45°C (113°F), to pain.

Thermal sensations are mediated by free nerve endings in the epidermis. The temperature ranges signaled by these nerve fibers are determined by the molecular composition of receptor molecules expressed in the distal nerve terminals and cell bodies of small-diameter DRG neurons. Studies by David Julius and his colleagues revealed that thermal stimuli activate specific classes of *transient receptor potential (TRP) channels* in these neurons (Figure 18–10). TRP channels are encoded by genes belonging to the same gene superfamily as the voltage-gated channels that give rise to the action potential (Chapter 8). They form nonselective cation channels that mediate inward depolarizing current. TRP channels comprise four identical protein subunits, each of which contains six transmembrane α-helices,

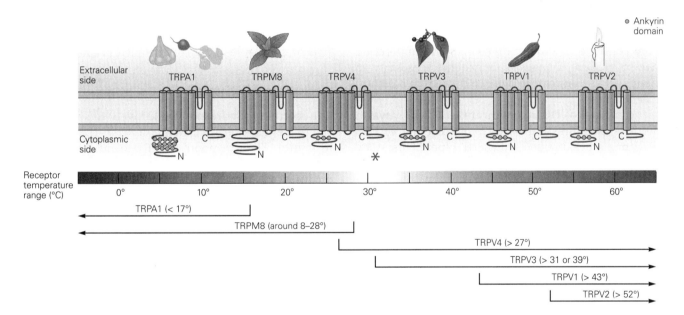

Figure 18–10 Transient receptor potential ion channels. TRP channels consist of membrane proteins with six transmembrane α-helices. A pore is formed between the fifth (S5) and sixth (S6) helices from the four subunits. Most of these receptors contain ankyrin repeats in the N-terminal domains and a common 25-amino acid motif adjacent to S6 in the C-terminal domain. Individual TRP channels are composed of four identical TRP proteins. All TRP channels are gated by temperature and various chemical ligands, but different types respond to different temperature ranges and have different activation thresholds. At least six types of TRP receptors have been identified in sensory neurons; the thermal sensitivity of a neuron is determined by the particular TRP receptors expressed in its nerve terminals. At 32°C (90°F), the resting skin temperature (**asterisk**), only TRPV4 and some TRPV3 receptors are

stimulated. TRPA1 and TRPM8 receptors are activated by cooling and cold stimuli. TRPM8 receptors also respond to menthol and various mints; TRPA1 receptors respond to allium-expressing plants such as garlic and radishes. TRPV3 receptors are activated by warm stimuli and also bind camphor. TRPV1 and TRPV2 receptors respond to heat and produce burning pain sensations. TRPV1 channels also respond to a variety of substances, temperatures, or forces that can elicit pain. Their sites of action on the receptor include binding sites for chili peppers' active ingredient (capsaicin), acids (lemon juice), spider venoms, and phosphorylation sites for second messenger-activated kinases. TRPV4 receptors are active at normal skin temperatures and respond to touch. (Adapted, with permission, from Jordt, McKemy, and Julius 2003; adapted from Dhaka, Viswanath, and Patapoutian 2006.)

with a pore-forming element between the fifth and sixth helices. Individual TRP receptors are distinguished by their sensitivity to heat or cold, showing sharp increases in conductance to cations when their thermal threshold is exceeded. Their names specify the genetic subfamily of TRP receptors and the member number. Examples include TRPV1 (for TRP vanilloid-1), TRPM8 (for TRP melastatin-8), and TRPA1 (for TRP ankyrin-1).

Two classes of TRP receptors are activated by cold temperatures and inactivated by warming. TRPM8 receptors respond to temperatures below 25°C (77°F); such temperatures are perceived as cool or cold. TRPA1 receptors have thermal thresholds below 17°C (63°F); this range is described as cold or frigid. Both TRPM8 and TRPA1 receptors are expressed in high-threshold cold receptor terminals, but only TRPM8 receptors are expressed in low-threshold cold receptor terminals.

Thermal signals from low-threshold cold receptors are transmitted by small-diameter, myelinated Aδ fibers with unmyelinated endings within the epidermis. These fibers express the transient receptor potential channel TRPM8 and respond to menthol applied to the skin. Cold receptors are approximately 100 times more sensitive to sudden drops in skin temperature than to gradual changes. This extreme sensitivity to change allows humans to detect a draft from a distant open window.

Four types of TRP receptors are activated by warm or hot temperatures and inactivated by cooling. TRPV3 receptors are expressed in warm type fibers; they respond to warming of the skin above 35°C (95°F) and generate sensations ranging from warm to hot. TRPV1 and TRPV2 receptors respond to temperatures exceeding 45°C (113°F) and mediate sensations of burning pain; they are expressed in heat nociceptors. TRPV4 receptors are active at temperatures above 27°C and signal normal skin temperatures.

Warm receptors are located in the terminals of C fibers that end in the dermis. Unlike the cold receptors, warm receptors act more like simple thermometers; their firing rates rise monotonically with increasing skin temperature up to the threshold of pain and then saturate at higher temperatures. Warm receptors are less sensitive to rapid changes in skin temperature than cold receptors. Consequently, humans are less responsive to warming than cooling; the threshold change for detecting sudden skin warming, even in the most sensitive subject, is about 0.1°C.

Heat nociceptors are activated by temperatures exceeding 45°C (113°F) and inactivated by skin cooling. The burning pain caused by high temperatures is transmitted by both myelinated Aδ fibers and unmyelinated C fibers.

The role of TRP receptors in thermal sensation was originally discovered by analyses of natural substances such as capsaicin and menthol that produce burning or cooling sensations when applied to the skin or injected subcutaneously. Capsaicin, the active ingredient in chili peppers, has been used extensively to activate nociceptive C fiber afferents that mediate sensations of burning pain. These studies indicate that the various TRP receptors also bind other molecules that induce painful sensations, such as toxins, venoms, and substances released by diseased or injured tissue. TRPA1 receptors bind pungent substances such as horseradish (wasabi), garlic, onions, and similar allium-expressing plants. These substances behave as irritants that may produce pain or itch through covalent modification of cysteines in the TRPA1 protein.

TRP channels are polymodal sensory integrators, because different sections of the protein respond directly to changes in temperature, pH, or osmolarity; to the presence of noxious substances such as capsaicin or toxins; or to phosphorylation by intracellular second messengers (see Figure 20–2). Their molecular structure and role in pain are detailed in Chapter 20.

Nociceptors Mediate Pain

The receptors that respond selectively to stimuli that can damage tissue are called *nociceptors* (Latin *nocere*, to injure). They respond directly to mechanical and thermal stimuli and indirectly to other stimuli by means of chemicals released from cells in the traumatized tissue. Nociceptors signal impending tissue injury, and more important, they provide a constant reminder of tissues that are already injured and must be protected.

Abnormal function in major organ systems resulting from disease or trauma evokes conscious sensations of pain. Much of our knowledge of the neural mechanisms of pain is derived from studies of cutaneous nociceptors because the mechanisms are easier to study in cutaneous nerves than in visceral nerves. Nevertheless, the neural mechanisms underlying visceral pain are similar to those for pain arising from the surface of the body.

Nociceptors in the skin, muscle, joints, and visceral receptors fall into two broad classes based on the myelination of their afferent fibers. Nociceptors innervated by thinly myelinated Aδ fibers produce short-latency pain that is described as sharp and pricking. The majority are called mechanical nociceptors or *high-threshold mechanoreceptors* (HTMRs) because they are excited by sharp objects that penetrate, squeeze, or pinch the skin

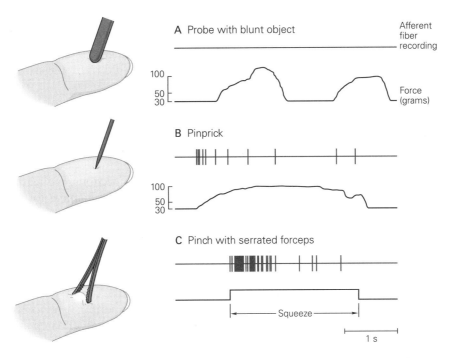

A Probe with blunt object

Afferent fiber recording

100
50
30

Force (grams)

B Pinprick

100
50
30

C Pinch with serrated forceps

Squeeze

1 s

Figure 18–11 Mechanical nociceptors respond to stimuli that puncture, squeeze, or pinch the skin. Sensations of sharp, pricking pain result from stimulation of Aδ fibers with free nerve endings in the skin. These receptors respond to sharp objects that puncture the skin (**B**), but not to strong pressure from a blunt probe (**A**). The strongest responses are produced by pinching the skin with serrated forceps that damage the tissue in the region of contact (**C**). (Adapted, with permission, from Perl 1968.)

(Figure 18–11) or by pulling hairs in hairy skin. Many of these fibers also respond to temperatures above 45°C (113°F) that burn the skin; these Aδ fibers also express the heat-sensitive TRPV2 channel.

Nociceptors innervated by C fibers produce dull, burning pain that is diffusely localized and poorly tolerated. The most common type encompasses polymodal nociceptors that respond to a variety of noxious mechanical, thermal, and chemical stimuli, such as pinch or puncture, noxious heat and cold, and irritant chemicals applied to the skin. As detailed in Chapter 20, most C-polymodal nociceptors express TRPV1 and/or TRPA1 receptors. Electrical stimulation of these fibers in humans evokes prolonged sensations of burning pain. In the viscera, nociceptors are activated by distension or swelling, producing sensations of intense pain.

Itch Is a Distinctive Cutaneous Sensation

Itch is a common sensory experience that is confined to the skin, the ocular conjunctiva, and the mucosa. It has some properties in common with pain and, until recently, was thought to result from low firing rates in nociceptive fibers. Like pain, itch is inherently unpleasant whatever its intensity; even at the expense of inducing pain, we attempt to eliminate it by scratching.

Recent studies by Diana Bautista and Sarah Wilson indicate that C fibers that express both TRPV1 and TRPA1 receptors mediate itch sensations evoked by pruritic (itch-producing) agents. Itch induced by intradermal injection of histamine or by procedures that release endogenous histamine activates a subset of TRPV1-expressing neurons that also contain the H1 histamine receptor; these itch sensations are blocked by antihistamines. Histamine-independent itch appears to be mediated by C fiber DRGs that express TRPA1 channels. Itch sensations in this pathway are triggered by dry skin or by pruritogens that bind to members of the Mas-related G protein–coupled receptor (Mrgpr) family, such as the antimalarial drug chloroquine.

How can TRPA1 receptors mediate itch when they are also involved in sensing noxious cold temperatures (<15°C)? Why do some TRPV1-expressing fibers mediate itch sensations rather than sensations of noxious heat? The answer lies in the use of *combinatorial* codes by small-diameter sensory nerve fibers. For example, noxious cold is sensed when both TRPA1 and TRPM8 receptors are excited, but itch is perceived when TRPM8 receptors are silent. Likewise, heat pain is sensed when TRPV1-, TRPV2-, and TRPV3-expressing fibers are co-activated, but itch may be perceived when only TRPV1-expressing fibers respond and TRPV2 and TRPV3 receptors are silent. Similar combinatorial codes using multiple receptors are commonly used by other chemical senses such as olfaction and taste.

Visceral Sensations Represent the Status of Internal Organs

Visceral sensations are important because they drive behaviors critical for survival, such as respiration, eating, drinking, and reproduction. The same molecular genetic strategies described earlier to study touch, pain, thermal senses, and proprioception in the dorsal root and trigeminal ganglia have been used to classify visceral afferents in the vagal sensory ganglia. Stephen Liberles and colleagues recently analyzed sensory responses in the vagal sensory ganglia (nodose/jugular complex) that receive mechanosensory or chemosensory information from the lungs, cardiovascular, immune, or digestive systems.

Vagal afferent fibers express a variety of G protein–coupled receptors (GPCRs) that have been labeled with fluorescent antibodies to identify their peripheral sensory receptor sites in specific viscera, as well as mark their distinctive central projections to specific zones in the nucleus of the solitary tract in the medulla. By expressing genetic markers of calcium transients (GCaMPs) in identified vagal ganglia neurons, Liberles and colleagues measured their physiological responses to mechanical stimuli such as stretch or their activation by nutrients or gastric hormones (serotonin, glucagon-like peptide 1, or cholecystokinin). The ability to label specific vagal afferents provides important tools for analyzing neural regulation of visceral function and tracing the pathways used to modulate these important bodily functions.

Although their cell bodies seem to be scattered randomly in the vagal nucleus, individual vagal neurons perform different sensory functions in specific organ systems. For example optogenetic stimulation of identified vagal sensory neurons reveals that there are at least two populations of vagal neurons controlling respiration. Neurons that express the GPCR *P2ry1* induce apnea, trapping the lung in expiration, while those expressing the GPCR *Npy2r* produce rapid shallow breathing. Stimulation of these neurons has no effect on heart rate or digestive function. Another set of GPCRs are used to label neurons that regulate gastrointestinal function. One set of gastric afferents are mechanoreceptors that sense distension of the stomach and upper intestine and modulate gastric motility, while other gastric afferents are chemoreceptors that sense specific nutrients in the gut and aid their absorption.

Action Potential Codes Transmit Somatosensory Information to the Brain

In the previous sections, we learned that a variety of stimuli, such as mechanical forces, temperature, and various chemicals, interact with receptor molecules at the distal axon terminals of DRG neurons to produce local depolarization of the sensory endings. As noted in Chapter 17, these receptor potentials are transformed into a digital pulse code of action potentials for transmission to the central nervous system.

The sensory terminal regions of peripheral nerve fibers are usually unmyelinated and do not express the voltage-gated Na⁺ and K⁺ channels that underlie action potential generation. For example, the lanceolate endings of hair follicle afferents are unmyelinated (Figure 18–2H). This design optimizes information gathering in the receptive field by dedicating the highly branched terminal membrane area to sensory transduction channels such as Piezo2 or TRP receptors.

The most distal action potential ion channels in myelinated fibers are usually located near the initial myelin segment (see Figure 3–10) or at the intersection of branches in unmyelinated fibers. This has important consequences for information transmission. Depolarizing sensory signals from multiple branches can summate more easily if channels involved in action potential generation are absent from receptive terminals, because of the regenerative properties of action potentials and the subsequent inactivation of the voltage-gated Na⁺ channels. Sensory messages arriving from later-activated receptors may be extinguished by collision with backward-propagating action potentials traveling along another branch of the fiber. Thus, the signals transmitted along a primary afferent axon may be a nonlinear reflection of the sensory stimulus, reflecting either spatial summation of excitation from multiple branches or winner-take-all suppression of late-generated activity. Sequential activation of different neurite branches can also aid detection of moving stimuli by generating long trains of action potentials if individual endings are stimulated at optimal rates so that their responses are not shunted by spikes generated earlier in other branches.

Action potential transmission along peripheral nerves depends on whether the axon is myelinated or unmyelinated and on the expression of specific sub-classes of voltage-dependent Na_V and K_V channels in each nerve fiber. Steven Waxman and colleagues reported that large-diameter Aα and Aβ fibers that innervate proprioceptors and low-threshold mechanoreceptors (LTMRs) express primarily $Na_V1.1$ and $Na_V1.6$ isoforms; these fibers generally fire action potentials at high rates, in part because they also express $K_V1.1$ and $K_V1.2$ channels that enable rapid repolarization of axons. Small-diameter peripheral nerves that mediate pain and itch sensations express $Na_V1.7$, $Na_V1.8$, and $Na_V1.9$ channels. The latter two Na_V subtypes have kinetic and voltage sensitivities that promote repetitive firing, thereby enhancing painful sensations: $Na_V1.8$ channels inactivate incompletely during action potentials and recover rapidly following them; $Na_V1.9$ channels activate at relatively negative potentials and undergo negligible inactivation, resulting in persistent inward currents that can amplify subthreshold stimuli.

Sensory Ganglia Provide a Snapshot of Population Responses to Somatic Stimuli

We conclude this survey of DRG neurons by examining the distribution of sensory responses within an individual mammalian somatosensory ganglion. Typically, peripheral nerve fibers have been studied one at a time, usually with optimal stimuli for particular receptor classes. However, even weak voices contribute to the neural chorale, and those have been largely ignored with classic single-cell recording techniques.

New in vivo functional imaging techniques provide useful tools for labeling, visualizing, and measuring ensemble responses to various types of somatosensory stimuli. For example, the Ca^{2+} currents evoked by sensory stimuli provide an alternative to electrophysiological recordings of spike trains in individual neurons. In the experiment illustrated in Figure 18–12, the genetically encoded Ca^{2+} sensor GCaMP6f was expressed in cells of the mouse trigeminal ganglia that also expressed the polymodal TRPV1 receptor, allowing researchers to visualize and quantify the activity of populations of neurons activated by a variety of somatosensory stimuli. Using a battery of tactile, noxious, and thermal stimuli first developed by William Willis to analyze somatosensory responses of neurons in the spinal cord, Nima Ghitani, Alexander Chesler, and colleagues recorded responses of 213 trigeminal neurons simultaneously. Their findings were quite remarkable. As shown in the heat map of Figure 18–12B1, neuronal responses are diverse, varying considerably in the intensity and duration of firing patterns to identical stimuli. Such ensemble recording techniques indicate that even at the receptor level there are no canonical responses to somatic stimuli, but rather common patterns of responses.

Furthermore, individual somatosensory neurons appear to be polysensory, responding to more than one modality, such as touch and pain. This study shows that individual trigeminal neurons distinguish noxious heat from mechanical pain (hair pull) and may respond, albeit weakly, to gentle touch or moderate thermal stimuli (Figure 18–12A). The most prevalent type of trigeminal ganglion neurons (49%) distinguish light touch (stroking the cheek) from thermal stimuli (Figure 18–12B). The next most common types are mechanical nociceptors (18%) or thermoreceptors (16%). Less common are polymodal types that respond to thermal and nociceptive stimuli (total 9%).

These new imaging techniques will enable neuroscientists to quantify sensory interactions in populations of somatosensory afferents, define combinatorial codes used by members of the active population, and thereby identify specific neural populations engaged in somatic sensation. Recording neurons simultaneously rather than one at a time is essential for decoding population activity and defining the circuits underlying diverse sensory modalities.

Lastly, we note that neurons in the dorsal root, trigeminal, and vagal ganglia do not appear to be spatially clustered or segregated functionally by modality such as mechanosensation or thermal or chemical events (Figure 18–12A). The principal organizational feature of these sensory ganglia is one of body topography: which particular area of skin or which muscle or visceral structure is innervated by particular sensory neurons. Such geographical specificity extends centrally to higher structures in the brain that analyze the sensory information and that organize specific behaviors.

Somatosensory Information Enters the Central Nervous System Via Spinal or Cranial Nerves

As the peripheral nerve fibers exit the dorsal root ganglia and approach the spinal cord, the large- and small-diameter fibers separate into medial and lateral divisions, to form the *spinal nerves* that project to distinct locations in the spinal cord and brain stem. The medial division includes large myelinated Aα and Aβ fibers, which transmit proprioceptive and tactile information from the innervated body region. The lateral division of a spinal nerve includes small, thinly myelinated Aδ fibers and unmyelinated C fibers, which transmit noxious, thermal, pruritic, and visceral

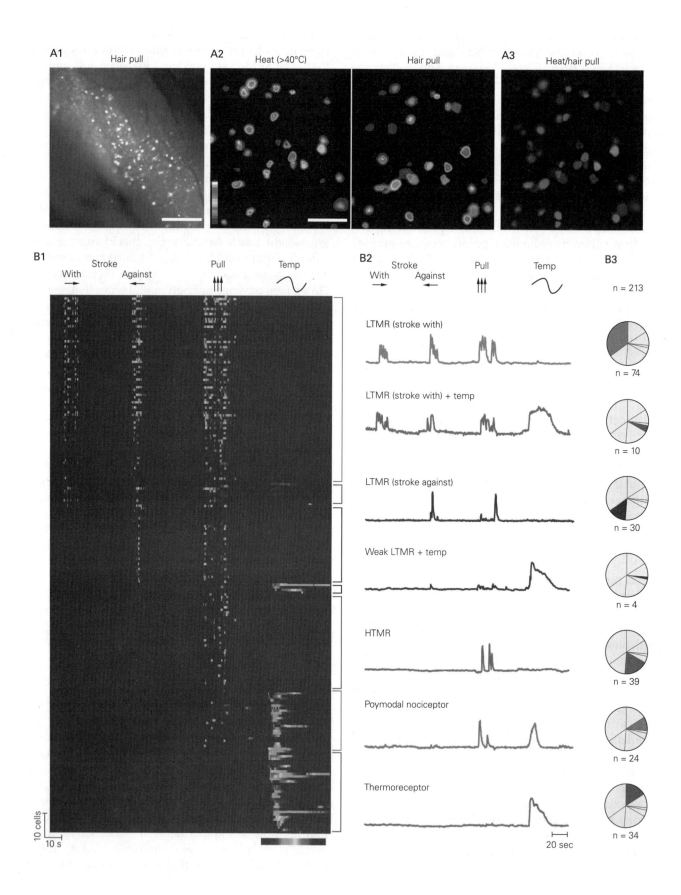

information from the same region of the body, as well as some tactile information.

Somatosensory information from the limbs and trunk reaches the central nervous system through the 31 spinal nerves, which enter the spinal cord through openings between the vertebrae of the spine. Individual spinal nerves are named for the vertebrae below the foramen through which they pass in cervical nerves or for the foramen above their entry point in thoracic, lumbar, and sacral nerves.

Somatosensory information from the head and neck is transmitted through the trigeminal, facial, glossopharyngeal, and vagus nerves, which enter through openings in the cranium. The trigeminal nerve conveys somatosensory information from the lips, mouth, cornea, and skin on the anterior half of the head, as well as the muscles of mastication. The facial and glossopharyngeal nerves innervate the taste buds of the tongue, the skin of the ear, and some of the skin of the tongue and pharynx. The glossopharyngeal and vagus nerves provide some cutaneous information, but their main sensory role is visceral. Vagal afferents regulating respiration and those regulating gastric motility project to distinct regions of the nucleus of the solitary tract.

Each spinal or cranial nerve receives sensory inputs from a particular region of the body called a *dermatome* (Figure 18–13); the muscles innervated by motor fibers in the corresponding peripheral nerve constitute a *myotome*. These are the skin and muscle regions affected by damage to peripheral nerves. Because the dermatomes overlap, three adjacent spinal nerves often have to be blocked to anesthetize a particular area of skin. The distribution of spinal nerves in the body forms the anatomical basis of the topographic maps of sensory receptors in the brain that underlie our ability to localize specific sensations.

Individual spinal or cranial nerve fibers terminate on neurons in specific zones of the spinal cord gray matter or the medullary dorsal horn (Figure 18–14). The spinal neurons that receive sensory input are either interneurons, which terminate upon other spinal neurons within the same or neighboring segments, or projection neurons, which serve as the cells of origin of major ascending pathways to higher centers in the brain.

The spinal gray matter is subdivided anatomically into 10 laminae (or layers), numbered I to X from dorsal to ventral, based on differences in cell and fiber composition. As a general rule, the largest fibers (Aα) terminate in or near the ventral horn, the medium-size fibers (Aβ) from the skin and muscle terminate in intermediate layers of the dorsal horn, and the smallest fibers (Aδ and C) terminate in the most dorsal portion of the spinal gray matter.

Lamina I consists of a thin layer of neurons capping the dorsal horn of the spinal cord and pars caudalis of the spinal trigeminal nucleus. Individual neurons of lamina I receive monosynaptic inputs from small myelinated fibers (Aδ) or unmyelinated C fibers of a

Figure 18–12 (Opposite) The distribution of somatosensory modalities among trigeminal ganglion neurons that innervate the hairy skin of the face. (Adapted, with permission, from Ghitani et al. 2017.)

A. In vivo epifluorescent imaging of a trigeminal ganglion in a TRPV1-GCaMP6f–expressing mouse. Calcium-sensitive dyes (GCaMP6f) fluoresce in response to Ca^{2+} entry through voltage-gated channels in individual trigeminal ganglion neurons. **A1.** Anatomical positions of 213 GCaMP6f-expressing neurons in the trigeminal ganglion of a mouse. Scale bar = 500 µm. These neurons are widely distributed within the trigeminal ganglion. **A2.** Higher magnification images of calcium signals in a subset of neurons that respond to heat pulses >40°C or to hair pull; the color bar in the left image indicates the strength of the calcium signal in each neuron. The strongest activity is shown in **white** or **red**; the weakest response in **blue**. Scale bar = 100 µm. **A3.** An overlay of the two population maps labeled in pseudocolor (**red** for heat, **green** for hair pulling) shows which neurons responded to each stimulus. These neurons were usually selective for heat or hair pull, but two responded to both modalities (**yellow**).

B. Quantification of the responses of all TRPV1-expressing neurons visualized in this trigeminal ganglion to various modes of tactile, noxious, and thermal stimuli. **B1.** Heatmap of the simultaneously recorded responses of all 213 labeled neurons to stroking the cheek, noxious mechanical (hair pull), and thermal stimuli. Each row illustrates the response of an individual neuron to these stimuli; the pixel color indicates the strength of each neuron's response (Δf/F). (Color range = 10%–60% Δf/F.) Neuronal responses are ordered vertically by the temporal onset of increased firing rates. The symbols above the heat map indicate the type and sequence of stimulation: stroking the cheek **with** or **against** the direction of hair growth, hair pull, and thermal stimuli ranging from 25°C to 47°C to 12°C. Although more than half of these neurons responded to gentle touch (stroking), they generally responded more vigorously to noxious mechanical stimuli (hair pull) than to stroking the skin. The strongest responses were observed to noxious heat, but such neurons composed only 30% of the population studied. At the end of the experiment, records from each neuron were sorted into one of the seven response categories identified in B2. **B2.** Averaged response amplitude and time course of Ca^{2+} signals for the seven categories of trigeminal sensory responses. Note the polymodal nature of responses using this objective mode of neuronal classification. (Abbreviations: **LTMR**, low-threshold mechanoreceptor; **HTMR**, high-threshold nociceptor-mechanoreceptor.) **B3.** Pie charts illustrate the number and fraction of neurons in each category.

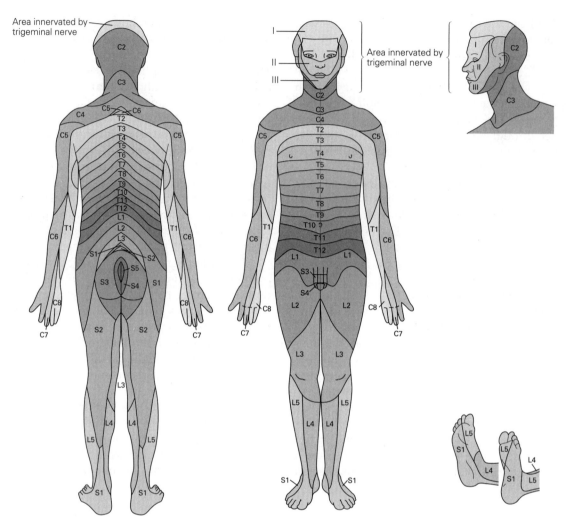

Figure 18–13 The distribution of dermatomes in the spinal cord and brain stem. A dermatome is the area of skin and deeper tissues innervated by a single dorsal root or branch of the trigeminal nerve. The dermatomes of the 31 pairs of dorsal root nerves are projected onto the surface of the body and labeled by the foramen through which each nerve enters the spinal cord. The 8 cervical (C), 12 thoracic (T), 5 lumbar (L), 5 sacral (S), and single coccygeal roots are numbered rostrocaudally for each division of the vertebral column. The facial skin, cornea, scalp, dura, and intraoral regions are innervated by the ophthalmic (I), maxillary (II), and mandibular (III) divisions of the trigeminal nerve (cranial nerve V). Level C1 has no dorsal root, only a ventral (or motor) root. Dermatome maps provide an important diagnostic tool for localizing the site of injury to the spinal cord and dorsal roots. However, the boundaries of the dermatomes are less distinct than shown here because the axons comprising a dorsal root originate from several different peripheral nerves, and each peripheral nerve contributes fibers to several adjacent dorsal roots.

single type (Figure 18–14) and therefore transmit information about noxious, thermal, or visceral stimuli. Inputs from warm, cold, itch, and pain receptors have been identified in lamina I, and some neurons have unique cellular morphologies that correlate with sensory modalities. Lamina I neurons generally have small receptive fields localized to one dermatome.

Neurons in lamina II are interneurons that receive inputs from Aδ and C fibers and make excitatory or inhibitory connections to neurons in laminae I, IV, and V that project to higher brain centers. The more superficial portion of lamina II receives input from peptidergic nociceptors that release substance P or CGRP together with glutamate at their central synapses. Fibers terminating in the deeper part of lamina II are purinergic; they release ATP at their central synapses and express the lectin IB4. Co-transmitters such as ATP provide useful immunostaining markers for identifying specific classes of sensory nerve fibers (Figure 18–2C,D).

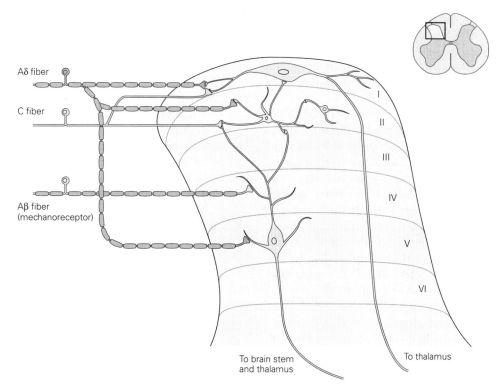

Figure 18–14 Touch and pain fiber projections to the spinal cord dorsal horn. The spinal gray matter in the dorsal horn and intermediate zone of the spinal cord is divided into six layers of cells (laminae I–VI), each with functionally distinct populations of neurons. Neurons in the marginal zone (lamina I) and in lamina II receive nociceptive or thermal inputs from receptors innervated by Aδ or C fibers. The zone for inputs from low-threshold mechanoreceptors (LTMR) is located below lamina II and spans laminae III to V, with the smallest fibers (C-LTMRs) located dorsally, and the largest fibers (Aβ LTMRs) terminating ventrally. LTMRs innervating a particular patch of skin are aligned to form a narrow cell column in the spinal dorsal horn, terminating on spinal interneurons or on projection neurons that send their axons to the brain stem. The medial-lateral arrangement of spinal nerves in the dorsal horn provides a somatotopic representation of adjacent skin areas in the body. The spinal nerve projections of Aβ LTMRs extend to multiple spinal segments along the rostrocaudal axis, whereas those of Aδ or C fibers are more localized to the immediate entry segment (not shown). Aβ LTMRs also send branches to the dorsal column nuclei in the brain stem (Chapters 19 and 20).

Neurons in laminae III to V are the main targets of LTMRs, particularly the large myelinated sensory (Aβ) fibers from cutaneous mechanoreceptors (Figure 18–14). Spinal cord circuits of the dorsal horn have been characterized anatomically and functionally by Victoria Abraira and David Ginty. These local spinal networks enable sensory integration of multiple modalities within a local zone of the body, enabling motoneuron pools to react rapidly to local sensory feedback. Large-diameter fibers mediating touch (Aβ) or proprioception (Aα) also send ascending branches to the medulla through the dorsal columns or dorsolateral funiculi.

Additionally, neurons of the cerebral cortex project to the dorsal horn, permitting direct cortical regulation of local sensorimotor circuits and thus coordinating purposeful behaviors. These higher-order, top-down pathways are supplemented by intraspinal circuits between dermatomes that enable coordinated movements of different fingers or distal and proximal joints.

Neurons in lamina V typically respond to more than one modality—low-threshold mechanical stimuli, visceral stimuli, or noxious stimuli—and are therefore named *wide-dynamic-range neurons*.

Many of the dorsal horn circuits also transmit somatosensory information directly to higher structures in the brain stem, such as the dorsal column, parabrachial, and raphe nuclei, and to the cerebellum or various thalamic nuclei.

Afferent C fibers from the viscera have widespread projections in the spinal cord that terminate ipsilaterally in laminae I, II, V, and X; some also cross the midline and terminate in lamina V and X of the contralateral gray matter. The extensive spinal distribution of visceral C fibers appears to be responsible for the poor localization of visceral pain sensations.

Afferent fibers from the pelvic viscera make important connections to cells in the central gray matter (lamina X) of spinal segments L5 and S1. Lamina X neurons in turn project their axons along the midline of the dorsal columns to the nucleus gracilis in a postsynaptic dorsal column pathway for visceral pain.

Primary afferent fibers that terminate in the deepest laminae in the ventral horn provide sensory information from proprioceptors (muscle spindles and Golgi tendon organs) that is required for somatic motor control, such as spinal reflexes (Chapter 32).

Somatosensory information is conveyed by several ascending pathways to higher centers in the brain, particularly the thalamus and cerebral cortex. The dorsal column–medial lemniscal system transmits tactile and proprioceptive information to the thalamus (Chapter 19), and the spinothalamic (anterolateral) tract conveys pain and thermal information to the midbrain parabrachial nucleus or to the thalamus (Chapter 20). A third pathway, the dorsolateral tract, conveys somatosensory information from the lower half of the body to the cerebellum. The anatomical and functional roles of these networks are described in detail in later chapters.

Highlights

1. The bodily senses mediate a wide range of experiences that are important for normal bodily function and for survival. Although diverse, they share common pathways and common principles of organization. The most important of those principles is *specificity*: Each of the bodily senses arises from specific types of receptors distributed throughout the body.

2. Dorsal root ganglion (DRG) neurons are the sensory receptor cells of the somatosensory system. The functional role of an individual DRG neuron is determined by the sensory receptor molecules expressed in its distal terminals in the body. Mechanoreceptors are sensitive to specific aspects of local tissue distortion, thermoreceptors to particular temperature ranges and shifts in temperature, and chemoreceptors to particular molecular structures. Recordings of physiological responses from these neurons reveal the cellular and molecular mechanisms underlying the senses of touch, pain, temperature, and proprioception, as well as visceral senses.

3. Mechanosensation is mediated by Piezo2 proteins that form ion channels in the axon terminals of DRG fibers sensitive to compression or stretch. These include touch fibers that innervate hair

follicles or specialized epithelia such as Merkel cells, Meissner and Pacinian corpuscles, or Ruffini endings. Muscle stretch is signaled by intramuscular spindle receptors and contractile force by Golgi tendon organs. These receptors transmit sensory information via rapidly conducting Aα and Aβ peripheral nerve fibers.

4. Thermoreceptors are excited by transient receptor potential (TRP) ion channels in the axon terminals that are gated in response to local temperature gradients and respond selectively to particular ranges of temperature: cold, cool, warm, or hot. Chemoreceptors change their conductance when binding specific chemicals, both natural and exogenous, giving rise to sensations of pain, itch, or visceral function. Thermosensory and chemosensory information is conveyed centrally via Aδ and C fiber pathways.

5. Activation of somatosensory receptors produces local depolarization of the distal nerve terminals, called the *receptor potential*, whose amplitude is proportional to the strength of the stimulus. Receptor potentials are converted near the distal nerve terminals to trains of action potentials whose frequency is linked to the strength of the stimulus, much as synaptic potentials at synapses produce complex firing patterns in postsynaptic neurons.

6. Individual DRG neurons have multiple sensory endings in the skin, muscle, or viscera, forming complex receptive fields with overlapping territories. The combination of divergent distal terminals and innervation of sense organs by multiple axons enables redundant, parallel pathways for information transmission to the brain.

7. The information transmitted from each type of somatosensory receptor in a particular part of the body is conveyed in discrete pathways to the spinal cord or brain stem by the axons of DRG neurons with cell bodies that generally lie in ganglia close to the point of entry. The axons are gathered together in peripheral nerves. Axon diameter and myelination, both of which determine the speed of action potential conduction, vary in different sensory pathways according to the need for rapid signaling.

8. When DRG axons enter the central nervous system, they separate to terminate in distinct layers of the spinal cord gray matter and/or project directly to higher centers in the brain stem. These circuits form the foundation of five separate sensory pathways with different properties. In three of those systems (the medial lemniscal, lamina I

spinothalamic, and solitary tract systems), the pathways for submodalities appear to be segregated until they reach the cerebral cortex.

9. Future studies of the peripheral nervous system will likely engage high-resolution optical methods for identification of specific receptor classes in the DRG that are labeled with genetic markers. Functional studies of these neurons will also employ optical imaging of entire sensory ganglia labeled with voltage-sensitive or calcium-sensitive fluorescent dyes that enable quantitative temporal monitoring of ensemble responses to specific somatosensory modalities. These receptor neurons will thereby be studied as identified physiological populations rather than one at a time in isolation.

Esther P. Gardner

Selected Reading

Abraira VE, Ginty DD. 2013. The sensory neurons of touch. Neuron 79:618–639.

Abraira VE, Kuehn ED, Chirila AM, et al. 2017. The cellular and synaptic architecture of the mechanosensory dorsal horn. Cell 168:295–310.

Bautista DM, Siemens J, Glazer JM, et al. 2007. The menthol receptor TRPM8 is the principal detector of environmental cold. Nature 448:204–208.

Delmas P, Hao J, Rodat-Despoix L. 2011. Molecular mechanisms of mechanotransduction in mammalian sensory neurons. Nat Rev Neurosci 12:139–153.

Dhaka A, Viswanath V, Patapoutian A. 2006. TRP ion channels and temperature sensation. Annu Rev Neurosci 29:135–161.

Iggo A, Andres KH. 1982. Morphology of cutaneous receptors. Annu Rev Neurosci 5:1–31.

Julius D. 2013. TRP channels and pain. Annu Rev Cell Dev Biol 29:355–384.

Kaas JH, Gardner EP (eds). 2008. *The Senses: A Comprehensive Reference*, Vol. 6, *Somatosensation*. Oxford: Elsevier.

LaMotte RH, Dong X, Ringkamp M. 2014. Sensory neurons and circuits mediating itch. Nat Rev Neurosci 15:19–31.

Lechner SG, Lewin GR. 2013. Hairy sensation. Physiology 28:142–150.

Li L, Rutlin M, Abraira VE, et al. 2011. The functional organization of cutaneous low-threshold mechanosensory neurons. Cell 147:1615–1627.

Patapoutian A, Tate S, Woolf CJ. 2009. Transient receptor potential channels: targeting pain at the source. Nat Rev Drug Discov 8:55–68.

Ranade SS, Syeda R, Patapoutian A. 2015. Mechanically activated ion channels. Neuron 87:1162–1179.

Vallbo ÅB, Hagbarth KE, Torebjörk HE, Wallin BG. 1979. Somatosensory, proprioceptive, and sympathetic activity in human peripheral nerves. Physiol Rev 59:919–957.

Vallbo ÅB, Hagbarth KE, Wallin BG. 2004. Microneurography: how the technique developed and its role in the investigation of the sympathetic nervous system. J Appl Physiol 96:1262–1269.

References

Bai L, Lehnert BP, Liu J, et al. 2015. Genetic identification of an expansive mechanoreceptor sensitive to skin stroking. Cell 163:1783–1795.

Bandell M, Macpherson LJ, Patapoutian A. 2007. From chills to chilis: mechanisms for thermosensation and chemesthesis via thermoTRPs. Curr Opin Neurobiol 17:490–497.

Bennett DL, Clark AJ, Huang J, Waxman SG, Dib-Hajj SD. 2019. The role of voltage-gated sodium channels in pain signaling. Physiol Rev 99:1079–1151.

Boyd IA, Davey MR. 1968. *Composition of Peripheral Nerves*. Edinburgh: Livingston.

Cao E, Liao M, Cheng Y, Julius D. 2013. TRPV1 structures in distinct conformations reveal activation mechanisms. Nature 504:113–118.

Chang RB, Strochlic DE, Williams EK, Umans BD, Liberles SD. 2015. Vagal sensory neuron subtypes that differentially control breathing. Cell 161:622–633.

Collins DF, Refshauge KM, Todd G, Gandevia SC. 2005. Cutaneous receptors contribute to kinesthesia at the index finger, elbow, and knee. J Neurophysiol 94:1699–1706.

Coste B, Xiao B, Santos JS, et al. 2012. Piezo proteins are pore-forming subunits of mechanically activated channels. Nature 483:176–181.

Cox CD, Bae C, Ziegler L, et al. 2016. Removal of the mechanoprotective influence of the cytoskeleton reveals PIEZO1 is gated by bilayer tension. Nat Commun 7:10366.

Darian-Smith I, Johnson KO, Dykes R. 1973. "Cold" fiber population innervating palmar and digital skin of the monkey: responses to cooling pulses. J Neurophysiol 36:325–346.

Darian-Smith I, Johnson KO, LaMotte C, Shigenaga Y, Kenins P, Champness P. 1979. Warm fibers innervating palmar and digital skin of the monkey: responses to thermal stimuli. J Neurophysiol 42:1297–1315.

Dib-Hajj SD, Cummins TR, Black JA, Waxman SG. 2010. Sodium channels in normal and pathological pain. Annu Rev Neurosci 33:325–347.

Edin BB, Vallbo AB. 1990. Dynamic response of human muscle spindle afferents to stretch. J Neurophysiol 63:1297–1306.

Erlanger J, Gasser HS. 1938. *Electrical Signs and Nervous Activity*. Philadelphia: Univ. of Pennsylvania Press.

Gandevia SC, McCloskey DI, Burke D. 1992. Kinaesthetic signals and muscle contraction. Trends Neurosci 15:62–65.

Gandevia SC, Smith JL, Crawford M, Proske U, Taylor JL. 2006. Motor commands contribute to human position sense. J Physiol 571:703–710.

Ghitani N, Barik A, Szczot M, et al. 2017. Specialized mechanosensory nociceptors mediating rapid responses to hair pull. Neuron 95:944–954.

Guharay F, Sachs F. 1984. Stretch-activated single ion channel currents in tissue-cultured embryonic chick skeletal muscle. J Physiol 352:685–701.

Guo YR, MacKinnon R. 2017. Structure-based membrane dome mechanism for Piezo mechanosensitivity. Elife 6:e33660.

Johansson RS, Vallbo ÅB. 1983. Tactile sensory coding in the glabrous skin of the human hand. Trends Neurosci 6:27–32.

Jordt S-E, McKemy DD, Julius D. 2003. Lessons from peppers and peppermint: the molecular logic of thermosensation. Curr Opin Neurobiol 13:487–492.

Liao M, Cao E, Julius D, Cheng Y. 2013. Structure of the TRPV1 ion channel determined by electron cryo-microscopy. Nature 504:107–112.

Lin S-Y, Corey DP. 2005. TRP channels in mechanosensation. Curr Opin Neurobiol 15:350–357.

Macefield VG. 2005. Physiological characteristics of low-threshold mechanoreceptors in joints, muscle and skin in human subjects. Clin Exp Pharmacol Physiol 32:135–144.

Macefield G, Gandevia SC, Burke D. 1990. Perceptual responses to microstimulation of single afferents innervating joints, muscles and skin of the human hand. J Physiol 429:113–129.

Maksimovic S, Nakatani M, Baba Y, et al. 2014. Epidermal Merkel cells are mechanosensory cells that tune mammalian touch receptors. Nature 509:617–621.

McGlone F, Wessberg J, Olausson H. 2014. Discriminative and affective touch: sensing and feeling. Neuron 82:737–755.

Murthy SE, Dubin AE, Patapoutian A. 2017. Piezos thrive under pressure: mechanically activated ion channels in health and disease. Nat Rev Mol Cell Biol 18:771–783.

Ochoa J, Torebjörk E. 1989. Sensations evoked by intraneural microstimulation of C nociceptor fibres in human skin nerves. J Physiol 415:583–599.

Ottoson D, Shepherd GM. 1971. Transducer properties and integrative mechanisms in the frog's muscle spindle. In: WR Lowenstein (ed). *Handbook of Sensory Physiology*, Vol. 1 *Principles of Receptor Physiology*, pp. 442–499. Berlin: Springer-Verlag.

Perl ER. 1968. Myelinated afferent fibers innervating the primate skin and their response to noxious stimuli. J Physiol (Lond) 197:593–615.

Perl ER. 1996. Cutaneous polymodal receptors: characteristics and plasticity. Prog Brain Res 113:21–37.

Ranade SS, Woo SH, Dubin AE, et al. 2014. Piezo2 is the major transducer of mechanical forces for touch sensation in mice. Nature 516:121–125.

Sachs F. 1990. Stretch-sensitive ion channels. Sem Neurosci 2:49–57.

Saotome K, Murthy SE, Kefauver JM, Whitwam T, Patapoutian A, Ward AB. 2018. Structure of the mechanically activated ion channel Piezo1. Nature 554:481–486.

Torebjörk HE, Vallbo ÅB, Ochoa JL. 1987. Intraneural microstimulation in man. Its relation to specificity of tactile sensations. Brain 110:1509–1529.

Wellnitz SA, Lesniak DR, Gerling GJ, Lumpkin EA. 2010. The regularity of sustained firing reveals two populations of slowly adapting touch receptors in mouse hairy skin. J Neurophysiol 103:3378–3388.

Wessberg J, Olausson H, Fernström KW, Vallbo ÅB. 2003. Receptive field properties of unmyelinated tactile afferents in the human skin. J Neurophysiol 89:1567–1575.

Williams EK, Chang RB, Strochlic DE, Umans BD, Lowell BB, Liberles SD. 2016. Sensory neurons that detect stretch and nutrients in the digestive system. Cell 166:209–221.

Wilson SR, Gerhold KA, Bifolck-Fisher A, et al. 2011. TRPA1 is required for histamine-independent, Mas-related G protein-coupled receptor-mediated itch. Nat Neurosci 14:595–602.

Wilson SR, Nelson AM, Batia L, et al. 2013. The ion channel TRPA1 is required for chronic itch. J Neurosci 33:9283–9294.

Woo SH, Lukacs V, de Nooij JC, et al. 2015. Piezo2 is the principal mechanotransduction channel for proprioception. Nat Neurosci 18:1756–1762.

Woo SH, Ranade S, Weyer AD, et al. 2014. Piezo2 is required for Merkel-cell mechanotransduction. Nature 509:622–626.

Zhao Q, Zhou H, Chi S, et al. 2018. Structure and mechanogating mechanism of the Piezo1 channel. Nature 554:487–492.

Zimmerman A, Bai L, Ginty DD. 2014. The gentle touch receptors of mammalian skin. Science 346:950–954.

19

Touch

I N THIS CHAPTER ON THE SENSE OF TOUCH, we focus on the hand because of its importance for this modality, in particular its role in the appreciation of object properties and in performance of skilled motor tasks. The human hand is one of evolution's great creations. The fine manipulative capacity provided by our fingers is possible because of their fine sensory capacity; if we lose tactile sensation in our fingers, we lose manual dexterity.

The softness and compliance of the glabrous skin play a major role in the sense of touch. When an object contacts the hand, the skin conforms to its contours, forming a mirror image of the object's surface. The resultant displacement and indentation of the skin stretches the tissue, thereby stimulating the sensory endings of mechanoreceptors at or near the region of contact.

These receptors are highly sensitive and are continually active as we manipulate objects and explore the world with our hands. They provide information to the brain about the object's position in the hand, its shape and surface texture, the amount of force applied at the contact points, and how these features change over time when the hand or the object moves. The fingertips are among the most densely innervated parts of the body, providing extensive and redundant somatosensory information about objects manipulated by the hand.

Moreover, the anatomical structure of the hand, with its multiple joints and apposable digits, enables humans to shape the hand in ways that mirror an object's overall shape, providing a hand-centered proprioceptive representation of the external world. This ability to internalize the shape of objects allows us to create tools that extend the abilities of our hands alone.

When we become skilled in the use of a tool, such as a scalpel or a pair of scissors, we feel conditions at the working surface of the tool as though our fingers were there because two groups of touch receptors monitor the vibrations and forces produced by those distant conditions. When we scan our fingers across a surface, we feel its form and texture because another group of mechanoreceptors has high spatial and temporal acuity. A blind person uses this capacity to read Braille at a hundred words per minute. When we grip and manipulate an object, we do so delicately, with only as much force as needed, because specific mechanoreceptors continually monitor slip and adjust our grip appropriately.

We are also able to recognize objects placed in the hand from touch alone. When we are handed a baseball, we recognize it instantly without having to look at it because of its shape, size, weight, density, and texture. We do not have to think about the information provided by each finger to deduce that the object must be a baseball; the information flows to memory and instantly matches previously stored representations of baseballs. Even if we have never previously handled a baseball, we perceive it as a single object, not as a collection of discrete features. The somatosensory pathways of the brain have the daunting task of integrating information from thousands of sensors in each hand and transforming it to a form suitable for cognition and action.

Sensory information is extracted for the purpose of motor control as well as cognition, and different kinds of information are extracted for those purposes. We can, for example, shift our attention from the baseball's shape to its location in the hand to readjust our grip for an effective throw or pitch. This selective attention to different aspects of the sensory information is brought about by cortical mechanisms.

Active and Passive Touch Have Distinct Goals

Touch is defined as direct contact between two physical bodies. In neuroscience, touch refers to the special sense by which contact with the body is perceived consciously. Touch can be active, as when you move your hand or some other part of the body against another surface, or passive, as when someone or something else touches you. Active touch is fundamentally a top-down process in which the subject has agency, seeks particular information, and controls what occurs. Subjects select relevant salient features of objects to determine subsequent behaviors. They choose which object to grasp and the most efficient hand shape needed to acquire it, and decide how to manipulate it to achieve particular goals. During active touch, somatosensory

information depicts the physical properties of objects as well as the motor actions of the subject's hand and arm, and their relation to the task goals. Importantly, active manipulation of objects is based upon the concept of touch as a three-dimensional modality designed to capture the volumetric, topographic, and elastic properties of objects, as first proposed by Roberta Klatzky and Susan Lederman. These three-dimensional qualities are best appreciated by active manipulation including grasping, rotation, and contour tracing by the hand.

Passive touch engages a bottom-up process in which subjects react to external stimuli specified by the experimenter or clinician. The experimenter selects and controls the location, amplitude, force, timing, duration, and spatial spread of stimuli delivered to the skin. Subsequent behaviors are guided by instructions provided in the paradigm. Tactile stimuli are classified into experimenter-selected categories and/or rated along an intensive or hedonic scale. Subjects therefore need to analyze all of the transmitted somatosensory information and select specific features guided in part by the task instructions.

Active and passive modes of tactile stimulation excite the same population of receptors in the skin and evoke similar responses in afferent fibers. They differ somewhat in cognitive features that reflect attention and behavioral goals during the period of stimulation. Passive touch is tested by naming objects or describing sensations; active touch is used when the hand manipulates objects. The sensory and motor components of touch are intimately connected anatomically in the brain and are important functionally in guiding motor behavior.

During active touch, descending fibers from motor centers of the cerebral cortex terminate on interneurons in the medial dorsal horn that receive tactile information from the skin. Similar fibers from cortical motor areas terminate in the dorsal column nuclei, providing an *efference copy* (or corollary discharge) of the motor commands that generate behavior (Chapter 30). In this manner, tactile signals from the hand resulting from active hand movements may be distinguished centrally from passively applied stimuli in the neurological exam or in psychophysical tests.

The distinction between active and passive touch is important clinically when patients have deficits in hand use. Motor deficits such as weakness, stiffness, or clumsiness may result from sensory loss, which is why passive sensory testing is important in the neurological examination. Common neurological tests for touch include measurements of detection thresholds, vibration sense, two-point or texture discrimination, and the ability to recognize form through touch (*stereognosis*). These tests measure the sensitivity and function of various receptors for touch. Deviations from expected values may help diagnose sensory deficits or lesions that underlie somatosensory dysfunction. The neural mechanisms underlying these tests are discussed in this chapter. Other common tests of somatosensory function—tendon reflexes, pinprick, and thermal tests—are discussed in other chapters.

The Hand Has Four Types of Mechanoreceptors

Tactile sensations in the human hand arise from four kinds of mechanoreceptors: Meissner corpuscles, Merkel cells, Pacinian corpuscles, and Ruffini endings

(Figure 19–1). Each receptor responds in a distinctive manner depending on its morphology, innervation pattern, and depth in the skin. The sense of touch can be understood as the combined result of the information provided by these four systems acting in concert.

Touch receptors are innervated by slowly adapting or rapidly adapting axons. Slowly adapting (SA) fibers respond to steady skin indentation with a sustained discharge, whereas rapidly adapting (RA) fibers stop firing when indentation becomes stationary (Figure 19–1 and Table 19–1). Sustained mechanical sensations from the hand must accordingly arise from the SA fibers; the sensation of motion on or across the skin is signaled primarily by RA fibers.

Touch receptors in the hand are further subdivided into two types based on size and location in the skin.

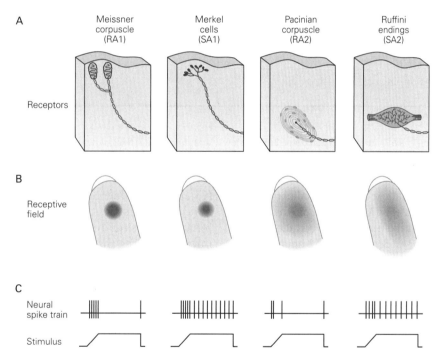

Figure 19–1 Four types of mechanoreceptors are responsible for the sense of touch in the human hand. The terminals of myelinated sensory nerves innervating the hand are surrounded by specialized structures that detect contact on the skin. The receptors differ in morphology, innervation patterns, location in the skin, receptive field size, and physiological responses to touch. (Adapted, with permission, from Johansson and Vallbo 1983.)

A. The superficial and deep layers of the glabrous (hairless) skin of the hand each contain distinct types of mechanoreceptors. The superficial layers contain small receptor cells: Meissner corpuscles (**RA1**, rapidly adapting type 1) and Merkel cells (**SA1**, slowly adapting type 1). The sensory nerve fibers that innervate these receptors have branching terminals that innervate multiple receptors of one type. The deep layers of the skin and subcutaneous tissue contain large receptors: Pacinian corpuscles (**RA2**,

rapidly adapting type 2) and Ruffini endings (**SA2**, slowly adapting type 2). Each of these receptors is innervated by a single nerve fiber, and each fiber innervates only one receptor.

B. The receptive field of a mechanoreceptor reflects the location and distribution of its terminals in the skin. Touch receptors in the superficial layers of the skin have smaller receptive fields than those in the deep layers.

C. The nerve fibers innervating each type of mechanoreceptor respond differently when activated. The schematic spike trains show responses of each type of nerve when its receptor is activated by slowly increasing and constant pressure against the skin. The rapidly adapting fibers respond to motion at the onset and end of a pressure stimulus and adapt rapidly to constant stimulation, whereas the slowly adapting fibers respond to both steady pressure and motion and adapt slowly.

Table 19–1 Cutaneous Mechanoreceptors in Glabrous Skin

	Type 1		Type 2	
	SA1	**RA1**[1]	**SA2**	**RA2**[2]
Receptor	Merkel cell/neurite complex (multiple endings)	Meissner corpuscle (multiple endings)	Ruffini ending (single ending)	Pacinian corpuscle (single ending)
Location	Base of intermediate ridge surrounding sweat duct	Dermal papillae (adjacent to limiting ridge)	Skin folds, skin over joints, nail bed	Dermis (deep tissue)
Axon diameter (μm)	7–11	6–12	6–12	6–12
Conduction velocity (ms)	40–65	35–70	35–70	35–70
Best stimulus	Edges, points	Lateral motion	Skin stretch	Vibration
Response to sustained indentation	Sustained with slow adaptation (irregular firing pattern)	Phasic at stimulus onset	Sustained with slow adaptation (regular firing rate)	Phasic at stimulus onset
Frequency range (Hz)	0–100	1–300		5–1,000
Best frequency (Hz)	5	50		200
Threshold for rapid indentation or vibration (best) (μm)	8	2	40	0.01

[1]Also called RA, QA, or FA1.
[2]Also called PC or FA2.
RA1, rapidly adapting type 1; **RA2**, rapidly adapting type 2; **SA1**, slowly adapting type 1; **SA2**, slowly adapting type 2.

Type 1 touch fibers terminate in clusters of small receptor organs (Meissner corpuscles or Merkel cells) in the superficial layers of the skin at the margin between the dermis and epidermis (Figure 19–2, Box 19–1.). RA1 fibers are the most numerous tactile afferents in the hand, reaching a density of approximately 150 per cm² at the fingertip in man and monkey; SA1 fibers are also widely distributed in the hand, at densities of 70 per cm² in the fingertips.

Type 2 fibers innervate the skin sparsely and terminate in single large receptors (Pacinian corpuscles and Ruffini endings) located in the dermis or in subcutaneous tissue. These receptors are larger and less numerous than the receptor organs of the type 1 fibers. The large size of type 2 receptors allows them to sense mechanical displacement of the skin at some distance from the sensory nerve endings. The density of RA2 fibers in human fingers is only 21 per cm²; SA2 fibers are the least abundant, providing only 9 fibers per cm².

A Cell's Receptive Field Defines Its Zone of Tactile Sensitivity

Individual mechanoreceptor fibers convey information from a limited area of skin called the *receptive field*

(Chapter 18). Tactile receptive fields in the human hand were first studied by Åke Vallbo and Roland Johansson using microneurography. They inserted microelectrodes through the skin into the median or ulnar nerves in the human forelimb and recorded the responses of individual afferent fibers. They found that in humans, as in other primates, there are important differences between touch receptors, both in their physiological responses and in the structure of their receptive fields.

Type 1 fibers have small, highly localized receptive fields with multiple spots of high sensitivity that reflect the branching patterns of their axon terminals in the skin (Figure 19–5). An RA1 axon typically innervates 10 to 20 Meissner corpuscles, integrating information from several adjacent fingerprint ridges. An SA1 fiber innervates approximately 20 Merkel cells in young adults (Figure 19–4B); the number of Merkel cells drops significantly as we age.

In contrast, type 2 fibers innervating the deep layers of skin are connected to only a single Pacinian corpuscle or Ruffini ending. As these receptors are large, they collect information from a broader area of skin. Their receptive fields typically contain a single

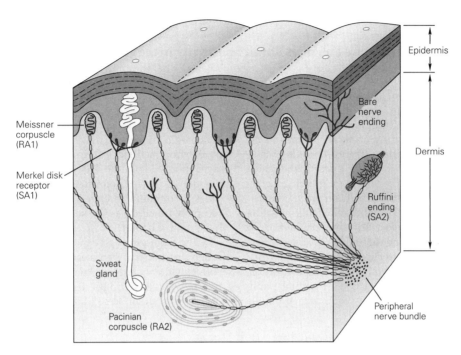

Figure 19–2 Tactile innervation of the glabrous skin in humans. A cross section of the glabrous skin shows the principal receptors for touch in the human hand. All of these receptors are innervated by large-diameter Aβ myelinated fibers. The Meissner corpuscles and Merkel cells lie in the superficial layers of the skin at the base of the epidermis, 0.5 to 1.0 mm below the skin surface. The Meissner corpuscles are located in the dermal papillae that border the edges of each papillary ridge. The Merkel cells form dense bands below the intermediate ridge surrounding the sweat gland ducts along the center of the papillary ridges. The RA1 and SA1 fibers that innervate these receptors branch at their terminals so that each fiber innervates several nearby receptor organs. The Pacinian and Ruffini corpuscles lie within the dermis (2–3 mm thick) and in deeper tissues. The RA2 and SA2 fibers that innervate these receptors each innervate only one receptor organ. (Abbreviations: **RA1**, rapidly adapting type 1; **RA2**, rapidly adapting type 2; **SA1**, slowly adapting type 1; **SA2**, slowly adapting type 2.)

"hot spot" where sensitivity to touch is greatest; this point is located directly above the receptor (Figure 19–5).

Receptive fields on the fingertips are the smallest on the body, averaging 11 mm^2 for SA1 fibers and 25 mm^2 for RA1 fibers. The small fields complement the high density of receptors in the fingertips. Receptive fields become progressively larger on the proximal phalanges and the palm, consistent with the lower density of mechanoreceptors in these regions. Importantly, the receptive fields of type 1 fibers are significantly smaller than most objects that contact the hand, and therefore signal the spatial properties of only a limited portion of an object. As in the visual system, the spatial features of objects are distributed across a population of stimulated receptors whose responses are integrated in the brain to form a unified percept.

Each RA2 axon terminates without branching in a single Pacinian corpuscle, and each Pacinian corpuscle receives but a single RA2 axon. Pacinian corpuscles are large onion-like structures in which successive layers of connective tissue are separated by fluid-filled spaces (see Figure 19–8A1). These layers surround the unmyelinated RA2 ending and its myelinated axon up to one or more nodes of Ranvier. The capsule amplifies high-frequency vibration, a role that is important for tool use. Estimates of the number of Pacinian corpuscles in the human hand range from 2,400 in the young to 300 in the elderly.

The SA2 fibers innervate Ruffini endings concentrated at the finger and wrist joints, the skin surrounding the fingernails, and along the skin folds in the palm. The Ruffini endings are elongated fusiform structures that enclose collagen fibrils extending from the subcutaneous tissue to folds in the skin at the joints, in the palm, or at the fingernail borders. The SA2 nerve endings are intertwined between the collagen fibers in the capsule, as in Golgi tendon organs (Box 32–4), and are excited by stimuli that stretch the skin along its long axis.

Two-Point Discrimination Tests Measure Tactile Acuity

The ability of humans to resolve spatial details of textured surfaces depends on which region of the body is

Box 19–1 Fingerprint Structure Enhances Touch Sensitivity in the Hand

The histological structure of glabrous skin—the smooth, hairless skin of the palm and fingertips—plays a crucial role in the hand's sensitivity to touch. The fingerprints are formed by a regular array of parallel ridges in the epidermis, the papillary ridges (Figure 19–3). Regularly spaced Merkel cells below sweat ducts that emerge from the center of each ridge provide a spatial grid that allows us to localize stimuli precisely on our fingertips.

Each ridge is bordered by epidermal folds—the limiting ridges—that are visible as thin lines on the fingers, palms, and feet. The limiting ridges increase the stiffness and rigidity of the skin, protecting it from damage when contacting objects or when walking barefooted. Meissner corpuscles are typically located in dermal papillae adjacent to the limiting ridges; each dermal papilla contains several Meissner corpuscles and is innervated by two to five RA1 axons (Figure 19–4A).

Merkel cells, innervated by an SA1 fiber, are densely clustered in the center of each papillary ridge, at the base of the intermediate ridge surrounding the epidermal sweat ducts (Figure 19–4A), placing them in an excellent position to detect deformation of the epidermis from pressure or lateral stretch. They perform similar tactile receptive functions as Merkel cells in the touch domes of hairy skin (Chapter 18).

The fingerprints give the glabrous skin a corrugated, rough structure that increases friction, allowing us to grasp objects without slippage. Frictional forces are augmented further when these ridges contact the textured surfaces of objects. Smooth surfaces slide easily underneath the fingers

Figure 19–3 The skin of the human fingertip.

A. Scanning electron micrograph of the fingerprints in the human index finger. The glabrous skin of the hand is structured as arrays of papillary ridges and intervening sulci (limiting ridges) that recur at regular intervals. Globules of sweat exude from ducts at the center of the papillary ridges, forming a regularly spaced grid-like pattern along the center of each ridge. The Merkel cells are located in dense clusters below the sweat ducts at the base of the epidermis along the center of the papillary ridges (see Figure 19–2). (Adapted, with permission, from Quilliam 1978.)

B. Histological section of the glabrous skin cut parallel to the skin surface. The Meissner corpuscles, here immunostained for cholinesterase, form regularly spaced chains along both sides of each papillary ridge adjacent to the limiting ridge. Thus, Meissner corpuscles and Merkel cells form alternating bands of rapidly adapting type 1 (RA1) and slowly adapting type 1 (SA1) touch receptors that span each fingerprint ridge. (Adapted, with permission, from Bolanowski and Pawson 2003.)

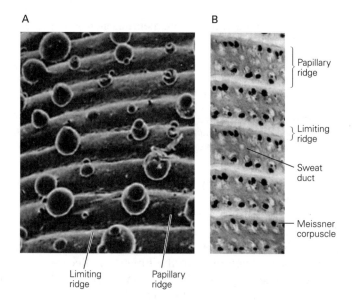

contacted. When a pair of probes is spaced several millimeters apart on the hand, each probe is perceived as a distinct point because it produces a separate dimple in the skin and stimulates nonoverlapping populations of receptors. As the probes are moved closer together, the two sensations become blurred because both probes are contained within the same receptive fields. The spatial interactions between tactile stimuli form the basis of neurological tests of *two-point discrimination* and texture recognition.

The threshold for *tactile acuity*—the separation that defines performance midway between chance and perfect discrimination—is approximately 1 mm on the fingertips of young adults, but declines in the elderly to about 2 mm. Tactile acuity is highest on the fingertips and the lips, where receptive fields are smallest.

A Glabrous skin

Meissner
corpuscle
terminals

RA1 fiber

Merkel cell
terminals

SA1 fiber

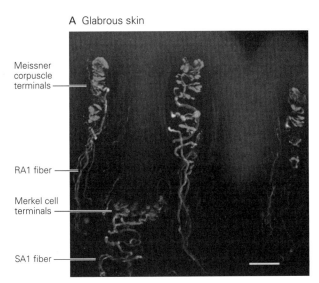

B Hairy skin

Merkel
cell

SA1 fiber

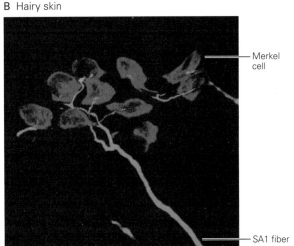

Figure 19–4 Innervation pattern of Meissner corpuscles and Merkel cells in glabrous and hairy skin.

A. A confocal transverse section of a papillary ridge in the human fingertip skin shows the innervation pattern of mechanoreceptors. Meissner corpuscles are located in dermal papillae just below the epidermis (**blue**) bordering the limiting ridge and are innervated by two or more rapidly adapting type 1 (**RA1**) fibers. The fibers lose their myelin sheaths (**orange**) when entering the receptor capsule, exposing broad terminal bulbs (**green**) at which sensory transduction occurs. Individual slowly adapting type 1 (**SA1**) fibers innervate groups of Merkel cells clustered at the base of the intermediate ridge, providing localized signals of pressure applied to that ridge. Scale bar = 50 µm. (Adapted, with permission, from Nolano et al. 2003. Copyright © 2003 American Neurological Association.)

B. A higher-magnification micrograph portrays keratin-8 antibody-labeled Merkel cells (**red**) innervated by an SA1 fiber (**green**) labeled with neurofilament heavy polypeptide (NFH$^+$). Each nerve fiber extends multiple branches parallel to the surface of the skin that allow it to integrate tactile information from multiple receptor cells in a small zone of skin. The diameter of each Merkel cell is approximately 10 µm. (Adapted, with permission, from Snider 1998. Copyright © 1998 Springer Nature.)

and thus require greater grip force to maintain stability in the hand; the screw caps on bottles are often ridged to make them easy to turn. Frictional forces between the limiting ridges and objects also amplify our sensations of surface features when we palpate objects, generating vibrations that allow us to detect small irregularities such as the grain of wood and threads of fabrics.

The regular spacing of the papillary ridges—and the precise localization of specific receptors within this grid—allows us to repeatedly scan surfaces with back-and-forth hand movements while preserving a constant spatial alignment of adjacent surface features. They also provide an anatomical grid for referencing the precise location of tactile stimuli.

Tactile acuity on proximal parts of the body decreases in parallel with the size of receptive fields of SA1 and RA1 fibers (Figure 19–6A).

When we grasp or touch an object, we can discriminate features of its surface separated by as little as 0.5 mm. Humans are able to distinguish horizontal and vertical orientations of gratings with remarkably narrow spacing of the ridges (Figure 19–6B).

Long edges, such as the ridges of a grating, evoke stronger responses from RA1 and SA1 afferents when they stimulate multiple sensory endings in the receptive field simultaneously, stressing the importance of multisensor receptive fields for tactile information processing. Roland Johansson and Andrew Pruszynski recently found that RA1 and SA1 fibers respond more intensely to edges that contact multiple sensory

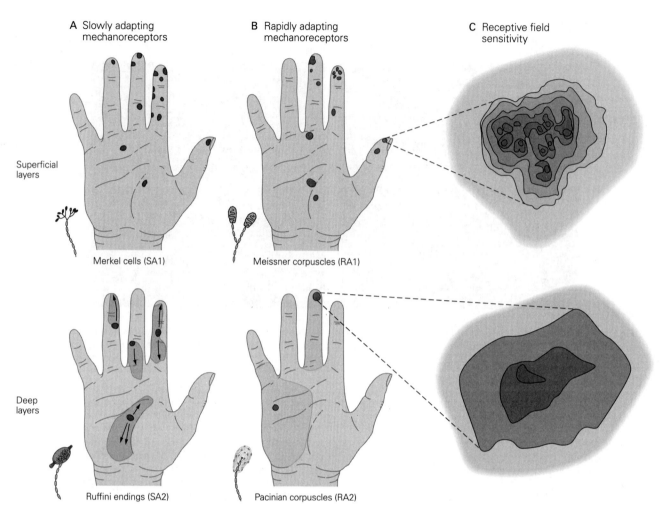

A Slowly adapting mechanoreceptors

B Rapidly adapting mechanoreceptors

C Receptive field sensitivity

Superficial layers

Merkel cells (SA1)

Meissner corpuscles (RA1)

Deep layers

Ruffini endings (SA2)

Pacinian corpuscles (RA2)

Figure 19–5 Receptive fields in the human hand are smallest at the fingertips. Each colored area on the hands indicates the receptive field of an individual sensory nerve fiber. (Adapted, with permission, from Johansson and Vallbo 1983.)

A–B. In the superficial layers of skin, the receptive fields of type 1 receptors encompass spot-like patches of skin. In the deep layers, type 2 receptive fields extend across wide regions of skin (**light shading**), but responses are strongest in the skin directly over the receptor (**dark spots**). The **arrows** indicate the directions of skin stretch that activate slowly adapting type 2 (**SA2**) fibers.

C. Pressure sensitivity throughout the receptive field is shown as a contour map. The most sensitive regions are indicated in **deep red** and the least sensitive areas in **pale pink**. The receptive field of a rapidly adapting type 1 (**RA1**) fiber (**above**) has many points of high sensitivity, marking the positions of the group of Meissner corpuscles innervated by the fiber. The receptive field of a rapidly adapting type 2 (**RA2**) fiber (**below**) has a single point of maximum sensitivity overlying the Pacinian corpuscle. The receptive field contour map of slowly adapting type 1 (**SA1**) fibers is similar to that of RA1 fibers. Likewise, the receptive field map of SA2 fibers resembles that of RA2 fibers.

endings, allowing these afferents to distinguish vertical, horizontal, or oblique orientations.

Tactile acuity is slightly greater in women than in men and varies between fingers but not between hands; the gender difference is related primarily to the smaller papillary ridge diameter in women, and the resultant higher density of SA1 fibers per cm^2 of skin. The distal pad of the index finger has the keenest sensitivity; spatial acuity declines progressively from the

index to the little finger and falls rapidly at locations proximal to the distal finger pads. Tactile spatial resolution is 50% poorer at the distal pad of the little finger and six to eight times coarser on the palm.

Blind individuals use the fine spatial sensitivity of SA1 and RA1 fibers to read Braille. The Braille alphabet represents letters as simple dot patterns that are easy to distinguish by touch. A blind person reads Braille by moving the fingers over the dot patterns. This hand

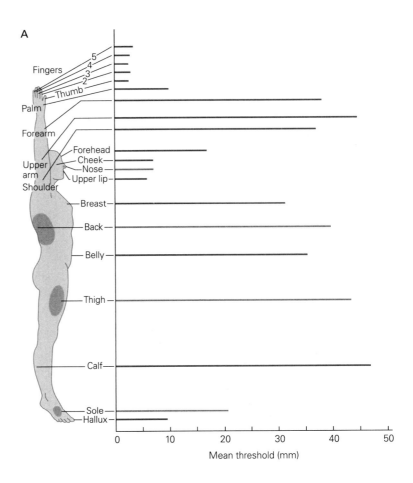

Figure 19–6 Tactile acuity in the human hand is highest on the fingertip.

A. The two-point threshold measures the minimum distance at which two stimuli are resolved as distinct. This distance varies for different body regions; it is approximately 2 mm on the fingers, but as much as 10 mm on the palm and 40 mm on the arm, thigh, and back. The mean two-point perceptual thresholds of different body parts, indicated by **pink** lines in the bar graph, match the mean receptive field diameters of the corresponding pink zones on the body. The greatest discriminative capacity is afforded in the fingertips, lips, and tongue, which have the smallest receptive fields. (Adapted, with permission, from Weinstein 1968. © Charles C. Thomas Publisher, Ltd.)

B. Spatial acuity is measured in psychophysical experiments by having a blindfolded subject touch a variety of textured surfaces. As shown here, the subject is asked to determine whether the surface of a wheel is smooth or contains a gap, whether the ridges of a grating are oriented across the finger or parallel to its long axis, or which letters appear on raised type used in letterpress printing. The tactile acuity threshold is defined as the groove width, ridge width, or font size that yields 75% correct performance (detectable midway between chance and perfect accuracy). The threshold spacing on the human fingertip is 1.0 mm in each of these tests. (Adapted, with permission, from Johnson and Phillips 1981.)

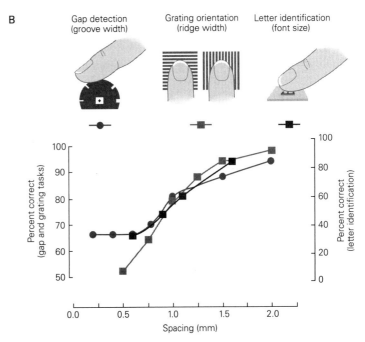

movement enhances the sensations produced by the dots. Because the Braille dots are spaced approximately 3 mm apart, a distance greater than the receptive field diameter of an SA1 fiber, each dot stimulates a different set of SA1 fibers. An SA1 fiber fires a burst of action potentials as a dot enters its receptive field and is silent once the dot leaves the field (Figure 19–7). Specific combinations of SA1 fibers that fire synchronously signal the spatial arrangement of the Braille dots. RA1 fibers also discriminate the dot patterns, enhancing the signals provided by SA1 fibers.

Although Pacinian corpuscles (RA2 fibers) respond to scanning Braille dots over the skin, their spike trains do not reflect the periodicity of dots in the Braille patterns. Instead, they signal the skin vibrations evoked by motion of the Braille dots over the skin. Sliman Bensmaia and colleagues recently found that when fine textures such as fabrics are tested with this method, RA2 afferents signal the periodicity of threads in the weave by generating their spike trains in phase with these surface features. SA1 fibers are less responsive to motion of textiles because the thread size is usually too small to indent the skin at sufficient amplitude. Nevertheless, all three types of tactile afferents contribute to human percepts of roughness and smoothness.

Slowly Adapting Fibers Detect Object Pressure and Form

The most important function of SA1 and SA2 fibers is their ability to signal skin deformation and pressure. The sensitivity of SA1 receptors to edges, corners, points, and curvature provides information about an object's compliance, shape, size, and surface texture. We perceive an object as hard or rigid if it indents the skin when we touch it, and soft if we deform the object.

Paradoxically, as an object's size and diameter increase, its surface curvature decreases. The responses of individual SA1 fibers are weaker and the resulting sensations feel less distinct. For example, the tip of a pencil pressed 1 mm into the skin feels sharp, unpleasant, and highly localized at the contact point, whereas a 1-mm indentation by the eraser feels blunt and broad. The weakest sensation is evoked by a flat surface pressed against the finger pad.

To understand why these objects evoke different sensations, we need to consider the physical events that occur when the skin is touched. When a pencil tip is pressed against the skin, it dimples the surface at the contact point and forms a shallow, sloped basin in the surrounding region (approximately 4 mm in radius). Although the indentation force is concentrated in the center, the surrounding region is also perturbed by

local stretch, called tensile strain. SA1 receptors at both the center and the surrounding "hillsides" of skin are stimulated, firing spike trains proportional to the degree of local stretch.

If a second probe is pressed close to the first one, more SA1 fibers are stimulated but the neural response of each fiber is reduced because the force needed to displace the skin is shared between the two probes. Ken Johnson and his colleagues have shown that as more probes are added within the receptive field, the response intensity at each sensory ending becomes progressively weaker because the displacement forces on the skin are distributed across the entire contact zone. Thus, the skin mechanics result in a case of "less is more." Individual SA1 fibers respond more vigorously to a small object than to a large one because the force needed to indent the skin is concentrated at a small contact point. In this manner, each SA1 fiber integrates the local skin indentation profile within its receptive field.

The sensitivity of SA1 receptors to local strain on the skin enables them to detect edges, the places where an object's curvature changes abruptly. SA1 firing rates are many times greater when a finger touches an edge than when it touches a flat surface because the force applied by an object boundary displaces the skin asymmetrically, beyond the edge as well as at the edge. This asymmetric distribution of force enhances responses from receptive fields located along the edges of an object. As edges are often perceived as sharp, we tend to grasp objects on flat or gently curved surfaces rather than by their edges.

The SA2 fibers that innervate Ruffini endings respond more vigorously to stretch of the skin than to indentation, because of their anatomical location along the palmar folds or at the finger joints. They provide information about the shape of large objects grasped with the entire hand, the "power grasp" in which an object is pressed against the palm.

The SA2 system may play a central role in stereognosis—the recognition of three-dimensional objects using touch alone—as well as other perceptual tasks in which skin stretch is a major cue. Benoni Edin has shown that SA2 innervation of the hairy skin on the dorsum of the hand plays a substantial role in the perception of hand shape and finger position. The SA2 fibers aid the perception of finger joint angle by detecting skin stretch over the knuckles, or in the webbing between the fingers. The Ruffini endings near these joints are aligned such that different groups of receptors are stimulated as the fingers move in specific directions (Figure 19–5A, bottom panel). In this manner, the SA2 system provides a neural representation

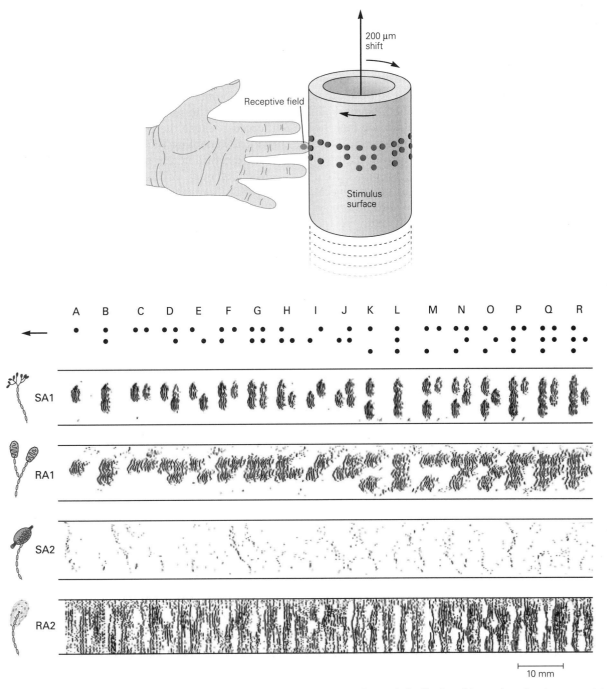

Figure 19–7 Responses of touch receptors to Braille dots scanned by the fingers. The Braille symbols for the letters A through R were mounted on a drum that was repeatedly rotated against the fingertip of a human subject. Following each revolution, the drum was shifted upward so that another portion of the symbols was scanned across the finger. Microelectrodes placed in the median nerve of this subject recorded the responses of the mechanoreceptive fibers innervating the fingertip. The action potentials discharged by the nerve fibers as the Braille symbols moved over the receptive field are represented in these records by small dots; each horizontal row of dots represents the responses of the fiber to a single revolution of the drum. The SA1 receptors register the sharpest image of the Braille symbols, representing each Braille dot with a series of action potentials and falling silent when the spaces between Braille symbols provide no stimulation. RA1 receptors provide a blurred image of the Braille symbols because their receptive fields are larger, but the individual dot patterns are still recognizable. Neither RA2 nor SA2 receptors are able to encode the spatial characteristics of the Braille patterns because their receptive fields are larger than the dot spacing. The high firing rate of the RA2 fibers reflects the keen sensitivity of Pacinian corpuscles to vibration. (Abbreviations: **RA1**, rapidly adapting type 1; **RA2**, rapidly adapting type 2; **SA1**, slowly adapting type 1; **SA2**, slowly adapting type 2.) (Reproduced, with permission, from Phillips, Johansson, and Johnson 1990. Copyright © 1990 Society for Neuroscience.)

of skin stretch over the entire hand, a proprioceptive rather than exteroceptive function.

The SA2 fibers also provide proprioceptive information about hand shape and finger movements when the hand is empty. If the fingers are fully extended and abducted, we feel the stretch in the palm and proximal phalanges as the glabrous skin is flattened. Similarly, if the fingers are fully flexed, forming a fist, we feel the stretch of the skin on the back of the hand, particularly over the metacarpal-phalangeal and proximal interphalangeal joints. Humans use this proprioceptive information to preshape their hand to grasp objects efficiently, opening the fingers just wide enough to clear the object and grasp it skillfully without too much force.

Rapidly Adapting Fibers Detect Motion and Vibration

Tests of vibration sense form an important component of the neurological exam. Touching the skin with a tuning fork that oscillates at a particular frequency evokes a periodic buzzing sensation because most touch receptors fire synchronized, periodic trains of action potentials in phase with the stimulus frequency (Figure 19–8A2). Vibration sense is a useful measurement of dynamic sensitivity to touch, particularly in cases of localized nerve damage.

The RA2 receptor, the Pacinian corpuscle, is the most sensitive mechanoreceptor in the somatosensory system. It is exquisitely responsive to high-frequency (30–500 Hz) vibratory stimuli and can detect vibration of 250 Hz in the nanometer range (Figure 19–8B2). The ability of Pacinian corpuscles to filter and amplify high-frequency vibration allows us to feel conditions at the working surface of a tool in our hand as if our fingers themselves were touching the object under the tool. The clinician uses this exquisite sensitivity to guide a needle into a blood vessel and to probe tissue stiffness. The auto mechanic uses vibratory sense to position wrenches on unseen bolts. We can write in the dark because we feel the vibration of the pen as it contacts the paper and transmits the frictional forces from the surface roughness to our fingers.

Although Pacinian corpuscles have the lowest vibration thresholds for frequencies greater than 40 Hz (Figure 19–8B2), vibratory stimuli of higher amplitude also excite SA1 and RA1 fibers, even if their evoked spike trains are weaker than those of Pacinian afferents. Figure 19–9A illustrates the evoked firing patterns of 15 different peripheral nerve fibers stimulated at 20 Hz at weak, moderate, and high amplitudes. Although these fibers differ in sensitivity to vibration,

their spike trains have certain important characteristics in common. First, each neuron fires at a particular phase of the vibratory cycle, usually when the probe indents the skin, and its phasic pattern of spikes replicates the vibratory frequency: when stimulated at 20 Hz, the spike bursts recur at intervals of approximately 50 ms. The patterning of the spike trains is further reinforced because the population of fibers fires synchronously, enabling the frequency information to be preserved centrally due to synaptic integration.

The total number of spikes per burst also increases as the stimulus amplitude rises, allowing each fiber to multiplex signals of vibratory frequency and intensity: the frequency information is conveyed by the temporal pattern of the spike train, and the vibratory amplitude is encoded by the total number of spikes fired per second by each fiber, as well as the total spike output of the ensemble of activated fibers. Finally, note that the spike trains of each neuron are very similar in time course and spike count from trial to trial for each condition, indicating the high reliability of sensory signaling provided by tactile afferent fibers. This reliability and predictability of sensory coding make vibration a particularly useful technique for assessing the sense of touch.

Both Slowly and Rapidly Adapting Fibers Are Important for Grip Control

In addition to their role sensing the physical properties of objects, touch receptors provide important information concerning hand actions during skilled movements. Roland Johansson and Gören Westling used microneurography to determine the role of touch receptors when objects are grasped in the hand. By placing microelectrodes in the median nerve, they were able to record the firing patterns of touch fibers as an object was initially contacted by the fingers, and when it was grasped between the thumb and index finger, lifted, held above a table, lowered, and returned to rest.

They found that all four classes of touch fibers respond to grasp and that each fiber type monitors a particular function. The RA1, RA2, and SA1 fibers are normally silent in the absence of tactile stimuli. They detect contact when an object is first touched (Figure 19–10). The SA1 fibers signal the amount of grip force applied by each finger, and the RA1 fibers sense how quickly the grasp is applied. The RA2 fibers detect the small shock waves transmitted through the object when it is lifted from the table and when it is returned. We know when an object makes contact with the table top because of these vibrations and therefore can

A Neural coding of vibration

1 Pacinian corpuscle

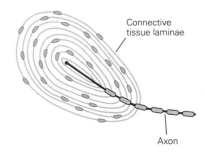

2 RA2 fiber

B Thresholds for detection of vibration

1 Human perceptual thresholds

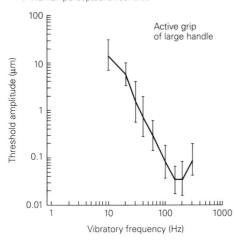

2 Neural thresholds

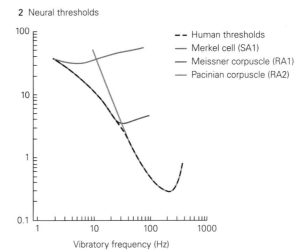

Figure 19–8 Rapidly adapting type 2 (RA2) fibers have the lowest threshold for vibration. Vibration is the sensation produced by sinusoidal stimulation of the skin, as by the hum of an electric motor, the strings of a musical instrument, or a tuning fork used in the neurological examination.

A. 1. The Pacinian corpuscle consists of concentric, fluid-filled lamellae of connective tissue that encapsulate the terminal of an RA2 fiber. This structure is uniquely suited to the detection of motion. Sensory transduction in the RA2 fiber occurs in stretch-sensitive cation channels linked to the inner lamellae of the capsule. **2.** When steady pressure is applied to the skin, the RA2 fiber fires a burst at the start and end of stimulation. In response to sinusoidal stimulation (vibration), the fiber fires at regular intervals such that each action potential signals one cycle of the stimulus. Our perception of vibration as a rhythmically repeating event results from the simultaneous activation

of many RA2 units, which fire in synchrony. (Adapted from Talbot et al. 1968.)

B. 1. Psychophysical thresholds for detection of vibration depend on the stimulation frequency. As shown here, humans can detect vibrations as small as 30 nm at 200 Hz when grasping a large object; the threshold is higher at other frequencies and when tested with small probes. (Adapted, with permission, from Brisben, Hsiao, and Johnson 1999.) **2.** Human thresholds for vibration, measured by a small probe tip indenting the skin, match those of the most sensitive touch fibers in each frequency range. Each type of mechanosensory fiber is most sensitive to a specific range of frequencies. Slowly adapting type 1 (**SA1**) fibers are the most sensitive population below 5 Hz, rapidly adapting type 1 (**RA1**) fibers between 10 Hz and 50 Hz, and RA2 fibers above 50 Hz and 400 Hz. (Adapted, with permission, from Mountcastle, LaMotte, and Carli 1972, and Johansson, Landström, and Lundström 1982.)

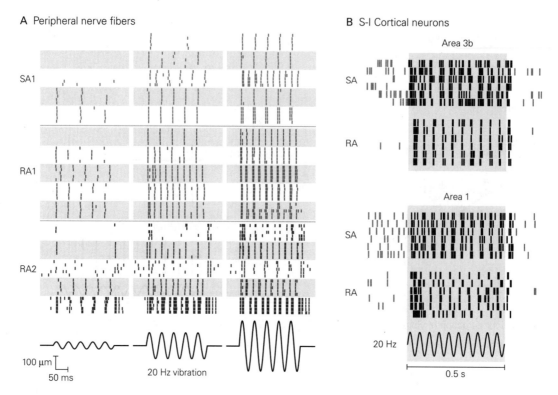

Figure 19–9 Suprathreshold vibration activates multiple classes of touch receptors.

A. Rasters of spike trains recorded from 15 different somatosensory fibers in macaque monkeys stimulated by 20-Hz vibratory stimuli with amplitudes of 35 (*left*), 130 (*center*), and 250 μm (*right*). The alternating **shaded and white bands** indicate the responses of individual slowly adapting type 1 (**SA1**), rapidly adapting type 1 (**RA1**), and rapidly adapting type 2 (**RA2**) touch fibers to five presentations of the same stimulus. Neural responses are grouped in bursts of one or more spikes that occur in phase with the indentation phase of each vibratory cycle. The total number of spikes per cycle in each fiber is correlated with the amplitude of the vibration; the total number of spikes fired across this population also reflects the vibratory amplitude. Although the individual neurons differ in the intensity of their responses, the spike

trains of each touch fiber are very similar from trial to trial and occur synchronously between neurons. (Adapted, with permission, from Muniak et al. 2007. Copyright © 2007 Society for Neuroscience.)

B. S-I cortical responses to 20-Hz vibration. Rasters of spike trains evoked in two neurons in area 3b (**top**) and two neurons in area 1 (**bottom**) of S-I cortex of a macaque monkey. The **shaded area** indicates the period of vibratory stimulation. As in the peripheral nerves, S-I cortical neurons respond to low-frequency vibration with bursts of impulses in phase with the stimulation rate. Note that the spike trains vary somewhat from trial to trial and are less periodic in area 1 than in area 3b. The periodicity of firing is even less pronounced in S-II cortex (see Figure 19–21) than in S-I. (Abbreviations: **RA**, rapidly adapting; **SA**, slowly adapting.) (Adapted, with permission, from Salinas et al. 2000. Copyright © 2000 Society for Neuroscience.)

manipulate the object without looking at it. The RA1 and RA2 fibers cease responding after grasp is established. The SA2 fibers signal flexion or extension of the fingers during grasp or release of the object and thereby monitor the hand posture as these movements proceed.

Signals from the hand that report on the shape, size, and texture of an object are important factors governing the application of force during grasping. Johansson and his colleagues found that we lift and manipulate an object with delicacy—with grip forces that just exceed the forces that result in overt slip—and that the grip force is adjusted automatically to compensate for differences in the frictional coefficient between the

fingers and object surface. Subjects predict how much force is required to grasp and lift an object and modify these forces based on tactile information provided by SA1 and RA1 afferents. Objects with smooth surfaces are grasped more firmly than those with rough textures, properties coded by RA1 afferents during initial contact of the hand with an object. The significance of the tactile information in grasping is seen in cases of nerve injury or during local anesthesia of the hand; patients apply unusually high grip forces, and coordination between the grip and load forces applied by the fingers is poor.

The information supplied by the RA1 receptors to monitor grasping actions is critical for grip control,

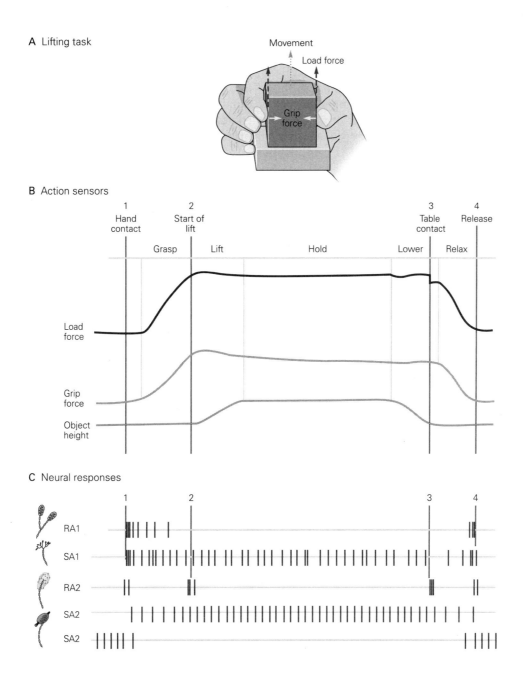

Figure 19–10 Sensory information from the hand during grasping and lifting. (Adapted, with permission, from Johansson 1996.)

A. The subject grasps and lifts a block between the thumb and fingertips, holds it above a table, and then returns it to the resting position. The normal (grip) force secures the object in the hand, and the tangential (load) force overcomes gravity. The grip force is adapted to the surface texture and weight of the object.

B. The grip and load forces are monitored with sensors in the object. These forces are coordinated following contact with

the object, stabilize as lift begins, and relax in concert after the object is returned to the table.

C. All four mechanoreceptors detect hand contact with the object, but each monitors a different aspect of the action as the task progresses. SA1 fibers encode the grip force and SA2 fibers the hand posture. RA1 fibers encode the rate of force application and movement of the hand on the object. RA2 fibers sense vibrations in the object during each task phase: at hand contact, lift-off, table contact, and release of grasp. (Abbreviations: **RA1**, rapidly adapting type 1; **RA2**, rapidly adapting type 2; **SA1**, slowly adapting type 1; **SA2**, slowly adapting type 2.)

allowing us to hold on to objects when perturbations cause them to slip unexpectedly. RA1 fibers are silent during steady grasp and usually remain quiet until the object is returned to rest and the grasp released. However, if the object is unexpectedly heavy or jolted by external forces and begins to slip from the hand, the RA1 fibers fire in response to the small tangential slip movements of the object. The net result of this RA1 activity is that grip force is increased by signals from the motor cortex.

Tactile Information Is Processed in the Central Touch System

Sensory afferent fibers innervating the hand transmit tactile and other somatosensory information to the central nervous system through the median, ulnar, and superficial radial nerves. These nerves terminate ipsilaterally in spinal segments C6 to T1; other branches of these fibers project through the ipsilateral dorsal columns directly to the medulla, where they make synaptic connections to neurons in the cuneate nucleus, the lateral division of the dorsal column nuclei (Figure 19–11).

Spinal, Brain Stem, and Thalamic Circuits Segregate Touch and Proprioception

Fibers in the dorsal columns, and neurons in the dorsal column nuclei, are organized topographically, with the upper body (including the hand) represented laterally in the cuneate fascicle and nucleus and the lower body represented medially in the gracile fascicle and nucleus. The somatosensory submodalities of touch and proprioception are also segregated functionally

in these regions, as individual spinal and brain stem neurons receive synaptic inputs from afferents of a single type, and neurons of distinct types are spatially separated. The rostral third of the dorsal column nuclei is dominated by neurons that process proprioceptive information from muscle afferents; tactile inputs predominate more caudally. Modality segregation is a consistent feature of the projection pathways to the primary somatosensory cortex.

Neurons in the dorsal column nuclei project their axons across the midline in the medulla to form the *medial lemniscus*, a prominent fiber tract that transmits tactile and proprioceptive information from the contralateral side of the body through the pons and midbrain to the thalamus. As a result of this crossing (or decussation) of sensory fibers, the left side of the brain receives somatosensory input from mechanoreceptors on the right side of the body, and vice versa. In transit, the somatotopic representation of the body in the medial lemniscus and within the thalamus becomes inverted; the topographic map of the body displays the face medially, the lower body laterally, and the upper body and hands in between.

Tactile and proprioceptive information from the hand and other regions of the body is processed in distinct subnuclei of the thalamus. Touch signals from the limbs and trunk are sent via the medial lemniscus to the ventral posterior lateral (VPL) nucleus, while those from the face and mouth are conveyed to the ventral posterior medial (VPM) nucleus. Proprioceptive information from muscles and joints, including those of the hand, is transmitted to the ventral posterior superior (VPS) nucleus. These nuclei send their outputs to different subregions of the parietal lobe of the cerebral cortex. The VPL and VPM nuclei transmit cutaneous information primarily to area 3b of the primary somatosensory

Figure 19–11 (Opposite) Somatosensory information from the limbs and trunk is conveyed to the thalamus and cerebral cortex by two ascending pathways. Brain slices along the neuraxis from the spinal cord to the cerebrum illustrate the anatomy of the two principal pathways conveying somatosensory information to the cerebral cortex. The two pathways are separated until they reach the pons, where they are juxtaposed.

Dorsal column—medial lemniscal system (orange). Touch and limb proprioception signals are conveyed to the spinal cord and brain stem by large-diameter myelinated nerve fibers and transmitted to the thalamus in this system. In the spinal cord, the fibers for touch and proprioception divide, one branch going to the ipsilateral spinal gray matter and the other ascending in the ipsilateral dorsal column to the medulla. The second-order fibers from neurons in the dorsal column nuclei cross the

midline in the medulla and ascend in the contralateral medial lemniscus toward the thalamus, where they terminate in the lateral and medial ventral posterior nuclei. Thalamic neurons in these nuclei convey tactile and proprioceptive information to the primary somatosensory cortex.

Anterolateral system (brown). Pain, itch, temperature, and visceral information is conveyed to the spinal cord by small-diameter myelinated and unmyelinated fibers that terminate in the ipsilateral dorsal horn. This information is conveyed across the midline by neurons within the spinal cord and transmitted to the brain stem and the thalamus in the contralateral anterolateral system. Anterolateral fibers terminating in the brain stem compose the spinoreticular and spinomesencephalic tracts; the remaining anterolateral fibers form the spinothalamic tract.

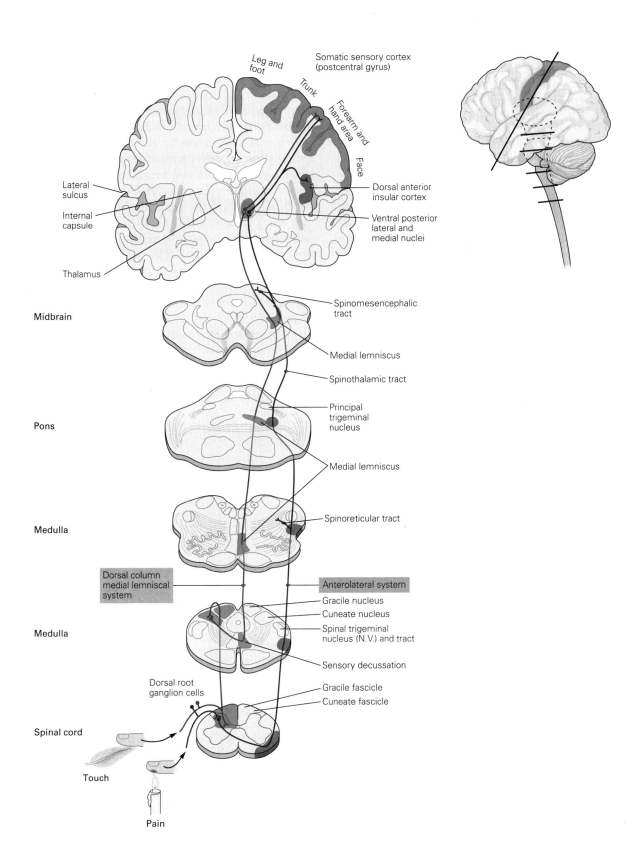

Somatic sensory cortex (postcentral gyrus)

Leg and foot

Trunk

Forearm and hand area

Face

Lateral sulcus

Internal capsule

Thalamus

Dorsal anterior insular cortex

Ventral posterior lateral and medial nuclei

Midbrain

Spinomesencephalic tract

Medial lemniscus

Spinothalamic tract

Pons

Principal trigeminal nucleus

Medial lemniscus

Medulla

Spinoreticular tract

Dorsal column medial lemniscal system

Anterolateral system

Gracile nucleus

Cuneate nucleus

Spinal trigeminal nucleus (N.V.) and tract

Sensory decussation

Medulla

Dorsal root ganglion cells

Gracile fascicle

Cuneate fascicle

Spinal cord

Touch

Pain

cortex (S-I), whereas the VPS nucleus conveys proprioceptive information principally to area 3a.

The Somatosensory Cortex Is Organized Into Functionally Specialized Columns

Conscious awareness of touch is believed to originate in the cerebral cortex. Tactile information enters the cerebral cortex through the primary somatosensory cortex (S-I) in the postcentral gyrus of the parietal lobe. The primary somatic sensory cortex comprises four cytoarchitectural areas: Brodmann's areas 3a, 3b, 1, and 2 (Figure 19–12). These areas are interconnected such that processing of sensory information in S-I involves both serial and parallel processing.

In a series of pioneering studies of the cerebral cortex, Vernon Mountcastle discovered that S-I cortex is organized into vertical columns or slabs. Each column is 300 to 600 μm wide and spans all six cortical layers from the pial surface to the white matter (Figure 19–13). Neurons within a column receive inputs from the same local area of skin and respond to the same class or classes of touch receptors. A column therefore comprises an elementary functional module of

the neocortex; it provides an anatomical structure that organizes sensory inputs to convey related information about location and modality.

The columnar organization of the cortex is a direct consequence of intrinsic cortical circuitry, the projection patterns of thalamocortical axons, and migration pathways of neuroblasts during cortical development. The pattern of connections within a column is oriented vertically, perpendicular to the cortical surface. Thalamocortical axons terminate primarily on clusters of stellate cells in layer IV, whose axons project vertically toward the surface of the cortex, as well as on star pyramid cells. Thus, thalamocortical inputs are relayed to a narrow column of pyramidal cells that are contacted by the layer IV cell axons. The apical dendrites and axons of cortical pyramidal cells in other cortical layers are also largely oriented vertically, parallel to the thalamocortical axons and stellate cell axons (Figure 19–14). This allows the same information to be processed by a column of neurons throughout the thickness of the cortex.

Pyramidal neurons form the principal excitatory class of somatosensory cortex; they compose approximately 80% of S-I neurons. Pyramidal neurons in

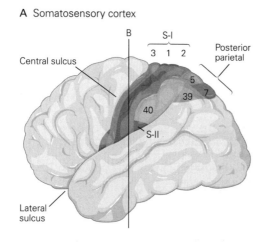

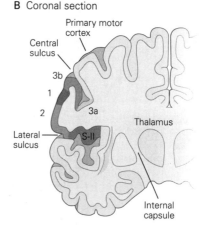

Figure 19–12 The somatosensory areas of the cerebral cortex in the human brain.

A. The somatosensory areas of cortex lie in the parietal lobe and consist of three major divisions. The *primary somatosensory cortex* (**S-I**) forms the anterior part of the parietal lobe. It extends throughout the postcentral gyrus beginning at the bottom of the central sulcus, extending posteriorly to the postcentral sulcus, and into the medial wall of the hemisphere to the cingulate gyrus (not shown). The S-I cortex comprises four distinct cytoarchitectonic regions: Brodmann's areas 3a, 3b, 1, and 2. The *secondary somatosensory cortex* (**S-II**) is located on the upper bank of the lateral sulcus (Sylvian fissure) and on the parietal operculum; it covers Brodmann's

area 43. The *posterior parietal cortex* surrounds the intraparietal sulcus on the lateral surface of the hemisphere, extending from the postcentral sulcus to the parietal-occipital sulcus and medially to the precuneus. The superior parietal lobule (Brodmann's areas 5 and 7) is a somatosensory area; the inferior parietal lobule (areas 39 and 40) receives both somatosensory and visual inputs.

B. A coronal section through the postcentral gyrus illustrates the anatomical relationship of S-I, S-II, and the primary motor cortex (area 4). S-II lies adjacent to area 2 in S-I and extends medially along the upper bank of the lateral sulcus to the insular cortex. The primary motor cortex lies rostral to area 3a within the anterior wall of the central sulcus.

A Sagittal section of monkey S-I cortex

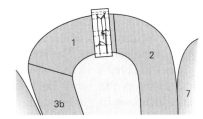

Figure 19–13 Organization of neuronal circuits within a column of somatosensory cortex. Sensory inputs from the skin or deep tissue are organized in columns of neurons that run from the surface of the brain to the white matter. Each column receives thalamic input primarily in layer IV from one part of the body. Excitatory neurons in layer IV send their axons vertically toward the surface of the cortex, contacting the dendrites of pyramidal neurons in layers II and III (supragranular layers) as well as the apical dendrites of pyramidal cells in the infragranular layers (layers V and VI). In this manner, tactile information from a body part such as a finger is distributed vertically within a column of neurons.

B Expanded view of cortical histology   C Schematic cortical circuits

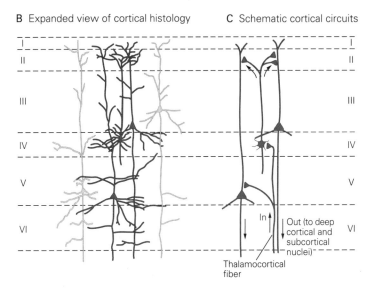

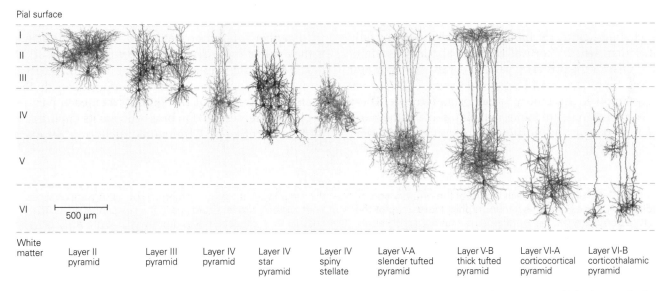

Figure 19–14 Columnar organization of the somatosensory cortex. Cortical excitatory neurons in the six layers have distinctive pyramidal-type shapes with large cell bodies, a single apical dendrite that projects vertically toward the cortical surface and arborizes in more superficial layers, and multiple basal dendrites that arborize close to the cell body. Pyramidal neurons differ in size, gene expression patterns, the length and thickness of their apical dendrite, and the projection targets of their axons.

All of these neurons synapse on targets within the cerebral cortex. Additionally, the pyramidal neurons in layer V project subcortically to the spinal cord, brain stem, midbrain, and basal ganglia. Corticothalamic neurons in layer VI project back to the afferent thalamic nucleus providing sensory input to that column. Spiny stellate neurons in layer IV are the only excitatory cells shown that are not pyramidal neurons. (Adapted, with permission, from Oberlaender et al. 2012.)

each of the six cortical layers project to specific targets (Figure 19–14). Recurrent horizontal connections link pyramidal neurons in the same or neighboring columns, allowing them to share information when activated simultaneously by the same stimulus. Neurons in layers II and III also project to layer V in the same column, to higher cortical areas in the same hemisphere, and to mirror-image locations in the opposite hemisphere. These feedforward connections to higher cortical areas allow complex signal integration, as described later in this chapter.

Pyramidal neurons in layer V provide the principal output from each column. They receive excitatory inputs from neurons in layers II and III in the same and adjacent columns as well as sparse thalamocortical inputs. Neurons in the superficial portion of layer V (layer V-A) send feedforward outputs bilaterally to layer IV of higher-order cortical areas (see Figure 19–17C) as well as to the striatum. Neurons deeper in layer V (layer V-B) project to subcortical structures, including the basal ganglia, superior colliculus, pontine and other brain stem nuclei, the spinal cord, and dorsal column nuclei. Layer VI neurons project to local cortical neurons, and back to the thalamus, particularly to regions of the ventral posterior nuclei providing inputs to that column.

In addition to feedforward signals of information from touch receptors, feedback signals from layers II and III of higher somatosensory cortical areas are provided to layer I in lower cortical areas, regulating their excitability. Such feedback signals originate not only in somatosensory cortical areas but also in sensorimotor areas of the posterior parietal cortex, frontal motor areas, limbic areas, and regions of the medial temporal lobe involved in memory formation and storage. These feedback signals are thought to play a role in the selection of sensory information for cognitive processing (by the mechanisms of attention) and in short-term memory tasks. Feedback pathways may also gate sensory signals during motor activity. Various local inhibitory interneurons within each column serve to focus columnar output.

Cortical Columns Are Organized Somatotopically

The columns within the primary somatic sensory cortex are arranged topographically such that there is a complete somatotopic representation of the body in each of the four areas of S-I (Figure 19–15). The cortical map of the body corresponds roughly to the spinal dermatomes (see Figure 18–13). Sacral segments are represented medially, lumbar and thoracic segments centrally, cervical segments more laterally, and the trigeminal representation of the face at the most lateral portion of S-I cortex. Knowledge of the neural map of the body in the brain is important for localizing damage to the cortex from stroke or head trauma.

The body surface is represented in at least 10 distinct neural maps in the parietal lobe: four in S-I, four in S-II, and at least two in the posterior parietal cortex. As a result, these regions mediate different aspects of tactile sensation. Neurons in areas 3b and 1 of S-I process details of surface texture, whereas those in area 2 represent the size and shape of objects. These attributes of somatic sensation are further elaborated in S-II and the posterior parietal cortex, where neurons are engaged in object discrimination and manipulation, respectively.

Another important feature of somatotopic maps is the amount of cerebral cortex devoted to each body part. The neural map of the body in the human brain, termed the *homunculus*, does not duplicate exactly the spatial topography of the skin. Rather, each part of the body is represented in proportion to its importance to the sense of touch. Disproportionately large areas are

Figure 19–15 (Opposite) Each region of the primary somatosensory cortex contains a topographic neural map of the entire body surface. (Adapted, with permission, from Nelson et al. 1980. Copyright © 1980 Alan R. Liss, Inc.)

A. The primary somatosensory cortex in the macaque monkey lies caudal to the central sulcus as in the human brain. The colored areas on the macaque cortex correspond to the homologous Brodmann's areas of the human brain in Figure 19–12. Area 5 in the macaque monkey is homologous to areas 5 and 7 in humans. Area 7 in macaques is homologous to areas 39 and 40 in humans.

B. The flat map diagram on the right shows the somatosensory cortex of the macaque monkey unfolded along the central sulcus (**dotted line** that parallels the border between areas 3b and 1). The upper part of the diagram includes cortex unfolded from the medial wall of the hemisphere. Body maps were obtained from microelectrode recordings in the postcentral gyrus. The body surface is mapped to columns within rostrocaudal bands arranged in the order of the spinal dermatomes. The body maps in areas 3b and 1 form mirror images of the distal-proximal or dorsal-ventral axes of each dermatome. Each finger (D5–D1) has its own representation along the medial-lateral axis of the cortex in areas 3b and 1, but inputs from several adjacent fingers converge in the receptive fields of neurons in areas 2 and 5.

C. Cortical magnification of highly innervated skin areas. Although the trunk (**violet**) is covered by a greater area of skin than the fingers (**red**), the number of cortical columns responding to touch on the fingers is nearly three times the number activated by touching the trunk because of the higher innervation density of the fingers.

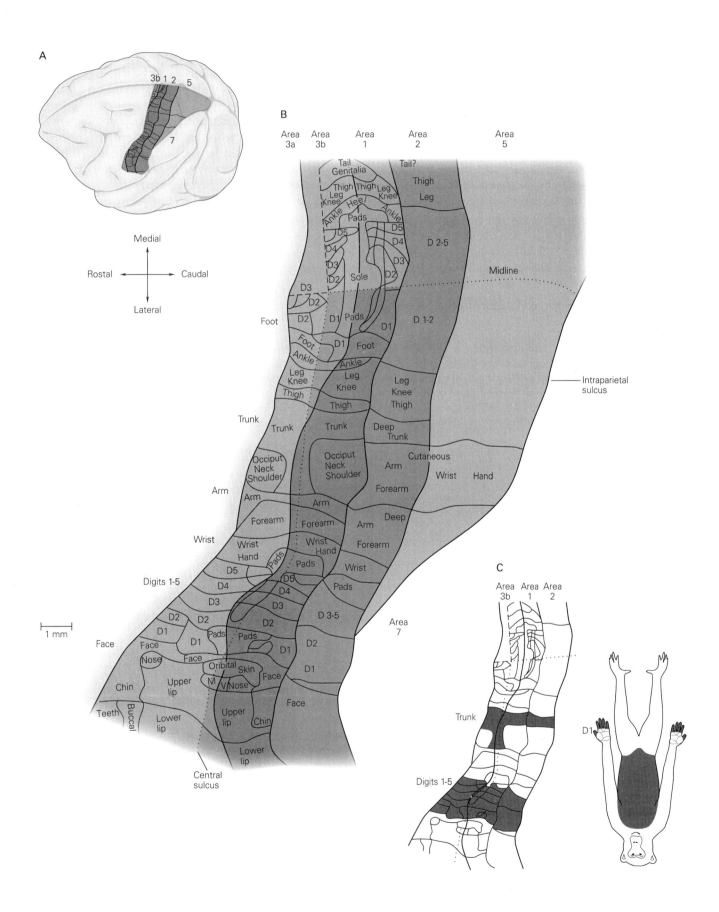

devoted to certain body regions, particularly the hand, foot, and mouth, and relatively smaller areas to more proximal body parts. In humans and monkeys, more cortical columns are devoted to the fingers than to the entire trunk (Figure 19–15C).

The amount of cortical area devoted to a unit area of skin—called the *cortical magnification*—varies by more than a hundredfold across different body surfaces. It is closely correlated with the innervation density and thus the spatial acuity of the touch receptors in an area of skin. The areas with greatest magnification in the human brain—the lips, tongue, fingers, and toes—have tactile acuity thresholds of 0.5, 0.6, 1.0, and 4.5 mm, respectively.

Rodents and other mammals that probe the environment with their whiskers have a large number of columns in S-I, named *barrels*, that receive inputs from individual vibrissae on the face (Box 19–2). Barrel cortex provides a widely used experimental preparation for studying cortical circuitry.

Box 19–2 The Rodent Whisker-Barrel System

The rodent whisker-barrel system is a widely used animal model in modern neuroscience. Most mammals and all primates except man possess specialized tactile hairs on their face called *vibrissae*. Distinct from other hairs on the skin, vibrissae grow from a follicle that is densely innervated by the trigeminal cranial nerve and surrounded by a blood-filled sinus.

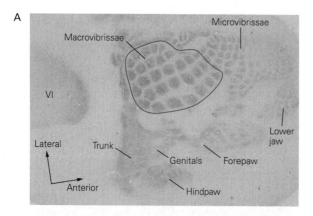

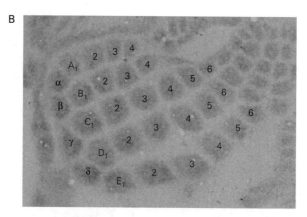

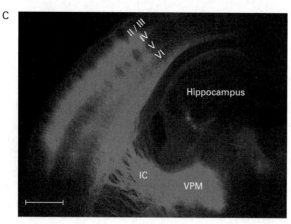

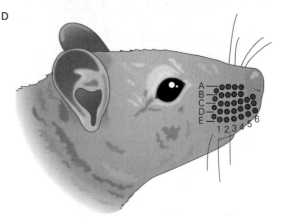

The Receptive Fields of Cortical Neurons Integrate Information From Neighboring Receptors

The neurons in S-I are at least three synapses beyond touch receptors in the skin. Their inputs represent information processed in the dorsal column nuclei, the thalamus, and the cortex itself. Each cortical neuron receives inputs arising from receptors in a specific area of the skin, and these inputs together are its receptive field. We perceive that a particular location on the skin is touched because specific populations of neurons in

the cortex are activated. This experience can be induced experimentally by electrical or optogenetic stimulation of the same cortical neurons.

The receptive fields of cortical neurons are much larger than those of somatosensory fibers in peripheral nerves. For example, the receptive fields of SA1 and RA1 fibers innervating the fingertip are tiny spots on the skin (Figure 19–5), whereas those of the cortical neurons receiving these inputs cover an entire fingertip or several adjacent fingers (Figure 19–17B). The receptive field of a neuron in area 3b represents a

---

Many mammalian species actively move these large facial whiskers using specialized muscles that wrap like slings around each individual follicle. Mice and rats, two of the most commonly used vertebrate model organisms, rely more heavily on their sense of whisker-mediated touch than on their other senses during exploration.

Rodents rhythmically sweep their whiskers across objects in much the same way that humans palpate objects with their fingertips. Despite their structural differences, vibrissae and fingertips afford similar psychophysical thresholds and discriminative sensitivities. Whiskers mediate diverse abilities, including localizing objects in space, discriminating textures and shapes, navigating the environment, interacting socially, and capturing prey.

The rodent somatosensory cortex has evolved proportional to this system's high ethological relevance. For

instance, the rat somatosensory cortex is thicker than the primary visual cortex of the cat, a highly visual animal.

The representation of the largest whiskers (macrovibrissae) in rodent S-I is enlarged relative to that of other parts of the body (Figure 19–16). In contrast to the continuous representations of the skin or retina, the cortical networks dedicated to processing information from individual whiskers are discrete and anatomically identifiable. Each whisker maps one-to-one onto a distinct cluster of excitatory neurons visible in cortical layer IV called a *barrel*.

Barrels are densely interconnected networks that are established during development by the interaction of thalamocortical axons with cortical neurons. This unique correspondence facilitates diverse studies of cortical microcircuits, development, experience-dependent plasticity, sensorimotor integration, tactile behavior, and disease.

Randy M. Bruno

Figure 19–16 (Opposite) The "barrel cortex" of rodents represents the vibrissae in topographic patterns. The barrel cortex, a subregion of the rodent primary somatosensory (S-I) cortex that represents the facial vibrissae, is a widely studied structure used to decipher cortical circuits. (Adapted from Bennett-Clarke et al. 1997 and Wimmer et al. 2010.)

A. Tangential histological section through layer IV of the somatosensory cortex of a juvenile rat stained for serotonin. The darker immunoreactive patches correspond to cortical representations of specific body parts. The largest part of the rodent somatosensory cortical map is devoted to the vibrissae.

B. Enlarged view of the macrovibrissae representation in S-I. The spatial pattern of the whiskers on the face is stereotyped from animal to animal, allowing each cortical "barrel"

to be identified by row with the letter, and by arc (column) with the number of the corresponding whisker. Neurons in each barrel are most responsive to motion of this principal whisker.

C. A rat brain section cut obliquely along the path axons travel from the ventroposterior medial (**VPM**) thalamic nucleus to S-I. Green fluorescent protein–labeled VPM axons project through the internal capsule (**IC**) to the subcortical white matter and travel parallel to the pial surface before entering the cortex. The axons densely innervate layer IV where they form discrete barrels and more sparsely and diffusely innervate the border of layers V and VI. Scale bar = 1 mm.

D. The topographic arrangement of the barrels in the cortex matches the spatial arrangement of vibrissae on the face in rows (letters) and arcs (numbers).

Figure 19–17 The hand area of S-I cortex.

A. This sagittal section through the hand representation illustrates the rostrocaudal anatomy of the four subregions of S-I (areas 3a, 3b, 1, and 2) in the human brain and the adjacent primary motor cortex (area 4) and posterior parietal cortex (area 5). Labels on the cortical surface indicate columns representing individual fingers (D2–D5); arrows to the right denote the section orientation in the brain. The four S-I regions process different types of somatosensory information indicated by color-matched rectangles below the cortical section. Neurons in area 5 respond mainly to goal-directed active hand movements. (Abbreviations: **RA1**, rapidly adapting type 1; **RA2**, rapidly adapting type 2; **SA1**, slowly adapting type 1.)

B. Typical receptive fields of neurons in each area of S-I of macaque monkeys are shown as colored patches on the hand icons. The fields are outlined by applying light touch to the skin or moving individual joints. Receptive fields are smallest in areas 3a and 3b, where tactile information first enters the cortex, and are progressively larger in areas 1, 2, and 5, reflecting convergent inputs from neurons in area 3b that are stimulated together when the hand is used. Neurons in area 5 and in S-II cortex often have bilateral receptive fields because they respond to touch at mirror-image locations on both hands. (Adapted from Gardner 1988; Iwamura et al. 1993; Iwamura, Iriki, and Tanaka 1994.)

C. Feedforward hierarchical connections between somatosensory cortical areas. The strength of thalamocortical and corticocortical connections is indicated by the thickness of arrows interconnecting these areas. Neurons in the thalamus send their axons mainly to areas 3a and 3b, but some also project to areas 1 and 2. In turn, neurons in cortical areas 3a and 3b project to areas 1 and 2. Information from the four areas of S-I is conveyed to neurons in the posterior parietal cortex (area 5) and in S-II. Many of these connections are bidirectional; neurons in higher order cortical areas project back to lower order regions, particularly to layer I. (**PR**, parietal rostroventral cortex; **PV**, parietal ventral cortex; **VPL**, ventral posterior lateral nuclei; **VPM**, ventral posterior medial nuclei; **VPS**, ventral posterior superior nuclei). (Adapted, with permission, from Felleman and Van Essen 1991. Copyright © 1991, Oxford University Press.)

A The hand area of primary somatosensory (S-I) cortex

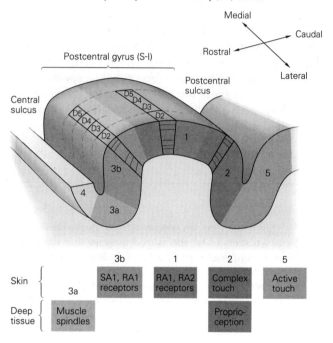

B Receptive fields

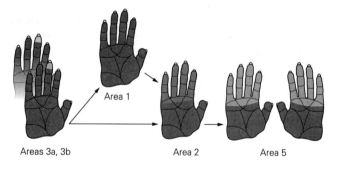

C Hierarchical connections to and from S-I

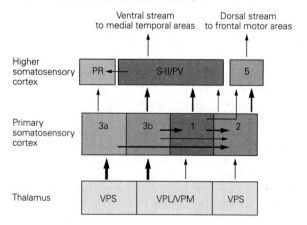

composite of inputs from 300 to 400 nerve fibers, and typically covers a single phalanx or palm pad. Inputs from SA1 and RA1 touch receptors in the same skin region converge on common neurons in area 3b.

Receptive fields in higher cortical areas are even larger, spanning functional regions of skin that are activated simultaneously during motor activity. These include the tips of several adjacent fingers, or an entire finger, or both the fingers and the palm. Neurons in areas 1 and 2 of S-I are concerned with information more abstract than just their innervation sites on the body. Neurons whose receptive fields include more than one finger fire at higher rates when several fingers are touched simultaneously and, in this way, signal the size and shape of objects held in the hand. These large receptive fields allow cortical neurons to integrate the fragmented information from individual touch receptors, enabling us to recognize the overall shape of an object. For example, such neurons may distinguish the handle of a screwdriver from its blade.

Convergent inputs from different sensory receptors in S-I may also allow individual neurons to detect the size and shape of objects. Whereas neurons in areas 3b and 1 respond only to touch and neurons in area 3a respond to muscle stretch, many of the neurons in area 2 receive both inputs. Thus, neurons in area 2 can integrate information about the hand shape used to grasp an object, the grip force applied by the hand, and the tactile stimulation produced by the object; this integrated information may be sufficient to recognize the object.

The receptive fields of cortical neurons usually have an excitatory zone surrounded by or superimposed upon inhibitory zones (Figure 19–18A). Stimulation of regions of skin outside the excitatory zone may reduce the neuron's responses to tactile stimulation within the receptive field. Similarly, repeated stimulation within the receptive field may also decrease neuronal responsiveness because the excitability of the pathway is diminished by longer lasting inhibition mediated by local interneurons.

Inhibitory receptive fields result from feedforward and feedback connections through interneurons in the dorsal column nuclei, the thalamus, and the cortex itself that limit the spread of excitation. Inhibition generated by strong activity in one circuit reduces the output of nearby neurons that are only weakly excited. The inhibitory networks ensure that the strongest of

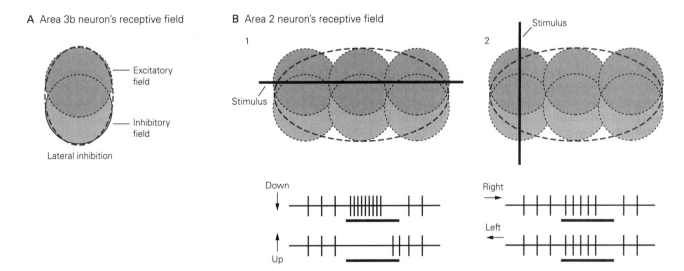

Figure 19–18 The spatial arrangement of excitatory and inhibitory inputs to a cortical neuron determines which stimulus features are encoded by the neuron.

A. A neuron in area 3b of the primary somatosensory cortex has overlapping excitatory and inhibitory zones within its receptive field. (Adapted, with permission, from DiCarlo et al. 1998; Sripati et al. 2006. Copyright © Society for Neuroscience.)

B. Convergence of three presynaptic neurons with the same arrangement of excitatory and inhibitory zones allows direction and orientation selectivity in a neuron in area 2. **1.** Downward motion of a horizontal bar across the receptive field of the postsynaptic cell produces a strong excitatory response because the excitatory fields of all three presynaptic neurons are contacted simultaneously. Upward motion of the bar strongly inhibits firing because it enters all three inhibitory fields first. The neuron responds poorly to upward motion through the excitatory field because the initial inhibition outlasts the stimulus. **2.** Motion of a vertical bar across the receptive field evokes a weak response because it simultaneously crosses the excitatory and inhibitory receptive fields of the input neurons. Motion to the left or right cannot be distinguished in this example.

several competing responses is transmitted, permitting a winner-take-all strategy. These circuits prevent blurring of tactile details such as texture when large populations of touch neurons are stimulated. In addition, higher centers in the brain use inhibitory circuits to focus attention on relevant information from the hand when it is used in skilled tasks, by suppressing unwanted, distracting inputs.

The size and position of receptive fields on the skin are not fixed permanently but can be modified by experience or injury to sensory nerves (Chapter 53). Cortical receptive fields appear to be formed during development and maintained by simultaneous activation of the input pathways. If a peripheral nerve is injured or transected, its cortical projection targets acquire new receptive fields from less effective sensory inputs that are normally suppressed by inhibitory networks, or from newly developed connections from neighboring skin areas that retain innervation. Likewise, extensive stimulation of afferent pathways through repeated practice may strengthen synaptic inputs, improving perception and thereby performance.

Touch Information Becomes Increasingly Abstract in Successive Central Synapses

Somatosensory information is conveyed in parallel from the four areas of S-I to higher centers in the cortex, such as the secondary somatosensory cortex (S-II), the posterior parietal cortex, and the primary motor cortex (Figure 19–17C). As information flows toward higher-order cortical areas, specific combinations of stimulus patterns are needed to excite individual neurons.

Signals from neighboring neurons are combined in higher cortical areas to discern global properties of objects such as their orientation on the hand, or the direction of motion (Figure 19–19). In general, cortical neurons in higher cortical areas are concerned with sensory features that are independent of the stimulus position in their receptive field, abstracting object properties common to a particular class of stimuli.

A cortical neuron is able to detect the orientation of an edge or the direction of motion because of the spatial arrangement of the presynaptic receptive fields. The receptive fields of the excitatory presynaptic neurons are typically aligned along a common axis that generates the preferred orientation of the postsynaptic neuron. In addition, the receptive fields of inhibitory presynaptic neurons at one side of the excitatory fields reinforce the orientation and direction selectivity of postsynaptic neurons (Figure 19–18B).

Cognitive Touch Is Mediated by Neurons in the Secondary Somatosensory Cortex

An S-I neuron's response to touch depends primarily on input from within the neuron's receptive field. This feedforward pathway is often described as a *bottom-up* process because the receptors in the periphery are the principal source of excitation of S-I cortical neurons.

Higher-order somatosensory areas not only receive information from peripheral receptors but are also strongly influenced by top-down cognitive processes, such as goal-setting and attentional modulation. Data obtained from a variety of studies—single-neuron studies in monkeys, neuroimaging studies in humans, and clinical observations of patients with lesions in higher-order somatosensory areas—suggest that the ventral and dorsal regions of the parietal lobe serve complementary functions in the touch system similar to the "what" and "where" pathways of the visual system (see Figure 17–13).

S-II is located on the upper bank and adjacent parietal operculum of the lateral sulcus in both humans and monkeys (Figures 19–12B and 19–20B). Like S-I, the S-II cortex contains four distinct anatomical subregions with separate maps of the body. The central zone—consisting of S-II proper and the adjacent parietal ventral area—receives its major input from areas 3b and 1, largely tactile information from the hand and face. A more rostral region, the parietal rostroventral area, receives information from area 3a about active hand movements as well as tactile information from areas 3b and 1 (Figure 19–20). The most caudal somatosensory region of the lateral sulcus extends onto the parietal operculum (Figure 19–12A). This region abuts the posterior parietal cortex and plays a role in integrating somatosensory and visual properties of objects.

Physiological studies indicate that S-II plays key roles in tactile recognition of objects placed in the hand (stereognosis), distinguishing spatial features, such as shape and texture, and temporal properties, such as vibratory frequency. The receptive fields of neurons in S-II are larger than those in S-I, covering the entire surface of the hand, and are often bilateral, representing symmetric, mirror-image locations on the contralateral and ipsilateral hands. Such large receptive fields enable us to sense the shape of an entire large object grasped in one hand, allowing us to integrate the overall contours of a tool as it contacts the palm and different fingers. Bilateral receptive fields enable us to perceive still larger objects with two hands, such as a watermelon or basketball, sharing the load between them.

The large receptive fields of S-II neurons also influence their physiological responses to motion and vibration. S-II neurons do not represent vibration as

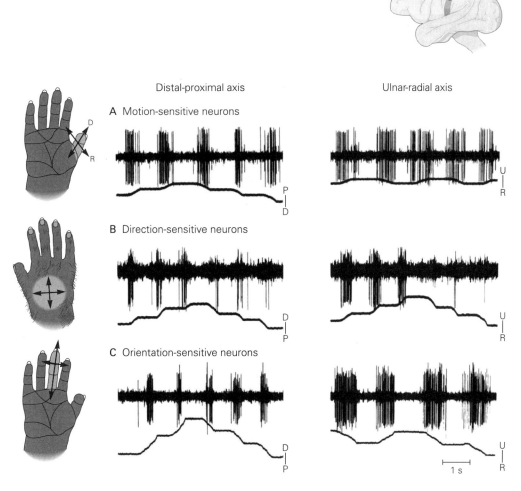

Distal-proximal axis   Ulnar-radial axis

A Motion-sensitive neurons

B Direction-sensitive neurons

C Orientation-sensitive neurons

1 s

Figure 19–19 Neurons in area 2 encode complex tactile information. These neurons respond to motion of a probe across the receptive field but not to touch at a single point. The lower trace indicates the direction of motion by upward and downward deflections. (Adapted, with permission, from Warren, Hämäläinen, and Gardner 1986.)

A. A motion-sensitive neuron responds to stroking the skin in all directions.

B. A direction-sensitive neuron responds strongly to motion toward the ulnar side of the palm but fails to respond to motion in the opposite direction. Responses to distal or proximal movements are weaker.

C. An orientation-sensitive neuron responds better to motion across a finger (ulnar-radial) than to motion along the finger (distal-proximal), but does not distinguish ulnar from radial or proximal from distal directions.

periodic spike trains linked to the oscillatory frequency, as do the sensory fibers from the skin or S-I neurons (Figure 19–9). Instead, S-II neurons abstract temporal or intensive properties of the vibratory stimulus, firing at different mean rates for different frequencies. A similar frequency-dependent transition from temporal- to rate-coding neurons underlies sound processing in primary auditory cortex (Chapter 28), a brain region juxtaposed to S-II cortex in the parietal operculum.

Importantly, the firing rates of S-II neurons depend on the behavioral context or motivational state of the subject. In elegant recent studies, Ranulfo Romo and his colleagues compared responses to vibratory stimuli of neurons in S-I, S-II, and various regions of the frontal lobe of monkeys while the animals performed a two-alternative forced-choice task. The animals were rewarded if they correctly recognized which of two vibratory stimuli was higher in frequency.

Neurons in S-I faithfully represent the vibratory cycles of each stimulus using a temporal code: they fire brief spike bursts in phase with each cycle (Figure 19–9B).

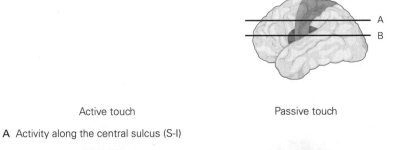

Figure 19–20 Responses in S-I and S-II to active touch are more complex than those evoked by passive touch. Cortical regions in the human brain stimulated by passive and active touch are localized using functional magnetic resonance imaging (fMRI). (Adapted, with permission, from Hinkley et al. 2007.)

A. Axial views of activity along the central sulcus during passive stroking of the right hand with a sponge (*right panel*) and during active touching of the sponge (*left panel*). Areas 3b and 1 are activated in the left hemisphere in both conditions. Active touch also engages the primary motor cortex (**M1**) in the left hemisphere, the anterior cingulate cortex (**ACC**), and evokes weak activity in the ipsilateral S-I (right hemisphere). These sites were confirmed independently using magnetoencephalography in the same subjects.

B. Axial views of activity along the Sylvian fissure in the same experiment. Bilateral activity occurs in S-II and the parietal ventral (**PV**) area during passive stroking and is stronger when the subject actively moves the hand. The parietal rostroventral area (**PR**) is active only during active touch. Magnetoencephalographic responses in S-II/PV and PR occur later than in S-I, reflecting serial processing of touch from S-I to S-II/PV and from S-II/PV to PR.

Active touch

Passive touch

A Activity along the central sulcus (S-I)

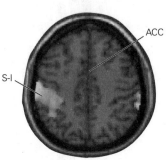

B Activity along the lateral sulcus (S-II)

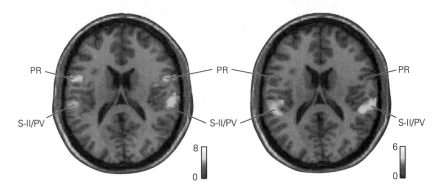

In contrast, S-II neurons respond to the first stimulus with nonperiodic spike trains in which their mean firing rates are directly or inversely correlated with the vibratory frequency (Figure 19–21A). Their responses to the second stimulus are even more abstract. S-II spike trains combine the frequencies of both stimuli (Figure 19–21B). In other words, S-II responses to vibration depend on the stimulus context: the same vibratory stimulus can evoke different firing rates depending on whether the preceding stimulus is higher or lower in frequency.

Even more interesting, Romo's group found that neurons in S-II send copies of the spike trains evoked by the first stimulus to the prefrontal and premotor cortex in order to preserve a memory of that response. Neurons in these frontal cortical areas continue to fire during the delay period after the first stimulus ends. Romo and colleagues proposed that these regions in

the frontal lobe send the memory signal back to S-II when the second stimulus occurs, thereby modifying the response of S-II neurons to the direct tactile signals from the hand. In this manner, sensorimotor memories of previous stimuli influence sensory processing in the brain, allowing subjects to make cognitive judgments about newly arriving tactile stimuli.

S-II is the gateway to the temporal lobe via the insular cortex. Regions of the medial temporal lobe, particularly the hippocampus, are vital to the storage of explicit memory (Chapter 53). We do not store in memory every scintilla of tactile information that enters the nervous system, only that which has some behavioral significance. In light of the demonstration that the firing patterns of S-II neurons are modified by selective attention, S-II could make the decision whether a particular bit of tactile information is remembered.

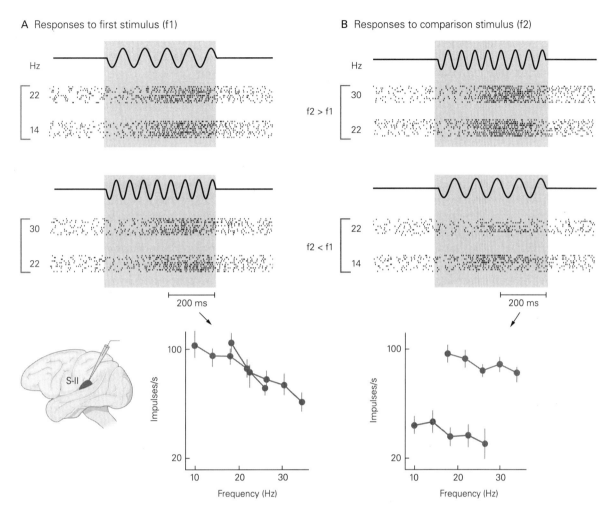

Figure 19–21 The sensitivity of an S-II neuron to vibratory stimuli is modulated by attention and behavioral conditions. A monkey was trained to compare two vibratory stimuli applied at a 3-second interval to the fingertips (**f1** and **f2**) and to indicate which had the higher frequency. The plots show the mean firing rates of the neuron during each of the two stimuli. The animal's decision about which frequency is higher can be predicted from the neural data during each type of trial. The mean firing rates of this neuron are significantly higher at each stimulation frequency when f2 is greater than f1 than when f2 is less than f1. (Adapted, with permission, from Romo et al. 2002. Copyright © 2002 Springer Nature.)

A. Raster plots show the responses of an S-II neuron to various sample stimuli (f1). The vertical tick marks in each row denote action potentials, and individual rows are separate trials of the

stimulus pairs. Trials are grouped according to the frequencies tested. The firing rate of the neuron encodes the vibratory frequency of the sample stimulus; it is higher for low-frequency vibration regardless of the subsequent events. Note that the firing patterns recorded in S-II are not phase-locked to the vibratory cycle as in S-I (see Figure 19–9B).

B. Each row in the raster plots illustrates responses to the comparison stimulus (f2) during the same trials shown in A. The neuron's response to f2 reflects the frequency of both f2 and f1. When f2 > f1, the neuron fires at high rates during f2 and the animal reports that f2 is the higher frequency. When f2 < f1, the neuron fires at low rates during f2 and the animal reports that f1 is the higher frequency. In this manner, the responses of S-II neurons reflect the animal's memory of an earlier event.

Active Touch Engages Sensorimotor Circuits in the Posterior Parietal Cortex

Studies in the mid-1970s by Vernon Mountcastle, Juhani Hyvärinen, and others demonstrated that regions of the posterior parietal cortex surrounding the intraparietal sulcus play an important role in the sensory guidance of movement rather than in discriminative touch. These regions include areas 5 and 7 in monkeys and the superior parietal lobule (Brodmann's areas 5 and 7) and inferior parietal cortex (areas 39 and 40) in humans. These and subsequent studies demonstrated that neural activity in the posterior parietal cortex during reaching and grasping coincides with activation of neurons in motor and premotor areas of the frontal cortex and precedes activity in S-I. Areas 5 and 7 are postulated to be engaged in the planning of hand actions, because the posterior parietal cortex receives convergent central and peripheral signals that allow it to compare central motor commands with somatosensory feedback during reaching and grasping behaviors. Sensory feedback from S-I to the posterior parietal cortex is used to confirm the goal of the planned action, thereby reinforcing previously learned skills or correcting those plans when errors occur.

Predicting the sensory consequences of hand actions is an important component of active touch. For example, when we view an object and reach for it, we predict how heavy it should be and how it should feel in the hand; we use such predictions to initiate grasping. Daniel Wolpert and Randy Flanagan have proposed that during active touch the motor system controls the afferent flow of somatosensory information to the brain so that subjects can predict when tactile information should arrive in S-I and reach consciousness. Convergence of central and peripheral signals allows neurons to compare planned and actual movements. Corollary discharge from motor areas to somatosensory regions of the cortex may play a key role in active touch. It provides posterior parietal cortex neurons with information on intended actions, allowing them to learn new skills and perform them smoothly.

Lesions in Somatosensory Areas of the Brain Produce Specific Tactile Deficits

Patients with lesions in S-I cortex have difficulty responding to simple tactile tests: touch thresholds, vibration and joint position sense, and two-point discrimination (Figure 19–22A). These patients also perform poorly on more complex tasks, such as texture discrimination, stereognosis, and visual–tactile matching tests.

Loss of tactile sensation in the hand produces significant motor as well as sensory deficits. Motor deficits are less pronounced than sensory losses, particularly during tests of force and position control. Exploratory movements and skilled tasks such as catching a ball or pinching small objects between the fingertips are also abnormal.

Local anesthesia of sensory nerve fibers in the hand provides a direct way to appreciate the sensorimotor role of touch. Under local anesthesia of the median and ulnar nerves, hand movements are clumsy and poorly coordinated, and force generation during grasping is abnormally slow. With the loss of tactile sensibility, one is completely reliant on vision for directing the hand. Loss of touch does not cause paralysis or weakness because much of skilled movement is predictive,

relying on sensory feedback for adjustment if necessary. The motor system in these subjects compensates for the absence of tactile information by generating more force than necessary.

These motor problems are exacerbated by long-term, chronic loss of tactile function because of injury to peripheral nerves or dorsal column lesions. Deafferentation produces major changes in the afferent connections in the brain, as do certain diseases. Myelinated afferent fibers in the dorsal columns degenerate in patients with demyelinating diseases, such as multiple sclerosis. In late-stage syphilis, the large-diameter neurons in the dorsal root ganglia are destroyed (tabes dorsalis). These patients have severe chronic deficits in touch and proprioception but often little loss of temperature perception and nociception. The somatosensory losses are accompanied by motor deficits: clumsy and poorly coordinated movements and dystonia. Similar impairments occur in patients with damage to S-I caused by stroke or head trauma, or following surgical excision of the postcentral gyrus.

Patients with lesions in the posterior parietal cortex usually have only mild difficulty with simple tactile tests. However, they have profound difficulty with complex tactile recognition tasks and use few exploratory and skilled movements (Figure 19–22B). They display kinematic deficits when interacting with objects, failing to shape and orient the hand properly to grasp them and misdirecting the arm during reaching. They typically use too much grip force when an object is placed in their hand and are unable to direct the fingers properly when asked to evaluate its size and shape. These deficits are described clinically as the "useless hand" syndrome (tactile apraxia).

Studies of sensory deficits in human patients are complicated by the fact that disease states or trauma rarely produce damage confined to one localized brain area. For this reason, analyses of experimentally controlled lesions in animals have been useful for understanding the etiology of the sensory deficits observed in human patients. For example, macaque monkeys with a lesion of the cuneate fascicle show chronic losses in tactile discrimination, such as higher touch thresholds, impaired vibration sense, and poor two-point discrimination. They also display major deficits in the control of fine finger movements during grooming, scratching, and object manipulation. A similar deficit in skilled movements can be produced experimentally in monkeys by inhibiting the neurons in the hand-representation region of area 2.

Experimental ablation of somatosensory areas of the cortex in monkeys has provided valuable information about the function of these areas. Small lesions

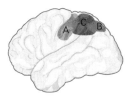

A Anterior parietal lesions

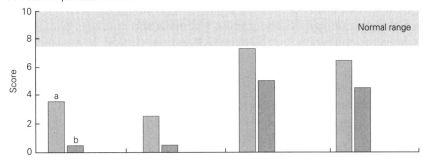

B Posterior parietal lesions

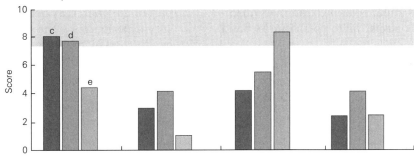

C Combined anterior and posterior parietal lesions

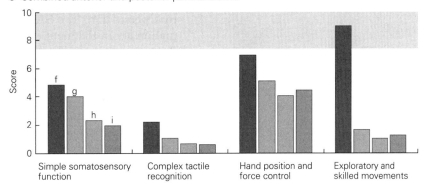

Simple somatosensory function   Complex tactile recognition   Hand position and force control   Exploratory and skilled movements

Figure 19–22 Lesions of anterior and posterior regions of the parietal lobe produce characteristic sensory and motor deficits of the hand. Bar graphs rank the performance of nine patients (a–i) with unilateral parietal cortex brain lesions on four sets of standardized tests of sensory and motor function of the contralateral hand. The behavioral scores are ranked from normal (10) to maximal deficit (0). The normal range shown is the performance score of these patients for the ipsilateral hand. Tests of *simple somatosensory function* include light touch from a 1-g force-calibrated probe, two-point discrimination on the finger and palm, vibration sense, and position sense of the index finger metacarpophalangeal joint. Tests of *complex tactile recognition* assess texture discrimination, form recognition, and size discrimination. Tests of *hand position and force control* measure grip force, tapping, and reaching to a target. Tests of *exploratory and skilled movements* evaluate insertion of pegs in slots, pincer grip of small objects, and exploratory movements

when palpating objects. (Adapted, with permission, from Pause et al. 1989. Copyright © 1989, Oxford University Press.)

A. Two patients with lesions to the anterior parietal lobe show severe impairment in both sets of tactile tests but only moderate impairment in the motor tasks.

B. Three patients with posterior parietal lesions show only minor deficits in simple somatosensory tests but severe impairment in complex tests of stereognosis and form. Motor deficits are greater in skilled tasks.

C. Four patients with combined lesions to anterior and posterior parietal cortex show severe impairment in all tests. Interestingly, the patient who showed the least impairment in this group (patient f) suffered brain damage at birth; the developing brain was able to compensate for the loss of major somatosensory areas. Lesions in the other patients resulted from strokes later in life.

limited to area 3b produce major deficits in touch sensation from a particular part of the body. Lesions in area 1 produce a defect in the assessment of the texture of objects, whereas lesions in area 2 alter the ability to differentiate the size and shape of objects. The damage to tactile function is less severe when such lesions are made in infant animals, apparently because in the developing brain S-II cortex may take over functions normally assumed by S-I.

Removal of S-II cortex in monkeys causes severe impairment in the discrimination of both shape and texture and prevents the animals from learning new tactile discriminations. Ablation or inhibition of areas 2 or 5 produces deficits in roughness discrimination but few other alterations in passive touch. However, motor performance is impaired as these animals misdirect reaching toward objects, fail to preshape the hand

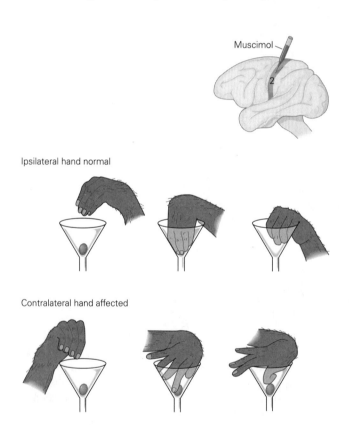

Ipsilateral hand normal

Contralateral hand affected

Figure 19–23 Finger coordination is disrupted when synaptic transmission in the somatic sensory cortex is inhibited in a monkey. Muscimol, a γ-aminobutyric acid (GABA) agonist that inhibits cortical cells, was injected into Brodmann's area 2 on the left side of a monkey's brain. Within minutes after injection, the finger coordination of the right hand (contralateral) was severely disrupted; the monkey was unable to pick up a grape from a funnel. The injection effects are shown to be specific to the injected hemisphere because the left hand (ipsilateral) continues to perform normally. (Adapted, with permission, from Hikosaka et al. 1985. Copyright © 1985 Elsevier B.V.)

to grasp objects skillfully, and have difficulty coordinating finger movements because tactile feedback is absent (Figure 19–23).

The similarity between impairments observed in humans and monkeys is an important basis for understanding clinical losses of somatosensory function. We shall learn in later chapters that lesioning studies of other cortical areas in monkeys have also provided insight into higher-order sensory and motor functions of the brain.

Highlights

1. When we explore an object with our hands, a large part of the brain may become engaged by the sensory experience, by the thoughts and emotions it evokes, and by motor responses to it. These sensations result from the parallel actions of multiple cortical areas engaged in feedforward and feedback networks.

2. At the first touch, the peripheral sensory apparatus deconstructs the object into tiny segments, distributed over a large population of approximately 20,000 sensory nerve fibers. The SA1 system provides high-fidelity information about the object's spatial structure that is the basis of form and texture perception. The SA2 system provides information about the hand conformation and posture during grasping and other hand movements. The RA1 system conveys information about motion of the object in the hand, which enables us to manipulate it skillfully. Together with RA2 receptors, they sense vibration of objects that allows us to use them as tools.

3. The information from touch receptors is conveyed to consciousness by the dorsal column fiber tracts of the spinal cord, relay nuclei in the brain stem and thalamus, and a hierarchy of intracortical pathways. By analyzing patterns of activity across the entire population, the brain constructs a neural representation of objects and actions of the hand.

4. Computations in central pathways are complex and accomplished serially, beginning in the dorsal column nuclei, progressing through the thalamus and several cortical stages, and terminating in regions of the medial temporal cortex concerned with memory and perception and in motor areas of the frontal lobe that mediate voluntary movements.

5. The brain's processing of touch is aided by the topographic, somatotopic organization of the neurons involved at each relay. Adjacent skin areas that

are stimulated together are linked anatomically and functionally in central relays. Body parts that are especially sensitive to touch—the hands, feet, and mouth—are represented in large areas of the brain, reflecting the importance of tactile information conveyed from these regions.

6. Another function of the central pathways is the transformation of the disaggregated representation of object properties among thousands of neurons to an integrated representation of complex object properties in a few neurons. Convergent excitatory connections between neurons representing neighboring skin areas and intracortical inhibitory circuits enable higher-order cortical cells to integrate global features of objects. In this manner, the somatosensory areas of the brain represent properties common to particular classes of objects.

7. A third function is regulating the afferent flow of somatosensory information. The peripheral fibers deliver much more information than can be handled at any one moment; the central neural pathways compensate by selecting information for delivery to the mechanisms of perception and memory. Recurrent pathways from higher brain areas modify the ascending information provided by touch receptors, thus fitting the stream of sensory information to previous experience and task goals.

8. Finally, the touch system provides information necessary for the control and guidance of movement. Interactions between sensory and motor areas of parietal and frontal cortex provide a neural mechanism for planning desired actions, for predicting the sensory consequences of motor behaviors, and for skill learning from repeated experience.

Esther P. Gardner

Selected Reading

Freund HJ. 2003. Somatosensory and motor disturbances in patients with parietal lobe lesions. Adv Neurol 93:179–193.

Harris KD, Shepherd GMG. 2015. The neocortical circuit: themes and variations. Nat Neurosci 18:170–181.

Johnson KO. 2001. The roles and functions of cutaneous mechanoreceptors. Curr Opin Neurobiol 11:455–461.

Jones EG. 2000. Cortical and subcortical contributions to activity-dependent plasticity in primate somatosensory cortex. Annu Rev Neurosci 23:1–37.

Jones EG, Peters A (eds). 1986. *Cerebral Cortex*. Vol 5, *Sensory-Motor Areas and Aspects of Cortical Connectivity*. New York: Plenum Press.

Kaas JH, Gardner EP (eds). 2008. *The Senses: A Comprehensive Reference*. Vol 6, *Somatosensation*. Oxford: Elsevier.

Milner AD, Goodale MA. 1995. *The Visual Brain in Action*. Oxford: Oxford Univ. Press.

Mountcastle VB. 1995. The parietal system and some higher brain functions. Cerebral Cortex 5:377–390.

Mountcastle VB. 2005. *The Sensory Hand: Neural Mechanisms of Somatic Sensation*. Cambridge, MA: Harvard Univ. Press.

Romo R, Salinas E. 2003. Flutter discrimination: neural codes, perception, memory and decision making. Nat Rev Neurosci 4:203–218.

Wing AM, Haggard P, Flanagan JR (eds). 1996. *Hand and Brain*. San Diego, CA: Academic Press.

References

Bennett-Clarke CA, Chiaia NL, Rhodes RW. 1997. Contributions of raphe-cortical and thalamocortical axons to the transient somatotopic pattern of serotonin immunoreactivity in rat cortex. Somatosens Mot Res 14:27–33.

Birznieks I, Macefield VG, Westling G, Johansson RS. 2009. Slowly adapting mechanoreceptors in the borders of the human fingernail encode fingertip forces. J Neurosci 29:9370–9379.

Bolanowski SJ, Pawson L. 2003. Organization of Meissner corpuscles in the glabrous skin of monkey and cat. Somatosens Mot Res 20:223–231.

Brisben AJ, Hsiao SS, Johnson KO. 1999. Detection of vibration transmitted through an object grasped in the hand. J Neurophysiol 81:1548–1558.

Brochier T, Boudreau M-J, Paré M, Smith AM. 1999. The effects of muscimol inactivation of small regions of motor and somatosensory cortex on independent finger movements and force control in the precision grip. Exp Brain Res 128:31–40.

Carlson M. 1981. Characteristics of sensory deficits following lesions of Brodmann's areas 1 and 2 in the postcentral gyrus of *Macaca mulatta*. Brain Res 204:424–430.

Chapman CE, Meftah el-M. 2005. Independent controls of attentional influences in primary and secondary somatosensory cortex. J Neurophysiol 94:4094–4107.

Connor C, Hsiao SS, Phillips J, Johnson KO. 1990. Tactile roughness: neural codes that account for psychophysical magnitude estimates. J Neurosci 10:3823–3836.

Costanzo RM, Gardner EP. 1980. A quantitative analysis of responses of direction-sensitive neurons in somatosensory cortex of alert monkeys. J Neurophysiol 43:1319–1341.

DiCarlo JJ, Johnson KO, Hsaio SS. 1998. Structure of receptive fields in area 3b of primary somatosensory cortex in the alert monkey. J Neurosci 18:2626–2645.

Edin BB, Abbs JH. 1991. Finger movement responses of cutaneous mechanoreceptors in the dorsal skin of the human hand. J Neurophysiol 65:657–670.

Felleman DJ, Van Essen DC. 1991. Distributed hierarchical processing in the primate cerebral cortex. Cereb Cortex 1:1–47.

Fitzgerald PJ, Lane JW, Thakur PH, Hsiao SS. 2006. Receptive field properties of the macaque second somatosensory cortex: representation of orientation on different finger pads. J Neurosci 26:6473–6484.

Flanagan JR, Vetter P, Johansson RS, Wolpert DM. 2003. Prediction precedes control in motor learning. Curr Biol 13:146–150.

Fogassi L, Luppino G. 2005. Motor functions of the parietal lobe. Curr Opin Neurobiol 15:626–631.

Gardner EP. 1988. Somatosensory cortical mechanisms of feature detection in tactile and kinesthetic discrimination. Can J Physiol Pharmacol 66:439–454.

Gardner EP. 2008. Dorsal and ventral streams in the sense of touch. In: JH Kaas, EP Gardner (eds). *The Senses: A Comprehensive Reference.* Vol. 6, *Somatosensation*, pp. 233–258. Oxford: Elsevier.

Gardner EP, Babu KS, Ghosh S, Sherwood A, Chen J. 2007. Neurophysiology of prehension: III. Representation of object features in posterior parietal cortex of the macaque monkey. J Neurophysiol 98:3708–3730.

Hikosaka O, Tanaka M, Sakamoto M, Iwamura Y. 1985. Deficits in manipulative behaviors induced by local injections of muscimol in the first somatosensory cortex of the conscious monkey. Brain Res 325:375–380.

Hinkley LB, Krubitzer LA, Nagarajan SS, Disbrow EA. 2007. Sensorimotor integration in S2, PV, and parietal rostroventral areas of the human Sylvian fissure. J Neurophysiol 97:1288–1297.

Hyvärinen J, Poranen A. 1978. Movement-sensitive and direction and orientation-selective cutaneous receptive fields in the hand area of the post-central gyrus in monkeys. J Physiol (Lond) 283:523–537.

Iwamura Y, Iriki A, Tanaka M. 1994. Bilateral hand representation in the postcentral somatosensory cortex. Nature 369:554–556.

Iwamura Y, Tanaka M, Sakamoto M, Hikosaka O. 1993. Rostrocaudal gradients in neuronal receptive field complexity in the finger region of the alert monkey's postcentral gyrus. Exp Brain Res 92:360–368.

Jenmalm P, Birznieks I, Goodwin AW, Johansson RS. 2003. Influence of object shape on responses of human tactile afferents under conditions characteristic of manipulation. Eur J Neurosci 18:164–176.

Johansson RS. 1996. Sensory control of dexterous manipulation in humans. In: AM Wing, P Haggard, JR Flanagan (eds). *Hand and Brain*, pp. 381–414. San Diego, CA: Academic Press.

Johansson RS, Flanagan JR. 2009. Coding and use of tactile signals from the fingertips in object manipulation tasks. Nat Rev Neurosci 10:345–359.

Johansson RS, Landström U, Lundström R. 1982. Responses of mechanoreceptive afferent units in the glabrous skin of the human hand to sinusoidal skin displacements. Brain Res 244:17–25.

Johansson RS, Vallbo ÅB. 1983. Tactile sensory coding in the glabrous skin of the human hand. Trends Neurosci 6:27–32.

Johnson KO, Phillips JR. 1981. Tactile spatial resolution: I. Two-point discrimination, gap detection, grating resolution and letter recognition. J Neurophysiol 46:1177–1191.

Jones EG, Powell TPS. 1969. Connexions of the somatic sensory cortex of the rhesus monkey. I. Ipsilateral cortical connexions. Brain 92:477–502.

Klatzky RA, Lederman SJ, Metzger VA. 1985. Identifying objects by touch: an "expert system." Percept Psychophys 37:299–302.

Koch KW, Fuster JM. 1989. Unit activity in monkey parietal cortex related to haptic perception and temporary memory. Exp Brain Res 76:292–306.

LaMotte RH, Mountcastle VB. 1979. Disorders in somesthesis following lesions of parietal lobe. J Neurophysiol 42:400–419.

Lederman SJ, Klatzky RL. 1987. Hand movements: a window into haptic object recognition. Cogn Psychol 19:342–368.

Lieber JD, Xia X, Weber AI, Bensmaia SJ. 2017. The neural code for tactile roughness in the somatosensory nerves. J Neurophysiol 118:3107–3117.

Manfredi LR, Saal, HP, Brown KJ, et al. 2014. Natural scenes in tactile texture. J Neurophysiol 111:1792–1802.

Mountcastle VB. 1997. The columnar organization of the neocortex. Brain 120:701–722.

Mountcastle VB, LaMotte RH, Carli G. 1972. Detection thresholds for stimuli in humans and monkeys: comparison with threshold events in mechanoreceptive afferent fibers innervating the monkey hand. J Neurophysiol 35:122–136.

Mountcastle VB, Lynch JC, Georgopoulos AP, Sakata H, Acuna C. 1975. Posterior parietal association cortex of the monkey: command functions for operations within extrapersonal space. J Neurophysiol 38:871–908.

Muniak MA, Ray S, Hsiao SS, Dammann JF, Bensmaia SJ. 2007. The neural coding of stimulus intensity: linking the population response of mechanoreceptive afferents with psychophysical behavior. J Neurosci 27:11687–11699.

Murray EA, Mishkin M. 1984. Relative contributions of SII and area 5 to tactile discrimination in monkeys. Behav Brain Res 11:67–83.

Nelson RJ, Sur M, Felleman DJ, Kaas JH. 1980. Representations of the body surface in postcentral parietal cortex of *Macaca fascicularis*. J Comp Neurol 192:611–643.

Nolano M, Provitera V, Crisci C, et al. 2003. Quantification of myelinated endings and mechanoreceptors in human digital skin. Ann Neurol 54:197–205.

Oberlaender M, de Kock CP, Bruno RM, et al. 2012. Cell type-specific three-dimensional structure of thalamocortical circuits in a column of rat vibrissal cortex. Cereb Cortex 22:2375–2391.

Pandya DN, Seltzer B. 1982. Intrinsic connections and architectonics of posterior parietal cortex in the rhesus monkey. J Comp Neurol 204:196–210.

Pause M, Kunesch E, Binkofski F, Freund H-J. 1989. Sensorimotor disturbances in patients with lesions of the parietal cortex. Brain 112:1599–1625.

Pei Y-C, Denchev P V, Hsiao SS, Craig JC, Bensmaia SJ. 2009. Convergence of submodality-specific input onto neurons in primary somatosensory cortex. J Neurophysiol 102:1843–1853.

Peters RM, Hackeman E, Goldreich D. 2009. Diminutive digits discern delicate details: fingertip size and the sex difference in tactile spatial acuity. J Neurosci 29:15756–15761.

Phillips JR, Johansson RS, Johnson KO. 1990. Representation of braille characters in human nerve fibres. Exp Brain Res 81:589–592.

Pons TP, Garraghty PE, Mishkin M. 1992. Serial and parallel processing of tactual information in somatosensory cortex of rhesus monkeys. J Neurophysiol 68:518–527.

Pons TP, Garraghty PE, Ommaya AK, Kaas JH, Taub E, Mishkin M. 1991. Massive cortical reorganization after sensory deafferentation in adult macaques. Science 252:1857–1860.

Pruszynski JA, Johansson RS. 2014. Edge-orientation processing in first-order tactile neurons. Nat Neurosci 17:1404–1409.

Quilliam TA. 1978. The structure of finger print skin. In: G Gordon (ed). *Active Touch*, pp. 1–18. Oxford: Pergamon Press.

Robinson CJ, Burton H. 1980. Somatic submodality distribution within the second somatosensory (SII), 7b, retroinsular, postauditory and granular insular cortical areas of *M. fascicularis*. J Comp Neurol 192:93–108.

Romo R, Hernandez A, Zainos A, Lemus L, Brody CD. 2002. Neuronal correlates of decision-making in secondary somatosensory cortex. Nat Neurosci 5:1217–1235.

Saal HP, Bensmaia SJ. 2014. Touch is a team effort: interplay of submodalities in cutaneous sensibility. Trends Neurosci 37:689–697.

Salinas E, Hernandez A, Zainos A, Romo R. 2000. Periodicity and firing rate as candidate neural codes for the frequency of vibrotactile stimuli. J Neurosci 20:5503–5515.

Snider WD. 1998. How do you feel? Neurotrophins and mechanotransduction. Nat Neurosci 1:5–6.

Srinivasan MA, Whitehouse JM, LaMotte RH. 1990. Tactile detection of slip: surface microgeometry and peripheral neural codes. J Neurophysiol 63:1323–1332.

Sripati AP, Yoshioka T, Denchev P, Hsiao SS, Johnson KO. 2006. Spatiotemporal receptive fields of peripheral afferents and cortical area 3b and 1 neurons in the primate somatosensory system. J Neurosci 26:2101–2114.

Talbot WH, Darian-Smith I, Kornhuber HH, Mountcastle VB. 1968. The sense of flutter-vibration: comparison of the human capacity with response patterns of mechanoreceptive afferents from the monkey hand. J Neurophysiol 31:301–334.

Vega-Bermudez F, Johnson KO. 1999. Surround suppression in the responses of primate SA1 and RA mechanoreceptive afferents mapped with a probe array. J Neurophysiol 81:2711–2719.

Warren S, Hämäläinen HA, Gardner EP. 1986. Objective classification of motion- and direction-sensitive neurons in primary somatosensory cortex of awake monkeys. J Neurophysiol 56:598–622.

Weber AI, Saal HP, Lieber JD, et al. 2013. Spatial and temporal codes mediate the tactile perception of natural textures. Proc Nat Acad Sci USA 110:17107–17112.

Weinstein S. 1968. Intensive and extensive aspects of tactile sensitivity as a function of body part, sex, and laterality. In: DR Kenshalo (ed). *The Skin Senses*, pp. 195–222. Springfield, IL: Thomas.

Westling G, Johansson RS. 1987. Responses in glabrous skin mechanoreceptors during precision grip in humans. Exp Brain Res 66:128–140.

Wimmer VC, Bruno RM, de Kock CP, Kuner T, Sakmann B. 2010. Dimensions of a projection column and architecture of VPM and POm axons in rat vibrissal cortex. Cereb Cortex 20:2265–2276.

20

Pain

ACCORDING TO THE INTERNATIONAL ASSOCIATION for the Study of Pain, pain is an unpleasant sensation and emotional experience associated with actual or potential tissue damage, or described in terms of such damage. Pricking, burning, aching, stinging, and soreness are among the most distinctive of all the sensory modalities. As with the other somatosensory modalities—touch, pressure, and position sense—pain serves an important protective function, alerting us to injuries that require evasion or treatment. In children born with insensitivity to pain, severe injuries often go unnoticed and can lead to permanent tissue damage. Yet pain is unlike other somatosensory modalities, or vision, hearing, and smell, in that it has an urgent and primitive quality, possessing a powerful emotional component.

The perception of pain is subjective and is influenced by many factors. An identical sensory stimulus can elicit quite distinct responses in the same individual under different conditions. Many wounded soldiers, for example, do not feel pain until they have been removed from the battlefield; injured athletes are often not aware of pain until a game is over. Simply put, there are no purely "painful" stimuli, sensory stimuli that invariably elicit the perception of pain in all individuals. The variability of the perception of pain is yet another example of a principle that we have encountered in earlier chapters: Pain is not the direct expression of a sensory event but rather the product of elaborate processing in the brain of a variety of neural signals.

When pain is experienced, it can be acute, persistent, or, in extreme cases, chronic. Persistent pain characterizes many clinical conditions and is usually the reason that patients seek medical attention. In contrast, chronic pain appears to have no useful purpose; it only makes patients miserable. Pain's highly individual and subjective nature is one of the factors that make it so difficult to define objectively and to treat clinically.

In this chapter, we discuss the neural processes that underlie the perception of pain in normal individuals and explain the origins of some of the abnormal pain states that are encountered clinically.

Noxious Insults Activate Thermal, Mechanical, and Polymodal Nociceptors

Many organs in the periphery, including skin and subcutaneous structures such as joints and muscles, possess specialized sensory receptors that are activated by noxious insults. Unlike the specialized somatosensory receptors for light touch and pressure, most of these *nociceptors* are simply the free nerve endings of primary sensory neurons. There are three main classes of

nociceptors—thermal, mechanical, and polymodal—as well as a more enigmatic fourth class, termed silent nociceptors.

Thermal nociceptors are activated by extremes in temperature, typically greater than 45°C (115°F) or less than 5°C (41°F). They include the peripheral endings of small-diameter, thinly myelinated Aδ axons that conduct action potentials at speeds of 5 to 30 m/s and unmyelinated C-fiber axons that conduct at speeds less than 1.0 m/s (Figure 20–1A). *Mechanical nociceptors* are activated optimally by intense pressure applied to the skin; they too are the endings of thinly myelinated Aδ axons. *Polymodal nociceptors* can be activated by high-intensity mechanical, chemical, or thermal (both hot and cold) stimuli. This class of nociceptors consists predominantly of unmyelinated C fibers (Figure 20–1A).

A Compound action potential

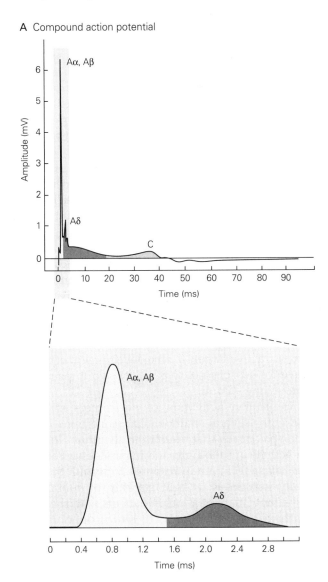

B First and second pain

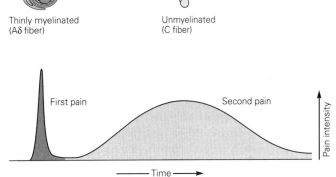

Figure 20–1 Propagation of action potentials in different classes of nociceptive fibers.

A. The speed at which action potentials are conducted is a function of each fiber's cross-sectional diameter. Wave peaks in the figure are labeled alphabetically in order of latency. The first peak and its subdivisions are the summed electrical activity of myelinated A fibers. A delayed (slowly conducting) deflection represents the summed action potentials of unmyelinated C fibers. The compound action potential of the A fibers is shown on a faster time-base to depict the summation of the action potentials of several fibers. (Adapted, with permission, from Perl 2007. Copyright © 2007 Springer Nature.)

B. First and second pain are carried by A delta and C fibers, respectively. (Adapted, with permission, from Fields 1987.)

These three classes of nociceptors are widely distributed in skin and deep tissues and are often co-activated. When a hammer hits your thumb, you initially feel a sharp pain ("first pain") followed by a more prolonged aching and sometimes burning pain ("second pain") (Figure 20–1B). The fast sharp pain is transmitted by Aδ fibers that carry information from damaged thermal and mechanical nociceptors. The slow dull pain is transmitted by C fibers that convey signals from polymodal nociceptors.

Silent nociceptors are found in the viscera. This class of receptors is not normally activated by noxious stimulation; instead, inflammation and various chemical agents dramatically reduce their firing threshold. Their activation is thought to contribute to the emergence of secondary hyperalgesia and central sensitization, two prominent features of chronic pain.

Noxious stimuli depolarize the bare nerve endings of afferent axons and generate action potentials that are propagated centrally. How is this achieved? The membrane of the nociceptor contains receptors that convert the thermal, mechanical, or chemical energy of noxious stimuli into a depolarizing electrical potential. One such protein is a member of a large family of so-called transient receptor potential (TRP) ion channels. This receptor-channel, TRPV1, is expressed selectively by nociceptive neurons and mediates the pain-producing actions of capsaicin, the active ingredient of hot peppers and many other pungent chemicals. The TRPV1 channel is also activated by noxious thermal stimuli, with a threshold for activation around 45°C, the temperature that provokes heat pain. Importantly, TRPV1-mediated membrane currents are enhanced by a reduction in pH, a characteristic of the chemical milieu of inflammation.

Other receptor-channels of the TRP channel family are expressed by nociceptors and underlie the perception of a wide range of temperatures, from cold to intense heat. Of particular interest is TRPM8, a menthol-responsive and cold-sensing channel that likely mediates the extreme cold hypersensitivity produced by many chemotherapeutic drugs (such as oxaliplatin). TRPA1 responds to a variety of irritants, from mustard oil to garlic and even air pollutants (Figure 20–2). Very recently, a family of mechanical transducers (Piezo1 and Piezo2) was described (Chapter 18). These channels may be important contributors to the mechanical hypersensitivity that is a prominent feature of many chronic pain conditions.

In addition to this constellation of TRP channels, sensory neurons express many other receptors and ion channels involved in the transduction of peripheral stimuli. Nociceptors selectively express many different voltage-gated Na^{2+} channels, which are the target of local anesthetics that so effectively block pain. (Think of the dentist who can completely eliminate tooth pain.) Nociceptors express Na^{2+} channels that are sensitive or resistant to tetrodotoxin (TTX). One type of TTX-sensitive channel, Nav1.7, is a key molecular mechanism in the perception of pain in humans, as revealed in the rare individuals who have a loss-of-function mutation in the corresponding *SCN9A* gene. These individuals are insensitive to pain but are otherwise healthy and exhibit normal sensory responses to touch, temperature, proprioception, tickle, and pressure. A second class of mutations in the *SCN9A* gene result in hyperexcitability of nociceptors; individuals with these mutations exhibit an inherited condition called erythromelalgia, in which there is intense, ongoing burning pain of the extremities, accompanied by profound redness (vasodilation). Since Nav1.7, unlike many other voltage-gated Na^+ channels, is not found in the central nervous system, pharmaceutical companies are developing antagonists that will hopefully provide a novel approach to regulating pain processing without the adverse side effects that can occur with systemic administration of lidocaine, which blocks all subtypes of voltage-gated Na^+ channels.

Nociceptors also express an ionotropic purinergic receptor, PTX3, which is activated by adenosine triphosphate (ATP) released from peripheral cells after tissue damage. In addition, they express members of the Mas-related G protein–coupled receptor (Mrg) family, which are activated by peptide ligands and serve to sensitize nociceptors to other chemicals released in their local environment (see Figure 20–7). Subsets of these unmyelinated afferents also include receptor-channels that respond to a variety of itch-provoking substances, including the pruritogens histamine and chloroquine. It follows that these receptors and channels are attractive targets for the development of drugs selective for sensory neurons responsive to pain and itch-provoking stimuli.

Uncontrolled activation of nociceptors is associated with several pathological conditions. Two common pain states that result from alterations in nociceptor activity are allodynia and hyperalgesia. Patients with *allodynia* feel pain in response to stimuli that are normally innocuous: a light stroking of sunburned skin, the movement of joints in patients with rheumatoid arthritis, and even the act of getting out of bed in the morning after a vigorous workout. Nevertheless, patients with allodynia do not feel pain constantly; in the absence of a peripheral stimulus, there is no pain. In contrast, patients with *hyperalgesia*—an exaggerated

A Thermosensitivity of TRP channels in *Xenopus* oocytes

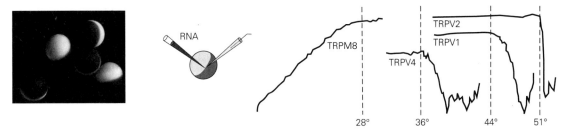

B Thermosensitivity of TRP channels in dorsal root ganglion cells

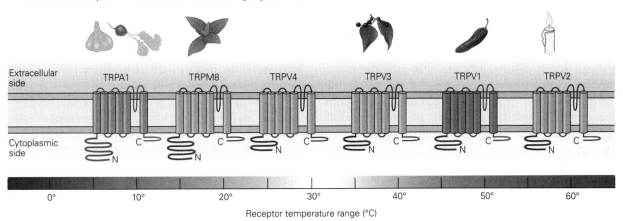

C Pathway to TRP channel opening

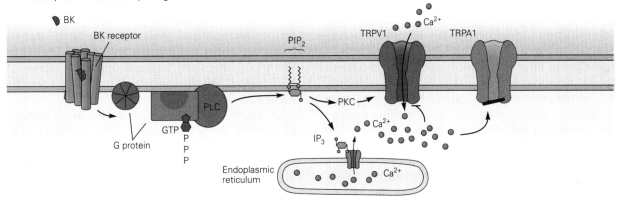

Figure 20–2 Transient receptor potential ion channels in nociceptive neurons.

A. Recordings from *Xenopus* oocytes injected with mRNA encoding transient receptor potential (**TRP**) channels reveal the thermosensitivity of the channels. The temperature (centigrade) at which a specific TRP channel is activated is shown by the downward deflection of the recording. (Photograph on left reproduced, with permission, from Erwin Siegel 1987; traces on the right reproduced, with permission, from Tominaga and Caterina 2004.)

B. Temperature response profiles of different TRP channels expressed by dorsal root ganglion neurons. (Adapted, with

permission, from Jordt, McKemy, and Julius 2003; Dhaka, Viswanath, and Patapoutian 2006.)

C. Bradykinin (**BK**) binds to G protein–coupled receptors on the surface of primary afferent neurons to activate phospholipase C (**PLC**), leading to the hydrolysis of membrane phosphatidylinositol bisphosphate (**PIP₂**), the production of inositol 1,4,5-trisphosphate (**IP₃**), and the release of Ca^{2+} from intracellular stores. Activation of protein kinase C (**PKC**) regulates TRP channel activity. The TRPV1 channel is sensitized, leading to channel opening and Ca^{2+} influx. (Source: Bautista et al. 2006.)

response to noxious stimuli—typically report persistent pain in the absence of sensory stimulation.

Persistent pain can be subdivided into two broad classes, nociceptive and neuropathic. *Nociceptive pain* results from the activation of nociceptors in the skin or soft tissue in response to tissue injury, and it usually occurs with inflammation. Sprains and strains produce mild forms of nociceptive pain, whereas arthritis or a tumor that invades soft tissue produce a much more severe nociceptive pain. Typically, nociceptive pain is treated with nonsteroidal anti-inflammatory drugs (NSAIDS; see later discussion) or, when severe, with opiates such as morphine.

Neuropathic pain results from direct injury to nerves in the peripheral or central nervous system, and is often accompanied by a burning or electric sensation. Neuropathic pains include complex regional pain syndrome, which can follow even very minor damage to a limb peripheral nerve; post-herpetic neuralgia, the severe pain experienced by many patients after a bout of shingles; or trigeminal neuralgia, an intense, shooting pain in the face that results from an as yet unknown pathology of the trigeminal nerve. Other neuropathic pains include phantom limb pain, which can occur after limb amputation (see Figure 20–14). In some instances, spontaneous, ongoing, often burning pain can even occur without a peripheral stimulus, a phenomenon termed *anesthesia dolorosa*. This syndrome can be triggered following attempts to treat trigeminal neuralgia by ablating trigeminal sensory neurons. Neuropathic pains do not respond to NSAIDS and are generally poorly responsive to opiates. Finally, lesions of the central nervous system, for example, in multiple sclerosis, after stroke, or after spinal cord injury, can also result in central neuropathic pain states. Since loss of inhibitory controls (as occurs in epilepsy) is an important contributor to neuropathic pain, the first-line therapy for neuropathic pain, not surprisingly, involves anticonvulsants, notably the gabapentinoids. (The reference to γ-aminobutyric acid [GABA] was based on a structural similarity of gabapentin to GABA. However, gabapentin in fact exerts its action by binding to the $\alpha_2\delta$-subunit of voltage-gated Ca^{2+} channels, ultimately decreasing neurotransmitter release.)

Signals From Nociceptors Are Conveyed to Neurons in the Dorsal Horn of the Spinal Cord

The sensation of noxious stimuli arises from signals in the peripheral axonal branches of nociceptive sensory neurons whose cell bodies are located in dorsal root ganglia. The central branches of these neurons terminate in the spinal cord in a highly orderly manner. Most terminate in the dorsal horn. Primary afferent neurons that convey distinct sensory modalities terminate in different laminae (Figure 20–3B) such that there is a tight link between the anatomical organization of dorsal horn neurons, their receptive properties, and their function in sensory processing.

Many neurons in the most superficial lamina of the dorsal horn, termed *lamina I* or the *marginal layer*, respond to noxious stimuli conveyed by Aδ and C fibers. Because they respond selectively to noxious stimulation, they have been called *nociceptive-specific neurons*. This set of neurons projects to the midbrain and thalamus. A second class of lamina I neurons receives input from C fibers that are activated selectively by cool stimuli. Other classes of lamina I neurons respond in a graded fashion to both innocuous and noxious mechanical stimulation and thus are termed *wide dynamic range neurons*.

Lamina II, the substantia gelatinosa, is a densely packed layer that contains many different classes of local interneurons, some excitatory and others inhibitory. Some of these interneurons respond selectively to pain-provoking inputs, whereas others are selectively activated by itch-provoking stimuli. Laminae III and IV contain a mixture of local interneurons and supraspinal projection neurons. Many of these neurons receive input from Aβ afferent fibers that respond to innocuous cutaneous stimuli, such as deflection of hairs and light pressure. Lamina V contains neurons that respond to a wide variety of noxious stimuli and project to the brain stem and thalamus. These neurons receive direct inputs from Aβ and Aδ fibers and, because their dendrites extend into lamina II, are also innervated by C-fiber nociceptors (Figure 20–3B).

Neurons in lamina V also receive input from nociceptors in visceral tissues. The convergence of somatic and visceral nociceptive inputs onto individual lamina V neurons provides one explanation for a phenomenon called "referred pain," a condition in which pain from injury to a visceral tissue is perceived as originating from a region of the body surface. Patients with myocardial infarction, for example, frequently report pain from the left arm as well as the chest (Figure 20–4). This phenomenon occurs because a single lamina V neuron receives sensory input from both regions, and thus a signal from this neuron does not inform higher brain centers about the source of the input. As a consequence, the brain often incorrectly attributes the pain to the skin, possibly because cutaneous inputs predominate. Another anatomical explanation for instances of referred pain is that the axons of nociceptive sensory

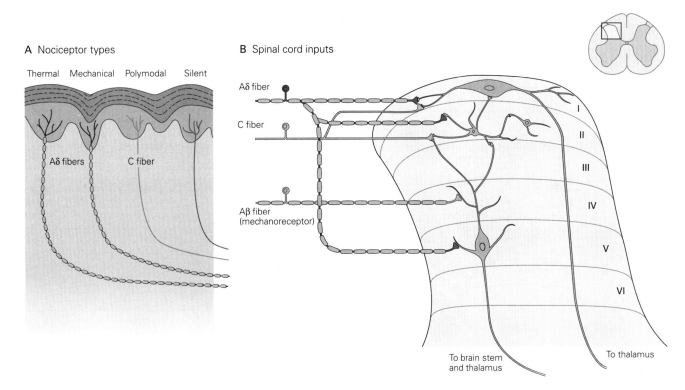

A Nociceptor types

Thermal   Mechanical   Polymodal   Silent

Aδ fibers   C fiber

B Spinal cord inputs

Aδ fiber

C fiber

Aβ fiber
(mechanoreceptor)

I
II
III
IV
V
VI

To brain stem
and thalamus

To thalamus

Figure 20–3 Nociceptive fibers terminate in different laminae of the dorsal horn of the spinal cord.

A. There are three main classes of peripheral nociceptor as well as the silent nociceptors, which are activated by inflammation and various chemical substances.

B. Neurons in lamina I of the dorsal horn receive direct input from myelinated (**Aδ**) nociceptive fibers and both direct and indirect input from unmyelinated (**C**) nociceptive fibers via interneurons in lamina II. Lamina V neurons receive low-threshold input from large-diameter myelinated Aβ mechanoreceptive fibers as well as inputs from nociceptive Aδ and C fibers. Lamina V neurons send dendrites to lamina IV, where they are contacted by the terminals of Aβ primary afferents. Axon terminals of lamina II interneurons can make contact with dendrites in lamina III that arise from cells in lamina V. Aα primary afferents contact motor neurons and interneurons in the ventral spinal cord (not shown). (Adapted, with permission, from Fields 1987.)

neurons branch in the periphery, innervating both skin and visceral targets.

Neurons in lamina VI receive inputs from large-diameter primary afferent fibers that innervate muscles and joints. These neurons are activated by innocuous joint movement and do not contribute to the transmission of nociceptive information. Many neurons in laminae VII and VIII, the intermediate and ventral regions of the spinal cord, do respond to noxious stimuli. These neurons typically have complex response properties because the inputs from nociceptors to these neurons are conveyed through many intervening synapses. Neurons in lamina VII often respond to stimulation of either side of the body, whereas most dorsal horn neurons receive unilateral input. The activation of lamina VII neurons is therefore thought to contribute to the diffuse quality of many pain conditions.

Nociceptive sensory neurons that activate neurons in the dorsal horn of the spinal cord release two major classes of neurotransmitters. Glutamate is the primary neurotransmitter of all primary sensory neurons, regardless of sensory modality. Neuropeptides are released as cotransmitters by many nociceptors with unmyelinated axons. These peptides include substance P, calcitonin gene–related peptide (CGRP), somatostatin, and galanin (Figure 20–5). Glutamate is stored in small, electron-lucent vesicles, whereas peptides are sequestered in large, dense-core vesicles at the central terminals of nociceptive sensory neurons (Figure 20–6). Separate storage sites permit these two classes of neurotransmitters to be selectively released under different physiological conditions.

Of the neuropeptide transmitters released by nociceptive sensory neurons, the actions of substance P, a member of the neurokinin peptide family, have been studied in most detail. Substance P is released from the central terminals of nociceptive afferents in response to tissue injury or after intense stimulation of peripheral

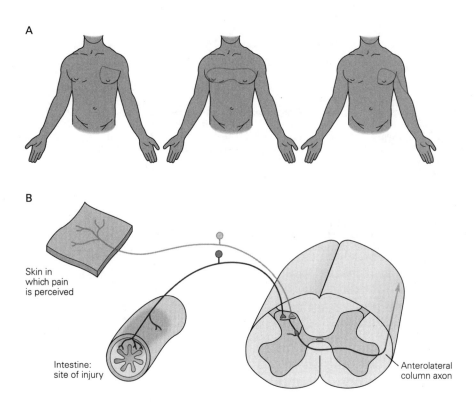

Figure 20–4 Signals from nociceptors in the viscera can be felt as "referred pain" elsewhere in the body.

A. Myocardial infarction and angina can be experienced as deep referred pain in the chest and left arm. The source of the pain cannot be readily predicted from the site of referred pain.

B. Convergence of visceral and somatic afferent fibers may account for referred pain. Nociceptive afferent fibers from

the viscera and fibers from specific areas of the skin converge on the same projection neurons in the dorsal horn. The brain has no way of knowing the actual site of the noxious stimulus and mistakenly associates a signal from a visceral organ with an area of skin. (Adapted, with permission, from Fields 1987.)

nerves. Its interaction with neurokinin receptors on dorsal horn neurons elicits slow excitatory postsynaptic potentials that prolong the depolarization elicited by glutamate. Although the physiological actions of glutamate and neuropeptides on dorsal horn neurons are different, these transmitters act coordinately to regulate the firing properties of dorsal horn neurons.

Details of the interaction of neuropeptides with their receptors on dorsal horn neurons have suggested strategies for chronic pain regulation. Infusion of substance P coupled to a neurotoxin into the dorsal horn of experimental animals results in selective destruction of neurons that express neurokinin receptors. Animals treated in this way fail to develop the central sensitization that is normally associated with peripheral injury. This method of neuronal ablation is more selective than traditional surgical interventions such as partial spinal cord transection (anterolateral cordotomy) and is being considered as a treatment for patients suffering from otherwise intractable chronic pain.

Hyperalgesia Has Both Peripheral and Central Origins

Up to this point, we have considered the conveyance of noxious signals in the normal physiological state. But the normal process of sensory signaling can be dramatically altered when peripheral tissue is damaged, resulting in an increase in pain sensitivity or hyperalgesia. This condition can be elicited by sensitizing peripheral nociceptors through repetitive exposure to noxious stimuli (Figure 20–7).

The sensitization is triggered by a complex mix of chemicals released from damaged cells that accumulate at the site of tissue injury. This cocktail contains peptides and proteins such as bradykinin, substance P, and nerve growth factor, as well as molecules such as ATP, histamine, serotonin, prostaglandins, leukotrienes, and acetylcholine. Many of these chemical mediators are released from distinct cell types, but together they act to decrease the threshold of nociceptor activation.

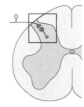

A Substance P

NK-1 receptor

B Enkephalin

μ-opioid receptor

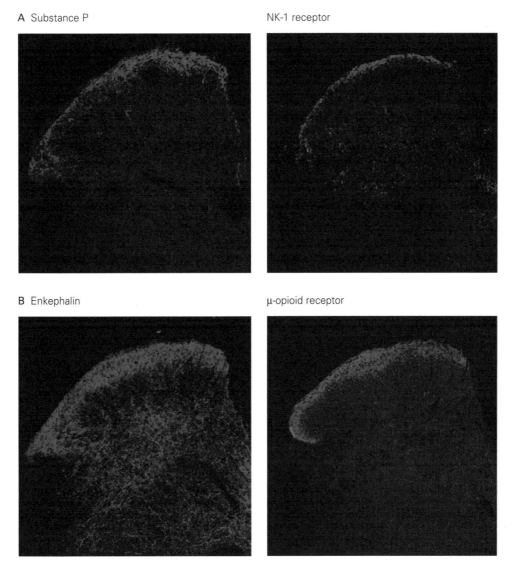

Figure 20–5 Neuropeptides and their receptors in the superficial dorsal horn of the rat spinal cord. (Images reproduced, with permission, from A. Basbaum.)

A. The terminals of unmyelinated primary sensory neurons are a major source of substance P in the superficial dorsal horn. Substance P activates the neurokinin-1 (**NK1**) receptor, which is

expressed by neurons in the superficial dorsal horn, the majority of which are projection neurons.

B. Enkephalin is localized in interneurons and found in the same region of the dorsal horn as terminals containing substance P. The μ-opioid receptor, which is targeted by enkephalins, is expressed by neurons in the superficial dorsal horn and also, presynaptically, on the terminals of sensory neurons.

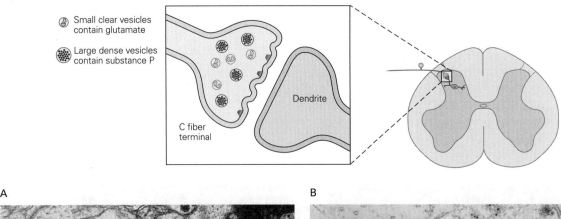

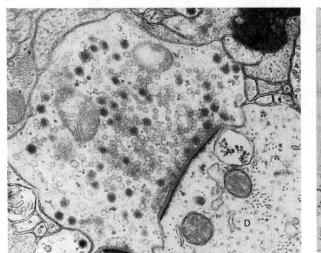

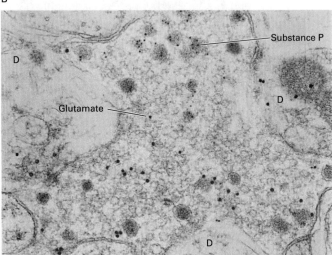

Figure 20–6 Transmitter storage in the synaptic terminals of primary nociceptive neurons in the dorsal spinal cord.

A. The terminal of a C fiber on the dendrite (D) of a dorsal horn neuron has two classes of synaptic vesicles that contain different transmitters. Small electron-lucent vesicles contain glutamate, whereas large dense-cored vesicles store neuropeptides. (Image reproduced, with permission, from H. J. Ralston III.)

B. Glutamate and the peptide substance P (marked by large and small gold particles, respectively) are scattered in the axoplasm of a sensory neuron terminal in lamina II of the dorsal horn. Dense core vesicles also store calcitonin gene–related peptide (CGRP). (Reproduced, with permission, from De Biasi and Rustioni 1990.)

Where do these chemicals come from, and what exactly do they do? Histamine is released from mast cells after tissue injury and activates polymodal nociceptors. The lipid anandamide, an endogenous cannabinoid agonist, is released under conditions of inflammation, activates the TRPV1 channel, and may trigger pain associated with inflammation. ATP, acetylcholine, and serotonin are released from damaged endothelial cells and platelets; they act indirectly to sensitize nociceptors by triggering the release of chemical agents such as prostaglandins and bradykinin from peripheral cells.

Bradykinin is one of the most active pain-producing agents. Its potency stems in part from the fact that it directly activates Aδ and C nociceptors and increases the synthesis and release of prostaglandins from nearby cells. Prostaglandins are metabolites of arachidonic acid that are generated through the activity of cyclooxygenase (COX) enzymes that cleave arachidonic acid (Chapter 14). The COX-2 enzyme is preferentially induced under conditions of peripheral inflammation, contributing to enhanced pain sensitivity. The enzymatic pathways of prostaglandin synthesis are targets of commonly used analgesic drugs. Aspirin and other nonsteroidal anti-inflammatory analgesics, such as ibuprofen and naproxen, are effective in controlling pain because they block the activity of the COX enzymes, reducing prostaglandin synthesis.

Activity of peripheral nociceptors can also produce all of the cardinal signs of inflammation, including heat

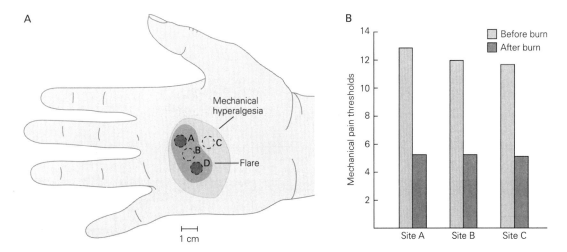

Figure 20–7 Hyperalgesia results from sensitization of nociceptors. (Reproduced, with permission, from Raja, Campbell, and Meyer 1984. Copyright © 1984, Oxford University Press.)

A. Mechanical thresholds for pain were recorded at sites A, B, and C before and after burns at sites A and D. The areas of reddening (flare) and mechanical hyperalgesia resulting from the burns

are shown on the hand of one subject. In all subjects, the area of mechanical hyperalgesia was larger than the area of flare. Mechanical hyperalgesia was present even after the flare disappeared.

B. Mean mechanical pain thresholds before and after burns. The mechanical threshold for pain is significantly decreased after the burn.

(calor), redness (rubor), and swelling (tumor). Heat and redness result from the dilation of peripheral blood vessels, whereas swelling results from plasma extravasation, a process in which proteins, cells, and fluids are able to penetrate postcapillary venules. Release of the neuropeptides substance P and CGRP from the peripheral terminals of C fibers provokes plasma extravasation and vasodilation, respectively. Because this form of inflammation depends on neural activity, it has been termed *neurogenic inflammation* (Figure 20–8). Importantly, as profound peripheral vasodilation is a critical trigger of many migraine headaches, the development of antibodies to CGRP, which counteract the vasodilation by scavenging CGRP, offers significant hope for a new migraine therapy.

The release of substance P and CGRP from the peripheral terminals of sensory neurons is also responsible for the *axon reflex*, a physiological process characterized by vasodilation in the vicinity of a cutaneous injury. Pharmacological antagonists of substance P are able to block neurogenic inflammation and vasodilation in humans; this discovery illustrates how knowledge of nociceptive mechanisms can be applied in improving clinical therapies for pain.

In addition to these small molecules and peptides, neurotrophins are causative agents in pain. Nerve growth factor (NGF) and brain-derived neurotrophic factor (BDNF) are particularly active in inflammatory pain states. The synthesis of BDNF is upregulated in many inflamed peripheral tissues (Figure 20–9). NGF-neutralizing molecules are effective analgesic agents in animal models of persistent pain. Indeed, inhibition of NGF function and signaling blocks pain sensation as effectively as COX inhibitors and opiates. Several promising clinical trials using antibodies to NGF for the management of knee osteoarthritis have been reported, once again demonstrating the translation of basic science to the clinic.

What accounts for the enhanced sensitivity of dorsal horn neurons to nociceptor signals? Under conditions of persistent injury, C fibers fire repetitively and the response of dorsal horn neurons increases progressively (Figure 20–10A). The gradual enhancement in the excitability of dorsal horn neurons has been termed "windup" and is thought to involve N-methyl-D-aspartate (NMDA)-type glutamate receptors (Figure 20–10B).

Repeated exposure to noxious stimuli therefore results in long-term changes in the response of dorsal horn neurons through mechanisms that are similar to those underlying the long-term potentiation of synaptic responses in many circuits in the brain. In essence, these prolonged changes in the excitability of dorsal horn neurons constitute a "memory" of the state of C-fiber input. This phenomenon has been termed *central sensitization* to distinguish it from sensitization at the peripheral terminals of the dorsal horn neurons, a process that involves activation of the enzymatic pathways of prostaglandin synthesis.

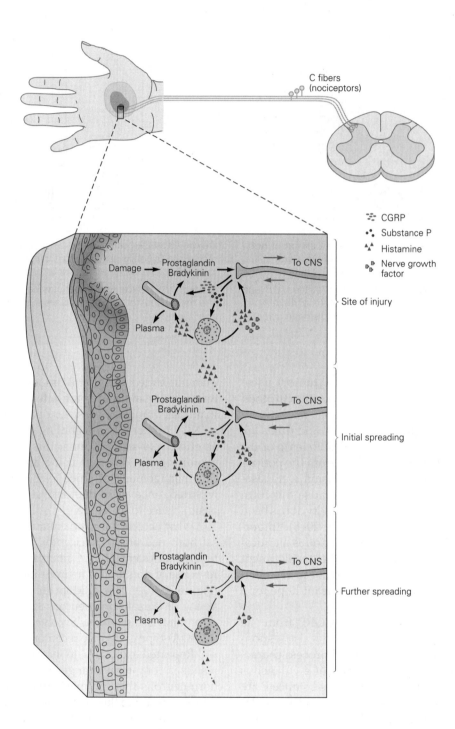

Figure 20–8 Neurogenic inflammation. Injury or tissue damage releases bradykinin and prostaglandins, which activate or sensitize nociceptors. Activation of nociceptors leads to the release of substance P and calcitonin gene–related peptide (**CGRP**). Substance P acts on mast cells (**light blue**) in the vicinity of sensory endings to evoke degranulation and the release of histamine, which directly excites nociceptors. Substance P also produces plasma extravasation and edema, and CGRP produces dilation of peripheral blood vessels (leading to reddening of the skin); the resultant inflammation causes additional liberation of bradykinin. These mechanisms also occur in healthy tissue, where they contribute to secondary or spreading hyperalgesia. (Abbreviation: **CNS**, central nervous system.)

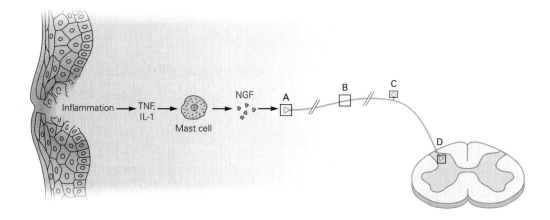

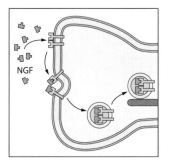

A Peripheral exposure to NGF

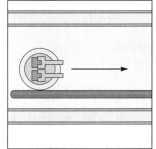

B Retrograde transport of signaling endosomes

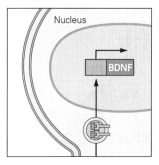

C Increased transcription of BDNF

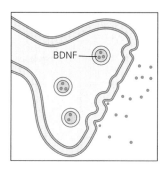

D Central release of BDNF

Figure 20–9 Neurotrophins are pain mediators. Local production of inflammatory cytokines such as interleukin-1 (**IL-1**) and tumor necrosis factor (**TNF**) promotes the synthesis and release of nerve growth factor (**NGF**) from several cell types in the periphery. Nerve growth factor binds to TrkA receptors on primary nociceptive terminals (**A**), triggering upregulation in expression of ion channels that increase nociceptor excitability. Retrograde transport of signaling endosomes to the cell body (**B**) results in enhanced expression of brain-derived neurotrophic factor (**BDNF**) (**C**), and its release from sensory terminals in the spinal cord (**D**) further increases excitability of dorsal horn neurons.

The sensitization of dorsal horn neurons also involves recruitment of second-messenger pathways and activation of protein kinases that have been implicated in memory storage in other regions of the central nervous system. One consequence of this enzymatic cascade is the expression of immediate-early genes that encode transcription factors such as *c-fos,* which are thought to activate effector proteins that sensitize dorsal horn neurons to sensory inputs. Most importantly, central sensitization of "pain" transmission circuitry in the dorsal horn is the process that can decrease pain thresholds (allodynia) and lead to *spontaneous pain* (ie, ongoing pain in the absence of peripheral stimulation).

Central sensitization is also a major contributor to neuropathic pain due to nerve injury. Here again, there is increased excitability of dorsal horn circuits mediated by NMDA receptors. There is also loss of inhibitory controls in the dorsal horn. Under normal conditions, GABAergic inhibitory interneurons in the dorsal horn are not only tonically active but are also turned on by activity of large-diameter, nonnociceptive Aβ fibers (Figure 20–11A). Peripheral nerve damage decreases the GABAergic controls, thus exacerbating the hyperactivity of these nociceptive pathways (Figure 20–11B). Recent studies also implicate nerve injury–induced activation of microglia and consequent reduced GABAergic inhibition in the central sensitization process (Figures 20–11C and 20–12). Together, these changes contribute to *mechanical allodynia* (ie, pain provoked by normally innocuous mechanical stimulation). Mechanical allodynia can also develop because of an inappropriate engagement of dorsal horn nociceptive pathway circuits by the Aβ myelinated afferents. In fact, spread of pain (secondary hyperalgesia) can occur because uninjured Aβ afferents outside of

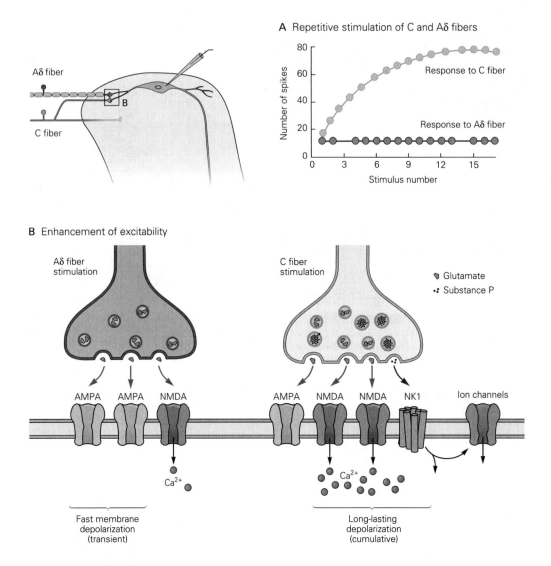

A Repetitive stimulation of C and Aδ fibers

B Enhancement of excitability

Figure 20–10 Mechanisms for enhanced excitability of dorsal horn neurons.

A. Typical responses of a dorsal horn neuron in the rat to electrical stimuli delivered transcutaneously at a frequency of 1 Hz. With repetitive stimulation, the long-latency component evoked by a C fiber increases gradually, whereas the short-latency component evoked by an A fiber remains constant.

B. Dorsal horn neurons receive mono- and polysynaptic input from Aδ and C fiber nociceptors. Elevation of residual Ca^{2+} in the presynaptic terminal leads to increased release of glutamate and substance P (and CGRP, not shown). *Left:* Activation of postsynaptic AMPA receptors by Aδ fibers causes a fast transient membrane depolarization, which relieves the Mg^{2+} block of the NMDA receptors. *Right:* Activation of the

postsynaptic NMDA receptors and neurokinin-1 (**NK1**) receptors by C fibers generates a long-lasting cumulative depolarization. The cytosolic Ca^{2+} concentration in the dorsal horn neuron increases because of Ca^{2+} entry through the NMDA receptor channels and voltage-sensitive Ca^{2+} channels. The elevated Ca^{2+} and activation by NK1 receptors of second-messenger systems enhance the performance of the NMDA receptors. Activation of NK1 receptors, cumulative depolarization, elevated cytosolic Ca^{2+}, and other factors regulate the behavior of voltage-gated ion channels responsible for action potentials, resulting in enhanced excitability, all of which contribute to the process of central sensitization. (Abbreviations: **AMPA**, α-amino-3-hydroxy-5-methylisoxazole-4-propionate; **NMDA**, *N*-methyl-D-aspartate)

Figure 20–11 Nerve injury triggers multiple dorsal horn central sensitization mechanisms that contribute to neuropathic pain.

A. Under normal conditions, nociceptors engage dorsal horn pain transmission circuits, via both monosynaptic and polysynaptic (excitatory) inputs to projection neurons of laminae I and V that transmit nociceptive information to the brainstem and thalamus. (See Figure 20–13.) The output of the projection neurons is regulated by GABAergic inhibitory interneurons, which can be activated by nonnociceptive, large-diameter, myelinated Aβ afferent fibers.

B. Peripheral nerve injury can result in a loss of the inhibitory control exerted by the Aβ afferents, via loss of GABAergic interneurons, reduced production of GABA, or reduced expression of GABAergic receptors by the projection neurons. Pathophysiological sprouting of Aβ afferents may also permit nonnociceptive inputs to directly engage the projection neurons (not shown), resulting in the condition of Aβ-mediated mechanical hypersensitivity/allodynia, a hallmark of neuropathic pain.

C. Peripheral nerve injury not only activates dorsal horn neurons directly but also activates microglia, which in turn release a host of mediators that enhance neuronal excitability and reduce the inhibitory controls exerted by GABAergic interneurons. Thus, targeting the mediators released from microglia introduces yet another potential approach to the pharmacotherapy of chronic pain.

A Normal pain control

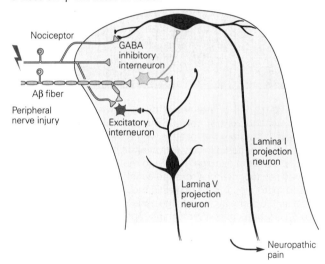

B Loss of Aβ–mediated inhibition

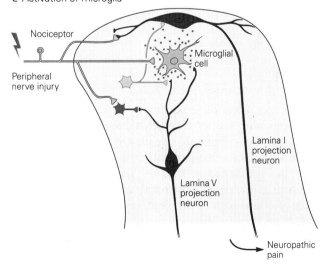

C Activation of microglia

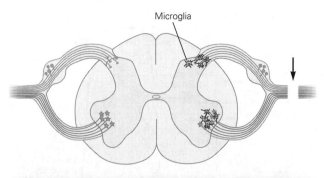

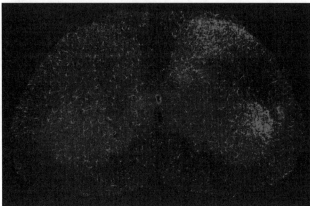

Figure 20–12 Peripheral nerve injury activates microglia in the dorsal and ventral horns. Schematic drawing and photomicrograph illustrate the location where microglia are activated after peripheral nerve injury. Activation of microglia in the dorsal horn results from damage (**arrow**) to the peripheral branch of primary sensory neurons (**orange cells**). Microglial activation around motor neuron cell bodies in the ventral horn occurs because the same injury damages efferent axons of the motor neurons (**green cells**). (Micrograph reproduced, with permission, from Julia Kuhn.)

the area of injury can inappropriately activate dorsal horn circuits that have undergone central sensitization.

Four Major Ascending Pathways Convey Nociceptive Information From the Spinal Cord to the Brain

Four major ascending pathways—the spinothalamic, spinoreticular, spinoparabrachial, and spinohypothalamic tracts—contribute sensory information to the central processes that generate pain.

The *spinothalamic tract* is the most prominent ascending nociceptive pathway in the spinal cord. It includes the axons of nociceptive-specific, thermosensitive, and wide-dynamic-range neurons in laminae I and V through VII of the dorsal horn. These axons cross the midline of the spinal cord near their segment of origin and ascend in the anterolateral white matter

before terminating in thalamic nuclei (Figure 20–13). The spinothalamic tract has a crucial role in the transmission of nociceptive information. Cells at the origin of this tract typically have discrete, unilateral receptive fields that underlie our ability to localize painful stimuli. Not surprisingly, electrical stimulation of the tract is sufficient to elicit the sensation of pain; conversely, lesioning this tract (anterolateral cordotomy), a procedure that is generally only used for intractable pain in terminal cancer patients, can result in a marked reduction in pain sensation on the side of the body contralateral to that of the lesion.

The *spinoreticular tract* contains the axons of projection neurons in laminae VII and VIII. This tract ascends in the anterolateral quadrant of the spinal cord with spinothalamic tract axons, and terminates in both the reticular formation and the thalamus. As neurons at the origin of the spinoreticular tract generally have large, often bilateral receptive fields, this pathway has been implicated more in the processing of diffuse, poorly localized pains.

The *spinoparabrachial tract* contains the axons of projection neurons in laminae I and V. Information transmitted along this tract is thought to contribute to the affective component of pain. This tract projects in the anterolateral quadrant of the spinal cord to the parabrachial nucleus at the level of the pons (Figure 20–13). This pathway has extensive collaterals to the mesencephalic reticular formation and periaqueductal gray matter. Parabrachial neurons project to the amygdala, a critical nucleus of the limbic system, which regulates emotional states (Chapter 42).

The *spinohypothalamic tract* contains the axons of neurons found in spinal cord laminae I, V, VII, and VIII. These axons project to hypothalamic nuclei that serve as autonomic control centers involved in the regulation of the neuroendocrine and cardiovascular responses that accompany pain syndromes (Chapter 41).

Several Thalamic Nuclei Relay Nociceptive Information to the Cerebral Cortex

The thalamus contains several relay nuclei that participate in the central processing of nociceptive information. Two of the most important regions of the thalamus are the lateral and medial nuclear groups. The *lateral nuclear group* comprises the ventroposterolateral (VPL), ventroposteromedial (VPM) and posterior/pulvinar nuclei. The VPL and VPM, respectively, receive inputs via the spinothalamic tract from nociception-specific and wide-dynamic-range neurons in laminae I and V of the dorsal horn and via the trigeminothalamic tract

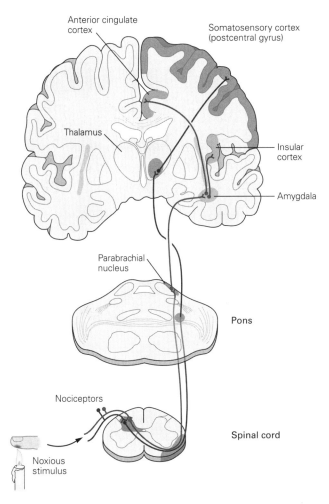

Figure 20–13 Major ascending pathways that transmit nociceptive information. Sensory discriminative features of the pain experience are transmitted from the spinal cord to the ventroposterolateral thalamus via the spinothalamic tract (**brown**). From there, information is transmitted predominantly to the somatosensory cortex. A second pathway, (the spinoparabrachial tract (**red**), carries information from the spinal cord to the parabrachial nucleus of the dorsolateral pons. These neurons in turn target limbic forebrain regions, including the insular and anterior cingulate cortex, which process emotional features of the pain experience.

from the trigeminal nucleus caudalis, the trigeminal homolog of the dorsal horn that processes nociceptive information from orofacial regions. The lateral thalamus processes information about the precise location of an injury, information usually conveyed to consciousness as acute pain. Consistent with this view, neurons in the lateral thalamic nuclei have small receptive fields, matching those of the presynaptic spinal neurons.

A cerebrovascular infarct that destroys the lateral thalamus can produce a central neuropathic pain condition called the Dejerine-Roussy (thalamic pain)

syndrome. Patients with this syndrome experience spontaneous burning pain as well as abnormal sensations (called dysesthesias) contralateral to the infarct. Electrical stimulation of the thalamus can also result in intense pain. In one dramatic clinical case, electrical stimulation of the thalamus rekindled sensations of angina pectoris that were so realistic that the anesthesiologist thought the patient was experiencing a heart attack. This and other clinical observations suggest that in chronic neuropathic pain conditions there is a fundamental change in thalamic and cortical circuitry. This hypothesis is consistent with studies demonstrating that the topographic map of the body in the thalamus and somatosensory cortex is not fixed, but can change with use and disuse. Loss of a limb can lead to shrinking and even disappearance of the cortical representation of the limb. Abnormal reorganization likely contributes to the phantom limb pain (Figure 20–14).

The *medial nuclear group* of the thalamus comprises the medial dorsal and central lateral nucleus of the thalamus and the intralaminar complex. Its major input is from neurons in laminae VII and VIII of the dorsal horn. The pathway to the medial thalamus was the first spinothalamic projection evident in the evolution of mammals and is therefore known as the *paleospinothalamic tract*. It is also sometimes referred to as the spinoreticulothalamic tract because it includes indirect connections through the reticular formation of the brain stem. The projection from the lateral thalamus to the ventroposterior lateral and medial nuclei is most developed in primates, and thus is termed the *neospinothalamic tract*. Many neurons in the medial thalamus respond optimally to noxious stimuli and project to many regions of the limbic system, including the anterior cingulate cortex.

The Perception of Pain Arises From and Can Be Controlled by Cortical Mechanisms

Anterior Cingulate and Insular Cortex Are Associated With the Perception of Pain

Imaging studies now show that no single area of the cortex is responsible for pain perception. Rather, many regions are activated when an individual experiences pain. In the somatosensory cortex, neurons typically have small receptive fields and may not contribute greatly to the diffuse perception of aches and pains that characterize most clinical syndromes. The anterior cingulate gyrus and insular cortex also contain neurons that are activated strongly and selectively by noxious somatosensory stimuli (Box 20–1).

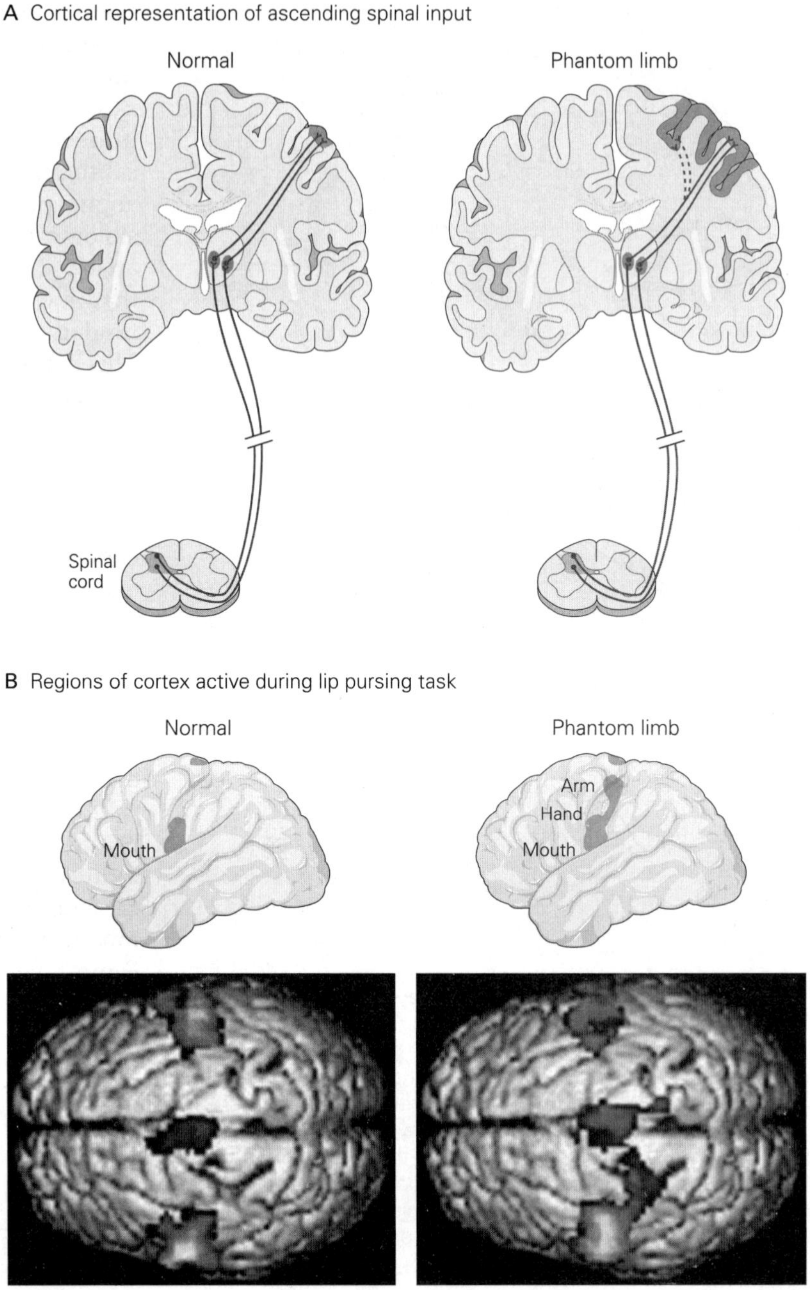

A Cortical representation of ascending spinal input

Normal

Phantom limb

Spinal
cord

Figure 20–14 Changes in neural activation in phantom limb pain.

A. The domain of cerebral cortex activated by ascending spinal sensory inputs is expanded in patients with phantom limb pain.

B. Functional magnetic resonance imaging (fMRI) of patients with phantom limb pain and healthy controls during a lip-pursing task. In amputees with phantom limb pain, cortical representation of the mouth has extended into the regions of the hand and arm. In amputees without pain, the areas of primary somatosensory and motor cortices that are activated are similar to those in healthy controls (image not shown). (Adapted, with permission, from Flor, Nokolajsen, and Jensen 2006. Copyright © 2006 Springer Nature.)

B Regions of cortex active during lip pursing task

Normal

Mouth

Phantom limb

Arm
Hand
Mouth

The anterior cingulate gyrus is part of the limbic system and is involved in processing emotional states associated with pain. The insular cortex receives direct projections from the thalamus as well as from the amygdala. Neurons in the insular cortex process information about the internal state of the body and contribute to the autonomic component of pain responses. Importantly, neurosurgical procedures that ablate the cingulate cortex or the pathway

from the frontal cortex to the cingulate cortex reduce the affective features of pain without eliminating the ability to recognize the intensity and location of the injury. Patients with lesions of the insular cortex present the striking syndrome of asymbolia for pain. They perceive noxious stimuli as painful and can distinguish sharp from dull pain but fail to display appropriate emotional responses. These observations implicate the insular cortex as an area in which

Box 20–1 Localizing Illusory Pain in the Cerebral Cortex

Thunberg's illusion, first demonstrated in 1896, is a strong, often painful heat felt after placing the hand on a grill of alternating warm and cool bars (Figure 20–15A).

One hypothesis proposes that this illusory sensation occurs as a result of differential grill responses of two classes of spinothalamic tract neurons, one sensitive to innocuous and another to noxious cold. This finding has led to a model of pain perception based on a central disinhibition or unmasking process in the cerebral cortex. The model predicts perceptual similarities between grill-evoked and cold-evoked pain, a prediction that has been verified psychophysically. The thalamocortical integration of pain and temperature stimuli may explain the burning sensation felt when nociceptors are activated by cold.

To identify the anatomical site of the unmasking phenomenon described above, positron emission tomography (PET) was used to compare the cortical areas activated by Thunberg's grill with those activated by cool, warm, noxious cold, and noxious heat stimuli separately. All thermal stimuli activate the insula and somatosensory cortices. The anterior cingulate cortex is activated by Thunberg's grill and by noxious heat and cold, but not by discrete warm and cool stimuli (Figure 20–15B).

A

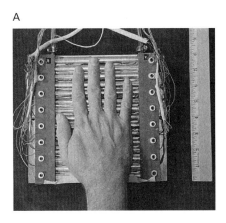

Figure 20–15A Thunberg's thermal grill. The stimulus surface (20 × 14 cm) is made of 15 sterling silver bars, each 1 cm wide, set approximately 3 mm apart. Underneath each bar are three longitudinally spaced thermoelectric (Peltier) elements (1 cm^2), and on top of each bar is a thermocouple. Alternate (even- and odd-numbered) bars can be controlled independently. (Adapted, with permission, from Craig and Bushnell 1994. Copyright © 1994 AAAS.)

Figure 20–15B Cortical areas activated by Thunberg's grill. The anterior cingulate and insula regions of the cerebral cortex are activated when the hand is placed on the grill but not when warm and cool stimuli are applied separately. (Reproduced, with permission, from Craig AD, Reiman EM, Evans A, et al. 1996. Functional imaging of an illusion of pain. Nature 384:258–260. Copyright © 1996 Springer Nature.)

B

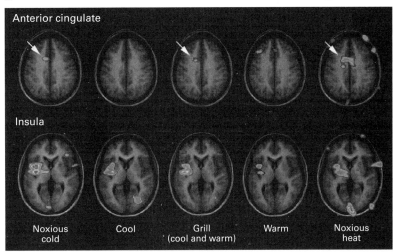

the sensory, affective, and cognitive components of pain are integrated.

Pain Perception Is Regulated by a Balance of Activity in Nociceptive and Nonnociceptive Afferent Fibers

Many projection neurons in the dorsal horn of the spinal cord respond selectively to noxious inputs, but others receive convergent inputs from both nociceptive and nonnociceptive afferents. The concept that the convergence of sensory inputs onto spinal projection neurons regulates pain processing first emerged in the 1960s.

Ronald Melzack and Patrick Wall proposed that the relative balance of activity in nociceptive and nonnociceptive afferents might influence the transmission and perception of pain. In particular, they proposed that activation of nonnociceptive sensory neurons, by engaging inhibitory interneurons in the dorsal horn, closes a "gate" for afferent transmission of nociceptive signals that can be opened by the activation of nociceptive sensory neurons. In the original and simplest form of this gate-control theory, the interaction between large and small fibers occurred at the first possible site of convergence on projection neurons in the dorsal horn of the spinal cord (Figure 20–16). We now know that such interactions can also occur at many supraspinal relay centers.

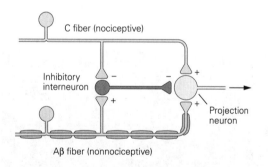

Figure 20–16 The gate control theory of pain. The gate-control hypothesis was proposed in the 1960s to account for the fact that activation of low-threshold primary afferent fibers can attenuate pain. The hypothesis focused on the interaction of neurons in the dorsal horn of the spinal cord: the nociceptive (C) and nonnociceptive (Aβ) sensory neurons, projection neurons, and inhibitory interneurons. In the original version of the model, as shown here, the projection neuron is excited by both classes of sensory neurons and inhibited by interneurons in the superficial dorsal horn. The two classes of sensory fibers also terminate on the inhibitory interneurons; the C fibers indirectly inhibit the interneurons, thus increasing the activity of the projection neurons (thereby "opening the gate"), whereas the Aβ fibers excite the interneurons, thus suppressing the output of the projection neurons (and "closing the gate").

The concept of convergence of different sensory modalities has provided an important basis for the design of new pain therapies. Viewed in its broadest sense, the convergence of high- and low-threshold inputs at spinal or supraspinal sites provided a plausible explanation for several empirical observations about the perception of pain. The shaking of the hand that follows a hammer blow or burn is a reflexive behavior and may alleviate pain by activating large-diameter afferent fibers that suppress the transmission of information about noxious stimuli.

The idea of convergence also helped to promote the use of transcutaneous electrical nerve stimulation (TENS) and spinal cord stimulation for the relief of pain. With TENS, stimulating electrodes placed at peripheral locations activate large-diameter afferent fibers that innervate areas that overlap but also surround the region of injury and pain. The region of the body in which pain is reduced maps to those segments of the spinal cord in which nociceptive and nonnociceptive afferents from that body region terminate. This makes intuitive sense: You do not shake your left leg to relieve pain in your right arm.

Electrical Stimulation of the Brain Produces Analgesia

Several sites of endogenous pain regulation are located in the brain. One effective means of suppressing nociception involves stimulation of the periaqueductal gray region, the area of the midbrain that surrounds the third ventricle and the cerebral aqueduct. In experimental animals, stimulation of this region elicits a profound and selective analgesia. This *stimulation-produced analgesia* is remarkably modality-specific; animals still respond to touch, pressure, and temperature within the body area that is not sensitive to pain. Stimulation-evoked analgesia has proved to be an effective way of relieving pain in a limited number of human pain conditions.

Stimulation of the periaqueductal gray matter blocks spinally mediated withdrawal reflexes that are normally evoked by noxious stimulation. Few of the neurons in the periaqueductal gray matter project directly to the dorsal horn of the spinal cord. Most make excitatory connections with neurons of the rostroventral medulla, including serotonergic neurons in a midline region called the nucleus raphe magnus. The axons of these serotonergic neurons project through the dorsal region of the lateral funiculus to the spinal cord, where they form inhibitory connections with neurons in laminae I, II, and V of the dorsal horn (Figure 20–17). Stimulation of the rostroventral

originates in the locus ceruleus and other nuclei of the medulla and pons (Figure 20–17). Through direct and indirect synaptic actions, these projections inhibit neurons in laminae I and V of the dorsal horn.

Opioid Peptides Contribute to Endogenous Pain Control

Since discovery of the opium poppy by the Sumerians in 3300 BC, the plant's active ingredients, opiates such as morphine and codeine, have been recognized as powerful analgesic agents. Over the past two decades, we have begun to understand many of the molecular mechanisms and neural circuits through which opiates exert their analgesic actions. In addition, we have come to realize that the neural networks involved in stimulation-produced and opiate-induced analgesia are intimately related.

Two key discoveries led to these advances. The first was the recognition that morphine and other opiates interact with specific receptors on neurons in the spinal cord and brain. The second was the isolation of endogenous neuropeptides with opiate-like activities at these receptors. The observation that the opiate antagonist, naloxone, blocks stimulation-produced analgesia provided the first clue that the brain contains endogenous opioids.

Endogenous Opioid Peptides and Their Receptors Are Distributed in Pain-Modulatory Systems

Opioid receptors fall into four major classes: mu (μ), delta (δ), kappa (κ), and orphanin FQ. The genes encoding each of these receptor types constitute a subfamily of G protein–coupled receptors. The μ receptors are particularly diverse; numerous μ receptor isoforms have been identified, many with different patterns of expression. This finding has prompted a search for analgesic drugs that target specific isoforms.

The opioid receptors were originally defined on the basis of the binding affinity of different agonist compounds. Morphine and other opioid alkaloids are potent agonists at μ receptors, and there is a tight correlation between the potency of an analgesic and its affinity of binding to μ receptors. Mice in which the gene for the μ receptor has been inactivated are insensitive to morphine and other opiate agonists. Many opiate antagonist drugs, such as naloxone, also bind to the μ receptor and compete with morphine for receptor occupancy without activating receptor signaling.

The μ receptors are highly concentrated in the superficial dorsal horn of the spinal cord, the ventral

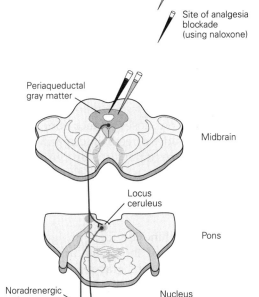

Site of opiate-induced analgesia

Site of analgesia blockade (using naloxone)

Periaqueductal gray matter

Midbrain

Locus ceruleus

Pons

Noradrenergic pathway

Nucleus raphe magnus

Medulla

Serotonergic pathway

Spinal cord

Figure 20–17 Descending monoaminergic pathways regulate nociceptive relay neurons in the spinal cord. A serotonergic pathway arises in the nucleus raphe magnus and projects through the dorsolateral funiculus to the dorsal horn of the spinal cord. A noradrenergic system arises in the locus ceruleus and other nuclei in the pons and medulla. (See Figure 40–11A for the locations and projections of monoaminergic neurons.) In the spinal cord, these descending pathways inhibit nociceptive projection neurons through direct connections as well as through interneurons in the superficial layers of the dorsal horn. Both the serotonergic nucleus raphe magnus and noradrenergic nuclei receive input from neurons in the periaqueductal gray region. Sites of opioid peptide expression and actions of exogenously administered opioids are shown.

medulla thus inhibits the firing of many classes of dorsal horn neurons, including projection neurons of the major ascending pathways that convey afferent nociceptive signals to the brain.

A second major monoaminergic descending system can also suppress the activity of nociceptive neurons in the dorsal horn. This noradrenergic system

Table 20-1 Four Major Classes of Endogenous Opioid Peptides

Propeptide	Peptide(s)	Preferential receptor
POMC	β-Endorphin	μ/δ
	Endomorphin-1	μ
	Endomorphin-2	μ
Proenkephalin	Met-enkephalin	δ
	Leu-enkephalin	δ
Prodynorphin	Dynorphin A	κ
	Dynorphin B	κ
Pro-orphanin FQ	Orphanin FQ	Orphan receptor

POMC, pro-opiomelanocortin.

medulla, and the periaqueductal gray matter—important anatomical sites for the regulation of pain. Nevertheless, like other classes of opioid receptors, they are also found at many other sites in the central and peripheral nervous systems. Their widespread distribution explains why systemically administered morphine influences many physiological processes in addition to the perception of pain.

The discovery of opioid receptors and their expression by neurons in the central and peripheral nervous systems led to the definition of four major classes of endogenous opioid peptides, each interacting with a specific class of opioid receptors (Table 20-1).

Three classes—the enkephalins, β-endorphins, and dynorphins—are the best characterized. These opioid peptides are formed from large polypeptide precursors by enzymatic cleavage (Figure 20-18) and encoded by distinct genes. Despite differences in amino acid sequence, each contains the sequence Tyr-Gly-Gly-Phe. β-Endorphin is a cleavage product of a precursor that also generates the active peptide adrenocorticotropic hormone (ACTH). Both β-endorphin and ACTH are synthesized by cells in the pituitary and are released into the bloodstream in response to stress. Dynorphins are derived from the polyprotein product of the *dynorphin* gene.

Members of the four classes of opioid peptides are distributed widely in the central nervous system, and individual peptides are located at sites associated with the processing or modulation of nociceptive information. Neuronal cell bodies and axon terminals containing enkephalin and dynorphin are found in the dorsal horn of the spinal cord, particularly in laminae I and II, as well as in the rostral ventral medulla and the periaqueductal gray matter. Neurons that synthesize

β-endorphin are confined primarily to the hypothalamus; their axons terminate in the periaqueductal gray region and on noradrenergic neurons in the brain stem. Orphanin FQ appears to participate in a broad range of other physiological functions.

Morphine Controls Pain by Activating Opioid Receptors

Microinjection of low doses of morphine, other opiates, or opioid peptides directly into specific regions of the rat brain produces a powerful analgesia. The periaqueductal gray region is among the most sensitive sites, but local administration of morphine into other regions, including the spinal cord, also elicits a powerful analgesia.

Systemic morphine-induced analgesia can be blocked by injection of the opiate antagonist naloxone into the periaqueductal gray region or the nucleus raphe magnus (Figure 20-17). In addition, bilateral transection of the dorsal lateral funiculus in the spinal cord blocks analgesia induced by central administration of morphine. Thus, the central analgesic actions of morphine involve the activation of descending pathways to the spinal cord, the same descending pathways that mediate the analgesia produced by electrical brain stimulation and morphine.

In the spinal cord, as elsewhere, morphine acts by mimicking the actions of endogenous opioid peptides. The superficial dorsal horn of the spinal cord contains interneurons that express enkephalin and dynorphin, and the terminals of these neurons lie close to synapses formed by nociceptive sensory neurons and spinal projection neurons (Figure 20-19A). Moreover, the μ, δ, and κ receptors are located on the terminals of the nociceptive sensory neurons as well as on the dendrites of dorsal horn neurons that receive afferent nociceptive input, thus placing endogenous opioid peptides in a strategic position to regulate sensory input. The C-fiber nociceptors, which mediate slow persistent pain or "second pain," have more μ receptors than the Aδ nociceptors, which mediate fast and acute pain or "first pain" (Figure 20-1). This may help to explain why morphine is more effective in the treatment of persistent rather than acute pains.

Opioids (both opiates and opioid peptides) regulate nociceptive transmission at synapses in the dorsal horn through two main mechanisms. First, they increase membrane K^+ conductances in dorsal horn neurons, hyperpolarizing the neurons and increasing their threshold for activation. Second, by binding to receptors on presynaptic sensory terminals, opioids block voltage-gated Ca^{2+} channels, which reduces Ca^+

A Precursor protein

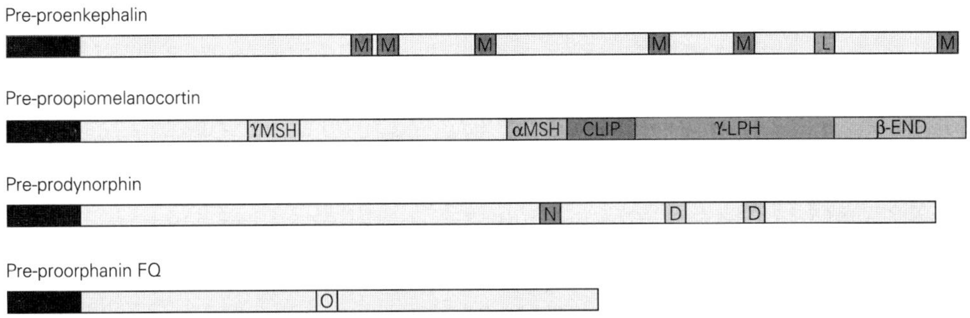

Pre-proenkephalin

Pre-proopiomelanocortin

Pre-prodynorphin

Pre-proorphanin FQ

B Proteolytically processed opioid peptides

Amino acid sequence

M	Methionine-enkephalin	**Tyr Gly Gly Phe** Met OH
L	Leucine-enkephalin	**Tyr Gly Gly Phe** Leu OH
β-END	β-Endorphin	**Tyr Gly Gly Phe** Met Thr Ser Glu Lys Ser Gln Thr Pro Leu Val Thr Leu Phe Lys Asn Ala Ile Val Lys Asn Ala His Lys Gly Gln OH
D	Dynorphin	**Tyr Gly Gly Phe** Leu Arg Arg Ile Arg Pro Lys Leu Lys Trp Asp Asn Gln OH
N	α-Neoendorphin	**Tyr Gly Gly Phe** Leu Arg Lys Tyr Pro Lys
O	Orphanin FQ	**Tyr Gly Gly Phe** Thr Gly Ala Arg Lys Ser Ala Arg Lys Leu Ala Asn Gln

Figure 20–18 Four families of endogenous opioid peptides arise from large precursor polyproteins.

A. Proteolytic enzymes cleave each of the precursor proteins to generate shorter, biologically active peptides, some of which are shown in this diagram. The proenkephalin precursor protein contains multiple copies of methionine-enkephalin (**M**), leucine-enkephalin (**L**), and several extended enkephalins. Proopiomelanocortin (POMC) contains β-endorphin (**β-END**, melanocyte-stimulating hormone (**MSH**), adrenocorticotropic

hormone (ACTH), and corticotropin-like intermediate-lobe peptide (**CLIP**). The prodynorphin precursor can produce dynorphin (**D**) and α-neoendorphin (**N**). The pro-orphanin precursor contains the orphanin FQ peptide (**O**), also called nociceptin. The **black domains** indicate a signal peptide.

B. Amino acid sequences of proteolytically processed bioactive peptides. The amino acid residues shown in **bold type** mediate interaction with opioid receptors. (Adapted, with permission, from Fields 1987.)

entry into the sensory nerve terminal (Figure 20–19B). This effect in turn inhibits the release of neurotransmitter and thereby decreases activation of postsynaptic dorsal horn neurons.

The wide distribution of opioid receptors within the brain and periphery accounts for the many side effects produced by opiates. Activation of opioid receptors expressed by muscles of the bowel and anal sphincter results in constipation. Similarly, opioid receptor–mediated inhibition of neuronal activity in the nucleus of the solitary tract underlies the respiratory depression and cardiovascular side effects. For this reason, direct spinal administration of opiates

has significant advantages. Morphine injected into the cerebrospinal fluid of the spinal cord subarachnoid space interacts with opioid receptors in the dorsal horn to elicit a profound and prolonged analgesia. Spinal administration of morphine is now commonly used in the treatment of postoperative pain, notably the pain associated with cesarean section during childbirth. In addition to producing prolonged analgesia, intrathecal morphine has fewer side effects because the drug does not diffuse far from its site of injection. Continuous local infusion of morphine to the spinal cord has also been used for the treatment of certain cancer pains.

A Nociceptor circuitry in the dorsal horn

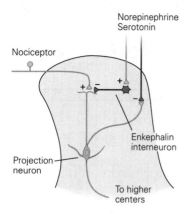

B Effects of opiates and opioids on nociceptor signal transmission

1 Sensory input alone

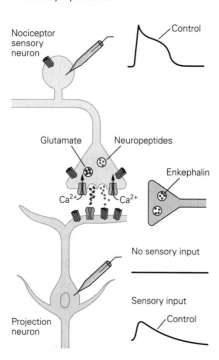

2 Sensory input + opiates/opioids

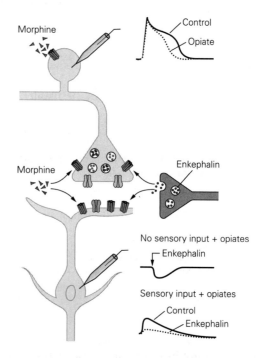

Figure 20–19 Local interneurons in the spinal cord integrate descending and afferent nociceptive pathways.

A. Nociceptive afferent fibers, local interneurons, and descending fibers interconnect in the dorsal horn of the spinal cord (see also Figure 20–3B). Nociceptive fibers terminate on second-order projection neurons. Local GABAergic and enkephalin-containing inhibitory interneurons exert both pre- and postsynaptic inhibitory actions at these synapses. Serotonergic and noradrenergic neurons in the brain stem activate the local interneurons and also suppress the activity of the projection neurons. Loss of these inhibitory controls contributes to ongoing pain and pain hypersensitivity.

B. Regulation of nociceptive signals at dorsal horn synapses. **1.** Activation of a nociceptor leads to the release of glutamate and neuropeptides from the primary sensory neuron, producing an excitatory postsynaptic potential in the projection neuron. **2.** Opiates decrease the duration of the postsynaptic potential, probably by reducing Ca^{2+} influx, and thus decrease the release of transmitter from the primary sensory terminals. In addition, opiates hyperpolarize the dorsal horn neurons by activating a K^+ conductance and thus decrease the amplitude of the postsynaptic potential in the dorsal horn neuron.

Opiates also act on receptors in the cerebral cortex. There is evidence, for example, that opiates can influence the affective component of the pain experience by an action in the anterior cingulate gyrus. Most interestingly, there is considerable evidence that placebo analgesia involves endorphin release and can be reversed by naloxone. This finding emphasizes that responses to a placebo do not indicate that the pain was somehow imaginary. Moreover, placebo analgesia is a component of the overall analgesic action of any pain-relieving drug, including morphine, provided that the patient believes that the treatment will be effective. On the other hand, some other psychological interventions to relieve pain, namely hypnosis, do not appear to involve release of endorphins.

Tolerance to and Dependence on Opioids Are Distinct Phenomena

The chronic use of morphine invites major problems, most notably tolerance and psychological dependence (addiction) (Chapter 43). The repeated use of morphine for pain relief can cause patients to develop resistance to the analgesic effects of the drug, such that progressively higher drug doses are required to achieve the same therapeutic effect. One theory holds that tolerance results from uncoupling of the opioid receptor from its G protein transducer. However, as the binding of naloxone to μ-opioid receptors can precipitate withdrawal symptoms in tolerant subjects, it appears that the opioid receptor is still active in the tolerance state. Tolerance may therefore also reflect a cellular response to the activation of opioid receptors, a response that counteracts the effects of the opiate and resets the system. It follows that when the opiate is abruptly removed or naloxone is administered, this compensatory response is unmasked and withdrawal results.

Such physiological tolerance differs from dependence/addiction, which is a psychological craving for the drug, one that is associated with its misuse and that contributes to opiate use disorders. Given the alarming increases in opiate-related deaths, either because of misuse and overdose of prescription opioids or a host of socioeconomic factors, further studies of the mechanisms that contribute to the development of and distinguish between tolerance and addiction are essential. Unquestionably, morphine and other opiate drugs are very useful in the management of postoperative pain. Whether they are equally effective for the management of chronic pain in noncancer patients remains controversial and needs further study.

Highlights

1. Peripheral nociceptive axons, with cell bodies in dorsal root ganglia, include small-diameter unmyelinated (C) and myelinated (Aδ) afferents. Larger diameter Aβ afferents respond only to innocuous stimulation but, following injury, can activate central nervous system pain circuitry.

2. All nociceptors use glutamate as their excitatory neurotransmitter; many also express an excitatory neuropeptide cotransmitter, such as substance P or CGRP.

3. Nociceptors are also molecularly distinguished by their expression of different receptors sensitive to temperature, plant products, mechanical stimuli, or ATP. As many of these molecules, including the Nav1.7 subtype of voltage-gated Na$^+$ channels, are exclusively expressed in sensory neurons, their selective pharmacological targeting suggests a novel approach to analgesic drug development.

4. Nociceptors terminate in the dorsal horn of the spinal cord where they excite interneurons and projection neurons. Neuropeptides are also released from the peripheral terminals of nociceptors and contribute to neurogenic inflammation, including vasodilatation of and extravasation from peripheral vessels. The development of antibodies to CGRP, to block vasodilation, is a new approach to managing migraine.

5. A major brain target of dorsal horn projection neurons is the ventroposterolateral thalamus, which processes location and intensity features of the painful stimulus. Other neurons target the parabrachial nucleus (PB) of the dorsolateral pons. PB neurons, in turn, project to limbic regions of the brain, which process affective/emotional features of the pain experience.

6. Allodynia, pain produced by an innocuous stimulus, results in part from peripheral sensitization of nociceptors. Peripheral sensitization occurs when there is tissue injury and inflammation and involves NSAID-sensitive production of prostaglandins, which lower the threshold for activating nociceptors. A great advantage of NSAIDs is that they act in the periphery, illustrating the importance of efforts to develop pharmacotherapies, such as antibodies to NGF, which cannot cross the blood–brain barrier, thus reducing their likelihood of having adverse side effects in the central nervous system.

7. Hyperalgesia (exacerbated pain in response to a painful stimuli) and allodynia also arise from

altered activity in the dorsal horn—a central sensitization process that contributes to spontaneous activity of pain-transmission neurons and amplification of nociceptive signals. Glutamate activation of spinal cord NMDA receptors and activation of microglia and astrocytes contribute, in particular, to the neuropathic pains that can occur after peripheral nerve injury. Understanding the consequences of central sensitization is critical to preventing the transition from acute to chronic pain.

8. Under normal conditions, input carried by large-diameter, nonnociceptive afferents can reduce the transmission of nociceptive information to the brain by engaging GABAergic inhibitory circuits in the dorsal horn. This inhibitory control is the basis of the pain relief produced by vibration and transcutaneous electrical stimulation. However, when injury induces central sensitization, Aβ input mediates mechanical allodynia.

9. Opiates are the most effective pharmacological tool for the management of severe pain. The inhibitory action of opiates and the related endogenous opioid peptides result from reduced neurotransmitter release or by hyperpolarization of postsynaptic neurons. All opioid actions can be blocked by the opiate receptor antagonist naloxone.

10. Endogenous opioids, including enkephalin and dynorphin, and their opioid receptor targets are not expressed only in pain-relevant areas of the brain. As a result, systemic administration of opiates is associated with many adverse side effects, including constipation, respiratory depression, and activation of the reward system. The latter can lead to psychological dependence and eventual misuse. Many of these adverse side effects limit opiate use for long-term pain control.

11. The brain not only receives nociceptive information leading to a perception of pain, but also regulates the output of the spinal cord to reduce pain by an endorphin-mediated pain control system. Electrical stimulation of the midbrain periaqueductal gray can engage a descending inhibitory control system, likely involving endorphins, which reduces the transmission of pain messages from the spinal cord to the brain.

12. The pain relief produced by some psychological manipulations (e.g., placebo analgesia) involves endorphin release; other manipulations, such as hypnosis, do not.

13. Tolerance and psychological dependence can arise after prolonged opiate use. Tolerance is manifested as a requirement for higher doses of the opiate to achieve the same physiological endpoint. Psychological dependence, in contrast, involves activation of the brain's reward system and the development of craving that can lead to misuse of opiates. Development of nonrewarding opioid analgesics, which can regulate the sensory-discriminative but not the emotional features of the pain experience, may significantly impact the ongoing opioid epidemic.

Allan I. Basbaum

Selected Reading

Basbaum AI, Bautista DM, Scherrer G, Julius D. 2009 Cellular and molecular mechanisms of pain. Cell 139:267–284.

Basbaum AI, Fields HL. 1984. Endogenous pain control systems: brainstem spinal pathways and endorphin circuitry. Annu Rev Neurosci 7:309–338.

Dib-Hajj SD, Geha P, Waxman SG. 2017. Sodium channels in pain disorders: pathophysiology and prospects for treatment. Pain 158(Suppl 1):S97–S107.

Grace PM, Hutchinson MR, Maier SF, Watkins LR. 2014. Pathological pain and the neuroimmune interface. Nat Rev Immunol 14:217–231.

Ji RR, Chamessian A, Zhang YQ. 2016. Pain regulation by non-neuronal cells and inflammation. Science 354:572–577.

Peirs C, Seal RP. 2016. Neural circuits for pain: recent advances and current views. Science 354:578–584.

Tracey I. 2017. Neuroimaging mechanisms in pain: from discovery to translation. Pain 158:S115–S122.

References

Akil H, Mayer DJ, Liebeskind JC. 1976. Antagonism of stimulation-produced analgesia by naloxone, a narcotic antagonist. Science 191:961–962.

Bautista DM, Jordt SE, Nikai T, et al. 2006. TRPA1 mediates the inflammatory actions of environmental irritants and proalgesic agents. Cell 124:1269–1282.

Bautista DM, Siemens J, Glazer JM, et al. 2007. The menthol receptor TRPM8 is the principal detector of environmental cold. Nature 44:204–208.

Benedetti F. 2014. Placebo effects: from the neurobiological paradigm to translational implications. Neuron 84:623–637.

Bliss TV, Collingridge GL, Kaang BK, Zhuo M. 2016. Synaptic plasticity in the anterior cingulate cortex in acute and chronic pain. Nat Rev Neurosci 17:485–496.

Caterina MJ, Schumacher MA, Tominaga M, Rosen TA, Levine JD, Julius D. 1997. The capsaicin receptor: a heat-activated ion channel in the pain pathway. Nature 389:816–824.

Colloca L, Ludman T, Bouhassira D, et al. 2017. Neuropathic pain. Nat Rev Dis Primers 3:17002.

Cox JJ, Reimann F, Nicholas AK, et al. 2006. An SCN9A channelopathy causes congenital inability to experience pain. Nature 444:894–898.

Craig AD, Bushnell MC. 1994. The thermal grill illusion: unmasking the burn of cold pain. Science 265:252–255.

Darland T, Heinricher MM, Grandy DK. 1988. Orphanin FQ/nociceptin: a role in pain and analgesia, but so much more. Trends Neurosci 21:215–221.

De Biasi S, Rustioni A 1990. Ultrastructural immunocytochemical localization of excitatory amino acids in the somatosensory system. J Histochem Cytochem 38: 1745–1754.

De Felice M, Eyde N, Dodick D, et al. 2013. Capturing the aversive state of cephalic pain preclinically. Ann Neurol 74:257–265.

Dejerine J, Roussy G. 1906. Le syndrome thalamique. Rev Neurol 14:521–532.

Dhaka A, Viswanath V, Patapoutian A. 2006. Trp ion channels and temperature sensation. Annu Rev Neurosci 29:135–161.

Fields H. 1987 *Pain*. New York: McGraw-Hill.

Flor H, Nikolajsen L, Jensen TS. 2006. Phantom limb pain: a case of maladaptive CNS plasticity? Nature Rev Neurosci 7:873–881.

Günther T, Dasgupta P, Mann A, et al. 2017. Targeting multiple opioid receptors—improved analgesics with reduced side effects? Br J Pharmacol 2018:2857–2868.

Han L, Ma C, Liu Q, et al. 2013. A subpopulation of nociceptors specifically linked to itch. Nat Neurosci 16:174–182.

Hosobuchi Y. 1986. Subcortical electrical stimulation for control of intractable pain in humans: report of 122 cases 1970–1984. J Neurosurg 64:543–553.

Jordt SE, Bautista DM, Chuang HH, et al. 2004. Mustard oils and cannabinoids excite sensory nerve fibres through the TRP channel ANKTM1. Nature 427:260–265.

Jordt SE, McKemy DD, Julius D. 2003. Lessons from peppers and peppermint: the molecular basis of thermosensation. Curr Opin Neurobiol 13:487–492.

Kelleher JH, Tewari D, McMahon SB. 2017. Neurotrophic factors and their inhibitors in chronic pain treatment. Neurobiol Dis 97:127–138.

Kuner R, Flor H. 2016. Structural plasticity and reorganisation in chronic pain. Nat Rev Neurosci 18:20–30.

Lane NE, Schnitzer TJ, Birbara CA, et al. 2010. Tanezumab for the treatment of pain from osteoarthritis of the knee. N Engl J Med 363:1521–1531.

Lenz FA, Gracely RH, Romanoski AJ, Hope EJ, Rowland LH, Dougherty PM. 1995. Stimulation in the human somatosensory thalamus can reproduce both the affective and sensory dimensions of previously experienced pain. Nat Med 1:910–913.

Mantyh PW, Rogers SD, Honore P, et al. 1997. Inhibition of hyperalgesia by ablation of lamina I spinal neurons expressing the substance P receptor. Science 278:275–279.

Matthes HW, Maldonado R, Simonin F, et al. 1996. Loss of morphine-induced analgesia, reward effect and withdrawal symptoms in mice lacking the μ-opioid-receptor gene. Nature 383:819–823.

McDonnell A, Schulman B, Ali Z, et al. 2016. Inherited erythromelalgia due to mutations in SCN9A: natural history, clinical phenotype and somatosensory profile. Brain 39:1052–1065.

Melzack R, Wall PD. 1965. Pain mechanisms: a new theory. Science 150:971–979.

Merzenich MM, Jenkins WM. 1993. Reorganization of cortical representations of the hand following alterations of skin inputs induced by nerve injury, skin island transfers, and experience. J Hand Ther 6:89–104.

Perl ER. 2007. Ideas about pain, a historical review. Nat Rev Neurosci 8:71–80.

Raja SN, Campbell JN, Meyer RA. 1984. Evidence for different mechanisms of primary and secondary hyperalgesia following heat injury to the glabrous skin. Brain 107:1179–1188.

Ross SE, Mardinly AR, McCord AE, et al. 2010. Loss of inhibitory interneurons in the dorsal spinal cord and elevated itch in Bhlhb5 mutant mice. Neuron 65:886–898.

Sorge RE, Mapplebeck JC, Rosen S, et al. 2015. Different immune cells mediate mechanical pain hypersensitivity in male and female mice. Nat Neurosci 18:1081–1083.

Talbot JD, Marrett S, Evans AC, Meyer E, Bushnell MC, Duncan GH. 1991. Multiple representations of pain in human cerebral cortex. Science 251:1355–1358.

Todd AJ. 2010. Neuronal circuitry for pain processing in the dorsal horn. Nat Rev Neurosci. 11:823–836.

Tominaga M, Caterina MJ. 2004. Thermosensation and pain. J Neurobiol 61:3–12.

Tracey I, Mantyh PW. 2007. The cerebral signature for pain perception and its modulation. Neuron 55:377–391.

Tso AR, Goadsby PJ. 2017. Anti-CGRP monoclonal antibodies: the next era of migraine prevention? Curr Treat Options Neurol 19:27.

Wercberger R. Basbaum AI. 2019. Spinal cord projection neurons: A superficial, and also deep analysis. Curr Opin Physiol 11:109–115.

Woo SH, Ranade S, Weyer AD, et al. 2014. Piezo2 is required for Merkel-cell mechanotransduction. Nature 509: 622–626.

Woolf CJ. 1983. Evidence for a central component of post-injury pain hypersensitivity. Nature 306:686–688.

Yaksh TL, Fisher C, Hockman T, Wiese A. 2017. Current and future issues in the development of spinal agents for the management of pain. Curr Neuropharmacol 15:232–259.

Zeilhofer HU, Wildner H, Yévenes GE. 2012. Fast synaptic inhibition in spinal sensory processing and pain control. Physiol Rev 92:193–235.

21

The Constructive Nature of Visual Processing

We are so familiar with seeing, that it takes a leap of imagination to realize that there are problems to be solved. But consider it. We are given tiny distorted upside-down images in the eyes and we see separate solid objects in surrounding space. From the patterns of stimulation on the retina we perceive the world of objects and this is nothing short of a miracle.

—Richard L. Gregory, *Eye and Brain*, 1966

MOST OF OUR IMPRESSIONS of the world and our memories of it are based on sight. Yet the mechanisms that underlie vision are not at all obvious. How do we perceive form and movement? How do we distinguish colors? Identifying objects in complex visual environments is an extraordinary computational achievement that artificial vision systems have yet to duplicate. Vision is used not only for object recognition but also for guiding our movements, and these separate functions are mediated by at least two parallel and interacting pathways.

The existence of parallel pathways in the visual system raises one of the central questions of cognition, the binding problem: How are different types of information carried by discrete pathways brought together into a coherent visual image?

Visual Perception Is a Constructive Process

Vision is often incorrectly compared to the operation of a camera. A camera simply reproduces point-by-point the light intensities in one plane of the visual field. The visual system, in contrast, does something fundamentally different. It interprets the scene and parses it into distinct components, separating foreground from background. The visual system is less accurate than a camera at certain tasks, such as quantifying the absolute level of brightness or identifying spectral color. However, it excels at tasks such as recognizing a charging animal (or a speeding car) whether in bright sunlight or at dusk, in an open field or partly occluded by trees (or other cars). And it does so rapidly to let the viewer respond and, if necessary, escape.

A potentially unifying insight reconciling the visual system's remarkable ability to grasp the bigger picture with its inaccuracy regarding details of the input is that vision is a biological process that has evolved in step with our ecological needs. This insight helps explain why the visual system is so efficient at extracting useful information such as the identities of objects independent of lighting conditions, while giving less importance to aspects like the exact nature of the ambient light. Moreover, vision does so using previously learned rules about the structure of the world. Some of these rules appeared to have become wired into our neural circuits over the course of evolution. Others are more plastic and help the brain guess at the scene presented to the eyes based on the individual's past experience. This complex, purposeful processing happens at all levels of the visual system. It starts even at the retina, which is specialized to pick out object boundaries rather than creating a point-by-point representation of uniform surfaces.

This *constructive* nature of visual perception has only recently been fully appreciated. Earlier thinking about sensory perception was greatly influenced by the British empiricist philosophers, notably John Locke, David Hume, and George Berkeley, who thought of perception as an atomistic process in which simple sensory elements, such as color, shape, and brightness, were assembled in an additive way, component by component. The modern view that perception is an active and creative process that involves more than just the information provided to the retina has its roots in the philosophy of Immanuel Kant and was developed in detail in the early 20th century by the German psychologists Max Wertheimer, Kurt Koffka, and Wolfgang Köhler, who founded the school of Gestalt psychology.

The German term *Gestalt* means configuration or form. The central idea of the Gestalt psychologists is that what we see about a stimulus—the perceptual interpretation we make of any visual object—depends not just on the properties of the stimulus but also on its context, on other features in the visual field. The Gestalt psychologists argued that the visual system processes sensory information about the shape, color, distance, and movement of objects according to computational rules inherent in the system. The brain has a way of looking at the world, a set of expectations that derives in part from experience and in part from built-in neural wiring.

Max Wertheimer wrote: "There are entities where the behavior of the whole cannot be derived from its individual elements nor from the way these elements fit together; rather the opposite is true: the properties of any of the parts are determined by the intrinsic structural laws of the whole." In the early part of the 20th century, the Gestalt psychologists worked out the laws of perception that determine how we group elements in the visual scene, including similarity, proximity, and good continuation.

We see a uniform six-by-six array of dots as either rows or columns because of the visual system's tendency to impose a pattern. If the dots in each row are similar, we are more likely to see a pattern of alternating rows (Figure 21–1A). If the dots in each column are closer together than those in the rows, we are more disposed to see a pattern of columns (Figure 21–1B). The principle of good continuation is an important basis for linking line elements into unified shapes (Figure 21–1C). It is also seen in the phenomenon of contour saliency, whereby smooth contours tend to pop out from complex backgrounds (Figure 21–1D). The Gestalt features that we are disposed to pick out are also ones that characterize objects in natural scenes. Statistical studies of natural scenes show that object boundaries are likely to contain visual elements that lie in close proximity, are continuous across intersections, or form smooth contours. It is tempting to speculate that the formal features of objects in natural scenes created evolutionary pressure on our visual systems to develop neural circuits that have made us sensitive to those features.

Separating the figure and background in a visual scene is an important step in object recognition. At different moments, the same elements in the visual field can be organized into a recognizable figure or serve as part of the background for other figures (Figure 21–2). This process of segmentation relies not only on certain geometric principles, but also on cognitive influences such as attention and expectation. Thus, a priming stimulus or an internal representation of object shape can facilitate the association of visual elements into a unified percept (Figure 21–3). This internal representation can take many different forms reflecting the wide range of time scales and mechanisms of neural encoding. It could consist of transient reverberating spiking activity selective to a shape or a decision, lasting a fraction of a second, or the selective modulation of synaptic weights during a particular context of a task or an expected shape, or circuit changes that could comprise a long-term memory.

The brain analyzes a visual scene at three levels: low, intermediate, and high (Figure 21–4). At the lowest level, which we consider in the next chapter (Chapter 22), visual attributes such as local contrast, orientation, color, and movement are discriminated. The intermediate level involves analysis of the layout of scenes and of surface properties, parsing the visual image into surfaces and global contours, and distinguishing foreground from background (Chapter 23). The highest level involves object recognition (Chapter 24). Once a scene has been parsed by the brain and objects recognized, the objects can be matched with memories of

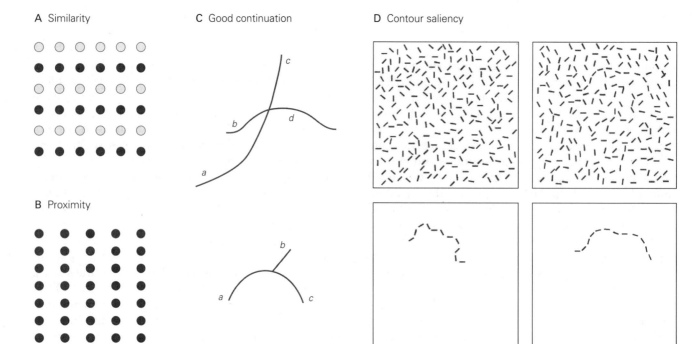

A Similarity

B Proximity

C Good continuation

D Contour saliency

Figure 21–1 Organizational rules of visual perception. To link the elements of a visual scene into unified percepts, the visual system relies on organizational rules such as similarity, proximity, and good continuation.

A. Because the dots in alternating rows have the same color, an overall pattern of blue and white rows is perceived.

B. The dots in the columns are closer together than those in the rows, leading to the perception of columns.

C. Line segments are perceptually linked when they are collinear. In the top set of lines, one is more likely to see line segment **a** as belonging with **c** rather than **d**. In the bottom set, **a** and **c** are perceptually linked because they maintain the same curvature, whereas **a** and **b** appear to be discontinuous.

D. The principle of good continuation is also seen in contour saliency. On the right, a smooth contour of line elements pops out from the background, whereas the jagged contour on the left is lost in the background. (Adapted, with permission, from Field, Hayes, and Hess 1993. Copyright © 1993 Elsevier Ltd.)

Figure 21–2 Object recognition depends on segmentation of a scene into foreground and background. Recognition of the white salamanders in this image depends on the brain "locating" the white salamanders in the foreground and the brown and black salamanders in the background. The image also illustrates the role of higher influences in segmentation: One can consciously select any of the three colors as the foreground. (Reproduced, with permission, from M.C. Escher's "Symmetry Drawing E56" © 2010 The M.C. Escher Company-Holland. All rights reserved. www.mcescher.com.)

Figure 21–3 Expectation and perceptual task play a critical role in what is seen. It is difficult to separate the dark and white patches in this figure into foreground and background without additional information. This figure immediately becomes recognizable after viewing the priming image on page 501. In this example, higher-order representations of shape guide lower-order processes of surface segmentation. (Reproduced, with permission, from Porter 1954. Copyright 1954 by the Board of Trustees of the University of Illinois. Used with permission of the University of Illinois Press.)

shapes and their associated meanings. Vision also has an important role in guiding body movement, particularly hand movement (Chapter 25).

In vision, as in other cognitive operations, various features—motion, depth, form, and color—occur together in a unified percept. This unity is achieved not by one hierarchical neural system but by multiple areas in the brain that are fed by parallel but interacting neural pathways. Because distributed processing is one of the main organizational principles in the neurobiology of vision, one must have a grasp of the anatomical pathways of the visual system to understand fully the physiological description of visual processing in later chapters.

In this chapter, we lay the foundation for understanding the neural circuitry and organizational principles of the visual pathways. These principles apply quite broadly and are relevant not only for the multiple areas of the brain concerned with vision but also for other types of sensory information processing by the brain.

Visual Processing Is Mediated by the Geniculostriate Pathway

The brain's analysis of visual scenes begins in the two retinas, which transform visual input using a strategy of parallel processing (Chapter 22). This important neural computation strategy is utilized at all stages of the visual pathway as well as in other sensory areas. The pixel-like bits of visual input falling on individual photoreceptors—rods and cones—are analyzed by retinal circuits to extract some 20 local features, such as the local contrasts of dark versus light, red versus

green, and blue versus yellow. These features are computed by different populations of specialized neural circuits forming independent processing modules that separately cover the visual field. Thus, each point in the visual field is processed in multiple channels that extract distinct aspects of the visual input simultaneously and in parallel. These parallel streams are then sent out along the axons of the retinal ganglion cells, the projection neurons of the retina, which form the optic nerves.

From the eye, the optic nerve extends to a midline crossing point, the optic chiasm. Beyond the chiasm, the fibers from each temporal hemiretina proceed to the ipsilateral hemisphere along the ipsilateral optic tract; fibers from the nasal hemiretinas cross to the contralateral hemisphere along the contralateral optic tract (Figure 21–5). Because the temporal hemiretina of one eye sees the same half of the visual field (hemifield) as the nasal hemiretina of the other, the partial decussation of fibers at the chiasm ensures that all the information about each hemifield is processed in the visual cortex of the contralateral hemisphere. The layout of the pathway also forms the basis for useful diagnostic information. As a consequence of the particular anatomy of this visual pathway, lesions at different points along the pathway lead to visual deficits with different geometric shapes (Figure 21–5) that can be distinguished reliably through clinical examination. The deficit could be entirely monocular; if present in both eyes, it could affect noncorresponding or corresponding parts of the visual field in the two eyes; it could be restricted to either the upper or the lower visual field or may extend into both, etc. Thus, the shape of the deficit could give valuable clues about type and location of the

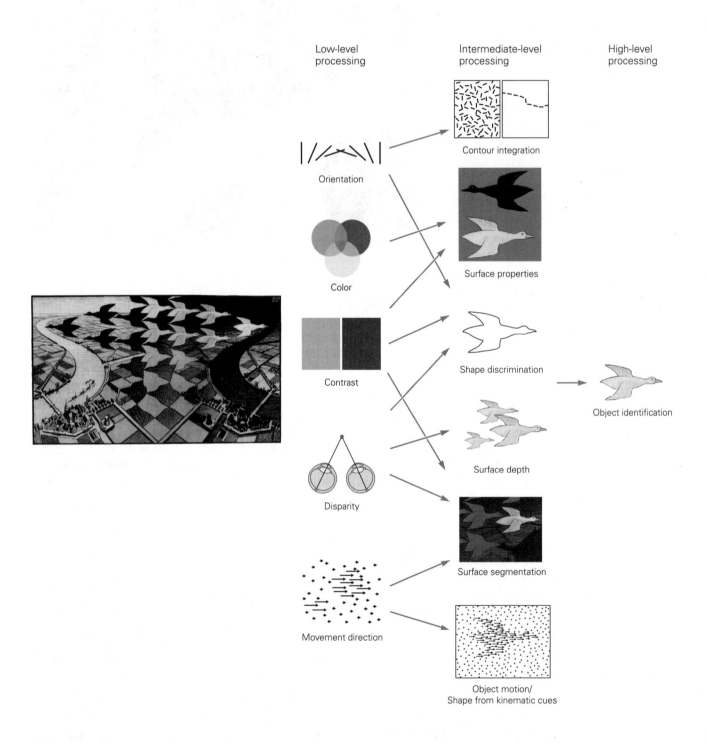

Figure 21–4 A visual scene is analyzed at three levels. Simple attributes of the visual environment are analyzed (low-level processing), and these low-level features are then used to parse the visual scene (intermediate-level processing): Local visual features are assembled into surfaces, objects are segregated from background (surface segmentation), local orientation is integrated into global contours (contour integration), and surface shape is identified from shading and kinematic cues. Finally, surfaces and contours are used to identify the object (high-level processing). (M.C. Escher's "Day and Night".
© 2020 The M.C. Escher Company—The Netherlands. All rights reserved. www.mcescher.com)

Priming image for Figure 21–3

underlying nerve damage or occlusion (ranging from optic nerve degeneration, such as due to multiple sclerosis, to tumors, strokes, or physical trauma).

Beyond the optic chiasm, the axons from nasal and temporal hemiretinas carrying input from one hemifield join in the optic tract, which extends to the lateral geniculate nucleus (LGN) of the thalamus. The LGN in primates consists of six primary layers: four parvocellular (Latin *Parvus*, small) and two magnocellular, each paired with a thin but dense intercalated or koniocellular (Greek *konio*, dust) layer (see Figure 21–14). The term "koniocellular" refers to the substantially smaller cell bodies in these layers relative to those of magnocellular or parvocellular layers. The parallel channels established in the retinas remain anatomically segregated through the LGN. Parvocellular layers get input from the midget retinal ganglion cells, which are the most numerous in the primate retina (~70%) and carry red-green opponent information (Chapter 22). Magnocellular layers get achromatic contrast information from the parasol ganglion cells (~10%). Koniocellular layers get input from the small and large bistratified ganglion cells, carrying blue-yellow information, that together make up the third most populous set of retinal projections to the LGN (~8%). Koniocellular layers also get inputs from a number of other numerically much smaller classes of retinal ganglion cells.

Each geniculate layer receives input from either the ipsilateral or the contralateral eye (see Figure 21–12) but is aligned so as to come from a matching region of the contralateral hemifield. Thus, they form a set of concordant maps stacked atop one another. The thalamic neurons then relay retinal information to

the primary visual cortex. But the LGN is not simply a relay; the retinal information it receives can be strongly modulated by attention and arousal through inhibitory connections to this brain region and by feedback from the visual cortex.

The primary visual pathway is also called the geniculostriate pathway (Figure 21–6A) because it passes through the LGN on its way to the primary visual cortex (V1), also known as the striate cortex because of the myelin-rich stripe that runs through its middle layers. A second pathway extends from the retina to the pretectal area of the midbrain, where neurons mediate the pupillary reflexes that control the amount of light entering the eyes (Figure 21–6B). A third pathway from the retina runs to the superior colliculus and is important in controlling eye movements. This pathway continues to the pontine formation in the brain stem and then to the extraocular motor nuclei (Figure 21–6C).

Each LGN projects to the primary visual cortex through a pathway known as the optic radiation. These afferent fibers form a complete neural map of the contralateral visual field in the primary visual cortex. Beyond the striate cortex lie the extrastriate areas, a set of higher-order visual areas that are also organized as neural maps of the visual field. The preservation of the spatial arrangement of inputs from the retina is called retinotopy, and a neural map of the visual field is described as retinotopic or having a retinotopic frame of reference.

The primary visual cortex constitutes the first level of cortical processing of visual information. From there, information is transmitted over two major pathways. A ventral pathway into the temporal lobe carries information about what the stimulus is, and a dorsal

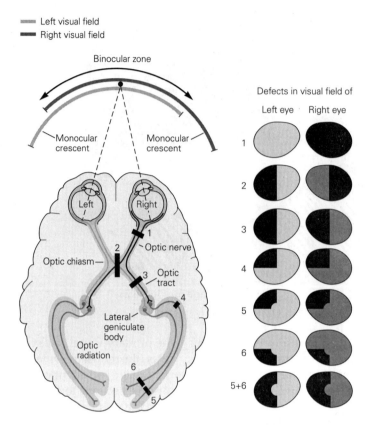

Figure 21–5 Representation of the visual field along the visual pathway. Each eye sees most of the visual field, with the exception of a portion of the peripheral visual field known as the monocular crescent. The axons of retinal neurons (ganglion cells) carry information from each visual hemifield along the optic nerve up to the optic chiasm, where fibers from the nasal hemiretina cross to the opposite hemisphere. Fibers from the temporal hemiretina stay on the same side, joining the fibers from the nasal hemiretina of the contralateral eye to form the optic tract. The optic tract carries information from the opposite visual hemifield originating in both eyes and projects into the lateral geniculate nucleus. Cells in this nucleus send their axons along the optic radiation to the primary visual cortex.

Lesions along the visual pathway produce specific visual field deficits, as shown on the *right*:

1. A lesion of an optic nerve causes a total loss of vision in one eye.

2. A lesion of the optic chiasm causes a loss of vision in the temporal half of each visual hemifield (bitemporal hemianopsia).

3. A lesion of the optic tract causes a loss of vision in the opposite half of the visual hemifield (contralateral hemianopsia).

4. A lesion of the optic radiation fibers that curve into the temporal lobe (Meyer's loop) causes loss of vision in the upper quadrant of the contralateral visual hemifield in both eyes (upper contralateral quadrantic anopsia).

5, 6. Partial lesions of the visual cortex lead to deficits in portions of the contralateral visual hemifield. For example, a lesion in the upper bank of the calcarine sulcus (**5**) causes a partial deficit in the inferior quadrant, while a lesion in the lower bank (**6**) causes a partial deficit in the superior quadrant. The central area of the visual field tends to be unaffected by cortical lesions because of the extent of the representation of the fovea and the duplicate representation of the vertical meridian in the hemispheres.

pathway into the parietal lobe carries information about where the stimulus is, information that is critical for guiding movement.

A major fiber bundle called the corpus callosum connects the two hemispheres, transmitting information across the midline. The primary visual cortex in each hemisphere represents slightly more than half the visual field, with the two hemifield representations overlapping at the vertical meridian. One of the functions of the corpus callosum is to unify the perception of objects spanning the vertical meridian by linking the cortical areas that represent opposite hemifields.

Form, Color, Motion, and Depth Are Processed in Discrete Areas of the Cerebral Cortex

In the late 19th and early 20th centuries, the cerebral cortex was differentiated into discrete regions by the anatomist Korbinian Brodmann and others using anatomical criteria. The criteria included the size, shape, and packing density of neurons in the cortical layers and the thickness and density of myelin. The functionally distinct cortical areas we have considered heretofore correspond only loosely to Brodmann's classification. The primary visual cortex (V1) is identical to Brodmann's area 17. In the extrastriate cortex, the secondary visual area (V2) corresponds to area 18. Beyond that, however, area 19 contains several functionally distinct areas that generally cannot be defined by anatomical criteria.

The number of functionally discrete areas of visual cortex varies between species. Macaque monkeys have more than 30 areas. Although not all visual areas in humans have yet been identified, the number is likely to be at least as great as in the macaque. If one includes oculomotor areas and prefrontal areas contributing to visual memory, almost half of the cerebral cortex is involved with vision. Functional magnetic resonance imaging (fMRI) has made it possible to establish homologies between the visual areas of the macaque and human brains (Figure 21–7). Based on pathway tracing studies in monkeys, we now appreciate that these areas are organized in functional streams (Figure 21–7B).

The visual areas of cortex can be differentiated by the functional properties of their neurons. Studies of

A Visual processing

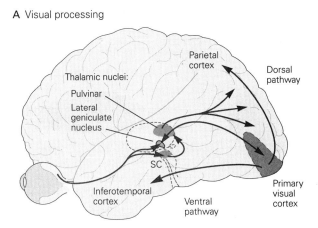

B Pupillary reflex and accommodation

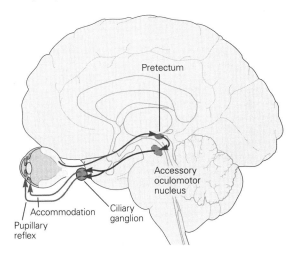

C Eye movement (horizontal)

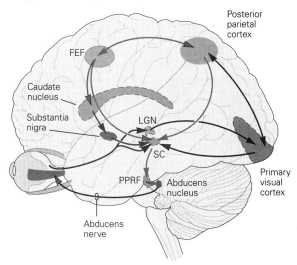

Figure 21–6 Pathways for visual processing, pupillary reflex and accommodation, and control of eye position.

A. *Visual processing.* The eye sends information first to thalamic nuclei, including the lateral geniculate nucleus and pulvinar, and from there to cortical areas. Cortical projections go forward from the primary visual cortex to areas in the parietal lobe (the dorsal pathway, which is concerned with visually guided movement) and areas in the temporal lobe (the ventral pathway, which is concerned with object recognition). The pulvinar also serves as a relay between cortical areas to supplement their direct connections. (Abbreviation: **SC**, superior colliculus).

B. *Pupillary reflex and accommodation.* Light signals are relayed through the midbrain pretectum, to preganglionic parasympathetic neurons in the accessory oculomotor (Edinger-Westphal) nucleus, and out through the parasympathetic outflow of the oculomotor nerve to the ciliary ganglion. Postganglionic neurons innervate the smooth muscle of the pupillary sphincter, as well as the muscles controlling the lens.

C. *Eye movement.* Information from the retina is sent to the superior colliculus (**SC**) directly along the optic nerve and indirectly through the geniculostriate pathway to cortical areas (primary visual cortex, posterior parietal cortex, and frontal eye fields) that project back to the superior colliculus. The colliculus projects to the pons (**PPRF**), which then sends control signals to oculomotor nuclei, including the abducens nucleus, which controls lateral movement of the eyes. (Abbreviations: **FEF**, frontal eye field; **LGN**, lateral geniculate nucleus; **PPRF**, paramedian pontine reticular formation.)

A Cortical visual areas in humans

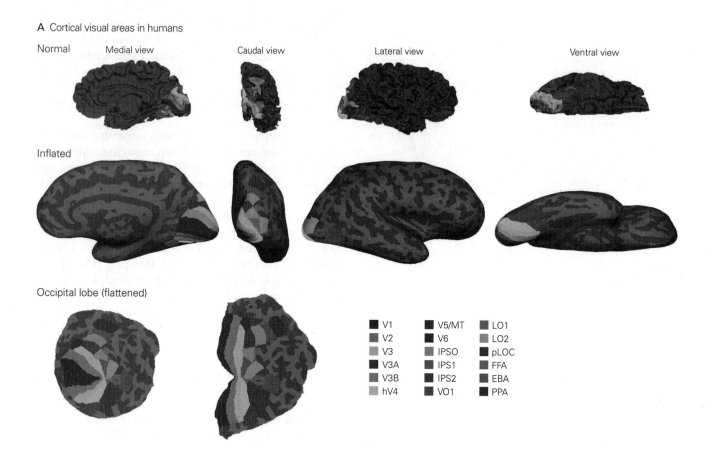

Normal   Medial view   Caudal view   Lateral view   Ventral view

Inflated

Occipital lobe (flattened)

■ V1	■ V5/MT	■ LO1
■ V2	■ V6	■ LO2
■ V3	■ IPSO	■ pLOC
■ V3A	■ IPS1	■ FFA
■ V3B	■ IPS2	■ EBA
■ hV4	■ VO1	■ PPA

B Visual pathways in the macaque monkey

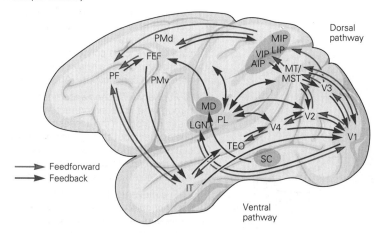

such functional properties have revealed that the visual areas are organized in two hierarchical pathways, a ventral pathway involved in object recognition and a dorsal pathway dedicated to the use of visual information for guiding movements. The ventral or object-recognition pathway extends from the primary visual cortex to the temporal lobe; it is described in detail in Chapter 24. The dorsal or movement-guidance pathway connects the primary visual cortex with the parietal lobe and then with the frontal lobes.

The pathways are interconnected so that information is shared. For example, movement information in the dorsal pathway can contribute to object recognition through kinematic cues. Information about movements in space derived from areas in the dorsal pathway is therefore important for the perception of object shape and is fed into the ventral pathway.

All connections between cortical areas are reciprocal—each area sends information back to the areas from which it receives input. These feedback connections provide information about cognitive functions, including spatial attention, stimulus expectation, and emotional content, to earlier levels of visual

processing. The pulvinar in the thalamus serves as a relay between cortical areas (see Figure 21–7B).

The dorsal pathway courses through the parietal cortex, a region that uses visual information to direct the movement of the eyes and limbs, that is, for visuomotor integration. The lateral intraparietal area, named for its location in the intraparietal sulcus, is involved in representing points in space that are the targets of eye movements or reaching. Patients with lesions of parietal areas fail to attend to objects on one side of the body, a syndrome called *unilateral neglect* (see Figure 59–1 in Chapter 59).

The ventral pathway extends into the temporal lobe. The inferior temporal cortex stores information about the shapes and identities of objects; one portion represents faces, for damage to that region results in the inability to recognize faces (*prosopagnosia*).

The dorsal and ventral pathways each comprise a hierarchical series of areas that can be delineated by several criteria. First, at many relays, the array of inputs forms a map of the visual hemifield. The boundaries of these maps can be used to demarcate the boundaries of visual areas. This is particularly useful at early levels of the pathway where the receptive fields of neurons

Figure 21–7 Visual pathways in the cerebral cortex.

A. Functional magnetic resonance imaging shows areas of the human cerebral cortex involved in visual processing. The **top row** shows areas on the gyri and sulci of a normal view of a brain; the **middle row** shows "inflated" views of the brain following a computational process that simulates inflating the brain like a balloon so as to stretch out the "wrinkles" of gyri and sulci into a smooth surface while minimizing local distortions. Light and dark gray regions identify gyri and sulci, respectively; the **bottom row** shows a two-dimensional representation of the occipital lobe (*left*) and a representation with less distortion by making a cut along the calcarine fissure. Different approaches are required for demarcating different functional areas. Retinotopic areas, by definition, contain continuous maps of visual space and are identified using stimuli such as rotating spirals or expanding circles that sweep through visual space. Maps in adjacent cortical areas run in opposite directions on the cortical surface and meet along boundaries of local mirror reversals. These mirror reversals can be used to identify area boundaries and thus demarcate each area. These retinotopic areas, including early visual areas V1, V2, and V3, and areas V3A, V3B, V6, hV4, VO1, LO1, LO2, and V5/ MT, share boundaries in pairs; these boundaries converge (at the representation of the fovea) at the occipital pole. A different approach, identifying loci of attention, is used to map areas IPS1 and IPS2. Yet further sets of approaches or responsiveness to specific attributes or classes of objects (such as faces) are used for less strictly retinotopic areas. Functional specificity has been demonstrated for a number of visual areas: VO1 is implicated in color processing, the lateral occipital complex (**LO2, pLOC**) codes object shape, fusiform face area (**FFA**)

codes faces, the parahippocampal place area (**PPA**) responds more strongly to places than to objects, the extrastriate body area (**EBA**) responds more strongly to body parts than objects, and V5/MT is involved in motion processing. Areas in the intraparietal sulcus (**IPS1** and **IPS2**) are involved in control of spatial attention and saccadic eye movements. (Images courtesy of V. Piech, reproduced with permission.)

B. In the macaque monkey, V1 is located on the surface of the occipital lobe and sends axons in two pathways. A dorsal pathway courses through a number of areas in the parietal lobe and into the frontal lobe and mediates attentional control and visually guided movements. A ventral pathway projects through V4 into areas of the inferior temporal cortex and mediates object recognition. In addition to feedforward pathways extending from primary visual cortex into the temporal, parietal, and frontal lobes (**blue arrows**), reciprocal or feedback pathways run in the opposite direction (**red arrows**). Feedforward and feedback can operate directly, between cortical areas, or indirectly, via the thalamus, in particular the pulvinar, which acts as a relay between cortical areas. The subcortical pathways involved include thalamic nuclei—the lateral geniculate nucleus (**LGN**), pulvinar nucleus (**PL**), and mediodorsal nucleus (**MD**)—and the superior colliculus (**SC**). (Abbreviations: **AIP**, anterior intraparietal area; **FEF**, frontal eye field; **IT**, inferior temporal cortex; **LIP**, lateral intraparietal area; **MIP**, medial intraparietal area; **MT**, middle temporal area; **PF**, prefrontal cortex; **PMd**, dorsal premotor cortex; **PMv**, ventral premotor cortex; **TEO**, posterior division of area IT; **V1**, primary visual cortex, Brodmann's area 17; **V2**, secondary visual area, Brodmann's area 18; **V3, V4**, third and fourth visual areas; **VIP**, ventral intraparietal area.)

are small and visuotopic maps are precisely organized (see the next section for the definition of receptive field). At higher levels, however, the receptive fields become larger, the maps less precise, and visuotopic organization is therefore a less reliable basis to delineate the boundaries of an area.

Another means to differentiate one area from another, as shown by experiments in monkeys, depends upon the distinctive functional properties exhibited by the neurons in each area. The clearest example of this is an area in the dorsal pathway, the middle temporal area (MT or V5), which contains neurons with a strong selectivity for the direction of movement across their receptive fields. Consistent with the idea that the middle temporal area is involved in the analysis of motion, lesions of this area produce deficits in the ability to track moving objects.

A classical view of the organization of visual cortical areas is a hierarchical one, where the areas at the bottom of the hierarchy, such as V1 and V2, represent the visual primitives of orientation, direction of movement, depth, and color. In this view, the top of the ventral pathway's hierarchy would represent whole objects, with the areas in between representing intermediate level vision. This idea of "complexification" along the hierarchy suggests a mapping between the levels of visual perception and stages in the sequence of cortical areas. But more recent findings indicate a more complex story, where even the primary visual cortex plays a role in intermediate-level vision, and neurons in the higher areas may process information on components of objects. Moreover, as shown in Figure 21–7, one also has to take into account the fact that there is a powerful reverse flow of information, or feedback, from the "higher" to the "lower" cortical areas. As will be described in Chapter 23, this reverse direction of information contains higher order "top-down" cognitive influences including attention, object expectation, perceptual task, perceptual learning, and efference copy. Top-down influences may play a role in scene segmentation, object relationships, and perception of object details, as well as object recognition itself.

The Receptive Fields of Neurons at Successive Relays in the Visual Pathway Provide Clues to How the Brain Analyzes Visual Form

In 1906, Charles Sherrington coined the term *receptive field* in his analysis of the scratch withdrawal reflex: "The whole collection of points of skin surface from which the scratch-reflex can be elicited is termed the receptive field of that reflex." When it became possible

to record from single neurons in the eye, H. Keffer Hartline applied the concept of the receptive field in his study of the retina of the horseshoe crab, *Limulus:* "The region of the retina which must be illuminated in order to obtain a response in any given fiber . . . is termed the receptive field of that fiber." In the visual system, a neuron's receptive field represents a small window on the visual field (Figure 21–8).

But responses to only one spot of light yielded a limited understanding of a cell's receptive field. Using two small spots of light, both Hartline and Stephen Kuffler, who studied the mammalian retina, found an inhibitory surround or lateral inhibitory region in the receptive field. In 1953, Kuffler observed that "not only the areas from which responses can actually be set up by retinal illumination may be included in a definition of the receptive field but also all areas which show a functional connection, by an inhibitory or excitatory effect on a ganglion cell." Kuffler thus demonstrated that the receptive fields of retinal ganglion cells have functionally distinct subareas. These receptive fields have a center-surround organization and fall into one of two categories: *on-center* and *off-center*. Later work demonstrated that neurons in the LGN have similar receptive fields.

The on-center cells fire when a spot of light is turned on within a circular central region. Off-center cells fire when a spot of light in the center of their receptive field is turned off. The surrounding annular region has the opposite sign. For on-center cells, a light stimulus anywhere in the annulus surrounding the center produces a response when the light is turned off, a response termed *on-center, off-surround*. The center and surround areas are mutually inhibitory (Figure 21–9). When both center and surround are illuminated with diffuse light, there is little or no response. Conversely, a light–dark boundary across the receptive field produces a brisk response. Because these neurons are most sensitive to borders and contours—to differences in illumination as opposed to uniform surfaces—they encode information about contrast in the visual field.

The size on the retina of a receptive field varies both according to the field's *eccentricity*—its position relative to the fovea, the central part of the retina where visual acuity is highest—and the position of neurons along the visual pathway. Receptive fields with the same eccentricity are relatively small at early levels in visual processing and become progressively larger at later levels. The size of the receptive field is expressed in terms of degrees of visual angle; the entire visual field covers nearly 180° (Figure 21–10A). In early relays of visual processing, the receptive fields near the fovea are the smallest. The receptive

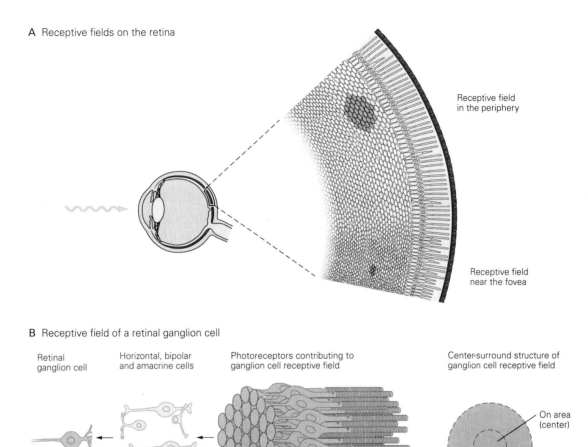

A Receptive fields on the retina

Receptive field
in the periphery

Receptive field
near the fovea

B Receptive field of a retinal ganglion cell

Retinal
ganglion cell

Horizontal, bipolar
and amacrine cells

Photoreceptors contributing to
ganglion cell receptive field

Center-surround structure of
ganglion cell receptive field

On area
(center)

Off area
(surround)

Light

Figure 21–8 Receptive fields of retinal ganglion cells in relation to photoreceptors.

A. The number of photoreceptors contributing to the receptive field of a retinal ganglion cell varies depending on the location of the receptive field on the retina. A cell near the fovea receives input from fewer receptors covering a smaller area, whereas a cell farther from the fovea receives input from many more receptors covering a larger area (see Figure 21–10).

B. Light passes through nerve cell layers to reach the photoreceptors at the back of the retina. Signals from the photoreceptors are then transmitted by neurons in the outer and inner nuclear layers to a retinal ganglion cell.

fields for retinal ganglion cells that monitor portions of the fovea subtend approximately 0.1°, whereas those in the visual periphery can be a couple of orders of magnitude larger.

The amount of cortex dedicated to a degree of visual space changes with eccentricity. More area of cortex is dedicated to the central part of the visual field, where the receptive fields are smallest and the visual system has the greatest spatial resolution (Figure 21–10C).

Receptive-field properties change from relay to relay along a visual pathway. By determining these properties, one can assay the function of each relay nucleus and how visual information is progressively analyzed by the brain. For example, the change in receptive-field structure that occurs between the LGN and cerebral cortex reveals an important mechanism in the brain's analysis of visual form. The key property of the form pathway is selectivity for the orientation of contours in the visual field. This is an emergent property of signal processing in primary visual cortex; it is not a property of the cortical input but is generated within the cortex itself.

Whereas retinal ganglion cells and neurons in the LGN have concentric center-surround receptive fields, those in the cortex, although equally sensitive to contrast, also analyze contours. David Hubel and Torsten Wiesel discovered this characteristic in 1958 while studying what visual stimuli provoked activity in neurons in the primary visual cortex. While showing an anesthetized animal slides containing a variety of images, they recorded extracellularly from individual neurons in the visual cortex. As they switched from one slide to another, they found a neuron that produced a brisk train of action potentials. The cell was responding not to the image on the slide but to the edge of the slide as it was moved into position.

The Visual Cortex Is Organized Into Columns of Specialized Neurons

The dominant feature of the functional organization of the primary visual cortex is the visuotopic organization of its cells: the visual field is systematically represented across the surface of the cortex (Figure 21–11A).

In addition, cells in the primary visual cortex with similar functional properties are located close together in columns that extend from the surface of the cortex to the white matter. The columns are concerned with the functional properties that are analyzed in any given cortical area and thus reflect the functional role of that area in vision. The properties that are developed in the primary visual cortex include orientation specificity and the integration of inputs from the two eyes, which is measured as the relative strength of input from each eye, or ocular dominance.

Ocular-dominance columns reflect the segregation of thalamocortical inputs arriving from different layers of the LGN. Alternating layers of this nucleus receive input from retinal ganglion cells located in either the ipsilateral or contralateral retina (Figure 21–12). This segregation is maintained in the inputs from the LGN to the primary visual cortex, producing the alternating left-eye and right-eye ocular dominance bands (Figure 21–11B).

Cells with similar orientation preferences are also grouped into columns. Across the cortical surface, there is a regular clockwise and counterclockwise cycling of orientation preference, with the full 180° cycle repeating every 750 μm (Figure 21–11C). Likewise, the left- and right-eye dominance columns alternate with a periodicity of 750 to 1,000 μm. One full cycle of orientation columns, or a full pair of left- and right-eye dominance columns, is called a *hypercolumn*. The orientation and ocular dominance columns at each point on the cortical surface are locally roughly orthogonal to each other. Thus, a cortical patch one hypercolumn in extent contains all possible combinations of orientation preference and left- and right-eye dominance.

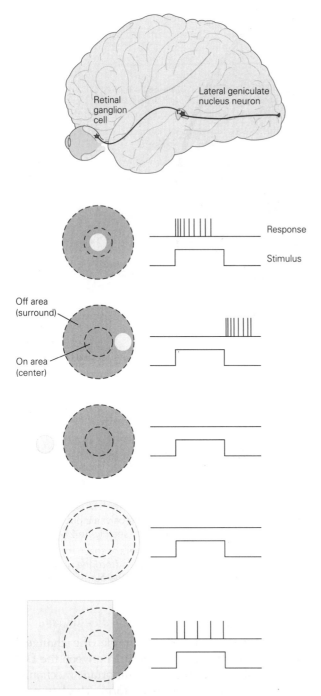

Figure 21–9 Receptive fields of neurons at early relays of visual pathways. A circular symmetric receptive field with mutually antagonistic center and surround is characteristic of retinal ganglion cells and neurons in the lateral geniculate nucleus of the thalamus. The center can respond to the turning on or turning off of a spot of light (**yellow**) depending on whether the receptied field belongs to an "on-center" or "off-center" class, respectively. The surround has the opposite response. Outside the surround, there is no response to light, thus defining the receptive field boundary. The response is weak when light covers both the center and surround, so these neurons respond optimally to contrast (a light–dark boundary) in the visual field.

A Map of retinal eccentricity

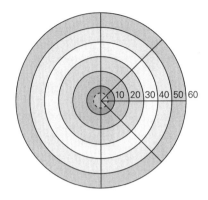

B Receptive field size varies systematically with eccentricity

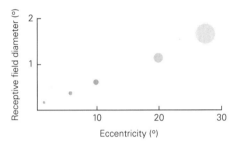

C Cortical magnification varies with eccentricity

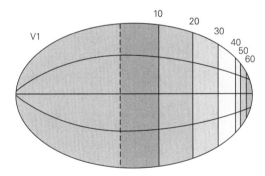

Figure 21–10 Receptive field size, eccentricity, retinotopic organization, and magnification factor. The color code refers to position in visual space or on the retina.

A. The distance of a receptive field from the fovea is referred to as the eccentricity of the receptive field.

B. Receptive field size varies with distance from the fovea. The smallest fields lie in the center of gaze, the fovea, where the visual resolution is highest; fields become progressively larger with distance from the fovea.

C. The amount of cortical area dedicated to inputs from within each degree of visual space, known as the magnification factor, also varies with eccentricity. The central part of the visual field commands the largest area of cortex. For example, in area V1, more area is dedicated to the central 10° of visual space than to all the rest. The map of V1 shows the cortical sheet unfolded.

Both types of columns were first mapped by recording the responses of neurons at closely spaced electrode penetrations in the cortex. The ocular-dominance columns were also identified by making lesions or tracer injections in individual layers of the LGN. More recently, a technique known as optical imaging has enabled researchers to visualize a surface representation of the orientation and ocular dominance columns in living animals. Developed for studies of cortical organization by Amiram Grinvald, this technique visualizes changes in surface reflectance associated with the metabolic requirements of active groups of neurons, known as intrinsic-signal optical imaging, or changes in fluorescence of voltage-sensitive dyes. Intrinsic-signal imaging depends on activity-associated changes in local blood flow and alterations in the oxidative state of hemoglobin and other intrinsic chromophores. These techniques are also now being complemented with imaging at cellular resolution using genetically encoded markers of neural activity.

An experimenter can visualize the distribution of cells with left or right ocular dominance, for example, by subtracting the image obtained while stimulating one eye from that acquired while stimulating the other. When viewed in a plane tangential to the cortical surface, the ocular dominance columns appear as alternating left- and right-eye stripes, each approximately 750 μm in width (Figure 21–11B).

The cycles of orientation columns form various structures, from parallel stripes to pinwheels. Sharp jumps in orientation preference occur at the pinwheel centers and "fractures" in the orientation map (Figure 21–11C).

Embedded within the orientation and ocular-dominance columns are clusters of neurons that have poor orientation selectivity but strong color preferences. These units of specialization, located within the superficial layers, were revealed by a histochemical label for the enzyme cytochrome oxidase, which is distributed in a regular patchy pattern of blobs and interblobs. In the primary visual cortex, these blobs are a few hundred micrometers in diameter and 750 μm apart (Figure 21–11D). The blobs correspond to clusters of color-selective neurons. Because they are rich in cells with color selectivity and poor in cells with orientation selectivity, the blobs are specialized to provide information about surfaces rather than edges.

In area V2, thick and thin dark stripes separated by pale stripes are evident with cytochrome oxidase labeling (Figure 21–11D). The thick stripes contain neurons selective for direction of movement and for binocular disparity as well as cells that are responsive to illusory contours and global disparity cues. The thin stripes hold cells specialized for color. The pale stripes contain orientation-selective neurons.

A Visuotopic map

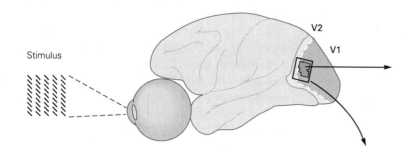

Stimulus

Pattern of excitation in response to striped stimulus

B Ocular dominance columns

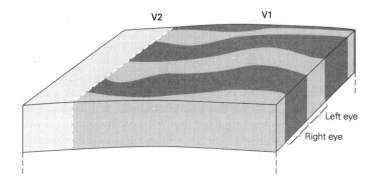

V2   V1

Left eye

Right eye

V2   V1

C Orientation columns

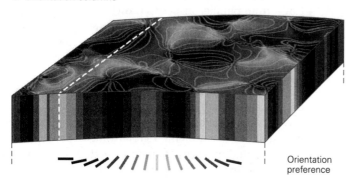

Orientation preference

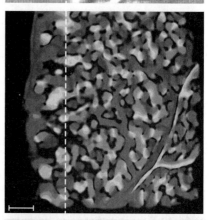

D Blobs, interblobs (V1), and stripes (V2)

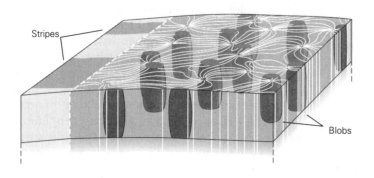

Stripes

Blobs

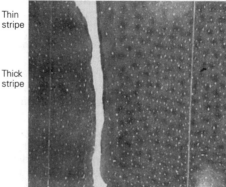

Thin stripe

Thick stripe

Figure 21–11 (Opposite) **Functional architecture of the primary visual cortex.** (Courtesy of M. Kinoshita and A. Das, reproduced with permission.)

A. The surface of the primary visual cortex is functionally organized as a map of the visual field. The elevations and azimuths of visual space are organized in a regular grid that is distorted because of variation in the magnification factor (see Figure 21–10). The grid is visible here in the dark stripes (visualized with intrinsic-signal optical imaging), which reflect the pattern of neurons that responded to a series of vertical candy stripes. Within this surface map, one finds repeated superimposed cycles of functionally specific columns of cells, as illustrated in B, C, and D.

B. The dark and light stripes represent the surface view of the left and right ocular dominance columns. These stripes

intersect the border between areas V1 and V2, the representation of the vertical meridian, at right angles.

C. Some columns contain cells with similar selectivity for the orientation of stimuli. The different colors indicate the orientation preference of the columns. The orientation columns in surface view are best described as pinwheels surrounding singularities of sudden changes in orientation (the center of the pinwheel). The scale bar represents 1 mm. (Surface image of orientation columns on the left courtesy of G. Blasdel, reproduced with permission.)

D. Patterns of blobs in V1 and stripes in V2 represent other modules of functional organization. These patterns are visualized with cytochrome oxidase.

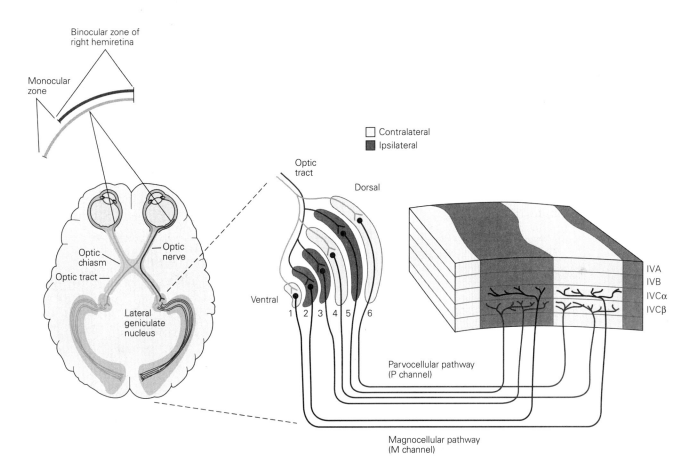

Figure 21–12 Projections from the lateral geniculate nucleus to the visual cortex. The lateral geniculate nucleus in each hemisphere receives input from the temporal retina of the ipsilateral eye and the nasal retina of the contralateral eye. The nucleus is a layered structure comprising four parvocellular layers (layers 3 to 6) and two magnocellular layers (layers 1 and 2). Each is paired with an intercalated koniocellular layer. (These layers are represented here by the gaps separating the primary layers. They are unlabeled to avoid clutter. See Figure 21–14.) The inputs from the two eyes terminate in different geniculate

layers: The contralateral eye projects to layers 1, 4, and 6, whereas the ipsilateral eye sends input to layers 2, 3, and 5. Neurons from these geniculate layers then project to different layers of cortex. The parvocellular geniculate neurons project to layer IVCβ, the magnocellular ones project to layer IVCα, and the koniocellular ones project to "blobs" in the upper cortical layers (see Figures 21–14 and 21–15). In addition, the afferents from the ipsilateral and contralateral layers of the lateral geniculate nucleus are segregated into alternating ocular-dominance columns.

For every visual attribute to be analyzed at each position in the visual field, there must be adequate tiling, or coverage, of neurons with different functional properties. As one moves in any direction across the cortical surface, the progression of the visuotopic location of receptive fields is gradual, whereas the cycling of columns occurs more rapidly. Any given position in the visual field can therefore be analyzed adequately in terms of the orientation of contours, the color and direction of movement of objects, and stereoscopic depth by a single computational module. The small segment of visual cortex that comprises such a module represents all possible values of all the columnar systems (Figure 21–13).

The columnar systems serve as the substrate for two fundamental types of connectivity along the visual pathway. *Serial processing* occurs in the successive connections between cortical areas, connections that run from the back of the brain forward. At the same time, *parallel processing* occurs simultaneously in subsets of fibers that process different submodalities such as form, color, and movement, continuing the neural processing strategy started in the retina.

Many areas of visual cortex reflect this arrangement; for example, functionally specific cells in V1 communicate with cells of the same specificity in V2. These pathways are not absolutely segregated, however, for there is some mixing of information between different visual attributes (Figure 21–14).

Columnar organization confers several advantages. It minimizes the distance required for neurons with similar functional properties to communicate with one another and allows them to share inputs from discrete pathways that convey information about particular sensory attributes. This efficient connectivity economizes on the use of brain volume and maximizes processing speed. The clustering of neurons into functional groups, as in the columns of the cortex, allows the brain to minimize the number of neurons required for analyzing different attributes. If all neurons were tuned for every attribute, the resultant combinatorial explosion would require a prohibitive number of neurons.

Intrinsic Cortical Circuits Transform Neural Information

Each area of the visual cortex transforms information gathered by the eyes and processed at earlier synaptic relays into a signal that represents the visual scene. This transformation is accomplished by local circuits comprising both excitatory and inhibitory neurons.

The principal input to the primary visual cortex comes from three parallel pathways that originate in the parvocellular, magnocellular, and the blue/yellow channels of koniocellular layers of the LGN (see Figure 21–12). Neurons in the parvocellular layers project to cortical layers IVCβ and 6, those in the

Figure 21–13 A cortical computational module. A chunk of cortical tissue roughly 1 mm square contains an orientation hypercolumn (a full cycle of orientation columns), one cycle of left- and right-eye ocular-dominance columns, and blobs and interblobs. This module would presumably contain all of the functional and anatomical cell types of primary visual cortex, which would be repeated hundreds of times to cover the visual field. (Adapted from Hubel 1988.)

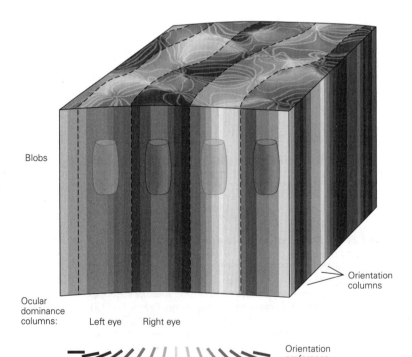

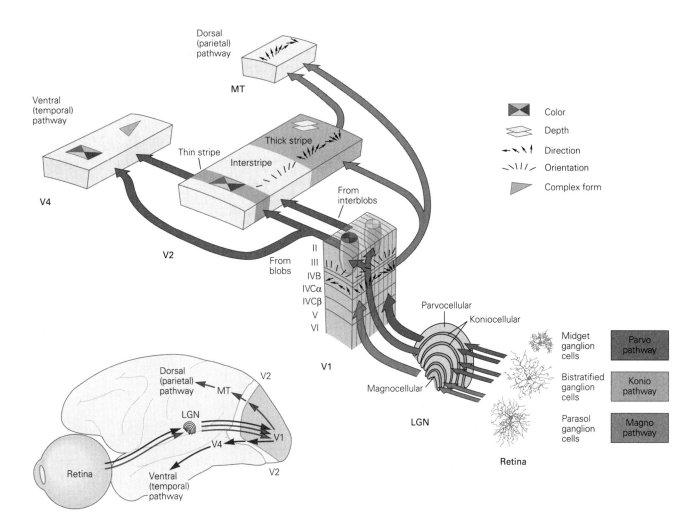

Figure 21–14 Parallel processing in visual pathways. The ventral stream is primarily concerned with object identification, carrying information about form and color. The dorsal pathway is dedicated to visually guided movement, with cells selective for direction of movement. These pathways are not strictly segregated, however, and there is substantial interconnection between them even in the primary visual cortex. (Abbreviations: **LGN**, lateral geniculate nucleus; **MT**, middle temporal area.) (Retinal ganglion cell images courtesy of Dennis Dacey, reproduced with permission.)

magnocellular layers project to layer IVCα and 6, while the koniocellular neurons project to layer 1 and to the cytochrome oxidase blobs in layers 2 and 3. From there, a sequence of interlaminar connections, mediated by the excitatory spiny stellate neurons, processes visual information over a stereotyped set of connections (Figure 21–15).

This characterization of parallel pathways is only an approximation, as there is considerable interaction between the pathways. This interaction is the means by which various visual features—color, form, depth, and movement—are linked, leading to a unified visual percept. One way this linkage, or binding, may be accomplished is through cells that are tuned to more than one attribute.

At each stage of cortical processing, pyramidal neurons extend output to other brain areas. Superficial-layer cells are responsible for connections to higher-order areas of cortex. Layer V pyramidal neurons project to the superior colliculus and pons in the brain stem. Layer VI cells are responsible for feedback projections, both to the thalamus and to lower-order cortical areas.

Neurons in different layers have distinctive receptive-field properties. Neurons in the superficial layer of V1 have small receptive fields, whereas neurons in deeper layers have large ones. The superficial-layer neurons are specialized for high-resolution pattern recognition. Neurons in the deeper layers, such as those in layer V that are selective for the direction of movement, are specialized for the tracking of objects in space.

A Distribution of cell types in the primary visual cortex

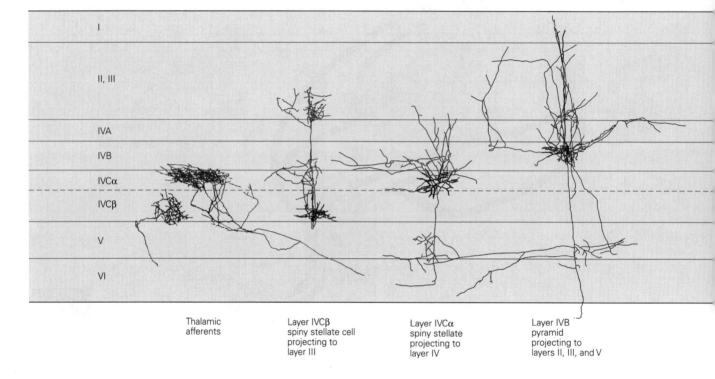

Thalamic afferents	Layer IVCβ spiny stellate cell projecting to layer III	Layer IVCα spiny stellate projecting to layer IV	Layer IVB pyramid projecting to layers II, III, and V

B Simplified diagram of intrinsic circuitry

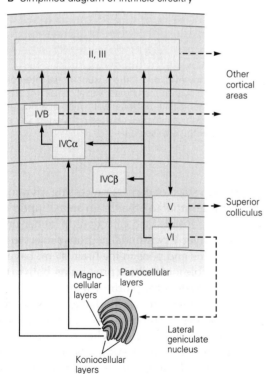

Figure 21–15 The intrinsic circuitry of the primary visual cortex.

A. Examples of neurons in different cortical layers responsible for excitatory connections in cortical circuits. Layer IV is the principal layer of input from the lateral geniculate nucleus of the thalamus. Fibers from the parvocellular layer terminate in layer IVCβ, whereas the magnocellular fibers terminate in layer IVCα. The intrinsic cortical excitatory connections are mediated by spiny stellate and pyramidal cells. A variety of γ-aminobutyric acid (GABA)-ergic smooth stellate cells (not shown) are responsible for inhibitory connections. Dendritic arbors are colored **blue**, and axonal arbors are shown in **brown**. (Cortical neurons courtesy of E. Callaway, reproduced with permission. Thalamic afferents adapted, with permission, from Blasdel and Lund 1983. Copyright © 1983 Society for Neuroscience.)

B. Diagram of excitatory connections within the primary visual cortex. Output to other regions of cortex is sent from every layer of visual cortex.

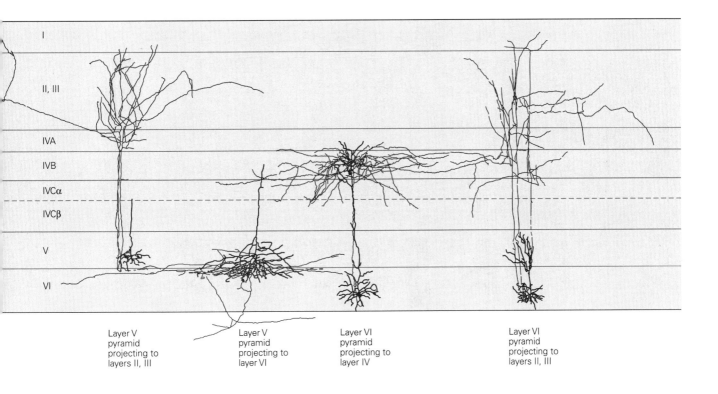

Layer V
pyramid
projecting to
layers II, III

Layer V
pyramid
projecting to
layer VI

Layer VI
pyramid
projecting to
layer IV

Layer VI
pyramid
projecting to
layers II, III

Feedback projections are thought to provide a means whereby higher centers in a pathway can influence lower ones. The number of neurons projecting from the cortex to the LGN is 10-fold the number projecting from the LGN to the cortex. Although this feedback projection is obviously important, its function is largely unknown.

The activity of the excitatory pyramidal and spiny stellate neurons that mediate information flow into or out of cortical regions is also tightly controlled by local networks of inhibitory interneurons. The spike rates of excitatory neurons are constantly nonlinearly balanced by matched inhibition that maintains the stability of the neural response to an input. Inhibitory interneurons come in multiple classes distinguished by their morphology and their coexpression of distinct peptides such as parvalbumin, somatostatin, or vasoactive intestinal polypeptide (VIP). Some of these interneurons form cascading circuits where interneurons of one class target interneurons of another class, which then target excitatory neurons. This leads to multistep control mechanisms in the neural circuit whereby increasing activity in the first class of inhibitory interneurons reduces activity in the second class, disinhibiting and increasing responses in the excitatory targets at the end

of the cascade. Such motifs of inhibitory control are likely to be common to multiple cortical sensory areas.

In addition to serial feedforward, feedback, and local recurrent connections, fibers that travel parallel to the cortical surface within each layer provide long-range horizontal connections (Figure 21–16). These connections and their role in the functional architecture of cortex were analyzed by Charles Gilbert and Torsten Wiesel, who used intracellular recordings and dye injection to correlate anatomical features with cortical function. Because the visual cortex is organized visuotopically, the horizontal connections allow target neurons to integrate information over a relatively large area of the visual field and are therefore important in assembling the components of a visual image into a unified percept.

Integration can also be achieved by other means. The considerable convergence and divergence of connections at the synaptic relays of the afferent visual pathway imply that the receptive fields of neurons are larger and more complex at each successive relay and thus have an integrative function. Feedback connections may also support integration, both because of their divergence and because they originate from cells with larger receptive fields.

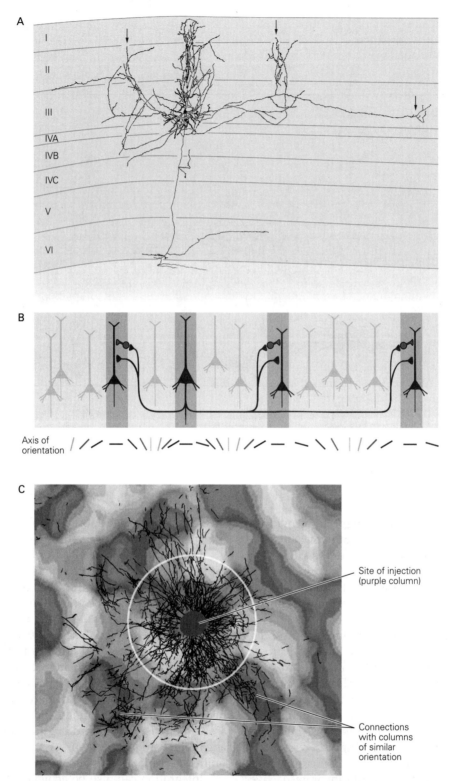

Axis of
orientation

Site of injection
(purple column)

Connections
with columns
of similar
orientation

Figure 21–16 Long-range horizontal connections in each layer of the visual cortex integrate information from different parts of the visual field.

A. The axons of pyramidal cells extend for many millimeters parallel to the cortical surface. Axon collaterals form connections with other pyramidal cells as well as with inhibitory interneurons. This arrangement enables neurons to integrate information over large parts of the visual field. An important characteristic of these connections is their relationship to the functional columns. The axon collaterals are found in clusters (**arrows**) at distances greater than 0.5 mm from the

cell body. (Reproduced, with permission, from Gilbert and Wiesel 1983. Copyright © 1983 Society for Neuroscience.)

B. Horizontal connections link columns of cells with similar orientation specificity.

C. The pattern of horizontal connections is visualized by injecting an adenoviral vector containing the gene encoding green fluorescent protein into one orientation column and superimposing the labeled image (**black**) on an optically imaged map of the orientation columns in the vicinity of the injection. (Diameter of white circle is 1 mm.) (Reproduced, with permission, from Stettler et al. 2002.)

Visual Information Is Represented by a Variety of Neural Codes

Individual neurons in a sensory pathway respond to a range of stimulus values. For example, a neuron in a color-detection pathway is not limited to responding to one wavelength but is instead tuned to a range of wavelengths. A neuron's response peaks at a particular value and tails off on either side of that value, forming a bell-shaped tuning curve with a particular bandwidth. Thus, a neuron with a peak response at 650 nm and a bandwidth of 100 nm might give identical responses at 600 nm and 700 nm.

To be able to determine the wavelength from neuronal signals, one needs at least two neurons representing filters centered at different wavelengths. Each neuron can be thought of as a *labeled line* in which activity signals a stimulus with a given value. When more than one such neuron fires, the convergent signals at the postsynaptic relay represent a stimulus with a wavelength that is the weighted average of the values represented by all the inputs.

A single visual percept is the product of the activity of a number of neurons operating in a specific combinatorial and interactive fashion called a *population code*. Population coding has been modeled in various ways. The most prevalent model is called *vector averaging*.

We can illustrate population coding with a population of orientation-selective cells, each of which responds optimally to a line with a specific orientation. Each neuron responds not just to the preferred stimulus but rather to any line that falls within a range of orientations described by a Gaussian tuning curve with a particular bandwidth. A stimulus of a particular orientation most strongly activates cells with tuning curves centered at that orientation; cells with tuning curves centered away from but overlapping that orientation are excited less strongly.

Each cell's preferred orientation, the line label, is represented as a vector pointing in the direction of that orientation. Each cell's firing is a "vote" for the cell's line label, and the cell's firing rate represents the weighting of the vote. The cell's signal can thus be represented by a vector pointing in the direction of the cell's preferred orientation with a length proportional to the strength of the cell's response. For all the activated cells, one can calculate a vector sum with a direction that represents the value of the stimulus (Figure 21–17).

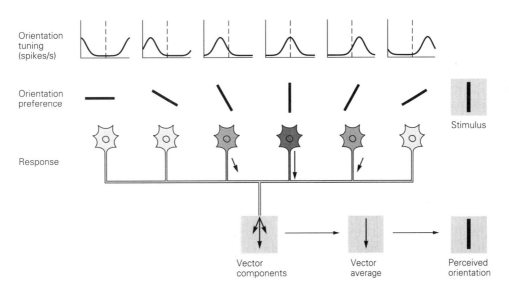

Figure 21–17 Vector averaging is one model for population coding in neural circuits. Vector averages describe the possible relationship between the responses in an ensemble of neurons, the tuning characteristics of individual neurons in the ensemble, and the resultant percept. Individual neurons respond optimally to a particular orientation of a stimulus in the visual field, but also respond at varying rates to a range of orientations. The stimulus orientation to which a neuron fires best can be thought of as a line label—when the cell fires briskly, its activity signifies the presence of a stimulus with that orientation. A number of neurons with different orientation preferences will respond to the same stimulus. Each neuron's response can be represented as a vector whose length indicates the strength of its response and whose direction represents its preferred orientation, or line label. (Adapted, with permission, from Kapadia, Westheimer, and Gilbert 2000.)

Another aspect of the population code is the variability of a neuron's response to the same stimulus. Repeated presentation of a stimulus to a neuron sensitive to that stimulus will elicit a range of responses. The most sensitive part of a neuron's tuning curve lies not at the peak but along the flanks, where the tuning curve is steepest. Here, small changes in the value of a stimulus produce the strongest change in response. Changes in stimulus value must, however, be sufficient to elicit a change in response that significantly exceeds the normal variability in the response of the neuron. One can compare that amount of change to the perceptual discrimination threshold. When many neurons contribute to the discrimination, the signal-to-noise ratio increases, a process known as probability summation, and the critical difference in stimulus value required for a significant change in neuronal response is less.

When the brain represents a piece of information, an important consideration is the number of neurons that participate in that representation. Although all information about a visual stimulus is present in the retina, the retinal representation is not sufficient for object recognition. At the other end of the visual pathway, some neurons in the temporal lobe are selective for complex objects, such as faces. Can an individual cell represent something as complex as a particular face? Such a hypothetical neuron has been dubbed a "grandmother cell" because it would represent exclusively a person's grandmother, or a "pontifical cell" because it would represent the apex of a hierarchical cognitive pathway.

The nervous system does not, however, represent entire objects by the activity of single neurons. Instead, some cells represent parts of an object, and an ensemble of neurons represents an entire object. Each member of the ensemble may participate in different ensembles that are activated by different objects. This arrangement is known as a *distributed code*. Distributed codes can involve a few neurons or many. In any case, a distributed code requires complex connectivity between the cells representing a face and those representing the name and experiences associated with that person.

The foregoing discussion assumes that neurons signal information by their firing rate and their line labels. An alternative hypothesis is that the timing of action potentials itself carries information, analogous to Morse code. The code might be read from the synchronous firing of different sets of neurons over time. At one instant, one group of cells might fire together followed by the synchronous firing of another group. Over a single train of action potentials, a single cell could participate in many such ensembles. Whether

sensory information is represented this way and whether the nervous system carries more information than that represented by firing rate alone are not known.

Highlights

1. Vision is a constructive process fundamentally different from the mere recording of visual input as in a camera. Rather, the brain uses visual input to infer information about the world around it, including information about objects, such as their sizes, shapes, distances, and identities and how rapidly they are moving.

2. The tuning of neural circuits for visual features such as contrast, orientation, and motion often matches the distribution of the feature in the natural environment. This suggests an evolutionary, ethologically driven origin for the neural circuitry.

3. Visual circuitry, and thus vision, are modulated by individual visual experience.

4. Vision makes extensive use of parallel processing. The higher visual centers form two distinct pathways. The dorsal pathway, located in parietal cortex, is involved in motion perception, attention, and visually guided action. The ventral pathway, located in temporal cortex, processes form and objects. Further subdivisions of the ventral pathway are specialized, for example, for recognizing faces. These pathways, although distinct, communicate with each other; this is likely important for the perception of objects as coherent wholes.

5. Parallel processing starts at the retina. Distinct retinal circuits analyze each point of the visual input for different local features including local contrasts of achromatic bright versus dark, red versus green, and blue versus yellow. The information is sent out through distinct classes of retinal ganglion cells (magnocelluar, parvocelluar, and koniocelluar, respectively, for the three features noted) whose axons form the optic nerves.

6. The optic nerves from the two eyes regroup at the optic chiasm such that all fibers from the left visual hemifield project to the right hemisphere of the brain, and vice versa. However, the parallel retinal channels remain anatomically segregated by eye and by visual feature, past a thalamic relay station, the lateral geniculate nucleus (LGN), up to primary visual cortex (V1).

7. The different channels enter V1 at different layers, although primarily they enter at the major input layers 4 and 6. The visual input is recombined to extract new sets of features. These include tuning for orientation, motion, and object depth (obtained by combining left- and right-eye inputs).

8. V1 neurons sharing basic properties such as spatial location or orientation preference form columns extending vertically from the pia to the white matter.

9. V1 neurons also form systematic horizontal maps of their response properties over cortex. The tuning for location forms a smooth "visuotopic" map of visual space, which changes gradually with distance, and is most finely resolved at the fovea, growing progressively coarser toward the periphery. Superimposed on the spatial map are locally smooth maps of orientation preference and left-versus right-eye preference, with interspersed columns that preferentially process color. These visual response features cycle over relatively short cortical distances, in effect completing one full cycle over each partial shift of the spatial map. Thus, V1 circuits effectively analyze each visual location, in parallel, for the full set of V1 visual features.

10. Neural processing in V1 reflects its architecture, with local vertical processing along columns and lateral processing across columns. In addition, there is long-range processing that spans multiple columns.

11. The output of V1 feeds into progressively higher visual areas comprising more than 30 centers distributed along the dorsal and ventral pathways. The connectivity is reciprocal, with higher loci sending dense feedback targeting lower areas including the LGN.

12. A useful measure of visual processing is provided by changes in neuronal "receptive fields" along the visual pathway. The receptive field is the region of visual space from which the neuron receives input; it is further characterized by the neuron's optimal visual stimulus. Receptive fields grow larger and more complex at successive stages along the visual pathway. Their optimal stimuli also increase in complexity from simple pixel-like dots for photoreceptors, to oriented lines for V1, to faces in higher face-selective centers of the ventral pathway.

13. Looking forward, one of the most important unsolved questions is the interaction between feedforward visual processing through progressively "higher" neural computations and feedback mediated via the dense plexus of connections from higher to lower levels. Understanding this interaction may be the key to understanding how the brain effortlessly forms complex visual percepts.

Charles D. Gilbert
Aniruddha Das

Selected Reading

Hubel DH, Wiesel TN. 1962. Receptive fields, binocular interaction and functional architecture in the cat's visual cortex. J Physiol 160:106–154.

Hubel DH, Wiesel TN. 1977. Functional architecture of macaque monkey visual cortex. Proc R Soc Lond B Biol Sci 198:1–59.

Hubener M, Shoham D, Grinvald A, Bonhoeffer T. 1997. Spatial relationships among three columnar systems in cat area 17. J Neurosci 17:9270–9284.

Isaacson JS, Scanziani M. 2011. How inhibition shapes cortical activity. Neuron 72:231–243.

Nassi JJ, Callaway EM. 2009. Parallel processing strategies of the primate visual system. Nat Rev Neurosci 10:360–372.

Orban GA, Van Essen D, Vanduffel W. 2004. Comparative mapping of higher visual areas in monkeys and humans. Trends Cogn Sci 8:315–324.

Stryker MP. 2014. A neural circuit that controls cortical state, plasticity, and the gain of sensory responses in mouse. Cold Spring Harb Symp Quant Biol 79:1–9.

Tsao DY, Moeller S, Freiwald WA. 2008. Comparing face patch systems in macaques and humans. Proc Natl Acad Sci U S A 105:19514–19519.

VanEssen DC, Anderson CH, Felleman DJ. 1992. Information processing in the primate visual system: an integrated systems perspective. Science 255:419–423.

Wertheimer M. 1938. *Laws of Organization in Perceptual Forms*. London: Harcourt, Brace & Jovanovitch.

Wiesel TN, Hubel DH. 1966. Spatial and chromatic interactions in the lateral geniculate body of the rhesus monkey. J Neurophysiol 29:1115–1156.

References

Blasdel GG, Lund JS. 1983. Termination of afferent axons in macaque striate cortex. J Neurosci 3:1389–1413.

Callaway EM. 1998. Local circuits in primary visual cortex of the macaque monkey. Annu Rev Neurosci 21:47–74.

Field DJ, Hayes A, Hess RF. 1993. Contour integration by the human visual system: evidence for a local "association field." Vision Res 33:173–193.

Gilbert CD, Li W. 2012. Adult visual cortical plasticity. Neuron 75:250–264.

Gilbert CD, Li W. 2013. Top-down influences on visual processing. Nat Rev Neurosci 14:350–363.

Gilbert CD, Wiesel TN. 1983. Clustered intrinsic connections in cat visual cortex. J Neurosci 3:1116–1133.

Hartline HK. 1941. The neural mechanisms of vision. Harvey Lect 37:39–68.

Hubel DH. 1988. *Eye, Brain and Vision*. New York: Scientific American Library.

Hubel DH, Wiesel TN. 1974. Uniformity of monkey striate cortex. A parallel relationship between field size, scatter and magnification factor. J Comp Neurol 158:295–306.

Kapadia MK, Westheimer G, Gilbert CD. 2000. Spatial distribution of contextual interactions in primary visual cortex and in visual perception. J Neurophysiol 84:2048–2062.

Kuffler SF. 1953. Discharge patterns and functional organization of mammalian retina. J Neurophysiol 16:37–68.

Porter PB. 1954. Another puzzle-picture. Am J Psychol 67:550–551.

Stettler DD, Das A, Bennett J, Gilbert CD. 2002. Lateral connectivity and contextual interactions in macaque primary visual cortex. Neuron 36:739–750.

22

Low-Level Visual Processing: The Retina

THE RETINA IS THE BRAIN'S WINDOW on the world. All visual experience is based on information processed by this neural circuit in the eye. The retina's output is conveyed to the brain by just one million optic nerve fibers, and yet almost half of the cerebral cortex is used to process these signals. Visual information lost in the retina—by design or deficiency—can never be recovered. Because retinal processing sets fundamental limits on what can be seen, there is great interest in understanding how the retina functions.

On the surface, the vertebrate eye appears to act much like a camera. The pupil forms a variable aperture, and the cornea and lens provide the refractive optics that project a small image of the outside world onto the light-sensitive retina lining the back of the eyeball (Figure 22–1). But this is where the analogy ends. The retina is a thin sheet of neurons, a few hundred micrometers thick, composed of five major cell types that are arranged in three cellular layers separated by two synaptic layers (Figure 22–2).

The photoreceptor cells, in the outermost layer, absorb light and convert it into a neural signal, a process known as phototransduction. These signals are

A Refraction of light onto the retina

B Focusing of light in the fovea

C Packing of photoreceptors in the fovea

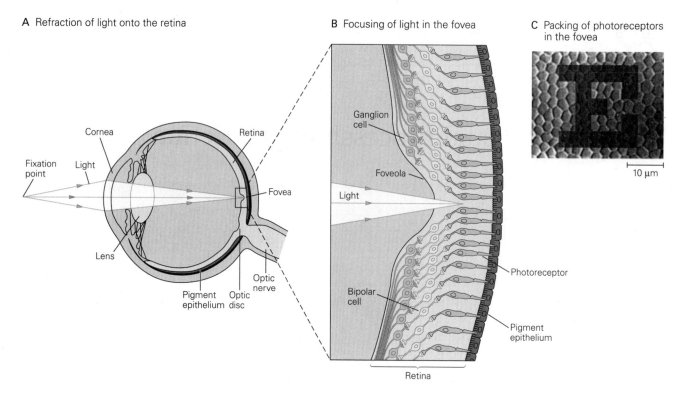

Figure 22–1 The eye projects the visual scene onto the retina's photoreceptors.

A. Light from an object in the visual field is refracted by the cornea and lens and focused onto the retina.

B. In the foveola, corresponding to the very center of gaze, the proximal neurons of the retina are shifted aside so light has direct access to the photoreceptors.

C. A letter from the eye chart used to assess normal visual acuity is projected onto the densely packed photoreceptors in the fovea. Although less sharply focused than shown here as a result of diffraction by the eye's optics, the smallest discernible strokes of the letter are approximately one cone diameter in width. (Adapted, with permission, from Curcio and Hendrickson 1991. Copyright © 1991 Elsevier Ltd.)

passed synaptically to bipolar cells, which in turn connect to retinal ganglion cells in the innermost layer. Retinal ganglion cells are the output neurons of the retina, and their axons form the optic nerve. In addition to this direct pathway from sensory to output neurons, the retinal circuit includes many lateral connections provided by horizontal cells in the outer synaptic layer and amacrine cells in the inner synaptic layer (Figure 22–3).

The retinal circuit performs low-level visual processing, the initial stage in the analysis of visual images. It extracts from the raw images in the eyes certain spatial and temporal features and conveys them to higher visual centers. The rules of this processing are adapted to changes in environmental conditions. In particular, the retina must adjust its sensitivity to ever-changing conditions of illumination. This adaptation allows our vision to remain more or less stable despite the vast range of light intensities encountered during the course of each day.

In this chapter, we discuss in turn the three important aspects of retinal function: phototransduction, preprocessing, and adaptation. We will illustrate both the neural mechanisms by which they are achieved and their consequences for visual perception.

The Photoreceptor Layer Samples the Visual Image

Ocular Optics Limit the Quality of the Retinal Image

The sharpness of the retinal image is determined by several factors: diffraction at the pupil's aperture, refractive errors in the cornea and lens, and scattering due to material in the light path. A point in the outside world is generally focused into a small blurred circle on the retina. As in other optical devices, this blur is smallest near the optical axis, where the image quality

A Section of retina   B Neurons in the retina

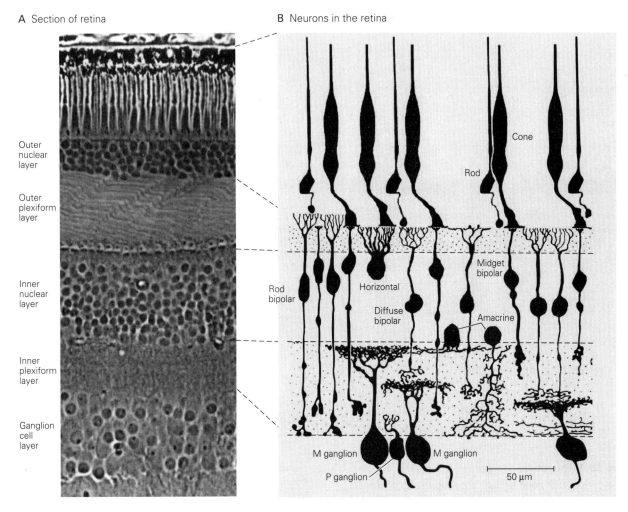

Figure 22–2 The retina comprises five distinct layers of neurons and synapses.

A. A perpendicular section of the human retina seen through the light microscope. Three layers of cell bodies are evident. The outer nuclear layer contains cell bodies of photoreceptors; the inner nuclear layer includes horizontal, bipolar, and amacrine cells; and the ganglion cell layer contains ganglion cells and some displaced amacrine cells. Two layers of fibers and synapses separate these:

the outer plexiform layer and the inner plexiform layer. (Reproduced, with permission, from Boycott and Dowling 1969. Permission conveyed through Copyright Clearance Center.)

B. Neurons in the retina of the macaque monkey based on Golgi staining. The cellular and synaptic layers are aligned with the image in part A. (Abbreviations: **M ganglion,** magnocellular ganglion cell; **P ganglion,** parvocellular ganglion cell.) (Reproduced, with permission, from Polyak 1941.)

approaches the limit imposed by diffraction at the pupil. Away from the axis, the image is degraded significantly owing to aberrations in the cornea and lens and may be degraded further by abnormal conditions such as light-scattering cataracts or refractive errors such as myopia.

The area of retina near the optical axis, the *fovea,* is where vision is sharpest and corresponds to the center of gaze that we direct toward the objects of our attention. The density of photoreceptors, bipolar cells, and ganglion cells is highest at the fovea (Figure 22–1B). The spacing between photoreceptors there is well matched to the size of the optical blur circle, and thus

the image is sampled in an ideal fashion. Light must generally traverse several layers of cells before reaching the photoreceptors, but in the center of the fovea, called the *foveola,* the other cellular layers are pushed aside to reduce additional blur from light scattering (Figure 22–1B). Finally, the back of the eye is lined by a black pigment epithelium that absorbs light and keeps it from scattering back into the eye.

The retina contains another special site, the *optic disc,* where the axons of retinal ganglion cells converge and extend through the retina to emerge from the back of the eye as the optic nerve (Figure 22–1A). By necessity, this area is devoid of photoreceptors and thus

A Cone signal circuitry

B Rod signal circuitry

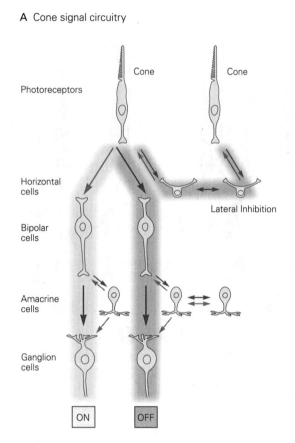

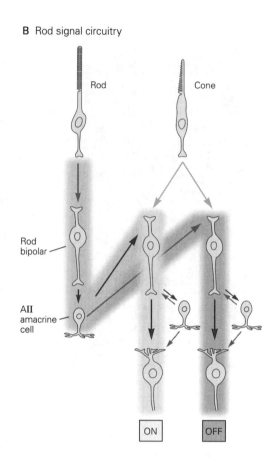

Figure 22–3 The retinal circuitry.

A. The circuitry for cone signals, showing the split into ON cell and OFF cell pathways (see Figure 22–10) as well as the pathway for lateral inhibition in the outer layer. **Red arrows** indicate sign-preserving connections through electrical or glutamatergic synapses. **Gray arrows** represent sign-inverting connections through GABAergic, glycinergic, or glutamatergic synapses.

B. Rod signals feed into the cone circuitry through AII amacrine cells, where the ON and OFF cell pathways diverge.

corresponds to a blind spot in the visual field of each eye. Because the disc lies nasal to the fovea of each eye, light coming from a single point never falls on both blind spots simultaneously, so that normally we are unaware of them. We can experience the blind spot by using only one eye (Figure 22–4). The blind spot demonstrates what blind people experience—not blackness, but simply nothing. This explains why damage to the peripheral retina often goes unnoticed. It is usually through accidents, such as bumping into an unnoticed object, or through clinical testing that a deficit of sight is revealed.

The blind spot is a necessary consequence of the inside-out design of the retina, which has puzzled and amused biologists for generations. The purpose of this organization may be to enable the tight apposition of photoreceptors with the retinal pigment epithelium, which plays an essential role in the turnover of retinal pigment and recycles photoreceptor membranes by phagocytosis.

There Are Two Types of Photoreceptors: Rods and Cones

All photoreceptor cells have a common structure with four functional regions: the outer segment, located at the distal surface of the neural retina; the inner segment, located more proximally; the cell body; and the synaptic terminal (Figure 22–5A).

Most vertebrates have two types of photoreceptors, rods and cones, distinguished by their morphology. A rod has a long, cylindrical outer segment within which the stacks of discs are separated from the plasma membrane, whereas a cone often has a shorter, tapered outer segment, and the discs are continuous with the outer membrane (Figure 22–5B).

Figure 22–4 The blind spot of the human retina. Locate the blind spot in your left eye by shutting the right eye and fixating the cross with the left eye. Hold the book about 12 inches from your eye and move it slightly nearer or farther until the circle on the left disappears. Now place a pencil vertically on the page and sweep it sideways over the circle. Note the pencil appears unbroken, even though no light can reach your retina from the region of the circle. Next, move the pencil lengthwise and observe what happens when its tip enters the circle. (Adapted, with permission, from Hurvich 1981.)

Rods and cones also differ in function, most importantly in their sensitivity to light. Rods can signal the absorption of a single photon and are responsible for vision under dim illumination such as moonlight. But as the light level increases toward dawn, the electrical response of rods becomes saturated and the cells cease to respond to variations in intensity. Cones are much less sensitive to light; they make no contribution to night vision but are solely responsible for vision in daylight. Their response is considerably faster than that of rods. Primates have only one type of rod but three kinds of cone photoreceptors, distinguished by the range of wavelengths to which they respond: the L (long-wave), M (medium-wave), and S (short-wave) cones (Figure 22–6).

The human retina contains approximately 100 million rods and 5 million cones, but the two cell types are differently distributed. The central fovea contains no rods but is densely packed with small cones. A few millimeters outside the fovea, rods greatly outnumber cones. All photoreceptors become larger and more widely spaced toward the periphery of the retina.

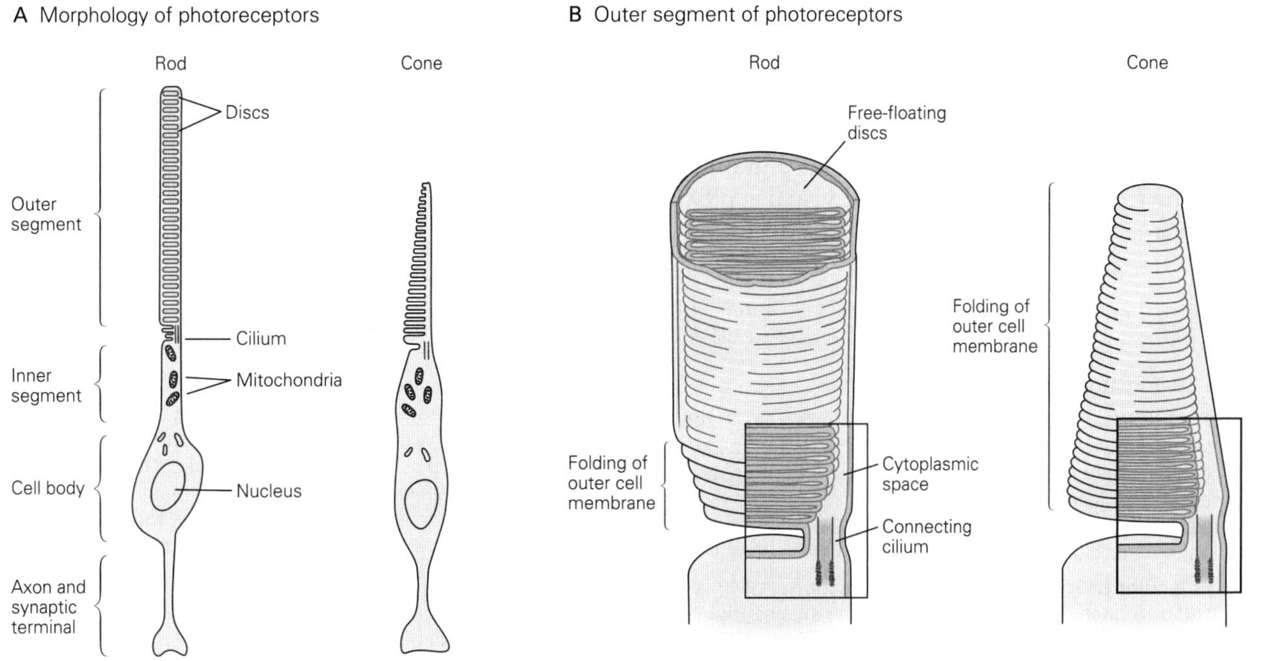

A Morphology of photoreceptors

B Outer segment of photoreceptors

Figure 22–5 Rod and cone photoreceptors have similar structures.

A. Both rod and cone cells have specialized regions called the outer and inner segments. The outer segment is attached to the inner segment by a cilium and contains the light-transducing apparatus. The inner segment holds mitochondria and much of the machinery for protein synthesis.

B. The outer segment consists of a stack of membranous discs that contain the light-absorbing photopigments. In both types of cells, these discs are formed by infolding of the plasma membrane. In rods, however, the folds pinch off from the membrane so that the discs are free-floating within the outer segment, whereas in cones, the discs remain part of the plasma membrane. (Adapted, with permission, from O'Brien 1982. Copyright © 1982 AAAS; Young 1970.)

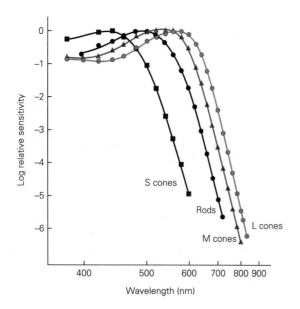

Figure 22–6 Sensitivity spectra for the three types of cones and the rod. At each wavelength, the sensitivity is inversely proportional to the intensity of light required to elicit a criterion response in the sensory neuron. Sensitivity varies over a large range and thus is shown on a logarithmic scale. The different classes of photoreceptors are sensitive to broad and overlapping ranges of wavelengths. (Reproduced, with permission, from Schnapf et al. 1988.)

The S cones make up only 10% of all cones and are absent from the central fovea.

The retinal center of gaze is clearly specialized for daytime vision. The dense packing of cone photoreceptors in the fovea sets the limits of our visual acuity. In fact, the smallest letters we can read on a doctor's eye chart have strokes whose images are just one to two cone diameters wide on the retina, a visual angle of about 1 minute of arc (Figure 22–1C). At night, the central fovea is blind owing to the absence of rods. Astronomers know that one must look just to the side of a dim star to see it at all. During nighttime walks in the forest, we nonastronomers tend to follow our daytime reflex of looking straight at the source of a suspicious sound. Mysteriously, the object disappears, only to jump back into our peripheral field of view as we avert our gaze.

Phototransduction Links the Absorption of a Photon to a Change in Membrane Conductance

As in many other neurons, the membrane potential of a photoreceptor is regulated by the balance of membrane conductances to Na^+ and K^+ ions, whose transmembrane gradients are maintained by metabolically active pumps (Chapter 9). In the dark, Na^+ ions flow into the photoreceptor through nonselective cation channels that are activated by the second messenger cyclic guanosine 3′-5′ monophosphate (cGMP).

Absorption of a photon by the pigment protein sets in motion a biochemical cascade that ultimately lowers the concentration of cGMP, thus closing the cGMP-gated channels and moving the cell closer to the K^+ equilibrium potential. In this way, light hyperpolarizes the photoreceptor (Figure 22–7). Here, we describe this sequence of events in detail. Most of this knowledge derives from studies of rods, but the mechanism in cones is very similar.

Figure 22–7 (Opposite) Phototransduction.

A. The rod cell responds to light. Rhodopsin molecules in the outer-segment discs absorb photons, which leads to the closure of cyclic guanosine 3′-5′ monophosphate (**cGMP**)-gated channels in the plasma membrane. This channel closure hyperpolarizes the membrane and reduces the rate of release of the neurotransmitter glutamate. (Adapted from Alberts 2008.)

B. 1. Molecular processes in phototransduction. cGMP is produced by a guanylate cyclase (**GC**) from guanosine triphosphate (**GTP**) and hydrolyzed by a phosphodiesterase (**PDE**). In the dark, the phosphodiesterase activity is low, the cGMP concentration is high, and the cGMP-gated channels are open, allowing the influx of Na^+ and Ca^{2+}. In the light, rhodopsin (**R**) is excited by absorption of a photon, then activates transducin (**T**), which in turn activates the PDE; the cGMP level drops, the membrane channels close, and less Na^+ and Ca^{2+} enter the cell. The transduction enzymes are all located in the internal membrane discs, and the soluble ligand cGMP serves as a messenger to the plasma membrane.

2. Calcium ions have a negative feedback role in the reaction cascade in phototransduction. Stimulation of the network by light leads to the closure of the cGMP-gated channels. This causes a drop in the intracellular concentration of Ca^{2+}. Because Ca^{2+} modulates the function of at least three components of the cascade—rhodopsin, GC, and the cGMP-gated channel—the drop in Ca^{2+} counteracts the excitation caused by light.

C. Voltage response of a primate rod and cone to brief flashes of light of increasing intensity. Higher numbers on the traces indicate greater intensities of illumination (not all traces are labeled). For dim flashes, the response amplitude increases linearly with intensity. At high intensities, the receptor saturates and remains hyperpolarized steadily for some time after the flash; this leads to the afterimages that we perceive after a bright flash. Note that the response peaks earlier for brighter flashes and that cones respond faster than rods. (Reproduced, with permission, from Schneeweis and Schnapf 1995. Copyright © 1995 AAAS.)

A Phototransduction and neural signaling

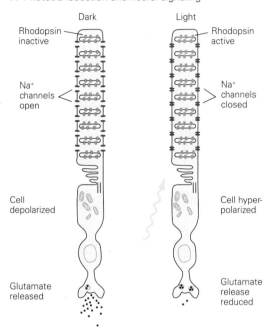

C Voltage response to light

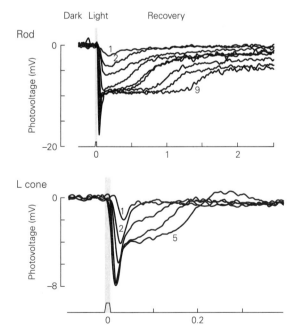

B₁ Molecular processes in phototransduction

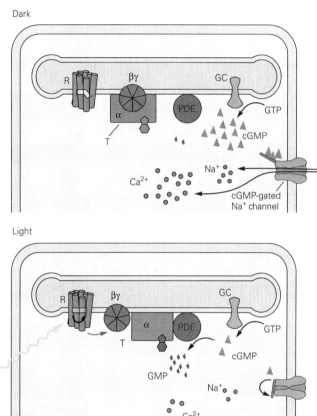

B₂ Reaction network in phototransduction

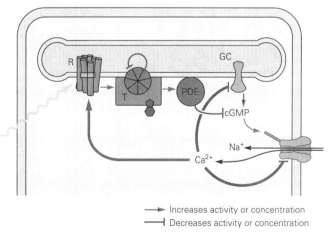

Light Activates Pigment Molecules in the Photoreceptors

Rhodopsin, the visual pigment in rod cells, has two components. The protein portion, *opsin*, is embedded in the disc membrane and does not by itself absorb visible light. The light-absorbing moiety, *retinal*, is a small molecule whose 11-*cis* isomer is covalently linked to a lysine residue of opsin (Figure 22–8A). Absorption of a photon by retinal causes it to flip from the 11-*cis* to the all-*trans* configuration. This reaction is the only light-dependent step in vision.

The change in shape of the retinal molecule causes a conformational change in the opsin to an activated state called *metarhodopsin II*, thus triggering the second

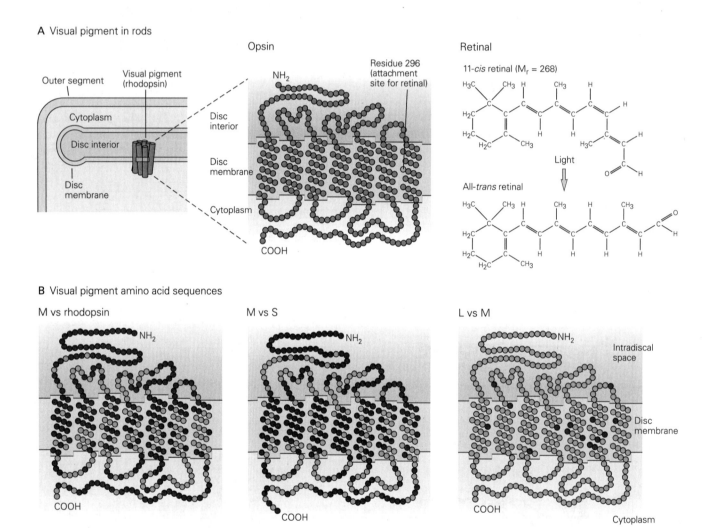

A Visual pigment in rods

B Visual pigment amino acid sequences

Figure 22–8 Structure of the visual pigments.

A. The visual pigment in rod cells, rhodopsin, is the covalent complex of two components. Opsin is a large protein with 348 amino acids and a molecular mass of approximately 40,000 daltons. It loops back and forth seven times across the membrane of the rod disc. Retinal is a small light-absorbing compound covalently attached to a side chain of lysine 296 in opsin's seventh membrane-spanning region. Absorption of light by 11-*cis* retinal causes a rotation around the double bond. As retinal adopts the more stable all-*trans* configuration, it causes a conformational change in opsin that triggers the subsequent events of visual transduction.

(Adapted, with permission, from Nathans and Hogness 1984.)

B. The **blue circles** denote identical amino acids; **black circles** denote differences. The forms of opsin in the three types of cone cells (L, M, and S) resemble each other as well as the rhodopsin in rod cells, suggesting that all four evolved from a common precursor by duplication and divergence. The L and M opsins are most closely related, with 96% identity in their amino acid sequences. They are thought to have evolved from a gene-duplication event approximately 30 million years ago, after Old World monkeys, which have three visual pigments, separated from New World monkeys, which generally have only two.

step of phototransduction. Metarhodopsin II is unstable and splits within minutes, yielding opsin and free all-*trans* retinal. The all-*trans* retinal is then transported from rods to pigment epithelial cells, where it is reduced to all-*trans* retinol (vitamin A), the precursor of 11-*cis* retinal, which is subsequently transported back to rods.

All-*trans* retinal is thus a crucial compound in the visual system. Its precursors, such as vitamin A, cannot be synthesized by humans and so must be a regular part of the diet. Deficiencies of vitamin A can lead to night blindness and, if untreated, to deterioration of receptor outer segments and eventually to blindness.

Each type of cone in the human retina produces a variant of the opsin protein. These three cone pigments are distinguished by their *absorption spectrum,* the dependence on wavelength of the efficiency of light absorption (see Figure 22–6). The spectrum is determined by the protein sequence through the interaction between retinal and certain amino acid side chains near the binding pocket. Red light excites L cones more than the M cones, whereas green light excites the M cones more. Therefore, the relative degree of excitation in these cone types contains information about the spectrum of the light, independent of its intensity. The brain's comparison of signals from different cone types is the basis for color vision.

In night vision, only the rods are active, so all functional photoreceptors have the same absorption spectrum. A green light consequently has exactly the same effect on the visual system as a red light of a greater intensity. Because a single-photoreceptor system cannot distinguish the spectrum of a light from its intensity, "at night all cats are gray." By comparing the sensitivity of a rod to different wavelengths of light, one obtains the absorption spectrum of rhodopsin. It is

a remarkable fact that one can measure this molecular property accurately just by asking human subjects about the appearance of various colored lights (Figure 22–9). The quantitative study of perception, or psychophysics, provides similar insights into other mechanisms of brain processing (Chapter 17).

Excited Rhodopsin Activates a Phosphodiesterase Through the G Protein Transducin

Activated rhodopsin in the form of metarhodopsin II diffuses within the disc membrane where it encounters transducin, a member of the G protein family (Chapter 14). As is the case for other G proteins, the inactive form of transducin binds a molecule of guanosine diphosphate (GDP). Interaction with metarhodopsin II promotes the exchange of GDP for guanosine triphosphate (GTP). This leads to dissociation of transducin's subunits into an active α-subunit carrying the GTP (Tα-GTP) and the β- and γ-subunits (T$\beta\gamma$). Metarhodopsin II can activate hundreds of additional transducin molecules, thus significantly amplifying the cell's response.

The active transducin subunit Tα-GTP forms a complex with a cyclic nucleotide phosphodiesterase, another protein associated with the disc membrane. This interaction greatly increases the rate at which the enzyme hydrolyzes cGMP to 5'-GMP. Each phosphodiesterase molecule can hydrolyze more than 1,000 molecules of cGMP per second, thus increasing the degree of amplification.

The concentration of cGMP controls the activity of the cGMP-gated channels in the plasma membrane of the outer segment. In darkness, when the cGMP concentration is high, a sizeable Na$^+$ influx through the open channels maintains the cell at a depolarized level

Figure 22–9 Absorption spectrum of rhodopsin. The absorption spectrum of human rhodopsin measured in a cuvette is compared with the spectral sensitivity of human observers to very dim light flashes. The psychophysical data have been corrected for absorption by the ocular media. (Reproduced, with permission, from Wald and Brown 1956. Copyright © 1956 Springer Nature.)

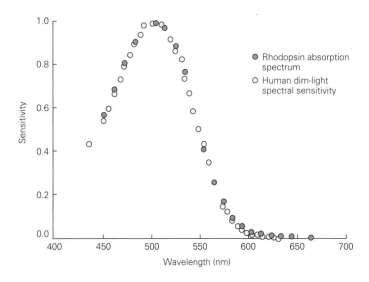

of approximately –40 mV. As a consequence, the cell's synaptic terminal continuously releases the transmitter glutamate. The light-evoked decrease in cGMP results in the closure of the cGMP-gated channels, thus reducing the inward flux of Na^+ ions and hyperpolarizing the cell (Figure 22–7B1). Hyperpolarization slows the release of neurotransmitter from the photoreceptor terminal, thereby initiating a neural signal.

Multiple Mechanisms Shut Off the Cascade

The photoreceptor's response to a single photon must be terminated so that the cell can respond to another photon. Metarhodopsin II is inactivated through phosphorylation by a specific rhodopsin kinase followed by binding of the soluble protein arrestin, which blocks the interaction with transducin.

Active transducin (Tα-GTP) has an intrinsic GTPase activity, which eventually converts bound GTP to GDP. Tα-GDP then releases phosphodiesterase and recombines with Tβγ, ready again for excitation by rhodopsin. Once the phosphodiesterase has been inactivated, the cGMP concentration is restored by a guanylate cyclase that produces cGMP from GTP. At this point, the membrane channels open, the Na^+ current resumes, and the photoreceptor depolarizes back to its dark potential.

In addition to these independent mechanisms that shut off individual elements of the cascade, an important feedback mechanism ensures that large responses are terminated more quickly. This is mediated by a change in the Ca^{2+} concentration in the cell. Calcium ions enter the cell through the cGMP-gated channels and are extruded by rapid cation exchangers. In the dark, the intracellular Ca^{2+} concentration is high, but during the cell's light response, when the cGMP-gated channels close, the Ca^{2+} level drops quickly to a few percent of the dark level.

This reduction in Ca^{2+} concentration modulates the biochemical reactions in three ways (Figure 22–7B2). Rhodopsin phosphorylation is accelerated through the action of the calcium-binding protein recoverin on rhodopsin kinase, thus reducing activation of transducin. The activity of guanylyl cyclase is accelerated by calcium-dependent guanylyl cyclase–activating proteins. Finally, the affinity of the cGMP-gated channel for cGMP is increased through the action of Ca^{2+}-calmodulin. All these effects promote the return of the photoreceptor to the dark state.

Defects in Phototransduction Cause Disease

Not surprisingly, defects in the phototransduction machinery have serious consequences. One prominent defect is color blindness, which results from loss or abnormality in the genes for cone pigments, as discussed later.

Stationary night blindness results when rod function has been lost but cone function remains intact. This disease is heritable, and mutations have been identified in many components of the phototransduction cascade: rhodopsin, rod transducin, rod phosphodiesterase, rhodopsin kinase, and arrestin. In some cases, it appears that the rods are permanently activated, as if exposed to a constant blinding light.

Unfortunately, many defects in phototransduction lead to *retinitis pigmentosa*, a progressive degeneration of the retina that ultimately results in blindness. The disease has multiple forms, many of which have been associated with mutations that affect signal transduction in rods. Why these changes in function lead to death of the rods and subsequent degeneration of the cones is not understood.

Ganglion Cells Transmit Neural Images to the Brain

The photoreceptor layer produces a relatively simple neural representation of the visual scene: Neurons in bright regions are hyperpolarized, whereas those in dark regions are depolarized. Because the optic nerve has only about 1% as many axons as there are receptor cells, the retinal circuit must edit the information in the photoreceptors before it is conveyed to the brain.

This step constitutes *low-level visual processing*, the first stage in deriving visual percepts from the pattern of light falling on the retina. To understand this process, we must first understand the organization of the retina's output and how retinal ganglion cells respond to various patterns of light.

The Two Major Types of Ganglion Cells Are ON Cells and OFF Cells

Many retinal ganglion cells fire action potentials spontaneously even in darkness or constant illumination. If the light intensity is suddenly increased, so-called ON cells fire more rapidly. Other ganglion cells, the OFF cells, fire more slowly or cease firing altogether. When the intensity diminishes again, the ON cells fire less and OFF cells fire more. The retinal output thus includes two complementary representations that differ in the polarity of their response to light.

This arrangement serves to communicate rapidly both brightening and dimming in the visual scene. If the retina had only ON cells, a dark object would be encoded by a decrease in firing rate. If the ganglion cell fired at a maintained rate of 10 spikes per second and then decreased its rate, it would take about 100 ms for the postsynaptic neuron to notice the change in frequency of action potentials. In contrast, an increase in firing rate to 200 spikes per second is noticeable within only 5 ms.

Many Ganglion Cells Respond Strongly to Edges in the Image

To probe the responses of a ganglion cell in more detail, one can test how the cell's firing varies with the location and time course of a small spot of light focused on different portions of the retina.

A typical ganglion cell is sensitive to light in a compact region of the retina near the cell body called the cell's *receptive field*. Within that area, one can often distinguish a *center* region and *surround* region where light produces opposite responses in the cell. An ON cell, for example, fires faster when a bright spot is focused in the cell's receptive field center but decreases its firing when the spot is focused on the surround. If light covers both the center and the surround, the response is much weaker than for center-only illumination. A bright spot on the center combined with a dark annulus covering the surround elicits very strong firing. For an OFF cell, these relationships are reversed; the cell is strongly excited by a dark spot and a bright annulus (Figure 22–10).

The output produced by a population of retinal ganglion cells thus enhances regions of spatial contrast in the input, such as an edge between two areas of different intensity, and gives less emphasis to regions of homogeneous illumination.

The Output of Ganglion Cells Emphasizes Temporal Changes in Stimuli

When an effective light stimulus appears, a ganglion cell's firing typically increases sharply from the resting level to a peak and then relaxes to an intermediate rate. When the stimulus turns off, the firing rate drops sharply then gradually recovers to the resting level.

The rapidity of decline from the peak to the resting level varies among ganglion cell types. *Transient neurons* produce a burst of spikes only at the onset of the stimulus, whereas *sustained neurons* maintain an almost steady firing rate for several seconds during stimulation (Figure 22–10).

In general, however, the output of ganglion cells favors temporal changes in visual input over periods of constant light intensity. In fact, when an image is

stabilized on the retina with an eye-tracking device, it fades from view within seconds. Fortunately, this never happens in normal vision; even when we attempt to fix our gaze, small automatic eye movements (saccades) continually scan the image across the retina and prevent the world from disappearing (Chapter 25).

Retinal Output Emphasizes Moving Objects

Based on these observations, we can understand more generally the response of ganglion cells to visual inputs. For example, the edges of a moving object elicit strong firing in the ganglion cell population because these are the only regions of spatial contrast and the only regions where the light intensity changes over time (Figure 22–11).

We can easily appreciate why the retina selectively responds to these features. The outline of an object is particularly useful for inferring its shape and identity. Similarly, objects that move or change suddenly are more worthy of immediate attention than those that do not. Retinal processing thus extracts low-level features of the scene that are useful for guiding behavior and selectively transmits those to the brain. In fact, the rejection of features that are constant either in space or in time accounts for the spatiotemporal sensitivity of human perception (Box 22–1).

Several Ganglion Cell Types Project to the Brain Through Parallel Pathways

Several different types of ganglion cells have been identified on the basis of their morphology and responses to light. The ON and OFF cells occur in every vertebrate retina, and in the primate retina, two major classes of cells, the P-cells and M-cells, each include ON and OFF types (see Figure 22–2B). At any given distance from the fovea, the receptive fields of M-cells (Latin *magno*, large) are much larger than those of P-cells (Latin *parvo*, small). The M-cells also have faster and more transient responses than P-cells. Some ganglion cells are intrinsically light-sensitive owing to expression of the visual pigment melanopsin.

In total, more than 20 types of ganglion cells have been described. The population of each type covers the retina in a tiled fashion, such that any point on the retina lies within the receptive field center of at least one ganglion cell. One can envision that the signals from each population together send a distinct neural representation of the visual field to the brain. In this view, the optic nerve conveys 20 or more neural representations that differ in polarity (ON or OFF), spatial resolution (fine or coarse), temporal responsiveness

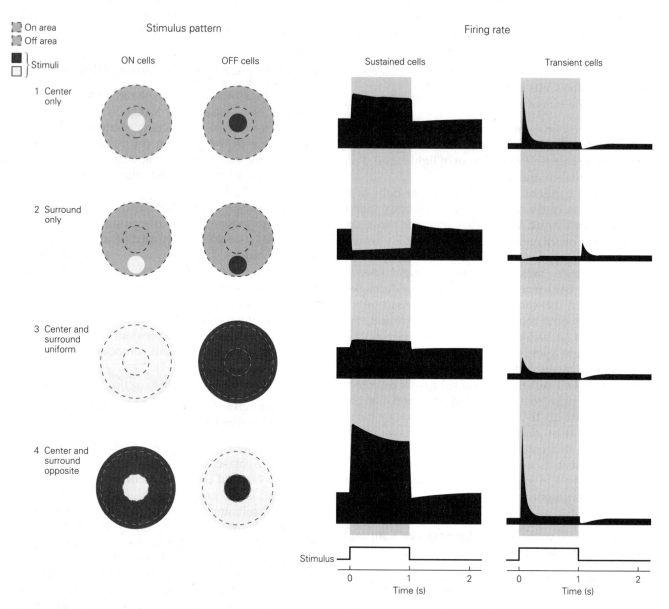

Figure 22–10 Responses of retinal ganglion cells with center-surround receptive fields. In these idealized experiments, the stimulus changes from a uniform gray field to the pattern of bright (**yellow**) and dark (**black**) regions indicated on the *left*. This leads to the firing rate responses shown on the *right*. **1.** ON cells are excited by a bright spot in the receptive field center, OFF cells by a dark spot. In *sustained cells*, the excitation persists throughout stimulation, whereas in *transient cells*, a brief burst of spikes occurs just after the onset of stimulation. **2.** If the same stimulus that excites the center is applied to the surround, firing is suppressed. **3.** Uniform stimulation of both center and surround elicits a response like that of the center, but much smaller in amplitude. **4.** Stimulation of the center combined with the opposite stimulus in the surround produces the strongest response.

A ON cell response

B Model prediction

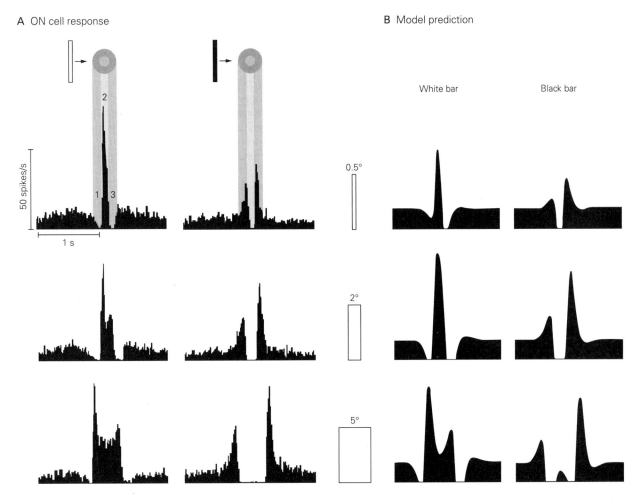

Figure 22–11 Responses of ganglion cells in the cat retina to moving objects.

A. The firing rate of an ON ganglion cell in response to a variety of bars (white or black, various widths) moving across the retina. Each bar moves at 10° per second; 1° corresponds to 180 µm on the retina. In response to the white bar, the firing rate first decreases as the bar passes over the receptive-field surround (1), increases as the bar enters the center (2), and decreases again as the bar passes through the surround on the opposite side (3). The dark bar elicits responses of the opposite sign. Because ganglion cells similar to this one are distributed throughout the retina, one can also interpret this curve as an instantaneous snapshot of activity in a population

of ganglion cells, where the horizontal axis represents location on the retina. In effect, this activity profile is the neural representation of the moving bar transmitted to the brain. A complementary population of OFF ganglion cells (not shown here) conveys another neural activity profile in parallel. In this way, both bright and dark edges can be signaled by a sharp increase in firing.

B. A simple model of retinal processing that incorporates center-surround antagonism and a transient temporal filter is used to predict ganglion-cell firing rates. The predictions match the essential features of the responses in part A. (Reproduced, with permission, from Rodieck 1965. Copyright © 1965 Elsevier Ltd.)

Box 22–1 Spatiotemporal Sensitivity of Human Perception

Although small spots of light are useful for probing the receptive fields of single neurons in visual pathways, different stimuli are needed to learn about human visual perception. Grating stimuli are commonly used to probe how our visual system deals with spatial and temporal patterns.

The subject views a display in which the intensity varies about the mean as a sinusoidal function of space (Figure 22–12). Then the contrast of the display—defined as the peak-to-peak amplitude of the sinusoid divided by the mean—is reduced to a threshold at which the grating is barely visible. This measurement is repeated for gratings of different spatial frequencies.

When the inverse of this threshold is plotted against the spatial frequency, the resulting *contrast sensitivity curve* provides a measure of sensitivity of visual perception to patterns of different scales (Figure 22–13A). When measured at high light intensity, sensitivity declines sharply at high spatial frequencies, with an absolute threshold at approximately 50 cycles per degree. This sensitivity is limited fundamentally by the quality of the optical image and the spacing of cone cells in the fovea (see Figure 22–1C).

Interestingly, sensitivity also declines at low spatial frequencies. Patterns with a frequency of approximately 5 cycles per degree are most visible. The visual system is said to have *band-pass* behavior because it rejects all but a band of spatial frequencies.

One can use the same techniques to measure the sensitivity of individual retinal ganglion cells in primates. The results resemble those for human subjects (Figure 22–13), suggesting that these basic features of visual perception are determined by the retina.

The band-pass behavior can be understood on the basis of spatial antagonism in center-surround receptive fields. A very fine grating presents many dark and bright stripes within the receptive-field center; their

effects cancel one another and thus provide no net excitation. A very coarse grating presents a single stripe to both the center and surround of the receptive field, and their antagonism again provides the ganglion cell little net excitation. The strongest response is produced by a grating of intermediate spatial frequency that just covers the center with one stripe and most of the surround with stripes of the opposite polarity (Figure 22–13B).

In dim light, the visual system's contrast sensitivity declines, but more so at high than at low spatial frequencies (Figure 22–13A). Thus, the peak sensitivity shifts to lower spatial frequencies, and eventually the curve loses its peak altogether. In this state, the visual system has so-called *low-pass* behavior, for it preferentially encodes stimuli of low spatial frequency. The fact that in dim light the receptive fields of ganglion cells lose their antagonistic surrounds explains the transition from band-pass to low-pass spatial filtering (Figure 22–13B).

Similar experiments can be done to test visual sensitivity to temporal patterns. Here, the intensity of a test stimulus flickers sinusoidally in time, while the contrast is gradually brought to the threshold level of detection. For humans, contrast sensitivity declines sharply at very high flicker frequencies, but declines also at very low frequencies (Figure 22–14A). Flicker at approximately 10 Hz is the most effective stimulus. One finds similar band-pass behavior in the flicker sensitivity of macaque retinal ganglion cells (Figure 22–14B).

Sensitivity to temporal contrast also depends on the mean light level. For human subjects, the optimum flicker frequency shifts downward at lower stimulus intensities and the peak in the curve becomes less and less prominent (Figure 22–14). The fact that primate retinal ganglion cells duplicate this behavior suggests that retinal processing limits the performance of the entire visual system in these simple tasks.

Low spatial frequency	High spatial frequency, high contrast	High spatial frequency, low contrast
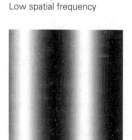 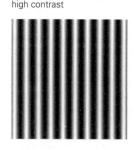 

Figure 22–12 Sinusoid grating displays used in psychophysical experiments with human subjects. Such stimuli were used in the experiments discussed in Figure 22–13.

A Sensitivity of humans and monkeys

1 Human subject

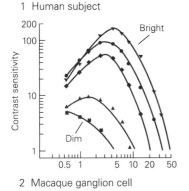

B Sensitivity of ganglion cell receptive field

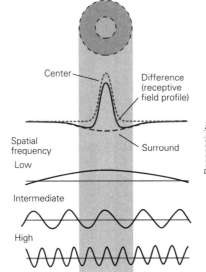

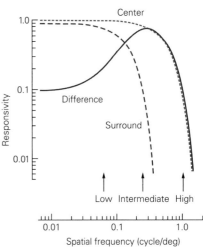

2 Macaque ganglion cell

Figure 22–13 Spatial contrast sensitivity.

A. 1. The contrast sensitivity of human subjects was measured using gratings with different spatial frequencies (see Figure 22–12). At each frequency, the contrast was increased to the threshold for detection, and the inverse of that contrast value was plotted against spatial frequency, as shown here. The curves were obtained at different mean intensities, decreasing by factors of 10 from the top to the bottom curve. (Reproduced, with permission, from De Valois, Morgan, and Snodderly 1974.) **2.** Contrast sensitivity of a P-type ganglion cell in the macaque retina measured at high intensity. At each spatial frequency, the contrast was gradually increased until it produced a detectable change in the neuron's firing rate. The inverse of that threshold contrast was plotted as in part **A-1**. The isolated dot at left marks the sensitivity at zero spatial frequency, a spatially uniform

field. (Reproduced, with permission, from Derrington and Lennie 1984.)

B. Stimulation of a center-surround receptive field with sinusoid gratings. The neuron's sensitivity to light at different points on the retina is modeled as a "difference-of-Gaussians" receptive field, with a narrow positive Gaussian for the excitatory center and a broad negative Gaussian for the inhibitory surround. Multiplying the profile of the grating stimulus (intensity vs position) with the profile of the receptive field (sensitivity vs position) and integrating over all space calculates the stimulus strength delivered by a particular grating. The resulting sensitivity of the receptive field to gratings of different frequency is shown in the plot on the right. At low spatial frequencies, the negative contribution from the surround cancels the contribution from the center, leading to a drop in the difference curve. (Reproduced, with permission, from Enroth-Cugell and Robson 1984.)

A Human subjects

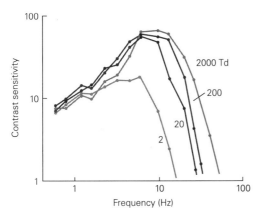

B Macaque ganglion cells

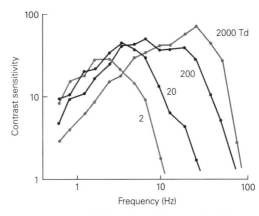

Figure 22–14 Temporal contrast sensitivity. (Reproduced, with permission, from Lee et al. 1990.)

A. The sensitivity of human subjects to temporal flicker was measured by methods similar to those in Figure 22–13A, but the stimulus was a large spot whose intensity varied sinusoidally in time rather than in space. The inverse of the threshold contrast required for detection is plotted against the frequency of the sinusoidal flicker. Sensitivity

declines at both high and low frequencies. The mean light level varied, decreasing by factors of 10 from the top to the bottom trace.

B. The flicker sensitivity of M-type ganglion cells in the macaque retina was measured by the same method applied to human subjects in part A. The detection threshold for the neural response was defined as a variation of 20 spikes per second in the cell's firing rate in phase with the flicker.

(sustained or transient), spectral filtering (broadband or dominated by red, green, or blue), and selectivity for other image features such as motion.

These neural representations are directed to various visual centers in the brain, including the lateral geniculate nucleus of the thalamus, a relay to the visual cortex; the superior colliculus, a midbrain region involved in spatial attention and orienting movements; the pretectum, involved in control of the pupil; the accessory optic system, which analyzes self-motion to stabilize gaze; and the suprachiasmatic nucleus, a central clock that directs circadian rhythm and whose phase can be set by light cues (Chapter 44). In many cases, the axons of one type of ganglion cell extend collaterals to multiple areas of the central nervous system. M-cells, for example, project to the thalamus and the superior colliculus.

A Network of Interneurons Shapes the Retinal Output

We now consider in more detail the retinal circuit and how it accounts for the intricate response properties of retinal ganglion cells.

Parallel Pathways Originate in Bipolar Cells

The photoreceptor forms synapses with bipolar cells and horizontal cells (see Figure 22–3A). In the dark, the photoreceptor's synaptic terminal releases glutamate continuously. When stimulated by light, the photoreceptor hyperpolarizes, less calcium enters the terminal, and the terminal releases less glutamate. Photoreceptors do not fire action potentials; like bipolar cells, they release neurotransmitter in a graded fashion using a specialized structure, the *ribbon synapse*. In fact, most retinal processing is accomplished with graded membrane potentials: Action potentials occur primarily in certain amacrine cells and in the retinal ganglion cells.

The two principal varieties of bipolar cells, ON and OFF cells, respond to glutamate at the synapse through distinct mechanisms. The OFF cells use ionotropic receptors, namely glutamate-gated cation channels of the AMPA-kainate variety (AMPA = α-amino-3-hydroxy-5-methylisoxazole-4-propionate). The glutamate released in darkness depolarizes these cells. The ON cells use metabotropic receptors that are linked to a G protein whose action ultimately closes cation channels. Glutamate activation of these receptors thus hyperpolarizes the cells in the dark.

Bipolar ON and OFF cells differ in shape and especially in the levels within the inner plexiform layer where their axons terminate. The axons of ON cells end in the proximal (lower) half, while those of OFF cells end in the distal (upper) half (Figure 22–15). There, they form specific synaptic connections on the dendrites of amacrine and ganglion cells. The ON bipolar cells excite ON ganglion cells, while OFF bipolar cells excite OFF ganglion cells (see Figure 22–3A). Thus, the two principal subdivisions of retinal output, the ON and OFF pathways, are already established at the level of bipolar cells.

Bipolar cells can also be distinguished by the morphology of their dendrites (Figure 22–15). In the central region of the primate retina, the *midget bipolar cell* receives input from a single cone and excites a P-type ganglion cell. This explains why the centers of P-cell receptive fields are so small. The *diffuse bipolar cell* receives input from many cones and excites an M-type ganglion cell. Accordingly, the receptive-field centers of M-cells are much larger. Thus, stimulus representations in the ganglion cell population originate in dedicated bipolar cell pathways that are differentiated by their selective connections to photoreceptors and postsynaptic targets.

Spatial Filtering Is Accomplished by Lateral Inhibition

Signals in the parallel on and off pathways are modified by interactions with horizontal and amacrine cells (see Figure 22–3A). Horizontal cells have broadly arborizing dendrites that spread laterally in the outer plexiform layer. Photoreceptors contact the tips of these arbors at glutamatergic terminals shared with bipolar cells. In addition, horizontal cells are electrically coupled to each other through gap junctions.

A horizontal cell effectively measures the average level of excitation of the photoreceptor population over a broad region. This signal is fed back to the photoreceptor terminal through an inhibitory synapse. Thus, the photoreceptor terminal is under two opposing influences: light falling on the receptor hyperpolarizes it, but light falling on the surrounding region depolarizes it through the sign-inverting synapses from horizontal cells. As a result, the bipolar cell has an antagonistic receptive-field structure.

This spatial antagonism in the receptive field is enhanced by lateral inhibition from amacrine cells in the inner retina. Amacrine cells are neurons whose processes ramify only in the inner plexiform layer. Approximately 30 types of amacrine cells are known, some with small arbors only tens of micrometers across, and others with processes that extend across the entire retina. Amacrine cells generally receive

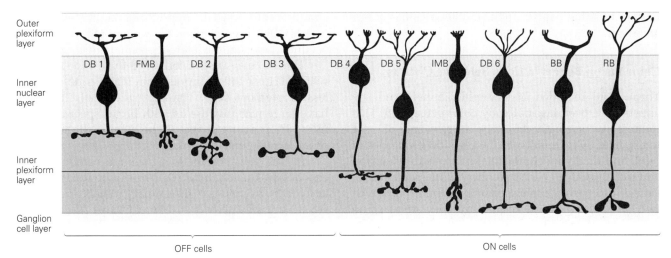

Figure 22–15 Bipolar cells in the macaque retina. The cells are arranged according to the depth of their terminal arbors in the inner plexiform layer. The horizontal line dividing the distal (**upper**) and proximal (**lower**) levels of this layer represents the border between the axonal terminals of OFF and ON type cells. Terminals in the upper half are presumed to be those of OFF cells, and those in the lower half ON cells. Cell types are diffuse bipolar cells (**DB**), ON and OFF midget bipolars (**IMB, FMB**), S-cone ON bipolar (**BB**), and rod bipolar (**RB**). (Reproduced, with permission, from Boycott and Wässle 1999.)

excitatory signals from bipolar cells at glutamatergic synapses. Some amacrine cells feed back directly to the presynaptic bipolar cell at a *reciprocal inhibitory synapse.* Some amacrine cells are electrically coupled to others of the same type, forming an electrical network much like that of the horizontal cells.

Through this inhibitory network, a bipolar cell terminal can receive inhibition from distant bipolar cells, in a manner closely analogous to the lateral inhibition of photoreceptor terminals (see Figure 22–3A). Amacrine cells also inhibit retinal ganglion cells directly. These lateral inhibitory connections contribute substantially to the antagonistic receptive field component of retinal ganglion cells.

Temporal Filtering Occurs in Synapses and Feedback Circuits

For many ganglion cells, a step change in light intensity produces a transient response, an initial peak in firing that declines to a smaller steady rate (see Figure 22–10). Part of this sensitivity originates in the negative-feedback circuits involving horizontal and amacrine cells. For example, a sudden decrease in light intensity depolarizes the cone terminal, which excites the horizontal cell, which in turn repolarizes the cone terminal (see Figure 22–3A). Because this feedback loop involves a brief delay, the voltage response of the cone peaks abruptly and then settles to a smaller steady level. Similar processing occurs at the reciprocal synapses between bipolar and amacrine cells in the inner retina.

In both cases, the delayed-inhibition circuit favors rapidly changing inputs over slowly changing inputs. The effects of this filtering, which can be observed in visual perception, are most pronounced for large stimuli that drive the horizontal and amacrine cell networks most effectively. For example, a large spot can be seen easily when it flickers at a rate of 10 Hz but not at a low rate (see Figure 22–14).

In addition to these circuit properties, certain cellular processes contribute to shaping the temporal response. For example, the AMPA-kainate type of glutamate receptor undergoes strong desensitization. A step increase in the concentration of glutamate at the dendrite of a bipolar or ganglion cell leads to an immediate opening of additional glutamate receptors. As these receptors desensitize, the postsynaptic conductance decreases again. The effect is to render a step response more transient.

Retinal circuits seem to go to great lengths to speed up their responses and emphasize temporal changes. One likely reason is that the very first cell in the retinal circuit, the photoreceptor, is exceptionally slow (see Figure 22–7C). Following a flash of light, a cone takes about 40 ms to reach the peak response, an intolerable delay for proper visual function. Through the various filtering mechanisms in retinal circuitry, subsequent neurons respond most vigorously during the rising phase of the cone's response. Indeed, some ganglion cells have a response peak only 20 ms after the flash. Temporal processing in the retina clearly helps to reduce visual reaction times, a life-extending trait as important in highway traffic as on the savannas of our ancestors.

Color Vision Begins in Cone-Selective Circuits

Throughout recorded history, philosophers and scientists have been fascinated by color perception. This interest was originally driven by the relevance of color to art, later by its relation to the physical properties of light, and finally by commercial interests in television and photography. The 19th century witnessed a profusion of theories to explain color perception, of which two have survived modern scrutiny. They are based on careful psychophysics that placed strong constraints on the underlying neural mechanisms.

Early experiments demonstrated that any given natural light could be color-matched by mixing together appropriate amounts of three primary lights. This led to the trichromatic theory of color perception based on absorption of light by three mechanisms, each with a different sensitivity spectrum. These correspond to the three cone types (see Figure 22–6), whose measured absorption spectra fully explain the color-matching results both in normal individuals and those with genetic anomalies in the pigment genes.

The so-called opponent-process theory was proposed to explain our perception of different hues. According to this theory, color vision involves three processes that respond in opposite ways to light of different colors: (y–b) would be stimulated by yellow and inhibited by blue light; (r–g) stimulated by red and inhibited by green; and (w–bk) stimulated by white and inhibited by black. We recognize some of these 19th century postulates in the postreceptor circuitry of the retina.

In the central 10° of the human retina, a single midget bipolar cell that receives input from a single cone excites each P-type ganglion cell. An L-ON ganglion cell, for example, has a receptive field center consisting of a single L cone and an antagonistic surround involving a mixture of L and M cones. When this neuron's receptive field is stimulated with a large uniform spot of light that extends over both the center and the surround, this neuron is depolarized by red light and hyperpolarized by green light. Similar antagonism holds for the three other P-cells: L-OFF, M-ON, and M-OFF. These P-cells send their signals to the parvocellular layers of the lateral geniculate nucleus.

A dedicated type of S-ON bipolar cell collects the signals of S-cones selectively and transmits them to ganglion cells of the small bistratified type. Because this ganglion cell also receives excitation from L-OFF and M-OFF bipolar cells, it is depolarized by blue light and hyperpolarized by yellow light. Another type of

ganglion cell shows the opposite signature: S-OFF and (L + M)-ON. These signals are transmitted to the koniocellular layers of the lateral geniculate nucleus.

The M-cells are excited by diffuse bipolar cells, which in turn collect inputs from many cones regardless of pigment type. These ganglion cells therefore have large receptive fields with broad spectral sensitivity. Their axons project to the magnocellular layers of the lateral geniculate nucleus.

In this way, chromatic signals are combined and encoded by the retina for transmission to the thalamus and cortex. In circuits of the primary visual cortex, these signals are recombined in different ways, leading to a great variety of receptive field layouts. Only about 10% of cortical neurons are preferentially driven by color contrast rather than luminance contrast. This likely reflects the fact that color vision—despite its great aesthetic appeal—makes only a small contribution to our overall fitness. As an illustration of this, recall that colorblind individuals, who in a sense have lost half of their color space, can grow up without ever noticing that defect.

Congenital Color Blindness Takes Several Forms

Few people are truly colorblind in the sense of being wholly unable to distinguish a change in color from a

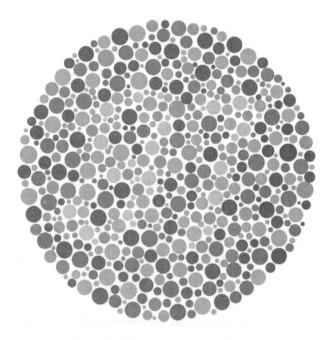

Figure 22–16 A test for some forms of color blindness. The numerals embedded in this color pattern can be distinguished by people with trichromatic vision but not by dichromats who are weak in red–green discrimination. If you don't see any numbers please have your vision tested. (Reproduced, with permission, from Ishihara 1993.)

change in the intensity of light, but many individuals have impaired color vision and experience difficulties in making distinctions that for most of us are trivial, for example between red and green. Most such abnormalities of color vision are congenital and have been characterized in detail; some other abnormalities result from injury or disease of the visual pathway.

Some people have only two classes of cones instead of three. These dichromats find it difficult or impossible to distinguish some surfaces whose colors appear distinct to trichromats. The dichromat's problem is that every surface reflectance function is represented by a two-value description rather than a three-value one, and this reduced description causes dichromats to confuse many more surfaces than do trichromats. Simple tests for color blindness exploit this fact (Figure 22–16).

Although there are three forms of dichromacy, corresponding to the loss of each of the three types of cones, two kinds are much more common than the third. The common forms correspond to the loss of the L cones or M cones and are called *protanopia* and *deuteranopia*, respectively. Protanopia and deuteranopia almost always occur in males, each with a frequency of about 1%. The conditions are transmitted by women who are not themselves affected, and so implicate genes on the X chromosome. A third form of dichromacy, *tritanopia*, involves loss or dysfunction of the S cone. It affects only about 1 in 10,000 people, afflicts women and men with equal frequency, and involves a gene on chromosome 7.

Because the L and M cones exist in large numbers, one might think that the loss of one or the other type would impair vision more broadly than just weakening color vision. In fact, this does not happen because the total number of L and M cones in the dichromat retina is not altered. All cells destined to become L or M cones are probably converted to L cones in deuteranopes and to M cones in protanopes.

In addition to the relatively severe forms of colorblindness represented by dichromacy, there are milder forms, again affecting mostly males. These so-called anomalous trichromats have cones whose spectral sensitivities differ from those in normal trichromats. Anomalous trichromacy results from the replacement of one of the normal cone pigments by an altered protein with a different spectral sensitivity. Two common forms, protanomaly and deuteranomaly, together affect about 7% of males and represent, respectively, the replacement of the L or M cones by a pigment with some intermediate spectral sensitivity.

The genetics of color vision defects are well understood. The genes for the L and M pigments reside on the X chromosome in a head-to-tail arrangement

(Figure 22–17A). The pigment proteins have very similar structures, differing in only 4% of their amino acids. People with normal color vision possess a single copy of the gene for the L pigment and from one to three—occasionally as many as five—nearly identical copies of the gene for the M pigment.

The proximity and similarity of these genes predisposes them to varied forms of recombination, leading either to the loss of a gene or to the formation of hybrid genes that account for the common forms of red–green defect (Figure 22–17B). Examination of these genes in dichromats reveals a loss of the L-pigment gene in protanopes and a loss of one or more M-pigment genes in deuteranopes. Anomalous trichromats have L-M or M-L hybrid genes that code for visual pigments with shifted spectral sensitivity; the extent of the shift

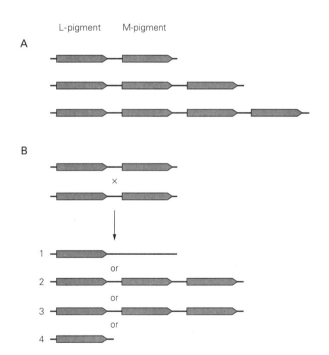

Figure 22–17 L- and M-pigment genes on the X chromosome.

A. The L- and M-pigment genes normally lie next to each other on the chromosome. The base of each arrow corresponds to the 5′ end of the gene, and the tip corresponds to the 3′ end. Males with normal color vision can have one, two, or three copies of the gene for the M pigment on each X chromosome. (Adapted, with permission, from Nathans, Thomas, and Hogness 1986. Copyright © 1986 AAAS.)

B. Recombinations of the L- and M-pigment genes can lead to the generation of a hybrid gene (3 and 4) or the loss of a gene (1), the patterns observed in colorblind men. Spurious recombination can also cause gene duplication (2), a pattern observed in some people with normal color vision. (Adapted from Streyer 1988. Used with permission from J. Nathans.)

depends on the point of recombination. In tritanopes, the loss of S-cone function arises from mutations in the S-pigment gene.

Rod and Cone Circuits Merge in the Inner Retina

For vision under low-light conditions, the mammalian retina has an ON bipolar cell that is exclusively connected to rods (see Figure 22–3B). By collecting inputs from up to 50 rods, this rod bipolar cell can pool the effects of dispersed single-photon absorptions in a small patch of retina. There is no corresponding OFF bipolar cell dedicated to rods.

Unlike all other bipolar cells, the rod bipolar cell does not contact ganglion cells directly but instead excites a dedicated neuron, the AII amacrine cell. This amacrine cell receives inputs from several rod bipolar cells and conveys its output to cone bipolar cells. It provides excitatory signals to ON bipolar cells through gap junctions as well as glycinergic inhibitory signals to OFF bipolar cells. These cone-bipolar cells in turn excite ON and OFF ganglion cells, as described earlier. Thus, the rod signal is fed into the cone system after a detour that produces the appropriate signal polarities for the ON and OFF pathways. The purpose of the added interneurons may be to allow greater pooling of rod signals than of cone signals.

Rod signals also enter the cone system through two other pathways. Rods can drive neighboring cones directly through electrical junctions, and they make connections with an OFF bipolar cell that services primarily cones. Once the rod signal has reached the cone bipolars through these pathways, it can take advantage of the same intricate circuitry of the inner retina. Thus, the rod system of the mammalian retina may have been an evolutionary afterthought added to the cone circuits.

The Retina's Sensitivity Adapts to Changes in Illumination

Vision operates under many different lighting conditions. The intensity of the light coming from an object depends on the intensity of the ambient illumination and the fraction of this light reflected by the object's surface, called the *reflectance*. The range of intensities encountered in a day is enormous, with variation spanning 10 orders of magnitude, but most of this variation is useless for the purpose of guiding behavior.

The illumination intensity varies by about nine orders of magnitude, mostly because our planet turns about its axis once a day, while the object reflectance varies much less, by about one order of magnitude in a typical scene. But this reflectance is the interesting quantity for vision, for it characterizes objects and distinguishes them from the background. In fact, our visual system is remarkably good at calculating surface reflectances independently of ambient illumination (Figure 22–18).

With an overall increase in ambient illumination, all points in the visual scene become brighter by the same factor. If the eye could simply reduce its sensitivity by that same factor, the neural representation of the image would remain unchanged at the level of the ganglion cells and could be processed by the rest of the brain in the same way as before the change in illumination. Moreover, the retinal ganglion cells would only need to encode the 10-fold range of image intensities owing to the different object reflectances, instead of the 10-billion-fold range that includes variations in ambient illumination. Some of this adjustment in sensitivity is performed by the pupil, which contracts in bright light, reducing retinal illumination by up to a factor of 10. In addition, the retina itself performs an automatic gain control, called *light adaptation*, that approaches the ideal normalization we have imagined here.

Light Adaptation Is Apparent in Retinal Processing and Visual Perception

When flashes of light of different intensity are presented with a constant background illumination, the responses of a retinal ganglion cell fit a sigmoidal curve (Figure 22–19A). The weakest flashes elicit no response, a graded increase in flash intensity elicits graded responses, and the brightest flashes elicit saturation. When the background illumination is increased, the response curve maintains the same shape but is shifted to higher flash intensities. Compensating for the increase in background illumination, the ganglion cell is now less sensitive to light variations: In the presence of a higher background, a larger change is needed to cause the same response. This lateral shifting of the stimulus–response relationship is a hallmark of light adaptation in the retina.

The consequences of this gain change for human visual perception are readily apparent in psychophysical experiments. When human subjects are asked to detect a flash in a background field of constant illumination, detection on a brighter background necessitates a brighter flash (Figure 22–19B). Under the ideal gain-control mechanism discussed earlier, two stimuli would produce the same response if they caused the same fractional change from the background intensity. In that case, the threshold flash intensity should

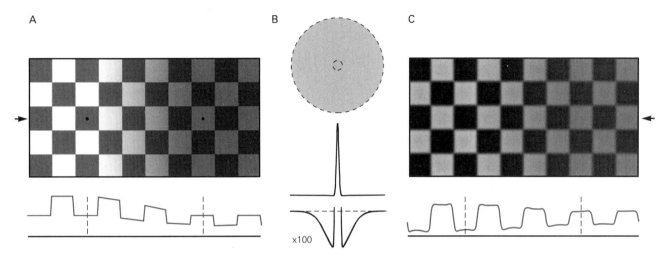

Figure 22–18 A brightness illusion.

A. The two tiles marked with small dots appear to have different color but actually reflect the same light intensity. (To see this, fold the page so they touch.) The trace underneath plots a profile of light intensity at the level of the **arrowheads**. Your visual system interprets this retinal image as a regular tile pattern under spatially varying illumination with a diffuse shadow in the right half. Under that interpretation, the right tile must have a lighter color than the left, which is what you perceive. This process is automatic and requires no conscious analysis.

B. Retinal processing contributes to the perception of "lightness" by discounting the shadow's smooth gradients of illumination and accentuating the sharp edges between

checkerboard fields. The receptive field for a visual neuron with an excitatory center and inhibitory surround is shown at the top. As shown in a hundredfold magnification at the bottom, the surround is weak but extends over a much larger area than the center.

C. The result when a population of visual neurons with receptive fields as in **B** processes the image in **A**. This operation—the convolution of the image in **A** with the profile in **B**—subtracts from each point in the visual field the average intensity in a large surrounding region. The neural representation of the object has largely lost the effects of shading, and the two tiles in question do indeed have different brightness values in this representation.

be proportional to the background intensity, a relationship known as *Weber's law of adaptation*, which we encountered in considering somatic receptor sensitivity (Chapter 17). The visual system follows Weber's law approximately: Over the entire range of vision, sensitivity decreases somewhat less steeply with increasing background intensity (Figure 22–19B).

Multiple Gain Controls Occur Within the Retina

The enormous change in gain required for light adaptation arises at multiple sites within the retina. In starlight, a single rod cell is stimulated by a photon only every few seconds, a rate insufficient to alter the cell's adaptation status. However, a retinal ganglion cell combines signals from many rods, thus receiving a steady stream of photon signals that can elicit a light-dependent gain change in the cell.

At somewhat higher light intensities, a rod bipolar cell begins to adapt, changing its responsiveness depending on the average light level. Next, we reach a light intensity at which the gain of individual rod cells gradually decreases. Beyond that, the rods saturate: All their cGMP-dependent channels are closed,

and the membrane potential no longer responds to the light stimulus. By this time, around dawn, the much less sensitive cone cells are being stimulated effectively and gradually take over from the rods. As the ambient light increases further, toward noon, light adaptation results principally from gain changes within the cones.

The cellular mechanisms of light adaptation are best understood in the photoreceptors. The calcium-dependent feedback pathways discussed earlier have a prominent role. Recall that when a light flash closes the cGMP-gated channels, the resulting decrease in intracellular Ca^{2+} accelerates several biochemical reactions that terminate the response to the flash (see Figure 22–7B). When illumination is continuous, however, the Ca^{2+} concentration remains low, and all these reactions are therefore in a steady state that both lowers the gain and accelerates the time course of the receptor's response to light (Figure 22–19C). As a result, the light-adapted photoreceptor can respond to rapid changes in intensity much more quickly. This has important consequences for human visual perception; the contrast sensitivity to high-frequency flicker increases with intensity, an effect observed in primate retinal ganglion cells as well (see Figure 22–14).

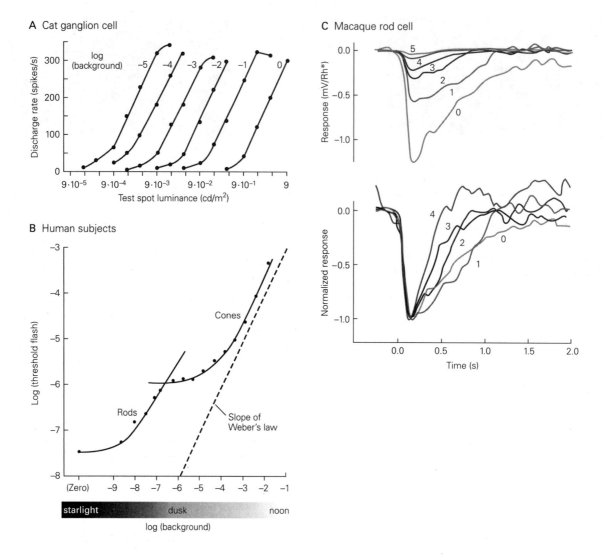

Figure 22–19 Light adaptation.

A. The receptive field of a cat retinal ganglion cell was illuminated uniformly at a steady background intensity, and a test spot was flashed briefly on the receptive field center. The peak firing rate following the flash was measured and plotted against the logarithm of the flash intensity. Each curve corresponds to a different background intensity, increasing by factors of 10 from left to right. (Reproduced, with permission, from Sakmann and Creutzfeldt 1969. Copyright © 1969 Springer.)

B. A small test spot was flashed briefly on a steadily illuminated background, and the flash intensity gradually increased to where a human subject could just detect it. The procedure was repeated at different background intensities. Here, the threshold flash intensity is plotted against the background intensity. The curve has two branches connected by a distinct kink: These correspond to the regimes of rod and cone vision. The slope

of Weber's law represents the idealization when the threshold intensity is proportional to the background intensity. (Adapted from Wyszecki and Stiles 1982.)

C. The top plot shows the responses of a macaque monkey's rod cell to flashes presented at varying background intensities. The cell's single-photon response was calculated from the recorded membrane potential divided by the number of rhodopsins (**Rh**) activated by the flash. The gain of the single-photon response decreases substantially with increasing background intensity. The background intensity, in photon/μm^2/s, is 0 for trace 0, 3.1 for trace 1, 12 for trace 2, 41 for trace 3, 84 for trace 4, and 162 for trace 5. In the bottom plot, the same data (except for the smallest response) are normalized to the same amplitude, showing that the time course of the single-photon response accelerates at high intensity. (Reproduced, with permission, from Schneeweis and Schnapf 2000.)

Light Adaptation Alters Spatial Processing

In addition to the sensitivity and speed of the retinal response, light adaptation also changes the rules of spatial processing. In bright light, many ganglion cells have a sharp center-surround structure in their receptive fields (see Figure 22–10). As the light dims, the antagonistic surround becomes broad and weak and eventually disappears. Under these conditions, the circuits of the retina function to simply accumulate the rare photons rather than computing local intensity gradients. These changes in receptive-field properties occur because of changes in the lateral inhibition produced by the networks of horizontal and amacrine cells (see Figure 22–3). An important regulator of these processes is dopamine, released in a light-dependent manner by specialized amacrine cells.

These retinal effects leave their signature on human perception. In bright light, our visual system prefers fine gratings to coarse gratings. But in dim light, we are most sensitive to coarse gratings: With the loss of center-surround antagonism, the low spatial frequencies are no longer attenuated (see Box 22–1 and Figure 22–13).

In conclusion, light adaptation has two important roles. One is to discard information about the intensity of ambient light while retaining information about object reflectances. The other is to match the small dynamic range of firing in retinal ganglion cells to the large range of light intensities in the environment. These large gain changes must be accomplished with graded neuronal signals before action potentials are produced in optic nerve fibers, because the firing rates of these fibers can vary effectively over only two orders of magnitude. In fact, the crucial need for light adaptation may be why this neural circuitry resides in the eye and not in the brain at the other end of the optic nerve.

Highlights

1. The retina transforms light patterns projected onto photoreceptors into neural signals that are conveyed through the optic nerve to specialized visual centers in the brain. Different populations of ganglion cells transmit multiple neural representations of the retinal image along parallel pathways.

2. The retina discards much of the stimulus information available at the receptor level and extracts certain low-level features of the visual field useful to the central visual system. Fine spatial resolution is maintained only in a narrow region at the center of gaze. Intensity gradients in the image, such as object edges, are emphasized over spatially uniform portions; temporal changes are enhanced over unchanging parts of the scene.

3. The retina adapts flexibly to the changing conditions for vision, especially the large diurnal changes in illumination. Information about the absolute light level is largely discarded, favoring the subsequent analysis of object reflectances within the scene.

4. The transduction of light stimuli begins in the outer segment of the photoreceptor cell when a pigment molecule absorbs a photon. This sets in motion an amplifying G protein cascade that ultimately reduces the membrane conductance, hyperpolarizes the photoreceptor, and decreases glutamate release at the synapse. Multiple feedback mechanisms, in which intracellular Ca^{2+} has an important role, serve to turn off the enzymes in the cascade and terminate the light response.

5. Rod photoreceptors are efficient collectors of light and serve nocturnal vision. Cones are much less sensitive and function throughout the day. Cones synapse onto bipolar cells that in turn excite ganglion cells. Rods connect to specialized rod bipolar cells whose signals are conveyed through amacrine cells to the cone bipolar cells.

6. The vertical excitatory pathways are modulated by horizontal connections that are primarily inhibitory. Through these lateral networks, light in the receptive-field surround of a ganglion cell counteracts the effect of light in the center. The same negative-feedback circuits also sharpen the transient response of ganglion cells.

7. The segregation of information into parallel pathways and the shaping of response properties by inhibitory lateral connections are pervasive organizational principles in the visual system.

Markus Meister
Marc Tessier-Lavigne

Selected Reading

Dowling JE. 2012. *The Retina: An Approachable Part of the Brain.* Cambridge, MA: Harvard Univ. Press.

Fain GL, Matthews HR, Cornwall MC, Koutalos Y. 2001. Adaptation in vertebrate photoreceptors. Physiol Rev 81:117–151.

Field GD, Chichilnisky EJ. 2007. Information processing in the primate retina: circuitry and coding. Ann Rev Neurosci 30:1–30.

Gollisch T, Meister M. 2010. Eye smarter than scientists believed: neural computations in circuits of the retina. Neuron 65:150–164.

Lamb TD. 2016. Why rods and cones? Eye (Lond) 30:179–185.

Masland RH. 2012. The tasks of amacrine cells. Vis Neurosci 29:3–9.

Meister M, Berry MJ. 1999. The neural code of the retina. Neuron 22:435–450.

Oyster CW. 1999. *The Human Eye: Structure and Function.* Sunderland, MA: Sinauer.

Roof DJ, Makino CL. 2000. The structure and function of retinal photoreceptors. In: DM Albert, FA Jakobiec (eds). *Principles and Practice of Ophthalmology,* pp. 1624–1673. Philadelphia: Saunders.

Shapley R, Enroth-Cugell C. 1984. Visual adaptation and retinal gain controls. Prog Retin Eye Res 3:223–346.

Wandell BA. 1995. *Foundations of Vision.* Sunderland, MA: Sinauer.

Wässle H. 2004. Parallel processing in the mammalian retina. Nat Rev Neurosci 5:747–757.

Williams DR. 2011. Imaging single cells in the living retina. Vision Res 51:1379–1396.

References

Alberts B, Johnson A, Lewis J, Raff M, Roberts K, Walter P. 2008. *Molecular Biology of the Cell,* 5th ed. New York: Garland Science.

Boycott BB, Dowling JE. 1969. Organization of the primate retina: light microscopy. Philos Trans R Soc Lond B Biol Sci 255:109–184.

Boycott B, Wässle H. 1999. Parallel processing in the mammalian retina: the Proctor Lecture. Invest Ophthalmol Vis Sci 40:1313–1327.

Curcio CA, Hendrickson A. 1991. Organization and development of the primate photoreceptor mosaic. Prog Retinal Res 10:89–120.

Derrington AM, Lennie P. 1984. Spatial and temporal contrast sensitivities of neurones in lateral geniculate nucleus of macaque. J Physiol 357:219–240.

De Valois RL, Morgan H, Snodderly DM. 1974. Psychophysical studies of monkey vision. 3. Spatial luminance contrast sensitivity tests of macaque and human observers. Vision Res 14:75–81.

Enroth-Cugell C, Robson JG. 1984. Functional characteristics and diversity of cat retinal ganglion cells. Basic characteristics and quantitative description. Invest Ophthalmol Vis Sci 25:250–227.

Hurvich LM. 1981. *Color Vision.* Sunderland, MA: Sinauer.

Ishihara S. 1993. *Ishihara's Tests for Colour-Blindness.* Tokyo: Kanehara.

Lee BB, Pokorny J, Smith VC, Martin PR, Valberg A. 1990. Luminance and chromatic modulation sensitivity of macaque ganglion cells and human observers. J Opt Soc Am A 7:2223–2236.

Nathans J, Hogness DS. 1984. Isolation and nucleotide sequence of the gene encoding human rhodopsin. Proc Natl Acad Sci U S A 81:4851–4855.

Nathans J, Thomas D, Hogness DS. 1986. Molecular genetics of human color vision: the genes encoding blue, green, and red pigments. Science 232:193–202.

O'Brien DF. 1982. The chemistry of vision. Science 218:961–966.

Polyak SL. 1941. *The Retina.* Chicago: Univ. Chicago Press.

Rodieck RW. 1965. Quantitative analysis of cat retinal ganglion cell response to visual stimuli. Vision Res 5:583–601.

Sakmann B, Creutzfeldt OD. 1969. Scotopic and mesopic light adaptation in the cat's retina. Pflügers Arch 313:168–185.

Schnapf JL, Kraft TW, Nunn BJ, Baylor DA. 1988. Spectral sensitivity of primate photoreceptors. Vis Neurosci 1:255–221.

Schneeweis DM, Schnapf JL. 1995. Photovoltage of rods and cones in the macaque retina. Science 228:1053–1056.

Schneeweis DM, Schnapf JL. 2000. Noise and light adaptation in rods of the macaque monkey. Vis Neurosci 17:659–666.

Solomon GS, Lennie P. 2007. The machinery of color vision. Nat Rev Neurosci 8:276–286.

Stryer L. 1988. *Biochemistry,* 3rd ed. New York: Freeman.

Wade NJ. 1998. *A Natural History of Vision.* Cambridge: MIT Press.

Wald G, Brown PK. 1956. Synthesis and bleaching of rhodopsin. Nature 177:174–176.

Wyszecki G, Stiles WS. 1982. *Color Science: Concepts and Methods, Quantitative Data and Formulas,* Chapter 7 "Visual Thresholds." 2nd ed. New York: Wiley.

Young RW. 1970. Visual cells. Sci Am 223:80–91.

23

Intermediate-Level Visual Processing and Visual Primitives

W E HAVE SEEN IN Chapters 21 and 22 that the eye is not a mere camera, but instead contains sophisticated retinal circuitry that decomposes the retinal image into signals representing contrast and movement. These data are conveyed through the optic nerve to the primary visual cortex, which uses this information to analyze the shape of objects. It first identifies the boundaries of objects, represented by numerous short line segments, each with a specific orientation. The cortex then integrates this information into a representation of specific objects, a process referred to as *contour integration.*

These two steps, local analysis of orientation and contour integration, exemplify two distinct stages of visual processing. Computation of local orientation is an example of low-level visual processing, which is concerned with identifying local elements of the light structure of the visual field. Contour integration is an example of intermediate-level visual processing, the first step in generating a representation of the unified visual field. At the earliest stages of analysis in the cerebral cortex, these two levels of processing are accomplished together.

A visual scene comprises many thousands of line segments and surfaces. Intermediate-level visual processing is concerned with determining which boundaries and surfaces belong to specific objects and which are part of the background (see Figure 21–4). It is also involved in distinguishing the brightness and color of a surface from the intensity and wavelength of light reflected from that surface. The physical characteristics of reflected light result as much from the intensity and color balance of the light that illuminates a surface as from the color of that surface. Determining the actual surface color of a single object requires comparison of the wavelengths of light reflected from multiple surfaces in a scene.

Intermediate-level visual processing thus involves assembling local elements of an image into a unified percept of objects and background. Although determining which elements belong together in a single object is a highly complex problem with an astronomical number of potential solutions, each relay in the visual circuitry of the brain has built-in logic that allows assumptions to be made about the likely spatial relationships between elements. In certain cases, these inherent rules can lead to the illusion of contours and surfaces that do not actually exist in the visual field (Figure 23–1).

Three features of visual processing help overcome ambiguity in the signals from the retina. First, the way in which a visual feature is perceived depends on everything that surrounds it. The perception of a point, line, or surface, for example, depends on the relationship between that feature and what else is present in the scene. That is, the response of a neuron in the visual cortex is context-dependent: It depends as much on the presence of contours and surfaces outside the cell's receptive field as on the attributes within it. Second, the functional properties of neurons in the visual cortex can be altered by visual experience or perceptual learning. Finally, visual processing in the cortex is subject to the influence of cognitive functions, specifically attention, expectation, and "perceptual task" (the active engagement in visual discrimination or detection). The interaction between these three factors— the context or entire set of signals representing a scene, experience-dependent changes in cortical circuitry,

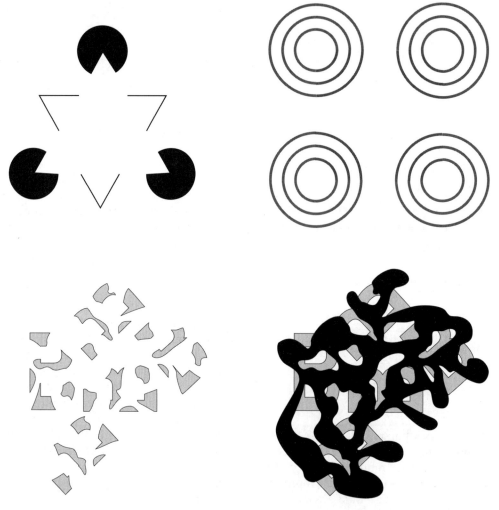

Figure 23–1 Illusory contours and perceptual fill-in. The visual system uses information about local orientation and contrast to construct the contours and surfaces of objects. This constructive process can lead to the perception of contours and surfaces that do not appear in the visual field, including those seen in illusory figures. **Top left:** In the Kanizsa triangle illusion, one perceives continuous boundaries extending between the apices of a white triangle, even though the only real contour elements are

those formed by the Pac-Man–like figures and the acute angles. **Top right:** The inside and outside of the illusory pink square are the same white color as the page, but a continuous transparent pink surface within the square is perceived. **Bottom:** Occluding surfaces can also facilitate contour integration and surface segmentation. The irregular shapes on the left appear to be unrelated, but when they are partially occluded by black shapes (right), they are easily seen as fragments of the letter B.

and expectation—is vital to the visual system's analysis of complex scenes.

In this chapter, we examine how the brain's analysis of the local features in a visual scene, or *visual primitives,* proceeds in parallel with the analysis of more global features. Visual primitives include contrast, line orientation, brightness, color, movement, and depth. Each type of visual primitive is subject to the integrative action of intermediate-level processing. Lines with particular orientations are integrated into object contours, local contrast information into surface brightness and surface segmentation, wavelength selectivity into color constancy, and directional selectivity into object motion.

The analysis of visual primitives begins in the retina with the detection of brightness and color and continues in the primary visual cortex with the analysis of orientation, direction of movement, and stereoscopic depth. Properties related to intermediate-level visual processing are analyzed together with visual primitives in the visual cortex starting in the primary visual cortex (V1), which plays a role in contour integration and surface segmentation. Other areas of the visual cortex specialize in different aspects of this task: V2 analyzes properties related to object surfaces, V4 integrates information about color and object shape, and V5—the middle temporal area or MT—integrates motion signals across space (Figure 23–2).

Internal Models of Object Geometry Help the Brain Analyze Shapes

A first step in determining an object's contour is identification of the orientation of local parts of the contour. This step commences in V1, which plays a critical role in both local and global analysis of form.

Neurons in the visual cortex respond selectively to specific local features of the visual field, including orientation, binocular disparity or depth, and direction of movement, as well as to properties already analyzed in the retina and lateral geniculate nucleus, such as contrast and color. Orientation selectivity, the first emergent property identified in the receptive fields of cortical neurons, was discovered by David Hubel and Torsten Wiesel in 1959.

Neurons in both the retina (Chapter 22) and the lateral geniculate nucleus (Chapter 21) have circular receptive fields with a center-surround organization. They respond to the light–dark contrasts of edges or lines in the visual field but are not selective for the orientations of those edges (see Figure 21–9). In the visual cortex, however, neurons respond selectively to lines of particular orientations. Each neuron responds to a narrow range of orientations, approximately 40°, and different neurons respond optimally to distinct orientations. Hubel and Wiesel proposed that this orientation selectivity reflects the arrangement of the inputs

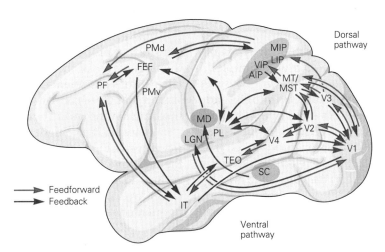

Figure 23–2 Cortical areas involved with intermediate-level visual processing. Many cortical areas in the macaque monkey, including V1, V2, V3, V4, and middle temporal area (**MT**), are involved with integrating local cues to construct contours and surfaces and segregating foreground from background. The shaded areas extend into the frontal and temporal lobes because cognitive output from these areas, including attention, expectation, and behavioral task, contributes to the process of scene segmentation. (Abbreviations: **AIP,** anterior intraparietal cortex; **FEF,** frontal eye fields; **IT,** inferior temporal cortex; **LGN,** lateral geniculate nucleus; **LIP,** lateral intraparietal cortex; **MD,** medial dorsal nucleus of thalamus; **MIP,** medial intraparietal cortex; **MST,** medial superior temporal cortex; **MT,** middle temporal cortex; **PF,** prefrontal cortex; **PL,** pulvinar; **PMd,** dorsal premotor cortex; **PMv,** ventral premotor cortex; **SC,** superior colliculus; **TEO,** occipitotemporal cortex; **VIP,** ventral intraparietal cortex; **V1, V2, V3, V4,** primary, secondary, third, and fourth visual areas.)

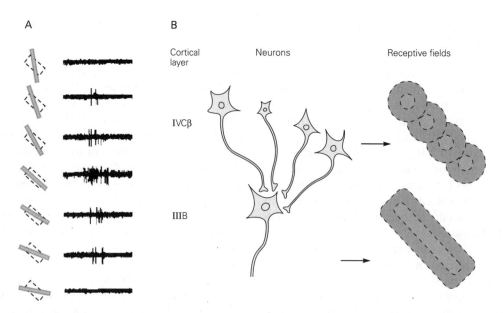

Figure 23-3 Orientation selectivity and mechanisms.

A. A neuron in the primary visual cortex responds selectively to line segments that fit the orientation of its receptive field. This selectivity is the first step in the brain's analysis of an object's form. (Reproduced, with permission, from Hubel and Wiesel 1968. Copyright © 1968 The Physiological Society.)

B. The orientation of the receptive field is thought to result from the alignment of the circular center-surround receptive fields of several presynaptic cells in the lateral geniculate nucleus. In the monkey, individual neurons in layer IVCβ of V1 have unoriented receptive fields. However, when several neighboring IVCβ cells project to a neuron in layer IIIB they create a receptive field with a specific orientation for that postsynaptic cell.

from the lateral geniculate nucleus, and there is now a body of supportive evidence for the idea. Each V1 neuron receives input from several neighboring geniculate neurons whose center-surround receptive fields are aligned so as to represent a particular axis of orientation (Figure 23-3). Two principal types of orientation-selective neurons, simple and complex, have been identified.

Simple cells have receptive fields divided into ON and OFF subregions (Figure 23-4). When a visual stimulus such as a bar of light enters the receptive field's ON subregion, the neuron fires; the cell also responds when the bar leaves the OFF subregion. Simple cells have a characteristic response to a moving bar; they discharge briskly when a bar of light leaves an OFF region and enters an ON region. The responses of these cells are therefore highly selective for the position of a line or edge in space.

Complex cells are less selective for the position of object boundaries. They lack discrete ON and OFF subregions (Figure 23-4) and respond similarly to light and dark at all locations across their receptive fields. They fire continuously as a line or edge stimulus traverses their receptive fields. Hubel and Wiesel proposed that the complex cells are a second stage of the elaboration of receptive fields after simple receptive fields and are built by overlapping simple receptive fields.

As one considers the range of receptive field properties that have been described in the early visual cortical areas, it is important to point out phylogenetic differences, with different species differing in the location in which these properties are first expressed and in the kinds of properties that are represented. In the cat, the target layer of the visual cortex for lateral geniculate neurons has oriented simple cells; it had been presumed that these cortical cells represent an obligatory first stage in the cortical processing of visual information, between the center-surround circularly symmetric receptive fields in the lateral geniculate nucleus and the receptive fields of complex cells in the superficial cortical layers. In primates, however, the geniculate target layers, 4Cα and β, have circularly symmetric, unoriented receptive fields. The postsynaptic target of the layer 4C cells, predominantly the superficial layers of the cortex, is populated with complex cells, therefore skipping a simple cell stage. In the mouse, orientation selectivity is seen in the lateral geniculate nucleus. The preceding comparison points out a few characteristics of the evolution of visual processing. One is the encephalization of function, where properties such as orientation are shifted to later stages of processing over stages of evolution. Another is the development of new pathways. It has been suggested that the

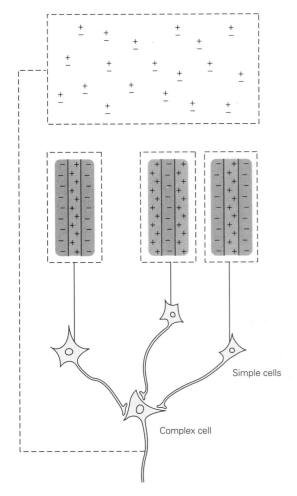

Figure 23–4 Simple and complex cells in the visual cortex.
The receptive fields of simple cells are divided into subfields
with opposite response properties. In an ON subfield (indicated
by +), the onset of a light triggers a response in the neuron;
in an OFF subfield (indicated by –), the extinction of a bar of
light triggers a response. Complex cells have overlapping ON
and OFF regions and respond continuously as a line or edge
traverses the receptive field along an axis perpendicular to the
receptive-field orientation.

Simple cells

Complex cell

magnocellular pathway in the monkey is equivalent to
the entire geniculostriate pathway in the cat, whereas
the parvocellular pathway, which mediates higher-
resolution vision and color vision, is new to the primate.

Moving stimuli are often used to study the receptive fields of visual cortex neurons, not only to simulate
the conditions under which an object moving in space
is detected but also to simulate the conditions produced by eye movements. As we scan the visual environment, the boundaries of stationary objects move
across the retina. In fact, visual perception requires eye
movement. Visual cortex neurons do not respond to an
image that is stabilized on the retina. These neurons

require transient stimulation (moving or flashing stimuli)
in order to be activated.

Some visual cortex neurons have receptive fields
in which an excitatory center is flanked by inhibitory regions. Inhibitory regions along the axis of orientation, a property known as *end-inhibition*, restrict
a neuron's responses to lines of a certain length
(Figure 23–5). End-inhibited neurons respond well to a
line that does not extend into the inhibitory flanks but
lies entirely within the excitatory part of the receptive
field. Because the inhibitory regions share the orientation preference of the central excitatory region, end-
inhibited cells are selective for line curvature and also
respond well to corners.

To define the shape of the object as a whole, the
visual system must integrate the information on local
orientation and curvature into object contours. The
way in which the visual system integrates contours
reflects the geometrical relationships present in the
natural world (Figure 23–6). As originally pointed out
by Gestalt psychologists early in the 20th century, contours that are immediately recognizable tend to follow
the rule of good continuation (curved lines maintain
a constant radius of curvature and straight lines stay
straight). In a complex visual scene, such smooth contours tend to "pop out," whereas more jagged contours
are difficult to detect.

The responses of a visual cortex neuron can be
modulated by stimuli that themselves do not activate
the cell and therefore lie outside the receptive field's
core. This *contextual modulation* endows a neuron with
selectivity for more complex stimuli than would be
predicted by placing the components of a stimulus at
different positions in and around the receptive field.
The same visual features that facilitate the detection of
an object in a complex scene (Figure 23–6A) also apply
to contextual modulation. The properties of the features that confer perception of contours, even illusory
ones, are reflected in the responses of neurons in the
primary visual cortex, which are sensitive to the global
characteristics of contours, even those that extend well
outside their receptive fields.

Contextual influences over large regions of visual
space are likely to be mediated by connections between
multiple columns of neurons in the visual cortex that
have similar orientation selectivity (Figure 23–6B). These
connections are formed by pyramidal-cell axons that run
parallel to the cortical surface (see Figure 21–16). The
extent and orientation dependency of these horizontal
connections provide the interactions that could mediate
contour saliency (see Figure 21–14).

Central to the process of contour integration is the
idea of the association field. The association field refers

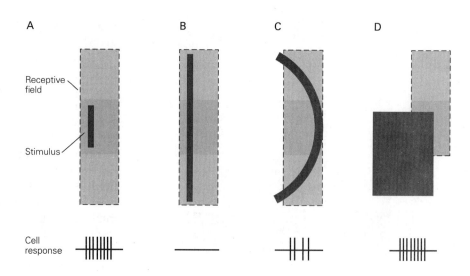

Figure 23–5 End-inhibited receptive fields. Some receptive fields have a central excitatory region flanked by inhibitory regions that have the same orientation selectivity. Thus, a short line segment or a long curved line will activate the neuron (**A** and **C**), but a long straight line will not (**B**). A neuron with a receptive field that displays only one inhibitory region in addition to the excitatory region can signal the presence of corners (**D**).

to the interactions across visual space required to perceptually link contour elements into global contours. It underlies the Gestalt principle of good continuation and the perceptual saliency of smooth contours embedded in complex scenes. Physiologically, it underlies the facilitation of neuronal responses by contour elements extending outside their "classical" receptive fields. Anatomically, it is mediated in part by the relationship between long-range horizontal connections and cortical functional architecture. Though it has been investigated most extensively in primary visual cortex, because of the ubiquity of horizontal connections across all areas of cortex, it is likely to be a strategy for associating bits of information that are mapped within every cortical area. The functional role of the association field in cortical areas outside of V1 depends on how information is mapped across the cortical surface and the relationship between these maps and the plexus of horizontal connections.

Depth Perception Helps Segregate Objects From Background

Depth is another key feature in determining the perceived shape of an object. An important cue for the perception of depth is the difference between the two eyes' views of the world, which must be computed and reconciled by the brain. The integration of binocular input begins in the primary visual cortex, the first level at which individual neurons receive signals from both eyes. The balance of input from the two eyes, a property known as ocular dominance, varies among cells in V1.

Binocular neurons in many visual cortical areas are also selective for depth, which is computed from

the relative retinal positions of objects placed at different distances from the observer. An object that lies in the *plane of fixation* produces images at corresponding positions on the two retinas (Figure 23–7). The images of objects that lie in front of or behind the plane of fixation fall on slightly different locations in the two eyes, a property known as binocular disparity. Individual neurons can be selective for a narrow range of disparities and therefore positions in depth. Some are selective for objects lying on the plane of fixation (tuned excitatory or inhibitory cells), whereas others respond only when objects lie in front of the plane of fixation (near cells) or behind that plane (far cells).

Depth plays an important role in the perception of object shape, in surface segmentation, and in establishing the three-dimensional properties of a scene. Objects that are placed near an observer can partially occlude those situated farther away. A surface passing behind an object is perceived as continuous even though its two-dimensional image on each retina represents two surfaces separated by the occluder. When the brain encounters a surface interrupted by gaps that have appropriate alignment and contrast, and lying in the near-depth plane, it fills in the gaps to create a continuous surface (Figure 23–8).

Although the depth of a single object can be established easily, determining the depths of multiple objects within a scene is a much more complex problem that requires linking the retinal images of all objects in the two eyes. The disparity calculation is therefore a global one: The calculation in one part of the visual image influences the calculation for other parts. When the assignment of depth is unambiguous in one part of an image, that information is applied to other parts of the image where there is insufficient information to

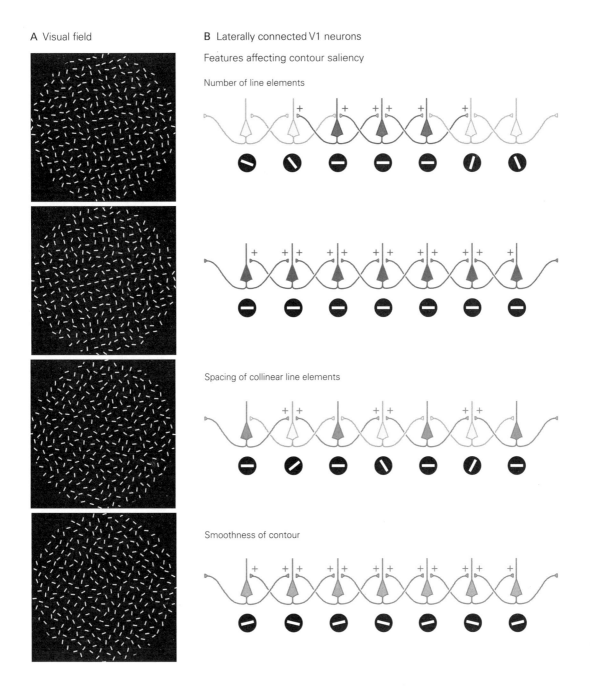

Figure 23–6 Contour integration reflects the perceptual rules of proximity and good continuation. (Adapted, with permission, from Li and Gilbert 2002.)

A. A straight line composed of one or more contour elements with the same oblique orientation appears in the center of each of the four images here. In some images, the line pops out more or less immediately, without searching. Factors that contribute to contour saliency include the number of contour elements (compare the first and second frames), the spacing of the elements (third frame), and the smoothness of the contour

(bottom frame). When the spacing between elements is too large or the orientation difference between them too great, one must search the image to find the contour.

B. These perceptual properties are reflected in the horizontal connections between columns of V1 neurons with similar orientation selectivity. As long as the visual elements are spaced sufficiently close together, excitation can propagate from cell to cell, thus facilitating the responses of V1 neurons. Each neuron in the network then augments the responses of neurons on either side, and the facilitated responses propagate across the network.

A Binocular disparity of retinal images

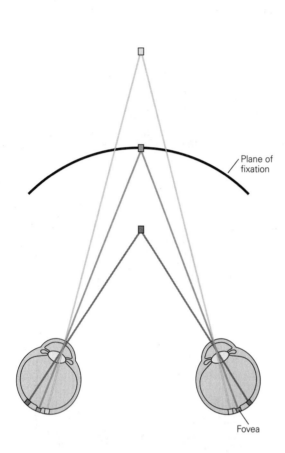

B Disparity-selective neurons

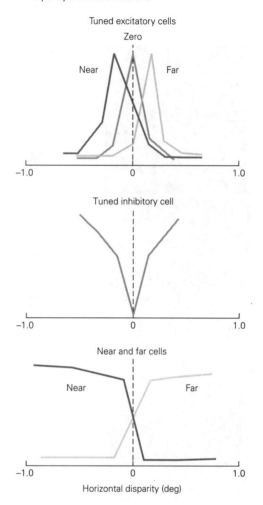

Figure 23–7 Stereopsis and binocular disparity.

A. Depth is computed from the positions at which images occur in the two eyes. The image of an object lying in the plane of fixation (**green**) falls on corresponding points on the two retinas. Images of objects lying in front of the plane of fixation (**blue**) or behind it (**yellow**) fall on noncorresponding locations on the two retinas, a phenomenon termed *binocular disparity*.

B. Neurons in many visual cortical areas are selective for particular ranges of disparity. Each plot shows the responses of a neuron to binocular stimuli with different disparities (abscissa). Some neurons are tuned to a narrow range of disparities and thus have particular disparity preferences (tuned excitatory or tuned inhibitory neurons), whereas others are tuned broadly for objects in front of the fixation plane (near cells) or beyond the plane (far cells). (Adapted, with permission, from Poggio 1995. Copyright © 1995 Oxford University Press.)

determine depth, a phenomenon known as disparity capture.

Random-dot stereograms provide a dramatic demonstration of the global scope of disparity analysis. The visual information presented to each eye appears to be incoherent, but when the stereogram is viewed binocularly, the disparity between the random array of dots in the two images allows an embedded shape to become visible (Figure 23–8C). The calculation underlying this percept is not simple, but requires determining which

features shown to the left eye correspond to features seen by the right eye and propagating local disparity information across the image.

Neurons in area V2 display sensitivity to global disparity cues. Distant depth cues can be used to link contour elements that belong to an object, and to separate them from the object background (Figure 23-B).

In addition to binocular disparity, the visual system also uses many monocular cues to discriminate depth. Depth determination through monocular

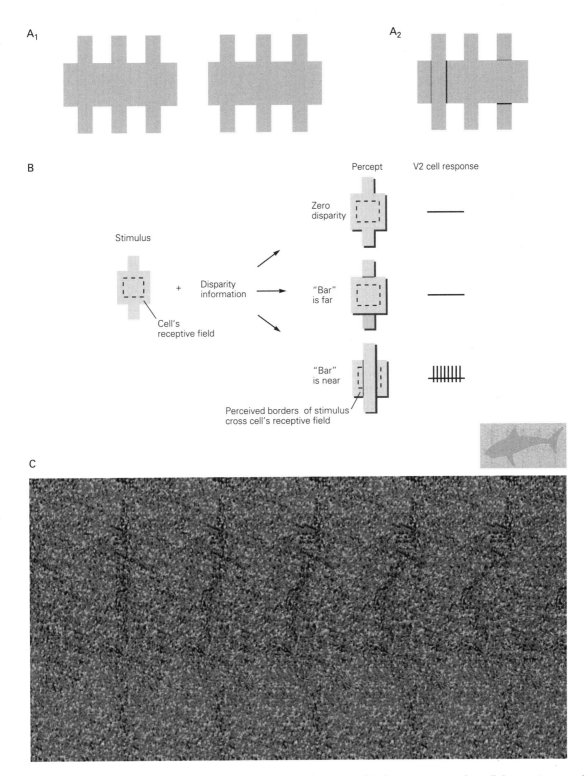

Figure 23–8 Global analysis of binocular disparity.

A. 1. Depth cues contribute to surface segmentation. If you view one of the images of three gray vertical bars crossing a gray horizontal rectangle, you see a uniform gray area within the rectangle. **2.** However, if you fuse the two rectangles with diverged eyes, the three vertical bars fall on the two retinas with near, zero, and far disparity. Seen this way, the bar at the left appears to hover in front of the rectangle with an illusory vertical edge crossing the rectangle, whereas the bar at the right appears to lie behind the edges of the horizontal rectangle.

B. A neuron in area V2 responds to illusory edges formed by binocular disparity cues. When the cell's receptive field is centered in the gray square, the cell does not respond to a vertical bar that has far disparity or the same disparity as the square. When the vertical bar has near disparity, the cell responds as the illusory vertical edge crosses its receptive field. (Reproduced, with permission, from Bakin, Nakayama, and Gilbert 2000. Copyright © 2000 Society for Neuroscience.)

C. A random-dot stereogram is seen as a random array of colored dots until you diverge or converge your eyes to bring the adjacent dark vertical stripes into register, producing a three-dimensional image of a shark that emerges from the background. This effect stems from systematic disparity for selected sets of dots. (© Fred Hsu/ Wikimedia Commons/CC-BY-SA-3.0.)

cues, such as size, perspective, occlusion, brightness, and movement, is not difficult. Another cue that originates outside the visual system is vergence, the angle between the optical axes of the two eyes for objects at varying distances. Yet another binocular cue, known as DaVinci stereopsis, is the presence of features visible to one eye but occluded in the other eye's view.

Neurons in areas V1 and V2 also signal foreground–background relationships. A cell with its receptive field in the center of a pattern within a larger surface may respond even when the boundary of that surface is distant from the receptive field. This response helps differentiate the object from its background. In making sense of an image, the brain must identify which edge belongs to which object and differentiate the edge of each object from the background. Some cells in area V2 have the property of "border ownership," firing only when a figure but not the background is to one side of the edge, even when the local edge information is identical in both instances (Figure 23–9).

Local Movement Cues Define Object Trajectory and Shape

The primary visual cortex determines the direction of movement of objects. Directional selectivity in neurons likely involves sequential activation of regions on different sides of the receptive field.

If an object moving at an appropriate velocity first encounters a region of a neuron's receptive field with long response latencies and then passes into regions with progressively shorter latencies, signals from throughout the receptive field will arrive at the cell simultaneously and the neuron will fire vigorously. If the object moves in the opposite direction, signals from the different regions will not summate and the cell may never reach the threshold for firing (Figure 23–10).

Early in the visual pathways, analysis of the movement of an object is limited by the size of the receptive fields of the sensory neurons. Even in the initial cortical areas V1 and V2, the receptive fields of neurons are small and might encompass only a fraction of an object. Eventually, however, information about the direction and speed of movement of discrete aspects of an object must be integrated into a computation of the movement of a whole object. This problem is more difficult than one might expect.

If one observes a complex shape moving through a small aperture, the part of the object's boundary

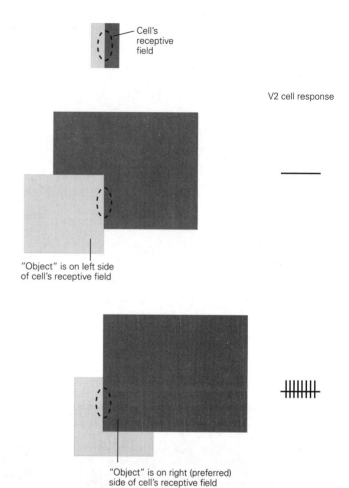

V2 cell response

"Object" is on left side of cell's receptive field

"Object" is on right (preferred) side of cell's receptive field

Figure 23–9 Border ownership. Cells in area V2 are sensitive to the boundaries of whole objects. Even though the local contrast is the same for the two rectangles within a cell's receptive field, the cell responds only when the boundary is part of the full rectangle that lies on the preferred side of the receptive field. (Adapted, with permission, from Zhou, Friedman, and von der Heydt 2000. Copyright © 2000 Society for Neuroscience.)

within the aperture appears to move in a direction perpendicular to the boundary's orientation (Figure 23–11A). One cannot detect a line's true direction of movement if the line's ends are not visible. The image of a line appears the same if it is moving slowly along an axis perpendicular to its orientation or more quickly along an oblique axis. This is the quandary presented by the receptive field of a V1 neuron. The visual system's solution is to assume that the movement of a contour is perpendicular to its orientation. Thus, an object is first presented to the visual system in countless small pieces with boundaries of different orientations, all of which appear to be moving in

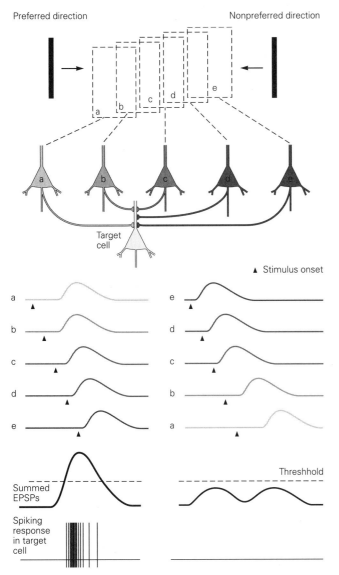

different directions and at different velocities (Figure 23–11A).

Determining the direction of motion of an object requires resolving multiple cues. This can be demonstrated readily by placing one grating on top of another and moving the two in different directions. The resulting checkerboard pattern appears to move in an intermediate direction between the trajectories of the individual gratings (Figure 23–11B). This percept depends on the relative contrast of the gratings and the area of grating overlap. With large relative contrasts, the gratings appear to slide across each other, moving in their individual directions rather than together in a common direction.

An important determinant of perceived direction is scene segmentation, the separation of moving elements into foreground and background. In a scene with moving objects, segmentation is not based on local cues of direction; instead, perception of direction depends on scene segmentation. The barber-pole illusion provides another example of the predominance of global relationships over the perception of simple attributes. The rotating stripes are perceived as moving vertically along the long axis of the pole (Figure 23–11C). The perception of motion in the visual field uses a complex algorithm that integrates the bottom-up analysis of local motion signals with top-down scene segmentation.

Integration of local motion signals in monkeys has been observed in the middle temporal area (area MT or V5), an area specializing in motion. The neurons in this area are selective for a particular direction of movement of an overall pattern, rather than individual components of the pattern. This dependency on the overall pattern is also seen in the correspondence of their responses with the perceived direction in the barber-pole effect.

Context Determines the Perception of Visual Stimuli

Brightness and Color Perception Depend on Context

The visual system measures the surface characteristics of objects by comparing the light arriving from different parts of the visual field. As a result, the perception of brightness and color is highly dependent on context. In fact, perceived brightness and color can be quite different from what is expected from the physical properties of an object. At the same time, perceptual constancies make objects appear similar even when the brightness and wavelength distribution of the light

Figure 23–10 Directional selectivity of movement. A neuron's selectivity for direction of movement depends on the response latencies of presynaptic neurons relative to the onset of a stimulus. The response latencies of presynaptic neurons *a* and *b* are somewhat longer than those of neurons *d* and *e*. When a stimulus moves from left to right, neurons *a* and then *b* are activated first, but because their response latencies are longer, their inputs arrive at the target neuron superimposed with the inputs from neurons *d* and *e*, and the summated inputs cause the neuron to fire. In contrast, stimuli moving leftward produce signals that arrive in the target neuron at different times and therefore do not reach the cell's threshold for firing. (Abbreviation: **EPSP**, excitatory postsynaptic potential.) (Adapted, with permission, from Priebe and Ferster 2008. Copyright © 2008 Elsevier.)

A

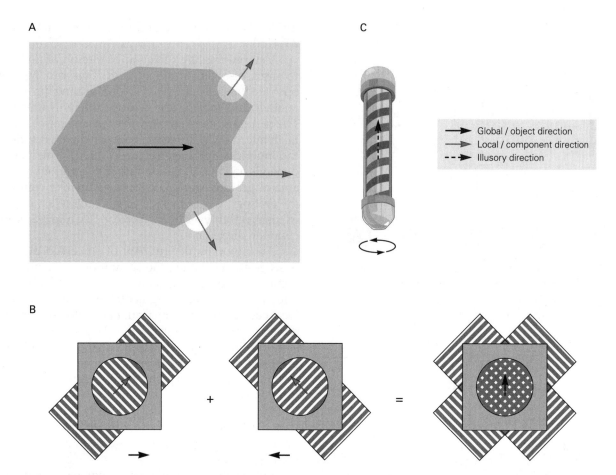

C

Global / object direction
Local / component direction
Illusory direction

B

+

=

Figure 23–11 The aperture problem and barber-pole illusion.
A. Although an object moves in one direction, each component edge when viewed through a small aperture appears to move in a direction perpendicular to its orientation. The visual system must integrate such local motion signals into a unified percept of a moving object.
B. Gratings are used to test whether a neuron is sensitive to local or global motion signals. When the gratings are superimposed and moved independently in different directions, one does not see the two gratings sliding past each other but rather a plaid pattern moving in a single, intermediate direction. Neurons in the middle temporal area of monkeys are responsive to such global motion rather than to local motion.

C. Motion perception is influenced by scene segmentation cues, as seen in the barber-pole illusion. Even though the pole rotates around its axis, one perceives the stripes as moving vertically, due to the global vertical rectangle surround of the barber pole enclosure.

that illuminates them changes from natural to artificial light, from sunlight to shadow, or from dawn to midday (Figure 23–12A).

As we move about or as the ambient illumination changes, the retinal image of an object—its size, shape, and brightness—also changes. Yet under most conditions, we do not perceive the object itself to be changing. As we move from a brightly lit garden into a dimly lit room, the intensity of light reaching the retina may vary a thousandfold. Both in the room's dim illumination and in the sun's glare, we nevertheless see a white shirt as white and a red tie as red. Likewise, as a friend walks toward you, she is seen as coming closer; you do not perceive her to be growing larger even though the

image on your retina does expand. Our ability to perceive an object's size and color as constant illustrates again a fundamental principle of the visual system: It does not record images passively, like a camera, but instead uses transient and variable stimulation of the retina to construct representations of a stable, three-dimensional world.

Another example of contextual influence is color induction, whereby the appearance of a color in one region shifts toward that in an adjoining region. Shape also plays an important role in the perception of surface brightness. Because the visual system assumes that illumination comes from above, gray patches on a folded surface appear very different when they lie on

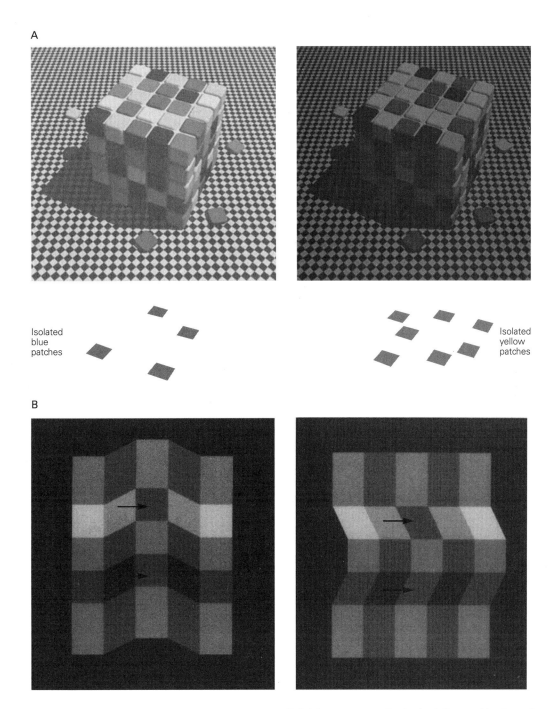

Figure 23–12 Color and brightness perception depend on contextual cues.

A. Perceived surface colors remain relatively stable under different illumination conditions and the consequent changes in wavelength of the light reflected from the surface. The **yellow squares** on the left and right cubes appear similar despite the fact that the wavelengths of light coming from the two sets of surfaces are very different. In fact, if the **blue squares** on the top of the left cube and the **yellow squares** on the top of the right cube are isolated from their contextual squares, their colors appear identical. (Reproduced, with permission, from www.lottolab.org.)

B. Brightness perception is also influenced by three-dimensional shape. The four gray squares indicated by **arrows** all have the same luminance. The apparent brightnesses are similar in the left illustration but different in the right illustration. This is because the visual system has an inherent expectation that illumination comes from above (the position of the sun relative to us), and thus the perception that the surface below the fold in the illustration on the right is brighter than the surface of the same luminance that lies above. (Reproduced, with permission, from Adelson 1993. Copyright © 1993 AAAS.)

the top or bottom of the surface, even when they are in fact the same shade of gray (Figure 23–12B).

The responses of some neurons in the visual cortex correlate with perceived brightness. Most visual neurons respond to surface boundaries; the center-surround structure of the receptive fields of retinal ganglion cells and geniculate neurons is suited to capturing boundaries. Most such cells do not respond to the interior parts of surfaces, for uniform interiors produce no contrast gradients across receptive fields. However, a small percentage of neurons do respond to the interiors of surfaces, signaling local brightness, texture, or color, and the responses of these neurons are influenced by context. The cell's response changes as the brightness of surfaces *outside* a cell's receptive field change, even when the brightness of the surface within the receptive field remains fixed.

Because most neurons respond to surface boundaries and not to areas of uniform brightness, the visual system calculates the brightness of surfaces from information about contrast at the edges of surfaces. The brain's analysis of surface qualities from boundary information is known as perceptual fill-in. If one fixates the boundary between a dark disk and a surrounding bright area for a few seconds, the disk will "fill in" with the same brightness as the surrounding area. This occurs because the cells that respond to edges fire only when the eye or stimulus moves. They gradually cease to respond to a stabilized image and no longer signal the presence of the boundary. Neurons with receptive fields within the disk gradually begin to respond in a fashion similar to those with receptive fields in the surrounding area, demonstrating short-term plasticity in their receptive-field properties.

An object's color always appears more or less the same despite the fact that under different conditions of illumination the wavelength distribution of light reflected from the object varies widely. To identify an object, we must know the properties of its surface rather than those of the reflected light, which are constantly changing. Computation of an object's color is therefore more complex than analyzing the spectrum of reflected light. To determine a surface's color, the wavelength distribution of the incident light must be determined. In the absence of that information, surface color can be estimated by determining the balance of wavelengths coming from different surfaces in a scene. Some neurons in V4 respond similarly to different illumination wavelengths if the perceived color remains constant. By being responsive to the light across an extensive surface, these neurons are selective for surface color rather than wavelength.

Receptive-Field Properties Depend on Context

The distinction between local and global effects—between stimuli that occur within a receptive field and those beyond—poses the problem of how the receptive field itself is defined. Because the original characterization of the receptive fields of visual cortex neurons did not take into account contextual influences, some investigators now distinguish between "classical" and "nonclassical" receptive fields.

However, even the earliest description of the sensory receptive field allowed for the possibility of influences from portions of the sensory surface outside the narrowly defined receptive field. In 1953, Steven Kuffler, in his pioneering observations on the receptive-field properties of retinal ganglion cells, noted that "not only the areas from which responses can actually be set up by retinal illumination may be included in a definition of the receptive field but also all areas which show a functional connection, by an inhibitory or excitatory effect on a ganglion cell. This may well involve areas which are somewhat remote from a ganglion cell and by themselves do not set up discharges."

A more useful distinction contrasts the response of a neuron to a simple stimulus, such as a short line segment, with its response to a stimulus with multiple components. Even in the primary visual cortex, neurons are highly nonlinear; their response to a complex stimulus cannot be predicted from their responses to a simple stimulus placed in different positions around the visual field. Their responses to local features are instead dependent on the global context within which the features are embedded. Contextual influences are pervasive in intermediate-level visual processing, including contour integration, scene segmentation, and the determination of object shape, object motion and surface properties.

Cortical Connections, Functional Architecture, and Perception Are Intimately Related

Intermediate-level visual processing requires sharing of information from throughout the visual field. The relationship of interconnections within the primary visual cortex to the functional architecture of this area suggests that this circuitry mediates contour integration.

Cortical circuits include a plexus of long-range horizontal connections formed by the axons of pyramidal neurons running parallel to the cortical surface. Horizontal connections exist in every area of the cerebral cortex, but their function varies from one area to the next depending on the functional architecture of each area. In the visual cortex, these connections mediate interactions between orientation columns of similar specificity, thus integrating information over a large area of visual cortex that represents a great expanse of the visual field (see Figure 21–16).

The fact that these horizontal connections link neurons similar in function but representing distant locations in the visual field suggests that these connections have a role in contour integration. Contour integration and the related property of contour saliency reflect the Gestalt principle of good continuation. Both are mediated by the horizontal connections in V1 (see Figure 23–6).

A final feature of cortical connectivity important for visuospatial integration is feedback projections from higher-order cortical areas. Feedback connections are as extensive as the feedforward connections that originate in the thalamus or at earlier stages of cortical processing. Little is known about the function of these feedback projections. They likely play a role in mediating the top-down influences of attention, expectation, and perceptual task, all of which are known to affect early stages in cortical processing.

Perceptual Learning Requires Plasticity in Cortical Connections

The synaptic connections in ocular-dominance columns are adaptable to experience only during a critical period in development (Chapter 49). This suggests that the functional properties of visual cortex neurons are fixed in adulthood. Nevertheless, many properties of cortical neurons remain mutable throughout life. For example, changes in the visual cortex can occur following retinal lesions.

When focal lesions occur in corresponding positions on the two retinas, the corresponding part of the cortical map, referred to as the lesion projection zone, is initially deprived of visual input. Over a period of several months, however, the receptive fields of cells within this region shift from the lesioned part of the retina to the functioning area surrounding the lesion. As a result, the cortical representation of the lesioned part of the retina shrinks while that of the surrounding region expands (Figure 23–13).

The plasticity of cortical maps and connections did not evolve as a response to lesions but as a neural mechanism for improving our perceptual skills. Many of the attributes analyzed by the visual cortex, including stereoscopic acuity, direction of movement, and orientation, become sharper with practice. Hermann von Helmholtz stated in 1866 that "the judgment of the senses may be modified by experience and by training derived under various circumstances, and may be adapted to the new conditions. Thus, persons may learn in some measure to utilize details of the sensation which otherwise would escape notice and not contribute to obtaining any idea of the object." This perceptual learning is a variety of implicit learning that does not involve conscious processes (Chapter 52).

Perceptual learning involves repeating a discrimination task many times and does not require error feedback to improve performance. Improvement manifests itself, for example, as a decrease in the threshold for discriminating small differences in the attributes of a target stimulus or in the ability to detect a target in a complex environment. Several areas of visual cortex, including the primary visual cortex, participate in perceptual learning.

An important aspect of perceptual learning is its specificity: Training on one task does not transfer to other tasks. For example, in a three-line bisection task, the subject must determine whether the centermost of three parallel lines is closer to the line on the left or the one on the right. The amount of offset from the central position required for accurate responses decreases substantially after repeated practice.

Learning of this task is specific to the location in the visual field and to the orientation of the lines. This specificity suggests that early stages of visual processing are responsible, for in the early stages, receptive fields are smallest, visuotopic maps are most precise, and orientation tuning is sharpest. The learning is also specific for the stimulus configuration. Training on three-line bisection does not transfer to a vernier discrimination task in which the context is a line that is collinear with the target line (Figure 23–14A).

The response properties of neurons in the primary visual cortex change during the course of perceptual learning in a way that tracks the perceptual improvement. An example of this is seen in contour saliency. With practice, subjects can more easily detect contours embedded in complex backgrounds. Detection improves with contour length, as do the responses of neurons in V1. With practice, subjects improve their ability to detect shorter contours and V1 neurons become correspondingly more sensitive to shorter contours (Figure 23–14B).

Visual Search Relies on the Cortical Representation of Visual Attributes and Shapes

The detectability of features such as color, orientation, and shape is related to the process of visual search. In a complex image, certain objects stand out or "pop out"

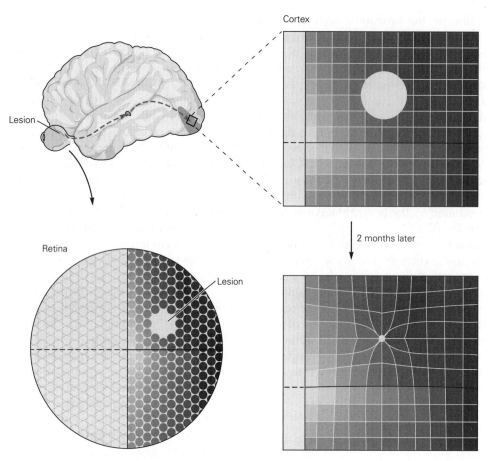

Figure 23–13 Adult cortical plasticity. When corresponding positions in both eyes are lesioned, the cortical area receiving input from the lesioned areas—the lesion projection zone—is initially silenced. The receptive fields of neurons in the lesion projection zone eventually shift from the area of the lesion to the surrounding, intact retina. This occurs because neurons surrounding the lesion projection zone sprout collaterals that form synaptic connections with neurons inside the zone. As a result, the cortical representation of the lesioned part of the retina shrinks while that of the surrounding retina expands.

because the visual system processes simultaneously, in parallel pathways, the features of the target and the surrounding distractors (Figure 23–15). When the features of a target are complex, the target can be identified only through careful inspection of an entire image or scene.

The pop-out phenomenon can be influenced by training. A stimulus that initially cannot be found without effortful searching will pop out after training. The neuronal correlate of such a dramatic change is not certain. Parallel processing of the features of an object and its background is possible because feature information is encoded in retinotopically mapped areas at multiple locations in the visual cortex. Pop-out probably occurs early in the visual cortex. The pop-out of complex shapes such as numerals supports the idea that early in visual processing neurons can represent, and be selective for, shapes more complex than line segments with a particular orientation.

Cognitive Processes Influence Visual Perception

Scene segmentation—the parsing of a scene into different objects—involves a combination of bottom-up processes that follow the Gestalt rule of good continuation and top-down processes that create object expectation.

One strong top-down influence is spatial attention, which can change focus without any movement of an observer's eyes. Spatial attention can be object-oriented in that the focus of attention is distributed over the area occupied by the attended object, allowing the visual cortex to analyze the shape and attributes of objects one at a time.

Attentional mechanisms can solve the superposition problem. Before we can recognize an object in a scene that includes many objects, we must determine which features correspond to which objects. Our sense that we identify all objects in the visual field simultaneously is illusory. Instead, we serially process objects in rapid succession by

A Perceptual learning is task-specific

Orientation discrimination

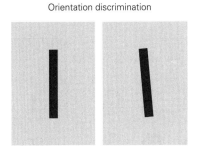

Three-line bisection task

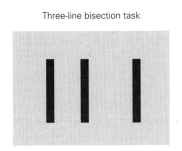

Vernier task

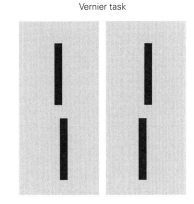

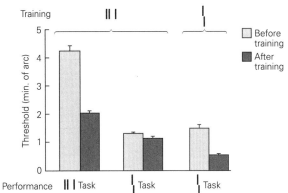

B Neuronal responsiveness changes during training

Contour detection task

5 collinear lines

9 collinear lines

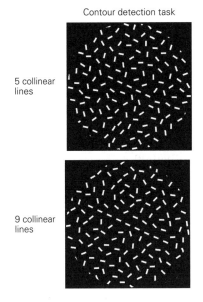

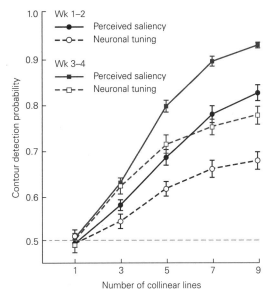

Figure 23–14 Perceptual learning. Perceptual learning is a form of implicit learning. With practice, one can learn to discriminate smaller differences in orientation, position, depth, and direction of movement of objects.

A. The improvement is seen as a reduction in the amount of change required to reliably detect a tilted line or one positioned to the left or right of a nearly collinear line (vernier task). Perceptual learning is highly specific, so that training on a three-line bisection task leads to substantial improvement in that task (*left pair of bars* in the bar graph) without affecting performance on the vernier discrimination task

(*central pair of bars*). However, training specifically on vernier discrimination does enhance performance on that task (*right pair of bars*).

B. Subjects can detect collinear line segments embedded in a random background more easily as the number of collinear segments is increased. The responses of neurons in V1 grow correspondingly stronger with the increase in the number of line segments. After practice, a line with fewer segments stands out more easily, and with this improvement, the responses in V1 also increase. (Reproduced, with permission, from Crist, Li, and Gilbert 2001; Li, Piech, and Gilbert 2008.)

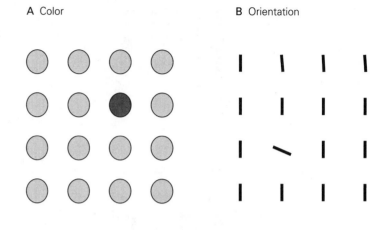

Figure 23–15 One object in a complex image stands out under certain conditions.

A. A differently colored object pops out.

B. A differently oriented line also pops out.

C. More complex shapes can pop out when they are very familiar, such as the numeral 2 embedded in a field of 5s. Rotating the image by 90° renders the elements of the figure less recognizable, making it more difficult to find the one figure that differs from the rest. (Reproduced, with permission, from Wang, Cavanagh, and Green 1994. Copyright © 1994 Springer Nature.)

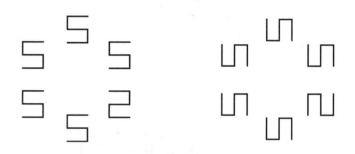

shifting attention from one to the next. The results of each analysis build up the perception of a complex environment populated with many distinct objects. A dramatic demonstration of the importance of attention in object recognition is *change blindness*. If a subject rapidly shifts between two slightly different views of the same scene, he will not be able to detect the absence of an important component of the scene in one view without considerable scrutiny (see Figure 25–8).

Another top-down influence is perceptual task. At early stages in visual processing, the properties of the same neuron vary with the type of visual discrimination being performed. Object identification involves a process of hypothesis testing in which information arriving from the retina is compared with internal representations of objects. This process is reflected in studies that have shown that early stages in processing, such as the primary visual cortex, are activated when scenes are imagined without visual input.

Highlights

1. Vision requires segregating objects from their backgrounds, a process involving contour integration and surface segmentation.

2. This process is simplified by relying on the statistical properties of natural forms. As recognized by the Gestalt psychologists early in the 20th century, we naturally link scene components based on grouping rules of similarity, proximity, and contour smoothness (referred to as "good continuation").

3. Neurons in visual cortical areas have properties consonant with Gestalt grouping rules. They perform a local and global analysis of scene properties in parallel. The local properties are the visual primitives, which include orientation selectivity, direction selectivity, contrast sensitivity, disparity selectivity, and color selectivity. The corresponding global properties include contour integration, object movement, border ownership, disparity capture, and color constancy.

4. Perception of visual features is dependent on context; similarly, neuronal responses are context dependent. The principle underlying these interactions is the association field, a pattern of interactions between bits of information that are mapped across each cortical area. The association field mediates contour integration in visual cortex but is likely to be a general feature of processing throughout the cerebral cortex. The anatomical

substrate for the association field includes a network of long-range horizontal connections formed by the axons of cortical pyramidal cells, which extend for long distances parallel to the cortical surface.

5. Different visual cortical areas contribute to the various global properties, and interactions between areas, including top-down influences, are required for their development. Though there has been considerable emphasis on selectivity for increasing stimulus complexity as one ascends a hierarchy of cortical areas through feedforward connections extending from the primary visual cortex to areas in the temporal (ventral pathway) and parietal (dorsal pathway) cortex, feedback connections are of equal importance.

6. Future studies will elucidate the relative contributions of intrinsic, feedforward, and feedback cortical connections, and the interactions between them, in cortical processing. Evidence is emerging that rather than having fixed functions, neurons are adaptive processors, taking on different functional roles under different behavioral contexts. Neurons may mediate this functional diversity by input selection, expressing task-relevant inputs and suppressing task-irrelevant inputs. When operating abnormally, these functional and connectivity dynamics may account for perceptual and behavioral phenomena associated with disorders such as autism and schizophrenia.

Charles D. Gilbert

Selected Reading

Albright TD, Stoner GR. 2002. Contextual influences on visual processing. Annu Rev Neurosci 25:339–379.

Gilbert CD, Sigman M. 2007. Brain states: top-down influences in sensory processing. Neuron 54:677–696.

Gilbert CD, Sigman M, Crist R. 2001. The neural basis of perceptual learning. Neuron 31:681–697.

Li W, Piech V, Gilbert CD. 2004. Perceptual learning and top-down influences in primary visual cortex. Nat Neurosci 7:651–657.

Li W, Piech V, Gilbert CD. 2006. Contour saliency in primary visual cortex. Neuron 50:951–962.

Priebe NJ, Ferster D. 2008. Inhibition, spike threshold, and stimulus selectivity in primary visual cortex. Neuron 57:482–497.

References

Adelson EH. 1993. Perceptual organization and the judgment of brightness. Science 262:2042–2044.

Bakin JS, Nakayama K, Gilbert CD. 2000. Visual responses in monkey areas V1 and V2 to three-dimensional surface configurations. J Neurosci 20:8188–8198.

Crist RE, Li W, Gilbert CD. 2001. Learning to see: experience and attention in primary visual cortex. Nat Neurosci 4:519–525.

Cumming BG, DeAngelis GC. 2001. The physiology of stereopsis. Annu Rev Neurosci 24:203–238.

Ferster D, Miller KD. 2000. Neural mechanisms of orientation selectivity in the visual cortex. Annu Rev Neurosci 23:441–471.

He ZJ, Nakayama K. 1994. Apparent motion determined by surface layout not by disparity or three-dimensional distance. Nature 367:173–175.

Hubel DH, Wiesel TN. 1968. Receptive fields and functional architecture of monkey striate cortex. J Physiol 195:215–243.

Li W, Gilbert CD. 2002. Global contour saliency and local colinear interactions. J Neurophysiol 88:2846–2856.

Li W, Piech V, Gilbert CD. 2008. Learning to link visual contours. Neuron 57:442–451.

Movshon JA, Adelson EH, Gizzi MS, Newsome WT. 1985. The analysis of moving visual patterns. In: C Chagas, R Gattass, CG Gross (eds). *Study Group on Pattern Recognition Mechanisms*, pp. 67–86. Vatican City: Pontifica Academia Scientiarum.

Nakayama K. 1996. Binocular visual surface perception. Proc Natl Acad Sci U S A 93:634–639.

Nakayama K, Joseph JS. 2000. Attention, pattern recognition and popout in visual search. In: R Parasuraman (ed). *The Attentive Brain*. Cambridge, MA: MIT Press.

Poggio GE. 1995. Mechanisms of stereopsis in monkey visual cortex. Cereb Cortex 5:193–204.

Purves D, Lotto RB, Nundy S. 2002. Why we see what we do. Am Sci 90:236–243.

Wang Q, Cavanagh P, Green M. 1994. Familiarity and pop-out in visual search. Percept Psychophys 56:495–500.

Zhou H, Friedman HS, von der Heydt R. 2000. Coding of border ownership in monkey visual cortex. J Neurosci 20:6594–6611.

24

High-Level Visual Processing: From Vision to Cognition

A S WE HAVE SEEN, LOW-LEVEL visual processing is responsible for detecting various types of contrasts in the patterns of light projected onto the retina. Intermediate-level processing is concerned with the identification of so-called visual primitives, such as contours and fields of motion, and the segregation of surfaces. High-level visual processing integrates information from a variety of sources and is the final stage in the visual pathway leading to visual perception.

High-level visual processing is concerned with identifying behaviorally meaningful features of the environment and thus depends on descending signals that convey information from short-term working memory, long-term memory, and executive areas of cerebral cortex.

High-Level Visual Processing Is Concerned With Object Recognition

Our visual experience of the world is fundamentally object-centered. We can recognize the same object even when the patterns of light it casts onto the retina vary greatly with viewing conditions, such as lighting, angle, position, and distance. And this is the case even for visually complex objects, those that include a large number of conjoined visual features.

Moreover, objects are not mere visual entities, but are commonly associated with specific experiences, other remembered objects, and sensations—such as the hum of the coffee grinder or the aroma of a lover's perfume—and a variety of emotions. It is the behavioral significance of objects that guides our action based on visual information. In short, object recognition establishes a nexus between vision and cognition (Figure 24–1).

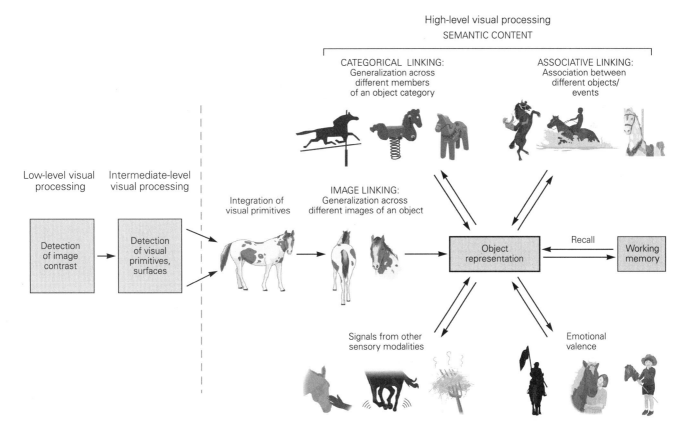

Figure 24–1 Representation of entire objects is central to high-level visual processing. Representation of entire objects involves integration of visual features extracted at earlier stages in the visual pathways. This integration is a generalization of the numerous retinal images generated by the same object and of different members of an object category. The representation also incorporates information from other sensory modalities, attaches emotional valence, and associates the object with the memory of other objects or events. Object representations can be stored in working memory and recalled in association with other memories.

The Inferior Temporal Cortex Is the Primary Center for Object Recognition

Primate studies implicate neocortical regions of the temporal lobe, principally the inferior temporal cortex, in object perception. Because the hierarchy of synaptic relays in the cortical visual system extends from the primary visual cortex to the temporal lobe, the temporal lobe is a site of convergence of many types of visual information.

Neuropsychological studies have found that damage to the inferior temporal cortex can produce specific failures of object recognition. Neurophysiological and functional imaging studies have, in turn, yielded remarkable insights into the ways in which the activity of inferior temporal neurons represents objects, how these representations relate to perceptual and cognitive events, and how they are modified by experience.

Visual signals originating in the retina are processed in the lateral geniculate nucleus of the thalamus before reaching the primary visual cortex (V1). Ascending visual pathways from V1 follow two main parallel and hierarchically organized streams: the ventral and dorsal streams (Chapter 21). The ventral stream extends ventrally and anteriorly from V1 through V2, via V4, into inferior temporal cortex, which, in macaque monkeys, comprises the lower bank of the superior temporal sulcus and the ventrolateral convexity of the temporal lobe (Figure 24–2). Neurons at each synaptic relay in this ventral stream receive convergent input from the preceding stage. At the top of the hierarchy, inferior temporal neurons are in a position to integrate a large and diverse quantity of visual information over a vast region of visual space.

The inferior temporal cortex is a large brain region. The patterns of anatomical connections to and from this region indicate that it comprises at least two main functional subdivisions—the posterior area temporo-occipital cortex and the anterior area temporal cortex—and functional evidence suggests further subdivisions

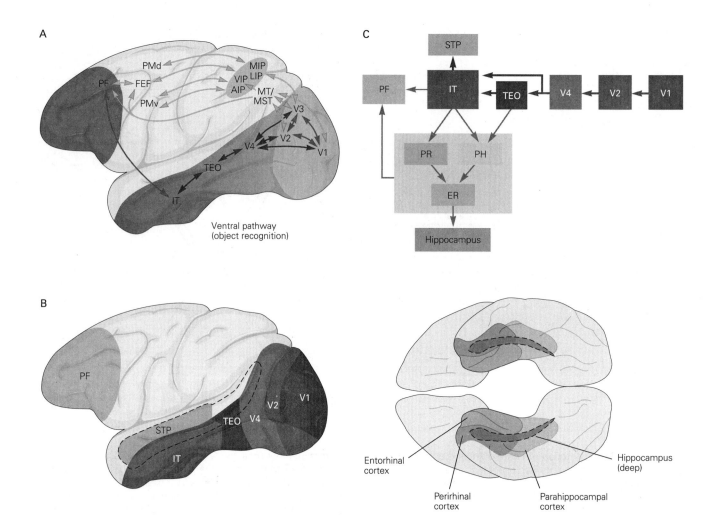

Figure 24–2 Cortical pathway for object recognition.

A. A lateral view of the macaque brain shows the major pathways involved in visual processing, including the pathway for object recognition (**red**). (Abbreviations: **AIP**, anterior intraparietal cortex; **FEF**, frontal eye fields; **IT**, inferior temporal cortex; **LIP**, lateral intraparietal cortex; **MIP**, medial intraparietal cortex; **MST**, medial superior temporal cortex; **MT**, middle temporal cortex; **PF**, prefrontal cortex; **PMd**, dorsal premotor cortex; **PMv**, ventral premotor cortex; **TEO**, temporo-occipital cortex; **VIP**, ventral intraparietal cortex.)

B. Lateral and ventral views of the macaque monkey brain show the cortical areas involved in object recognition.

(Abbreviations: **IT**, inferior temporal cortex; **PF**, prefrontal cortex; **STP**, superior temporal polysensory area; **TEO**, temporo-occipital cortex.)

C. The inferior temporal cortex (**IT**) is the end stage of the ventral stream (**red arrows**) and is reciprocally connected with neighboring areas of the medial temporal lobe and prefrontal cortex (**gray arrows**). This chart illustrates the main connections and predominant direction of information flow. (Abbreviations: **ER**, entorhinal cortex; **PF**, prefrontal cortex; **PH**, parahippocampal cortex; **PR**, perirhinal cortex; **STP**, superior temporal polysensory area; **TEO**, temporo-occipital cortex.)

into multiple functionally specialized areas. As we shall see, the distinction between anterior and posterior parts of the inferior temporal cortex is supported by both neuropsychological and neurophysiological evidence.

Clinical Evidence Identifies the Inferior Temporal Cortex as Essential for Object Recognition

The first clear insight into the neural pathways mediating object recognition was obtained in the late 19th century when the American neurologist Sanger Brown

and the British physiologist Edward Albert Schäfer found that experimental lesions of the temporal lobe in primates abolished the ability to recognize objects. Unlike the deficits that accompany lesions of occipital cortical areas, temporal lobe lesions do not impair sensitivity to basic visual attributes, such as color, motion, and distance. Because of the unusual type of visual loss, the impairment was originally called psychic blindness, but this term was later replaced by *visual agnosia* ("without visual knowledge"), a term coined by Sigmund Freud.

In humans, there are two basic categories of visual agnosia, apperceptive and associative, the description of which led to a two-stage model of object recognition in the visual system. With apperceptive agnosia, the ability to match or copy complex visual shapes or objects is impaired (Figure 24–3). This impairment results from disruption of the first stage of object recognition: integration of visual features into sensory representations of entire objects. With associative agnosia, the ability to match or copy complex objects remains intact, but the ability to identify objects is impaired. This impairment results from disruption of the second stage of object recognition: association of the sensory representation of an object with knowledge of the object's meaning or function.

Consistent with this functional hierarchy, apperceptive agnosia is most common following damage to the posterior inferior temporal cortex, whereas associative agnosia, a higher-order perceptual deficit, is more common following damage to the anterior inferior temporal cortex. Neurons in the anterior subdivision exhibit a variety of memory-related properties not seen in the posterior area.

More focal lesions within temporal cortex can lead to specific deficits. Damage to a small region of the human temporal lobe results in an inability to

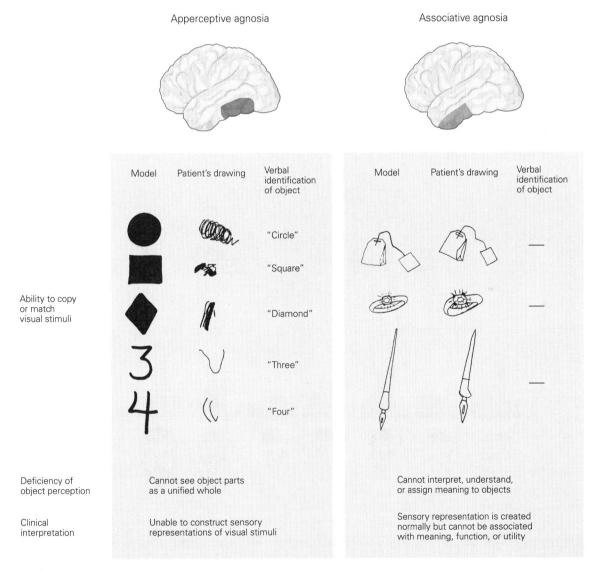

Figure 24–3 Neurons in the temporal lobe of humans are involved in object recognition. Damage to the inferior temporal cortex impairs the ability to recognize visual objects, a condition known as visual agnosia. There are two major categories of visual agnosia: Apperceptive agnosia results from damage to the posterior region, and associative agnosia results from damage of the anterior region. (Reproduced, with permission, from Farah 1990. © 1990 Massachusetts Institute of Technology.)

recognize faces, a form of associative agnosia known as *prosopagnosia*. Patients with prosopagnosia can identify a face as a face, recognize its parts, and even detect specific emotions expressed by the face, but they are unable to identify a particular face as belonging to a specific person.

Prosopagnosia is an example of a *category-specific agnosia*, in which patients with temporal lobe damage fail to recognize particular items belonging to a specific semantic category. Category-specific agnosias for living things, fruits, vegetables, tools, or animals have also been reported. Owing to the pronounced behavioral significance of faces and the normal ability of people to recognize an extraordinarily large number of faces, prosopagnosia may simply be the most commonly diagnosed variety of category-specific agnosia.

Neurons in the Inferior Temporal Cortex Encode Complex Visual Stimuli and Are Organized in Functionally Specialized Columns

The coding of visual information in the temporal lobe has been studied extensively using electrophysiological techniques, beginning with the work of Charles Gross and colleagues in the 1970s. Neurons in this region have distinctive response properties. They are relatively insensitive to simple stimulus features such as orientation and color. Instead, the vast majority possess large, centrally located receptive fields and encode complex stimulus features. These selectivities often appear somewhat arbitrary. An individual neuron might, for example, respond strongly to a crescent-shaped pattern of a particular color and texture. Cells with such unique selectivities likely provide inputs to higher-order neurons that respond to specific meaningful objects.

In fact, within the inferior temporal cortex, several small subpopulations of neurons are activated by highly meaningful objects, such as faces and hands (Figure 24–4), as Charles Gross discovered. For cells that respond to the sight of a hand, individual fingers are particularly critical. Among cells that respond to faces, the most effective stimulus for some cells is a frontal view of the face, whereas for others it is a side view. Although some neurons respond preferentially to faces in general, others respond only to specific

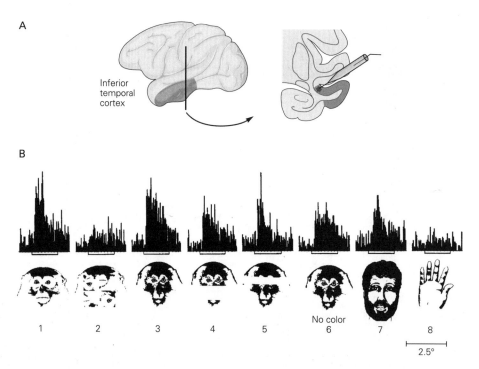

Figure 24–4 Neurons in the inferior temporal cortex of the monkey are involved in face recognition. (Reproduced, with permission, from Desimone et al. 1984. Copyright © 1984 Society for Neuroscience.)

A. The location of the inferior temporal cortex of the monkey is shown in a lateral view and coronal section. The colored area is the location of the recorded neurons.

B. Peristimulus time histograms illustrate the frequency of action potentials in a single neuron in response to different images (shown below the histograms). This neuron responded selectively to faces. Masking of critical features, such as the mouth or eyes (**4, 5**), led to a substantial but not complete reduction in response. Scrambling the parts of the face (**2**) nearly eliminated the response.

facial expressions. It seems likely that such cells contribute directly to face recognition.

In initial relays in the cortical visual system, neurons that respond to the same stimulus features, such as orientation or direction of motion, but from different parts of the visual field are organized in columns. Cells within the inferior temporal cortex are similarly organized. Columns of neurons representing the same or similar stimulus properties commonly extend throughout the cortical thickness and over a range of approximately 400 μm. The columns are arranged such that different stimuli that possess some similar features are represented in partially overlapping columns (Figure 24–5). Thus, one stimulus can activate multiple columns. Horizontal connections can span many millimeters and may facilitate the formation of distributed networks for encoding objects.

The Primate Brain Contains Dedicated Systems for Face Processing

Prosopagnosia often occurs in the absence of any other form of agnosia. Such a highly specific perceptual deficit could be explained by focal lesions of face-selective neurons located in exclusive clusters. This idea was strengthened by the discovery of face-selective regions in the human brain by Nancy Kanwisher and colleagues using functional magnetic resonance imaging (fMRI) and by Gregory McCarthy and colleagues using direct electrophysiological recordings from the surface of the human brain. Kanwisher and colleagues found that during the presentation of pictures of faces and other objects one area in the human temporal lobe, the fusiform face area, responded significantly more during the presentation of faces compared to other objects.

Subsequently, several more face-selective areas were found, primarily in temporal but also in prefrontal cortex. Early studies of these areas provided circumstantial evidence for clustering of face-selective neurons. In later studies, Doris Tsao, Winrich Freiwald, and colleagues directly demonstrated such clustering and showed that face processing might be performed by a dedicated face-processing network spanning from the posterior part of inferior temporal cortex to prefrontal cortex. Using fMRI, they found six areas in temporal cortex and three in prefrontal cortex of the macaque monkey that responded more selectively to faces than to other objects. These areas, called face patches, are found at highly consistent locations across individuals and thus are named based on their location. Each face patch is a few millimeters in diameter and thus differs organizationally from the inferior temporal columns. Intracellular recordings from the face patches

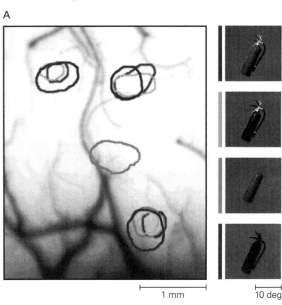
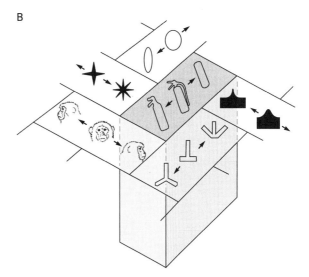

Figure 24–5 Neurons in the anterior portion of the inferior temporal cortex that respond to complex visual stimuli are organized into columns. (Reproduced, with permission, from Tanaka 2003. Copyright © 2003 Oxford University Press.)

A. Optical images of the surface of the anterior inferior temporal cortex illustrate regions selectively activated by the objects shown at the right.

B. Neurons of the inferior temporal cortex are organized in functionally specialized columns that extend from the surface of the cortex. According to this model, each column includes neurons that respond to a specific visually complex object. Columns of neurons that represent variations of an object, such as different faces or different fire extinguishers, constitute a hypercolumn.

revealed that the vast majority of cells respond selectively more to faces than to other objects. Thus, millions of face cells are clustered into a fixed number of small areas. These areas are directly connected to each other, thus forming a face-processing network. Within this network, each node appears to be functionally specialized. From posterior to anterior locations within the temporal lobe, the initial face patches respond to particular views of the face, and then face patches become gradually more selective to identity and less selective for angle of view. Furthermore, dorsal face areas within the temporal lobe exhibit a selectivity for natural facial motion, which ventral areas lack. Thus, a highly specialized network, located primarily in temporal cortex, processes the multiple dimensions of information conveyed by a face (Figure 24–6).

The Inferior Temporal Cortex Is Part of a Network of Cortical Areas Involved in Object Recognition

Object recognition is intimately intertwined with visual categorization, visual memory, and emotion, and the outputs of the inferior temporal cortex contribute to these functions (see Figure 24–2). Among the principal projections are those to the perirhinal and parahippocampal cortices, which lie medially adjacent to the ventral surface of the inferior temporal cortex (Figure 24–2C). These regions project, in turn, to the entorhinal cortex and the hippocampal formation, both of which are involved in long-term memory storage and retrieval. A second major projection from the inferior temporal cortex is to the prefrontal cortex, an important site for high-level visual processing. As we shall see, prefrontal neurons play important roles in object categorization, visual working memory, and memory recall.

The inferior temporal cortex also provides input—directly and indirectly via the perirhinal cortex—to the amygdala, which is believed to apply emotional valence to sensory stimuli and to engage the cognitive and visceral components of emotion (Chapter 42). Finally, the inferior temporal cortex is a major source of input to multimodal sensory areas of cortex such as the superior temporal polysensory area (Figure 24–2B), which lies dorsally adjacent to the inferior temporal cortex.

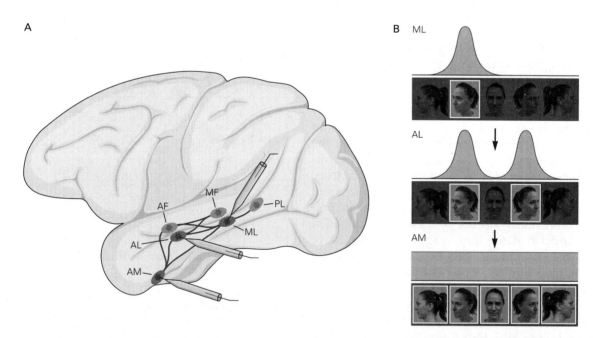

Figure 24–6 The temporal lobe contains a network of face-selective areas.

A. Functional magnetic resonance imaging of macaque monkeys watching pictures of faces and other objects identified six face-selective areas in the temporal lobe, inside and around the superior temporal sulcus. These areas occur at the same locations across subjects and have been given names based on their anatomical location (**PL**, posterior lateral; **ML**, medial lateral; **MF**, medial fundus of the superior temporal sulcus;

AL, anterior lateral; **AF**, anterior fundus; **AM**, anterior medial). These areas are interconnected to form a face-processing network.

B. Single-neuron recordings from areas ML, AL, and AM show tuning to head orientation. ML cells are tuned to specific head orientations, many AL cells are tuned to multiple orientations that are mirror-symmetric versions of each other, and AM cells are broadly and more weakly tuned to head orientation. These three representations in interconnected areas can be thought of as transformations of each other (**arrows**).

Object Recognition Relies on Perceptual Constancy

The ability to recognize objects as the same under different viewing conditions, despite the sometimes markedly different retinal images, is one of the most functionally important requirements of visual experience. The invariant attributes of an object—for example, spatial and chromatic relationships between image features or characteristic features such as the stripes of a zebra—are cues to the identity and meaning of the objects.

For object recognition to take place, these invariant attributes must be represented independently of other image properties. The visual system does this with proficiency, and its behavioral manifestation is termed *perceptual constancy.* Perceptual constancy has many forms ranging from invariance across simple transformations of an object, such as changes of size or position, to more difficult ones, such as rotation in depth or changes in lighting, and even to the sameness of objects within a category: All zebras look alike.

One of the best examples is *size constancy.* An object placed at different distances from an observer is perceived as having the same size, even though the object produces images of different absolute size on the retina. Size constancy has been recognized for centuries, but only in the past several decades has it been possible to identify the neural mechanisms responsible. An early study found that lesions of the inferior temporal cortex lead to failures of size constancy in monkeys, suggesting that neurons in this area play a critical role in size constancy. Indeed, one of the most striking properties of individual inferior temporal neurons is the invariance of their shape selectivity even to very big changes in stimulus size (Figure 24–7A).

Another type of perceptual constancy is *position constancy,* in which objects are recognized as the same regardless of their location in the visual field. The pattern of selective response of many inferior temporal neurons does not vary when an object changes position within their large receptive fields (Figure 24–7B). *Form-cue invariance* refers to the constancy of a form when the cues that define the form change. The silhouette of Abraham Lincoln's head, for example, is readily recognizable whether it is black on white, white on black, or red on green. The responses of many inferior temporal neurons do not change with changes in contrast polarity (Figure 24–7C), color, or texture.

Viewpoint invariance refers to the perceptual constancy of three-dimensional objects observed from different angles. Because most objects we see are three-dimensional and opaque, when looked at from different viewpoints, some parts become invisible, while others are revealed, and all others change in appearance. Yet despite the limitless range of retinal images that might be cast by a familiar object, an observer can readily recognize an object independently of the angle at which it is viewed. There are notable exceptions to this rule, which generally occur when an object is viewed from an angle that yields an uncharacteristic retinal image, such as a bucket viewed from directly above.

Thus, object recognition mechanisms must infer the identity of objects from apparent complex shapes. Many neurons in inferior temporal cortex do not exhibit viewpoint invariance. In fact, many are systematically tuned to viewing angle. Yet at more anterior locations, neurons are not only more size and position invariant, but they also exhibit greater invariance to viewpoint. The face-processing system is a case in point. Neurons in posterior face patches are tuned to viewing angle, while neurons in anterior face patches exhibit great robustness to changes in viewpoint. Thus, population responses in posterior face areas contain more information about head orientation than those in anterior areas, while the anterior face patches provide more information about face identity across head orientations compared to posterior face areas. The degree of viewpoint invariance achieved in anterior inferior temporal cortex, by individual neurons and populations of neurons, might be sufficient to account for perceptual viewpoint invariance. But this has not been directly shown yet. Alternatively, viewpoint invariance may be achieved at a higher stage of cortical processing, such as the prefrontal cortex.

Studies of the conditions under which viewpoint invariance fails may lead to insights into the neural mechanisms of the behavior. One such condition is presentation of mirror images. Although mirror images are not identical, they are frequently perceived as such, a confusion reflecting a false-positive identification by the system for viewpoint invariance. Carl Olson and colleagues examined the responses of neurons in a particular region of the inferior temporal cortex to stimuli that were mirror images. Consistent with the perceptual confusion, many inferior temporal neurons responded similarly to both images. Similarly, in one face area between the posterior and anterior ones described earlier, profile-selective cells respond similarly to the left and right profile of a face. These results reinforce the conclusion that activity in the inferior temporal cortex reflects perceptual invariance, albeit incorrectly in this case, rather than the actual features of a stimulus.

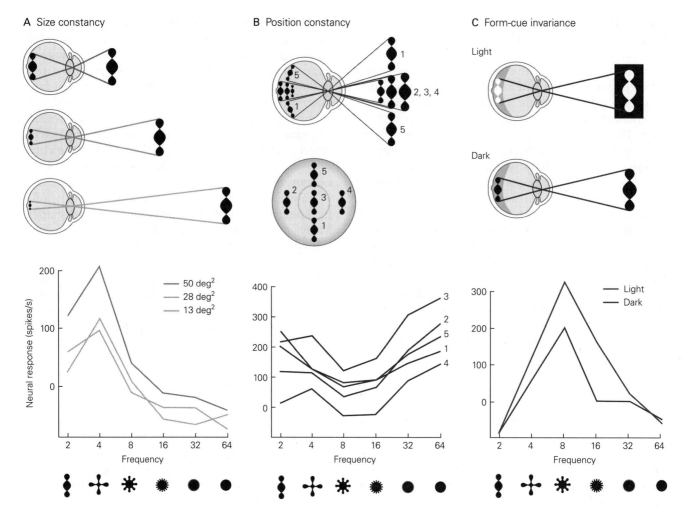

Figure 24–7 Perceptual constancy is reflected in the behavior of neurons in the inferior temporal cortex. The responses of many inferior temporal neurons are selective for stimuli with a particular frequency (number) of lobes but invariant to object size, position, and reflectance. (Reproduced, with permission, from Schwartz et al. 1983.)

A. *Size constancy.* An object is perceived to be the same even when the retinal image size decreases with the distance of the object in the visual field. The response of the vast majority of inferior temporal neurons to substantial changes in retinal image size is invariant, as illustrated here by the record of a single cell.

B. *Position constancy.* An object is perceived to be the same despite changes in position in the retinal image. Almost all inferior temporal neurons respond similarly to the same stimulus in different positions in the visual field, as illustrated here by the record of a single neuron.

C. *Form-cue invariance.* An object is perceived to be the same despite changes in reflectance. Most inferior temporal neurons respond similarly to the two images illustrated, as shown in the record of an individual neuron.

Categorical Perception of Objects Simplifies Behavior

All forms of perceptual constancy are the product of the visual system's attempts to generalize across different retinal images generated by a single object. A still more general type of constancy is the perception of individual objects as belonging to the same semantic category. The apples in a basket or the many appearances of the letter *A* in different fonts, for example, are physically distinct but are effortlessly perceived as *categorically* identical.

Categorical perception is classically defined as the ability to distinguish objects of different categories better than objects of the same category. For example, it is more difficult to discriminate between two red lights that differ in wavelength by 10 nm than to discriminate between red and orange lights with the same wavelength difference.

Categorical perception simplifies behavior. For example, it usually does not matter whether an apple is completely spherical or slightly mottled on the left side or whether the seat we are offered is a Windsor or a Chippendale side chair. Similarly, reading ability requires that one be able to recognize the alphabet in a broad variety of type styles. Like the simpler forms of perceptual constancy, categorical perception relies on the brain's ability to extract invariant features of objects seen.

Is there a population of neurons that respond uniformly to objects within a category and differentially to objects of different categories? To test this, David Freedman and Earl Miller and colleagues created a set of images in which features of dogs and cats were merged; the proportions of dog and cat in the composite images varied continuously from one extreme to the other. Monkeys were trained to identify these stimuli reliably as either dog or cat. Miller and colleagues then recorded from visually responsive neurons in the dorsolateral prefrontal cortex, a region that receives direct input from the inferior temporal cortex. Not only did these neurons exhibit the predicted category-selective responses—responding well to cat but not dog, or vice versa—but the neuronal category boundary also corresponded to the behaviorally learned boundary (Figure 24–8). By contrast, neurons in inferior temporal cortex represented similarity of features, not categories.

The fact that category-specific agnosias sometimes follow damage to the temporal lobe suggests there are neurons in the inferior temporal cortex that are category-selective similar to those of neurons in the prefrontal cortex. Face-selective cells in the temporal cortex appear to meet this criterion, because their responses to a range of faces are often similar. Yet, these may constitute a special case, whereas for most stimulus conditions, category-selective responses may be characteristic of neurons in the prefrontal cortex, where visual responses are more commonly linked to the behavioral significance of the stimuli.

Visual Memory Is a Component of High-Level Visual Processing

Visual experience can be stored as memory, and visual memory influences the processing of incoming visual information. Object recognition in particular relies on the observer's previous experiences with objects. Thus, the contributions of the inferior temporal cortex to object recognition must be modifiable by experience.

Studies of the role of experience in visual perception have focused on two distinct types of experience-dependent plasticity. One stems from repeated exposure or practice, which leads to improvements in visual discrimination and object recognition ability. These experience-dependent changes constitute a form of implicit learning known as perceptual learning (Chapter 23). The other occurs in connection with the storage of explicit learning, the learning of facts or events that can be recalled consciously (Chapter 54).

Implicit Visual Learning Leads to Changes in the Selectivity of Neuronal Responses

The ability to discriminate complex visual stimuli is highly modifiable by experience. For example, individuals who attend to fine differences between automobile models improve their ability to recognize such differences.

In the inferior temporal cortex, neuronal selectivity for complex objects can undergo change that parallels change in the ability to distinguish objects. For example, in a study by Logothetis and colleagues, monkeys were trained to identify novel three-dimensional objects, such as randomly bent wire forms, from two-dimensional views of the objects. Extensive training led to pronounced improvements in the ability to recognize the objects from two-dimensional views. After training, a population of neurons was found that exhibited marked selectivity for the views seen earlier but not for other two-dimensional views of the same objects (Figure 24–9).

Other studies with monkeys have shown that familiarization with novel faces alters the tuning of face-selective neurons in the inferior temporal cortex. Similarly, when an animal has experience with novel objects formed from simple features, inferior temporal neurons become selective for those objects. Such neuronal changes have been found as a consequence of the animal engaging in active discrimination or simply passive viewing of visual stimuli, and they are often manifested as a sharpening of neural selectivity rather than changes in absolute firing rate. Sharpening is precisely the sort of neuronal change that could underlie improvements in perceptual discrimination of visual stimuli.

The Visual System Interacts With Working Memory and Long-Term Memory Systems

Object recognition and learning are intricately linked. In fact, learning can generate entire areas of functional specialization within inferior temporal cortex. For example, monkeys who learn at a young age to associate specific shapes (eg, a number symbol) with particular reward magnitudes develop specialized brain areas that process these specific shapes. These brain regions

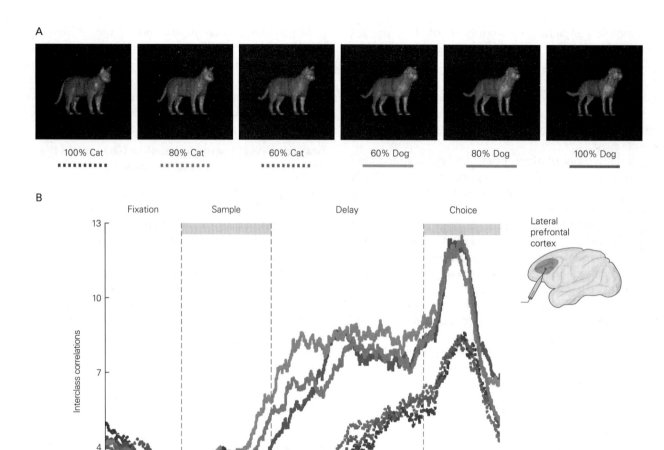

Figure 24–8 Neural coding for categorical perception.
(Reproduced, with permission, from Freedman et al. 2002.)

A. The images combine cat and dog features in varying proportions. Monkeys were trained to categorize an image as cat or dog if it had 50% or more features of that animal.

B. Peristimulus time histograms illustrate the responses of a prefrontal cortex neuron to the images shown in part A.

The neuron responded much more weakly to images of cats (100%, 80%, and 60%) than to images of dogs (60%, 80%, and 100%). Responses to images from the same category were very similar despite variations in retinal images that were as large as or even larger than the differences in retinal images between categories. Thus, the cell was category-specific. Such category-specific responses were common among visual neurons of the lateral prefrontal cortex.

develop close to the temporal lobe face patches discussed earlier.

Two issues concerning interaction between vision and memory have been investigated. First, how is visual information maintained in short-term working memory? Working memory has a limited capacity, acting like a buffer in a computer operating system, and consolidation into long-term memory is susceptible to interference (Chapter 54). Second, how are long-term visual memories and the associations between them stored and recalled?

In a visual delayed-response task requiring access to stimulus information beyond the duration of the stimulus (Box 24–1), many vision-related neurons in both the inferior temporal and prefrontal cortices continue firing during the delay. This delay-period activity is thought to maintain information in short-term working memory (Figure 24–11). Delay-period activity in the inferior temporal and prefrontal cortices differs in a number of ways. For one, activity in the inferior temporal cortex is associated with the short-term storage of visual patterns and color information, whereas

A Familiar object

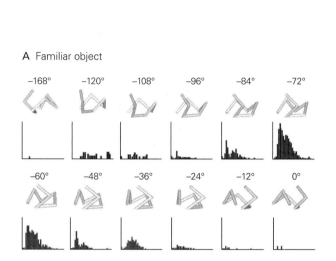

B Unfamiliar object

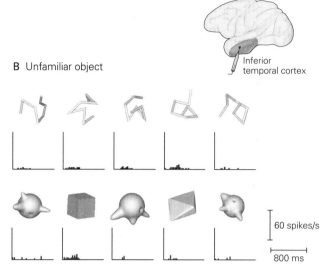

60 spikes/s

800 ms

Figure 24–9 Familiarity with particular complex objects leads inferior temporal neurons to respond selectively for those objects. (Reproduced, with permission, from Logothetis and Pauls 1995. Copyright © 1995 Oxford University Press.)

A. Monkeys were trained to recognize a randomly bent wire from a set of two-dimensional views of the wire. The wire form was rotated 12° in successive views. Once recognition performance was stable at a high level, recordings were made from

neurons in the inferior temporal cortex while each view was presented. Peristimulus time histograms show the responses of a typical neuron to each view. This neuron responded selectively to views that represented a small range of rotation of the object.

B. When the same neuron was tested with two sets of stimuli that were unfamiliar to the monkey, it failed to respond to any of these stimuli.

activity in the prefrontal cortex encodes visuospatial information as well as information received from other sensory modalities. Delay-period activity in the inferior temporal cortex also appears to be closely attuned to visual perception, for it encodes the sample image, but can be eliminated by the appearance of another image.

In the prefrontal cortex, by contrast, delay-period activity depends more on task requirements and is not terminated by intermittent sensory inputs, suggesting that it may play a role in the recall of long-term memories. Experiments by Earl Miller and colleagues support this view. In these experiments, monkeys were trained to associate multiple pairs of objects. They were then tested on whether they had learned these pairwise associations, using the following procedure. First, a single (sample) object was presented; then, after a brief delay, a second (test) object appeared. The monkey was instructed to indicate whether the test object was the object paired with the sample during previous training.

There are two possible ways to solve this task. During the delay, the animal could use a sensory code and keep a representation of the sample object online until the appearance of the test object, or it could remember the sample object's associate and keep information

about the associate object online in a "prospective code" of what might appear as the test object. Remarkably, neuronal activity appears to transition from one to the other during the delay. Neurons in the prefrontal cortex initially encode the sensory properties of the sample object—the one just seen—but later begin to encode the expected (associated) object. As we shall see, such prospective coding in the prefrontal cortex may be the source of top-down signals to the inferior temporal cortex, activating neurons that represent the expected object and thus giving rise to conscious recall of that object.

The relation between long-term declarative memory storage and visual processing has been explored extensively in the context of remembered associations between visual stimuli. Over a century ago, William James, a founder of the American school of experimental psychology, suggested that learning visual associations might be mediated by enhanced connectivity between the neurons encoding individual stimuli. To test this hypothesis, Thomas Albright and colleagues trained monkeys to associate pairs of objects that had no prior physical or semantic relatedness. The monkeys were later tested while extracellular recordings of neurons in the inferior temporal cortex were made. Objects that had been paired often elicited similar neuronal

Box 24–1 Investigating Interactions Between Vision and Working Memory

The relationship between vision and memory can be studied by combining a neuropsychological approach with single-cell electrophysiological methods.

One behavioral paradigm used to study memory is the *delayed-response task*. A subject is required to make a specific response based on information remembered during a brief delay. In one form of this task, known as *delayed match-to-sample*, the subject must indicate whether a visual stimulus is the same or different from a previously viewed cue stimulus (sample) (Figure 24–10A).

When used in conjunction with single-cell recording, this task allows the experimenter to isolate three key components of a neuronal response: (1) the sensory component, the response elicited by the cue stimulus; (2) the short-term or working-memory component, the response that occurs during the delay between the cue

and the match; and (3) the recognition-memory or familiarity component, the difference between the response elicited by the match stimulus and the earlier response to the cue stimulus.

A second behavioral paradigm, the *visual paired-association task*, has been used in conjunction with electrophysiology to explore the cellular mechanisms underlying the long-term storage and recall of associations. This task differs from the delayed match-to-sample task in that the match and cue are two different stimuli (Figure 24–10B).

The sample stimulus might consist of the letter *A* and the match stimulus the letter *B*. Through repeated temporal pairing and conditional reinforcement, subjects learn that *A* and *B* are predictive of one another: They are associated.

A

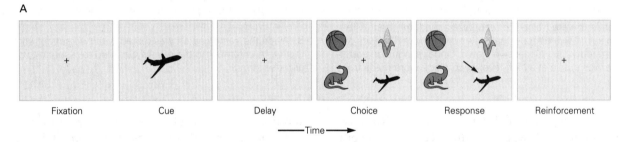

Fixation   Cue   Delay   Choice   Response   Reinforcement

⟶Time⟶

Figure 24–10A Delayed match-to-sample task. In this paradigm, a trial begins with the appearance of a fixation spot that directs the subject's attention and gaze to the center of the computer screen. A cue stimulus (the "sample") then appears briefly, typically for 500 ms, followed by a delay in which the display is blank. The delay can be varied to fit the experimental goals. Following the delay, the choice display appears, which contains several images, one of which is the cue (the "match"). The subject must respond by choosing

the cue stimulus, typically either by pressing a button or by a saccade to the stimulus. In the task illustrated here, all of the test images appear at once (a simultaneous match-to-sample task). They can also be presented sequentially (a sequential match-to-sample task). Although the trial's duration may be longer for the sequential task, this paradigm can be advantageous for electrophysiological studies by limiting the visual stimuli present at any time.

B

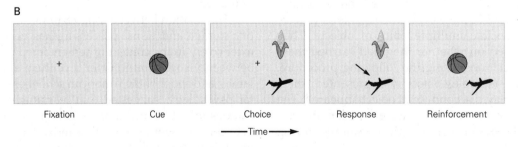

Fixation   Cue   Choice   Response   Reinforcement

⟶Time⟶

Figure 24–10B The paired-association task. This paradigm resembles the match-to-sample paradigm except that the cue and match are different stimuli. In the illustrated example, the basketball is the cue stimulus and the airplane is the experimenter-designated match stimulus. Because these stimuli have no inherent association, the

subject must discover the designated association through trial-and-error learning. The task is thus to establish an association between nonidentical stimuli. The paired-association task can also incorporate a delay between presentation of the sample and test stimuli, and it can be used in both simultaneous (shown) and sequential forms.

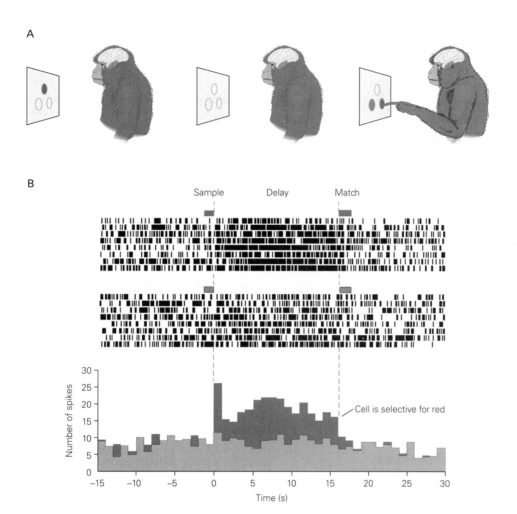

Figure 24–11 Neural activity representing an object is sustained while the object is held in working memory. (Reproduced, with permission, from Fuster and Jervey 1982. Copyright © 1982 Society for Neuroscience.)

A. Monkeys were trained to perform a color match-to-sample task. For example, a red stimulus was first presented and the animal later had to choose a red stimulus from among many colored stimuli. The task incorporated a brief delay (1–2 seconds) between display of the sample and the match, during which information about the correct target color had to be maintained in working memory. The **purple** area in the monkey's brain indicates the inferior temporal cortex.

B. Peristimulus time histograms and raster plots of action potentials illustrate responses of a single neuron in the inferior temporal cortex during the delayed match-to-sample task. The upper record is from trials in which the sample was red, and the lower record is from trials in which it was green (shown here as **blue**). The recordings show that the cell responds preferentially to red stimuli. In trials with a green sample, the activity of the neuron does not change, whereas in trials with a red sample, the cell exhibited a brief burst of activity following presentation of the sample and continued firing throughout the delay. Many visual neurons in the inferior temporal and prefrontal cortices exhibit this kind of behavior.

responses, as one would expect if functional connections had been enhanced, whereas responses elicited by unpaired objects were unrelated. Recordings from individual inferior temporal neurons while monkeys were learning new visual associations showed that a cell's responses to paired objects became more similar over the course of training (Figure 24–12). Most importantly, the changes in neuronal activity occurred on the same timescale as the changes in behavior, and the changes in neural activity depended on successful learning.

These learning-dependent changes in the stimulus selectivity of inferior temporal cortex neurons are long-lasting, suggesting that this cortical region is part of the neural circuitry for associative visual memories. The experimental results also support the view that learned associations are implemented rapidly by changes in the strength of synaptic connections between neurons representing the associated stimuli.

We know that the hippocampus and neocortical areas of the medial temporal lobe—the perirhinal, entorhinal, and parahippocampal cortices—are

A Animals learn to associate pairs of stimuli

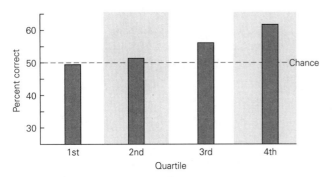

B After training neurons respond similarly to paired stimuli

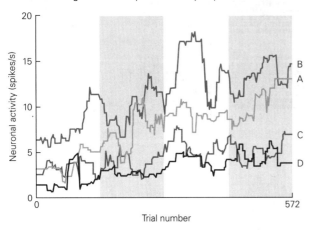

Figure 24–12 Object recognition is linked to associative memory. Monkeys learned associations between pairs of visual stimuli while activity was recorded from a neuron in the inferior temporal cortex. (Reproduced, with permission, from Messinger et al. 2001. © 2001 National Academy of Sciences.)

A. Behavioral performance on a paired-association task is plotted for each quartile of a single training session (572 trials). The animal was presented with four novel stimuli (A, B, C, D) and was required to learn two paired associations (A–B, C–D). As expected, performance began at chance (50% correct) and gradually climbed as the animal learned the associations.

B. Mean firing rates of an inferior temporal neuron recorded during the behavioral task described in part A. Each trace represents the firing rate during presentation of one of the four stimuli (A, B, C, or D). The responses to all stimuli were of similar magnitude at the outset. As the paired associations were learned, the neuronal responses to the paired stimuli A and B began to cluster at a different level from responses to the paired stimuli C and D. The neuron's activity thus corresponded to the learned associations between the two pairs.

essential both for the acquisition of associative visual memories and for the functional plasticity of the inferior temporal cortex. In fact, work by Yasushi Miyashita and colleagues showed that the aforementioned pair-association neurons are much more prevalent in perirhinal cortex than in anterior inferior temporal cortex.

Thus, although learning changes the stimulus selectivity of neurons in both areas, the association between visually associated pairs grows stronger from inferior temporal to perirhinal cortex (Figure 24–2C). The hippocampus and medial temporal lobe may facilitate the reorganization of local neuronal circuitry in the inferior temporal cortex necessary to store associative visual memories. The reorganization itself may be a form of Hebbian plasticity (Chapter 49) initiated by the temporal coincidence of the associated visual stimuli.

Associative Recall of Visual Memories Depends on Top-Down Activation of the Cortical Neurons That Process Visual Stimuli

One of the most intriguing features of high-level visual processing is the fact that the detection of an image in one's visual field and the recall of the same image are subjectively similar. The former depends on the bottom-up flow of visual information and is what we traditionally regard as vision. The latter, by contrast, is a product of top-down information flow. This distinction is anatomically accurate but obscures the fact that under normal conditions afferent and descending signals collaborate to yield visual experience.

The study of associative visual memory has provided valuable insights into the cellular mechanisms underlying visual recall. As we have seen, visual associative memories are stored in the visual cortex through changes in the functional connectivity between neurons that independently represent the associated stimuli. The practical consequence of this change is that a neuron that responded only to stimulus *A* prior to learning will respond to both *A* and *B* after these stimuli have been associated (Figure 24–13). Activation of an *A*-responsive neuron by stimulus *B* can be viewed as the neuronal correlate of top-down recall of stimulus *A*.

Neurons in the inferior temporal cortex exhibit precisely this behavior. The activity correlated with cued recall is nearly identical to the bottom-up activation by the stimulus. These neurophysiological findings are supported by a number of brain imaging studies that have identified selective activity in the visual cortex during cued and spontaneous recall of objects.

Although learned associations between images are likely to be stored through circuit changes in the inferior temporal cortex, activation of these circuits for conscious recall depends on input from the prefrontal cortex. The afferent signal for one of a pair of images might be received by the inferior temporal cortex and relayed to prefrontal cortex, where the information

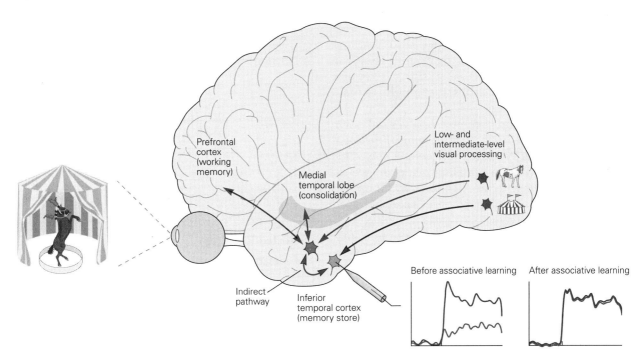

Figure 24–13 Circuits for visual association and recall. Bottom-up signals—afferent signals conveying information about objects in the observer's visual field—are combined into object representations in the inferior temporal cortex. Before associative learning, a neuron (**blue**) responds well to the circus tent but not to the horse. Learned associations between objects are mediated in the inferior temporal cortex by strengthening connections between neurons representing each of the paired objects (the indirect pathway in the figure). Thus, recall of the circus tent following presentation of the horse is achieved by activating the indirect pathway. Indirect activation can also be triggered by the contents of working memory (feedback from the prefrontal cortex). Under normal conditions, visual perception is the product of a combination of direct and indirect inputs to inferior temporal neurons.

would be maintained in working memory. As we have seen, the continued firing of many prefrontal neurons during the delay period of a delayed match-to-sample task initially represents information about the sample image but changes to the associated image that is expected to follow. Signals from prefrontal cortex to the inferior temporal cortex would selectively activate neurons representing the associated image, and that activation would constitute the neural correlate of visual recall.

Highlights

1. A key function of high-level vision is object recognition. Object recognition imbues visual perception with meaning. As the eminent neuropsychologist Hans-Lukas Teuber once wrote, failure of object recognition "would appear in its purest form as a normal percept that has somehow been stripped of its meaning."

2. Object recognition is difficult, primarily because of changes in appearance with changes in position, distance, orientation, or lighting conditions, possibly rendering different objects similar in appearance. Building computer models mimicking primate object recognition capabilities is a major challenge for current and future research.

3. Object recognition relies on a region of the temporal lobe called inferior temporal cortex. Visual information reaching inferior temporal cortex has already been processed through mechanisms of low- and mid-level vision.

4. Lesions to inferior temporal cortex cause visual agnosia, a loss in the ability to recognize objects. Apperceptive agnosia, the inability to match or copy complex objects, is distinguished from associative agnosia, the impairment of the ability to recognize an object's meaning or function. Predicting the exact nature of an agnosia from the pattern of lesioned or inactivated areas, and thus to go from understanding the *correlates* to the *causes* of neural object representations, is a major goal for the field of object recognition and neurology.

5. Individual cells in inferior temporal cortex can be highly shape-selective and respond selectively, eg, to a hand or a face. They can maintain selectivity across position, size, and even rotation—properties that might explain perceptual constancy.

6. Inferotemporal cortex comprises a yet-unknown number of areas with very different functional specializations. While the functional logic of the overall organization remains unknown, we do know that cells with similar selectivity group into cortical columns and that face cells are organized into larger units called face areas.

7. Face recognition is supported by multiple face areas, each with a unique functional specialization. Face areas are selectively coupled to form a face-processing network, which has emerged as a model system for high-level vision.

8. Inferotemporal cortex is interconnected with perirhinal and parahippocampal cortices for memory formation, with the amygdala for the assignment of emotional valence to objects, and with prefrontal cortex for object categorization and visual working memory. If associative memories are stored as patterns of connections between neurons, what then are the specific contributions of hippocampus and neocortical structures of the medial temporal lobe, and by what cellular mechanisms do they exert their influences? The confluence of molecular-genetic, cellular, neurophysiological, and behavioral approaches promises to solve these and other problems.

9. Objects are perceived as members of a category. This simplifies the selection of appropriate behaviors, which often do not depend on stimulus details. Neurons with categorical selectivity are found in dorsolateral prefrontal cortex, a main projection site of inferior temporal cortex.

10. Object recognition relies on past experience. Perceptual learning can improve the ability to discriminate between complex objects and refine neural selectivity in inferior temporal cortex.

11. Visual information can be held in short-term working memory to be available beyond the duration of a sensory stimulus. Neurons in temporal and prefrontal cortex can exhibit delay-period activity after the disappearance of a stimulus. How these networks establish the ability to keep information online is an open question.

12. High-level visual information processing changes with top-down modulation. The sensory experience of an image in view and the recall of the same stimulus from memory are subjectively similar. Neurons in inferior temporal cortex exhibit similar activity during bottom-up activation and cued recall.

Thomas D. Albright
Winrich A. Freiwald

Selected Reading

Freedman DJ, Miller EK. 2008. Neural mechanisms of visual categorization: insights from neurophysiology. Neurosci Biobehav Rev 32:311–329.

Gross CG. 1999. *Brain, Vision, Memory: Tales in the History of Neuroscience.* Cambridge, MA: MIT Press.

Kanwisher N, McDermott J, Chun MM. 1997. The fusiform face area: a module in human extrastriate cortex specialized for face perception. J Neurosci 17:4302–4311.

Logothetis NK, Sheinberg DL. 1996. Visual object recognition. Annu Rev Neurosci 19:577–621.

McCarthy G, Puce A, Gore J, Allison T. 1997. Face-specific processing in the human fusiform gyrus. J Cog Neurosci 9:605–610.

Messinger A, Squire LR, Zola SM, Albright TD. 2005. Neural correlates of knowledge: stable representation of stimulus associations across variations in behavioral performance. Neuron 48:359–371.

Miller EK, Li L, Desimone R. 1991. A neural mechanism for working and recognition memory in inferior temporal cortex. Science 254:1377–1379.

Miyashita Y. 1993. Inferior temporal cortex: where visual perception meets memory. Annu Rev Neurosci 16:245–263.

Schlack A, Albright TD. 2007. Remembering visual motion: neural correlates of associative plasticity and motion recall in cortical area MT. Neuron 53:881–890.

Squire LR, Zola-Morgan S. 1991. The medial temporal lobe memory system. Science 253:1380–1386.

Ungerleider LG, Courtney SM, Haxby JV. 1998. A neural system for human visual working memory. Proc Natl Acad Sci U S A 95:883–890.

References

Baker CI, Behrmann M, Olson CR. 2002. Impact of learning on representation of parts and wholes in monkey inferotemporal cortex. Nat Neurosci 5:1210–1216.

Brown S, Schafer ES. 1888. An investigation into the functions of the occipital and temporal lobes of the monkey's brain. Philos Trans R Soc Lond B Biol Sci 179:303–327.

Damasio AR, Damasio H, Van Hoesen GW. 1982. Prosopagnosia: anatomic basis and behavioral mechanisms. Neurology 32:331–341.

Desimone R, Albright TD, Gross CG, Bruce CJ. 1984. Stimulus selective properties of inferior temporal neurons in the macaque. J Neurosci 8:2051–2062.

Desimone R, Fleming J, Gross CG. 1980. Prestriate afferents to inferior temporal cortex: an HRP study. Brain Res 184:41–55.

Farah MJ. 1990. *Visual Agnosia: Disorders of Object Recognition and What They Tell Us About Normal Vision.* Cambridge, MA: MIT Press.

Felleman DJ, Van Essen DC. 1991. Distributed hierarchical processing in the primate cerebral cortex. Cereb Cortex 1:1–47.

Freedman DJ, Riesenhuber M, Poggio T, Miller EK. 2002. Visual categorization and the primate prefrontal cortex: neurophysiology and behavior. J Neurophysiol 88:929–941.

Freiwald WA, Tsao DY. 2010. Functional compartmentalization and viewpoint generalization within the macaque face-processing system. Science 330:845–851.

Fujita I, Tanaka K, Ito M, Cheng K. 1992. Columns for visual features of objects in monkey inferotemporal cortex. Nature 360:343–346.

Fuster JM, Jervey JP. 1982. Neuronal firing in the inferotemporal cortex of the monkey in a visual memory task. J Neurosci 2:361–375.

Gross CG, Bender DB, Rocha-Miranda CE. 1969. Visual receptive fields of neurons in inferotemporal cortex of the monkey. Science 166:1303–1306.

Kosslyn SM. 1994. *Image and Brain.* Cambridge, MA: MIT Press.

Leibo JZ, Liao Q, Anselmi F, Freiwald WA, Poggio T. 2017. View-tolerant face recognition and Hebbian learning imply mirror-symmetric neural tuning to head orientation. Curr Biol 27:62–67.

Logothetis NK, Pauls J. 1995. Psychophysical and physiological evidence for viewer-centered object representations in the primate. Cereb Cortex 5:270–288.

Messinger A, Squire LR, Zola SM, Albright TD. 2001. Neuronal representations of stimulus associations develop in the temporal lobe during learning. Proc Natl Acad Sci U S A 98:12239–12244.

Miyashita Y, Chang HS. 1988. Neuronal correlate of pictorial short-term memory in the primate temporal cortex. Nature 331:68–70.

Rainer G, Rao SC, Miller EK. 1999. Prospective coding for objects in primate prefrontal cortex. J Neurosci 19:5493–5505.

Rollenhagen JE, Olson CR. 2000. Mirror-image confusion in single neurons of the macaque inferotemporal cortex. Science 287:1506–1508.

Sakai K, Miyashita Y. 1991. Neural organization for the long-term memory of paired associates. Nature 354:152–155.

Schwartz EL, Desimone R, Albright TD, Gross CG. 1983. Shape recognition and inferior temporal neurons. Proc Natl Acad Sci U S A 80:5776–5778.

Suzuki WA, Amaral DG. 2004. Functional neuroanatomy of the medial temporal lobe memory system. Cortex 40:220–222.

Tanaka K. 2003. Columns for complex visual object features in the inferotemporal cortex: clustering of cells with similar but slightly different stimulus selectivities. Cereb Cortex 13:90–99.

Teuber HL. 1968. Disorders of memory following penetrating missile wounds of the brain. Neurology 18:287–288.

Tomita H, Ohbayashi M, Nakahara K, Hasegawa I, Miyashita Y. 1999. Top-down signal from prefrontal cortex in executive control of memory retrieval. Nature 401:699–703.

Tsao DY, Freiwald WA, Tootell RB, Livingstone MS. 2006. A cortical region consisting entirely of face-selective cells. Science 311:670–674.

Wheeler ME, Petersen SE, Buckner RL. 2000. Memory's echo: vivid remembering reactivates sensory-specific cortex. Proc Natl Acad Sci U S A 97:11125–11129.

25

Visual Processing for Attention and Action

The Brain Compensates for Eye Movements to Create a Stable Representation of the Visual World

    Motor Commands for Saccades Are Copied to the Visual System

    Oculomotor Proprioception Can Contribute to Spatially Accurate Perception and Behavior

Visual Scrutiny Is Driven by Attention and Arousal Circuits

The Parietal Cortex Provides Visual Information to the Motor System

Highlights

Tʜᴇ ʜᴜᴍᴀɴ ʙʀᴀɪɴ ʜᴀs ᴀɴ ᴀᴍᴀᴢɪɴɢ ability to direct action to objects in the visual world—a baby reaching for an object, a tennis player hitting a ball, an artist looking at a model. This ability requires that the visual system solve three problems: making a spatially accurate analysis of the visual world, choosing the object of interest from the welter of stimuli in the visual world, and transferring information on the location and details of the object to the motor system.

The Brain Compensates for Eye Movements to Create a Stable Representation of the Visual World

Although the visual system produces vivid representations of our visual world, as described in preceding chapters, a visual image is not like an instantaneous photographic record but is dynamically constructed from information conveyed in several discrete neural pathways from the eyes. When we look at a painting, for example, we explore it with a series of quick eye movements (saccades) that redirect the fovea to different objects of interest in the visual field. The brain must take into account these eye movements in the course of producing an interpretable visual image from the light stimuli in the retina.

As each saccade brings a new object onto the fovea, the image of the entire visual world shifts on the fovea. These shifts occur several times per second, such that after several minutes the record of movement is a jumble (Figure 25–1). With such constant movement, visual images should resemble an amateur video in which the image jerks around because the camera operator is not skilled at holding the camera steady. In fact, however, our vision is so stable that we are ordinarily unaware of the visual effects of saccades. This is so because the brain makes continual adjustments to the images falling on the retina after each saccade.

A simple laboratory experiment, shown in Figure 25–2, illustrates the biological challenge to the brain.

Motor Commands for Saccades Are Copied to the Visual System

The first insight into the brain mechanisms underlying visual stability came from an observation by Hermann von Helmholtz in the 19th century. He saw a patient who could not move his eye horizontally toward his ear because of a paralysis of the lateral rectus muscle. Whenever the patient attempted to look toward his ear, the entire visual world jumped in the opposite direction and then returned to the center of gaze.

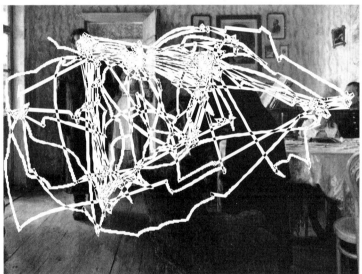

Figure 25–1 Eye movements during vision. A subject viewed this painting (*An Unexpected Visitor* by Ilya Repin) for several minutes, making saccades to selected fixation points, primarily to faces. Lines indicate saccades, and spots indicate points of fixation. (Reproduced, with permission, from Yarbus 1967).

Helmholtz postulated that a copy of the motor command for each saccade was fed to the visual system so that the representation of the visual world could be adjusted to compensate for eye movement. This adjustment would lead to a stable image of the visual world. In the 19th century, Helmholtz called such a copy a "sense of effort," and in the 20th century, it was named an efference copy or corollary discharge.

The corollary discharge solves the problem of the double-step saccade. In order for a corollary discharge to affect visual perception across eye movements, motor information has to affect the activity of visual neurons.

This is precisely what happens to neurons in the parietal cortex, frontal eye field, prestriate visual cortex, and superior colliculus when a monkey makes a saccade. Each saccade can be considered a vector with two dimensions—direction and amplitude. Although the retinal image is different after each saccade, the brain can use the vector of each saccade to reconstruct the whole visual scene from the sequence of retinal images.

The corollary discharge can be seen at the level of a single cell. Physiological studies in the Rhesus monkey, an animal whose oculomotor and visual systems resemble those of humans, have illuminated the

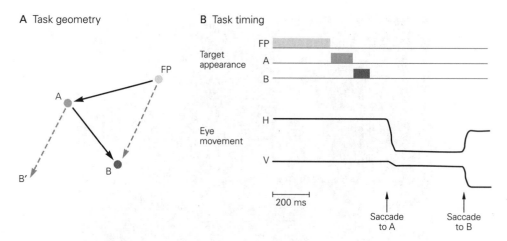

A Task geometry

B Task timing

Figure 25–2 The double-step task illustrates how the brain stabilizes images during saccades.

A. A subject starts by looking at a fixation point (**FP**) that disappears, after which two saccade targets **A** and **B** appear and disappear sequentially before the subject can make the saccade. The first saccade (to target A) is simple. The retinal vector (FP→A) and the saccade vectors are the same. After the first saccade, the subject is looking at A. The retinal vector is A→B', but the monkey must make a saccade whose vector is A→B. The brain must adjust the retinal vector to compensate for the first saccade.

B. Timing. The upper records show when the targets appear (**colored bars**). (Abbreviations: **H**, horizontal; **V**, vertical.)

problem. Every time a monkey makes a saccade, a stimulus currently not in the receptive field of a neuron in the lateral intraparietal area, and therefore incapable of exciting the neuron, will excite the neuron if the impending saccade will bring the stimulus into the receptive field, even before the saccade occurs (Figure 25–3). Thus, a corollary discharge of the impending saccade affects the visual responsiveness of the parietal neuron.

This transient remapping of the receptive field explains how subjects can perform the double-step task. Consider the diagram in Figure 25–2A. The task begins with the monkey directing gaze to the fixation point (FP). After the monkey makes the first saccade, the retinal vector A→B' is no longer useful for making the A→B saccade. However, the FP→A saccade remaps the activity of the cell describing the vector A→B, so it responds to the target at the retinal location of B, which was not in its receptive field when the monkey was looking at FP. Remapping is found in a number of cortical and subcortical areas, including lateral intraparietal area, frontal eye field, medial intraparietal area, intermediate layers of the superior colliculus, and prestriate areas V4, V3a, and V2. As we shall see, remapping facilitates both visual perception around the time of a saccade and the accuracy of visually guided movement.

The first question this raises is: How does the brain obtain the vector of the saccade that it feeds back to the visual system? We know from decades of research that the motor command for the vector is represented in the superior colliculus on the roof of the midbrain (Chapter 35). Each neuron in the superior colliculus

is tuned to saccades of a given vector, such that the neurons collectively provide a map of the vectors of all possible saccades. Inactivation of the superior colliculus affects the monkey's ability to make saccades. Electrical stimulation of the superior colliculus evokes saccades of the vector described by the neurons at the stimulation site. But this provides the vectors that actually drive the eye, not the vectors that inform perception about the vector of the saccade. How does the vector information used to move the eye become available to brain processes that do not move the eye but do require information about how it moved?

Since the vectors for moving the eye have been identified in the superior colliculus, it is reasonable to expect that this also might be the source of a corollary discharge. Indeed, it is. The superior colliculus has both descending pathways for generating the saccades and ascending pathways to the cerebral cortex that could carry the corollary discharge of the impending movement (Figure 25–4). The pathways to the cortex pass through the thalamus, as does all internal and almost all external information reaching the cerebral cortex.

The motor signal in the thalamus is not necessarily a corollary discharge; it could also be a movement command that simply passes through the cerebral cortex. That is not the case, however, because inactivation of this pathway in the thalamus does not alter the amplitude and direction of saccades. It is not driving saccades. It is more likely to be a corollary discharge. After inactivation of the thalamic pathway, monkeys cannot accurately perform the second saccade of the double-step task. In addition, inactivation disrupts the

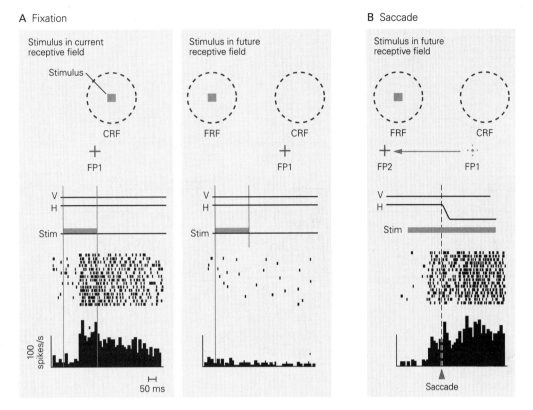

Figure 25–3 Remapping of the receptive field of a visual neuron in the parietal cortex of a monkey in conjunction with saccadic eye movements. (**A** [*left*] and **B** adapted, with permission, from Duhamel, Colby, and Goldberg 1992. **A** [*right*] reproduced with permission, from M.E. Goldberg.)

A. *Left:* The monkey looks at fixation point 1 (**FP1**), and the cell responds to the abrupt onset of a task-irrelevant stimulus in the current receptive field (**CRF**). Successive trials are synchronized on the appearance of the stimulus. (Abbreviations: **H**, horizontal

eye position; **V**, vertical eye position.) *Right:* The monkey looks at FP1, and the cell does not respond to a stimulus flashed in the future receptive field (**FRF**).

B. The monkey makes a saccade from FP1 to FP2, which will bring the cell's receptive field onto the stimulus in the FRF. Now the cell fires even before the saccade begins, which means that a corollary discharge of the saccade plan remapped the area of the retina to which the cell responds.

receptive field remapping described earlier (Figure 25–3B). Because disrupting the corollary discharge disrupts both receptive field remapping and the behavioral compensation for eye movements, it is likely that the corollary discharge is essential for solving the problem of spatial accuracy for action.

To determine whether the corollary discharge also provides the information that allows the visual system to perceive the location of objects that appeared before a saccade, the monkey is trained to indicate where it thinks its eyes are directed at the end of the saccade. We can measure where the motor system moved the eye, but what we want to know is the monkey's perception of the change in its eye direction with each saccade. This can be determined using a task developed for humans by Heiner Deubel and his colleagues and adapted for monkeys. In this task, the monkey looks at a fixation point and then makes a saccade to a target (Figure 25–5A). During the saccade, the target

temporarily disappears; when it reappears, it has been displaced to a location left or right of the original target. After the trial, the monkey moves a bar to the right or left to indicate the direction of the displacement (Figure 25–5A).

Over a series of trials, the monkey's responses are plotted to generate a psychometric curve (Figure 25–5B). This curve show the actual intrasaccadic target displacement (horizontal axis) in the same (forward) or opposite (backward) direction as the initial saccade, and how frequently the monkey reports that it was moved forward (vertical axis). The monkey responded that the target had moved forward 100% of the time when the target was 3° to the right. When the target moved 3° to the left, the monkey responded that it had never moved forward. The point on the psychometric curve where the monkey reported forward and backward displacements with equal frequency (the 50% horizontal line) was taken as the perceptual null point.

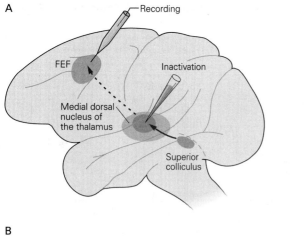

A

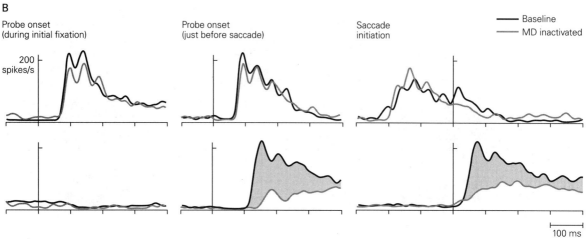

B

Probe onset
(during initial fixation)

Probe onset
(just before saccade)

Saccade
initiation

—— Baseline
—— MD inactivated

200
spikes/s

100 ms

Figure 25–4 A corollary discharge from the motor program for saccades directs a shift in location of the receptive field of frontal eye field neurons prior to the saccade. (Adapted, with permission, from Sommer and Wurtz 2008. Copyright © 2008 by Annual Reviews.)

A. One possible pathway for the corollary discharge originates in saccade-generating neurons in the superior colliculus, passes through the medial dorsal nucleus of the thalamus, and terminates in the frontal eye field (**FEF**) in the frontal cortex.

B. When the medial dorsal nucleus (**MD**) is inactivated, the response of a frontal eye field neuron to a stimulus probe in the cell's current receptive field is unaffected (**upper records**), whereas the response to a stimulus in a future (post-saccade) receptive field is severely impaired (**lower records**). This result demonstrates that a corollary discharge from the saccade motor program directs the shift in the neuron's receptive field properties.

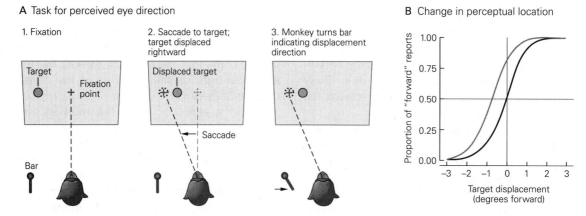

A Task for perceived eye direction

1. Fixation

Target

Fixation
point

Bar

2. Saccade to target;
target displaced
rightward

Displaced target

Saccade

3. Monkey turns bar
indicating displacement
direction

B Change in perceptual location

Figure 25–5 Perceived saccade direction changes with disruption of corollary discharge.

A. At the start of each trial, the monkey fixates a target on a screen (**1**). When the fixation point is turned off, the monkey makes a saccade to the target; during the saccade, the target is displaced randomly (up to 3°) either to the left or to the right (**2**). After the saccade to the original target, the monkey receives a reward for manually moving a bar in the direction of the target displacement (**3**).

B. Psychometric curves before (**black**) and after (**purple**) inactivation of the medial dorsal nucleus of the thalamus, which contains the relay neurons for the corollary discharge in its pathway between the superior colliculus and the frontal cortex. The curve shows the proportion of forward (in the direction of the saccade) judgments (y-axis) for each target displacement (x-axis). The post-saccadic target location at which the monkey perceived no displacement is defined as the perceptual null location. (Adapted, with permission, from Cavanaugh et al. 2016.)

We take this point to be the monkey's perception of the original target location. If the target were not perceived to move, it must be in the same location as before the saccade; in a normal monkey, that point is close to zero (Figure 25–5B).

We now have a corollary discharge that can provide the vector for each saccade and a task for a monkey that allows us to determine where it perceives the target to be at the end of the saccade. If the corollary discharge contributes to the monkey's perception, then inactivating the corollary discharge should change the animal's perception of target location. It does. The purple curve in Figure 25–5B represents the perceived location after corollary discharge inactivation; the curve shifts to the left after inactivation of the medial dorsal nucleus of the thalamus. The conclusion is that the corollary discharge does provide the vector of the saccade, which is necessary for the monkey to perceive that the target had moved. With each saccade, corollary discharge information provides perceptual information for determining the amplitude and direction of the current saccade, and it does so with machine-like precision several times per second.

The corollary discharge provides the vector information available before the saccade is made, but it is not the only source of information. Two other types of information must be evaluated after the saccade has taken place: visual cues and eye muscle proprioception. Visual cues are unlikely to be a factor in the perceptual experiment described (Figure 25–5) because the experiment was done in total darkness except for light scattered from the very dim fixation point and saccade target. In the light, however, could visual cues be a factor? In fact, repeating the experiment in the light did not improve the monkey's judgment and frequently made it worse.

Oculomotor proprioception is unlikely to provide the vector information at the end of the saccade because, on average, the metrics of the saccades before and during inactivation do not change, so there is little reason to expect that the muscle proprioception will have changed. In addition, while the corollary discharge begins at least 100 ms before the saccade, neuronal activity from oculomotor proprioception reaches the lateral intraparietal area about 150 ms after the saccade. As we will see in the next section, the role of proprioception in perception might be to provide information long after the saccade ends.

Finally, there is a second potential disruption of vision produced by saccades: a blur as the saccade sweeps the visual scene across the retina. The blur is not seen, however, because neuronal activity in a number of visual areas is suppressed around the time of every saccade. This so-called saccadic suppression was first seen in the superior colliculus and has subsequently been seen in the thalamus and areas of visual cortex beyond primary visual cortex.

A corollary discharge contributes to this neuronal activity suppression because the suppression occurs even in total darkness (no vision) and even if eye movement is blocked (no proprioception). Suppression can also be produced by visual masking, which occurs when one stimulus reduces the perception of a following or preceding stimulus. If a saccade starts in total darkness, and an object is then flashed and extinguished before the saccade ends, a blur can be seen during the saccade. If a mask is flashed after the saccade, the blur is suppressed. A correlate of such a masking effect is clearly seen in neurons in primary visual cortex. The suppression resulting from a corollary discharge is relatively weak but is present with all saccades; that from visual masking is much stronger but is present only in the light.

Oculomotor Proprioception Can Contribute to Spatially Accurate Perception and Behavior

Charles Sherrington suggested that the way the brain compensates for a moving eye is to measure directly where the eyes are in the orbit and adjust the visual signal for changes in position. Richard Andersen and Vernon Mountcastle discovered that the responses of parietal visual neurons with retinotopic receptive fields are modulated by the position of the eye in the orbit in a linear fashion called the *gain field* (Figure 25–6). From this relationship, the position of an object in head-centered (craniotopic) coordinates can easily be calculated.

Where does the eye position signal that creates the gain fields come from? It could come from a corollary discharge of eye position, or it could come from a proprioceptive mechanism. Human eye muscles have two structures that could contribute to oculomotor proprioception: muscle spindles and myotendinous cylinders, or palisade endings, an eye-specific structure. Area 3a, the region of somatosensory cortex to which skeletal muscle spindles project, has a representation of the position of the eye, which arises from proprioceptors in the contralateral orbit (Figure 25–7).

However, the proprioceptive measurement of eye position lags changes in eye position by 60 ms, and for 150 ms after a saccade, the gain fields modulate the visual response as if the monkey were still looking at the presaccadic target, long after the corollary discharge has remapped the visual response. Therefore, the eye position signal creating the gain fields probably arises from a proprioceptive mechanism. The possibility

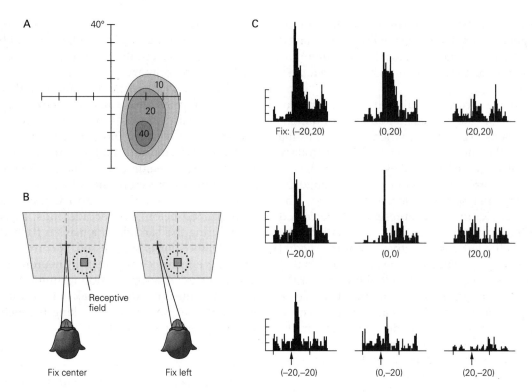

Figure 25–6 The position of the eye in the orbit affects the responses of parietal visual neurons with retinotopic receptive fields.

A. Receptive field relative to the fovea. Contour plot indicates spike rates for different spatial locations. Numbers are spikes per second for each contour at the maximum position.

B. The receptive field moves in space with the eye. On the left the monkey is fixating the center of the screen. On the right the same monkey is fixating 20° to the left of center. For the

recordings in C, the stimulus (**blue square**) is always presented in the center of the receptive field.

C. Responses to a stimulus at the optimum location in the receptive field change as a function of the position of the eye in the orbit, from a maximum when the monkey fixates a point at –20°,20° to a minimum when the monkey fixates a point at 20°,–20°. **Arrows** indicate onset of stimulus flash. Trial duration, 1.5 sec; ordinate, 25 spikes/division. (Adapted, with permission, from Andersen, Essick, and Siegel 1985. Copyright © 1985 AAAS.)

exists that the brain calculates the spatial location of an object that appeared before an eye movement using two mechanisms: a corollary discharge that is rapid and a proprioceptive signal that is slow but can be more accurate than the corollary discharge. The proprioceptive signal can also be used to calibrate the corollary discharge.

Visual Scrutiny Is Driven by Attention and Arousal Circuits

In the 19th century, William James described attention as "the taking possession by the mind in clear and vivid form, of one out of what seem several simultaneously possible objects or trains of thought. It implies withdrawal from some things in order to deal effectively with others." James went on to describe two different kinds of attention: "It is either passive, reflex, non-voluntary, effortless or active and voluntary. In

passive immediate sensorial attention the stimulus is a sense-impression, either very intense, voluminous, or sudden … big things, bright things, moving things … blood."

Your attention to this page as you read it is an example of voluntary attention. If a bright light suddenly flashed, your attention would probably be pulled away involuntarily from the page. Large changes in the visual scene that occur outside the focus of attention are often missed until the subject directs attention to them, a phenomenon referred to as change blindness (Figure 25–8).

Voluntary attention is closely linked to saccadic eye movements because the fovea has a much denser array of cones than the peripheral retina (Chapter 17) and moving the fovea to an attended object permits a finer-grain analysis than is possible with peripheral vision. Attention that selects a point in space, whether or not it is accompanied by a saccade, is called spatial attention. Searching for a specific kind of object, for

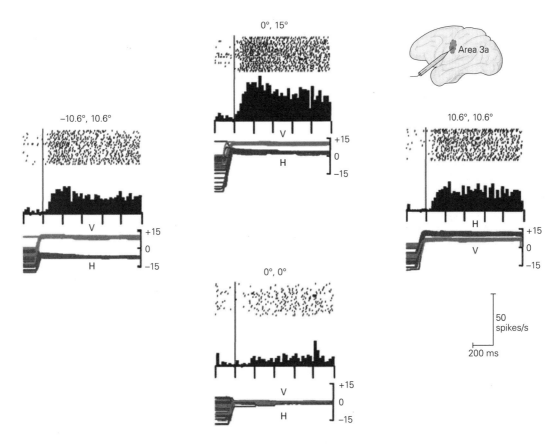

Figure 25–7 Eye position neuron in somatosensory cortex area 3a. Each panel shows horizontal (**H**) and vertical (**V**) eye position and the activity of the neuron after the monkey made a saccade to the eye position indicated above each raster. The neuron responds much more briskly when the eye is at 0°,15° than when it is at 0°,0°. (Reproduced, with permission, from Wang et al. 2007.)

example a red O among red and green Qs, involves a second kind of attention, feature attention: In your search, you ignore the green letters and attend only to the red ones.

Attention, both voluntary and involuntary, shortens reaction time and makes visual perception more sensitive. This increased sensitivity includes the abilities to detect objects at a lower contrast and ignore distracters close to an attended object. The abrupt appearance of a behaviorally irrelevant cue, such as a light flash, reduces the reaction time to a test stimulus presented 300 ms later in the same place. Conversely, when the cue appears away from the test stimulus, the reaction time is increased. The light flash draws involuntary attention to its location, thus accelerating the visual response to the test stimulus. Similarly, when a subject plans a saccade to a particular part of the visual field, the contrast threshold at which any object there can be detected is improved 50% by a cue.

Clinical studies have long implicated the parietal lobe in visual attention. Patients with lesions of the right parietal lobe have normal visual fields. When their visual perception is studied with a single stimulus in an uncomplicated visual environment, their responses are normal. However, when presented with a more complicated visual environment, with objects in the left and right visual hemifields, these patients tend to report less of what lies in the left hemifield (contralateral to their lesion) than in the right hemifield (ipsilateral to their lesion). This deficit, known as *neglect* (Chapter 59), arises because attention is focused on the visual hemifield ipsilateral to the lesion. Even when patients are presented with only two stimuli, one in each hemifield, they report seeing only the stimulus in the ipsilateral hemifield. When attention is focused on one stimulus in the affected hemifield and a second stimulus is presented in the unaffected hemifield, patients do not have the ability to shift attention to the new stimulus, even though the sensory pathway from the eye to the striate and prestriate cortex is intact.

This neglect of the contralateral visual hemifield extends to the neglect of the contralateral half of

Blank (80 ms)

Blank (80 ms)

Blank (80 ms)

Figure 25–8 Change blindness. In a test for change blindness, one picture is presented followed by a blank screen for 80 ms, followed by the second picture, another blank screen, and a repeat of the cycle (*left*). The subject is asked to report what changed in the scene. Although there is a substantial difference between the two pictures, it takes multiple repetitions for most observers to detect the difference. (Reproduced, with permission, from Ronald Rensink.)

individual objects (Figure 25–9). Patients with right parietal lobe deficits often have difficulty reproducing drawings. When asked to draw a clock, for example, they may force all of the numbers into the right side of the clock's face, or when asked to bisect a line, they may place the midline well to the right of the line's actual center.

The process of attentional selection is evident at the level of single parietal neurons in the monkey. The responses of neurons in the lateral intraparietal area to a visual stimulus depend not only on the physical properties of the stimulus but also on its importance

Figure 25–9 Drawing of a candlestick by a patient with a lesion of the right parietal lobe. The patient neglects the left side of the candlestick, drawing only its right half. (Reproduced, with permission, from Halligan and Marshall 2001. Copyright © 2001 Academic Press.)

to the monkey. Thus, the responses to a behaviorally irrelevant stimulus are much smaller than for any event that evokes attention, such as the abrupt onset of a visual stimulus in the receptive field or the planning of a saccade to the receptive field of the neuron.

Although neurons in the lateral intraparietal area collectively represent the entire visual hemifield, the neurons active at any one moment represent only the important objects in the hemifield, a priority map of the visual field. The lateral intraparietal area acts as a summing junction for a number of different signals: saccade planning, abrupt stimulus onset, and the cognitive aspects of a searched-for feature.

The absolute value of the neuronal response evoked by an object does not by itself determine whether that animal is attending to that object. When a monkey plans a saccade to a stimulus in the visual field, attention is on the goal of the saccade, and the activity evoked by the saccade plan lies at the peak of the priority map. However, if a bright light appears elsewhere in the visual field, attention is involuntarily drawn to the bright light, which evokes more neuronal activity than does the saccade plan. Thus, the locus of attention can be identified only by examining the entire priority map and choosing its peak; it cannot be identified by monitoring activity at any one point alone (Box 25–1).

Box 25–1 The Priority Map in Parietal Cortex

Neurons in the lateral intraparietal area of the monkey represent only those objects of potential importance to the monkey, a priority map of the visual field. This selectivity for objects of behavioral importance can be demonstrated by recording from neurons in a monkey while the animal makes eye movements across a stable array of objects.

Stable objects in the visual world are rarely the objects of attention. In the lateral intraparietal area, as in most other visual centers of the brain, neuronal receptive fields are retinotopic; that is, they are defined relative to the center of gaze. As a monkey scans the visual field, fixed objects enter and leave the receptive fields of neurons with every eye movement without disrupting the monkey's attention (Figure 25–10).

The abrupt appearance of a visual stimulus involuntarily evokes attention. When a task-irrelevant light flashes in the receptive field of a lateral intraparietal

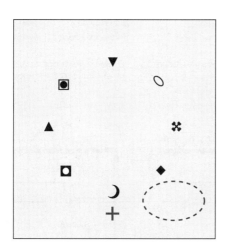

Figure 25–10 Exploring a stable array of objects. The monkey views a screen with a number of objects that remain in place throughout the experiment. The monkey's gaze can be positioned so that none of the objects are included in the receptive field of a neuron (*left*), or the monkey can make a saccade that brings one of the objects into the receptive field (*right*). (Reproduced, with permission, from Kusunoki, Gottlieb, and Goldberg 2000.)

(continued)

The Parietal Cortex Provides Visual Information to the Motor System

Vision interacts with the supplementary and premotor cortices to prepare the motor system for action. For example, when you pick up a pencil, your fingers are separated from your thumb by the width of the pencil; when you pick up a drink, your fingers are separated from your thumb by the width of the glass. The visual system helps to adjust the grip width before your hand arrives at the object. Similarly, when you insert a letter into a mail slot, your hand is aligned to place the letter in the slot. If the slot is tilted, your hand tilts to match.

Patients with lesions of the parietal cortex cannot adjust their grip width or wrist angle using visual information alone, even though they can verbally describe the size of the object or the orientation of the slot. Conversely, patients with intact parietal lobes and deficits in the ventral stream cannot describe the size of an object or its orientation but can adjust their grip width and orient their hands as well as normal subjects can. Neurons in parietal cortex are a critical source of information needed to manipulate or move objects.

Box 25–1 The Priority Map in Parietal Cortex (continued)

neuron, that cell responds briskly (Figure 25–11A). In contrast, a stable, task-irrelevant stimulus evokes little response when eye movement brings it into the neuron's receptive field (Figure 25–11B).

It is possible that the saccade that brings the stable object into the receptive field suppresses the visual response. This is not the case. A second experiment uses a similar array, except there is no stimulus at the location to which the saccade had brought the receptive field in

the stable array experiment. The monkey fixates so that no member of the array is in the receptive field, and then the task-irrelevant stimulus suddenly appears at the post-saccade location of the receptive field. Now the monkey makes a saccade to the center of the array, bringing the recently appeared stimulus into the receptive field, and the cell fires intensely (Figure 25–11C). When the monkey makes the saccade, the two arrays are identical. However, the stable stimulus is presumably unattended,

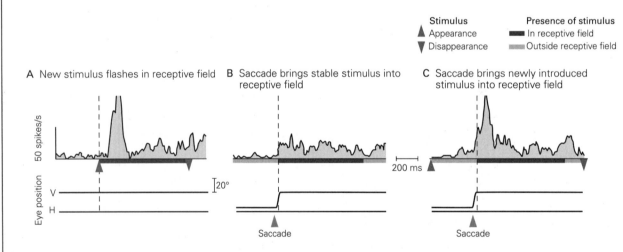

Figure 25–11 A neuron in the lateral intraparietal area fires only in response to salient stimuli. In each panel, neuronal activity and eye positions are plotted across time.

A. A stimulus flashes in the receptive field while the monkey fixates.

B. The monkey makes a saccade that brings a stable, task-irrelevant stimulus into the receptive field.

C. The monkey makes a saccade that brings the location of the recent light flash into the receptive field.

The neural operations behind visually guided movements involve identifying targets, specifying their qualities, and ultimately generating a motor program to accomplish the movement. Neurons in the parietal cortex provide the visual information necessary for independent movement of the fingers.

The representation of space in the parietal cortex is not organized into a single map like the retinotopic map in primary visual cortex. Instead, it is divided into at least four areas (LIP, MIP, VIP, AIP) that analyze the visual world in ways appropriate for individual motor systems. These four areas

project visual information to the areas of premotor and frontal cortex that control individual voluntary movements (Figure 25–13).

Neurons in the medial intraparietal area describe the targets for reaching movements and project to the premotor area that controls reaching movements. The anterior intraparietal cortex has neurons that signal the size, depth, and orientation of objects that can be grasped. Neurons in this area respond to stimuli that could be the targets for a grasping movement, and these neurons are also active when the animal makes the movement (Figure 25–14). Neurons in the lateral

---

whereas the recently flashed stimulus evokes attention and a much larger response. Stable objects can evoke enhanced responses when they become relevant to the animal's current behavior.

A stable object can also be made behaviorally important. In that case, the neurons increase their firing rate when the monkey has to attend to the stable object brought into the receptive field by the saccade (Figure 25–12).

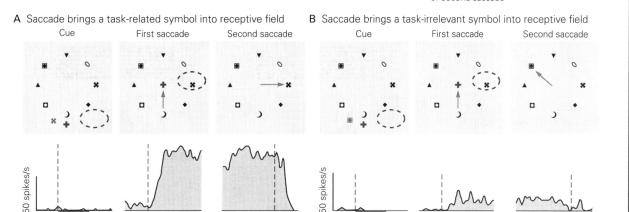

Figure 25–12 A neuron in the lateral intraparietal area fires before a saccade to a significant stable object. On each trial, one object in a stable array becomes significant to the monkey because the monkey must make a saccade to it. The monkey fixates a point outside the array, and a cue that matches an object in the array appears outside the neuron's receptive field. The monkey must then make a saccade to the center of the array and a second saccade to the object that matches the cue. Two experiments are shown (in parts **A** and **B**). The *left* panel shows the neuron's response to the appearance of the cue outside the receptive field, the *center* panel shows the response after the first saccade brings the cued object into the receptive

field, and the *right* panel shows the response just before the second saccade to the cued object. The cues are shown here in **green** for clarity but were black in the experiment. The visual scene at the time of the saccade is identical in both experiments.

A. The monkey is trained to make the second saccade to the cued object; the cell fires intensely when the first saccade brings the object into the receptive field.

B. The monkey is trained to make the second saccade to an object outside the receptive field; the cell fires much less when the saccade brings the task-irrelevant stimulus into the receptive field.

Figure 25–13 Pathways involved in visual processing for action. The dorsal visual pathway (**blue**) extends to the posterior parietal cortex and then to the frontal cortex. The ventral visual pathway (**pink**) is considered in Chapter 24. There are bidirectional projections from the inferior temporal cortex to the prefrontal cortex. (Abbreviations: **AIP**, anterior intraparietal cortex; **FEF**, frontal eye field; **IT**, inferior temporal cortex; **LIP**, lateral intraparietal cortex; **MIP**, medial intraparietal cortex; **MST**, medial superior temporal cortex; **MT**, middle temporal cortex; **PF**, prefrontal cortex; **PMd, PMv**, dorsal and ventral premotor cortices; **TEO**, occipitotemporal cortex; **VIP**, ventral intraparietal cortex; **V1–V4**, areas of visual cortex.)

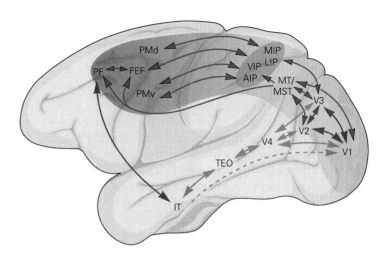

Figure 25–14 Neurons in the anterior intraparietal cortex respond selectively to specific shapes. The neuron shown here is selective for a rectangle, whether viewing the object or reaching for it. The neuron is not responsive to the cylinder in either case. (Reproduced, with permission, from Murata et al. 2000.)

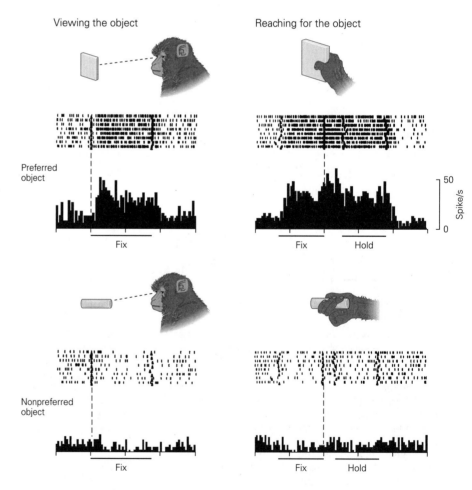

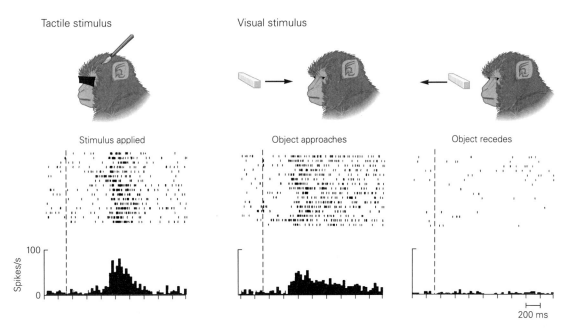

Figure 25–15 Bimodal neurons in the ventral intraparietal cortex of a monkey respond to both visual and tactile stimuli. The neuron shown here responds to tactile stimulation of the monkey's head or to a visual stimulus coming toward the head, but not to the same stimulus moving away from the head. (Reproduced, with permission, from Duhamel, Colby, and Goldberg 1998.)

intraparietal area specify the targets for saccades, and project to the frontal eye field.

Because a monkey cannot see its mouth, the ventral intraparietal area has bimodal neurons that respond to tactile stimuli on the face (Figure 25–15) and to objects in the visual world that are approaching the tactile receptive field, allowing the brain to estimate that an object is near the mouth. The ventral intraparietal area projects to the face area of premotor cortex.

Highlights

1. The image of the world enters the brain via the eye, which is constantly moving in the head. The visual system must compensate for changes in eye position to calculate spatial locations from retinal locations. Helmholtz postulated that the brain solves this problem by feeding back the motor signal that drives the eye to the visual system, to compensate for the effect of the eye movement. This motor feedback to the visual system is called corollary discharge.

2. Neurons in the lateral intraparietal area, which provides visual information to the oculomotor system, show evidence of this corollary discharge. Neurons that ordinarily do not respond to a particular stimulus in space will respond to it if an impending saccade will bring that stimulus into its receptive field.

3. This receptive field remapping depends on a pathway that goes from the intermediate layers of the superior colliculus to the medial dorsal nucleus of the thalamus to the frontal eye field. Medial dorsal nucleus inactivation impairs monkeys' ability to identify where their eyes land after a saccade, suggesting the corollary discharge has a perceptual as well as a motor role.

4. Sherrington postulated that the brain uses eye position to calculate the spatial location of objects from the position of their images on the retina. There is a representation of eye position in somatosensory cortex. Eye position modulates the visual responses of parietal neurons, and target position in space is simple to calculate from this modulation.

5. An unanswered question is how the brain chooses between the eye position and corollary discharge mechanisms to determine spatial position. Because corollary discharge precedes the change in eye position and proprioception follows it, could the brain use both positions at different times?

6. Attention is the ability of the brain to select objects in the world for further analysis. Without attention, spatial perception is severely limited. For

example, humans have great difficulty noticing a change in the visual world unless their attention is drawn to the spatial location of a change.

7. The activity of neurons in the parietal cortex predicts a monkey's locus of spatial attention as measured by their perceptual thresholds. The parietal cortex sums a number of different signals—motor, visual, cognitive—to create a priority map of the visual field. The motor system uses this map to choose targets for movement. The visual system uses the same map to find the locus of visual attention.

8. Lesions in the parietal cortex cause a neglect of the contralateral visual world.

9. Visual information provided by the parietal cortex enables the motor system to adjust hand grip to match the size of the object to which it reaches before the hand actually lands on the target. By contrast, patients with perceptual deficits caused by lesions in inferior temporal cortex adjust their grip perfectly well even though they cannot describe the nature or size of the object to which they reach perfectly.

10. There are at least four different visual maps in the intraparietal sulcus, each of which corresponds to a particular motor workspace.

11. Neurons in the anterior intraparietal area respond to targets for grasping, respond even when monkeys make grasping movements in total darkness, and project to the grasp region of premotor cortex.

12. Neurons in the ventral intraparietal area respond to objects coming toward the mouth, have tactile receptive fields on the face, and project to the mouth area of premotor cortex.

13. Neurons in the medial intraparietal area have a representation of arm position and respond to targets for reaching.

14. Neurons in the lateral intraparietal area respond to targets for eye movements and objects of visual attention, discharge before eye movements, and have a representation of eye position. Activity of these neurons is modulated by the position of the eyes in the orbit.

15. Neurons in the face region of area 3a in the somatosensory cortex have a representation of the position of the eye in the orbit that arises from the contralateral eye.

Michael E. Goldberg
Robert H. Wurtz

Selected Reading

Bisley JW, Goldberg ME. 2010. Attention, intention, and priority in the parietal lobe. Annu Rev Neurosci 33:1–21.

Cohen YE, Andersen RA. 2002. A common reference frame for movement plans in the posterior parietal cortex. Nat Rev Neurosci 3:553–562.

Colby CL, Goldberg ME. 1999. Space and attention in parietal cortex. Annu Rev Neurosci 23:319–349.

Henderson JM, Hollingworth A. 1999. High-level scene perception. Annu Rev Psychol 50:243–271.

Milner AD, Goodale MA. 1996. *The Visual Brain in Action*. Oxford: Oxford Univ. Press.

Rensink RA. 2002. Change detection. Annu Rev Psychol 53:245–277.

Ross J, Ma-Wyatt A. 2004. Saccades actively maintain perceptual continuity. Nat Neurosci 7:65–69.

Sommer MA, Wurtz RH. 2008. Brain circuits for the internal monitoring of movements. Annu Rev Neurosci 31:317–338.

Sun LD, Goldberg ME. 2016. Corollary discharge and oculomotor proprioception: cortical mechanisms for spatially accurate vision. Annu Rev Vis Sci 2:61–84.

Wurtz RH. 2008. Neuronal mechanisms of visual stability. Vision Res 48:2070–2089.

Yarbus AL. 1967. *Eye Movements and Vision*. New York: Plenum.

References

Andersen RA, Essick GK, Siegel RM. 1985. Encoding of spatial location by posterior parietal neurons. Science 230:456–458.

Bisley JW, Goldberg ME. 2003. Neuronal activity in the lateral intraparietal area and spatial attention. Science 299:81–86.

Cavanaugh J, Berman RA, Joiner WM, Wurtz RH. 2016. Saccadic corollary discharge underlies stable visual perception. J Neurosci 36:31–42.

Cohen YE, Andersen RA. 2002. A common reference frame for movement plans in the posterior parietal cortex. Nat Rev Neurosci 3:553–562.

Deubel H, Schneider WX, Bridgeman B. 1996. Postsaccadic target blanking prevents saccadic suppression of image displacement. Vision Res 36:985–996.

Duhamel J-R, Colby CL, Goldberg ME. 1992. The updating of the representation of visual space in parietal cortex by intended eye movements. Science 255:90–92.

Duhamel J-R, Colby CL, Goldberg ME. 1998. Ventral intraparietal area of the macaque: congruent visual and somatic response properties. J Neurophysiol 79:126–136.

Duhamel J-R, Goldberg ME, FitzGibbon EJ, Sirigu A, Grafman J. 1992. Saccadic dysmetria in a patient with a right frontoparietal lesion: the importance of corollary discharge for accurate spatial behavior. Brain 115:1387–1402.

Goodale MA, Meenan JP, Bulthoff HH, Nicolle DA, Murphy KJ, Racicot CI. 1994. Separate neural pathways for the visual analysis of object shape in perception and prehension. Curr Biol 4:604–610.

Hallett PE, Lightstone AD. 1976. Saccadic eye movements to flashed targets. Vision Res 16:107–114.

Halligan PW, Marshall JC. 2001. Graphic neglect—more than the sum of the parts. Neuro Image 14:S91–S97.

Henderson JM, Hollingworth A. 2003. Global transsaccadic change blindness during scene perception. Psychol Sci 14:493–497.

Kusunoki M, Gottlieb J, Goldberg ME. 2000. The lateral intraparietal motion, and task relevance. Vision Res 40:1459–1468.

Morrone MC, Ross J, Burr DC. 1997. Apparent position of visual targets during real and simulated saccadic eye movements. J Neurosci 17:7941–7953.

Murata A, Gallese V, Luppino G, Kaseda M, Sakata H. 2000. Selectivity for the shape, size, and orientation of objects for grasping in neurons of monkey parietal area AIP. J Neurophysiol 83:2580–2601.

Nakamura K, Colby CL. 2002. Updating of the visual representation in monkey striate and extrastriate cortex during saccades. Proc Natl Acad Sci U S A 99:4026–4031.

Perenin MT, Vighetto A. 1988. Optic ataxia: a specific disruption in visuomotor mechanisms. I. Different aspects of the deficit in reaching for objects. Brain 111:643–674.

Rensink RA. 2002. Change detection. Annu Rev Psychol 53:245–277.

Rizzolatti G, Luppino G, Matelli M. 1998. The organization of the cortical motor system: new concepts. Electroencephalogr Clin Neurophysiol 106:283–296.

Snyder LH, Batista AP, Andersen RA. 1997. Coding of intention in the posterior parietal cortex. Nature 386:167–170.

Thiele A, Henning P, Kubischik M, Hoffmann KP. 2002. Neural mechanisms of saccadic suppression. Science 295:2460–2462.

Umeno MM, Goldberg ME. 1997. Spatial processing in the monkey frontal eye field. I. Predictive visual responses. J Neurophysiol 78:1373–1383.

Walker MF, Fitzgibbon EJ, Goldberg ME. 1995. Neurons in the monkey superior colliculus predict the visual result of impending saccadic eye movements. J Neurophysiol 73:1988–2003.

Wang X, Zhang M, Cohen IS, Goldberg ME. 2007. The proprioceptive representation of eye position in monkey primary somatosensory cortex. Nat Neurosci 10:640–646.

Xu B, Karachi C, Goldberg M. 2012. The postsaccadic unreliability of gain fields renders it unlikely that the motor system can use them to calculate target position in space. Neuron 76:1201–1209.

26

Auditory Processing by the Cochlea

Human experience is enriched by the ability to distinguish a remarkable range of sounds—from the intimacy of a whisper to the warmth of a conversation, from the complexity of a symphony to the roar of a stadium. Hearing begins when the sensory cells of the cochlea, the receptor organ of the inner ear, transduce sound energy into electrical signals and forward them to the brain. Our ability to recognize small differences in sounds stems from the cochlea's capacity to distinguish among frequency components, their amplitudes, and their relative timing.

Hearing depends on the remarkable properties of hair cells, the cellular microphones of the inner ear. Hair cells transduce mechanical vibrations elicited by sounds into electrical signals, which are then relayed to the brain for interpretation. The hair cells can measure motions of atomic dimensions and transduce stimuli ranging from static inputs to those at frequencies of tens of kilohertz. Remarkably, hair cells can also serve as mechanical amplifiers that augment auditory sensitivity. Each of the paired cochleae contains approximately 16,000 of these cells. Deterioration of hair cells and their innervation accounts for most of the hearing loss that afflicts about 10% of the population in industrialized countries.

The Ear Has Three Functional Parts

Sound consists of alternating compressions and rarefactions propagated by an elastic medium, the air, at a speed of approximately 340 m/s. This wave of pressure changes carries mechanical energy that stems from the work produced on air by our vocal apparatus or some other sound source. The mechanical energy is captured and transmitted to the receptor organ, where it is transduced into electrical signals suitable for neural analysis. These three tasks are associated with the external ear, the middle ear, and the cochlea of the inner ear, respectively (Figure 26–1).

The most obvious component of the human external ear is the auricle, a prominent fold of cartilage-supported skin. The auricle acts as a reflector to capture sound efficiently and focus it into the external auditory meatus, or ear canal. The ear canal ends at the tympanum, or eardrum, a diaphragm approximately 9 mm in diameter and 50 μm in thickness.

The external ear is not uniformly effective at capturing sound from all directions; the auricle's corrugated surface collects sounds best when they originate at different, but specific, positions with respect to the head. Our capacity to localize sounds in space, especially along the vertical axis, depends critically on these sound-gathering properties. Each auricle has a unique topography; its effect on sound reflections at different frequencies is learned by the brain early in life.

The middle ear is an air-filled pouch connected to the pharynx by the Eustachian tube. Airborne sound traverses the middle ear as vibrations of the auditory ossicles, three tiny bones that are linked together: the malleus (hammer), incus (anvil), and stapes (stirrup; Figure 26–1). A long extension of the malleus is attached to the tympanic membrane; its other extreme makes a ligamentous connection to the incus, which is similarly connected to the stapes. The flattened base of the stapes, the footplate, is seated in an opening—the oval window—in the bony covering of the cochlea. The auditory ossicles are relics of evolution. The stapes was originally a component of the gill support of ancient fish; the malleus and incus were components of the primary jaw joint in reptilian ancestors.

The inner ear includes the auditory sensory organ, the cochlea (Greek *cochlos*, snail), a coiled structure of progressively diminishing diameter wound around a conical bony core (Figure 26–1). In humans, the cochlea is approximately 9 mm across, the size of a chickpea, and is embedded within the temporal bone. The interior of the cochlea consists of three parallel liquid-filled compartments termed *scalae*. In a cross section of the cochlea at any position along its spiral course, the top compartment is the *scala vestibuli* (Figure 26–2). At the broad, basal end of this chamber is the oval window, the opening that is sealed by the footplate of the stapes. The bottom compartment is the *scala tympani*; it too has a basal aperture, the round window, which is closed by

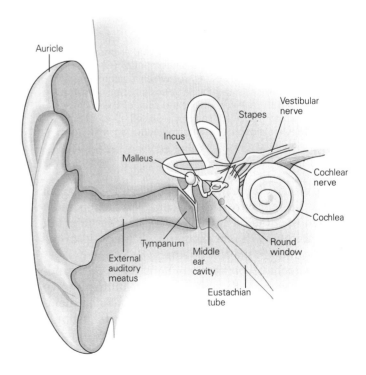

Figure 26–1 The structure of the human ear. The external ear, especially the prominent auricle, focuses sound into the external auditory meatus. Alternating increases and decreases in air pressure vibrate the tympanum. These vibrations are conveyed across the air-filled middle ear by three tiny, linked bones: the malleus, the incus, and the stapes. Vibration of the stapes stimulates the cochlea, the hearing organ of the inner ear.

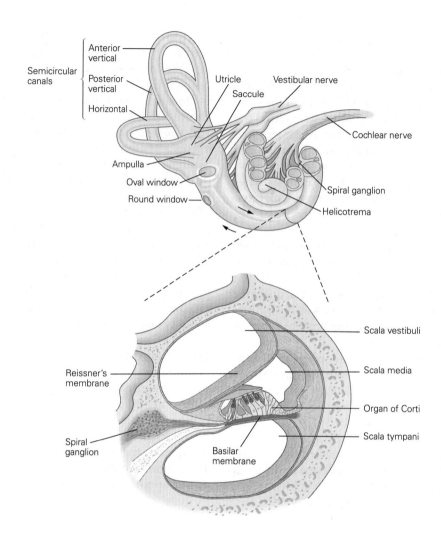

Figure 26–2 The structure of the cochlea. A cross section of the cochlea shows the arrangement of the three liquid-filled ducts or scalae, each of which is approximately 33 mm long. The scala vestibuli and scala tympani communicate through the helicotrema at the apex of the cochlea. At the base, each duct is closed by a sealed aperture. The scala vestibuli is closed by the oval window, against which the stapes pushes in response to sound; the scala tympani is closed by the round window, a thin, flexible membrane. Between these two compartments lies the scala media, an endolymph-filled tube whose epithelial lining includes the 16,000 hair cells in the organ of Corti surmounting the basilar membrane (**blue**). The hair cells are covered by the tectorial membrane (**green**). The cross section in the lower diagram has been rotated so that the cochlear apex is oriented toward the top.

a thin, elastic diaphragm beyond which lies the air of the middle-ear cavity. The two chambers are separated along most of their length by the cochlear partition but communicate with one another at the very tip of the cochlea, through the helicotrema.

The cochlear partition contains the third liquid-filled cavity, the *scala media*, and is delimited by two membranes. The thin Reissner's, or vestibular, membrane divides the scala media from the scala vestibuli. The basilar membrane separates the cochlear partition from the scala tympani and supports the complex sensory structure involved in auditory transduction, the organ of Corti (Figure 26–2).

Hearing Commences With the Capture of Sound Energy by the Ear

Psychophysical experiments have established that we perceive an approximately equal increment in loudness for each 10-fold increase in the magnitude of a sound stimulus. This type of relation is characteristic of many of our senses and is the basis of the Weber-Fechner law (Chapter 17). A logarithmic scale is therefore useful in relating the magnitude of sound pressure to perceived loudness. Sound pressure corresponds to the sound-evoked modulation of the air pressure with respect to the mean atmospheric pressure; the louder the sound, the larger is the modulation. The sound-pressure level, L, of any sound may be expressed in decibels (dB) as

$$L = 20 \cdot \log_{10}(P / P_{\text{REF}}),$$

in which P, the magnitude of the stimulus, is the root-mean-square sound pressure (in units of pascals, abbreviated Pa, or newtons per square meter). For a sinusoidal stimulus, the amplitude exceeds the root-mean-square value by a factor of $\sqrt{2}$. The arbitrary reference level on this scale, 0 dB sound-pressure level (SPL), corresponds to a root-mean-square sound pressure, P_{REF}, of 20 µPa. This level represents the approximate threshold of human hearing at 1 to 4 kHz, the frequency range in which our ears are most sensitive.

Sound consists of very small alternating changes in the local air pressure. The loudest sound tolerable to humans, approximately 120 dB SPL, transiently alters the local atmospheric pressure by only ±0.01%. In contrast, a sound at the threshold level causes a change in the local pressure of much less than one part in a billion. From the faintest sounds that can be detected to sounds so intense that they hurt, the sound pressure increases by one millionfold, which correspond to a trillionfold range in stimulus power. The dynamic range of hearing is enormous.

Despite their small magnitude, sound-induced increases and decreases in air pressure move the tympanum inward and outward (Figure 26–3A,B). Near threshold, the amplitude of vibration is in the picometer range, which is comparable to the tympanum's own thermal fluctuations. Even loud sounds elicit vibrations of the tympanum that do not exceed 1 µm in amplitude. The resulting motions of the ossicles are essentially like those of two interconnected levers (the malleus and incus) and a piston (the stapes). The vibration of the incus alternately drives the stapes deeper into the oval window and retracts it, like a piston that pushes and pulls cyclically upon the liquid in the scala vestibuli. In humans, the area of the eardrum is about 20-fold larger than that of the stapes footplate. As a result, pressure changes applied on the liquid of scala vestibuli by the stapes footplate are larger than those pushing and pulling the tympanum. Pressures are further magnified by the lever operating between the malleus and the incus, the incus in humans being only about 70% of the length of the malleus.

The action of the stapes produces pressure changes that propagate through the liquid of the scala vestibuli at the speed of sound in water. Because liquids are virtually incompressible, however, the primary effect of the stapes's motion is to displace the liquid in the scala vestibuli in the one direction that is not restricted by a rigid boundary: toward the elastic cochlear partition (Figure 26–3B). The deflection of the cochlear partition downward increases the pressure in the scala tympani, displacing a liquid mass that causes outward bowing of the round window. Each cycle of a sound stimulus thus evokes a cycle of up-and-down movement of a minuscule volume of liquid in each of the cochlea's three chambers, thus displacing the sensory organ.

By increasing the magnitude of pressure changes by up to 30-fold, the overall effect of the middle ear is to match the low impedance of the air outside the ear to the higher impedance of the cochlear partition, thus ensuring the efficient transfer of sound energy from the first medium to the second. The pressure gain afforded by the middle ear depends on sound frequency, which determines the U-shape tuning curve of auditory threshold.

Changes of the middle ear's normal structure that reduce its displacement amplitudes can lead to *conductive hearing loss,* of which two forms are especially common. First, scar tissue caused by middle-ear infection (*otitis media*) can immobilize the tympanum or ossicles. Second, a proliferation of bone in the ligamentous attachments of the ossicles can reduce their normal freedom of motion. This chronic condition of unknown origin, termed *otosclerosis,* can lead to severe deafness.

A clinician may test for conductive hearing loss by the simple Rinné test. A patient is asked to assess the loudness of a vibrating tuning fork under two conditions: when the tuning fork is held in the air or when it is pressed against the head just behind the ear. For the second stimulus, sound is conducted through bone to the cochlea. If the second stimulus is perceived to be louder, the patient's conductive pathway through the middle ear may be damaged, but the inner ear is likely to be intact. In contrast, if bone conduction is not more efficient than airborne stimulation, the patient may have inner-ear damage, that is, sensorineural hearing loss. The diagnosis of conductive hearing loss is important, because surgical intervention can be highly effective: Removal of scar tissue or reconstitution of the conductive pathway with a prosthesis may restore excellent hearing.

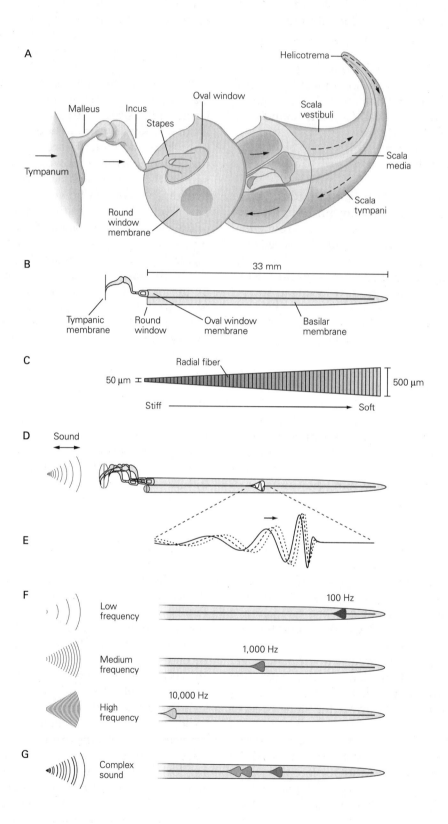

The Hydrodynamic and Mechanical Apparatus of the Cochlea Delivers Mechanical Stimuli to the Receptor Cells

The Basilar Membrane Is a Mechanical Analyzer of Sound Frequency

The continuous variation of the mechanical properties of the basilar membrane along the cochlea's length, approximately 33 mm, is key to the cochlea's operation. The basilar membrane at the base of the human cochlea is less than one-fifth as broad as at the apex. Thus, although the cochlear chambers become progressively smaller from the organ's base toward its apex, the basilar membrane *increases* in width (Figure 26–3C). Moreover, the basilar membrane is relatively thick toward the base of the cochlea but thinner at the apex. Both morphological gradients contribute to a base-to-apex decrease in basilar-membrane stiffness. Radial collagen fibers within the membrane determine most of its elasticity. The basilar membrane may schematically be regarded as a set of weakly coupled radial segments of increasing length along the longitudinal axis of the cochlea, with the shortest segment at the base and the longest segment at the apex, analogous to the multiple strings of a piano.

Stimulation with a pure tone evokes a complex and elegant movement of the basilar membrane. Over one complete cycle of a tone, each affected segment along the basilar membrane undergoes a single cycle of vibration (Figure 26–3D,E). The various segments of the membrane do not, however, oscillate in phase with one another. As first demonstrated by Georg von Békésy using stroboscopic illumination, each segment reaches its maximal amplitude of motion slightly later than its basal neighbor. The normalized sinusoidal movement of the basilar membrane reproduces that of the stapes, but with a time delay that increases with the distance from the cochlear base.

The overall pattern of motion of the membrane is that of a traveling wave that traverses the cochlea from the stiff base toward the floppier apex. As each wave advances toward the apex, the amplitude of vibration grows to a maximum and then declines rapidly. The position at which the traveling wave reaches its maximal amplitude depends on sound frequency. The basilar membrane at the base of the cochlea responds best to the highest audible frequencies—in humans approximately 20 kHz. At the cochlear apex, the membrane responds to frequencies as low as 20 Hz. The intervening frequencies are represented along the basilar

Figure 26–3 (Opposite) Motion of the basilar membrane.

A. An uncoiled cochlea, with its base displaced to show its relation to the scalae, indicates the flow of stimulus energy. Sound vibrates the tympanum, which sets the three ossicles of the middle ear in motion. The piston-like action of the stapes, a bone inserted partially into the elastic oval window, produces oscillatory pressure differences that rapidly propagate along the scala vestibuli and scala tympani. Low-frequency pressure differences are shunted through the helicotrema, where the two ducts communicate.

B. The functional properties of the cochlea are conceptually simplified if the cochlea is viewed as a linear structure with only two liquid-filled compartments separated by the elastic basilar membrane.

C. The basilar membrane, here represented in a surface view, increases in width from approximately 50 μm near the base to 500 μm near the apex of the cochlea. Radial collagen fibers run from the neural to the abneural edge of the membrane. As the result of its morphological gradients, the basilar membrane's mechanical properties vary continuously along its length.

D. The oscillatory stimulation of a sound causes a traveling wave on the basilar membrane, shown here within the envelope of maximal displacement over an entire cycle. The magnitude of movement is grossly exaggerated in the vertical direction; the loudest tolerable sounds move the basilar membrane by only ±150 nm, a scaled distance less than one-hundredth the width of the lines representing the basilar membrane in these figures.

E. An enlargement of the active region in **D** demonstrates the motion of the basilar membrane in response to stimulation with sound of a single frequency. The continuous curve depicts a traveling wave at one instant; the vertical scale of basilar-membrane deflection is exaggerated about one-millionfold. The **dashed** and **dotted curves** portray the traveling wave at successively later times as it progresses from the cochlear base (*left*) toward the apex (*right*). As the wave approaches the characteristic place for the stimulus frequency, it slows and grows in amplitude. The stimulus energy is then transferred to hair cells at positions within the wave's peak.

F. Each frequency of stimulation excites maximal motion at a particular position along the basilar membrane. Low-frequency sounds produce basilar-membrane motion near the apex, where the membrane is relatively broad and soft. Mid-frequency sounds excite the membrane in its middle. The highest frequencies that we can hear excite the basilar membrane at its narrow, stiff base. The mapping of sound frequency onto the basilar membrane is approximately logarithmic.

G. The basilar membrane performs spectral analysis of complex sounds. In this example, a sound with three prominent frequencies, such as the three formants of a vowel sound, excites basilar-membrane motion in three regions, each of which represents a particular frequency. Hair cells in the corresponding positions transduce the basilar-membrane oscillations into receptor potentials, which in turn excite the nerve fibers that innervate these particular regions.

membrane in a continuous array (Figure 26–3F). In the 19th century, the German physiologist Hermann von Helmholtz was the first to appreciate that the basilar membrane's operation is essentially the inverse of a piano's. The piano synthesizes a complex sound by combining the pure tones produced by numerous vibrating strings; the cochlea, by contrast, deconstructs a complex sound by isolating the component tones at the appropriate segments of the basilar membrane.

For any frequency within the auditory range, there is a characteristic place along the basilar membrane at which the magnitude of vibration is maximal. Although the morphological gradients of the basilar membrane are key to the process, the actual dispersion of a sound's frequency components along the cochlea's longitudinal axis depends on the mechanical properties of the cochlear partition as a whole. In particular, as we shall detail later, the hair cells within the organ of Corti provide active mechanical feedback that sharpens mechanical tuning of the basilar membrane and enhances its sensitivity to sound. The arrangement of vibration frequencies along the basilar membrane is an example of a *tonotopic map*. The relationship between frequency and position along the basilar membrane varies monotonically, but is not linear; the logarithm of the frequency decreases roughly in proportion to the distance from the cochlea's base. The frequencies from 20 kHz to 2 kHz, those between 2 kHz and 200 Hz, and those spanning 200 Hz to 20 Hz are each represented by approximately one-third of the basilar membrane's extent.

Analysis of the response to a complex sound illustrates how the basilar membrane operates in daily life. A vowel sound in human speech, for example, ordinarily comprises three dominant frequency components termed formants. Each frequency component of the stimulus establishes a traveling wave that, to a first approximation, is independent of the waves evoked by the others (Figure 26–3G) and reaches its peak excursion at a point on the basilar membrane appropriate for that frequency component. The basilar membrane thus acts as a mechanical frequency analyzer by distributing the energies associated with the different frequency components of the stimulus to hair cells arrayed along its length. In doing so, the basilar membrane begins the encoding of the frequencies in a sound.

The Organ of Corti Is the Site of Mechanoelectrical Transduction in the Cochlea

The organ of Corti, a ridge of epithelium extending along the basilar membrane, is the receptor organ of the inner ear. Each organ of Corti contains approximately 16,000 hair cells that are innervated by approximately 30,000 *afferent* nerve fibers; these are fibers that carry information into the brain along the eighth cranial nerve. Like the basilar membrane itself, each hair cell is most sensitive to a particular frequency, and these frequencies are logarithmically mapped in descending order from the cochlea's base to its apex. Thus, the information transmitted by these sensory cells to their innervating nerve fibers is also tonotopically organized.

The organ of Corti includes a variety of cells, some of unknown function, but four types have obvious importance. First, there are two types of hair cells. The *inner hair cells* form a single row of approximately 3,500 cells, whereas approximately 12,000 *outer hair cells* lie in three rows farther from the central axis of the cochlear spiral (Figure 26–4). The space between the inner and outer hair cells is delimited and mechanically supported by pillar cells. The outer hair cells are supported at their bases by Deiters's (phalangeal) cells.

A second epithelial ridge adjacent to the organ of Corti, but nearer the cochlea's central axis, gives rise to the tectorial membrane, a gelatinous shelf that covers the organ of Corti (Figure 26–4). The tectorial membrane is anchored at its base, and its tapered distal edge forms a fragile connection with the organ of Corti.

Hair cells are not neurons; they lack both dendrites and axons (Figure 26–5A). A special saline solution, the endolymph that fills scala media, bathes the cell's apical aspect. Tight junctions between hair cells and supporting cells separate this liquid from the standard extracellular fluid, or perilymph, that contacts the basolateral surface of the cell. Immediately below the tight junctions, a desmosomal junction provides a strong mechanical attachment for the hair cell to its neighbors.

The hair bundle, which serves as a receptive antenna for mechanical stimuli, projects from the flattened apical surface of the hair cell. Each bundle comprises a few tens to a few hundred cylindrical processes, the *stereocilia*, arranged in 2 to 10 parallel rows and extending several micrometers from the cell surface. Successive stereocilia across a cell's surface vary monotonically in height; a hair bundle is beveled like the tip of a hypodermic needle (Figure 26–5B). The inner hair-cell bundles of the mammalian cochlea, when viewed from above, have a roughly linear form. Outer hair-cell bundles, in contrast, have a V or W shape (Figure 26–6).

Each stereocilium is a rigid cylinder whose core consists of a fascicle of actin filaments that are heavily cross-linked by the proteins plastin (fimbrin), fascin, and epsin. Cross-linking renders a stereocilium

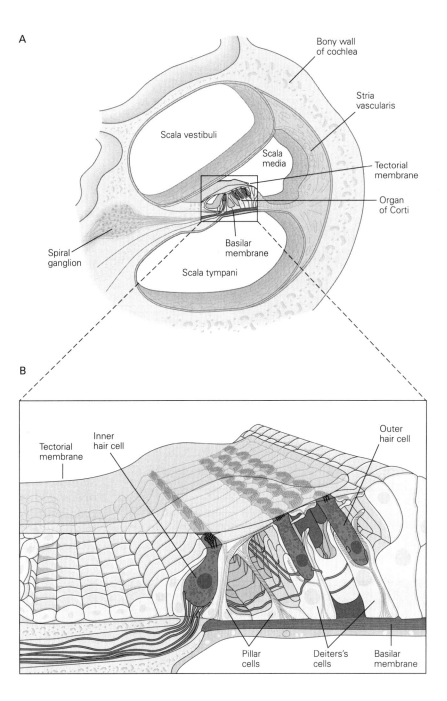

Figure 26–4 Cellular architecture of the human organ of Corti. Although there are differences among species, the basic plan is similar for all mammals.

A. The organ of Corti, the inner ear's receptor organ, is an epithelial strip that surmounts the elastic basilar membrane. The organ contains some 16,000 hair cells arrayed in four rows: a single row of inner hair cells and three rows of outer hair cells. The mechanically sensitive hair bundles of these receptor cells protrude into endolymph, the liquid within the scala media. Reissner's membrane, which provides the upper boundary of the scala media, separates the endolymph from the perilymph in the scala vestibuli. The hair bundles of outer hair cells are attached at their tops to the lower surface of the tectorial membrane, a gelatinous shelf that extends the full length of the basilar membrane.

B. The hair cells are separated and supported by pillar cells and Deiters's cells. One hair cell has been removed from the middle row of outer hair cells to reveal the three-dimensional relationship between supporting cells and hair cells. Afferent and efferent nerve endings are colored in **red** and **green**, respectively.

much more rigid than would be expected for a bundle of unconnected actin filaments. The actin core of the stereocilium is covered by a tubular sheath of plasma membrane. Although a stereocilium is of constant diameter along most of its length, it tapers just above its basal insertion (see Figure 25–5B). Correspondingly, the number of actin filaments diminishes from several hundred to only a few dozen. This thin cluster of microfilaments anchors the stereocilium in the cuticular plate, a thick mesh of interlinked actin filaments beneath the apical cell membrane. Because of this tapered structure, a mechanical force applied at the tip causes the stereocilium to pivot around its basal insertion. Horizontal top connectors interconnect adjacent stereocilia near their tips. These extracellular filaments restrict the bundle to move as a unit during stimulation at low frequencies. At high frequencies, the viscosity of the liquid between the stereocilia also opposes their separation and thus ensures the unitary motion of the hair bundle.

During its early development, every hair bundle includes at its tall edge a single true cilium, the *kinocilium* (Figure 26–5). Like other cilia, this structure possesses at its core an axoneme, or array of nine paired microtubules, and often an additional central pair of microtubules. The kinocilium is not essential for mechanoelectrical transduction, for in mammalian cochlear hair cells, it degenerates around the time of birth.

Hair Cells Transform Mechanical Energy Into Neural Signals

Deflection of the Hair Bundle Initiates Mechanoelectrical Transduction

Just as in vestibular organs (Chapter 27), mechanical deflection of the hair bundle is the stimulus that excites hair cells of the cochlea. Stimuli elicit an electrical response, the receptor potential, by opening or closing—a process termed "gating"—mechanically sensitive ion channels. The hair cell's response depends on the direction and magnitude of the stimulus.

In an unstimulated cell, 10% to 50% of the channels involved in stimulus transduction are open. As a result, the cell's resting potential, which lies within a range of approximately –70 to –30 mV, is determined in part by the influx of cations through these channels. A stimulus that displaces the bundle toward its tall edge opens additional channels, thereby depolarizing the cell (Figure 26–7). In contrast, a stimulus that displaces the bundle toward its short edge shuts transduction channels that are open at rest, thus hyperpolarizing

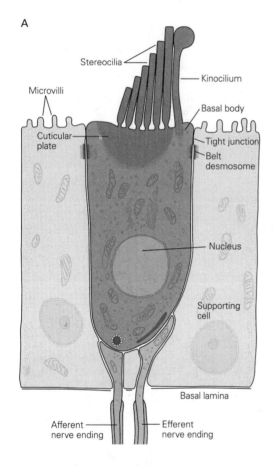

A

Stereocilia

Microvilli

Kinocilium

Basal body

Cuticular plate

Tight junction

Belt desmosome

Nucleus

Supporting cell

Basal lamina

Afferent nerve ending

Efferent nerve ending

B

Figure 26–5 (Left) Structure of a vertebrate hair cell.

A. The epithelial character of the hair cell is evident in this drawing of the sensory epithelium from a frog's internal ear. The cylindrical hair cell is joined to adjacent supporting cells by a junctional complex around its apex. The hair bundle, a mechanically sensitive organelle, extends from the cell's apical surface. The bundle comprises some 60 stereocilia arranged in stepped rows of varying length. At the bundle's tall edge stands the single kinocilium, an axonemal structure with a bulbous swelling at its tip; in the mammalian cochlea, this organelle degenerates around the time of birth. Deflection of the hair bundle's top to the right depolarizes the hair cell; movement in the opposite direction elicits hyperpolarization. The hair cell is surrounded by supporting cells, whose apical surfaces bear a stubble of microvilli. Afferent and efferent synapses contact the basolateral surface of the plasma membrane.

B. This scanning electron micrograph of a hair cell's apical surface reveals the hair bundle protruding approximately 8 μm into the endolymph. (Image reproduced, with permission, from A.J. Hudspeth.)

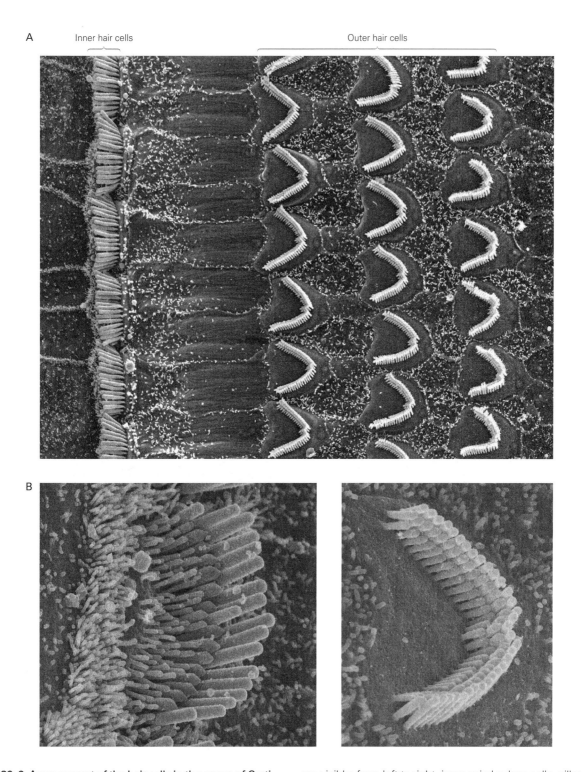

Figure 26–6 Arrangement of the hair cells in the organ of Corti. (Images reproduced, with permission, from D. Furness, Keele University, United Kingdom.)

A. Inner hair cells form a single row, and the stereocilia of each cell are arranged linearly. In contrast, outer hair cells are distributed in three rows, and the stereocilia of each cell are arranged in a V configuration. The apical surfaces of several other cells

are visible: from left to right, inner spiral sulcus cells, pillar cells, Deiters's cells, and Hensen's cells (see Figure 26–4).

B. Higher magnification shows the linear configuration of the hair bundle atop an inner hair cell (*left*) and the V configuration of an outer hair-cell bundle (*right*), as well as the arrangement of the stereocilia in rows of increasing heights.

the cell. Hair cells respond most to stimuli parallel to the hair bundle's axis of morphological mirror symmetry: Stimuli at right angles to the axis produce little change from the resting potential. An oblique stimulus elicits a response proportional to its vectorial projection along the axis of sensitivity.

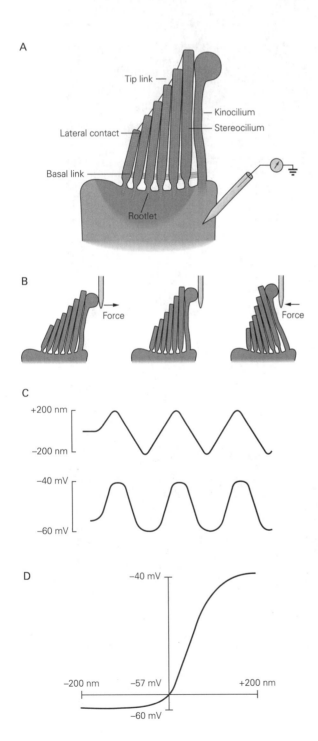

A hair cell's receptor potential is graded. As the stimulus amplitude increases, the receptor potential grows progressively larger up to the point of saturation. The receptor potential of an inner hair cell can be as great as 25 mV in peak-to-peak magnitude. The relation between a bundle's deflection and the resulting electrical response is sigmoidal (Figure 26–7D). A displacement of only ±100 nm represents approximately 90% of the response range. During normal stimulation, a hair bundle moves through an angle of ±1° or so, that is, by much less than the diameter of one stereocilium.

When observed in vitro, a hair bundle exhibits Brownian motion of approximately ±3 nm, whereas the threshold of hearing corresponds to basilar-membrane movements of as little as ±0.3 nm. There are at least three mechanisms that explain how the hair bundle may respond to motion smaller than its own noise. First, because the cochlear partition does not move as a rigid body, the movement of the hair bundle is larger than that of the basilar membrane. Second, frequency-selective amplification of low stimuli actively pulls the signal out of the noise. Finally, mechanical coupling to a group of neighbors results in synchronization that effectively reduces noise. At hearing threshold, a stimulus evokes a receptor potential near 100 µV in amplitude.

The ion channels in hair cells that mediate mechanoelectrical transduction are relatively nonselective, cation-passing pores with a conductance near 100 pS. From the known size of small organic cations and fluorescent molecules that can traverse the channel, the transduction channel's pore must be about 1.3 nm in diameter. Most of the transduction current is carried by K^+, the cation with the highest concentration in the endolymph bathing the hair bundle. Although

Figure 26–7 (Left) Mechanical sensitivity of a hair cell.

A. A recording electrode is inserted into a frog hair cell.

B. A probe attached to the bulbous tip of the stereocilium is moved by a piezoelectric stimulator, deflecting the elastic hair bundle from its resting position. The actual deflections are generally only one-tenth as large as those portrayed.

C. When the top of a hair bundle is displaced back and forth (**upper trace**), the opening and closing of mechanically sensitive ion channels produce an oscillatory receptor potential (**lower trace**) that—as here—may saturate in both the depolarizing and the hyperpolarizing directions.

D. The relation between hair-bundle deflection (abscissa) and receptor potential (ordinate) is sigmoidal. The entire operating range is only approximately 100 nm, less than the diameter of an individual stereocilium. At rest, the hair bundle operates within the steep region of the sigmoid, which ensures significant receptor potentials in response to weak stimuli.

endolymph is relatively poor in Ca^{2+}, a small fraction of the transduction current is carried by this ion. Fluorescent indicators indicate that Ca^{2+} entry, and thus mechanoelectrical transduction, occurs precisely at the stereociliary tips of a deflected hair bundle. Single-channel recordings, together with the observation that the magnitude of the transduction current is roughly proportional to the number of functional stereocilia remaining in a microdissected bundle, indicate that there are probably only two active transduction channels per stereocilium.

The large diameter and poor selectivity of the pore permit transduction channels to be blocked by aminoglycoside antibiotics such as streptomycin, gentamicin, and tobramycin. When used in large doses to counter bacterial infections, these drugs have a toxic effect on hair cells; the antibiotics damage hair bundles and eventually kill hair cells. These drugs pass through transduction channels at a low rate and thus cause long-term toxic effects by interfering with protein synthesis on the mitochondrial ribosomes, which resemble bacterial ribosomes. Consistent with this hypothesis, human sensitivity to aminoglycosides is maternally inherited, as are the mitochondria, and in many instances reflects a single base change in the 12S ribosomal RNA gene of the mitochondrion.

Mechanical Force Directly Opens Transduction Channels

The mechanism for gating of transduction channels in hair cells differs fundamentally from the mechanisms used for electrical signals in neurons such as the action potential or postsynaptic potential. Many ion channels respond to changes in membrane potential or to specific ligands (Chapters 8, 10, and 12–14). In contrast, two lines of evidence suggest that the mechanoelectrical transduction channels in the hair cell are activated by mechanical strain.

First, a bundle is stiffer along its axis of mechanical sensitivity than at a right angle. This observation suggests that a portion of the work done in deflecting a bundle goes into elastic elements, termed *gating springs*, which pull on the molecular gates of the transduction channels. Because the gating springs contribute over half of a hair bundle's stiffness, the transduction channels efficiently capture the energy supplied when a bundle is deflected. In addition, the mechanical properties of a hair bundle vary during channel gating: When channels open or close, stiffness decreases and friction increases. Both phenomena are expected if the channels are gated directly through a mechanical linkage to the hair bundle.

A second indication that transduction channels are directly controlled by gating springs is the rapidity with which hair cells respond. The response latency is so brief, only a few microseconds, that gating is more likely to be direct than to involve a second messenger (Chapter 14). Moreover, the electrical responses of hair cells to a series of step stimuli of increasing magnitude become both larger and faster. This behavior favors a kinetic scheme in which mechanical force controls the rate constant for channel gating.

The *tip link* is a probable component of the gating spring. A tip link is a fine molecular braid joining the distal end of one stereocilium to the side of the longest adjacent process (Figure 26–8A). Deflection of a hair bundle toward its tall edge tenses the tip link and promotes channel opening; movement in the opposite direction slackens the link and allows the associated channels to close (Figure 26–8B).

Three experimental results suggest that the tip links are components of the gating springs. First, tip links are universal features of hair bundles and are situated at the site of transduction. The transduction channels are indeed located at the stereociliary tips, thus near the lower insertion point of the tip link. Second, the orientation of the links is consistent with the vectorial sensitivity of transduction. The links invariably interconnect stereocilia in a direction parallel with the hair bundle's axis of mechanosensitivity. Finally, when tip links are disrupted by exposing hair cells to Ca^{2+} chelators, transduction vanishes. As the tip links regenerate over the course of approximately 12 hours, a hair cell regains mechanosensitivity. It remains unclear whether the elasticity of gating springs resides primarily in the tip links or in the structures at their two insertions.

In the mammalian cochlea, hair bundles are deflected through their linkage to the tectorial membrane. When the basilar membrane oscillates up and down in response to a sound, the organ of Corti and the overlying tectorial membrane move with it. Because the basilar and tectorial membranes pivot about different lines of insertion, however, their up-and-down motion is accompanied by a back-and-forth shearing motion between the upper surface of the organ of Corti and the lower surface of the tectorial membrane. This is the motion that is detected by hair cells (Figure 26–9).

The hair bundles of outer hair cells, whose tips are firmly attached to the tectorial membrane, are directly deflected by this movement. The hair bundles of inner hair cells, which do not contact the tectorial membrane, are deflected by movement of the liquid beneath the membrane. This mode of stimulation affords some mechanical magnification of the signals reaching hair

Figure 26–8 Mechanoelectrical transduction by hair cells.

A. A tip link connects each stereocilium to the side of the longest adjacent stereocilium, as seen in a scanning electron micrograph (*left*) and a transmission electron micrograph (*right*) of a hair bundle's top surface. Each tip link is only 3 nm in diameter and 150 to 200 nm in length. The links appear stouter in the illustration on the left because of metallic coating during specimen preparation. (Reproduced, with permission, from Assad, Shepherd, and Corey 1991; reproduced, with permission, from Hudspeth and Gillespie 1994.)

B. Top: Ion flux through the channel that underlies mechanoelectrical transduction in hair cells is regulated by a molecular gate. The opening and closing of the gate are controlled by the tension in an elastic element, the gating spring, which senses hair-bundle displacement. (Adapted, with permission, from Howard and Hudspeth 1988.)

Bottom: When the hair bundle is at rest, each transduction channel fluctuates between closed and open states, spending most of its time shut. Displacement of the bundle in the positive direction increases the tension in the gating spring, here assumed to be in part a tip link, attached to each channel's molecular gate. The enhanced tension promotes channel opening and the influx of cations, thereby producing a depolarizing receptor potential. (Adapted, with permission, from Hudspeth 1989.)

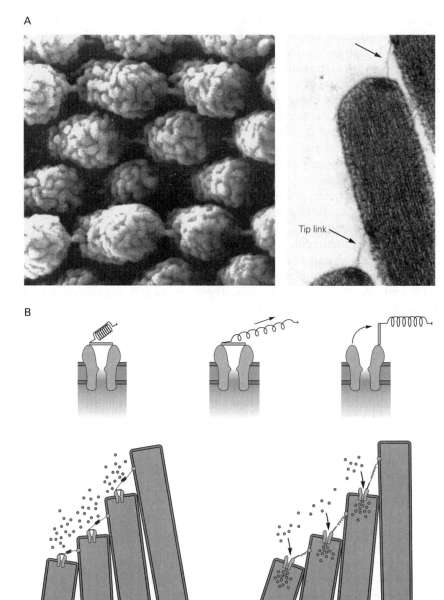

A

B

Tip link

bundles. At least for high-frequency stimuli, the movements of hair bundles are thought to be severalfold greater than that of the basilar membrane.

Direct Mechanoelectrical Transduction Is Rapid

Hair cells operate much more quickly than do other sensory receptor cells of the vertebrate nervous system

and, indeed, more quickly than neurons themselves. To deal with the frequencies of biologically relevant sounds, transduction by hair cells must be rapid. Given the behavior of sound in air and the dimensions of sound-emitting and sound-absorbing organs such as vocal cords and eardrums, optimal auditory communication occurs in the frequency range of 10 Hz to 100 kHz. Much higher frequencies propagate

A

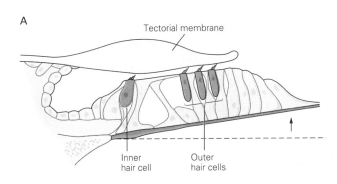

Tectorial membrane

Inner
hair cell

Outer
hair cells

B

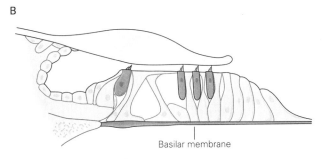

Basilar membrane

C

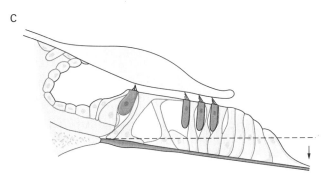

Figure 26–9 Forces acting on cochlear hair cells. Hair cells in the cochlea are stimulated when the basilar membrane is driven up and down by differences in the pressure between the scala vestibuli and scala tympani. This motion is accompanied by shearing movements between the tectorial membrane and organ of Corti. These motions deflect the hair bundles of outer hair cells, which are attached to the lower surface of the tectorial membrane. The hair bundles of inner hair cells, which are not attached to the tectorial membrane, are deflected by the movement of liquid in the space beneath that structure. In both instances, the deflection initiates mechanoelectrical transduction of the stimulus.

A. When the basilar membrane is driven upward, shear between the hair cells and the tectorial membrane deflects hair bundles in the excitatory direction, toward their tall edge.

B. At the midpoint of an oscillation, the hair bundles resume their resting position.

C. When the basilar membrane moves downward, the hair bundles are driven in the inhibitory direction.

poorly through air; much lower frequencies are inefficiently produced and poorly captured by animals of moderate size. Even in animals sensitive to relatively low frequencies, such as frogs, the in vitro transduction current in response to a step stimulus of moderate intensity rises with a time constant of only 80 µs at room temperature. For mammals to be able to respond to frequencies greater than 100 kHz, the hair cells evidently display gating times that are an order of magnitude smaller. Locating sound sources, one of the most important functions of hearing, sets even more stringent limits on the speed of transduction (Chapter 28). A sound from a source directly to one side of a person reaches the nearer ear somewhat sooner than the farther, by at most 700 µs in humans. An observer can locate sound sources on the basis of much smaller delays, about 10 µs. For this to occur, hair cells must be capable of transducing acoustic waveforms with microsecond-level resolution.

Deafness Genes Provide Components of the Mechanotransduction Machinery

Genetic studies of deafness in both humans and mouse models have provided entry points into the molecular composition of the mechanotransduction machinery of the hair cell. In particular, the upper two-thirds of the tip link consist of two parallel molecules of cadherin-23, whereas the lower third comprises two parallel molecules of protocadherin-15 (Figure 26–10). The two components are joined at their tips in a Ca^{2+}-sensitive manner; lowering the extracellular Ca^{2+} concentration below approximately 1 µM disrupts their association. In humans, mutations in the genes coding for cadherin 23 (*USH1D*) and protocadherin 15 (*USH1F*) lead to the most severe form of the Usher syndrome, an autosomal recessive disorder that associates severe-to-profound congenital deafness, constant vestibular dysfunction, and retinitis pigmentosa with a prepubertal onset. The study of other genes involved in this type of Usher syndrome has revealed that the upper end of the tip link is anchored to the actin core of a stereocilium by a protein complex that includes the scaffolding proteins sans (*USH1G*) and harmonin (*USH1C*), as well as the molecular motor myosin 7a (*USH1B*).

The small number of channels in a hair cell, along with the lack of high-affinity ligands with which to label them, explains why the biochemical identity of the transduction channels has long remained uncertain. However, recent genetic, biochemical, and biophysical experiments indicate that four integral transmembrane proteins are intimately related to the transduction channel: transmembrane channel-like proteins 1 and

A

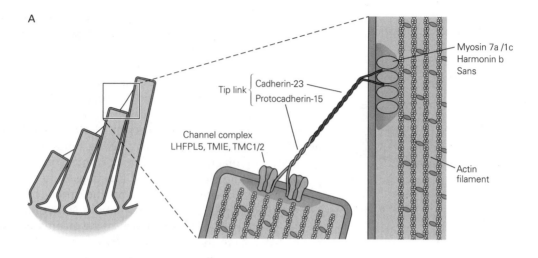

B   Model of the transduction-channel complex

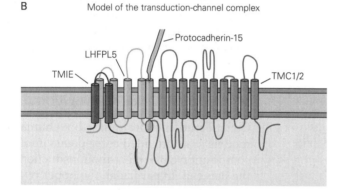

Figure 26–10 Molecular composition of the transduction machinery.

A. The tip link is composed of the heterophilic association of protocadherin 15 and cadherin 23. Two transduction channels are localized near the lower insertion point of the tip link at the tip of the shorter stereocilium. Each channel is part of a molecular complex that includes the proteins TMC1/2, LHFPL5, and TMIE. At the upper insertion point on the flank of the longer stereocilium, cadherin 23 interacts with harmonin b and the molecular motor myosin 7a, which both bind to actin and thus anchor the tip link. The protein sans serves as a scaffolding protein. In vestibular hair cells, myosin 1c may set the tip link under tension, but the presence of this motor protein is uncertain in cochlear hair cells.

B. Model of the transduction-channel complex. TMC1/2, LHFPL5, and TMIE interact with protocadherin 15 and, thus, with the lower end of the tip link. TMIE also interacts with LHFPL5. The detailed arrangement of these proteins within the transduction apparatus is still unknown. Unlike what the figure suggests, TMC1 has been proposed to assemble as a dimer, with each TMC1 molecule contributing a permeation pathway for cations. (Adapted, with permission, from Wu and Müller 2016 and Pan et al. 2018.)

2 (TMC1 and TMC2), tetraspan membrane protein in hair-cell stereocilia (TMHS; official nomenclature LHFPL5), and transmembrane inner-ear-expressed gene (TMIE; Figure 26–10). Mechanotransduction is abolished in mouse hair cells lacking TMIE, even though all other known components of the transduction machinery appear to be properly in place. However, because TMIE contains only two predicted transmembrane domains, it seems highly unlikely that this protein alone constitutes an ion channel. In the absence of LHFPL5, the conductance of the transduction channel is reduced, but significant transduction currents can still be measured, suggesting that this protein is not an essential part of the channel pore.

Multiple lines of evidence advocate for TMC1 and TMC2 as components of the transduction channel. Both proteins are localized near the lower insertion point of the tip link, where the transduction current enters the hair cell, interact with the tip-link constituent protocadherin 15, and their onset of expression coincides with that of mechanoelectrical transduction. In addition, transduction channels in a mouse carrying a single point mutation in the *Tmc1* gene show lower conductance and Ca^{2+} permeability, indicating that TMC1 is very close to the channel's pore. Individual cysteine substitutions made at sites that are predicted to be in or near the channel pore confirm that TMC1 belongs to the main conductive pathway. Indeed, covalent modification of the cysteine residues with a positively charged reagent leads to a reduction of the single-channel conductance in several TMC1 mutants. The cysteine-modifying reagent has no effect when the reagent is applied after the hair bundle has been deflected towards its short edge (to close the transduction channels) or when access to the channel pore is prevented by a channel blocker. TMC1 is thus unlikely to constitute an accessory channel subunit that forms a vestibule to the pore-forming protein(s). Instead, the evidence is strong that TMC1 forms at least part of the pore of the transduction channel. In cochlear hair cells, TMC1 and TMC2 are coexpressed during neonatal development, but only TMC1 expression is maintained through adulthood.

Dynamic Feedback Mechanisms Determine the Sensitivity of the Hair Cells

Hair cells must cope with acoustic stimuli that have a very low energy content. If the stimulus consists of a periodic signal, such as the sinusoidal pressure of a pure tone, a detection system can increase the signal-to-noise ratio by enhancing selectively the response to a relevant frequency. Hair cells respond best at a characteristic frequency of acoustic stimulation. The frequency selectivity of a given hair cell results in part from passive extrinsic filtering of its mechanical input, in particular as a result of the tonotopic arrangement of the mammalian basilar membrane. In addition, when it is appropriate that low-frequency inputs be disregarded, hair cells possess a unique mechanism of adaptation that acts as a high-pass filter. Hair cells also employ mechanical amplification that enhances and further tunes their mechanosensitivity.

Hair Cells Are Tuned to Specific Stimulus Frequencies

Every cochlear hair cell is most sensitive to stimulation at a specific frequency, termed its characteristic, natural, or best frequency. On average, the characteristic frequencies of adjacent inner hair cells differ by approximately 0.2%; adjacent piano strings, in comparison, are tuned to frequencies some 6% apart. Because the traveling wave evoked even by a pure sinusoidal stimulus spreads somewhat along the basilar membrane, the sensitivity of a cochlear hair cell extends within a limited range above and below its characteristic frequency, the more so with a greater level of stimulation. At low levels, a pure tone recruits

approximately 100 hair cells. The frequency sensitivity of a hair cell may be displayed as a tuning curve. To construct a tuning curve, an experimenter stimulates the ear with pure tones at numerous frequencies below, at, and above the cell's characteristic frequency. The level of stimulation is adjusted for each frequency until the cell's response reaches a predefined criterion magnitude. The tuning curve is then a graph of sound level, presented logarithmically in decibels SPL, as a function of stimulus frequency.

The tuning curve for an inner hair cell is typically V-shaped (Figure 26–11). The curve's tip represents the cell's characteristic frequency, the frequency that produces the criterion response for the lowest level of the stimulus. Sounds of greater or lesser frequencies require higher levels to excite the cell to the criterion response. As a consequence of the traveling wave's shape, the slope of a tuning curve is far steeper on its high-frequency flank than on its low-frequency flank.

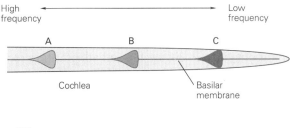

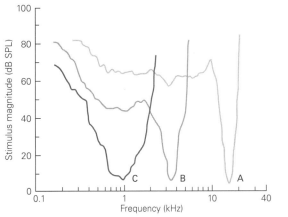

Figure 26–11 Tuning curves for cochlear hair cells. To construct a curve, the experimenter presents sound at several frequencies. At each frequency, the stimulus intensity is adjusted until the cell produces a criterion response, here 1 mV. The curve thus reflects the threshold of the cell for stimulation over a range of frequencies. Each cell is most sensitive to a specific frequency, its characteristic frequency. The threshold rises briskly—the sensitivity falls rapidly—as the stimulus frequency is raised or lowered. The characteristic frequency depends on the position of the hair cell along the longitudinal axis of the cochlea. (Reproduced, with permission, from Kiang 1980. Copyright © 1980 Acoustical Society of America.)

In the same way as a tuning fork's resonant frequency depends on the size of its tines, the heights of the hair bundles vary systematically along the tonotopic axis. Hair cells that respond to low-frequency stimuli have the tallest bundles, whereas those that respond to the highest-frequency signals possess the shortest bundles. In the human cochlea, for example, an inner hair cell with a characteristic frequency of 20 kHz bears a 4-µm hair bundle. At the opposite extreme, a cell sensitive to a 20-Hz stimulus has a bundle more than 7 µm high. A similar morphological gradient is observed with outer hair cells, supplementing the extrinsic tuning accomplished by the basilar membrane.

Hair Cells Adapt to Sustained Stimulation

Despite the precision with which a hair bundle grows, it cannot develop in such a way that the sensitive transduction apparatus is always perfectly poised at its position of greatest mechanosensitivity. Some mechanism must compensate for developmental irregularities, as well as for environmental changes, by adjusting the gating springs so that transduction channels are responsive to weak stimuli at the bundle's resting position. To ensure this, an adaptation process continuously resets the hair bundle's range of mechanical sensitivity. As a result of adaptation, a hair cell can maintain a high sensitivity to transient stimuli while rejecting static inputs a million times as large.

Adaptation manifests itself as a progressive decrease in the receptor potential during protracted deflection of the hair bundle (Figure 26–12). The process is not one of desensitization, for the responsiveness of the receptor persists. Instead, during a prolonged step stimulus, the sigmoidal relationship between the initial receptor potential and the bundle's position shifts in the direction of the applied stimulus. As a result, the membrane potential of the hair cell progressively returns to near its resting value. Adaptation is incomplete, however; the relation between the membrane potential and the bundle position shifts by approximately 80% of the deflected position.

How does adaptation occur? Because the mechanical force exerted by a hair bundle changes as adaptation proceeds, the process evidently involves an adjustment in the tension borne by the gating springs. It appears likely that the structure anchoring the upper end of each tip link, the *insertional plaque*, is repositioned during adaptation by an active molecular motor (Figure 26–12). The transduction channels are inherently more stable in a closed state, for they close when the tip links are disrupted. A motor is thus also required to maintain a significant fraction (10%–50%) of the transduction

channels open at rest by continuously pulling on the gating springs. Several dozen myosin molecules associated with the upper end of each tip link are thought to maintain tension by ascending the actin core of the stereocilium and pulling the link's insertion with them.

When a stimulus step increases the tension in a gating spring, the associated transduction channel opens, permitting an influx of cations. As Ca^{2+} ions accumulate in the stereociliary cytoplasm, they reduce the upward force of the myosin molecules, thereby shortening the gating spring. When the spring reaches its resting tension, closure of the channel reduces the Ca^{2+} influx to its original level, restoring a balance between the upward force of myosin and the downward tension in the spring.

Hair bundles contain at least five isoforms of myosin, the motor molecule associated with motility along actin filaments (Chapter 31). In vestibular hair cells, immunohistochemical studies and site-directed mutagenesis implicate myosin 1c in adaptation. In cochlear hair cells, the role of myosin 1c in adaptation has remained elusive. Another motor protein, myosin 7a, is present near the upper insertion point of the tip link and mutations in the corresponding gene (*USH1B*) are associated with deafness. Hair-cell bundles defective for myosin 7a are disorganized, suggesting that this motor is involved at least in hair-bundle development.

If it were only to set the operating point of the transduction apparatus, adaptation could afford to operate on much slower timescales than the period of acoustic stimuli. This is the case in response to large deflections of the hair bundle, for which the time constant of adaptation is approximately 20 ms or more when endolymph bathes the hair bundle. This slow adaptation is compatible with the activity of a myosin-based motor driven by the cyclical hydrolysis of adenosine triphosphate (ATP). Yet, after being pulled open by an excitatory step stimulus of small magnitude, transduction channels reclose with typical timescales of less than 1 ms and thus short enough to be compatible with auditory frequencies. Current models posit that Ca^{2+} ions entering a hair cell through a transduction channel bind to or near the channel's pore, thereby energetically favoring channel closure. The kinetics of this fast adaptation varies systematically along the tonotopic axis of auditory organs, indicating that adaptation may help in setting the hair cell's characteristic frequency of maximal responsiveness. In addition, the reciprocal relationship between channel gating and tip-link tension means that adaptive channel rearrangements evoke internal forces that drive active hair-bundle movements. The mechanical correlate of adaptation

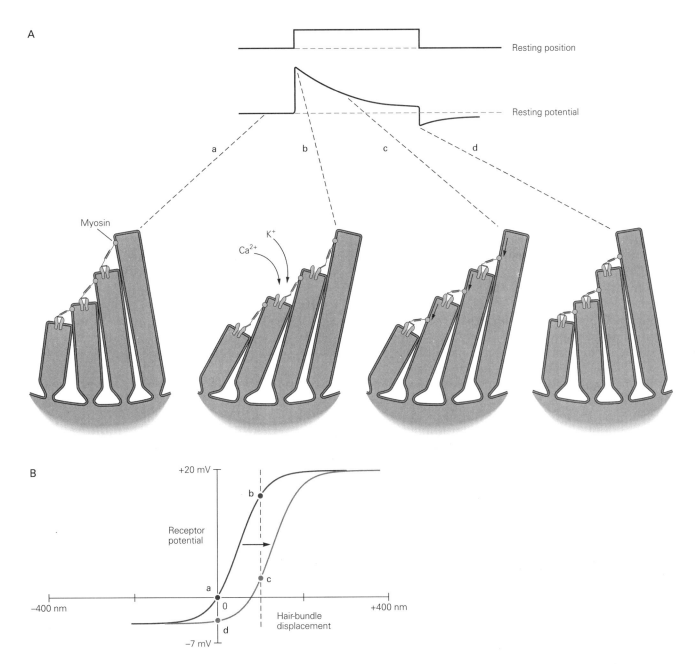

Figure 26–12 Adaptation of mechanoelectrical transduction in hair cells. A. Prolonged deflection of the hair bundle in the positive direction (**upper trace**) elicits an initial depolarization followed by a decline to a plateau and an undershoot at the cessation of the stimulus (**lower trace**). The four schematic hair-cell bundles illustrate the states of the transduction channels before (**a**) and during the illustrated phases of the adaptation (**b–d**). Initially, the stimulation increases tension in the tip link (second bundle), thus opening transduction channels. As stimulation continues, however, a tip link's upper attachment is thought to slide down the stereocilium, allowing each channel to close during adaptation (third bundle). Prolonged deflection of the hair bundle in the negative direction elicits a complementary response. The cell is slightly hyperpolarized at first but shows a rebound depolarization at the end of stimulation; tension is restored to the initially slack tip link as myosin molecules actively pull up the link's upper insertion.

B. As adaptation proceeds, the sigmoidal relation between hair bundle displacement and the receptor potential of the hair cell shifts to the right along the abscissa, in the direction of the new hair-bundle position (**dashed line**), without substantial changes in the curve's shape or amplitude. The shift explains why the receptor potential decreases over time as shown in **A** between states **b** and **c**, restoring the membrane potential of the hair cell to near its value when there is no stimulus. This result implies that adaptation restores mechanical sensitivity to small rapid deflections of the hair bundle in the presence of a protracted stimulus that would otherwise saturate mechanoelectrical transduction. The four states of the transduction channels shown in **A** are marked (**a–d**). (Adapted, with permission, from Hudspeth and Gillespie 1994.)

thus provides feedback that can enhance the stimulus to the hair cell.

Sound Energy Is Mechanically Amplified in the Cochlea

The inner ear faces an important obstacle to efficient operation: A large portion of the energy in an acoustic stimulus goes into overcoming the damping effects of cochlear liquids on hair-cell and basilar-membrane motion rather than into excitation of hair cells. The sensitivity of the cochlea is too great, and auditory frequency selectivity too sharp, to result solely from the inner ear's passive mechanical properties. The cochlea must therefore possess some means of actively amplifying sound energy.

One indication that amplification occurs in the cochlea comes from measurements of the basilar membrane's movements with sensitive laser interferometers. In a preparation stimulated with low-level sound, the basilar-membrane motion is highly dependent on frequency. The movement is maximal at the appropriate frequency for the position at which the measurement is made—the characteristic frequency—but drops abruptly at higher or lower frequencies. As the sound level is increased, however, the frequency selectivity of the vibration becomes less sharp; the peak in the relationship between amplitude and frequency broadens. In addition, the membrane's sensitivity to sound, defined as the vibration amplitude per unit of sound pressure, declines precipitously. When stimulated at the characteristic frequency, the sensitivity of basilar-membrane motion to stimulation at 80 dB SPL is less than 1% of that for 10 dB SPL excitation. The basilar membrane displays a compressive nonlinearity that accommodates the millionfold variation of sound pressure that characterizes audible sounds (0–120 dB SPL) into only two to three orders of magnitude of vibration amplitude (±0.3–300 nm). The sensitivity and frequency selectivity predicted in modeling studies of a passive cochlea correspond to those observed with high-level stimuli. This result implies that the motion of the basilar membrane is augmented more than 100-fold during low-level stimulation at the characteristic frequency, but that amplification diminishes progressively as the stimulus grows in strength. Consequently, amplification lowers the threshold of hearing by more than 40 to 50 dB SPL.

In addition to this circumstantial evidence, experimental observations support the idea that the cochlea contains a mechanical amplifier. When a normal human ear is stimulated with a click, that ear emits one to several measurable pulses of sound within

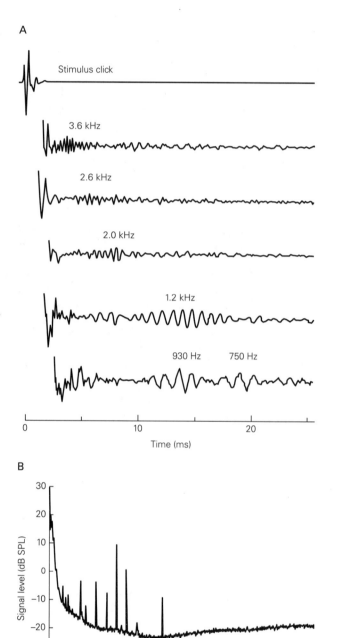

Figure 26–13 The cochlea actively emits sounds.

A. The records display evoked otoacoustic emissions from the ears of five human subjects. A brief click (**top trace**) was played into each ear through a miniature speaker. A few milliseconds later, a tiny microphone in the external auditory meatus detected one or more bursts of sound emission from the ear. (Adapted, with permission, from Wilson 1980. Copyright © 1980 Elsevier B.V.)

B. Under suitably quiet recording conditions, spontaneous otoacoustic emissions occur in most normal human ears. This spectrum displays the acoustic power of six prominent emissions and several smaller ones from one ear. (Reproduced, with permission, from Murphy et al. 1995. Copyright © 1995, Acoustical Society of America.)

milliseconds (Figure 26–13A). Because they can carry more energy than the stimulus, these so-called *evoked otoacoustic emissions* cannot simply be echoes; they represent the emission of mechanical energy by the cochlea, triggered by acoustic stimulation. In accordance with the compressive nonlinearity associated with cochlear amplification, the relative level of the emissions decreases with the stimulus level.

A still more compelling manifestation of the cochlea's active amplification is *spontaneous otoacoustic emission*. When a suitably sensitive microphone is used to measure sound pressure in the ear canals of subjects in a quiet environment, at least 70% of normal human ears continuously emit one or more pure tones (Figure 26–13B). Although these sounds are generally too faint to be directly audible by others, physicians have reported actually hearing sounds emanating from the ears of newborns!

What is the source of evoked and spontaneous otoacoustic emissions, and thus presumably of cochlear amplification as well? Several lines of evidence implicate outer hair cells as the elements that enhance cochlear sensitivity and frequency selectivity and hence act as the motors for amplification. The afferent nerve fibers that extensively innervate the inner hair cells make only minimal contacts with the outer hair cells (Figure 26–4). Instead, the outer hair cells receive an extensive efferent innervation that, when activated, decreases cochlear sensitivity and frequency discrimination. In addition, when stimulated electrically, an isolated outer hair cell displays the unique phenomenon of electromotility: The cell body shortens by up to several micrometers when depolarized and elongates when hyperpolarized (Figure 26–14). This response can occur at frequencies exceeding 80 kHz, an attractive feature for a process postulated to assist high-frequency hearing.

The energy for these movements is drawn from the experimentally imposed electrical field rather than from hydrolysis of an energy-rich substrate such as ATP. Movement occurs when changes in the electric field across the membrane reorient molecules of the protein prestin. The concerted movement of several million of these molecules, which are packed in the lateral cell membranes of outer hair cells, changes the membrane's area and thus the cell's length. When an outer hair cell transduces mechanical stimulation of its hair bundle into receptor potentials, cochlear amplification might then occur when voltage-induced movement of the cell body augments basilar-membrane motion. Consistent with this hypothesis, mutation of certain amino acid residues required for the voltage sensitivity of prestin abolishes the active process in mice.

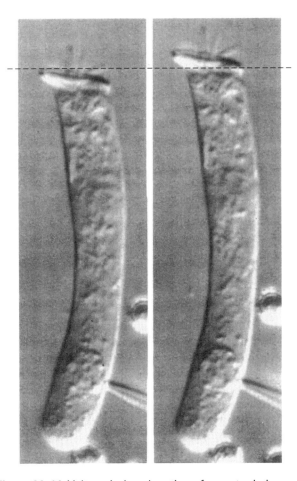

Figure 26–14 Voltage-induced motion of an outer hair cell. Depolarization of an isolated outer hair cell through the electrode at its base causes the cell body to shorten (*left*); hyperpolarization causes it to lengthen (*right*). The oscillatory motions of outer hair cells may provide the mechanical energy that amplifies basilar-membrane motion and thus enhances the sensitivity of human hearing. (Reproduced, with permission, from Holley and Ashmore 1988.)

Because sharp frequency selectivity, high sensitivity, and otoacoustic emissions are also observed in animal species that lack outer hair cells and lack high concentrations of prestin, electromotility cannot be the only form of mechanical amplification by hair cells. In addition to detecting stimuli, hair bundles are also mechanically active and contribute to amplification. Hair bundles can make spontaneous back-and-forth movements that have been shown in some nonmammals to underlie spontaneous otoacoustic emissions. Under experimental conditions, bundles can exert force against stimulus probes, performing mechanical work and thereby amplifying the input. In vitro experiments indicate that active hair-bundle motility contributes to the cochlear active process even in the mammalian ear.

Active hair-bundle movements can be fast enough to mediate otoacoustic emissions at sound frequencies at least as high as a few kilohertz. However, it remains uncertain whether bundles can generate forces at the very high frequencies at which sharp frequency selectivity and otoacoustic emissions are observed in the mammalian cochlea. Active hair-bundle motility and somatic electromotility may function synergistically, with the former serving metaphorically as a tuner and preamplifier and the latter as a power amplifier. Alternatively, hair-bundle motility may dominate at relatively low frequencies but be superseded by electromotility at higher frequencies.

Cochlear Amplification Distorts Acoustic Inputs

When stimulated by two tones at nearby frequencies f_1 and f_2 ($f_1 < f_2$), the basilar membrane vibrates not only at these frequencies but also at additional frequencies—the distortion products—that are not present in the acoustic stimulus. As reported by the Italian violinist Giuseppe Tartini in the 18th century, distortion products can be heard as phantom tones in the auditory percept. Remarkably, the cubic difference tone $2f_1 - f_2$ is heard even at very low sound levels, and its magnitude grows in proportion to the stimulus. Correspondingly, the relative level of distortion remains practically constant over a broad range of sound levels. This phenomenon is explained by the particular form of the compressive nonlinearity associated with cochlear amplification. Clearly, the cochlea does not work as a high-fidelity sound receiver. Distorted cochlear vibrations are strong enough to be reemitted from the ear canal as *distortion-product otoacoustic emissions*. Because they are a property of healthy ears, these emissions are extensively used to screen hearing in newborns.

The Hopf Bifurcation Provides a General Principle for Sound Detection

Detailed in vivo and in vitro studies have revealed four cardinal features of auditory responsiveness. First, an active amplification process lowers the detection threshold. Second, because amplification operates only near a characteristic frequency, the input to the sensory system is actively filtered, which sharpens frequency selectivity. Third, for stimulation near the characteristic frequency, the response displays a compressive nonlinearity that represents a wide range of stimulus levels by a much narrower range of vibration amplitudes. Finally, even in the absence of a stimulus, mechanical activity can produce self-sustained oscillations that result in otoacoustic emissions.

These features have been recognized as signatures of an active dynamical system—a critical oscillator—that operates on the verge of an oscillatory instability termed the Hopf bifurcation (Box 26–1). They are generic: They do not depend on the subcellular and molecular details of the candidate mechanism that brings the system to the brink of spontaneous oscillation. The fact that active hair-bundle motility demonstrates a Hopf bifurcation in vitro provides further evidence that this mechanism contributes to cochlear amplification.

Within this framework, the characteristic frequency is set by that of the critical oscillator. The cochlear partition may be viewed as a set of active oscillatory modules that are hydrodynamically coupled by the cochlear fluids and with characteristic frequencies tonotopically distributed along the longitudinal axis of the cochlea. The hypothesis of critical oscillation facilitates modeling of the traveling wave and cochlear amplification. This is because the generic behaviors of a critical oscillator can be described by a single equation termed the "normal form" (Box 26–1). A critical oscillator is ideally suited for auditory detection, even if its inherent nonlinearity yields pronounced distortions in response to complex sound stimuli. Nonlinear interference between the frequency components of complex stimuli in fact appears as a necessary price to pay for the exquisite sensitivity, sharp frequency selectivity, and wide dynamic range of auditory detection afforded by a critical oscillator. The Hopf bifurcation provides generic properties that account for numerous disparate experimental observations at the level of a single hair bundle, of the basilar membrane, and even in psychoacoustics. This physical principle of auditory detection thus greatly simplifies our understanding of hearing. Although this chapter focusses primarily on mammalian hearing, the common necessity to hear with high sensitivity and sharp frequency selectivity poses similar physical constraints to the ears of all land vertebrates. These constraints have led to the independent evolution of ears that share similar structural features and whose operation is based on similar physical principles, including the use of critical oscillators to amplify sound (Box 26–2).

Hair Cells Use Specialized Ribbon Synapses

Being sensory receptors, hair cells form synapses with sensory neurons. The basolateral membrane of each cell contains several presynaptic active zones at which

Box 26–1 Generic Properties Near a Hopf Bifurcation

A dynamical system displays a Hopf bifurcation when it abruptly transits from quiescence to a state of spontaneous oscillation while subject to continuous variation of a control parameter C. If the system is poised in the vicinity of the critical point, at which $C = C_C$, its steady-state response to sinusoidal forces can be described by a single equation—the "normal form"—of a complex variable Z:

$$\Lambda \frac{dZ}{dt} \cong -\Lambda(C_C - C - 2i\pi f_C)Z - B\,|Z|^2 Z + F \quad \textbf{(26–1)}$$

Here, the real part of Z may represent the position of the basilar membrane or of the hair bundle, Λ is a friction coefficient, and F is the external force provided by a sound stimulus. In the absence of an external force ($F = 0$), spontaneous oscillations emerge when the control parameter becomes larger than the critical value C_C (Figure 26-15A); the parameter f_C corresponds to the frequency of spontaneous oscillation at the critical point. A Hopf bifurcation must be driven by an active process, whereby the system mobilizes internal resources of energy to power spontaneous movements. A system operating precisely at the critical point is called a critical oscillator. The response of a critical oscillator to sinusoidal stimuli is endowed with generic properties (Figure 26–15B,C) that are characteristic of sound detection in the ear.

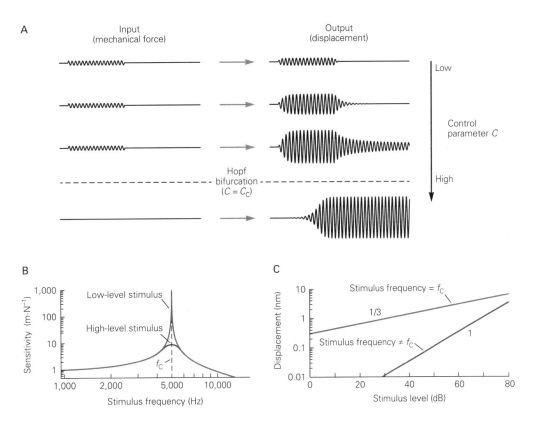

Figure 26–15 Frequency-selective amplification near a Hopf bifurcation.

A. As the control parameter C increases to approach the critical value C_C, the response to a sinusoidal stimulus of constant amplitude increases: The system gets more sensitive. For $C > C_C$, the system oscillates spontaneously at a constant amplitude and frequency, even if no stimulus is applied. The Hopf bifurcation corresponds to $C = C_C$. (Reproduced, with permission, from Hudspeth 2014.)

B. When the system is poised at the bifurcation ($C = C_C$), the sensitivity to a low-level stimulus is greatly enhanced near the characteristic frequency f_C (here 5000 Hz; **dashed line**), but drops rapidly if the stimulus is detuned from this frequency (**red line**). For a stimulus 60 dB more intense, the maximal sensitivity is 100-fold lower, and the peak in sensitivity is 100-fold broader (**blue line**): A critical oscillator

enhances the weakest stimuli much more than the strongest and with much sharper frequency selectivity. (Adapted, with permission, from Hudspeth, Jülicher, and Martin 2010.)

C. At the characteristic frequency f_C, the response—here displacement—displays a compressive growth that corresponds to a line of slope 1/3 in this doubly logarithmic plot (**red line**): A large range of stimulus levels is represented by a much narrower range of response amplitudes. In contrast, when the frequency of the stimulus departs significantly from the characteristic frequency (**blue line**), the response is proportional to the input, corresponding to a line of slope unity. By amplifying weak inputs near its characteristic frequency, the critical oscillator lowers the stimulus level required to elicit a threshold vibration, here by 60 dB for a threshold of 0.3 nm. (Adapted, with permission, from Hudspeth, Jülicher, and Martin 2010.)

Box 26–2 The Evolutionary History of Hearing Resulted in Similarities Between Groups

Mammals are not alone in possessing sensitive and frequency-selective hearing. Amphibians and reptiles, including birds, also do. It is a remarkable fact that these various groups of land vertebrates actually acquired their good hearing systems largely independently. The small, dedicated auditory receptor organ was present in the inner ear of their common ancestor. Much later, the ancestors of modern lizards, birds, and mammals each independently evolved middle-ear systems with eardrums collecting sound from the outside world. Some species, such as birds and their relatives, even evolved two groups of sensory hair cells that have a division of labor similar to that of mammalian inner and outer hair cells. Comparison of middle-ear and inner-ear structures and functions across all living vertebrate groups has revealed that they share many common features and that hearing performance is largely comparable between them. Sound amplification associated with active hair-bundle motility was already present in the very first hair cells that evolved, even before the first fishes. This amplifying system was inherited by all groups and, as described earlier, plays a critical role in improving hearing sensitivity and sharpening frequency selectivity. The greatest difference between mammals and the other groups is that the upper frequency limit of hearing is generally higher in mammals. Nonmammalian ears are limited in response to

frequencies lower than about 12 to 14 kHz, whereas some mammals can hear beyond 100 kHz.

In addition to active hair-bundle motility, the second mechanism that tunes individual hair cells to specific frequencies in many nonmammalian ears is electrical in nature. In many fishes, amphibians, and birds, the membrane potential of each hair cell resonates at a particular frequency. Several factors, including alternative splicing of the mRNA encoding cochlear K^+ channels and expression of these channels' auxiliary β subunit, tune the characteristic frequency of the resonance along the tonotopic axis of the auditory organ. Whether electrical resonance contributes to frequency tuning in the ears of mammals, including humans, remains uncertain. It is plausible that mammalian hair cells use instead an interplay between somatic electromotility, which seems absent in nonmammalian species, and the micromechanical environment, including hair-bundle motility, to actively amplify and filter their inputs.

The key signatures of a Hopf bifurcation have been recognized in spontaneous mechanical oscillations of the hair bundle, in electrical oscillations of the membrane potential, and in sound-evoked vibration of the basilar membrane. It is likely that the parallel evolution of hearing organs in different groups of vertebrates resulted in several ways of benefitting from the generic properties of critical oscillation.

chemical neurotransmitter is released. An active zone is characterized by four prominent morphological features (Figure 26–16).

A presynaptic dense body or synaptic ribbon lies in the cytoplasm adjacent to the release site. This fibrillar structure may be spherical, ovoidal, or flattened, and usually measures a few hundred nanometers across. The dense body resembles the synaptic ribbon of a photoreceptor cell and represents a specialized elaboration of the smaller presynaptic densities found at many other synapses. In addition to molecular components shared with conventional synapses, ribbon synapses contain large amounts of the protein ribeye.

The presynaptic ribbon is surrounded by clear synaptic vesicles, each 35 to 40 nm in diameter, which are attached to the dense body by tenuous filaments. Between the dense body and the presynaptic cell membrane lies a striking presynaptic density that comprises

several short rows of fuzzy-looking material. Within the cell membrane, rows of large particles are aligned with the strips of presynaptic density. These particles include the Ca^{2+} channels involved in the release of transmitter as well as the K^+ channels that participate in electrical resonance in nonmammalian vertebrates.

Studies of nonmammalian experimental models show that, as with most other synapses (Chapter 15), the release of transmitter by hair cells is evoked by presynaptic depolarization and requires influx of Ca^{2+} from the extracellular medium. Hair cells lack synaptotagmins 1 and 2, however, and the role of those proteins as rapid Ca^{2+} sensors has probably been assumed by the protein otoferlin, which also promotes the replenishment of synaptic vesicles. Although glutamate is the principal afferent neurotransmitter, other substances are released as well.

The presynaptic apparatus of hair cells has several unusual features that underlie the signaling abilities of

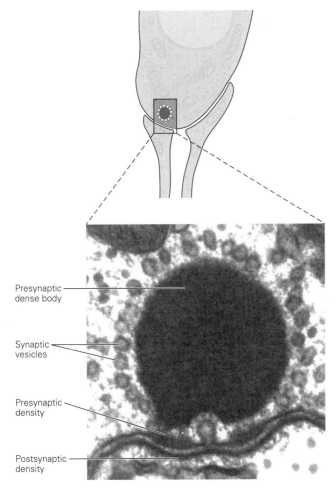

Presynaptic
dense body

Synaptic
vesicles

Presynaptic
density

Postsynaptic
density

Figure 26–16 The presynaptic active zone of a hair cell. This transmission electron micrograph shows the spherical presynaptic dense body or synaptic ribbon that is characteristic of the hair cell's presynaptic active zone. It is surrounded by clear synaptic vesicles. Beneath the ribbon lies a presynaptic density, in the middle of which one vesicle is undergoing exocytosis. A modest postsynaptic density lies along the inner aspect of the plasmalemma of the afferent terminal. (Reproduced, with permission, from Jacobs and Hudspeth 1990.)

these cells. At rest, inner hair cells continuously release synaptic transmitter. The rate of transmitter release can be modulated upward or downward, depending on whether the hair cell is respectively depolarized or hyperpolarized. Consistent with this observation, some Ca^{2+} channels of hair cells are activated at the resting potential, providing a steady leak of Ca^{2+} that evokes transmitter release from unstimulated cells. Another unusual feature of the hair cell's synapses is that, like those of photoreceptors, they must be able to release neurotransmitter reliably in response to a threshold receptor potential of only 100 μV or so. This

feature, too, depends on the fact that the presynaptic Ca^{2+} channels are activated at the resting potential.

Outer hair cells receive inputs from neurons in the brainstem in the form of large boutons on their basolateral surfaces (Figure 26–4). This efferent system desensitizes the cochlea by hyperpolarizing outer hair cells, which turns down the active process. The efferent terminals contain numerous clear synaptic vesicles about 50 nm in diameter, as well as a smaller number of larger, dense-core vesicles. The principal transmitter at these synapses is acetylcholine (ACh); calcitonin gene–related peptide (CGRP) also occurs in efferent terminals and may be co-released with ACh. ACh binds to nicotinic ionotropic receptors consisting of α9 and α10 subunits that have a substantial permeability to Ca^{2+} as well as to Na^+ and K^+. The Ca^{2+} that enters through these channels activates small-conductance Ca^{2+}-sensitive K^+ channels (SK channels), whose opening leads to a protracted hyperpolarization. The cytoplasm of a hair cell immediately beneath each efferent terminal holds a single cisterna of smooth endoplasmic reticulum. This structure may be involved in the reuptake of the Ca^{2+} that enters the cytoplasm in response to efferent stimulation, thus accelerating the return to the cell's resting potential.

Auditory Information Flows Initially Through the Cochlear Nerve

Bipolar Neurons in the Spiral Ganglion Innervate Cochlear Hair Cells

Information flows from cochlear hair cells to neurons whose cell bodies lie in the cochlear ganglion. The central processes of these bipolar neurons form the cochlear division of the vestibulocochlear nerve (eighth cranial nerve). Because this ganglion follows a spiral course around the bony core of the cochlea, it is also called the *spiral ganglion*. Approximately 30,000 ganglion cells innervate the hair cells of each inner ear.

The afferent pathways from the human cochlea reflect the functional distinction between inner and outer hair cells. At least 90% of the spiral ganglion cells terminate on inner hair cells (Figure 26–17). Each axon contacts only a single inner hair cell, but each cell directs its output to many nerve fibers, on average nearly 10. This arrangement has three important consequences.

First, the neural information from which hearing arises originates almost entirely at inner hair cells. Second, because the output of each inner hair cell is

Figure 26–17 Innervation of cochlear hair cells. The great majority of sensory axons (orange) in the cochlea carry signals from inner hair cells, each of which constitutes the sole input to an average of 10 axons. A few sensory axons of small caliber transmit information from the outer hair cells. Efferent axons (green) largely innervate the outer hair cells and do so directly. In contrast, efferent innervation of inner hair cells is sparse and occurs on the sensory axon terminals. (Adapted, with permission, from Spoendlin 1974.)

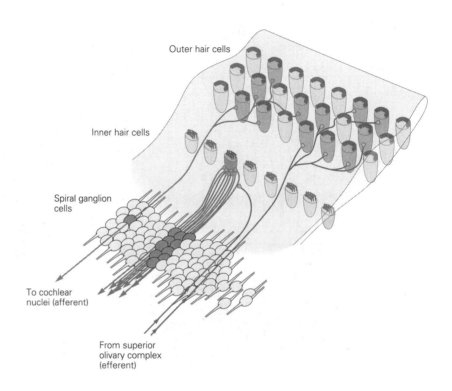

sampled by many afferent nerve fibers, the information from one receptor is encoded independently in parallel channels. Third, at any point along the cochlear spiral, or at any position within the spiral ganglion, each ganglion cell responds best to stimulation at the characteristic frequency of the presynaptic hair cell. The tonotopic organization of the auditory neural pathways thus begins at the earliest possible site, immediately postsynaptic to inner hair cells.

Relatively few cochlear ganglion cells contact outer hair cells, and each such neuron extends branching terminals to numerous outer hair cells. Although the ganglion cells that receive input from outer hair cells are known to project into the central nervous system, these neurons are so few that it is not certain whether their projections contribute significantly to the analysis of sound.

The patterns of efferent and afferent connections of cochlear hair cells are complementary. Mature inner hair cells do not receive efferent input; just beneath these cells, however, are extensive axo-axonic synaptic contacts between efferent axon terminals and the endings of afferent nerve fibers. In contrast, other efferent nerves have extensive connections with outer hair cells on their basolateral surfaces. Each outer hair cell receives input from several large efferent terminals, which fill most of the space between the cell's base and the associated supporting cell, leaving little space for afferent terminals.

Cochlear Nerve Fibers Encode Stimulus Frequency and Level

The acoustic sensitivity of axons in the cochlear nerve mirrors the connection pattern of the spiral ganglion cells to the hair cells. Each axon is most responsive to a characteristic frequency. Stimuli of lower or higher frequency also evoke responses, but only when presented at greater levels. An axon's responsiveness may be characterized by a frequency selectivity, or tuning, curve which is V-shaped like the curves for basilar-membrane motion and hair-cell sensitivity (Figure 26–11). The tuning curves for nerve fibers with different characteristic frequencies resemble one another but are shifted along the frequency axis.

The relationship between sound level in decibels SPL and firing rate in each fiber of the cochlear nerve is approximately linear. Because of the dependence of decibel level on sound pressure, this relation implies that sound pressure is logarithmically encoded by neuronal activity. At the upper end of a fiber's dynamic range, very loud sounds saturate the response. Because an action potential and the subsequent refractory period each last almost 1 ms, the greatest sustainable firing rate is about 500 spikes per second.

Even among nerve fibers with the same characteristic frequency, the threshold of responsiveness varies from axon to axon. The most sensitive fibers, whose response thresholds extend down to approximately

0 dB SPL, characteristically have high rates of spontaneous activity and produce saturating responses for stimulation at moderate intensities, approximately 30 dB SPL. At the opposite extreme, the least sensitive afferent fibers have very little spontaneous activity and much higher thresholds, but respond in a graded fashion to levels even in excess of 100 dB SPL. The activity patterns of most fibers range between these extremes.

The afferent neurons of lowest sensitivity contact the surface of an inner hair cell nearest the axis of the cochlear spiral. The most sensitive afferent neurons, on the other hand, contact the hair cell's opposite side. The multiple innervation of each inner hair cell is therefore not redundant. Instead, the output from a given hair cell is directed into parallel channels of differing sensitivity and dynamic range.

The firing pattern of fibers in the eighth cranial nerve exhibits both phasic and tonic components. Brisk firing occurs at the onset of a tone but, as adaptation occurs, the firing rate declines to a plateau level over a few tens of milliseconds. When stimulation ceases, there is usually a transitory cessation of activity with a similar time course to that of adaptation, before gradual resumption of the spontaneous firing rate (Figure 26–18).

When a periodic stimulus such as a pure tone is presented, the firing pattern of a cochlear nerve fiber encodes information about the periodicity of the stimulus. For example, a relatively low-frequency tone at a moderate intensity might produce one spike in a nerve fiber during each cycle of stimulation. The phase of firing is also stereotyped. Each action potential might occur, for example, during the compressive phase of the stimulus. As the stimulation frequency rises, the stimuli eventually become so rapid that the nerve fiber can no longer produce action potentials on a cycle-by-cycle basis. Up to a frequency in excess of 3 kHz, however, phase-locking persists; a fiber may produce an action potential only every few cycles of the stimulus, but its firing continues to occur at a particular phase in the stimulus cycle.

Periodicity in neuronal firing enhances the information about the stimulus frequency. Any pure tone of sufficient level evokes firing in numerous cochlear nerve fibers. Those fibers whose characteristic frequency coincides with the frequency of the stimulus respond at the lowest stimulus level, but respond still more briskly for stimuli of moderate intensity. Other nerve fibers with characteristic frequencies further from the stimulus also respond, although less vigorously. Regardless of their characteristic frequencies, however, all the responsive fibers may display phase

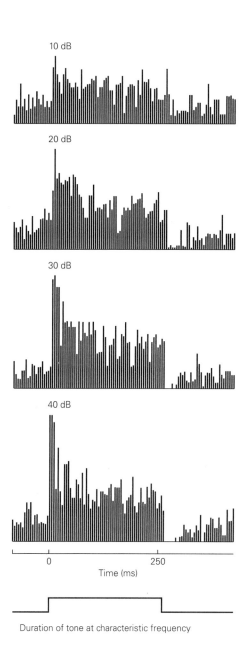

Figure 26–18 The firing pattern of a cochlear nerve fiber. A cochlear nerve fiber is stimulated for somewhat more than 250 ms with a tone burst at about 5 kHz, the cell's characteristic frequency. After a quiet period, the stimulus is repeated. Histograms show the average response patterns of the fiber as a function of stimulus level. The sample period is divided into discrete temporal bins, and the number of spikes occurring in each bin is displayed. An initial, phasic increase in firing is correlated with the onset of the stimulus. The discharge continues during the remainder of the stimulus during adaptation, but decreases following termination. This pattern is evident when the stimulus is 20 dB or more above threshold. Activity gradually returns to baseline during the interval between stimuli. (Adapted, with permission, from Kiang 1965.)

locking: Each tends to fire during a particular part of the stimulus cycle.

The central nervous system can therefore gain information about stimulus frequency in two ways. First, there is a *place code*: The fibers are arrayed in a tonotopic map in which the position is related to characteristic frequency. Second, there is a *frequency code*: The phase-locked firing of the fiber provides information about the frequency of the stimulus, at least for frequencies below 3 kHz.

Sensorineural Hearing Loss Is Common but Is Amenable to Treatment

Whether mild or profound, most deafness falls into the category of *sensorineural hearing loss,* often misnamed "nerve deafness." Although hearing loss can result from direct damage to the eighth cranial nerve, for example from an acoustic neuroma, deafness stems primarily from the loss of cochlear hair cells and their afferent fibers.

The 16,000 hair cells in each human cochlea are not replaced by cell division but must last a lifetime. However, in amphibians and birds, supporting cells can be induced to divide and their progeny to produce new hair cells. In the zebrafish and in birds, some hair cell populations are regenerated continually by the activity of stem or supporting cells. Researchers have recently succeeded in replenishing mammalian hair cells in vitro. Until we understand how hair cells can be restored to the organ of Corti, however, we must cope with hearing loss.

The past few decades have brought remarkable advances in our ability to treat deafness. For the majority of patients who have significant residual hearing, hearing aids can amplify sounds to a level sufficient to activate the surviving hair cells. A modern aid is custom-tailored to compensate for each individual's hearing loss, so that the device amplifies sounds at frequencies to which the wearer is least sensitive, while providing little or no enhancement to those that can still be heard well.

When most or all of a person's cochlear hair cells have degenerated, no amount of amplification can assist hearing. However, a degree of hearing can be restored by bypassing the damaged organ of Corti with a cochlear prosthesis or implant. A user wears a compact unit that picks up sounds, separates their frequency components, and forwards electronic signals representing these constituents along separate wires to small antennae situated just behind the auricle. The signals are then transmitted transdermally to receiving antennae implanted in the temporal bone. From there, fine wires bear the signals to appropriate electrodes implanted as an array in the cochlea at various positions along the scala tympani. Activation of the electrodes excites action potentials in any nearby axons that have survived the degeneration of the hair cells (Figure 26–19).

The cochlear prosthesis takes advantage of the tonotopic representation of stimulus frequency along the cochlea—the *place code* (Figure 26–11 and Chapter 28). The axons innervating each segment of the cochlea are concerned with a specific, narrow range of frequencies. Each electrode in a prosthesis can excite a cluster of nerve fibers that represent similar frequencies. The stimulated neurons then forward their outputs along the eighth nerve to the central nervous system, where these signals are interpreted as a sound of the frequency represented at that position on the basilar membrane. An array of approximately 20 electrodes can mimic a complex sound by appropriately stimulating several clusters of neurons.

The number of implanted cochlear prostheses worldwide is now approaching 350,000. Their effectiveness, however, varies widely from person to person. In the best outcome, an individual can, under quiet conditions, understand speech nearly as well as a normally hearing person and can even conduct telephone conversations. At the other extreme are patients who derive little benefit from prostheses, presumably because of extensive degeneration of the nerve fibers near the electrode array. Most patients find their prostheses of great value. Even if hearing is not completely restored, the devices help in lip reading and alert patients to noises in the environment.

Hearing loss is often accompanied by another distressing symptom, *tinnitus,* or "ringing in the ears." By interfering with concentration and disrupting sleep, tinnitus can exasperate, depress, and even madden its victims. Because on rare occasions tinnitus stems from lesions to the auditory pathways, such as acoustic neuromas, it is important in neurological diagnosis to exclude such causes. Most tinnitus, however, is idiopathic: Its cause is uncertain. More and more studies implicate stress as an important factor. Some drugs also trigger the condition; antimalarial drugs related to quinine and aspirin at the high dosages used in the treatment of rheumatoid arthritis are notorious for this. Often, however, tinnitus occurs at high frequencies to which a damaged ear is no longer sensitive. In these instances, tinnitus may reflect hypersensitivity in the deafferented central nervous system, a phenomenon analogous to phantom limb pain (Chapter 20).

Figure 26–19 A cochlear prosthesis.
(Reproduced, with permission, from Loeb et al. 1983.)

A. Transmitting antennas receive electrical signals from a sound processor, located behind the subject's auricle or on the frame of his eyeglasses, and transmit them across the skin to receiving antennas implanted subdermally behind the auricle. The signals are then conveyed in a fine cable (**dark purple**) to an electrode array in the cochlea.

B. This cross section of the cochlea shows the placement of pairs of electrodes in the scala tympani. A portion of the extracellular current passed between an electrode pair is intercepted by nearby cochlear nerve fibers, which are thus excited and send action potentials to the brain.

A Sound transmission to cochlea

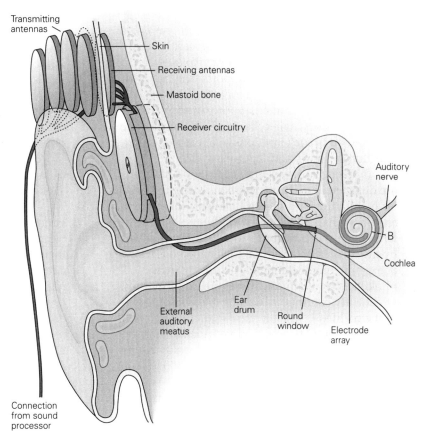

B Electrode array in cochlea

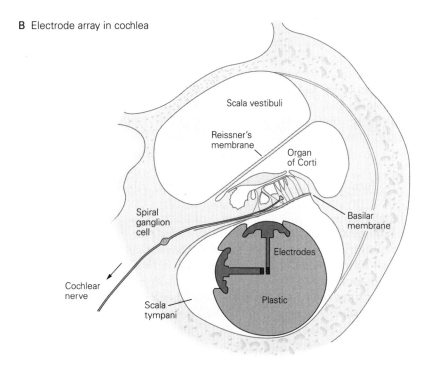

Highlights

1. Hearing begins with capture of sound by the ear. Mechanical energy captured by the outer ear flows through the middle ear to the cochlea, where it causes the elastic basilar membrane to oscillate.

2. The basilar membrane supports the receptor organ of the inner ear—the organ of Corti, an epithelial strip that contains approximately 16,000 mechanosensory hair cells. Hair cells transduce basilar-membrane vibrations into receptor potentials that cause sensory neurons to fire.

3. The frequency components of a sound stimulus are detected at different locations along the basilar membrane by different hair cells, following a tonotopic map. Mechanical gradients of the basilar membrane contribute to frequency analysis by the cochlea. In addition, each hair cell is tuned at a characteristic frequency according to its morphological, mechanical, and electrical properties, which vary continuously along the tonotopic axis of the cochlea.

4. Hair cells operate much more quickly than do other sensory receptors, which allows them to respond to sound frequencies beyond 100 kHz in some mammalian species. Accordingly, the mechanoelectrical transduction channels in the hair cell are activated directly by mechanical strain.

5. Each hair cell projects from its apical surface a tuft of cylindrical stereocilia—the hair bundle, which works as a mechanical antenna that vibrates in response to sound stimuli. The transduction channels occur at the stereociliary tips. Their open probability is modulated by tension changes in tip links that interconnect neighboring stereocilia.

6. Uniquely among sensory receptors, hair cells amplify their inputs to enhance their sensitivity, sharpen their frequency selectivity, and widen the range of stimulus levels that they can detect. Two forms of cellular motility contribute to this active process. First, receptor potentials evoke length changes of the somata of outer hair cells, a biological analog of piezoelectricity called electromotility. Second, the hair bundle—the mechanosensory antenna of the hair cell—can vibrate autonomously.

7. The ear not only receives sound but also emits sound called otoacoustic emissions. Spontaneous and evoked otoacoustic emissions result from the cochlea's active amplification processes.

8. The cochlea does not work as a high-fidelity sound receiver; instead, it introduces conspicuous distortions that contribute to sound perception. The auditory nonlinearity originates in the cochlea, which amplifies preferentially weak sound stimuli, and constitutes a hallmark of sensitive hearing that is used to screen hearing deficits in newborns.

9. A large variety of experimental observations at the level of a single hair bundle, of the basilar membrane, and in psychoacoustics are readily explained if the cochlea contains active mechanical modules that each operate on the verge of an oscillatory instability—the Hopf bifurcation. The Hopf bifurcation provides a general principle of auditory detection that simplifies our understanding of hearing.

10. The evolutionary history of hearing reveals that the various groups of land vertebrates acquired their hearing systems largely independently, but that their sensitivity and frequency selectivity are similar. In particular, both mammalian and non-mammalian ears benefit from mechanical amplification of sound inputs and show otoacoustic emissions. Mammals most notably differ from other groups in that their hearing range extends to frequencies beyond 12 to 14 kHz.

11. Analysis of the genetic forms of deafness has provided information on dozens of proteins key to the function of the hair cell, in particular those responsible for mechanoelectrical transduction and for synaptic transmission between hair cells and fibers of the auditory nerve. Although these genes may serve as potential targets for future therapies, sensorineural hearing loss is currently treated mostly with hearing aids or cochlear prostheses. New strategies, such as hair-cell replenishment via stem-cell differentiation or optogenetic stimulation of the spiral ganglion, provide promising avenues for research on hearing restoration.

Pascal Martin
Geoffrey A. Manley

Selected Reading

Hudspeth AJ. 1989. How the ear's works work. Nature 341:397–404.

Hudspeth AJ. 2014. Integrating the active process of hair cells with cochlear function. Nat Rev Neurosci 15:600–614.

Hudspeth AJ, Jülicher F, Martin P. 2010. A critique of the critical cochlea: Hopf—a bifurcation—is better than none. J Neurophysiol 104:1219–1229.

Kazmierczak P, Sakaguchi H, Tokita J, et al. 2007. Cadherin 23 and protocadherin 15 interact to form tip-link filaments in sensory hair cells. Nature 449:87–91.

Loeb GE. 1985. The functional replacement of the ear. Sci Am 252:104–111.

Pickles JO. 2008. *An Introduction to the Physiology of Hearing,* 3rd ed. New York: Academic.

Robbles L. Ruggero MA. 2001. Mechanics of the mammalian cochlea. Physiol Rev 81:1305–1352.

Zheng J, Shen W, He DZZ, Long KB, Madison LD, Dallos P. 2000. Prestin is the motor protein of cochlear outer hair cells. Nature 405:149–155.

References

Art JJ, Crawford AC, Fettiplace R, Fuchs PA. 1985. Efferent modulation of hair cell tuning in the cochlea of the turtle. J Physiol 360:397–421.

Ashmore JF. 2008. Cochlear outer-hair-cell motility. Physiol Rev 88:173–210.

Assad JA, Shepherd GM, Corey DP. 1991. Tip-link integrity and mechanical transduction in vertebrate hair cells. Neuron 7:985–994.

Avan P, Buki B, Petit C. 2013. Auditory distortions: origins and functions. Physiol Rev 93:1563–1619.

Barral J, Dierkes K, Lindner B, Jülicher F, Martin P. 2010. Coupling a sensory hair-cell bundle to cyber clones enhances nonlinear amplification. Proc Natl Acad Sci USA 107:8079–8084.

Barral J, Martin P. 2012. Phantom tones and suppressive masking by active nonlinear oscillation of the hair-cell bundle. Proc Natl Acad Sci USA 109:E1344–E1351.

Beurg M, Fettiplace R, Nam J-H, Ricci AJ. 2009. Localization of inner hair cell mechanotransducer channels using high-speed calcium imaging. Nat Neurosci 12:553–558.

Chan DK, Hudspeth AJ. 2005. Ca²⁺ current-driven nonlinear amplification by the mammalian cochlea in vitro. Nat Neurosci 8:149–155.

Corey D, Hudspeth AJ. 1983. Kinetics of the receptor current in bullfrog saccular hair cells. J Neurosci 3:962–976.

Crawford AC, Fettiplace R. 1981. An electrical tuning mechanism in turtle cochlear hair cells. J Physiol 312:377–412.

Fettiplace R, Kim KX. 2014. The physiology of mechano-electrical transduction channels in hearing. Physiol Rev 94:951–986.

Frolenkov GI, Atzori M, Kalinec F, Mammano F, Kachar, B. 1998. The membrane-based mechanism of cell motility in cochlear outer hair cells. Mol Biol Cell 9:1961–1968.

Glowatzki E, Fuchs PA. 2002. Transmitter release at the hair cell ribbon synapse. Nat Neurosci 5:147–154.

Helmholtz HLF. [1877] 1954. *On the Sensations of Tone as a Physiological Basis for the Theory of Music.* New York: Dover.

Holley MC, Ashmore JF. 1988. On the mechanism of a high-frequency force generator in outer hair cells isolated from the guinea pig cochlea. Proc R Soc Lond B Biol Sci 232:413–429.

Howard J, Hudspeth AJ. 1988. Compliance of the hair bundle associated with gating of mechanoelectrical transduction channels in the bullfrog's saccular hair cell. Neuron 1:189–199.

Hudspeth AJ, Gillespie PG. 1994. Pulling springs to tune transduction: adaptation by hair cells. Neuron 12:1–9.

Jacobs RA, Hudspeth AJ. 1990. Ultrastructural correlates of mechanoelectrical transduction in hair cells of the bullfrog's internal ear. Cold Spring Harbor Symp Quant Biol 55:547–561.

Johnson SL, Beurg M, Marcotti W, Fettiplace R. 2011. Prestin-driven cochlear amplification is not limited by the outer hair cell membrane time constant. Neuron 70:1143–1154.

Kemp DT. 1978. Stimulated acoustic emissions from within the human auditory system. J Acoust Soc Am 64:1386–1391.

Kiang NY-S. 1965. *Discharge Patterns of Single Fibers in the Cat's Auditory Nerve.* Cambridge, MA: MIT Press.

Kiang NY-S. 1980. Processing of speech by the auditory nervous system. J Acoust Soc Am 68:830–835.

Liberman MC. 1982. Single-neuron labeling in the cat auditory nerve. Science 216:1239–1241.

Loeb GE, Byers CL, Rebscher SJ, et al. 1983. Design and fabrication of an experimental cochlear prosthesis. Med Biol Eng Comput 21:241–254.

Manley GA. 2012. Evolutionary paths to mammalian cochleae. J Assoc Res Otolaryngol 13:733–743.

Manley GA, Köppl C. 1998. Phylogenic development of the cochlea and its innervation. Curr Opin Neurobiol 8:468–474.

Martin P, Hudspeth AJ. 1999. Active hair-bundle movements can amplify a hair cell's response to oscillatory mechanical stimuli. Proc Natl Acad Sci USA 96:14306–14311.

Michalski N, Petit C. 2015. Genetics of auditory mechanoelectrical transduction. Pflugers Arch 467:49–72.

Murphy WJ, Tubis A, Talmadge CL, Long GR. 1995. Relaxation dynamics of spontaneous otoacoustic emissions perturbed by external forces. II. Suppression of interacting emissions. J Acoust Soc Am 97:3711–3720.

Oshima K, Shin K, Diensthuber M, Peng AW, Ricci AJ, Heller S. 2010. Mechanosensitive hair cell-like cells from embryonic and induced pluripotent stem cells. Cell 141:704–716.

Pan B, Akyuz N, Liu XP, et al. 2018. TMC1 forms the pore of mechanosensory transduction channels in vertebrate inner ear hair cells. Neuron 99:736–753.

Probst R, Lonsbury-Martin BL, Martin GK. 1991. A review of otoacoustic emissions. J Acoust Soc Am 89:2027–2067.

Reichenbach T, Hudspeth AJ. 2014. The physics of hearing: fluid mechanics and the active process of the inner ear. Rep Prog Phys 77:0706601.

Ricci AJ, Crawford AC, Fettiplace R. 2003. Tonotopic variation in the conductance of the hair cell mechanotransducer channel. Neuron 40:983–990.

Sotomayor M, Weihofen WA, Gaudet R, Corey DP. 2012. Structure of a force-conveying cadherin bond essential for the inner-ear mechanotransduction. Nature 492:128–132.

Spoendlin H. 1974. Neuroanatomy of the cochlea. In: E Zwicker, E Terhardt (eds). *Facts and Models in Hearing*, pp. 18–32. New York: Springer-Verlag.

Stauffer EA, Scarborough JD, Hirono M, et al. 2005. Fast adaptation in vestibular hair cells requires myosin-1c activity. Neuron 47:541–553.

Tinevez JY, Jülicher F, Martin P. 2007. Unifying the various incarnations of active hair-bundle motility by the vertebrate hair cell. Biophys J 93:4053–4067.

von Békésy G. 1960. *Experiments in Hearing*. EG Wever (ed, transl). New York: McGraw-Hill.

Wilson JP. 1980. Evidence for a cochlear origin for acoustic re-emissions, threshold fine-structure and tonal tinnitus. Hear Res 2:233–252.

Wu Z, Müller U. 2016. Molecular identity of the mechanotransduction channel in hair cells: not quiet there yet. J Neurosci 36:10927–10934.

27

The Vestibular System

MODERN VEHICULAR TRAVEL ON EARTH and through extraterrestrial space relies upon sophisticated guidance systems that integrate acceleration, velocity, and positional information through transducers, computational algorithms, and satellite triangulation. Yet the principles of inertial guidance are ancient: Vertebrates have used analogous systems for 500 million years and invertebrates for even longer. In these animals, the inertial guidance system, termed the vestibular system, serves to detect and interpret motion through space as well as orientation relative to gravity.

Through extensive research over many decades, it is apparent that most, if not all, organisms on Earth have evolved to sense one of the most prevalent "forces" in our universe, gravity. The mechanisms for the sensory transduction are as diverse as nature could devise. Gravity is most precisely referenced as gravito-inertial acceleration (GIA), a distinct form of linear acceleration directed toward the core of our planet. In truth, gravity varies systematically by as much as 0.5% between the equator and the poles; it increases over mineral-dense regions and decreases over mineral-light regions of the Earth's surface. Yet every single behavior that animals perform is referenced to the GIA, and all of our actions and cognitive directives depend upon knowledge of our motion and orientation relative to it. The first developments of what we refer to as a vestibular system were actually gravity sensors; as behavior became increasingly mobile, sensory organs evolved to process rotational accelerations as well.

In this chapter we will concentrate on the vestibular system of vertebrates, which has remained highly conserved across many species. Vestibular signals originate in the labyrinths of the internal ear (Figure 27–1B). The *bony labyrinth* is a hollow structure within the petrous portion of the temporal bone. Within it lies the *membranous labyrinth,* which contains sensors for both the vestibular and auditory systems.

The vestibular receptors consist of two parts: two otolith organs, the utricle and saccule, which measure linear accelerations, and three semicircular canals, which measure angular accelerations. Rotational motion (angular acceleration) is experienced during head turns, whereas linear acceleration occurs during walking, falling, vehicular travel (ie, translations), or head tilts relative to gravity. These receptors send vestibular information to the brain, where it is integrated into an appropriate signal regarding direction and speed of motion, as well as the position of the head relative to the GIA. Many of the central vestibular neurons at the first junction with receptor afferent fibers also

Figure 27–1 The vestibular apparatus of the inner ear.

A. The orientations of the vestibular and cochlear divisions of the inner ear are shown with respect to the head.

B. The inner ear is divided into bony and membranous labyrinths. The bony labyrinth is bounded by the petrosal portion of the temporal bone. Lying within this structure is the membranous labyrinth, which contains the receptor organs for hearing (the cochlea) and equilibrium (the utricle, saccule, and semicircular canals). The space between bone and membrane is filled with perilymph, whereas the membranous labyrinth is filled with endolymph. Sensory cells in the utricle, saccule, and ampullae of the semicircular canals respond to motion of the head. (Adapted from Iurato 1967.)

A

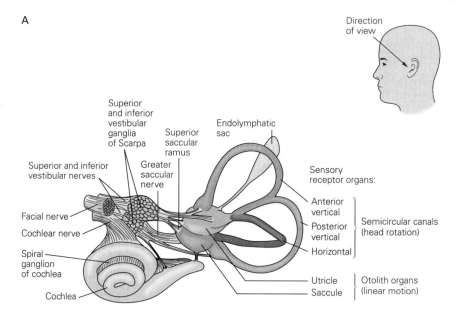

B

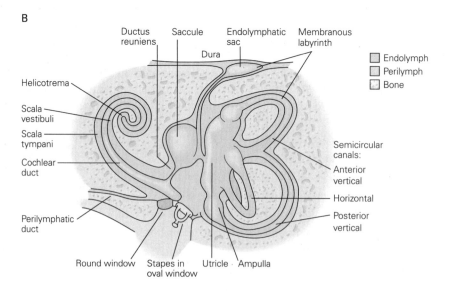

receive convergent signals from other systems such as proprioceptors, visual signals, and motor commands. Central processing of these multimodal signals occurs very rapidly to ensure adequate coordination of visual gaze and postural responses, autonomic responses, and awareness of spatial orientation.

The Vestibular Labyrinth in the Inner Ear Contains Five Receptor Organs

The membranous labyrinth is supported within the bony labyrinth by a filamentous network of connective tissue. The vestibular portion of the membranous labyrinth lies lateral and posterior to the cochlea. Vestibular receptors are contained in specialized enlarged regions of the membranous labyrinth, termed the ampullae for the semicircular canals and maculae for the otolith organs (Figure 27–1B). Both of the otolith organs lie in a central compartment of the membranous labyrinth, the vestibule, which is surrounded by the bony labyrinth of the same name.

The membranous labyrinth is filled with endolymph, a K$^+$-rich (150 mM) and Na$^+$-poor (16 mM) fluid whose composition is maintained by the action of ion pumps in specialized cells. Endolymph bathes the surface of the vestibular receptor cells. Surrounding the membranous labyrinth, in the space between the membranous labyrinth and the wall of the bony labyrinth, is *perilymph*. Perilymph is a high-Na$^+$ (150 mM), low-K$^+$ (7 mM) fluid similar in composition to cerebrospinal fluid, with which it is in communication through the cochlear duct. Perilymph bathes the basal surface of the receptor epithelia and the vestibular nerve fibers. Two fluid-tight partitions in the bony labyrinth, the oval and round windows (Figure 27–1B), connect the perilymphatic space to the middle ear cavity. The oval window is connected to the tympanic membrane by the middle ear ossicles. These windows are important for sound transduction (Chapter 26). The endolymph and perilymph are kept separate by a junctional complex of support cells that surrounds the apex of each receptor cell. Disruption of the balance between these two fluids (by trauma or disease) can result in vestibular dysfunction, leading to dizziness, vertigo, and spatial disorientation.

During development, the labyrinth progresses from a simple sac to a complex set of interconnected sensory organs, but retains the same fundamental topological organization. Each organ originates as an epithelium-lined pouch that buds from the otic cyst, and the endolymphatic spaces within the several organs remain continuous in the adult. The endolymphatic spaces of the vestibular labyrinth are also connected to the cochlear duct through the ductus reuniens (Figure 27–1B). In addition, the membranous labyrinth contains a small tube, the endolymphatic duct, which extends through a space in the sigmoid bone, the vestibular aqueduct, to end in a blind sac adjacent to the dura in the epidural space of the posterior cranial fossa. It is thought that the endolymphatic sac has both absorptive and excretive functions to maintain the ionic composition of the endolymphatic fluid.

Hair Cells Transduce Acceleration Stimuli Into Receptor Potentials

Each of the five receptor organs has a cluster of hair cells responsible for transducing head motion into vestibular signals. Hair cells are so named due to an array of nearly 100 staggered height stereocilia. The shortest stereocilia are at one end of the cell and the tallest at the other, ending with the only true cilium of the hair cell, termed the kinocilium. The kinocilium is typically the tallest of all stereocilia. Angular or linear acceleration of the head leads to a deflection of the stereocilia, which together compose the hair bundle (Figure 27–2).

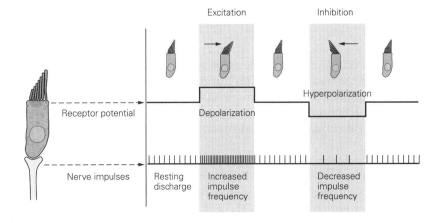

Figure 27–2 Hair cells in the vestibular labyrinth transduce mechanical stimuli into neural signals. At the apex of each cell are the stereocilia, which increase in length toward the single kinocilium. The membrane potential of the receptor cell depends on the direction in which the stereocilia are bent. Deflection toward the kinocilium causes the cell to depolarize and thus increases the rate of firing in the afferent fiber. Bending away from the kinocilium causes the cell to hyperpolarize, thus decreasing the afferent firing rate. (Adapted, with permission, from Flock 1965.)

Specialized ion channels in the tips of the hair bundle stereocilia allow K⁺ to enter or be blocked from the surrounding endolymph (Chapter 26). This action allows hair cells to act as mechanoreceptors, where deflection of the stereocilia produces a depolarizing or hyperpolarizing receptor potential depending on which direction the hair bundle moves (Figure 27–2). These depolarizations and hyperpolarizations of the receptor membrane lead to excitation and inhibition, respectively, in the firing rate of the innervating afferent (Figure 27–2). In each vestibular receptor organ, hair cells are arranged so that movement directional specificity is defined by excitation in some cells and inhibition in other cells.

Vestibular signals are carried from the hair cells to the brain stem by branches of the vestibulocochlear nerve (cranial nerve VIII), which enter the brain stem and terminate in the ipsilateral vestibular nuclei, cerebellum, and reticular formation. Cell bodies of the vestibular nerve are located in Scarpa's ganglia within the internal auditory canal (Figure 27–1A). The *superior vestibular nerve* innervates the horizontal and anterior canals and the utricle, whereas the *inferior vestibular nerve* innervates the posterior canal and the saccule. The labyrinth's vascular supply, which arises from the anterior inferior cerebellar artery, travels with nerve VIII. The anterior vestibular artery supplies the structures innervated by the superior vestibular nerve, and the posterior vestibular artery supplies the structures innervated by the inferior vestibular nerve.

All vertebrate receptor hair cells receive efferent inputs from the brain stem. The function of the efferent innervation of vestibular receptors is still a subject of debate. Stimulation of the efferent fibers from the brain stem changes the sensitivity of the afferent axons from the hair cells. It increases the excitability of some afferents and hair cells while inhibiting others, and varies across species.

The Semicircular Canals Sense Head Rotation

An object undergoes angular acceleration when its rate of rotation about an axis changes. Therefore, the head undergoes angular acceleration when it turns or tilts, when the body rotates, and during active or passive locomotion. The three semicircular canals of each vestibular labyrinth detect these angular accelerations and report their magnitudes and motion directions to the brain.

Each semicircular canal is a semicircular tube of membranous labyrinth extending from the vestibule. One end of each canal is open to the vestibule, whereas at the other end, the ampulla, the entire lumen of the canal is traversed by a fluid-tight gelatinous diaphragm, the cupula. The stereocilia and the kinocilium protrude into the gelatinous cupula, while the hair cells are located below in a receptor epithelium, the crista, along with the innervating afferent terminals (Figure 27–3).

The vestibular organs detect accelerations of the head because the inertia of endolymph and cupula results in forces acting on the stereocilia. Consider the simplest situation, a rotation in the plane of a semicircular canal. When the head begins to rotate, the membranous and bony labyrinths move along with it. Because of its inertia, however, the endolymph lags behind the surrounding membranous labyrinth, thus pushing the cupula in a direction opposite that of the head (Figure 27–3B).

The motion of endolymph in a semicircular canal can be demonstrated with a cup of coffee. While gently twisting the cup about its vertical axis, observe a particular bubble near the fluid's outer boundary. As the cup begins to turn, the coffee tends to maintain its initial orientation in space and thus counter-rotates in the cup. If you continue rotating the cup at the same speed, the coffee (and the bubble) eventually catches up to the cup and rotates with it. When the cup decelerates and stops, the coffee keeps rotating, moving in the opposite direction relative to the cup.

In the ampulla, this relative motion of the endolymph creates pressure on the cupula, bending it toward or away from the adjacent vestibule, depending on the direction of endolymph flow. The resulting deflection of the stereocilia alters the membrane potential of the hair cells, thereby changing the firing rates of the associated sensory fibers. Each semicircular canal is maximally sensitive to rotations in its plane. The horizontal canal is oriented approximately 30° elevated above the naso-occipital axis (roughly in the horizontal plane as a person walks and looks at the ground ahead) and thus is most sensitive to rotations in the horizontal plane. The stereocilia are arranged so that leftward rotational motion is excitatory for the left horizontal canal and inhibitory for the right horizontal canal. The anterior and posterior canals are oriented more vertically in the head, at an angle of approximately 45 degrees from the sagittal plane (Figure 27–4). Similar rotational motion downward in the plane of the anterior canals is excitatory for anterior canal hair cells, while upward head motion is excitatory for posterior canals.

Because there is approximate mirror symmetry of the left and right labyrinths, the six canals effectively operate as three coplanar pairs. The two horizontal canals form one pair; each of the other pairs consists of one anterior canal on one side of the head and the contralateral posterior canal. Further, the three

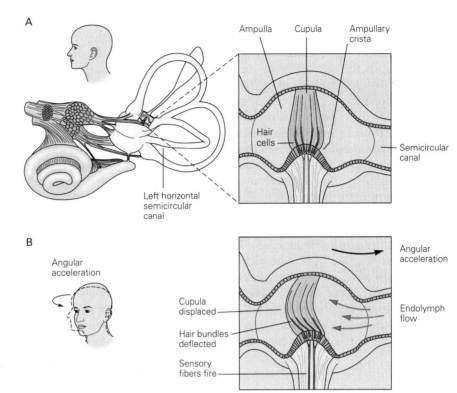

Figure 27–3 The ampulla of a semicircular canal.

A. A thickened zone of epithelium, the ampullary crista, contains the hair cells. The stereocilia and the kinocilia of the hair cells extend into a gelatinous diaphragm, the cupula, which stretches from the crista to the roof of the ampulla.

B. The cupula is displaced by the relative movement of endolymph when the head turns. As a result, the hair bundles are also displaced. Their movement is greatly exaggerated in the diagram.

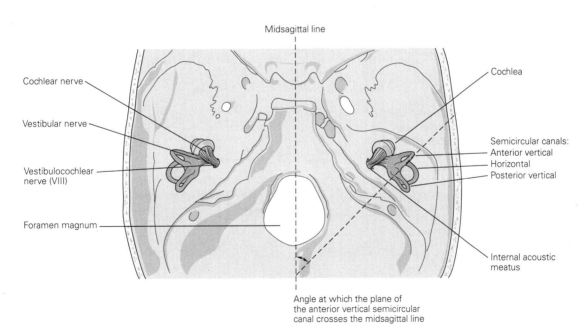

Figure 27–4 The bilateral symmetry of the semicircular canals. The horizontal canals on both sides lie in approximately the same plane and therefore are functional pairs. The bilateral vertical canals have a more complex relationship. The anterior canal on one side and the posterior canal on the opposite side lie in parallel planes and therefore constitute a functional pair. The vertical semicircular canals lie nearly 45° from the midsagittal plane. Each of the semicircular canals on one side of the head lie in approximately orthogonal planes to each other.

semicircular canals on each side of the head lie roughly orthogonal to each other (Figure 27–4). When the head moves toward the receptor hair cells (eg, leftward head turns for the left horizontal semicircular canal), the stereocilia are bent toward the tall kinocilium, thus exciting (depolarizing) the cell. Head motion in the opposite direction causes bending away from the kinocilium and toward the smallest stereocilia, thus closing the channels and inhibiting (hyperpolarizing) the cell.

The left and right ear semicircular canals have opposite polarity; thus, when you turn your head to the left, the receptors in the left horizontal semicircular canal will be excited (increased firing rate), whereas right horizontal canal receptors will be inhibited (decreased firing rate; Figure 27–5). The same relationship is true for the vertical semicircular canals. The canal planes are also roughly aligned to the pulling planes of specific eye muscles. The pair of horizontal canals lies in the pulling plane of the lateral and medial rectus muscles. The left anterior and right posterior canal pair lie in the pulling plane of the left superior

and inferior rectus and right superior and inferior oblique muscles. The right anterior and left posterior pair occupies the pulling plane of the left superior and inferior oblique and right superior and inferior rectus muscles.

The Otolith Organs Sense Linear Accelerations

The vestibular system must compensate not only for head rotations but also for linear motion. The two otolith organs, the utricle and saccule, detect linear motion as well as the static orientation of the head relative to gravity, which is itself a linear acceleration. Each organ consists of a sac of membranous labyrinth approximately 3 mm in the longest dimension. The hair cells of each organ are arranged in a roughly elliptical patch called the *macula*. The human utricle contains approximately 30,000 hair cells, whereas the saccule contains some 16,000.

The hair bundles of the otolithic hair cells extend into a gelatinous sheet, the *otolithic membrane*, which

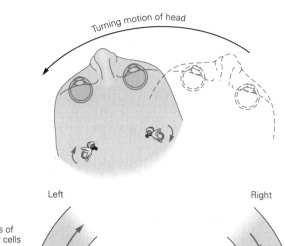

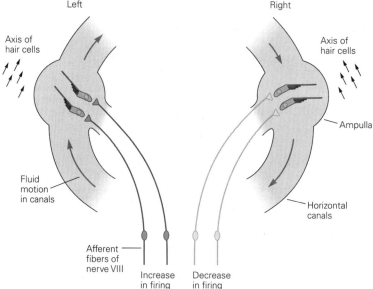

Figure 27–5 The left and right horizontal semicircular canals work together to signal head movement. Because of inertia, rotation of the head in a counterclockwise direction causes endolymph to move clockwise with respect to the canals. This deflects the stereocilia in the left canal in the excitatory direction, thereby exciting the afferent fibers on this side. In the right canal, the afferent fibers are hyperpolarized so that firing decreases.

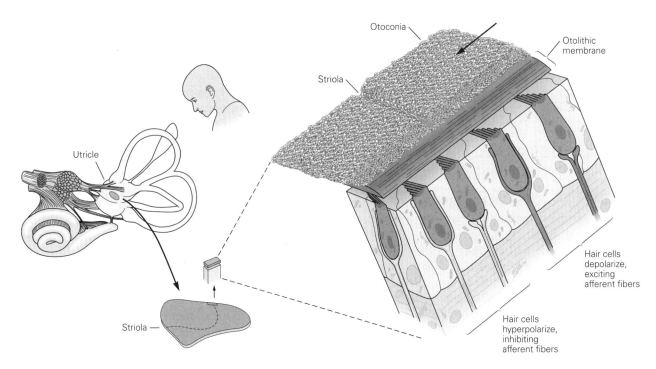

Figure 27–6 The utricle detects tilt of the head. Hair cells in the epithelium of the utricle have apical hair bundles that project into the otolithic membrane, a gelatinous material that is covered by millions of calcium carbonate particles (otoconia). The hair bundles are polarized but are oriented in different directions. The directional polarity of each hair cell is organized relative to a reversal region running through the center of the utricle, termed the striola (see Figure 27–7). Thus, when the head is tilted, the gravitational force on the otoconia bends each hair bundle in a particular direction. When the head is tilted in the direction of a hair cell's axis of polarity, that cell depolarizes and excites the afferent fiber. When the head is tilted in the opposite direction, the same cell hyperpolarizes and inhibits the afferent fiber. (Adapted from Iurato 1967.)

covers the entire macula (Figure 27–6). Embedded on the surface of this membrane are fine, dense particles of calcium carbonate called *otoconia* (Greek root translates to "ear dust"), which give the otolith ("ear stone") organs their name. Otoconia are typically 0.5 to 30 μm long; thousands of these particles are attached to the otolithic membranes of the utricle and saccule.

Gravity and other linear accelerations exert shear forces on the otoconial matrix and the gelatinous otolithic membrane, which can move relative to the membranous labyrinth. This results in a deflection of the hair bundles, altering activity in the vestibular nerve to signal linear acceleration owing to translational motion or gravity. The orientations of the otolith organs and the directional sensitivity of individual hair cells are such that a linear acceleration along any axis can be sensed. For example, with the head in its normal position, the macula of each utricle is raised above the naso-occipital axis by approximately 30°, similar to the horizontal semicircular canal. In normal resting head position, the utricle is deviated to bring the utricle approximately equal to an Earth horizontal plane. Any acceleration in the horizontal plane excites some hair cells in each utricle and inhibits others, according to their orientations (Figures 27–6 and 27–7).

The operation of the paired saccules resembles that of the utricles. The hair cells represent all possible

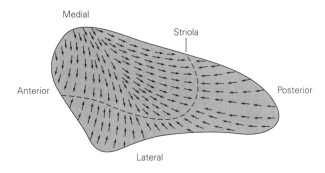

Figure 27–7 The axis of mechanical sensitivity of each hair cell in the utricle is oriented toward the striola. The striola curves across the surface of the macula containing the hair cells, resulting in a characteristic variation in the axes of mechanosensitivity (**arrows**) in the population of hair cells. Because of this arrangement, tilt in any direction depolarizes some cells and hyperpolarizes others, while having no effect on the remainder. (Adapted, with permission, from Spoendlin 1966.)

orientations within the plane of each saccular macula, but the maculae are oriented vertically in nearly parasagittal planes. The saccules are therefore especially sensitive to vertical accelerations. Certain saccular hair cells also respond to accelerations in the horizontal plane, in particular those along the anterior–posterior axis.

Central Vestibular Nuclei Integrate Vestibular, Visual, Proprioceptive, and Motor Signals

The vestibular nerve projects ipsilaterally from the vestibular ganglion mainly to four vestibular nuclei (medial, lateral, superior, and descending) in the dorsal part of the pons and medulla, in the floor of the fourth ventricle. Many vestibular nerve fibers also bifurcate, sending a direct projection to the fastigial nucleus, the nodulus and uvula, and the reticular formation (Figure 27–8A). These nuclei integrate signals from the vestibular organs with signals from the spinal cord, cerebellum, and visual system.

The vestibular nuclei project, in turn, to many central targets, including the oculomotor nuclei, reticular and spinal centers concerned with gaze and postural movement, and the thalamus (Figure 27–9). Many vestibular nuclei neurons have reciprocal connections with the cerebellum, primarily in the floculo-nodular lobe, that form important regulatory mechanisms for eye movements, head movements, and posture (Figures 27–8 and 27–9). The vestibular nuclei receive inputs from the premotor cortex, the accessory optic system (nucleus of the optic tract), the neural integrator nuclei (nucleus prepossitus hypoglossi and interstitial nucleus of Cajal), and the reticular formation (Figure 27–8). Further projections from the vestibular nuclei reach the rostral and caudal lateral medulla nuclei that are involved in regulation of blood pressure, heart rate, respiration, and bone remodeling, as well as the parabrachial nucleus for homeostasis modulation. Finally, there are projections from the vestibular nuclei to the medial geniculate (auditory) nuclei, as well as the supragenual nucleus and dorsal tegmental nucleus, which contribute to spatial orientation (Figure 27–9).

The superior and medial vestibular nuclei receive fibers predominantly from the semicircular canals in the medial regions and some otolith input in the lateral regions (Figure 27–8). They send fibers predominantly to the cerebellum, reticular formation, thalamus, oculomotor centers, and spinal cord (Figure 27–9). Oculomotor center outputs include the three oculomotor nuclei (abducens, oculomotor, trochlear), as well as the neural integrators for converting head velocity into head position signals in the nucleus hypoglossi

(horizontal eye movements) and interstitial nucleus of Cajal (vertical eye movements). These nuclei are described in some detail later.

Another major output pathway concerned with gaze control arises from the medial vestibular nucleus (as well as lesser projections from the descending and lateral vestibular nuclei) and projects bilaterally to the cervical spinal cord through the medial vestibulospinal tract (Figure 27–9; see Chapter 35). There are two categories of medial vestibulospinal fibers. Vestibulospinal neurons project only to the spinal cord to control neck musculature. Vestibulo-ocular neurons project to both the spinal cord and the oculomotor nuclei and are involved in coordinated eye and head movements to maintain gaze stability.

The lateral vestibular nucleus (Deiters' nucleus) receives fibers from the semicircular canals medially and the otolith organs laterally. There is a major output to all levels of the ipsilateral spinal cord through the lateral vestibulospinal tract that is concerned principally with postural reflexes through modulation of limb and axial musculature (Figure 27–9). Lateral vestibular nuclei neurons also project heavily to the reticular formation. The descending vestibular nucleus receives predominantly otolithic input, but also receives semicircular canal fibers medially, and projects to the cerebellum, reticular formation, and spinal cord (medial vestibulospinal tract). The primary neurotransmitters for excitatory vestibular nuclear projections include glutamate, whereas the inhibitory projections are either glycine or γ-aminobutyric acid (GABA). Vestibular projections to the spinal systems are discussed in more detail in Chapter 36.

The Vestibular Commissural System Communicates Bilateral Information

Many of these vestibular nuclei neurons receive convergent motion information from the opposite ear through an inhibitory commissural pathway that uses GABA as a neurotransmitter (Figure 27–8B). The commissural pathway is highly organized according to the type of receptor from which information is received. For example, cells receiving signals from the ipsilateral horizontal excitatory canal will also receive signals from the contralateral horizontal canal through an inhibitory interneuron. Due to the directional selectivity of the receptors in each ear, the contralateral horizontal canal input will always be decreased during an ipsilateral head turn, in effect "disinhibiting" the inhibitory input from the contralateral side.

The effect of the commissural system is to increase the response of the vestibular nuclei neuron and

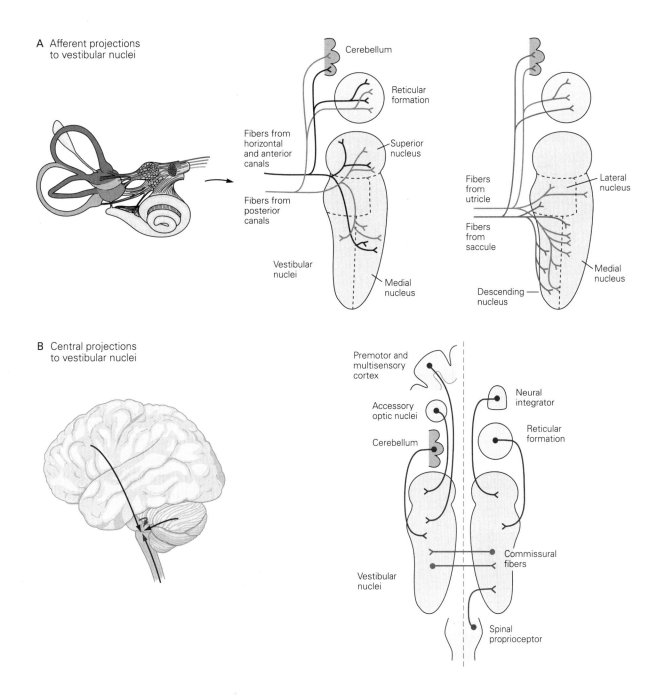

Figure 27–8 Afferent fiber and central projections to the vestibular nuclei.

A. Afferent fibers from vestibular receptors terminate in the brain stem and cerebellum. Fibers from semicircular canals project primarily to the medial portions of the superior and medial vestibular nuclei, the descending vestibular nucleus, the cerebellum (nodulus and uvula), and the reticular formation. Fibers from the otoliths primarily project to the lateral portions of

all vestibular nuclei, the nodulus and uvula, and the reticular formation. (Adapted, with permission, from Gacek and Lyon 1974.)

B. Central projections to the vestibular nuclei arise from a number of cortical, brain stem, and spinal cord regions. These include the premotor and multisensory cortices, accessory optic nuclei, cerebellum, neural integrator nuclei, reticular formation, spinal cord, and commissural fibers from the contralateral vestibular nuclei.

decrease noise from the incoming afferent signal, giving rise to a "push-pull" vestibular function. From an engineering point of view, the "push-pull" set point in the nuclei neurons constantly updates canal signals from the opposing ear to act as a comparator junction and can explain the relatively high spontaneous firing rate of canal afferents at nearly 100 spikes/s. For example, during a leftward head turn, left brain stem nuclei neurons receive high firing rate signals from the left horizontal canal and low firing rate signals from the right horizontal canal. The comparison of activity is interpreted as a left head turn (Figure 27–5). Similar comparisons between signals also occur for inputs from the anterior semicircular canal on one side and the posterior semicircular canal on the opposite ear side. Thus, for rotational motion in any head plane, the comparator is able to determine the direction of movement with great specificity.

Any disruption of the normal balance between left and right ear canal inputs (eg, from trauma or disease in the receptor organs or nerve) will be interpreted by the brain as a head rotation, even though the head is stationary. These effects often lead to illusions of spinning or rotating that can be quite upsetting and may produce nausea or vomiting. However, over time, the commissural fibers provide for vestibular compensation, a process by which the loss of unilateral vestibular receptor function is partially restored centrally and behavioral responses such as the vestibulo-ocular reflex mostly recover.

Combined Semicircular Canal and Otolith Signals Improve Inertial Sensing and Decrease Ambiguity of Translation Versus Tilt

In some instances, the vestibular input from a single receptor may be ambiguous. For example, Einstein (1908) showed that linear accelerations are equivalent whether they arise from translational motion or tilts of the head relative to gravity. The otolith receptors cannot

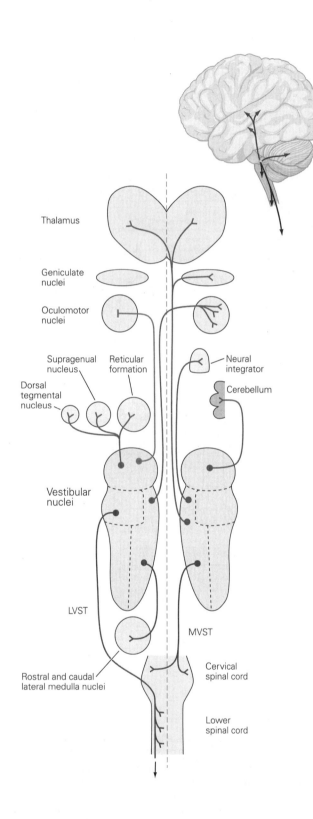

Thalamus

Geniculate nuclei

Oculomotor nuclei

Supragenual nucleus

Reticular formation

Dorsal tegmental nucleus

Neural integrator

Cerebellum

Vestibular nuclei

LVST

MVST

Rostral and caudal lateral medulla nuclei

Cervical spinal cord

Lower spinal cord

Figure 27–9 (Left) Output projections from the vestibular nuclei. The vestibular nuclei project to a number of brain regions below the cortical level. Two separate descending pathways project through the lateral and medial vestibulospinal tracts (**LVST, MVST**) to terminate in the spinal cord. The vestibular nuclei also project to the reticular formation and the lateral medullary nuclei in the brain stem. Ascending projections to the supragenual nucleus, the dorsal tegmental nucleus, the oculomotor nuclei (abducens, oculomotor, and trochlear), and the neural integrator nuclei are very prominent (**red line**, excitatory; **gray line**, inhibitory), as are projections to the cerebellum (nuclei, nodulus, and uvula). Other prominent vestibular projections terminate in the geniculate nuclei and the thalamus (ventral lateral, posterior, and intralaminar thalamic regions).

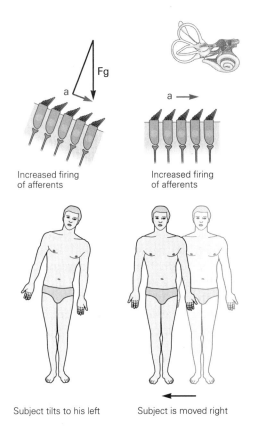

Figure 27–10 Vestibular inputs signaling body posture and motion can be ambiguous. The postural system cannot distinguish between tilt and linear acceleration of the body based on otolithic inputs alone. The same shearing force acting on vestibular hair cells can result from tilting of the head (*left*), which exposes the hair cells to a portion of the acceleration (**a**) owing to gravity (**Fg**), or from horizontal linear acceleration of the body (*right*).

passively generated head movements. Specifically, in contrast to vestibular afferents, some neurons in the vestibular nuclei and cerebellum well known for responding to vestibular stimuli during passive movement lose or reduce their sensitivity during self-generated movement. The preferential response to passive motion, or to the passive components of combined active and passive motion, has been interpreted as sensory prediction error signals: The brain predicts how self-generated motion activates the vestibular organs and subtracts these predictions from afferent signals. Such error signals are important for the on-line control of head movement, as well as head movement estimation.

Computationally, these properties have been interpreted quantitatively using concepts common to all sensorimotor systems; that is, active and passive motion signals are processed by internal models of the motion sensor (ie, the canals, otolith organs, and neck proprioceptors). The brain uses an internal representation of the laws of physics and sensory dynamics (which can be elegantly modeled as forward internal models of the sensors) to process vestibular signals. Without such error signals, accurate self-motion estimation would be severely compromised. These computational insights suggest that, unlike early interpretations, vestibular signals remain critically important when coupled to self-motion estimation and head movement control during actively generated head movements.

discriminate between the two: So how is it that we can tell the difference between translating rightward and tilting leftward, where the linear acceleration signaled by the otolith afferents is the same (Figure 27–10)?

It is now well established that convergent vestibular nuclei and cerebellar neurons use combined signals from both the semicircular canals and the otolith receptors and some simple computations to discriminate between tilt and translation. As a result, some central vestibular and cerebellar cells encode head tilt, whereas other cells encode translational motion, which, as we will see, is extremely important for the control of head and eye movements.

Vestibular Signals Are a Critical Component of Head Movement Control

An important discovery is the differing responses in some vestibular nuclei neurons to actively versus

Vestibulo-Ocular Reflexes Stabilize the Eyes When the Head Moves

In order to see clearly and maintain focus on visual objects during head motion, the eyes maintain foveal fixation through a series of vestibulo-ocular reflexes (VORs). If you shake your head back and forth while reading, you can still discern words because of the VORs. If instead you move the book at a similar speed while holding your head steady, you can no longer read the words.

In the latter instance, vision provides the brain with the only corrective feedback for stabilizing of the image on the retina, and visual processing in vertebrates is much slower (around 100 ms latency) and less effective than vestibular processing (around 10 ms) for image stabilization. The vestibular apparatus signals how fast the head is rotating, and the oculomotor system uses this information to stabilize the eyes to fix visual images on the retina.

There are two components of VORs. The *rotational VOR* compensates for head rotation and receives its input predominantly from the semicircular canals. The *translational VOR* compensates for linear head movement. These two VOR responses arise from connections from vestibular nuclei neurons to the abducens, oculomotor, and trochlear nuclei (Figure 27–9).

The Rotational Vestibulo-Ocular Reflex Compensates for Head Rotation

When the semicircular canals sense head rotation in one direction, the eyes rotate in the opposite direction at equal velocity in the orbits (Figure 27–11). This compensatory eye rotation is called the vestibular slow phase, although it is not necessarily slow: The eyes may reach speeds of more than 200 degrees per second if the head's rotation is fast. During fast head movements, the VOR must act quickly to maintain stable gaze. A trisynaptic pathway, the three-neuron arc, connects each semicircular canal to the appropriate eye muscle (Figure 27–11).

The rotational VOR represents a phylogenetically old reflex. Many invertebrates and all vertebrate species, from amphibians, reptiles, fish, and birds to nonhuman primates, have the ability to reflexively rotate their eyes opposite to the direction of head rotation, thus keeping the visual world stable on the retina. Primary afferents from the horizontal semicircular canals send excitatory signals through the vestibular nuclei and the medial longitudinal fasciculus to the contralateral abducens nucleus (Figure 27–11). Abducens motor neurons send impulses via cranial nerve VI to excite the ipsilateral lateral rectus muscle. At the same time, abducens interneurons send excitatory signals to motor neurons in the contralateral oculomotor nucleus, which innervates the medial rectus muscle (see Chapter 35 for details on other projections).

The three-synapse pathway illustrated in Figure 27–11 is not sufficient to elicit appropriate compensatory eye movements. This is because the afferent signal from the semicircular canals is proportional to head velocity, while the compensatory eye movement requires eye position changes. To convert velocity to position requires temporal integration (simple calculus) that occurs through neural networks in the brain stem nuclei for most head motion speeds. However, at high rotation frequencies, the viscoelastic properties of the eyeball, eye muscles, and surrounding tissues provide an additional integration step. Thus, the rotational VOR is thought to consist of two parallel processes.

The first process consists of the direct neural pathway known as the three-neuron arc (Figure 27–11).

The second neural integrator process consists of additional parallel pathways that ensure that the correct proportion of velocity and position commands are delivered to the oculomotor nuclei to move the eye appropriately (Figure 27–9 and see Chapter 35). Without this second indirect integrator pathway, the response to a head rotation would initially bring the eye to the correct position, but the eye would drift away from that position since the oculomotor neurons would lack the tonic input to compensate for the elastic restoring forces of the eyeball (Chapter 35). This is exactly what happens after lesions of brainstem and cerebellar structures that are thought to participate in this neural integration (eg, the prepositus hypoglossi and the interstitial nucleus of Cajal; Figure 27–9). It is generally thought that the integrator pathway is shared by all conjugate eye movement systems (saccades, smooth pursuit, and the VOR), although the direct pathway is at least partly segregated for different types of eye movements (ie, VOR, smooth pursuit, saccades).

With continued head rotation, the eyes eventually reach the limit of their orbital range and stop moving. To prevent this, a rapid saccade-like movement called a quick phase displaces the eyes to a new point of fixation in the direction of head rotation.

If rotation is prolonged, the eyes execute alternating slow and quick phases called *nystagmus* (Figure 27–12). Although the slow phase is the primary response of the rotational VOR, the direction of nystagmus is defined in clinical practice by the direction of its quick phase. Since prolonged rightward rotation excites the right horizontal canal and inhibits the left horizontal canal, leftward slow phases and a *right-beating nystagmus* result.

If the angular velocity of the head remains constant, the inertia of the endolymph is eventually overcome, as in the earlier coffee cup example. The cupula relaxes and vestibular nerve discharge returns to its baseline rate. As a consequence, slow-phase velocity decays and the nystagmus stops, although the head is still rotating.

In fact, the nystagmus lasts longer than would be expected based on cupular deflection. By a process called *velocity storage*, a brain stem network provides a velocity signal to the oculomotor system, although the vestibular nerve no longer signals head movement. Eventually, however, the nystagmus does decay and the sense of motion vanishes in darkness. Further, the same rotation in the presence of a visual surround activates the optokinetic reflex (Chapter 35) and elicits a steady-state nystagmus pattern that is sustained indefinitely. The interactions between canal and optokinetic

A Excitatory connections

B Inhibitory connections

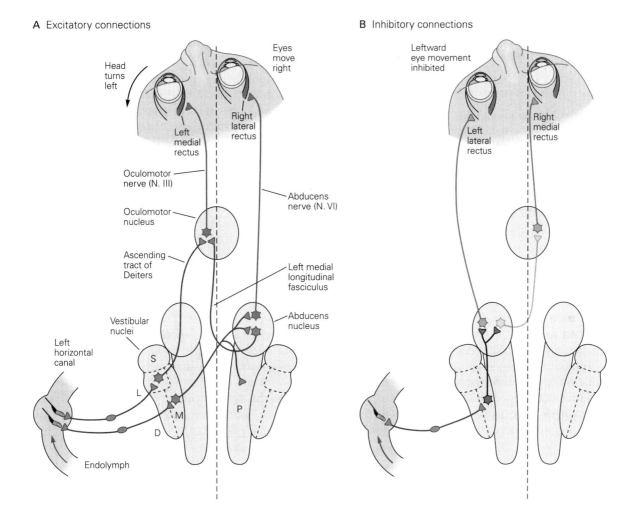

Figure 27–11 The horizontal vestibulo-ocular reflex. Similar pathways connect the anterior and posterior canals to the vertical recti and oblique muscles.

A. Leftward head rotation excites hair cells in the left horizontal canal, thus exciting neurons that evoke rightward eye movement. The vestibular nuclei include two populations of first-order neurons. One lies in the medial vestibular nucleus (**M**); its axons cross the midline and excite neurons in the right abducens nucleus and nucleus prepositus hypoglossi (**P**). The other population is in the lateral vestibular nucleus (**L**); its axons ascend ipsilaterally in the tract of Deiters and excite neurons in the left oculomotor nucleus, which project in the oculomotor nerve to the left medial rectus muscle.

The right abducens nucleus has two populations of neurons. A set of motor neurons projects in the abducens nerve and excites the right lateral rectus muscle. The axons of a set of interneurons cross the midline and ascend in the left medial longitudinal fasciculus to the oculomotor nucleus, where they

excite the neurons that project to the left medial rectus muscle. These connections facilitate the rightward horizontal eye movement that compensates for leftward head movement. Other nuclei shown are the superior (**S**) and descending (**D**) vestibular nuclei.

B. During counterclockwise head movement, leftward eye movement is inhibited by sensory fibers from the left horizontal canal. These afferent fibers excite neurons in the medial vestibular nucleus that inhibit motor neurons and interneurons in the left abducens nucleus. This action reduces the excitation of the motor neurons for the left lateral and right medial rectus muscles. The same head movement results in a decreased signal in the right horizontal canal (not shown), which has similar connections. The weakened signal results in decreased inhibition of the right lateral and left medial rectus muscles and decreased excitation of the left lateral and right medial rectus muscles. (Adapted from Sugiuchi et al. 2005.)

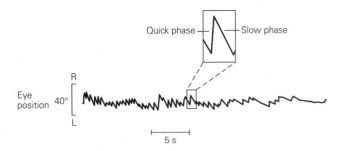

Figure 27–12 Vestibular nystagmus. The trace shows the eye position of a subject in a chair rotated counterclockwise at a constant rate in the dark. At the beginning of the trace, the eye moves slowly at the same speed as the chair (slow phase) and occasionally makes rapid resetting movements (quick phase). The speed of the slow phase gradually decreases until the eye no longer moves regularly. (Reproduced, with permission, from Leigh and Zee 2015.)

signals during rotation occurs through the velocity storage network.

If head rotation stops abruptly, the endolymph continues to be displaced in the same direction that the head had formerly rotated. With rightward rotation, this inhibits the right horizontal canal and excites the left horizontal canal, resulting in a sensation of leftward rotation and a corresponding left-beating nystagmus. However, this occurs only in darkness. In the light, optokinetic reflexes suppress postrotatory nystagmus since there is no visual motion stimulus.

The Translational Vestibulo-Ocular Reflex Compensates for Linear Motion and Head Tilts

When the head rotates, all images move with the same velocity on the retina. When the head moves sideways, however, the image of a close object moves more rapidly across the retina than does the image of a distant object. This can be understood easily by considering what happens when a person looks out the side window of a moving car. Objects near the side of the road move out of view almost with the speed of the car, whereas distant objects disappear more slowly. To compensate for linear head movement, the vestibular system must take into account the distance to the object being viewed—the more distant the object, the smaller the needed eye movement. During linear movements that do not involve head rotation, an appropriate translational VOR is elicited, driven by input from the otolith organs. Neurons in the vestibular nuclei, including some different from those providing the main drive to the rotational VOR, carry this signal to the extraocular motor neuron pools.

Side-to-side head movements result in a horizontal eye movement in a direction opposite to the head movement. Vertical displacements of the body, such as during walking or running, elicit oppositely directed vertical eye movements to stabilize gaze. However, in contrast to the rotational VOR where a head rotation is compensated by an equal but opposite eye rotation, horizontal displacement must be compensated by an eye rotation that depends on the viewed object distance, a nontrivial computation. For example, during a lateral head displacement, nearby objects move on the retina more rapidly than distant ones. So, in order to stabilize a nearby object on the retina, the eyes need to rotate by a larger amount than is needed for a distant object. Thus, the horizontal compensatory eye movements that are elicited during lateral motion scale with target distance; the closer the target, the larger is the compensatory eye movement. Similarly, as in the rotational VOR, compensatory responses to translation occur at relatively short latency (10–12 ms).

Fore-aft translations produce converging and diverging eye movements that bring the eyes together or move them apart. The amount of convergence or divergence is also dependent upon visual target distance, such that close visual objects produce large eye movements and distant visual objects produce little eye movements. Further, the amount of relative left and right eye movement is dependent upon visual object eccentricity relative to straight ahead. Unlike the rotational VOR that is a full-field image stabilization reflex, the goal of the translational VOR is to selectively stabilize visual objects on the fovea. In general, the two eyes move disjunctively, consisting of either a pure vergence movement or a combination of vergence and conjugate eye movements. In practice, although the direction of the evoked eye movement is typically consistent with geometrical predictions, the primate/human translational VOR typically undercompensates for near-target viewing, with gains of only about 0.5.

The translational VOR differs from the rotational VOR in the ability to generate compensatory eye movements during translation that optimize visual acuity on the central retina. These abilities appear to be specific to frontal-eyed animals, such as primates. Many lateral-eyed species, like the rabbit, do not generate eye movements that compensate for the visual consequences of translation during self-motion.

Because gravity exerts a constant linear acceleration force on the head, the otolith organs also sense the orientation of the head relative to gravity. When the head tilts away from the vertical in the roll plane—around the axis running from the occiput to the nose—the eyes rotate in the opposite direction to reduce the tilt of the retinal image. This ocular counter-rolling reflex—the ability to use a gravity-sensing mechanism to maintain gaze relative to the horizon—is of paramount importance for lateral-eyed, afoveate species that typically lack a well-developed saccadic system. But such functional utility for these tilt responses has lost its advantage in the primate oculomotor system, where static ocular counter-rolling and counter-pitching in humans have a gain of less than 0.1.

Vestibulo-Ocular Reflexes Are Supplemented by Optokinetic Responses

The VORs compensate for head movement imperfectly. They are best at sensing the onset or abrupt change of motion; they compensate poorly for sustained motion at constant speed during translation or constant angular velocity during rotation. In addition, they are insensitive to very slow rotations or low-amplitude linear accelerations.

Thus, vestibular responses during prolonged motion in the light are supplemented by visual stabilization reflexes that maintain nystagmus when vestibular input ceases: optokinetic nystagmus, a full-field stabilization system, and ocular following, a foveal stabilization system. Although the two classes of reflexes are distinct, their pathways overlap.

The Cerebellum Adjusts the Vestibulo-Ocular Reflex

As we have seen, the VOR keeps the gaze constant when the head moves. There are times, however, when the reflex is inappropriate. For example, when you turn your head while walking, you want your gaze to follow. The rotational VOR, however, would prevent your eyes from turning with your head. To prevent this sort of biologically inappropriate response, the VOR is under the control of the cerebellum and cortex, which suppress the reflex during volitional head movements.

In addition, the VOR must be continuously calibrated to maintain its accuracy in the face of changes within the motor system (fatigue, injury to vestibular organs or pathways, eye-muscle weakness, or aging) and differing visual requirements (wearing corrective lenses). Indeed, the VOR is a highly modifiable reflex. The brain continuously monitors its performance by evaluating the clarity of vision during head movements. When head turns are consistently associated with image motion across the retina, the VOR undergoes gain changes in the direction appropriate to improve the compensatory ability of the reflex. For example, when viewing the world through spectacles that magnify or miniaturize the visual scene, the rotational VOR gain (in darkness) increases or decreases accordingly. The reflex behavior can adapt over several minutes, hours, and days. This is accomplished by sensory feedback that modifies the motor output. If the reflex is not working properly, the image moves across the retina. The motor command to the eye muscles must be adjusted until the gaze is again stable, rotational retinal image motion is zero, and there is no error.

Anyone who wears eyeglasses depends on this plasticity of the VOR. Because lenses for nearsightedness shrink the visual image, a smaller eye rotation is needed to compensate for a given head rotation, and the gain of the VOR must be reduced. Conversely, glasses for farsightedness magnify the image, so the VOR gain must increase during their use. More complicated is the instance of bifocal or progressive spectacles, in which the reflex must use different gains for the different magnifications. In the laboratory, the reflex can be conditioned by altering the visual consequences of head motion. For example, if a subject is rotated for a period of time while wearing magnifying glasses, the reflex gain gradually increases (Figure 27–13A).

This process requires changes in synaptic transmission in both the cerebellum and the brain stem. If the flocculus and paraflocculus of the cerebellum are lesioned, the gain of the VOR can no longer be modulated. Mossy fibers carry vestibular, visual, and motor signals from the pontine and vestibular nuclei to the cerebellar cortex; the granule cells, with their parallel fiber axons, relay these signals to the Purkinje cells (Figure 27–13B). The synaptic efficacy of parallel fiber input to a Purkinje cell could be modified by the concurrent action of climbing fiber input. Indeed, the climbing fiber input to the cerebellum carries a retinal error signal, thought to serve as a "teaching" signal enabling the cerebellum to correct the error in the VOR. This adaptation requires long-term plasticity of multiple mechanisms through multiple sites (Chapter 37).

In addition to the Purkinje cell, plasticity is also found in the vestibular nuclei, in a particular class of neurons known as flocculus target neurons, which receive GABAergic inhibitory input from Purkinje cells in the flocculus as well as direct inputs from vestibular sensory fibers. During adaptation of the VOR, these neurons change their sensitivity to the vestibular inputs in the appropriate way, and after adaptation, they can maintain those changes without further input from the cerebellum. The importance of the cerebellum in calibrating eye movements is also evident in patients with cerebellar disease, who are often characterized by a VOR response of abnormal amplitude or direction.

A Adaptability of the vestibulo-ocular reflex

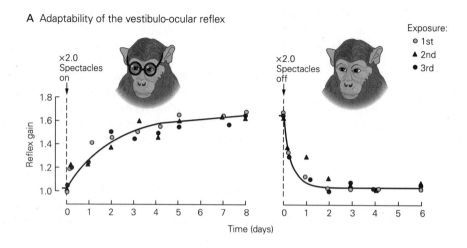

B Sites of adaptive learning

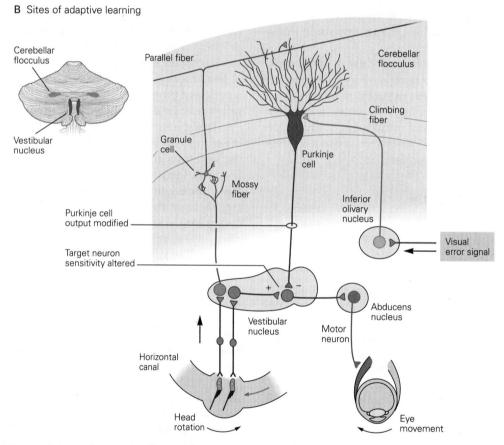

Figure 27–13 The vestibulo-ocular reflex is adaptable.

A. For several days, the monkey continuously wears magnifying spectacles that double the speed of the retinal-image motion evoked by head movement. Each day, the gain of the vestibulo-ocular reflex—the amount the eyes move for a given head movement—is tested in the dark so that the monkey cannot use retinal motion as a clue to modify the reflex. Over a period of 4 days, the gain increases gradually (*left*). It quickly returns to normal when the spectacles are removed (*right*). (Adapted, with permission, from Miles and Eighmy 1980.)

B. Adaptation of the vestibulo-ocular reflex occurs in cerebellar and brain stem circuits. A visual error signal, triggered by motion of the retinal image during head movement, reaches the inferior olivary nucleus. The climbing fiber transmits this error signal to the Purkinje cell, affecting the parallel fiber–Purkinje cell synapse. The Purkinje cell transmits changed information to the floccular target cell in the vestibular nucleus, changing its sensitivity to the vestibular input. After the reflex has been adapted, the Purkinje cell input is no longer necessary.

The Thalamus and Cortex Use Vestibular Signals for Spatial Memory and Cognitive and Perceptual Functions

For decades, vestibular function has been studied primarily in relation to reflexes, both vestibulo-ocular and vestibulospinal. Yet, in the past decade, it has become increasingly clear that the function of the vestibular system is as important for cognitive processes as for reflexes. The difficulty in understanding the vestibular system's role in spatial cognition stems from the fact that these functions are inherently multisensory, arising through convergence of vestibular, visual, somatosensory, and motor cues, following principles that remain poorly understood. Some of these perceptual functions of the vestibular system include tilt perception, visual-vertical perception, and visuospatial constancy.

Tilt perception. Vestibular information is critical for spatial orientation—the perception of how our head and body are positioned relative to the outside world. Nearly all species orient themselves using gravity, which provides a global, external reference. Thus, spatial awareness is governed by our orientation relative to gravity, collectively typically referred to as tilt.

Visual-vertical perception. We commonly experience the visual scene as perceptually oriented relative to earth-vertical orientation, regardless of our spatial orientation in the world. This ability has been studied psychophysically in humans and monkeys using tasks in which a subject is turned ear-down in the dark and asked to orient a dimly lit bar vertically in space (to align it with gravity). The results suggest that the neural representation of the visual scene is modified by static vestibular and proprioceptive signals that indicate the orientation of the head and body.

Visuospatial constancy. Vestibular signals are also important for the perception of a stable visual world despite constantly changing retinal images caused by movement of the eyes, head, and body. The projection of the scene onto the retina continuously changes because of these movements. Despite the changing retinal image, the percept of the scene as a whole remains stable; this stability is critical not only for vison but also for sensorimotor transformations (eg, to update the motor goal of an eye or arm movement).

Vestibular Information Is Present in the Thalamus

Vestibular projections to the thalamus are complicated and overall less clear, partly because of the strong multisensory nature of the responses in these cells and the difficulty in comparing thalamic regions and nomenclature across studies and species. Some neurons in all vestibular nuclei and likely the fastigial cerebellar nuclei project bilaterally to the thalamus, but most fibers terminate in the contralateral thalamic nuclei (Figure 27–9).

Several major thalamic regions receive vestibular projections, including the ventral posterolateral and ventral lateral thalamic nuclei and, to a lesser extent, the ventral posteroinferior nuclei, the posterior group, and the anterior pulvinar. These nuclei are traditionally thought to also receive somatosensory input and project to the primary and secondary somatosensory cortices, as well as the posterior parietal cortex (areas 5 and 7) and the insula of the temporal cortex.

Vestibular Information Is Widespread in the Cortex

A number of cortical areas receiving short-latency vestibular signals either alone or more commonly in concert with proprioceptive, tactile, oculomotor, visual, and auditory signals have been identified (Figure 27–14). Although vestibular signals are widely distributed to a number of cortical regions, all such regions are multimodal and none seems to represent a purely vestibular cortex, similar to other modalities such as vision, proprioception, and audition.

Vestibular modulation has been established in the lateral sulcus (parietoinsular vestibular cortex), somatosensory cortex (areas 3a and 2v), oculomotor cortex (frontal and supplementary eye fields), extrastriate visual motion cortex (dorsal medial superior temporal area), and parietal cortex (ventral intraparietal area and area 7a). In the primary somatosensory cortex, area 2v lies at the base of the intraparietal sulcus just posterior to the areas of the postcentral gyrus representing the hand and mouth. Electrical stimulation of area 2v in humans produces sensations of whole-body motion. Area 3a lies at the base of the central sulcus, adjacent to the motor cortex. Many cells in the parietoinsular vestibular cortex are multisensory, responding to body motion, somatosensory, proprioceptive, and visual motion stimuli. Patients with lesions in this region report episodes of vertigo, unsteadiness, and a loss of perception for visual vertical. Neurons in the medial intraparietal and medial superior temporal areas respond to both visual (optic flow) and vestibular signals. These cells utilize multisensory cue integration (Bayesian) frameworks to assist in the cognitive perception of motion through space.

Imaging studies reveal an even larger portion of cerebral cortex involved in processing vestibular information, including the temporoparietal cortex and the insula, the superior parietal lobe, the

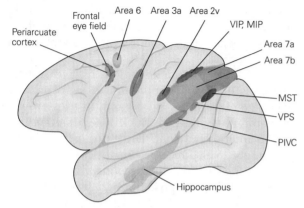

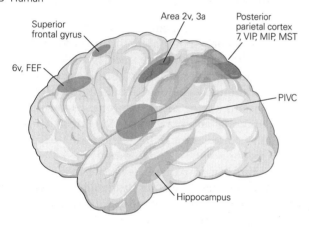

Figure 27–14 The vestibular cortex.

A. This lateral view of a monkey's brain shows the areas of cerebral cortex in which vestibular responses have been recorded. Areas in monkey cortex include periarcuate cortex, area 6, frontal eye fields, areas 3a and 2v, ventral intraparietal area (**VIP**), medial intraparietal area (**MIP**), area 7, visual posterior sylvian area (**VPS**), medial superior temporal area (**MST**), parieto-insular vestibular cortex (**PIVC**), and the hippocampal formation.

B. In the human cortex, areas recording vestibular activity include 6v, frontal eye fields (**FEF**), superior frontal gyrus, 2v, 3a, posterior parietal cortex, PIVC, and the hippocampal formation.

pre- and postcentral gyri, anterior cingulate and posterior middle temporal gyri, premotor and frontal cortices, inferior parietal lobule, putamen, and hippocampal regions. Using electrical stimulation of the vestibular nerve in patients activates the prefrontal lobe and anterior portion of the supplementary motor area at relatively short latencies. However, imaging and, to a lesser extent, single-cell recording studies may overstate the range of vestibular representations. In particular, vestibular stimuli often co-activate the

somatosensory and proprioceptive systems, as well as evoke postural and oculomotor responses, which might in turn result in increased cortical activations.

Vestibular Signals Are Essential for Spatial Orientation and Spatial Navigation

Our ability to move about depends on a stable directional orientation. Certain cells in the thalamus, hippocampal region, entorhinal cortex, and subiculum are involved in navigation tasks. Damage to these areas impairs a variety of spatial and directional abilities. At least six cell types contributing to spatial orientation have been identified, including place cells, grid cells, head direction cells, border cells, speed cells, and conjunctive cells. In the hippocampus, place cells discharge relative to the animal's location in the environment (Chapter 54). Head direction cells in the dorsal thalamus, parahippocampal regions, and several regions of the cortex indicate the animal's heading direction like a compass. Grid cells in the entorhinal cortex respond to multiple spatial locations in a unique triangular grid pattern. Border cells in the entorhinal cortex signal environmental boundaries, speed cells discharge in proportion to the animal's running speed, and conjunctive cells exhibit a combination of several of these properties.

These regions are intimately connected and appear to work together in a "navigation network" to provide for spatial orientation, spatial memory, and our ability to move through our surroundings. Think of walking through your house, driving to the store, or knowing which direction to go in a new city. Lesions of central vestibular networks disrupt head direction, place, and grid responses. Patients with disease or trauma to the vestibular system, hippocampus, and anterior thalamus regions often exhibit severe deficits in their ability to orient in familiar environments or even find their way home.

All of these cells depend on a functioning vestibular system to maintain their spatial orientation properties. The pathway by which vestibular signals reach the navigation network and the computational principles determining how vestibular cues influence these spatially tuned cells is not well understood. We know that there are at least three different influences: Semicircular canal signals contribute to the estimate of head direction; gravity signals influence the three-dimensional properties of head direction cells; and translation signals influence the estimate of linear speed, which controls both grid cell properties and the magnitude and frequency of theta oscillations in the hippocampal network. What is clear is that there is no evidence linking vestibular nuclei response properties directly to head direction or other spatially tuned cell types, and no direct projections from the vestibular nuclei to the brain areas thought to house these spatially tuned neurons have been identified. Furthermore, vestibular nuclei responses are inappropriate for driving these spatially tuned cells, as these signals need to encompass the total head movement, rather than individual components during active or passive head movement.

It has long been recognized that proprioceptive and motor efference cues should participate, together with vestibular signals, to track head direction over time. It has been proposed that internally generated information from vestibular, proprioceptive, and motor efference cues can be utilized to keep track of changes in directional heading. More recent insights have started to shed light on how each of these cues contributes to the final self-motion estimate that can be precisely predicted and quantitatively estimated based on a Bayesian framework. Although as yet difficult to define, quantitative internal models govern the relationship of vestibular and other multisensory self-motion cues for computing the spatial properties of navigation circuit cells.

Clinical Syndromes Elucidate Normal Vestibular Function

As we have seen, rotation excites hair cells in the semicircular canal whose hair bundles are oriented in the direction of motion and inhibits those in the canals oriented away from the motion direction. This imbalance in vestibular signals is responsible for the compensatory eye movements and the sensation of rotation that accompanies head movement. It can also originate from disease of one labyrinth or vestibular nerve, which results in a pattern of afferent vestibular signaling analogous to that stemming from rotation away from the side of the lesion, that is, more discharge from the intact side. There is accordingly a strong feeling of spinning, called vertigo.

Caloric Irrigation as a Vestibular Diagnostic Tool

Nystagmus can be used as a diagnostic indicator of vestibular system integrity. In patients complaining of dizziness or vertigo, the function of the vestibular labyrinth is typically assessed by a caloric test (Figure 27–15). Either warm (44°C) or cold (30°C) water is introduced into the external auditory canal. In normal persons, warm water induces nystagmus that beats toward the ear into which the water has been introduced, whereas cold water induces nystagmus that beats away from the ear into which the water has been introduced. This relationship is encapsulated in the mnemonic COWS: Cold water produces nystagmus beating to the Opposite side; Warm water produces nystagmus beating to the Same side. In normal persons, the two ears give equal responses. If there is a unilateral lesion in the vestibular pathway, however, nystagmus will be induced and directed toward the side opposite the lesion.

The vertigo and nystagmus resulting from an acute vestibular lesion typically subside over several days, even if peripheral function does not recover. This is because central compensatory mechanisms restore the balance in vestibular signals in the brain stem, even when peripheral input is permanently lost or unbalanced.

The loss of input from one labyrinth also means that all vestibular reflexes must be driven by a single labyrinth. For the VOR, this condition is quite effective at low speeds because the intact labyrinth can be both excited and inhibited. However, during rapid, high-frequency rotations, inhibition is not sufficient, such that the gain of the reflex is reduced when the head rotates toward the lesioned side. This is the basis of an important clinical test of canal function, the head-impulse test. In this test, the head is moved rapidly one time along the axis of rotation of a single canal. If there is a significant decrease in gain owing to canal dysfunction, the movement of the eyes will lag behind that of the head, and there will be a visible catch-up saccade.

Bilateral Vestibular Hypofunction Interferes With Normal Vision

Vestibular function is sometimes lost simultaneously on both sides, for example, from ototoxicity owing to aminoglycoside antibiotics such as gentamicin or cancer treatment medications such as cisplatin. The symptoms of bilateral vestibular hypofunction are different from those of unilateral loss. First, vertigo is absent because there is no imbalance in vestibular signals; input is reduced equally from both sides. For the same reason, there is no spontaneous nystagmus. In fact, these patients may have no symptoms when they are at rest and the head is still.

In humans, receptor and nerve fiber loss due to disease, trauma, or ototoxicity is permanent. However, in other animal classes such as amphibians, reptiles, and birds, spontaneous regeneration does occur over time. Although the differences in regeneration between animal groups is not yet understood, recent

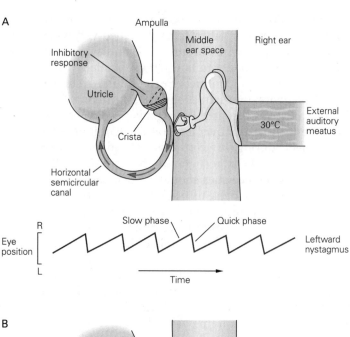

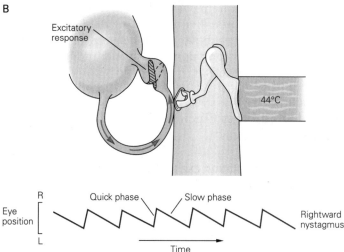

Figure 27–15 Bithermal caloric test of the vestibulo-ocular reflex. The vestibular caloric test remains the primary test used today in clinics around the world to determine if there is system dysfunction. The head is elevated 30° to align the horizontal semicircular canals with gravity.

A. Cold water or air introduced into the right ear causes a downward convection current in the endolymph, producing an inhibitory response in the right ear hair cells and afferent fibers. The result is a leftward (opposite side) beating nystagmus (as determined by fast phase direction).

B. Warm water or air introduced into the right ear produces an upward endolymph movement, producing an excitatory response in the hair cells and afferents. The result is a rightward (same side) beating nystagmus.

research shows promise for the future development of regenerative treatments in humans.

For the present, the loss of vestibular reflexes is devastating. A physician who lost his vestibular hair cells because of a toxic reaction to streptomycin wrote a dramatic account of this loss. Immediately after the onset of streptomycin toxicity, he could not read without steadying his head to keep it motionless. Even after partial recovery, he could not read signs or recognize friends while walking in the street; he had to stop to see clearly. Some patients may even "see" their heartbeat if the VOR fails to compensate for the miniscule head movements that accompany each arterial pulse.

Highlights

1. The vestibular system provides the brain with a rapid estimate of head movement. Vestibular signals are used for balance, visual stability, spatial orientation, movement planning, and motion perception.
2. Vestibular receptor hair cells are mechanotransducers that sense rotational and linear accelerations. Through kinematic and neural processing mechanisms, movements are transformed into acceleration, velocity, and position signals. These signals are used throughout the brain efficiently and quickly to guide behavior and cognition.

3. Receptor cells are polarized to detect the direction of motion. Three semicircular canals in each inner ear detect rotational motion and work in bilateral synergistic pairs through convergent commissural pathways in the vestibular nuclei. Two otolith organs in each ear detect linear translations and tilts relative to gravity.

4. Vestibular nuclei neurons receive converging multisensory and motor signals from visual, proprioceptive, cerebellar, and cortical sources. The multisensory integration allows for discrimination between active and passive body motion, as well as appropriate motor responses for reactive or volitional behavior.

5. Projections from the vestibular nuclei to the oculomotor system allow eye muscles to compensate for head movement through the vestibulo-ocular reflex to hold the image of the external world motionless on the retina. Cortical projections to the vestibular and oculomotor nuclei allow volitional eye movements to be separated from reflex eye movements but work through a final common pathway. Motor learning through vestibulocerebellar networks provides compensatory changes in eye movement responses to changing visual conditions through the use of spectacles, disease, or aging.

6. Projections from the vestibular nuclei to motor areas and the spinal cord facilitate postural stability. Gaze stability coordinates eye and neck movements through the medial vestibulospinal pathway. Postural control is exerted through the lateral vestibulospinal pathway.

7. Projections from the vestibular nuclei to the rostral and caudal medulla nuclei are involved in regulation of blood pressure, heart rate, respiration, bone remodeling, and homeostasis.

8. Projections from the vestibular nuclei to thalamus and cortex ensure spatial orientation and influence spatial perception more generally.

9. Vestibular signals processed in the hippocampal regions are crucial for spatial location and navigation functions.

10. Vestibular signals are combined with visual signals in several cortical regions through Bayesian cue integration to provide motion perception.

11. Disease or trauma to the vestibular system can produce nausea, vertigo, dizziness, balance disorders, visual instability, and spatial confusion.

12. We are only beginning to appreciate the role of the vestibular system in cognition. However, it is clear that vestibular signals contribute to our perception of self, conception of body presence, and memory.

13. New approaches in computation and theory promise to provide the lapidary keys needed to unlock our understanding of how vestibular signals contribute to the essence of brain function.

J. David Dickman
Dora Angelaki

Selected Reading

Baloh RW, Honrubia V. 1990. *Clinical Neurology of the Vestibular System*, 2nd ed. Philadelphia: FA Davis.

Beitz AJ, Anderson JH. 2000. *Neurochemistry of the Vestibular System*. Boca Raton, FL: CRC Press.

Goldberg JM, Wilson VJ, Cullen KE, et al. 2012. *The Vestibular System: A Sixth Sense*. New York: Oxford Univ. Press.

Leigh RJ, Zee DS. 2015. *The Neurology of Eye Movements*, 5th ed. New York: Oxford Univ. Press.

References

Angelaki DE, Cullen KE. 2008. Vestibular system: the many facets of a multimodal sense. Ann Rev Neurosci 31:125–150.

Angelaki DE, Shaikh AG, Green AM, Dickman JD. 2004. Neurons compute internal models of the physical laws of motion. Nature 430:560–564.

Brandt T, Dieterich M. 1999. The vestibular cortex. Its locations, functions, and disorders. Ann N Y Acad Sci 871:293–312.

Clark BJ, Taube JS. 2012. Vestibular and attractor network basis of the head direction cell signal in subcortical circuits. Front Neural Circuits 6:1–12.

Crèmer PD, Halmagyi GM, Aw ST, et al. 1998. Semicircular canal plane head impulses detect absent function of individual semicircular canals. Brain 121:699–716.

Cullen KE, Roy JE. 2004. Signal processing in the vestibular system during active versus passive head movements. J Neurophys 91:1919–1933.

Curthoys IS, Halmagyi GM. 1992. Behavioral and neural correlates of vestibular compensation. Behav Clin Neurol 1:345–372.

Dickman JD, Angelaki DE. 2002. Vestibular convergence patterns in vestibular nuclei neurons of alert primates. J Neurophys 88:3518–3533.

Dieterich M, Brandt T. 1995. Vestibulo-ocular reflex. Curr Opin Neurol 8:83–88.

Distler C, Mustari MJ, Hoffmann KP. 2002. Cortical projections to the nucleus of the optic tract and dorsal terminal nucleus and to the dorsolateral pontine nucleus in macaques: a dual retrograde tracing study. J Comp Neurol 444:144–158.

Einstein A. 1908. Über das Relativitätsprinzip und die aus demselben gezogenen Folgerungen. Jahrbuch Radioaktiv Electronik 4:411–462.

Fernandez C, Goldberg JM. 1971. Physiology of peripheral neurons innervating semicircular canals of the squirrel monkey. II. Response to sinusoidal stimulation and dynamics of peripheral vestibular system. J Neurophysiol 34:661–675.

Fernandez C, Goldberg JM. 1976a. Physiology of peripheral neurons innervating otolith organs of the squirrel monkey. I. Response to static tilts and to long-duration centrifugal force. J Neurophysiol 39:970–984.

Fernandez C, Goldberg JM. 1976b. Physiology of peripheral neurons innervating otolith organs of the squirrel monkey. II. Directional selectivity and force-response relations. J Neurophysiol 39:985–995.

Flock Å. 1965. Transducing mechanisms in the lateral line canal organ receptors. Cold Spring Harbor Symp Quant Biol 30:133–145.

Fukushima K. 1997. Corticovestibular interactions: anatomy, electrophysiology, and functional considerations. Exp Brain Res 117:1–16.

Gacek RR, Lyon M. 1974. The localization of vestibular efferent neurons in the kitten with horseradish peroxidase. Acta Otolaryngol (Stockh) 77:92–101.

Goldberg JM, Fernández C. 1971. Physiology of peripheral neurons innervating semicircular canals of the squirrel monkey. I. Resting discharge and response to constant angular accelerations. J Neurophysiol 34:635–660.

Goldberg ME, Colby CL. 1992. Oculomotor control and spatial processing. Curr Opin Neurobiol 2:198–202.

Grüsser OJ, Pause M, Schreiter U. 1990. Localization and responses of neurons in the parieto-insular vestibular cortex of awake monkeys (Macaca fascicularis). J Physiol (Lond) 430:537–557.

Gu Y, Watkins PV, Angelaki DE, DeAngelis GC. 2006. Visual and nonvisual contributions to three-dimensional heading selectivity in the medial superior temporal area. J Neurosci 26:73–85.

Hillman DE, McLaren JW. 1979. Displacement configuration of semicircular canal cupulae. Neuroscience 4:1989–2000.

Hudspeth AJ, Corey DP. 1977. Sensitivity, polarity, and conductance change in the response of vertebrate hair cells to controlled mechanical stimuli. Proc Nat Acad Sci 74:2407–2411.

Iurato S. 1967. Submicroscopic Structure of the Inner Ear. Oxford: Pergamon Press.

Laurens J, Kim H, Dickman JD, Angelaki DE. 2016. Gravity orientation tuning in macaque anterior thalamus. Nat Neurosci 19:1566–1568.

Miles FA, Eighmy BB. 1980. Long-term adaptive changes in primate vestibuloocular reflex. I. Behavioral observations. J Neurophysiol 43:1406–1425.

Moser EI, Moser MB. 2008. A metric for space. Hippocampus 18:1142–1156.

Mustari MJ, Fuchs AF. 1990. Discharge patterns of neurons in the pretectal nucleus of the optic tract (NOT) in the behaving primate. J Neurophysiol 64:77–90.

Newlands SD, Vrabec JT, Purcell IM, Stewart CM, Zimmerman BE, Perachio AA. 2003. Central projections of the saccular and utricular nerves in macaques. J Comp Neurol 466:31–47.

O'Keefe J. 1976. Place units in the hippocampus of the freely moving rat. Exp Neurol 51:78–109.

Raymond JL, Lisberger SG. 1998. Neural learning rules for the vestibulo-ocular reflex. J Neurosci 18:9112–9129.

Spoendlin H. 1966. Ultrastructure of the vestibular sense organ. In: RJ Wolfson (ed). The Vestibular System and Its Diseases, pp. 39–68. Philadelphia: Univ. of Pennsylvania Press.

Sugiuchi Y, Izawa Y, Ebata S, Shinoda Y. 2005. Vestibular cortical areas in the periarcuate cortex: its afferent and efferent projections. Ann N Y Acad Sci 1039:111–123.

Taube JS. 1995. Head direction cells recorded in the anterior thalamic nuclei of freely moving rats. J Neurosci 15:70–86.

Waespe W, Henn V. 1977. Neuronal activity in the vestibular nuclei of the alert monkey during vestibular and optokinetic stimulation. Exp Brain Res 27:523–538.

Watanuki K, Schuknecht HF. 1976. A morphological study of human vestibular sensory epithelia. Arch Otolaryngol Head Neck Surg 102:583–588.

28

Auditory Processing by the Central Nervous System

HEARING IS CRUCIAL FOR LOCALIZING and identifying sound; for humans, it is particularly important because of its role in the understanding and production of speech. The auditory system has several noteworthy features. Its subcortical pathway is longer than that of other sensory systems. Unlike the visual system, sounds can enter the auditory system from all directions, day and night, when we are asleep as well as when we are awake. The auditory system processes not only sounds emanating from outside the body (environmental sounds, sounds generated by others) but also self-generated sounds (vocalizations and chewing sounds). The location of sound stimuli in space is not conveyed by the spatial arrangement of sensory afferent neurons but is instead computed by the auditory system from representations of the physical cues.

Sounds Convey Multiple Types of Information to Hearing Animals

Hearing helps to alert animals to the presence of unseen dangers or opportunities and, in many species, also serves as a means for communication. Information about where sounds arise and what they mean must be extracted from the representations of the physical characteristics of sound at each of the ears. To understand how animals process sound, it is useful first to consider which cues are available.

Most vertebrates take advantage of having two ears for localizing sounds in the horizontal plane. Sound sources at different positions in that plane affect the two ears differentially: Sound arrives earlier and is more intense at the ear nearer the source (Figure 28–1A). Interaural time and intensity differences carry information about where sounds arise.

The size of the head determines how interaural time delays are related to the location of sound sources; the neuronal circuitry determines the precision with which time delays are resolved. Because air pressure waves travel at roughly 340 m/s in air, the maximal interaural delay in humans is approximately 600 µs; in small birds, the greatest delay is only 35 µs. Humans can resolve the location of a sound source directly ahead to within approximately 1 degree, corresponding to an interaural time difference of 10 µs. Interaural time differences are particularly well conveyed by neurons that encode relatively low frequencies. These neurons can fire at the same position in every cycle of the sound and in this way encode the interaural time difference as an interaural phase difference. Sounds of high frequencies produce *sound shadows* or intensity differences between the two ears. For many mammals with small heads, high-frequency sounds provide the primary cue for localizing sound in the horizontal plane.

Mammals can localize sounds in the vertical plane and with a single ear using spectral filtering. High-frequency sounds, with wavelengths that are close to or smaller than the dimensions of the head, shoulders, and external ears, interact with those parts of the body to produce constructive and destructive interference, introducing broad spectral peaks and deep, narrow spectral notches whose frequency changes with the location of the sound (Figure 28–1B). High-frequency sounds from different origins are filtered differently

because in mammals the shape of the external ear differs back-to-front as well as top-to-bottom. Animals learn to use these spectral cues to locate sound sources. If the shape of the ear is experimentally altered, even adult humans can learn to make use of a new pattern of spectral cues. If animals lose hearing in one ear, they lose interaural timing and intensity cues and must depend completely on spectral cues for localizing sounds.

How do we make sense of the complex and changing sounds that we hear? Most natural sounds contain energy over a wide range of frequencies and change rapidly with time. The information used to recognize sounds varies among animal species, and depends on listening conditions and experience. Human speech, for example, can be understood in the midst of noise, over electronic devices that distort sounds, and even through cochlear implants. One reason for its robustness is that speech contains redundant cues: The vocal apparatus produces sounds in which multiple parameters covary. At the same time, this makes the task of understanding how animals recognize patterns a complicated one. It is not clear which cues are used by animals under various conditions.

Music is a source of pleasure to human beings. Musical instruments and human voices produce sounds that have energy at the fundamental frequency that corresponds to its perceived pitch, as well as at multiples of that frequency, giving sounds a quality that allows us, for example, to distinguish a flute from a violin when their pitch is the same. Musical pitches are largely in the low-frequency range in which auditory nerve fibers fire in phase with sounds. In music, sounds are combined simultaneously to produce chords and successively to produce melodies. Euphonious, pleasant chords elicit regular, periodic firing in cochlear nerve fibers. In dissonant sounds, there is less regularity both in the sound itself and in the firing of auditory nerve fibers; the component frequencies are so close that they interfere with one another instead of periodically reinforcing one another.

The Neural Representation of Sound in Central Pathways Begins in the Cochlear Nuclei

The neural pathways that process acoustic information extend from the ear to the brain stem, through the midbrain and thalamus, to the cerebral cortex (Figure 28–2). Acoustic information is conveyed from cells in the cochlear ganglion (see Figure 26–17) to the cochlear nuclei in the brain stem. There information is received by several different types of neurons, most of which are arranged tonotopically.

A Sound localization using interaural difference

B Sound localization using spectral filtering

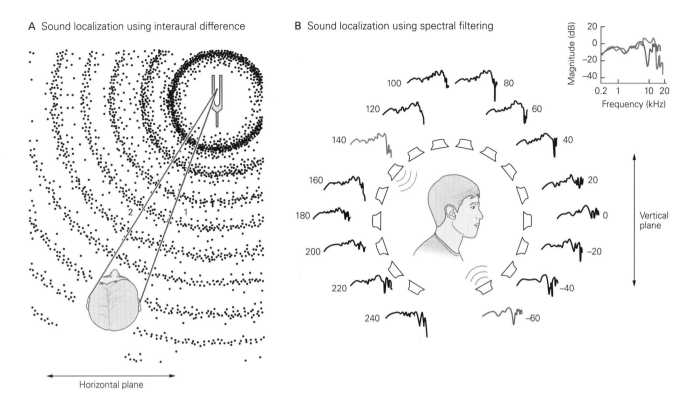

Horizontal plane

Figure 28–1 Cues for localizing sound sources in the horizontal plane.

A. Interaural time and intensity differences are cues for localizing sound sources in the horizontal plane, or azimuth. A sound arising in the horizontal plane arrives differently at the two ears: Sounds arrive earlier and are louder at the ear nearer the source. A sound that arises directly in the front or back travels the same distance to the right and left ears and thus arrives at both ears simultaneously. Interaural time and intensity do not vary with the movement of sound sources in the vertical plane, so it is impossible to localize a pure sinusoidal tone in the vertical plane. In humans, the maximal interaural time difference is approximately 600 µs. High-frequency sounds, with short wavelengths, are deflected by the head, producing a sound shadow on the far side. (Adapted, with permission, from Geisler 1998.)

B. Mammals can localize broadband sounds in both the vertical and horizontal planes on the basis of spectral filtering. When a noise that has equal energy at all frequencies over the human hearing range (*white noise*) is presented through a speaker, the ear, head, and shoulders cancel energy at some frequencies and enhance others. The white noise that is emitted from the speaker has a flat power spectrum, but by the time the noise

has reached the bottom of the ear canal, its spectrum is no longer flat.

In the figure, the sound energy at each frequency at the eardrum relative to that of the white noise is shown by the traces beside each speaker; these traces plot the relative sound magnitude in decibels against spectral frequency (*head-related transfer function*). The small plot in the upper right compares two head-related transfer functions: one for a noise that arises low and in front of a listener (**blue**) and one for a noise from behind the listener's head (**brown**). Head-related transfer functions have deep notches at frequencies greater than 8 kHz, whose frequencies vary depending on where the sounds arose. Sounds that lack energy at high frequencies and narrowband sounds are difficult to localize in the vertical plane. Since spectral filtering also varies in the horizontal plane, it provides the only location cue to animals that have lost hearing in one ear.

You can test the salience of these spectral cues with a simple experiment. Close your eyes as a friend jingles keys directly in front of you at various elevations. Compare your ability to localize sounds under normal conditions and when you distort the shape of both ears by pushing them with your fingers from the back. (Data from D. Kistler and F. Wightman.)

The axons of the different types of neurons take different routes to the brain stem and midbrain, where they terminate on separate targets. Some of the pathways from the cochlear nuclei to the contralateral inferior colliculus are direct; others involve one or two synaptic stages in brain stem auditory nuclei. From the bilateral inferior colliculi, acoustic information

flows two ways: to the ipsilateral superior colliculus, where it participates in orienting the head and eyes in response to sounds, and to the ipsilateral thalamus, the relay to auditory areas of the cerebral cortex. The afferent auditory pathways from the periphery to higher brain regions include efferent feedback at many levels.

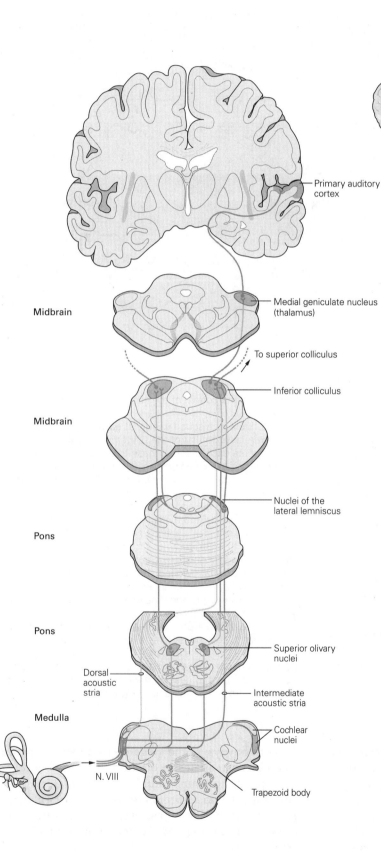

Midbrain

Primary auditory
cortex

Medial geniculate nucleus
(thalamus)

To superior colliculus

Inferior colliculus

Midbrain

Nuclei of the
lateral lemniscus

Pons

Pons

Superior olivary
nuclei

Dorsal
acoustic
stria

Intermediate
acoustic stria

Medulla

Cochlear
nuclei

N. VIII

Trapezoid body

**Figure 28–2 The central auditory pathways
extend from the brain stem through the
midbrain and thalamus to the auditory
cortex.** The fibers in the cochlear nerve
(cranial nerve VIII) terminate in the cochlear
nuclei of the brain stem. The neurons of these
nuclei project in several parallel pathways to
the inferior colliculus. Their axons exit through
the trapezoid body, intermediate acoustic stria,
or dorsal acoustic stria. Some cells terminate
directly in the inferior colliculus. Others contact
cells in the superior olivary complex and in the
nuclei of the lateral lemniscus, which in turn
project to the inferior colliculus. Neurons of
the inferior colliculus project to the superior
colliculus and to the medial geniculate nucleus
of the thalamus. Thalamic neurons project to
the auditory cortex. The cochlear nuclei and the
ventral nuclei of the lateral lemniscus are the
only central auditory neurons that receive monaural input. (Adapted, with permission, from
Brodal 1981.)

The Cochlear Nerve Delivers Acoustic Information in Parallel Pathways to the Tonotopically Organized Cochlear Nuclei

The afferent nerve fibers from cochlear ganglion cells are bundled in the cochlear or auditory component of the vestibulocochlear nerve (cranial nerve VIII) and terminate exclusively in the cochlear nuclei. The cochlear nerve in mammals contains two groups of fibers: a large number (95%) of myelinated fibers that receives input from inner hair cells and a small number (5%) of unmyelinated fibers that receive input from outer hair cells.

The larger, more numerous, myelinated fibers are much better understood than the unmyelinated fibers. Each type detects energy over a narrow range of frequencies; the tonotopic array of cochlear nerve fibers thus carries detailed information about how the frequency content of sounds varies from moment to moment. The unmyelinated fibers terminate both on the large neurons in the ventral cochlear nuclei and also on the small granule cells that surround the ventral cochlear nuclei. Because it is difficult to record from these tiny fibers, the information they convey to the brain is not well understood. The unmyelinated fibers integrate information from a relatively wide region of the cochlea but are not responsive to sound. It has been suggested that these fibers respond to cochlear damage and contribute to hyperacusis—pain after exposure to loud sounds that damages the cochlea.

Two features of the cochlear nuclei are important. First, these nuclei are organized tonotopically. Fibers that carry information from the apical end of the cochlea, which detects low frequencies, terminate ventrally in the ventral and dorsal cochlear nuclei; those that carry information from the basal end of the cochlea, which detects high frequencies, terminate dorsally (Figure 28–3). Second, each cochlear nerve fiber innervates several different areas within the cochlear nuclei, contacting various types of neurons that have distinct projection patterns to higher auditory centers. As a result, the auditory pathway comprises at least four parallel ascending pathways that simultaneously extract different acoustic information from the signals carried by cochlear nerve fibers. Parallel circuits are a general feature of vertebrate sensory systems.

The Ventral Cochlear Nucleus Extracts Temporal and Spectral Information About Sounds

The principal cells of the unlayered ventral cochlear nucleus sharpen temporal and spectral information and convey it to higher centers of the auditory pathway. Three types of neurons are intermingled and form separate pathways through the brain stem (Figure 28–4).

Bushy cells project bilaterally to the superior olivary complex. This pathway has two parts. One courses through the medial superior olive and compares the time of arrival of sounds at the two ears; the other travels through the medial nucleus of the trapezoid body and the lateral superior olive and compares interaural intensity. Large spherical bushy cells sense low frequencies and project bilaterally to the medial superior olive, forming a circuit that detects interaural time delay and contributes to the localization of low-frequency sounds in the horizontal plane. The small spherical bushy cells and globular bushy cells sense higher frequencies. Small spherical bushy cells excite the lateral superior olive ipsilaterally. The globular bushy cells, through calyceal endings, excite neurons in the contralateral medial nucleus of the trapezoid body that in turn inhibit principal cells of the lateral superior olive. Neurons in the lateral superior olive integrate the ipsilateral excitation and contralateral inhibition to measure interaural intensity and to localize sources of high-frequency sounds in the horizontal plane (see Figure 28–6).

Stellate cells terminate widely. They excite neurons in the ipsilateral dorsal cochlear nucleus, the medial olivocochlear efferent neurons in the ventral nucleus of the trapezoid body, the periolivary nuclei in the vicinity of the ipsilateral lateral superior olive, and the contralateral ventral nucleus of the lateral lemniscus, inferior colliculus, and thalamus. The tonotopic array of stellate cells encodes the spectrum of sounds.

Octopus cells excite targets in the contralateral paraolivary nucleus and terminate in large excitatory calyceal endings on neurons of the ventral nucleus of the lateral lemniscus, which in turn provide sharply timed glycinergic inhibition to the inferior colliculus. Octopus cells detect onsets of sounds that allow animals to detect brief gaps. They mark the spectral components that come from one source that necessarily start together.

The differences in the integrative tasks performed by these pathways through the ventral cochlear nucleus are reflected in cell morphology. The shapes of their dendrites reflect the way they collect information from cochlear nerve fibers. The dendrites of the sharply tuned bushy and stellate cells receive input from relatively few cochlear nerve fibers, whereas those of the broadly tuned octopus cells, in contrast, lie perpendicular to the path of cochlear nerve fibers, poised to receive input from many cochlear nerve fibers. Many of the inputs to bushy cells are from unusually large

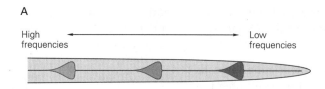

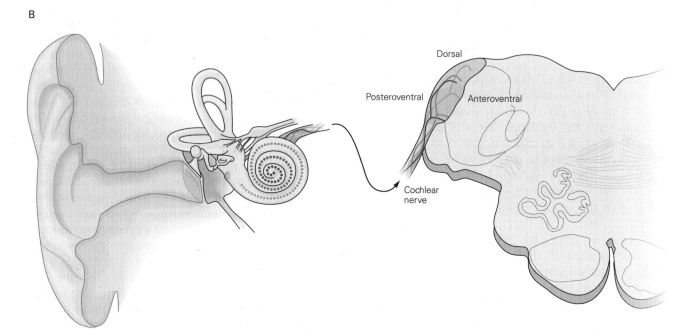

Figure 28–3 The dorsal and ventral cochlear nuclei.

A. Stimulation with three frequencies of sound vibrates the schematically uncoiled basilar membrane at three positions, exciting distinct populations of hair cells and their afferent nerve fibers.

B. Cochlear nerve fibers project in a tonotopic pattern to the cochlear nuclei. Those encoding the lowest frequencies (**red**) terminate most ventrally, whereas those encoding higher frequencies (**yellow**) terminate more dorsally. The cochlear nuclei include the ventral and dorsal nuclei. Each afferent fiber enters at the nerve root and splits into branches that run anteriorly (the ascending branch) and posteriorly (the descending branch). The ventral cochlear nucleus is thus divided functionally into anteroventral and posteroventral divisions.

terminals that envelop the bushy cell bodies, meeting their need for large synaptic currents. The need for large synaptic currents in octopus cells is met by summing inputs from large numbers of small terminals.

The biophysical properties of neurons determine how synaptic currents are converted to voltage changes and over how long a time synaptic inputs are integrated. Octopus and bushy cells in the ventral cochlear nucleus are able to respond with exceptionally rapid and precisely timed synaptic potentials. These neurons have a prominent, low-voltage-activated K^+ conductance that confers a low input resistance and rapid responsiveness and prevents repetitive firing (Figure 28–4C). The large synaptic currents that are required to trigger action potentials in these leaky cells are delivered through rapidly gated, high-conductance,

AMPA-type (α-amino-3-hydroxy-5-methylisoxazole-4-propionate) glutamate receptors at many synaptic release sites. In contrast, stellate cells, in which even relatively small depolarizing currents produce large protracted voltage changes, generate slower excitatory postsynaptic potentials (EPSPs) in response to synaptic currents, and *N*-methyl-D-aspartate (NMDA)-type glutamate receptors enhance those responses.

The Dorsal Cochlear Nucleus Integrates Acoustic With Somatosensory Information in Making Use of Spectral Cues for Localizing Sounds

Among vertebrates, only mammals have dorsal cochlear nuclei. The dorsal cochlear nucleus receives input from two systems of neurons that project to different layers (Figure 28–4A,B). Its principal cells, fusiform cells, integrate those two systems of inputs and convey the result directly to the contralateral inferior colliculus.

The outermost molecular layer is the terminus of a system of parallel fibers, the unmyelinated axons of granule cells that are scattered in and around the cochlear nuclei. This system transmits somatosensory, vestibular, and auditory information from widespread regions of the brain to the molecular layer.

The deep layer receives acoustic information. Not only cochlear nerve fibers but also stellate cells of the ventral cochlear nucleus terminate in the deep layer. Acoustic inputs are tonotopically organized in isofrequency laminae that run at right angles to parallel fibers.

Fusiform cells, the principal cells of the dorsal cochlear nucleus, integrate the two systems of inputs. Parallel fibers in the molecular layer excite fusiform cells through spines on apical dendrites in the molecular layer. Parallel fibers also terminate on spines of dendrites of cartwheel cells, interneurons that bear a strong resemblance to cerebellar Purkinje cells, which in turn inhibit fusiform cells. Cochlear nerve fibers and stellate cells in the ventral cochlear nucleus excite fusiform cells and inhibitory interneurons via synapses on the smooth basal dendrites in the deep layer.

Recent experiments suggest that the circuits of the dorsal cochlear nucleus distinguish between unpredictable and predictable sounds. An animal's own chewing or licking sounds, for example, are predictable and canceled through these circuits. The changes in spectral cues that arise when animals move their heads or ears or shoulders, changing the angle of incidence of sounds to the ears, are unpredictable, especially when an external sound source is moving. Somatosensory and vestibular information about the position of the head and ears, as well as descending information from higher levels of the nervous system about the animal's own movements, pass through the molecular layer to modulate acoustic information that arrives in the deep layer.

The Superior Olivary Complex in Mammals Contains Separate Circuits for Detecting Interaural Time and Intensity Differences

In many vertebrates, including mammals and birds, neurons in the superior olivary complex compare the activity of cells in the bilateral cochlear nuclei to locate sound sources. Separate circuits detect interaural time and intensity differences and project to the inferior colliculi.

The Medial Superior Olive Generates a Map of Interaural Time Differences

Differences in arrival times at the ears are not represented at the cochlea. Instead, they are first represented in the medial superior olive where a map of interaural phase is created by a comparison of the timing of action potentials in the responses to sounds from the two ears. Sounds arrive at the near ear before they arrive at the far ear, with interaural time differences being directly related to the location of sound sources in the horizontal plane (Figure 28–5A).

Cochlear nerve fibers tuned to frequencies below 4 kHz and their bushy cell targets encode sounds by firing in phase with the pressure waves. This property is known as *phase-locking*. Although individual neurons may fail to fire at some cycles, some set of neurons fires with every cycle. In so doing, these neurons carry information about the timing of inputs with every cycle of the sound. Sounds arriving from one side evoke phase-locked firing that is consistently earlier at the near ear than at the far ear, resulting in consistent interaural phase differences (Figure 28–5A).

In 1948, Lloyd Jeffress suggested that an array of detectors of coincident inputs from the two ears, transmitted through *delay lines* comprised of axons with systematically differing lengths, could form a map of interaural time differences and thus a map of the location of sound sources (Figure 28–5B). In such a circuit, conduction delays compensate for the earlier arrival at the near ear. Interaural time delays increase systematically as sounds move from the midline to the side, resulting in coincident firing further toward the edge of the neuronal array.

Such neuronal maps have been found in the barn owl in the homolog of the medial superior olivary nucleus. Mammals and chickens use a variant of this input arrangement. The principal neurons of the medial superior olive form a sheet of one or a few cells' thickness on each side of the midline. Each neuron has two tufts of dendrites, one extending to the lateral face of the sheet, and the other projecting to the medial face of the sheet (Figure 28–5C). The dendrites at the lateral face are contacted by the axons of large spherical bushy cells from the ipsilateral cochlear nucleus, whereas the dendrites at the medial face are contacted by large spherical bushy cells of matching best frequency from the contralateral cochlear nucleus. The axons of bushy cells terminate in the contralateral

A

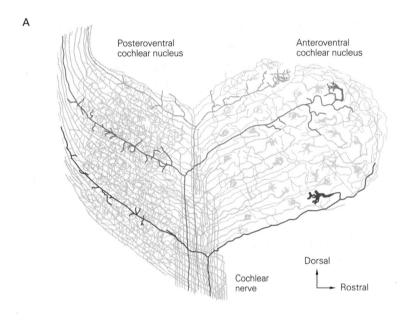

Posteroventral
cochlear nucleus

Anteroventral
cochlear nucleus

Cochlear
nerve

Dorsal

Rostral

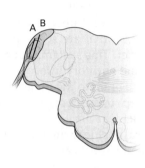

A B

B

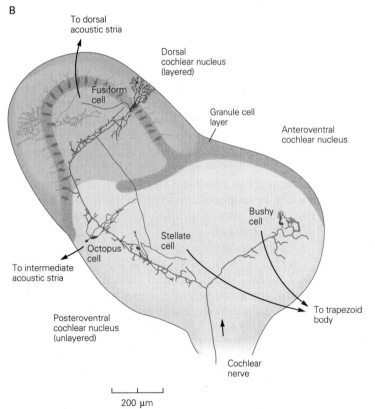

To dorsal
acoustic stria

Dorsal
cochlear nucleus
(layered)

Fusiform
cell

Granule cell
layer

Anteroventral
cochlear nucleus

Bushy
cell

Stellate
cell

Octopus
cell

To intermediate
acoustic stria

To trapezoid
body

Posteroventral
cochlear nucleus
(unlayered)

Cochlear
nerve

200 µm

C

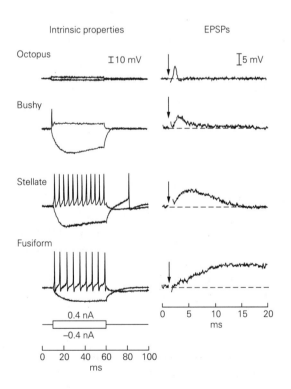

Intrinsic properties

EPSPs

Octopus

Bushy

Stellate

Fusiform

I 10 mV

I 5 mV

0.4 nA

−0.4 nA

0   20   40   60   80 100
ms

0   5   10   15   20
ms

medial superior olive with delay lines just as Jeffress had suggested, but the branches that terminate in the ipsilateral medial superior olive are of equal length (see Figure 28–5C).

The conduction delays are such that each medial superior olive receives coincident excitatory inputs from the two ears only when sounds come from the contralateral half of space. As sound sources move from the midline to the most lateral point on the contralateral side of the head, the earlier arrival of sounds at the contralateral ear needs to be compensated by successively longer delay lines. This results in inputs from the two ears coinciding at successively more posterior and lateral regions of the medial superior olive. Inhibition superimposed on these excitatory inputs plays a significant role in sharpening the map of interaural phase.

In encoding interaural phase, individual neurons in the medial superior olive provide ambiguous information about interaural time differences. Phase ambiguities are resolved when sounds have energy at multiple frequencies, as natural sounds almost always do. The sheet of neurons of the medial superior olive forms a representation of interaural phase along the rostrocaudal and lateromedial dimensions. The array of bushy cell inputs also imposes a tonotopic organization in the dorsoventral dimension. Sounds that contain energy at multiple frequencies evoke maximal coincident firing in a single dorsoventral column of neurons that localizes sound sources unambiguously. The beauty of using interaural phase to encode interaural time disparities is that the brain receives information about interaural time differences not just at the beginning and end of the sound but with every cycle of an ongoing sound.

Principal cells of the medial superior olive also receive sharply timed inhibition driven by sounds from both the ipsilateral and contralateral sides through the lateral and medial nuclei of the trapezoid body, respectively. Remarkably, the inhibition through pathways from both sides precedes the arrival of excitation and sharpens the summation of excitation even though inhibition is mediated through a pathway that has an additional synapse. The great conduction speed through the disynaptic pathway through the medial nucleus of the trapezoid body is made possible by the large axons of globular bushy cells and the large calyceal terminals of Held that activate neurons in the medial nucleus of the trapezoid body with short and consistently timed delays. The pathway that brings ipsilateral inhibition through the lateral nucleus of the trapezoid body is less well understood.

Each medial superior olive thus forms a map of the location of sound sources in the contralateral hemifield. The striking difference between this spatial representation of stimuli and those in other sensory systems is that it is not the result of the spatial arrangement of inputs, like retinotopic or somatosensory maps, but is inferred by the brain from computations made in the afferent pathways.

The Lateral Superior Olive Detects Interaural Intensity Differences

Sounds with wavelengths that are similar to or smaller than the head are deflected by the head, causing the

---

Figure 28–4 (Opposite) Different types of cells in the cochlear nuclei extract distinct types of acoustic information from cochlear nerve fibers.

A. The differing sizes and shapes of terminals along the length of each cochlear nerve fiber in the ventral cochlear nucleus of a newborn dog reflect differences in their postsynaptic targets. The large end bulbs form synapses on bushy cells; smaller boutons contact stellate and octopus cells. The nerve fibers shown here are color-coded as in Figure 28–3: the **yellow** fiber encodes the highest frequencies and the **red** fiber the lowest. (Adapted, with permission, from Cajal 1909.)

B. A layer of mouse granule cells (**light brown**) separates the unlayered ventral cochlear nucleus (**pink**) from the layered dorsal nucleus (**tan and light brown**). In the dorsal cochlear nucleus, the cell bodies of fusiform and granule cells are intermingled in a region between the outermost molecular layer and the deep layer. Cochlear nerve fibers, color-coded for frequency as in Figure 28–3, terminate in both nuclei but with different patterns of convergence on the principal cells. Each bushy,

stellate, and fusiform cell receives input from a few auditory nerve fibers and is sharply tuned, whereas individual octopus cells are contacted by many auditory nerve fibers and are broadly tuned.

C. Differences in the intrinsic electrical properties of the principal cells of mouse cochlear nuclei are reflected in the patterns of voltage change in the cells. When steadily depolarized, stellate and fusiform cells fire repetitive action potentials, whereas repetitive firing in bushy and octopus cells is prevented by low-voltage-activated conductances. The low input resistance of bushy and octopus cells in the depolarized voltage range makes depolarizing voltage changes rapid but also small; the rise and fall of voltage changes in stellate and fusiform cells is slower. Synaptic potentials, too, are different. The brief synaptic potentials in bushy and octopus cells require larger synaptic currents but encode the timing of auditory nerve inputs more faithfully than do the longer-lasting synaptic potentials in stellate or fusiform cells. (Reproduced, with permission, from N. Golding.)

A Phase-locked firing in bushy cells

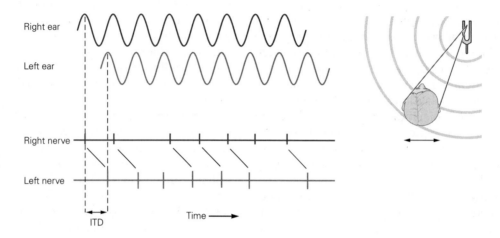

B Mapping of ITD onto array of neuronal
   coincidence neurons

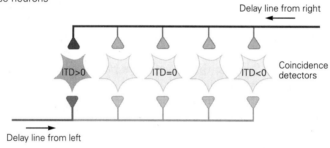

C Bilateral medial superior olivary nuclei

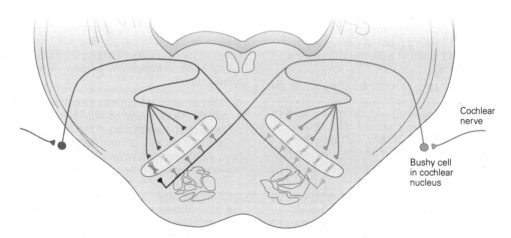

intensity at the near ear to be greater than that at the far ear. In humans, interaural intensities can differ in sounds that have frequencies greater than about 2 kHz. Interaural intensity differences produced by such *head shadowing* are detected by a neuronal circuit that includes the medial nucleus of the trapezoid body and the lateral superior olive.

Although the lateral superior olive does not form a map of the location of sounds in the horizontal plane, it performs the first of several integrative steps that use interaural intensity differences to localize sounds. Neurons in this nucleus balance ipsilateral excitation with contralateral inhibition. Excitation comes from small spherical bushy cells and stellate cells in the ipsilateral ventral cochlear nucleus. Inhibition comes from a disynaptic pathway that includes globular bushy cells in the contralateral ventral cochlear nucleus and principal neurons of the ipsilateral medial nucleus of the trapezoid body (Figure 28–6A). Sounds that arise ipsilaterally generate relatively strong excitation and relatively weak inhibition, whereas those that arise contralaterally generate stronger inhibition than excitation. Neurons in the lateral superior olive are activated more strongly by sounds from the ipsilateral than from the contralateral hemifield. The firing of lateral superior olivary neurons is a function of the location of the sound source and thus carries information about where sounds arise in the horizontal plane (Figure 28–6B).

In order to balance excitation and inhibition stimulated by one sound, the ipsilateral excitation and contralateral inhibition must arrive at neurons in the lateral superior olive at the same time. Thus, excitation that arises monosynaptically from the ipsilateral ventral cochlear nucleus must arrive at the same time as inhibition that arises disynaptically from the contralateral ventral cochlear nucleus. The inhibition comes from the medial nucleus of the trapezoid body whose inputs through large axons of globular bushy cells and large calyces of Held produce synaptic responses

Figure 28–5 (Opposite) Interaural differences in the arrival of a sound help localize sound in the horizontal plane.

A. When a sound such as a pure tone arises from the right, the right ear detects the sound earlier than the left ear. The difference in the time of arrival at the two ears is the interaural time delay (**ITD**). Cochlear nerve fibers and their bushy cell targets fire in phase with pressure changes. Although individual bushy cells may fail to fire at some cycles, a set of cells will encode the timing of a low-frequency sound and its frequency with every cycle. Comparison of the onset of action potentials in the bushy cells at the two sides reveals the ITDs (**slanted black lines**).

B. Interaural time differences can be measured by an array of neurons whose inputs from the two ears are delay lines as proposed by Lloyd Jeffress (1948). Action potentials propagate to reach the nearest terminals before they reach the farthest ones; thus, in the delay line from the right, terminals will generate synaptic potentials sequentially from right to left, and in the delay line from the left, terminals will generate synaptic potentials sequentially from left to right. Suppose that such postsynaptic neurons are coincidence detectors, firing only when they receive excitatory postsynaptic potentials (EPSPs) simultaneously from the right and left. Sounds that arise at the midline reach the right and left ears simultaneously with no interaural time disparity (ITD = 0). The neuron in the middle of the array that receives input from equally long axons from the two sides will thus receive simultaneous EPSPs from the two sides. When sounds come from the right, signals from the right ear arrive at the central nervous system earlier than those from the left ear (ITD >0). Sound from the right generates synchronous EPSPs in the (**yellow**) neuron because the earlier arrival of sound from the right (**red**) is compensated by a longer conduction delay relative to that from the left (**blue**). Likewise, when sound arises from the left, the ITD <0 and conduction delays

from the left (**blue**) compensate for the early arrival at the left. Such a neuronal circuit produces a map of interaural time disparities in the coincidence detectors; as sounds move from the right to left, they activate coincidence detectors sequentially from left to right. Such an arrangement of delay lines has been found in the nucleus laminaris of the barn owl, the homolog of the mammalian medial superior olivary nucleus.

C. Mammals use delay lines only in the nucleus contralateral to a sound source to form a map of interaural time differences. The bitufted neurons of the medial superior olivary nucleus form a sheet that is contacted on its lateral face by bushy cells from the ipsilateral cochlear nucleus and on the medial face by bushy cells from the contralateral cochlear nucleus. (Although it is depicted here schematically in a coronal section of the brain stem, the encoding of interaural disparities is in a sheet of neurons that also has a rostrocaudal dimension.) On the ipsilateral side, the branches of the bushy cell axon are of equal length and thus initiate synaptic currents in their targets in the medial superior olive simultaneously. On the contralateral side, the branches deliver synaptic currents sequentially first to the regions closest to the midline, and then to progressively more lateral regions. Neurons of the medial superior olive detect synchronous excitation from the two ears only when sounds arise from the contralateral half of space. When sounds arise from the right side, their early arrival at the right ear is compensated by progressively longer conduction delays to activate neurons more and more toward the lateral end of the left medial superior olive (the **yellow cell** is activated by a sound from the far right, as in part **B**). When sounds arise from the front and there is no interaural time difference, neurons in the anterior end of the medial superior olive are activated synchronously from both sides. Each medial superior olive forms a map of where sounds arise in the contralateral hemifield. (Adapted, with permission, from Yin 2002.)

Figure 28–6 Interaural differences in the intensity of a sound also help localize sound in the horizontal plane.

A. Principal cells of the lateral superior olivary nucleus (**LSO**) receive excitatory input from the ipsilateral cochlear nucleus (**CN**) and inhibitory input from the contralateral cochlear nucleus. A coronal section through the brain stem of a cat illustrates the anatomical connections. Small spherical bushy cells and stellate cells in the ipsilateral ventral cochlear nucleus provide direct excitation. Globular bushy cells in the contralateral ventral cochlear nucleus project across the midline and excite neurons in the medial nucleus of the trapezoid body (**MNTB**) via large terminals, the calyces of Held. Cells of the medial nucleus of the trapezoid body inhibit neurons in the lateral superior olive as well as in the medial superior olive (**MSO**). For neurons of the lateral superior olive to compare intensities of the same sound, the timing of the ipsilateral excitatory input must be matched with the timing of the contralateral inhibitory input. To this end, globular bushy cells have particularly large axons that terminate in a calyx of Held in the medial nucleus of the trapezoid body where synaptic transmission is strong and thus the synaptic delay is short and invariant in its timing.

B. The firing of neurons in the lateral superior olive reflects a balance of ipsilateral excitation and contralateral inhibition. When sounds arise from the ipsilateral side, excitation is relatively stronger and inhibition is relatively weaker than when sounds arise from the contralateral side. The transition between the dominance of excitation and inhibition reflects the location of the sound source.

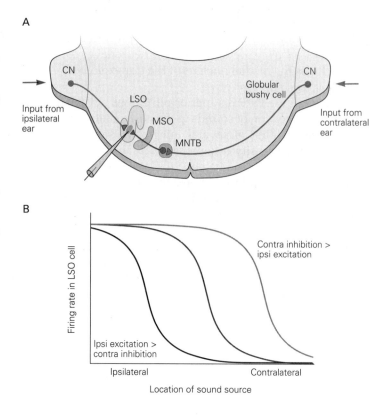

with short and consistently timed delays. The axons of small spherical bushy cells and stellate cells that carry ipsilateral excitation conduct more slowly than those of globular bushy cells.

The terminals of the globular bushy cells, the calyces of Held, engulf the cell bodies of trapezoid-body neurons so dramatically that they caught the attention of early anatomists and modern biophysicists. A single somatic terminal releases neurotransmitter at numerous release sites and generates large synaptic currents. The reliability of pre- and postsynaptic recordings at this synapse makes the site ideal for detailed studies of the mechanisms of synaptic transmission (Chapter 15).

The Superior Olivary Complex Provides Feedback to the Cochlea

Although sensory systems are largely afferent, bringing sensory information to the brain, recent studies have led to an appreciation of the importance of efferent signaling at many levels of the auditory system.

Olivocochlear neurons form a feedback loop from the superior olivary complex to hair cells in the cochlea. Their cell bodies lie around the major dense clusters of cell bodies in the olivary nuclei. In mammals, two groups of olivocochlear neurons have been functionally distinguished. The medial olivocochlear neurons' axons terminate on the outer hair cells bilaterally; the lateral olivocochlear neuron axons terminate ipsilaterally on the afferent fibers associated with inner hair cells.

Most medial olivocochlear neurons, with cell bodies that lie ventral and medial within the olivary complex, send their axons to the contralateral cochlea (Figure 28–7), but many also project to the ipsilateral cochlea. These cholinergic neurons act on hair cells through a special class of nicotinic acetylcholine receptor-channels formed from $\alpha9$ and $\alpha10$ subunits. The influx of Ca^{2+} through these channels leads to the opening of K^+ channels that hyperpolarize outer hair cells. These neurons thus mediate tuned negative feedback and are binaural, being driven predominantly but not exclusively by stellate cells of the contralateral

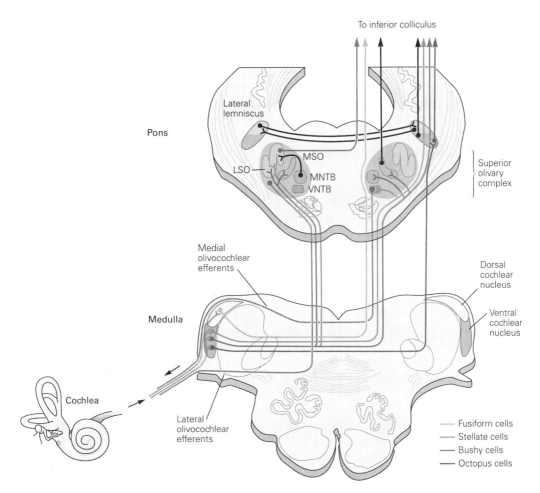

Figure 28–7 Major components of the ascending and descending auditory pathways. The auditory pathway is bilaterally symmetrical; the major connections among the nuclei that form the early auditory pathway are shown. The ascending pathway begins in the cochlea and progresses through several parallel pathways through the cochlear nuclei: the cochlear nuclei, the superior olivary nuclei, and the ventral and dorsal nuclei of the lateral lemniscus. These signals converge in the inferior colliculus, which projects to the medial geniculate body of the thalamus and thence to the cerebral cortex (see Figure 28–2). Some of the connections are through excitatory pathways (**colored lines**) and others through inhibitory pathways (**black lines**). These same nuclei are also interconnected through descending pathways (**blue lines**) and bilaterally through commissural projections. (**LSO**, lateral superior olivary nucleus; **MNTB**, medial nucleus of the trapezoid body; **MSO**, medial superior olive; **VNTB**, ventral nucleus of the trapezoid body).

ventral cochlear nucleus. Activity in these efferent fibers reduces the sensitivity of the cochlea and protects it from damage by loud sounds. Collateral branches of medial olivocochlear neurons terminate on stellate cells in the cochlear nucleus, acting on conventional nicotinic and muscarinic acetylcholine receptors, forming an excitatory feedback loop.

Lateral olivocochlear neurons, with cell bodies that lie in and around the lateral superior olive, send their axons exclusively to the ipsilateral cochlea, where they terminate on the afferent fibers from inner hair cells. Charles Liberman and his colleagues demonstrated that these efferents balance the excitability of cochlear nerve fibers at the two ears.

Ventral and Dorsal Nuclei of the Lateral Lemniscus Shape Responses in the Inferior Colliculus With Inhibition

Fibers from the cochlear and superior olivary nuclei run in a band, or lemniscus, along the lateral edge of the brain as they ascend from the brainstem to the inferior colliculus. Along this band of fibers are groups of neurons that form the dorsal and ventral nuclei of the lateral lemniscus. Neurons in the ventral nuclei of the lateral lemniscus receive input from all major groups of principal cells of the ventral cochlear nuclei and respond predominantly to monaural input, driven by the contralateral ear, while neurons in the dorsal nucleus receive input from the lateral and medial superior olivary nuclei and respond to inputs from both ears. Neurons in both subdivisions are inhibitory and project to the inferior colliculus. Their roles are intriguing but not fully understood.

Since understanding the meaning of sounds is not greatly compromised by the loss of one ear, it would make sense that the largely monaural functions of the ventral nuclei of the lateral lemniscus involve the processing of the meaning of sounds. Furthermore, mammals vary in the information they extract from their acoustic environments, which may account for differences between species in the structure and function of the ventral nuclei of the lateral lemniscus.

A border that is more distinct in some mammalian species than in others separates the ventral and intermediate nuclei and the subdivisions of the ventral nucleus of the lateral lemniscus. Neurons differ in their shapes, biophysical properties, and pattern of convergence of cochlear nuclear inputs. One group of glycinergic neurons is innervated by large calyceal terminals from octopus cells. These could generate inhibitory temporal reference signals in the inferior colliculus. Some broadly tuned neurons fire almost exclusively at the onset of tones with sharply timed action potentials but convey periodicity in complex sounds, raising the question of whether these neurons might have a role in encoding pitch in music and speech. Others respond by firing as long as a tone is present; these neurons track the fluctuations in intensity or the envelopes of sounds, a feature that is useful for understanding the meaning of sounds including speech. Tuning curves of the neurons are variable, with many being broad or W-shaped.

Neurons in the dorsal nucleus are predominantly binaural, receiving input from the ipsilateral medial superior olive and from the lateral superior olive, primarily from the contralateral side. These neurons are GABAergic, targeting the inferior colliculi on both sides and also targeting the contralateral dorsal nucleus of the lateral lemniscus. Excitation in neurons of the dorsal nucleus is amplified by NMDA-type glutamate receptors so that the inhibition they generate in their targets outlasts sound stimuli for tens of milliseconds and thus has been termed persistent inhibition. To localize sounds accurately, animals must ignore the reflections of sounds from surrounding surfaces that arrive after the initial direct wave front. Psychophysical

experiments have shown that mammals suppress all but the first-arriving sound, a phenomenon termed the *precedence effect*. It has been proposed that persistent inhibition in the inferior colliculus from the dorsal nucleus of the lateral lemniscus serves to suppress spurious localization cues such as echoes and thus that it contributes to the precedence effect.

Afferent Auditory Pathways Converge in the Inferior Colliculus

The inferior colliculus occupies a central position in the auditory pathway of all vertebrate animals because all auditory pathways ascending through the brain stem converge there (Figure 28–7). The most important sources of excitation are stellate cells from the contralateral ventral cochlear nucleus, fusiform cells from the contralateral dorsal cochlear nucleus, principal cells of the ipsilateral medial superior olive and of the contralateral lateral superior olive, principal cells of ipsilateral and contralateral dorsal nuclei of the lateral lemniscus, commissural connections from the contralateral inferior colliculus, and pyramidal cells in layer V of the auditory cortex. Important sources of inhibition include the nuclei of the lateral lemniscus, the ipsilateral lateral superior olive, the superior paraolivary nucleus, and the contralateral inferior colliculus.

The inferior colliculus of mammals is subdivided into the central nucleus, dorsal cortex, and external cortex. The central nucleus is tonotopically organized. Low frequencies are represented dorsolaterally and high frequencies ventromedially in laminae that have similar best frequencies. Fine mapping has shown that the tonotopic organization is discontinuous; the separation between best frequencies corresponds to psychophysically measured critical bands of approximately one-third octave. Although the central nucleus is organized tonotopically, the spectral range of inputs to these neurons is broader than at earlier stages in the auditory pathway. Inhibition can be broad and narrows the responses of excitatory neurons. Furthermore, tuning can be modulated by descending inputs from the cortex.

Many neurons in the central nucleus carry information about the location of sound sources. The majority of these cells are sensitive to interaural time and intensity differences, essential cues for localizing sounds in the horizontal plane. Neurons are also sensitive to spectral cues that localize sounds in the vertical plane. Physiological correlates of the precedence effect have been measured in the inferior colliculus, where inhibition suppresses simulated reflections of sounds.

The inferior colliculus is not only a convergence point but also a branch point for ascending or outflow pathways. Neurons of the central nucleus project to the external cortex of the inferior colliculus and also to the thalamus and the nucleus of the brachium of the inferior colliculus, both of which then project to the superior colliculus (or the optic tectum in birds).

Sound Location Information From the Inferior Colliculus Creates a Spatial Map of Sound in the Superior Colliculus

The inferior colliculus is not only a convergence point but also a branch point for ascending or outflow pathways. Central nucleus neurons project to the thalamus and also to the external cortex of the inferior colliculus and the nucleus of the brachium of the inferior colliculus, both of which then project to the superior colliculus (or the optic tectum in birds).

The superior colliculus is critical for reflexive orienting movements of the head and eyes to acoustic and visual cues in space. By the time the binaural sound cues and the monaural spectral cues that underlie mammalian sound localization reach the superior colliculus, they have been merged to create a spatial map of sound in which neurons are unambiguously tuned to specific sound directions. This convergence is critical since binaural differences in level and timing alone cannot unambiguously code for a single position in space. The spectral cues that provide information about vertical location must be taken into account, as different locations in the vertical plane can give rise to identical interaural differences in time or intensity. Such unambiguous spatial mapping occurs both in birds and in some mammals (Figure 28–8). In ferrets and guinea pigs, it occurs in the external cortex and the nucleus of the brachium of the inferior colliculus.

Within the superior colliculus, the auditory map is aligned with maps of visual space and the body surface. Unlike the visual and somatosensory spatial maps, the auditory spatial map does not reflect the peripheral receptor surface; instead, it is computed from a combination of cues that identify the specific position of a sound source in space.

Auditory, visual, and somatosensory neurons in the superior colliculus all converge on output pathways in the same structure that controls orienting movements of the eyes, head, and external ears. The motor circuits of the superior colliculus are mapped with respect to motor targets in space and are aligned with the sensory maps. Such sensory-motor correspondence facilitates the sensory guiding of movements.

The Inferior Colliculus Transmits Auditory Information to the Cerebral Cortex

Auditory information ascends from the inferior colliculus to the medial geniculate body of the thalamus and from there to the auditory cortex. The pathways from the inferior colliculus include a lemniscal or core pathway and extralemniscal or belt pathways. Descending projections from the auditory cortex to the medial geniculate body are prominent both anatomically and functionally.

Stimulus Selectivity Progressively Increases Along the Ascending Pathway

A marked feature of auditory neurons at structures along the ascending pathway is their progressively increased stimulus selectivity. An auditory nerve fiber is primarily selective to one stimulus dimension, the frequency of a pure tone. The stimulus selectivity of neurons in the central auditory system may be multidimensional, such as frequency, spectral bandwidth, sound intensity, modulation frequency, and spatial location. In this multidimensional acoustic space, neurons become more selective at successive auditory areas along the ascending pathway.

Many neurons in the auditory cortex (especially those in upper cortical layers) are highly selective to acoustic stimuli, such that the preferred (nearly optimal) stimulus of a neuron occupies only a small region of its receptive field in the multidimensional acoustic space. The region of the preferred stimulus becomes increasingly smaller at structures along the path to the auditory cortex (Figure 28–9A). Pure tones and broadband noises are two extreme cases of a wide range of acoustic stimuli that could preferentially drive auditory cortex neurons. The majority of neurons in the auditory cortex are preferentially driven by stimuli with greater spectral and temporal complexity than pure tones and broadband noises.

The increased stimulus selectivity is also accompanied by changes in a neuron's firing pattern. When neurons are driven by their preferred stimuli, they respond not only with higher firing rates but also with sustained firing throughout the stimulus duration (Figure 28–9B). The receptive field of a cortical neuron contains a "sustained firing region" (corresponding to preferred stimuli) within a larger "onset firing region" (corresponding to nonpreferred stimuli). This explains why it is common for experimenters to observe onset (phasic) responses in auditory cortex when a continuous sound is played.

A Directional tuning of neurons in the ferret

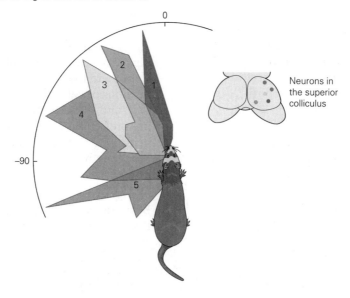

Neurons in
the superior
colliculus

B Directional tuning of a neuron in the barn owl

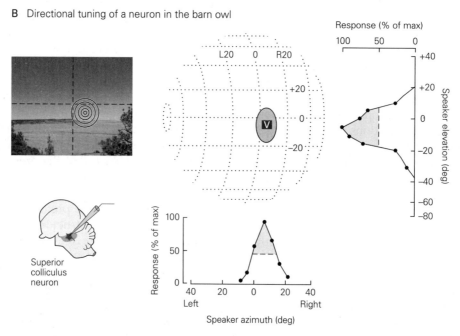

Figure 28–8 A spatial map of sound is formed in the superior colliculus.

A. Neurons in the ferret's superior colliculus are directionally tuned to sound in the horizontal plane. The illustration shows the firing rate profiles of collicular neurons 1 through 5 as a function of where the sounds are located, plotted in polar coordinates centered on the head. The drawing on the right shows the location of the recorded neurons in the colliculus. Note that neuron 1 responds best to sounds in front of the animal, whereas neurons that are located progressively more caudally in the colliculus gradually shift their responses to sounds that originate farther contralaterally. (Adapted, with permission, from King 1999.)

B. The normalized responses of a neuron in the superior colliculus of a barn owl to noise bursts presented at various

locations along the horizon are plotted below (*bottom right*). The **yellow areas** in these tuning curves indicate where responses exceed 50% of the maximum. The sensitivity of the neuron to a particular location along the horizon or a particular elevation (*top right*) creates a discrete best auditory area in space for this neuron (*top middle*), shown as the colored ellipse on a plot of spatial locations with respect to a point straight in front of the owl. The neuron also responds to visual cues from the same area (the box labeled **V**). The photo illustrates the neuron's best area in space with respect to the position of the head (the intersection of the vertical and horizontal dotted lines indicates where the owl's head is pointing). The recording site for this neuron is also shown. (Adapted, with permission, from Cohen and Knudsen 1999. Copyright © 1999 Elsevier Science.)

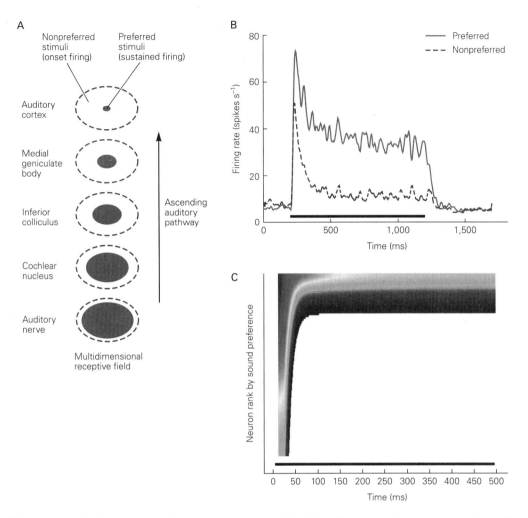

Figure 28–9 Stimulus selectivity increases along the ascending auditory pathway.

A. Stimulus selectivity and the relationship between sustained and onset firings along the ascending auditory pathway. Each open ellipse represents the multidimensional receptive field (RF) of a neuron illustrated on a two-dimensional plane. The filled ellipse represents the "sustained firing region" (corresponding to preferred stimuli) of a neuron's RF. The rest of the area within the RF is the "onset firing region" (corresponding to nonpreferred stimuli). A neuron exhibits sustained or onset firing depending on which region of the RF is stimulated. The neuron does not fire if stimuli fall outside the RF. (Adapted, with permission, from Wang 2018.)

B. Population-averaged firing rate in response to each neuron's preferred and nonpreferred stimuli from primary auditory cortex

(A1). Extracellular recordings were made in awake marmoset monkeys. **Thick bar** = stimulus duration. (Adapted, with permission, from Wang et al. 2005. Copyright © 2005 Springer Nature.)

C. Distribution of activity among A1 neurons in response to a sound burst. On the *y*-axis, all A1 neurons are ranked according to their preference for a particular stimulus. The **blue-to-red color gradient** represents increasing firing rate. The neuron with the highest firing rate is located at the top end of the *y*-axis. **Black bar** = stimulus duration. Most neurons show a brief phasic response to the onset of the sound, but only those particularly tuned to the sound maintain their response until the end of the sound. (Adapted, with permission, from Middlebrooks 2005. Copyright © 2005 Springer Nature.)

The discovery of how sustained firing is evoked in the auditory cortex is important because it provides a direct link between neural firing and the perception of a continuous acoustic event. Such sustained firing by auditory cortex neurons has been observed only in awake animals. In contrast, an auditory nerve fiber typically shows sustains firing in response to a wide range of acoustic signals as long as the spectral energy of the stimulus falls within the neuron's receptive field, under either anesthetized or awake conditions. When David Hubel and his colleagues ventured into the auditory cortex more than half a century ago, they were puzzled by how difficult it was to drive neurons in the auditory cortex of awake cats. Now we know it was because they were probably recording from highly selective neurons and using nonpreferred stimuli. The availability of digital technology since then has made it possible to create and test a large battery of acoustic stimuli in search of the preferred stimulus of a highly selective neuron in auditory cortex. The overall picture elucidated by experimenters is that when a sound is heard, the auditory cortex first responds with transient discharges (encoding the onset of a sound) across a relatively large population of neurons. As the time passes, the activation becomes restricted to a smaller population of neurons that are preferentially driven by the sound (Figure 28–9C), which results in a selective representation of the sound within the neuronal population and over time. Because each neuron has its own preferred stimulus that differs from preferred stimuli of other neurons, neurons in the auditory cortex collectively cover the entire acoustic space with their sustained firing regions. Therefore, any particular sound can evoke sustained firing throughout its duration in a particular population of neurons in the auditory cortex. In other words, the region of auditory cortex activated by acoustic stimulation in whole-brain imaging (eg, functional magnetic resonance imaging [fMRI], positron emission tomography [PET]) comprises neurons that are preferentially driven by the acoustic stimulus.

The Auditory Cortex Maps Numerous Aspects of Sound

The auditory cortex includes multiple distinct functional areas on the dorsal surface of the temporal lobe. The most prominent projection is from the ventral division of the medial geniculate nucleus to the primary auditory cortex (A1, or Brodmann's area 41). As in the subcortical structures, the neurons in this cytoarchitectonically distinct region are arranged tonotopically. In monkeys, neurons tuned to low frequencies are found at the rostral end of A1, while those responsive to high frequencies are in the caudal region (Figure 28–10). Thus, like the visual and somatosensory cortices, A1 contains a map reflecting the sensory periphery.

Because the cochlea encodes discrete frequencies at different points along the basilar membrane, however, a one-dimensional frequency map from the periphery is spread across the two-dimensional surface of the cortex, with a smooth frequency gradient in one direction and isofrequency contours along the other direction. In many species, subregions of the auditory cortex that represent biologically significant frequencies are larger than others because of extensive

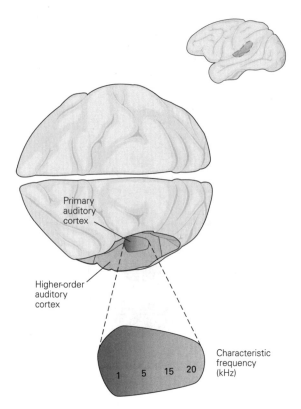

Figure 28–10 The auditory cortex of primates has multiple primary and secondary areas. The expanded figure of the primary auditory cortex shows its tonotopic organization. The primary areas are surrounded by higher-order areas (see Figure 28–11).

inputs, similar to the large area in the primary visual cortex devoted to inputs from the fovea.

In addition to frequency, other features of auditory stimuli are mapped in the primary auditory cortex, although the overall organization is less clear and precise than for vision. Auditory neurons in A1 are excited either by input from both ears (EE neurons), with the contralateral input usually stronger than the ipsilateral contribution, or by a unilateral input (EI). The EI neurons are inhibited by stimulation of the opposite ear.

Certain neurons in A1 also seem to be organized according to bandwidth, that is, according to their responsiveness to a narrow or broad range of frequencies. Neurons near the center of the isofrequency contours are tuned more narrowly to bandwidth or frequency than those located away from the center. Distinct subregions of A1 form clusters of cells with narrow or broadband tuning within individual isofrequency contours. Within intracortical circuits, neurons receive input primarily from neurons with similar bandwidths and characteristic frequencies.

This modular organization of bandwidth selectivity may allow redundant processing of incoming signals through neuronal filters of varying bandwidths as well as center frequencies, which could be useful for the analysis of spectrally complex sounds such as species-specific vocalizations, including speech.

Several other parameters are represented in A1. These include neuronal response latency, loudness, modulation of loudness, and the rate and direction of frequency modulation. Although it remains to be seen how these various maps intersect, this array of parameters clearly endows each neuron and each location in A1 with the ability to represent many independent variables of sound and thus allows for a great diversity of neuronal selectivity.

As is true for visual and somatosensory areas of the cortex, sensory representation in A1 can change in response to alterations in input pathways. After peripheral hearing loss, tonotopic mapping in A1 can be altered so that neurons that were previously responsive to sounds within the lost range of hearing will begin to respond to adjacent frequencies. The work of Michael Merzenich and others has shown that behavioral training of adult animals can also result in large-scale reorganization of the auditory cortex, so that the most behaviorally relevant frequencies—those specifically associated with attention or reinforcement—come to be overrepresented.

The auditory areas of young animals are particularly plastic. In rodents, the frequency organization of A1 emerges gradually during development from an early, crude frequency map. Raising animals in acoustic environments in which they are exposed to repeated tone pulses of a particular frequency results in a persistent expansion of cortical areas devoted to that frequency, accompanied by a general deterioration and broadening of the tonotopic map. This result not only suggests that the development of A1 is experience-dependent but also raises the possibility that early exposure to abnormal sound environments can create long-term disruptions of high-level sensory processing. A greater understanding of how this happens and whether it is also true for human fetuses and infants may provide insights into the origin and remediation of disorders in which central auditory processing is impaired, such as many forms of dyslexia. Moreover, the ability to induce plasticity in the auditory cortex of adults by engaging attention or reward raises new hopes for brain repair even in adulthood.

The primary auditory area of mammals is surrounded by multiple distinct regions, some of which are tonotopic. Adjacent tonotopic fields have mirror-image tonotopy: The direction of tonotopy reverses at the boundary between fields. In monkeys, as many as 7 to 10 secondary (belt) areas surround the three or four primary or primary-like (core) areas (see Figure 28–11). The secondary areas receive input from the core areas of the auditory cortex and, in some cases, from thalamic nuclei. Electrophysiological and imaging studies have confirmed that A1 in humans lies on Heschl's gyrus, in the temporal lobe, medial to the Sylvian fissure. In addition, recent fMRI studies have revealed that in humans, just as in monkeys, pure tones activate primarily core areas, whereas the neurons of belt areas prefer complex sounds such as narrowband noise bursts.

A Second Sound-Localization Pathway From the Inferior Colliculus Involves the Cerebral Cortex in Gaze Control

Many neurons in the auditory cortex have broad spatial tuning, but neurons with narrow spatial tuning are also found when studied in awake animals. In monkeys, auditory cortex neurons are tuned to both frontal space and rear space (outside the coverage of vision), as well as the space above and below the horizontal plane. In contrast to the auditory midbrain, however, there is yet no evidence for a spatially organized map of sound in any of the cortical areas sensitive to sound location.

The sound-localization pathways in the cortex originate in the central nucleus of the inferior colliculus and ascend through the auditory thalamus and the primary and secondary cortical areas, eventually reaching the frontal eye fields involved in gaze control. Eye or head movements can be elicited by stimulating the frontal eye fields, which connect directly to brain stem tegmentum premotor nuclei that mediate gaze changes as well as to the superior colliculus. But why should there be this second sound-localization pathway connected to gaze control circuitry when the midbrain pathway from location-sensitive neurons in the inferior colliculus to the superior colliculus to gaze control circuitry directly controls orientation movements of the head, eyes, and ears?

Behavioral experiments shed light on this question. Although lesions of A1 can result in profound sound-localization deficits, no deficiency is seen when the task is simply to indicate the side of the sound source by pushing a lever. The deficit becomes apparent only when the animal must approach the location of a brief sound source; that is, when the task is the more complex one of forming an image of the source, remembering it, and moving toward it.

Experiments in barn owls have produced particularly compelling evidence. The ability of owls to orient to sounds in space is unaffected by inactivation of the avian equivalent of the frontal eye fields. Similarly, when the midbrain sound localization pathway is disrupted by pharmacological inactivation of the superior colliculus, the probability of an accurate head turn is decreased, but animals still respond correctly more than half of the time. In contrast, when both structures are inactivated, animals are completely unable to orient accurately to acoustic stimuli on the contralateral side. Thus, cortical and subcortical sound-localization pathways have parallel access to gaze control centers, perhaps providing some redundancy. Moreover, when only the frontal eye fields are inactivated, birds lose their ability to orient their gaze toward a target that has been extinguished and must be remembered, just as is seen with mammalian A1 lesions. Thus, in both mammals and birds, cortical pathways are required for more complex sound-localization tasks.

This appears to be a general difference between cortical and subcortical pathways. Subcortical circuits are important for rapid and reliable performance of behaviors that are critical to survival. Cortical circuitry allows for working memory, complex recognition tasks, and selection of stimuli and evaluation of their significance, resulting in slower but more differentiated performance. Examples of this also exist in auditory pathways not involved in localization. Conditioned fear responses to simple auditory stimuli are mediated by direct rapid pathways from the auditory thalamus to the amygdala; they can still be elicited after cortical inactivation. However, fear responses that require more complex discrimination of auditory stimuli require pathways through the cortex and are accordingly slower but more specific.

Auditory Circuits in the Cerebral Cortex Are Segregated Into Separate Processing Streams

In the visual system, the output from the primary visual cortex is segregated into separate dorsal and ventral streams concerned respectively with object location in space and object identification. A similar division of labor is thought to exist in the somatosensory cortex, and recent evidence suggests that the auditory cortex also follows this plan.

Anatomical tracing studies of the three most accessible belt areas in monkeys show that the more rostral and ventral areas connect primarily to the more rostral and ventral areas of the temporal lobe, whereas the more caudal area projects to the dorsal and caudal temporal lobe. In addition, these belt areas and their

temporal lobe targets both project to largely different areas of the frontal lobes (Figure 28–11).

The frontal areas receiving anterior auditory projections are generally implicated in nonspatial functions, whereas those that are targets of posterior auditory areas are implicated in spatial processing. Electrophysiological and imaging studies provide support for this. Caudal and parietal areas are more active when a stimulus must be localized or moves, and ventral areas are more active during identification of the same stimulus or analysis of its pitch. Thus anterior-ventral pathways may identify auditory objects by analyzing spectral and temporal characteristics of sounds, whereas the more dorsal-posterior pathways may specialize in sound-source location, detection of sound-source motion, and spatial segregation of sources.

Although the idea that all sensory areas of the cerebral cortex initially segregate object identification and location is attractive, it is likely an oversimplification. It is clear that the medial-belt areas of the auditory cortex project to both dorsal and ventral frontal cortices, and neurons with broad spatial responsiveness are distributed throughout caudal and anterior areas. Nonetheless, although the details may differ between systems, the basic concept holds that sensory systems deconstruct stimuli into features and analyze each type in discrete pathways.

The Cerebral Cortex Modulates Sensory Processing in Subcortical Auditory Areas

An intriguing feature of all mammalian cortical areas, and one shared by the auditory system, is the massive projection from the cortex back to lower areas. There are almost 10 times as many corticofugal fibers entering the sensory thalamus as there are axons projecting from the thalamus to the cortex. Projections from the auditory cortex also innervate the inferior colliculus, olivocochlear neurons, some basal ganglionic structures, and even the dorsal cochlear nucleus.

Insights into possible functions of this feedback have come from the bat's auditory system. Silencing of frequency-specific cortical areas leads to decreased responses in thalamus and inferior colliculus in the corresponding frequency-specific areas, whereas activation of cortical projections increases and sharpens the responses of some neurons. The auditory cortex can therefore actively adjust and improve auditory signal processing in subcortical structures. A variety of evidence suggests that cortical feedback also occurs in other mammals. This challenges the view of ascending

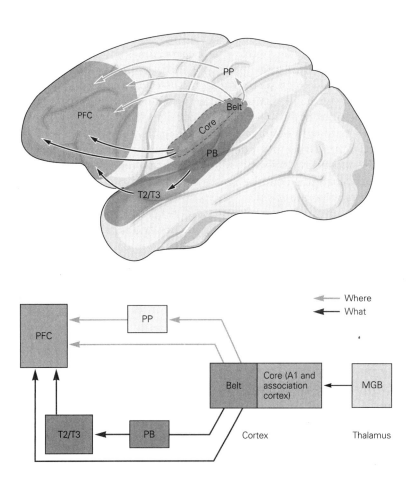

Figure 28–11 The "what" and "where" streams in the auditory cortical system of primates. The ventral "what" stream and dorsal "where" stream originate in different parts of primary and belt cortex and ultimately project to distinct regions of prefrontal cortex through independent paths. (**MGB**, medial geniculate body of the thalamus; **PB**, parabelt cortex; **PFC**, prefrontal cortex; **PP**, posterior parietal cortex; **T2/T3**, areas of temporal cortex.) (Adapted, with permission, from Rauschecker and Tian 2000. Copyright 2000 National Academy of Sciences; adapted from Romanski and Averbeck 2009.)

sensory pathways as purely feedforward circuits and suggests that we should regard the thalamus and cortex as reciprocally and highly interconnected circuits in which the cortex exercises some top-down control of perception.

The Cerebral Cortex Forms Complex Sound Representations

The Auditory Cortex Uses Temporal and Rate Codes to Represent Time-Varying Sounds

An important function of the auditory system is to represent time-varying sounds across multiple time scales, from a few milliseconds to tens and hundreds of milliseconds or even longer. In the auditory nerve, firing patterns largely mirror the temporal structure of sounds, firing in phase with sounds to the limit of the phase-locking. The precision of this temporally based neural representation gradually decreases as information ascends toward the auditory cortex due to synaptic integration at the soma and dendrites.

The upper limit of the phase-locking to periodic sounds progressively decreases along the ascending auditory pathway from approximately 3,000 Hz in the auditory nerve to less than approximately 300 Hz in the medial geniculate body in the thalamus and less than 100 Hz in A1. The upper limit of the phase-locking in A1 is similar to that found in the primary visual and somatosensory areas of cortex. In the auditory cortex, the temporal firing pattern alone is inadequate to represent the entire range of time-varying sounds that are perceived by humans and animals.

Cortical neurons use an alternative method to represent time-varying sounds that change more rapidly than the upper limit of the phase-locking in A1. When an animal listens to a sequence of periodic clicks, two types of neural responses are observed in A1. One population of neurons displays phase-locked periodic firing in response to click trains with long intervals between clicks or slowly varying sounds, but not to click trains with short intervals between clicks or rapidly varying sounds (Figure 28–12A). The second population of neurons does not respond to click trains at long interclick intervals, but instead fires increasingly rapidly as the interclick interval becomes shorter (Figure 28–12B). These two populations of A1 neurons, referred to as *synchronized* and *nonsynchronized*, respectively, have complementary response properties. Neurons of the synchronized population *explicitly* represent slowly occurring sound events by synchronized neural firing (a temporal code), whereas neurons of the nonsynchronized population *implicitly* represent rapidly changing sound events by changes in average firing rates (a rate code).

The nonsynchronized neurons have been observed in the auditory cortex of awake primates and rodents. In A1, neural representation changes from a temporal code to a rate code at the interclick interval of about 25 ms, corresponding to a repetition rate of approximately 40 Hz (Figure 28–12A,B). This is near the boundary where our perception of a periodic click train changes from being "discrete" to "continuous."

The combination of temporal and rate codes to represent the whole range of time-varying sounds is the consequence of a progressive transformation beginning in the auditory nerve, where only a temporal code (phase-locking) is available. The progressive reduction in the upper limit of the phase-locking along the ascending auditory pathway is accompanied by the emergence of firing-rate-based representations. In the medial geniculate body of the thalamus, the intersection between temporal and rate codes is at a shorter interclick interval than in A1 (Figure 28–12C). This indicates that neurons in the medial geniculate body can phase-lock to more rapidly time-varying sounds than A1 neurons, but still utilize a rate code to represent rapidly changing sounds beyond their phase-locking limit.

The prevalence of rate-coding neurons in A1 has important functional implications. It shows that a considerable transition from temporal to rate coding has taken place by the time auditory signals reach the auditory cortex. The importance of the nonsynchronized neural responses is that they represent transformed instead of preserved temporal information. It suggests that cortical processing of sound streams operates on a segment-by-segment basis rather than on a moment-by-moment basis, as found in the auditory nerve. This is necessary for complex integration because higher-level processing tasks require temporal integration over a time window. The reduction in A1 of the upper limit of phase-locking is a prerequisite for multisensory integration in the cerebral cortex. Auditory information is encoded at the periphery at a much faster temporal modulation rate than visual or tactile information, but phase-locking is similar across

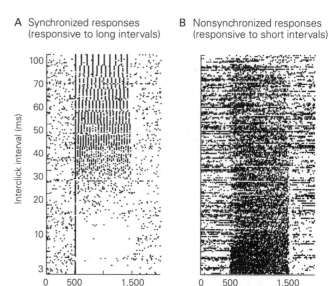

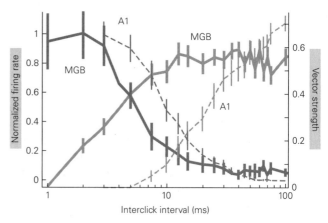

Figure 28–12 Temporal and rate coding of time-varying sounds.

A. Stimulus-synchronized responses of a neuron to periodic click trains recorded from A1 of an awake marmoset. The horizontal bar below the *x*-axis indicates the duration of the stimulus. (Adapted, with permission, from Lu, Liang, and Wang 2001. Copyright © 2001 Springer Nature.)

B. Nonsynchronized responses of a neuron to periodic click trains recorded from A1 in an awake marmoset. (Adapted, with permission, from Lu, Liang, and Wang 2001. Copyright © 2001 Springer Nature.)

C. Comparison of temporal response properties between primary auditory cortex (**A1**) and medial geniculate body of the thalamus (**MGB**). Stimulus-synchronized responses are quantified by vector strength, a measure of the strength of phase-locking. Nonsynchronized responses are quantified by the normalized firing rate (data curves identified as A1 rate and MGB rate). Error bars represent standard error of the mean. (Adapted, with permission, from Bartlett and Wang 2007.)

primary sensory areas of the cortex. The slowing of the phase-locking limit along the ascending auditory pathway and accompanying transition from a temporal code to a rate code are necessary for auditory information to be integrated in the cerebral cortex with information from other sensory modalities that are intrinsically slower.

Primates Have Specialized Cortical Neurons That Encode Pitch and Harmonics

Pitch perception is crucial for perceiving speech and music and for recognizing auditory objects in a complex acoustic environment. Pitch is the percept that allows harmonically structured periodic sounds to be perceived and ordered on a musical scale. Pitch carries crucial linguistic information in tonal languages such as Chinese and prosodic information in European languages. We use pitch to identify a particular voice from a noisy background in a cocktail party. When listening to an orchestra, we hear the melody of the soloist over the background of accompanying instruments.

An important phenomenon for understanding pitch is the perception of "missing fundamental," also referred to as the residue pitch. When the harmonics of a fundamental frequency are played together, the pitch is perceived as the fundamental frequency even if the fundamental frequency is missing. For example, the harmonics of the fundamental frequency of 200 Hz are at 400, 600, 800 Hz, and so on. Playing the frequencies 400, 600, and 800 Hz together will generate a pitch perception of 200 Hz, even though a distinct frequency component of 200 Hz is not physically present in the sound. We encounter this phenomenon routinely when we listen to music over speakers that are too small to generate sounds at low frequencies.

Many combinations of frequencies can give rise to a common fundamental frequency or pitch, making it a particularly valuable auditory cue. This is especially useful when pitch conveys behaviorally important information, as in the case of human speech or animal vocalizations. Sounds propagated through the environment can become spectrally degraded, losing high or low frequencies. While such spectral filtering distorts spectral information, the perception of the missing fundamental is robust despite the loss of some harmonic components.

The ability to perceive pitch is not unique to humans; birds, cats, and monkeys can also pick out pitch. Monkeys are capable of spectral pitch discrimination, melody recognition, and octave generalization, each of which requires the perception of pitch. Marmoset monkeys (*Callithrix jacchus*), a highly vocal New

World primate species whose hearing range is similar to that of humans, exhibit human-like pitch perception. Marmosets are able to discriminate a missing fundamental in harmonic sounds with a precision as small as one semitone for the periodicity above 440 Hz.

Given that both humans and some animals experience a pitch that generalizes across a variety of sounds with the same periodicity (including harmonic sounds with a missing fundamental), it is reasonable to expect that some neurons extract pitch from complex sounds. Xiaoqin Wang and his colleagues discovered a decade ago that a small region in the auditory cortex of marmoset monkeys contains "pitch-selective neurons." These neurons are tuned to pure tones with a best frequency and respond to harmonic complexes with a fundamental frequency near its best frequency even when the harmonics lay outside the neuron's excitatory-frequency response area (Figure 28–13A).

A pitch-selective neuron responds to pitch-evoking sounds (eg, harmonic sounds, click trains) when the pitch is near the neuron's preferred best frequency. Pitch-selective neurons increase their firing rates as the behavioral salience of pitch increases and prefer sounds with periodicity over aperiodic sounds. It is important to note that the pitch-selective neurons in marmoset monkeys, which extract and code for pitch embedded in harmonic sounds (a highly nonlinear computation), are distinctly different from neurons in subcortical areas or A1 that merely "reflect" information on pitch in their firing patterns.

The region containing the pitch-selective neurons in marmoset monkeys is confined to the low-frequency border of A1, the rostral auditory cortex (area R), and lateral belt areas (Figure 28–13B). Human imaging studies have identified a restricted region at the lateral end of Heschl's gyrus anterolateral to A1 that extracts pitch of harmonic complex sounds and is sensitive to changes in pitch salience. The location of this region mirrors the location of the pitch center in marmoset monkeys (Figure 28–13B).

The core regions of auditory cortex in marmosets also contain a class of harmonic template neurons that respond weakly or not at all to pure tones or two-tone combinations but respond strongly to particular combinations of multiple harmonics. The harmonic template neurons show stronger responses to harmonic sounds than inharmonic sounds and selectivity for particular harmonic structures. In contrast to the pitch-selective neurons that are localized within a small cortical region lateral to the low-frequency border between A1 and R and have best frequencies

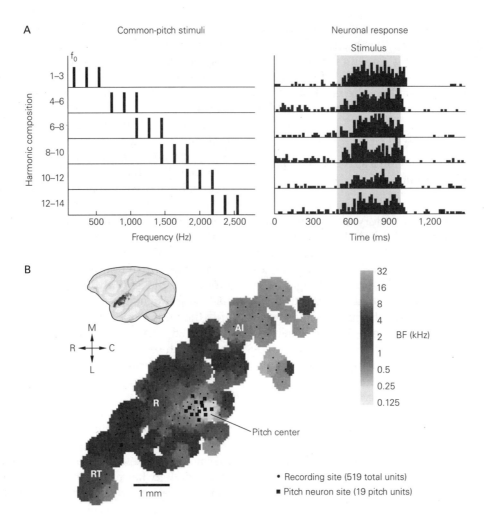

Figure 28–13 Pitch is encoded by specialized neurons in primate auditory cortex.

A. An example of a pitch-selective neuron recorded from marmoset auditory cortex. *Left*: Frequency spectra of a series of harmonic stimuli that share the same fundamental frequency (f_0). *Right*: Peristimulus time histogram of the neuron's response to the stimuli (stimulus duration indicated by the shaded region). (Adapted, with permission, from Bendor and Wang 2005. Copyright © 2005 Springer Nature.)

B. Anatomical organization of the marmoset auditory cortex and the location of a pitch center. **Top:** Side view of the marmoset brain. **Bottom:** Tonotopic map of the left auditory cortex characterized in one marmoset. Pitch-selective neurons (**black squares**) are clustered near the low-frequency border between A1 and area R (rostral auditory cortex). Frequency reversals indicate the borders between A1/R and R/RT (rostrotemporal auditory cortex). (BF: best frequency.) (Adapted from Bendor and Wang 2005. Copyright © 2005 Springer Nature.)

less than 1,000 Hz, the harmonic template neurons are distributed across A1 and R and have best frequencies ranging from approximately 1 kHz to approximately 32 kHz, a range that covers the entire hearing range of marmosets.

Whereas in the periphery single auditory nerve fibers encode individual components of harmonic sounds, the properties of the harmonic template neurons reveal harmonically structured receptive fields for extracting harmonic patterns. The change in neural representation of harmonic sounds from auditory nerve fibers to the auditory cortex reflects a principle of neural coding in sensory systems. Neurons in sensory pathways transform the representation of physical features, such as the frequency of sounds in hearing or luminance of images in vision, into a representation of perceptual features, such as pitch in hearing or curvature in vision. Such features lead to the formation of auditory or visual percepts. The harmonic template neurons in the auditory cortex are key to processing sounds with harmonic structures such as animal vocalizations, human speech, and music.

Insectivorous Bats Have Cortical Areas Specialized for Behaviorally Relevant Features of Sound

Although it is generally assumed that upstream auditory areas perform increasingly specialized functions related to hearing, much less is known about the functions of serial relays in the auditory system compared to the visual system. In humans, one of the most important aspects of audition is its role in processing language, but we know relatively little about how speech sounds are analyzed by neural circuits. New techniques for imaging the human brain are gradually providing insights into the functional specialization of cortical areas associated with language (Chapter 55).

Evidence for specialized analysis of complex auditory signals in the cerebral cortex comes from studies of insectivorous bats. These animals find their prey almost entirely through *echolocation*, emitting ultrasonic pulses of sound that are reflected by flying insects. Bats analyze the timing and structure of the echoes to help locate and identify the targets, and discrete auditory areas are devoted to processing different aspects of the echoes.

Many bats, such as the mustached bat studied by Nobuo Suga and his collaborators, emit echolocating pulses with two components. An initial *constant-frequency* (CF) component consists of several harmonically related sounds. These harmonics are emitted stably for tens to hundreds of milliseconds, akin to human vowel sounds. The constant-frequency component is followed by a sound that decreases steeply in frequency, the *frequency-modulated* (FM) component, which resembles the rapidly changing frequency of human consonants (Figure 28–14A).

The FM sounds are used to determine the distance to the target. The bat measures the interval between the emitted sound and the returning echo, which corresponds to a particular distance, based on the relatively constant speed of sound. Neurons in the FM-FM area of auditory cortex (Figure 28–14B) respond preferentially to pulse-echo pairs separated by a specific delay. Moreover, these neurons respond better to particular combinations of sounds than to the individual sounds in isolation; such neurons are called *feature detectors* (Figure 28–14C). The FM-FM area contains an array of such detectors, with preferred delays systematically ranging from 0.4 to 18 ms, corresponding to target ranges of 7 to 280 cm (Figure 28–14B). These neurons are organized in columns, each of which is responsive to a particular combination of stimulus frequency and delay. In this way, the bat, like the barn owl in its inferior colliculus, is able to represent an acoustic feature that is not directly represented by sensory receptors.

The CF components of bat calls are used to determine both the speed of the target relative to the bat and the acoustic image of the target. When an echolocating bat is flying toward an insect, the sounds reflected from the insect are Doppler-shifted to a higher frequency at the bat's ear, for the bat is moving toward the returning sound waves from the target, causing a relative speeding up of these waves at its ear. Similarly, a receding insect yields reflections of lowered frequency at the bat's ear. Neurons in the CF-CF area (Figure 28–14B) are sharply tuned to a combination of frequencies close to the emitted frequency or its harmonics. Each neuron responds best to a combination of a pulse of a particular fundamental frequency with an echo corresponding to the first or second harmonic of the pulse, Doppler-shifted to a specific extent. As in the FM-FM area, neurons do not respond to the pulse or echo alone, but rather to the combination of the two CF signals.

CF-CF neurons are arranged in columns, each encoding a particular combination of frequencies. These columns are arranged regularly along the cortical surface, with the fundamental frequency along one axis and the echo harmonics along a perpendicular axis. This dual-frequency coordinate system creates a map wherein a specific location corresponds to a particular Doppler shift and thus a particular target velocity, ranging systematically from –2 m/s to 9 m/s.

The CF components of returning echoes are also used for detailed frequency analysis of the acoustic image, presumably important in its identification. The Doppler-shifted constant-frequency area (DSCF) of the mustached bat is a dramatic expansion of the primary auditory cortex's representation of frequencies between 60 kHz and 62 kHz, corresponding well to the set of returning echoes from the major CF component of the bat's call (Figure 28–14B). Within the DSCF area, individual neurons are extremely sharply tuned to frequency, so that the tiny changes in frequency created by fluttering moth wings are easily detected.

Transient inactivation of some of these specialized cortical areas while the bat performs a discrimination task strikingly supports the importance of their functional specialization in behavior. Silencing of the DSCF selectively impairs fine frequency discrimination while leaving time perception intact. Conversely, inactivation of the FM-FM area impairs the bat's ability to detect small differences in the time of arrival of two echoes, while leaving frequency perception unchanged.

Investigation of this auditory system was greatly facilitated by knowledge of the stimuli relevant to bats. It remains to be seen whether these cortical areas are functionally or anatomically analogous to particular

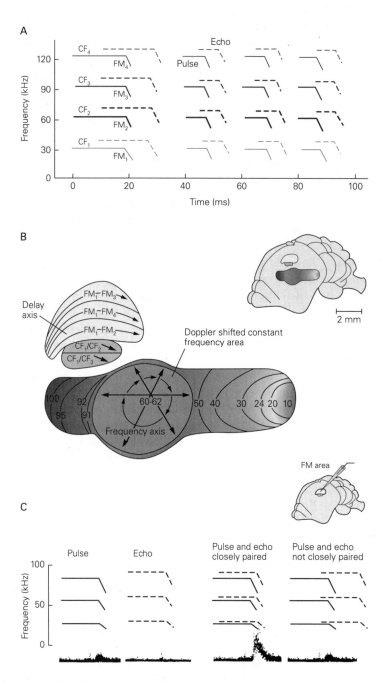

Figure 28–14 The auditory system of the bat has specialized areas for locating sounds.

A. A sonogram of an animal's calls (**solid lines**) and the resultant echoes (**dashed lines**) illustrates the two components of the call: the protracted, harmonically related constant-frequency (**CF**) signal and the briefer frequency-modulated (**FM**) signal. The duration of the calls declines as the animal approaches its target. (Adapted, with permission, from Suga 1984.)

B. A view of the cerebral hemisphere of the mustached bat shows three of the functional areas within the auditory cortex. The FM area is where the distance from the target is computed; the CF area is where the velocity of the target is computed; and the Doppler-shifted CF area is specialized for the identification of small fluttering objects. The expanded cortical representation of Doppler-shifted CF signals near the second harmonic of the call frequency (60–62 kHz) forms the acoustic "fovea." (Adapted, with permission, from Suga 1984.)

C. The FM-FM combination-sensitive neuron shown does not respond significantly to either pulses or echoes alone, but responds very strongly to a closely paired pulse-echo. However, the neuron is also sensitive to the time difference between the pulse and echo, as seen in the record on the right, where the neuron fails to respond to a pulse-echo combination that is not closely paired. (Adapted, with permission, from Suga et al. 1983.)

fields in cats, monkeys, and humans. Regardless, the choice of appropriate stimuli is likely to be as important in studying these other species as it has been in studies of bats.

The Auditory Cortex Is Involved in Processing Vocal Feedback During Speaking

Vocal communication involves both speaking and hearing, often taking place concurrently. When we speak, the sound of our voice is delivered not only to the intended listener but also back to our own ears. Such feedback to our auditory system during vocal production is conducted not only through the air but also through bone and can be loud as a result of the proximity of the mouth and the ears.

The auditory system must distinguish an auditory percept as being self-generated or externally generated. To monitor external sounds from the acoustic environment during speaking, self-generated sounds have to be masked. At the same time, the auditory system must also monitor our own voice in order to detect errors in vocal production. An accurate representation of one's own voice through vocal feedback is crucial to maintaining desired vocal production and to the learning of a new language. In humans and animals, perturbations of the vocal feedback can lead to alterations in vocal production, and interruptions or blockages of the vocal feedback can result in degradation in vocal learning.

The evidence for the involvement of the auditory cortex in processing vocal feedback comes from both human and animal studies. Responses in the auditory cortex of human subjects to their own voice while speaking are smaller than the responses to the playback of the same sounds. This reduction can be observed in electrocorticographical (ECoG) recordings (Figure 28–15A) or with a variety of imaging methods (eg, fMRI, PET, magnetoencephalography [MEG]).

Single-neuron recordings from the auditory cortex of vocalizing monkeys have shown that self-initiated vocalizations result in suppression of cortical responses to monkeys' own vocalizations, of external sounds heard during vocalization, and also spontaneous activity (Figure 28–15B). Because in many instances firing rates are suppressed to below spontaneous activity, the suppression is likely caused by inhibition. Neurons suppressed by self-initiated vocalizations show frequency and intensity tuning, as is typical of auditory cortical neurons, and respond to the playback of vocalizations.

The vocalization-induced suppression begins several hundred milliseconds prior to the onset of vocalization (Figure 28–15B), suggesting that these neurons receive modulatory signals originating in vocal production circuits. In humans, vocal production is carried out by cortical areas in the frontal lobe, from Broca's area to premotor and motor cortex. In humans and monkeys, axons from the premotor cortex to auditory regions of the superior temporal gyrus have been described, and presumably, they mediate the vocalization-induced suppression. This modulatory connection is not active when humans or monkeys simply listen to vocal sounds played to them.

Why do we suppress our auditory cortex when we speak? A simple answer is that this suppression helps reduce the masking effect of our own voice, which can be very loud. A more interesting answer is that this suppression results from a vocal feedback-monitoring network in auditory cortex. In humans, there is less or no suppression of auditory cortex if vocal feedback is experimentally altered through earphones, for example, when the pitch of the voice is shifted (Figure 28–15A). In marmoset monkeys, neurons suppressed by self-initiated vocalizations may become less suppressed or even excited when the animal hears its own frequency-shifted vocalizations (Figure 28–15C). This sensitivity to feedback perturbations suggests that neurons exhibiting vocalization-induced suppression are part of a network responsible for monitoring vocal feedback signals. The presence of vocal feedback-related neural activity in the auditory cortex of both humans and monkeys suggests that the auditory cortex combines both internal modulation and vocal feedback responses, rather than merely responding to sensory signals coming through the ears.

Not all neurons in the auditory cortex are suppressed by speaking or vocalizing. A smaller proportion (~30%) of neurons in marmoset A1 increase their responses during self-initiated vocalizations, consistent with their auditory response characteristics. In contrast to vocalization-induced suppression, vocalization-related excitation begins after the onset of vocalization and is likely the result of feedback through the ascending auditory pathway. The vocalization-related excitation may help maintain the sensitivity of the auditory cortex to the external acoustic environment during speaking or vocalizing.

Vocalization-induced suppression of auditory responses has been observed in several mammalian subcortical structures, including the brain stem and inferior colliculus. Such suppression begins a few milliseconds before or is synchronized with vocal production. In contrast, cortical suppression begins several hundred milliseconds before the vocal onset. It is possible that subcortical suppression of auditory responses during speaking or vocalizing is initiated by cortical commands.

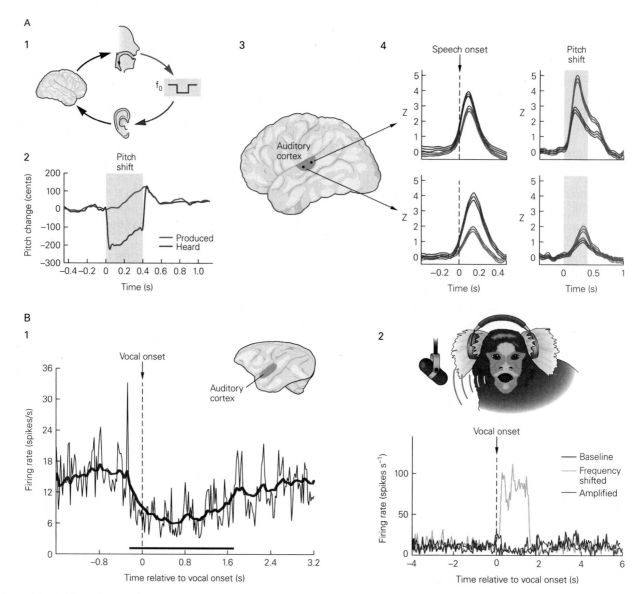

Figure 28–15 Vocal feedback processing in auditory cortex.

A. Examples of vocalization-induced suppression and sensitivity to pitch perturbation in human cerebral cortex. **1.** A subject's vocalizations (**red arrow**) went through a digital signal processor that shifted pitch and delivered the distorted auditory feedback (**blue arrow**) to the subject's earphones. **2.** Pitch track of an example trial shows the pitch recorded by the microphone (produced) and the pitch delivered to the earphones (heard). Shaded region indicates the time interval when the signal processor shifted the pitch by −200 cents (1 cent = 1/1200 octave). **3.** The locations of electrodes that recorded from two sites in the auditory cortex on the surface of the superior temporal gyrus. **4.** The Z variable represents the power in the 50 to 150 Hz (high-γ) range of cortical activity, which has been shown to correlate well with neuronal spiking activity. It was extracted from the signals recorded at each electrode in the speaking (**red**) and listening (**blue**) conditions. Vertical lines in the left column of plots indicate vocalization onset, and shaded regions in the right column of plots indicate the onset and offset of perturbation. The response of a subject's auditory cortex to his or her own self-produced vocalization is generally smaller than the response seen when the subject passively listens to playback of the same vocalization (*left column*). The response of auditory

cortex to the perturbation during active phonation (speaking) is enhanced (*right column*). (Adapted, with permission, from Houde and Chang, 2015.)

B. 1. Vocalization-induced suppression of neural activity in marmoset monkey auditory cortex. Population-averaged firing rate of all vocalization-suppressed responses are aligned by vocal onset (a "Phee" call). The **blue line** is a moving average (100 ms window) and shows that suppression begins prior to vocalization (indicated by **arrow**). The **thick bar** indicates the period over which suppression is continuously significant ($P < 0.05$). (Adapted, with permission, from Eliades and Wang 2003.)

2. Neurons subject to vocalization-induced suppression are sensitive to vocal feedback perturbations. **Top:** Self-produced vocalizations with or without feedback alterations were delivered to the marmoset through a customized headphone. **Bottom:** This auditory cortical neuron was suppressed during normal vocalization (**dark blue**) but showed a large increase in firing rate when the auditory feedback of the vocalization was shifted in the frequency domain (**light blue**). Amplifying auditory feedback alone did not generate firing rate changes (**black**). (Adapted, with permission, from Eliades and Wang 2008; Crapse and Sommer 2008.)

Highlights

1. Sound impinging on two ears carries information that the brain uses to compute where sounds arise and what they mean. Sounds are characterized by the amount of energy at one or more frequencies. To determine where sounds arise in the horizontal plane, many mammals compute differences in the time of arrival at the two ears for sounds less than approximately 3,000 Hz. To determine where sounds arise in the vertical dimension and whether they arise from the front or the back, mammals use spectral filtering of sounds greater than approximately 6,000 Hz by the head, shoulders, and external ears.

2. Acoustic information is brought to the brain from the cochlea by auditory nerve fibers, each sharply tuned to a narrow range of frequencies and together representing the entire hearing range of the animal. Auditory nerve fibers terminate in the ventral and dorsal cochlear nuclei, distributing acoustic information to four major groups of principal cells that form parallel ascending pathways through the brain stem. The topographic organization of the auditory nerve inputs imparts a tonotopic organization to the ipsilateral cochlear nuclei that is preserved all along the auditory pathway, including auditory cortex.

3. A marked feature of auditory neurons at processing stations along the ascending pathway is their progressively increasing stimulus selectivity.

4. The ventral cochlear nucleus extracts three features of sounds: (a) The monaural pathways through octopus cells of the ventral cochlear nucleus, the superior paraolivary nucleus, and ventral nucleus of the lateral lemniscus detect coincident firing of auditory nerve fibers that is useful for detecting onsets and gaps in sounds. (b) Stellate cells detect and sharpen the encoding of spectral peaks and valleys and convey that spectral information to the dorsal cochlear nucleus, olivocochlear neurons in the ventral nucleus of the lateral lemniscus, ventral nucleus of the lateral lemniscus, inferior colliculus, and thalamus. Spectral information is used for understanding the meaning of sounds and for localizing their sources. (c) Bushy cells sharpen and convey information about the fine structure of sounds, which is used in the binaural pathways through the medial and lateral superior olivary nuclei to make the interaural comparisons of timing and intensity of sounds

at the two ears, which are used to localize sound sources along the azimuth.

5. The dorsal cochlear nucleus integrates acoustic signals with somatosensory information in its principal cells. Somatosensory information helps distinguish the spectral cues generated by an animal's own movements, which are biologically uninteresting, from those that arise from the environment.

6. Auditory brainstem pathways converge in the inferior colliculus. The inferior colliculus feeds acoustic information through the medial geniculate body of the thalamus to auditory cortex.

7. A projection from the inferior colliculus carries information about the location of sounds to the superior colliculus, a part of the brain that controls reflexive orienting movements of the head and eyes.

8. Within auditory cortex, auditory neurons continue to become more selective to the stimuli to which they respond. Subregions of the auditory cortex represent different biologically significant features such as pitch of tones that form harmonic complexes. Auditory cortex also transforms rapidly varying features of sounds into firing-rate-based representations, while representing slowly varying sounds using spike timing.

9. Auditory circuits in the cerebral cortex are segregated into separate processing streams, with dorsal and ventral streams concerned respectively with sound location in space and sound identification.

10. The cerebral cortex modulates processing in subcortical auditory areas. Projections from the auditory cortex innervate the thalamus, inferior colliculus, olivocochlear neurons, some basal ganglionic structures, and even the dorsal cochlear nucleus.

11. Auditory cortex is involved in processing vocal feedback signals during speaking. Speaking induces suppression of neural activity in auditory cortex that begins several hundred milliseconds prior to the vocal onset. This suppression results from a vocal feedback-monitoring network that functions to guide vocal production and learning.

Donata Oertel
Xiaoqin Wang

Selected Reading

Bendor DA, Wang X. 2005. The neuronal representation of pitch in primate auditory cortex. Nature 436:1161–1165.

Chase SM, Young ED. 2006. Spike-timing codes enhance the representation of multiple simultaneous sound-localization cues in the inferior colliculus. J Neurosci 26:3889–3898.

Eliades SJ, Wang X. 2008. Neural substrates of vocalization feedback monitoring in primate auditory cortex. Nature 453:1102–1106.

Gao E, Suga N. 2000. Experience-dependent plasticity in the auditory cortex and the inferior colliculus of bats: role of the corticofugal system. Proc Natl Acad Sci U S A 97:8081–8086.

Hofman PM, Van Riswick JG, Van Opstal AJ. 1998. Relearning sound localization with new ears. Nat Neurosci 1:417–421.

Joris PX, Smith PH, Yin TC. 1998. Coincidence detection in the auditory system: 50 years after Jeffress. Neuron 21:1235–1238.

Joris PX, Yin TCT. 2007. A matter of time: internal delays in binaural processing. Trends Neurosci 30:70–78.

Oertel D, Young ED. 2004. What's a cerebellar circuit doing in the auditory system? Trends Neurosci 27:104–110.

Schneider DM, Mooney R. 2018. How movement modulates hearing. Annu Rev Neurosci 41:553–572.

Schreiner CE, Read HL, Sutter ML. 2000. Modular organization of frequency integration in primary auditory cortex. Annu Rev Neurosci 23:501–529.

Suga N. 1990. Cortical computational maps for auditory imaging. Neural Netw 3:3–21.

Wang X. 2018. Cortical coding of auditory features. Annu Rev Neurosci 41:527–552.

Zhang LI, Bao S, Merzenich MM. 2001. Persistent and specific influences of early acoustic environments on primary auditory cortex. Nat Neurosci 4:1123–1130.

References

Bartlett EL, Wang X. 2007. Neural representations of temporally-modulated signals in the auditory thalamus of awake primates. J Neurophysiol 97:1005–1017.

Bendor DA, Wang X. 2006. Cortical representations of pitch in monkeys and humans. Curr Opin Neurobiol 16:391–399.

Brodal A. 1981. *Neurological Anatomy in Relation to Clinical Medicine.* New York: Oxford Univ. Press.

Cajal SR. 1909. *Histologie du Systeme Nerveux de l'Homme et des Vertebres.* Paris: A. Maloine.

Cariani PA, Delgutte B. 1996. Neural correlates of the pitch of complex tones. I. Pitch and pitch salience. J Neurophysiol 76:1698–1716.

Cohen YE, Knudsen EI. 1999. Maps versus clusters: different representations of auditory space in the midbrain and forebrain. Trends Neurosci 22:128–135.

Crapse TB, Sommer MA. 2008. Corollary discharge circuits in the primate brain. Curr Opin Neurobiol 18:552–557.

Darrow KN, Maison SF, Liberman MC. 2006. Cochlear efferent feedback balances interaural sensitivity. Nat Neurosci 9:1474–1476.

Eliades SJ, Wang X. 2003. Sensory-motor interaction in the primate auditory cortex during self-initiated vocalizations. J Neurophysiol 89:2194–2207.

Feng L, Wang X. 2017. Harmonic template neurons in primate auditory cortex underlying complex sound processing. Proc Natl Acad Sci U S A 114:E840–E848.

Gao L, Kostlan K, Wang Y, Wang X. 2016. Distinct subthreshold mechanisms underlying rate-coding principles in primate auditory cortex. Neuron 91:905–919.

Geisler CD. 1998. *From Sound to Synapse, Physiology of the Mammalian Ear.* New York: Oxford Univ. Press.

Houde JF, Chang EF 2015. The cortical computations underlying feedback control in vocal production. Curr Opin Neurobiol 33:174–181.

Hubel DH, Henson CO, Rupert A, Galambos R. 1959. Attention units in the auditory cortex. Science 129:1279–1280.

Jeffress LA. 1948. A place theory of sound localization. J Comp Physiol Psychol 41:35–39.

Kanold PO, Young ED. 2001. Proprioceptive information from the pinna provides somatosensory input to cat dorsal cochlear nucleus. J Neurosci 21:7848–7858.

King AJ. 1999. Sensory experience and the formation of a computational map of auditory space in the brain. BioEssays 21:900–911.

King AJ, Bajo VM, Bizley JK, et al. 2007. Physiological and behavioral studies of spatial coding in the auditory cortex. Hear Res 229:106–115.

Liberman MC. 1978. Auditory-nerve response from cats raised in a low-noise chamber. J Acoust Soc Am 63:442–455.

Lu T, Liang L, Wang X. 2001. Temporal and rate representations of time-varying signals in the auditory cortex of awake primates. Nature Neurosci 4:1131–1138.

Merzenich MM, Knight PL, Roth GL. 1975. Representation of cochlea within primary auditory cortex in the cat. J Neurophysiol 38:231–249.

Mesgarani N, Cheung C, Johnson K, Chang EF. 2014. Phonetic feature encoding in human superior temporal gyrus. Science 343:1006–1010.

Middlebrooks JC. 2005. Auditory cortex cheers the overture and listens through the finale. Nature Neurosci 8:851–852.

Musicant AD, Chan JCK, Hind JE. 1990. Direction-dependent spectral properties of cat external ear: new data and cross-species comparisons. J Acoust Soc Am 87:757–781.

Oertel D, Bal R, Gardner SM, Smith PH, Joris PX. 2000. Detection of synchrony in the activity of auditory nerve fibers by octopus cells of the mammalian cochlear nucleus. Proc Nat Acad Sci U S A 97:11773–11779.

Palmer AR, King AJ. 1982. The representation of auditory space in the mammalian superior colliculus. Nature 299:248–249.

Penagos H, Melcher JR, Oxenham AJ. 2004. A neural representation of pitch salience in nonprimary human auditory cortex revealed with functional magnetic resonance imaging. J Neurosci 24:6810–6815.

Raman IM, Zhang S, Trussell LO. 1994. Pathway-specific variants of AMPA receptors and their contribution to neuronal signaling. J Neurosci 14:4998–5010.

Rauschecker JP, Tian B. 2000. Mechanisms and streams for processing of "what" and "where" in auditory cortex. Proc Nat Acad Sci U S A 97:11800–11806.

Rauschecker JP, Tian B, Hauser M. 1995. Processing of complex sounds in the macaque nonprimary auditory cortex. Science 268:111–114.

Recanzone GH, Schreiner CE, Merzenich MM. 1993. Plasticity in the frequency representation of primary auditory cortex following discrimination training in adult owl monkeys. J Neurosci 13:87–103.

Remington ED, Wang X. 2019. Neural representations of the full spatial field in auditory cortex of awake marmoset (*Callithrix jacchus*). Cereb Cortex 29:1199–1216.

Riquimaroux H, Gaioni SJ, Suga N. 1991. Cortical computational maps control auditory perception. Science 251: 565–568.

Romanski LM, Averbeck BB. 2009. The primate cortical auditory system and neural representation of conspecific vocalizations. Annu Rev Neurosci 32:315–346.

Sadagopan S, Wang X. 2009. Nonlinear spectrotemporal interactions underlying selectivity for complex sounds in auditory cortex. J Neurosci 29:11192–11202.

Schreiner CE, Winer JA. 2007. Auditory cortex mapmaking: principles, projections, and plasticity. Neuron 56: 356–365.

Scott LL, Mathews PJ, Golding NL. 2005. Posthearing developmental refinement of temporal processing in principal neurons of the medial superior olive. J Neurosci 25: 7887–7895.

Song X, Osmanski MS, Guo Y, Wang X. 2016. Complex pitch perception mechanisms are shared by humans and a New World monkey. Proc Natl Acad Sci U S A 113:781–786.

Spirou GA, Young ED. 1991. Organization of dorsal cochlear nucleus type IV unit response maps and their relationship to activation by bandlimited noise. J Neurophysiol 66:1750–1768.

Suga N, O'Neill WE, Kujirai K, Manabe T. 1983. Specificity of combination-sensitive neurons for processing of complex biosonar signals in auditory cortex of the mustached bat. J Neurophysiol 49:1573–626.

Suga N. 1984. Neural mechanisms of complex-sound processing for echolocation. Trends Neurosci 7:20–27.

Tollin DJ, Yin TC. 2002. The coding of spatial location by single units in the lateral superior olive of the cat. II. The determinants of spatial receptive fields in azimuth. J Neurosci 22:1468–1479.

Wang X, Lu T, Snider RK, Liang L. 2005. Sustained firing in auditory cortex evoked by preferred stimuli. Nature 435:341–346.

Warr WB. 1992. Organization of olivocochlear efferent systems in mammals. In: DB Webster, AN Popper, RR Fay (eds). *The Mammalian Auditory Pathway: Neuroanatomy*, pp. 410–448. New York: Springer.

Winer JA, Saint Marie RL, Larue DT, Oliver DL. 1996. GABAergic feedforward projections from the inferior colliculus to the medial geniculate body. Proc Natl Acad Sci U S A 93:8005–8010.

Yin TCT. 2002. Neural mechanisms of encoding binaural localization cues in the auditory brainstem. In: D Oertel, RR Fay, AN Popper (eds). *Integrative Functions in the Mammalian Auditory Pathway*, pp. 238–288. New York: Springer.

29

Smell and Taste: The Chemical Senses

THROUGH THE SENSES OF SMELL and taste, we are able to perceive a staggering number and variety of chemicals in the external world. These chemical senses inform us about the availability of foods and their potential pleasure or danger. Smell and taste also initiate physiological changes required for the digestion and utilization of food. In many animals, the olfactory system also serves an important social function by detecting pheromones that elicit innate behavioral or physiological responses.

Although the discriminatory ability of humans is somewhat limited compared with that of many other animals, odor chemists estimate that the human olfactory system may be capable of detecting more than 10,000 different volatile chemicals. Perfumers who are highly trained to discriminate odorants can distinguish as many as 5,000 different types of odorants, and wine tasters can discern more than 100 different components of taste based on combinations of flavor and aroma.

In this chapter, we consider how odor and taste stimuli are detected and how they are encoded in patterns of neural signals transmitted to the brain. In recent years, much has been learned about the mechanisms underlying chemosensation in a variety of animal species. Certain features of chemosensation have been conserved through evolution, whereas others are specialized adaptations of individual species.

A Large Family of Olfactory Receptors Initiate the Sense of Smell

Odorants—volatile chemicals that are perceived as odors—are detected by olfactory sensory neurons in the nose. The sensory neurons are embedded in a specialized olfactory epithelium that lines part of the nasal cavity, approximately 5 cm^2 in area in humans (Figure 29–1), and are interspersed with glia-like supporting cells (Figure 29–2). They are relatively short lived, with a life span of only 30 to 60 days, and are continuously replaced from a layer of basal stem cells in the epithelium.

The olfactory sensory neuron is a bipolar nerve cell. A single dendrite extends from the apical end to the epithelial surface, where it gives rise to numerous

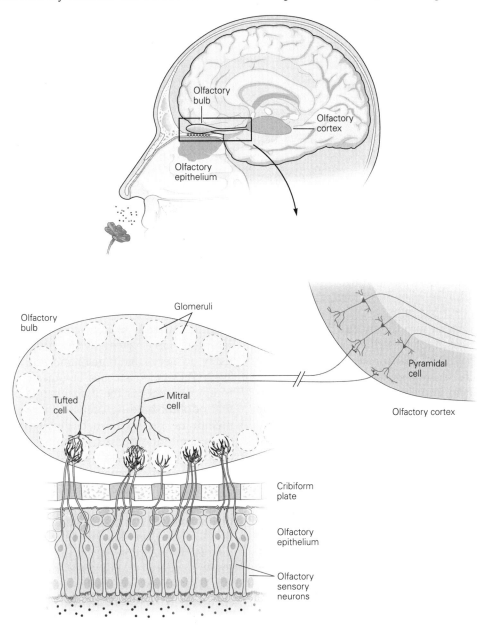

Figure 29–1 The olfactory system. Odorants are detected by olfactory sensory neurons in the olfactory epithelium, which lines part of the nasal cavity. The axons of these neurons project to the olfactory bulb, where they terminate on the dendrites of mitral and tufted cell relay neurons within glomeruli. In turn, the axons of the relay neurons project to the olfactory cortex, where they terminate on the dendrites of pyramidal neurons whose axons project to other brain areas.

A

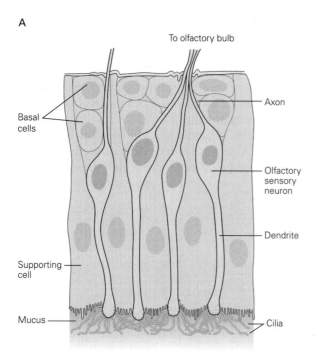

To olfactory bulb

Basal
cells

Axon

Olfactory
sensory
neuron

Dendrite

Supporting
cell

Mucus

Cilia

B

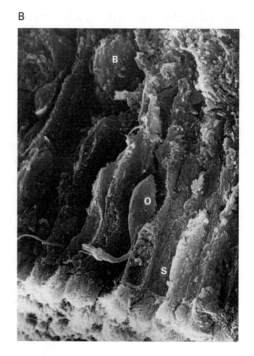

Figure 29–2 The olfactory epithelium.

A. The olfactory epithelium contains sensory neurons interspersed with supporting cells as well as a basal layer of stem cells. A single dendrite extends from the apical end of each neuron; sensory cilia sprout from the end of the dendrite into the mucus lining the nasal cavity. An axon extends from the basal end of each neuron to the olfactory bulb.

B. A scanning electron micrograph of the olfactory epithelium shows the dense mat of sensory cilia at the epithelial surface. Supporting cells (**S**) are columnar cells that extend the full depth of the epithelium and have apical microvilli. Interspersed among the supporting cells is an olfactory sensory neuron (**O**) with its dendrite and cilia, and a basal stem cell (**B**). (Reproduced, with permission, from Morrison and Costanzo 1990. Copyright © 1990 Wiley-Liss, Inc.)

thin cilia that protrude into the mucus that coats the nasal cavity (Figure 29–2). The cilia contain the odorant receptors as well as the transduction machinery needed to amplify sensory signals from the receptors and transform them into electrical signals in the neuron's axon, which projects from the basal pole of the neuron to the brain. The axons of olfactory sensory neurons pass through the cribriform plate, a perforated region in the skull above the nasal cavity, and then terminate in the olfactory bulb (see Figure 29–1).

Mammals Share a Large Family of Odorant Receptors

Odorant receptors are proteins encoded by a multigene family that is evolutionarily conserved and found in all vertebrate species. Humans have approximately 350 different odorant receptors, whereas mice have approximately 1,000. Although odorant receptors belong to the G protein–coupled receptor superfamily, they share sequence motifs not seen in other superfamily members. Significantly, the odorant receptors vary considerably in amino acid sequence (Figure 29–3A).

Like other G protein–coupled receptors, odorant receptors have seven hydrophobic regions that are likely to serve as transmembrane domains (Figure 29–3A). Detailed studies of other G protein–coupled receptors, such as the β-adrenergic receptor, suggest that odorant binding occurs in a pocket in the transmembrane region formed by a combination of the transmembrane domains. The amino acid sequences of odorant receptors are especially variable in several transmembrane domains, providing a possible basis for variability in the odorant binding pocket that could account for the ability of different receptors to recognize structurally diverse ligands.

A second, smaller family of chemosensory receptors is also expressed in the olfactory epithelium. These receptors, called trace amine-associated receptors (TAARs), are G protein–coupled, but their protein sequence is unrelated to that of odorant receptors. They are encoded by a small family of genes present in humans and mice as well as fish. Studies in mice, which have 14 different olfactory TAARs, indicate that TAARs recognize volatile amines, one of which is present in high concentrations in the urine of male

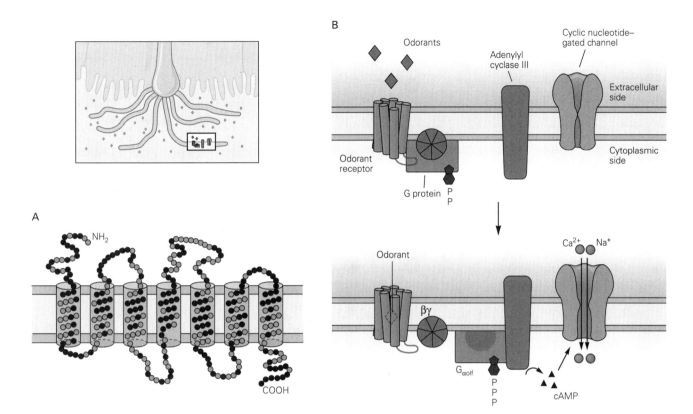

Figure 29–3 Odorant receptors.

A. Odorant receptors have the seven transmembrane domains characteristic of G protein–coupled receptors. They are related to one another but vary in amino acid sequence (positions of highest variability are shown here as **black balls**). (Reproduced, with permission, from Buck and Axel 1991.)

B. Binding of an odorant causes the odorant receptor to interact with $G\alpha_{olf}$, the α-subunit of a heterotrimeric G

protein. This causes the release of a guanosine triphosphate (GTP)-coupled $G\alpha_{olf}$, which stimulates adenylyl cyclase III, leading to an increase in cyclic adenosine monophosphate (**cAMP**). The elevated cAMP in turn induces the opening of cyclic nucleotide–gated cation channels, causing cation influx and a change in membrane potential in the ciliary membrane.

mice and another in the urine of some predators. It is possible that this small receptor family has a function distinct from that of the odorant receptor family, perhaps one associated with the detection of animal cues. Another family of 12 receptors, called MS4Rs, is also found in mice, where it may be involved in the detection of pheromones and certain food odors.

The binding of an odorant to its receptor induces a cascade of intracellular signaling events that depolarize the olfactory sensory neuron (Figure 29–3B). The depolarization spreads passively to the cell body and then the axon, where action potentials are generated that are actively conducted to the olfactory bulb.

Humans and other animals rapidly accommodate to odors, as seen for example in the weakening of detection of an unpleasant odor that is continuously present. The ability to sense an odorant rapidly recovers when the odorant is temporarily removed. The adaptation to odorants is caused in part by modulation

of a cyclic nucleotide–gated ion channel in olfactory cilia, but the mechanism by which sensitivity is speedily restored is not yet understood.

Different Combinations of Receptors Encode Different Odorants

To be distinguished perceptually, different odorants must cause different signals to be transmitted from the nose to the brain. This is accomplished in two ways. First, each olfactory sensory neuron expresses only one odorant receptor gene and therefore one type of receptor. Second, each receptor recognizes multiple odorants, and conversely, each odorant is detected by multiple different receptors (Figure 29–4). Importantly, however, each odorant is detected, and thereby encoded, by a unique combination of receptors and thus causes a distinctive pattern of signals to be transmitted to the brain.

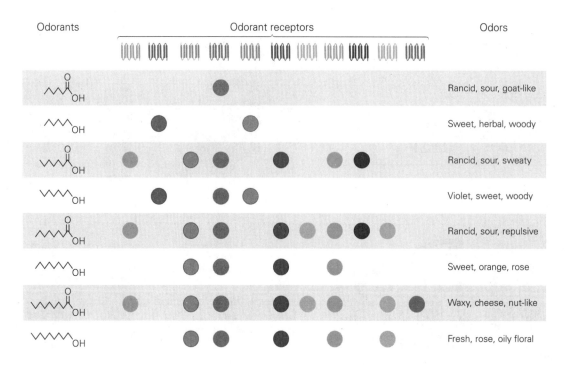

| Odorants | Odorant receptors | Odors |

Figure 29–4 Each odorant is recognized by a unique combination of receptors. A single odorant receptor can recognize multiple odorants, but different odorants are detected, and thus encoded, by different combinations of receptors. This combinatorial coding explains how mammals can distinguish odorants with similar chemical structures as having different scents. The data

in the figure were obtained by testing mouse olfactory sensory neurons with different odorants and then determining the odorant receptor gene expressed by each responsive neuron. The perceived qualities of these odorants in humans shown on the right illustrate how highly related odorants can have different scents. (Adapted, with permission, from Malnic et al. 1999.)

The combinatorial coding of odorants greatly expands the discriminatory power of the olfactory system. If each odorant were detected by only three different receptors, this strategy could in theory generate millions of different combinatorial receptor codes—and an equivalently vast number of different signaling patterns sent from the nose to the brain. Interestingly, even odorants with nearly identical structures are recognized by different combinations of receptors (Figure 29–4). The fact that highly related odorants have different combinatorial receptor codes explains why a slight change in the chemical structure of an odorant can alter its perceived odor. In some cases, the result is dramatic, for example, changing the perception of a chemical from rose to sour.

A change in concentration of an odorant can also change the perceived odor. For example, a low concentration of thioterpineol smells like tropical fruit, whereas a higher concentration smells like grapefruit and an even higher concentration smells putrid. As the concentration of an odorant is increased, additional receptors with lower affinity for the odorant are recruited into the response and thus change the combinatorial receptor code, providing an explanation for the effects of odorant concentration on perception.

Olfactory Information Is Transformed Along the Pathway to the Brain

Odorants Are Encoded in the Nose by Dispersed Neurons

How are signals from a large array of different odorant receptors organized in the nervous system to generate diverse odor perceptions? This question has been investigated in rodents. Studies in mice have revealed that olfactory information undergoes a series of transformations as it travels from the olfactory epithelium to the olfactory bulb and then to the olfactory cortex.

The olfactory epithelium has a series of spatial zones that express different olfactory receptors. Each receptor type is expressed in approximately 5,000 neurons that are confined to one zone (Figure 29–5). (Recall that each neuron expresses only one odorant receptor gene.) Neurons with the same receptor are randomly scattered within a zone so neurons with different receptors are interspersed. All zones contain a variety of receptors, and a specific odorant may be recognized by receptors in different zones. Thus, despite a rough organization of odorant receptors into spatial zones, information provided by the odorant receptor family is highly distributed in the epithelium.

Because each odorant is detected by an ensemble of neurons widely dispersed across the epithelial sheet, receptors in one part of the epithelium will be able to detect a particular odorant even when those in another part are impaired by respiratory infection.

Sensory Inputs in the Olfactory Bulb Are Arranged by Receptor Type

The axons of olfactory sensory neurons project to the ipsilateral olfactory bulb, whose rostral end lies just above the olfactory epithelium. The axons of olfactory sensory neurons terminate on the dendrites of olfactory bulb neurons within bundles of neuropil called glomeruli that are arrayed over the bulb's surface (Figure 29–1). In each glomerulus, the sensory axons make synaptic connections with three types of neurons: mitral and tufted projection (relay) neurons, which project axons to the olfactory cortex, and periglomerular interneurons, which encircle the glomerulus (Figure 29–6).

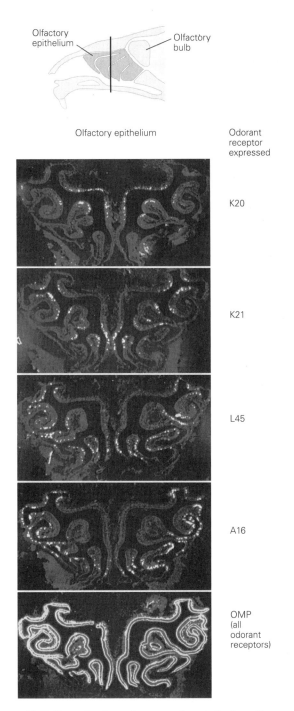

Figure 29–5 Organization of sensory inputs in the olfactory epithelium. The olfactory epithelium has different spatial zones that express different sets of odorant receptor genes. Each sensory neuron expresses only one receptor gene and thus one type of receptor. Neurons with the same receptor are confined to one zone but randomly scattered within that zone, such that neurons with different receptors are interspersed. The micrographs show the distribution of neurons labeled by four different receptor probes in sections through the mouse nose. An olfactory marker protein (**OMP**) probe labels all neurons expressing odorant receptors. (Adapted, with permission, from Ressler, Sullivan, and Buck 1993; Sullivan et al. 1996.)

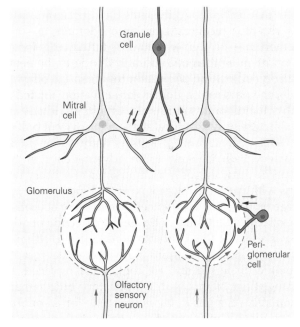

Figure 29–6 Olfactory bulb interneurons. In addition to excitatory mitral and tufted relay neurons, the olfactory bulb contains inhibitory interneurons. Within each glomerulus, the dendrites of GABAergic periglomerular cells receive excitatory input from olfactory sensory neurons and have reciprocal synapses with the primary dendrites of mitral and tufted relay neurons, suggesting a possible role in signal modification. The dendrites of GABAergic granule cells deeper in the bulb have reciprocal excitatory-inhibitory synapses with the secondary dendrites of the relay neurons and are thought to provide negative feedback to relay neurons that shapes the odor response. (Adapted from Shepherd and Greer 1998.)

The axon of an olfactory sensory neuron as well as the primary dendrite of each mitral and tufted relay neuron terminate in a single glomerulus. In each glomerulus, the axons of several thousand sensory neurons converge on the dendrites of approximately 40 to 50 relay neurons. This convergence results in approximately a 100-fold decrease in the number of neurons transmitting olfactory signals.

The organization of sensory information in the olfactory bulb is dramatically different from that of the epithelium. Whereas olfactory sensory neurons with the same odorant receptor are randomly scattered in one epithelial zone, their axons typically converge in two glomeruli at specific locations, one on either side of the olfactory bulb (Figure 29–7C). Each glomerulus, and each mitral and tufted relay neuron connected to it, receives input from just one type of odorant receptor. The result is a precise arrangement of sensory inputs from different odorant receptors, one that is similar between individuals.

Because each odorant is recognized by a unique combination of receptor types, each also activates a particular combination of glomeruli in the olfactory bulb (Figure 29–7B). At the same time, just as one odorant receptor recognizes multiple odorants, a single glomerulus—or a given mitral or tufted cell—is activated by more than one odorant. Owing to the nearly stereotyped pattern of receptor inputs in the olfactory bulb, the patterns of glomerular activation elicited by individual odorants are similar in all individuals and are bilaterally symmetrical in the two adjacent bulbs.

This organization of sensory information in the olfactory bulb is likely to be advantageous in two respects. First, signals from thousands of sensory neurons with the same odorant receptor type always converge on the same few glomeruli, and relay neurons in the olfactory bulb may optimize the detection of odorants present at low concentrations. Second, although olfactory sensory neurons with the same receptor type are dispersed and are continually replaced, the arrangement of inputs in the olfactory bulb remains unaltered. As a result, the neural code for an odorant in the brain is maintained over time, assuring that an odorant encountered previously can be recognized years later.

One mystery that remains unsolved is how all the axons of olfactory sensory neurons with the same type of receptor are directed to the same glomeruli. Studies using transgenic mice indicate that the odorant receptor itself somehow determines the target of the axon, but how it does so is not yet understood.

Sensory information is processed and possibly refined in the olfactory bulb before it is forwarded to the olfactory cortex. Each glomerulus is encircled by periglomerular interneurons that receive excitatory input from sensory axons and form inhibitory dendrodendritic synapses with mitral and tufted cell dendrites in that glomerulus and perhaps adjacent glomeruli. The periglomerular interneurons may therefore have a role in signal modulation. In addition, granule cell interneurons deep in the bulb provide negative feedback onto mitral and tufted cells. The granule cell interneurons are excited by the basal dendrites of mitral and tufted cells and in turn inhibit those relay neurons and others with which they are connected. The lateral inhibition afforded by these connections is thought to dampen signals from glomeruli and relay neurons that respond to an odorant only weakly, thereby sharpening the contrast between important and irrelevant sensory information before its transmission to the cortex.

Other potential sources of signal refinement are the retrograde projections to the olfactory bulb from the olfactory cortex, basal forebrain (horizontal limb of the diagonal band), and midbrain (locus ceruleus and raphe nuclei). These connections may modulate olfactory bulb output according to the physiological or behavioral state of an animal. When the animal is hungry, for example, some centrifugal projections might heighten the perception of the aroma of foods.

The Olfactory Bulb Transmits Information to the Olfactory Cortex

The axons of the mitral and tufted relay neurons of the olfactory bulb project through the lateral olfactory tract to the olfactory cortex (Figure 29–8 and see Figure 29–1). The olfactory cortex, defined roughly as that portion of the cortex that receives a direct projection from the olfactory bulb, comprises multiple anatomically distinct areas. The six major areas are the anterior olfactory nucleus, which connects the two olfactory bulbs through a portion of the anterior commissure; the anterior and posterior-lateral cortical nuclei of the amygdala; the olfactory tubercle; part of the entorhinal cortex; and the piriform cortex, the largest and considered the major olfactory cortical area.

The functions of the different olfactory cortical areas are largely unknown. However, the piriform cortex is thought to be important for odor learning. Recent studies indicate that the posterior-lateral cortical amygdala may have a role in innate attraction and fear behaviors, and the amygdalo-piriform transition area, a minor olfactory cortical area, a role in stress hormone responses to predator odors detected in the nose.

In the piriform cortex, the axons of olfactory bulb mitral and tufted cells leave the lateral olfactory tract

A Axons of neurons with the same odorant receptor converge on a few glomeruli

B One odorant can activate many glomeruli

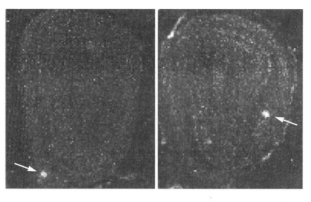
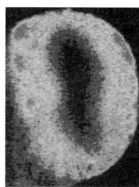

C The olfactory bulb has a precise map of odorant receptor inputs

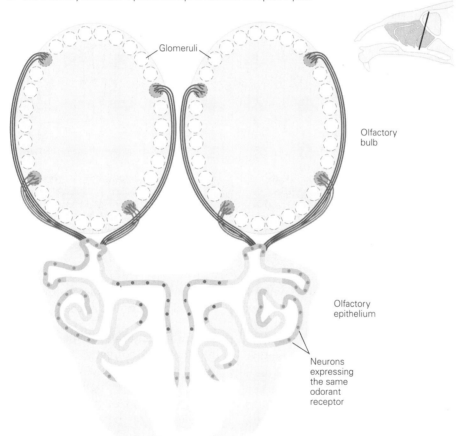

Glomeruli

Olfactory bulb

Olfactory epithelium

Neurons expressing the same odorant receptor

Figure 29–7 Odor responses in the olfactory bulb.

A. The axons of sensory neurons with the same odorant receptor type usually converge in only two glomeruli, one on each side of the olfactory bulb. Here, a probe specific for one odorant receptor gene labeled a glomerulus on the medial side (*left*) and lateral side (*right*) of a mouse olfactory bulb. The probe hybridized to receptor messenger RNAs present in sensory axons in these coronal sections. (Adapted, with permission, from Ressler, Sullivan, and Buck 1994.)

B. A single odorant often activates multiple glomeruli with input from different receptors. This section of a rat olfactory bulb shows the uptake of radiolabeled 2-deoxglucose

at multiple foci (**red**) following exposure of the animal to the odorant methyl benzoate. The labeled foci correspond to numerous glomeruli at different locations in the olfactory bulb. (Reproduced, with permission, from Johnson, Farahbod, and Leon 2005. Copyright © 2005 Wiley-Liss, Inc.)

C. The olfactory bulb has a precise map of odorant receptor inputs because each glomerulus is dedicated to only one type of receptor. The maps in the two olfactory bulbs are bilaterally symmetrical and are nearly identical across individuals. The maps on the medial and lateral sides of each bulb are similar, but slightly displaced along the dorsal-ventral and anterior-posterior axes.

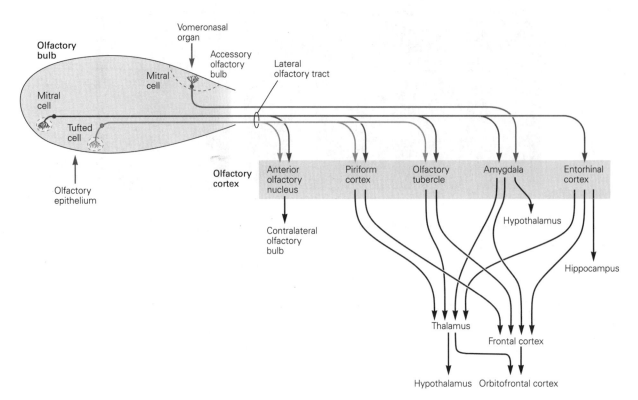

Figure 29–8 Afferent pathways to olfactory cortex. The axons of mitral and tufted relay neurons of the olfactory bulb project through the lateral olfactory tract to the olfactory cortex. The olfactory cortex consists of a number of distinct areas, the largest of which is the piriform cortex. From these areas, olfactory information is transmitted to other brain areas directly as well as indirectly via the thalamus. Targets include frontal and orbitofrontal areas of the neocortex, which are thought to be important for odor discrimination, and the amygdala and hypothalamus, which may be involved in emotional and physiological responses to odors. Mitral cells in the accessory olfactory bulb project to specific areas of the amygdala that transmit signals to the hypothalamus.

to form excitatory glutamatergic synapses with pyramidal neurons, the projection neurons of the cortex. Pyramidal neuron activity appears to be modulated by inhibitory inputs from local GABAergic interneurons as well as by excitatory inputs from other pyramidal neurons in the same and other olfactory cortical areas and the contralateral piriform cortex. The piriform cortex also receives centrifugal inputs from modulatory brain areas, suggesting that its activity may be adjusted according to physiological or behavioral state. Finally, the olfactory cortex projects to the olfactory bulb, providing yet another possible means of signal modulation.

As with the olfactory bulb relay neurons, individual pyramidal neurons can be activated by more than one odorant. However, the pyramidal neurons activated by a particular odorant are scattered across the piriform cortex, an arrangement different from that of the olfactory bulb. Mitral cells in different parts of the olfactory bulb can project axons to the same subregion of the piriform cortex, further indicating that the highly organized map of odorant receptor inputs in the olfactory bulb is not recapitulated in the cortex.

Output From the Olfactory Cortex Reaches Higher Cortical and Limbic Areas

Pyramidal neurons in the olfactory cortex transmit information indirectly to the orbitofrontal cortex through the thalamus and directly to the frontal cortex. These pathways to higher cortical areas are thought to be important in odor discrimination. In fact, people with lesions of the orbitofrontal cortex are unable to discriminate odors. Interestingly, recordings in the orbitofrontal cortex suggest that some individual neurons in that area receive multimodal input, responding, for example, to the smell, sight, or taste of a banana.

Many areas of the olfactory cortex also relay information to nonolfactory areas of the amygdala, which is linked to emotions, and to the hypothalamus, which controls basic drives, such as appetite, as well as a number of innate behaviors. These limbic areas are thought to play a role in the emotional and motivational aspects of smell as well as many of the behavioral and physiological effects of odorants. In animals, they may be important in the generation of stereotyped behavioral and physiological responses to odors of predators or to pheromones that are detected in the olfactory epithelium.

Olfactory Acuity Varies in Humans

Olfactory acuity can vary as much as 1,000-fold among humans, even among people with no obvious abnormality. The most common olfactory aberration is *specific anosmia*. An individual with a specific anosmia has lowered sensitivity to a specific odorant even though sensitivity to other odorants appears normal. Specific anosmias to some odorants are common, with a few occurring in 1% to 20% of people. For example, 12% of individuals tested in one study exhibited a specific anosmia for musk. Recent studies indicate that specific anosmias can be caused by mutations in particular odorant receptor genes.

Far rarer abnormalities of olfaction, such as *general anosmia* (complete lack of olfactory sensation) or *hyposmia* (diminished sense of smell), are often transient and can derive from respiratory infections. Chronic anosmia or hyposmia can result from damage to the olfactory epithelium caused by infections; from particular diseases, such as Parkinson disease; or from head trauma that severs the olfactory nerves passing through holes in the cribriform plate, which then become blocked by scar tissue. Olfactory hallucinations of repugnant smells (*cacosmia*) can occur as a consequence of epileptic seizures.

Odors Elicit Characteristic Innate Behaviors

Pheromones Are Detected in Two Olfactory Structures

In many animals, the olfactory system detects not only odors but also pheromones, chemicals that are released from animals and influence the behavior or physiology of members of the same species. Pheromones play important roles in a variety of mammals, although they have not been demonstrated in humans. Often contained in urine or glandular secretions, some pheromones modulate the levels of reproductive hormones or stimulate sexual behavior or aggression. Pheromones are detected by two separate structures: the nasal olfactory epithelium, where odorants are detected, and the vomeronasal organ, an accessory olfactory organ thought to be specialized for the detection of pheromones and other animal cues.

The vomeronasal organ is present in many mammals, although not in humans. It is a tubular structure in the nasal septum that has a duct opening into the nasal cavity and one inner wall lined by a sensory epithelium. Signals generated by sensory neurons in the epithelium of the vomeronasal organ follow a distinct pathway. They travel through the accessory olfactory bulb primarily to the medial amygdala and posterior-medial cortical amygdala and from there to the hypothalamus.

Sensory detection in the vomeronasal organ differs from that in the olfactory epithelium. The vomeronasal organ has two different families of chemosensory receptors, the V1R and V2R families. In the mouse, each family has more than 100 members. Variation in amino acid sequence between members of each receptor family suggests that each family may recognize a variety of different ligands. Like odorant receptors, V1R and V2R receptors have the seven transmembrane domains typical of G protein–coupled receptors. The V2R receptor differs from both V1R and odorant receptors in having a large extracellular domain at the N-terminal end (Figure 29–9A). By analogy with receptors with similar structures, ligands may bind V1R receptors in a membrane pocket formed by a combination of transmembrane domains, whereas binding to V2R receptors may occur in the large extracellular domain. Although the V1R receptors are thought to recognize volatile chemicals, at least some V2Rs are thought to recognize proteins. These include a protein pheromone present in tears, mouse urinary proteins that stimulate aggression, and predator proteins from cats and rats that stimulate fear in mice.

The V1R and V2R families are expressed in different spatial zones in the vomeronasal organ that express different G proteins (Figure 29–9B,C). Each *V1R* or *V2R* gene is expressed in a small percentage of neurons scattered throughout one zone, an arrangement similar to that of odorant receptors in the olfactory epithelium. Similar to the main olfactory bulb, vomeronasal neurons with the same receptor type project to the same glomeruli in the accessory olfactory bulb, although the glomeruli for each receptor type are more numerous and their distribution less stereotyped than in the main olfactory bulb. In addition to V1R and V2R receptors, the vomeronasal organ has a family of five formyl peptide-related receptors (FPRs). These receptors are related to immune system FPRs that detect bacterial proteins, raising speculation that they might play a role in detecting diseased animals of the same species.

Invertebrate Olfactory Systems Can Be Used to Study Odor Coding and Behavior

Because invertebrates have simple nervous systems and often respond to olfactory stimuli with stereotyped behaviors, they are useful for understanding the relationship between the neural representation of odor and behavior.

Certain features of chemosensory systems are highly conserved in evolution. First, all metazoan animals can detect a variety of organic molecules using specialized chemosensory neurons with cilia or microvilli that contact the external environment. Second, the initial events of odor detection are mediated by families of transmembrane receptors with specific expression patterns in peripheral sensory neurons. Other features of the olfactory system differ between species, reflecting selection pressures and evolutionary histories of the animals.

The primary sensory organs of insects are the antennae and appendages known as maxillary palps near the mouth (Figure 29–10A). Whereas mammals have millions of olfactory neurons, insects have a much smaller number. There are approximately 2,600 olfactory neurons in the fruit fly *Drosophila* and approximately 60,000 in the honeybee.

The insect odorant receptors were discovered by finding multigene receptor families in the *Drosophila* genome, and these genes have now been examined in other insect genomes as well. Remarkably, they have little similarity to mammalian odorant receptors save for the presence of many transmembrane domains. Indeed, insect receptors appear to have an independent evolutionary origin from mammalian receptors and may not even be G protein–coupled receptors—an extreme example of the fast evolutionary change observed across all olfactory receptor systems. In *Drosophila*, the main odorant receptor family has only 60 genes, rather than the hundreds characteristic of vertebrates. The malaria mosquito *Anopheles gambiae* and the honeybee have similar numbers (85–95 genes), whereas leaf-cutter ants have more than 350 odorant

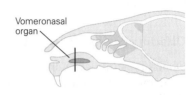

A Receptor structure

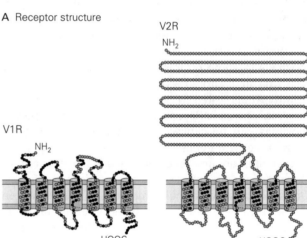

B Receptor distribution

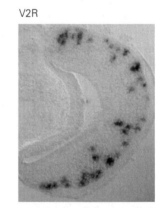

C Receptor and G protein distribution

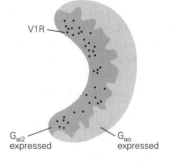
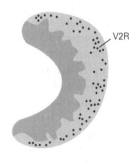

Figure 29–9 (Right) Candidate pheromone receptors in the vomeronasal organ.

A. The V1R and V2R families of receptors are expressed in the vomeronasal organ. In the mouse, each family has more than 100 members, which vary in protein sequence. Members of both families have the seven transmembrane domains of G protein–coupled receptors, but V2R receptors also have a large extracellular domain at the N-terminal end that may be the site of ligand binding.

B. Sections through the vomeronasal organ show individual V1R and V2R probes hybridized to subsets of neurons in two distinct zones. (Reproduced, with permission, from Dulac and Axel 1995; Matsunami and Buck 1997.)

C. The two zones express high levels of different G proteins, $G\alpha_{\alpha i2}$ and $G\alpha_{\alpha o}$.

A Olfactory pathways

Arista

Antenna

Lateral
protocerebrum

Maxillary
palp

Mushroom
body

Antennal
lobe

C, D

B1

B2, B3

Figure 29–10 Olfactory pathways from the antenna to the brain in *Drosophila*.

A. The axons of olfactory neurons with cell bodies and dendrites in the antenna and maxillary palp project axons to the antennal lobe. Projection neurons in the antennal lobe then project to two regions of the fly brain, the mushroom body and lateral protocerebrum. (Reproduced, with permission, from Takaki Komiyama and Liqun Luo.)

B. The neurons that express one type of olfactory receptor gene, detected by RNA in situ hybridization, are scattered in the maxillary palp (**1**) or antenna (**2, 3**).

C. All neurons that express the olfactory receptor gene *OR47* converge on a glomerulus in the antennal lobe. (Reproduced, with permission, from Vosshall et al. 1999; Vosshall, Wong, and Axel 2000.)

D. Each odorant elicits a physiological response from a subset of glomeruli in the antennal lobe. Two-photon calcium imaging was used to detect odor-evoked signals. (Reproduced, with permission, from Wang et al. 2003. Copyright © 2003 Elsevier.)

B Organization of receptor expression

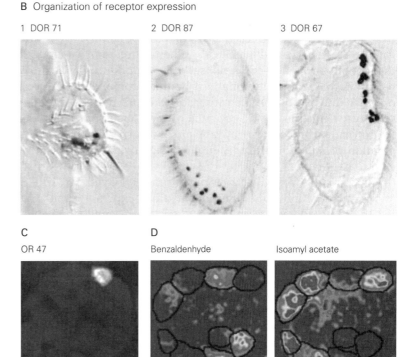

1 DOR 71 2 DOR 87 3 DOR 67

C OR 47

D Benzaldenhyde Isoamyl acetate

receptor genes, suggesting a wide variation in receptor number in insects.

Despite molecular differences in receptors, the anatomical organization of the fly's olfactory system is quite similar to that of vertebrates. Each olfactory neuron expresses one or sometimes two functional odorant receptor genes. The neurons expressing a particular gene are loosely localized to a region of the antenna but interspersed with neurons expressing other genes (Figure 29–10B). This scattered distribution is not the case at the next level of organization, the antennal lobe. Axons from sensory neurons that express one type of receptor converge on two invariant glomeruli in the antennal lobe, one each on the left and right sides of the animal (Figure 29–10C). This organization is strikingly similar to that of the first sensory relay in the vertebrate olfactory bulb and is also found in the moth, honeybee, and other insects.

Because there are only a few dozen receptor genes in *Drosophila*, it is possible to characterize the entire repertory of odorant-receptor interactions, a goal that is not yet attainable in mammals. Sophisticated genetic methods can be used to label and record from a *Drosophila* neuron expressing a single known odorant receptor gene. By repeating this experiment with many receptors and odors, the receptive fields of the odorant receptors have been defined and shown to be quite diverse.

In insects, individual odorant receptors can detect large numbers of odorants, including odorants with very different chemical structures. This broad recognition of odorants by "generalist" receptors is necessary if only a small number of receptors is available to detect all biologically significant odorants. A single insect receptor protein that detects many odors can be stimulated by some odors and inhibited by others, often with distinct temporal patterns. A subset of insect odorant receptors that convey information about pheromones or other unusual odors like carbon dioxide are more selective. Thus, the coding potential of each olfactory neuron can be broad or narrow and arises from a combination of stimulatory and inhibitory signals delivered to its receptors.

Information from the olfactory neurons is relayed to the antennal lobe where sensory neurons expressing the same odorant receptor converge onto a small number of projection neurons in one glomerulus (Figure 29–10A). Because *Drosophila* glomeruli are stereotyped in position and have one type of odorant receptor input, the transformation of information across the synapse can be described. Convergence of many olfactory sensory axons onto a few projection neurons leads to a great increase in the signal-to-noise ratio of olfactory signals, so projection neurons are much more sensitive to odor than individual olfactory neurons. Within the antennal lobe, excitatory interneurons distribute signals to projection neurons at distal locations, and inhibitory interneurons feed back onto the olfactory sensory neurons to dampen their input. Thus, while activity of an individual olfactory neuron is conveyed to one glomerulus. its activity is also distributed across the entire antennal lobe, as it is processed by excitatory and inhibitory local interneurons that connect many glomeruli.

The projection neurons from the antennal lobe extend to higher brain centers called mushroom bodies and lateral protocerebrum (Figure 29–10A). These structures may represent insect equivalents of the olfactory cortex. The mushroom bodies are sites of olfactory associative learning and multimodal associative learning; the lateral protocerebrum is important for innate olfactory responses. At this stage, projection neurons form complex connections with a large number of downstream neurons. Neurons in higher brain centers in *Drosophila* have the potential to integrate information from many receptors.

Olfactory Cues Elicit Stereotyped Behaviors and Physiological Responses in the Nematode

The nematode roundworm *Caenorhabditis elegans* has one of the simplest nervous systems in the animal kingdom, with only 302 neurons in the entire animal. Of these, 32 are ciliated chemosensory neurons. Because *C. elegans* has strong behavioral responses to a wide variety of chemicals, it has been a useful experimental animal for relating olfactory signals to behavior. Each chemosensory neuron detects a specific set of chemicals, and activation of the neuron is required for the behavioral responses to those substances. The neuron for a particular response, such as attraction to a specific odor, occurs in the same position in all individuals.

The molecular mechanisms of olfaction in *C. elegans* were elucidated through genetic screens for mutant worms lacking the ability to detect odors (anosmia). The G protein–coupled receptor for the volatile odorant diacetyl emerged from these screens (Figure 29–11). This receptor is one of approximately 1,700 predicted G protein–coupled chemoreceptor genes in *C. elegans*, the largest number of chemoreceptors among known genomes. Other kinds of chemosensory receptors are also present; for example, *C. elegans* senses external oxygen levels indirectly by detecting soluble guanylate cyclases that bind directly to oxygen. With so many chemoreceptors, nematodes are able to recognize a large variety of odors with great sensitivity. Some chemosensory neurons use G proteins to regulate cyclic guanosine $3',5'$-monophosphate (cGMP) and a cGMP-gated channel, a signal transduction pathway like that of vertebrate photoreceptors. Other chemosensory neurons signal through a transient receptor potential vanilloid (TRPV) channel, like vertebrate nociceptive neurons.

The "one neuron, one receptor" principle observed in vertebrates and insects does not operate in nematodes because the number of neurons is much smaller

A

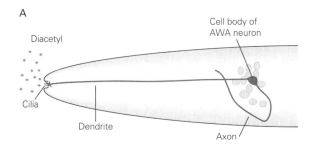

Diacetyl

Cell body of
AWA neuron

Cilia

Dendrite

Axon

B

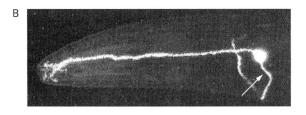

Figure 29–11 The receptor for diacetyl in the *Caenorhabditis elegans* worm.

A. A lateral view of the worm's anterior end shows the cell body and processes of the AWA chemosensory neuron. A dendrite terminates in cilia that are exposed to environmental chemicals. The neuron detects the volatile chemical diacetyl; animals with a mutation in the *odr-10* gene are unable to sense diacetyl.

B. The *odr-10* gene is active only in the AWA neurons. The micrograph here shows the gene product marked with fusion to a fluorescent reporter protein; the **arrow** indicates the neuron's axon. (Reproduced, with permission, from Sarafi-Reinach and Sengupta 2000.)

than the number of receptors. Each chemoreceptor gene is typically expressed in only one pair of chemosensory neurons, but each neuron expresses many receptor genes. The small size of the *C. elegans* nervous system limits olfactory computations. For example, a single neuron responds to many odors, but odors can be distinguished efficiently only if they are sensed by different primary sensory neurons.

The relationship between odor detection and behavior has been explored in *C. elegans* through genetic manipulations. For example, diacetyl is normally attractive to worms, but when the diacetyl receptor is experimentally expressed in an olfactory neuron that normally senses repellents, the animals are instead repelled by diacetyl. This observation indicates that specific sensory neurons encode the hardwired behavioral responses of attraction or repulsion and that a "labeled line" connects specific odors to specific behaviors. Similar ideas have emerged from genetic manipulations of taste systems in mice and flies, where sweet and bitter preference pathways are encoded by different sets of sensory cells.

Olfactory cues are linked to physiological responses as well as behavioral responses in nematodes. Food and pheromone cues that regulate development are detected by specific sensory neurons through G protein–coupled receptors. With low pheromone levels and plentiful food, animals rapidly develop to adulthood, whereas with high pheromone levels and scarce food, animals arrest in a long-lived larval stage called *dauer larvae* (Figure 29–12). Activation of these sensory neurons ultimately regulates the activity of an insulin signaling pathway that controls physiology and growth as well as the life span of the nematode. It is an open question whether the chemosensory systems and physiological systems of other animals are as entangled as they are in nematodes.

Strategies for Olfaction Have Evolved Rapidly

Why have independent families of odorant receptors evolved in mammals, nematodes, and insects? And why have the families changed so rapidly compared to genes involved in other important biological processes? The answer lies in a fundamental difference between olfaction and other senses such as vision, touch, and hearing.

Most senses are designed to detect physical entities with reliable physical properties: photons, pressure, or sound waves. By contrast, olfactory systems are designed to detect organic molecules that are infinitely variable and do not fit into a simple continuum of properties. Moreover, the organic molecules that are detected are produced by other living organisms, which evolve far more rapidly than the world of light, pressure, and sound.

An ancient olfactory system was present in the common ancestor of all animals that exist today. That ancestor lived in the ocean, where it gave rise to different lineages for mammals, insects, and nematodes. Those three phyla of animals came onto land hundreds of millions of years after the phyla diverged. Each phylum independently modified its olfactory system to detect airborne odors, leading to diversification of the receptors.

A consideration of the natural history of dipteran and hymenopteran insects, which have evolved in the last 200 million years, helps explain the rapid diversification of the odorant receptors. These insects include honeybees that pollinate flowers, fruit flies that feed on rotting fruit, flesh flies that arrive within minutes of death, and mosquitoes that prey on living animals. The odorants important for the survival of these insects are radically different, and receptor genes tuned to those odorants have evolved accordingly.

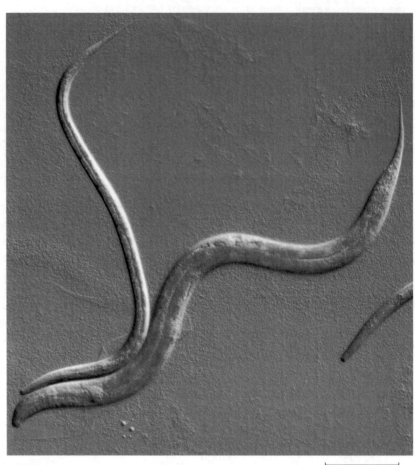

Figure 29–12 Chemosensory cues regulate the development of *C. elegans*. When exposed to different chemosensory cues, two larvae of the same age follow different development paths. A dauer larva, which forms under stressful conditions of low food and high population density, develops into a small slender adult (*left*). It is a nonfeeding, nonreproducing, stress-resistant form of the worm. In contrast, a larva in a rich environment favoring reproductive growth develops into a normal adult (*right*). (Reproduced, with permission, from Manuel Zimmer.)

100 μm

The Gustatory System Controls the Sense of Taste

Taste Has Five Submodalities That Reflect Essential Dietary Requirements

The gustatory system is a specialized chemosensory system dedicated to evaluating potential food sources. It is the only sensory system that detects sugars and harmful compounds present in foods, and it serves as a main driver of feeding decisions. Unlike the olfactory system, which distinguishes millions of odors, the gustatory system recognizes just a few taste categories.

Humans and other mammals can distinguish five basic taste qualities: sweet, bitter, salty, sour, and umami, a Japanese word meaning delicious and associated with the "savory" taste of amino acids. This limited palate detects all essential dietary requirements of animals: A sweet taste invites consumption of energy-rich foods; bitter taste warns against the ingestion of toxic, noxious chemicals; salty taste promotes a diet

that maintains proper electrolyte balance; sour taste signals acidic, unripened, or fermented foods; and umami indicates protein-rich foods.

Consistent with the nutritional importance of carbohydrates and proteins, both sweet and umami tastants elicit innately pleasurable sensations in humans and are attractants for animals in general. In contrast, bitter and sour tastants elicit innately aversive responses in humans and animals.

Taste is often thought to be synonymous with flavor. However, taste refers strictly to the five qualities encoded in the gustatory system, whereas flavor, with its rich and varied qualities, stems from the multisensory integration of inputs from the gustatory, olfactory, and somatosensory systems (eg, texture and temperature).

Tastant Detection Occurs in Taste Buds

Tastants are detected by taste receptor cells clustered in taste buds. Although the majority of taste buds in humans are located on the tongue surface, some can also be found on the palate, pharynx, epiglottis, and upper third of the esophagus.

Taste buds on the tongue occur in structures called papillae, of which there are three types based on morphology and location. *Fungiform papillae*, located on the anterior two-thirds of the tongue, are peg-like structures that are topped with taste buds. Both the *foliate papillae*, situated on the posterior edge of the tongue, and the *circumvallate papillae*, of which there are only a few in the posterior area of the tongue, are structures surrounded by grooves lined with taste buds (Figure 29–13A). In humans, each fungiform papilla contains one to five taste buds, whereas each foliate and circumvallate papilla may contain hundreds to thousands of taste buds, respectively.

The taste bud is a garlic-shaped structure embedded in the epithelium. A small opening at the epithelial surface, the taste pore, is the point of contact with tastants (Figure 29–13B). Each taste bud contains approximately 100 taste receptor cells (taste cells), elongated cells that stretch from the taste pore to the basal area of the bud. The taste bud also contains other elongated cells that are thought to serve a supporting function, as well as a small number of round cells at the base, which are thought to serve as stem cells. Each taste cell extends microvilli into the taste pore, allowing the cell to contact chemicals dissolved in saliva at the epithelial surface.

At its basal end, the taste cell contacts the afferent fibers of gustatory sensory neurons, whose cell bodies reside in specific sensory ganglia (see Figure 29–17). Although taste cells are nonneural, their contacts with the gustatory sensory neurons have the morphological characteristics of chemical synapses, including clustered presynaptic vesicles. Taste cells also resemble neurons in that they are electrically excitable; they have voltage-gated Na^+, K^+, and Ca^{2+} channels and are capable of generating action potentials. Taste cells are very short-lived (days to weeks) and are continually replaced from the stem cell population. This turnover requires that newborn taste cells differentiate to detect one of the five taste qualities and connect to the terminals of appropriate gustatory sensory neurons, such that a sweet taste cell connects to sweet sensory neurons and a bitter taste cell to bitter sensory neurons.

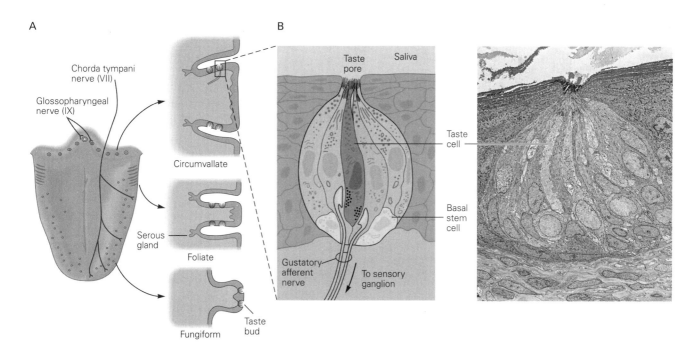

Figure 29–13 Taste buds are clustered in papillae on the tongue.

A. The three types of papillae—circumvallate, foliate, and fungiform—differ in morphology and location on the tongue and are differentially innervated by the chorda tympani and glossopharyngeal nerves.

B. Each taste bud contains 50 to 150 elongated taste receptor cells, as well as supporting cells and a small population of basal stem cells. The taste cell extends microvilli into the taste pore, allowing it to detect tastants dissolved in saliva. At its basal end, the taste cell contacts gustatory sensory neurons that transmit stimulus signals to the brain. The scanning electron micrograph shows a taste bud in a foliate papilla in a rabbit. (Reproduced, with permission, from Royer and Kinnamon 1991. Copyright © 1991 Wiley-Liss, Inc.)

Each Taste Modality Is Detected by Distinct Sensory Receptors and Cells

The five taste qualities are detected by sensory receptors in the microvilli of different taste cells. There are two general types of receptors: Bitter, sweet, and umami tastants interact with G protein–coupled receptors, whereas salty and sour tastants interact directly with specific ion channels (Figure 29–14). These interactions depolarize the taste cell, leading to the generation of action potentials in the afferent gustatory fibers.

Sweet Taste Receptor

Compounds that humans perceive as sweet include sugars, artificial sweeteners such as saccharin and aspartame, a few proteins such as monellin and thaumatin, and several D-amino acids. All of these sweet-tasting compounds are detected by a heteromeric receptor composed of two members of the T1R taste receptor family, T1R2 and T1R3 (Figure 29–15). The

T1R receptors are a small family of three related G protein–coupled receptors that participate in sweet and umami detection.

Receptors of the TIR family have a large N-terminal extracellular domain (Figure 29–14) that serves as the main ligand-binding domain, similar to the V2R receptor of vomeronasal neurons. This domain recognizes many different sugars with low-affinity binding in the millimolar range. This ensures that only high sugar concentrations of nutritive value are detected. Changing a single amino acid in this domain in mice can alter an animal's sensitivity to sweet compounds. Indeed, T1R3 was initially discovered by examining genes at the mouse saccharin preference (Sac) locus, a chromosomal region that governs sensitivity to saccharin, sucrose, and other sweet compounds.

In mice, taste cells with T1R2 receptors are found mostly in palate, foliate, and circumvallate papillae; almost invariably, those cells also possess T1R3 receptors (Figure 29–16A). Gene knockout experiments in mice indicate that the T1R2/T1R3 complex mediates

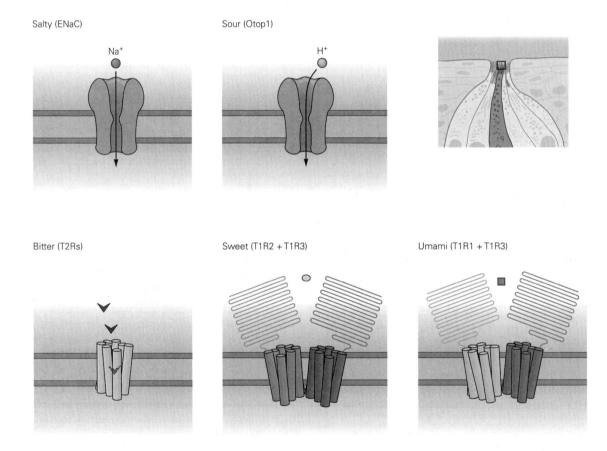

Salty (ENaC)   Sour (Otop1)

Bitter (T2Rs)   Sweet (T1R2 + T1R3)   Umami (T1R1 + T1R3)

Figure 29–14 Sensory transduction in taste cells. Different taste qualities involve different detection mechanisms in the apical microvilli of taste cells (see Figure 29–13B). Salty and sour tastants directly activate ion channels, whereas tastants perceived as bitter, sweet, or umami activate G protein–coupled receptors. Bitter tastants are detected by T2R receptors, whereas sweet tastants are detected by a combination of T1R2 and T1R3, and umami tastants by a combination of T1R1 and T1R3.

A T1R2 + T1R3 recognizes sweet tastants

C T2R receptors recognize bitter tastants

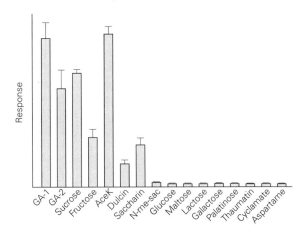

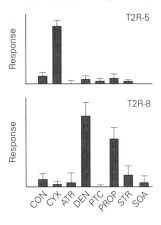

B T1R1 + T1R3 recognizes umami tastants

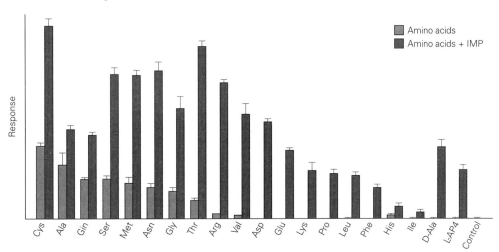

Figure 29–15 Tastants recognized by T1R and T2R receptors. A calcium-sensitive dye was used to test whether T1R and T2R receptors expressed in a tissue culture cell line could detect tastants.

A. Cells expressing both rat T1R2 and rat T1R3 responded to a number of sweet compounds. (Reproduced, with permission, from Nelson et al. 2001.)

B. Cells expressing mouse T1R1 and mouse T1R3 responded to numerous L-amino acids (umami taste). Responses were potentiated by inosine monophosphate (**IMP**). (Reproduced,

with permission, from Nelson et al. 2002. Copyright © 2002 Springer Nature.)

C. Cells expressing different T2R receptors responded selectively to different bitter compounds. Cells expressing mouse T2R5 responded most vigorously to cycloheximide (**CYX**), whereas cells expressing mouse T2R8 responded preferentially to denatonium (**DEN**) and 6-n-propyl-2-thiouracil (**PROP**). (**ATR**, atropine; **CON**, control; **PTC**, phenyl thiocarbamide; **SOA**, sucrose octaacetate; **STR**, strychnine.) (Reproduced, with permission, from Chandrashekar et al. 2000.)

the detection of all sweet compounds except for high concentrations of sugars, which may also be detected by T1R3 alone.

Umami Taste Receptor

Umami is the name given to the savory taste of monosodium glutamate, an amino acid widely used as a

flavor enhancer. It is believed that the pleasurable sensation associated with umami taste encourages the ingestion of proteins and is thus evolutionarily important for nutrition.

The receptor for umami taste is a complex of two T1R receptor subunits: T1R1, specific to the umami receptor, and T1R3, present in both sugar and umami receptors (Figure 29–14). In mice, the T1R1/T1R3 complex can interact with all L-amino acids (Figure 29–15B), but in humans it is preferentially activated by glutamate. Purine nucleotides, such as inosine 5′-monophosphate (IMP), are often added to monosodium glutamate to enhance its pleasurable umami taste. Interestingly, in vitro studies demonstrated that IMP potentiates the responsiveness of T1R1/T1R3 to L-amino acids, acting as a strong positive allosteric modulator of the receptor (Figure 29–15B).

Taste cells with both T1R1 and T1R3 are concentrated in fungiform papillae (Figure 29–16A). Studies in genetically engineered mice in which individual T1R genes have been deleted indicate that the T1R1/T1R3 complex is solely responsible for umami taste, whereas T1R2/T1R3 is solely responsible for sweet taste. As expected, a genetic knockout of T1R1 selectively abolishes umami taste, a knockout of T1R2 specifically abolishes sweet taste, while a knockout of T1R3 eliminates both sweet and umami taste (exactly as predicted, given that it is a common subunit of both the umami and sweet taste receptors).

Sweet and umami receptors differ significantly among different species. Most interestingly, different T1R subunits have been lost in some species, likely reflecting their evolutionary niche and diet. For example, the giant panda, which feeds almost exclusively on a bamboo diet, lacks a functional umami receptor. On the other hand, domestic cats, tigers, and cheetahs do not have a functional sweet receptor, whereas vampire bats that feed on a blood diet have mutations that have eliminated both sweet and umami functional receptors.

Figure 29–16 (Right) Expression of T1R and T2R receptors on the tongue. Sections of mouse or rat tongue were hybridized to probes that label T1R or T2R mRNAs to detect their sites of expression in taste cells.

A. The T1R3 receptor is expressed in taste cells of all three types of papillae. However, T1R1 is found mostly in fungiform papillae, whereas T1R2 is located predominantly in circumvallate (and foliate) papillae. Overlap between sites of expression appears as yellow cells in the micrographs at the top. The T1R1-T1R3 umami receptor is more frequently found in fungiform papillae, whereas the T1R2-T1R3 sweet receptor is more frequently found in circumvallate and foliate papillae. (Reproduced, with permission, from Nelson et al. 2001.)

B. A taste cell that detects bitter tastants can express several variants of T2R receptors. Here, probes for T2R3 and T2R7 labeled the same taste cells in circumvallate papillae. (Reproduced, with permission, from Adler et al. 2000.)

C. The T1R and T2R receptors are expressed in different taste cells. Taste cells labeled by a T1R3 probe or mixed T1R probes (**green**) did not overlap with cells labeled by a mixture of T2R probes (**red**). (Reproduced, with permission, from Nelson et al. 2001.)

Bitter Taste Receptor

Bitter taste is thought to have evolved as an aversive signal of toxic molecules. Bitter taste sensation is elicited by a variety of compounds, including caffeine, nicotine, alkaloids, and denatonium, the most bitter-tasting chemical known (this compound is sometimes added to toxic products that are odorless and tasteless to prevent their ingestion).

Bitter tastants are detected by a family of approximately 30 G protein–coupled receptors called

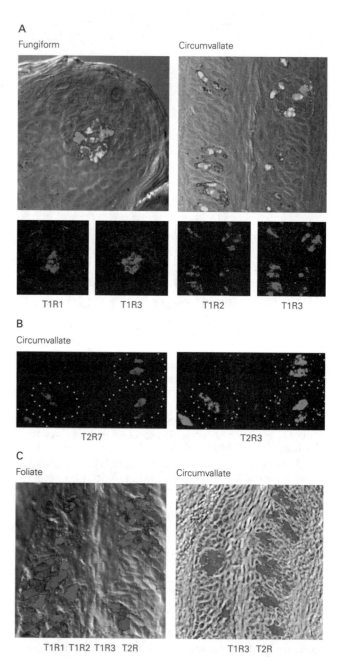

T2Rs (Figure 29–14). However, different animal species contain different numbers of bitter receptors (varying from just a handful in the chicken genome to over 50 in the western clawed frog; humans have 28 T2R genes). These receptors recognize bitter compounds that have diverse chemical structures, with each T2R tuned to detect a small number of bitter compounds (Figure 29–15C). The T2R receptors recognize chemicals with high-affinity binding in the micromolar range, allowing detection of minute quantities of harmful compounds. A single taste cell expresses many, probably most, types of T2R receptors (Figure 29–16B). This arrangement implies that information about different bitter tastants is integrated in individual taste cells. Because different bitter compounds are detected by the same cells, all these compounds elicit the same perceptual bitter taste quality. The degree of bitterness might be caused by a compound's effectiveness in activating bitter taste cells.

Interestingly, genetic differences in the ability to perceive specific bitter compounds have been identified in both humans and mice. For example, humans are either super-tasters, tasters, or taste-blind to the bitter chemical 6-n-propylthiouracil. It was by mapping variation in this trait to specific chromosomal loci, and then by searching for novel G protein–coupled receptor genes within that chromosomal interval, that the T2R receptors were first identified. In the case of 6-n-propylthiouracil detection, the gene responsible for the genetic difference has proven to be a particular T2R gene. Thus, some bitter compounds may be recognized predominantly by only one of the approximately 30 T2R receptor types.

Taste cells expressing T2R receptors are found in both foliate and circumvallate papillae in mice (Figure 29–16C). A given taste cell expresses either T2R or T1R receptors (ie, one taste cell–one receptor class), but a single taste bud can contain taste cells of all types (eg, sweet, umami, bitter). Such mixing of cells accords with the observation that a single taste bud can be activated by more than one class of tastant; for example, sweet as well as bitter.

Salty Taste Receptor

Salt intake is critical to maintaining electrolyte balance. Perhaps because electrolytes must be maintained within a stringent range, the behavioral response to salt is concentration dependent: Low salt concentrations are appetitive, whereas high salt concentrations are aversive. How does the response to salt change based on concentration? It turns out that multiple taste cells detect salt. The essential salt taste receptor cell

uses the epithelial sodium channel ENaC (see Figure 29–14). These specialized salt taste receptors are distinct from sweet, bitter, or umami receptors. At much higher salt concentrations, some bitter and sour taste cells also respond to salt, although the molecular details of detection have not been determined. Therefore, appetitive concentrations of salt drive responses via the ENaC salt taste receptor in the salt-sensing cells, whereas high salt concentrations activate the bitter and sour cells and thus trigger behavioral aversion.

Sour Taste Receptor

Sour taste is associated with acidic or fermented foods or drink. As with bitter compounds, animals are innately averse to sour substances, suggesting that the adaptive advantage of sour taste is avoidance of spoiled foods. Sour, like the other 4 taste qualities, is also detected by its own type of taste receptor cells (Figure 29–14). The ion channel Otopetrin-1 (Otop1), a proton-selective channel normally involved in the sensation of gravity in the vestibular system, is the sour-sensing ion channel in the taste system. As expected, a knockout of Otop1 in mice eliminated acid responses from sour taste receptor cells. Furthermore, mice engineered to express Otop1 in sweet taste receptor cells now have sweet cells that also respond to sour stimuli, demonstrating that this channel is sufficient to confer acid sensing.

Molecular-genetic studies have demonstrated that the different taste modalities are detected by distinct subsets of taste cells. As we have seen, a combination of T1R1 and T1R3 is responsible for all umami taste, and a combination of T1R2 and T1R3 is needed for all sweet taste detection except for the detection of high concentrations of sugars, which can be mediated by T1R3 alone. The T1R1 and T1R2 receptors are expressed by separate subsets of taste cells, indicating that the detection of sweet and umami tastants is segregated. Similarly, receptors and molecular markers uniquely define bitter, low salt, and sour taste cells.

A dramatic demonstration that each taste quality is detected by a different category of taste cells comes from studies of mice lacking a specific taste receptor gene or cell type. These studies showed that the loss of one taste modality did not affect the others. For example, mice in which sweet cells have been genetically ablated do not detect sugars but still detect amino acids, bitter compounds, salts, and sour compounds. Similarly, mice engineered to lack specific taste receptors cannot detect the corresponding tastants. For instance, mice lacking selective bitter receptors are not responsive to the corresponding bitter tastants, and mice lacking ENaC cannot detect the taste of salt. These types of studies have shown that different tastes are detected by different receptors expressed in different classes of taste cells that drive specific behaviors.

Studies in mice further indicate that it is the taste cells rather than the receptors that determine the animal's response to a tastant. The human bitter receptor T2R16 recognizes a bitter tastant that mice cannot detect. When this receptor was expressed in mouse taste cells that normally express T2R bitter receptors, the ligand caused strong taste aversion. However, when that receptor was expressed in cells that express the T1R2/T1R3 sweet complex (ie, sweet cells), the bitter ligand elicited strong taste acceptance. These findings showed that innate responses of mice to different tastants (sweet and bitter in this example) operate via labeled lines that link the activation of different subsets of taste cells to different behavioral outcomes.

Gustatory Information Is Relayed From the Periphery to the Gustatory Cortex

Each taste cell is innervated at its base by the peripheral branches of the axons of primary sensory neurons (Figure 29–13). Each sensory fiber branches many times, innervating several taste cells within taste buds. The release of neurotransmitter from taste cells onto the sensory fibers induces action potentials in the fibers and the transmission of signals to the sensory cell body.

The cell bodies of gustatory sensory neurons lie in the geniculate, petrosal, and nodose ganglia. The peripheral branches of these neurons travel in cranial nerves VII, IX, and X, while the central branches enter the brain stem, where they terminate on neurons in the gustatory area of the nucleus of the solitary tract (Figure 29–17). In most mammals, neurons in this nucleus transmit signals to the parabrachial nucleus of the pons, which in turn sends gustatory information to the ventroposterior medial nucleus of the thalamus. In primates, however, these neurons transmit gustatory information directly to the taste area of the thalamus.

From the thalamus, taste information is transmitted to the gustatory cortex, a region of the cerebral cortex located along the border between the anterior insula and the frontal operculum (Figure 29–17). The gustatory cortex is believed to mediate the conscious perception and discrimination of taste stimuli. The taste areas of the thalamus and cortex also transmit information both directly and indirectly to the hypothalamus, which controls feeding behavior and autonomic responses.

Large-scale calcium imaging revealed that some neurons in the gustatory cortex respond preferentially to one taste modality, such as bitter or sweet. These neurons are localized in segregated cortical fields or hot spots. Interestingly, using a light-activated ion channel to activate neurons in the sweet hot spot elicits innately attractive responses. In contrast, activation of the bitter hot spot evokes suppression of licking and strong aversive orofacial responses, mimicking what is often seen in response to bitter tastants. These experiments showed that direct control of primary taste cortex can evoke specific, reliable, and robust behaviors that mimic responses to natural tastants. They also illustrated that the gustatory pathway can activate innate, immediate responses to sweet and bitter chemicals. To demonstrate that these cortically triggered behaviors are innate (ie, independent of learning or experience), similar stimulation experiments were performed in mutant mice that had never tasted sweet or bitter chemicals (the mutation abolished all sweet and bitter signal transduction). Even in these animals, activation of the corresponding cortical fields triggered the expected behavioral response, thus substantiating the predetermined nature of the sense of taste.

Perception of Flavor Depends on Gustatory, Olfactory, and Somatosensory Inputs

Much of what we think of as the flavor of foods derives from information provided by the integration of the taste and olfactory systems. Volatile molecules released from foods or beverages in the mouth are pumped into the back of the nasal cavity ("retronasal passage") by the tongue, cheek, and throat movements that accompany chewing and swallowing. Although the olfactory epithelium of the nose clearly makes a major contribution to sensations of flavor, such sensations are localized in the mouth rather than in the nose.

The somatosensory system is also thought to be involved in this localization of flavors. The coincidence between taste, somatosensory stimulation of the tongue, and the retronasal passage of odorants into the nose is assumed to cause odorants to be perceived as flavors in the mouth. Sensations of flavor also frequently have a somatosensory component that includes the texture of food as well as sensations evoked by spicy or minty foods and by carbonation.

Insects Have Modality-Specific Taste Cells That Drive Innate Behaviors

Insects have a specialized gustatory system that evaluates potential nutrients and toxins in food. Taste neurons are found on the proboscis, internal mouthparts, legs, wings, and ovipositor, allowing insects to sample

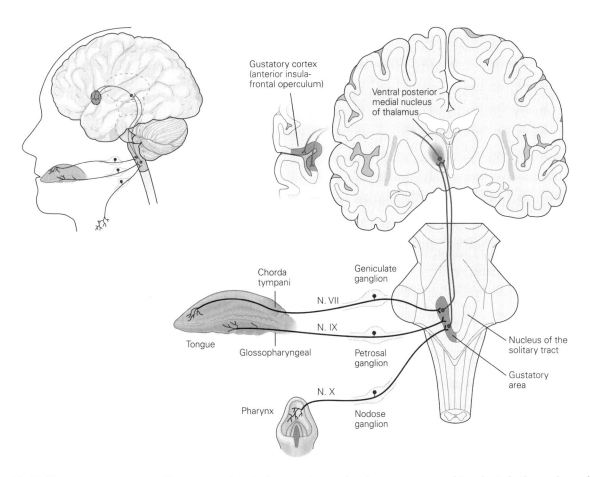

Figure 29–17 The gustatory system. Tastants are detected in taste buds in the oral cavity. Taste buds on the tongue and pharynx are innervated by the peripheral fibers of gustatory sensory neurons, which travel in the glossopharyngeal, chorda tympani and vagus nerves and terminate in the nucleus of the solitary tract in the brain stem. From there, taste information is relayed through the thalamus to the gustatory cortex as well as to the hypothalamus.

the local chemical environment prior to ingestion. As in mammals, only a few different types of taste cells detect different tastes. In the *Drosophila* fly, the different taste cell classes include those that sense sugars, bitter compounds, water, and pheromones. As in mammals, activation of these different taste cells drives different innate behaviors; for example, activation of sugar cells drives food acceptance behavior, whereas activation of bitter cells drives food rejection. Thus, the basic organization of taste detection is remarkably similar in insects and mammals, despite divergent evolutionary histories.

The taste receptors in insects are not related to vertebrate receptors. Members of the gustatory receptor (GR) gene family participate in the detection of sugars and bitter compounds. The GRs are membrane-spanning receptors that are distantly related to the odorant receptors of the fly. The fly has approximately 70 GR genes, a surprisingly large number considering

it has approximately 60 olfactory receptor genes. Different GRs are found in sugar cells versus bitter cells, with many GRs present in a single neuron. In addition to GRs, other gene families participate in insect taste, including variants of ionotropic glutamate receptors and other ion channel classes. Similar to olfactory detection, the gene families involved in taste recognition differ across phyla, demonstrating that the gene families for chemical recognition have evolved independently.

Highlights

1. Odor detection in the nose is mediated by a large family of odorant receptors that number approximately 350 in humans and 1,000 in mice. These receptors vary in protein sequence, consistent with an ability to detect structurally diverse odorants.

2. Individual odorant receptors can detect multiple odorants, and different odorants activate different combinations of receptors. This combinatorial strategy explains how we can discriminate a multitude of odorants and how nearly identical odorants can have different scents.

3. Each olfactory sensory neuron in the nose expresses a single type of receptor. Thousands of neurons with the same receptor are dispersed in the olfactory epithelium and intermingled with neurons expressing other receptors.

4. In the olfactory bulb, the axons of the sensory neurons expressing the same receptor converge in a few receptor-specific glomeruli, generating a map of odorant receptor inputs that is similar among individuals.

5. The axons of olfactory bulb projection neurons project broadly to multiple areas of the olfactory cortex, generating a highly distributed organization of cortical neurons responsive to individual odorants. The olfactory cortex transmits information to many other brain areas.

6. In mice, pheromones can be detected in the nose or in the vomeronasal organ, a structure absent in humans. Signals from the nose and vomeronasal organ travel through different neural pathways in the brain.

7. The olfactory system of the fruit fly *Drosophila melanogaster* resembles that of mammals in many aspects. It uses a large number of diverse olfactory receptors, with one or a few olfactory receptors expressed by each olfactory sensory neuron. Moreover, neurons with the same receptor synapse in a few specific glomeruli in the antennal lobe of the brain. From there, olfactory signals are transmitted to two major brain areas involved in innate versus learned odor responses. The ease of using genetic approaches in fruit flies has enabled rapid study of mechanisms underlying odor coding and behavior.

8. The gustatory system detects five basic tastes: sweet, sour, bitter, salty, and umami (amino acids). Tastants that activate these taste qualities are detected by taste receptor cells located primarily in taste buds on the tongue and palate epithelium. The detection of the five different taste modalities is mediated by different taste receptor cells, each dedicated to one modality.

9. Sweet tastants are detected by a single type of receptor, which is composed of two subunits, T1R2 and T1R3. Umami receptors are related but comprise a combination of T1R1 and T1R3 subunits.

10. Bitter taste receptors constitute a family of approximately 30 related but diverse receptors that vary in ligand specificity. Individual taste receptor cells express many or all bitter receptors.

11. In contrast to sweet, umami, and bitter receptors, which are all G protein–coupled receptors, salty and sour tastes are detected by ion channels: ENaC for salt taste and Otopetrin-1 for sour taste.

12. Taste signals travel from taste buds through cranial nerves from gustatory sensory neurons in the geniculate, petrosal, and nodose ganglia via labeled lines (sweet taste receptor cells to sweet neurons, bitter taste cells to bitter neurons, etc.). They then travel to the gustatory area of the nucleus of the solitary tract and parabrachial nucleus, and from there to the taste area of the thalamus and then the gustatory cortex. The gustatory cortex, in turn, projects to many brain areas, including those involved in motor control, feeding, hedonic value, learning, and memory.

13. The gustatory cortex contains hot spots for sweet and bitter taste, which, when directly stimulated, can elicit behavioral responses similar to those obtained with tastants applied to the tongue.

14. The fruit fly *Drosophila* also has a specialized gustatory system that evaluates potential nutrients and toxins in food. Different classes of taste cells sense sugars, bitter compounds, pheromones, or water. Activation of these different peripheral sensors drives different innate behaviors, such as food acceptance or rejection.

Linda Buck
Kristin Scott
Charles Zuker

Selected Reading

Bargmann CI. 2006. Comparative chemosensation from receptors to ecology. Nature 444:295–301.

Giessel AJ, Datta SR. 2014. Olfactory maps, circuits and computations. Curr Opin Neurobiol 24:120–132.

Stowers L, Kuo TH. 2015. Mammalian pheromones: emerging properties and mechanisms of detection. Curr Opin Neurobiol 34:103–109.

Touhara K, Vosshall LB. 2009. Sensing odorants and pheromones with chemosensory receptors. Annu Rev Physiol 71:307–332.

Wilson DA, Sullivan RM. 2011. Cortical processing of odor objects. Neuron 72:506–519.

Yarmolinsky DA, Zuker CS, Ryba NJ. 2009. Common sense about taste: from mammals to insects. Cell 139: 234–244.

References

Adler E, Hoon MA, Mueller KL, Chandrashekar J, Ryba NJ, Zuker CS. 2000. A novel family of mammalian taste receptors. Cell 100:693–702.

Bachmanov AA, Bosak NP, Lin C, et al. 2014. Genetics of taste receptors. Curr Pharm Des 20:2669–2683.

Buck L, Axel R. 1991. A novel multigene family may encode odorant receptors: a molecular basis for odor recognition. Cell 65:175–187.

Chandrashekar J, Kuhn C, Oka Y, et al. 2010. The cells and peripheral representation of sodium taste in mice. Nature 464:297–301.

Chandrashekar J, Mueller KL, Hoon MA, et al. 2000. T2Rs function as bitter taste receptors. Cell 100:703–711.

Chen X, Gabitto M, Peng Y, Ryba NJ, Zuker CS. 2011. A gustotopic map of taste qualities in the mammalian brain. Science 333:1262–1266.

Dulac C, Axel R. 1995. A novel family of genes encoding putative pheromone receptors in mammals. Cell 83:195–206.

Dulac C, Torello AT. 2003. Molecular detection of pheromone signals in mammals: from genes to behaviour. Nat Rev Neurosci 7:551–562.

Glusman G, Yanai I, Rubin I, Lancet D. 2001. The complete human olfactory subgenome. Genome Res 1:685–702.

Godfrey PA, Malnic B, Buck LB. 2004. The mouse olfactory receptor gene family. Proc Natl Acad Sci U S A 101:2156–2161.

Greer PL, Bear DM, Lassance JM, et al. 2016. A family of non-GPCR chemosensors defines an alternative logic for mammalian olfaction. Cell 165:1734–1748.

Hallem EA, Carlson JR. 2006. Coding of odors by a receptor repertoire. Cell 125:143–160.

Herrada G, Dulac C. 1997. A novel family of putative pheromone receptors in mammals with a topographically organized and sexually dimorphic distribution. Cell 90:763–773.

Hoon MA, Adler E, Lindemeier J, Battey JF, Ryba NJ, Zuker CS. 1999. Putative mammalian taste receptors: a class of taste-specific GPCRs with distinct topographic selectivity. Cell 96:541–551.

Johnson BA, Farahbod H, Leon M. 2005. Interactions between odorant functional group and hydrocarbon structure influence activity in glomerular response modules in the rat olfactory bulb. J Comp Neurol 483:205–216.

Keller A, Zhuang H, Chi Q, Vosshall LB, Matsunami H. 2007. Genetic variation in a human odorant receptor alters odour perception. Nature 449:468–472.

Kondoh K, Lu Z, Ye X, Olson DP, Lowell BB, Buck LB. 2016. A specific area of olfactory cortex involved in stress hormone responses to predator odours. Nature 532:103–106.

Liberles SD, Buck LB. 2006. A second class of chemosensory receptors in the olfactory epithelium. Nature 442:645–650.

Liberles SD, Horowitz LF, Kuang D, et al. 2009. Formyl peptide receptors are candidate chemosensory receptors in the vomeronasal organ. Proc Natl Acad Sci U S A 106:9842–9847.

Malnic B, Godfrey PA, Buck LB. 2004. The human olfactory receptor gene family. Proc Natl Acad Sci U S A 101:2584–2589.

Malnic B, Hirono J, Sato T, Buck LB. 1999. Combinatorial receptor codes for odors. Cell 96:713–723.

Matsunami H, Buck LB. 1997. A multigene family encoding a diverse array of putative pheromone receptors in mammals. Cell 90:775–784.

Matsunami H, Montmayeur JP, Buck LB. 2000. A family of candidate taste receptors in human and mouse. Nature 404:601–604.

Mombaerts P, Wang F, Dulac C, et al. 1996. Visualizing an olfactory sensory map. Cell 87:675–686.

Montmayeur JP, Liberles SD, Matsunami H, Buck LB. 2001. A candidate taste receptor gene near a sweet taste locus. Nat Neurosci 4:492–498.

Morrison E, Constanzo R. 1990. Morphology of the human olfactory epithelium. J Comp Neurol 297:1–13.

Mueller KL, Hoon MA, Erlenbach I, Chandrashekar J, Zuker CS, Ryba NJ. 2005. The receptors and coding logic for bitter taste. Nature 434:225–229.

Nelson G, Chandrashekar J, Hoon MA, et al. 2002. An amino-acid taste receptor. Nature 416:199–202.

Nelson G, Hoon MA, Chandrashekar J, et al. 2001. Mammalian sweet taste receptors. Cell 106:381–390.

Neville KR, Haberly LB. 2004. The olfactory cortex. In: GM Shepherd (ed). *The Synaptic Organization of the Brain*, pp. 415–454. New York: Oxford University Press.

Niimura Y, Nei M. 2005. Evolutionary changes of the number of olfactory receptor genes in the human and mouse lineages. Gene 14:23–28.

Northcutt RG. 2004. Taste buds: development and evolution. Brain Behav Evol 64:198–206.

Oka Y, Butnaru M, von Buchholtz L, Ryba NJ, Zuker CS. 2013. High salt recruits aversive taste pathways. Nature 494:472–475.

Peng Y, Gillis-Smith S, Jin H, Tränkner D, Ryba NJ, Zuker CS. 2015. Sweet and bitter taste in the brain of awake behaving animals. Nature 527:512–515.

Ressler KJ, Sullivan SL, Buck LB. 1993. A zonal organization of odorant receptor gene expression in the olfactory epithelium. Cell 73:597–609.

Ressler KJ, Sullivan SL, Buck LB. 1994. Information coding in the olfactory system: evidence for a stereotyped and highly organized epitope map in the olfactory bulb. Cell 79:1245–1255.

Riviere S, Challet L, Fluegge D, Spehr M, Rodriguez I. 2009. Formyl peptide receptor-like proteins are a novel family of vomeronasal chemosensors. Nature 459:574–577.

Root CM, Denny CA, Hen R, Axel R. 2014. The participation of cortical amygdala in innate, odour-driven behaviour. Nature 515:269–273.

Royer SM, Kinnamon JC. 1991. HVEM Serial-section analysis of rabbit foliate taste buds. I. Type III cells and their synapses. J Comp Neurol 306:49–72.

Sarafi-Reinach TR, Sengupta P. 2000. The forkhead domain gene *unc*-130 generates chemosensory neuron diversity in *C. elegans*. Genes Dev 14:2472–2485.

Shepherd GM, Chen WR, Greer CA. 2004. The olfactory bulb. In: GM Shepherd (ed). *The Synaptic Organization of the Brain*, 5th ed., pp. 164–216. New York: Oxford Univ. Press.

Shepherd GM, Greer CA. 1998. The olfactory bulb. In: GM Shepherd (ed). *The Synaptic Organization of the Brain*, 4th ed., pp. 159–203. New York: Oxford Univ. Press.

Stettler DD, Axel R. 2009. Representations of odor in the piriform cortex. Neuron 63:854–864.

Sullivan SL, Adamson MC, Ressler KJ, Kozak CA, Buck LB. 1996. The chromosomal distribution of mouse odorant receptor genes. Proc Natl Acad Sci U S A 93:884–888.

Teng B, Wilson CE, Tu YH, Joshi NR, Kinnamon SC, Liman ER. 2019. Cellular and neural responses to sour stimuli require the proton channel Otop1. Curr Biol 4:3647–3656.

Troemel ER, Kimmel BE, Bargmann CI. 1997. Reprogramming chemotaxis responses: sensory neurons define olfactory preferences in *C. elegans*. Cell 91:161–169.

Vassar R, Ngai J, Axel R. 1993. Spatial segregation of odorant receptor expression in the mammalian olfactory epithelium. Cell 74:309–318.

Vosshall L, Amrein H, Morozov PS, Rzhetsky A, Axel R. 1999. A spatial map of olfactory receptor expression in the *Drosophila* antenna. Cell 96:725–736.

Vosshall LB, Stocker RF. 2007. Molecular architecture of smell and taste in *Drosophila*. Annu Rev Neurosci 30:505–533.

Vosshall LB, Wong AM, Axel R. 2000. An olfactory sensory map in the fly brain. Cell 102:147–159.

Wang JW, Wong AM, Flores J, Vosshall LB, Axel R. 2003. Two-photon calcium imaging reveals an odor-evoked map of activity in the fly brain. Cell 112:271–282.

Wilson RI. 2013. Early olfactory processing in *Drosophila*: mechanisms and principles. Annu Rev Neurosci 36:217–241.

Zhang X, Firestein S. 2002. The olfactory receptor gene superfamily of the mouse. Nat Neurosci 5:124–133.

Zhang Y, Hoon MA, Chandrashekar J, et al. 2003. Coding of sweet, bitter, and umami tastes: different receptor cells sharing similar signaling pathways. Cell 112:293–301.

Zhang J, Jin H, Zhang W, et al. 2019. Sour Sensing from the Togue to the Brain. Cell 179:392–402.

Zhao GQ, Zhang Y, Hoon MA, et al. 2003. The receptors for mammalian sweet and umami taste. Cell 115:255–266.